Smart Sensors for Structural Health Monitoring and Nondestructive Evaluation

Smart Sensors for Structural Health Monitoring and Nondestructive Evaluation

Editor

Zenghua Liu

MDPI

Basel • Beijing • Wuhan • Barcelona • Belgrade • Novi Sad • Cluj • Manchester

Editor
Zenghua Liu
Beijing University of
Technology
Beijing
China

Editorial Office
MDPI
St. Alban-Anlage 66
4052 Basel, Switzerland

This is a reprint of articles from the Special Issue published online in the open access journal *Gels* (ISSN 2310-2861) (available at: https://www.mdpi.com/journal/sensors/special_issues/SHM_NDE).

For citation purposes, cite each article independently as indicated on the article page online and as indicated below:

Lastname, A.A.; Lastname, B.B. Article Title. *Journal Name* **Year**, *Volume Number*, Page Range.

ISBN 978-3-7258-0225-8 (Hbk)
ISBN 978-3-7258-0226-5 (PDF)
doi.org/10.3390/books978-3-7258-0226-5

Cover image courtesy of Zenghua Liu

Contents

About the Editor

Zenghua Liu

Zenghua Liu is currently a professor with the Faculty of Information Technology, Beijing University of Technology, Beijing, China. He received a Ph.D. from Beijing University of Technology, Beijing, China, in 2006. He spent 2010 and 2011 at University College London, United Kingdom, as a visiting scholar. He has authored and co-authored over 200 papers, 20 patents and 1 book. He currently serves as the chairman for the Ultrasonic Section of the Chinese Society for Nondestructive Testing (CHSNDT), as a deputy director for the Equipment Structural Health Monitoring and Prognostics Institution of the China Instrument and Control Society (CIS), and as a Member and Deputy Secretary-General of the National Working Group on Standardization of Equipment Structural Health Monitoring of China. His research interests include nondestructive testing and evaluation, structural health monitoring, material characterization, ultrasonic imaging, ultrasonic guided waves, electromagnetic acoustic transducers, signal processing and THz inspection.

sensors

Editorial

Smart Sensors for Structural Health Monitoring and Nondestructive Evaluation

Zenghua Liu

Faculty of Information Technology, Beijing University of Technology, Beijing 100124, China; liuzenghua@bjut.edu.cn

1. Introduction

During manufacturing, processing, and usage, various types of damage may be caused in structures. For example, corrosion and fatigue cracks are common defects in metal plates, while the main defects in composite plates are delamination, debonding, etc. Thus, it is important to develop defect detection and monitoring techniques to ensure the integrity of structures [1–3]. Structural health monitoring (SHM) and nondestructive evaluation (NDE) technologies can be used to identify defects or damages and evaluate the health status of components and systems to avoid structural failure or catastrophes. NDE is usually a technique for detecting offline defects without destroying structures using removable sensors. In contrast, SHM is a potential alternative whereby the global and local security state of a structure is permanently monitored through embedded or attached sensors or sensor arrays. By monitoring the condition, or change in conditions, of a structure, the detection of damage at any location within a structure becomes possible [4–6]. SHM and NDE both aim to assess the integrity of a structure nondestructively and these fields often overlap, whether that be to a greater or lesser extent [7].

With complex working conditions of structures, including harsh and high-temperature environments, increasing structure scale, and increasing detection requirements, SHM and NDE are progressively becoming more challenging to implement [8–11]. A lot of techniques with different working mechanisms, such as the piezoelectric effect [12], magnetostriction effect [13], and magnetic effect [14], can be used for SHM and NDE. In order to detect how safe a structure is, it is crucial to use appropriate sensors and sensor arrays to obtain effective information to try to evaluate a safety status using various signal processing methods. A major development in the fields of SHM and NDE over the past few decades has been the development and use of smart sensors.

In short, the development and application of sensors are key research topics in the areas of SHM and NDE. This Special Issue "Smart Sensors for Structural Health Monitoring and Nondestructive Evaluation" has collected the most recent original contributions relating to all facets of smart sensors utilized in structural health monitoring and nondestructive evaluation. The call for papers for this Special Issue included topics such as sensors and sensor arrays, sensor modeling and simulation, SHM systems and technology, nondestructive testing and evaluation, structural diagnosis and performance evaluation, signal processing, artificial intelligence applications in SHM and NDE, system and instrument development, and field applications of SHM and NDE.

Citation: Liu, Z. Smart Sensors for Structural Health Monitoring and Nondestructive Evaluation. *Sensors* **2024**, *24*, 603. https://doi.org/10.3390/s24020603

Received: 2 January 2024
Accepted: 11 January 2024
Published: 17 January 2024

2. Overview of Published Papers

In this context, this Special Issue includes 33 papers focused on the latest advancements in the field of smart sensors for structural health monitoring and nondestructive evaluation. Each of the 33 original contributions (2 review papers and 31 research papers) accepted for publication have undergone a rigorous review process by a minimum of two expert reviewers across at least two rounds of revision. The papers published in the current Special Issue are briefly summarized as follows:

In contribution 1, the authors present an overview of recent progress in the field of piezoelectric materials and sensors for structural health monitoring. The article commences with a brief introduction to the fundamental physical science of the piezoelectric effect. Emphases are placed on piezoelectric materials engineered using various strategies and applications of piezoelectric sensors for structural health monitoring. Finally, challenges, along with opportunities for future research and the development of high-performance piezoelectric materials and sensors for structural health monitoring, are highlighted.

In contribution 2, the authors present a systematic literature review on the state-of-the-art electrochemical methods and physical methods used so far for corrosion monitoring that are compatible with low-cost sensors and data acquisition devices for metallic and concrete structures. In addition, special attention is paid to the use of these devices for corrosion monitoring and detection for in situ applications in different industries. This analysis demonstrates the possible applications of low-cost sensors in the corrosion monitoring sector. In addition, this study provides scholars with preferred techniques and the most common microcontrollers to overcome corrosion monitoring difficulties in the construction industry.

In contribution 3, the authors propose a two-step physics- and machine learning (ML)-based electromechanical impedance (EMI) measurement data evaluation approach for sandwich face-layer debonding detection and size estimation in SHM applications. This approach was shown to be robust against unknown artificial disturbances and it outperformed a previous method for debonding size estimation.

In contribution 4, the authors propose a novel baseline-free method for damage localization using Lamb waves based on a hyperbolic algorithm. This method employs a special array with a relatively small number of transducers and only one branch of the hyperbola. The novel symmetrical array was arranged on plate structures to eliminate the direct waves. The imaging results showed that both the damages outside and inside the diamond-shaped arrays could be localized, and the positioning error reached its maximum for the diamond-shaped array using the minimum size.

In contribution 5, the authors discuss the need for surface monitoring of municipal solid waste (MSW) landfills. A properly 3D-mapped landfill mass was the basis for ensuring the geotechnical safety of the restored landfill. Based on archival data and current measurements of Radiowo landfill (Poland), this study compared the advantages and limitations of the following measurement techniques: linear and angular measurements, satellite measurements, TLS, and UAV scanning and photogrammetry, considering specific conditions regarding the location and vegetation of the landfill. Solutions for long-term monitoring were proposed, considering the cost and time resolution necessary for creating a differential model of landfill geometry changes.

In contribution 6, the authors present a novel adaptive filtering approach to enhance the signal-to-noise ratio (SNR) of a measured ultrasonic signal from the inspection of a stainless steel component, enabling the detection of hidden flaws under strong noise. The performance of the proposed method for SNR enhancement was evaluated by both the simulated and experimental signal and its effectiveness was successfully demonstrated.

In contribution 7, the authors propose a method to simultaneously estimate the mechanical parameters of vehicles, bridges, and road unevenness with only a few constraints. Signals from acceleration sensors attached to vehicles traveling on bridges were processed. Methods were proposed to estimate the modal parameters of bridges and road unevenness from vehicle vibrations on a case-by-case basis. The road surface irregularities estimated by the proposed method were compared with the measured values, which were somewhat smaller than the measured values.

In contribution 8, the authors propose a deep neural network to detect failures and investigate sensor sensitivity to damage classification for an orbiting satellite hosting a distributed network of accelerometers on a large mesh reflector antenna. The proposed deep learning framework and sensor configuration were shown to accurately detect failures in the most critical area or the structure while opening new investigation possibilities regarding geometrical properties and sensor distribution.

In contribution 9, the authors investigate the effect of the cycle period of a stimulation wave on amplitude-phase results by performing various numeric simulations and laboratory tests. The recommended cycle period from the prediction surfaces was experimentally validated using two samples. After various laboratory experiments, the temperature, amplitude, and phase results validated the previous equations and prediction surfaces relating to the poly(methyl methacrylate) (PMMA) and carbon fiber reinforced polymer (CFRP) sample.

In contribution 10, the authors develop a wireless impedance monitoring system to have cheap, mobile, and handy practical features compared to wired commercial impedance analyzers. A Raspberry Pi platform sensor node was designed to acquire impedance signals using a PZT interface technique. The software scheme was designed to operate the Raspberry Pi platform and impedance sensor node. The calibration procedure was designed for the impedance sensor node. The feasibility of the proposed Raspberry Pi platform SSeL-Pi system was experimentally evaluated for PZT interfaces that were subjected to various compressive loadings.

In contribution 11, the authors design a sandwich piezoelectric ceramic transducer and analyze the vibration of each part of the transducer, making use of elastic mechanics and piezoelectric theory. The vibration characteristics of the transducer under different parameters such as voltage and frequency were analyzed, and the accuracy of the vibration model was verified. The results showed that the equivalent simplified model established in this study could effectively be used to design the inherent frequency of the transducer, and the operation of the first-order inherent frequency met the one-dimensional assumptions of this study.

In contribution 12, the authors present research on the performance of composite and monolithic sensors for distributed fiber optic sensing (DFOS). The performance of each DFOS nondestructive tool was investigated in the close vicinity of the cracks—both the new ones, opening within the tension zone, and the existing ones, closing within the compression zone. Qualitative (detection) and quantitative (width estimation) crack analyses were performed and discussed. Finally, examples of actual applications within concrete structures, including bridges, were presented, with some examples of in situ results.

In contribution 13, the authors develop a laser ultrasonic system to indirectly evaluate the preload force of different-frequency piezoelectric bolts and achieve the goal of non-contact excitation and synchronous reception of laser-induced ultrasonic signals using a combination of a smart piezoelectric sensor and a magnetically mounted transducer connector. The results indicate that the proposed system based on a surface-mounted piezoelectric sensor is a promising system for evaluating axial preload change in a connector and fastener and is an additional potential laser ultrasonic system for nondestructive tests.

In contribution 14, the authors propose a new configuration of angled shear vertical (SV) wave EMAT with horizontal magnetization to reduce the influence of a head wave. The results from simulations and experiments showed that the proposed EMAT had a larger signal amplitude and significantly reduced interference in large-incidence angle scenarios. Moreover, it was found that an incidence angle of an SV wave of up to 60 degrees could be achieved, which will help to improve the performance and capability of nondestructive testing. In contribution 15, the authors present an experimental field study using an unattended corrosion sensor developed on the basis of ultrasound technology to assess the thickness loss caused by general atmospheric corrosion on land close to the sea (coastal regions). The system described here used FPGA, a low-power microcontroller, analog front-end devices in the sensor node, and a Beaglebone black wireless board to post data to a server. Over the course of 5 months, the proposed experiment continuously monitored the corrosion rate in an equivalent corrosion process, showing an average thickness loss rate of 0.134 mm/year.

In contribution 16, the authors propose an improved RAPID (reconstruction algorithm for probabilistic inspection of defect) imaging method based on machine learning (ML) to precisely visualize the location and features of defects in a composite plate. The simulation results showed that the proposed method can intuitively characterize the location and related feature information of a defect and effectively improve the accuracy of defect imaging.

In contribution 17, the authors investigate directional magnetic incremental permeability in an iron–cobalt magnetic sheet. The study revealed that an angle of $\pi/2$ between DC (Hsurf DC) and AC (Hsurf AC) magnetic excitations with a flux density Ba at Hsurf DC = 10 $kA{\cdot}m^{-1}$ constituted the ideal experimental situation and the highest correlated parameter in homogeneous imposed tensile stress. Magnetic incremental permeability was linked to the magnetic domain wall bulging magnetization mechanism; this study thus provided insights for understanding such a mechanism.

In contribution 18, the authors present a measurement system involving the application of 3D cameras for a wind tunnel test. A specific application of displacement measurement using a 3D shape was shown and an analysis of acquired data allowed the authors to obtain the displacement of each side of each structure, for different sections. This proposed system can be seen as a complete system for measuring displacement, with higher accuracy than relying on the uncertainty of 3D sensors, and a system for measuring the 3D geometry of structures in harsh environments.

In contribution 19, the authors propose an efficient semi-analytical method capable of modeling the propagation of flexural waves on cracked plate structures with any form of excitation, based on the same group of vibration characteristics and validated by a non-contact scanning Laser Doppler Vibrometer (LDV) system. The proposed modeling method was based on the superposition of the vibrational normal modes of the detected structure, which could be applied to analyze long-time and full-field transient wave propagations. The experimental results indicated that the proposed semi-analytical method can model flexural waves, and in doing so, crack information can be revealed.

In contribution 20, the authors present the effect of different curing methods on gas permeability with the help of laboratory and on-site tests, showing that inadequate curing leads to increased permeability in the near-surface area of concrete. The measurement results of concrete samples and components with the same composition but varying curing treatments were compared and evaluated. Influences on the quality of concrete, such as concrete composition and environmental factors, were observed.

In contribution 21, the authors deduce the theoretical process of calculating the propagation characteristics of Lamb waves in functionally graded material (FGM) sandwich plates by combining the FGM volume fraction curve and Legendre polynomial series expansion method. For comparison purposes, the Lamb wave dispersion curve of the sliced layer model for the FGM sandwich plate was obtained using the global matrix method. Meanwhile, the FGM sandwich plate was subjected to finite element simulation, which was also based on the layered-plate model. The acoustic characteristic detection experiment was performed via simulation using a defocusing measurement. Thus, the Lamb wave dispersion curves were obtained through V(f, z) analysis.

In contribution 22, the authors thoroughly investigate pulse modulation eddy current (PMEC) for the imaging and assessment of ILC through theoretical simulations and experiments. A semi-analytical model of PMEC evaluation of ILC occurring at the interlayer of two conductor layers was established based on extended truncated region eigenfunction expansion (ETREE) along with an efficient algorithm for the numerical computation of eigenvalues for reflection coefficients of the stratified conductor under inspection. The theoretical and experimental results revealed the feasibility of PMEC for the imaging and evaluation of ILC in stratified conductors.

In contribution 23, the authors investigate the validation of a numerical method from the literature which was used to simulate an H-N source on a complex plate. The location results of the numerical delta-t map technique were compared with those of traditional TOA techniques and an experimental delta-t map technique. The viability of using the FE method was demonstrated to decrease the time and labor required for manually collecting and processing training data whilst maintaining a reasonable degree of source location accuracy with an average error of 3.88 mm. With such high source location accuracy, the specific area of concern requiring inspection using other NDE techniques can be greatly reduced.

In contribution 24, the authors propose an adaptive signal truncation method based on signal difference analysis to improve the defect location accuracy of the probabilistic elliptic imaging algorithm. The experimental results showed that the probabilistic ellipse imaging algorithm, based on the new adaptive signal truncation method, could effectively locate a single defect on a pressure vessel.

In contribution 25, the authors derive an analytical expression for the response of piezoelectric transducers under the action of stress waves to obtain an overall mathematical model of an acoustic emission signal from generation to reception. By comparing the finite element simulations, experimental validation, and analytical modeling, it was found that they were almost consistent in the time and frequency domains, and the presence of the Lamb wave S0 mode was clearly observed. All of this validated the accuracy of the analytical modeling predictions.

In contribution 26, the authors propose a novel time reversal (TR)-based localization method for pipeline leakage. In the proposed method, so-called TR self-adaptive cancellation was developed to improve leak localization resolution. The experimental results showed that the leak positions could be accurately revealed by using the proposed approach. Furthermore, the resolution of the proposed approach was shown to be ten times that of the conventional TR localization method.

In contribution 27, the authors measure the acoustic parameters of guided wave propagation in different cement tailing ratios and different curing times of cemented paste backfills (CPBs). Combined with the uniaxial compression strength of CPB, the relationships between CPB strength and the guided wave parameters were established. Based on the relationship curves between CPB strength and guided wave velocity and attenuation, the guided wave technique was proven to be feasible in monitoring the strength development of CPB.

In contribution 28, the authors propose a method using only two piezoelectric transducers and based on orthogonal matching pursuit (OMP) decomposition to detect damage with the fewest sensors and high resolution. The experimental results showed that the OMP-based algorithm was beneficial for resolution improvement and transducer usage reduction.

In contribution 29, the authors propose a new method for identifying many different types of rail defects. Segment principal component analysis (S-PCA) was utilized to extract characteristics from signals collected by sensors located at different positions. The Support Vector Machine (SVM) model was used to identify different defects depending on the features extracted. The results showed that the defect classification accuracy rates were 96.29% and 96.15% for combining multiple signals, such as the method of single-point excitation and multi-point reception or the method of multi-point excitation and reception at a single point.

In contribution 30, the authors propose a new concept of nonlinearity parameter grouping with multi-frequency excitation as an early failure parameter. An analytical solution of the one-dimensional wave equation was derived with four fundamental frequencies and a total of 64 individual and 30 group nonlinearity parameters. Experimental validation of the approach was conducted on metal plates with different types of cracks and on turbine blades where cracks originated under service conditions. The results showed that the use of multi-frequency excitation offered advantages in detecting cracks.

In contribution 31, the authors present a metamaterial lens design and its use in far-field microwave imaging for subwavelength defect detection in nondestructive evaluation (NDE). Theoretical formulation and numerical studies of the metamaterial lens design were presented, followed by experimental demonstration and characterization of metamaterial behavior. A microwave homodyne receiver-based system was used in conjunction with the metamaterial lens to develop a far-field microwave NDE sensor system. The system was shown to be sensitive to a defect of size $0.17\lambda \times 0.06\lambda$ in a Teflon sample. Consecutive positions of the defect with a separation of 0.23λ were resolvable using the proposed system.

In contribution 32, the authors present a coherent optical fiber sensor with adequate sensitivity for detecting acoustic emission (AE) during the propagation of a crack in a ferrous material. The proposed fiber optic sensor was successfully compared, in terms of the SNR (signal-to-noise ratio) and detectable AE energy levels, to commercially available AE piezo-transducers and was proven to be an effective and advantageous alternative for sensing and monitoring fatigue damage in structural applications.

In contribution 33, the authors propose a new triangle eddy current sensor array to increase the level of quantifying hole-edge crack parameters, especially the crack angle. This new sensor array consists of triangular coils instead of planar rectangular coils. The configuration of the novel sensor array, including the excitation current directions and the excitation winding shape, was optimized through simulation. The ability of the proposed sensing film to identify the crack parameters was verified using finite element simulations and experiments. The results showed that triangular coils with the same current directions in circumferentially adjacent coils and opposite current directions in axially adjacent coils achieved better performance in terms of sensor linearity and resolution compared to rectangular coils.

3. Conclusions

In this Special Issue, we selected 33 papers that address different topics related to smart sensors for structural health monitoring and nondestructive evaluation in order to delineate the state of the art and the future of this field. We hope that the selected papers may provide useful insights into the research areas of smart sensors for structural health monitoring and nondestructive evaluation, inspiring future work in these rapidly developing research fields.

We would like to thank all of the authors who contributed their work to this Special Issue, as well as all of the reviewers of the submitted papers who dedicated their time and expertise to providing high-quality suggestions and comments that allowed us to finalize a successful Special Issue.

Funding: This work was supported in part by the National Key R&D Program of China under grant no. 2022YFC3005003 and the National Natural Science Foundation of China under grant no. 12172015.

Conflicts of Interest: The author declares no conflicts of interest.

List of Contributions

1. Ju, M.; Dou, Z.; Li, J.-W.; Qiu, X.; Shen, B.; Zhang, D.; Yao, F.-Z.; Gong, W.; Wang, K. Piezoelectric Materials and Sensors for Structural Health Monitoring: Fundamental Aspects, Current Status, and Future Perspectives. *Sensors* **2023**, *23*, 543. https://doi.org/10.3390/s23010543.
2. Komary, M.; Komarizadehasl, S.; Tošić, N.; Segura, I.; Lozano-Galant, J.A.; Turmo, J. Low-Cost Technologies Used in Corrosion Monitoring. *Sensors* **2023**, *23*, 1309. https://doi.org/10.3390/s23031309.
3. Kralovec, C.; Lehner, B.; Kirchmayr, M.; Schagerl, M. Sandwich Face Layer Debonding Detection and Size Estimation by Machine-Learning-Based Evaluation of Electromechanical Impedance Measurements. *Sensors* **2023**, *23*, 2910. https://doi.org/10.3390/s23062910.
4. Xu, J.; Zhu, W.; Qiu, X.; Xiang, Y. A Novel Baseline-Free Method for Damage Localization Using Guided Waves Based on Hyperbola Imaging Algorithm. *Sensors* **2023**, *23*, 2050. https://doi.org/10.3390/s23042050.
5. Pasternak, G.; Zaczek-Peplinska, J.; Pasternak, K.; Jóźwiak, J.; Pasik, M.; Koda, E.; Vaverková, M.D. Surface Monitoring of an MSW Landfill Based on Linear and Angular Measurements, TLS, and LIDAR UAV. *Sensors* **2023**, *23*, 1847. https://doi.org/10.3390/s23041847.
6. Wu, B.; Yang, H.; Huang, Y.; Zhou, W.; Liu, X. A Novel Adaptive Time-Frequency Filtering Approach to Enhance the Ultrasonic Inspection of Stainless Steel Structures. *Sensors* **2023**, *23*, 1030. https://doi.org/10.3390/s23021030.
7. Shin, R.; Okada, Y.; Yamamoto, K. Discussion on a Vehicle–Bridge Interaction System Identification in a Field Test. *Sensors* **2023**, *23*, 539. https://doi.org/10.3390/s23010539.
8. Angeletti, F.; Iannelli, P.; Gasbarri, P.; Panella, M.; Rosato, A. A Study on Structural Health Monitoring of a Large Space Antenna via Distributed Sensors and Deep Learning. *Sensors* **2023**, *23*, 368. https://doi.org/10.3390/s23010368.

9. Ramos Silva, A.; Vaz, M.; Leite, S.; Mendes, J. Lock-In Thermal Test Simulation, Influence, and Optimum Cycle Period for Infrared Thermal Testing in Non-Destructive Testing. *Sensors* **2023**, *23*, 325. https://doi.org/10.3390/s23010325.
10. Pham, Q.-Q.; Ta, Q.-B.; Park, J.-H.; Kim, J.-T. Raspberry Pi Platform Wireless Sensor Node for Low-Frequency Impedance Responses of PZT Interface. *Sensors* **2022**, *22*, 9592. https://doi.org/10.3390/s22249592.
11. Lu, Y.; Xu, C.; Pan, Q.; Yu, Q.; Xiao, D. Research on Inherent Frequency and Vibration Characteristics of Sandwich Piezoelectric Ceramic Transducer. *Sensors* **2022**, *22*, 9431. https://doi.org/10.3390/s22239431.
12. Bednarski, Ł.; Sieńko, R.; Howiacki, T.; Zuziak, K. The Smart Nervous System for Cracked Concrete Structures: Theory, Design, Research, and Field Proof of Monolithic DFOS-Based Sensors. *Sensors* **2022**, *22*, 8713. https://doi.org/10.3390/s22228713.
13. Ren, G.; Zhan, H.; Liu, Z.; Jiang, W.; Li, R.; Liu, S. Evaluation of Axial Preload in Different-Frequency Smart Bolts by Laser Ultrasound. *Sensors* **2022**, *22*, 8665. https://doi.org/10.3390/s22228665.
14. Qu, Z.; Li, Z.; Yang, R.; Hu, S.; Wang, S. Extending the Incidence Angle of Shear Vertical Wave Electromagnetic Acoustic Transducer with Horizontal Magnetization. *Sensors* **2022**, *22*, 8589. https://doi.org/10.3390/s22228589.
15. Thibbotuwa, U.C.; Cortés, A.; Irizar, A. Small Ultrasound-Based Corrosion Sensor for Intraday Corrosion Rate Estimation. *Sensors* **2022**, *22*, 8451. https://doi.org/10.3390/s22218451.
16. Deng, F.; Zhang, X.; Yu, N.; Zhao, L. An Improved RAPID Imaging Method of Defects in Composite Plate Based on Feature Identification by Machine Learning. *Sensors* **2022**, *22*, 8413. https://doi.org/10.3390/s22218413.
17. Toutsop, B.; Ducharne, B.; Lallart, M.; Morel, L.; Tsafack, P. Characterization of Tensile Stress-Dependent Directional Magnetic Incremental Permeability in Iron-Cobalt Magnetic Sheet: Towards Internal Stress Estimation through Non-Destructive Testing. *Sensors* **2022**, *22*, 6296. https://doi.org/10.3390/s22166296.
18. Marchisotti, D.; Schito, P.; Zappa, E. 3D Measurement of Large Deformations on a Tensile Structure during Wind Tunnel Tests Using Microsoft Kinect V2. *Sensors* **2022**, *22*, 6149. https://doi.org/10.3390/s22166149.
19. Wang, D.-F.; Chuang, K.-C.; Liu, J.-J.; Liao, C.-Y. Modeling Full-Field Transient Flexural Waves on Damaged Plates with Arbitrary Excitations Using Temporal Vibration Characteristics. *Sensors* **2022**, *22*, 5958. https://doi.org/10.3390/s22165958.
20. Ptacek, L.; Strauss, A.; Bos, C.; Peyerl, M.; Torrent, R. Concrete Curing Performance Assessment Based on Gas Permeability Testing in the Lab and on Site. *Sensors* **2022**, *22*, 4672. https://doi.org/10.3390/s22134672.
21. Gao, J.; Zhang, J.; Lyu, Y.; Song, G.; He, C. Lamb Waves Propagation Characteristics in Functionally Graded Sandwich Plates. *Sensors* **2022**, *22*, 4052. https://doi.org/10.3390/s22114052.
22. Liu, Z.; Li, Y.; Ren, S.; Ren, Y.; Abidin, I.M.Z.; Chen, Z. Pulse-Modulation Eddy Current Evaluation of Interlaminar Corrosion in Stratified Conductors: Semi-Analytical Modeling and Experiments. *Sensors* **2022**, *22*, 3458. https://doi.org/10.3390/s22093458.
23. Yang, H.; Wang, B.; Grigg, S.; Zhu, L.; Liu, D.; Marks, R. Acoustic Emission Source Location Using Finite Element Generated Delta-T Mapping. *Sensors* **2022**, *22*, 2493. https://doi.org/10.3390/s22072493.
24. Li, Q.; Luo, Z.; Hu, G.; Zhou, S. A New Probabilistic Ellipse Imaging Method Based on Adaptive Signal Truncation for Ultrasonic Guided Wave Defect Localization on Pressure Vessels. *Sensors* **2022**, *22*, 1540. https://doi.org/10.3390/s22041540.
25. Mu, W.; Gao, Y.; Wang, Y.; Liu, G.; Hu, H. Modeling and Analysis of Acoustic Emission Generated by Fatigue Cracking. *Sensors* **2022**, *22*, 1208. https://doi.org/10.3390/s22031208.
26. Mo, Y.; Bi, L. TR Self-Adaptive Cancellation Based Pipeline Leakage Localization Method Using Piezoceramic Transducers. *Sensors* **2022**, *22*, 696. https://doi.org/10.3390/s22020696.
27. He, W.; Zheng, C.; Li, S.; Shi, W.; Zhao, K. Strength Development Monitoring of Cemented Paste Backfill Using Guided Waves. *Sensors* **2021**, *21*, 8499. https://doi.org/10.3390/s21248499.
28. Mu, W.; Gao, Y.; Liu, G. Ultrasound Defect Localization in Shell Structures with Lamb Waves Using Spare Sensor Array and Orthogonal Matching Pursuit Decomposition. *Sensors* **2021**, *21*, 8127. https://doi.org/10.3390/s21238127.
29. Deng, F.; Li, S.-Q.; Zhang, X.-R.; Zhao, L.; Huang, J.-B.; Zhou, C. An Intelligence Method for Recognizing Multiple Defects in Rail. *Sensors* **2021**, *21*, 8108. https://doi.org/10.3390/s21238108.

30. Mevissen, F.; Meo, M. Nonlinear Ultrasound Crack Detection with Multi-Frequency Excitation—A Comparison. *Sensors* **2021**, *21*, 5368. https://doi.org/10.3390/s21165368.
31. Datta, S.; Mukherjee, S.; Shi, X.; Haq, M.; Deng, Y.; Udpa, L.; Rothwell, E. Negative Index Metamaterial Lens for Subwavelength Microwave Detection. *Sensors* **2021**, *21*, 4782. https://doi.org/10.3390/s21144782.
32. Di Luch, I.; Ferrario, M.; Fumagalli, D.; Carboni, M.; Martinelli, M. Coherent Fiber-Optic Sensor for Ultra-Acoustic Crack Emissions. *Sensors* **2021**, *21*, 4674. https://doi.org/10.3390/s21144674.
33. Fan, S.; Yi, J.; Sun, H.; Yun, F. Quantifying Hole-Edge Crack of Bolt Joints by Using an Embedding Triangle Eddy Current Sensing Film. *Sensors* **2021**, *21*, 2567. https://doi.org/10.3390/s21072567.

References

1. Liu, Z.H.; Chen, H.L. Application and challenges of signal processing techniques for Lamb waves structural integrity evaluation: Part A—Lamb Waves Signals Emitting and Optimization Techniques. In *Structural Health Monitoring from Sensing to Processing;* Wahab, M.A., Zhou, Y.L., Maia, N.M.M., Eds.; Intechopen Limited: London, UK, 2018; pp. 61–86.
2. Liu, Z.H.; Chen, H.L. Application and challenges of signal processing techniques for Lamb waves structural integrity evaluation: Part B—Defects imaging and recognition techniques. In *Structural Health Monitoring from Sensing to Processing;* Wahab, M.A., Zhou, Y.L., Maia, N.M.M., Eds.; Intechopen Limited: London, UK, 2018; pp. 87–115.
3. Momeni, H.; Ebrahimkhanlou, A. High-dimensional data analytics in structural health monitoring and non-destructive evaluation: A review paper. *Smart Mater. Struct.* **2022**, *31*, 043001. [CrossRef]
4. Croxford, A.J.; Wilcox, P.D.; Drinkwater, B.W.; Konstantinidis, G. Strategies for guided-wave structural health monitoring. *Proc. R. Soc. A* **2007**, *463*, 2961–2981. [CrossRef]
5. Jia, J.F.; Zhang, L.G.; Ou, J.P.; Chen, X.Z. Nondestructive Testing and Health Monitoring Techniques for Structural Effective Prestress. *Struct. Control Health* **2023**, *2023*, 8940008. [CrossRef]
6. Yang, Z.Y.; Yang, H.J.; Tian, T.; Deng, D.S.; Hu, M.T.; Ma, J.T.; Gao, D.Y.; Zhang, J.Q.; Ma, S.Y.; Yang, L.; et al. A review on guided-ultrasonic-wave-based structural health monitoring: From fundamental theory to machine learning techniques. *Ultrasonics* **2023**, *133*, 107014. [CrossRef] [PubMed]
7. Cawley, P. Structural health monitoring: Closing the gap between research and industrial deployment. Struct. *Health Monit.* **2018**, *17*, 1225–1244. [CrossRef]
8. Dutta, C.; Kumar, J.; Das, T.K.; Sagar, S.P. Recent advancements in the development of sensors for the structural health monitoring (SHM) at high-temperature environment: A review. *IEEE Sens. J.* **2021**, *21*, 15904–15916. [CrossRef]
9. Vaghefi, K.; Ahlborn, T.M.; Harris, D.K. Combined imaging technologies for concrete bridge deck condition assessment. *J. Perform. Constr. Facil.* **2015**, *29*, 04014102. [CrossRef]
10. Bae, J.Y.; Lee, K.S.; Hur, H.; Nam, K.H.; Hong, S.J.; Lee, A.Y.; Chang, K.S.; Kim, G.H.; Kim, G. 3D defect localization on exothermic faults within multi-layered structures using lock-in thermography: An experimental and numerical approach. *Sensors* **2017**, *17*, 2331. [CrossRef] [PubMed]
11. Towsyfyan, H.; Biguri, A.; Boardman, R.; Blumensath, T. Successes and challenges in non-destructive testing of aircraft composite structures. *Chin. J. Aeronaut.* **2020**, *33*, 771–791. [CrossRef]
12. Giurgiutiu, V. *Structural Health Monitoring with Piezoelectric Wafer Active Sensors*, 2nd ed.; Academic Press: Columbia, NC, USA, 2014; pp. 13–32.
13. Hirao, M.; Ogi, H. *Electromagnetic Acoustic Transducers: Noncontacting Ultrasonic Measurements Using EMATs*, 2nd ed.; Springer: Tokyo, Japan, 2017; pp. 15–37.
14. Wang, Z.D.; Gu, Y.; Wang, Y.S. A review of three magnetic NDT technologies. *J. Magn. Magn. Mater.* **2012**, *324*, 382–388. [CrossRef]

sensors MDPI

Review

Low-Cost Technologies Used in Corrosion Monitoring

Mahyad Komary [1], Seyedmilad Komarizadehasl [1], Nikola Tošić [1], I. Segura [1], Jose Antonio Lozano-Galant [2] and Jose Turmo [1,*]

1 Department of Civil and Environment Engineering, Universitat Politècnica de Catalunya, BarcelonaTech, C/ Jordi Girona 1-3, 08034 Barcelona, Spain
2 Department of Civil Engineering, Universidad de Castilla-La Mancha, Av. Camilo Jose Cela s/n, 13071 Ciudad Real, Spain
* Correspondence: jose.turmo@upc.edu

Abstract: Globally, corrosion is the costliest cause of the deterioration of metallic and concrete structures, leading to significant financial losses and unexpected loss of life. Therefore, corrosion monitoring is vital to the assessment of structures' residual performance and for the identification of pathologies in early stages for the predictive maintenance of facilities. However, the high price tag on available corrosion monitoring systems leads to their exclusive use for structural health monitoring applications, especially for atmospheric corrosion detection in civil structures. In this paper a systematic literature review is provided on the state-of-the-art electrochemical methods and physical methods used so far for corrosion monitoring compatible with low-cost sensors and data acquisition devices for metallic and concrete structures. In addition, special attention is paid to the use of these devices for corrosion monitoring and detection for in situ applications in different industries. This analysis demonstrates the possible applications of low-cost sensors in the corrosion monitoring sector. In addition, this study provides scholars with preferred techniques and the most common microcontrollers, such as Arduino, to overcome the corrosion monitoring difficulties in the construction industry.

Keywords: corrosion monitoring; electrochemical sensors; data acquisition; data analysis; physical methods; structural health monitoring; low-cost sensors; atmospheric corrosion

Citation: Komary, M.; Komarizadehasl, S.; Tošić, N.; Segura, I.; Lozano-Galant, J.A.; Turmo, J. Low-Cost Technologies Used in Corrosion Monitoring. *Sensors* **2023**, 23, 1309. https://doi.org/10.3390/s23031309

Academic Editor: Zenghua Liu

Received: 15 December 2022
Revised: 17 January 2023
Accepted: 19 January 2023
Published: 23 January 2023

1. Introduction

Continuous maintenance and repair work are two vital factors to guarantee the necessary conditions of safety, economy, and functionality during the life cycle of infrastructures. However, the lack or inefficiency of these actions can significantly accelerate the deterioration of structures leading to more costly reparations. One harmful phenomenon that might jeopardize the structural integrity of structures is corrosion, which is a thermodynamically spontaneous process. In fact, the cost incurred from corrosion contributes more than the total damage caused by all natural disasters (such as earthquakes, floods, and storms) combined. The World Corrosion Organization (WCO) estimates that the annual cost of problems caused by corrosion worldwide reaches $US2.4 trillion, which is 3% of the world's gross domestic product (GDP) [1]. In many developed nations, corrosion can cause structural damages with costs approaching 3.5% to 4.5% of the gross national product (GNP) [2]. For example, on 22 November 2013, corrosion was the main reason for the oil pipeline explosion in the city of Qingdao in China's eastern Shandong province, as it thinned the pipeline's wall, leading to the break and causing a substantial economic loss of $US123.9 million [3]. In addition to economic losses, corrosion might lead to structural collapses with consequent casualties. For instance, the Guadalajara sewer explosion in Mexico in April 1992 claimed the lives of 215 people and injured another 1500 [4]. In addition, a $US75 million financial loss was estimated. The incident was linked to a gas line's degradation, which led to a gas leak into the nearby sewage main.

Moreover, corrosion is also responsible for the leakage of devastating pollution into the environment. For instance, on 12 December 1999; the tanker Erika sank off of France's Brittany coast [4]. This type of catastrophe also caused a heavy oil spill globally in 1998, about 19,000 tons [5]. Another recent accident occurred during a severe rainfall event on 14 August 2018, when a 210 m portion of Ponte Morandi collapsed. Corrosion of the concrete reinforcements was listed as one of the primary causes of the collapse [6].

These disasters could have been avoided if regular corrosion detections had been conducted. Corrosion detection, in general, is the process of evaluating the relevant types of corrosion in a particular environment and identifying the critical variables impacting the rate of propagation, which is divided into corrosion inspection and corrosion monitoring [7]. Inspection typically refers to short-term "once-off" measurements performed in accordance with maintenance and inspection schedules, whereas corrosion monitoring refers to the measurement of corrosion damage or variables that accelerate the corrosion rate over a longer time period. Corrosion monitoring frequently includes an attempt to gain a deeper understanding of how and why the corrosion rate fluctuates over time. Therefore, nowadays, corrosion monitoring has been attracting considerable attention among different industries, such as oil and gas, civil infrastructures, and the marine industry as a means to guarantee their structures' safe operation. Figure 1 illustrates the distribution of papers focused on corrosion monitoring from the SCOPUS database from the years 2004 to 2021. Analysis of this graph shows a constant interest in this field among researchers.

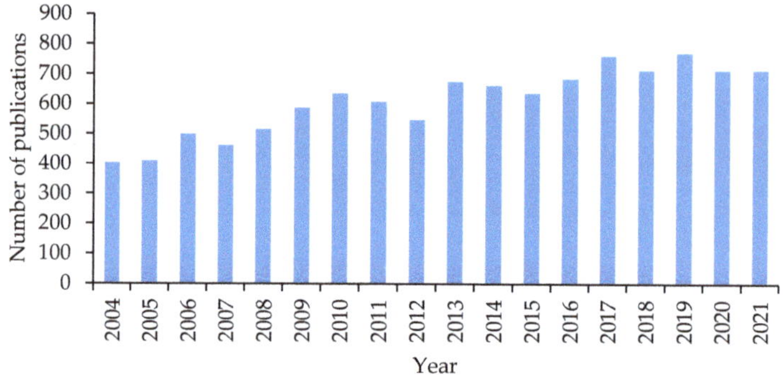

Figure 1. Distribution of publications on corrosion monitoring over time (2004–2021) in the SCOPUS database.

Corrosion monitoring is carried out using a set of detectors, data storage and transmission systems. The implemented system analyzes the detected data and assesses the corrosion state of the structure and its components [4]. Among different types of detectors, metal coupons are one of the most widely used and reliable methods for corrosion detection in different industries. Metal coupons are exposed to corrosive conditions in order to calculate the mean corrosion rate from the mass loss, mass gain, or coulometric decrease of corrosion products. However, this technique does not provide real-time measurements, and corrosion monitoring usually relies on electronic corrosion sensors or probes [8]. These electronic corrosion sensors or probes continuously or semi-continuously transmit corrosion information to the system. Afterwards, by using the proper corrosion management algorithm in the corresponding data acquisition system, not only the performance data but also a basis for life prediction of the under-study system are provided. Therefore, continued development of these electronic devices has a high level of importance, as corrosion monitoring is one of the most essential components in corrosion prevention and corrosion control. Figure 2 illustrates the scheme of a corrosion monitoring system, and, as can be seen, it contains three main parts: (1) a sensor or probe that is connected to a corrosive area in order to measure the events and changes of the environmental parameters; (2) a data acquisition system which is used to collect and transfer the data; and (3) a data processing and storing component where the collected data are analyzed and stored.

Figure 2. Scheme of a corrosion monitoring system.

Corrosion is a highly complex process involving at least two phases. For instance, solid and liquid; solid and gas; solid, liquid, and gas; or even a solid and a first and second liquid might be involved in the corrosion process. Therefore, corrosion monitoring is a multidisciplinary task, and very often, two or more methods are needed in order to adequately manage the monitoring needs in a given situation. However, because almost all corrosion processes in nature take place in the aqueous phase, electrochemical detection and monitoring procedures are considered to be more efficient compared to physical methods [9]. In addition, electrochemical techniques have several advantages, including sensitivity to low corrosion rates, a short experimental period, and a well-established theoretical understanding [10]. Nevertheless, due to their high sensitivity and accuracy in corrosion measurements, monitoring for field applications is more complex than laboratory experiments. Electrochemical techniques that are appropriate for field applications should have two key properties. On the one hand, the methods must be dependable and straightforward, which is crucial for long-term corrosion monitoring. Numerous complex electrochemical techniques exist that are effective for understanding the kinetics and mechanism of corrosion. Nevertheless, they may not be applicable to field evaluation, because they are either too slow or too complex for the typical operator to utilize in the field. On the other hand, in order to minimize the negative impact on the materials being studied, the techniques used should be non-destructive.

Depending on different natural environments, such as corrosion in soils, atmospheric corrosion, fresh-water corrosion and seawater corrosion, corrosion monitoring instances are usually considered separately, and different electrochemical or physical techniques are used. For instance, in the case of corrosion in different industries (i.e., chemical processing, oil and gas production, power generation, civil infrastructures), different electronic systems are implemented [11]. Lately, with the development of technology the demand for adopting sensors in civil infrastructures and buildings for static or dynamic structural system identification (SSI) has increased. However, in order to implement sensors to monitor the corrosion in these structures, the atmospheric conditions are normally monitored. Although there are many methods available for monitoring metal corrosion, only some of them can be applied in the atmosphere [12]. Among all of the methods available, atmospheric corrosion monitoring (ACM), electrochemical impedance spectroscopy (EIS), electrochemical noise (EN), and electrical resistance (ER) probes stand out for their unique ability to monitor the real-time data of environmental corrosiveness [13]. ACM is also known as a galvanic corrosion sensor, which is composed of two different metal electrodes insulated from each other. One of them is fashioned from a noble metal, and the other one is formed from the metal of interest in order to create galvanic corrosion, and by measuring the galvanic current using a galvanometer or a zero-resistance ammeter (ZRA), the corrosion current and corrosion rate are measured [14]. EIS is a non-destructive monitoring technique that provides information about the corrosion progress and condition of protective coatings. EIS measures the amplitude and phase of the surface impedance at different frequencies.

Traditionally, measurements are performed by employing a conventional electrochemical cell filled with an electrolytic solution so that, if a corrosion product layer coats the metallic surface, the current gets through owing to both an electronic and an ionic contribution. Since corrosion products, such as metallic oxides and hydroxides, have low electronic conductivity, their ionic contribution can become significant. Therefore, measuring the impedance at different frequencies enables the study of electrochemical reactions on the metallic surface and permits highlighting of the presence of cracks and porosity on the surface layer, both of which can act as starting points for localized corrosion attacks [15–17]. Although the use of EIS in atmospheric circumstances is constrained by the requirement for an electrolyte, it has recently been used in a number of atmospheric corrosion experiments [18,19]. Generally speaking, EN refers to the current or potential fluctuation that occurs on a corroding electrode. In this method, the current between two identical electrodes (ECN) and the potential difference between the two identical electrodes compared with the potential of a reference electrode (EPN) are measured simultaneously. The potential measurement (EPN) is related to the corrosion mechanism, whereas the current data (ECN) indicates the corrosion rate [20]. EN is the only electrochemical approach that is ideal for in situ measurements [21], straightforward to set up, and does not affect the system while conducting experiments [22]. ER probes have been widely used since the 1950s to track corrosion in soils, solutions, concrete, and atmosphere [23]. In short, a metallic sensor crafted from the same element as the materials of the structure under study is exposed to the corrosive environment and its resistance is monitored by detecting a potential drop while applying a small current. Primarily thin films are used for atmospheric systems due to their high sensitivity. The principle relies on the fact that the resistance of an element is a function of its resistivity and geometrical parameters (length, width, and thickness). Therefore, by tracking the resistance values, the values of the thickness loss due to corrosion and the corrosion depth (CD) will be acquired, and consequently, the corrosion rate, which is the slope of CD with respect to time, is monitored [5].

Table 1 provides an overview of the techniques mentioned for monitoring corrosion in the atmosphere. For a quick orientation, the sensitivity is indicated in general categories. Moreover, the main benefits and drawbacks of each method are presented.

Table 1. Comparison of atmospheric corrosion methods.

Method	Sensitivity	Benefits	Drawbacks
Coupons	Medium, unless there are long exposure times	Standardized method Simple to analyze the data	Does not have real-time updates Requires a long time for measurements
Galvanic	Medium	Not affected by temperature changes Suitably equipped for harsh outdoor conditions	Galvanic corrosion increases the corrosion rate Electrolyte connection required
EIS	Medium	Details of the mechanism of corrosion Non-destructive examination of coatings	Data interpretation requires deep knowledge of EIS Unevaluable data in the presence of a thick corrosion product Electrolyte connection required
EN	Medium	Localized detection Identification of the corrosion mechanism	Complicated data interpretation Electrolyte connection required

Table 1. *Cont.*

Method	Sensitivity	Benefits	Drawbacks
ER	High	Standardized method Very sensitive Data interpretation and operation are simple. The best option for consistent corrosion monitoring	Temperature dependent Monitoring limitations for non-uniform corrosion

The cost of commercial corrosion systems is a crucial determinant to their suitability for use in given services. Therefore, Table 2 presents a list of common corrosion monitoring devices with real-time updates available on the market for each technique, including their names, measurement ranges, and their retail prices in 2022.

Table 2. List of commercial corrosion sensors in the market for real-time update.

Technique	Device	Corrosion Rate Range (mpy)	Price ($US)	Ref.
ER	ZK9800	0.00394–39.34	1250	[24]
EIS	CST1808	0.00394–0.3934	5511	[25]
Galvanic	CST480MS	0.00394–3.934	3150	[26]
LPR	AquaMate	0.01–200	2925	[27]

As can be seen from Table 2, commercial corrosion monitoring devices are characterized by their high precision and their high price. In addition to these prices, the cost of data acquisition equipment (such as cables, a corrosion data logger, an inductance transmitter, and a corrosion management system) must be added. These needed auxiliary elements can increase the prices of the aforementioned sensors by up to an additional $US7000. Moreover, it is well established that corrosion detection and monitoring in outdoor applications is more complicated in comparison to laboratory experiments. Some areas that require ongoing development are probe/sensor design, data processing, and electrochemical instrumentation. In some circumstances, environmental noise is mixed with the observed data, thereby necessitating the use of appropriate signal processing technology to filter or even analyze this noise. In fact, because many corrosion systems are not steady-state, and as the corrosion potential may change, it is possible to apply transient techniques like electrochemical noise analysis (ENA) or other physical methods (image analysis [28–30], acoustic emission [31–34]). However, these systems are generally quite large and expensive, because they need to be equipped with complex auxiliary assets.

The high price tag on corrosion monitoring systems leads to their exclusive use with limited measuring points. These systems are typically used only in areas with the highest probability of corrosion, for example, inside petrochemical pipelines. However, in conventional structures, which are those in which many pathologies occur, the provision of corrosion monitoring systems is not usual, and insufficient attention has been paid to evaluating the progress of corrosion, especially in the case of atmospheric corrosion attacks in civil infrastructure. Additionally, this high price tag is a barrier to the long-term monitoring of structures on a low structural monitoring budget.

Therefore, the demand to use low-cost probes and open-source platforms for data acquisition systems by implementing various techniques to improve their accuracy compared to high-cost precision systems has attracted considerable attention for structural assessments [35–37]. This includes improvement such as increasing the density of measurement points as well as making it possible to implement long-term monitoring systems in order to reduce the uncertainty of the obtained data [38]. Low-cost sensor (LCS) usage for many aspects of structural health monitoring is evident in the literature due to the rapid advancement of micro sensor technology and microcontrollers (such as Arduino and Raspberry Pi) [39,40]. The monitoring of acceleration [41–43], tilt [44], temperature [45,46], humidity [47], and corrosion [19] are a few examples of this research. Komarizadehsasl et al. [48] reported that, with a proper code and the right mode to exploit the features and potential of inexpensive electronics, it is possible to obtain valuable data that can provide useful insights for structural health monitoring applications. In this work, a high-precision low-cost inclinometer was developed through the fusion of multiple microelectromechanical systems (MEMS). There are also various review papers in the literature concerning the use of low-cost sensors (LCSs) in structural health monitoring applications. Mobaraki et al. [49] reviewed the application of some low-cost sensors for the measurement of temperature, humidity, and airflow for building monitoring.

For corrosion monitoring systems, there are several literature reviews concerning different electrochemical methods used both for in situ and laboratory assessments [50], corrosion monitoring in atmospheric conditions [13], different techniques for monitoring the corrosion of steel rebar in reinforced concrete structures [51], using fiber optic sensors for the assessment of rebar corrosion in RC structures [52], and the advancement of a sensor capable of corrosion detection [53]. However, a systematic literature review on the use of LCSs and inexpensive technologies as low-cost solutions for corrosion monitoring applications is lacking. Therefore, in order to fill this gap, the current review offers a thorough, up-to-date analysis of the state of the art for using low-cost digital technologies for sustainable corrosion monitoring by examining the SCOPUS database. Additionally, the authors pay close attention to the presentation of many elements of the main fields of corrosion monitoring (such as techniques and ambient conditions) and the most popular sensors used in each technique.

The current review is divided into four sections. The primary topic of discussion in this study is the research on in situ measurements of atmospheric corrosion. The background and research significance of corrosion detection and monitoring are briefly introduced in Section 1. The rest of the paper is organized as follows. The data sampling technique for low-cost corrosion technologies is described in Section 2. In addition, the approach for processing data from inexpensive monitoring equipment is presented. The results are presented in Section 3. Specifically, Section 3.1 provides a thorough discussion of the many domains of corrosion monitoring as well as the reviewed articles. The distribution of the publications among different countries and industries are also reviewed in this section. Section 3.2 reviews the use of low-cost corrosion monitoring systems using electrochemical methods. In addition, the open-source microcontrollers discussed in the literature are evaluated, and the ones recommended for use in the installation of inexpensive monitoring devices are described. Section 3.3 describes and compares the use of low-cost technologies to detect and monitor the physical parameters induced by corrosion. The conclusions of the current literature review are presented in the final section.

2. Materials and Methods

This section describes the sources of the data, the evaluation techniques, and the classifications of low-cost technologies used for corrosion monitoring. Two main fields of corrosion monitoring in the industry are associated with (1) electrochemical methods and (2) physical methods. These two fields could provide and cover a structure's corrosion monitoring for atmospheric corrosion. Measurements made in the field are the next primary discipline of this review. This paper addresses the publications in the SCOPUS database between the years 1986 and 2021. The current review was carried out in November 2022. The initial search technique used for this systematic review paper consists of three major components. The first stage is meant to provide an initial set of articles found by employing a certain search algorithm using the Boolean operator AND (Figure 3). The obtained articles in the first stage were then filtered further in the next step in order to eliminate duplicates and irrelevant ones. Following this process 209 articles were deemed inappropriate because they lacked information on corrosion monitoring or did not clearly cover low-cost features. It shall be highlighted that in this study, the authors only considered peer-reviewed papers written in English. This exclusion was accomplished using the "AND NOT" command, and the final outcome was 38 using electrochemical methods and 52 articles implementing physical methods to track and monitor the essential parameters of corrosion across all industries.

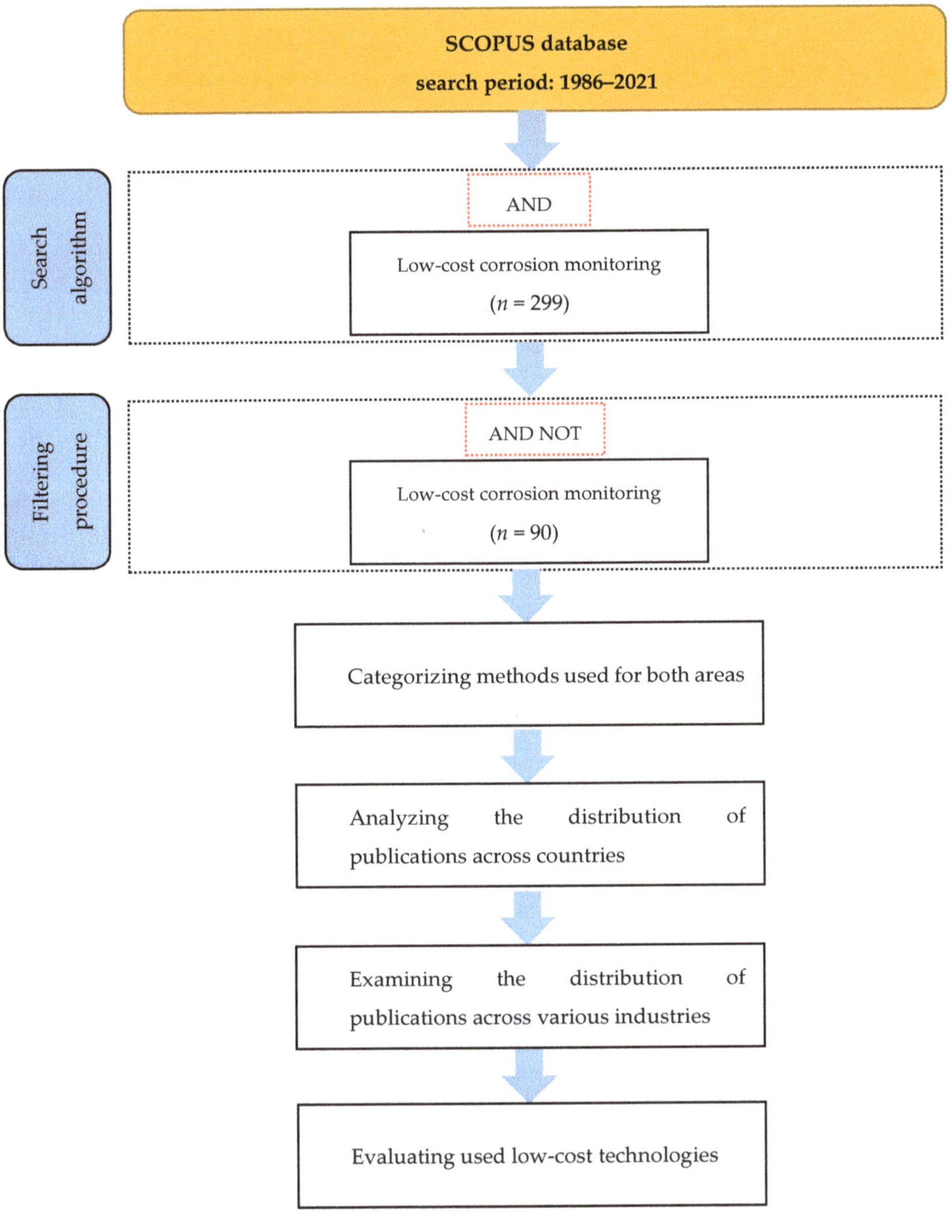

Figure 3. Flowchart depicting the systematic data sampling procedure.

In the final step, a detailed review of the selected articles was conducted in terms of defining the main methods featured in the selected articles, assessing their distribution in different countries around the world, analyzing their contribution in different industries, and finally, evaluating the types of sensors and technologies used for the installation of the low-cost corrosion monitoring system.

There are several methods available in the literature to detect corrosion and monitor its essential parameters for different type of structures. However, the principal aim of this article is to present comprehensive information about low-cost technologies implementing electrochemical methods in this field, and to do so, only corrosion detection methods that are compatible with the intended inexpensive electronics are reviewed in this paper. Therefore, this review can be used as a thorough reference for scholars interested in low-cost corrosion monitoring. Moreover, this review provides comprehensive information about low-cost technologies measuring physical parameters to track corrosion progress as well as the preferred data acquisition systems and their important parameters for use in low-cost corrosion monitoring applications. Table 3 provides a summary of the corrosion monitoring methods introduced so far in the literature and their designated categories.

Table 3. Shows methods for the two main fields of electrochemical and physical techniques.

Fields of Corrosion Monitoring	Electrochemical					Physical				
Methods used	EIS	LPR	Galvanic	EN	ER	Optical fiber	Radio frequency	Image sensor	Ultrasonic	Electromagnetic induction

Recent advances in data processing technology, along with an ever-increasing demand for online structural health monitoring, have resulted in the development of more efficient monitoring systems. One of the most essential features of current low-cost monitoring systems is that they run open-source software. This enables the users to access the library source of the low-cost sensors, thereby giving them the chance to improve and develop their own algorithms in the post-processing stage [54]. However, the vast majority of the commercial corrosion monitoring devices on the market have already been programmed by the associated companies, and they are based on closed source platforms. This implies that users have no access to inspect and develop the software, which represents a huge drawback for researchers interested in this field. Therefore, this paper provides insight to the scholars for choosing the most desirable low-cost open-source monitoring system for their project by reviewing the most-preferred options in the literature.

3. Results

3.1. Overview of the Results

The first study found in the SCOPUS database dealing with the idea of low-cost corrosion monitoring is dated to 1984 [55]. However, this paper is not included in the statistical analysis of this research, as it only introduced the use of various types of probes with portable devices to reduce the risk and the cost of the corrosion monitoring procedure in the oil and gas industry. In fact, the actual first paper that focused on the use of a low-cost corrosion monitoring device was dated to 1986. Knuutila et al. were the first researchers to describe the use of a low-cost microcomputer compatible with electrochemical systems for polarization and cyclic voltammogram measurements [56]. Since then, the number of yearly publications on the subject was limited until 2008, when a significant increase was observed (Figure 4). A second relative rebound can be seen between 2008 and 2012. It was observed that more than 54% of all the publications over a 36-year period (1986–2021) were condensed between 2012 and 2021, highlighting the importance of technology's growth on the development of new low-cost electronics. It can be seen from Figure 4 that the majority of the authors investigated low-cost approaches using physical methods (almost 58%), whereas only 42% of the articles introduced economic developments by employing electrochemical methods. In addition, it was observed that from 1986 until 2007, the number of papers associated with the use of electrochemical methods for corrosion monitoring was quite few (23.7%), and more than 76.3% of the articles were published in the last 14 years (from 2008 to 2021). This rebound in projects concentrating on the use of physical methods for corrosion monitoring applications occurred within the same amount of time as the number of papers that contributed more than 78.9% in the last 14 years, and only 21.1% of them were published in the years 1986 to 2007.

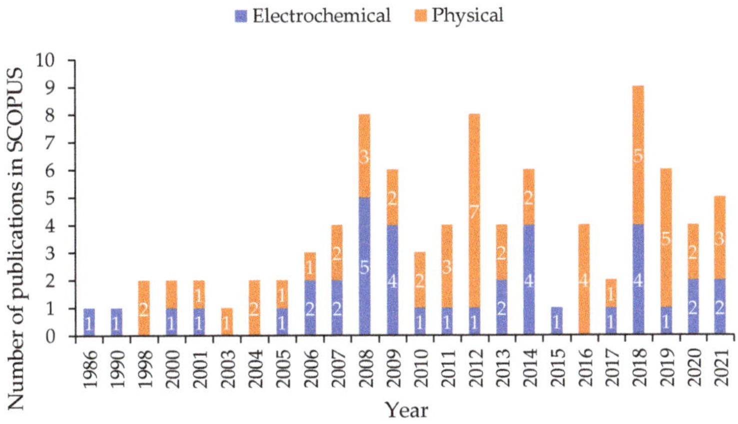

Figure 4. Distribution of publications over time (1986–2021) from the SCOPUS database.

Figure 5 illustrates the number of papers by country. It an be observed that, in terms of the publication numbers, the countries that put forth the largest contribution to the use of low-cost technologies for corrosion monitoring are the United states (43 papers), the United Kingdom (7 papers), Italy and China (6 papers each), Canada (5 papers), and Australia and Portugal (3 papers each), followed by Spain, Mexico, India, and Germany (2 papers each). In total, researchers from 26 countries around the world contributed to investigating and introducing the use of new low-cost systems in different industries in order to overcome corrosion monitoring's high price tag problem in all 90 of the reviewed publications.

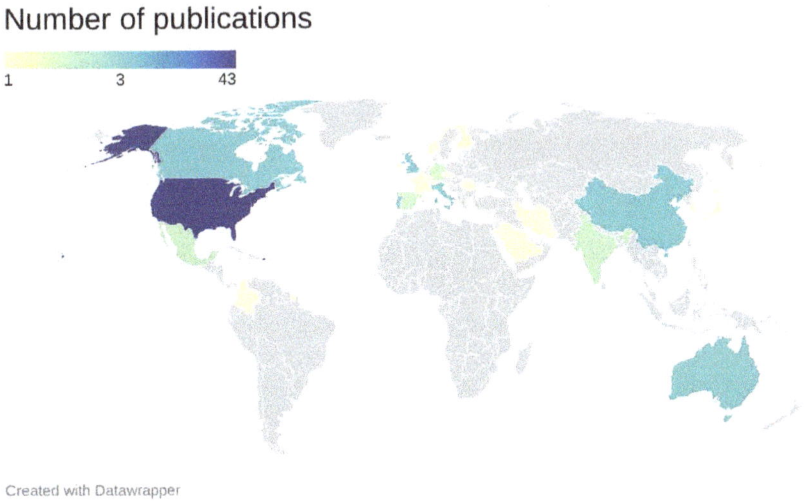

Figure 5. Distribution of the publications on low-cost corrosion monitoring by country.

Figure 6 shows the contribution of the reviewed publications to different industries. It can be seen that 51% of the researchers devoted their studies to developing low-cost solutions for monitoring the corrosion parameters of civil structures. The second most-interesting industry for corrosion monitoring using low-cost technologies is the oil and gas industry, for which the most pipeline pathologies have occurred, and 18% of the publications implemented low-cost solutions to overcome the high price tag of conventional corrosion monitoring systems in this industry. The analysis in Figure 6 shows that the monitoring of artifacts is considered to be a vital area among researchers for developing low-cost monitoring techniques for tracking and controlling the corrosion process of these valuable assets. The rest of the industries for which researchers developed low-cost corrosion monitoring systems are shown in Figure 6.

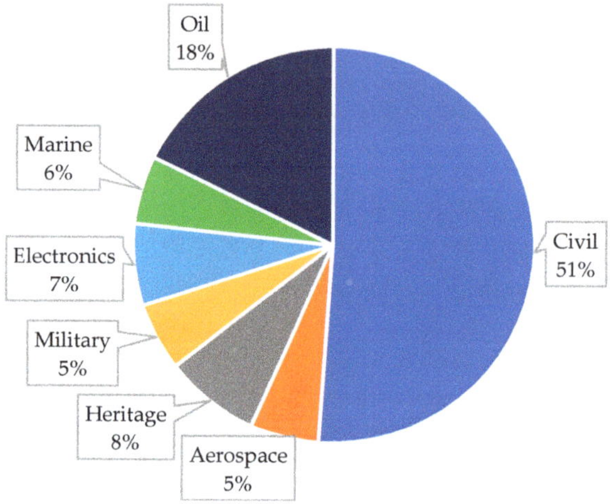

Figure 6. Contribution of the publications on low-cost corrosion monitoring by industry.

3.2. Electrochemical Methods Used for Corrosion Monitoring

In this section, the reviewed articles associated with the use of electrochemical methods for low-cost corrosion monitoring techniques are presented. This first group of papers can be classified into four separate categories. The first category is a group of publications that devoted their research to the application of EIS in the corrosion field and especially to the development of low-cost systems compatible with this method. In the second category, authors utilized linear polarization resistance (LPR)-based devices to remotely track corrosion potential in their studies. In Sections 3 and 4, Galvanic and EN methods were respectively used to measure and monitor electrochemical parameters for low-cost corrosion monitoring. Figure 7 shows the distribution of the reviewed publications in SCOPUS associated with the above-mentioned methods.

It can be seen that among the four groups, the highest number of papers dealt with the application of the EIS method for the measuring and monitoring of electrochemical parameters involved in the corrosion process. This category contained 58% of the publications from 1986 to 2021. Analysis of Figure 7 proves that EIS is considered among authors to be the most reliable method of monitoring the coating condition and the corrosion progress for their projects [57–59]. It also shows that the distribution of the publications dealing with LPR as their main method was almost constant during the years (from 2007 to 2018) with no significant changes, representing 34% of the total publications in the field of low-cost corrosion monitoring. Although EN and galvanic are considered to be useful electrochemical methods in the corrosion industry, they have not attracted the attention of researchers for low-cost monitoring applications, and they only contributed 3% and 5% (respectively) of the entire number of publications. The main reason for this low participation is their complicated data interpretation and the short life span of their sensors, respectively [13].

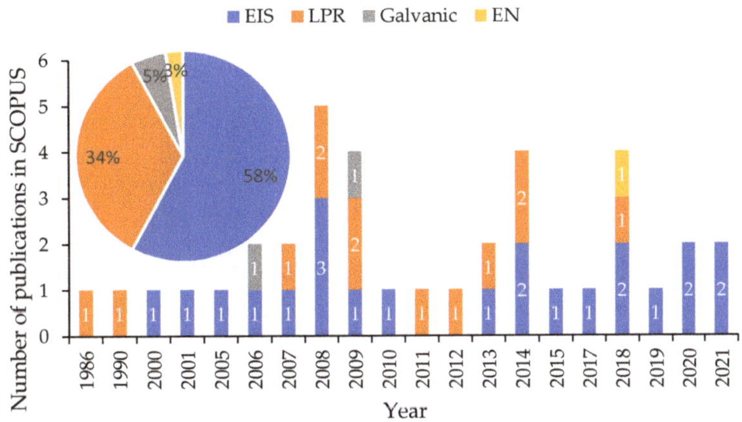

Figure 7. Distribution of the reviewed articles associated with the application of electrochemical methods for corrosion monitoring from 1986 to 2021.

3.2.1. Electrochemical Impedance Spectroscopy

This subsection provides detailed information on the reviewed publications that introduced low-cost corrosion monitoring systems based on the EIS principle. EIS is one of the most commonly used methods in the industry that enables researchers to study and monitor the coating conditions of metallic structures. However, the corrosion rate measurements produced by EIS are not so simple, and it is normally considered to be a time-consuming method due to the requirement of recording data at various frequencies [60]. In addition, EIS measurements require 3 electrodes as well as an electrolyte connection, which makes mostly limits this method to laboratory use. Therefore, attempts to develop low-cost portable EIS-based devices are considered to be highly valuable among various industries. These attempts are focused either on the development of the sensing component (low-cost electrodes or sensors) or the data acquisition implement (use of a low-cost microcontroller).

Licina et al. [61], Thien et al. [62], Park et al. [63], Yu et al. [64], Thien et al. [65], Abdur Rahman et al. [66], Mejía-Aguilar et al. [67], Azhari et al. [68], Friedersdorf et al. [69], Andrews et al. [70], Nazir et al. [57], Bansal et al. [71], Li et al. [72], Corva et al. [73], Chowdhury et al. [74], and Strachotová et al. [75] all employed new technologies (such as piezoelectric materials and advanced nanocomposites) for the sensing section of their EIS-based corrosion monitoring systems in order to obtain low-cost and low-power features. For instance, Yu et al. [64] used a piezoelectric wafer active sensor (PWAS) for the detection of coating thickness and the monitoring of the corrosion condition of metallic plates and pipes with a price of $US10 for each sensor.

Thien et al. [62] developed a low-cost active-sensing-based diagnostic system using flexible micro-fiber composite (MFC) patches as both the sensors and the actuators in order to detect cracks or corrosion damage along the pipeline structures by recording the impedance responses. Figure 8 illustrates that only 27% of the authors—such as Carullo et al. [76], Davis et al. [77], Angelini et al. [19], Matsiev [78], Grassini et al. [79], and Sebar et al. [80]—focused on the use of open-source microcontrollers (such as Arduino and AD5933) in the development of their reliable low-cost EIS-based data acquisition systems. Among the reviewed microcontrollers that are able to perform impedance measurement through various frequencies, Arduino board—which costs less than $US50—stands out for its unique benefits, such as ease of use, active user community and various available libraries due its open-source feature. Angelini et al. [19] and Grassini et al. [79] proposed a simple low-cost solution based on a commercial Arduino board coupled with a logarithmic amplifier to perform impedance measurements in the frequency range of 0.01 Hz to 100 kHz. Sebar et al. [80] proposed a novel low-cost device that measured EIS using a simple

Teensyduino board in the frequency range of 0.01 to 10,000 Hz without need of an additional analog amplifier.

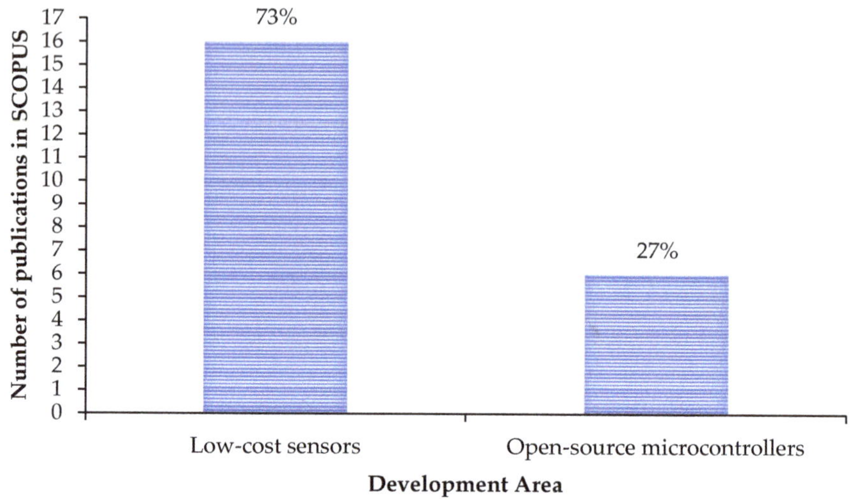

Figure 8. Two areas of development in the SCOPUS database for the application of low-cost corrosion monitoring solutions using the EIS technique: Area 1 focusing on the development of the sensing elements, and Area 2 focusing on the use of open-source microcontrollers as the low-cost data acquisition system.

3.2.2. Linear Polarization Resistance

LPR is one of the most commonly used methods for corrosion monitoring in field applications. In this method, a very small polarization potential (usually $\pm$ 30 mV) above and below the corrosion potential is applied to the sample. The current response recorded over this short range near the corrosion potential is linear. As a result, the polarization resistance (R_P) is defined as the slope of this current–potential curve, which is constant. Under certain conditions R_P is inversely related to the instantaneous corrosion rate, according to the Stern–Geray equation [81]. In fact, polarization techniques such as EIS and LPR are considered highly accurate with easy data interpretation. Hence, many investigators have tried to develop low-cost monitoring solutions through the use of either open-source microcontrollers or developed sensing elements. In Figure 9 it is shown that 54% of the publications that implemented LPR method focused the development of their data acquisition system on using open-source microcontrollers, while 46% of the articles addressed the use of low-cost sensors.

For instance, Niblock et al. [82], Wood et al. [83], Abu Yosef et al. [84], Rafezi et al. [85], Perveen et al. [86], and Ahmad et al. [87] employed low-cost passive wireless sensors such as piezoelectric sensors for the detection of corrosion in pipelines and in concrete by taking advantage of the LPR technique. At the same time, Knuutila et al. [56], Arpaia et al. [88], Mugambi et al. [89], Kwan et al. [90,91], Samoilă et al. [92], and Degrigny et al. [93] devoted the main objective of their research to developing open-source hardware systems with low-cost and easy-to-use instrumentation for in-field measurements. For example, Samoilă et al. [92] presented virtual instrument technology made up of a low-cost but powerful Cypress PSoC microcontroller based on the polarization resistance technique.

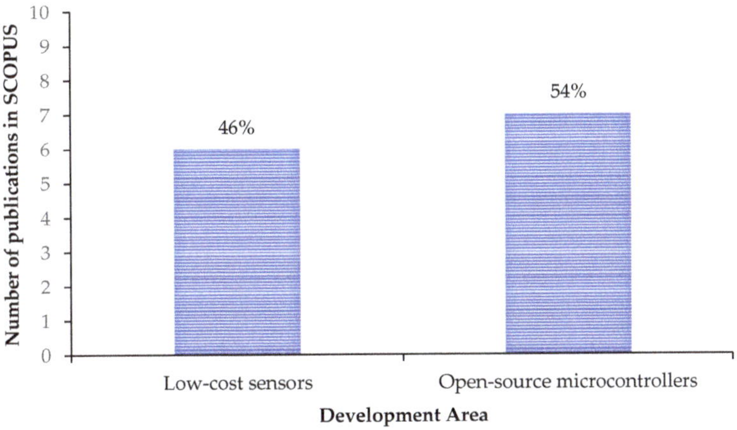

Figure 9. Two areas of development in the SCOPUS database on the application of low-cost corrosion monitoring solutions using the LPR technique: Area 1 focusing on the development of the sensing elements, and Area 2 focusing on the use of open-source microcontrollers as the low-cost data acquisition system.

3.2.3. Galvanic or Atmospheric Corrosion Monitoring

In the late 1950s, the first attempts to quantify atmospheric corrosion using galvanic sensors were made. An ACM system is composed of two different metal electrodes insulated from each other. One of them formed from is a noble metal, while the other is crafted from the metal of interest in order to create galvanic corrosion, and by measuring the galvanic current using a galvanometer or a zero-resistance ammeter (ZRA), the corrosion current and corrosion rate are measured. Figure 10 shows a schematic illustration of an ACM sensor used to monitor automotive corrosion. However, due to the fact that galvanic coupling accelerates the corrosion process, which results in a shorter life span of the sensors and unclear data interpretation in harsh environments, researchers have not found this technique interesting for the development of low-cost monitoring solutions. Nevertheless, Castro-Borges et al. [94] proposed a reasonable, low-cost solution for the detection and monitoring of the corrosion performance of concrete columns after localized repairs using galvanic sensors to measure the electrochemical parameters. Mierau [95] proposed a remote monitoring solution for corrosion monitoring at airports, solving the access and field-diagnostic issues. The proposed low-cost PC-based control system, which employs galvanic sensors, provides corrosion engineers with a remote monitoring system with real-time updates.

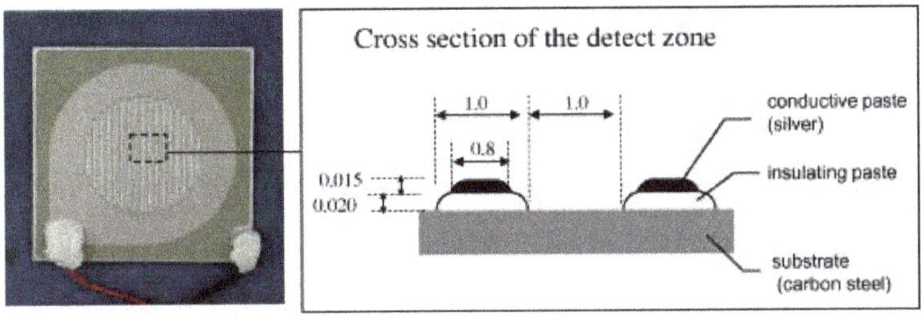

Figure 10. Fe–Ag-type atmospheric corrosion monitor (ACM) sensor (reproduced with permission from Reference [26]).

3.2.4. Electrochemical Noise

EN is a useful electrochemical method for investigating and monitoring corrosion processes such as pitting and stress corrosion cracks' growth, microbial corrosion, and uniform corrosion. The measuring mode, the surface area of the working electrodes, the electrolyte resistance, and the symmetry of the electrode system all have a significant impact on the EN results. In addition, depending on the measurement mode, this technique can be used for in situ measurements employing either three-electrode or two-electrode setups [96]. Although the EN method has numerous benefits, such as localized corrosion detection because of its complex and unclear data interpretation, very little attention has been paid to using it to implement new technologies for low-cost monitoring developments. However, Arellano-Pérez et al. [27] developed a portable device for measuring and monitoring corrosion parameters such as electrochemical potential (EP) and electrochemical current (EC) based on EN signals. In this research, the corrosion type and rate, a probe made of three identical electrodes (6061-T6 aluminum) was constructed in order to evaluate. Furthermore, the proposed low-cost portable device includes a signal amplifier, a physical filter, and an analog-to-digital convertor (ADC), which enables it to measure weak signals and analyze the corrosion variables in different materials. Figure 11 illustrates both the three-electrode probe and the proposed low-cost portable device.

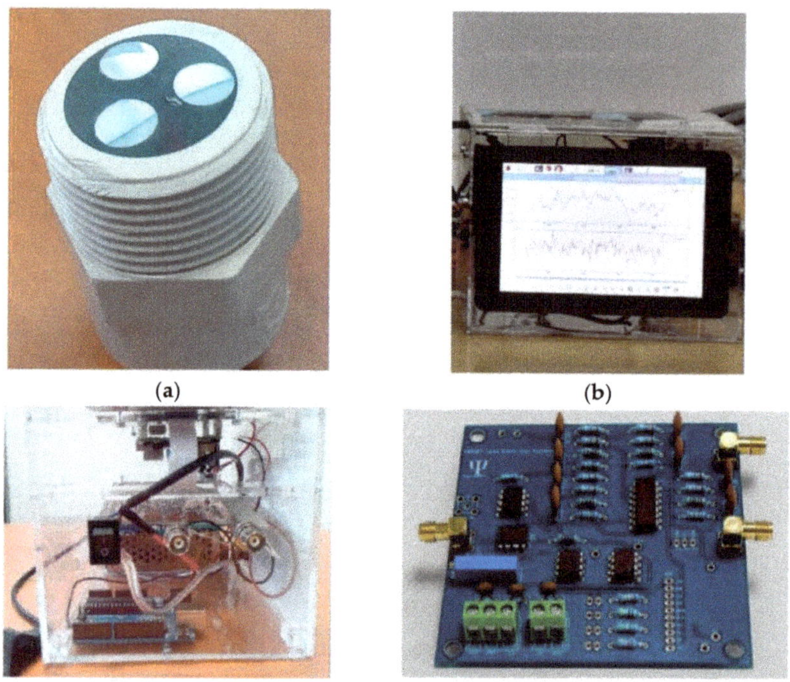

Figure 11. The developed low-cost EN-based corrosion monitoring system: (**a**) three-electrode probe made of aluminum; (**b**) touchscreen display for communication and graphical interface; (**c**) on–off switch and connectors of the portable device; (**d**) the electronic circuit board for the signal amplification (reproduced with permission from Reference [27]).

3.3. *Physical Methods Used for Corrosion Monitoring*

In this section low-cost approaches for corrosion monitoring implementing physical methods are reviewed. Controlling and tracking the physical parameters of the corrosion progress—such as metal thickness loss, crack propagation, and physical pathologies—are the main features of this method [97]. Unlike the electrochemical methods, these techniques do not directly measure the corrosion parameters such as corrosion potential or corrosion current. However, depending on the requirements of the projects, these techniques provide the possibility of measuring and tracking the physical changes caused by corrosion, and furthermore, calculating the corrosion rate. Therefore, in this section, the existing low-cost technologies for the measurement of the physical changes to structures caused by corrosion are reviewed. The reviewed publications are categorized into six distinct groups. The first category is a group of researchers that used ER as their principal method of measuring and monitoring corrosion rate using low-cost electronics. The second group of authors profited from radio frequency sensors (RF) and their unique features in their respective projects aimed at developing inexpensive remote corrosion monitoring systems. In the third group are authors who employed optical fiber (OF) sensors as their principal physical detection method for low-cost corrosion monitoring applications. The fourth group of authors focused on adapting ultrasonic waves (UW) to introduce their novel low-cost remote technologies for monitoring the coating and internal corrosion condition of structures in different industries. Authors from Group 5 implemented the electromagnetic induction (EMI) method to track physical changes using passive wireless sensors. Finally, in Group 6, authors addressed a novel technique taking advantage of image recognition (IR) technology to develop low-cost image sensors for tracking the progression of corrosion in their projects.

Figure 12 illustrates the rise in the number of publications associated with low-cost corrosion monitoring using physical methods that were published over the period from 1998 to 2021 and found in the SCOPUS database. It can be seen that researchers chose a more balanced distribution of physical methods suitable for low-cost corrosion monitoring applications as compared to the electrochemical approaches mentioned before. Nevertheless, among the six groups, the majority of the authors employed ER sensors and RF-based methods to detect physical changes (such as strain and thickness loss) and monitor them using inexpensive wireless technologies. These two groups contain 25% and 23% of the publications from 1998 to 2021, respectively. Analysis of Figure 12 shows that 75% of the articles focusing on the use of RF-based sensors were published in the past 10 years. At the same time, the least attention was paid to the use of IR for detecting and tracking structural pathologies. This technique has only been investigated in publications in recent years, with 100% of the articles having been published in the last 6 years. In fact, the developments in camera sensors and artificial intelligence (AI) technology in recent years have made them more accessible and thus more affordable for the development of low-cost corrosion monitoring systems [98]. Figure 12 also illustrates that the distribution of publications employing OF sensors and UW for low-cost corrosion monitoring solutions has been almost constant through across the years 1998 to 2021. However, low-cost monitoring solutions based on the EMI technique have been only reviewed by authors over a 10-year period between 2006 and 2016.

3.3.1. Electrical Resistance

Low-cost sensors based on the ER principle have been widely utilized to monitor corrosion in various industries since 1998. As shown in Figure 13, these sensors generally contain two parts: a reference element, which is protected by a coating layer, and an exposed element. The resistance of the sensing element is measured by comparing its potential drop to the reference element, which is proportional to the loss in metal thickness experienced by the sensing element for a uniform corrosion mechanism. Furthermore, these values are used to calculate the accurate corrosion rate over time. Due to the simplicity and high precision of this technique, many scholars such as Taylor [99], Hautefeuille et al. [100], McCarter et al. [101], Holst et al. [102], Materer et al. [103], Halabe et al. [104], Hurley et al. [105], and Corva et al. [106] have introduced the use of low-cost ER sensors in their projects. Stromment et al. [107] introduced FSM (the field signature method) for the first time in 1998, which is an economical solution based on the ER principle for internal corrosion and crack monitoring in pipelines and vessels. In that study the authors combined non-destructiveness, high sensitivity, and real-time updateability with the ability to cover large areas and no maintenance requirements as the main features of their technique in order to develop their low-cost corrosion monitoring system. Andringa et al. [108,109] and Pasupathy et al. [110,111] demonstrated in their publications the use of sacrificial transducers as sensing elements for monitoring corrosion in concrete. This passive wireless sensor, which is ideal for in situ measurements in inaccessible locations, contains a resonant inductor–capacitor circuit with a resistive transducer that is exposed to the corrosive ambient in order to perform resistance measurement based on the ER technique.

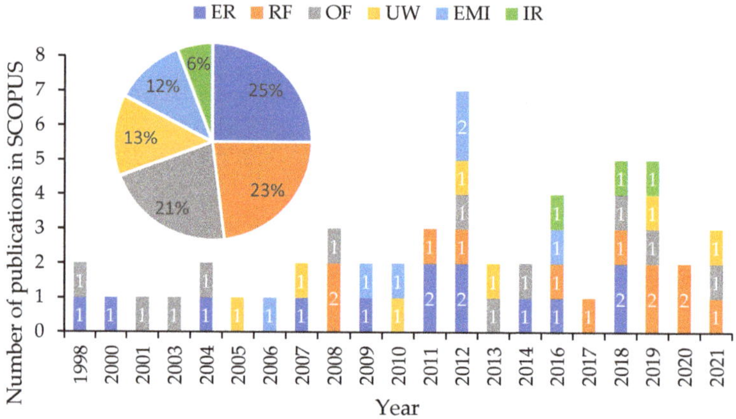

Figure 12. Distribution of the reviewed articles associated with the application of physical methods for corrosion monitoring from 1986 to 2021.

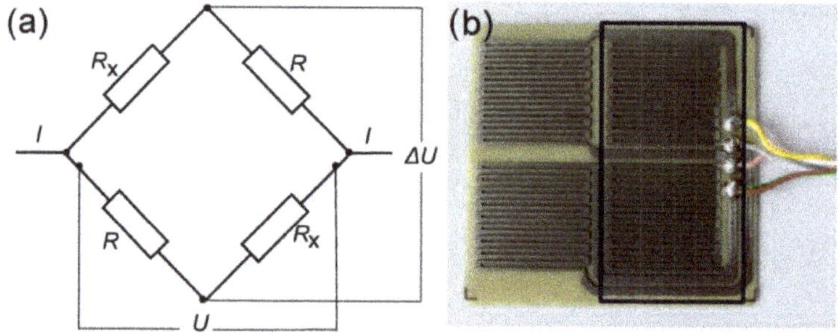

Figure 13. An ER sensor made of high-strength steel with four elements: (**a**) a schematic illustration of the sensor; (**b**) a photo of the sensor, protected by a transparent epoxy coating (reproduced with permission from Reference [24]).

3.3.2. Radio Frequency

Radio-frequency identification (RFID) sensors are generally employed for inventory management and product tracking, and their application for corrosion monitoring is a relatively new concept established in the previous decade. However, in recent years numerous authors have addressed applications of this method in the field of corrosion monitoring. Bouzaffour et al. [112] introduced an innovative embedded UHF RFID (ultra-high-frequency RFID) sensor in concrete in order to detect the mass loss of the steel and monitor its corrosion process. El Masri et al. [113] addressed the sensitivity feature of the UHF RFID sensors and investigated their potential as indoor environmental corrosiveness detectors sensitive to variation of the metal thickness loss in the range of tens of nanometers. One of the controversial issues for corrosion monitoring sensors is the interpretation of the output data when localized corrosion phenomena appear. Yasri et al. [114] proposed a novel RF-based technique for atmospheric corrosion monitoring, the results of which showed the ability to distinguish between uniform and localized corrosion mechanisms. Another important usage for this technique within corrosion monitoring applications is in tracking the condition of the coating layer and monitor its degradation process [115]. This ability allows for the improvement of infrastructure maintenance operations and can result in large financial savings in many industrial fields. Generally, the basic components of an RFID system are a reader and a tag sensor. The tag is a transponder with integrated circuits and an antenna that can be activated by the reader's electromagnetic radiation, and it communicates stored information back. Figure 14 shows a detailed view of a commercial RFID system´s scheme for crack detection. Depending on the power supply, tags can be active or passive. Active tags can convey data and have their own power sources (such as a battery or solar cell), making them larger and more expensive. Passive tags convey data in response to reader inquiry, since their only power source is the electromagnetic field transmitted by the reader. These tags are simpler, less expensive, more dependable, and more widely utilized. Therefore, for low-cost corrosion monitoring applications, passive wireless RFID sensors are predominantly used. In the publications reviewed in this study, Lippincott [116], Dante et al. [117], Palmer et al. [118], He et al. [119], Zhang et al. [120], Sunny et al. [121], Bruciati et al. [122], and Marindra et al. [123] addressed the use of low-cost passive wireless RFID sensors (tags) in order to detect physical parameters such as coating thickness and crack propagation which are essential corrosion elements in pipelines and concrete structures. It is important to mention that in the aforementioned studies, real-time corrosion monitoring and corrosion rate calculation were not performed. However, Zhang et al. [124], Zarifi et al. [125], Deif et al. [126], and Rioual et al. [127] developed RFID-based low-cost corrosion monitoring systems with real-time update ability.

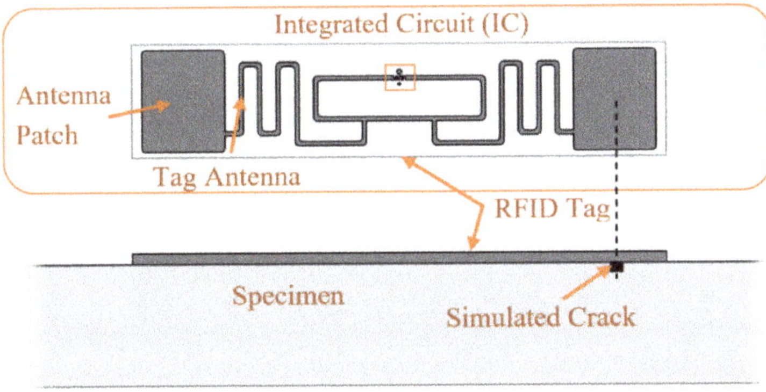

Figure 14. The commercial UHF RFID tag ALN-9662 (reproduced with permission from Reference [122]).

3.3.3. Optical Fiber

Over the last few decades, optical fibers have been developed and widely used for a variety of applications in the oil and gas sector; civil engineering construction; structural health monitoring; and cracks, strain, temperature, and pH monitoring. Essentially, the approach employs optical fibers composed of glass or plastic that convey light along their length and are thus sensitive to any parameter that can change the intensity, frequency, polarity, or phase of the light. For corrosion monitoring, optical fiber sensors such as fiber Bragg gratings (FBG), interferometers, surface plasmon resonances (SPR), distributed sensing, or optical intensity modulations are used. As can be seen in Figure 15, optical fiber sensors for low-cost corrosion monitoring applications are categorized into four types based on how they detect the presence of corrosion in the structure. The first group employs the direct measurement of corrosion effects. For that purpose, sensors measure parameters such as corrosion-induced strain, deformation, and displacements within the structure. FBGs, white light interferometry, and optical time domain interferometry (OTDR)—a technology within optical distributed sensing—are among the optical sensors used in this category. These sensors are implanted in the building's structure in order to monitor the corrosion in concrete structures as well as in civil and military airframes where corrosion is a worry because they are working beyond their projected lifetime in some circumstances.

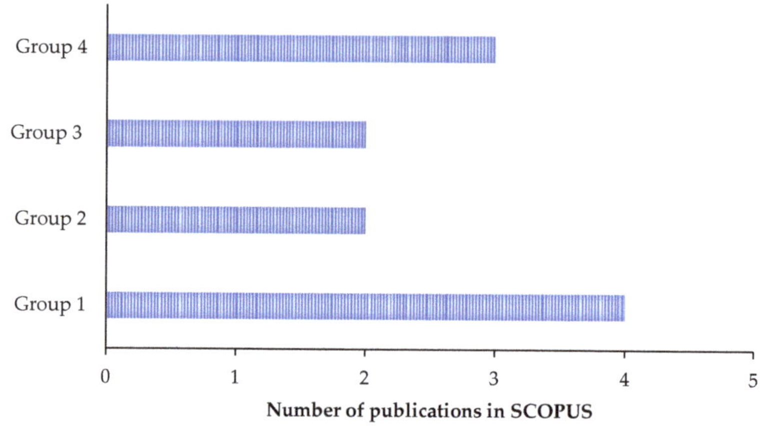

Figure 15. Distribution of publications in the SCOPUS database between four areas of corrosion measurement using optical fiber sensors.

Among the authors who developed low-cost corrosion monitoring solutions based on optical fiber sensors, Zhao et al. [97] implemented a white light interferometer, Ramani et al. [128] addressed the use of lens-based plastic optical fibers (LPOF), and Dos Santos et al. [129] and Chen et al. [130] employed long-period grating (LPG) sensors to detect and monitor parameters such strain and corrosion crack propagation in their projects. In the second group, the corrosion of the structure of interest was estimated rather than directly measured by monitoring the corrosion of a similar substance deposited onto an optical fiber. Obviously, this sensor must be situated near the structure whose corrosion is being investigated in order to experience identical conditions. Wan et al. [131] and Guan et al. [132] employed this approach to monitor corrosion in metals such as steel and aluminum as well as for the monitoring of coating layer thickness. In terms of corrosion precursors, optical fiber sensors can detect ions that cause a higher corrosion rate, with waterborne chlorides being the most relevant. Elster et al. [133] and Chen et al. [134] used fluorescence-based optical fiber sensors to detect chlorides ions inside of concrete structures.

Environmental parameters such as humidity and pH play an important role in corrosion, therefore optical fiber sensors such as fluorescence or colorimetry are widely used for monitoring these elements. Cooper et al. [135], Casas et al. [136], and Kiremidjian et al. [137] developed novel low-cost systems based on the combination of type 1 and type 4 measurements using LPG sensors in order to detect both the strain and environmental parameters and therefore obtain better corrosion data on the analyzed structure.

3.3.4. Ultrasonic Wave

Ultrasonic corrosion monitoring is a type of non-destructive testing (NDT) that is used to track the corrosion process. High-frequency acoustic waves (sound waves) are used in this monitoring technique to measure or map the interior structure, thickness, and other properties of the material being monitored. There are numerous commercial ultrasound instruments available that provide useful information regarding various types of degradation (i.e., cracks, deformations, thinning, and corrosion). They are, however, generally classified as inspection equipment rather than monitoring devices. Therefore, in this this paper, publications that presented low-cost corrosion monitoring solutions for various industries based on ultrasound technology are reviewed. For instance, in the aerospace industry, the use of ultrasonic sensors is very common. Zhao et al. [138], Meyendorf et al. [139], and Perelli et al. [140] introduced novel low-cost monitoring systems based on ultrasound technology for aircraft´s health monitoring applications. In the oil industry, in order to track the health of the pipelines, Wan et al. [141] and Barshinger et al. [142] developed low-cost wireless ultrasound-based monitoring devices. Harper et al. [143] and Jalali et al. [144] proposed remote ultrasound-based techniques for corrosion monitoring in civil structures. The aforementioned corrosion monitoring systems take highly frequent readings in order to detect minor variations, which in this case are connected to corrosion phenomena. The accuracy of ultrasound signals enables very frequent thickness measurements that can be used to monitor corrosion.

3.3.5. Electromagnetic Induction

Over the last few years, a novel corrosion control method has been created for protecting damaged, painted surfaces in contact with atmospheric conditions, and it employs electromagnetically induced surface currents. In addition, this technique has attracted considerable attention for the initial monitoring of corrosion in concrete structures. Several scholars have introduced monitoring sensors based on Faradays law of electromagnetic induction (EMI) for the early assessment of corrosion in reinforced concrete (RC) structures. In this review, authors such as Andringa et al. [145], Vinogradov [146], Chen et al. [147,148], and Kranz et al. [149] introduced low-cost corrosion monitoring solutions based on this technique. The monitoring devices used in the aforenamed experiments are made up of multiple loop coils (MLCs) which operate as sensors and have a receiver and a transmitter coil. The positioning of the two coils on concrete structures measures the potential difference, which is then used to analyze the extent of the corrosion. Gallo et al. [150] introduced a novel corrosion monitoring method based on the magnetic field induced by the electrochemical activity from a uniform corrosion mechanism of a metal sample. Figure 16 shows a scheme of the monitoring system containing a two-unit set of commercial giant magneto-resistive (GMR) sensors for tracking the magnetic fields from the active corrosion of metal plates.

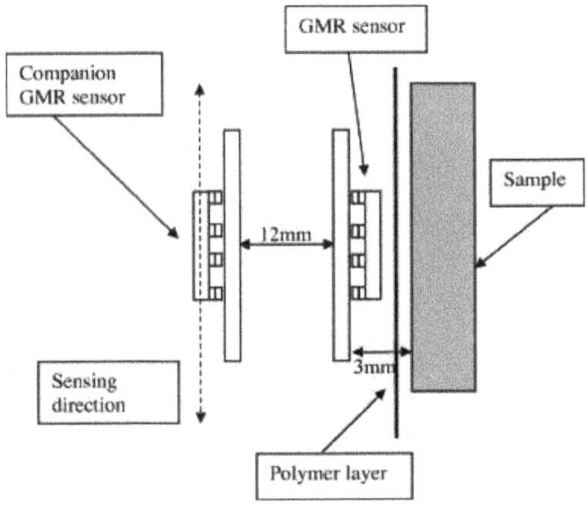

Figure 16. Scheme of the side view of the sensor setup (reproduced with permission from Reference [150]).

3.3.6. Image Recognition

Human-based visual inspection, which is time-consuming, costly, and even dangerous, has so far been one of the most-used techniques for structural health monitoring applications. Concerns about safety, high costs, as well as recent developments in image detection sensors and AI technology have highlighted the urge for the development of a reliable, low-cost, safe, and quantitative solution known as image recognition (IR) or image analysis technology. In fact, the phenomenon of corrosion leads to changes in the surface structure, morphology, and composition of materials [151]. These changes can be captured by a number of image sensors, and by implementing image recognition technologies, information about the type and extent of corrosion can be obtained [152]. Figure 17 illustrates a simple image-collection setup using a portable CCD camera. Xia et al. [153] presented a comprehensive review of the digital image analysis systems as well as the developed algorithms used to assess and process the corrosion images. In addition, some physical and electrochemical methods that can be employed to support the IR results are summarized in this study. However, this technique is still considered to be new for low-cost corrosion monitoring applications, and it has not yet been widely investigated. Among the reviewed publications, Igoe et al. [154] addressed the use of smartphone cameras to detect and monitor the corrosion process on the surface of a clean iron sample. Smartphone technology offers numerous prospects for increased participation in scientific and technological research. In fact, in this research, congruent validation tests and errors of less than 5% demonstrated the efficiency of this method for the characterization of a smartphone image sensor response to the degree of steel corrosion. These findings show that a smartphone can be used to evaluate surface corrosion efficiently and at a low cost. Moreover, Jindal et al. [155] proposed the use of a panoramic image transmission and refinement technique in order to overcome difficulties such high time consumption, blurred data, and color variations that are common problems in underwater pipeline monitoring applications. The proposed method was validated efficiently in terms of energy consumption and communication, and it showed maximum data delivery at a low-cost compared to conventional techniques. A more recent development in the field of corrosion monitoring using image sensors was produced by Benkhoui et al. [156]. In this study the authors introduced the use of an unmanned aerial vehicle (UAV) for crack detection in a bridge and investigated the adoption of stereo cameras based on passive and active depth calculation for the assessment of bridge structural integrity.

Figure 17. Collection of corrosion images of metallic materials in the field using a portable CCD camera: (**a**) image collection, (**b**,**c**) corrosion images of 304 SS exposed to a marine atmosphere (reproduced with permission from Reference [153]).

4. Conclusions

In this paper, a systematic literature review of low-cost technologies for corrosion monitoring applications has been presented. After the filtering process, 90 articles published since 1986 were examined that address the use of low-cost technologies such as sensors and microcontrollers in the corrosion monitoring field. According to the reviewed articles, this paper could provide the readers with an overview of the low-cost technologies that have been implemented so far for the monitoring of corrosion parameters using electrochemical methods (EIS, LPR, galvanic, and EN) and physical methods (ER, RF, OF, UW, EMI, and IR). In the case of using electrochemical methods as the principal measurement technique, it was found that the majority of the authors (58%) focused their studies on the use of EIS, as it turned out to be the most reliable and feasible method for corrosion monitoring applications. In the case of physical methods, ER has proven to be the most reliable method, indicating that 25% of the scholars were focused on development of their respective low-cost corrosion monitoring systems based on this technique. Regarding the architecture of the low-cost data acquisition system, it has been discovered that the Arduino microcontroller has attracted a lot of attention for EIS and ER measurements. Moreover, among the physical methods, it was found that RFID and OF sensors were the two second-most-reliable areas in which the majority of the researchers focused the development of their respective low-cost corrosion monitoring systems (23% and 21%, respectively). Furthermore, it has been observed that most of the reviewed articles were published in the US (43 out of 90), and the majority of the projects were established in the civil structure industry (46 out of 90).

This article shows that, despite the fact that a significant number of studies have been reviewed with the goal of employing low-cost technologies for corrosion monitoring, there are still some gaps in the literature that must be addressed in future research. The same systematic literature research should be conducted using other databases (such as IEEE explore or Google Scholar) in order to give a full analysis of the used low-cost solutions for corrosion monitoring. It is also necessary to provide an overview of the existing low-cost sensors for corrosion monitoring in the market and compare their features with the those in the literature.

Author Contributions: Conceptualization, M.K., S.K. and J.A.L.-G.; methodology, M.K. and S.K.; software, M.K.; validation, M.K., J.A.L.-G., N.T. and J.T.; formal analysis, M.K.; investigation, M.K.; resources, J.T.; data curation, M.K.; writing—original draft preparation, M.K.; writing—review and editing, N.T., J.A.L.-G. and J.T.; visualization, M.K.; supervision, J.T. and I.S.; project administration, M.K.; funding acquisition, J.T., I.S. and J.A.L.-G. All authors have read and agreed to the published version of the manuscript.

Funding: The authors are indebted to the projects PID2021-126405OB-C31 and PID2021-126405OB-C32 funded by FEDER funds—A Way to Make Europe and Spanish Ministry of Economy and Competitiveness MICIN/AEI/10.13039/501100011033/, project PID2019-106555RB-I00 and project IDEAS 2.14 from Ports 4.0. It should also be noted that funding for this research was provided for Seyedmilad Komarizadehasl by the European Social Fund and the Spanish Agencia Estatal de Investigación del Ministerio de Ciencia Innovación y Universidades, grant (PRE2018-083238).

Institutional Review Board Statement: Not applicable.

Informed Consent Statement: Not applicable.

Data Availability Statement: Not applicable.

Conflicts of Interest: The authors declare no conflict of interest.

References

1. Scheuer, C.; Boot, E.; Carse, N.; Clardy, A.; Gallagher, J.; Heck, S.; Marron, S.; Martinez-Alvarez, L.; Masarykova, D.; Mcmillan, P.; et al. Statistical modelling of pitting corrosion: Extrapolation of the maximum pit depth-growth. *Int. J. Electrochem. Sci.* **2014**, *9*, 4129–4143. [CrossRef]
2. Zajec, B.; Bajt Leban, M.; Kosec, T.; Kuhar, V.; Legat, A.; Lenart, S.; Fifer Bizjak, K.; Gavin, K. Corrosion Monitoring of Steel Structure Coating Degradation. *Teh. Vjesn.* **2018**, *25*, 1348–1355. [CrossRef]
3. Ding, H.; Lan, J.; Yao, S.; Zhang, D.; Han, B.; Pan, G.; Li, X. Evolution of polycyclic aromatic hydrocarbons in the surface sediment of southern Jiaozhou Bay in northern China after an accident of oil pipeline explosion. *Mar. Pollut. Bull.* **2022**, *183*, 114039. [CrossRef]
4. Yang, L. *Techniques for Corrosion Monitoring*, 2nd ed.; Elsevier: Amsterdam, The Netherlands, 2020; ISBN 9780081030042.
5. Li, S.Y.; Kim, Y.G.; Jung, S.; Song, H.S.; Lee, S.M. Application of steel thin film electrical resistance sensor for in situ corrosion monitoring. *Sens. Actuators B Chem.* **2007**, *120*, 368–377. [CrossRef]
6. Romano, N.D. *Applying FIDIC Contracts in Italy*; Informa Law from Routledge: Boca Raton, FL, USA, 2022; ISBN 9781003201236.
7. Roberge, P.R. *Corrosion Inspection and Monitoring*; John Wiley & Sons: Hoboken, NJ, USA, 2007; ISBN 9780470099766.
8. Prosek, T.; Kouril, M.; Dubus, M.; Taube, M.; Hubert, V.; Scheffel, B.; Degres, Y.; Jouannic, M.; Thierry, D. Real-time monitoring of indoor air corrosivity in cultural heritage institutions with metallic electrical resistance sensors. *Stud. Conserv.* **2013**, *58*, 117–128. [CrossRef]
9. Song, S.Z.; Zhao, W.X.; Wang, J.H.; Li, J.; Gao, Z.M.; Xia, D.H. Field Corrosion Detection of Nuclear Materials Using Electrochemical Noise Techinique. *Prot. Met. Phys. Chem. Surf.* **2018**, *54*, 340–346. [CrossRef]
10. Yang, L. (Ed.) *Techniques for Corrosion Monitoring*; Woodhead Publishing: Cambridge, UK, 2008; ISBN 9781845691875.
11. Cramer, S.D.; Covino, B.S., Jr. *Corrosion: Environments and Industries*; Cramer, S.D., Covino, B.S., Eds.; ASM International: Almere, The Netherlands, 2006; Volume 13C, ISBN 978-1-62708-184-9.
12. Xia, D.-H.; Song, S.; Qin, Z.; Hu, W.; Behnamian, Y. Electrochemical Probes and Sensors Designed for Time-Dependent Atmospheric Corrosion Monitoring: Fundamentals, Progress, and Challenges. *J. Electrochem. Soc.* **2020**, *167*, 037513. [CrossRef]
13. Qiao, Y.; Wang, Z.; Popova, K.; Prošek, T. Corrosion Monitoring in Atmospheric Conditions: A Review. *Metals* **2022**, *12*, 171. [CrossRef]
14. Ahn, J.H.; Jeong, Y.S.; Kim, I.T.; Jeon, S.H.; Park, C.H. A Method for Estimating Time-Dependent Corrosion Depth of Carbon and Weathering Steel Using an Atmospheric Corrosion Monitor Sensor. *Sensors* **2019**, *19*, 1416. [CrossRef]
15. Kendig, M.; Scully, J. Basic Aspects of Electrochemical Impedance Application for the Life Prediction of Organic Coatings on Metals. *Corrosion* **1990**, *46*, 22–29. [CrossRef]
16. Bonora, P.L.; Deflorian, F.; Fedrizzi, L. Electrochemical impedance spectroscopy as a tool for investigating underpaint corrosion. *Electrochim. Acta* **1996**, *41*, 1073–1082. [CrossRef]
17. Mansfeld, F. Use of electrochemical impedance spectroscopy for the study of corrosion protection by polymer coatings. *J. Appl. Electrochem.* **1995**, *25*, 187–202. [CrossRef]
18. Davis, G.D.; Krebs, L.A.; Dacres, C.M. Coating evaluation and validation of accelerated test conditions using an in-situ corrosion sensor. *J. Coat. Technol.* **2002**, *74*, 69–74. [CrossRef]
19. Angelini, E.; Corbellini, S.; Parvis, M.; Ferraris, F.; Grassini, S. An Arduino-based EIS with a logarithmic amplifier for corrosion monitoring. In Proceedings of the 2014 IEEE International Instrumentation and Measurement Technology Conference (I2MTC), Montevideo, Uruguay, 11–14 May 2014; pp. 905–910. [CrossRef]
20. Ma, C.; Wang, Z.; Behnamian, Y.; Gao, Z.; Wu, Z.; Qin, Z.; Xia, D.H. Measuring atmospheric corrosion with electrochemical noise: A review of contemporary methods. *Measurement* **2019**, *138*, 54–79. [CrossRef]
21. Xia, D.-H.; Ma, C.; Song, S.; Xu, L. Detection of Atmospheric Corrosion of Aluminum Alloys by Electrochemical Probes: Theoretical Analysis and Experimental Tests. *J. Electrochem. Soc.* **2020**, *167*, 089004. [CrossRef]
22. Ma, C.; Song, S.; Gao, Z.; Wang, J.; Hu, W.; Behnamian, Y.; Xia, D.H. Electrochemical noise monitoring of the atmospheric corrosion of steels: Identifying corrosion form using wavelet analysis. *Corros. Eng. Sci. Technol.* **2017**, *52*, 432–440. [CrossRef]
23. *ASTM G96-90(2008)*; Standard Guide for Online Monitoring of Corrosion in Plant Equipment (Electrical and Electrochemical Methods). ASTM: West Conshohocken, PA, USA, 2008.
24. Kosec, T.; Kuhar, V.; Kranjc, A.; Malnarič, V.; Belingar, B.; Legat, A. Development of an Electrical Resistance Sensor from High Strength Steel for Automotive Applications. *Sensors* **2019**, *19*, 1956. [CrossRef]
25. Nishikata, A.; Zhu, Q.; Tada, E. Long-term monitoring of atmospheric corrosion at weathering steel bridges by an electrochemical impedance method. *Corros. Sci.* **2014**, *87*, 80–88. [CrossRef]
26. Mizuno, D.; Suzuki, S.; Fujita, S.; Hara, N. Corrosion monitoring and materials selection for automotive environments by using Atmospheric Corrosion Monitor (ACM) sensor. *Corros. Sci.* **2014**, *83*, 217–225. [CrossRef]
27. Arellano-Pérez, J.H.; Ramos Negrón, O.J.; Escobar-Jiménez, R.F.; Gómez-Aguilar, J.F.; Uruchurtu-Chavarín, J. Development of a portable device for measuring the corrosion rates of metals based on electrochemical noise signals. *Meas. J. Int. Meas. Confed.* **2018**, *122*, 73–81. [CrossRef]
28. Lee, S. *Digital Color Image Processing System for Civil Infrastructure Health Assessment and Monitoring: Steel Bridge Coating Case*; Purdue University: Purdue, IN, USA, 2005.

29. Pereira, F.M.V.; Bueno, M.I.M.S. Image evaluation with chemometric strategies for quality control of paints. *Anal. Chim. Acta* **2007**, *588*, 184–191. [CrossRef]
30. Shen, H.K.; Chen, P.H.; Chang, L.M. Automated steel bridge coating rust defect recognition method based on color and texture feature. *Autom. Constr.* **2013**, *31*, 338–356. [CrossRef]
31. Kovač, J.; Legat, A.; Zajec, B.; Kosec, T.; Govekar, E. Detection and characterization of stainless steel SCC by the analysis of crack related acoustic emission. *Ultrasonics* **2015**, *62*, 312–322. [CrossRef] [PubMed]
32. Etiemble, A.; Idrissi, H.; Roué, L. Effect of Ti and Al on the pulverization resistance of MgNi-based metal hydride electrodes evaluated by acoustic emission. *Int. J. Hydrogen Energy* **2013**, *38*, 1136–1144. [CrossRef]
33. Kovač, J.; Leban, M.; Legat, A. Detection of SCC on prestressing steel wire by the simultaneous use of electrochemical noise and acoustic emission measurements. *Electrochim. Acta* **2007**, *52*, 7607–7616. [CrossRef]
34. Wu, K.; Byeon, J.W. Morphological estimation of pitting corrosion on vertically positioned 304 stainless steel using acoustic-emission duration parameter. *Corros. Sci.* **2019**, *148*, 331–337. [CrossRef]
35. Komary, M.; Komarizadehasl, S.; Ramos Schneider, G.; Tošić, N.; Turmo Coderque, J.; Torralba Mendiola, V. Full review of low-cost electronics implemented in structural health monitoring applications for bridges. In Proceedings of the Elevated International Converence on Bridge Maintenance, Safety and Management (IMBAS 2022); Bridge Safety, Maintenance, Management, Life-Cycle, Resilience and Sustainability, Barcelona, Spain, 11–15 July 2022; pp. 980–985. [CrossRef]
36. Komarizadehasl, S.; Lozano, F.; Komary, M.; Lozano-Galant, J.A.; Turmo, J. Development of an accurate low-cost device for structural vibration acquisition. In Proceedings of the IABSE Symposium Prague, 2022: Challenges for Existing and Oncoming Structures—Report, Online, 25–27 May 2022.
37. Komarizadehasl, S.; Komary, M.; Turmo Coderque, J.; Lozano Galant, F.; Lozano Galant, F.; Lozano Galant, J.A. Using few accelerometer for improving the resolution and accuracy of low-cost accelerometers. In Proceedings of the Elevated International Converence on Bridge Maintenance, Safety and Management (IMBAS 2022); Bridge Safety, Maintenance, Management, Life-Cycle, Resilience and Sustainability, Barcelona, Spain, 11–15 July 2022; pp. 1575–1579. [CrossRef]
38. Soojin, C.; Yun, C.B.; Lynch, J.P.; Zimmerman, A.T.; Spencer, B.F., Jr.; Nagayama, T. Smart Wireless Sensor Technology for Structural Health Monitoring of Civil Structures. *Int. J. Steel Struct.* **2008**, *8*, 267–275.
39. Komarizadehasl, S.; Huguenet, P.; Lozano, F.; Lozano-Galant, J.A.; Turmo, J. Operational and Analytical Modal Analysis of a Bridge Using Low-Cost Wireless Arduino-Based Accelerometers. *Sensors* **2022**, *22*, 9808. [CrossRef]
40. Mobaraki, B.; Komarizadehasl, S.; Castilla Pascual, F.J.; Lozano-Galant, J.A.; Porras Soriano, R. A Novel Data Acquisition System for Obtaining Thermal Parameters of Building Envelopes. *Buildings* **2022**, *12*, 670. [CrossRef]
41. Komarizadehasl, S.; Lozano, F.; Komary, M.; Lozano-Galant, J.A.; Turmo, J. Resolution improvement of Low-Cost MEMS accelerometer by aligning Simulations sensors. In Proceedings of the IABSE Symposium Prague, 2022: Challenges for Existing and Oncoming Structures—Report, Online, 25–27 May 2022.
42. Komarizadehasl, S.; Komary, M.; Turmo, J.; Torralba, V.; Lozano, F.; Lozano-Galant, J.A. Low-cost accurate acceleration acquisition sensor. In *Bridge Safety, Maintenance, Management, Life-Cycle, Resilience and Sustainability*; CRC Press: Boca Raton, FL, USA, 2022; pp. 803–810. [CrossRef]
43. Komarizadehasl, S.; Lozano, F.; Lozano-Galant, J.A.; Ramos, G.; Turmo, J. Low-Cost Wireless Structural Health Monitoring of Bridges. *Sensors* **2022**, *22*, 5725. [CrossRef]
44. Komary, M.; Komarizadehasl, S.; Ramos Schneider, G.; Torralba Mendiola, V. Developing and validation of an inclinometer sensor based on fusion of a magnetometer, an accelerometer and a gyroscope sensor for SHM applications. In Proceedings of the Elevated International Converence on Bridge Maintenance, Safety and Management (IMBAS 2022); Bridge Safety, Maintenance, Management, Life-Cycle, Resilience and Sustainability, Barcelona, Spain, 11–15 July 2022; pp. 1607–1611. [CrossRef]
45. Mobaraki, B.; Komarizadehasl, S.; Castilla Pascual, F.J.; Lozano Galant, J.A. Determination of environmental parameters based on Arduino based low-cost sensors. In Proceedings of the Recent Trends in Construction and Education RTCEE International Conference, Brisbane, Australia, 10–11 September 2020; pp. 103–108.
46. Mobaraki, B.; Komarizadehasl, S.; Castilla Pascual, F.J.; Lozano-Galant, J.A. Application of Low-Cost Sensors for Accurate Ambient Temperature Monitoring. *Buildings* **2022**, *12*, 1411. [CrossRef]
47. Mobaraki, B.; Komarizadehasl, S.; Pascual, F.J.C.; Lozano-Galant, J.A. Environmental monitoring system based on low-cost sensors. In Proceedings of the Current Topics and Trends on Durability of Building Materials and Components: Proceedings of the XV edition of the International Conference on Durability of Building Materials and Components (DBMC 2020), Barcelona, Spain, 20–23 October 2020; pp. 1679–1686. [CrossRef]
48. Komarizadehasl, S.; Komary, M.; Alahmad, A.; Lozano-Galant, J.A.; Ramos, G.; Turmo, J. A Novel Wireless Low-Cost Inclinometer Made from Combining the Measurements of Multiple MEMS Gyroscopes and Accelerometers. *Sensors* **2022**, *22*, 5605. [CrossRef]
49. Mobaraki, B.; Lozano-Galant, F.; Soriano, R.P.; Pascual, F.J.C. Application of Low-Cost Sensors for Building Monitoring: A Systematic Literature Review. *Buildings* **2021**, *11*, 336. [CrossRef]
50. Xia, D.H.; Deng, C.M.; Macdonald, D.; Jamali, S.; Mills, D.; Luo, J.L.; Strebl, M.G.; Amiri, M.; Jin, W.; Song, S.; et al. Electrochemical measurements used for assessment of corrosion and protection of metallic materials in the field: A critical review. *J. Mater. Sci. Technol.* **2022**, *112*, 151–183. [CrossRef]
51. Fan, L.; Shi, X. Techniques of corrosion monitoring of steel rebar in reinforced concrete structures: A review. *Struct. Health Monit.* **2022**, *21*, 1879–1905. [CrossRef]

52. Tang, F.; Zhou, G.; Li, H.N.; Verstrynge, E. A review on fiber optic sensors for rebar corrosion monitoring in RC structures. *Constr. Build. Mater.* **2021**, *313*, 125578. [CrossRef]
53. Reddy, M.S.B.; Ponnamma, D.; Sadasivuni, K.K.; Aich, S.; Kailasa, S.; Parangusan, H.; Ibrahim, M.; Eldeib, S.; Shehata, O.; Ismail, M.; et al. Sensors in advancing the capabilities of corrosion detection: A review. *Sensors Actuators A Phys.* **2021**, *332*, 113086. [CrossRef]
54. Saini, J.; Dutta, M.; Marques, G. Indoor Air Quality Monitoring Systems Based on Internet of Things: A Systematic Review. *Int. J. Environ. Res. Public Health* **2020**, *17*, 4942. [CrossRef]
55. Strmmen, R. Internal Corrosion Monitoring. 1984. Available online: https://www.scopus.com/record/display.uri?eid=2-s2.0-85040802642&origin=resultslist&sort=plf-f (accessed on 16 January 2023).
56. Knuutila, K.; Aromaa, J.; Forsen, O.; Ylasaari, S. Low-Cost Microcomputer System for Controlling and Measuring Electrochemical Experiments. *Dechema Monogr. (Dtsch. Ges. Fuer Chem. Appar.)* **1986**, *101*, 339–348.
57. Nazir, M.H.; Khan, Z.A.; Saeed, A. A Novel Non-Destructive Sensing Technology for On-Site Corrosion Failure Evaluation of Coatings. *IEEE Access* **2017**, *6*, 1042–1054. [CrossRef]
58. Yu, L.; Giurgiutiu, V.; Chao, Y.; Pollock, P. In-situ multi-mode sensing with embedded piezoelectric wafer active sensors for critical pipeline health monitoring. *ASME Int. Mech. Eng. Congr. Expo. Proc.* **2008**, *10 Pt A*, 687–696. [CrossRef]
59. Simmers, G.E., Jr.; Sodano, H.A.; Park, G.; Inman, D.J. Impedance based corrosion detection. *SPIE* **2005**, *5767*, 328–339. [CrossRef]
60. Khan, A.; Qurashi, A.; Badeghaish, W.; Noui-Mehidi, M.N.; Aziz, M.A. Frontiers and challenges in electrochemical corrosion monitoring; surface and downhole applications. *Sensors* **2020**, *20*, 6583. [CrossRef] [PubMed]
61. Licina, G.J.; Carney, C.S. Monitoring biofilm formation and incipient MIC in real time. In Proceedings of the CORRISION 99, San Antonio, TX, USA, 25–30 April 1999.
62. Thien, A.B.; Chiamori, H.C.; Ching, J.T.; Wait, J.R.; Park, G. Piezoelectric active sensing for damage detection in pipeline structures. *Conf. Proc. Soc. Exp. Mech. Ser.* **2005**.
63. Park, S.; Yun, C.B.; Inman, D.J. A Self-contained active sensor system for health monitoring of civil infrastructures. In Proceedings of the IEEE Sensors, Daegu, Republic of Korea, 22–25 October 2006; pp. 798–802. [CrossRef]
64. Yu, L.; Giurgiutiu, V.; Pollock, P. A multi-mode sensing system for corrosion detection using piezoelectric wafer active sensors. *Sensors Smart Struct. Technol. Civil Mech. Aerosp. Syst.* **2008**, *6932*, 69322H. [CrossRef]
65. Thien, A.B.; Chiamori, H.C.; Ching, J.T.; Wait, J.R.; Park, G. The use of macro-fibre composites for pipeline structural health assessment. *Struct. Control Health Monit.* **2008**, *15*, 43–63. [CrossRef]
66. Abdur Rahman, A.R.; Bhat, S.; Bhansali, S. Design, Fabrication, and Impedance Characterization of a Capacitance-Based Salinity Sensor for Marine Applications. *J. Electrochem. Soc.* **2008**, *155*, J355. [CrossRef]
67. Mejía-Aguilar, A.; Pallàs-Areny, R. Electrical impedance measurement using voltage/current pulse excitation. In Proceedings of the XIX Imeko World Congress: Fundamental and Applied Metrology, Lisbon, Portugal, 6–11 September 2009; Volume 1, pp. 591–596.
68. Azhari, F.; Banthia, N. Structural health monitoring using piezoresistive cementitious composites. In Proceedings of the 2nd International Conference on Sustainable Construction Materials and Technologies, Ancona, Italy, 23–28 June 2010; pp. 335–345.
69. Friedersdorf, F.J.; Ridder, K.B.; Andrews, C. Tank coating damage assessment using an artificial neural network model. In Proceedings of the CORROSION 2013, Orlando, FL, USA, 17 March 2013.
70. Andrews, C.; Bunget, G.; Farinholt, K.M.; Friedersdorf, F.J. Low-cost autonomous condition monitoring system for tank coatings. In Proceedings of the CORRISION 2015, Dallas, TX, USA, 15–19 March 2015.
71. Bansal, T.; Talakokula, V.; Bhalla, S. Rebar corrosion assessment comparison of different piezo configurations in blended concrete. In Proceedings of the 7th Asia-Pacific Workshop on Structural Health Monitoring, Hong Kong, China, 12–15 November 2018; pp. 914–925.
72. Li, W.; Liu, T.; Wang, J.; Zou, D.; Gao, S. Finite-Element Analysis of an Electromechanical Impedance–Based Corrosion Sensor with Experimental Verification. *J. Aerosp. Eng.* **2019**, *32*, 04019012. [CrossRef]
73. Corva, D.M.; Hosseini, S.S.; Collins, F.; Adams, S.D.; Gates, W.P.; Kouzani, A.Z. Miniature resistance measurement device for structural health monitoring of reinforced concrete infrastructure. *Sensors* **2020**, *20*, 4313. [CrossRef]
74. Chowdhury, T.; D'souza, N.; Dahotre, N. Low-cost reliable corrosion sensors using zno-pvdf nanocomposite textiles. *Sensors* **2021**, *21*, 4147. [CrossRef]
75. Strachotová, K.C.; Kouřil, M. High-Sensitivity Sensors For Monitoring of Lead Atmospheric Corrosion. In Proceedings of the 30th Anniversary International Conference on Metallurgy and Materials, Brno, Czech Republic, 26–28 May 2021; pp. 936–941. [CrossRef]
76. Carullo, A.; Ferraris, F.; Parvis, M.; Vallan, A.; Angelini, E.; Spinelli, P. Low-cost electrochemical impedance spectroscopy system for corrosion monitoring of metallic antiquities and works of art. *IEEE Trans. Instrum. Meas.* **2000**, *49*, 371–375. [CrossRef]
77. Davis, G.D.; Ross, R.A.; Raghu, S. Coating health monitoring system for army ground vehicles. In Proceedings of the CORROSION 2007, Nashville, TN, USA, 11–15 March 2007; pp. 072301–0723011.
78. Matsiev, L. Improving performance and versatility of systems based on single-frequency dft detectors such as AD5933. *Electronics* **2014**, *4*, 1–34. [CrossRef]
79. Grassini, S.; Corbellini, S.; Parvis, M.; Angelini, E.; Zucchi, F. A simple Arduino-based EIS system for in situ corrosion monitoring of metallic works of art. *Meas. J. Int. Meas. Confed.* **2018**, *114*, 508–514. [CrossRef]

80. Sebar, L.E.; Angelini, E.; Grassini, S.; Iannucci, L.; Parvis, M. An op amp-less electrochemical impedance spectroscopy system. In Proceedings of the 2020 IEEE International Instrumentation and Measurement Technology Conference (I2MTC), Dubrovnik, Croatia, 25–28 May 2020. [CrossRef]
81. Nishikata, A.; Suzuki, F.; Tsuru, T. Corrosion monitoring of nickel-containing steels in marine atmospheric environment. *Corros. Sci.* **2005**, *47*, 2578–2588. [CrossRef]
82. Niblock, T.; Mian, M.; Clements, R.; Darr, D.; O'Day, J.; Morse, J.; Laskowski, B.C.; Sims, M.; Ooi, T.K.; Corder, A. Advanced sensing, degradation detection, diagnostic and prognostic capabilities for structural health management. In Proceedings of the SPIE Smart Structures and Materials + Nondestructive Evaluation and Health Monitoring, San Diego, CA, USA, 11–15 March 2009; pp. 777–786. [CrossRef]
83. Wood, S.L.; Neikirk, D.P. Passive sensors for infrastructure monitoring. In Proceedings of the Sensors and Smart Structures Technologies for Civil, Mechanical, and Aerospace Systems, Denver, CO, USA, 5–8 March 2009; Volume 7292. [CrossRef]
84. Abu Yosef, A.E.; Pasupathy, P.; Wood, S.L.; Neikirk, D.P. Low-cost passive sensors for monitoring corrosion in concrete structures. In Proceedings of the Nondestructive Characterization for Composite Materials, Aerospace Engineering, Civil Infrastructure, and Homeland Security 2011, San diego, CA, USA, 7–10 March 2011; Volume 7983. [CrossRef]
85. Rafezi, H.; Rahmani, B. Corrosion detection in pipes by piezoelectric sensors using Artificial Neural Network. *Adv. Mater. Res.* **2012**, *403–408*, 748–752. [CrossRef]
86. Perveen, K.; Bridges, G.E.; Bhadra, S.; Thomson, D.J. Compact passive wireless reinforced concrete corrosion initiation sensor that can be installed in existing steel. In Proceedings of the EWSHM-7th European Workshop on Structural Health Monitoring, Nantes, France, 8–11 July 2014; pp. 1017–1024.
87. Ahmad, S.; Jibran, M.A.A.; Azad, A.K.; Maslehuddin, M. A simple and reliable setup for monitoring corrosion rate of steel rebars in concrete. *Sci. World J.* **2014**, *2014*, 525678. [CrossRef]
88. Arpaia, P.; Landi, C. Microcontroller-based system for pipe-to-soil potential measurement. *Meas. Control* **1990**, *23*, 237–241. [CrossRef]
89. Mugambi, E.M.; Kwan, K.; Laskowski, B.C.; Ooi, T.K.; Corder, A. MEMS based strain and corrosion sensors for structural health monitoring. In Proceedings of the 6th International Workshop on Structural Health Monitoring, Stanford, CA, USA, 11–13 September 2007; pp. 730–737.
90. Kwan, K.; Caldwell, B.; Morse, J.; O'Day, J.; Laskowski, B.C.; Ooi, T.K.; Corder, A. Wireless sensors with advanced detection and prognostic capabilities for corrosion health management. *Adv. Mater. Res.* **2008**, *38*, 123–131. [CrossRef]
91. Kwan, K.; Morse, J.; Laskowski, B.C. Optimal condition based maintenance monitoring by using corrosion degradation AI modeling for Structural Health Monitoring. In Proceedings of the 32nd Congress of International Council of the Aeronautical Sciences, Shanghai, China, 6-10 September 2008; pp. 551–559.
92. Samoilă, C.; Nascov, V.; Ursuţiu, D. Virtual instrumentation in corrosion measurements. In Proceedings of the 2013 10th International Conference on Remote Engineering and Virtual Instrumentation (REV), Sydney, Australia, 6–8 February 2013. [CrossRef]
93. Degrigny, C.; Menart, E.; Erny, G. Easy-to-use, low-cost electrochemical open-source hardware to analyse heritage metals: Possibilities and limits. *Curr. Top. Electrochem.* **2018**, *20*, 15–23.
94. Castro-Borges, P.; Ordaz, J.M. Corrosion performance of concrete columns after localized repairs in a tropical coastal environment. *Corrosion* **2009**, *65*, 118–125. [CrossRef]
95. Mierau, T. Remote monitoring & control systems as a solution for corrosion monitoring at airports with elevated security standards. In Proceedings of the CORROSION 2006, San Diego, CA, USA, 12–16 March 2006; pp. 063091–0630912.
96. Xia, D.-H.; Song, S.; Behnamian, Y.; Hu, W.; Cheng, Y.F.; Luo, J.-L.; Huet, F. Electrochemical Noise Applied in Corrosion Science: Theoretical and Mathematical Models towards Quantitative Analysis. *J. Electrochem. Soc.* **2020**, *167*, 081507. [CrossRef]
97. Zhao, X.; Cui, Y.; Wei, H.; Kong, X.; Zhang, P.; Sun, C. Research on corrosion detection for steel reinforced concrete structures using the fiber optical white light interferometer sensing technique. *Smart Mater. Struct.* **2013**, *22*, 065014. [CrossRef]
98. Wen, F.; Pray, J.; McSweeney, K.; Gu, H. Emerging inspection technologies—Enabling remote surveys/inspections. In Proceedings of the Annual Offshore Technology Conference 2019, Houston, TX, USA, 6–9 May 2019. [CrossRef]
99. Taylor, S.R. Electrical resistance changes as an alternate method for monitoring the corrosion of steel in concrete and mortar. *Corrosion* **2000**, *56*, 951–959. [CrossRef]
100. Hautefeuille, M.; O'Flynn, B.; Peters, F.H.; O'Mahony, C. Development of a microelectromechanical system (MEMS)-Based multisensor platform for environmental monitoring. *Micromachines* **2011**, *2*, 410–430. [CrossRef]
101. McCarter, W.J.; Chrisp, T.M.; Starrs, G.; Adamson, A.; Owens, E.; Basheer, P.A.M.; Nanukuttan, S.V.; Srinivasan, S.; Holmes, N. Developments in Performance Monitoring of Concrete Exposed to Extreme Environments. *J. Infrastruct. Syst.* **2012**, *18*, 167–175. [CrossRef]
102. Holst, A.; Budelmann, H. Low-cost sensor for integrated durability monitoring and life-cycle assessment of reinforced concrete structures. In Proceedings of the Life-Cycle and Sustainability of Civil Infrastructure Systems: Proceedings of the Third International Symposium on Life-Cycle Civil Engineering (IALCCE'12), Vienna, Austria, 3–6 October 2012; pp. 1414–1421.
103. Materer, N.; Field, P.; Ley, N.; Soufiani, A.R.; Scott, D.; Ley, T.; Apblett, A. Passive Wireless Detection of Corrosive Salts in Concrete Using Wire-Based Triggers. *J. Mater. Civ. Eng.* **2014**, *26*, 918–922. [CrossRef]

104. Halabe, U.B.; Kavi, J.; GangaRao, H.V.S. Innovative Sensors for Monitoring the Corrosion of Steel Embedded in Concrete Structural Components. In Proceedings of the Geotechnical and Structural Engineering Congress 2016, Phoenix, AZ, USA, 14–17 February 2016; pp. 1016–1031. [CrossRef]
105. Hurley, M.F.; Cunningham, A.; Lysne, D.; Acharya, S.; Jaques, B.J.; Butt, D.P. A condition monitor for atmospheric induced stress corrosion cracking. In Proceedings of the CORROSION 2018, Phoenix, AZ, USA, 15–19 April 2018.
106. Corva, D.M.; Hosseini, S.S.; Adams, S.; Gates, W.; Collins, F.; Kouzani, A.Z. A miniature device for in-situ measurement of concrete corrosion. In Proceedings of the 2018 IEEE International Conference on Mechatronics and Automation (ICMA), Changchun, China, 5–8 August 2018; pp. 2185–2190. [CrossRef]
107. Strommen, R.; Gartland, P.O.; Wold, K. *FSM (Field Signature Method)—The New Technology for Internal Corrosion Monitoring of Pipelines, Vessels and Pressure Equipment*; American Society of Mechanical Engineers: New York, NY, USA, 1998.
108. Andringa, M.M.; Neikirk, D.P.; Wood, S.L. Unpowered wireless analog resistance sensor. In *Smart Structures and Materials 2004: Sensors and Smart Structures Technologies for Civil, Mechanical, and Aerospace Systems*; SPIE: Bellingham, DC, USA, 2004; Volume 5391, p. 356. [CrossRef]
109. Andringa, M.M.; Puryear, J.M.; Neikirk, D.P.; Wood, S.L. In situ measurement of conductivity and temperature during concrete curing using passive wireless sensors. In *Smart Structures and Materials: Technologies for Civil, Mechanical, and Aerospace Systems*; SPIE: Bellingham, DC, USA, 2007; Volume 6529. [CrossRef]
110. Pasupathy, P.; Abu Yousef, A.; Neikirk, D.P.; Wood, S.L. Wireless Electronic Structural Surveillance sensors using inductively coupled sacrificial transducers. *Prog. Electromagn. Res. Symp.* **2011**, 756–760.
111. Pasupathy, P.; Munukutla, S.; Neikirk, D.P.; Wood, S.L. Versatile wireless sacrificial transducers for electronic structural surveillance sensors. In Proceedings of the IEEE Sensors 2009, Christchurch, New Zealand, 25–28 October 2009; pp. 979–983. [CrossRef]
112. Bouzaffour, K.; Lescop, B.; Talbot, P.; Gallee, F.; Rioual, S. Development of an Embedded UHF-RFID Corrosion Sensor for Monitoring Corrosion of Steel in Concrete. *IEEE Sens. J.* **2021**, *21*, 12306–12312. [CrossRef]
113. El Masri, I.; Lescop, B.; Talbot, P.; Vien, G.N.; Becker, J.; Thierry, D.; Rioual, S. Development of a RFID sensitive tag dedicated to the monitoring of the environmental corrosiveness for indoor applications. *Sens. Actuators B Chem.* **2020**, *322*, 128602. [CrossRef]
114. Yasri, M.; Lescop, B.; Diler, E.; Gallée, F.; Thierry, D.; Rioual, S. Monitoring uniform and localised corrosion by a radiofrequency sensing method. *Sens. Actuators B Chem.* **2018**, *257*, 988–992. [CrossRef]
115. Khalifeh, R.; Lescop, B.; Gallée, F.; Rioual, S. Development of a radio frequency resonator for monitoring water diffusion in organic coatings. *Sensors Actuators A Phys.* **2016**, *247*, 30–36. [CrossRef]
116. Lippincott, C. Automating Pipeline Integrity Monitoring. *Fierce Electronics*, 1 August 2008. Available online: https://www.fierceelectronics.com/components/automating-pipeline-integrity-monitoring (accessed on 16 January 2023).
117. Dante, J.F.; Steiner, A.; Friedersdorf, F. RF system for corrosivity monitoring. In Proceedings of the CORROSION 2008, New Orleans, LA, USA, 16–20 March 2008; pp. 082041–0820411.
118. Palmer, C.; Gutterman, A.; Argenna, G.; Inclan, V.; Zyuzin, A. Wireless, batteryless distributed strain sensing for structural health monitoring. *Structure Health Monit.* **2011**, *1*, 386–393.
119. He, Y.; Lo, J.; McLaughlin, S.; Vincelli, A. Low-cost corrosion monitoring sensors based on Radio Frequency Identification (RFID) technology. In Proceedings of the Aterials Science and Technology Conference and Exhibition 2012, Pittsbourg, PA, USA, 7–11 October 2012; pp. 606–613.
120. Zhang, H.; Yang, R.; He, Y.; Wu, R. Characterisation of steel corrosion using high frequency RFID. In Proceedings of the 2017 Far East NDT New Technology & Application Forum (FENDT), Xiamen, China, 15–17 December 2018; pp. 127–132. [CrossRef]
121. Sunny, A.I.; Zhang, J.; Tian, G.Y.; Tang, C.; Rafique, W.; Zhao, A.; Fan, M. Temperature independent defect monitoring using passive wireless RFID sensing system. *IEEE Sens. J.* **2019**, *19*, 1525–1532. [CrossRef]
122. Bruciati, B.; Jang, S.; Fils, P. RFID-based crack detection of ultra high-performance concrete retrofitted beams. *Sensors* **2019**, *19*, 1573. [CrossRef] [PubMed]
123. Marindra, A.M.J.; Tian, G.Y. Chipless RFID sensor for corrosion characterization based on frequency selective surface and feature fusion. *Smart Mater. Struct.* **2020**, *29*, 125010. [CrossRef]
124. Zhang, J.; Tian, G.Y. UHF RFID Tag Antenna-Based Sensing for Corrosion Detection & Characterization Using Principal Component Analysis. *IEEE Trans. Antennas Propag.* **2016**, *64*, 4405–4414. [CrossRef]
125. Zarifi, M.H.; Deif, S.; Daneshmand, M. Wireless passive RFID sensor for pipeline integrity monitoring. *Sens. Actuators A Phys.* **2017**, *261*, 24–29. [CrossRef]
126. Deif, S.; Daneshmand, M. Long Array of Microwave Sensors for Real-Time Coating Defect Detection. *IEEE Trans. Microw. Theory Tech.* **2020**, *68*, 2856–2866. [CrossRef]
127. Rioual, S.; Lescop, B.; Pellé, J.; Radicchi, G.D.A.; Chaumat, G.; Bruni, M.D.; Becker, J.; Thierry, D. Monitoring of the environmental corrosivity in museums by rfid sensors: Application to pollution emitted by archeological woods. *Sustainability* **2021**, *13*, 6158. [CrossRef]
128. Ramani, V.; Kuang, K.S.C. Monitoring of rebar corrosion in concrete structures using a lens-based plastic optical fiber (LPOF) sensor. *Constr. Build. Mater.* **2021**, *276*, 122129. [CrossRef]
129. Dos Santos, P.S.S.; Jorge, P.A.S.; De Almeida, J.M.M.M.; Coelho, L. Low-cost interrogation system for long-period fiber gratings applied to remote sensing. *Sensors* **2019**, *19*, 1500. [CrossRef]

130. Chen, Y.; Xue, X. Advances in the structural health monitoring of bridges using piezoelectric transducers. *Sensors* **2018**, *18*, 4312. [CrossRef] [PubMed]
131. Wan, K.T.; Leung, C.K.Y.; Chen, L. A novel optical fiber sensor for steel corrosion in concrete structures. *Sensors* **2008**, *8*, 1960–1976. [CrossRef]
132. Guan, S.; Ripley, G. Monitoring technology for integrity management of the hull structure of floating production, storage and offloading (FPSO) vessel. *Mater. Sci. Technol. Conf. Exhib.* **2014**, 2, 761–767.
133. Elster, J.L.; Greene, J.A.; Jones, M.E.; Bailey, T.A.; Lenahan, S.M.; Velander, W.H.; VanTassell, R.; Hodges, W.; Perez, I.M. Optical-fiber-based chemical sensors for detection of corrosion precursors and by-products. *Chem. Biochem. Environ. Fiber Sensors X* **1999**, *3540*, 251–257. [CrossRef]
134. Chen, L.; Xu, Q.; Wu, M.; Hu, J.; Song, H.; Yang, H.; Han, J.; Chen, Y. Corrosion monitoring for marine concrete based on a first-order RC model. In Proceedings of the Oceans 2012 MTS/IEEE Harnessing Power Ocean, Virginia Beach, VA, USA, 14–19 October 2012. [CrossRef]
135. Cooper, K.R.; Elster, J.; Jones, M.; Kelly, R.G. Optical fiber-based corrosion sensor systems for health monitoring of aging aircraft. In Proceedings of the IEEE Systems Readiness Technology Conference, Valley Forge, PA, USA, 20–23 August 2002; pp. 847–856. [CrossRef]
136. Casas, J.R.; Cruz, P.J.S. Fiber Optic Sensors for Bridge Monitoring. *J. Bridg. Eng.* **2003**, *8*, 362–373. [CrossRef]
137. Kiremidjian, G.K.; Kiremidjian, A.S.; Lynch, J.P. Wireless structural monitoring for homeland security applications. *Nondestruct. Detect. Meas. Homel. Secur. II* **2004**, *5395*, 82–90. [CrossRef]
138. Zhao, X.; Gao, H.; Zhang, G.; Ayhan, B.; Yan, F.; Kwan, C.; Rose, J.L. Active health monitoring of an aircraft wing with embedded piezoelectric sensor/actuator network: I. Defect detection, localization and growth monitoring. *Smart Mater. Struct.* **2007**, *16*, 1208–1217. [CrossRef]
139. Meyendorf, N.; Frankenstein, B.; Schubert, L. Structural health monitoring for aircraft, ground transportation vehicles, wind turbines and pipes-prognosis. In *Emerging Technologies in Non-Destructive Testing*; CRC Press: Boca Raton, FL, USA, 2012; pp. 15–22. [CrossRef]
140. Perelli, A.; Harput, S.; De Marchi, L.; Freear, S. Frequency warping compressive sensing for structural monitoring of aircraft wing. In Proceedings of the 2013 18th International Conference on Digital Signal Processing (DSP), Thera, Greece, 1–3 July 2013. [CrossRef]
141. Wan, C.; Mita, A. Recognition of potential danger to buried pipelines based on sounds. *Struct. Control Health Monit.* **2010**, *17*, 317–337. [CrossRef]
142. Barshinger, J.; Feydo, M.; Strachan, S. Development of a non-intrusive, wireless, corrosion monitoring device. In Proceedings of the CORROSION 2019, Warsaw, Poland, 27–29 May 2019.
143. Harper, M.F.; Thompson, M. Health monitoring of a complex structure using a sparse array of modally selective transducers. *Nondestruct. Test. Eval.* **2005**, *20*, 159–173. [CrossRef]
144. Jalali, H.; Rizzo, P. A Remote Solitary Wave-Based Technique For Monitoring Corrosion in Steel Structures: Numerical Analysis and Experimental Validation. In Proceedings of the 48th Annual Review of Progress in Quantitative Nondestructive Evaluatio, Online, 28–30 July 2021. [CrossRef]
145. Andringa, M.M.; Puryear, J.M.; Neikirk, D.P.; Wood, S.L. Low-cost wireless corrosion and conductivity sensors. In *Smart Structures and Materials 2006: Sensors and Smart Structures Technologies for Civil, Mechanical, and Aerospace Systems*; SPIE: Bellingham, DC, USA, 2006; Volume 6174. [CrossRef]
146. Vinogradov, S.A. Magnetostrictive transducer for torsional guided waves in pipes and plates. *Mater. Eval.* **2009**, *67*, 333–341.
147. Chen, Y.; Munukutla, S.; Pasupathy, P.; Neikirk, D.P.; Wood, S.L. Magneto-inductive waveguide as a passive wireless sensor net for structural health monitoring. In *Sensors and Smart Structures Technologies for Civil, Mechanical, and Aerospace Systems 2010*; SPIE: Bellingham, DC, USA, 2010; Volume 7647. [CrossRef]
148. Chen, Y.; Pasupathy, P.; Trivedi, T.; Neikirk, D.P.; Wood, S.L. Improved magneto-inductive waveguide as wireless sensor net for structural health monitoring. In *Nondestructive Characterization for Composite Materials, Aerospace Engineering, Civil Infrastructure, and Homeland Security 2012*; SPIE: Bellingham, DC, USA, 2012; Volume 8347. [CrossRef]
149. Kranz, M.S.; English, B.A.; Whitley, M.R. RFID-inspired wireless microsensors for structural health monitoring. In Proceedings of the 2016 IEEE Aerospace Conference, Big Sky, MT, USA, 5–12 March 2016. [CrossRef]
150. Gallo, G.E.; Popovics, J.S. Monitoring active corrosion of metals in natural environments with magnetometry. *Corros. Sci.* **2012**, *63*, 1–4. [CrossRef]
151. Tao, L.; Song, S.; Zhang, X.; Zhang, Z.; Lu, F. Image analysis of atmospheric corrosion of field exposure high strength aluminium alloys. *Appl. Surf. Sci.* **2008**, *254*, 6870–6874. [CrossRef]
152. Xia, D.H.; Ma, C.; Song, S.; Jin, W.; Behnamian, Y.; Fan, H.; Wang, J.; Gao, Z.; Hu, W. Atmospheric corrosion assessed from corrosion images using fuzzy Kolmogorov–Sinai entropy. *Corros. Sci.* **2017**, *120*, 251–256. [CrossRef]
153. Xia, D.H.; Song, S.; Tao, L.; Qin, Z.; Wu, Z.; Gao, Z.; Wang, J.; Hu, W.; Behnamian, Y.; Luo, J.L. Review-material degradation assessed by digital image processing: Fundamentals, progresses, and challenges. *J. Mater. Sci. Technol.* **2020**, *53*, 146–162. [CrossRef]
154. Igoe, D.; Parisi, A.V. Characterization of the corrosion of iron using a smartphone camera. *Instrum. Sci. Technol.* **2016**, *44*, 139–147. [CrossRef]

155. Jindal, H.; Kasana, S.S.; Saxena, S. Underwater pipelines panoramic image transmission and refinement using acoustic sensors. *Int. J. Wavelets Multiresolut. Inf. Process.* **2018**, *16*, 18500133. [CrossRef]
156. Benkhoui, Y.; El Korchi, T.; Reinhold, L. UAS-based crack detection using stereo cameras: A comparative study. In Proceedings of the 2019 International Conference on Unmanned Aircraft Systems (ICUAS), Atlanta, GA, USA, 11–14 June 2019; pp. 1031–1035. [CrossRef]

sensors
MDPI

Review

Piezoelectric Materials and Sensors for Structural Health Monitoring: Fundamental Aspects, Current Status, and Future Perspectives

Min Ju [1], Zhongshang Dou [1,*], Jia-Wang Li [1], Xuting Qiu [1], Binglin Shen [1], Dawei Zhang [1], Fang-Zhou Yao [1,2], Wen Gong [1] and Ke Wang [1,3,*]

1 Research Center for Advanced Functional Ceramics, Wuzhen Laboratory, Jiaxing 314500, China
2 Center of Advanced Ceramic Materials and Devices, Yangtze Delta Region Institute of Tsinghua University, Jiaxing 314500, China
3 State Key Laboratory of New Ceramics and Fine Processing, School of Materials Science and Engineering, Tsinghua University, Beijing 100084, China
* Correspondence: douzs@wuzhenlab.com (Z.D.); wang-ke@tsinghua.edu.cn (K.W.)

Abstract: Structural health monitoring technology can assess the status and integrity of structures in real time by advanced sensors, evaluate the remaining life of structure, and make the maintenance decisions on the structures. Piezoelectric materials, which can yield electrical output in response to mechanical strain/stress, are at the heart of structural health monitoring. Here, we present an overview of the recent progress in piezoelectric materials and sensors for structural health monitoring. The article commences with a brief introduction of the fundamental physical science of piezoelectric effect. Emphases are placed on the piezoelectric materials engineered by various strategies and the applications of piezoelectric sensors for structural health monitoring. Finally, challenges along with opportunities for future research and development of high-performance piezoelectric materials and sensors for structural health monitoring are highlighted.

Keywords: piezoelectric materials; sensors; structural health monitoring

Citation: Ju, M.; Dou, Z.; Li, J.-W.; Qiu, X.; Shen, B.; Zhang, D.; Yao, F.-Z.; Gong, W.; Wang, K. Piezoelectric Materials and Sensors for Structural Health Monitoring: Fundamental Aspects, Current Status, and Future Perspectives. *Sensors* **2023**, *23*, 543. https://doi.org/10.3390/s23010543

Academic Editor: Zenghua Liu

Received: 30 November 2022
Revised: 30 December 2022
Accepted: 30 December 2022
Published: 3 January 2023

1. Introduction

Structural health monitoring (SHM) is a ubiquitous technology to evaluate the status and integrity of structures, and even predict their lifetime by constantly collecting and analyzing the data acquired from the sensors integrated in the structures [1,2]. SHM is of particular importance for complex engineering structures, which require costly maintenance, to significantly lower the maintenance cost and guarantee the safety and reliability thereof [3]. It is estimated that the market of SHM reached up to USD 2 billion in 2022 [4]. In addition, driven by the growing demands of internet of things, the SHM market is predicted to expand at a high compound annual growth rate of 14.6 in the following 5 years.

Sensors with high sensitivity, good reliability, and low cost are the cornerstone for the structural health monitoring (SHM). Various kinds of sensors have been developed to realize the SHM, such as strain gages [5], accelerometers [6], fiber optical sensors [7,8], displacement sensors [9], piezoelectric sensors [10], and laser Doppler vibrometers [11]. Piezoelectric materials are capable of becoming electrically polarized upon the application of external stress or deform in response to electrical stimuli. Therefore, sensors based on piezoelectric effect could be used as multipurpose sensors to realize the SHM using a variety of methods, including electromechanical impedance technology [12,13], ultrasonic propagation monitoring [14], acoustic emission [15], and stress monitoring [16]. Compared with other monitoring sensors or techniques, piezoelectric sensors have numerous advantages, such as small size, light weight, low cost, availability in a variety of formats, high sensitivity, and so on.

To meet the growing demands for high-performance piezoelectric sensors for SHM, there has been considerable research interest in this domain. In the recent years, lots of work has been focused on the piezoelectric sensors, such as piezoelectric transducers, smart aggregates, direct deposition piezoelectric sensors on structure, flexible smart sensors, and so on [17–21]. Numerous review articles have been published on the applications of piezoelectric SHM, for instance, bonded structures [22], polymer-matrix composites [23,24], aircraft applications [14], wind turbine blades [25,26], bridge applications [27], or other engineering structures [10]. Moreover, some reviews were focused on the monitoring techniques, such as impedance-based SHM [12,13], ultrasonic Lamb, or both [28].

In this review, we provide an overview of the currently available piezoelectric materials and sensors for SHM. In particular, focuses are placed on the high-performance piezoelectric materials, covering organic piezoelectrics, inorganic piezoelectrics and piezoelectric composites, engineered by various strategies, and piezoelectric sensors operated in active and passive modes for SHM. We conclude by highlighting some challenges and opportunities for future developments.

2. Piezoelectric Effect

The working principle of a piezoelectric sensor depends on "piezoelectric effect" of piezoelectric materials, first discovered by the Curie brothers in 1880 [29]. They found that, when an external force (pressure or tension) is applied in a specific direction of some dielectric crystals, the surface of both ends of the crystal will generate positive and negative bound charges of equal amount of electricity, and the density of bound charges is proportional to the magnitude of the applied stress, which is called the "positive piezoelectric effect". Subsequently, G. Lippman and the Curie brothers predicted and confirmed the existence of inverse piezoelectric effect in theory and experiment, respectively, that is, the material with piezoelectric effect will produce corresponding deformation under a certain electric field, and the deformation of the material will be restored when the applied electric field is removed. The positive piezoelectric effect and the inverse piezoelectric effect are reciprocal inverse effects, which jointly characterize the ability of piezoelectric materials to realize the conversion of mechanical energy and electric energy. The schematic diagram is shown in Figure 1a [30].

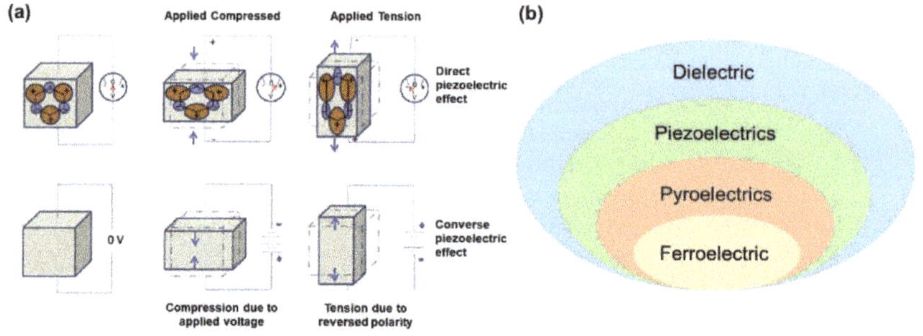

Figure 1. (**a**) Schematic of direct and converse piezoelectric effect [30]; (**b**) the affiliations of dielectric, piezoelectric, pyroelectric, and ferroelectric.

Whether the piezoelectric effect exists in a crystal depends on the symmetry of its crystal structure. Neumann principle requires that the symmetry of any macroscopic physical properties of a crystal should include the symmetry of the point group to which the crystal belongs. Therefore, it is possible for crystals without symmetry centers to have piezoelectric effects. Strict mathematical derivation shows that there are 32 kinds of macroscopic symmetries of crystals, that is, 32 kinds of point groups without translation operation. According to symmetry, these 32 point groups can be divided into two categories: central symmetry and noncentral symmetry. Of these, 11 are centrally symmetric, so only 21 are likely to have piezoelectric effects. However, although the point group 432 (O) has no symmetry center, its symmetry is very high and it does not have the piezoelectric effect. In addition, the spontaneous polarization of a part of the pyroelectric body can be reversed under the action of an applied electric field. Such crystals are called ferroelectrics. It must be pointed out that piezoelectric material must first be dielectric. Secondly, some piezoelectric materials are pyroelectric or ferroelectric. The affiliations of dielectric, piezoelectric, pyroelectric, and ferroelectric are shown in Figure 1b.

3. Piezoelectric Materials

3.1. Inorganic Piezoelectric Materials

Inorganic piezoelectric materials include piezoelectric single crystals (e.g., quartz), piezoelectric ceramics, and piezoelectric films, and their development process is the process of improving their piezoelectric properties. The piezoelectric properties of piezoelectric materials are mainly controlled by element doping. The development of an inorganic piezoelectric material system has gone through two stages: single component (such as $BaTiO_3$ [31] and $PbTiO_3$ [32]) and morphotropic phase boundary (such as lead zirconate titanate (PZT) [33,34], (1-x)$Pb(Mg_{1/3}Nb_{2/3})O_3$-x$PbTiO_3$ (PMN-PT) [35], and $Ba(Zr_{0.2}Ti_{0.8})O_3$-x$(Ba_{0.7}Ca_{0.3})TiO_3$ (BCT-BZT) [36]). With the development of material systems, the properties of inorganic piezoelectric materials have been greatly improved. Among them, PZT is one of the piezoelectric materials with outstanding comprehensive performance. As shown in Figure 2a, Lin et al. prepared Sb_2O_3-doped PZT-based piezoelectric ceramics that possess simultaneously enhanced piezoelectric coefficient d_{33} and large mechanical quality factor Q_m value [37].

However, with the increasing application requirements and scenarios of piezoelectric materials, the demand for high-performance piezoelectric materials is increasingly urgent. The existing material systems developed based on traditional methods have reached a bottleneck stage, which is gradually difficult to meet the requirements of precision sensing. In recent years, studies have shown that the doping of new complex multi-component components (such as Sm-doped PMN-PT [38]) can greatly improve the piezoelectric properties of piezoelectric materials, and the piezoelectric coefficient can be more than twice that of the existing piezoelectric system. The new ultra-high performance piezoelectric materials have the characteristics of multi-rare-earth element doping and multi-scale complex structures, which make the traditional methods, such as composition traversal preparation methods and single-scale material structure characterization methods, have a huge workload. To meet the increasing demand for high-end electrical materials, the development of new ultra-high-performance piezoelectric materials and their devices based on artificial intelligence methods, such as machine learning or high-throughput computation and experiments, has become an inevitable trend in the future of the sensing field.

In addition, due to the toxicity of Pb, environmentally benign lead-free piezoelectrics have received intensive attention, such as $BaTiO_3$ [39–41], sodium bismuth titanate (BNT) [42–44], and potassium sodium niobate (KNN) [45–52]. In Figure 2b, Jong et al. demonstrated a $BaTiO_3$ hybrid film by adopting a simple and facile inkjet-printing process [53]. By optimization of the ceramic particle movement in the flow that occurred by solvent evaporation in a droplet of ink, they successfully formed the $BaTiO_3$ ceramic layer, whose packing density of ceramic particles was over 55% in volume. Wang et al. proposed the engineering of oxygen vacancy, aiming to solve the antagonistic relationship between Q_m and d_{33} in hard piezoelectrics of KNN by the hot-pressing combined with a post-annealing process, leading to an enhancement of Q_m by 60% (Figure 2c) [46].

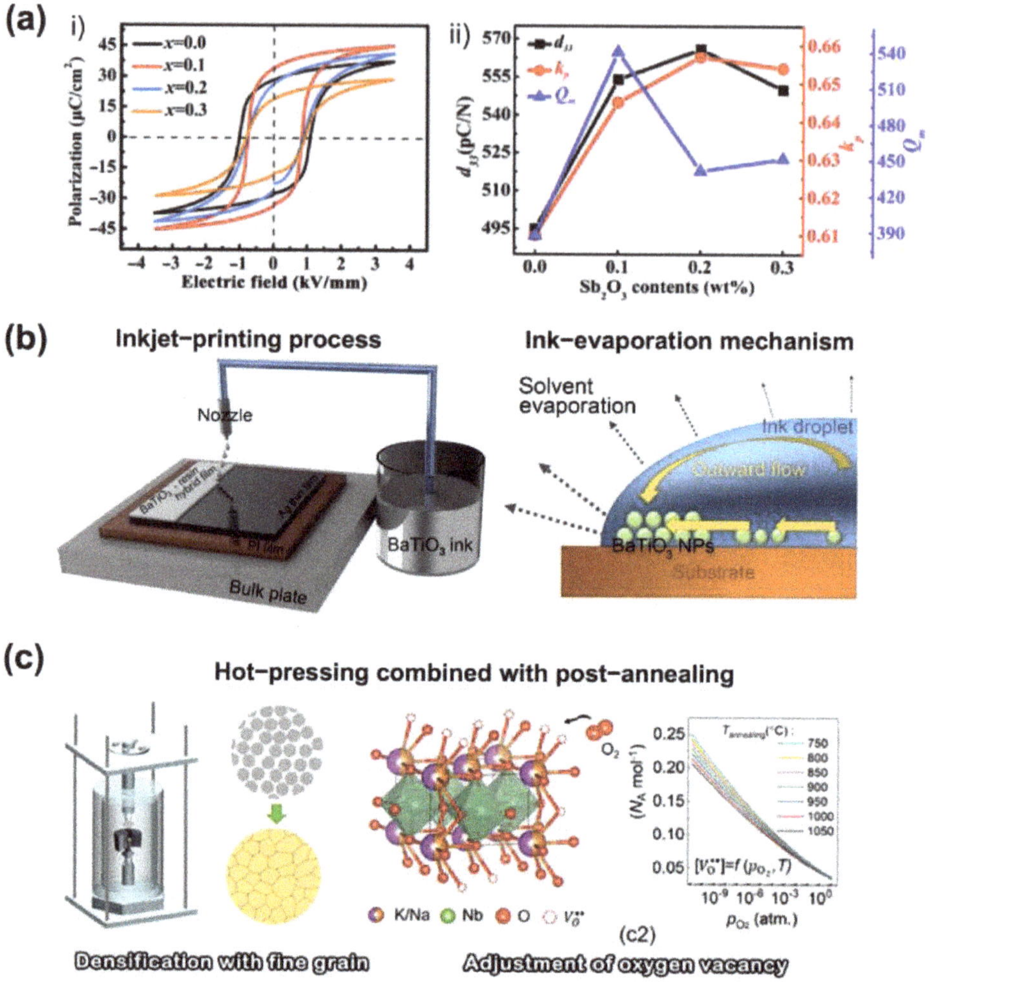

Figure 2. (**a**) P(E) loops of unpoled (xSb, Mn)-doped PZT ceramics at 1 Hz (i) and d_{33}, k_p, and Q_m of poled (xSb, Mn)-doped PZT ceramics (ii) [37]. (**b**) Schematic illustrations for a facile inkjet-printing process and ink solvent evaporation mechanism in the printed droplet [53]. (**c**) Hot-pressing combined with a post-annealing process for the KNN preparation [46].

3.2. Organic Piezoelectric Materials

Organic piezoelectric materials are mainly piezoelectric polymers, which are widely used in flexible sensors due to their good mechanical properties, including poly(L-lactide) (PLLA) [54], poly(vinylidene fluoride) (PVDF) [55], poly(vinylidene fluoride trifluoroethylene) (P(VDF-TrFE)) [56], polyimides (PI) [57], polyacrylonitrile (PAN) [58], etc. As shown in Figure 3a, Sohini et al. first reported the direct observation of shear piezoelectricity in highly crystalline and oriented PLLA nanowires grown by a novel template-wetting method [59].

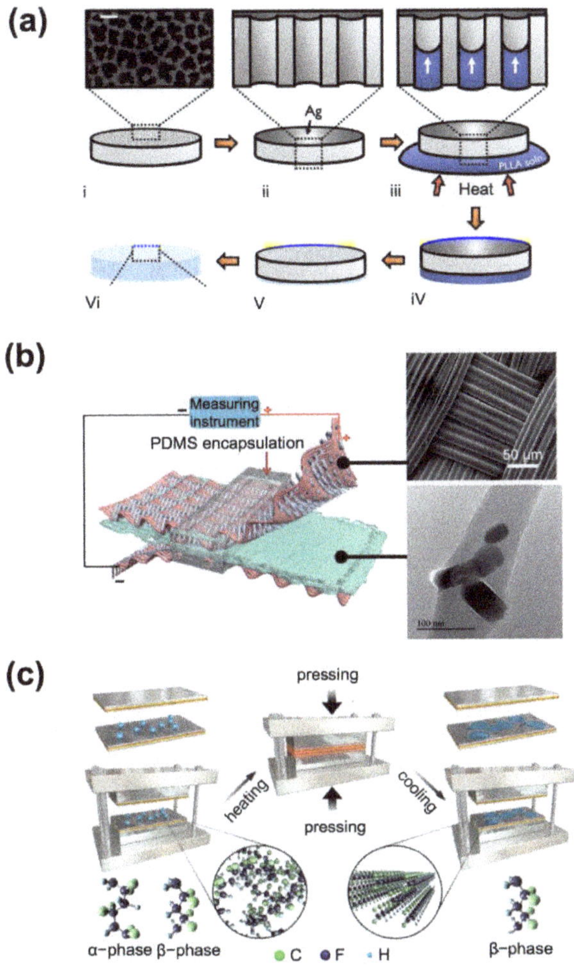

Figure 3. (**a**) A schematic of the temperature controlled capillary infiltration process used to grow the PLLA nanowires [59]; (**b**) schematic diagram of ZnO-nanoparticle-reinforced PVDF nanofibers by electrospinning [60]; (**c**) a typical hot-pressing method adopted to fabricate ultrathin PVDF nanoflakes [55].

PVDF is the most typical organic piezoelectric material, which is versatile and light in weight in comparison to piezoelectric ceramics. In consequence, thin films of any desired form can be drawn into them, giving them an advantage over piezoceramics in various applications involving complex designs of sensors. Despite organic piezoelectric materials having lower electromechanical coupling compared to inorganic counterparts, other characteristics that make the piezoelectric polymers attractive are their low electrical permittivity, low acoustic impedance, high voltage sensitivity, and relatively lower cost. Mandal et al. designed ZnO-nanoparticle-reinforced PVDF nanofibers by electrospinning (Figure 3b) [60]. Nanofibers act as the active layer and interlocked conducting microfiber composite mats as electrodes to convert the mechanical energy into electrical power. Taking advantage of high flexibility and easy processability of PVDF, Wang et al. fabricated ultrathin PVDF nanoflakes with thicknesses down to 7 nm by using a hot-pressing method (Figure 3c) [55]. This thermo-mechanical strategy simultaneously induces robust thermodynamic α to electroactive β phase transformation, with β fraction as high as 92.8% in sub-10 nm flakes.

3.3. Composite Piezoelectric Materials

As mentioned above, piezoelectric polymers have good mechanical properties. However, their low piezoelectric coefficients limit the performance of sensors. Therefore, many typical inorganic piezoelectric materials, such as PZT [61] and ZnO [62], have been used to fabricate piezoelectric composites to improve the performance of piezoelectric polymer sensors. At the same time, the piezoelectric composites obtained by doping the inorganic fillers in the polymer also show better physical properties than the single component. Niu et al. prepared stretchable ceramic/polymer piezoelectric composite by mixing PZT with solid silicone rubber, with 92% filler mass ratio and 30% deformation (Figure 4a) [61]. In another work, Gu et al. prepared a novel flexible ZnO/PVDF nanocomposite fibrous membrane by electrospinning (Figure 4b) [62]. The study showed that the synergistic effect between the rod-like piezoelectric nanofillers and electroactive β-crystals of PVDF plays an important role in enhancing the piezoelectric behaviors of a ZnO/PVDF nanocomposite.

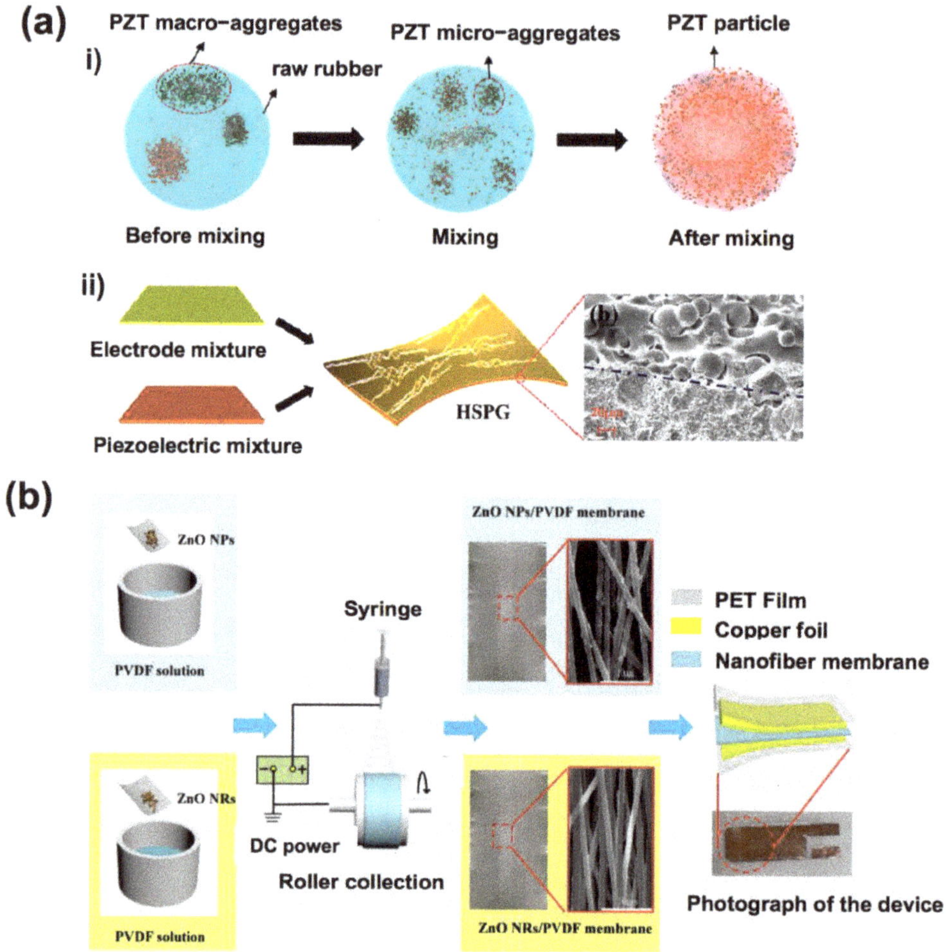

Figure 4. (**a**) Fabrication process and characterizations of the high performance PZT-based stretchable piezoelectric composite [61]; (**b**) schematic illustration of ZnO/PVDF nanocomposite fibrous membrane [62].

4. Piezoelectric Sensors for Structural Health Monitoring

The following section presents an updated literature review of piezoelectric sensors for the structural health monitoring in the recent years. In order to elaborate clearly, different technologies are simply introduced and recent relevant progress on piezoelectric sensors are reviewed in detail. The piezoelectric SHM can act in both active and passive modes to realize the SHM. Active monitoring sensors can be utilized based on electro-mechanical impedance, guided wave propagation, or ultrasonic propagation, while passive monitoring sensors based on acoustic emission or stress wave propagation.

4.1. Sensors Based on Electro-Mechanical Impedance (Active Mode)

The electromechanical impedance-based (EMI) monitoring is one of the innovative and powerful structure health monitoring techniques. The electromechanical impedance-based SHM utilizes the electromechanical property of piezoelectric materials and the coupling of piezoelectric sensors and target structure. The EMI sensors can be attached to the surface of the structure or embedded into the structure, such as smart aggregates. In this active monitoring method, the piezoelectric EMI sensors can work as actuators, converting the electric voltage signal into a mechanical stress solicitation. Meanwhile, the piezoelectric sensors can act as the sensors, converting the structure's mechanical response to an electric signal. The feedback electric signals can be further dealt, using algorithms to detect, localize, and characterize damages in the structure. The EMI methods are used for continuous monitoring and early detection of structural defects, such as joint looseness, debonding, and crack detection. Liang et al. were the first to propose the concept of the electromechanical impedance for the EMI system and suggested 1-DOF free-body diagram of a PZT–structure system to explain the interaction between the PZT and the host structure [63,64]. In addition, for the last decade, EMI sensors have been used to detect the occurrence, location, and characterization of damage in the concrete structures [65,66], composite structures [67], rotary blades in turbomachine [68,69], and bolted or adhesive joints [70,71]. Due to the limitation of damage-detecting area, EMI sensors are always attached to the positions that are important and prone to damage, such as bolted joints, adhesive joints, blades, etc. In the development of EMI sensors, the sensitivity and self-diagnosis of piezoelectric sensors have been investigated to guarantee the adequate application of EMI sensors.

In order to ensure immunity to ambient noise and vibrations commonly present in practical applications, variable high frequencies (typically larger than 30 kHz) are preferred. This makes the EMI sensors sensitive to the minor damage; however, it also brings the issue of weak signals from far away, which limits the EMI sensor to the local damage monitoring. Some investigations on the design of EMI sensors have been conducted to improve the sensitivity of the monitoring system [72–74]. The sensitivity of detecting damage through EMI is closely related to the selected frequency band/wavelength of the excitation signal, which is emitted by the EMI sensor. The size of PZT piezoelectric sensor affects ZS/ZT (ratio of host structures' mechanical impedance to the PZT transducer's mechanical impedance), which is equivalent to the sensitivity. For using the frequencies below 125 kHz, the sizes (length and width) of sensors should fall into the range of 5 mm to 20 mm and thickness of sensors of 0.1 mm to 0.3 mm [72]. Hire et al. optimized the size of the piezoelectrical patch for optimum corrosion detection in reinforced concrete by combining theoretical and experimental studies [74]. They utilized the model developed by Giurgiutiu et al. to evaluate the impact of sensor patch sizes to the sensitivity range. They also measured the impedance of actual steel with sensor patch in air and in concrete. As shown in Figure 5, the experiment results are consistent with the theoretical results. In addition, it demonstrates that the appropriate design of the piezoelectric patch can improve its sensitivity for damage detection.

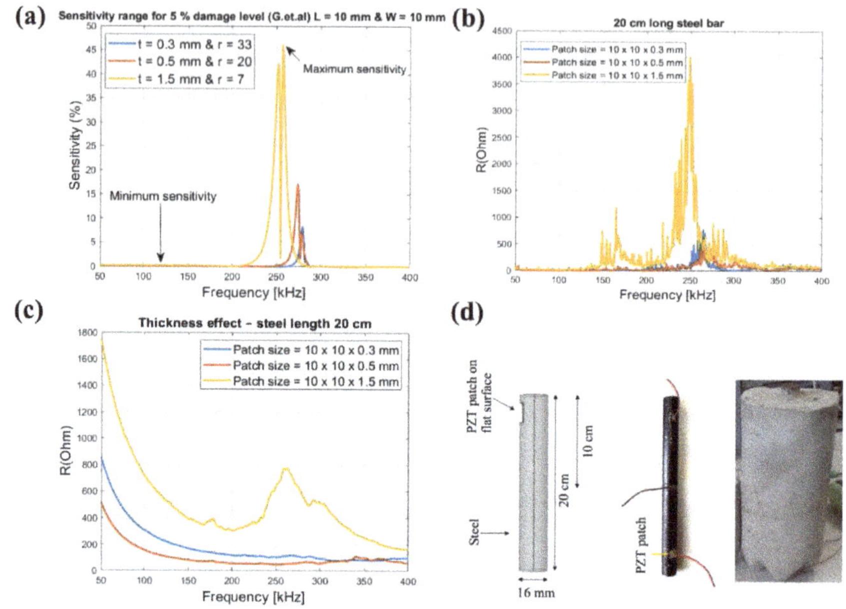

Figure 5. (**a**) Sensitivity levels with respect to thickness variations using Giurgiutiu's model; (**b**) the real part of impedance in free air over frequency ranges from 50 to 400 kHz related to different patch size; (**c**) the real part of the impedance in reinforced concrete with varying patch thickness; (**d**) diagram of steel dimensions, actual steel, and actual concrete used in the experiments [74].

Recently, the investigation of sensors with the capability of self-diagnosis for electromechanical health monitoring has become an essential issue [75–78]. Distinguishing sensor faults and functional degradations from structure damage will directly affect the effectiveness and accuracy of SHM. Jiang et al. introduced a K-means clustering analysis and artificial neural network to realize the self-diagnosis of piezoelectric active sensor for electromechanical impedance monitoring [75]. Three principal components, including the average change of conductance peak, the RMSD of susceptance, and the RMSD of conductance, were extracted by principal component analysis from the impedance signals. Then, the K-means algorithm was used to cluster different cases of sensor damages represented by the principal components. Finally, the analysis and artificial neural network were used to identify degree of the sensor damages. Four kinds of sensor damages, namely, pseudo-soldering, debonding, wear, and breakage, can be distinguished from the structure damage using the K-means clustering analysis based on admittance characteristics. Nguyen et al. established a finite element model corresponding to an experimental model based on a bolted steel girder connection to investigate the EMI response characteristics of a degraded piezoelectric-based smart interface [76]. Four common degradation types, including shear lag effect, transducer debonding, transducer breakage, and interface detaching, are simulated and their effects on EMI response are comprehensively analyzed. Figure 6 shows the photo and schematic picture of smart interface of PZT sensors. It was found that the transducer breakage occurring in the smart interface can result in unique shifts in the imaginary admittance and they can be feasibly diagnosed and differentiated from the structural damage through a diagnosis process, as shown in Figure 6.

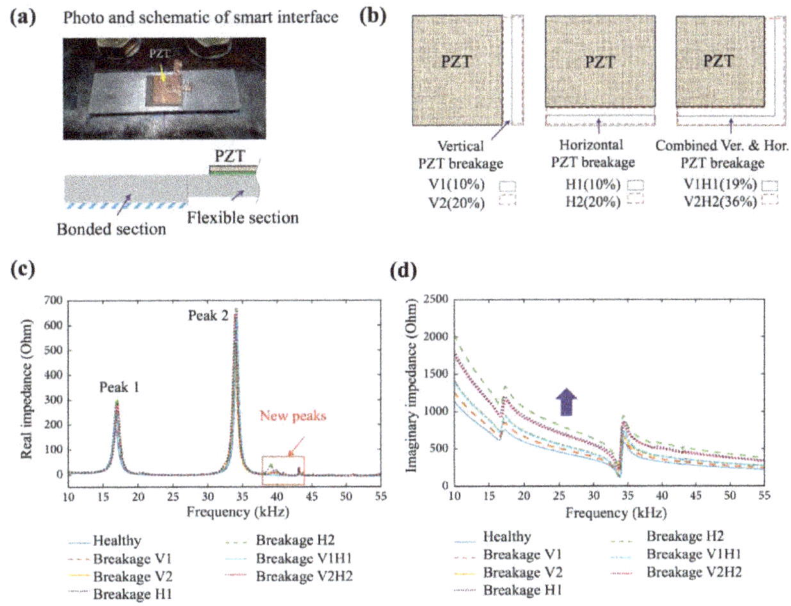

Figure 6. (**a**) Photo and schematic of smart interface with PZT transducer; (**b**) schematic of PZT breakage; (**c**,**d**) numerical EMI responses under PZT debonding defects, PZT breakage, and interface debonding defect [76].

4.2. Sensors Based on Guided Wave or Ultrasonics Propagation (Active Mode)

In this section, a typical structural health monitoring technique based on a specific wave, which can be guided wave or ultrasonics, is introduced. Guided wave, such as Lamb and Rayleigh waves, have the characteristics of surface propagation, low energy loss, and long propagation range. It determines that sensors based on guided wave propagation are technology of particular importance in SHM.

Guided wave-propagation-based SHM is most widely used for damage detection in metallic and composite structures, existing in aircraft, pressure vessels, missiles, pipelines, and steel bridges [14]. Piezoelectric components can be utilized as actuators or sensors in the guided wave-propagation-based SHM. According to the functionality of piezoelectric sensors, there are mainly four modes of the guided wave propagation, including pitch–catch mode, pulse–echo mode, thickness mode, and impact detection mode, as shown in Figure 7 [79]. Taking the typical pitch–catch mode as an example, a pair of piezoelectric transducers are attached on the plate-like structures. Firstly, ultrasonic guided waves are induced by the piezo-actuator attached to the surface of a flat plate-like structure. Secondly, ultrasonic disturbances occur and propagate radially to the around in the structure. Finally, the piezo-sensor around receives the electric charge signal, owing to the induced mechanical strains and output voltage signals (sensing waveform). While there is an even damage existing in the structure, the guided wave (such as Lamb wave) would incur dispersion and energy would attenuate during the propagation in the pitch–catch mode, pulse–echo mode, and thickness mode. The thickness mode can be used for the detection of corrosion thickness loss. In the impact detection mode, the piezoelectric sensors would receive a signal of acoustic guided wave, while impact events on the structure and advancing cracks occur. This is specific in acoustic emission, which will be introduced in detail in Section 4.3.

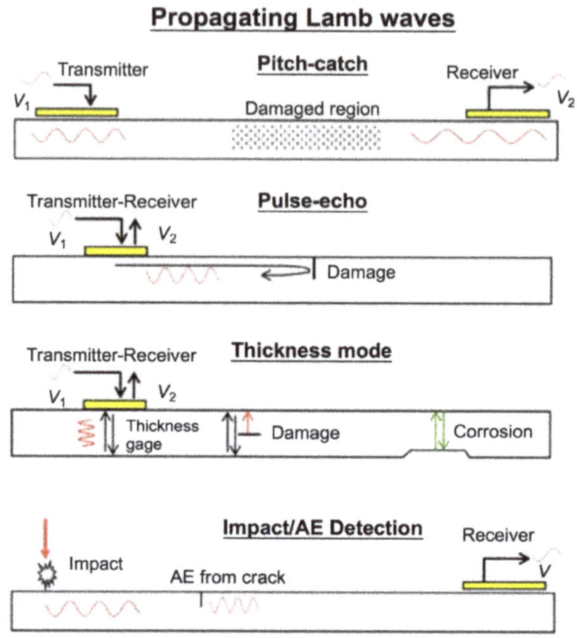

Figure 7. Modes of operation of piezoelectric transducers using propagating Lamb waves [79].

Compared with the local electromechanical impedance-based SHM, the guided wave monitoring technology can both realize local damage monitoring and global monitoring. In the case of local damage monitoring, the guided waves could be used to monitor the hybrid bonded joints. Jahanbin [80] utilized the ultrasonic interface guided waves that were propagating on the boundaries of bonded joints to inspect the disbond and delamination by recording the change in wave form, energy attenuation, and time of flight. For global monitoring, the selection or design of the piezoelectric sensors/transducers is of significant importance. To realize the global monitoring using a lower quantity of transducers, piezoelectric transducers exciting high purity of shear horizontal (SH0) wave are preferred. Boivin [81] optimized the geometry of the transducer to obtain a 23.0 dB SH0 wave using both simulations and experiments, and Zennaro [82] revised the transducer design (removed the wrap-around electrode) to eliminate the out-of-plane wave well in the propagation direction of fundamental SH0 mode. Ochoa et al. proposed a systematic multiparameter design methodology for the piezoelectric transducers used in SHM, and the multiparameter design includes the transducer shape, piezoelectric material, and transducer size [83].

Apart from the transducers themselves, transducer/sensor network optimization offers an opportunity to obtain robust and low-cost active SHM systems. Two reviews had focused on the sensor networks for SHM by Ostachowicz [84] and Mustapha [85]. Sensor networks should be well-designed as they affect the SHM system integration, system performance, and accuracy of assessment. In the sensor network, deciding the number and locations of the sensors are the primary work for accurate damage detection and localization, then followed by the data transmission, data processing, etc. Many researchers have focused on determining the minimum number of piezoelectric transducers and achieving full or high coverage of damage monitoring by developing algorithms. A lot of algorithms, such as iterative optimization, combinatorial optimization, genetic algorithms (GA), and artificial neural network (ANN) techniques, have been developed to optimize distribution of sensors [85]. Recently, Ismail et al. proposed an approach of transforming any complex or closed structure surface, and then used a genetic algorithm to optimize the deployment of piezoelectric sensors [86]. The coverage of the optimized sensor network increases from 85 to 99% for a pipe-like structure, as shown in Figure 8a. Experimental validation was performed on a circular section (pipe) and the artificial damage can be accurately located within 18 mm from the actual location in Figure 8b. They also developed a similar model for sensor network optimization based on genetic algorithms, which was further validated on a large cargo door of an A330 airplane [87].

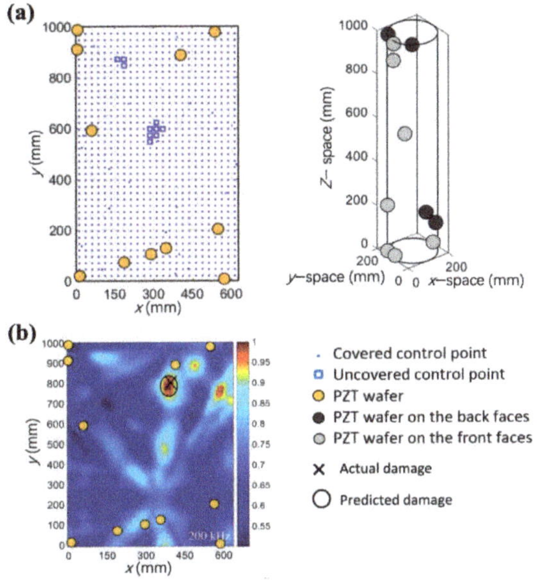

Figure 8. (**a**) A sensor network mounted on a pipe-like structure with 99% detecting coverage; (**b**) the location of damage identified on the pipe surface [86].

In addition to the optimization of sensor network for global monitoring, there are some interesting works on the topic of large area monitoring. Nonlinear lamb waves have attracted intensive attentions to be used for contact-type damages (e.g., disbands, delamination, and micro-crack) because of its high sensitivity to these damages [88,89]. Ju et al. proposed a nonlinear ultrasonic testing method for large-area monitoring of practical structures with arbitrary complexity using multi-mode guided waves [88]. When the multi-mode guided waves diffusely propagate through the structure, all available guided wave modes are automatically down-selected by the medium through attenuation or dispersion and the remaining modes efficiently transfer energy (for example, to their second harmonic modes) after encountering the micro-cracks. This method assures the success in detecting the crack that is close to the middle of actuator and sensor with a distance of 0.95 m. When large-scale monitoring is demanded, smart sensor networks are proposed to overcome the problems regarding large number of sensors, the complex cable, and placement efficiency. Many investigators developed the smart sensor networks combing the piezoelectric materials and flexible circuit technology [90–92]. The smart sensor networks have the advantages of light weight, high placement efficiency, and being suitable for complex structural forms. Ren et al. developed a large-scale PZT network layer (LPNL) design method based on FPC technology and one of the LPNLs connecting 37 PZT transducers has a dimension of 565 mm × 500 mm [91]. A small section of LPNL is shown in Figure 9a and one of the LPNL network is attached to the composite plate with the detail placement in Figure 9b,c. The maximum localization error of its damage diagnosis using this LPNL network is 0.41 cm, as shown in Figure 9d.

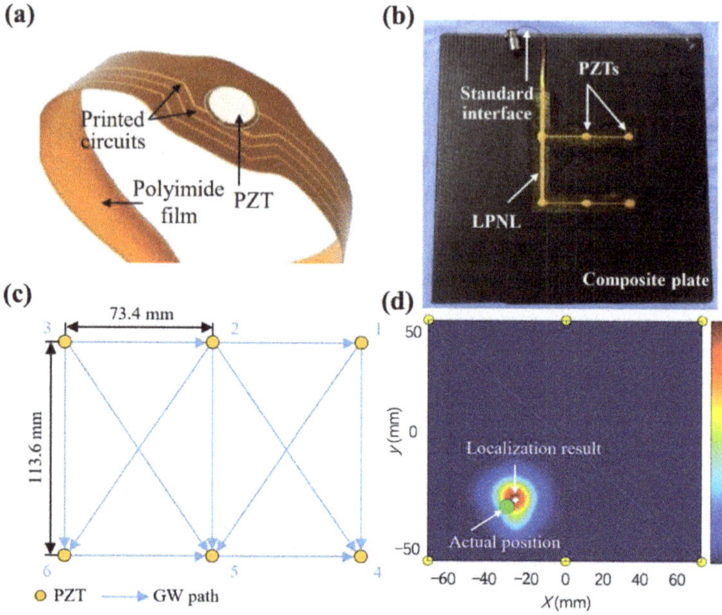

Figure 9. (**a**) Small section of LPNL and (**b**) the adopted LPNL, (**c**) its placement, (**d**) its imaging result of damage [91].

4.3. Sensors Based on Acoustic Emission and Stress Wave (Passive Mode)

While the structures are subjected to impact by external force or irreversible damage with ultra-limited internal stress, the structures would generate transient elastic waves to release strain energy. Structure health monitoring based on this kind of elastic wave is also called acoustic-emission-based SHM. This typical acoustic-emission-based SHM is preferred to be introduced separately from the foregoing topics, as this method is a specific passive monitoring. Acoustic emission method is available only for damage initiation and propagation, such as impact event, crack initiation, fiber breakage, debonding, and delamination. For example, when a sudden crack occurs in the structure, piezoelectric sensors can catch the signal of acoustic emission from the crack. The acoustic-emission-based SHM is a local monitoring. Piezoelectric sensors should be placed near the key objects where damages or defects are prone to occurring.

The acoustic-emission-based SHM has been deeply developed after its initial trail for damage monitoring. In the early research, the acoustic emission method was only used for the detection of the damage occurrence [93,94]. Then, this method was improved to localize the damage in metallic plates and composite structures [95–97]. Capineri et al. have reviewed the acoustic emission sensors and advanced methods for impact detection and localization [98]. Many papers have been published to improve the reliability and accuracy of detection and localization of damages, such as Akaike Information Criterion (AIC) for the accurate estimation of measured differential time of arrival [95], artificial neural network (ANN) [99], and theoretical modeling based on the phase velocity analysis [100]. In the recent years, much more advanced methods were evaluated to determine the type, magnitude, and severity of the impact or defects in the structures [101]. Garrett et al. proposed an artificial intelligence approach to estimate the fatigue crack length in thin metallic plates using acoustic-emission-based SHM [101]. Finite element modeling was firstly conducted to establish the simulated frequency spectra of calculated PWAS responses for different fatigue-crack-generated AE signals. Then, the Choi–Williams transform (CWT) result of the experimental inspected structure could be obtained from the raw acoustic emission signals following the process flow, as shown in Figure 10. It shows that a convolutional neural network (CNN) was successfully used for the artificial intelligence processing of the AE signals to predict the crack length.

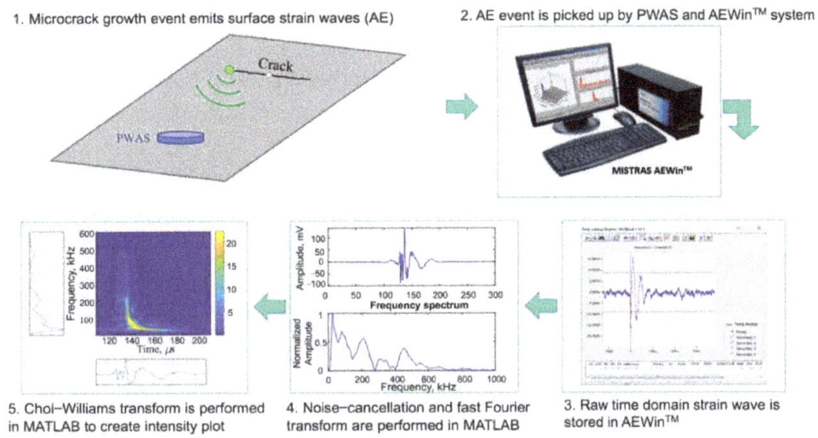

Figure 10. Flowchart of acoustic emission signal processing from microcrack growth event to the time–frequency intensity plot of Choi–Williams transform [101].

The piezoelectric stress/strain sensing technique is a promising approach of SHM techniques for dynamic loading to the structures, such as railway bridge [102] and aircraft wings [103]. When the structures are under low-frequency dynamic loadings, elastic waves would generate and propagate to the piezoelectric materials. The stress variation of structures can be reflected by analyzing the output voltage of piezoelectric sensors based on the direct piezoelectric effect. The application of piezoelectric stress/strain sensing technique for SHM has triggered worldwide interest after the first investigation by Krueger et al. [104]. Sha et al. developed an embedded smart piezoelectric sensor for concrete SHM [16]. The ratio of encapsulation materials is optimized to have proper mechanical performance. Furthermore, the mechanical sensing property of the embedded sensors with dynamic compressive loadings were studied in concrete, which demonstrates the great potentials of applying the piezoelectric sensors for concrete SHM.

4.4. Integrated Passive and Active Sensors

In some practical applications, only passive monitoring is not enough. For example, the impact events on airplane or aerospace structures might cause damage to the structures and the damage could also worsen or become extended with time due to the operational fatigue. In this scenario, the active monitoring is needed to continuously monitor the damage progression. In analogy, purely active monitoring also has its drawbacks. Piezoelectric transducers need to send out inspect waves continuously, resulting in useless work when they are not demanded. Therefore, it is desirable to integrate the passive and active sensors to overcome their drawbacks.

Bulletti et al. developed an integrated acoustic/ultrasonic SHM system for composite pressure vessels (massively used as fuel tanks) using the same piezoelectric transducers [105]. Two flexible arrays of PVDF interdigital transducers were designed. The transducers have two functionalities: passive detection of impacts and active damage assessment using guided Lamb waves. Guo et al. proposed a piezoelectric transducer-based integrated SHM system for impact monitoring and impedance measurement [106]. A "scheduling module" method is utilized to schedule the commands to the PZT sensors and transfer their signals to different preprocessing units for impact detecting or EMI measurement. This designed system is deployed in a supporting structure of a sailplane. Gayakwad et al. developed smart sensing units (SSU) to improve the effectiveness of the monitoring system, which contain PZT patches to detect both near-field and far-field damage in concrete structures through EMI and wave propagation techniques [107]. The PZT patch in the SSU-1 is used as an EMI admittance sensor for local damage identification and, meanwhile, the same EMI sensor is used to acquire elastic waves generated by another PZT patch in SSU-2 to monitor damages outside the EMI admittance sensor's sensing area, as shown in Figure 11. Figure 11b,c show the schematic diagram of a concrete cube with embedded SSU and an experimental specimen with embedded PZT patches and crack, respectively.

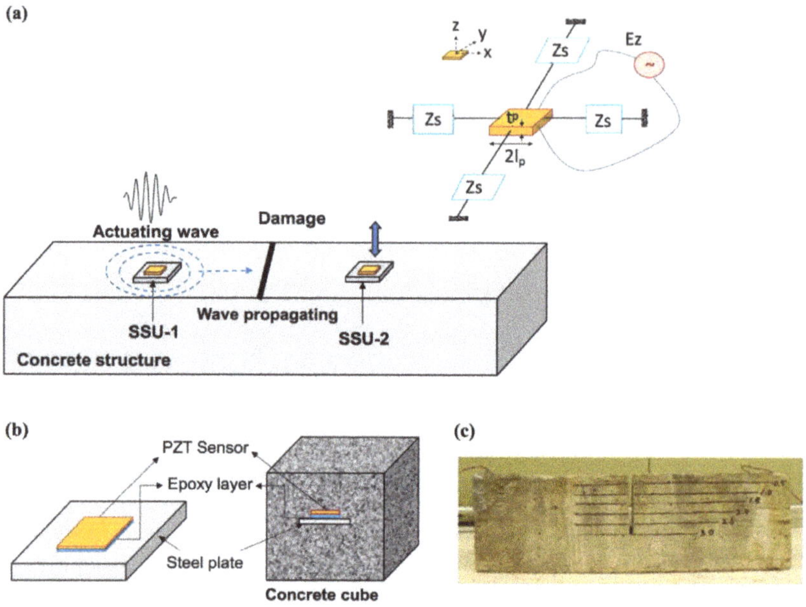

Figure 11. (**a**) Schematic representation of the combined EMI-WP technique for damage detection in concrete structure utilizing two SSUs; (**b**) the schematic diagram of concrete cube with embedded SSU and (**c**) an experimental specimen with embedded PZT patches and crack [107].

5. Challenges, Opportunities, and Future Prospects

Piezoelectric sensors are indispensable for structural health monitoring. The last decades have witnessed many exciting developments of piezoelectric materials and sensors for structural health monitoring. Despite these breakthroughs, there is still plenty of room for further improvement, as exemplified by the following challenges.

Piezoelectric materials are the key component of piezoelectric sensors for structural health monitoring. The development of high-performance piezoelectric materials is essential to enable high-end piezoelectric sensors with exceptional sensitivity and further promote their practical applications in structural health monitoring. The combination of experimental synthesis, comprehensive characterization and the concept of materials genome, high-throughput calculations, and machine learning is expected to substantially accelerate the discovery of piezoelectric materials with unprecedented piezoelectrical properties [108]. As mentioned above, inorganic piezoelectric materials usually have good piezoelectric property but poor flexibility. In contrast, stretchable organic piezoelectric materials have low piezoelectric coefficients. Composite piezoelectric materials and sensors integrated with simultaneously high piezoelectricity and decent flexibility show unprecedented opportunity to improve the performance of SHM systems, such as working range, complex mechanical loadings, etc.

Piezoelectric materials always work as stacked wafers in the piezoelectric sensors and SHM systems, as shown in many reviewed examples. However, it increases the difficulty of placement and decreases the reliability, such as debonding fault. Piezoelectric films can be attached to the surface of the structure or embedded in the composite structure, which make them highly flexible and adaptable for deployment on complex configurations [92]. Advanced material processing technologies represented by additive manufacturing and three-dimensional printing offer new platforms to manufacture and implement piezoelectric sensors for structural health monitoring [109]. Piezoelectric materials, including piezoceramics, polymers, or composites, can be facilely processed into sensor components by these technologies.

So far, most of the work focuses on the research to address the deployment problems in SHM techniques in the category of experiment. From the experiment result to the practical applications, many problems need to be resolved, such as sensors self-detection, influence of environmental change on the monitoring signals, fundamental structure failure mode, and so on. For a reliable SHM system, especially in engineering system, accurate sensor self-diagnostics is one of the major issues. The sensor failures without self-identification will lead to false result in damage detection. For the long-term monitoring in the engineering SHM, sensor failure or debonding failure tends to occur. Different sensor faults need to be investigated to increase reliability of SHM systems [110]. Self-diagnostics for piezoelectric transducers have been investigated by much research based on the change in wave propagation signature or the change in EMI signature [77,111,112]. Sensor self-diagnostics would improve the reliability in the engineering SHM application in the future. The change in environment condition would bring error to the SHM systems. Lots of properties of piezoelectric materials would vary when the operating temperature changes, such as piezoelectric properties, dielectric constant, coupling constant, and Young's modulus. For the guided-wave-based SHM systems, the temperature would make an impact on the guided wave baseline comparison and an optimal baseline selection method for the environment temperature range should be adopted [113,114]. For the EMI-based SHM technique, the dielectric constant exhibits the most significant effect on the electrical impedance of PZT sensor, which results in the shift of frequency and amplitude of impedance signatures [12]. Furthermore, the environmental effect of ambient induced noises, vibrations, and external loads also should be dealt with appropriately in the practical engineering SHM systems.

The material, sensor, and structural designs play an important role in the SHM systems. This phenomenon is particularly obvious in the EMI-based monitoring techniques. The working frequency range is decided not only by the design of piezoelectric sensor, but also by the structure properties. Traditional EMI methodology for indicating a failure is not enough. Developing a technique to determine the impact of damage on structural properties is demanding. Aabid et al. proposed some open research areas, such as packaging piezoelectric materials to generate a high working frequency range, integrating a network of sensors with local artificial intelligence (AI)/machine learning (ML) data-processing platforms and so on [10]. As reviewed above, only sensors based on guided wave propagation or acoustic emission have the opportunity to realize the global damage monitoring. However, structure health monitoring in the practical engineering applications is always preferred to cover the large area, such as in the bridge, aircraft, airplane, etc. It would make the SHM systems contain plenty of piezoelectric transducers and a complex line layout. The wireless sensor network to collect and process the information becomes the priority way to optimize the SHM system. Generally, bulky batteries are needed to provide power for the wireless sensor network, and these batteries should be frequently replaced due to the limited capacity. Therefore, a self-powered wireless sensor network is highly demanded, particularly for the case of structural health monitoring. The direct piezoelectric effect enables the piezoelectric devices to harvest electrical power from ambient mechanical and vibrational energies, such as structure vibration, airflow, etc. [115]. The piezoelectric energy harvesters can be integrated to the wireless sensor network for structural health monitoring to provide an unbounded power source for the system.

6. Conclusions

In this contribution, the recent progress in piezoelectric materials and sensors for structural health monitoring has been systematically reviewed. A brief introduction of the fundamental physical science of piezoelectric effect was introduced. Emphases are placed on the piezoelectric materials engineered by various strategies and the applications of piezoelectric sensors for structural health monitoring. Finally, challenges along with opportunities for future research and development of high-performance piezoelectric materials and sensors for structural health monitoring are highlighted. It is expected that the contribution could accelerate the development of high-end piezoelectric materials and sensors for structural health monitoring.

Author Contributions: Conceptualization, K.W. and F.-Z.Y.; writing—original draft preparation and writing—review and editing, M.J., Z.D. and F.-Z.Y.; Validation and visualization, J.-W.L., X.Q., B.S., D.Z. and W.G. All authors have read and agreed to the published version of the manuscript.

Funding: This research was funded by National Key Research and Development Program of China (No.2021YFB2012502) and National Nature Science Foundation of China (No. 52032005, U22A20254).

Institutional Review Board Statement: Not applicable.

Informed Consent Statement: Not applicable.

Data Availability Statement: Not applicable.

Conflicts of Interest: The authors declare no conflict of interest.

References

1. Sivasuriyan, A.; Vijayan, D.S.; Gorski, W.; Wodzynski, L.; Vaverkova, M.D.; Koda, E. Practical implementation of structural health monitoring in multi-story buildings. *Buildings* **2021**, *11*, 263. [CrossRef]
2. He, Z.G.; Li, W.T.; Salehi, H.D.; Zhang, H.; Zhou, H.Y.; Jiao, P.C. Integrated structural health monitoring in bridge engineering. *Autom. Constr.* **2022**, *136*, 104168. [CrossRef]
3. Das, S.; Saha, P. A review of some advanced sensors used for health diagnosis of civil engineering structures. *Measurement* **2018**, *129*, 68–90. [CrossRef]
4. Yin, J.; Chen, S.T.; Wong, V.K.; Yao, K. Thermal sprayed lead-free piezoelectric ceramic coatings for ultrasonic structural health monitoring. *IEEE Trans. Ultrason. Ferroelectr. Freq. Control* **2022**, *69*, 3070–3080. [CrossRef] [PubMed]
5. Glisic, B. Concise historic overview of strain sensors used in the monitoring of civil structures: The first one hundred years. *Sensors* **2022**, *22*, 2397. [CrossRef]
6. Ragam, P.; Sahebraoji, N.D. Application of MEMS-based accelerometer wireless sensor systems for monitoring of blast-induced ground vibration and structural health: A review. *Iet Wirel. Sens. Syst.* **2019**, *9*, 103–109. [CrossRef]
7. Garcia, I.; Zubia, J.; Durana, G.; Aldabaldetreku, G.; Illarramendi, M.A.; Villatoro, J. Optical fiber sensors for aircraft structural health monitoring. *Sensors* **2015**, *15*, 15494–15519. [CrossRef]
8. Sakiyama, F.I.H.; Lehmann, F.; Garrecht, H. Structural health monitoring of concrete structures using fibre-optic-based sensors: A review. *Mag. Concr. Res.* **2021**, *73*, 174–194. [CrossRef]
9. Bonopera, M. Fiber-bragg-grating-based displacement sensors: Review of recent advances. *Materials* **2022**, *15*, 5561. [CrossRef]
10. Aabid, A.; Parveez, B.; Raheman, M.A.; Ibrahim, Y.E.; Anjum, A.; Hrairi, M.; Parveen, N.; Mohammed Zayan, J. A review of piezoelectric material-based structural control and health monitoring techniques for engineering structures: Challenges and opportunities. *Actuators* **2021**, *10*, 101. [CrossRef]
11. Dilek, A.U.; Oguz, A.D.; Satis, F.; Gokdel, Y.D.; Ozbek, M. Condition monitoring of wind turbine blades and tower via an automated laser scanning system. *Eng. Struct.* **2019**, *189*, 25–34. [CrossRef]
12. Huynh, T.-C. Advances and challenges in impedance-based structural health monitoring. *Struct. Monit. Maint.* **2017**, *4*, 301–329.
13. Na, W.S.; Baek, J. A review of the piezoelectric electromechanical impedance based structural health monitoring technique for engineering structures. *Sensors* **2018**, *18*, 1307. [CrossRef]
14. Qing, X.; Li, W.; Wang, Y.; Sun, H. Piezoelectric transducer-based structural health monitoring for aircraft applications. *Sensors* **2019**, *19*, 545. [CrossRef]
15. Nasir, V.; Ayanleye, S.; Kazemirad, S.; Sassani, F.; Adamopoulos, S. Acoustic emission monitoring of wood materials and timber structures: A critical review. *Constr. Build. Mater.* **2022**, *350*, 128877–128893. [CrossRef]
16. Sha, F.; Xu, D.; Cheng, X.; Huang, S. Mechanical sensing properties of embedded smart piezoelectric sensor for structural health monitoring of concrete. *Res. Nondestruct. Eval.* **2021**, *32*, 88–112. [CrossRef]
17. Miao, H.; Li, F. Shear horizontal wave transducers for structural health monitoring and nondestructive testing: A review. *Ultrasonics* **2021**, *114*, 106355–106381. [CrossRef]

18. Song, S.; Hou, Y.; Guo, M.; Wang, L.; Tong, X.; Wu, J. An investigation on the aggregate-shape embedded piezoelectric sensor for civil infrastructure health monitoring. *Constr. Build. Mater.* **2017**, *131*, 57–65. [CrossRef]
19. Li, Y.; Feng, W.; Meng, L.; Tse, K.M.; Li, Z.; Huang, L.; Su, Z.; Guo, S. Investigation on in-situ sprayed, annealed and corona poled PVDF-TrFE coatings for guided wave-based structural health monitoring: From crystallization to piezoelectricity. *Mater. Design* **2021**, *199*, 109415–109432. [CrossRef]
20. Guo, S.; Chen, S.; Zhang, L.; Liew, W.H.; Yao, K. Direct-write piezoelectric ultrasonic transducers for pipe structural health monitoring. *NDT E Int.* **2019**, *107*, 102131–102137. [CrossRef]
21. Zhang, X.; Wang, Y.; Shi, X.; Jian, J.; Wang, X.; Li, M.; Ji, Y.; Qian, F.; Fan, J.; Wang, H.; et al. Heteroepitaxy of flexible piezoelectric $Pb(Zr_{0.53}Ti_{0.47})O_3$ sensor on inorganic mica substrate for lamb wave-based structural health monitoring. *Ceram. Int.* **2021**, *47*, 13156–13163. [CrossRef]
22. Tenreiro, A.F.G.; Lopes, A.M.; da Silva, L.F.M. A review of structural health monitoring of bonded structures using electromechanical impedance spectroscopy. *Struct. Health Monit.* **2021**, *21*, 228–249. [CrossRef]
23. Tuloup, C.; Harizi, W.; Aboura, Z.; Meyer, Y.; Khellil, K.; Lachat, R. On the use of in-situ piezoelectric sensors for the manufacturing and structural health monitoring of polymer-matrix composites: A literature review. *Compos. Struct.* **2019**, *215*, 127–149. [CrossRef]
24. Tuloup, C.; Harizi, W.; Aboura, Z.; Meyer, Y. Integration of piezoelectric transducers (PZT and PVDF) within polymer-matrix composites for structural health monitoring applications: New success and challenges. *Int. J. Smart Nano Mater.* **2020**, *11*, 343–369. [CrossRef]
25. Kaewniam, P.; Cao, M.; Alkayem, N.F.; Li, D.; Manoach, E. Recent advances in damage detection of wind turbine blades: A state-of-the-art review. *Renew. Sustain. Energy Rev.* **2022**, *167*, 112723–112746. [CrossRef]
26. Le, T.C.; Luu, T.H.T.; Nguyen, H.P.; Nguyen, T.H.; Ho, D.D.; Huynh, T.C. Piezoelectric impedance-based structural health monitoring of wind turbine structures: Current status and future perspectives. *Energies* **2022**, *15*, 5459. [CrossRef]
27. Chen, Y.; Xue, X. Advances in the structural health monitoring of bridges using piezoelectric transducers. *Sensors* **2018**, *18*, 4312. [CrossRef]
28. Jiao, P.; Egbe, K.-J.I.; Xie, Y.; Matin Nazar, A.; Alavi, A.H. Piezoelectric sensing techniques in structural health monitoring: A state-of-the-art review. *Sensors* **2020**, *20*, 3730. [CrossRef]
29. Curie, J.; Curie, P. Développement par compression de l'électricité polaire dans les cristaux hémièdres à faces inclinées. *Bull. Minéralogie* **1880**, *3*, 90–93. [CrossRef]
30. Tandon, B.; Blaker, J.J.; Cartmell, S.H. Piezoelectric materials as stimulatory biomedical materials and scaffolds for bone repair. *Acta Biomater.* **2018**, *73*, 1–20. [CrossRef]
31. Park, K.I.; Xu, S.; Liu, Y.; Hwang, G.T.; Kang, S.J.; Wang, Z.L.; Lee, K.J. Piezoelectric $BaTiO_3$ thin film nanogenerator on plastic substrates. *Nano Lett.* **2010**, *10*, 4939–4943. [CrossRef]
32. Gavrilyatchenko, V.; Semenchev, A.; Fresenko, E. Dielectric, elastic and piezoelectric constants of $PbTiO_3$ single crystals. *Ferroelectrics* **2011**, *158*, 31–35. [CrossRef]
33. Park, K.I.; Son, J.H.; Hwang, G.T.; Jeong, C.K.; Ryu, J.; Koo, M.; Choi, I.; Lee, S.H.; Byun, M.; Wang, Z.L.; et al. Highly-efficient, flexible piezoelectric PZT thin film nanogenerator on plastic substrates. *Adv. Mater.* **2014**, *26*, 2514–2520. [CrossRef]
34. Chen, B.; Li, H.; Tian, W.; Zhou, C. PZT based piezoelectric sensor for structural monitoring. *J. Electron. Mater.* **2019**, *48*, 2916–2923. [CrossRef]
35. Hur, S.; Lee, S.Q.; Choi, H.S. Fabrication and characterization of PMN-PT single crystal cantilever array for cochlear-like acoustic sensor. *J. Mech. Sci. Technol.* **2010**, *24*, 181–184. [CrossRef]
36. Shi, P.; Li, T.; Lou, X.; Yu, Z.; Zhu, X.; Zhou, C.; Liu, Q.; He, L.; Zhang, X.; Yang, S. Large electric-field-induced strain and energy storage properties in $Bi_{0.5}Na_{0.5}TiO_3$-$(0.5Ba_{0.7}Ca_{0.3}TiO_3$-$0.5BaTi_{0.8}Zr_{0.2}O_3)$ lead-free relaxor ferroelectric ceramics. *J. Alloys Compd.* **2021**, *860*, 158369–158378. [CrossRef]
37. Fei, X.; Lin, X.; Wang, C.; Li, W.; Yu, F.; Yang, C.; Huang, S. Sb_2O_3-modified lead zirconate titanate piezoelectric ceramics with enhancing piezoelectricity and low loss. *J. Am. Ceram. Soc.* **2023**, *106*, 501–512. [CrossRef]
38. Fang, Z.; Tian, X.; Zheng, F.; Jiang, X.; Ye, W.; Qin, Y.; Wang, X.; Zhang, Y. Enhanced piezoelectric properties of Sm^{3+}-modified PMN-PT ceramics and their application in energy harvesting. *Ceram. Int.* **2022**, *48*, 7550–7556. [CrossRef]
39. Song, R.; Zhao, Y.; Li, W.; Yu, Y.; Sheng, J.; Li, Z.; Zhang, Y.; Xia, H.; Fei, W.-D. High temperature stability and mechanical quality factor of donor-acceptor co-doped $BaTiO_3$ piezoelectrics. *Acta Mater.* **2019**, *181*, 200–206. [CrossRef]
40. Zhao, C.; Gao, S.; Yang, T.; Scherer, M.; Schultheiß, J.; Meier, D.; Tan, X.; Kleebe, H.J.; Chen, L.Q.; Koruza, J.; et al. Precipitation hardening in ferroelectric ceramics. *Adv. Mater.* **2021**, *33*, 2102421–2102430. [CrossRef]
41. Liu, W.; Ren, X. Large piezoelectric effect in Pb-free ceramics. *Phys. Rev. Lett.* **2009**, *103*, 257602–257604. [CrossRef] [PubMed]
42. Song, T.K.; Cho, J.H.; Lee, S.C.; Wang, L.; Yeo, H.G.; Sung, Y.S.; Kim, M.H.; Kim, S.S.; Choi, B.C. Dielectric properties and phase transitions in hetero-valent ions substituted $(Bi_{0.5}Na_{0.5})TiO_3$ ceramics. *J. Korean Phys. Soc.* **2010**, *56*, 457–461. [CrossRef]
43. Yin, J.; Shi, X.; Tao, H.; Tan, Z.; Lv, X.; Ding, X.; Sun, J.; Zhang, Y.; Zhang, X.; Yao, K.; et al. Deciphering the atomic-scale structural origin for large dynamic electromechanical response in lead-free $Bi_{0.5}Na_{0.5}TiO_3$-based relaxor ferroelectrics. *Nat. Commun.* **2022**, *13*, 6333–6341. [CrossRef] [PubMed]
44. Zhang, J.; Pan, Z.; Guo, F.-F.; Liu, W.-C.; Ning, H.; Chen, Y.B.; Lu, M.-H.; Yang, B.; Chen, J.; Zhang, S.-T.; et al. Semiconductor/relaxor 0–3 type composites without thermal depolarization in $Bi_{0.5}Na_{0.5}TiO_3$-based lead-free piezoceramics. *Nat. Commun.* **2015**, *6*, 6615–6624. [CrossRef] [PubMed]
45. Cheng, M.L.; Ren, W.L.; Li, H.X.; Liu, X.G.; Bandaru, S.; Zhang, J.; Zhang, X.F. Multiscale collaborative coupling of wood-derived porous carbon modified by three-dimensional conductive magnetic networks for electromagnetic interference shielding. *Compos. Part B Eng.* **2021**, *224*, 109169–109180. [CrossRef]

46. Liu, Y.X.; Qu, W.; Thong, H.C.; Zhang, Y.; Zhang, Y.; Yao, F.Z.; Nguyen, T.N.; Li, J.W.; Zhang, M.H.; Li, J.F.; et al. Isolated-oxygen-vacancy hardening in lead-free piezoelectrics. *Adv. Mater.* **2022**, *34*, 2202558–2202567. [CrossRef]
47. Liu, Y.-X.; Thong, H.-C.; Cheng, Y.-Y.-S.; Li, J.-W.; Wang, K. Defect-mediated domain-wall motion and enhanced electric-field-induced strain in hot-pressed $K_{0.5}Na_{0.5}NbO_3$ lead-free piezoelectric ceramics. *J. Appl. Phys.* **2021**, *129*, 024102–024110. [CrossRef]
48. Zhang, M.H.; Shen, C.; Zhao, C.; Dai, M.; Yao, F.Z.; Wu, B.; Ma, J.; Nan, H.; Wang, D.; Yuan, Q.; et al. Deciphering the phase transition-induced ultrahigh piezoresponse in (K,Na)NbO_3-based piezoceramics. *Nat. Commun.* **2022**, *13*, 3434–3445. [CrossRef]
49. Liu, H.; Liu, Y.X.; Song, A.Z.; Li, Q.; Yin, Y.; Yao, F.Z.; Wang, K.; Gong, W.; Zhang, B.P.; Li, J.F. (K, Na)NbO_3-based lead-free piezoceramics: One more step to boost applications. *Natl. Sci. Rev.* **2022**, *9*, nwac101–nwac103. [CrossRef]
50. Gao, X.; Cheng, Z.; Chen, Z.; Liu, Y.; Meng, X.; Zhang, X.; Wang, J.; Guo, Q.; Li, B.; Sun, H.; et al. The mechanism for the enhanced piezoelectricity in multi-elements doped (K,Na)NbO_3 ceramics. *Nat. Commun.* **2021**, *12*, 881–889. [CrossRef]
51. Wu, H.; Ning, S.; Waqar, M.; Liu, H.; Zhang, Y.; Wu, H.H.; Li, N.; Wu, Y.; Yao, K.; Lookman, T.; et al. Alkali-deficiency driven charged out-of-phase boundaries for giant electromechanical response. *Nat. Commun.* **2021**, *12*, 2841–2848. [CrossRef]
52. Li, P.; Zhai, J.W.; Shen, B.; Zhang, S.J.; Li, X.L.; Zhu, F.Y.; Zhang, X.M. Ultrahigh Piezoelectric Properties in Textured (K,Na)NbO_3-Based Lead-Free Ceramics. *Adv. Mater.* **2018**, *30*, 1705171–1705178. [CrossRef]
53. Lim, J.; Jung, H.; Baek, C.; Hwang, G.-T.; Ryu, J.; Yoon, D.; Yoo, J.; Park, K.-I.; Kim, J.H. All-inkjet-printed flexible piezoelectric generator made of solvent evaporation assisted $BaTiO_3$ hybrid material. *Nano Energy* **2017**, *41*, 337–343. [CrossRef]
54. Curry, E.J.; Le, T.T.; Das, R.; Ke, K.; Santorella, E.M.; Paul, D.; Chorsi, M.T.; Tran, K.T.M.; Baroody, J.; Borges, E.R.; et al. Biodegradable nanofiber-based piezoelectric transducer. *Proc. Natl. Acad. Sci. USA* **2020**, *117*, 214–220. [CrossRef]
55. Hussain, N.; Zhang, M.H.; Zhang, Q.; Zhou, Z.; Xu, X.; Murtaza, M.; Zhang, R.; Wei, H.; Ou, G.; Wang, D.; et al. Large piezoelectric strain in sub-10 nanometer two-dimensional polyvinylidene fluoride nanoflakes. *ACS Nano* **2019**, *13*, 4496–4506. [CrossRef]
56. Katsouras, I.; Asadi, K.; Li, M.; van Driel, T.B.; Kjaer, K.S.; Zhao, D.; Lenz, T.; Gu, Y.; Blom, P.W.; Damjanovic, D.; et al. The negative piezoelectric effect of the ferroelectric polymer poly(vinylidene fluoride). *Nat. Mater.* **2016**, *15*, 78–84. [CrossRef]
57. Gonzalo, B.; Vilas, J.L.; San Sebastián, M.; Breczewski, T.; Pérez-Jubindo, M.Á.; de la Fuente, M.R.; Rodríguez, M.; León, L.M. Electric modulus and polarization studies on piezoelectric polyimides. *J. Appl. Polym. Sci.* **2012**, *125*, 67–76. [CrossRef]
58. Shao, H.; Wang, H.; Cao, Y.; Ding, X.; Fang, J.; Wang, W.; Jin, X.; Peng, L.; Zhang, D.; Lin, T. High-performance voice recognition based on piezoelectric polyacrylonitrile nanofibers. *Adv. Electron. Mater.* **2021**, *7*, 2100206–2100215. [CrossRef]
59. Smith, M.; Calahorra, Y.; Jing, Q.; Kar-Narayan, S. Direct observation of shear piezoelectricity in poly-l-lactic acid nanowires. *APL Mater.* **2017**, *5*, 074105–074112. [CrossRef]
60. Mahanty, B.; Ghosh, S.K.; Jana, S.; Mallick, Z.; Sarkar, S.; Mandal, D. ZnO nanoparticle confined stress amplified all-fiber piezoelectric nanogenerator for self-powered healthcare monitoring. *Sustain. Energy Fuels* **2021**, *5*, 4389–4400. [CrossRef]
61. Niu, X.; Jia, W.; Qian, S.; Zhu, J.; Zhang, J.; Hou, X.; Mu, J.; Geng, W.; Cho, J.; He, J.; et al. High-performance PZT-based stretchable piezoelectric nanogenerator. *ACS Sustain. Chem. Eng.* **2018**, *7*, 979–985. [CrossRef]
62. Li, J.; Chen, S.; Liu, W.; Fu, R.; Tu, S.; Zhao, Y.; Dong, L.; Yan, B.; Gu, Y. High performance piezoelectric nanogenerators based on electrospun ZnO nanorods/poly(vinylidene fluoride) composite membranes. *J. Phys. Chem. C* **2019**, *123*, 11378–11387. [CrossRef]
63. Liang, C.; Sun, F.P.; Rogers, C.A. Coupled electro-mechanical analysis of adaptive material systems—Determination of the actuator power consumption and system energy transfer. *J. Intell. Mater. Syst. Struct.* **1994**, *5*, 12–20. [CrossRef]
64. Liang, C.; Sun, F.P.; Rogers, C.A. Electro-mechanical impedance modeling of active material systems. *Smart Mater. Struct.* **1996**, *5*, 171–186. [CrossRef]
65. Voutetaki, M.E.; Naoum, M.C.; Papadopoulos, N.A.; Chalioris, C.E. Cracking diagnosis in fiber-reinforced concrete with synthetic fibers using piezoelectric transducers. *Fibers* **2022**, *10*, 5. [CrossRef]
66. Perera, R.; Torres, L.; Diaz, F.J.; Barris, C.; Baena, M. Analysis of the impact of sustained load and temperature on the performance of the electromechanical impedance technique through multilevel machine learning and FBG sensors. *Sensors* **2021**, *21*, 5755. [CrossRef]
67. Sikdar, S.; Singh, S.K.; Malinowski, P.; Ostachowicz, W. Electromechanical impedance based debond localisation in a composite sandwich structure. *J. Intell. Mater. Syst. Struct.* **2021**, *33*, 1487–1496. [CrossRef]
68. Hoshyarmanesh, H.; Abbasi, A.; Moein, P.; Ghodsi, M.; Zareinia, K. Design and implementation of an accurate, portable, and time-efficient impedance-based transceiver for structural health monitoring. *IEEE/ASME Trans. Mechatron.* **2017**, *22*, 2809–2814. [CrossRef]
69. Hoshyarmanesh, H.; Abbasi, A. Structural health monitoring of rotary aerospace structures based on electromechanical impedance of integrated piezoelectric transducers. *J. Intell. Mater. Syst. Struct.* **2018**, *29*, 1799–1817. [CrossRef]
70. Shao, J.; Wang, T.; Yin, H.; Yang, D.; Li, Y. Bolt looseness detection based on piezoelectric impedance frequency shift. *Appl. Sci.* **2016**, *6*, 298. [CrossRef]
71. Tenreiro, A.F.G.; Lopes, A.M.; da Silva, L.F.M.; Amorim, J.D.P. Effect of mechanical properties and geometric dimensions on electromechanical impedance signatures for adhesive joint integrity monitoring. *Mech. Adv. Mater. Struct.* **2022**. [CrossRef]
72. Baptista, F.G.; Filho, J.V.; Inman, D.J. Sizing PZT transducers in impedance-based structural health monitoring. *IEEE Sens. J.* **2011**, *11*, 1405–1414. [CrossRef]
73. Baptista, F.G.; Filho, J.V. Optimal frequency range selection for PZT transducers in impedance-based SHM systems. *IEEE Sens. J.* **2010**, *10*, 1297–1303. [CrossRef]
74. Hire, J.H.; Hosseini, S.; Moradi, F. Optimum PZT patch size for corrosion detection in reinforced concrete using the electromechanical impedance technique. *Sensors* **2021**, *21*, 3903. [CrossRef]

75. Jiang, X.; Zhang, X.; Zhang, Y.; Richiedei, D. Piezoelectric active sensor self-diagnosis for electromechanical impedance monitoring using K-means clustering analysis and artificial neural network. *Shock Vib.* **2021**, *2021*, 5574898–5574910. [CrossRef]
76. Nguyen, B.-P.; Tran, Q.H.; Nguyen, T.-T.; Pradhan, A.M.S.; Huynh, T.-C.; Lo Iudice, F. Understanding impedance response characteristics of a piezoelectric-based smart interface subjected to functional degradations. *Complexity* **2021**, *2021*, 5728679–5728702. [CrossRef]
77. Ai, D.; Luo, H.; Zhu, H. Diagnosis and validation of damaged piezoelectric sensor in electromechanical impedance technique. *J. Intell. Mater. Syst. Struct.* **2016**, *28*, 837–850. [CrossRef]
78. Taylor, S.G.; Park, G.; Farinholt, K.M.; Todd, M.D. Diagnostics for piezoelectric transducers under cyclic loads deployed for structural health monitoring applications. *Smart Mater. Struct.* **2013**, *22*, 025024–025034. [CrossRef]
79. Giurgiutiu, V.; Soutis, C. Enhanced composites integrity through structural health monitoring. *Appl. Compos. Mater.* **2012**, *19*, 813–829. [CrossRef]
80. Jahanbin, M. Application of interface guided waves for structural health monitoring of hybrid bonded joints. *IOP Conf. Series Mater. Sci. Eng.* **2021**, *1060*, 012006. [CrossRef]
81. Boivin, G.; Viens, M.; Belanger, P. Plane wave SH_0 piezoceramic transduction optimized using geometrical parameters. *Sensors* **2018**, *18*, 542. [CrossRef] [PubMed]
82. Zennaro, M.; O'Boy, D.J.; Lowe, P.S.; Gan, T.H. Characterization and design improvement of a thickness-shear lead zirconate titanate transducer for low frequency ultrasonic guided wave applications. *Sensors* **2019**, *19*, 1848. [CrossRef] [PubMed]
83. Ochôa, P.; Groves, R.M.; Benedictus, R. Systematic multiparameter design methodology for an ultrasonic health monitoring system for full-scale composite aircraft primary structures. *Struct. Control Health Monit.* **2019**, *26*, 2340–2362. [CrossRef]
84. Ostachowicz, W.; Soman, R.; Malinowski, P. Optimization of sensor placement for structural health monitoring: A review. *Struct. Health Monit.* **2019**, *18*, 963–988. [CrossRef]
85. Mustapha, S.; Lu, Y.; Ng, C.-T.; Malinowski, P. Sensor networks for structures health monitoring: Placement, implementations, and challenges—A review. *Vibration* **2021**, *4*, 551–584. [CrossRef]
86. Ismail, Z.; Mustapha, S.; Tarhini, H. Optimizing the placement of piezoelectric wafers on closed sections using a genetic algorithm—Towards application in structural health monitoring. *Ultrasonics* **2021**, *116*, 106523–106545. [CrossRef]
87. Ismail, Z.; Mustapha, S.; Fakih, M.A.; Tarhini, H. Sensor placement optimization on complex and large metallic and composite structures. *Struct. Health Monit.* **2019**, *19*, 262–280. [CrossRef]
88. Ju, T.; Findikoglu, A.T. Large area detection of microstructural defects with multi-mode ultrasonic signals. *Appl. Sci.* **2022**, *12*, 2082. [CrossRef]
89. Jhang, K.-Y. Nonlinear ultrasonic techniques for nondestructive assessment of micro damage in material: A review. *Int. J. Precis. Eng. Manuf.* **2009**, *10*, 123–135. [CrossRef]
90. Qiu, L.; Yuan, S.F.; Shi, X.L.; Huang, T.X. Design of piezoelectric transducer layer with electromagnetic shielding and high connection reliability. *Smart Mater. Struct.* **2012**, *21*, 075032–075045. [CrossRef]
91. Ren, Y.; Tao, J.; Xue, Z. Design of a large-scale piezoelectric transducer network layer and its reliability verification for space structures. *Sensors* **2020**, *20*, 4344. [CrossRef]
92. Bekas, D.G.; Sharif-Khodaei, Z.; Aliabadi, M.H.F. An innovative diagnostic film for structural health monitoring of metallic and composite structures. *Sensors* **2018**, *18*, 2084. [CrossRef]
93. Morton, T.M.; Harrington, R.M.; Bjeletich, J.G. Acoustic emissions of fatigue crack growth. *Eng. Fract. Mech.* **1973**, *5*, 691–697. [CrossRef]
94. Berkovits, A.; Fang, D. Study of fatigue crack characteristics by acoustic emission. *Eng. Fract. Mech.* **1995**, *51*, 401–416. [CrossRef]
95. De Simone, M.E.; Ciampa, F.; Boccardi, S.; Meo, M. Impact source localisation in aerospace composite structures. *Smart Mater. Struct.* **2017**, *26*, 125026–125038. [CrossRef]
96. Ebrahimkhanlou, A.; Salamone, S. Acoustic emission source localization in thin metallic plates: A single-sensor approach based on multimodal edge reflections. *Ultrasonics* **2017**, *78*, 134–145. [CrossRef]
97. Seno, A.H.; Aliabadi, M.H.F. Impact localisation in composite plates of different stiffness impactors under simulated environmental and operational conditions. *Sensors* **2019**, *19*, 3659. [CrossRef]
98. Capineri, L.; Bulletti, A. Ultrasonic guided-waves sensors and integrated structural health monitoring systems for impact detection and localization: A review. *Sensors* **2021**, *21*, 2929. [CrossRef]
99. Hesser, D.F.; Kocur, G.K.; Markert, B. Active source localization in wave guides based on machine learning. *Ultrasonics* **2020**, *106*, 106144–106154. [CrossRef]
100. Hakoda, C.; Lissenden, C.J. Using the partial wave method for wave structure calculation and the conceptual interpretation of elastodynamic guided waves. *Appl. Sci.* **2018**, *8*, 966. [CrossRef]
101. Garrett, J.C.; Mei, H.; Giurgiutiu, V. An artificial intelligence approach to fatigue crack length estimation from acoustic emission waves in thin metallic plates. *Appl. Sci.* **2022**, *12*, 1372. [CrossRef]
102. Kołakowski, P.; Szelążek, J.; Sekuła, K.; Świercz, A.; Mizerski, K.; Gutkiewicz, P. Structural health monitoring of a railway truss bridge using vibration-based and ultrasonic methods. *Smart Mater. Struct.* **2011**, *20*, 35016–35026. [CrossRef]
103. Chen, Y.X.; Xu, M.H.; Li, X.Y.; Liu, J.H.; Zhu, N.; Zhang, M.H.; Ma, W.M.; Shi, Z.; Zhang, J.; Lu, X.X.; et al. Concurrently improved breakdown strength and storage energy capacitance in the core-shell-structured aromatic polythiourea@$BaTiO_3$ polymer nanocomposites induced by the nature of interfacial polarization and crystallization. *ACS Appl. Energy Mater.* **2021**, *4*, 470–481. [CrossRef]

104. Krueger, H.H.A.; Berlincourt, D. Effects of high static stress on the piezoelectric properties of transducer materials. *J. Acoust. Soc. Am.* **1961**, *33*, 1339–1344. [CrossRef]
105. Bulletti, A.; Giannelli, P.; Calzolai, M.; Capineri, L. An integrated acousto/ultrasonic structural health monitoring system for composite pressure vessels. *IEEE Trans. Ultrason. Ferroelectr. Freq. Control.* **2016**, *63*, 864–873. [CrossRef]
106. Guo, Z.; Huang, T.; Schröder, K.-U. Development of a piezoelectric transducer-based integrated structural health monitoring system for impact monitoring and impedance measurement. *Appl. Sci.* **2020**, *10*, 2062. [CrossRef]
107. Gayakwad, H.; Thiyagarajan, J.S. Structural damage detection through EMI and wave propagation techniques using embedded PZT smart sensing units. *Sensors* **2022**, *22*, 2296. [CrossRef]
108. Xue, D.Z.; Balachandran, P.V.; Yuan, R.H.; Hu, T.; Qian, X.; Dougherty, E.R.; Lookman, T. Accelerated search for $BaTiO_3$-based piezoelectrics with vertical morphotropic phase boundary using Bayesian learning. *Proc. Natl. Acad. Sci. USA* **2016**, *113*, 13301–13306. [CrossRef]
109. Cui, H.C.; Yao, D.S.; Hensleigh, R.; Lu, H.T.; Calderon, A.; Xu, Z.P.; Davaria, S.; Wang, Z.; Mercier, P.; Tarazaga, P.; et al. Design and printing of proprioceptive three-dimensional architected robotic metamaterials. *Science* **2022**, *376*, 1287–1293. [CrossRef]
110. Mueller, I.; Fritzen, C.P. Inspection of piezoceramic transducers used for structural health monitoring. *Materials* **2017**, *10*, 71. [CrossRef]
111. Liang, D.; Wu, L.N.; Fan, Z.F.; Xu, Y. Self-diagnosis and self-reconfiguration of piezoelectric actuator and sensor network for large structural health monitoring. *Int. J. Distrib. Sens. Netw.* **2015**, *11*, 207303–207318. [CrossRef]
112. Kang, M.; Wang, T.; Pant, S.; Genest, M.; Liu, Z. Fault detection and diagnosis for PZT sensors with electro-mechanical impedance technique by using one-dimensional convolutional autoencoder. In *Health Monitoring of Structural and Biological Systems XV*; SPIE: Bellingham, WA, USA, 2021; p. 1159309.
113. Konstantinidis, G.; Drinkwater, B.W.; Wilcox, P.D. The temperature stability of guided wave structural health monitoring systems. *Smart Mater. Struct.* **2006**, *15*, 967–976116. [CrossRef]
114. Salmanpour, M.S.; Khodaei, Z.S.; Aliabadi, M.H.F. Impact damage localisation with piezoelectric sensors under operational and environmental conditions. *Sensors* **2017**, *17*, 1178. [CrossRef]
115. Feng, R.F.; Wei, T.; Zhong, L.W. Flexible nanogenerators for energy harvesting and self-powered electronics. *Adv. Mater.* **2016**, *28*, 4283–4305.

sensors

MDPI

Article

Sandwich Face Layer Debonding Detection and Size Estimation by Machine-Learning-Based Evaluation of Electromechanical Impedance Measurements

Christoph Kralovec [1,*,†], Bernhard Lehner [2,†], Markus Kirchmayr [1] and Martin Schagerl [1]

1 Institute of Structural Lightweight Design, Johannes Kepler University Linz, 4040 Linz, Austria
2 Silicon Austria Labs GmbH, 4040 Linz, Austria
* Correspondence: christoph.kralovec@jku.at
† These authors contributed equally to this work.

Abstract: The present research proposes a two-step physics- and machine-learning(ML)-based electromechanical impedance (EMI) measurement data evaluation approach for sandwich face layer debonding detection and size estimation in structural health monitoring (SHM) applications. As a case example, a circular aluminum sandwich panel with idealized face layer debonding was used. Both the sensor and debonding were located at the center of the sandwich. Synthetic EMI spectra were generated by a finite-element(FE)-based parameter study, and were used for feature engineering and ML model training and development. Calibration of the real-world EMI measurement data was shown to overcome the FE model simplifications, enabling their evaluation by the found synthetic data-based features and models. The data preprocessing and ML models were validated by unseen real-world EMI measurement data collected in a laboratory environment. The best detection and size estimation performances were found for a One-Class Support Vector Machine and a K-Nearest Neighbor model, respectively, which clearly showed reliable identification of relevant debonding sizes. Furthermore, the approach was shown to be robust against unknown artificial disturbances, and outperformed a previous method for debonding size estimation. The data and code used in this study are provided in their entirety, to enhance comprehensibility, and to encourage future research.

Keywords: sandwich debonding; detection and size estimation; electromechanical impedance method; machine learning; feature engineering; physics-based

Citation: Kralovec, C.; Lehner, B.; Kirchmayr, M.; Schagerl, M. Sandwich Face Layer Debonding Detection and Size Estimation by Machine-Learning- Based Evaluation of Electromechanical Impedance Measurements. *Sensors* **2023**, *23*, 2910. https://doi.org/10.3390/s23062910

Academic Editor: Zenghua Liu

Received: 3 February 2023
Revised: 24 February 2023
Accepted: 3 March 2023
Published: 7 March 2023

1. Introduction

Modern products face increasing demand for sustainable manufacturing and operation. These demands are addressed by research in a variety of technological fields, amongst which, lightweighting of the mechanical structure of a product is often applied. Applying high-performance materials, such as fiber-reinforced polymer composites, and modern numerical design methods, such as the finite element (FE) method, enables the highest optimization of mechanical structures. However, high optimization always implies the possibility of critical failure, e.g., when loads are underestimated, when unexpected manufacturing defects exist, or when damages occur during operation. These uncertainties are a significant issue in industries where the highest structural reliability is requested, e.g., transportation industries. To address these uncertainties, e.g., in aviation engineering, damage tolerance design philosophy is used together with maintenance programs: this produces high operational costs, due to downtime and labor. An upcoming approach to reducing maintenance costs, while reliably obtaining integrity, is structural health monitoring (SHM) [1,2].

SHM is the continuous on-board monitoring of the condition of a mechanical structure, during operation, by online and integrated systems of sensors, and can be classified into five levels: (i) damage detection; (ii) damage localization; (iii) damage quantification;

(iv) damage typification; and (v) structural integrity assessment. Numerous active and passive SHM methods readily exist, achieving up to level four in laboratory conditions [3–5]. These apply systems of sensors to measure different physical effects, and continuously evaluate the collected data for potential damage. Among the various methods, acoustic methods—such as the passive acoustic emission (AE) method or the active guided waves (GW) and electromechanical impedance (EMI) method—are of the highest research interest, due to their sensitivity to structural change, and their potential for damage identification up to level four [2,3,5–11]. Furthermore, the often-used piezoelectric wafer active sensors (PWAS) are lightweight and easy to apply; however, due to the high sensitivity of acoustic methods, they are also very prone to environmental influences, measurement equipment, etc. This makes their application in realistic operational conditions extremely challenging, and has thus, to date, prevented their widespread industrial application. This is particularly true for the EMI method, which evaluates the dynamic response of a structure of interest (part of an electromechanical system) to continuous harmonic excitation, i.e., it is a vibration-based method. Consequently, the evaluation of potential damages is extremely complicated, due to reflections from more distant structural regions, e.g., boundaries and correlated disturbances [12]. Classical statistical damage metrics typically do not allow reliable damage identification [9]. However, recently, with the development and increasing distribution of novel statistical methods like machine learning (ML), and the availability of large computational power, new potential to solve the issue has been identified by the scientific community [13]. This has been similarly reported by Avci [14] for "vibration-based methods", by a review of a number of selected journals. Furthermore, numerous recent review articles have highlighted the application and the advances of ML-based measurement data evaluation for vibration-based damage monitoring methods, such as the EMI [13–16]. Fan [13] classified EMI measurement data evaluation methods into data-based and physics-based methods.

Significant advances in data-based methods, which rely solely on real-world measurement data, are currently being achieved, in light of the increasing number of damage indicators, via the application of novel ML algorithms. Lopes [17] used an artificial neural network (ANN) to evaluate EMI measurements for monitoring the loosening of bolted joints of a scaled steel bridge section. The algorithm evaluated the data of four PWAS in two steps, and succeeded in detecting and locating loose joints, and identifying the severity of the loosening (one, two, or three bolts loose per joint). The training was done according to the same structure, by systematically loosening and tightening bolts. Giurgiutiu [11] demonstrated, in laboratory conditions, the superiority of a probabilistic neural network (PNN) to conventional damage-metric-based EMI measurement data evaluations, by means of artificially introduced cracks in an aging aircraft panel. He applied a number of the largest resonance frequencies as features, and classified unknown data into pristine and non-pristine (i.e., damaged). For training, only pristine state data was used. While the classical damage metric was only capable of detecting a crack in a very narrow region, the PPN-based evaluation also identified all cracks, at a distance of 100 mm. Another demonstration of PNN-based EMI measurement data evaluation for damage was reported by Palomino [18], who investigated a number of different damage scenarios in laboratory conditions, and was able to detect, localize, and classify cracks and rivet losses in large aluminum aircraft fuselage components by means of a sensor array consisting of eight PWAS. Park [19] demonstrated the potential of principal component analysis (PCA) to extract relevant features for damage monitoring from EMI measurement data: he used a bolted metal–metal single-lap shear joint, and used bolt loosening and bolt re-tightening as damage. He clearly showed the improvement of the feature extraction by the PCA, for differentiating between a loose and a re-tightened joint by a simple damage metric, and he also demonstrated the applicability of the k-means clustering algorithm to damage identification, using only two principal components. Min [20] addressed the sensitivity of EMI signal damage metrics to the considered frequency ranges, and solved the issue by training an ANN with damage metric values of discretized frequency spectra, to identify five different damage scenarios at a building structure (zero to four loose bolts at a joint). The same approach also performed well in an experimental study on the detection of loose bolts and cracks induced on a steel bridge. Oliveira [21] first proposed a convolutional-neural-network(CNN)-based evaluation method for damage detection, by the use of EMI measurements, and demonstrated its superiority to other ML methods on an aluminum plate equipped with three PWAS and with nuts adhesively bonded to it as artificial damages. Rezende [22] used aluminum beams, with weight added as damage, at varying temperature conditions, and demonstrated CNN-based evaluation also for temperature compensation. A similar demonstration was performed by Li [23] for the more complex use case of a concrete cube with cracks at varying temperature conditions, using CNN in combination with Pearson Correlation Coefficient indices from measured electromechanical admittance spectra sub-ranges, to identify both crack length and applied temperature. Furthermore, Li [23] used the Orthogonal Matching Pursuit algorithm to synthetically generate data based on measurement data, to reduce training costs for the CNN. Other successful applications of CNN models were reported by Ai [24,25], who demonstrated the identification of concurrent compressive stresses and damage (crack number and width), and minor mass loss of a concrete specimen. The application of a deep residual network to the EMI data evaluation for damage, and its potential to overcome the necessity of programmer-dependent preprocessing, was demonstrated by Alazzawi [26], who managed to directly evaluate time domain impedance data for crack quantification and localization in a steel beam. This potential for the evaluation of electromechanical admittance raw data was also found for a deep neural network by Nguyen [27], who applied it to the EMI data-based prestress loss monitoring of a post-tensioned reinforced concrete girder, in laboratory conditions. Singh [28] successfully used unsupervised self-organizing maps (SOM) to fuse multi-sensor data for the identification of hole sizes in a large aluminum plate. Furthermore, he found that data reduction of measured EMI spectra by PCA is not only more efficient but also improves the quality of the data for the realized experiments. The latter was also supported by the findings of Lopes [17], who stated that feature engineering, i.e., the selection of meaningful features of the raw data, often (i) highlights dynamic characteristics of the evaluation objective, and (ii) strongly reduces the required amount of training data. For a summary of the key contributions to data-driven and ML-based EMI measurement data, see Fan [13].

Physics-based methods use physical models of the monitored structure of interest and the applied sensor system, to evaluate real-world EMI measurement data. Analytical or numerical models are used to, e.g., find sensitive frequency ranges or specific sensitive features and their responses to a considered damage and its location and properties [29,30]. Another physics-based approach is model updating, whereby predefined parameters of a model, which represent damage, temperature changes, etc., are optimized until the EMI measurement data is matched by the model's simulation results. Albakri [31], e.g., used a length-varying spectral element mesh to efficiently model a beam with a single crack-like opening, the opening parameters of which were subsequently optimized to fit the EMI measurement data of the equivalent, but physical, beam with a single unknown opening, thereby enabling the identification of the opening's location, width, and severity. However, due to the large parameter space for more complex mechanical structures, and their large number of possible damage modes, the computational effort for model updating is large, and fast data evaluation is challenging. Ezzat [32] addressed this issue by increasing the efficiency of the model updating evaluation method by the application of surrogate models and a multi-stage statistical calibration framework. The framework used a pre-screening step to reduce the parameter space, by identifying probable damage locations before the model updating and signal matching steps and, therefore, strongly reduced the overall computational effort for damage identification. Another approach to accelerating data evaluation is to pre-simulate impedance spectra for varying damage parameters of the structure of interest. The typically large database is computationally expensive; however, its calculation can be done prior to the operational life of the monitored structure. To reduce this large computational effort, the application of numerical surrogate models (e.g., ML models) can be used to lower the number of required simulations by interpolation. An approach was recently demonstrated for the accurate interpolation of simulated guided wave signals by Humer [33]. The available simulated signals (and their correlated and known damage parameters) enable very fast damage identification, by simply comparing their similarity to measured EMI spectra. Shuai [34] and Kralovec [9] demonstrated this approach by the identification of the crack location and severity in a simple metallic beam, and the debonding size in an aluminum sandwich panel, respectively. However, physics-based methods are still limited to a low-frequency range or simple mechanical structures (mostly beams are investigated), as computational accuracy and efficiency are still issues [13].

Nevertheless, ML methods are generally found to be superior to traditional evaluation methods, in which the vibration signal is fuzzy and noise-contaminated [14,21]. To date, ML algorithms applied to vibration-based monitoring methods, such as EMI, are predominantly supervised learning algorithms based on real-world measurement data. Avci [14] concludes that the latest ML data evaluation approaches to potential damage increasingly apply signal features other than modal parameters, as these are often very sensitive to environmental conditions, e.g., temperature. Furthermore, the curation of experimental data is typically very costly. Thus, approaches using synthetic data (e.g., simulated by an FE model) are utilized, to reduce the cost of data acquisition [35,36]. Similarly, Fan [13] highlighted the significant potential of physics-based (i.e., based on synthetic data from physical models) EMI data evaluation, and the need for efficient numerical modeling.

The present research contributes to the field of EMI data evaluation for structural damage identification, by proposing a novel two-step physics- and ML-based EMI data evaluation approach to sandwich debonding detection and size estimation. The proposed novel approach addresses the above-mentioned research shortcomings by (i) very fast and accurate damage identification by supervised ML models, (ii) the replacement of costly real-world measurement data for training, by synthetic FE model-based data, and (iii) the engineering of EMI signal features that are robust against simplifications of the FE model (used to generate training data), environmental disturbances and small variations of the debonding shape (in reality, debondings are not perfectly circular). As a case example, a comparatively large and complex mechanical structure—a circular sandwich with aluminum face layers and an aluminum honeycomb core—was used. A single PWAS located at the center of the sandwich directly above the considered idealized face layer debonding was used for the EMI measurements. In contrast to many other EMI studies, the FE model used to simulate the artificial EMI data was strongly simplified, and was not intended to perfectly reflect all features of the real structural response: this, on the one hand, enabled a very efficient simulation of the artificial EMI data, which may be very relevant for larger structural components or more complex damage modes, and on the other hand, it represented a test of the proposed method's robustness against incomplete training data. The considered idealized face layer debonding was increased stepwise by milling and EMI measurements, which were evaluated by the proposed physics-based ML approach. The approach was tested for the consistency, accuracy, reliability, and robustness of its evaluation results, and was benchmarked.

2. Electromechanical Impedance Data Generation

In the present research, we used synthetic and real-world EMI spectra as data. The synthetic EMI spectra were generated by a multi-physics FE model. The real-world EMI spectra were generated by measurements from a PWAS attached to a physical test specimen. To fully understand the EMI data generation, the present section presents (i) the fundamentals of the EMI method, (ii) the electromechanical system of the used case example of a circular sandwich plate with idealized face layer debonding and adhesively attached PWAS, (iii) the FE model and the conducted parameter study, and the generated data, and (iv) the real-world measurement setup and procedure, and the collected data.

2.1. Electromechanical Impedance Method

The EMI method is a vibration-based SHM method, which evaluates the dynamic response of a mechanical structure of interest to harmonic excitation. A single PWAS is usually used for both excitation and dynamic response measurement. The PWAS is adhesively bonded to the mechanical structure, resulting in the evaluated electromechanical system. For evaluation, the PWAS—and, thus, the whole electromechanical system—is excited over a wide frequency range by a harmonic voltage signal $U(f)$, and the system's dynamic response is measured by the PWAS's electrical impedance $Z(f) = U(f)/I(f)$ or its reciprocal, the admittance, where $I(f)$ is the electrical current and f the frequency. Consequently, changes in the mechanical structure are also reflected in the measured impedance signal, and allow conclusions on its health state. The underlying correlation is given for one-dimensional strain and neglected PWAS dynamics by [2]

$$Z(f) = \frac{1}{j2\pi fC}\left[1 - \kappa_{31}^2 \frac{k_{st}(f)}{k_{st}(f) + k_p}\right]^{-1}, \quad (1)$$

where C is the capacity, k_p the static stiffness, and k_{31} the non-dimensional coupling coefficient of the PWAS. The behavior of the structure enters the equation by the frequency-dependent dynamic stiffness $k_{st}(f)$. The imaginary unit is denoted by j.

2.2. Test Specimen

The investigated case example was the identification of sandwich debonding. A single circular aluminum honeycomb sandwich panel, with an idealized debonding in the center, was used in the present study. Figure 1 presents the used test specimen. The circular panel had the dimensions ⌀500 × 20 mm^2. The face layers were produced from aluminum ENAW5754-H22, and had a thickness of 1 mm; thus, the aluminum honeycomb core (3/8–3000–0.0025) had a thickness of 18 mm. The in-plane orientation of the orthotropic honeycomb core was arbitrary, and was not considered in the present study. The face layer and core were adhesively bonded. To enable the physical introduction of debonding initiation and propagation, circular debonding was idealized by removing the core and bottom face layer in the debonded area. For the EMI measurements, a single PWAS of dimensions ⌀10 × 0.25 mm^2 (PIC 255 material provided by PI Ceramic GmbH) was adhesively bonded (by LOCTITE EA 9466) to the center of the top face layer, i.e., directly above the debonded area. The material properties are presented later in Section 2.3.

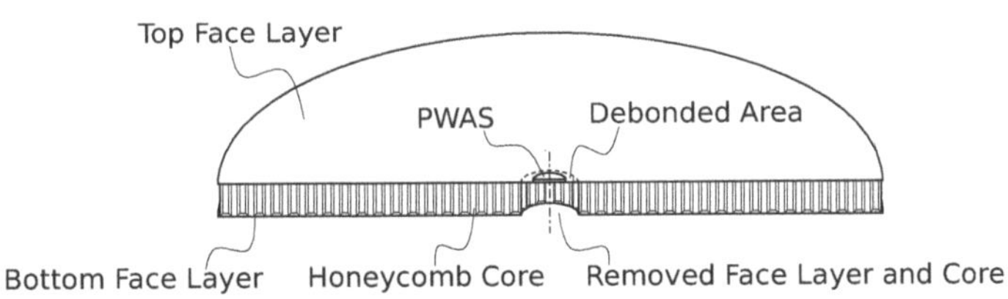

Figure 1. Circular sandwich panel with central face layer debonding and applied PWAS (cut and non-scaled visualization).

2.3. Synthetic EMI Data Generation

The synthetic data were EMI spectra that were simulated by the variation of the material parameters of a number of multi-physics FE models that included different face layer debonding sizes.

2.3.1. Simulation Model

The FE model was set up by the commercially available FE analysis software, Abaqus, provided by Dassault Systemes. The direct-solution steady-state dynamics analysis of

Abaqus/Standard was used to simulate the dynamic response to harmonic excitation for the same 801 frequencies as measured on the physical test specimen (see Section 2.4). For efficiency, and to test the influence of unreflected features (e.g., all asymmetric modes and local resonances within core cells;, cf., [9]) in the training data, the FE model of the (in reality, orthotropic) sandwich structure was strongly simplified to an axisymmetric, and thus two-dimensional FE model. Consequently, the orthotropic honeycomb structure of the core was neglected, and was modeled by volume elements with homogenized transverse isotropic material properties ($\overline{G}_{12} = \overline{G}_{23} = (G_{12} + G_{23})/2$). Figure 2 presents the employed FE model.

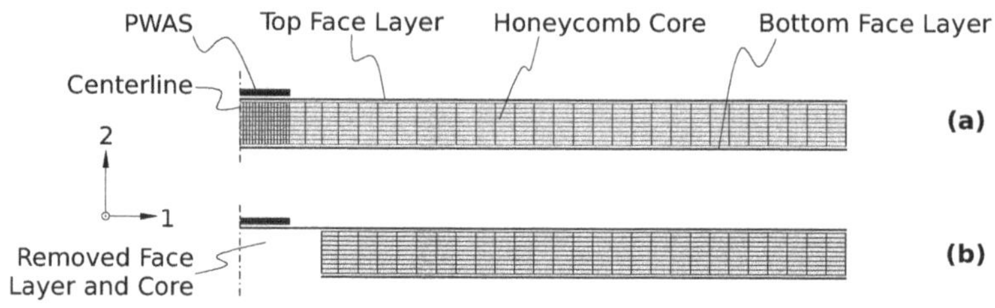

Figure 2. Axisymmetric FE model of sandwich panel in (**a**) pristine condition, and (**b**) with face layer debonding of a radius of 8.33 mm (non-scaled visualization).

The dimensions of the FE model were according to the considered test specimen (see Section 2.2). The face layers were modeled by quadratic, axisymmetric shell elements (Abaqus: SAX2). The core was modeled by quadratic, axisymmetric, and reduced integrated elements (Abaqus: CAX8R). The PWAS was modeled by quadratic, axisymmetric, piezoelectric, and reduced integrated elements (Abaqus: CAX8RE). The face layers were assumed to be perfectly bonded to the core and PWAS, and thus connected by tie constraints. The basic model discretization was based on a previous study that experimentally validated the model on an identical sandwich structure with idealized debonding damage. For further details of the FE model, see Kralovec [9]. The idealized debonding was modeled by simply removing the elements from the bottom face layer and the core, i.e., every damage size had its own FE model, where the remaining mesh was unchanged in respect of the pristine condition. Consequently, the simulated debonding sizes were modeled perfectly circularly, and were predefined by the model discretization. For the present work, the considered debonding sizes were given by their radius. Simulated debonding radii are $\{0, 0.3125, \ldots\ 5, 6.6670, \ldots\ 10, 12, \ldots\ 40\}$ mm: thus, 35 different FE models were used (including pristine condition). The material properties used for the FE models were typically from manufacturer data sheets; however, these always differed from the true properties, thus resulting in deviations between simulated and measured results. This could be addressed by, e.g., model updating during the initial pristine condition of a monitored structure of interest. However, due to, e.g., environmental conditions—such as temperature, moisture, or surface pollution—the properties of the structure could also vary during operation. To overcome these issues, we included possible variations of the material properties, via a large FE-simulation-based parameter study implemented by the software OptiSLang 8.0.0, provided by ANSYS Inc. Table 1 presents the considered material parameter ranges. The reference values were taken from manufacturer data sheets, or were typical values from the literature [2,9]. The parameter ranges represented extreme values found in data sheets and the literature, or were simply estimated within a spacious but meaningful range. The parameter range, in percent, was an additive scaling factor for all parameters in the table row, i.e., these parameters were assumed to change relative to one other, which (i) seemed reasonable, and (ii) strongly reduced the parameter space. The following are some arguments for the parameter range selection. The density of aluminum is exacted to be very constant, as it is defined by the alloy elements. However, the aluminum honeycomb core can be easily stretched or compressed during manufacturing: thus, its density and stiffness may vary together in a larger range. The Young's modulus of aluminum typically varies in a narrow range. The review of different data sheets yielded the assumed parameter range. The Poisson's ratio was found to be very constant for aluminum. Typical values for structural damping of mechanical structures are 0–0.05 [2]. The structural damping of aluminum is low. The aluminum honeycomb core was assumed to have stronger damping properties, due to the enclosed air. For the PWAS material parameter range, no adequate literature was found: thus, a small scattering was assumed to include the influence of these parameters.

Table 1. Range for material parameter variations considered for the FE parameter study.

Parameter	Unit	Reference Value	Range
face layer			
density	kg/m^3	2660	–
Young's modulus	GPa	70	±3%
Poisson's ratio	1	0.33	–
structural damping		0.01	0.0–0.05
honeycomb			
density	kg/m^3	59.3	±18%
Young's moduli E_{11},E_{22},E_{33}	GPa	0.634, 0.001, 0.001	±18%
shear moduli G_{12},G_{23},G_{23}	GPa	0.137, 0.001, 0.275	±18%
structural damping	1	0.02	0.0–0.05
PWAS			
density	kg/m^3	7800	–
Poisson's ratio	1	0.34	–
elastic compliance coefficients s^E_{11}, s^E_{22}	m^2/N	$\{16.1, 20.7\} \times 10^{-12}$	±5%
coupling factor for transverse oscillation	1	0.35	–
relative permitivity ε^T_{22}, ε^T_{11}	1	$\{1750, 1650\}\ \varepsilon_0$	±5%
piezoelectric large-signal deformation coefficient d_{21}, d_{22}, d_{15}	C/N	$\{-180, 400, 550\} \times 10^{-12}$	±5%
dielectric loss factor	$\tan(\delta)$	0.02	–

To cover the material parameter space, space-filling Latin hypercube sampling was applied, with a sample size of 200: thus, a total of 35 damage states × 200 parameter variations = 7000 simulation results were generated (with 801 discrete dynamic frequency responses each). The overall computation time on a desktop machine was about 29 days. Using piezoelectric elements to model the PWAS in Abaqus, the dynamic frequency response was given by the frequency-dependent complex nodal charge output, $Q_k(f)$, where $k \in N_e$ were the nodes of the PWAS electrode supplied by the harmonic excitation voltage signal with constant amplitude U_{fe}. The nodal charges could be used to calculate the complex impedance of the PWAS for the $i = 1 \ldots 801$ frequencies f_i, and thus, for the whole electromechanical system, by [9,29]

$$Z(f_i) = U_{fe}\left[j2\pi f_i \left(\underbrace{\sum_{k=1}^{N_e} Q_k(f_i)}_{\text{FE result}} - \underbrace{j\delta\epsilon^T_{33}\frac{U_{fe}}{t_p}A_p}_{\substack{\text{electrical}\\ \text{damping}}} \right)\right]^{-1}, \quad (2)$$

where A_p, t_p and δ were the electrode area, the thickness, and the dielectric loss factor of the PWAS, respectively. The latter could also vary in a range of ±20%. Thus, varying the dielectric loss factor arbitrarily within this range enabled us to further multiply the synthetic data generated for training. In the present work, five arbitrarily selected values for the dielectric loss factor were used, thus yielding a theoretical number of 35,000 calculated EMI spectra. However, some of the generated parameter combinations did not give meaningful results, and were thus sifted out, yielding a final number of used synthetic EMI spectra of 34,152.

2.3.2. Simulated Data

The FE-based simulation results were 34,152 complex EMI spectra of different perfectly circular debonding sizes and varying material parameters (cf., Section 2.3). Each of the simulated EMI spectra consisted of 801 complex values, which were evenly distributed over the measured frequency range of 0.004 kHz to 60 kHz (i.e., the simulated and measured results were at the same frequencies). For the selection of the material parameters, space-filling Latin hypercube sampling was used: thus, there was no sequence of simulation results in which the effect of a single parameter might be illustrated. However, Figure 3 shows the real part of a number of simulated EMI spectra, which reflect the influence of different debonding sizes (cf., Figure 3a) and the influence of the varied material properties for a constant debonding size (cf., Figure 3b).

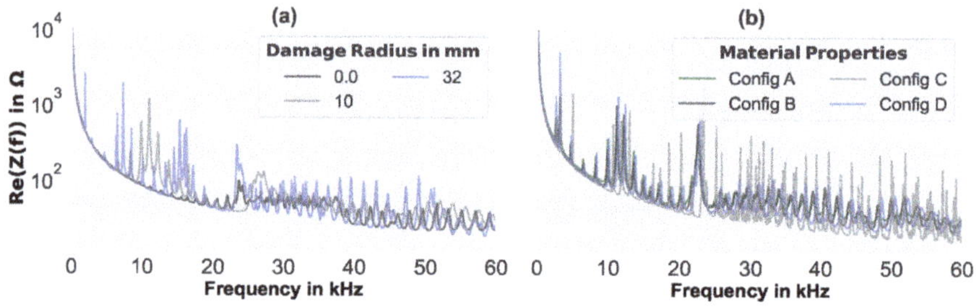

Figure 3. Examples of synthetic EMI spectra simulated by the FE model of the sandwich plate, with different arbitrarily selected material properties and (**a**) different debonding sizes, (**b**) debonding radius of 24 mm.

For the circular plate with debonding, clearly higher and more resonance peaks can be observed than in the pristine case (i.e., damage radius 0; cf., Figure 3a). Furthermore, with debonding size increase, more resonance peaks appear at lower frequencies. Some of the first significant resonance peaks could be assigned to axisymmetric resonances of the debonded face layer, as demonstrated in [9]. Generally, it is difficult to conclude on specific effects of the debonding size on the simulated EMI spectra, as the material properties varied, and strongly influenced the baseline and resonance peak heights and frequencies (cf., Figure 3b). However, the intended representation of a wide range of material parameters in the simulated EMI spectra seems to have been achieved.

2.4. Real-World EMI Data Generation

The real-world data were EMI spectra that were measured on the considered test specimen with debonding sizes increased stepwise, and different artificial disturbances applied.

2.4.1. Measurement Setup and Procedure

The experimental setup for the real-world EMI measurements is presented in Figure 4.

The aluminum sandwich panel was placed on foamed polymer, to allow free vibration. The PWAS was electrically connected, via soldered 400 mm-long AWG 20 cables, to the four-terminal probe, type L2000, of the impedance analyzer, type IM3570—both provided by the Hioki E.E. Corporation. The excitation voltage amplitude of $5\sqrt{2}$ was constant over the whole considered frequency range of 0.004 kHz to 60 kHz. The frequency range was discretized by 800 evenly distributed steps, thus resulting in 801 complex impedance values per measurement. The excitation speed of the impedance analyzer was set to medium, and the averaging was set to three, to get a good balance between measurement noise and efficiency. In total, 225 EMI measurements were conducted, including both pristine and damaged states. The debonding was manufactured successively by removing, stepwise, the bottom face layer and the core, by milling. Figure 5 shows the clamped sandwich panel after tooling different debonding sizes in the milling machine.

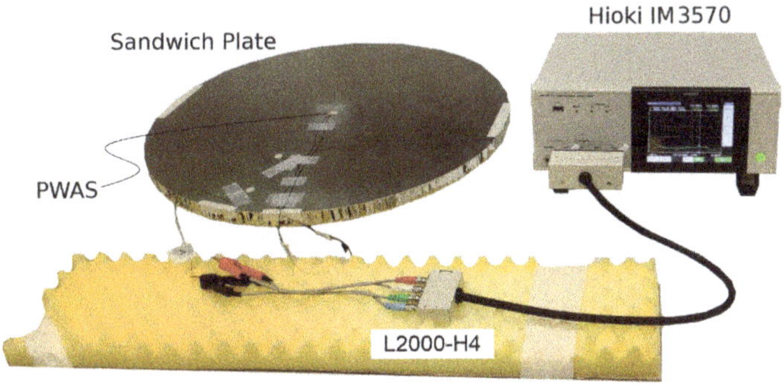

Figure 4. Setup for real-world EMI measurements.

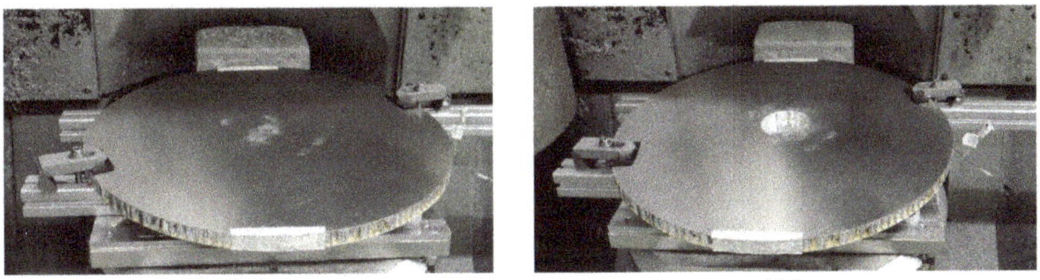

Figure 5. Circular sandwich plate clamped in the milling machine, with a debonding of the radius of (**a**) 2.5 mm, and (**b**) 40 mm.

The manufactured and measured 25 different debonding sizes are summarized in Table 2. The debonding shapes were mostly perfectly circular. Five debonding sizes had an elongated shape, which was manufactured by simply elongating the readily existing hole symmetrically along one arbitrary direction, until its largest extent had the value given in Table 2.

Table 2. Experimentally measured debonding sizes in mm.

0	1.5	2.5	3.5	4.5	5.5	7 *	8	10 *	12	14	16 *	17
19	21	23 *	25	27	29	31 *	33	35	37	39	40	

* elongated shape.

Furthermore, eight artificial environmental disturbances were used to influence the EMI measurements. The used disturbances were local forces and damping applied by human fingers, by a whole human hand, and by small modeling compound rolls. These disturbances were found, in previous experimental tests, to influence the measured EMI spectrum, and are summarized in Table 3.

Figure 6 shows four examples of the application of these disturbances.

For every debonding size, one EMI spectrum was measured without any disturbance, sequentially followed by EMI measurements with the disturbances presented in Table 3.

Thus, nine measurements were done per debonding size, yielding a total number of 25 damage states × 9 disturbance states = 225 experimentally measured EMI spectra.

Table 3. Artificial environmental disturbances of EMI measurements.

Type	Approximate Total Load	Approximate Distance Plate Center to First Contact
whole hand	10 N	25 mm
one finger	10 N	25 mm
two finger	10 N	25 mm
three finger	10 N	25 mm
one roll	-	25 mm
two rolls	-	25 mm
three rolls	-	25 mm
four rolls	-	25 mm

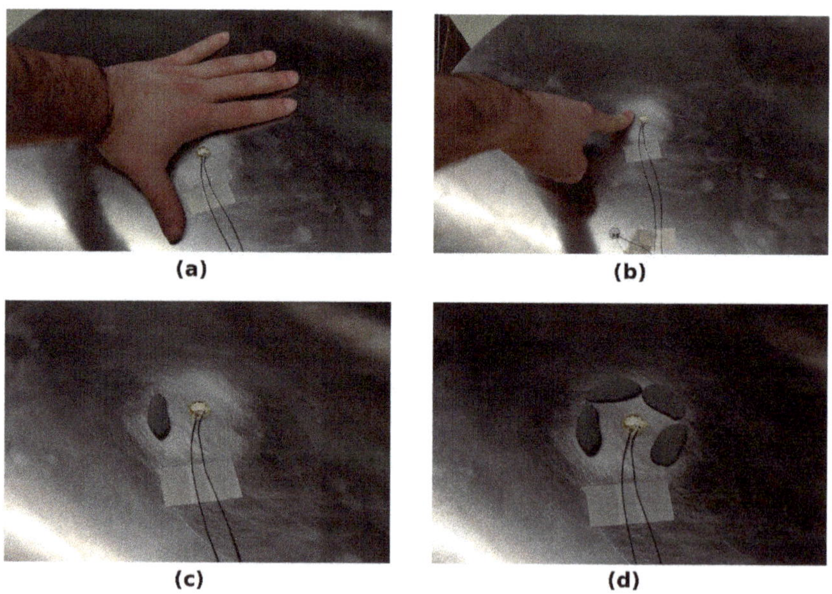

Figure 6. Examples of artificial disturbances: (**a**) hand next to PWAS; (**b**) one finger next to PWAS; (**c**) one modeling compound roll next to PWAS; (**d**) four modeling compound rolls next to PWAS.

2.4.2. Measured Data

The results of the real-world EMI measurements were 225 complex impedance spectra of different debonding states and additional artificial environmental disturbances. Each of the EMI spectra consisted of 801 complex values, which were evenly distributed over the measured frequency range of 0.004 kHz to 60 kHz. Figure 7 presents the real part of a number of measured EMI spectra for (a) comparatively small, and (b) middle-range debonding sizes.

Readily small debonding sizes clearly changed the EMI spectra from the pristine state (debonding radius 0, cf., Figure 7a). With increasing debonding size, single very sharp resonance peaks appeared in the low-frequency range of the spectra (below 10 kHz). Furthermore, resonance frequency accumulations seem to have taken place around specific frequencies (e.g., at 17 kHz and 50 kHz for a debonding radius of 8 mm), yielding significant deviations from the pristine spectra. Moreover, the values of the presented EMI spectra generally increased with debonding size, and the resonances tended to lower in their frequencies. Similar trends may be seen also for middle-range debonding sizes; however, they are not clear anymore, as can be seen in Figure 7b. Similar experimental results were also found by the authors, in a previous work, for EMI measurements on identical circular sandwich plates with artificial debonding of a thickness of 20 mm and 40 mm, where the correlation between dominant resonance frequencies in the low-frequency range and the first eigenmodes of the debonded face layer was analytically, numerically, and experimentally demonstrated [9].

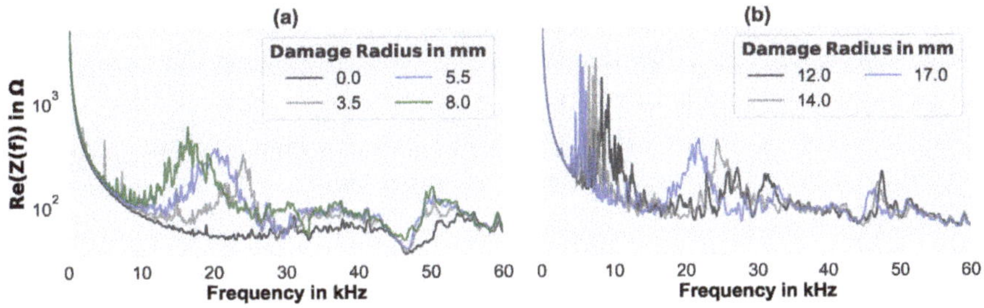

Figure 7. Examples of measured EMI spectra of the sandwich plate without any artificial disturbance, and with different perfectly circular debondings of (**a**) small size, and (**b**) middle-range size.

In addition to the debonding size, other effects—such as measurement noise, ambient noise, pollution, and environmental temperature changes—may also have an influence on real-world EMI measurements [37,38]. Therefore, in the present work, strong artificial disturbances were applied, to test the robustness of the proposed evaluation method, and also to test some possible environmental influences during the operational life of a structural component. Figure 8 presents the real part of a number of EMI spectra measured on the circular sandwich plate, with perfectly circular debonding of the radius of 29 mm, and different applied artificial disturbances (cf., Section 2.4).

The effects of the artificial disturbances could be clearly observed in the measured EMI spectra. As expected, all applied disturbances resulted in a damping effect, that reduced the resonance peaks and smoothed out the spectra. For modeling compound rolls, the resonances additionally shifted towards lower frequencies, and the baseline of the EMI spectra appeared slightly increased. This may also have been due to the added mass. All the applied artificial disturbances clearly influenced the measured EMI spectra. Generally, the influence increased with the application of more fingers or more compound rolls. Furthermore, the influence of the artificial disturbances on the EMI spectra increased with the debonding size, in particular, when the disturbing fingers, compound rolls, etc., were readily located within the debonded area.

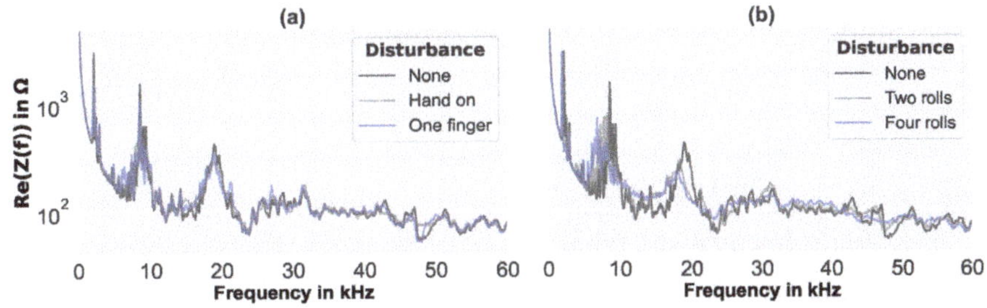

Figure 8. Examples of measured EMI spectra of the sandwich plate with perfectly circular debonding of the 29 mm radius, and (**a**) hand and fingers, (**b**) compound rolls applied as artificial disturbances.

3. Data Preprocessing

A three-step approach was implemented for preprocessing the input data for the ML-based evaluation, in order to decrease its dimensionality and, at the same time, improve the robustness of the overall method: firstly, relevant frequencies were identified through analysis of the frequency spectrum, allowing for the elimination of unnecessary information; secondly, a log-scaled filterbank was applied to the data, leading to more robust representation, in terms of slight variations in resonance frequencies; lastly, a discrete cosine transform (DCT) was applied to the processed data, for further data compression and smoothing of the spectrum. After each step, it was important to confirm that enough relevant information was still present in the data. To ensure this, a well-known data exploration technique called t-SNE (t-distributed stochastic neighbor embedding) was applied. T-SNE is a non-linear, unsupervised technique for visualizing high-dimensional data in 2D or 3D maps, which preserves the local data structure: that is, data points close to each other in high-dimensional space will end up close to each other in low-dimensional space. This technique allows for the visualization of high-dimensional data in a lower-dimensional space, making it easier to identify patterns, and to confirm the preservation of relevant information. This way, the original data representation, with a dimensionality of 801, was reduced to only 31.

The features obtained through the preprocessing steps were then extracted from both the synthetic and the real-world EMI data, before conducting the ML-based damage evaluation experiments described in Section 4. The real-world EMI data were additionally calibrated to the synthetic data, to compensate for the strong simplifications of the FE model (see Section 2.3) used to generate the synthetic data. The finally defined preprocessing for the ML-based damage evaluation of real-world EMI data is depicted in Figure 9, for a better overview. The preprocessing for synthetic EMI data was the same, but without the calibration.

3.1. Feature Engineering

The goal of feature engineering is to improve the performance and robustness of ML-based algorithms: to this end, the dimensionality of data can be reduced for denoising purposes, and features can be developed to extract relevant information, while at the same time suppressing irrelevant information. The features used in this analysis were obtained through the process of feature engineering using exclusively synthetic data.

We started with an exploration of the raw synthetic spectra, by looking at damage size-specific characteristics, to identify the most informative range of frequencies. In Figure 10, we can see the mean and standard deviation values of each damage size individually.

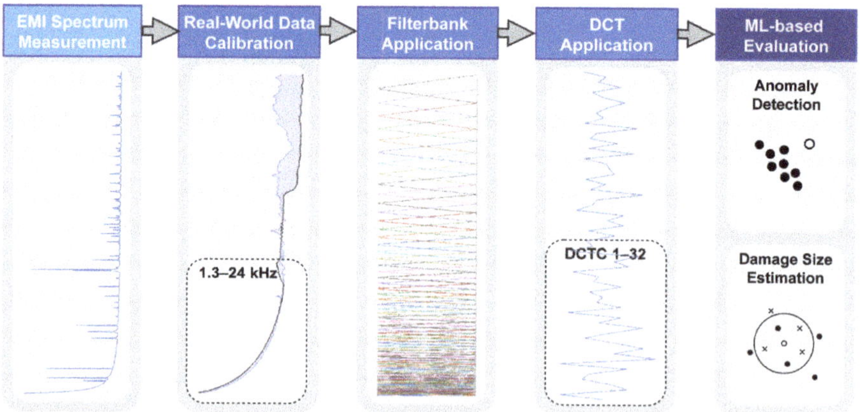

Figure 9. Flowchart of the preprocessing for real-world EMI data, where the original data representation,

with 801 spectral bins (real part of impedance), was reduced step-by-step to only 31 DCT coefficients. The calibration process involved the adjustment of pristine real-world spectra to match pristine synthetic spectra. In the case that we only propagated a subset of the signal to the next step, it is outlined accordingly by a dashed window.

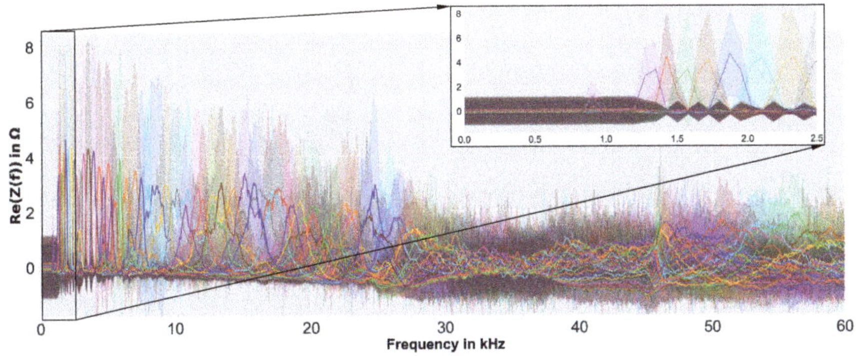

Figure 10. Summary of raw synthetic EMI spectra, scaled to have zero mean and unit standard deviation, for better clarity. The centerline is the mean value, and the surrounding translucent range represents the standard deviation per frequency of all spectra with the same damage size (approximately 1000 each).

Note that we scaled the spectra to have zero mean and unit standard deviation, for better clarity. In general, it can be seen that different damage sizes caused peaks (i.e., resonances) at different frequencies. However, frequencies below 1.3 kHz were found to be less informative (see Figure 10 detail), and were therefore excluded. Furthermore, the size-specific damage peaks were less pronounced, between 28 kHz and 46 kHz. Given our knowledge, from previous work, about the precision decrease of the simulation results with increasing frequency, and the clear objective to find an efficient evaluation method, we defined a frequency range between 1.3 kHz and 28 kHz to be considered, thereby readily shrinking the dimensionality of the data to less than half. With the help of the previously mentioned t-SNE plots, it was possible to further optimize the frequency range from 1.3 kHz to 24 kHz, i.e., only 38% of the available data. Figure 11 shows the clear structure of the t-SNE plot of the finally used frequency range.

The central cluster consisted of comparatively small damage sizes up to a radius of 5 mm. Furthermore, there was some overlap for very small damage sizes (debonding radius < 2 mm) in the lower part.

After identifying the relevant frequencies, a filterbank was applied, which consisted of triangular filters that were 50% overlapping, and uniformly spaced on a logarithmic frequency scale. The overlapping of the filters allowed for a more fine-grained representation of the frequency spectrum, while the logarithmic spacing ensured that the filterbank was more sensitive to lower frequencies than to higher frequencies. This led to a more robust representation in terms of slight variations in resonance frequencies, while at the same time compressing the frequency axis of the spectrum. Additionally, log-scaling the magnitude compressed the dynamic range of the spectrum, which can be useful for handling signals with wide variations in amplitude. Considering the range of frequencies obtained in the previous step, and the filterbank configuration, a log-scaled spectrum with 66 bins was computed.

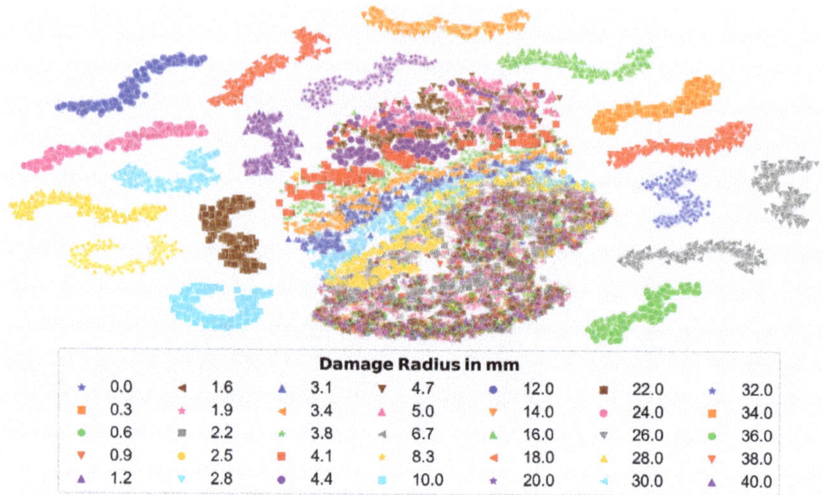

Figure 11. 2D T-SNE embedding of synthetic data of the final frequency range from 1.3 kHz to 24 kHz, with a clear structure. The big cluster in the center comprises smaller defects up to 5 mm, and only very small defects (<2 mm) seem to overlap a lot (lower part of the center cluster).

Finally, the log-scaled spectrum obtained through the previous steps was transformed, using the DTC [39]. The DCT is a mathematical technique used to decorrelate and transform data into a new coordinate system: that is, orthogonal cosine basis vectors with specific frequencies. The resulting transformed data will be a set of uncorrelated basis vectors, each representing a different frequency component of the original data. This step can further improve the robustness and efficiency of the data representation, by eliminating the dependencies between the data, furthermore only keeping the lower-order coefficients of the DCT results, in a signal that is more compact, and less affected by irrelevant information. This is based on the assumption that a signal can be separated—at least to a certain extent—into low-order components comprising relevant information, and high-order components that are irrelevant to the ML model, to perform the task at hand. In other words, the lower DCT coefficients represent a low-pass filtered spectrum that converges to its envelope. Furthermore, ignoring the 0th DCT coefficient increases the robustness of the method even further, by eliminating the dependency on the overall energy level of the signal.

The final feature vector consisted of 31 DCT coefficients that were computed on a log-scaled spectrum that exhibited a frequency range from 1.3 kHz to 24 kHz. This feature vector, which represented a robust and compact representation of the input signal, captured the essential spectral information relevant to damage identification. In the case of real-world data, there was an additional calibration step required, which is discussed in the next Section.

3.2. *Calibration of Real-World EMI Data*

The real-world data (i.e., the measured EMI spectra) were calibrated to the synthetic data (i.e., the simulated EMI spectra), to increase comparability. Therefore, it was possible to evaluate real-world EMI spectra by synthetic data-based features and ML models. This data preprocessing step was required, as the FE model used for the simulation of the synthetic EMI spectra was strongly simplified, to increase efficiency and robustness. The applied calibration was a simple shift of the real-world EMI spectra Z_i^r by the difference of the frequency-dependent mean values of the real-world pristine data $\mu_i^{Z,r,p}$ and the synthetic pristine data $\mu_i^{Z,s,p}$, thereby aligning the pristine spectra of the real-world measurements and the synthetic simulations. For illustration, Figure 12 shows the mean real-world EMI spectra $\mu_i^{Z,r,p}$ for all measurements at the pristine sandwich panel, and the mean synthetic EMI spectra for all simulations by the pristine FE model $\mu_i^{Z,s,p}$.

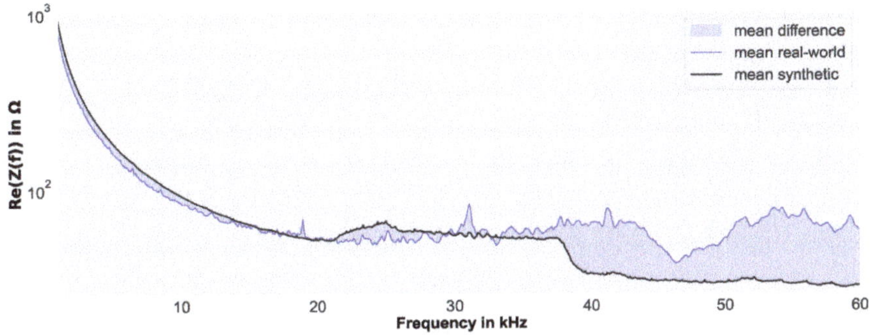

Figure 12. Difference between mean EMI spectra of all pristine real-world EMI spectra and mean EMI spectra of all pristine synthetic EMI spectra used to calibrate single real-world EMI spectra before evaluation.

Furthermore, the difference between the two mean spectra used to shift, and thus calibrate, the real-world data is highlighted. This difference was used to calculate the calibrated real-world EMI spectra by

$$Z_i^{r,C} = Z_i^r - (\mu_i^{Z,r,p} - \mu_i^{Z,s,p}), \tag{3}$$

where i represents the bin of the spectrum (i.e., the 801 discrete frequency steps).

4. Data Evaluation Experiments

The data evaluation experiments tested a number of frequently used ML models for evaluating the existence and the size of a potential sandwich debonding. For a fair and meaningful comparison of the ML models, with respect to their performance, the available EMI spectra were split into arbitrary subsets, and the experiments were repeated at each of the subsets, thereby also enabling the evaluation of the consistency of the results, and thus the robustness of each ML model when repeated.

4.1. Debonding Detection by Anomaly Detection

In this Section, the results of the evaluation experiments for damage detection by anomaly detection methods are presented: firstly, the setup of the experiments is described; secondly, the results of synthetic EMI spectra are presented, i.e., the models are trained and tested by synthetic data; thirdly, the results of real-world EMI spectra are presented, i.e., the models are trained exclusively by synthetic data, but tested by calibrated real-world data.

4.1.1. Setup of Experiments

The problem of anomaly detection was formulated as a novelty detection problem. The training data were not polluted by anomalies, and whether or not a new observation was an anomaly was evaluated. This could be done with the following models, which only needed normal data to learn their characteristics and identify deviations from this pattern as anomalies [40–42]:

1. Isolation Forest is an ensemble of binary trees specifically trained to isolate anomalies; decisions are made based on depth, where anomalous data points have shorter average depths in the tree structure;
2. One-Class Support Vector Machine (SVM) learns a boundary around the normal data points, which can be used to identify data points that deviate significantly from the normal data;

3. Elliptic Envelope assumes that the data are normally distributed, and fits an ellipse around them; the ellipse is determined by the mean and covariance of the data, and the assumption that anomalies are located in low-density regions.

For a fair and meaningful comparison, a cross-validation setting, which produced consistent results when experiments were repeated, was set up. The following setting was found to be adequate: the cross-validation used 128 splits, and 95% randomly picked EMI spectra from the pristine FE simulation model (synthetic data) as training data. For the experiments with synthetic data, the remaining 5% of the pristine, and all the anomalous EMI spectra examples, were used as test data. For the experiments with real-world data, only the real-world data were used as test data. All the models used the same cross-validation splits for training and testing.

4.1.2. Results on Synthetic Data

Figure 13 presents the results of the anomaly detection methods on synthetic data.

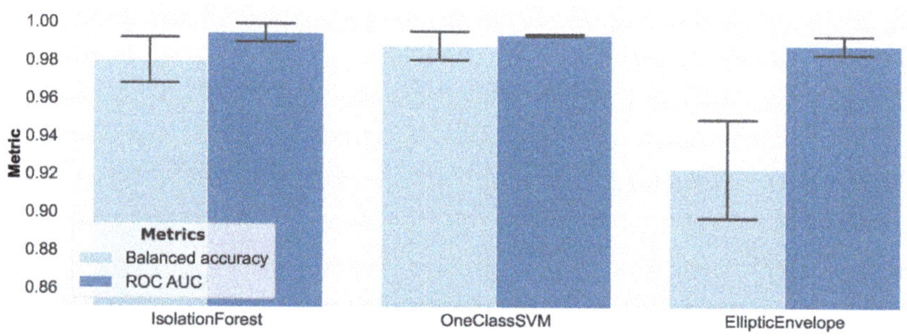

Figure 13. Anomaly detection results on synthetic data, by different ML models. According to balanced accuracy, the Isolation Forest and One-Class SVM seemed to perform reasonably well. The Elliptic Envelope exhibited the worst performance, due to a relatively high false positive rate.

The height of the bars refers to the mean value, and the error bars correspond to one standard deviation. According to balanced accuracy, the Isolation Forest and One-Class SVM seemed to perform reasonably well, while the Elliptic Envelope exhibited the worst performance. The ROC AUC metric did not reflect this as clearly, and should therefore be considered less appropriate in this setting. On closer inspection, the bad performance of the Elliptic Envelope was caused by a relatively high false positive rate on small damage sizes. A false positive example represented an anomalous example that was classified as normal (i.e., pristine), and a true negative example represented a correctly classified anomaly (i.e., damaged). Therefore, the false positive and true negative rates covered the performance of the complete set of anomalous examples, and always summed up to 1. In Figure 14, the false positive and true negative rates from the same experiment, with a focus on the comparatively small debonding radii of 2.2 mm and 2.5 mm, are shown.

In that regard, the Isolation Forest was the best-performing model, while the Elliptic Envelope had the poorest performance. Debonding sizes larger than a radius of 2.5 mm were accurately and consistently detected by all models.

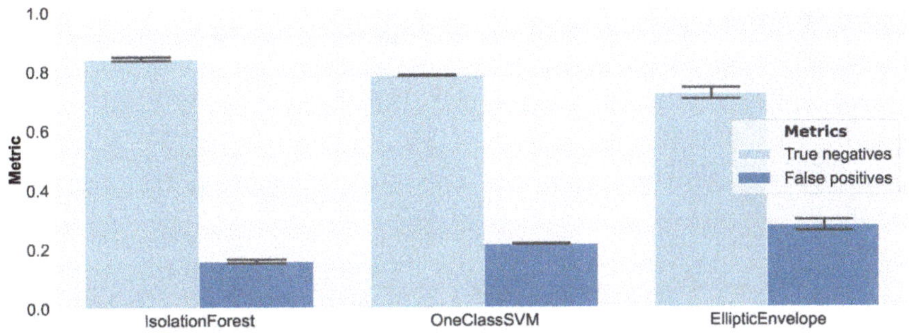

Figure 14. Anomaly detection results, on synthetic data, of small debonding radii of 2.2 mm and 2.5 mm. Larger damage was consistently detected accurately by all models.

4.1.3. Results on Real-World Data

Figure 15 presents the evaluation results of the real-world EMI spectra measurements.

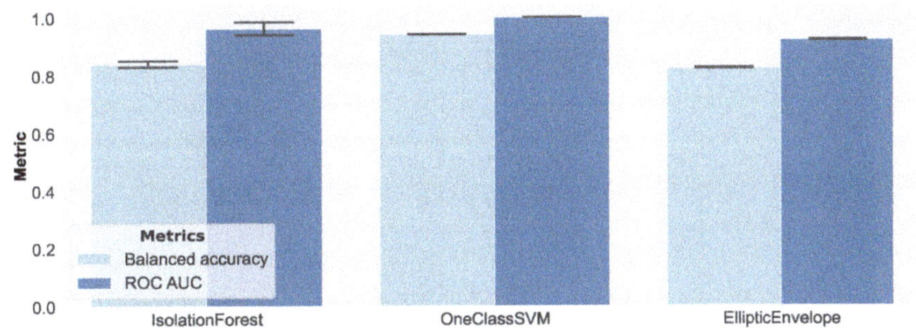

Figure 15. Anomaly detection results on real-world data.

Interestingly, compared to the results of synthetic data, the ranking of the performance of the models was preserved, where the One-Class SVM was the best, and the Elliptic Envelope was the worst performing model. However, this time there was a greater difference in performance, with respect to the considered damage sizes. Figure 16 shows the false positive rate of each model, with respect to damage size.

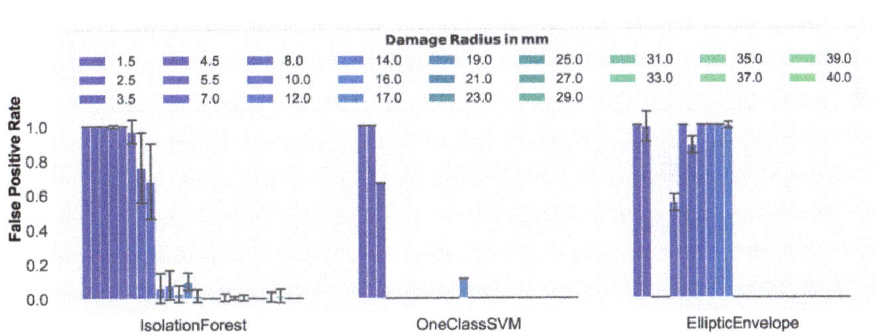

Figure 16. False positive rates of real-world data.

As can be seen, smaller damages could not be correctly classified as anomalous. This was to be expected, given the geometric structure of the honeycomb core of the considered sandwich panel. The core material consisted of hexagonal cells with thin aluminum walls, whose core cells were oriented normally to the face layers, and had a size of 3/8″ (i.e., they enclosed a circle with a 4.75 mm radius). Furthermore, it can be confirmed that only the One-Class SVM consistently detected damage > 3.5 mm, which was the reason for its best performance among the models. The Elliptic Envelope exhibited unexpectedly good performance on damage with 3.5 mm and 4.5 mm, but the reasons remain unclear at this time, and are still under investigation. According to these results, the One-Class SVM turned out to be the best-suited model, as the good performance of the synthetic data was well-aligned to the performance of the real-world data.

4.2. Debonding Radius Estimation by Regression

In this Section, the results of the evaluation experiments for debonding radius estimation by regression methods are presented: firstly, the setup of the experiments is described; secondly, the results of the synthetic EMI spectra are presented (i.e., the models trained and tested by synthetic data); thirdly, the results of real-world EMI spectra are presented (i.e., the models trained exclusively by synthetic data, but tested by calibrated real-world data), and are compared to a baseline previously published by Kralovec [9].

4.2.1. Setup of Experiments

The regression problem, where the debonding size had to be estimated based on the EMI measurement data, was realized by a baseline algorithm (BASE; a simple nearest-neighbor-based approach proposed by Kralovec [9], which works directly on the spectra of the measurements) and the following frequently used algorithms [43–45]:

1. K-Nearest Neighbor (KNN), which selects the K-Nearest Neighbors, based on a chosen distance (we used Euclidean distance and $K = 1$), and outputs the corresponding target value;
2. Support Vector Regression (SVR), which is the counterpart of the nonlinear SVM algorithm for regression problems;
3. Multilayer Perceptron (MLP), which is a fully connected, feed-forward multi-layer neural network. We implemented one with just a single hidden layer of 48 neurons (MLP_48), and another with two hidden layers of 64 and 32 neurons (MLP_64.32).

The nearest-neighbor-based method (KNN) is interesting, insofar as the only difference to the baseline (BASE) is the representation of the measurement data: that is, the baseline takes the raw spectra as input, and the novel method takes the features computed from the raw spectra (data preprocessing) as input. Therefore, we could evaluate the impact of our feature extraction pipeline in the most meaningful way.

As with the previous debonding detection task (see Section 4.1), a cross-validation setting that produced consistent results when repeated was needed, to compare the models fairly. This time, the cross-validation used 32 splits, and 90% randomly picked examples as training data. For the experiments with synthetic data, the remaining 10% of the examples were used as test data. For the experiments with real-world data, only the real-world data were used as test data. All the models used the same cross-validation splits for training and testing.

4.2.2. Results on Synthetic Data

Figure 17 presents the regression results for the synthetic data set.

Compared to the baseline (BASE), all methods gave substantially better results, whereas the nearest-neighbor-based method (KNN) yielded the smallest overall Mean Squared Error (MSE). The MLP with higher capacity (MLP_64.32) predictably gave better results than the simpler MLP architecture with fewer neurons (MLP_48), and was on a par with the performance of the SVR.

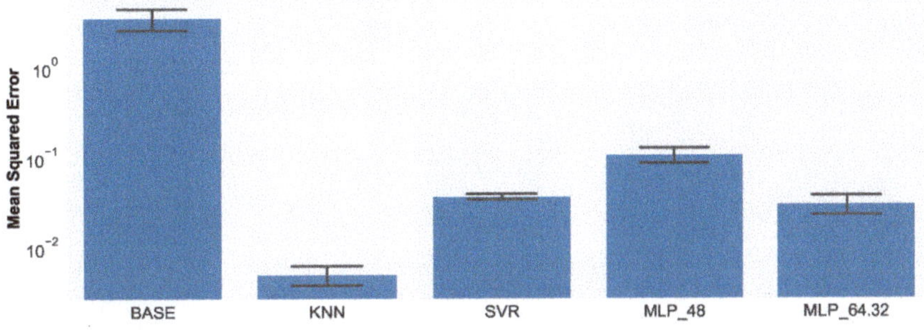

Figure 17. Regression results on synthetic data. Note the logarithmic scale on the y-axis. Compared to the baseline (BASE), all methods gave substantially better results, whereas the nearest-neighbor-based method (KNN) yielded the smallest overall MSE.

In Figure 18, a closer look at the results is given.

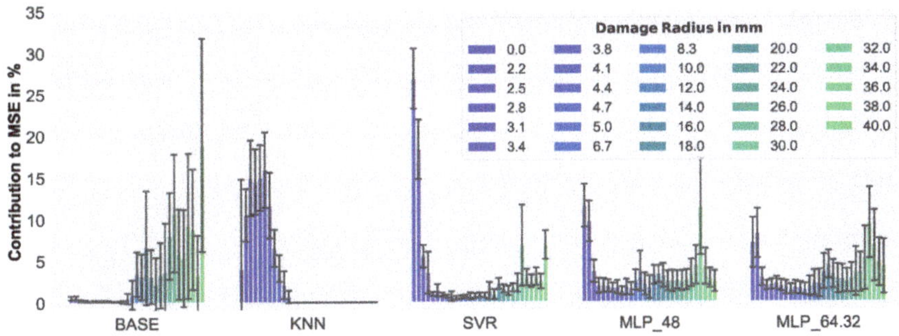

Figure 18. Relative contribution to the overall MSE, for each defect size, of synthetic data.

Here, the relative contribution to the MSE of each damage size is presented. Note that in such a visualization, the performance of the models cannot be compared in terms of MSE, but it sheds light on the behavior with respect to damage size. As can be seen, three different characteristics of the models can be distinguished: firstly, an increased error-proneness with increasing damage size (BASE); secondly, perfect handling of damage with ≥ 5 mm radius, and higher error-proneness with damage < 5 mm radius (KNN); thirdly, higher error-proneness only with small (0–2.5 mm radius) and large damage (>30 mm radius) (SVR, MLP_48, MLP_64.32).

4.2.3. Results on Real-World Data

Figure 19 presents the regression results of the real-world EMI measurements.

The MSE increased compared to the results on the synthetic data. This was expected, as no real-world data were used for training any of the models. All methods, except the baseline (BASE) and the smallest neural network (MLP_48), seemed to generalize reasonably well to unseen data. Similar to the anomaly detection task, the ranking between models, compared to the results of the synthetic data, was preserved, and the nearest-neighbor-based method (KNN) yielded the smallest MSE. Additionally, the results of KNN and SVR seemed to exhibit very small variances, which was also consistent with the results of the synthetic data. MLP_48 yielded the highest MSE, matching approximately the performance of the baseline (BASE).

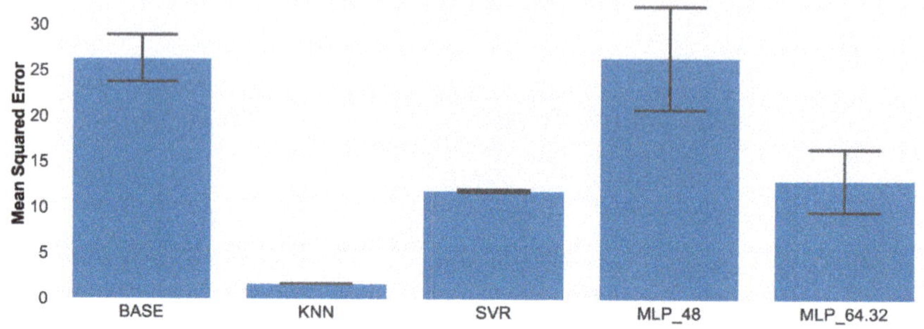

Figure 19. Regression results on real-world data. All methods except the baseline (BASE) and the smallest neural network (MLP_48) seemed to generalize reasonably well to unseen data.

In Figure 20, the relative contribution to the MSE of each damage size is presented.

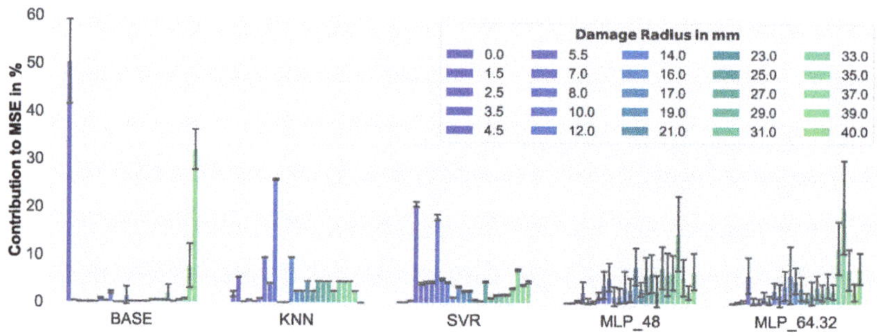

Figure 20. Relative contribution to the overall MSE, for each defect size, of real-world data.

Compared to the results of the synthetic data, the behavior of the models was quite different. To give an example, the contribution to MSE increased for the baseline (BASE) with defect size for synthetic data, but for real-world data, the highest contributions came from the normal examples with no defect, and the largest damage with 40.0 mm. On the other hand, the increased error-proneness with increasing damage size, which was noticed for the baseline (BASE) (see Figure 19), could now be observed in the MLP-based algorithms (MLP_48, MLP_64.32). The low variance that had already been observed with respect to MSE for KNN and SVR (see Figure 19) could also be observed here. This clearly indicates that the weaknesses of these models did not change, even when slightly different training data were used.

The erratic behavior of the baseline (BASE), especially for normal measurements—where often large damage sizes were predicted—rendered it unreliable. Given the overall low MSE and consistency of the results, the KNN algorithm seemed to be the best choice for damage size estimation, followed by the SVR. The MLP-based algorithms might perform better if more architectures were evaluated, but from those considered, only the more complex one (MLP_64.32) could outperform even the baseline.

5. Discussion

5.1. Evaluation Experiments

In the present research, great care was taken to meticulously select an appropriate cross-validation setup for each task individually, recognizing that this crucial step is overlooked in many studies. Thus, each experimental setup with cross-validation produced consistent results when repeated. Interestingly, the ranking of models was preserved between synthetic and real-world data for both tasks: this demonstrated that the synthetic data, in combination with the experimental setup, were sufficiently accurate for developing methods for a specific task, and for determining their relative performance in real-world settings.

5.2. Preprocessing

It is important to understand that although exclusively synthetic data were used to set up the complete preprocessing pipeline and ML models for evaluation, real-world data were still needed for calibration purposes. However, this was not a significant issue in practice, as the calibration process only required pristine baseline samples, which could be easily collected during the installation of the sensor network and the corresponding data acquisition system. Despite this, the value is recognized of further investigating more advanced preprocessing methods that could eliminate the need for calibration altogether.

5.3. Data

There are numerous disturbing influences on EMI measurements of operating mechanical structures, and thus challenges to the robustness of damage evaluation methods. The present research does not reflect all of these. Among them, measurement disturbances by ambient structural vibrations should be highlighted, as the results presented for the proposed two-step evaluation approach used comparatively low frequencies, and recent studies clearly show its relevance [38,46].

Thus, future research will address the approach's robustness against ambient structural vibrations. Further frequently reported and not directly addressed influences on EMI measurements are temperature and moisture, as these are not expected to significantly affect aluminum sandwich panels. However, other important lightweight design materials, e.g., fiber-reinforced polymer composite, are well-known to be sensitive to temperature and moisture. The authors expect that an FE parameter study that also considers the density change of the face layers will reflect all parameters that are affected by temperature and moisture, and thus also allow damage evaluation in composite materials. This is also supported, as the real-world EMI data used for model testing already include measurements of unknown elongated debondings and artificial disturbances, which may have effects similar to temperature and moisture on composite properties. However, extended studies should also include composite materials and the examination of other non-idealized types of damages, such as delamination and cracks. The proposed physics-based two-step approach is expected to be applicable to all these cases, but may require inefficiently detailed FE models.

The considered debonding in the FE model is a strong idealization, and the results of the real-world measurements may also be interpreted as the identification of a hole in the back side of the sandwich panel. However, unpublished simulations with the same FE model, but with the remaining core and back-side face layer, show very similar characteristics of the EMI spectra. This is also supported by previous results of the authors [9], who found a clear correlation between the resonances in the EMI spectra and the debonded face layer area. Furthermore, damages distant from the sensor are expected to be very challenging to the proposed approach, due to the strongly increased parameter space. Nevertheless, today's computational power, and the fusion of the data of a sensor array, are considered promising for solving this issue.

It is noteworthy that, while the real-world data used in this study provided valuable insights, the sample size was limited; therefore, the conclusions drawn from it should be considered within the scope of this limitation. Further research, with a larger and more diverse set of real-world data, is necessary, to fully validate the results and conclusions of this study.

5.4. Debonding Detection by Anomaly Detection

Depending on the material properties, small damages may only be detectable if a relatively high rate of false negatives (i.e., normal examples classified as anomalous) is accepted. However, by monitoring over a longer period of time, and evaluating summary statistics, satisfactory results may still be obtained. In regard to the real-world specimen used in this study, it might not be considered practical to investigate debonding damage sizes that are smaller than the size of the honeycomb core cells. However, the ability of the proposed method to identify even smaller damages could be useful in certain applications, such as predictive maintenance. These applications may also be supported by the short data evaluation times of the considered ML models, which are in the magnitude of $1/10^{-3}$ (for damage detection) and $1/10^{-5}$ (for damage size estimation) of the duration of a real-world EMI measurement.

Overall, the results are promising, and provide valuable insights into the use of anomaly detection models for identifying damage in sandwich panels.

5.5. Debonding Size Estimation by Regression

One of the main disadvantages of using the KNN approach is that it requires the traversal of the entire training data set during inference; therefore, the running time of a KNN is directly proportional to the number of examples in the data set, which means that the larger the data set, the more time-consuming the process will be. This can lead to increased computational time and memory usage, particularly when dealing with large data sets. In cases where real-time or near-real-time predictions are needed, the KNN approach may not be the most efficient choice, due to this requirement. If the computational efficiency is less important, the KNN results may be further improved by increasing K, the number of considered NNs. However, as most of the time the NNs correspond to the same target value, little improvement is expected. The ML models considered in the present research showed short data evaluation times in the magnitude of $1/10^{-3}$ of the duration of a real-world EMI measurement, and thus were hardly constrained in their possible applications. This efficiency may even question the need for the previous anomaly detection step. However, anomaly detection is (i) believed to be more robust in the evaluation of data that strongly deviate from the used training data, and (ii) 100-times faster, which appears meaningful for applications where normal data (data from the pristine structure) are predominantly evaluated.

6. Conclusions

The presented results clearly demonstrate the applicability of the proposed two-step physics- and ML-based EMI spectra evaluation approach for sandwich debonding detection and size estimation. The approach is capable of evaluating measured real-world EMI spectra exclusively based on synthetic, FE-simulated EMI spectra for debonding detection (step one) and debonding size estimation (step two). In addition, the feature engineering and the evaluation model development are conducted exclusively based on synthetic, FE-simulated EMI spectra. Therefore, the approach is very economic, compared to training with measured real-world data. The ML models applied for damage evaluation are, once trained, approximately 10^5-times (for damage detection) and 10^3-times (for damage size estimation) faster than the EMI measurement itself. This also enables quasi-real-time applications and edge computing at small and lightweight local processing units.

The best performance for damage detection by real-world EMI data was given by the combination of the developed data preprocessing with a One-Class SVM anomaly detection model. Misclassifications were only given for radial debonding sizes below 2.5 mm, i.e., debondings smaller than the honeycomb core's cell size. The best performance for damage size estimation by real-world EMI data (1.7 mm MSE and 0.01 mm standard deviation) was given by the combination of the developed data preprocessing with a KNN regression model, which clearly outperformed a previous method used as a benchmark. Both the damage detection and the size estimation models were cross-validated, and showed high consistency in their results. Furthermore, the physics-based two-step approach demonstrated robustness against unknown artificial disturbances that were applied during real-world data acquisition. Data preprocessing was identified as the key factor in the performance of the proposed two-step evaluation approach, which—after comprehensive data exploration—found the consideration of a frequency range between 1.3 kHz and 24 kHz to be sufficient. This strongly reduced the ML model complexity, and allowed for a less discretized FE model, leading to a significant reduction in the simulation costs of the synthetic data.

However, the application of synthetic data-based features to the evaluation of real-world EMI measurement data still requires calibration to baseline measurements, to overcome the strong simplifications of the used FE model—a drawback that shall be addressed in future research. Future research must also investigate the proposed approach's applicability to other important materials (e.g., composites) and their peculiar damage types, and damage evaluation with distant sensors. In addition, damage assessment with more remote sensors, and the extension of the approach to data fusion of a sensor network, need to be addressed.

The present work is also intended to promote future research in the field, by providing the complete EMI spectra data and the full code for data exploration, data preprocessing, and data evaluation (access see: Data Availability Statement).

Author Contributions: Conceptualization, C.K. and B.L.; methodology, C.K. and B.L.; software, B.L.; validation, C.K., B.L. and M.S.; formal analysis, C.K., B.L. and M.K.; investigation, C.K., B.L. and M.K.; resources, M.S.; data curation, B.L. and M.K.; writing—original draft preparation, C.K. and B.L.; writing—review and editing, C.K., B.L. and M.S.; visualization, C.K. and B.L. All authors have read and agreed to the published version of the manuscript.

Funding: This research received no external funding.

Institutional Review Board Statement: Not applicable.

Informed Consent Statement: Not applicable.

Data Availability Statement: Data can be found at https://zenodo.org/record/6758723 (accessed on 23 December 2022). Code is available at https://github.com/berni-lehner/structural_health_monitoring/ (accessed on 23 December 2022).

Acknowledgments: This work was supported by Silicon Austria Labs (SAL), owned by the Republic of Austria, the Styrian Business Promotion Agency (SFG), the federal state of Carinthia, the Upper Austrian Research (UAR), and the Austrian Association for the Electric and Electronics Industry (FEEI).

Conflicts of Interest: The authors declare no conflict of interest.

References

1. Viechtbauer, C.; Schagerl, M.; Schröder, K.-U. From NDT over SHM to SHC—The future for wind turbines. In Proceedings of the 6th International Conference on Structural Health Monitoring of Intelligent Infrastructure, Hong Kong, China, 9–11 December 2013.
2. Giurgiutiu, V. *Structural Health Monitoring with Piezoelectric Wafer Active Sensors*, 2nd ed.; Academic Press Inc.: Cambridge, MA, USA, 2014.
3. Rytter, A. Vibration based inspection of Civil Engineering. Ph.D. Thesis, Department of Building Technology and Structural Engineering, Aalborg University, Aalborg, Denmark, 1993.
4. Viechtbauer, C. A Novel Approach to Monitor and Assess Damages in Lightweight Structures. Ph.D. Thesis, Institute of Structural Lightweight Design, JKU Linz, Linz, Austria, 2015.
5. Kralovec, C.; Schagerl, M. Review of structural health monitoring methods regarding a multi-sensor approach for damage assessment of metal and composite structures. *Sensors* **2020**, *20*, 826. [CrossRef] [PubMed]
6. Scala, C.M.; Bowles, S.J.; Scott, I.G. *The Development of Acoustic Emission for Structural Integrity Monitoring of Aircraft*; Technical Report; Aeronautical Research Labs Melbourne: Melbourne, Australia, 1988.
7. Humer, C.; Kralovec, C.; Schagerl, M. Testing the scattering analysis method for guided waves by means of artificial disturbances. In Proceedings of the 7th APWSHM, Hong Kong, China, 12–15 November 2018. Available online: https://www.ndt.net/search/docs.php3?id=24082 (accessed on 23 December 2022).
8. Yeasin Bhuiyan, M.; Shen, Y.; Giurgiutiu, V. Interaction of Lamb waves with rivet hole cracks from multiple directions. *Proc. Inst. Mech. Eng. C J. Mech. Eng. Sci.* **2017**, *231*, 2974–2987. [CrossRef]
9. Kralovec, C.; Schagerl, M. Electro-mechanical impedance measurements as a possible SHM method for sandwich debonding detection. In Proceedings of the 21st Symposium on Composites, Bremen, Germany, 5–7 July 2017; pp. 763–777. [CrossRef]

10. Kralovec, C.; Schagerl, M.; Mayr, M. Localization of damages by model-based evaluation of electro-mechanical impedance measurements. In Proceedings of the 9th EWSHM, Manchester, UK, 10–13 July 2018. Available online: http://www.ndt.net/?id=23338 (accessed on 23 December 2022).
11. Giurgiutiu, V.; Zagrai, A. Damage detection in thin plates and aerospace structures with the electro-mechanical impedance method. *Struct. Health Monit.* **2005**, *4*, 99–118. [CrossRef]
12. Li, W.; Palardy, G. Damage Monitoring Methods for Fiber-Reinforced Polymer Joints: A Review. *Comp. Struct.* **2022**, *299*, 116043. [CrossRef]
13. Fan, X.; Li, J.; Hao, H. Review of piezoelectric impedance based structural health monitoring: Physics-based and data-driven methods. *Adv. Struct. Eng.* **2021**, *24*, 3609–3626. [CrossRef]
14. Avci, O.; Abdeljaber, O.; Kiranyaz, S.; Hussein, M.; Gabbouj, M.; Inman, D.J. A review of vibration-based damage detection in civil structures: From traditional methods to Machine Learning and Deep Learning applications. *Mech. Syst. Signal Process.* **2021**, *147*, 107077. [CrossRef]
15. Tenreiro, A.; Francisco, G.; António, M.L.; Da Silva, L.F.M. A review of structural health monitoring of bonded structures using electromechanical impedance spectroscopy. *Struct. Health Monit.* **2022**, *21*, 228–249. [CrossRef]
16. Aabid, A.; Parveez, B.; Raheman, M.A.; Ibrahim, Y.E.; Anjum, A.; Hrairi, M.; Parveen, N.; Mohammed Zayan, J. A review of piezoelectric material-based structural control and health monitoring techniques for engineering structures: Challenges and opportunities. *Actuators* **2022**, *10*, 101. [CrossRef]
17. Lopes, V., Jr.; Park, G.; Cudney, H.H.; Inman, D.J. Impedance-based structural health monitoring with artificial neural networks. *J. Intell. Mater. Syst. Struct.* **2000**, *11*, 206–214. [CrossRef]
18. Palomino, L.V.; Steffen, V.; Finzi Neto, R.M. Probabilistic neural network and fuzzy cluster analysis methods applied to impedance-based SHM for damage classification. *Shock Vib.* **2014**, *2014*, 40194. [CrossRef]
19. Park, S.; Lee, J.J.; Yun, C.B.; Inman, D.J. Electro-mechanical impedance-based wireless structural health monitoring using PCA-data compression and k-means clustering algorithms. *J. Intell. Mater. Syst. Struct.* **2008**, *19*, 509–520. [CrossRef]
20. Min, J.; Park, S.; Yun, C.B. Impedance-based structural health monitoring using neural networks for autonomous frequency range selection. *Smart Mater. Struct.* **2010**, *19*, 125011. [CrossRef]
21. De Oliveira, M.A.; Monteiro, A.V.; Vieira Filho, J. A new structural health monitoring strategy based on PZT sensors and convolutional neural network. *Sensors* **2018**, *18*, 2955. [CrossRef] [PubMed]
22. De Rezende, S.W.F.; de Moura, J.D.R.V.; Neto, R.M.F.; Gallo, C.A.; Steffen, V. Convolutional neural network and impedance-based SHM applied to damage detection. *Eng. Res. Express* **2020**, *2*, 035031. [CrossRef]
23. Li, H.; Ai, D.; Zhu, H.; Luo, H. Integrated electromechanical impedance technique with convolutional neural network for concrete structural damage quantification under varied temperatures. *Mech. Syst. Signal Process.* **2021**, *152*, 107467. [CrossRef]
24. Ai, D.; Mo, F.; Han, Y.; Wen, J. Automated identification of compressive stress and damage in concrete specimen using convolutional neural network learned electromechanical admittance. *Eng. Struct.* **2022**, *259*, 114176. 2022.114176. [CrossRef]
25. Ai, D.; Cheng, J. A deep learning approach for electromechanical impedance based concrete structural damage quantification using two-dimensional convolutional neural network. *Mech. Syst. Signal Process.* **2023**, *183*, 109634. 2022.109634. [CrossRef]
26. Alazzawi, O.; Wang, D. Damage identification using the PZT impedance signals and residual learning algorithm. *J. Civ. Struct. Health Monit.* **2021**, *11*, 1225–1238. [CrossRef]
27. Nguyen, T.T.; Phan, T.T.V.; Ho, D.D.; Pradhan, A.M.S.; Huynh, T.C. Deep learning-based autonomous damage-sensitive feature extraction for impedance-based prestress monitoring. *Eng. Struct.* **2022**, *259*, 114172. [CrossRef]
28. Singh, S.K.; Sikdar, S.; Malinowski, P.H. An optimized data fusion strategy for structural damage assessment using electromechanical impedance. *Smart Mater. Struct.* **2021**, *30*, 035012. [CrossRef]
29. Winklberger, M.; Kralovec, C.; Humer, C.; Heftberger, P.; Schagerl, M. Crack identification in necked double shear lugs by means of the electro-mechanical impedance method. *Sensors* **2021**, *21*, 44. [CrossRef] [PubMed]
30. Naidu, A.S.; Soh, C.K.; Pagalthivarthi, K.V. Bayesian network for E/M impedance-based damage identification. *J. Comput. Civ. Eng.* **2006**, *20*, 227–236. [CrossRef]
31. Albakri, M.I.; Tarazaga, P.A. Electromechanical impedance–based damage characterization using spectral element method. *J. Intell. Mater. Syst. Struct.* **2017**, *28*, 63–77. [CrossRef]
32. Ezzat, A.A.; Tang, J.; Ding, Y. A model-based calibration approach for structural fault diagnosis using piezoelectric impedance measurements and a finite element model. *Struct. Health Monit.* **2020**, *19*, 1839–1855. [CrossRef]
33. Humer, C.; Höll, S.; Kralovec, C.; Schagerl, M. Damage identification using wave damage interaction coefficients predicted by deep neural networks. *Ultrasonics* **2022**, *124*, 106743. [CrossRef]
34. Shuai, Q.; Zhou, K.; Zhou, S.; Tang, J. Fault identification using piezoelectric impedance measurement and model-based intelligent inference with pre-screening. *Smart Mater. Struct.* **2017**, *26*, 045007. Available online: https://iopscience.iop.org/article/10.1088/1361-665X/aa5d41/meta (accessed on 23 December 2022). [CrossRef]
35. Kovács, P.; Lehner, B.; Thummerer, G.; Mayr, G.; Burgholzer, P.; Huemer, M. A Hybrid Approach for Thermographic Imaging With Deep Learning. In Proceedings of the ICASSP 2020—2020 IEEE International Conference on Acoustics, Speech and Signal Processing (ICASSP), Barcelona, Spain, 4–8 May 2020; pp. 4277–4281. [CrossRef]
36. Kovács, P.; Lehner, B.; Thummerer, G.; Mayr, G.; Burgholzer, P.; Huemer, M. Surfing Virtual Waves to Thermal Tomography: From model- to deep learning-based reconstructions. *IEEE Signal Process. Mag.* **2022**, *39*, 55–67. [CrossRef]

37. Park, G. Assessing Structural Integrity Using Mechatronic Impedance Transducers with Applications in Extreme Environments. Doctoral Dissertation, Virginia Tech, Blacksburg, VA, USA, 2000.
38. De Castro, B.A.; Baptista, F.G.; Ciampa, F. A comparison of signal processing techniques for impedance-based damage characterization in carbon fibers under noisy inspections. *Mater. Today Proc.* **2021**, *34*, 275–279. [CrossRef]
39. Makhoul, J. A fast cosine transform in one and two dimensions. *IEEE Trans. Acoust. Speech Signal Process.* **1980**, *28*, 27–34. Available online: https://ieeexplore.ieee.org/document/1163351 (accessed on 23 December 2022). [CrossRef]
40. Liu, F.T.; Ting, K.M.; Zhou, Z.H. Isolation forest. In Proceedings of the eighth IEEE International Conference on Data Mining, Pisa, Italy, 15–19 December 2008; pp. 413–422. [CrossRef]
41. Schölkopf, B.; Platt, J.C.; Shawe-Taylor, J.; Smola, A.J.; Williamson, R.C. Estimating the support of a high-dimensional distribution. *Neural Comput.* **2001**, *13*, 1443–1471. [CrossRef]
42. Hubert, L.; Arabie, P. Comparing partitions. *J. Classif.* **1985**, *2*, 193–218. [CrossRef]
43. Rosenblatt, F. The perceptron: A probabilistic model for information storage and organization in the brain. *Psychol. Rev.* **1958**, *65*, 386–408. [CrossRef]
44. Cortes, C.; Vapnik, V. Support-vector networks. *Mach. Learn.* **1995**, *20*, 273–297. [CrossRef]
45. Fix, E.; Hodges, J.L. Discriminatory analysis. Nonparametric discrimination: Consistency properties. *Int. Stat. Rev. Int. Stat.* **1989**, *57*, 238–247. [CrossRef]
46. De Castro, B.A.; Baptista, F.G.; Ciampa, F. New imaging algorithm for material damage localisation based on impedance measurements under noise influence. *Measurement* **2020**, *163*, 107953. [CrossRef]

sensors

MDPI

Article

A Novel Baseline-Free Method for Damage Localization Using Guided Waves Based on Hyperbola Imaging Algorithm

Jichao Xu, Wujun Zhu, Xunlin Qiu and Yanxun Xiang *

Key Laboratory of Safety Science of Pressurized System of MOE, School of Mechanical and Power Engineering, East China University of Science and Technology, Shanghai 200237, China
* Correspondence: yxxiang@ecust.edu.cn

Abstract: Most imaging methods based on ultrasonic Lamb waves in structural health monitoring requires reference signals, recorded in the intact state. This paper focuses on a novel baseline-free method for damage localization using Lamb waves based on a hyperbolic algorithm. This method employs a special array with a relatively small number of transducers and only one branch of the hyperbola. The novel symmetrical array was arranged on plate structures to eliminate the direct waves. The time difference between the received signals at symmetrical sensors was obtained from the damage-scattered waves. The sequence of time difference for constructing the hyperbolic trajectory was calculated by the cross-correlation method. Numerical simulation and experimental measurements were implemented on an aluminum plate with a through-thickness hole in the current state. The imaging results show that both the damages outside and inside the diamond-shaped arrays can be localized, and the positioning error reaches the maximum for the diamond-shaped array with the minimum size. The results indicate that the position of the through-hole in the aluminum plate can be identified and localized by the proposed baseline-free method.

Keywords: Lamb waves; baseline-free; hyperbola; cross-correlation; damage localization

Citation: Xu, J.; Zhu, W.; Qiu, X.; Xiang, Y. A Novel Baseline-Free Method for Damage Localization Using Guided Waves Based on Hyperbola Imaging Algorithm. *Sensors* **2023**, *23*, 2050. https://doi.org/10.3390/s23042050

Academic Editor: Zenghua Liu

Received: 14 January 2023
Revised: 8 February 2023
Accepted: 9 February 2023
Published: 11 February 2023

1. Introduction

Plate-like structures are widely used in the fields of mechanical engineering, civil engineering, aerospace engineering and so on. During the fabrication and service process, plate-like structures inevitably undergo various degrees of damage with the possibility of failure under extreme conditions and environment. Structural health monitoring (SHM) based on Lamb waves is an important means to effectively monitor the structural safety and predict the service life of structures, due to its outstanding advantages of low attenuation, long propagation distance, large detection area and, most importantly, high sensitivity to structural damage and material inhomogeneity [1–4]. However, phenomena like dispersion, multimodes and the mode conversion (change in wave velocity when the wave interacts with structural discontinuities or boundaries) of Lamb waves make reliable damage localization a challenging task, because it is difficult to effectively extract damage features from the received signals.

In the process of damage localization, the typical detection technology is to refer to the baseline signals, namely, to compare the signals in the current state with those in the healthy state. A scattered signal containing some damage-related information (location, size, orientation, type) can be obtained by performing reference signal subtraction. The time of flight (ToF), which corresponds to the time taken by the wave packet to travel from an actuator to a sensor along a certain path, is a straightforward feature of the damage-scattered signal for damage localization. The ellipse and hyperbola algorithms become the two most important methods for locating defects based on the ToF in plate-like structures [5–10].

This technology will be very effective and accurate if damage is the only factor causing the change in the baseline signals. In practice, operational and environmental conditions, such as stress, load, temperature, etc., are constantly changing, which may lead to significant changes in the baseline signals and can easily overwhelm the signal changes caused by damage [11,12]. This could cause significant errors and may even corrupt the damage localization results.

To overcome such drawbacks, baseline-free damage detection algorithms have become a focus in the development of SHM. An instantaneous baseline can be obtained through a comparison between different sensor pairs that have a similar path [13], but the distance of each pair must be kept identical. A large number of undamaged wave-propagating paths are also required, which may not be satisfied simultaneously in practice. The time reversal method (TRM) is widely used as a promising candidate for baseline-free damage detection in thin-walled plate/shell-like structures [14]. It is based on the concept of spatial reciprocity and time-reversal invariance, originally observed in acoustic waves [15]. This is due to an input signal, which can be reconstructed at the source transducer, when the output signal measured at the receiver transducer is time-reversed and emitted back. Liu et al. [16] combined the probability imaging algorithm with the virtual time reversal method to study the damage in a composite board by air-coupled sensors; the effects of the damage type and size on the probability imaging algorithm were discussed to achieve the baseline-free detection. Huang et al. [17] proposed a reciprocity index-based probabilistic imaging algorithm to achieve the baseline-free identification of delamination on the path of a piezoelectric piece. Kannusamy et al. [18] presented a refined time reversal method in conjunction with the probability imaging algorithm for baseline-free damage localization in thin plates, using a sparse array arrangement of piezoelectric wafer transducers.

However, the aforementioned methods need a lot of actuators/sensors to form enough monitoring paths for damage localization. Damage in the survey areas, but not covered by the monitoring paths, is hard to detect accurately. To circumvent these limitations, damage can be detected and localized anywhere in the monitoring areas by using the ToF of the damage-scattered signals. Wang and Yuan [19] applied the time reversal method on the scattered signals so as to focus on the damage location and achieve the baseline-free damage imaging using a delay and sum algorithm. Jeong et al. [20] obtained the ToF information using the time reversal of a scattered Lamb wave and constructed a baseline-free localization image by a beamforming technique. Jun and Lee [21] proposed a new baseline-free technique, which combined a hybrid time reversal method with a beamforming method using the ToF of the wave packets generated by the damage. In general, TRM often requires a programmable transducer array to synthesize time-reverse signals simultaneously, which limits its application due to the complexity and cost of the hardware system. Therefore, a baseline-free method without complex manipulation, such as time-reversal based on the ToF information, is urgently needed.

In addition to the ToF of the scattered signals, the time difference of arrival (TDOA) between two scattered signals can also be used for damage localization. This method utilizes the difference in the ToF to localize damage based on the intersection of hyperbolas. Yelve et al. [22] introduced a baseline-free method based upon the TDOA of nonlinear Lamb waves to localize barely visible impact damage in composite laminates. Li et al. [23] introduced a probability-based hyperbola diagnostic imaging method based on the different ToFs of the damage signals to achieve the baseline-free damage localization and imaging. This method required at least twice as many transducers as in the general hyperbola method to form sufficient pairs of two sensors. This study aims to achieve baseline-free damage localization by the hyperbola algorithm using far fewer transducers. A new array composed of these transducers can not only eliminate the direct wave, but can also make the boundary reflection signal, and other noise, cancel each other out. The proposed baseline-free method is expected to be more convenient and effective in engineering practice.

According to the aforementioned research, a crucial step of the baseline-free damage imaging, based on the ToF, is to extract the damage-scattered signals accurately. In this paper, a novel baseline-free method for damage localization was presented, including a peculiar symmetrical array of PZT actuators/sensors to eliminate the direct waves from the received signals. Meanwhile, the cross-correlation method (CCM) was introduced to determine the sequence of time difference accurately from the damage-scattered waves. The damage was identified and localized by a hyperbolic algorithm with only one branch of the hyperbolic curve. To demonstrate the feasibility and the effectiveness of the proposed strategy, numerical simulation and experimental validation were carried out to localize a through-thickness hole in an aluminum plate. The influence of different damage locations and array forms on the damage localization results was discussed.

2. Methodology for Baseline-Free Damage Localization

Lamb wave based damage localization is based on the fundamental idea that a traveling wave in a plate structure will be scattered by the damage. Note that the damage location is usually determined by the group velocity and the ToF of the scattered Lamb waves. While the group velocity of a certain Lamb wave mode is often determined in a homogenous and isotropic plate structure, the ToF information becomes a key factor in damage localization. There are many imaging and localizing methods based on the ToF, including ellipse and hyperbola algorithms.

Whereas the ellipse algorithm considers two transducers to be acting in pairs, the hyperbola algorithm mainly considers combinations of one actuator (transmitter), n, and pairs of sensors (receivers), i and j, in a spatially distributed array, as shown in Figure 1a. Without the loss of generality, a received signal on the sensing path consists, at least, of a direct signal and a scattered signal. If a damage is positioned at the point $D(x,y)$, the difference in arrival time of the scattered signals at the two sensors is defined as:

$$\Delta t_{ij}(x,y) = \frac{l_{n-D-i} - l_{n-D-j}}{V_g}, \quad (1)$$

where l_{n-D-i} and l_{n-D-j} are the distances from the actuator to the damage, and then to the two sensors, respectively. Further, V_g is the velocity of the signals. Since the distance from the actuator to the damage is the same for both sensors (Figure 1a), it does not affect the difference in the arrival time. Equation (1) can then be simplified as [24]:

$$\Delta t_{ij}(x,y) = \frac{\sqrt{(x_i - x)^2 + (y_i - y)^2} - \sqrt{(x_j - x)^2 + (y_j - y)^2}}{V_g}, \quad (2)$$

where (x_i, y_i) and (x_j, y_j) are the coordinates of the two sensors, respectively. The solution of Equation (2) is a hyperbola with its foci on the two sensors. By performing this for all the available nij groups of sensors, the damage location in the sensor network is obtained as the intersection point of all the hyperbolas. The time difference of the scattered signals is a robust feature of the wave signals, for which it is not necessary to know the time origin of the excitation.

In general, the scattered signals are always subtracted by the corresponding baselines in the intact state, depicted by:

$$f_{SC}^{i}(t) = f_S^i(t) - f_C^i(t), \quad (3)$$

where $f_{SC}^{i}(t)$ is the scattered signal for the sensor i, $f_S^i(t)$ is the received signal for the sensor i in the intact state, which is the state without damage in the structure, and $f_C^i(t)$ is the received signal for the sensor i in the current state. The same processing is carried out for the sensor j. To realize baseline-free localization and imaging, a novel symmetrical sensor arrangement is presented to eliminate the direct propagating waves and extract the damage-scattered waves in this paper. As shown in Figure 1b, one actuator is placed in the perpendicular bisector of the line segment connecting the two sensors, while one actuator and two sensors are located in random positions in the typical sensor array. This form of array can be in a diamond or square shape, as long as the diagonals of the array are perpendicular to each other. Since the distances between the actuator and the two sensors are equal, according to vertical theorem in mathematics, the direct propagating signals on the two sensors from the actuator are almost identical. Under this situation, the damage-scattered signals can be extracted by subtracting the two received signals in the current state, which is described by:

$$f_{ij}(t) = f_i(t) - f_j(t), \tag{4}$$

where $f_{ij}(t)$ is the damage-scattered signal; $f_i(t)$ and $f_j(t)$ are the received signals for the sensors i and j in the current state, respectively. At the same time, boundary reflection signals, and other noise, can also cancel each other out because of the symmetrical positioning of the two sensors in the structure. Therefore, the time difference $\Delta t_{ij}(x,y)$ can be obtained according to the characteristics of the damage-scattered signals. Here, Hilbert transform [12] is used to estimate the ToF by extracting the time corresponding to the maximum of the envelope of the signal component. This is described by:

$$H_{ij}(T) = \frac{1}{\pi}\int_{-\infty}^{+\infty} \frac{f_{ij}(t)}{T-t}dt, \tag{5}$$

where $H_{ij}(T)$ is the Hilbert transform of the damage-scattered signal $f_{ij}(t)$.

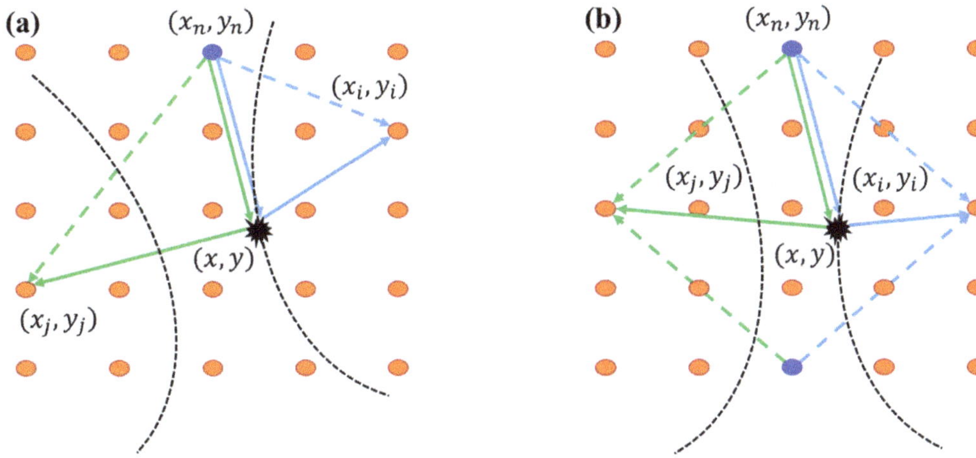

Figure 1. Schematic diagram of hyperbolic positioning with: (**a**) a traditional array; and (**b**) a new symmetrical array for exciting and receiving signals.

In addition, one should also pay further attention to the sign of the time difference $\Delta t_{ij}(x,y)$, because the hyperbolic loci usually have two branches by definition, which may bring pseudo-intersection points. To overcome such a problem, a cross-correlation method (CCM) is used to determine the sequence of time difference with a positive or negative sign. The time delays are calculated by means of the CCM between the damage-scattered signals and the two received signals, respectively, in which the cross-correlation function exhibits a notable characteristic value at $\Delta t_{ij}(x,y)$. This can be expressed by [25]:

$$R_{iji}(T) = \int_{-\infty}^{+\infty} f_{ij}(t)f_i(t-T)dt = \int_{-\infty}^{+\infty} f_{ij}(t+T)f_i(t)dt \tag{6}$$

and

$$R_{ijj}(T) = \int_{-\infty}^{+\infty} f_{ij}(t)f_j(t-T)dt = \int_{-\infty}^{+\infty} f_{ij}(t+T)f_j(t)dt. \tag{7}$$

Since the damage-scattered signals are subtracted from the two received signals, the cross-correlation function will give a peak value at time 0, simultaneously. If the peak value appears at $\Delta t_{ij}(x,y)$ positive relative to 0, the scattered signal in another received signal will be behind that in the received signal. Otherwise, it is in advance. Once the sign of $\Delta t_{ij}(x,y)$ has been determined, whether it is positive or negative, it can specify only one branch of the hyperbola, so as to achieve damage localization precisely.

In this imaging method, a cumulative distribution function (CDF) $F(z_{ij})$ is introduced to determine a specific pixel at each node when the inspection area of the structure is virtually and evenly meshed, which is defined as [26]:

$$F(z_{ij}) = \int_{-\infty}^{z_{ij}} f(z)dz. \tag{8}$$

where $f(z) = \frac{1}{\sigma\sqrt{2\pi}} \exp\left[-\frac{z}{2\sigma}\right]$ is the Gaussian distribution function that relates the probability density of damage occurrence at the meshed node (x,y). Further, σ is the standard deviation of the relevant damage feature as a tolerance factor in the imaging process, and z_{ij} is the upper limit of integral function, which means the shortest distance from the node to the hyperbolic loci. Thus, the probability value $D_{ij}(x,y)$ on each node constructed by sensor i and j is then defined as:

$$D_{ij}(x,y) = 1 - \left[F(z_{ij}) - F(-z_{ij})\right]. \tag{9}$$

For simplicity, a general overview of the proposed method in the symmetrical array is given in Figure 2.

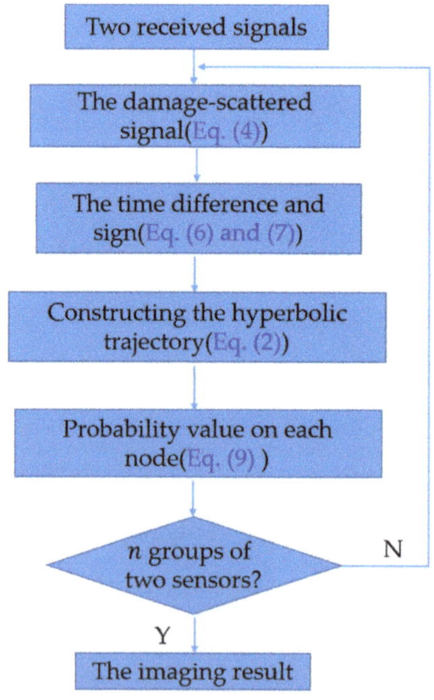

Figure 2. A flowchart of the proposed method in the symmetrical array.

3. Validation of Numerical Simulation

In order to verify the effectiveness of the proposed method, a three-dimensional finite element (FE) model was firstly developed in conjunction with a commercial FEM software (ABAQUS/CAE). The FE model was considered as an effective method to study the characteristics of wave propagation in plate structures.

3.1. Simulation Model

The dimensions of the aluminum plate employed for the simulation were 300 mm × 300 mm × 1 mm. A circular through-thickness hole with a diameter of 5 mm was dug on the aluminum plate as a damage source, as shown in Figure 3. Additionally, in order to realize very minimal or no wave reflection from the boundary, a gradually damping artificial boundary was introduced to simulate a nonreflecting boundary condition [27]. The damping coefficients of the individual layers were gradually increased from the inner layer to the finite boundary. Both the boundary length and the damping coefficient of each layer were dependent on the excitation frequency. The Young's modulus, Poisson's ratio and the density of the aluminum plate were 69 GPa, 0.33 and 2700 kg/m^3, respectively.

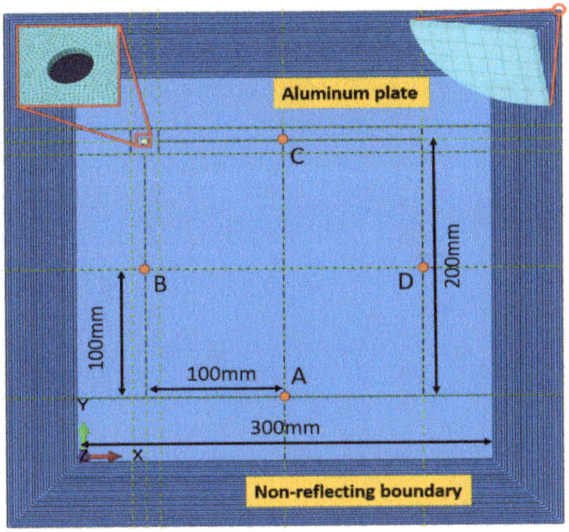

Figure 3. A schematic diagram of the simulation model setup.

Since piezoelectric elements were not available in ABAQUS/explicit, concentrated forces were applied to simulate the Lame wave excitation for round PZT transducers. Hanning window modulated five-cycle sinusoidal tone bursts, at a central frequency of 250 kHz, were excited by applying a pair of symmetric forces on two points, which were symmetrically positioned on the upper and lower surfaces of the plate. This kind of point displacement modeling to generate a single S0 mode Lamb wave was described in detail in [28]. The S0 mode Lamb wave was selected as an incident wave because it is more sensitive to the through-hole than an A0 mode. The Lamb waves were received at monitored points on the upper surface of the plate.

A global cartesian coordinate was introduced, in which the abscissa axis was parallel to the lower edge of the plate, and the bottom left corner of the plate was at the origin. The points for exciting and receiving signals, forming a diamond-shaped array, were labelled as A, B, C and D at different coordinates, respectively. The coordinates of the diamond-shaped array, and different damage positions, were shown in Table 1; the coordinates of the diamond-shaped arrays with different sizes and the damage location, were shown in Table 2. These diamond-shaped arrays were numbered from 1 to 6, both in the case of numerical simulation and experimental testing.

Table 1. The diamond-shaped array with the through-hole of different positions. (Unit: mm).

Position of Array	Center Position of the Hole	Localization Result (Simulation)	Error	Localization Result (Experiment)	Error	Number
Diagonal distance was 200, A (150, 50), B (50, 150), C (150, 250), D (250, 150).	(50, 250)	(60.1, 240.1)	3.3%	(35.6, 266)	5.3%	1
	(100, 150)	(101.2, 146.8)	1.07%	(92.9, 149.8)	2.36%	2

Table 2. The locations of the through-hole in the diamond-shaped arrays with different sizes (Unit: mm).

Number	Position of Array	Center Position of the Hole	Localization Result (Simulation)	Error	Localization Result (Experiment)	Error
3	Diagonal distance was 80, A (150, 110), B (110, 150), C (150, 190), D (190, 150)	(230, 230)	(211.1, 211.1) (211.3, 211.3)	6.3% 6.2%	(244.6, 253.1)	7.7%
4	Diagonal distance was 120, A (150, 90), B (90, 150), C (150, 210), D (210, 150)		(218.6, 220.7)	3.8%	(251.9, 249.1)	7.3%
5	Diagonal distance was 160, A (150, 70), B (70, 150), C (150, 230), D (230, 150)		(220.4, 219.8)	3.4%	(244.1, 243.1)	4.7%
6	Diagonal distance was 200, A (150, 50), B (50, 150), C (150, 250), D (250, 150)		(221.6, 223.4)	2.8%	(241.4, 239.3)	3.8%

Eight-noded 3D reduced integrated linear brick elements, C3D8R, were adopted to complete the structured mesh division. In order to obtain sufficient accuracy and high efficiency, the dimension of the elements was around 0.5 mm × 0.5 mm × 0.5 mm, so that there were at least 20 elements per wavelength of the Lamb waves. A time step of 20 ns was used to ensure the accuracy of the wave propagation modeling [29].

3.2. Signal Analysis

According to the above-mentioned diamond-shaped array for excitation and reception, all the received signals were obtained from the pair of two sensors. As shown in Figure 4, there are two time-domain received signals from the sensing path AB (blue line) and AD (red line) in the diamond-shaped array with number 1, including a direct signal and a scattered signal, respectively. The direct signals are almost the same in the two received signals. By taking the difference between the two received signals (Equation (4)), a damage-scattered signal was obtained by eliminating the direct signals from each other.

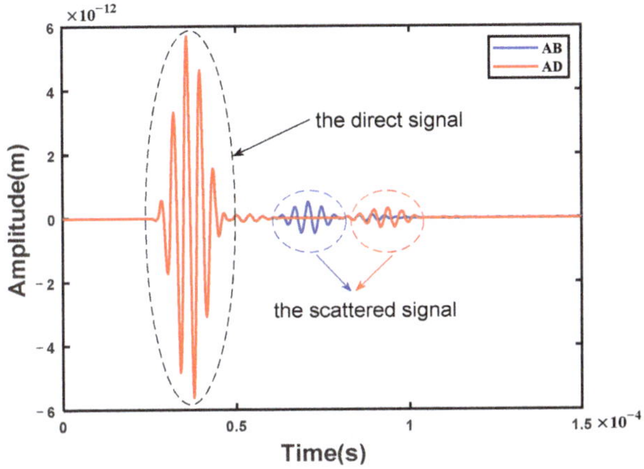

Figure 4. A schematic diagram of two received signals from the sensing path AB (blue line) and AD (red line) in the diamond-shaped array with the damage location of number 1.

As shown in Figure 5a, the two wave packets in the damage-scattered signal represent the scattered signals in the received signals, respectively. Since the direct propagation distances from one actuator to pairs of two sensors are equal theoretically, the direct signals are canceled by each other. Moreover, since the two receiving sensors are geometrically symmetrical in the structure, other noise and boundary reflection signals are also eliminated in the subtraction process, so as to highlight the damage-scattered signals. The envelope curve (blue line in Figure 5a) was constructed by Hilbert transform (Equation (5)) from the damage-scattered signal, hence the relative time difference Δt was determined from the corresponding times (two red dots of the envelope curve in Figure 5a). According to the Δt and group velocity of the S0 mode, the hyperbolic trajectory is constructed, and the points on the trajectory indicate the possible locations of the damage. Obviously, the group velocity of the S0 mode can be calculated from the received signals with different propagation distances, which is about 5164.7 m/s in this numerical simulation.

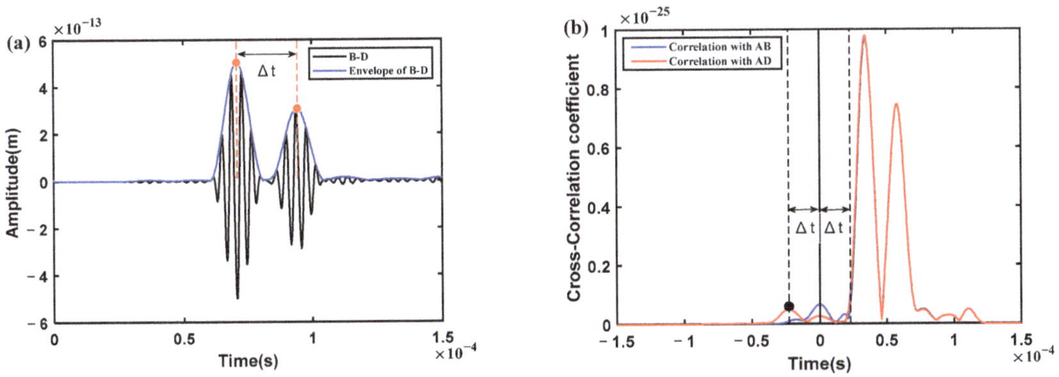

Figure 5. (**a**) A damage-scattered signal obtained by subtracting two received signals; (**b**) the cross-correlation curves between the damage-scattered signal and the two received signals, respectively. (Red dots denote the amplitude corresponding to the time, black dot denotes the peak corresponding to the time difference).

At the same time, the sequence of Δt was determined by means of the CCM (Equations (6) and (7)) between the damage-scattered signal and the two received signals, respectively. As shown in Figure 5b, there are two envelope curves based on the cross-correlation function, in which a peak always appears at time zero. Considering time zero as the reference, and the relative time difference Δt as the interval, the corresponding peaks at Δt are found on the left- or right hand sides of the time zero, respectively. Therefore, the sign of Δt is determined when the time sequence is judged from the time lag in the cross-correlation curve. For instance, the red envelope curve was calculated from the cross-correlation between the damage-scattered signal and the received signal AD. Obviously, the peak at Δt is on the left hand side of the time zero, which means that the scattered signal in the received signal AB is in advance of that in the received signal AD. On the contrary, it is behind. Similar results can be obtained from the blue envelope curve. Unfortunately, the peak may not be easy to find in the blue envelope curve because of the interference of the other components of the cross-correlation function. Nevertheless, any of the envelope curves can be used to determine the sequence of the time difference Δt, and there is no need to calculate the value of the relative time difference. This may be a relatively robust feature for damage localization.

After obtaining the time difference, the damage location can be realized based on the hyperbolic trajectory; the imaging results of the through-hole location in the simulation is shown in Figure 6a. The maximum probability position of the imaging result is at 60.1, 240.1, while the center position of the through-hole is at 50, 250. It is shown that the proposed method can be used to achieve baseline-free damage localization. At the same time, the imaging result of the traditional hyperbolic method in the symmetrical array is shown in Figure 6b. The coordinates of the maximum pixel value are 56.5, 243.5. It shows that the traditional hyperbolic method can also achieve damage positioning, and is more accurate than the proposed method. The reason is probably due to the use of more groups of signals in the traditional hyperbolic method; there are 12 pairs of signals in total, while the proposed method includes four groups of signals. It should be noted that the more signals, the longer the processing time of localization, and the traditional method needs the reference signals in the intact state.

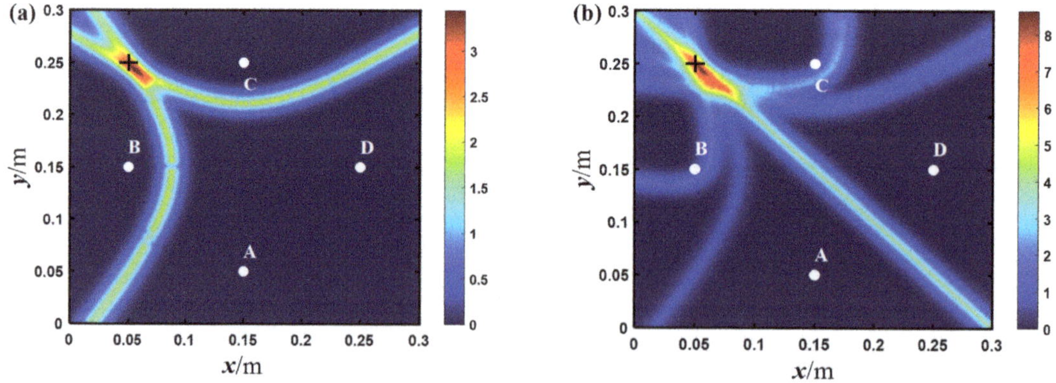

Figure 6. The imaging results of: (**a**) the proposed method; and (**b**) the traditional hyperbolic method in the diamond-shaped array with number 1 in the simulation (white points denotes the exciting and receiving position, + indicates the center position of the circular hole).

4. Experimental Verification

The proposed method based on Lamb waves was then corroborated by the ultrasonic experimental testing of a through-hole in an aluminum plate.

4.1. Experimental Setup

The experimental testing system, as shown in Figure 7, consisted of a dual channel arbitrary/function generator (Textronix AFG 3022C), a power amplifier (EPA 104), a mixed domain oscilloscope (Textronix MDO 3012) and a personal computer (PC) connected with the oscilloscope. The specimen was a common aluminum plate with a dimension of 500 mm × 500 mm ×1 mm, and a through-hole with a diameter of 5 mm was used as a damage source, which was the same as that in the numerical simulations. Four PZT piezoelectric transducers, each 10 mm in diameter and 1 mm in thickness, were permanently bonded to the surface of the aluminum plate by means of a thin cyanoacrylate adhesive layer.

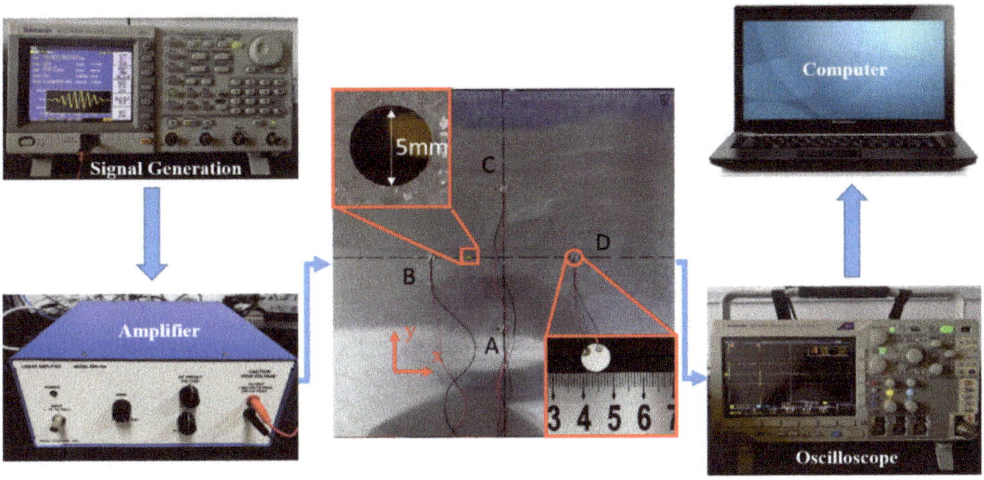

Figure 7. A schematic of the experimental testing system.

A global cartesian coordinate was also introduced for convenience, as in the simulation model. The PZT transducers were labelled as A, B, C and D; damage locations also had the same coordinates as in the numerical simulation. It should be noted that the bottom left corner of the aluminum plate was not at the origin of the coordinate axis.

A five-cycle Hanning windowed sine wave with a central frequency of 250 kHz, generated by the arbitrary function generator, was applied as the excitation tone burst. The burst was then amplified by means of the power amplifier before it was sent to one of the PZT transducers. The PZT transducers, at two symmetrical positions, were used to receive the Lamb wave signals. For example, the PZT transducers B and D, which were considered to be symmetrical in the diamond-shaped array, were used to receive the signals when PZT transducer A was excited. The signals were captured by the oscilloscope at a sampling rate of 25 MHz with 256-time averaging, and was then transmitted to the computer. The excitation-acquisition procedure was implemented in turn on each PZT transducer, so that in total eight received signals were collected from these sensing paths. It is worth noting that the received signals were always measured in the current state, but not in the intact state.

4.2. Signal Processing

The experiments and simulations have some similar results because of the same excitation and reception process with the diamond-shaped array. Firstly, the group velocity of the S0 mode is about 5451.2 m/s, as calculated from the direct waves with different propagation distances. It is considered that the PZT transducers can effectively excite the S0 mode Lamb wave. Then, the received time-domain signals of the sensing paths AB and AD, in the diamond-shaped array with number 1, are represented in Figure 8, including the direct waves and the scattered waves. Meanwhile, there is an electromagnetic interference signal at the front of the received signals, which is mainly introduced by the transducer and circuit imperfections.

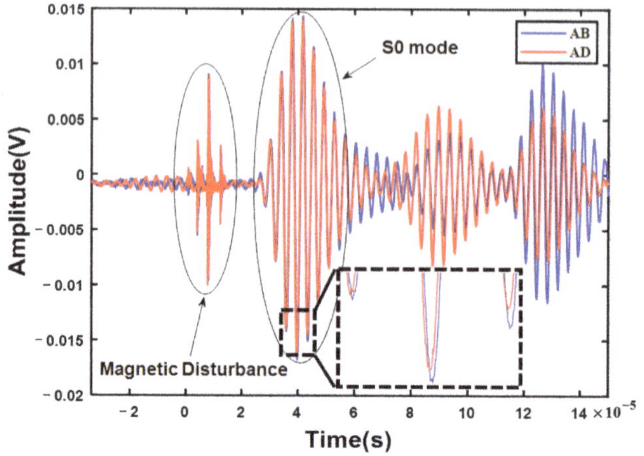

Figure 8. A schematic diagram of two received signals from the sensing path AB (blue line) and AD (red line) in the diamond-shaped array with the damage location of number 1.

After the electromagnetic signal was filtered out, the two received signals were compared, according to Equation (4). A damage-scattered wave was obtained by eliminating the direct waves via subtraction of the received signals. However, as shown in Figure 9a, the damage-scattered wave still has an obvious wave packet at the position of the direct wave, which means that the direct waves are not fully consistent with each other and not completely eliminated, as in the numerical simulation. This is because the direct waves are not exactly the same in phase and amplitude, as shown in the inset of Figure 8. This is due to a slight difference in the performance of the adhesive and the geometric position of the PZT transducers, which cannot be avoided in the experiments. Fortunately, the difference between the two direct waves can be reduced considerably by phase compensation and standard normalization to obtain the damage-scattered wave, as shown in Figure 9b. The next steps are consistent with those in the simulation, including calculating the time difference Δt and constructing the hyperbolic trajectory. It is noted here that the first wave packet in the damage-scattered wave (the first red dot shown in Figure 9b) may be the A0 mode direct wave. This is because the A0 mode could also be excited in the above process of the effective S0 mode excitation, in which the amplitude of the A0 mode is seemingly small. Although the S0 mode direct wave can be eliminated through the above manipulation, the A0 mode direct wave may not be completely excluded.

At the same time, the sequence of Δt was determined by means of the CCM (Equations (6) and (7)). As shown in Figure 10, the red envelope curve is the cross-correlation curve between the damage-scattered wave and the received signal AD. A characteristic point corresponding to the time difference Δt is found on the left hand side of the time zero, shown as a black circle in Figure 10. It indicates that the scattered wave in the received signal AB is in advance of that in the received signal AD. On the other hand, the blue envelope is the cross-correlation curve between the damage-scattered wave and the received signal AB, and it does not show the characteristic point at the time difference Δt. Regardless, the time sequence could be obtained as long as a feature point in any of the cross-correlation curves is found. It should be noted that the peak of the cross-correlation curve does not appear at the time zero. This is because there is a time offset in the process of phase compensation. In addition, the point corresponding to the time difference Δt is actually in the valley of the signals, not at the peak. This could be attributed to the fact that the first red dot corresponding to the time difference Δt of the damage-scattered signal is also in the trough of the waves, which is caused by the failure to completely eliminate the A0 mode direct wave.

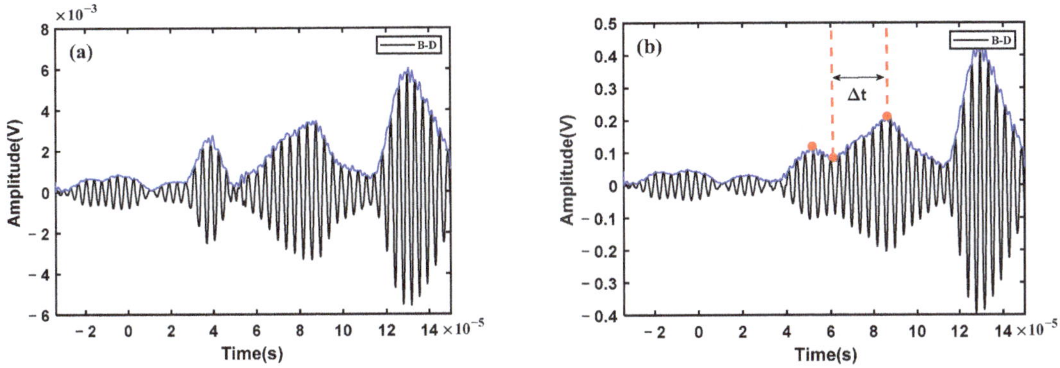

Figure 9. A damage-scattered signal obtained by subtracting two received signals: (**a**) without phase compensation; and (**b**) with phase compensation. (Red dots denote the amplitude corresponding to the time).

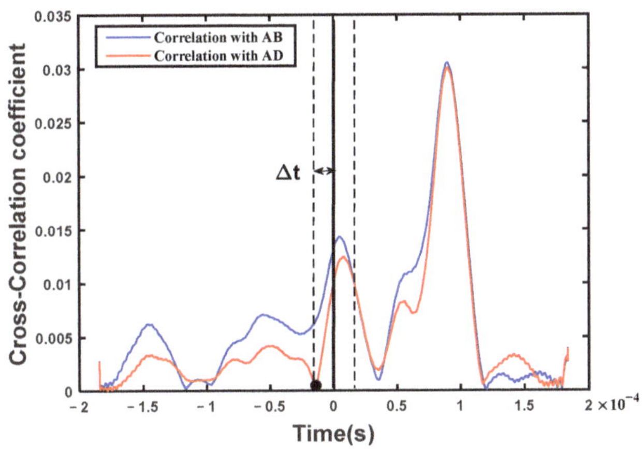

Figure 10. The cross-correlation curves between the damage-scattered signal and the two received signals, respectively. (Black dot denotes the characteristic point corresponding to the time difference).

5. Results and Discussion

In the aforementioned diamond-shaped array, the time difference Δt is obtained from the two received signals in the current state. Then, using the proposed baseline-free method, the imaging results of damage locations in the experiments are shown in Figures 11 and 12.

As shown in Figure 11a, there are four intersections of the hyperbolic trajectories in the diamond-shaped array, resulting in artifacts in the imaging result, and it is not able to achieve damage localization accurately. Obviously, the hyperbola itself has two branches using the relative time difference of each pair of the two received signals. Therefore, it is still necessary to determine the sign of the relative time difference (namely, the time sequence), for precise damage localization. By using the cross-correlation function, only one branch of the hyperbola can be determined and the damage location can be achieved precisely, as shown in Figure 11b. The maximum probability position of the imaging result considered to be the most likely location for the damage is at 35.6, 266, while the center position of the through-hole is at 50, 250. According to the error calculation method in [30], the relative error of the imaging result is about 5.3%. It shows that the probability position of the imaging result is very close to the actual location of the through-hole, and the method gives the damage location fairly well.

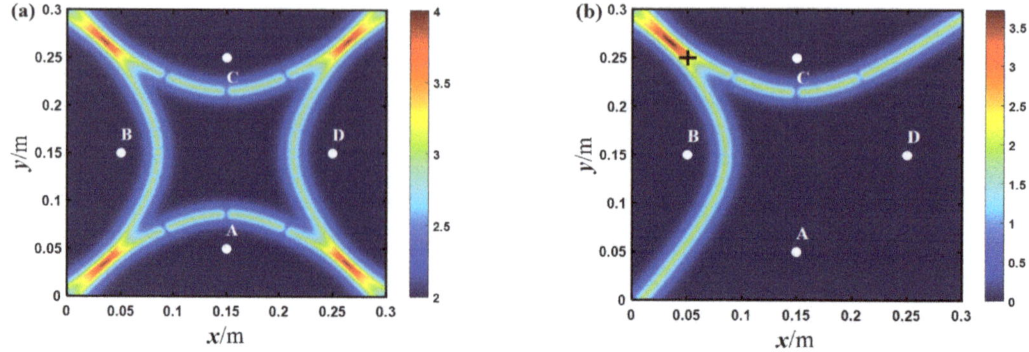

Figure 11. The imaging results of the experiments with: (**a**) two branches; and (**b**) one branch of hyperbola in the diamond-shaped array with number 1 (white points denotes the exciting and receiving position, + indicates the center position of the circular hole).

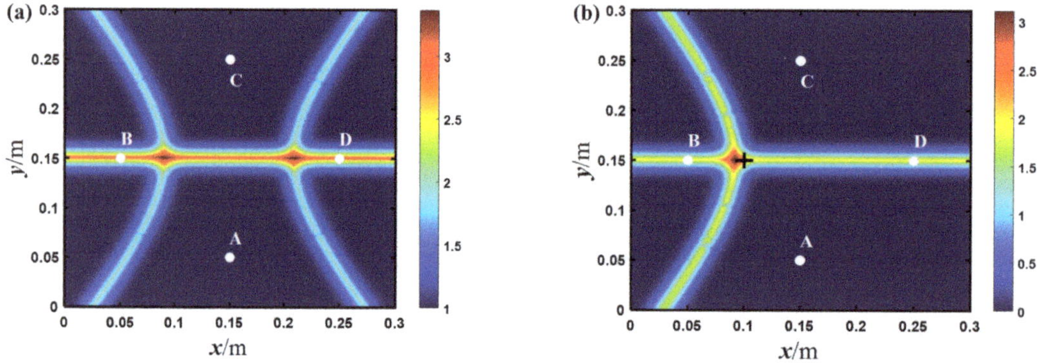

Figure 12. The imaging results of the experiments with: (**a**) two branches; and (**b**) one branch of hyperbola in the diamond-shaped array with number 2 (white points denotes the exciting and receiving position, + indicates the center position of the circular hole).

The imaging result in Figure 12 is similar to that in Figure 11, while the most obvious difference is that in Figure 12 the damage location is inside the diamond-shaped array and on the perpendicular bisector of certain line segment(s) of the pairs of two receiving sensors. Here, a correlation coefficient of two received signals is used to prejudge whether the damage location is on some of the perpendicular bisectors. If the correlation coefficient exceeds a certain threshold, it means that the damage is most likely on the perpendicular bisector, and vice versa. As shown in Figure 12b, the maximum probability position of the imaging result is at 92.9, 149.8, which is almost at the actual center position of the through-hole at 100, 150, with a relative error of only about 2.36%. The above results show that no matter whether the damage location is inside or outside the diamond-shaped array, or even in the position symmetrical with respect to the pairs of two receiving sensors, the proposed method can be used to achieve baseline-free damage localization.

In order to further study the influence of the array forms on the imaging results, diamond-shaped arrays with different sizes were used for the baseline-free damage localization of the aluminum plate. As shown in Figure 13, the imaging results in the experiments are obtained for different arrays. Meanwhile, the size of the arrays, the damage location and relative errors are shown in Table 2. As shown in Figure 13a, when the diagonal distance of the diamond-shaped array is about 80 mm, the maximum relative error is 7.7%. Compared with the array with maximum size, the array with minimum size has not only a higher relative error, but also two maximum probability locations in the simulated imaging results. It is worth mentioning that these imaging results come from some intersecting areas, not trajectory intersections; this is because this size of array has a small time difference, which makes the hyperbolic trajectory infinitely approach their asymptotes. Consequently, the maximum probability position becomes the intersection area of the asymptotes. Based on the above research, it is believed that when the diagonal distance of the diamond-shaped array is less than 80 mm, it will deteriorate the positioning accuracy and cause noticeable positioning errors, no matter where the damage location is. However, as the size of the array (namely, the diagonal distance) increases, the maximum probability position approaches the actual locations of damage and the accuracy of damage localization also increases correspondingly. As shown in Figure 13d, when the diagonal distance of the diamond-shaped array is about 200 mm, the maximum probability location of the imaging result is at 241.4, 239.3, while the actual center position of the through-hole is at 230, 230. The relative error is about 3.8%, which means that the imaging result is close enough to the damage location. Therefore, the positioning accuracy can be significantly improved when the diagonal distance is larger than 80 mm, especially up to 200 mm. In summary, by using the diamond-shaped arrays with appropriate size, and the hyperbola algorithm, it is possible to effectively achieve damage localization with the proposed baseline-free method.

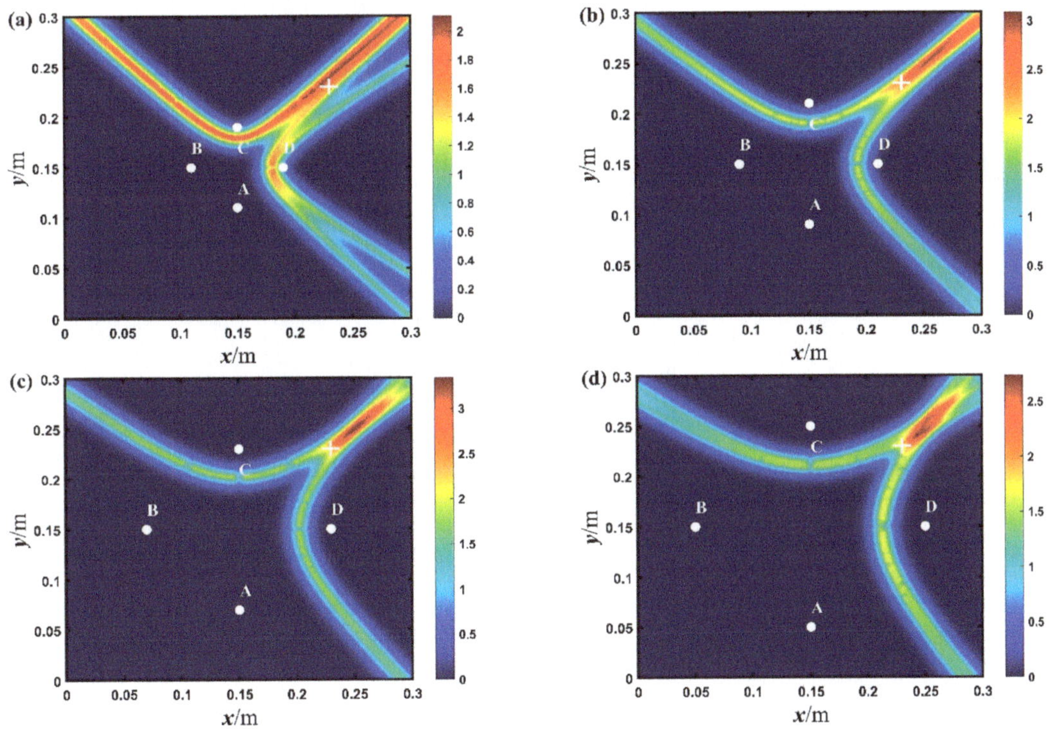

Figure 13. The imaging results of the through-hole for different diamond-shaped arrays with a diagonal distance of: (**a**) 80 mm; (**b**) 120 mm; (**c**) 160 mm; and (**d**) 200 mm, in the experiments (white points denotes the exciting and receiving position, + indicates the center position of the circular hole).

The simulations and experiments have similar imaging results because of the same excitation and reception process in the diamond-shaped arrays. The positioning errors of arrays with different sizes in the experimental testing and numerical simulations are shown in Table 2. The results show that the accuracy of damage localization in the simulations is higher than that in the experiments. The reason may be that it is easy to excite a single S0 mode in the simulations, and the wave propagation is relatively simple, while the A0 mode can also be excited in the experiments, and the imaging results are probably affected by the multimode characteristics of Lamb waves. It can be seen that the damage localization is realized through simulations and experiments, which indicates that the proposed method is an effective baseline-free method for damage localization.

In further study, in order to verify the effectiveness of the proposed method, a comparative analysis was made of the baseline-free method presented in Ref. [31]. The array form in the baseline-free method is shown in Figure 14a. Although it is somewhat similar to the array form mentioned above, the difference is that two transducers are closely placed as a pair in the special transducer arrangement. Thus, the direct wave and damage-scattered wave can be separated effectively because of the large difference of their time of flight. The damage-scattered wave is easily extracted, and the imaging result of through-hole location in the experiment is shown in Figure 14b. The maximum probability position of the imaging result is at 94.3, 146.3; it shows that the baseline-free method can also achieve damage positioning. Compared with the coordinate 92.9, 149.8 of the maximum value in Figure 12b, it could be concluded that the positioning accuracy of the two methods is almost the same, even though it seems that the baseline-free method is more accurate than the proposed method, according to the positioning error here. Thus, these two kinds of baseline-free method can achieve damage localization by the special arrangement. It is worth noting that there are four groups of signals in these methods, while only four transducers are used in the proposed method. Further, the baseline-free method in Ref. [31] has a high resolution and can detect multi defects due to the employed single mode nondispersive SH0 wave.

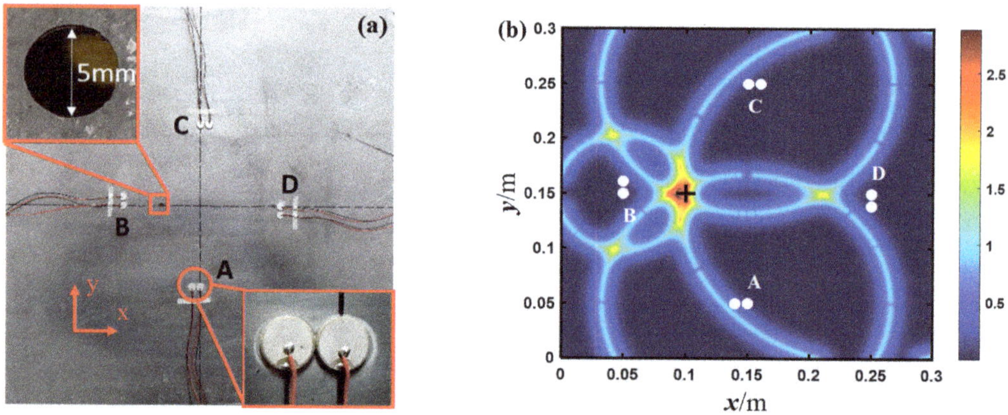

Figure 14. (**a**) A special transducer arrangement and (**b**) the imaging result of the baseline-free method [31] in the experiment (white points denotes the exciting and receiving position, + indicates the center position of the circular hole).

6. Conclusions

Due to the changing of operational and environmental conditions, typical detection methods involving baseline signals have large damage localization error. To address this problem, a novel baseline-free method for damage localization using Lamb waves based on a hyperbolic algorithm, was proposed in this study.

To eliminate the direct signals, diamond-shaped arrays were arranged on the plate structures to receive signals at two symmetrical positions. The time difference between the received signals was obtained from the damage-scattered signals. The sequence of time difference calculated by the cross-correlation method determined only one branch of the hyperbola. Damage localization was realized by a hyperbolic algorithm with the damage probability of each point in the structure. Numerical simulations and experimental measurements were implemented on an aluminum plate with a through-thickness hole in the current state. The imaging results show that the damage localization can be achieved no matter whether the position of the defect is outside or inside the diamond-shaped arrays. Meanwhile, because of the intersection of the asymptote of the hyperbola, the diamond-shaped array with the smallest size has the biggest positioning error of about 7.7%. Nevertheless, the relative error can be reduced to 3.8% when the diagonal distance of the diamond-shaped array increases to 200 mm. The consistency between the numerical and experimental results indicates that the position of the through-hole in the aluminum plate can be accurately identified and localized by the proposed baseline-free method. Ultimately, the baseline-free method proposed here may largely improve the practicability of the Lamb wave damage localization technique.

Author Contributions: Conceptualization, W.Z.; Methodology, J.X.; Validation, J.X.; Formal analysis, W.Z.; Investigation, J.X.; Data curation, J.X.; Writing—original draft, J.X.; Writing—review & editing, X.Q. and Y.X.; Project administration, X.Q.; Funding acquisition, Y.X. All authors have read and agreed to the published version of the manuscript.

Funding: This work was supported by the National Key Research and Development Program of China (2021YFC3001802) and the National Natural Science Foundation of China (Grant Nos. 12025403, U1930202, 12004114 and 12174102).

Institutional Review Board Statement: Not applicable.

Informed Consent Statement: Not applicable.

Data Availability Statement: The data presented in this study are available on request from the corresponding author.

Conflicts of Interest: The authors declare that they have no known competing financial interest or personal relationships that could have appeared to influence the work reported in this paper.

References

1. Su, Z.; Ye, L. *Identification of Damage Using Lamb Waves: From Fundamentals to Applications*; Springer Science & Business Media: Berlin/Heidelberg, Germany, 2009; Volume 48.
2. Rose, J.L. *Ultrasonic Guided Waves in Solid Media*; Cambridge University Press: Cambridge, UK, 2014.
3. Nokhbatolfoghahai, A.; Navazi, H.M.; Groves, R.M. Evaluation of the sparse reconstruction and the delay-and-sum damage imaging methods for structural health monitoring under different environmental and operational conditions. *Measurement* **2021**, *169*, 108495. [CrossRef]
4. Xu, C.; Wang, J.; Yin, S.; Deng, M. A focusing MUSIC algorithm for baseline-free Lamb wave damage localization. *Mech. Syst. Signal Process.* **2022**, *164*, 108242. [CrossRef]
5. Soman, R.; Golestani, A.; Balasubramaniam, K.; Karpiński, M.; Malinowski, P.; Ostachowicz, W. Application of ellipse and hyperbola methods for guided waves based structural health monitoring using fiber Bragg grating sensors. In *Health Monitoring of Structural and Biological Systems XV*; SPIE: Washington, DC, USA, 2021; Volume 11593, pp. 65–73.
6. Wu, Y.; Shen, X.; Li, D. Numerical and experimental research on damage shape recognition of aluminum alloy plate based on Lamb wave. *J. Intell. Mater. Syst. Struct.* **2021**, *32*, 2273–2288. [CrossRef]
7. Feng, B.; Pasadas, D.J.; Ribeiro, A.L.; Ramos, H.G. Locating defects in anisotropic CFRP plates using ToF-based probability matrix and neural networks. *IEEE Trans. Instrum. Meas.* **2019**, *68*, 1252–1260. [CrossRef]
8. Huo, H.; He, J.; Guan, X. A Bayesian fusion method for composite damage identification using Lamb wave. *Struct. Health Monit.* **2021**, *20*, 2337–2359. [CrossRef]
9. Chen, H.; Liu, Z.; Wu, B.; He, C. An intelligent algorithm based on evolutionary strategy and clustering algorithm for Lamb wave defect location. *Struct. Health Monit.* **2021**, *20*, 2088–2109. [CrossRef]
10. Mu, W.; Gao, Y.; Liu, G. Ultrasound defect localization in shell structures with Lamb waves using spare sensor array and orthogonal matching pursuit decomposition. *Sensors* **2021**, *21*, 8127. [CrossRef]
11. Sohn, H. Effects of environmental and operational variability on structural health monitoring. *Philos. Trans. R. Soc. A Math. Phys. Eng. Sci.* **2007**, *365*, 539–560. [CrossRef]
12. Michaels, J.E. Detection, localization and characterization of damage in plates with an in situ array of spatially distributed ultrasonic sensors. *Smart Mater. Struct.* **2008**, *17*, 035035. [CrossRef]
13. Anton, S.R.; Inman, D.J.; Park, G. Reference-free damage detection using instantaneous baseline measurements. *AIAA J.* **2009**, *47*, 1952–1964. [CrossRef]
14. Wang, J.; Shen, Y.; Rao, D.; Xu, W. Physical-virtual time reversing of nonlinear Lamb waves for fatigue crack detection and quantification. *Mech. Syst. Signal Process.* **2021**, *160*, 107921. [CrossRef]

15. Ing, R.K.; Fink, M. Time-reversed Lamb waves. *IEEE Trans. Ultrason. Ferroelectr. Freq. Control.* **1998**, *45*, 1032–1043. [CrossRef] [PubMed]
16. Liu, Z.; Yu, H.; Fan, J.; Hu, Y.; He, C.; Wu, B. Baseline-free delamination inspection in composite plates by synthesizing non-contact air-coupled Lamb wave scan method and virtual time reversal algorithm. *Smart Mater. Struct.* **2015**, *24*, 045014. [CrossRef]
17. Huang, L.; Zeng, L.; Lin, J. Baseline-free damage detection in composite plates based on the reciprocity principle. *Smart Mater. Struct.* **2017**, *27*, 015026. [CrossRef]
18. Kannusamy, M.; Kapuria, S.; Sasmal, S. Accurate baseline-free damage localization in plates using refined Lamb wave time-reversal method. *Smart Mater. Struct.* **2020**, *29*, 055044. [CrossRef]
19. Qiang, W.; Shenfang, Y. Baseline-free imaging method based on new PZT sensor arrangements. *J. Intell. Mater. Syst. Struct.* **2009**, *20*, 1663–1673. [CrossRef]
20. Jeong, H.; Cho, S.; Wei, W. A baseline-free defect imaging technique in plates using time reversal of Lamb waves. *Chin. Phys. Lett.* **2011**, *28*, 064301. [CrossRef]
21. Jun, Y.; Lee, U. Computer-aided hybrid time reversal process for structural health monitoring. *J. Mech. Sci. Technol.* **2012**, *26*, 53–61. [CrossRef]
22. Yelve, N.P.; Mitra, M.; Mujumdar, P.M.; Ramadas, C. A hybrid method based upon nonlinear Lamb wave response for locating a delamination in composite laminates. *Ultrasonics* **2016**, *70*, 12–17. [CrossRef]
23. Li, D.; Jing, Z.; Jin, M. Plate-like structure damage location identification based on Lamb wave baseline-free probability imaging method. *Adv. Mech. Eng.* **2017**, *9*, 1–10. [CrossRef]
24. Gao, W.; Huo, L.; Li, H.; Song, G. An embedded tubular PZT transducer based damage imaging method for two-dimensional concrete structures. *IEEE Access* **2018**, *6*, 30100–30109. [CrossRef]
25. Gao, F.; Wang, L.; Hua, J.; Lin, J.; Mal, A. Application of Lamb wave and its coda waves to disbond detection in an aeronautical honeycomb composite sandwich. *Mech. Syst. Signal Process.* **2021**, *146*, 107063. [CrossRef]
26. Hong, M.; Su, Z.; Lu, Y.; Sohn, H.; Qing, X. Locating fatigue damage using temporal signal features of nonlinear Lamb waves. *Mech. Syst. Signal Process.* **2015**, *60*, 182–197. [CrossRef]
27. Niu, X.; Zhang, J.; Croxford, A.; Drinkwater, B. Efficient finite element modelling of guided wave scattering from a defect in three dimensions. *Nondestruct. Test. Eval.* **2022**, 1–21. [CrossRef]
28. Wang, K.; Fan, Z.; Su, Z. Orienting fatigue cracks using contact acoustic nonlinearity in scattered plate waves. *Smart Mater. Struct.* **2018**, *27*, 09LT01. [CrossRef]
29. Zhu, W.; Xu, Z.; Xiang, Y.; Liu, C.; Deng, M.; Qiu, X.; Sun, D.; Xuan, F. Nonlinear ultrasonic detection of partially closed cracks in metal plates using static component of lamb waves. *NDT E Int.* **2021**, *124*, 102538. [CrossRef]
30. Wei, N.; Xu, J.; Xiang, Y. Research on Damage Imaging Accuracy of the Guided Wave. In Proceedings of the 2018 IEEE Far East NDT New Technology & Application Forum (FENDT), Xiamen, China, 6–8 July 2018; IEEE: Piscataway, NJ, USA, 2018; pp. 165–169.
31. Huan, Q.; Li, F. A baseline-free SH wave sparse array system for structural health monitoring. *Smart Mater. Struct.* **2019**, *28*, 105010. [CrossRef]

sensors

MDPI

Article

Surface Monitoring of an MSW Landfill Based on Linear and Angular Measurements, TLS, and LIDAR UAV

Grzegorz Pasternak [1], Janina Zaczek-Peplinska [2], Klaudia Pasternak [3], Jacek Jóźwiak [1], Mariusz Pasik [2], Eugeniusz Koda [1] and Magdalena Daria Vaverková [1,4,*]

1 Institute of Civil Engineering, Warsaw University of Life Sciences—SGGW, Nowoursynowska 159, 02-776 Warsaw, Poland
2 Faculty of Geodesy and Cartography, Warsaw University of Technology, Pl. Politechniki 1, 00-661 Warsaw, Poland
3 Department of Imagery Intelligence, Faculty of Civil Engineering and Geodesy, Military University of Technology (WAT), 00-908 Warsaw, Poland
4 Department of Applied and Landscape Ecology, Faculty of AgriSciences, Zemědělská 1, Mendel University in Brno, 613 00 Brno, Czech Republic
* Correspondence: magdalena_vaverkova@sggw.edu.pl

Abstract: Surface monitoring of landfills is crucial not only during their operation but also for later land restoration and development. Measurements concern environmental factors, such as leachate, migration of pollutants to water, biogas, and atmospheric emissions, and geotechnical factors, such as stability and subsidence. Landfill subsidence can be measured using modern surveying techniques. Modern measurement methods for landfill body displacement monitoring and their control after restoration and adaptation as recreational areas include terrestrial laser scanning (TLS), and scanning and low-altitude photogrammetric measurements from an unmanned aerial vehicle (UAV). The acquired measurement data in the form of 3D point clouds should be referenced to the local control network to enable a comprehensive analysis of data acquired using various techniques, including geotechnical sensors such as benchmarks, piezometers, and inclinometers. This study discusses the need for surface monitoring of municipal solid waste (MSW) landfills. A properly 3-D mapped landfill mass is the basis for ensuring the geotechnical safety of the restored landfill. Based on archival data and current measurements of the Radiowo landfill (Poland), this study compares the advantages and limitations of the following measurement techniques: linear and angular measurements, satellite measurements, TLS, and UAV scanning and photogrammetry, considering specific conditions of the location and vegetation of the landfill. Solutions for long-term monitoring were proposed, considering the cost and time resolution necessary for creating a differential model of landfill geometry changes.

Keywords: point cloud; landfill; unmanned aerial vehicle; terrestrial laser scanner; aerial laser scanning; geotechnical structure

Citation: Pasternak, G.; Zaczek-Peplinska, J.; Pasternak, K.; Jóźwiak, J.; Pasik, M.; Koda, E.; Vaverková, M.D. Surface Monitoring of an MSW Landfill Based on Linear and Angular Measurements, TLS, and LIDAR UAV. *Sensors* **2023**, *23*, 1847. https://doi.org/10.3390/s23041847

Academic Editor: Zenghua Liu

Received: 18 January 2023
Revised: 1 February 2023
Accepted: 5 February 2023
Published: 7 February 2023

1. Introduction

Mass movements and failures of municipal and post-production landfills, landslides in open-pit mines, or bulk product storage areas cause serious economic losses and safety hazards for workers during landfilling and for the public after landfill restoration [1–3]. Potential damage resulting from insufficient compaction of waste layers and inadequate slope protection requires continuous monitoring of the surface of municipal solid waste (MSW) landfills. Early detection of displacement of slowly creeping material is an important part of early warning against landslides. Stability assessment and prediction of large deformations and failures, including in open-pit mines and slopes during construction works, are usually performed using surveying techniques [4,5].

The requirement for geodetic monitoring for aforementioned facilities in Poland is set forth in the Regulation of the Minister of Environment of 30 April 2013, on landfills [6]. The regulation specifies the minimum frequency of geodetic measurements at a landfill every 3 months during the operational stage and every 12 months during the post-operational stage. The scope of monitoring should include control of the subsidence of the surface of the structure using geodetic methods based on measurements of displacements at control points, and evaluation of slope stability using geotechnical methods.

The stability of slopes can be achieved by using retaining walls, weight-bearing embankments, local slope mitigation, local replacement and recompaction of waste in the road substructure, and horizontal reinforcement with geogrid and car tire mattresses [7,8]. Accurate mapping of the surface of the restored landfill provides the starting material for the design of the target land use, such as forestry, agriculture, construction, parks, and recreation. In the case of the Radiowo landfill (Poland) selected for this study, the plan is to develop part of the area for sports and recreation and another part (slopes) for the installation of photovoltaic panels (PP) and wind turbines (RES). Given the large size of the landfill, steep slopes, and deformable material prone to landslides, it is necessary to measure the current shape and deformation of the mass over time to ensure geotechnical safety, preferably covering the entire landfill surface [9–11].

Landslide monitoring uses a number of surveying techniques and remote sensing tools to detect surface deformation at various scales and magnitudes. These include the Global Navigation Satellite System (GNSS), total station surveying [12], ground-based interferometric synthetic aperture radar (GB-InSAR), thermal infrared imagery [13], terrestrial laser scanning (TLS) [14], and techniques using UAVs [15–18].

The classic surveying techniques (GNSS, leveling, total station) allow obtaining results with high accuracy, but only at certain points, in a discrete manner, without depicting changes over the entire area. In contrast to these surveying techniques, laser scanning and photogrammetric techniques cover the entire monitored area with a dense grid of points, while being fast and efficient [19–21]. These techniques are widely used in various fields of geology/geomorphology, in issues related to landslide monitoring [22] and landslide risk assessment [16,17,23–26], for acquiring a high accuracy terrain model or making precise volumetric measurements.

Unlike traditional methods, monitoring using 3D laser scanning allows for non-contact point measurement, so it can complement traditional monitoring methods, improve the efficiency of slope monitoring, and provide data support/supplementation for site protection [4]. For example, Huang et al. [14] proposed a method for monitoring slow-moving landslides using long-range TLS. The proposed method features several key steps of TLS point cloud processing, including ground point extraction based on multidimensional classification, hybrid-weighted iterative closest point (HWICP) algorithm, and ALC2M-based deformation calculation [14]. Numerous literature examples indicate that TLS is an effective solution for providing accurate information on the movement of inaccessible areas characterized by landslide hazards [27–30]. In view of the expectations of planners and engineers designing both sports and recreational facilities, as well as those related to renewable energy sources, the use of appropriate/efficient measurement techniques provides the starting material for further work and enables the monitoring of phenomena occurring inside the landfill and on its surface.

The purpose of this paper is to analyze the possibility of using modern measurement techniques in monitoring the restored Radiowo landfill near Warsaw (Poland). The landfill was turned into a recreational and winter park with a slope for skiers and tobogganers, and in the summer for cyclists, runners, modelers, etc. The current geodetic monitoring of the site is based on point measurements of specially stabilized points (benchmarks) using GNSS methods and does not fully reflect the changes occurring in the entire area of the landfill, especially in the area of the ski run. The inventory measurement of this slope, a method that can serve as a baseline measurement for monitoring using point cloud analysis methods, is the focus of the article. Due to the specific conditions of the landfil's location (proximity to the airport and protected areas), the existing vegetation, and the large size of the landfill (area of 17.3 hectares, height of more than 60 m), the paper proposes three alternative measurement methods. Two UAV-based survey methods, i.e., LIDAR and low-altitude photogrammetry, were first proposed. However, the close proximity to the airport imposes certain limitations and requirements for permits issued by Airport Flight Information Service (AFIS) may pose a serious problem in the context of long-term monitoring of the landfill. For this reason, TLS has been proposed as an alternative. The paper compares the measurement results of these methods in the form of point clouds with reference data obtained from linear and angular measurements (trigonometric leveling).

2. Study Area

The study concerned the Radiowo landfill site in Warsaw in July 2022. The site is located near the northwestern border of Warsaw at the junction of the Bemowo district and the Stare Babice municipality (Figure 1). The area of the landfill is about 17.3 hectares, and its height relative to ground level is about 60 m [31].

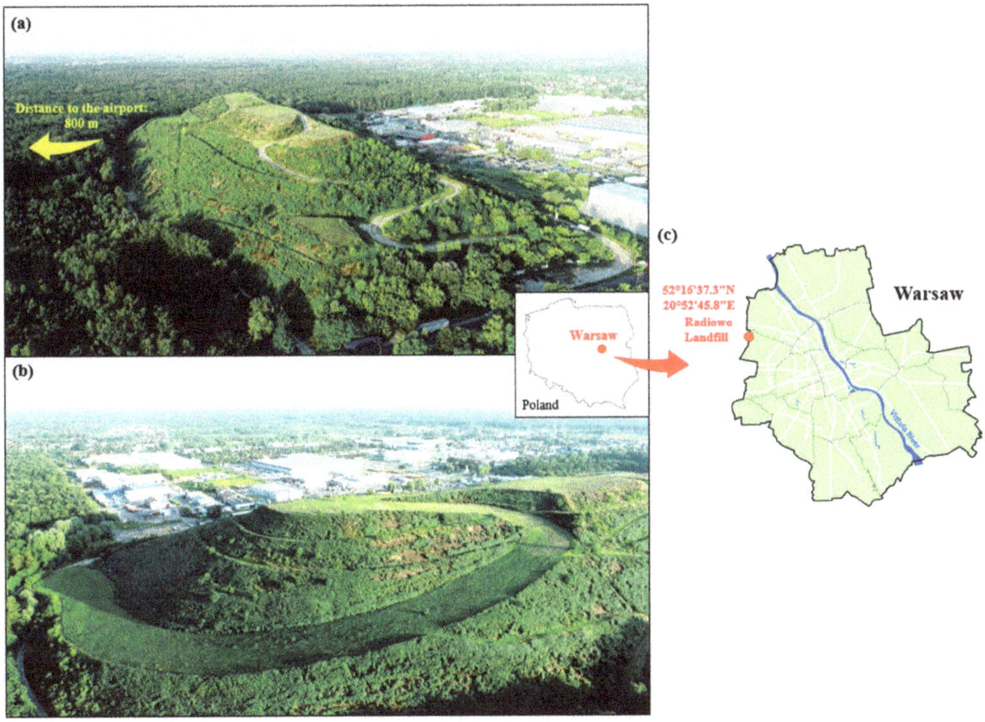

Figure 1. Location of the Radiowo landfill in Warsaw: (**a**) view from the access road (northeast slope), (**b**) view from the ski run (southeast slope), and (**c**) location of the landfill within the administrative borders of Warsaw.

Initially, the Radiowo landfill received only unsegregated MSW (1961–1991). Due to the unfavorable location in terms of geotechnical conditions (wetlands) and the occurrence of numerous slope landslides, from 1992 the landfill began to accept only solid waste generated from compost produced at the Radiowo composting plant (foils, tires, textiles, scrap metal), and, from 2012 to 2018, stabilized compost produced by the mechanical-biological processing of MSW. Since 1993, to reduce the slopes and thus improve their stability, a number of reclamation works have been carried out, which were completed in 2018. Since then, cyclic geodetic monitoring on installed benchmarks has made it possible to assess the condition of the structure's geometry in terms of horizontal displacements and subsidence. This process was aimed at preparing the landfill, which was closed in

2017, for its transformation into a public facility, i.e., a ski run for skiers, snowboarders, and tobogganers, as well as runners and cyclists [7,32]. The regulation [6] allows building constructions on the crown of a landfill after fifty years from the date of its closure. However, it allows a shorter time based on geotechnical and sanitary expert opinions in cases where this does not endanger human life and health or cause a risk to the environment.

The lump of the Radiowo landfill was monitored for many years by measuring controlled points using traditional surveying methods (more than 100 benchmarks were installed), which over time were supplemented by satellite measurements using GNSS receivers. The survey work at the Radiowo landfill used a survey set consisting of two GNSS Real Time Kinematic (RTK) receivers operating in a Base and Rover mode [8]. In such a configuration, one of the receivers serves as a stationary base and sends corrections via radio to the other mobile receiver (Rover) performing the measurement. The advantage of such a solution is that a single base station can cooperate with several mobile receivers which serve as the equipment for controlling construction machines, such as dozers, excavators, and graders used in earthmoving works at the landfill [7].

3. Materials and Methods

3.1. Periodic Archival Measurements

In the past, the studied landfill experienced frequent landslides caused by improper compaction of waste layers and inadequate protection of steep slopes. Accordingly, landslide areas have been rebuilt in accordance with earthwork rules and considering slope stability conditions (horizontal reinforcements, retaining structures), and geodetic monitoring has been introduced. Periodic geodetic monitoring of control points allows early detection of potential hazards and early response to the occurring changes in the shape of the landfill, especially in the areas of slopes with landslide hazards.

During the 30 years of geodetic monitoring, a total of 105 control points (benchmarks) were established, 59 of which were monitored for more than 5 years. Some of them have existed since the beginning of the site's monitoring, i.e., since 1993, and are located on the northern and western slopes of the landfill. Long-term monitoring has allowed the assessment of the mechanical parameters of the waste (deformation and strength) and the prediction of the displacements occurring at the landfill [8]. Figure 2 shows the location of the control points and the values of their measured subsidence over a one-year period: December 2020–December 2021.

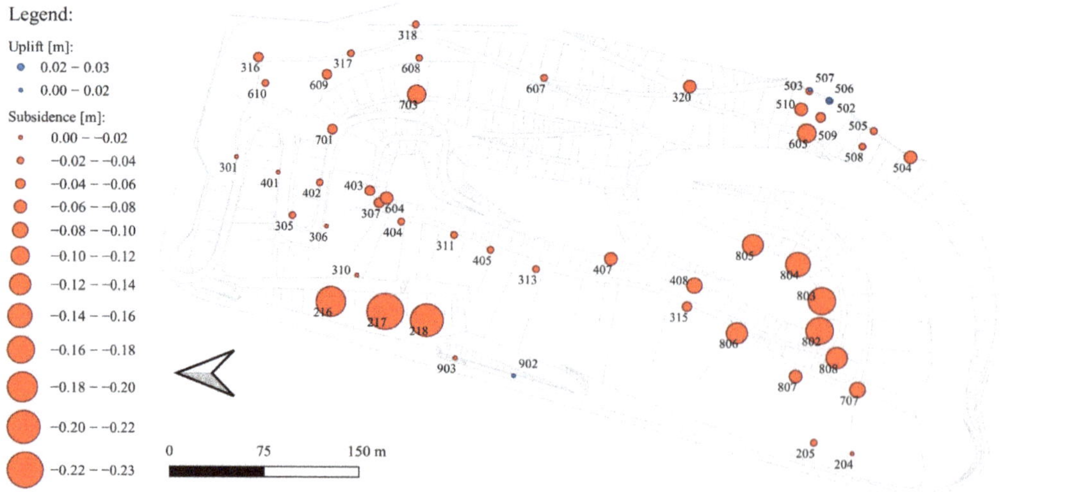

Figure 2. Map of annual subsidence and uplift of control points (December 2020–December 2021) [7].

As of 2017, i.e., after the formation of the target landfill body, the largest overall subsidence has been about 1.8 m and applies to the southern slope formed along the designed ski run. This area is still subject to increased subsidence, which is visualized in Figure 2 in the form of circles with a diameter corresponding to local displacement. The value of displacements per year reaches about 0.20 m [7]. The subsidence is expected to slowly disappear over time; however, it may locally increase following the installation of facilities on the surface of the landfill (buildings, PP, pole structures, etc.).

3.2. Data Sets

In the area of the landfill in 2022, the periodic measurements were made using four methods: (i) Method 1 (set 1)—photogrammetric measurements from an unmanned aerial vehicle (UAV); (ii) Method 2 (set 2)—aerial laser scanning (ALS); (iii) Method 3 (set 3)—terrestrial laser scanning (TLS); and (iv) Method 4 (set 4)—linear and angular measurements.

As a result of the measurement using methods 1–3, point clouds were obtained, differing in density and resolution from the recorded data. Figure 3 summarizes the general technical and accuracy characteristics of the instruments and reference points used.

	AERIAL MEASUREMENT		GROUND MEASUREMENT	
Set / Method	1 (Photogrammetry UAV)	2 (ALS)	3 (TLS)	4 (Angular–Linear Measurements)
Platform	 Phantom 4 Pro V2	 Matrice M600	 Imager Z + F	 Leica Flex Line TS03
Features	• Sony CMOS 1" matrix has a resolution of 20MP • FOV 84° • F2.8 optics with a focal length of 24mm • ISO range 100 – 12,800 • Max. Flight time 30 min • Positioning GPS / GLONASS	• Scanning range up to 70 [m] • System accuracy ±5 [cm] • Scanning angle (longitudinal) between −15° and +15° • Max. Flight time 20 min	• Min. range: 0.4 m • Resolution range: 0.1 mm • Max. data acquisition rate: 1,016,727 px/sec • Linearity error up to 50 m: ≤ 1 mm • Resolution in high mode: ± 6.3 mm / 10 m	• Single prism accuracy: (precise) + / Once: 1 mm + 1.5 ppm • Non-prism accuracy: 0 m – 500 m: 2 mm + 2 ppm • Accuracy Hz, V: 2"/3"/5" • Prism range: 1.5 – 3500 m • Long prism range: > 10,000m
Reference target				

Figure 3. Overview of the research measurements. Developed based on [33–35].

Set 1: Digital images were obtained from a photogrammetric flight taken using a Phantom 4 Pro platform, which is equipped with a 1-inch, 20-megapixel CMOS sensor with a 24-mm FOV lens. The measurements were taken from a height of 60 m above the landfill crown. The BSP measurement was divided into five flights. During the first flight, vertical (nadir) images were taken. In the next four, flights were carried out with oblique photos in four different directions (north, south, east, and west). A total of 18 longitudinal and 18 transverse series were taken. Coverage of longitudinal and transverse images was 80% for nadir images and 50% for oblique images with an off-nadir angle of 30°. A total of 1768 images were acquired. In the field, before the flights, 13 ground control points (GCPs) were marked and their XYZ coordinates were measured. Their shape, color, and size enabled unambiguous identification of the points in the field. The point clouds from imaging data were generated using the Pix4D Mapper software.

Set 2: Data acquired with an airborne laser scanning system using a DJI Matrice M600 platform. The flight was performed at an altitude of 60 m above the crown of the landfill. The distance between each flight path was 20 m. A total of four overflights were performed. The S50 LiDAR system mounted on the platform consists of a scanner: Velodyne's VLP-16 and a Sony A6000 RGB camera. The survey can be performed at a distance range of up to 70 m with an accuracy of ±0.05 m and a scanning angle of −15° to 15° (longitudinal). In the field, 10 control points were marked in the form of dedicated reference mats prior to the flights (Figure 3). During the flight, a GNSS reference station was used to reference the data. The average density of the resulting point cloud was about 160 pts/m^2. In areas of double coverage, the point cloud density was 200–250 pts/m^2.

Set 3: Point clouds (9 scans) acquired with the Z + F Imager 5006H laser scanner. The clouds were recorded in XYZI format (where I denotes the intensity of reflection of the laser beam from the measured structure). Benchmarks were the spheres with known diameters situated on geodetic tripods. The measurement was performed with high resolution reaching ±6.3 mm/10 m. The data acquisition process included several steps: (i) planning the measurement stations; (ii) arrangements of reference spheres for surveying tripods; and (iii) landfill scanning. The oriented point cloud was developed (filtering, referring the point cloud into a global coordinate system) in Faro Scene software.

Set 4: Reference data: In order to reference measurements made by alternative methods (set 1, set 2, and set 3), the position of control points located on the landfill was determined using trigonometric leveling. The measurement was made using the Leica Flex Line TS03 total station. To ensure the highest possible accuracy of the measurement, the so-called "three tripod method" was used. The measurement was referred to reference points located outside the area of the structure's influence.

3.3. Data Processing

3.3.1. Method 1

The processing of the images acquired from the photogrammetric flights was performed in Pix 4D Mapper software. It consists of calculating the internal and external orientation elements of each photo by generating tie points. This stage as well as the process of generating a dense point cloud is based on multi-image correlation (Dense Matching).

The alignment of the photo block was carried out based on the signaled matrix measured in the field—the ground control points (GCPs). After measuring the photopoints, a dense point cloud was generated. A GSD value of 0.03 m was obtained. The average density of the point cloud was 185 points per 1 m^2.

An important step in the processing of point clouds was their filtering, resulting from the intensive coverage of slopes with vegetation. Figure 4 shows an example of vegetation filtering of photogrammetric data. The landfill area is largely covered with vegetation, especially in the western part, as indicated by the red color in Figure 4b.

Figure 4. Point cloud generated from photos showing (**a**) green vegetation visible on the landfill, (**b**) the process of vegetation filtration, and (**c**) the surface of the site after filtration.

3.3.2. Method 2

The processing of ALS data was performed using LIDAR 360 software, which consisted of georeferencing the point cloud based on the position of the GNSS ground antenna and applying appropriate data filtering. Points that were recorded during the UAV's turnaround, ascent, and descent were removed from the dataset. Redundant points covering data areas recorded in overflights on adjacent paths were eliminated. Points recorded by beams emitted at angles greater than 25° from the nadir were removed. Vegetation filtering was also performed, as well as correction of the mutual position of fragments of the point cloud registered during the flight on two adjacent parallel paths (elimination of the so-called "broadside error").

3.3.3. Method 3

The acquired scans were recorded in local coordinate systems. The first step in processing the TLS data was to orient the scans. This process was performed based on reference spheres and points homologously located on adjacent scans using Faro Scene software. The accuracy of the scan orientation results is affected by (i) the incidence angle of the laser beam; (ii) surface color; (iii) the type of surface material; and (iv) the number and geometry of overlapping points on several scans.

Based on the measured GCPs, the resulting point cloud was transformed into the coordinate system in which the other measurements were made (EPSG:2178) and was filtered to remove vegetation from the image.

3.3.4. Method 4 (Reference Method)

The results of the angle-line measurements were used to calculate the elevation and topographic position of the control points. Calculations of the angular–linear network were made using the WinKalk program using the least squares method. Separate calculations were made for the altimetric network and horizontal geodetic network for topographic measurements. The average error in the calculation of the height of the points was 4 mm, while the average error in the topographic position was 25 mm.

4. Methodology

The study used four data sets (sets 1–4), acquired using different measurement systems. The measurements focused on the areas of the access road to the crown of the landfill (northern part), the crown of the landfill, and the ski run located in the southeastern part of the landfill. The alternative measurement methods proposed in the paper (sets 1 and 2) were compared with classical methods (set 4). Due to the large area of the landfill, TLS measurements were used only to image the western, less vegetated part of the landfill which is, however, the main area exposed to the landslide associated with the use of the slope for recreation in winter (ski run). The comparison of measurement methods was performed in the vertical plane, keeping in mind the results of archival measurements and the trends of change determined based on the analysis of behavioral models of similar objects, indicating changes in elevation as the main components of the displacement of control points on the site.

The study proposed three approaches to compare the methods used: (i) Approach 1 based on comparing the altitude of homologous points; (ii) Approach 2 based on the direct comparison of point clouds using the cloud-to-cloud (C2C) method; and (iii) Approach 3 based on the comparison of cross-sections created from point clouds.

Approach 1: To carry out a comprehensive analysis of the point clouds in the landfill area, the height differences determined at the control points based on the heights measured on the point clouds (obtained from methods 1 and 2) and the heights measured by method 4 (the reference data) were compared. The analyzed points are located at different locations in the landfill area (Figure 5). These include points located on the concrete road (the northern part of the landfill), among the vegetation along the road, on the slopes (mainly on the western side of the site) and on the crown of the landfill. Figure 5 shows the location of the control points measured using method 4.

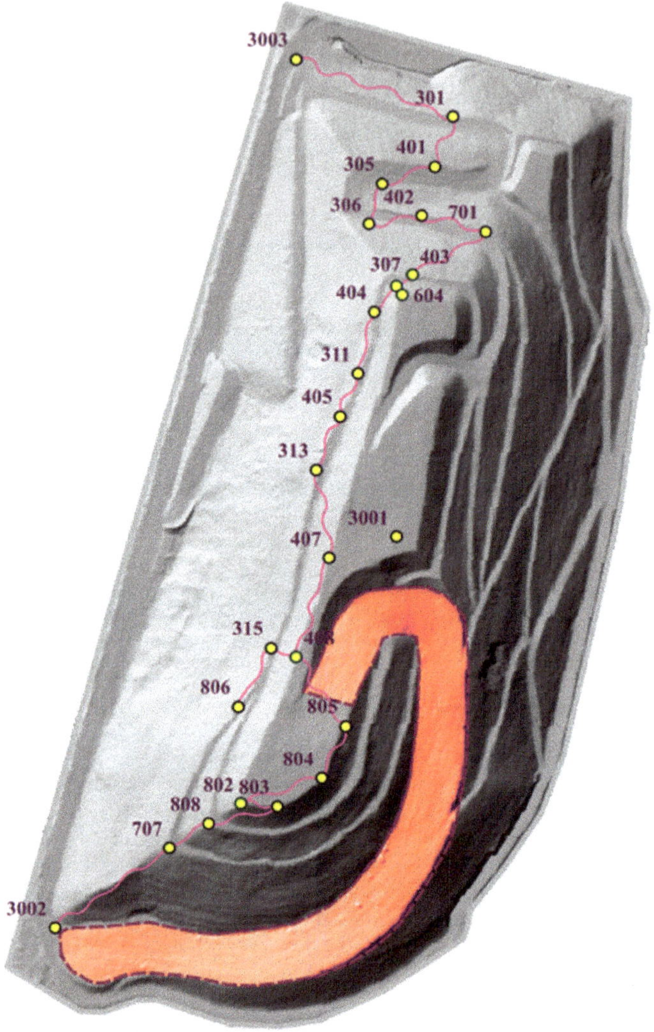

Figure 5. Location of the analyzed homologous points and the ski run.

Approach 2: TLS measurement was limited to the ski run located in the southeastern part of the landfill. The range of measurements is marked in red in Figure 5 (bounded by a purple dashed line). Only cloud-to-cloud distance analyses obtained from methods 1–3 were performed in this area.

Approach 3: When studying the quality of point clouds, attention should also be paid to their detail level related to their resolution [36]. The assessment of their detail levels was based on a comparative analysis of the profiles created from the measurements. In this approach, cross-sections along the ski run were extracted and their geometry was compared for each of the three-point clouds (sets 1–3).

5. Results and Discussion

Approach 1: This approach presented a comparison of point clouds obtained by methods 1 and 2 with reference data (set 4). The analyses performed included a comparison of height differences on the control points and a comparison of the difference in height between adjacent points, which are marked with a purple line in Figure 5. The results are shown in Tables 1 and 2.

Table 1. Summary of height differences at the control points.

Point No.	Height Differences (m)		
	ΔH_{4-2}	ΔH_{4-1}	$\Delta H_{4-2}-\Delta H_{4-1}$
3003	0.143	−0.067	0.210
301	1.225	0.996	0.229
401	0.507	0.295	0.212
305	0.699	0.648	0.051
306	2.009	1.908	0.101
402	0.858	0.778	0.080
701	0.326	0.173	0.153
604	−0.292	−0.018	−0.274
403	−0.019	−0.041	0.022
307	−0.665	−0.502	−0.163
404	0.031	−0.102	0.133
311	−0.184	−0.140	−0.044
405	−6.226	−6.392	0.166
313	−0.131	−0.188	0.057
407	−0.325	−0.341	0.016
408	−0.259	−0.103	−0.156
315	−1.135	−1.068	−0.067
806	0.098	0.109	−0.011
802	0.004	−0.067	0.071
803	−0.022	0.036	−0.058
804	0.241	0.231	0.010
805	−0.142	−0.043	−0.099
3001	0.033	−0.037	0.070
808	−0.875	−0.783	−0.092
707	−0.278	−0.280	0.002
3002	−0.212	−0.198	−0.014

The largest height differences (0.20–0.27 m) were obtained for points 3003, 301, 401, and 604 located in the northern part of the landfill on the road or in its close location to the slopes amidst dense vegetation. In other areas of the landfill, the differences were smaller. In general, the recorded height differences did not exceed 0.28 m. The standard deviation was 0.12 m. The results obtained for the area excluding the northern slope were sufficient and in accordance with the requirements for the accuracy of topographic-elevation measurements for earth structures.

An analysis of elevation differences was also conducted for selected sections located between control points, and the results are shown in Table 2.

Table 2. Summary of elevation differences on selected sections between control points.

Number of Section	Height Differences (m)					
	ΔH_4	ΔH_2	ΔH_1	ΔH_{4-2}	ΔH_{4-1}	ΔH_{4-2}–ΔH_{4-1}
3003–301	−11.866	−10.784	−10.803	−1.082	−1.063	−0.019
301–401	−8.460	−9.179	−9.161	0.719	0.701	0.018
401–305	−2.823	−2.631	−2.470	−0.192	−0.353	0.161
305–306	−4.893	−3.583	−3.633	−1.310	−1.260	−0.050
306–402	−2.072	−3.223	−3.202	1.151	1.130	0.021
402–701	−3.620	−4.152	−4.225	0.532	0.605	−0.073
701–403	−14.560	−14.904	−14.774	0.344	0.214	0.130
403–307	−2.290	−2.937	−2.751	0.647	0.461	0.186
307–404	2.113	2.809	2.513	−0.696	−0.400	−0.296
404–311	−0.285	−0.501	−0.323	0.216	0.038	0.178
311–405	0.882	−5.160	−5.370	6.042	6.252	−0.210
405–313	0.267	6.362	6.471	−6.095	−6.204	0.109
313–407	−8.591	−8.785	−8.744	0.194	0.153	0.041
407–3001	−0.668	−0.311	−0.364	−0.357	−0.304	−0.053
407–408	0.758	0.824	0.996	−0.066	−0.238	0.172
408–315	9.722	8.846	8.757	0.876	0.965	−0.089
315–806	0.259	1.492	1.436	−1.233	−1.177	−0.056
408–805	0.119	0.236	0.179	−0.117	−0.060	−0.057
805–804	−1.088	−0.705	−0.814	−0.383	−0.274	−0.109
804–802	0.500	0.263	0.202	0.237	0.298	−0.061
802–803	0.087	0.061	0.190	0.026	−0.103	0.129
803–808	10.314	9.461	9.495	0.853	0.819	0.034
808–707	9.965	10.563	10.468	−0.598	−0.503	−0.095
707–3002	35.242	35.307	35.324	−0.065	−0.082	0.017

Maximum height differences were obtained for sections 307–404 and 311–405. These sections are located in the northern part of the landfill crown in an area covered with vegetation. The value of the standard deviation was 0.12 m. The obtained values of height differences did not exceed 0.30 m.

Approach 2: Cloud-to-cloud distance analyses were performed to calculate the average distance between point clouds derived from methods 1–3. This approach adopted the principle of comparing the results side by side. The differences in the distances of the resulting point clouds are shown in Figures 6–8.

Table 3 shows the statistical data calculated using the Gaussian elimination for the mean value of the height differences obtained from the comparison of the different methods, as well as their standard deviation (mean value, standard deviation).

Table 3. Results of statistical analysis using the Gaussian elimination.

	Sets 1 vs. 2	Sets 2 vs. 3	Sets 3 vs. 1
Mean (m)	0.201	−0.024	−0.209
Standard deviation (m)	0.153	0.321	0.329

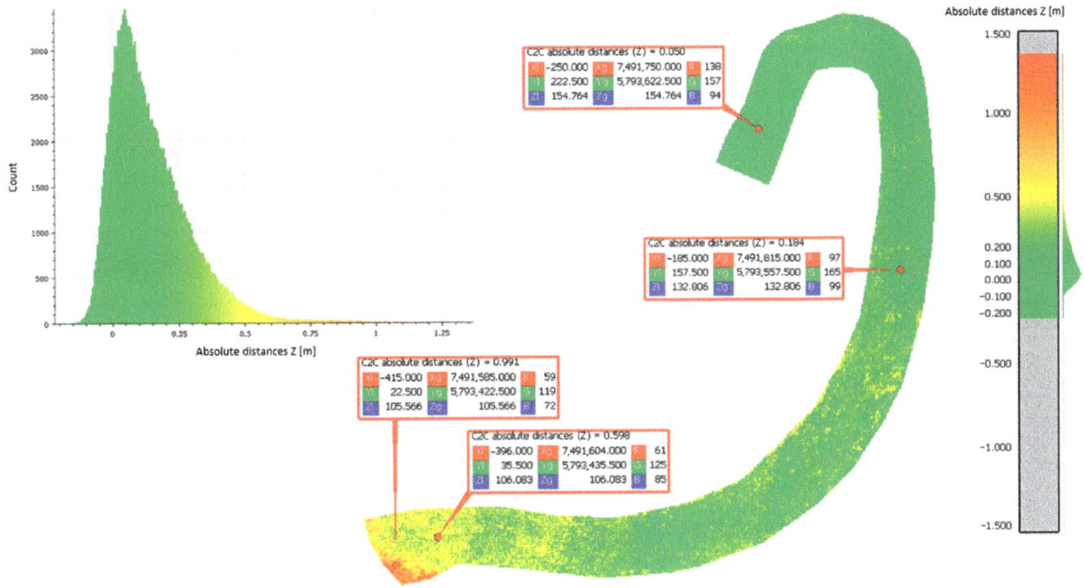

Figure 6. C2C distance analyses for set 1 compared to set 2.

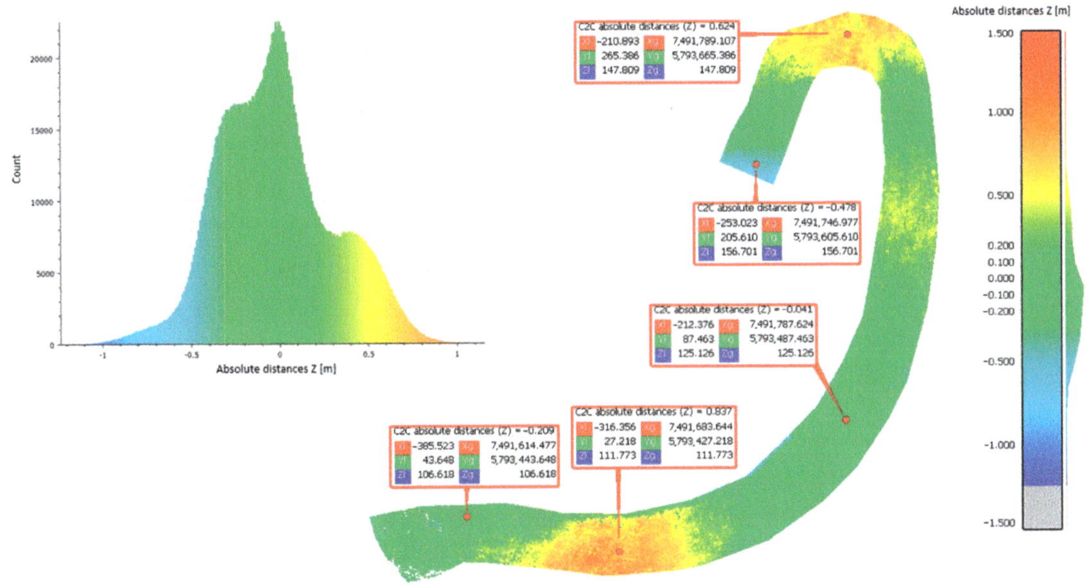

Figure 7. C2C distance analyses for set 2 compared to set 3.

Based on the obtained results of measurements and approximations, the mean value of the height difference for the compared methods was 0.20 m (set 1 vs. set 2), −0.02 m (set 2 vs. set 3), and −0.21 m (sets 3 vs. set 1), respectively. The largest average differences were observed in the comparison of set 1 (photogrammetric method). Methods 2 (ALS) and 3 (TLS) are based on laser scanning to penetrate vegetation and better reach the ground surface, while method 1 is based on surface measurement. Therefore, the difference may be due to the vegetation cover lushly growing on the landfill during the summer. The standard deviation values were 0.15 m (set 1 vs. set 2), 0.32 m (set 2 vs. set 3), and 0.33 m (set 3 vs. set 1), respectively. The largest standard deviations were found when comparing set 3 acquired by using method 3 (TLS). This parameter indicates how widely the values of a given size are scattered around its mean. Its larger value for set 3 may indicate errors made in the mutual orientation of the measurement stations, as shown in Figures 7 and 8.

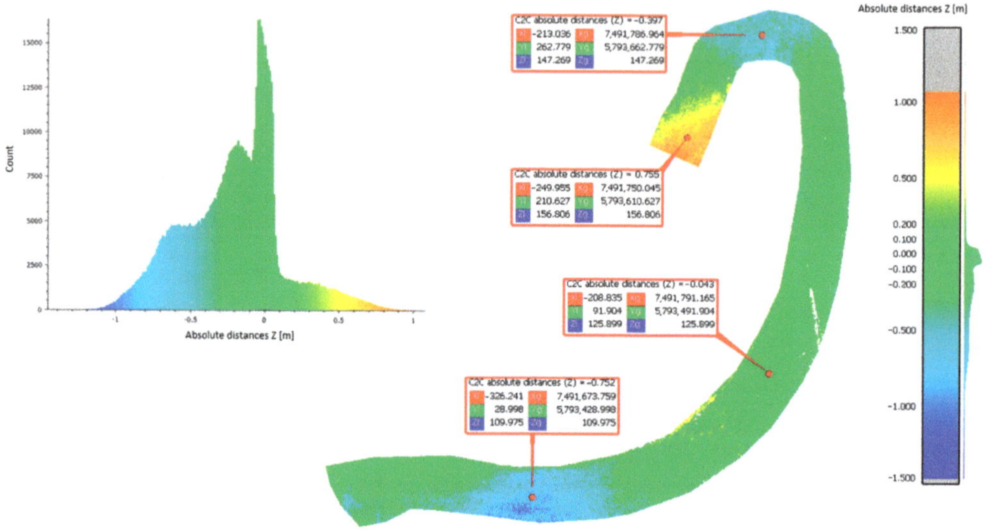

Figure 8. C2C distance analyses for set 3 compared to set 1.

Approach 3: Point cloud analysis was carried out based on profiles generated in the landfill area. Figure 9 shows three selected profiles generated in the northern, central, and southern areas of the ski run axis. The profiles were generated from the acquired point clouds (sets 1–3). Strips of 0.5 m width were used in profile extraction.

Analyzing the generated land surface profiles it can be concluded that the greatest differences in elevation occurred between the data acquired using method 1 and method 2. This is due to the greater penetration of vegetation in method 2 and the surface measurement in method 1.

The highest accuracy of measurements at the Radiowo landfill was obtained in discrete linear and angular measurements (trigonometric leveling) of controlled points, which is due to the accuracy of the method itself, the precise measuring equipment used and the high regime of measurement execution (adequate repeatability of measurements of leveling sections—both in forward and backward leveling).

The quasi-continuous methods of measurement that produce point clouds can be an excellent complement to classical measurements. However, due to the accuracies of the methods themselves and the data filtering necessary due to vegetation, methods 1, 2, and 3 were consistent with each other in terms of accuracy (height differences of 0.20 m) but inferior compared to the GNSS RTK, classical linear and angular methods.

On the other hand, the main advantage of methods 1–3 over method 4 is the representation of the whole body of the geotechnical structure and not just the coordinates of selected individual points. Data in the form of a point cloud can provide a basis for modeling the shape of the solid of the entire landfill, not just periodic measurements of coordinates of control points.

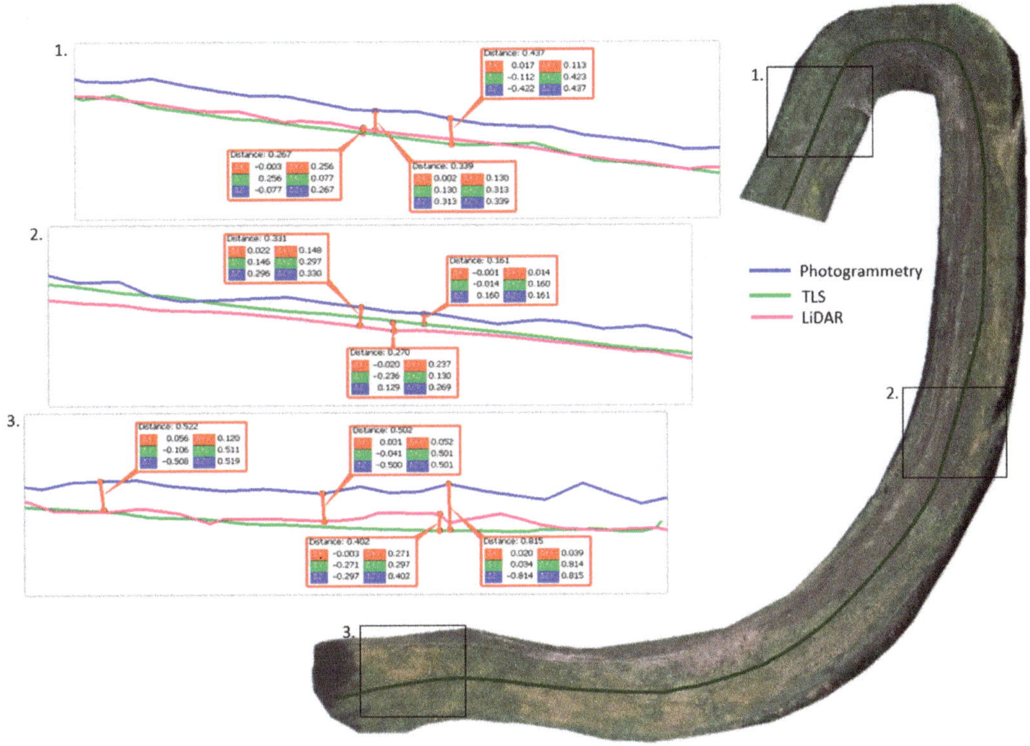

Figure 9. Height differences on selected profiles between point clouds extracted based on methods 1–3.

However, it is necessary to compare not only the area coverage or economic context, but, above all, the accuracy and resolution of the acquired data and their further use in the post-restoration management of the landfill. In particular, the selection of the method of periodic and inventory measurement must consider the expected values of change and the purpose of the measurement. In particularly fortified areas that are the basis of the stability of the landfill (steep slope, access road, crown), periodic control measurements of displacements should be carried out discreetly in the network of control points using linear and angular methods, precision leveling or GNSS RTK or STATIC methods, which will allow achieving repeatability and accuracy in the range of single millimeters and centimeters (depending on the method). ALS and TLS methods are suitable in areas intended for intensive use and annual renewal of geometry (artificial ski run) to control displacements in decimeter ranges.

The use of selected measurement methods to inventory the landfill was compared from an economic point of view, considering the selected two aspects, i.e., time (to perform field work, conduct an intimate study) and the number of people required to perform the measurement, the results of the comparison of the proposed measurement methods are shown in Figure 10.

The analysis of the aforementioned data (Figure 10) shows a clear advantage of methods 2 and 1, both in terms of the time taken to perform the measurement, the indoor analysis, and the number of people required to perform the field measurements. The most accurate results were obtained using the trigonometric leveling method, despite only obtaining the coordinates of single points.

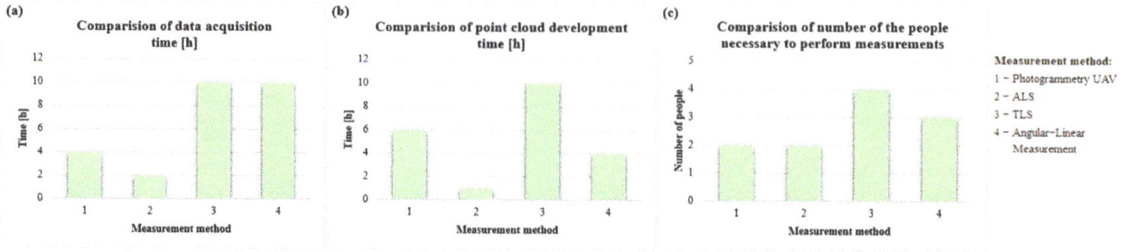

Figure 10. Comparison of the applied methods in terms of (**a**) data acquisition time, (**b**) point cloud development time, and (**c**) the number of people necessary to perform the measurement.

The photogrammetric method [16–18], the ALS method [15], and the TLS method [4,5,13,14,19,22,23] are successfully used in the periodic geodetic monitoring of facilities similar in their construction to MSW landfills (large-scale, soil-built facilities). These include open-pit mines, natural landslide slopes, post-production landfills, and various types of bulk product storage areas. On this basis, it can be concluded that these methods will also find application in the periodic monitoring of MSW landfills. The paper attempts to validate the above-mentioned methods in the area of the reclaimed MSW landfill Radiowo, which is a modern approach to a paper. The research is a prelude to a better understanding of modern measurement methods in the context of periodic monitoring of such facilities.

6. Conclusions

The analyses and comparisons of the measurement methods to control and inventory periodic changes at landfills described in the paper allowed identifying of their advantages and limitations in the context of long-term geodetic monitoring. In structures such as landfills, the expected and recorded annual displacements reach decimeters in the course of a year, and in multi-year cycles even several meters, depending on the way the slopes are reinforced and how they are used. The use of proposed remote sensing methods is justified mainly because of the magnitude of predicted displacements of the earth. The recorded height differences between the data set from the photogrammetric method and the data set from ALS did not exceed 0.28 m and the standard deviation was 0.12 m. The subsidence of the Radiowo landfill surface is variable. Its values depend on the morphology and age of the waste. In the last five years, the values of subsidence on the crown of the landfill have ranged from 0.6 to 2.1 m, giving an average annual increase range from 0.12 to 0.40 m. In the next ten years, average annual increments in subsidence are forecast to be between 0.02 and 0.10 m.

TLS-based monitoring allows for high-resolution data but is labor-intensive and not very economical in the context of monitoring the entire landfill. This method can be used to monitor selected portions of the landfill that are particularly vulnerable to displacement and in areas where high accuracy is required, for example, in the ski run or slopes with exceeded critical conditions. When performing such measurements, it is important to keep in mind the close location of measurement stations and as many benchmarks as possible, due to the vast area of the landfill and the lack of terrain details that hinder the process of orienting the scans.

Monitoring using UAVs (photogrammetry and ALS) can be an excellent complementation of the classical methods in monitoring periodic changes on a landfill, especially in areas where it is not possible to stabilize controlled points, or due to dense vegetation (used as slope reinforcement) that prevents good visibility between points. On the other hand, it cannot completely replace classical measurement methods based on cyclic measurement of control points, due to the required accuracy of displacement determination and current legal regulations in Poland (requirement to install and periodic measurement benchmarks).

Due to the growing seasons in which the quality of data and the need for filtration change significantly, and also due to changing atmospheric conditions (humidity: precipitation and haze), geodetic monitoring of landfills should be carried out in late autumn (late October/early November) or early spring (late March/early April). Comparative analysis of the acquired point cloud data shows the greatest differences in areas with high and dense vegetation. This difference can be seen especially in the results of the photogrammetric method compared to the ALS and TLS methods. The mean value of the height difference for the compared methods was 0.20 m (photogrammetric method vs. ALS), −0.02 m (ALS vs. TLS), and −0.21 m (TLS vs. photogrammetric method). High and dense vegetation makes it impossible in the photogrammetric method to directly measure the area of ground cover of the landfill.

The presented results of the 3-D measurements will be used during the design of facilities at the restored landfill, including sports and recreational equipment and renewable energy sources (PP, RES, and a landfill gas collection system). The obtained imaging will also be useful in the periodic assessment of the geotechnical safety of the structure and in designing appropriate strengthening solutions on the slopes of the landfill. Finally, they may be used as references for future inventory studies and visualizations of changes.

Author Contributions: Conceptualization, G.P. and J.Z.-P.; methodology, G.P. and J.Z.-P.; software, G.P.; validation, J.Z.-P., E.K. and M.D.V.; formal analysis, G.P. and K.P.; investigation, G.P. and K.P.; resources, G.P., J.J., M.P., J.Z.-P. and K.P.; writing—original draft preparation, G.P. and K.P.; writing—review and editing, J.Z.-P., E.K. and M.D.V.; visualization, G.P. and K.P.; supervision, J.Z.-P.; funding acquisition, M.D.V. All authors have read and agreed to the published version of the manuscript.

Funding: This research received no external funding.

Institutional Review Board Statement: Not applicable.

Informed Consent Statement: Not applicable.

Data Availability Statement: The data presented in this study are available on request from the co-author Grzegorz Pasternak, e-mail: grzegorz_pasternak@sggw.edu.pl.

Conflicts of Interest: The authors declare no conflict of interest.

References

1. Jahanfar, A.; Dubey, B.; Gharabaghi, B.; Movahed, S.B. Landfill failure mobility analysis: A probabilistic approach. *Int. J. Environ. Ecol. Eng.* **2016**, *10*, 568–576.
2. Koelsch, F.; Fricke, K.; Mahler, C.; Damanhuri, E. Stability of Landfill—The Bandung Dumpsite Deseaster. In Proceedings Sardinia. 2005. Available online: http://dr-koelsch.de/assets/images/sardinia_2005.pdf (accessed on 15 December 2022).
3. Blight, G. Slope failures in municipal solid waste dumps and landfills: A review. *Waste Manag. Res.* **2008**, *26*, 448–463. [CrossRef] [PubMed]
4. Guo, Y.; Li, X.; Ju, S.; Lyu, Q.; Liu, T. Utilization of 3D Laser Scanning for Stability Evaluation and Deformation Monitoring of Landslides. *J. Environ. Public Health* **2022**, *2022*, 8225322. [CrossRef] [PubMed]
5. Weidner, L.; Walton, G. Monitoring the Effects of Slope Hazard Mitigation and Weather on Rockfall along a Colorado Highway Using Terrestrial Laser Scanning. *Remote Sens.* **2021**, *13*, 4584. [CrossRef]
6. Regulation of the Minister of Environment of 30 April 2013 on Landfills (Journal of Laws 2013, Item 523). Available online: https://isap.sejm.gov.pl/isap.nsf/DocDetails.xsp?id=WDU20130000523 (accessed on 15 December 2022).
7. Pasternak, G. Geodetic monitoring of geotechnical structures displacements: A case study of Radiowo landfill site in Warsaw. *Acta Sci. Pol. Archit.* **2022**, *21*, 75–83. [CrossRef]
8. Koda, E.; Kiersnowska, A.; Kawalec, J.; Osinski, P. Landfill slope stability improvement incorporating reinforcements in reclamation process applying Observational Method. *Appl. Sci.* **2020**, *10*, 1572. [CrossRef]
9. Adamcová, D.; Barton, S.; Osinski, P.; Pasternak, G.; Podlasek, A.; Vaverková, M.; Koda, E. Analytical modelling of MSW landfill surface displacement based on GNSS monitoring. *Sensors* **2020**, *20*, 5998. [CrossRef] [PubMed]
10. Koda, E.; Żakowicz, S. Physical and hydraulics properties of the MSW for water balance of the landfill. In Proceedings of the 3rd International Congress on Environmental Geotechnics, Lisbon, Portugal, 7–11 September 1998; A.A. Balkema Publishers: Rotterdam, The Netherlands, 1998; Volume 1, pp. 217–222.
11. Mohammad, A.; Osinski, P.; Koda, E.; Singh, D.N. A Case Study on Establishing the State of Decomposition of Municipal Solid Waste in a Bioreactor Landfill in India. *Waste Manag. Res.* **2021**, *39*, 1375–1388. [CrossRef] [PubMed]

12. Rose, N.D.; Hungr, O. Forecasting potential rock slope failure in open pit minesusing the inverse-velocity method. *Int. J. Rock Mech. Min. Sci.* **2007**, *44*, 308–320. [CrossRef]
13. Ma, J.; Niu, X.; Liu, X.; Wang, Y.; Wen, T.; Zhang, J. Thermal Infrared Imagery Integrated with Terrestrial Laser Scanning and Particle Tracking Velocimetry for Characterization of Landslide Model Failure. *Sensors* **2019**, *20*, 219. [CrossRef]
14. Huang, R.; Jiang, L.; Shen, X.; Dong, Z.; Zhou, Q.; Yang, B.; Wang, H. An efficient method of monitoring slow-moving landslides with long-range terrestrial laser scanning: A case study of the Dashu landslide in the Three Gorges Reservoir Region, China. *Landslides* **2019**, *16*, 839–855. [CrossRef]
15. Štroner, M.; Urban, R.; Línková, L. A New Method for UAV Lidar Precision Testing Used for the Evaluation of an Affordable DJI ZENMUSE L1 Scanner. *Remote Sens.* **2021**, *13*, 4811. [CrossRef]
16. Baiocchi, V.; Napoleoni, Q.; Tesei, M.; Servodio, G.; Alicandro, M.; Costantino, D. UAV for monitoring the settlement of a landfill. *Eur. J. Remote Sens.* **2019**, *52*, 41–52.
17. Mello, C.C.S.; Salim, D.H.C.; Simões, G.F. UAV-based landfill operation monitoring: A year of volume and topographic measurements. *Waste Manag.* **2022**, *137*, 253–263. [CrossRef]
18. Incekara, A.H.; Delen, A.; Seker, D.Z.; Goksel, C. Investigating the Utility Potential of Low-Cost Unmanned Aerial Vehicles in the Temporal Monitoring of a Landfill. *Int. Soc. Photogramm. Remote Sens.* **2019**, *8*, 22. [CrossRef]
19. Lefu, N.; Nokwe, V. Use of laser scanner technology as part of the slope stability risk management strategy at Letšeng diamond mine. In *Slope Stability 2020: Proceedings of the 2020 International Symposium on Slope Stability in Open Pit Mining and Civil Engineering*; Australian Centre for Geomechanics: Perth, Australia, 2020; pp. 241–254.
20. Mikrut, S. Classical Photogrammetry and UAV-Selected Ascpects. *Int. Arch. Photogramm. Remote Sens. Spat. Inf. Sci.* **2016**, *XLI-B1*, 947–952. [CrossRef]
21. Kedzierski, M.; Fryskowska, A. Methods of laser scanning point clouds integration in precise 3D building modeling. *Measurement* **2015**, *74*, 221–232. [CrossRef]
22. Chen, X.; Ban, Y.; Hua, X.; Lu, T.; Tao, W.; An, Q. A method for the calculation of Detectable Landslide using Terrestrial Laser Scanning data. *Measurement* **2020**, *160*, 107852. [CrossRef]
23. Yin, C.; Li, H.; Hu, Z.; Li, Y. Application of the terrestrial laser scanning in slope deformation monitoring: Taking a highway slope as an example. *Appl. Sci.* **2020**, *10*, 2808. [CrossRef]
24. Li, H.B.; Qi, S.C.; Yang, X.G.; Li, X.W.; Zhou, J.W. Geological survey and unstable rock block movement monitoring of a post-earthquake high rock slope using terrestrial laser scanning. *Rock Mech. Rock Eng.* **2020**, *53*, 4523–4537. [CrossRef]
25. DiFrancesco, P.M.; Bonneau, D.; Hutchinson, D.J. The implications of M3C2 projection diameter on 3D semi-automated rockfall extraction from sequential terrestrial laser scanning point clouds. *Remote Sens.* **2020**, *12*, 1885. [CrossRef]
26. Berlinghieri, M.; Pandolfi, O. Remote monitoring of natural slopes: Combining use of Terrestrial Laser Scanning and innovative rockfall alert system. *IOP Conf. Ser. Earth Environ. Sci.* **2021**, *833*, 012153. [CrossRef]
27. Ge, Y.; Tang, H.; Gong, X.; Zhao, B.; Lu, Y.; Chen, Y.; Lin, Z.; Chen, H.; Qiu, Y. Deformation Monitoring of Earth Fissure Hazards Using Terrestrial Laser Scanning. *Sensors* **2019**, *19*, 1463. [CrossRef] [PubMed]
28. Spreafico, M.C.; Perotti, L.; Cervi, F.; Bacenetti, M.; Bitelli, G.; Alena, V.; Mandanici, E.; Tini, M.A.; Borgatti, L. Terrestrial remote sensing techniques to complement conventional geomechanical surveys for the assessment of landslide hazard: The San Leo case study (Italy). *Eur. J. Remote Sens.* **2015**, *48*, 639–660. [CrossRef]
29. Stumvoll, M.J.; Schmaltz, E.M.; Glade, T. Dynamic characterization of a slow-moving landslide system—Assessing the challenges of small process scales utilizing multi-temporal TLS data. *Geomorphology* **2021**, *389*, 107803. [CrossRef]
30. Haerani, N.; Abidin, H.Z.; Gumilar, I.; Sadarviana, V.; Wijaya, D.D. On the Performance of Terrestrial Laser Scanner for Volcanic and Landslide Hazard Assessment in Indonesia 2013 (8131). In Proceedings of the FIG Working Week 2016 Rocovery from Disaster Christchurch, Christchurch, New Zealand, 2–6 May 2016.
31. Podlasek, A.; Jakimiuk, A.; Vaverková, M.D.; Koda, E. Monitoring and Assessment of Groundwater Quality at Landfill Sites: Selected Case Studies of Poland and the Czech Republic. *Sustainability* **2021**, *13*, 7769. [CrossRef]
32. Koda, E.; Rybak-Niedziółka, K.; Winkler, J.; Černý, M.; Osinski, P.; Podlasek, A.; Kawalec, J.; Vaverková, M.D. Space Redevelopment of Old Landfill Located in the Zone between Urban and Protected Areas: Case Study. *Energies* **2022**, *15*, 146. [CrossRef]
33. DJI Website. Available online: https://www.dji.com (accessed on 15 December 2022).
34. Available online: https://www.mertind.com (accessed on 15 December 2022).
35. Available online: https://leica-geosystems.com (accessed on 15 December 2022).
36. Fryskowska, A.; Kedzierski, M.; Walczykowski, P.; Wierzbicki, D.; Delis, P.; Lada, A. Effective detection of sub-surface archaeological features from laser scanning point clouds and imagery data. *Int. Arch. Photogramm. Remote Sens. Spat. Inf. Sci.* **2017**, *42*, 245–251. [CrossRef]

sensors

Article

A Novel Adaptive Time-Frequency Filtering Approach to Enhance the Ultrasonic Inspection of Stainless Steel Structures

Biao Wu [1], Haitao Yang [1], Yong Huang [2], Wensong Zhou [2] and Xiaohui Liu [3,*]

1 College of Civil Engineering, Nanjing Tech University, Nanjing 211816, China
2 Key Laboratory of Structural Dynamic Behavior and Control of the Ministry of Education, School of Civil Engineering, Harbin Institute of Technology, Harbin 150090, China
3 Institute of Engineering Mechanics, China Earthquake Administration, Langfang 065201, China
* Correspondence: liuxh1998@yeah.net

Abstract: Ultrasonic nondestructive testing (NDT) provides a valuable insight into the integrity of stainless steel structures, but the noise caused by the scattering of stainless steel microstructure often limits the effectiveness of inspection. This work presents a novel adaptive filtering approach to enhance the signal-to-noise ratio (SNR) of a measured ultrasonic signal from the inspection of a stainless steel component, enabling the detection of hidden flaws under strong noise. After the spectral modeling of the noisy ultrasonic NDT signal, the difference between the spectral characteristics of a flaw echo and that of grain noise is highlighted, and a reference spectrum model to estimate the frequency spectrum of the echo reflected by any possible flaw is developed. Then, the signal is segmented and the similarity between the spectra of data segments and the reference spectra is evaluated quantitatively by the spectral similarity index (SSI). Based on this index, an adaptive time-frequency filtering scheme is proposed. Each data segment is processed by the filtering to suppress the energy of noise. The processed data segments are recombined to generate the de-noised signal after multiplying weighting coefficients, which again is determined by the SSI. The performance of the proposed method for SNR enhancement is evaluated by both the simulated and experimental signal and the effectiveness has been successfully demonstrated.

Keywords: stainless steel; ultrasonic nondestructive testing; grain noise; noise suppression; time-frequency filtering

Citation: Wu, B.; Yang, H.; Huang, Y.; Zhou, W.; Liu, X. A Novel Adaptive Time-Frequency Filtering Approach to Enhance the Ultrasonic Inspection of Stainless Steel Structures. *Sensors* **2023**, *23*, 1030. https://doi.org/10.3390/s23021030

Academic Editor: Zenghua Liu

Received: 2 December 2022
Revised: 10 January 2023
Accepted: 11 January 2023
Published: 16 January 2023

1. Introduction

Excellent mechanical properties such as toughness, high strength as well as resistance to corrosion and heat have made stainless steel important building material widely applied in civil and marine engineering. For example, introducing stainless steel tubes to replace the carbon steel tube in concrete-filled-steel-tube systems can fundamentally solve the corrosion problem [1]. Stainless steel is also used as reinforcement for concrete structures in the coastal environment [2] as well as storage canisters and piping systems in the nuclear industry [3]. During long-term service, properties of stainless steel such as strength, ductility and corrosion resistance deteriorate due to aging [4]. Among the factors responsible for the performance deterioration, chloride-induced stress corrosion cracking of welded heat affected zones is of special concern as a well-documented mode of attack for stainless steels [5]. Manufacturing flaws (inclusions, voids or lack of fusion) are another factor of performance deterioration that could also exist in the welded joints. To ensure the safety and integrity of stainless steel structures, ultrasonic nondestructive testing techniques would be very helpful to detect, locate and size potential in-service cracks and manufacturing flaws [6,7]. Especially, state-of-the-art additive manufacturing (AM) requires advanced ultrasonic NDT techniques to evaluate the integrity of AM components [8] and to design an inspection scheme based on the reliability and cost analysis of AM components [9].

However, stainless steels are coarse-grained, heterogeneous and anisotropic. The interaction between incident ultrasonic waves and the grain microstructure of stainless steel results in a type of complex acoustic noise, often called grain noise, leading to low SNR of the measured signal and poor detectability of flaws. The instrumental noise of the NDT system could further lower the SNR.

For the effective inspection of stainless steel, SNR enhancement of the measured ultrasonic signal is of great importance, and numerous methods have been proposed for it. Widely explored methods include wavelet-based processing [10–13], split spectrum processing [14–16] and deconvolution-based approaches [17–19]. In more recent years, efforts based on sparse signal representation [20–22] and empirical mode decomposition [19] also contributed to the enhancement of SNR for ultrasonic NDT.

When an ultrasonic pulse propagates inside a stainless steel component, the pulse will interact with the grain microstructure, resulting in the distortion of the pulse due to scattering and attenuation. In stainless steel, as reported in [23], the ratio of the wavelength λ to the average grain diameter $\overline{D}$ satisfies that $\lambda/\overline{D} \gg 1$. The scattering of the incident ultrasonic wave by the grain microstructure falls in the Rayleigh scattering region where attenuation is proportional to the fourth power of frequency. As a result, the high-frequency component of the ultrasonic wave attenuates more significantly than the low-frequency component, leading to the waveform distortion of the ultrasonic pulse that the center frequency decreases while the duration dilates. Meanwhile, the interaction between the incident ultrasonic pulse and a flaw is different from that of a grain. Generally, the size of a flaw is comparable or larger than the ultrasonic wavelength λ and thus can be viewed as a geometrical reflector, and the reflected signal can be regarded as frequency independent [24]. On the contrary, the backscattering of the incident ultrasonic pulse by a grain is frequency dependent on the amplitude of scattering being proportional to the second power of frequency [25]. The difference between the two interaction mechanisms can be useful for noise suppression in the ultrasonic NDT of stainless steel structures.

Based on the above analysis, this work presents a novel methodology to enhance the ultrasonic NDT of stainless steel materials by exploiting the time-frequency characteristics of the ultrasonic signal. Specifically, the commonly used pulse-echo testing mode will be considered here. It is important to note that the echo reflected by a defect has a finite-time duration and the distortion of the waveform is dependent on the traveling distance. Taking these characteristics into consideration, it is desirable to exploit the time-frequency features of the signal rather than analyzing the signal as a whole. We start with the spectral modeling of grain noise and a typical flaw echo, and how the two spectra evolve with traveling distance is established. The difference between the spectral characteristics of a flaw echo and that of noise is highlighted. A reference spectrum model based on this difference is developed for the echo reflected by any possible flaw. Next, the signal is segmented and the similarity between the spectra of data segments and the corresponding reference spectra is evaluated quantitatively by a term defined as the spectral similarity index (SSI) in a moving window fashion. For each data segment, its SSI will determine the shape of the filter as well as the weighting coefficient which is used in the recombination stage where all the filtered data segments are recombined to generate the de-noised signal.

The rest of this paper is organized as follows. In Section 2, detailed spectral analysis and modeling of the components of a typical ultrasonic NDT signal are conducted. In Section 3, the proposed method is presented in detail. In Section 4, we demonstrate our method by synthetic ultrasonic signals and signals from an experimental ultrasonic inspection of a stainless steel block that contains prefabricated flaws, respectively. In the final section, some conclusive remarks are drawn.

2. Modeling of Ultrasonic NDT Signal

When conducting ultrasonic NDT on a stainless steel component in pulse-echo mode, the measured signal $y(t)$ can be written as $y(t) = e(t) + g(t) + w(t)$ where $e(t)$ refers to the echoes reflected by defects in the component, $g(t)$ and $w(t)$ are backscattered grain noise and electrical noise caused by the instruments, respectively.

In pulse-echo mode where the transducer is functioning as both a transmitter and a receiver, the scattered wave by a grain scatterer in the time domain can be viewed as the convolution of the impulse response of a grain scatterer $r(t,\lambda)$ and the impinging ultrasonic wave $h(t,\lambda)$ on the scatterer:

$$g(t) = h(t,\lambda) * r(t,\lambda) \tag{1}$$

where $*$ denotes the convolution operator. The impulse response function $r(t,\lambda)$ mathematically describes the backscattered wave as a function of wavelength λ (or equivalently, frequency) in the time domain. Meanwhile, the impinging ultrasonic pulse $h(t,\lambda)$ is a convolution of the transmitted pulse from the transducer $u(t,\lambda)$ and the attenuation function $a(t,\lambda)$, since the transmitted pulse will experience frequency-dependent attenuation during its propagation from the transmitter to the scatterer. Therefore, the backscattered signal by a grain scatterer received by the receiver can be expressed as:

$$g(t) = u(t,\lambda) * a(t,\lambda) * r(t,\lambda) \tag{2}$$

In the frequency domain, the above expression can be written as:

$$G(f) \propto |U(f)||S(f)||R(f)||A(f)| \tag{3}$$

where $|U(f)|$ is the impulse response of the transducer which is usually modeled as a bandpass Gaussian-shaped spectrum. $|S(f)|$ is the frequency modulation function due to the sum of grain scatterers with random orientations and phases [26]. $|R(f)|$ is the frequency-dependent scattering function which is proportional to the second power of frequency in the Rayleigh scattering region. $|A(f)|$ is the transfer function corresponding to the frequency-dependent attenuation characteristics of the ultrasonic wave during its propagation.

Similar to grain noise, the echo reflected by a flaw can be viewed as a convolution between the impinging ultrasonic pulse and the impulse response function of a flaw, considering the effect of frequency-dependent attenuation:

$$e(t) = u(t,\lambda) * a(t,\lambda) * p_f \tag{4}$$

where p_f is the impulse response function of a flaw and is independent on frequency. In the frequency domain, a flaw echo can be expressed as:

$$E(f) \propto |U(f)||A(f)| \tag{5}$$

By comparing Equations (3) and (5), it is obvious that the spectrum of a flaw differs from that of grain noise, represented by $|S(f)|$ and $|R(f)|$. The inherent randomness property of $|S(f)|$ will make the spectrum of grain noise take on a grass-like pattern; while $|R(f)|$ means that the spectrum of grain noise shifts upward compared to the spectrum of a flaw, due to the frequency dependence scattering property of $|R(f)|$ which is proportional to the second power of frequency in the Rayleigh scattering region.

From Equation (5), it is seen that once the attenuation function $A(f)$ is determined, one can estimate the frequency spectrum of a flaw echo at any given depth. According to earlier studies, the attenuation function can be written as [23,24]:

$$A(f) \propto exp[-2\int_0^z \alpha(z,f)\mathrm{d}z] = exp\{-2\int_0^z [\alpha_a(z,f) + \alpha_s(z,f)]\mathrm{d}z\} \tag{6}$$

where z is the distance of a reflector from the transducer. $\alpha(z,f)$ is the frequency-dependent attenuation coefficient defined as the sum the scattering term $\alpha_s(z,f)$ and the absorption term $\alpha_a(z,f)$. In the Rayleigh scattering region, the former varies with the third power of grain diameter and the fourth power of frequency while the latter increases linearly with frequency. However, the absorption term is usually insignificant and negligible when compared to the scattering term; therefore, the attenuation model can be expressed as:

$$A(f,t) = exp[-\alpha(2\pi f)^4 vt] \tag{7}$$

where v is the velocity of the sound in stainless steel and t is the time delay of the pulse. In practice, the attenuation coefficient α can be estimated beforehand [27]. Generally, the spectrum of the transmitted pulse $U(f)$ possesses a Gaussian shape with the following form:

$$U(f) = exp[-\frac{(f-f_c)^2}{2s^2}] \tag{8}$$

where f_c is the center frequency of the transmitted pulse, s is a parameter that describes the frequency bandwidth of the transmitted pulse. Combining Equations (7) and (8), the spectrum of a flaw echo can be estimated as:

$$E(f) = exp[-\frac{(f-f_c)^2}{2s^2}]exp[-\alpha(2\pi f)^4 vt] \tag{9}$$

3. The Proposed Methodology

The procedure of the proposed method is illustrated in Figure 1. First, the measured signal is segmented successively along the time axis at a step of one sample. Here, the segment length is set to be slightly longer than the duration of the transmitted pulse because the duration of the pulse increases during propagation due to frequency-dependent attenuation, which results in the decrease of the frequency bandwidth of the pulse. Next, a fast Fourier transform (FFT) operation is performed on the ith segmented data of which the time delay is t_i, after multiplying a Hamming window function, to extract its spectrum P_i. At the same time, the reference spectrum T_i is obtained by substituting t_i into Equation (9). The similarity between P_i and T_i is calculated, yielding the SSI value S_i. For each data segment, based on its S_i, a time-frequency filter can be designed employing the Tukey window. The filter is adaptive because, for each data segment, one filter is specifically tailored whose width and shape are determined by the spectral characteristics of the data segment. The spectrum P_i is then fed to the filter to produce the spectrum of the de-noised data segment. After inverse FFT operation and recombination, the final de-noised signal can be obtained. The calculation of SSI and the designing of the filter for each data segment are elaborated as follows.

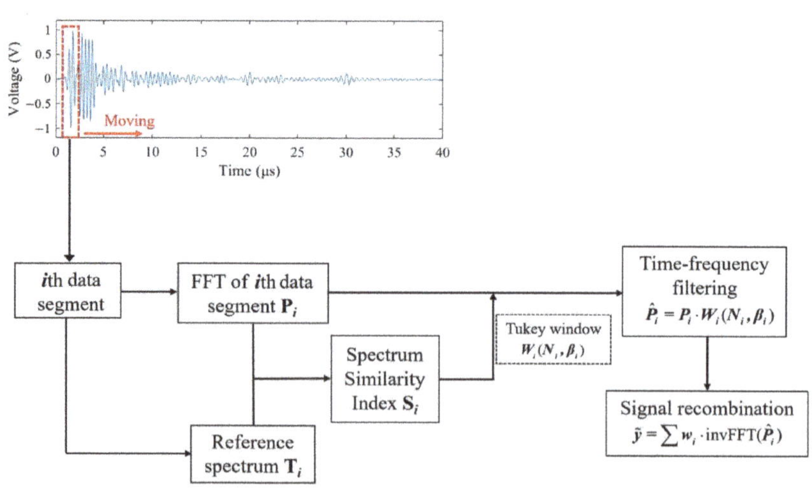

Figure 1. Illustration of the proposed method.

3.1. Calculation of Spectral Similarity Index

To evaluate the degree of similarity between P_i and T_i, an appropriate similarity measure is designed first. The correlation coefficient is one of the most used similarity measures. For the two vectors P_i and T_i, the correlation coefficient is calculated by:

$$C = \frac{P_i \times T_i}{\sqrt{\|P_i\|^2 \cdot \|T_i\|^2}} = \frac{\langle P_i, T_i \rangle}{\sqrt{\langle P_i, P_i \rangle \cdot \langle T_i, T_i \rangle}} \tag{10}$$

where $\langle T_i, P_i \rangle = {T_i}^T \cdot P_i$ is the inner product of P_i and T_i. It is worthwhile noting that the calculation of the correlation coefficient is a linear operation and is appropriate for data that are best described by second-order statistics. To evaluate the similarity between two spectra that have complex patterns, second-order statistics may not be sufficient. As reported in [28], by mapping the two variables into high-dimensional space before calculating the correlation coefficient, high-order statistics of the variables can be exploited, thus better performance can be achieved. In this study, a nonlinear technique employing a kernel function is used in the evaluation of similarity between P_i and T_i. First, both P_i and T_i are nonlinearly mapped into a potentially higher dimensional space, denoted as $\widetilde{P_i} = \Phi(P_i)$ and $\widetilde{T_i} = \Phi(T_i)$, respectively. With the help of a kernel function, one can compute the inner product in the high dimensional space without explicitly computing the nonlinear mapping Φ.

A kernel function κ computes the inner product of mapped P_i and T_i as follows:

$$\kappa(P_i, T_i) = \langle \Phi(P_i), \Phi(T_i) \rangle \tag{11}$$

Substituting normalized variables into Equation (11), one has

$$\widetilde{\kappa}(P_i, T_i) = \left\langle \frac{\Phi(P_i)}{\|\Phi(P_i)\|}, \frac{\Phi(T_i)}{\|\Phi(T_i)\|} \right\rangle = \frac{\kappa(P_i, T_i)}{\sqrt{\kappa(P_i, P_i) \cdot \kappa(T_i, T_i)}} \tag{12}$$

From Equations (10)–(12), it is easily noticed that Equation (12) is equivalent to computing the correlation coefficient between $\Phi(P_i)$ and $\Phi(T_i)$. In this study, a Gaussian kernel function is employed which has the following form:

$$\kappa(P_i, T_i) = exp(-\frac{\|P_i - T_i\|^2}{2\sigma^2}) \tag{13}$$

where $\sigma > 0$ is a parameter that controls the flexibility of the kernel. Note that the denominator in Equation (13) equals 1. Therefore, the SSI between P_i and T_i can be calculated by

$$S_i = \widetilde{\kappa}(P_i, T_i) = exp(-\frac{\|P_i - T_i\|^2}{2\sigma^2}) \tag{14}$$

Note that the calculation of SSI above combines two commonly used similarity criteria, i.e., the correlation coefficient and the sum of squares of deviations.

3.2. SSI-Ssi-Based Adaptive Time-Frequency Filtering

Essentially, the reference spectrum T_i estimates the frequency band and spectrum shape of any possible flaw echo of time delay t_i. Therefore, a time-frequency filtering scheme can be devised based on the SSI calculated above, which evaluates the similarity between the spectrum of the ith data segment and its corresponding reference spectrum. The proposed time-frequency filtering is expressed as follows:

$$\hat{P_i} = P_i \cdot W_i(N_i, \beta_i) \tag{15}$$

where $\hat{P}_i$ is the frequency spectrum of the ith data segment after the filtering process, $W_i(N_i, \beta_i)$ is the Tukey window function that has the following form [29]:

$$W(n) = \begin{cases} \frac{1}{2}[1 - \cos(\frac{2\pi n}{\beta N})], & 0 \leq n \leq \frac{\beta N}{2} \\ 1, & \frac{\beta N}{2} \leq n \leq \frac{N}{2} \\ W(N-n), & 0 \leq n \leq N \end{cases} \tag{16}$$

where N is the width of the Tukey window, β is a flexible parameter that controls the shape of the window so that at $\beta = 0$ the window becomes rectangular and at $\beta = 1$ the window becomes a Hann window. The relation between window shape and β is illustrated in Figure 2.

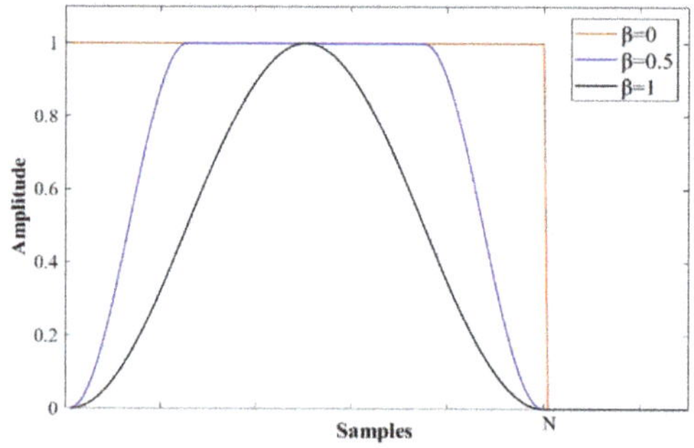

Figure 2. Shapes of Tukey window with various β values.

Because the spectrum of a flaw echo is assumed to process a Gaussian shape where the energy of the signal is distributed in a finite bandwidth centered at the center frequency, the width of the Tukey window N can be determined by the bandwidth of significant energy distribution, which is set to be the -20 dB bandwidth in this study. For example, for the data segment centered at t_i, its corresponding reference spectrum T_i is estimated using Equation (9), the width of the Tukey window N_i for this data segment is equal to the -20 dB bandwidth of the reference spectrum, and the shape parameter is determined as $\beta_i = 1 - S_i$.

From Equation (15), it can be observed that if the spectrum of a data segment P_i highly resembles its reference spectrum T_i, the SSI between the two spectra will approach 1, the Tukey window is almost rectangular and the designed filter is equivalent to a band-pass filter and has no impact on P_i. Conversely, if P_i is very different from T_i, the SSI between the two spectra will be near 0, the Tukey window is now a Hann window that modulates the spectrum P_i so that the power of signal outside the main frequency band will be reduced to zero and at the meantime keeps remained power concentrated at the predicted center frequency and decayed toward both sides. By doing this, the power of noise is significantly reduced.

After the filtering stage, inverse FFT is applied to $\hat{P}_i$, and the final de-noised signal can be obtained by the recombination process:

$$\tilde{y} = \sum w_i \cdot \text{invFFT}(\hat{P}_i) \tag{17}$$

where w_i is a weighting coefficient which is determined by SSI as follows:

$$w_i = \begin{cases} 1, & \text{if } S_i \geq thres \\ 0, & \text{if } else \end{cases} \tag{18}$$

where *thres* is the threshold value to dismiss the data segments having insignificant SSI values. In this study, $thres = 0.3max(S_i)$.

The implementation of the proposed method is summarized in Algorithm 1.

Algorithm 1. Procedure of the proposed method.

Input: noisy signal y
Output: de-noised signal $\widetilde{y}$

Step 1: Data segmentation.
Successive segmentation of the signal, such that:
(1) Segment length L is slightly larger than the length of transmitted pulse;
(2) The overlapped length of adjacent segments is $L - 1$.
Step 2: Filtering of each data segment.
For the ith segment whose time center is t_i:
(1) Calculate its frequency spectrum P_i by FFT;
(2) Obtain the reference spectrum: $T_i = exp[-\frac{(f-f_c)^2}{2s^2}]exp[-\alpha(2\pi f)^4 v t_i]$;
(3) Calculate its SSI value S_i: $S_i = \widetilde{\kappa}(P_i, T_i) = exp(-\frac{\|P_i - T_i\|^2}{2\sigma^2})$;
(4) Let $\beta = 1 - S_i$, calculate the Tukey window:

$$W(n) = \begin{cases} \frac{1}{2}[1 - \cos(\frac{2\pi n}{\beta N})], & 0 \le n \le \frac{\beta N}{2} \\ 1, & \frac{\beta N}{2} \le n \le \frac{N}{2} \\ W(N-n), & 0 \le n \le N \end{cases}$$

(5) Calculate the filtered spectrum $\hat{P}_i$ by: $\hat{P}_i = P_i \cdot W_i(N_i, \beta_i)$;
(6) Calculate the weighting coefficient w_i: $w_i = \begin{cases} 1, & \text{if } S_i \ge thres \\ 0, & \text{if } else \end{cases}$.

Step 3: Signal reconstruction.
Reconstruct the de-noised signal $\widetilde{y}$: $\widetilde{y} = \sum w_i \cdot \mathrm{invFFT}(\hat{P}_i)$.

4. Results

4.1. Simulated Signals

Considering the typical inspection scenario shown in Figure 3, simulation studies are conducted to evaluate the SNR enhancement performance of the proposed method. The ultrasonic transducer used here has a center frequency of 5 MHz and full width at half maximum (FWHM) of 2 MHz. The specimen is a stainless steel block containing two flaws located 60 mm and 72 mm below the surface, respectively. The transmitted pulse is assumed to be a Gabor pulse, which is characterized by:

$$g(t) = \frac{A}{\sqrt{2\pi}s} exp[-(t-u)^2 / \left(2s^2\right)] exp[i\omega(t-u)] \tag{19}$$

where A is an energy normalization parameter, s determines the duration of the pulse, u is the time delay of the Gabor pulse, ω is the angular frequency. The frequency spectrum and waveform of the transmitted pulse are given in Figure 4.

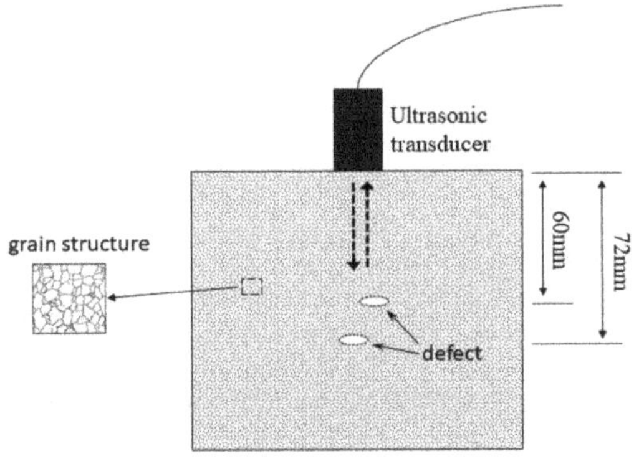

Figure 3. The ultrasonic inspection scenario for the simulation study.

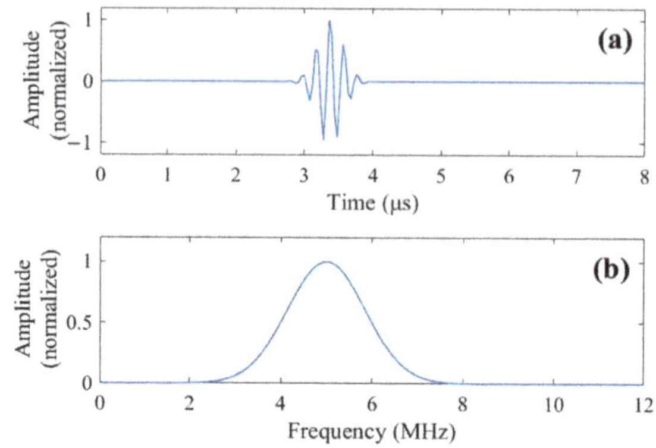

Figure 4. (**a**) Waveform of the transmitted pulse from the transducer. (**b**) Frequency spectrum of the transmitted pulse.

For the inspection scenario above, the received signal contains two flaw echoes and grain noise backscattered by the grain microstructure of stainless steel as well as white Gaussian noise from the circuits of instruments. Employing the model in [30], the frequency spectrum of grain noise can be expressed as:

$$G(\omega) = Z(\omega) \cdot \sum_{j=1}^{K_g} \gamma_j \frac{\omega^2}{d_j} exp(-2\alpha d_j \omega^4) exp(-2i\omega d_j / v) \tag{20}$$

where $Z(\omega)$ is the frequency spectrum of the transmitted pulse from the transducer, K_g is the number of grain scatterers in the ensonified area, γ_j is the scattering coefficient of the jth grain scatterer which is d_j away from the transducer, α is the frequency-dependent attenuation coefficient, v is longitudinal wave velocity in stainless steel. In this study, the parameters are chosen as follows: $K_g = 2500$, $\alpha = 1 \times 10^{-28}$, $v = 6000$ m/s. The frequency spectrum of a simulated grain noise by Equation (20) is shown in Figure 5a. Applying inverse Fourier transform to $G(\omega)$, the time domain waveform of the grain noise $g(t)$ can be generated as shown in Figure 5b. Similarly, the frequency spectrum of an echo reflected by a flaw at depth d_{flaw} can be expressed as:

$$E(\omega) = Z(\omega) \cdot exp(-2\alpha d_{flaw} \omega^4) \cdot exp(-\frac{2i\omega d_{flaw}}{v}) \tag{21}$$

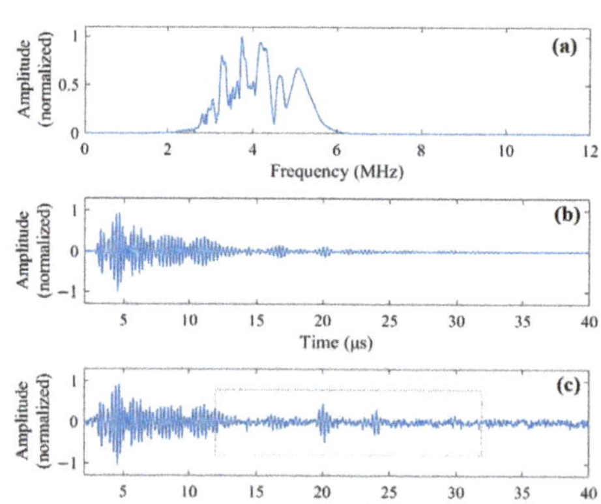

Figure 5. (**a**) Frequency spectrum of simulated grain noise. (**b**) The time-domain waveform of the simulated grain noise. (**c**) The simulated noisy signal.

The simulated noisy signal is the superposition of grain noise $g(t)$ and flaw echoes $e(t)$ plus some white Gaussian noise. In Figure 5c, the simulated noisy signal is presented.

To evaluate the SNR enhancement performance of the proposed method, normalized signal-to-noise ratio (NSNR) is introduced which is calculated as [31]:

$$\text{NSNR} = \frac{\sum_{u-l_p/2}^{u+l_p/2} y(n)^2}{\sum_1^{N_s} y(n)^2} \tag{22}$$

where u denotes the time delay of a flaw echo, l_p is the width of the flaw echo, and N_s is the length of signal $y(t)$. It can be observed that a noise-free ultrasonic signal has a NSNR = 1, and if only noise is present, NSNR approaches 0. Therefore, the enhancement of SNR can be evaluated by comparing NSNR values before and after signal de-noising.

For the highlighted signal segment in Figure 5c, which contains two flaw echoes as shown in Figure 6a, the NSNR is 0.5946. After signal processing by the proposed method, the de-noised signal is shown in Figure 6b along with the original noise-free flaw echoes. After de-noising, the NSNR increases to 0.9488, meaning that the SNR is significantly increased. Meanwhile, by comparing the de-noised signal and the original noise-free echoes, it can be seen that the waveform of the two echoes can be preserved with high fidelity. For comparison, the four-level discrete wavelet transform (DWT) using the sym25 wavelet proposed in [11] is employed to process the same noisy signal, and the de-noised signal, which has NSNR = 0.7378, is given in Figure 6c. According to [11], after an extensive study, the authors concluded that a four-level sym25 wavelet yields the best result for signal-to-noise ratio enhancement. The comparison clearly demonstrates a better performance of our method. Although the second reconstructed echo by our method has a slightly lower amplitude, the majority of noise is effectively removed. On the contrary, the de-noised signal by four-level sym25 DWT still contains notable residual noise, resulting in a lower NSNR value.

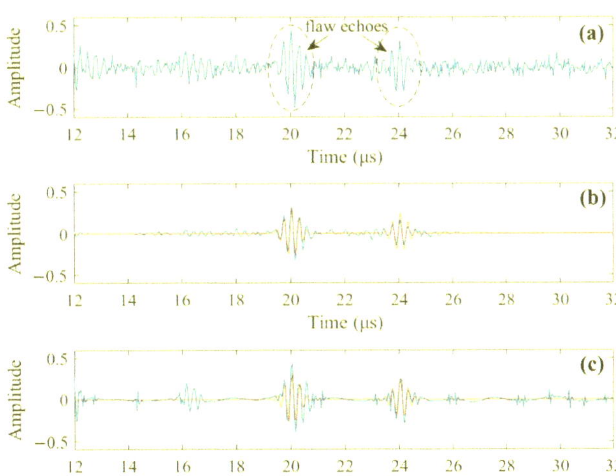

Figure 6. (**a**) The analyzed signal segment containing two flaw echoes, NSNR = 0.5946. (**b**) De-noised signal by our method along with noise-free flaw echoes, NSNR = 0.9488 (blue line: reconstructed signal; red line: original noise-free echoes). (**c**) De-noised signal by DWT (level-4 sym25) along with noise-free flaw echoes, NSNR = 0.7378 (blue line: reconstructed signal; red line: original noise-free echoes).

4.2. Experimental Signals

In this section, the proposed methodology is employed to process signals obtained from ultrasonic inspection of a stainless steel block in the laboratory to demonstrate its effectiveness. The specimen being tested is a heat-treated Type 304 stainless steel block with two fabricated flaws (side drilled holes with a diameter of 2 mm) as shown in Figure 7a. The dimensions of the specimen are 150 mm × 82 mm × 59 mm, and the two holes are about 31 mm and 51 mm below the top surface, respectively. Heat treatment is conducted at a temperature of 1250 °C for 8 h, followed by quenching in a water bath. Empirically, grain sizes around $\overline{D} \approx 120$ μm will be formed. The ultrasonic DNT system is shown in Figure 7b. The transducer used is type 5P20 made by Shantou Institute of Ultrasonic Instruments Co., Ltd. (Shantou, China), with a center frequency of 5 MHz. The ultrasonic pulse transmitted by the transducer is given in Figure 8a and the frequency spectra of the transmitted pulse is shown in Figure 8b. The ultrasonic pulse/receiver, Type 5072PR, is from Olympus Co., Ltd.(Tokyo, Japan). A-scan signals are obtained in pulse-echo mode and sampled at 50 MHz using DPO2024 oscilloscope produced by Tektronix, Inc. (Beaverton, OR, USA).

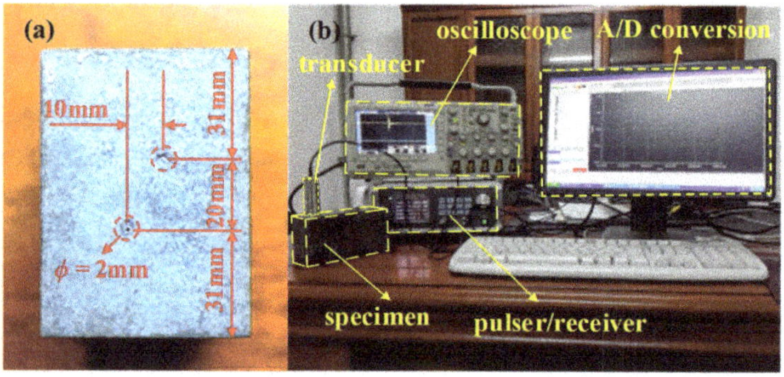

Figure 7. (**a**) A stainless steel specimen with two flaws. (**b**) The setup of the experiment system.

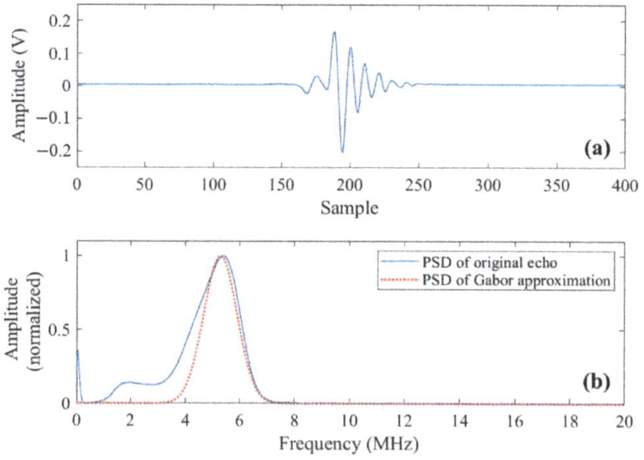

Figure 8. (**a**) The measured pulse transmitted by the transducer. (**b**) Frequency spectrum (F.S.) of the transmitted pulse and its Gaussian fitting.

Longitudinal sound velocity in the block is measured first on the time-of-flight principle. The thickness of the specimen is measured ten times using a Vernier caliper, and the average thickness of the specimen is about 81.4 mm. The round-trip travel time of the ultrasonic pulse reflected by the bottom of the specimen is 28.4 μs; therefore, the longitudinal sound velocity equals $v = 5735$ m/s. Now one can notice that the wavelength of the ultrasonic pulse is $\lambda \approx 1$ mm, which means $\lambda/\overline{D} \gg 1$, confirming that the scattering is in the Rayleigh region.

In Figure 9a, an experimental signal obtained when the transducer is positioned between the two flaws so the received signal contains two flaw echoes is presented. The signal segment to be analyzed is highlighted in Figure 9a and shown in Figure 9b, which has an NSNR = 0.2864. By using the proposed method, the de-noised signal is shown in Figure 9c. Notable SNR enhancement is observed, as its NSNR now increases to 0.9596. From the de-noised signal, the time delays of the two flaw echoes are easily identified, which are 11.28 µs and 18.35 µs, respectively. Based on the time-of-flight principle, the depths of the flaws are obtained and compared with their actual values, as tabulated in Table 1. The detection results reveal good detection accuracy. For comparison, the DWT approach is implemented to process the same noisy signal, revealing NSNR = 0.2685. The low value of NSNR is due to the inability of the DWT approach to remove the strong grain noise on the left side. The de-noised signal by the DWT approach (4-level sym25) is shown in Figure 9d. Although the two flaw echoes are successfully revealed and the yielded time delays are in line with our method, the waveform of the two flaw echoes in the de-noised signal is not sound, because theoretically, the duration of the second flaw echo will be longer than the first one due to frequency-dependent attenuation.

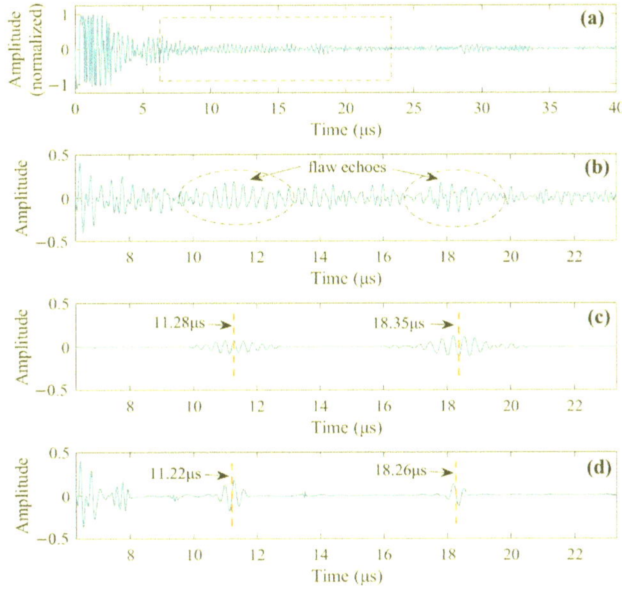

Figure 9. (**a**) A noisy signal measured from the experiment. (**b**) The signal to be analyzed containing two flaw echoes, NSNR = 0.2864. (**c**) The signal de-noised by our method, NSNR = 0.9596. (**d**) The signal de-noised by DWT (4-level sym25), NSNR = 0.2685.

Table 1. Flaw detection results using the proposed method.

	Upper Flaw	Lower Flaw
Detected depth (mm)	32.3	52.6
Actual depth (mm)	31.0	51.0
Absolute error (mm)	1.3	1.6

5. Conclusions

In this work, a novel time-frequency filtering approach is proposed to enhance the ultrasonic inspection performance of stainless steel structures by suppressing noise induced by the grain microstructure of stainless steel; thus, increasing the SNR of the measured signal. Spectral modeling of a flaw echo and grain noise shows that, in the same time frame, the spectrum of a flaw echo and that of grain noise exhibits different characteristics which can be exploited to suppress the energy of noise. As such, the spectral similarity index is introduced, which is used to determine the shape of the Tukey window in the filtering stage and to determine the weighting coefficients in the signal recombination stage. This approach has been validated both in the processing of simulated signals and signals measured from experiments. In both cases, the enhancement of SNR is significant. The high fidelity of the de-noised signal enables accurate flaw depth detection, showing good performance of the proposed method. Moreover, the method is also effective for scenarios where multiple flaws exist.

Author Contributions: Conceptualization, B.W. and X.L.; methodology, B.W., W.Z. and X.L.; investigation, B.W., H.Y., Y.H. and X.L.; writing—original draft preparation, B.W., H.Y., Y.H. and X.L.; writing—review and editing, B.W., W.Z. and X.L. All authors have read and agreed to the published version of the manuscript.

Funding: This work was supported by the National Natural Science Foundation of China (grant number 52008205), the Natural Science Foundation for College and University in Jiangsu Province (grant number 20KJB560013) and Key Lab of Structural Dynamic Behavior and Control of the Ministry of Education (grant number HITCE202002).

Institutional Review Board Statement: Not applicable.

Informed Consent Statement: Not applicable.

Data Availability Statement: Not applicable.

Conflicts of Interest: The authors declare no conflict of interest.

References

1. Dai, P.; Yang, L.; Wang, J.; Lin, M.; Fan, J. Bond stress-slip relationship in concrete-filled square stainless steel tubes. *Constr. Build. Mater.* **2022**, *326*, 127001. [CrossRef]
2. Hasan, M.A.; Akiyama, M.; Kashiwagi, K.; Kojima, K.; Peng, L. Flexural behaviour of reinforced concrete beams repaired using a hybrid scheme with stainless steel rebars and CFRP sheets. *Constr. Build. Mater.* **2020**, *265*, 120296. [CrossRef]
3. Remillieux, M.C.; Kaoumi, D.; Ohara, Y.; Geesey, M.A.S.; Xi, L.; Schoell, R.; Bryan, C.R.; Enos, D.G.; Summa, D.A.; Ulrich, T.; et al. Detecting and imaging stress corrosion cracking in stainless steel, with application to inspecting storage canisters for spent nuclear fuel. *NDT E Int.* **2020**, *109*, 102180. [CrossRef]
4. Joseph, M.J.; Jabbar, M.A. Effect of aging process on the microstructure, corrosion resistance and mechanical properties of stainless steel AISI 204. *Case Stud. Constr. Mater.* **2019**, *11*, e00253. [CrossRef]
5. Kain, R.M. Marine atmospheric stress corrosion cracking of austenitic stainless steels. *Mater. Perform.* **1990**, *29*, 60–62.
6. Hwang, Y.-I.; Seo, M.-K.; Oh, H.G.; Choi, N.; Kim, G.; Kim, K.-B. Detection and Classification of Artificial Defects on Stainless Steel Plate for a Liquefied Hydrogen Storage Vessel Using Short-Time Fourier Transform of Ultrasonic Guided Waves and Linear Discriminant Analysis. *Appl. Sci.* **2022**, *12*, 6502. [CrossRef]
7. Malmström, M.; Jansson, A.; Hutchinson, B. Application of Laser-Ultrasonics for Evaluating Textures and Anisotropy. *Appl. Sci.* **2022**, *12*, 10547. [CrossRef]
8. Coro, A.; Macareno, L.M.; Aguirrebeitia, J.; López de Lacalle, L.N. A Methodology to Evaluate the Reliability Impact of the Replacement of Welded Components by Additive Manufacturing Spare Parts. *Metals* **2019**, *9*, 932. [CrossRef]
9. Pereira, J.C.; Zubiri, F.; Garmendia, M.J.; Tena, M.; Gonzalez, H.; de Lacalle, L.N.L. Study of laser metal deposition additive manufacturing, CNC milling, and NDT ultrasonic inspection of IN718 alloy preforms. *Int. J. Adv. Manuf. Technol.* **2022**, *120*, 2385–2406. [CrossRef]
10. Bettayeb, F.; Haciane, S.; Aoudia, S. Improving the time resolution and signal noise ratio of ultrasonic testing of welds by the wavelet packet. *NDT E Int.* **2005**, *5*, 478–484. [CrossRef]
11. Manjula, K.; Vijayarekha, K.; Venkatramen, B. Quality Enhancement of Ultrasonic TOFD Signals from Carbon Steel Weld Pad with Notches. *Ultrasonics* **2018**, *84*, 264–271. [CrossRef]
12. Praveen, A.; Vijayarekha, K.; Abraham, S.T.; Venkatraman, B. Signal quality enhancement using higher order wavelets for ultrasonic TOFD signals from austenitic stainless steel welds. *Ultrasonics* **2013**, *53*, 1288–1292. [CrossRef] [PubMed]
13. Matz, V.; Smid, R.; Starman, S.; Kreidl, M. Signal-to-noise ratio enhancement based on wavelet filtering in ultrasonic testing. *Ultrasonics* **2009**, *49*, 752–759. [CrossRef] [PubMed]
14. Rodriguez, A.; Miralles, R.; Bosch, I.; Vergara, L. New analysis and extensions of split-spectrum processing algorithms. *NDT E Int.* **2012**, *45*, 141–147. [CrossRef]

15. Bosch, I.; Vergara, L. Normalized split-spectrum: A detection approach. *Ultrasonics* **2008**, *48*, 56–65. [CrossRef]
16. Bouden, T.; Djerfi, F.; Nibouche, M. Adaptive split spectrum processing for ultrasonic signal in the pulse echo test. *Russ. J. Nondestr. Test.* **2015**, *51*, 245–257. [CrossRef]
17. Boßmann, F.; Plonka, G.; Peter, T.; Nemitz, O.; Schmitte, T. Sparse Deconvolution Methods for Ultrasonic NDT. *J. Nondestruct. Eval.* **2012**, *31*, 225–244. [CrossRef]
18. Abdessalem, B.; Farid, C. Resolution Improvement of Ultrasonic Signals Using Sparse Deconvolution and Variational Mode Decomposition Algorithms. *Russ. J. Nondestr. Test.* **2020**, *56*, 479–489. [CrossRef]
19. Sun, X.; Lin, L.; Jin, S.J. Improving Time Resolution of Ultrasonic Signals with Frequency-Domain Sparse Blind Deconvolution (FSBD) Method. *J. Nondestruct. Eval.* **2022**, *41*, 37. [CrossRef]
20. Zhang, G.M.; Zhang, C.Z.; Harvey, D.M. Sparse signal representation and its applications in ultrasonic NDE. *Ultrasonics* **2012**, *52*, 351–363. [CrossRef]
21. Wu, B.; Huang, Y.; Krishnaswamy, S. A Bayesian approach for sparse flaw detection from noisy signals for ultrasonic NDT. *NDT E Int.* **2017**, *85*, 76–85. [CrossRef]
22. Mor, E.; Aladjem, M.; Azoulay, A. A Sparse Approximation Method for Ultrasonic Monitoring the Degradation of Adhesive Joints. *NDT E Int.* **2018**, *98*, 17–26. [CrossRef]
23. Papadakis, E.P. Ultrasonic Attenuation Caused by cattering in Polycrystalline Metals. *J. Acoust. Soc. Am.* **1965**, *37*, 711–717. [CrossRef]
24. Saniie, J.; Jin, X.M. Spectral analysis for ultrasonic nondestructive evaluation applications using autoregressive, Prony, and multiple signal classification methods. *J. Acoust. Soc. Am.* **1996**, *100*, 3165–3171. [CrossRef]
25. Goebbels, K.; Hirsekorn, S.; Willems, H. The use of ultrasound in the determination of microstructure: A review. In Proceedings of the IEEE 1984 Ultrasonics Symposium, Dallas, TX, USA, 14–16 November 1984; pp. 841–846.
26. Saniie, J.; Wang, T.; Bilgutay, N.M. Analysis of homomorphic processing for ultrasonic grain signal characterizations. *IEEE Trans. Ultrason. Ferroelectr. Freq. Contr.* **1989**, *36*, 365–375. [CrossRef]
27. Wu, B.; Huang, Y. Flaw Detection in Highly Scattering Materials Using a Simple Ultrasonic Sensor Employing Adaptive Template Matching. *Sensors* **2022**, *22*, 268. [CrossRef]
28. Liu, R.M.; Lu, Y.H.; Gong, C.L.; Liu, Y. Infrared point target detection with improved template matching. *Infrared. Phys. Techn.* **2012**, *55*, 380–387. [CrossRef]
29. Bloomfield, P. *Fourier Analysis of Time Series: An Introduction*; Wiley-Interscience: New York, NY, USA, 2000.
30. Gustafsson, M.G.; Stepinski, T. Studies of split spectrum processing, optimal detection, and maximum likelihood amplitude estimation using a simple clutter model. *Ultrasonics* **1997**, *35*, 31–52. [CrossRef]
31. Izquierdo, M.A.G.; Hernández, M.G.; Graullera, O.; Ullate, L. Time–frequency Wiener filtering for structural noise reduction. *Ultrasonics* **2002**, *40*, 259–261. [CrossRef]

sensors

MDPI

Article

Discussion on a Vehicle–Bridge Interaction System Identification in a Field Test

Ryota Shin [1], Yukihiko Okada [2,3] and Kyosuke Yamamoto [2,3,*]

1 Graduate School of Systems and Information Engineering, University of Tsukuba, 1-1-1 Tennodai, Tsukuba 305-8573, Ibaraki, Japan
2 Institute of Systems and Information Engineering, University of Tsukuba, 1-1-1 Tennodai, Tsukuba 305-8573, Ibaraki, Japan
3 Center for Artificial Intelligence Research, University of Tsukuba, 1-1-1 Tennodai, Tsukuba 305-8577, Ibaraki, Japan
* Correspondence: yamamoto_k@kz.tsukuba.ac.jp; Tel.: +81-29-853-5146

Abstract: For infrastructures to be sustainable, it is essential to improve maintenance and management efficiency. Vibration-based monitoring methods are being investigated to improve the efficiency of infrastructure maintenance and management. In this paper, signals from acceleration sensors attached to vehicles traveling on bridges are processed. Methods have been proposed to individually estimate the modal parameters of bridges and road unevenness from vehicle vibrations. This study proposes a method to simultaneously estimate the mechanical parameters of the vehicle, bridge, and road unevenness with only a few constraints. Numerical validation examined the effect of introducing the Kalman filter on the accuracy of estimating the mechanical parameters of vehicles and bridges. In field tests, vehicle vibration, bridge vibration, and road unevenness were measured and verified, respectively. The road surface irregularities estimated by the proposed method were compared with the measured values, which were somewhat smaller than the measured values. Future studies are needed to improve the efficiency of vehicle vibration preprocessing and optimization methods and to establish a methodology for evaluating accuracy.

Keywords: drive-by bridge monitoring; vehicle–bridge interaction; system identification; field test

Citation: Shin, R.; Okada, Y.; Yamamoto, K. Discussion on a Vehicle–Bridge Interaction System Identification in a Field Test. *Sensors* **2023**, *23*, 539. https://doi.org/10.3390/s23010539

Academic Editor: Zenghua Liu

Received: 27 October 2022
Revised: 25 December 2022
Accepted: 28 December 2022
Published: 3 January 2023

1. Introduction

Transportation and logistics are supported by civil structures. Civil structures deteriorate due to daily use and natural disasters. If the deterioration of infrastructures is left unattended, it will lead to structural damage. Structural damage sometimes causes serious accidents, with economic consequences and loss of life. Therefore, infrastructure development is essential, and structure health monitoring (SHM) is attracting attention in many countries [1–3]. Infrastructures are maintained and managed through detailed inspections, but inspection costs are high. Therefore, methods focusing on vibration have been proposed as low-cost bridge inspection methods. The suggested methods can be divided into two categories. One is direct monitoring, in which sensors are installed directly on the bridges. This method uses multiple sensors on each bridge to measure vibration data. Based on the measured data, the modal parameters of the bridge are estimated, and the bridge's state is monitored. However, this method requires the installation of multiple sensors on each bridge, which is costly.

Indirect monitoring has been proposed as an alternative to costly direct monitoring methods [4]. Indirect monitoring uses mounting sensors on vehicles; the data are measured when vehicles pass over a bridge. The vehicle acts as a bridge exciter and vibration receiver in this method. When a vehicle passes over a bridge, it is shaken by the vehicle. At the same time, the vehicle is also shaken by the bridge. This interaction is called vehicle–bridge interaction (VBI). The vehicle vibration includes the response due to this interaction. Therefore, it is possible to estimate the structural parameters of the bridge from the measured vehicle vibration. Once the structural parameters of the bridge are estimated, priorities for bridge inspection can be determined. Indirect monitoring can be implemented quickly and inexpensively because it is sufficient to mount sensors only on the vehicle. A conceptual diagram of the efficiency of detailed inspections with drive-by monitors is shown in Figure 1.

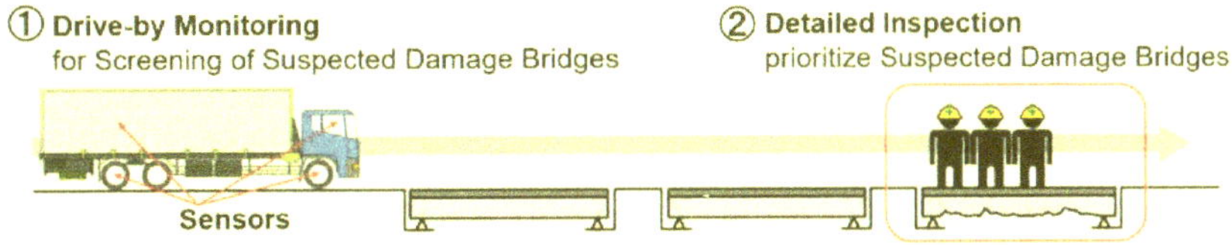

Figure 1. Conceptual diagram of drive-by monitoring.

Existing drive-by-bridge monitoring techniques are affected by road surface irregularities, limited VBI time, and the environment [5]. Therefore, crowdsourcing information collection and the integrated analysis of multiple data runs are considered [6]. Integrated data analysis can eliminate influences unrelated to bridge damage, such as measurements and factors that may affect the behavior of bridges. However, existing methods separately estimate the mass, damping, and stiffness of bridges and road unevenness. Therefore, an analytical method capable of integrated analysis is needed to detect bridge damage from vehicle vibration.

The authors proposed the VBI system identification (VBISI) method [7–9]. The VBISI method can simultaneously estimate vehicles' and bridges' mechanical parameters and road unevenness from vehicle vibration and location information, under the condition that both the wheelbase and total weight of the vehicle are known. The VBISI method is inspired by technology that simultaneously estimates the vehicles' mechanical parameters and road unevenness [10,11]. The numerical simulation of bridge vibration is combined with methods proposed in previous studies. The numerical validation of the VBISI method [7–9] has already shown the feasibility of this method. However, it is not verified in a field test.

This study aims to identify issues based on the results of field tests of the VBISI method. A Kalman filter is used to mitigate the effects of measurement noise in vehicle vibration.

The contribution of this research is as follows.

- The Kalman filter can be used to estimate road unevenness, but estimating mechanical parameters is challenging.
- The applicability of the VBISI method was verified using experimental data from a 14-t truck and an actual concrete bridge with a 30 m span.
- Through the VBISI method in the field study, a vehicle response analysis problem was proposed.

This paper is organized as follows: Section 2 introduces the related techniques of the VBISI method. Section 3 presents the identification method for the coupled vehicle–bridge system. Section 4 describes the basic mathematics of vehicle–bridge interaction system identification. Next, Section 5 verifies the effectiveness of applying the findings of previous studies to this technology through numerical simulations. Section 6 analyzes the field test data, and Section 7 discusses this study's limitations and future challenges. Section 8 summarizes the conclusions of this paper.

2. Related Work

This section reviews previous studies. Fortunately, there is an extensive review article [5,6,12–15]. Malekjafarian et al. [5] provided the first comprehensive review of bridge monitoring using vehicle response. Yang et al. [12] summarized methods for estimating the bridge mode shapes and detecting damage from vehicle vibration. Sholravi, H. et al. [13] outlined the broad framework of vehicle-assisted monitoring, which considers the vehicle classification-based SHM from bridge vibration. Hou et al. [14] comprehensively summarized SHM for bridges from 2010 to 2019 and touched on the challenges of damage detection by drive-by monitoring. Based on the previous review paper [5], Malekjafarian et al. [6] discussed drive-by monitoring by dividing it into individual and multiple passes.

2.1. Estimating the Bridge Modal Parameter

The estimation of the bridge's modal parameters is a central topic in indirect monitoring. Modal parameters of bridges are mainly natural frequency, damping, and mode shape. Since indirect monitoring was proposed, identifying bridge frequencies [16–27] from vehicle vibrations has been an exciting topic. Yang et al. [4] proposed the first method to estimate the natural frequencies of bridges from vehicle vibrations. After that, Lin and Yang [16] showed applicability in a field test, and Yang and Lin [17] also investigated the estimation of higher-order frequencies. Jian et al. [25] focused on coupled vibration in a 3D model of the vehicle–bridge interaction system. By taking the difference in the acceleration of the front and rear wheels in the frequency domain, it is possible to suppress the influence of road unevenness and identify the bridge's natural frequency. In addition, estimations of mode shapes [28–36] and damping [37–44] were also performed. Yang et al. [35] proposed a method for estimating the bridge mode shapes from the acceleration responses on towed vehicles. Next, the bridge stiffness is calculated from the estimated bridge mode shape. From the estimated bridge stiffness, the bridge deflection was calculated using the finite element method and compared to the measured values [36]. Other methods have been proposed that use machine learning to identify bridge damage from multiple data [45] and estimate bridge vibration from vehicles traveling at high speeds [46]. Yamamoto and Takahashi [47] proposed a damage index that can detect minor damage, such as bolt dropout. Shin et al. [48] proposed a model to discriminate whether the vehicle vibration data were obtained from driving on a bridge or not to implement VBI technology in society.

2.2. Estimating the Road Unevenness

Road unevenness also shakes vehicles. If the road pavement is rough, the luggage will deteriorate due to the shaking of the vehicle, and the passenger's satisfaction will be low. Road roughness affects logistics efficiency, so the World Bank adopted it as an investment decision index. Therefore, it is crucial to manage pavement conditions. Road unevenness reduces the accuracy of estimating bridge modal parameters and damage states from vehicle vibration. Therefore, estimating road unevenness from vehicle vibration is essential. McGetrick et al. [49] proposed a method to estimate road unevenness by calculating dynamic vehicle forces from vehicle vibrations. A study of bridge span lengths and vehicle speeds was conducted to verify the method's robustness. He and Yang [50] proposed a method for estimating road unevenness on a bridge from a single vehicle using a Kalman filter. The proposed method showed robust results against VBI, vehicle speed, noise contained in vehicle vibration, and bridge damping. Yang et al. [51] estimated the displacement input to the vehicle system (input profile in this study) using the Kalman filter from the measured vehicle vibration. The input profile is the sum of bridge vibration and road unevenness at the axle position. The bridge vibration component is obtained from the difference between the input profile's front and rear axle positions. Road unevenness can be estimated by subtracting the bridge vibration from the input profile. Hasegawa et al. [52] proposed a road unevenness estimation method using regularized least squares minimization by dynamic programming. Compared to the Kalman filter, it is helpful to use fewer hyperparameters for optimization. The methods described above aim at estimating road unevenness from vehicle vibration.

2.3. Estimating the Mechanical Parameters and the Road Unevenness

On the other hand, methods for estimating mechanical vehicle parameters simultaneously with estimating road unevenness have been proposed. Keenahan et al. [53] proposed a method for estimating road unevenness from the vibrations of multiple vehicles traveling on the same route. Their research is interesting because it proposes a method that can integrally process multiple vehicles and driving data. The proposed method was verified by numerical simulations and can also estimate vehicle parameters. Previous research [10,11] estimated vehicles' mechanical parameters and road unevenness using a Kalman filter from vehicle vibration data measured by smartphones. The smartphone is installed in only one vehicle, and in field tests, road unevenness has been estimated with high accuracy. Techniques for simultaneously identifying road unevenness and vehicles' mechanical parameters from vehicle vibrations are susceptible to noise. Therefore, it is necessary to reduce the influence of noise using a Kalman filter or the like. The noise reduction method means it is possible to estimate the flatness of a road surface even when the vehicle vibration contains high noise [10,11]. However, when the Kalman filter is used, the accuracy of estimating the mechanical parameters of a vehicle decreases. In particular, the VBISI method can simultaneously estimate a bridge's mass, damping, and stiffness parameters. However, papers have yet to investigate the relationship between the Kalman filter and the accuracy of estimating the mechanical parameters of vehicles and bridges. In the context of comparison with the VBISI method, the methods for estimating road surface irregularities from vehicle vibrations are summarized in Table 1. The first three columns in Table 1 depict the targets estimated by the methods proposed in each paper. They also distinguish whether the model includes bridges, and whether the estimation process incorporates the Kalman filter. If the relevant condition is satisfied, it is indicated by a symbol ○.

Table 1. Comparison with previous studies.

	Estimated Target			Including Bridges	Kalman Filter
	Road	Vehicle	Bridge		
McGetrick et al. [49]	○			○	
He and Yang [50]	○			○	○
Yang et al. [51]	○			○	○
Hasegawa et al. [52]	○				
Xue et al. [10] Nagayama et al. [11]	○	○			○
Keenahan et al. [53]	○	○	○	○	
VBISI method [7–9]	○	○	○	○	○ [7,9]

3. Preliminaries

Before explaining the VBISI method, this section mentions what a VBI system is.

3.1. Vehicle–Bridge Interaction System

VBI systems consist of bridge and vehicle systems. The bridge system takes the contact-point force of a vehicle as an input and returns bridge vibration as an output. Bridge vibration as an output is the response at a fixed point of a bridge. Now, vibrations at fixed points are converted into vibrations at moving points using interpolation. In this paper, the bridge vibration at a moving point is called the bridge profile. On the other hand, the vehicle system receives the input profile and returns the vehicle vibration as an output. The input profile is the sum of the road profile and the bridge profile. Here, the road profile represents road unevenness at the moving point. The vehicle's contact-point force is obtained from the output vehicle vibration, which is used as the input of the bridge system. In summary, the vehicle and bridge systems have mutual input and output relationships. The VBI system is nonlinear because of repeated interactions as vehicles travel over bridges.

3.2. Vehicle–Bridge Interaction System Identification Method

Next, an overview of the VBISI method to be verified in this study is provided. The VBISI method first assumes random vehicle and bridge mechanical parameters. The VBISI method is divided into two processes. The first is the vehicle system's IEP (input estimation problem) from the vehicle vibration. In this research, the input profile is estimated using the Kalman filter from the equation of motion of the vehicle. The second is the bridge system's DRS (dynamic response simulation). The input is the contact-point force calculated from the measured vehicle responses. In DRS, the dynamic response of the bridge is calculated by numerical schemes. The bridge profile is obtained from the obtained bridge vibration. The road profile is given by subtracting the bridge profile from the input profile. The obtained road profile can also be converted back to road unevenness. Assuming the vehicle's pathway is straight, the front and rear wheels run on the same road. However, because the mechanical parameters of the vehicle and bridge are assumed randomly, the estimated road unevenness usually does not match. The objective function is the residual of the road surface roughness estimated for the front and rear wheels. The dynamic parameters of the vehicle and bridge are estimated by solving an optimization problem to minimize this objective function.

4. Methodology

The VBISI method verified in this study simultaneously estimates road unevenness and all mechanical parameters (mass, damping, stiffness) both of vehicle and bridge only from the position and vibration data of a traveling vehicle. The formulas of the VBISI method have already been published [7–9].

4.1. Overview of Vehicle–Bridge Interaction

Figure 2 shows a conceptual diagram of the VBI system used in this study. A half-car model is adopted for the vehicle, and the bridge is a simple one-dimensional beam model that considers only bending. The half-car model considers four independent degrees of freedom: the translation and rotation of the car body and the translation of the half-car axle. As the VBI system is a non-linear system and the input/output of the vehicle is the output/input of the bridge, convergence calculation is required to reproduce the vehicle vibration. First, input displacement is given to the vehicle model to obtain the vehicle vibration, and the contact-point force calculated from the vehicle vibration is input to the bridge model to obtain the bridge vibration. The input displacement of the vehicle is updated from the obtained bridge vibration, and the vehicle vibration is calculated again. This process is repeated until the vehicle vibration converges.

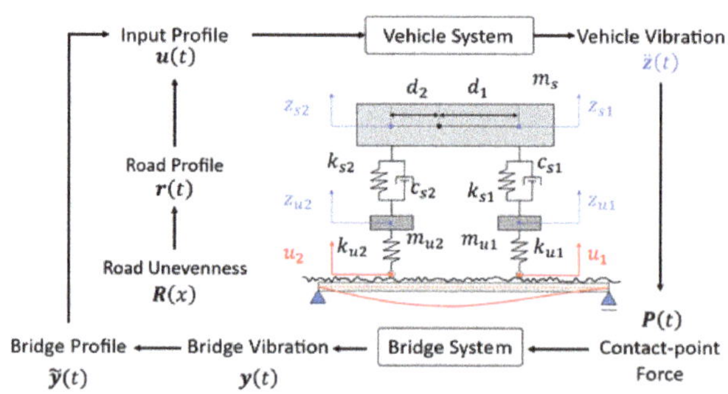

Figure 2. Conceptual diagram of vehicle–bridge interaction [9].

4.2. Vehicle

For the half-car model shown in Figure 2, m_s is the mass of the vehicle body and c_{si}, k_{si}, d_i, m_{ui}, and k_{ui} represent the suspension damping, the suspension stiffness, the distance from the gravity point, the unspring-mass, and the tire stiffness of the i-th axle, respectively. Let z_{si} be the vertical displacement of sprung-mass vibration at the axle, z_{si} be the unsprung mass, u_i be the input profile at the axle. The sprung mass models the vehicle body, and the unsprung mass models the tires and axles. The subscript i corresponds to the axle, with

1 being the front wheel and 2 being the rear wheel. The equation of motion of the vehicle is expressed as [7]

$$\mathbf{M}_v \ddot{z}(t) + \mathbf{C}_v \dot{z}(t) + \mathbf{K}_v z(t) = \boldsymbol{F}_v \tag{1}$$

where t represents the time. $\ddot{z}(t)$ and $\dot{z}(t)$ represent the second and first derivatives of $z(t)$, which are the velocity vibration and the acceleration vibration. $z(t)$ and $\boldsymbol{F}_v(t)$ can be represented as

$$z(t) = \begin{Bmatrix} z_{s1}(t) \\ z_{s2}(t) \\ z_{u1}(t) \\ z_{u2}(t) \end{Bmatrix} \tag{2}$$

$$\boldsymbol{F}_v(t) = \begin{Bmatrix} 0 \\ 0 \\ k_{u1}u_1(t) \\ k_{u2}u_2(t) \end{Bmatrix} \tag{3}$$

$$\mathbf{M}_v = \begin{bmatrix} \frac{d_2 m_s}{d_1+d_2} & \frac{d_1 m_s}{d_1+d_2} & & \\ \frac{I}{d_1+d_2} & -\frac{I}{d_1+d_2} & & \\ & & m_{u1} & \\ & & & m_{u2} \end{bmatrix} \tag{4}$$

$$\mathbf{C}_v = \begin{bmatrix} c_{s1} & c_{s2} & -c_{s1} & -c_{s2} \\ d_1 c_{s1} & -d_2 c_{s2} & -d_1 c_{s1} & d_2 c_{s2} \\ -c_{s1} & 0 & c_{s1} & 0 \\ 0 & -c_{s2} & 0 & c_{s2} \end{bmatrix} \tag{5}$$

$$\mathbf{K}_v = \begin{bmatrix} k_{s1} & k_{s2} & -k_{s1} & -k_{s2} \\ d_1 k_{s1} & -d_2 k_{s2} & -d_1 k_{s1} & d_2 k_{s2} \\ -k_{s1} & 0 & k_{s1} + k_{u1} & 0 \\ 0 & -k_{s2} & 0 & k_{s2} + k_{u2} \end{bmatrix} \tag{6}$$

$\mathbf{M}_v$, $\mathbf{C}_v$, and $\mathbf{K}_v$ are the mass, damping, and stiffness matrices of the vehicle. If the center of rotation coincides with the gravity point, it is known that

$$I = m_s d_1 d_2 \tag{7}$$

4.3. Bridge

Let flexural rigidity and mass per unit length of bridge be $EI(x)$ and $\rho A(x)$, respectively; the equation of motion of the bridge system can be expressed as

$$\rho A \ddot{y}(x,t) + \frac{\partial^2}{\partial x^2} EI\left(\frac{\partial^2}{\partial x^2} y(x,t)\right) = p(x,\ t) \tag{8}$$

where $y(x,t)$ denotes the deflection and x represents the position. The external force p consists of the contact-point force $P_i(t)$ of the vehicle and the reaction forces R_A and R_B at both supports. $x = 0$ indicates the bridge entrance, and $x = L$ indicates the exit. The bridge span length is L. Let the positions of the fulcrums be also $x_A = 0$ and $x_B = L$, respectively, and the position of the i-th wheel be $x_i(t)$. The function $\delta(x)$ represents Dirac's Delta function. The external force p is expressed as follows [7].

$$p = \sum_{i=1}^{2} \delta(x - x_i) P_i(t) + \delta(x) R_A + \delta(x - L) R_B \tag{9}$$

This study applies the finite element method to solve Equation (8) numerically. The finite element formulation is derived by the WRM (weighted residual method). The weighted residual formula in Equation (8) is

$$\int_0^L \omega\left(\rho A\frac{\partial^2 y}{\partial t^2} + EI\frac{\partial^4 y}{\partial x^4} - p\right)\mathrm{d}x = 0 \tag{10}$$

Let ω be the weight. The weak form of Equation (10) is given by Equation (11).

$$\int_0^L \left(\rho A\omega\frac{\partial^2 y}{\partial t^2} + EI\frac{\partial^2 \omega}{\partial x^2}\frac{\partial^2 y}{\partial x^2} - p\right)\mathrm{d}x = 0 \tag{11}$$

A one-dimensional finite element model discretizes the bridge vibration $y(x,t)$ with a Hermite basis.

$$\begin{cases} \phi_1(s) = \frac{1}{4}(s-1)(s-1)(s+2) \\ \phi_2(s) = \frac{\Delta x}{8}(s-1)(s-1)(s+1) \\ \phi_3(s) = -\frac{1}{4}(s+1)(s+1)(s-2) \\ \phi_4(s) = \frac{\Delta x}{8}(s-1)(s+1)(s+1) \end{cases} \tag{12}$$

where s represents the normalized local coordinate in each element. Assuming that the j-th and $(j+1)$-th nodes compose the j-th beam elements, $X = -1$ indicates the position of the j-th node, and $X = 1$ indicates the position of the $(j+1)$-th node. $\Delta x = x_j - x_{j+1}$ when the whole system is inside beam element j, which consists of node x_j and node x_{j+1}. Define a basis function vector $N(x)$ whose components are

$$\begin{cases} N_{2j-1}(x) = \phi_1(s) \\ N_{2j+0}(x) = \phi_2(s) \\ N_{2j+1}(x) = \phi_3(s) \\ N_{2j+2}(x) = \phi_4(s) \end{cases} \tag{13}$$

All components outside the element are set to zero. Using the bridge deflection $y(x_j,t)$ and deflection angle $\theta(x_j,t)$ at the nodes, the deformation vector $\boldsymbol{y}(t)$ is

$$\begin{cases} y_{2(j-1)+1}(t) = y(x_j,t) \\ y_{2(j-1)+2}(t) = \theta(x_j,t) \end{cases} \tag{14}$$

Then, the approximate solution of $y(x,t)$ is

$$y(x,t) = \mathbf{N}(x)\cdot \boldsymbol{y}(t) \tag{15}$$

Similarly, by setting the weights to $\omega(x) = \mathbf{N}(x)\cdot \boldsymbol{\omega}$ and substituting them into Equation (11), we obtain the following.

$$\boldsymbol{\omega}^{\mathrm{T}}(\mathbf{M}_b\ddot{\boldsymbol{y}}(t) + \mathbf{K}_b\boldsymbol{y}(t) - \boldsymbol{F}(t)) = 0 \tag{16}$$

$\mathbf{M}_b$, $\mathbf{C}_b$, and $\mathbf{K}_b$ are the mass, damping, and stiffness matrices of the bridge.

$$\mathbf{M}_b = \int_0^L \boldsymbol{N}\boldsymbol{N}^{\mathrm{T}}\mathrm{d}x \tag{17}$$

$$\mathbf{K}_b = \int_0^L \frac{\partial^2 \mathbf{N}}{\partial x^2}\frac{\partial^2 \mathbf{N}^{\mathrm{T}}}{\partial x^2}\mathrm{d}x \tag{18}$$

$F(t)$ is an external force vector whose components are the external forces (concentrated load and moment of force) at each node. Considering Rayleigh damping,

$$\mathbf{C}_b = \alpha\mathbf{M}_b + \beta\mathbf{K}_b \tag{19}$$

Solving for the integration condition in Equation (16) for any ω gives the following finite element equation.

$$\mathbf{M}_b\ddot{\boldsymbol{y}}(t) + \mathbf{C}_b\dot{\boldsymbol{y}}(t) + \mathbf{K}_b\boldsymbol{y}(t) = \boldsymbol{F}(t) \tag{20}$$

4.4. Vehicle–Bridge Interaction

In general, the responses of vehicles and bridges are modeled by interactions with each other's outputs as inputs. In order to realize a numerical simulation considering this interaction, the following steps are performed. First, the vehicle vibration is calculated by inputting only the road profile. Then, the contact-point force to the bridge is obtained from the obtained vehicle vibration, and the bridge vibration is calculated. Adding this bridge profile to the road profile creates a new input profile, and the vehicle vibration is obtained again. By repeating this process, the displacement vibration of the vehicle and the bridge is obtained. The input profile and contact-point force, which are the inputs of the vehicle and bridge, are explained below [7].

4.4.1. Input Profile

The input profile $\boldsymbol{u}(t)$, which is the input of the vehicle system, is given by the sum of the road profile $\boldsymbol{r}(t)$ and the bridge profile $\widetilde{\boldsymbol{y}}(t)$ and is expressed by Equation (21).

$$\boldsymbol{u}(t) = \boldsymbol{r}(t) + \widetilde{\boldsymbol{y}}(t) \tag{21}$$

Here, the road profile represents the road at the axle position. When the road unevenness is $R(x)$ and the axle position is $x_i(t)$,

$$r_i(t) = R(x_i(t)) \tag{22}$$

$r_i(t)$ is the component of $\boldsymbol{r}(t)$. On the other hand, the bridge profile is the bridge vibration $y(x,t)$ at the axle position $x_i(t)$. In other words, the bridge vibration is the deformation vector of each node fixed on the bridge. Therefore, converting to bridge displacement at the axle position is necessary. The same basis used in the discretization was used. The transformation matrix is

$$\mathbf{L}(t) = \begin{bmatrix} \boldsymbol{N}(x_1(t)) & \boldsymbol{N}(x_2(t)) \end{bmatrix} \tag{23}$$

The bridge profile is

$$\widetilde{\boldsymbol{y}}(t) = \mathbf{L}^{\mathrm{T}}(t)\boldsymbol{y}(t) \tag{24}$$

4.4.2. Contact-Point Force

The contact-point force, which is the input to the bridge, corresponds to the restoring force acting on the tire. However, as the equation of motion of the vehicle (Equation (1)) is based on the equilibrium position, it should be noted that the gravity term has disappeared. In calculating the restoring force, the effect of gravity is considered based on the equilibrium length. Considering that the center of gravity of the vehicle body is the center of gravity, the contact-point force between the front and rear wheels is

$$\begin{aligned} V_1(t) &= -\tfrac{d_2 m_s}{d_1+d_2}(g+\ddot{z}_{s1}) - m_{u1}(g+\ddot{z}_{u1}) \\ V_2(t) &= -\tfrac{d_1 m_s}{d_1+d_2}(g+\ddot{z}_{s2}) - m_{u2}(g+\ddot{z}_{u2}) \end{aligned} \tag{25}$$

The external force vector acting on the bridge is

$$\boldsymbol{F}(t) = \mathbf{L}(t)\begin{bmatrix} V_1(t) & V_2(t) \end{bmatrix} + \boldsymbol{H}(t) \tag{26}$$

$\boldsymbol{H}(t)$ represents the fulcrum reaction force.

4.5. System Identification

It is assumed that vehicle vibration is obtained as measurement data. Here, the road profile can be estimated by substituting the vehicle vibration data and the vehicle and bridge parameters whose initial values are randomly assumed in the VBI system. By positionally synchronizing the obtained road profile, the estimated road unevenness $R_1(x)$ and $R_2(x)$ for the front and rear wheels can be obtained. When the vehicle travels straight, the front and rear wheels are considered to run on the same road unevenness. Therefore, the estimated road unevenness $R_1(x)$ and $R_2(x)$ should also match. However, as the mechanical parameters are given randomly, they do not match. Therefore, the optimization problem is to update the dynamic parameters to minimize the error between the estimated road unevenness $R_1(x)$ and $R_2(x)$. In other words, the VBISI method is a search method for mechanical parameters where $R_1(x) = R_2(x)$.

4.5.1. Kalman Filter

In order to estimate the road unevenness from the vehicle vibration, this study also introduced the discrete-time extended state-space model proposed by Xu et al. [10]. The following equations give the state vector and observation vector.

$$\mathbf{Z}_k = \begin{Bmatrix} z(k\Delta t) \\ \dot{z}(k\Delta t) \\ u(k\Delta t) \\ \dot{u}(k\Delta t) \end{Bmatrix} \tag{27}$$

$$s_k = \begin{Bmatrix} \ddot{z}(k\Delta t) \\ z(k\Delta t) \end{Bmatrix} \tag{28}$$

The state vector is given by the vehicle's vertical displacement vibration, velocity vibration, input profile, and velocity vibration. The discrete-time extended state-space model in this study can be written as follows:

$$\mathbf{Z}_k = \bar{\mathbf{V}}\mathbf{Z}_{k-1} + \boldsymbol{\omega}_k \tag{29}$$

$$s_k = \mathbf{H}\mathbf{Z}_k + \epsilon_k \tag{30}$$

The matrix index $\overline{\boldsymbol{V}}$ is given by Equation (31), where ω_k, ϵ_k represent system noise and observation noise, and $\omega \sim N(\mathbf{0}, \boldsymbol{Q}), \epsilon \sim N(\mathbf{0}, \boldsymbol{R})$.

$$\bar{\mathbf{V}} = \text{expm}[\mathbf{V}\Delta t] = \mathbf{U}\text{diag}(\exp(\mathbf{D}))\mathbf{U}^{-1} \tag{31}$$

$$\mathbf{V} = \begin{bmatrix} \mathbf{O}^{4\times4} & \mathbf{I}^{4\times4} & \mathbf{O}^{4\times2} & \mathbf{O}^{2\times2} \\ -\mathbf{M}_V^{-1}\mathbf{K}_V & -\mathbf{M}_V^{-1}\mathbf{C}_V & \mathbf{M}_V^{-1}\mathbf{F}_V & \mathbf{O}^{2\times2} \\ \mathbf{O}^{2\times4} & \mathbf{O}^{2\times4} & \mathbf{I}^{2\times2} & \mathbf{O}^{2\times2} \\ \mathbf{O}^{2\times4} & \mathbf{O}^{2\times4} & \mathbf{O}^{2\times2} & \mathbf{O}^{2\times2} \end{bmatrix} \tag{32}$$

$\boldsymbol{U}$ and $\boldsymbol{D}$ are the modal matrices and diagonal matrix when $\mathbf{V}\Delta t$ is diagonalized and are expressed as follows.

$$\mathbf{V}\Delta t = \mathbf{U}\mathbf{D}\mathbf{U}^{-1} \tag{33}$$

Additionally, $\mathbf{H}$ is given by

$$\mathbf{H} = \begin{bmatrix} -\frac{k_{s1}}{m_s} & 0 & \frac{k_{s1}}{m_s} & 0 & -\frac{c_{s1}}{m_s} & 0 & \frac{c_{s1}}{m_s} & 0 & 0 & 0 & 0 & 0 \\ 0 & -\frac{k_{s2}}{m_s} & 0 & \frac{k_{s2}}{m_s} & 0 & -\frac{c_{s2}}{m_s} & 0 & \frac{c_{s1}}{m_s} & 0 & 0 & 0 & 0 \\ \frac{k_{s1}}{m_{u1}} & 0 & -\frac{(k_{s1}+k_{u1})}{m_{u1}} & 0 & \frac{c_{s1}}{m_{u1}} & 0 & -\frac{c_{s1}}{m_{u1}} & 0 & \frac{k_{u1}}{m_{u1}} & 0 & 0 & 0 \\ 0 & \frac{k_{s1}}{m_{u2}} & 0 & -\frac{(k_{s2}+k_{u2})}{m_{u2}} & 0 & \frac{c_{s2}}{m_{u2}} & 0 & -\frac{c_{s1}}{m_{u2}} & 0 & \frac{k_{u2}}{m_{u2}} & 0 & 0 \\ 1 & 0 & 0 & 0 & 0 & 0 & 0 & 0 & 0 & 0 & 0 & 0 \\ 0 & 1 & 0 & 0 & 0 & 0 & 0 & 0 & 0 & 0 & 0 & 0 \\ 0 & 0 & 1 & 0 & 0 & 0 & 0 & 0 & 0 & 0 & 0 & 0 \\ 0 & 0 & 0 & 1 & 0 & 0 & 0 & 0 & 0 & 0 & 0 & 0 \end{bmatrix} \quad (34)$$

Let $\boldsymbol{Q}$ and $\boldsymbol{R}$ be the variance–covariance matrices of system noise $\boldsymbol{\omega}_k$ and observation noise $\boldsymbol{\epsilon}_k$.

$$\mathbf{Q} = \mathrm{E}\left[\boldsymbol{\omega}_k \boldsymbol{\omega}_k^{\mathrm{T}}\right] \quad (35)$$

$$\mathbf{R} = \mathrm{E}\left[\boldsymbol{\epsilon}_k \boldsymbol{\epsilon}_k^{\mathrm{T}}\right] \quad (36)$$

Apply the Kalman filter [54] to the obtained discrete-time extended state-space model. According to Xue et al. [10], it is possible to estimate the input profile even in a noisy environment. In addition, the Robbins–Monro algorithm [55] can be used to dynamically estimate the variance–covariance matrix of process noise and observation noise. Xue et al. [10] also used Rauch-Tung-Striebel smoothing [56], which was also applied in this study. The Kalman filter [55] is an efficient method for obtaining state vectors based on observed data and a dynamic model. Let the estimated state vector be $\hat{\boldsymbol{Z}}_k$. Now, when $\hat{\boldsymbol{Z}}_{k-1}$ is obtained, the candidate X of $\hat{\boldsymbol{Z}}_k$ from Equation (29) is $X \sim N(\mu_a, \boldsymbol{\Sigma}_a)$. On the other hand, when $\boldsymbol{s}_k$ is obtained, the candidate Y of $\hat{\boldsymbol{Z}}_k$ derived from Equation (30) follows $Y \sim N(\mu_b, \boldsymbol{\Sigma}_b)$.

$$\mu_a = \bar{\mathbf{V}}\hat{\mathbf{Z}}_{k-1} \quad (37)$$

$$\boldsymbol{\Sigma}_{a,k} = \bar{\mathbf{V}}\mathbf{P}_{k-1}\bar{\mathbf{V}}^{\mathrm{T}} + \mathbf{Q} \quad (38)$$

$$\mu_b = \mathbf{H}^{-1}\boldsymbol{s}_k \quad (39)$$

$$\boldsymbol{\Sigma}_b = \mathbf{H}^{-1}\mathbf{R}\mathbf{H}^{-\mathrm{T}} \quad (40)$$

After applying the Kalman filter, the state vector $\hat{\boldsymbol{Z}}_k$ is estimated as the maximum likelihood value, assuming that the two candidates X and Y follow a normal distribution. That is, $\hat{\boldsymbol{Z}}_k$ is derived as follows.

$$\begin{aligned} \hat{Z}_k &= \left(\boldsymbol{\Sigma}_{a,k}^{-1} + \boldsymbol{\Sigma}_b^{-1}\right)^{-1}\left(\boldsymbol{\Sigma}_{a,k}^{-1}\mu_a + \boldsymbol{\Sigma}_b^{-1}\mu_b\right) \\ &= [\mathbf{I} - \mathbf{G}_k\mathbf{H}]\left\{\bar{\mathbf{V}}\hat{Z}_{k-1}\right\} + [\mathbf{G}_k\mathbf{H}]\left\{\mathbf{H}^{-1}s_k\right\} \\ &= [\mathbf{I} - \mathbf{G}_k\mathbf{H}]\left\{\bar{\mathbf{V}}\hat{Z}_{k-1}\right\} + \mathbf{G}_k s_k \end{aligned} \quad (41)$$

where $\boldsymbol{G}_k$ and $\boldsymbol{P}_k$ are calculated as follows.

$$\mathbf{G}_k = \boldsymbol{\Sigma}_{a,k}\mathbf{H}^{\mathrm{T}}\left(\mathbf{H}\boldsymbol{\Sigma}_{a,k}\mathbf{H}^{\mathrm{T}} + \mathbf{R}\right)^{-1} \quad (42)$$

$$\mathbf{P}_k = \left(\boldsymbol{\Sigma}_{a,k}^{-1} + \boldsymbol{\Sigma}_b^{-1}\right)^{-1} = [\mathbf{I} - \mathbf{G}\mathbf{H}]\boldsymbol{\Sigma}_a \quad (43)$$

Subsequently, the Robbins–Monro algorithm [55] dynamically estimates the variance–covariance matrix of the process noise $\boldsymbol{\omega}_k$ and the observation noise $\boldsymbol{\epsilon}_k$. Correctly estimating the order of noise is important because it affects the practicality of Kalman filter-based road profile estimation. Here, $\alpha_{Q,k}$ and $\alpha_{R,k}$ are positive real numbers smaller than one.

$$\mathbf{Q}_k = (1 - \alpha_{Q,k})\mathbf{Q}_{k-1} + \alpha_{Q,k}\mathbf{G}_k\left(s_k - \mathbf{H}\hat{\mathbf{Z}}_k\right)\left(s_k - \mathbf{H}\hat{\mathbf{Z}}_k\right)^{\mathbf{T}}\mathbf{G}_k^{\mathrm{T}} \quad (44)$$

$$\mathbf{R}_k = (1 - \alpha_{R,k})\mathbf{R}_{k-1} + \alpha_{R,k}\left(s_k - \mathbf{H}\hat{\mathbf{Z}}_k\right)\left(s_k - \mathbf{H}\hat{\mathbf{Z}}_k\right)^{\mathrm{T}} \tag{45}$$

4.5.2. Object Function

First, $\dot{z}(t)$ and $z(t)$ are obtained by applying the Newmark-β method to vehicle vibration $\ddot{z}(t)$, which is the measurement data. $\boldsymbol{M}_v$, $\boldsymbol{C}_v$, and $\boldsymbol{K}_v$ can also be obtained by randomly assuming the system parameters of the vehicle. At this time, u_1 and u_2 can be obtained by estimating the state vector using the Kalman filter. Next, the vehicle vibration data $\ddot{z}(t)$ and the assumed vehicle system parameters are substituted into Equation (25) to obtain contact-point forces $V_1(t)$ and $V_2(t)$. Suppose the system parameters of the bridge are also assumed randomly. In that case, the bridge vibration $\boldsymbol{y}(t)$ can be obtained using the equation of motion of the bridge and the Newmark-β method, as in numerical simulation. By substituting into Equation (24), the bridge profile $\widetilde{\boldsymbol{y}}(t)$ can be obtained. Then, $\mathbf{r}$(t) can be estimated by subtracting $\widetilde{\boldsymbol{y}}(t)$ from u(t) obtained earlier. Here, $R_1(x)$ is obtained by synchronizing $r_1(t)$ with $x_1(t)$, and $R_2(x)$ is obtained by synchronizing $r_2(t)$ with $x_2(t)$. Based on the assumption that road unevenness $R_1(x)$ and $R_2(x)$ should be equal, the problem of estimating mechanical parameters is treated as an optimization problem that minimizes the squared error of $R_1(x)$ and $R_2(x)$. The objective function of this optimization problem is

$$J(x) = \sum |R_1(x) - R_2(x)|^2 \tag{46}$$

If the parameters of the vehicle and bridge are all correct values, the two road unevenness calculations match. Therefore, if the parameters can be updated so that the calculated road unevenness matches, it can be expected that the parameters will eventually approach the correct values. However, the equations of motion are equivalent when all parameters are multiplied by the same factor. Therefore, at least one parameter must be known. This research assumes that the gross vehicle weight $M = m_s + m_{u1} + m_{u2}$ and the distance between axles $D = d_1 + d_2$, which are easy to measure, are known parameters [7].

4.5.3. Optimization Method

This research adopts the Nelder–Mead method [57,58] as the parameter search method. The method creates an initial simplex (a simulated triangle in high-dimensional space). The simplex is then iteratively modified to approach the minimum or maximum value of the function. The simplex is then modified using the reflection, expansion, contraction, and shrink methods. The Nelder–Mead method is independent of the slope of the objective function and searches for the optimal solution relatively quickly. This study uses the faster adaptive Nelder-Mead method [58]. In addition, there are non-negative conditions and constraints on mechanical parameters. Therefore, the objective function of the VBISI method is changed from a constrained objective function to an unconstrained objective function using a penalty function. The penalty function sets the value of the objective function to infinity if the parameters do not satisfy the following conditions.

1. One of the parameters is negative.
2. The sum of axle weights m_{u1} and m_{u2} exceeds the vehicle weight.
3. The center of gravity position d_1 exceeds the wheelbase value.

For the initial value, give 0.8 to 1.2 times the value assumed in advance.

4.6. Implementation of Numerical Simulation

Based on the model shown in Figure 2, the coupled vibration of the vehicle and bridge is reproduced by numerical simulation. Here, the vehicle model is separated from the bridge model, and the computation is repeated until the vehicle vibration converges [7].

4.6.1. Newmark-β Method

Vehicle and bridge vibrations are obtained by applying the Newmark-β method to the respective equations of motion. The arbitrary equations of motion are shown below.

$$\boldsymbol{M}\ddot{\boldsymbol{\eta}}(t) + \boldsymbol{C}\dot{\boldsymbol{\eta}}(t) + \boldsymbol{K}\boldsymbol{\eta}(t) = \boldsymbol{\xi}(t) \tag{47}$$

Discretize the time function $\boldsymbol{\eta}(t)$, and let $\boldsymbol{\eta}_k$ be the displacement response of the vehicle or bridge, where Δt is the time increment

$$\boldsymbol{\eta}_k = \boldsymbol{\eta}(k\Delta t). \tag{48}$$

In the Newmark-β method,

$$\begin{aligned} \dot{\boldsymbol{\eta}}_k &= \dot{\boldsymbol{\eta}}_{k-1} + \Delta t((1-\gamma)\ddot{\boldsymbol{\eta}}_{k-1} + \gamma\ddot{\boldsymbol{\eta}}_{k-1}) \\ \boldsymbol{\eta}_k &= \boldsymbol{\eta}_{k-1} + \Delta t\dot{\boldsymbol{\eta}}_{k-1} + \Delta t^2\left(\left(\tfrac{1}{2}-\beta\right)\ddot{\boldsymbol{\eta}}_{k-1} + \beta\ddot{\boldsymbol{\eta}}_k\right) \end{aligned} \tag{49}$$

From the above equation, the equation can be written as follows

$$\boldsymbol{M}\ddot{\boldsymbol{\eta}}_k + \boldsymbol{C}\dot{\boldsymbol{\eta}}_k + \boldsymbol{K}\boldsymbol{\eta}_k = \boldsymbol{\xi}_k \tag{50}$$

Applying the Newmark-β method, we receive

$$\boldsymbol{A}\ddot{\boldsymbol{\eta}}_k = \boldsymbol{b}_k. \tag{51}$$

From this, the following can be derived

$$\boldsymbol{A} = \left[\boldsymbol{M} + \Delta t\gamma\boldsymbol{C} + \Delta t^2\beta\boldsymbol{K}\right] \tag{52}$$

$$\boldsymbol{b}_k = \left\{\boldsymbol{\xi}_k - \boldsymbol{C}(\dot{\boldsymbol{\eta}}_{k-1} + \Delta t(1-\gamma)\ddot{\boldsymbol{\eta}}_{k-1}) - \boldsymbol{K}\left(\boldsymbol{\eta}_{k-1} + \Delta t\dot{\boldsymbol{\eta}}_{k-1} + \Delta t^2\left(\frac{1}{2}-\beta\right)\ddot{\boldsymbol{\eta}}_{k-1}\right)\right\} \tag{53}$$

4.6.2. Iterative Computation

Iterative calculations reproduce the coupled vibration of the vehicle and bridge. First, vehicle vibration is calculated using only the road unevenness as an input. The data matrix $\boldsymbol{Z} = [\cdots z_k \cdots]$ represents the discretized vehicle vibration. The obtained vehicle vibration $\boldsymbol{Z}$ is set to $\boldsymbol{Z}^0$, as convergence calculations are performed, and using $\boldsymbol{Z}^0$, the ground forces can be obtained. Replacing Equation (47) with the equation of motion of the bridge (Equation (20), the bridge vibration is obtained in the same way. Let this be $\boldsymbol{Y}^0$ from $\boldsymbol{Y}^0$; $\boldsymbol{Z}^1$ is obtained. This process is repeated to obtain $\boldsymbol{Y}^l$ from $\boldsymbol{Z}^l$ and $\boldsymbol{Z}^{l+1}$ from $\boldsymbol{Y}^l$. The convergence condition for this iterative calculation is

$$\varepsilon = \frac{\left|\boldsymbol{Z}^{l+1} - \boldsymbol{Z}^l\right|}{\left|\boldsymbol{Z}^{l+1}\right|} \leq \varepsilon_{max}. \tag{54}$$

The computation is terminated when the update ratio of $\boldsymbol{Z}^l$ is less than the threshold value ε_{max}. Additionally, $|\ |$ denotes the quadratic norm.

The time increment is set to $1.0 \times 10^{-3}[s]$, and the threshold for convergence judgment is set to 1.0×10^{-6}. The Newmark-β method employs the average acceleration method, with $\gamma = \frac{1}{2}$ and $\beta = \frac{1}{4}$.

5. Numerical Validation

5.1. Setting

First, this study verifies whether road unevenness can be estimated from vehicle vibration calculated by numerical simulation. To see how the Kalman filter's application changes the parameter estimation accuracy. The mechanical parameters of the vehicle were determined from the vehicle inspection certificate. The vehicle is assumed to be a 14-t truck that can sufficiently vibrate the bridge. The mechanical parameters of the bridge were determined concerning the bridge ledger (Table 2). The roughness of the road profile used in the numerical simulation was determined based on the values calculated by the road profiler. The vehicle displacement vibration is estimated from the acceleration using the *Newmark-β* method. The estimated vehicle displacement vibration is high-pass filtered at 0.1 Hz. The filtering process is to remove trends due to numerical integration.

Table 2. The assumed mechanical parameters of vehicle and bridge.

Vehicle Parameter		Assumed		Bridge Parameter		Assumed	
d_1	Distance between wheels and center of gravity	2.18	[m]	L	Length	30.0	[m]
d_2		2.67	[m]		Number of Elements	7	
m_s	Mass	13,060	[kg]	EI_i	Bending stiffness	5.50×10^{10}	[Nm^2]
m_{ui}	Mass (sprung)	3.20×10^3	[kg]	ρA	Mass per unit	1.80×10^4	[kg/m]
c_{si}	Damping (sprung)	3.00×10^4	[kg/s]	α	Rayleigh damping coefficient	0.7024	
k_{si}	Stiffness (sprung)	4.00×10^3	[N/m]	β		0.0052	
k_{ui}	Stiffness (unsprung)	4.00×10^5	[N/m]				

The authors assumed values for the variance–covariance matrix based on previously measured vehicle vibration data. Table 3 summarizes the initial value diagonal components of the variance–covariance matrix of the process and observation noise. Even if the model parameters corresponding to an actual vehicle are estimated, there will consistently be modeling errors because the half-car model is a simplified linear system with limited degrees of freedom. It is generally difficult to estimate the degree of the first through eighth components of **Q** corresponding to this modeling error [10]. Therefore, the values were determined by trial and error with reference to [10]. On the other hand, the observed noise value can be determined by referring to the noise level of the measurement equipment and the installation method. However, it is difficult to accurately estimate these values for each measurement vehicle and installation method. Therefore, **R** was also determined by trial and error. In this study, the values were set based on the RMS values of vehicle vibration measured when the vehicle was stationary; the off-diagonal elements of **Q** and **R** are zero.

Table 3. Variance–Covariance Matrix Settings.

(i,i)	(1, 1)	(2, 2)	(3, 3)	(4, 4)	(5, 5)	(6, 6)
Q_{ii}	1.35×10^{-4}	1.47×10^{-4}	1.84×10^{-4}	1.85×10^{-4}	5.11×10^{-4}	5.73×10^{-4}
R_{ii}	4.50×10^{-3}	5.20×10^{-3}	3.34×10^{-3}	3.34×10^{-3}	1.25×10^{-4}	1.37×10^{-4}
(i,i)	(7, 7)	(8, 8)	(9, 9)	(10, 10)	(11, 11)	(12, 12)
Q_{ii}	1.20×10^{-3}	1.20×10^{-3}	1.00×10^{-9}	1.00×10^{-9}	1	1
R_{ii}	1.74×10^{-4}	1.75×10^{-4}				

5.2. Result and Discussion

Parameter identification and road unevenness estimation are performed using vehicle vibrations. The effect of the Kalman filter on parameter identification was confirmed through comparison with the case where the Kalman filter was not used. The diagonal elements of the variance–covariance matrices R and Q of the Kalman filter without consideration of noise are assumed to be 1.00×10^{-9} for all one to eight elements. In addition, the case where noise is included in the vehicle vibration is also considered for field tests.

The noise was set based on the value measured by an acceleration sensor installed in the vehicle in a static state. Figure 3 shows the estimation of road unevenness by the VBISI method without a Kalman filter. Figure 4 shows the results using the Kalman filter. The upper and lower rows show the estimated road unevenness when noise is not considered and when noise is considered, respectively. The set vehicle speed is 8.1 [m/s], and the data for about 5 s is used. The Nelder–Mead method was adopted to minimize the objective function, and the number of updates was set to 1000 times. The input profile without the Kalman filter is obtained from Equations (1) and (3).

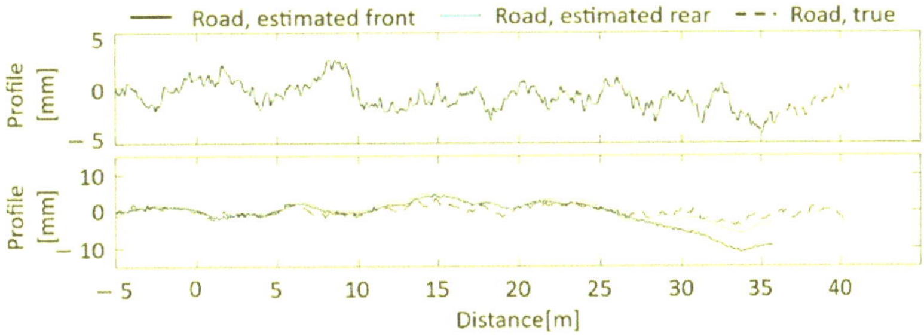

Figure 3. Road unevenness estimation results without Kalman filter [upper: without noise, lower: with noise].

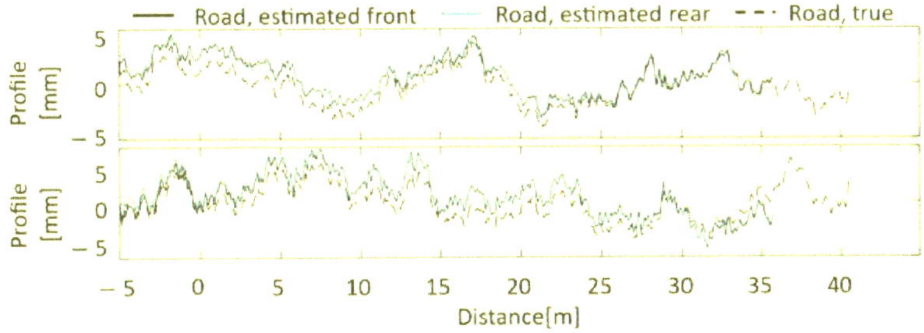

Figure 4. Road unevenness estimation results with Kalman filter [upper: without noise, lower: with noise].

When the Kalman filter is not used, the road unevenness estimated for the front and rear wheels almost always coincide with the absence of noise. In addition, the road unevenness calculated for the front and rear wheels almost always overlaps with the correct value. However, when noise is considered, the estimation accuracy is significantly reduced. On the other hand, when the Kalman filter is used, the decrease in the accuracy of road unevenness is slight regardless of the presence or absence of noise. However, the road unevenness estimated for the front and rear wheels differs from the correct ones. There are several possible reasons for this. One is that the parameters of the variance–covariance matrix of the Kalman filter need to be better adjusted, which may have been affected by the noise processing. It is also known that the accuracy of the Kalman filter in estimating road unevenness decreases as the distance from the center of gravity increases [7]. There is a possibility that the influence of the estimation error of the front and rear wheels cannot be ignored. Table 4 summarizes the parameter estimation results. The two columns on the left in Table 4 indicate the presence or absence of the Kalman filter in the estimation process and the consideration or lack thereof of noise to vehicle vibration. If the conditions are met, a symbol of ○ is given; if not, a symbol of × is assigned.

Table 4. Variance–Covariance Matrix Settings.

KF	Noise	d_1	c_{s1}	c_{s2}	k_{s1}	k_{s2}	m_{u1}	m_{u2}	k_{u1}	k_{u2}	ρA	α	bC	EI_1	EI_4	EI_7
×	×	1.00	1.01	1.01	1.05	1.04	1.02	1.02	1.01	1.02	1.00	0.99	1.06	1.00	0.94	0.96
×	○	0.00	0.20	0.13	0.07	3.02	2.32	0.37	2.69	1.19	0.33	1.73	2.01	0.62	0.36	0.02
○	×	0.96	1.22	1.05	0.93	0.94	0.92	1.02	1.05	1.07	1.09	1.06	0.94	1.05	0.89	0.94
○	○	1.13	1.82	2.11	0.70	1.41	1.05	0.93	0.68	0.85	1.60	1.86	0.84	1.07	1.08	0.38

Estimated parameters were divided by their correct values and normalized. Only the bridge endpoints and the central part are represented as bridge stiffness values. The highest accuracy is obtained when the Kalman filter is not used and noise is not considered. Even if the Kalman filter is used, it is possible to estimate the mechanical parameters of the vehicle and bridge without considering noise. However, the accuracy of the estimated parameters is slightly lower than before using the Kalman filter. This phenomenon occurs even though the front and rear wheels estimate almost the same road surface. In other words, the difference between the estimated road unevenness and the correct road unevenness is the cause. In addition, when noise is considered, the parameter estimation accuracy decreases in both methods regardless of the introduction of the Kalman filter.

6. Field Test

6.1. Bridge Description

Matsumi Bridge is a one-span bridge constructed in 1973 across Kaede Street in Tsukuba City, Ibaraki Prefecture, Japan (Figure 5). The bridge's total length is about 30.88 m, and the width is about 12.98 m. The main girder is a PC box post-tension girder. It is paved with asphalt and integrated with the main girder.

(a) **(b)**

Figure 5. The view of the bridge: (**a**) full view; (**b**) expansion joint [48].

To provide a baseline for comparison with drive-by measurements, the natural frequency of Matsumi Bridge is directly measured by installing accelerometers on the bridge. A three-axis wireless MEMS accelerometer was used to record the forced vibration of the bridge. Two sensors were installed at the center of each bridge for bridge measurement. A wireless accelerometer driven by a mobile battery was attached using double-sided tape for construction. Vertical acceleration was measured at 300 Hz.

The first natural frequency of Matsumi Bridge was calculated by the fast Fourier transform of free vibration and forced vibration data. Figure 6 shows an example of the free vibration and FFT of the bridge after passing vehicles measured at two locations. Additionally, forced vibration is shown in Figure 6. In this example, a passing measurement vehicle excites a bridge. Comparing the free vibration and vehicle vibration, the latter has more frequencies in the lower frequency range than the fundamental frequency of the bridge (including vehicle frequencies). As shown in Figure 6, the first peak in the frequency domain, corresponding to the fundamental frequency of the bridge, was evident in the data from all two sensors. The averages of five measurements of free vibration and forced vibration due to the passage of the measurement vehicle were 18.6 Hz and 18.33 Hz. The lower frequency of forced oscillation than that of free oscillation can be attributed to the mass increase by the measurement vehicle [59]. As there are only two sensors in this study, it is impossible to determine which bridge mode is responsible for the observed peaks.

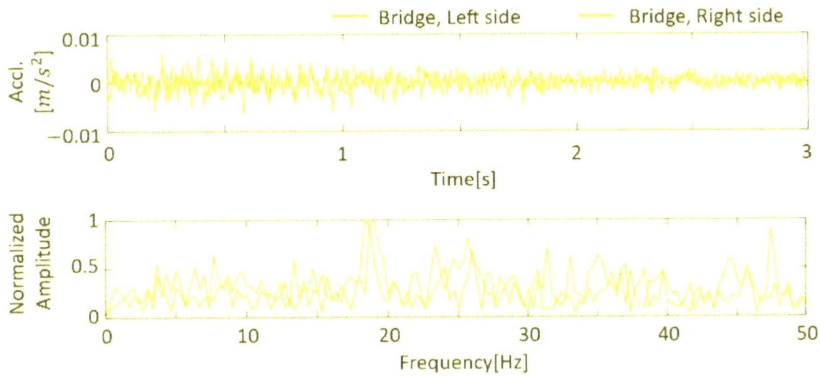

Figure 6. Free-vibration response of the Matsumi Bridge in both the time and frequency domain.

6.2. Measured Data

Measurements were taken with multiple accelerometers and multiple GPSs for a two-axis vehicle. The layout of the installed sensors is shown in Figure 7. The instrumentation and testing details are presented in [48]. Over 18 and 19 May 2022, the vertical acceleration was measured when the measurement vehicle crossed Matsumi Bridge 100 times. In VBI studies, the measured acceleration vibration is often in the vertical direction. However, it is difficult to obtain data only in the vertical direction when accelerometers are installed in vehicles, depending on the location. Therefore, the acceleration data were corrected using a rotation matrix before conducting this analysis. The accelerometers used in this study were triaxial, and the vertical direction can be inferred in the post-process by correcting the data with a rotation matrix. The sensor tilt was estimated from the vehicle vibration measured while the vehicle was stopped and corrected. The obtained rotation matrix is applied to the vehicle vibration data for analysis to obtain the mean value of the acceleration signal in the vertical direction, and the process of average zeroing is performed. Figure 8 shows an example of the vertical acceleration of a vehicle and the forced vibration of a bridge and its FFT results. The upper, middle, and lower rows represent the vehicle's front, rear, and bridge vibrations. For vehicle vibration, the vibration of the vehicle body and the vibration of the axle are plotted simultaneously. In common with front and rear vehicle vibrations, the amplitude of axle vibration is larger than vehicle body vibration. In the frequency domain, both the vehicle and bridge vibrations have a peak of about 3 Hz. By VBI, a peak was observed that did not exist during forced vibration (Figure 8b). Some peaks are seen around 13 Hz at the front of the vehicle. These are due to the effects of the vehicle's natural frequency, engine vibration, and road unevenness. In addition, the FFT of vehicle vibration does not show a peak near 18 Hz. Therefore, it is difficult to identify the fundamental frequency of a bridge simply by the FFT of the measurement data. The Nelder–Mead method was adopted to minimize the objective function, and the number of updates was set to 10,000 times.

6.3. Results

The results of the VBISI method run using the data measured in the field test are summarized in Figures 9 and 10 and Table 4. Figure 9a shows the acceleration vibration of the bridge vibration estimated by the VBISI method, its FFT result, and the estimated road profile. In the frequency domain of Figure 6, a peak around 18 Hz can be observed for the bridge vibration. This peak could not be observed from the estimated bridge vibration, as shown in Figure 9a. Figure 9b shows the road profile estimated by the VBISI method. Large amplitudes can be seen around 0 m and 30 m of the estimated road unevenness. These amplitudes can be attributed to the expansion joint in front of and behind the bridge. The road unevenness estimated for the front and rear wheels show similar characteristics, but they did not match. Figure 10 compares the power spectral density function (PSD) of the VBISI-estimated and measured road unevenness. The estimated road unevenness was assessed at the vehicle's front axle position. The road unevenness was measured using a road profiler. The estimated road unevenness is lower than the measured road unevenness. Table 5 summarizes the vehicle and mechanical bridge parameters estimated by the VBISI method. The correct values of the modal parameters for vehicles and bridges in a field test have yet to be discovered. Therefore, evaluating the parameters estimated by this method is a technical issue.

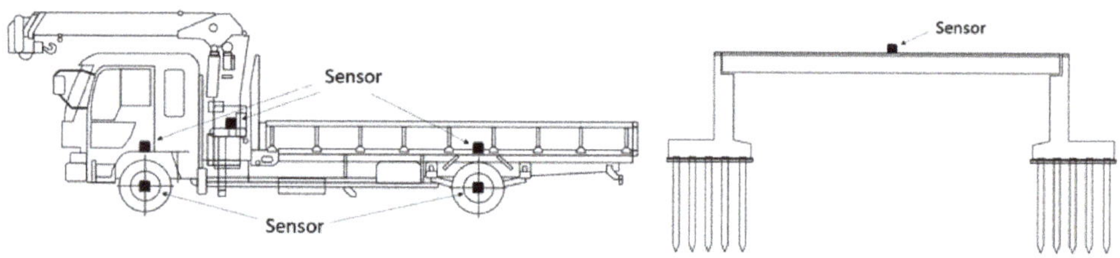

(**a**) (**b**)

Figure 7. Layout of sensor installation: (**a**) vehicle; (**b**) bridge.

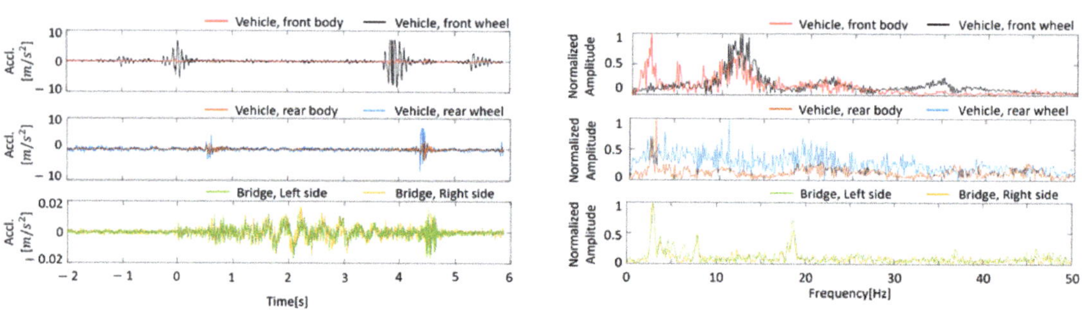

(**a**) (**b**)

Figure 8. Vibration response of vehicle and bridge: (**a**) in time domain; (**b**) in frequency domain.

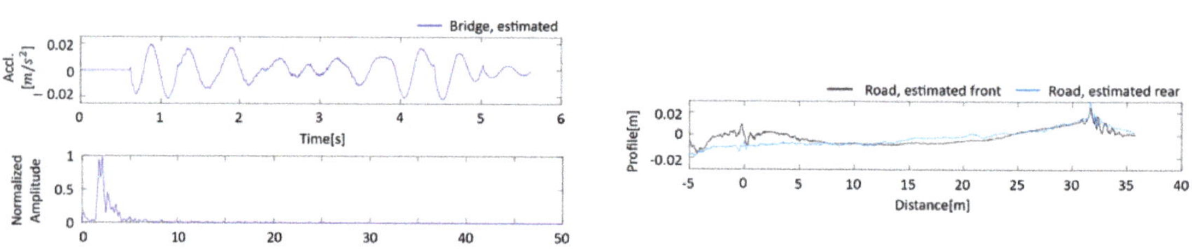

(**a**) (**b**)

Figure 9. The VBISI method estimation result: (**a**) bridge vibration in time and domain; (**b**) road profile.

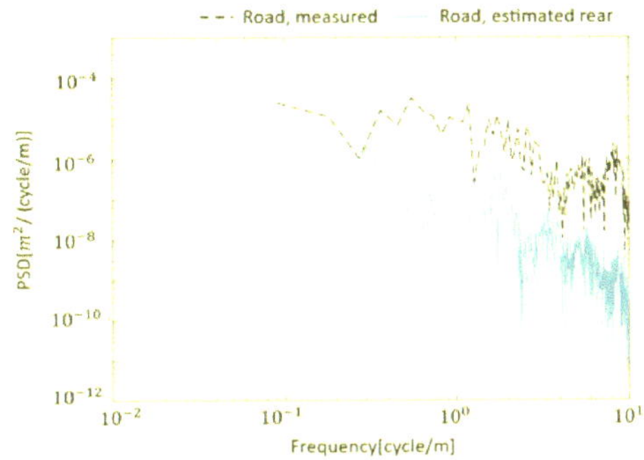

Figure 10. Comparison of estimated and measured PSD functions of road unevenness.

Table 5. The estimated mechanical parameters of vehicle and bridge.

Parameter		Estimated	Parameter		Estimated
d_1	[m]	2.18	EI_1	[Nm^2]	5.48×10^{10}
d_2	[m]	2.67	EI_2	[Nm^2]	5.35×10^{10}
m_{u1}	[kg]	3.25×10^2	EI_3	[Nm^2]	5.75×10^{10}
m_{u2}	[kg]	3.27×10^2	EI_4	[Nm^2]	5.19×10^{10}
c_{s1}	[kg/s]	3.02×10^4	EI_5	[Nm^2]	5.80×10^{10}
c_{s2}	[kg/s]	3.04×10^4	EI_6	[Nm^2]	5.38×10^{10}
k_{s1}	[N/m]	4.22×10^3	EI_7	[Nm^2]	5.25×10^{10}
k_{s2}	[N/m]	4.17×10^3	α		0.6921
k_{u1}	[N/m]	4.04×10^5	β		0.0055
k_{u2}	[N/m]	4.07×10^5	ρA	[kg]	1.79×10^4

7. Discussion

When the estimated road unevenness matches the correct values [8,9], the mechanical parameters of the vehicle and bridge also mostly match the correct values. Therefore, the accurate estimation of road unevenness is essential. To improve the estimation accuracy, actions are taken to improve the accuracy of the road unevenness estimation. In this study, several possible factors may reduce the accuracy of road unevenness estimation.

One is the problem of measured data. In this study, vehicle acceleration vibration and displacement vibration are used as observation data. However, vehicle displacement vibration is calculated by the numerical integration of vehicle acceleration vibration. The acceleration vibration of a vehicle contains various noises. Vibrations included in vehicle vibration may include engine vibration. Therefore, the numerically integrated displacement vibration of the vehicle has a trend. Because it is difficult to remove this effect, the accuracy of the road unevenness estimation is reduced. It is possible to use only acceleration vibration as the observed data. However, displacement vibration is essential to satisfy observability in the state space model. In a model that does not satisfy observability, it is difficult to determine the state variables from the observed data uniquely. On the other hand, the vehicle is also equipped with a GPS sensor to measure vertical displacement. The model's accuracy can be improved if the GPS's vertical displacement can compensate for the accelerometer's displacement oscillations.

Optimization also poses challenges. Many nonlinear optimization problems have multiple local solutions. As the objective function of the VBISI method is also a nonlinear function, there are local solutions. Therefore, even if the optimal solution is obtained, there may be a different value. Moreover, in the absence of blueprints, it is necessary to increase the range of parameters to be explored. Therefore, a more efficient parameter search method is needed for future verification. There is also potential for improvement in the objective function. Although this study used road unevenness estimated from the front and rear wheel positions, the synchronization of the positions takes work.

It is also necessary to consider how to verify the accuracy of this method. Therefore, it is practical to conduct a laboratory experiment to examine the applicability of this method. In laboratory experiments, it is relatively easy to grasp the mechanical parameters of the vehicle and bridge. In addition, it may be possible to estimate mechanical vehicle parameters from free vibration tests using humps and compare them with the results. Mechanical bridge parameters are also being considered to use findings from direct monitoring of bridges. A sensor is installed directly on the bridge, and the excitation test estimates the modal parameters of the bridge. Nikkhoo et al. [60] proposed a method for estimating the natural frequencies and dynamic response of various beams subjected to excitation by a moving mass. If these direct bridge monitoring techniques can be utilized, more efficient and reliable parameter estimation may be possible. Additionally, Yang et al. [36] proposed a method to evaluate the stiffness of a bridge estimated from vehicle vibration. Based on the estimated bridge stiffness, the deflection of the bridge, when given a specific load, is calculated by the finite element method. A similar load is applied to an actual bridge and compared with its displacement to verify accuracy. In addition, the VBISI method does not examine vehicle speed restrictions, which have been pointed out in previous studies [61]. Therefore, it is necessary to verify the practicality of this technique through parametric studies.

8. Conclusions

A field study of an indirect bridge health monitoring method for single-span concrete bridges is described. Acceleration responses extracted from sensor-mounted sensors were used. The study examined methods for estimating (i) mechanical vehicle parameters, (ii) mechanical bridge parameters, and (iii) road unevenness. Direct measurements using sensors installed on the bridges confirmed that the vehicles shook the bridge. Road profilers were also used to measure road unevenness. It is difficult to identify the fundamental frequencies of bridges from the estimated bridge vibrations. In addition, the estimated road unevenness needs to be more accurate when compared to the measured road unevenness, although they show some of the same trends. This paper is the first to examine the simultaneous identification of vehicle and bridge parameters and road unevenness in a field test. The VBISI method is a new approach that expands the possibilities of indirect monitoring.

A limitation of this study is that it took a lot of work to find a combination of parameters to estimate the road unevenness by the studied method between the front and rear wheels. The VBISI method had to consider the effects of noise fully, and the optimization method needed to be revised. Therefore, it is conceivable to adopt a less susceptible technique to engine vibration and measurement noise in future research. Establishing a methodology to evaluate the estimated mechanical parameters of the vehicle and bridge is also essential.

A summary of the results is as follows:

1. The VBISI method can estimate the mechanical parameters of the vehicle and bridge and road unevenness from vehicle vibration and position information. The only information required for the estimation is the vehicle's total weight and the wheelbase.
2. The Kalman filter improves the accuracy of estimating road unevenness but reduces the accuracy of estimating the mechanical parameters of the vehicle and bridge.
3. The method was validated with vehicle vibrations measured during field tests. The values estimated by the proposed method are compared to the directly measured vibrations.
4. It is not easy to estimate the natural frequencies of bridges from the bridge vibrations estimated by the proposed method. On the other hand, some of the estimated road unevenness showed similar trends.

5. To improve the accuracy of the VBISI method, the vibration preprocessing and optimization methods need to be improved. In addition, a method for evaluating the estimated mechanical parameters needs to be established.

Author Contributions: Conceptualization, K.Y.; methodology, R.S. and K.Y.; validation, R.S.; resources, K.Y.; data curation, R.S. and K.Y.; writing—original draft preparation, R.S.; writing—review and editing, Y.O. and K.Y.; visualization, R.S.; supervision, Y.O. and K.Y.; funding acquisition, R.S., Y.O. and K.Y. All authors have read and agreed to the published version of the manuscript.

Funding: This study is supported by (1) MEXT/JSPS KAKENHI Grant Number JP19H02220, JP22J10994 and (2) Shoreki Commemorative Foundation 2022.

Institutional Review Board Statement: Not applicable.

Informed Consent Statement: Not applicable.

Data Availability Statement: Data is contained within the article.

Acknowledgments: The authors are grateful to M. Ono (University of Tsukuba) for developing the sensor for this experiment. The authors would also thank K. Tsukada, M. Eugene, M. Sakai, N. Kaneko, and T. Miyagi (University of Tsukuba) and Y. Takahashi (Yachiyo Engineering Co., Ltd.) for experimental cooperation. The authors would thank K. Murakami (University of Tsukuba) for organizing the formulas.

Conflicts of Interest: The authors declare no conflict of interest.

References

1. Moreu, F.; Li, X.; Li, S.; Zhang, D. Technical specifications of structural health monitoring for highway bridges: New Chinese structural health monitoring code. *Front. Built Environ.* **2018**, *4*, 10. [CrossRef]
2. Fujino, Y.; Siringoringo, D.M. Recent research and development programs for infrastructures maintenance, renovation and management in Japan. *Struct. Infrastruct. Eng.* **2020**, *16*, 3–25. [CrossRef]
3. Rizzo, P.; Enshaeian, A. Bridge health monitoring in the United States: A review. *Struct. Monit. Maint.* **2021**, *8*, 1–50.
4. Yang, Y.B.; Lin, C.W.; Yau, J.D. Extracting bridge frequencies from the dynamic response of a passing vehicle. *J. Sound Vib.* **2004**, *272*, 471–493. [CrossRef]
5. Malekjafarian, A.; McGetrick, P.J.; Obrien, E.J. A review of indirect bridge monitoring using passing vehicles. *Shock Vib.* **2015**, *2015*, 286139. [CrossRef]
6. Malekjafarian, A.; Corbally, R.; Gong, W. A review of mobile sensing of bridges using moving vehicles: Progress to date, challenges and future trends. *Structures* **2022**, *44*, 1466–1489. [CrossRef]
7. Yamamoto, K.; Murakami, K.; Shin, R.; Okada, Y. Application of Particle Swarm Optimization method to On-going Monitoring for estimating vehicle-bridge interaction system. *arXiv* **2021**, arXiv:2201.08014.
8. Yamamoto, K.; Fujiwara, S.; Tsukada, K.; Shin, R.; Okada, Y. Numerical Studies on Bridge Inspection using Data obtained from sensors on vehicle. In Proceedings of the 13th IWSHM, Stanford University, Stanford, CA, USA, 7–9 December 2021; pp. 756–763.
9. Shin, R.; Okada, Y.; Yamamoto, K. A Numerical Study of Parameter Estimation Accuracy in System Identification Using Mobile Sensing. *J. Struct. Eng. A* **2022**, *68*, 298–309.
10. Xue, K.; Nagayama, T.; Zhao, B. Road profile estimation and half-car model identification through the automated processing of smartphone data. *Mech. Syst. Signal Process.* **2020**, *142*, 106722. [CrossRef]
11. Nagayama, T.; Zhao, B.; Xue, K. Half CAR Model Identification and Road Profile Estimation Using Vibration Responses of a running vehicles. *J. Jpn. Soc. Civ. Eng. Ser. E1 (Pavement Eng.)* **2019**, *75*, 1–16. [CrossRef]
12. Yang, Y.B.; Yang, J.P. State-of-the-art review on modal identification and damage detection of bridges by moving test vehicles. *Int. J. Struct. Stab. Dyn.* **2018**, *18*, 1850025. [CrossRef]
13. Shokravi, H.; Shokravi, H.; Bakhary, N.; Heidarrezaei, M.; Rahimian Koloor, S.S.; Petrů, M. Vehicle-assisted techniques for health monitoring of bridges. *Sensors* **2020**, *20*, 3460. [CrossRef] [PubMed]
14. Hou, R.; Xia, Y. Review on the new development of vibration-based damage identification for civil engineering structures: 2010–2019. *J. Sound Vib.* **2021**, *491*, 115741. [CrossRef]
15. Yang, Y.B.; Yang, J.P.; Wu, Y.; Zhang, B. *Vehicle Scanning Method for Bridges*; John Wiley and Sons, Ltd.: Chichester, UK, 2019.
16. Lin, C.W.; Yang, Y.B. Use of a passing vehicle to scan the fundamental bridge frequencies: An experimental verification. *Eng. Struct.* **2005**, *27*, 1865–1878. [CrossRef]
17. Yang, Y.B.; Lin, C.W. Vehicle–bridge interaction dynamics and potential applications. *J. Sound Vib.* **2005**, *284*, 205–226. [CrossRef]
18. Yang, Y.B.; Chang, K.C. Extraction of bridge frequencies from the dynamic response of a passing vehicle enhanced by the EMD technique. *J. Sound Vib.* **2009**, *322*, 718–739. [CrossRef]

19. Yang, Y.B.; Chang, K.C. Extracting the bridge frequencies indirectly from a passing vehicle: Parametric study. *Eng. Struct.* **2009**, *31*, 2448–2459. [CrossRef]
20. Siringoringo, D.M.; Fujino, Y. Estimating bridge fundamental frequency from vibration response of instrumented passing vehicle: Analytical and experimental study. *Adv. Struct. Eng.* **2012**, *15*, 417–433. [CrossRef]
21. Cerda, F.; Chen, S.; Bielak, J.; Garrett, J.H.; Rizzo, P.; Kovacevic, J. Indirect structural health monitoring of a simplified laboratory-scale bridge model. *Smart Struct. Syst.* **2014**, *13*, 849–868. [CrossRef]
22. He, W.Y.; Zhu, S. Moving load-induced response of damaged beam and its application in damage localization. *J. Vib. Control* **2016**, *22*, 3601–3617. [CrossRef]
23. Yang, Y.B.; Zhang, B.; Qian, Y.; Wu, Y. Contact-point response for modal identification of bridges by a moving test vehicle. *Int. J. Struct. Stab. Dyn.* **2018**, *18*, 1850073. [CrossRef]
24. Corbally, R.; Malekjafarian, A. Examining changes in bridge frequency due to damage using the contact-point response of a passing vehicle. *J. Struct. Integr. Maint.* **2021**, *6*, 148–158. [CrossRef]
25. Jian, X.; Xia, Y.; Sun, L. Indirect identification of bridge frequencies using a four-wheel vehicle: Theory and three-dimensional simulation. *Mech. Syst. Signal Process.* **2022**, *177*, 109155. [CrossRef]
26. Jin, N.; Dertimanis, V.K.; Chatzi, E.N.; Dimitrakopoulos, E.G.; Katafygiotis, L.S. Subspace identification of bridge dynamics via traversing vehicle measurements. *J. Sound Vib.* **2022**, *523*, 116690. [CrossRef]
27. Zhang, J.; Yang, D.; Ren, W.X.; Yuan, Y. Time-varying characteristics analysis of vehicle-bridge interaction system based on modified S-transform reassignment technique. *Mech. Syst. Signal Process.* **2021**, *160*, 107807. [CrossRef]
28. Malekjafarian, A.; OBrien, E.J. Identification of bridge mode shapes using short time frequency domain decomposition of the responses measured in a passing vehicle. *Eng. Struct.* **2014**, *81*, 386–397. [CrossRef]
29. Yang, Y.B.; Li, Y.C.; Chang, K.C. Constructing the mode shapes of a bridge from a passing vehicle: A theoretical study. *Smart Struct. Syst.* **2014**, *13*, 797–819. [CrossRef]
30. Oshima, Y.; Yamamoto, K.; Sugiura, K. Damage assessment of a bridge based on mode shapes estimated by responses of passing vehicles. *Smart Struct. Syst.* **2014**, *13*, 731–753. [CrossRef]
31. OBrien, E.J.; Malekjafarian, A. A mode shape-based damage detection approach using laser measurement from a vehicle crossing a simply supported bridge. *Struct. Control Health Monit.* **2016**, *23*, 1273–1286. [CrossRef]
32. OBrien, E.J.; Malekjafarian, A.; González, A. Application of empirical mode decomposition to drive-by bridge damage detection. *Eur. J. Mech. -A/Solids* **2017**, *61*, 151–163. [CrossRef]
33. Yang, Y.; Zhu, Y.; Wang, L.L.; Jia, B.Y.; Jin, R. Structural damage identification of bridges from passing test vehicles. *Sensors* **2018**, *18*, 4035. [CrossRef] [PubMed]
34. Nie, Z.; Shen, Z.; Li, J.; Hao, H.; Lin, Y.; Ma, H. Output-only complete mode shape identification of bridges using a limited number of sensors. *Mech. Syst. Signal Process.* **2022**, *178*, 109246. [CrossRef]
35. Yang, Y.; Lu, H.; Tan, X.; Chai, H.K.; Wang, R.; Zhang, Y. Fundamental mode shape estimation and element stiffness evaluation of girder bridges by using passing tractor-trailers. *Mech. Syst. Signal Process.* **2022**, *169*, 108746. [CrossRef]
36. Yang, Y.; Tan, X.; Lu, H.; Xue, S.; Wang, R.; Zhang, Y. Indirect Approach to Identify Girder Bridge Element Stiffness Based on Blind Source Separation. *Symmetry* **2022**, *14*, 1963. [CrossRef]
37. McGetrick, P.J.; Gonzlez, A.; OBrien, E.J. Theoretical investigation of the use of a moving vehicle to identify bridge dynamic parameters. *Insight-Non-Destr. Test. Cond. Monit.* **2009**, *51*, 433–438. [CrossRef]
38. González, A.; OBrien, E.J.; McGetrick, P.J. Identification of damping in a bridge using a moving instrumented vehicle. *J. Sound Vib.* **2012**, *331*, 4115–4131. [CrossRef]
39. Keenahan, J.; OBrien, E.J.; McGetrick, P.J.; Gonzalez, A. The use of a dynamic truck–trailer drive-by system to monitor bridge damping. *Struct. Health Monit.* **2014**, *13*, 143–157. [CrossRef]
40. Yang, Y.B.; Zhang, B.; Chen, Y.; Qian, Y.; Wu, Y. Bridge damping identification by vehicle scanning method. *Eng. Struct.* **2019**, *183*, 637–645. [CrossRef]
41. Yang, Y.B.; Shi, K.; Wang, Z.L.; Xu, H.; Zhang, B.; Wu, Y.T. Using a single-DOF test vehicle to simultaneously retrieve the first few frequencies and damping ratios of the bridge. *Int. J. Struct. Stab. Dyn.* **2021**, *21*, 2150108. [CrossRef]
42. Eshkevari, S.S.; Pakzad, S.N.; Takáč, M.; Matarazzo, T.J. Modal identification of bridges using mobile sensors with sparse vibration data. *J. Eng. Mech.* **2020**, *146*, 04020011.
43. Yang, Y.B.; Li, Z.; Wang, Z.L.; Shi, K.; Xu, H.; Qiu, F.Q.; Zhu, J.F. A novel frequency-free movable test vehicle for retrieving modal parameters of bridges: Theory and experiment. *Mech. Syst. Signal Process.* **2022**, *170*, 108854. [CrossRef]
44. Gong, F.; Han, F.; Wang, Y.; Xia, Y. Bridge Damping Extraction Method from Vehicle–Bridge Interaction System Using Double-Beam Model. *Appl. Sci.* **2021**, *11*, 10304. [CrossRef]
45. Malekjafarian, A.; Golpayegani, F.; Moloney, C.; Clarke, S. A machine learning approach to bridge-damage detection using responses measured on a passing vehicle. *Sensors* **2019**, *19*, 4035. [CrossRef] [PubMed]
46. Elhattab, A.; Uddin, N.; OBrien, E. Drive-by bridge frequency identification under operational roadway speeds employing frequency independent underdamped pinning stochastic resonance (FI-UPSR). *Sensors* **2018**, *18*, 4207. [CrossRef] [PubMed]
47. Yamamoto, K.; Takahashi, Y. Experimental Validation of Bridge Screening Method based on Vehicle Response Analysis. *Proc. World Congr. Eng.* **2017**, *2*, 928–933.

48. Shin, R.; Okada, Y.; Yamamoto, K. Application of C-LSTM Networks to Automatic Labeling of Vehicle Dynamic Response Data for Bridges. *Sensors* **2022**, *22*, 3486. [CrossRef]
49. McGetrick, P.; Kim, C.W.; González, A. Dynamic axle force and road profile identification using a moving vehicle. *Int. J. Archit. Eng. Constr.* **2013**, *2*, 1–16. [CrossRef]
50. He, Y.; Yang, J.P. Using Kalman filter to estimate the pavement profile of a bridge from a passing vehicle considering their interaction. *Acta Mech.* **2021**, *232*, 4347–4362. [CrossRef]
51. Yang, Y.B.; Wang, B.; Wang, Z.; Shi, K.; Xu, H. Scanning of Bridge Surface Roughness from Two-Axle Vehicle Response by EKF-UI and Contact Residual: Theoretical Study. *Sensors* **2022**, *22*, 3410. [CrossRef]
52. Hasegawa, S.; Kim, C.W.; Chang, K.C. Road Profile Identification by Means of Regularized Least Square Minimization with Dynamic Programming Utilizing Accelerations of a Moving Vehicle. *J. Jpn. Soc. Civ. Eng. Ser. A1 (Struct. Eng. Earthq. Eng. (SE/EE))* **2022**, *78*, 78–93. [CrossRef]
53. Keenahan, J.; Ren, Y.; OBrien, E.J. Determination of road profile using multiple passing vehicle measurements. *Struct. Infrastruct. Eng.* **2020**, *16*, 1262–1275. [CrossRef]
54. Kalman, R.E. A new approach to linear filtering and prediction problems. *J. Basic Eng.* **1960**, *82*, 35–45. [CrossRef]
55. Robbins, H.; Monro, S. A stochastic approximation method. *Annu. Math. Stat.* **1951**, *22*, 400–407. [CrossRef]
56. Rauch, H.E.; Tung, F.; Striebel, C.T. Maximum likelihood estimates of linear dynamic systems. *AIAA J.* **1965**, *3*, 1445–1450. [CrossRef]
57. Nelder, J.A.; Mead, R. A simplex method for function minimization. *Comput. J.* **1965**, *7*, 308–313. [CrossRef]
58. Gao, F.; Han, L. Implementing the Nelder-Mead simplex algorithm with adaptive parameters. *Comput. Optim. Appl.* **2012**, *51*, 259–277. [CrossRef]
59. Malekjafarian, A.; Khan, M.A.; OBrien, E.J.; Micu, E.A.; Bowe, C.; Ghiasi, R. Indirect Monitoring of Frequencies of a Multiple Span Bridge Using Data Collected from an Instrumented Train: A Field Case Study. *Sensors* **2022**, *22*, 7468. [CrossRef]
60. Nikkhoo, A.; Farazandeh, A.; Ebrahimzadeh, H.M. On the computation of moving mass/beam interaction utilizing a semi-analytical method. *J. Braz. Soc. Mech. Sci. Eng.* **2016**, *38*, 761–771. [CrossRef]
61. Nikkhoo, A.; Rofooei, F.R.; Shadnam, M.R. Dynamic behavior and modal control of beams under moving mass. *J. Sound Vib.* **2007**, *306*, 712–724. [CrossRef]

sensors

MDPI

Article

A Study on Structural Health Monitoring of a Large Space Antenna via Distributed Sensors and Deep Learning

Federica Angeletti [1,2,*], Paolo Iannelli [3,*], Paolo Gasbarri [1], Massimo Panella [4] and Antonello Rosato [4]

1 School of Aerospace Engineering, Sapienza University of Rome, Via Salaria 851, 00138 Rome, Italy
2 Since February 1st 2022—OHB System AG, Manfred-Fuchs-Straße 1, 82234 Weßling, Germany
3 Department of Mechanical and Aerospace Engineering (DIMA), Sapienza University of Rome, Via Eudossiana 18, 00184 Rome, Italy
4 Department of Information Engineering, Electronics and Telecommunications (DIET), Sapienza University of Rome, Via Eudossiana 18, 00184 Rome, Italy
* Correspondence: federica.angeletti@ohb.de (F.A.); paolo.iannelli@uniroma1.it (P.I.)

Abstract: Most modern Earth and Universe observation spacecraft are now equipped with large lightweight and flexible structures, such as antennas, telescopes, and extendable elements. The trend of hosting more complex and bigger appendages, essential for high-precision scientific applications, made orbiting satellites more susceptible to performance loss or degradation due to structural damages. In this scenario, Structural Health Monitoring strategies can be used to evaluate the health status of satellite substructures. However, in particular when analysing large appendages, traditional approaches may not be sufficient to identify local damages, as they will generally induce less observable changes in the system dynamics yet cause a relevant loss of payload data and information. This paper proposes a deep neural network to detect failures and investigate sensor sensitivity to damage classification for an orbiting satellite hosting a distributed network of accelerometers on a large mesh reflector antenna. The sensors-acquired time series are generated by using a fully coupled 3D simulator of the in-orbit attitude behaviour of a flexible satellite, whose appendages are modelled by using finite element techniques. The machine learning architecture is then trained and tested by using the sensors' responses gathered in a composite scenario, including not only the complete failure of a structural element (structural break) but also an intermediate level of structural damage. The proposed deep learning framework and sensors configuration proved to accurately detect failures in the most critical area or the structure while opening new investigation possibilities regarding geometrical properties and sensor distribution.

Keywords: space structures; space antenna; damage classification; deep learning; structural health monitoring

Citation: Angeletti, F.; Iannelli, P.; Gasbarri, P.; Panella, M.; Rosato, A. A Study on Structural Health Monitoring of a Large Space Antenna via Distributed Sensors and Deep Learning. *Sensors* **2023**, *23*, 368. https://doi.org/10.3390/s23010368

Academic Editor: Zenghua Liu

Received: 7 November 2022
Revised: 24 December 2022
Accepted: 26 December 2022
Published: 29 December 2022

1. Introduction

Modern large space structures (LSS) have a higher level of structural flexibility and larger dimensions than those experienced in past spacecraft. Due to the constraints imposed by the launch vehicle on the lift-off mass, the size of such flexible appendages (i.e., antennas, telescopes, solar panels, etc.) comes at the expense of the overall stiffness of their structural elements. Therefore, many large structures exhibit a multitude of closely spaced and lightly damped elastic modes, thus making the Control/Structure Interaction (CSI) phenomenon a very critical issue that can significantly affect the satellite's behaviour and degrade its performance. In fact, external and internal disturbances such as thermal loads, reaction wheels/mechanisms exported forces and vibrations, among others, or the control inputs for attitude stabilisation and orbital manoeuvres, can easily excite the vibrational responses of the satellites' flexible appendages [1]. Moreover, flexible appendages are generally subjected to narrow pointing and instrument requirements [2–4]. However, during their operative life, such parts are potentially exposed to several sources of damage and degradation, as well as sudden stiffness loss, due to debris impact or fatigue and material ageing or wear phenomena. When a failure or deterioration occurs, a change in the overall system properties takes place. Even minor local structural damages may induce a relevant change in the stiffness and dimensional stability of spacecraft scientific instruments, possibly compromising the acquisition of data and images.

Spacecrafts such as the Soil Moisture Active Passive (SMAP) [5,6], launched by NASA in 2015, or the European Copernicus Imaging Microwave Radiometer (CIMR) [7], currently under development by Thales Alenia Space, can be mentioned as representative cases of such large flexible systems. Both the cited missions consist of a main non-spun platform and a rotating part composed of an antenna boom, a deployable large mesh reflector model (LMRM), and a rotation mechanism. The interest in such larger and structurally complex missions imposes the precise analysis of how, where, and when failures can be detected with sufficient accuracy. To this end, we analyse in this paper a satellite hosting an LMRM payload with properties and characteristics comparable to the SMAP and CIMR missions. In particular, the LMRM is deemed a case of interest given its design maturity, promising current and future space applications, and also its high-precision pointing and balance requirements [8], which make it particularly susceptible to performance degradation in case of local structural damages. The goal is to study the sensitivity of a distributed network of sensors to structural damages when using a deep learning approach.

On account of this, the current and next LSS and related missions are calling for new de-risking technologies, aimed at minimising the impact of damages directly on orbiting satellites. A possible approach is to implement a set of sensors—whose amount and location will depend on the specific application—to promptly identify the location and the entity of the damage, and to assess the impact on the mission and relevant mitigation strategies [9]. In such cases, Structural Health Monitoring (SHM) strategies have hence assumed an increasing importance on the consideration of improved performance, safety, and costs, especially referring to modern space lightweight structures [10,11]. An important aspect of the SHM of the LSS is related to the use of in situ sensing to detect the structural damage of the structures and their related components [12–14]. Different sensing technologies can be used to identify the damages of on-board satellites: the solutions range from cameras allowing image processing (such as the debris hit detection on Sentinel 1-A's solar panel) to distributed devices—such as accelerometers, piezoelectrics, or optical fibre sensors—at the structural level, for both a time response acquisition and strain sensing [15].

Cameras are used to visually identify the faults by acquiring pictures and/or extracting the relevant information, such as displacement measurements. A vision-based method for in-plane displacement acquisition of cantilever beams was applied in [16], while Ye et al. [17] combined a vision-based system with digital image processing and proposed a hybrid approach using a multipoint pattern-matching algorithm. Although the vision systems generally prove to capture good results when compared to other types of in situ sensors, it is found that illumination has a critical effect on the overall performance. Vision in space faces even more challenges in respect to lighting conditions, exposition, and the accessibility of partially hidden areas. According to the orbit selected, there is the chance of detectors being saturated or blinded by direct sun light, as well as interference and a poorer performance due to the noisy environment for the light background.

As this paper focuses on identifying local failures that could be then difficultly observed via cameras [18], the focus is given to the possibility of implementing a network of distributed sensors. In particular, accelerometers are selected as the sensing technology to acquire structural nodal acceleration time histories, as they are space qualified, easily installable, and have a low cost/mass impact, as demonstrated by the Honeywell, InnaLabs, and PCB catalogues [19]. Additionally, the presence of a distributed set of sensors, e.g., accelerometers and piezoelectrics, is of particular interest as being crucial to pave the road to another key technology, such as active structural control for spacecraft [20–22], especially in terms of failure-tolerant strategies.

Over the past decades, several techniques have been studied to perform a diagnosis and failure analysis for space systems [23,24]. Such strategies are still mostly based on traditional methods, such as lamb wave-based approaches [25], used to detect cracks or flaws which perturb the waves propagation, or a comparison between healthy and damaged structural parameters, as in dynamic strain measurements [26] and optimisation methods for model updates [27] or Transmissibility Functions (TF) [28]. Another class of techniques successfully assessed and tested in a real-time SHM is based on the eigenvalue perturbation theory [29,30], including both first- and higher-order perturbation methods [31]. Recent advancements in singular spectral analysis [32,33] proved that not only can this approach allow to detect degradation patterns and model features in real time (even by using a single sensor output), but it can also relevantly filter out the noise component [34], i.e., without any external filters. Moreover, data-driven approaches can be used in space applications focusing on in-orbit sensed acceleration profiles. The research in this category traditionally adopts modern signal processing methods and tries to catch a sudden change in signals caused by the occurrence of damage [35,36]. Recently, methodologies based on data analysis and information extraction, in the broad field of machine learning, are being increasingly used to address damage/failure identification problems to achieve a wider range of applicability [37,38]. In order to overcome the limitations associated to traditional neural networks solutions [39], such as real-world noise, more complex deep learning (DL) models and techniques, with higher generalisation capabilities, have been introduced as data extractors, classifiers, and predictors [40–42]. Such models can include also recurrent neural networks (RNN) [43] to efficiently obtain the information from time-series data. In fact, a state-of-the-art recurrent DL network, known as the Long Short-Term Memory NN (LSTM-NN), was applied to failure identification for an LSS by the authors [9] with promising results.

The purpose of this study is then to expand and test the limits of the approach proposed in the authors' previous research [9]. Indeed, the present study focuses on a much more complex study case from the structural point of view, different from both the geometrical and dynamic point of view, and on the possibility to detect diverse failure types (not only in the antenna attachment area with the satellite but also on the supporting structure) that are quite difficult to detect due to a neglectable impact on the satellite dynamics. Moreover, the concept of "damage" is now extended to include not only complete failures of structural elements but also partial damages that are notoriously more difficult to classify. Indeed, one of the main objectives of this paper is to propose a more generalisable deep neural network (DNN) system for the SHM and to investigate the architecture sensitivity to different damages both in terms of the locations and structural type. In this research, a data acquisition process is carried out based on attitude manoeuvres simulations, the sensor readings are correlated to the damage presence and location, and the processing is performed by using Long Short-Term Memory neural networks (LSTM-NN) layers combined with non-linear activation layers and fully connected blocks. The resultant deep learning model is then trained to solve the sequence-to-label classification issue of structural damage detection.

The paper is divided as follows. Firstly, the fully coupled 3D mathematical model of a flexible spacecraft is implemented in a simulator for carrying out a wide set of in-orbit attitude manoeuvres (Section 2). Then, the spacecraft test case model is described, including the network of distributed sensors for the SHM and the damage configurations addressed herein (Section 3). In Section 4, the implemented deep neural architecture for the damage classification is described, based on an LSTM variant. Section 5 reports the main results and analyses the performance of the trained classification network. Finally, Section 6 discusses the relevant findings while Section 7 highlights some of conclusions and provides the indications about further steps in this research effort.

2. Dynamics of a Flexible Spacecraft

This section deals with the general equations of a flexible spacecraft in a gravitational field. The main steps for the description of the motion of a flexible body are:

- The definition of kinematic parameters for flexible structures;
- The definition of the functionals: kinetic, elastic, and gravitational;
- The definition of the Lagrangian;
- The writing of the equilibrium equations through the Hamilton's principle.

To write the above-mentioned quantities, the position of a generic point P belonging to the spacecraft in the inertial reference frame $\mathcal{R}_i$ can be expressed as the superposition of a rigid motion plus a combination of structural modes:

$$\mathbf{x}_P^I = \mathbf{x}_O^I + \mathbf{P}_{B/I}\left(\xi + \sum_{k=1}^{N} A_k(t)\phi_k(\xi)\right) \tag{1}$$

where $\mathbf{x}_O^I$ is the position of a reference point O of the body in $\mathcal{R}_i$, ξ is the rigid component of the position vector of point P with respect to the reference point O in the body reference frame $\mathcal{R}_b$ while its flexible component is expressed via the classical modal decomposition through N modal amplitudes A_k and their relevant mode shapes $\phi_k(\xi)$. Finally, $\mathbf{P}_{B/I} = \mathbf{P}_{I/B}^T$ is the rotation matrix that transforms vectors from $\mathcal{R}_b$ to $\mathcal{R}_i$.

Using Hamilton's principle [44], the equilibrium equations of a flexible spacecraft can be obtained by evaluating the functionals associated to: the kinetic energy, the elastic potential of the system, the gravitational field, and the control actions (if any). Only the final equations of motion, Equations (2)–(4), are reported in this work for brevity's sake while the methods leading to their derivation are covered in [44–46]. Hence, the governing translational, rotational, and modal base flexible equations of a flexible body under gravitational force, respectively, can be written as:

$$m\ddot{\mathbf{x}}_O^I + \omega \times (\omega \times \tilde{p}) + 2\omega \times \sum_{k=1}^{N} \mathbf{\Lambda}_k \dot{A}_k + \dot{\omega} \times \tilde{p} + \sum_{k=1}^{N} \mathbf{\Lambda}_k \ddot{A}_k = f_g \tag{2}$$

$$\tilde{p} \times \ddot{\mathbf{x}}_O^I + \tilde{\mathbf{J}}\dot{\omega} + \sum_{k=1}^{N}\left(\tilde{\mathbf{J}}_1^k \dot{A}_k\right)\omega + \sum_{k=1}^{N}\sum_{t=1}^{N} \Xi^{kt} \dot{A}_k \dot{A}_t + \sum_{k=1}^{N} \tilde{\mathbf{\Gamma}}_k \ddot{A}_k + \omega \times \tilde{\mathbf{J}}\omega + \omega \times \sum_{k=1}^{N} \tilde{\mathbf{\Gamma}}_k \dot{A}_k = C_g + C_c \tag{3}$$

$$\mathbf{\Lambda}_k^T \ddot{\mathbf{x}}_O^I + \tilde{\mathbf{\Gamma}}_k^T \dot{\omega} + \ddot{A}_k + \omega_k^2 A_k + 2\zeta_k \omega_k \dot{A}_k - \frac{1}{2}\omega^T \tilde{\mathbf{J}}^k \omega + 2\omega^T \sum_{t=1}^{N} \Xi^{kt} \dot{A}_t = \tilde{f}_{g,k} \quad k = 1, \ldots, N \tag{4}$$

where the symbol $\times$ indicates the cross product operation between two vectors; ω is the angular velocity of the spacecraft defined with respect to $\mathcal{R}_i$; m, $\tilde{p}$, and $\tilde{\mathbf{J}}$, respectively, are the total mass of the spacecraft, the static moment, and its inertia tensor. The translational and rotational modal participation factors, accounting for the most significant contribution to the coupling between rigid and flexible dynamics, are, respectively, expressed as $\mathbf{\Lambda}_k$ and $\tilde{\mathbf{\Gamma}}_k$. ω_k and ζ_k are the natural frequency and damping factor associated to the k-th mode of vibration. Further parameters, such as Ξ^{kt} and $\tilde{\mathbf{J}}_1^k$ (first variation in the inertia tensor due to the flexibility), appear in higher-order terms in Equations (3) and (4).

Moreover, the gravitational forces f_g (in the inertial reference frame $\mathcal{R}_i$), the gravity gradient torques C_g (in the body reference frame $\mathcal{R}_b$), the generalised gravitational forces projected on the k-th elastic mode $\tilde{f}_{g,k}$ are reported as follows:

$$f_g = -m\mu_{\oplus}\frac{\hat{s}}{|\mathbf{x}_O^I|^2} \quad \text{with } \hat{s} = \frac{\mathbf{x}_O^I}{|\mathbf{x}_O^I|} \tag{5}$$

$$C_g = -\frac{\mu_{\oplus}}{|\mathbf{x}_O^I|^2}(\tilde{p} \times \hat{s}) + \frac{3\mu_{\oplus}}{|\mathbf{x}_O^I|^3}(\hat{s} \times \tilde{\mathbf{J}}\hat{s}) \tag{6}$$

$$\widetilde{f}_{g,k} = -\frac{\mu_\oplus}{|\mathbf{x}_O^I|^2}\left(\widehat{\mathbf{s}}^T \boldsymbol{\Lambda}_k\right) \quad (7)$$

being $\widehat{\mathbf{s}}$ the unitary vector of the position of the origin of $\mathcal{R}_b$ in the inertial frame $\mathcal{R}_i$ and $\mu_\oplus$ the gravitational constant of Earth. Finally, $\boldsymbol{C}_c$ is the vector of the applied control torques.

For more details about the symbols reported in the above equations, please refer to [44,47].

3. Spacecraft Model and Dataset Generation

To test the damage classification DL architecture, the case of a spacecraft equipped with an LMRM is considered. Such a model, representative of a realistic Earth Observation (EO) satellite, is composed of a central platform of prismatic shape, two solar arrays (3×8 m), and a large mesh reflector (diameter ~ 12 m).

In detail, the LMRM is attached to a central platform—considered to be rigid with respect to the flexible structure—in correspondence to an attachment point P_1 defined in the S/C reference frame, whose origin O is located at the centre of the launch vehicle payload adapter, while the two large solar panels are connected to the bus at points P_2 and P_3 (see Figure 1). Each array is reinforced via a supporting structure in Carbon-Fibre-Reinforced Polymers (CFRP) and a Yoke is added to reproduce the attachment of the panel to the platform. Because the central hub is assumed rigid, each flexible substructure is directly assembled in MSC Nastran by connecting them via rigid body element connections to a single node coinciding with the point O. Then, the relevant data concerning the inertial properties, modal participation factors, and the natural frequencies of the flexible structure computed with respect to O are imported in the Matlab environment, implementing the dynamics of the flexible spacecraft described in Section 2. The inertial properties of the platform, without considering the appendages, are listed in Table 1.

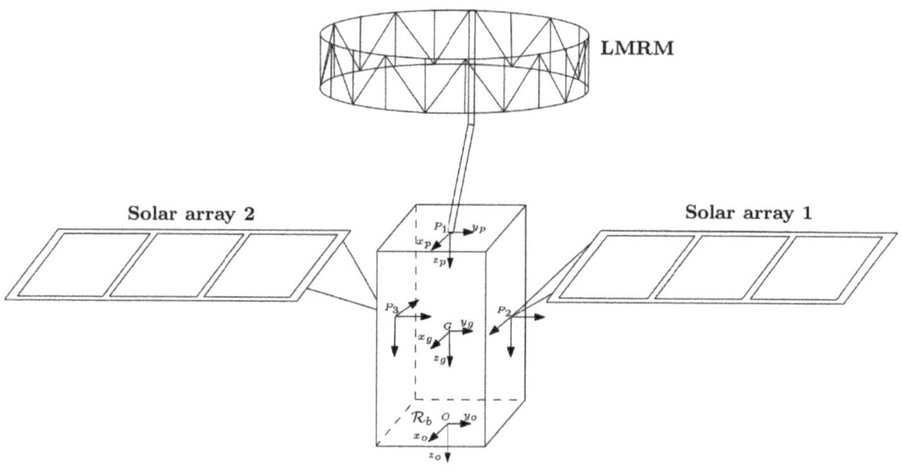

Figure 1. Schematic of the spacecraft model.

Table 1. Platform inertial properties.

Mass (kg)	Inertia Wrt CoM (kgm^2)						CoM Wrt $\mathcal{R}_b$ (m)		
	Jxx	Jyy	Jzz	Jxy	Jxz	Jyz	X	Y	Z
1200	903.6	985.2	885.6	2	4	−7	0	0	−2

3.1. Large Mesh Reflector Structural Model

This section briefly introduces a representative case of a realistic LMRM, which was designed using MSC Nastran FEM tool based on information available in the literature [48]. The LMRM is here composed of a circular truss backbone structure, a mesh reflector, support cables, and a deployable boom. In detail, the truss structure counts 30 unit bays, while the parabolic surface (with an areal density of about 0.3 kg/m^2) is sustained by a net of rod elements representing the reflector cable mesh. Concerning the LMRM, the horizontal and vertical truss tubes are designed to be 2.5 m long (external radius 0.021 m, internal radius 0.02 m), while the diagonal ones 2.81 m (external radius 0.01 m, internal radius 0.009 m). The selected material for the truss is Carbon-Fibre-Reinforced Polymer (CFRP), with density equal to 1550 kg/m^3 and Young's modulus 125 GPa. The cable mesh is represented by rod elements with area equal to 1 mm^2, in CFRP material. The reflector's main structural properties are listed in Table 2. An 8 m long, 20 kg, extendable boom is also modelled to complete the structural model of the LMRM.

Table 2. Reflector properties.

Property	LMRM
Mass (kg)	Truss: 38.34 Mesh: 4.25
	Total: 42.6
1st Mode (Hz)	Truss: 0.20 Truss + Mesh: 0.53

Generally, LMRM-deployed structures show three different constrained modes, namely yaw, pitch, and roll modes. They correspond to torsion and bending with respect to the main coordinated axes. When mounted on a satellite hosting other flexible appendages, the flexible modes of the payload interact with the structural dynamics of the additional substructures. In particular, the presented study case is composed of a central bus, equipped with both symmetric solar panels and the LMRM. The set of modes of the assembled spacecraft is illustrated in Figure 2, along with an overview of the complete system. It can be noticed how the system modal behaviour is characterised by both symmetric/asymmetric bending and torsion modes of the solar panels, interacting with the three traditional modes of the LMRM payload.

3.2. Training Set Generation and Processing

With respect to the location of damage on the structure of the spacecraft, failures are expected to lead most likely to major consequences in the attachment element of the mesh reflector to the satellite. In the following, as reported in Figure 3, six possible structural elements of the reflector are supposed to be susceptible to a failure. In particular, failures are placed in those areas where issues may occur, such as connection areas, where, for instance, hinges or similar components are located due to the structure deployment strategy. Moreover, they are considered at both the attachment area of the antenna with the satellite and on the supporting truss to test the failure/damage of elements with completely different impact on the overall satellite dynamics. Furthermore, for each element, two damage scenarios are considered: the first one consists of the complete failure of the structural element ("Broken element" in Table 3) while the second models a partial damage of the element itself ("Damaged element" in Table 3). To mathematically reproduce these behaviours, the "Broken elements", associated to the fully damaged location of the structure, are erased from the finite element model (in other words, the structural stiffness is considered null, and the order of the stiffness matrix is changed), while the partial structural failure, on "Damaged elements", is introduced into the structure considering a 50% reduction in the element's Young's modulus, which is representative of the mechanical properties of the material of the damaged element. Table 3 also reports the associated label to each elements and damage type used to train the classification DL network.

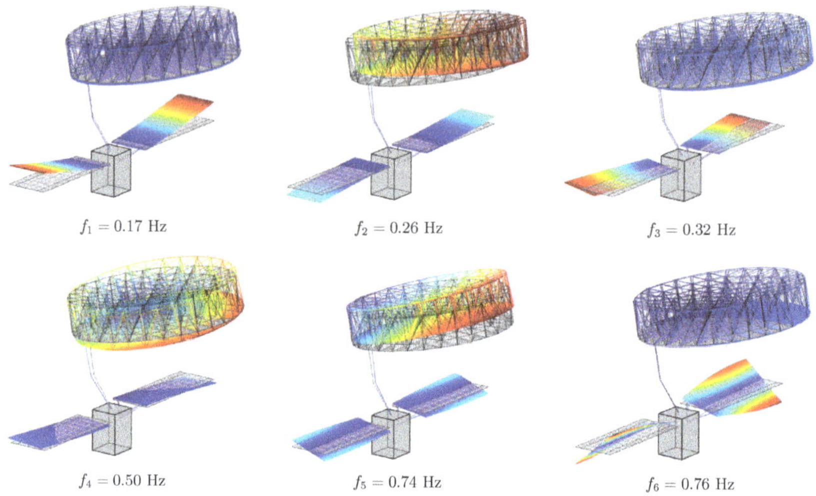

Figure 2. Mode shapes and natural frequency of the assembled spacecraft.

Table 3. Damaged element ID, associated damage configuration (broken or partial failure), and associated label.

Damaged Element ID	Damage Configuration	Associated Label
-	Undamaged system	0
Elm 4115	Failure 1—Broken element Failure 2—Damaged element	1 2
Elm 4110	Failure 3—Broken element Failure 4—Damaged element	3 4
Elm 3872	Failure 5—Broken element Failure 6—Damaged element	5 6
Elm 3920	Failure 7—Broken element Failure 8—Damaged element	7 8
Elm 3823	Failure 9—Broken element Failure 10—Damaged element	9 10
Elm 3972	Failure 11—Broken element Failure 12—Damaged element	11 12

The behaviour of the LMRM during its operational life is observed by a network of sensors that has been designed in order to measure the time response of the structure in some critical points, namely Nodes of Interest (NOIs), of the FEM (see Figure 4). In detail, a set of 12 tri-axial accelerometers is considered, with the majority of them localised in proximity to critical areas, such as the connection of the reflection with the boom.

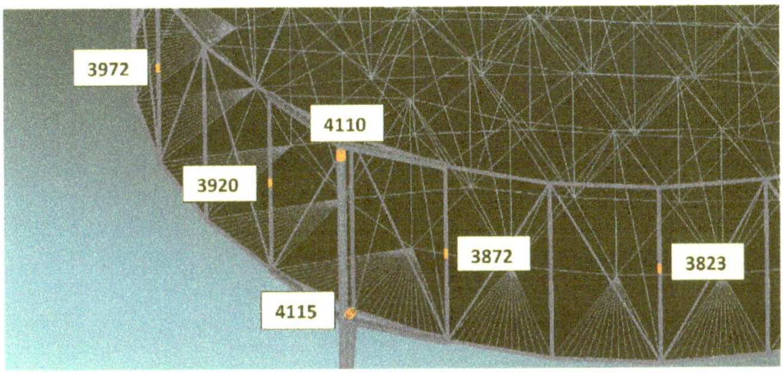

Figure 3. Possible damage location on the reflector structure (labels are used to identify the element in which the failure can occur).

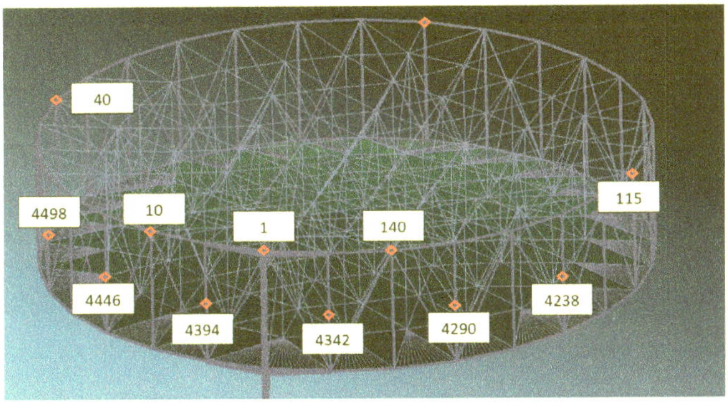

Figure 4. Network of distributed sensors (labels are used to identify the node on which the sensor is placed).

Once the damaged structural sub-models, deriving from the original undamaged one, and the network of sensors are defined, the dataset generation for the training of the DNN can be set up. This process, summarised in Figure 5, consists of the following steps:

- The set N_d structural models extrapolated from the finite element suite (MSC Nastran), as previously described, are imported in Matlab to perform further analysis. At this point, the non-linear simulator of the flexible spacecraft (built on the basis of the equations reported in Section 2) is used to carry out N_m attitude manoeuvres and produce the measurements from the sensors network. A quaternion-based PD controller [49,50] is implemented to produce the desired control torque C_c:

$$C_c = -K_p \bar{q}_e^I \operatorname{sign}(q_0) - K_d \omega \tag{8}$$

  with K_p and K_d as proportional and derivative gain matrices, respectively, $\bar{q}_e^I$ the error quaternion [50], q_0 scalar component of the quaternion, and ω the angular rate of the spacecraft. The dataset of possible manoeuvres is built by varying properly both the desired final attitude angles of the manoeuvre (considering also one- and two-axis manoeuvres to improve the variation in the data collection) and the gains of the controller.
- s quantities of interest (i.e., $s = 36$ because 12 tri-axial accelerometers are mounted on the structure) are then collected by the sensors network with a sampling frequency of 10 Hz. At this stage, the raw data, extrapolated from the simulator, consists of a multidimensional array $X_1 \in R^{s \times k \times q}$ with $k = N_d \cdot N_m$ and q the number of time samples. Moreover, a Gaussian noise equal to the 2% of measured values is considered when acquiring the data from the accelerometers to simulate a realistic condition.

- The raw multidimensional array X_1 is not directly fed to the network for the training, but it passes through two intermediate steps of pre-processing. Indeed, data-driven classification algorithms need a relevant amount of training data, which require to be properly processed to avoid introducing biases and ill-conditioning in the results, while being used in a computationally and time-expensive process (in particular for recurrent neural networks). In this paper, firstly, each time sequence is truncated ($\boldsymbol{X}_1 \rightarrow \boldsymbol{X}_R \in R^{s \times k \times q_r}$) in order to preserve only parts of the signals where relevant dynamical content is detected. In the current study, the sequences are truncated to 25s for a total of 251 samples. Such a reduction is supported by the fact that responses below a certain threshold do not improve classification accuracy during network training or operation. Conversely, it was also observed in [9] that feeding the full-time sequence to the network rather jeopardises the performance of the training process (and subsequently also the performance of damage identification in real-time condition). The second step consists of normalising the measurements so as to ensure proper dynamic range of the variables in the learning space of the NN model. Therefore, in this study, each time sequence is normalised according to their mean and standard deviation as reported here:

$$\boldsymbol{X}_{\sigma_{ijl}} = \frac{\boldsymbol{X}_{R_{ijl}} - \mu_x}{\sigma_x} \quad \text{with} \quad \begin{array}{l} i = 1, \ldots, s \\ j = 1, \ldots, k \\ l = 1, \ldots, q_r \end{array} \tag{9}$$

where μ_x and σ_x are, respectively, the mean value and standard deviation of the measured time history $\left[\boldsymbol{X}_{R_{ij1}}\ \boldsymbol{X}_{R_{ij2}}\ \cdots\ \boldsymbol{X}_{R_{ijq_r}}\right]$.

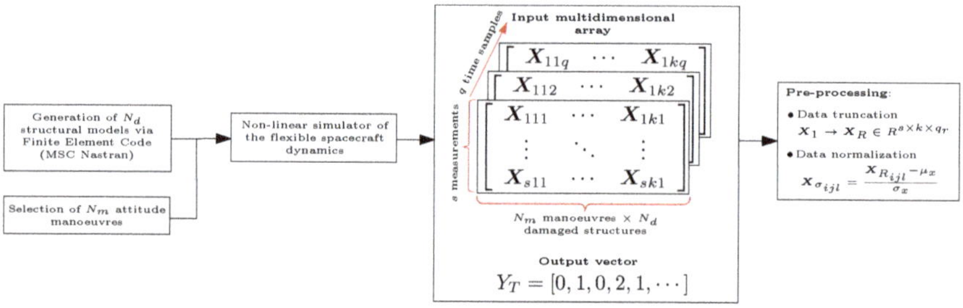

Figure 5. Dataset generation and processing prior to network training.

For the sake of illustration, we show, in Figure 6, the behaviour of the truncated and normalised 36-sensor time series associated with one observation having class label "0" (undamaged system). The picture aims at pointing out that the most relevant dynamic content of the acceleration time sequences is detected in the first-time instants of the manoeuvre. Indeed, after the rotation of the satellite has occurred, the residual naturally damped oscillations fade out and the system returns to a stationary condition while the measurements go to zero. Therefore, the time histories have been cut to avoid including in the dataset the components (equal to zero or under a certain threshold) which would contribute to flatten the dataset. The final dataset used for the training consists of an input multidimensional array $\boldsymbol{X}_R$ of size $36 \times 3003 \times 251$ while the output vector contains the labels of the failures reported in Table 3 and it has dimensions equal to 3003×1, where each simulation is associated with the label of the corresponding failure.

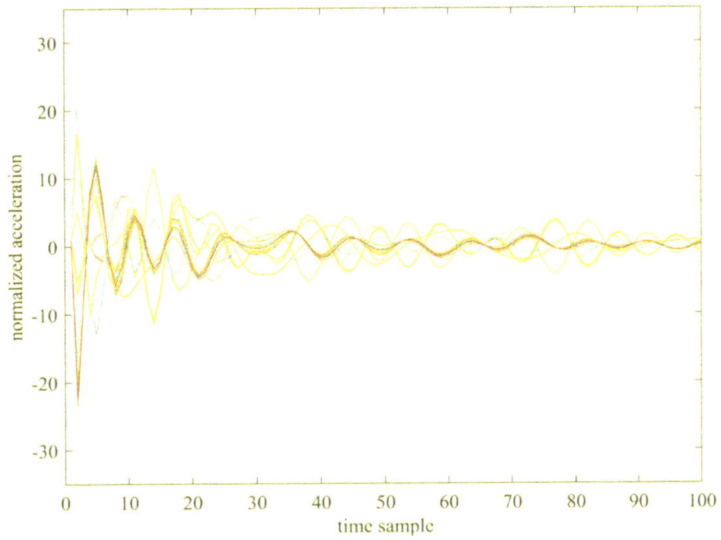

Figure 6. Time−series behaviour (first 100 time samples) of the 36-sensor data relevant to one of the 231 observations of the dataset with the undamaged system (class label "0").

4. Bidirectional Long-Short Term Memory Network

In recent times, deep recurrent neural networks have been the staple for time-series classification problems. In particular, because most of the real-world use cases revolve around sequences that are in the hundreds or thousands of observations, it was deemed necessary to utilise learning models that are able to connect information found far in the past to the present task. In this regard, LSTM networks are a well-known solution and have been applied to a multitude of problems. While the LSTM model is indeed reliable, there are some cases in which a more robust information extraction is needed, such as the classification of long sequences. To this end, bidirectional Long Short-Term Memory (Bi-LSTM) networks have been introduced and applied to many practical contexts [51].

Unlike the classic, unidirectional LSTM model, the Bi-LSTM applies a first LSTM on the input sequence in the given order, then it reverses the sequence and feeds the second LSTM. It was proved [52] that the Bi-LSTM architecture works better in a multivariate time-series analysis and classification, which is the application studied in this work. A detailed scheme of the workings of the Bi-LSTM model is reported in Figure 7. The input size is equal to N and the hidden state $\mathbf{h}^{i}_{j}$ refers to the j-th time step of the i-th (backward or forward) LSTM.

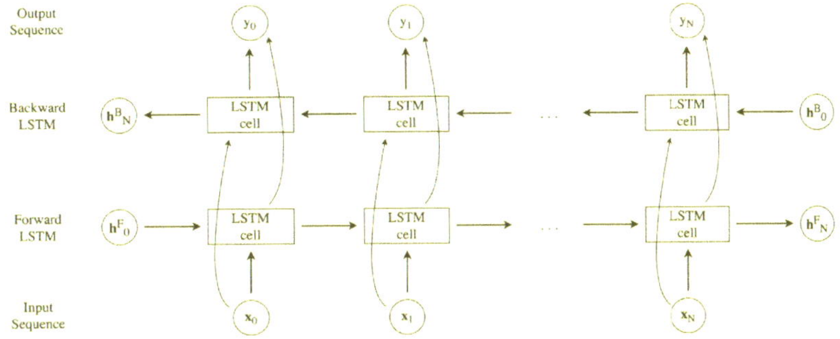

Figure 7. Unrolled network structure of a bidirectional LSTM.

It is important to underline that the Bi-LSTM is trained twice, once for both directions (forward and backward). The bidirectional application of this more complex LSTM cell structure is performed to robustly save information from both past and future samples, improving the accuracy of the model [53].

In the present work, we employ a multivariate deep classification model which we called deep Bi-LSTM (DBLSTM); it is composed of an input layer (in the form of a sequence) and two stacked Bi-LSTM layers, interspersed with a dropout layer, a dense layer, a Softmax layer, and a classification layer. A qualitative trade-off was carried out between the different network architectures to find the best compromise between the computational costs and accuracy. Therefore, the network proposed in this paper was selected as the simplest stack that could be applied to this configuration.

The sequence input layer creates the proper input sequence to feed the network. The two Bi-LSTM layers are the ones charged with learning the temporal dependencies among the data; two stacked layers are preferred over a single layer to enhance the predictive power of the model [54]. A dropout layer is inserted between the Bi-LSTM layers as a regularisation factor to avoid overfitting. A fully connected (FC) layer is stacked on the last Bi-LSTM unit to map the output of the Bi-LSTM layer to a desired output size. The Softmax layer turns the output of the FC layer into probability values summing to one so that the final classification layer is able to perform the binary classification. The architecture of the network is visualised in Figure 8.

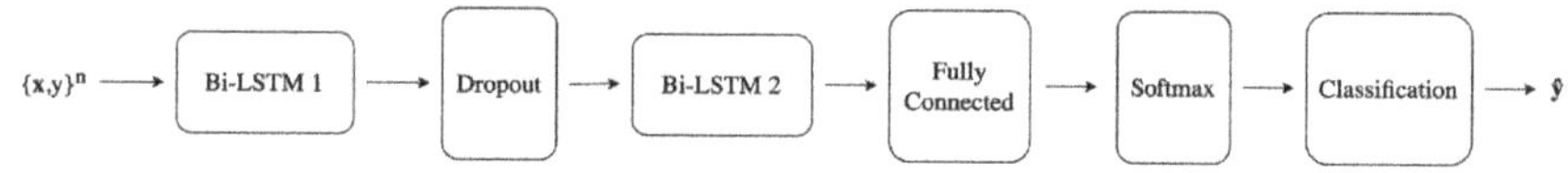

Figure 8. Network configuration of the proposed DBLSTM.

The output of the model is a binary vector $\hat{\mathbf{y}}$, whose dimension is equal to n and which contains the predicted outcome.

In the DBLSTM scheme, the number of hidden units in the Bi-LSTMs is denoted as $N_h^{(1)}$ for the first layer and $N_h^{(2)}$ for the second layer; these are the recurrently connected blocks (i.e., the computational units) and their number should be optimised based on the application and data purposes.

In fact, based on the dataset defined earlier, it is possible to empirically set the hyperparameters for the DBLSTM network structure; the first Bi-LSTM layer has 36 features and two or three classes in the FC and Softmax layers, depending on the specific problem, as explained in Section 5.

5. Results

The present study investigates the sensitivity and accuracy of the structural damage recognition by using an ensemble of sensors distributed across the structure to be monitored. We dealt with this problem by using a suited DNN architecture for the multivariate time-series classification, namely the one introduced in Section 4, and the dataset obtained by the measures through the simulations on the spacecraft model defined in Section 3.

We carried out three different types of analysis; in every problem, all the sensor data are used to build a 36-element vector associated with each time sample of the multivariate time series that feed the input of the adopted DNN:

- First, six binary classification problems were considered, one for each structural element reported in Table 3. Two classes only are selected among the observations: 231 time series having the class label "0" (undamaged system) and 462 times series having a unique class label associated with that element, either broken or damaged (i.e., a unique class named "1-2" for "Elm 4115", a unique class named "3-4" for "Elm 4110", and so on). This meets the normal logic of damage detection to firstly determine whether damage has occurred or not and then to determine the degree of the damage.

- Successively, twelve binary classification problems were considered, two of them for each structural element reported in Table 3. Two classes only are selected among the observations: 231 time series having the class label "0" (undamaged system) and 231 times series having one class label (between "1" and "12"), which is associated with a specific failure of the considered structural element. Here, the rationale is to find whether, by analysing the data coming from all the sensors, is it possible to recognise one specific failure at a time, either a broken or damaged element, with respect to an undamaged condition.
- The third analysis pertains to six three-class classification problems, one for each structural element reported in Table 3. Three classes are considered in this case: 231 time series having the class label "0" (undamaged system); 231 times series having an odd class label (between "1" and "11"), which is associated with the broken failure of the considered structural element; and 231 times series having the even class label (between "2" and "12") associated with the damaged failure of the same element, which is the previous odd class label increased by one. In this situation, we want to study the capability of the proposed multivariate deep learning approach to identify both a broken and a same damaged element with respect to the undamaged condition.

The optimal model performance and selection of the considered deep neural architecture were obtained by means of a nested cross-validation procedure. Namely, an outer 10-fold cross-validation was carried out for a model performance evaluation. In each of these folds, using the training data only, a grid search procedure was applied to obtain the best setting of the hyperparameters for an optimal model selection. For each hyperparameters' setting (i.e., for each point of the grid to be searched), an inner 3-fold cross-validation was carried out for obtaining different partitions into training and validation sets in order to actually train the neural network with the ADAM algorithm [55].

Because of the randomness in any k-fold partitioning and in neural network parameters' initialisation, both the outer and inner cross-validations were repeated for 10 and 5 runs, respectively, using a different seed for the random number generation. Accordingly, the statistical average of the classification accuracy on the validation set, over the different inner folds and runs, was used for the best hyperparameters' selection during the grid search; the statistical average of the classification accuracy on the test set, over the different outer folds and runs, was used to evaluate the final model performance, and it is reported in the following tables.

The hyperparameters optimised in each grid search were the number of units $N_h^{(1)}$ and $N_h^{(2)}$ in the first and second Bi-LSTM layer of the adopted network, respectively. The possible values for both of them were searched in the range from 5 to 50 with a step of 5 units. We note that the optimal values of the hyperparameters, which are obtained at the end of every grid search procedure, generally differ among the several outer folds and hence only a statistical average of the optimal setting can be reported in the following tables. The other main hyperparameters set in advance were a dropout rate of 0.1; 150 maximum epochs; five iterations per epoch (which affects the mini batch size according to the actual training size); an ADAM learning rate of 1.0 (0.75% drop factor every 10 epochs); a regularisation factor of 0.0001; a gradient decay factor of 0.9; and the validation patience of 50 epochs before early stopping.

All the experiments were performed using Matlab® R2022a on a machine provided with an AMD Ryzen Threadripper™ 3970X CPU (32 cores at 4.5 GHz), 128 GB of DDR4 RAM (at 3.2 GHz), and an NVIDIA® RTX A6000 GPU (10752 cores at 1.8 GHz) with 48 GB of GDDR6 RAM (bus width 384 bits at 16 GB/s).

The numerical results for the two-class recognition problems described before, both the six and twelve arrangements, are reported in Table 4. As explained, the accuracy is the average on the several test sets while the optimal values of the Bi-LSTM slightly differ among the different optimisation tests. It is evident that the structural element "Elm 4110" can be well recognised both in the case of explicit breaking (class "3") or damage (class "4"); also, the prior discrimination as to whether general damage occurred (class "3-4") or not is obtained with a maximum accuracy close to 100%. The element "Elm 4115" is well classified in the case of explicit breaking only (class "1"), while the prior discrimination of general damage (class "1-2") is less accurate but good enough around 75%.

The other failures are recognised at a poor rate around 50% in the case of either broken or damage recognition. In these cases, the prior analysis as to whether general damage has occurred or not seems more accurate around 66.7%, but this is due only to the unbalanced datasets used in such situations, as the undamaged time series are 231 while the ones associated with general damage (either the "broken" or "damaged" class) are doubled to 462. Nonetheless, a discussion in this regard is reported in the next section. We also note that the optimal values of the Bi-LSTM units are always considerably smaller than 50, which was the largest value considered during the grid search; thus, the overfitting of the tested networks was correctly prevented.

Table 4. Classification Results of the 2-class Structural Health Recognition Problems.

Failure	Class Labels	Accuracy (%)	Bi-LSTM Units 1	Bi-LSTM Units 2
Elm 4115 broken–damaged	0, 1-2	75.8 ± 4.9	20 ± 10.0	11 ± 4.2
Elm 4115 broken	0, 1	100.0 ± 0.0	30 ± 12.2	12 ± 2.7
Elm 4115 damaged	0, 2	47.3 ± 8.5	14 ± 5.5	13 ± 2.7
Elm 4110 broken–damaged	0, 3-4	99.9 ± 0.3	22 ± 4.5	6 ± 2.2
Elm 4110 broken	0, 3	99.8 ± 0.5	16 ± 5.5	5 ± 0.0
Elm 4110 damaged	0, 4	100.0 ± 0.0	22 ± 11.0	6 ± 2.2
Elm 3872 broken–damaged	0, 5-6	66.7 ± 0.3	16 ± 8.9	14 ± 5.5
Elm 3872 broken	0, 5	58.8 ± 21.3	18 ± 11.0	16 ± 6.5
Elm 3872 damaged	0, 6	50.4 ± 1.0	32 ± 17.9	10 ± 5.0
Elm 3920 broken–damaged	0, 7-8	66.7 ± 0.3	20 ± 17.3	10 ± 7.1
Elm 3920 broken	0, 7	54.3 ± 7.5	26 ± 21.9	15 ± 6.1
Elm 3920 damaged	0, 8	50.0 ± 0.4	26 ± 16.7	8 ± 4.5
Elm 3823 broken–damaged	0, 9-10	67.1 ± 1.1	28 ± 20.5	10 ± 6.1
Elm 3823 broken	0, 9	49.8 ± 0.3	18 ± 8.4	11 ± 8.2
Elm 3823 damaged	0, 10	50.0 ± 0.4	16 ± 8.9	6 ± 2.2
Elm 3972 broken–damaged	0, 11-12	66.7 ± 0.3	18 ± 7.9	8 ± 4.5
Elm 3972 broken	0, 11	50.0 ± 0.4	32 ± 17.9	6 ± 2.2
Elm 3972 damaged	0, 12	49.8 ± 0.3	30 ± 14.1	9 ± 6.5

It is interesting to show which kind of mistakes are made in wrong situations. For instance, in Figure 9, the confusion matrix in the case of "Elm 4115" damaged (class "2") is reported. In this case, the number of undamaged observations that are misclassified as "Elm 4115" damaged (i.e., false positive) is 70, which is less than the 149 observations of "Elm 4115" damaged that are misclassified as undamaged (i.e., false negative). Conversely, as shown in Figure 10 in the case of "Elm 3823" broken (class "9"), the number of false positive observations is 139, which is larger than the 93 false negative observations. So, no general rules can be proved to exist about the majority of false positive or false negative observations.

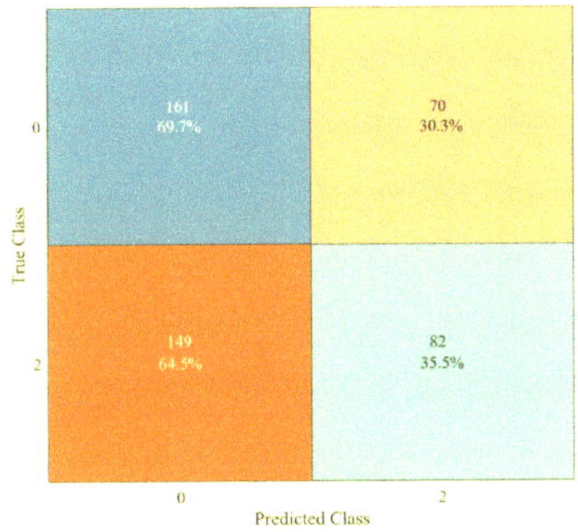

Figure 9. Confusion matrix for the binary classification between undamaged system (class "0") and "Elm 4115" damaged (class "2").

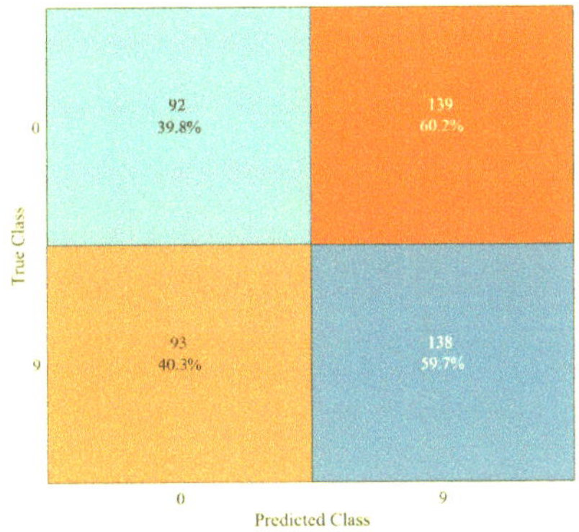

Figure 10. Confusion matrix for the binary classification between undamaged system (class "0") and "Elm 3823" broken (class "9").

The numerical results for the three-class recognition problem are reported in Table 5. It is confirmed that, also in this case, element "Elm 4110" is well-classified in both situations; in fact, the accuracy is close to 100%. Similarly, "Elm 4115" is well-classified only in part of the situations, therefore achieving an accuracy close to 72%. The failures of the other elements are poorly recognised, staying always around 33% (i.e., 1 out of 3). Moreover, in this case, the optimal values of the Bi-LSTM units are far from the largest possible value of 50, thus showing that the optimal networks are not overfitted.

Table 5. Classification Results of the 3-class Structural Health Recognition Problem.

Failure	Class Labels	Accuracy (%)	Bi-LSTM Units 1	Bi-LSTM Units 2
Elm 4115 broken, damaged	0, 1, 2	72.7 ± 3.0	32 ± 11.0	11 ± 4.2
Elm 4110 broken, damaged	0, 3, 4	99.8 ± 0.3	18 ± 8.4	8 ± 2.8
Elm 3872 broken, damaged	0, 5, 6	33.8 ± 0.5	20 ± 7.1	8 ± 4.5
Elm 3920 broken, damaged	0, 7, 8	34.5 ± 6.0	22 ± 11.0	16 ± 6.5
Elm 3823 broken, damaged	0, 9, 10	33.2 ± 0.1	22 ± 8.4	8 ± 6.7
Elm 3972 broken, damaged	0, 11, 12	33.3 ± 0.3	32 ± 17.9	14 ± 6.5

By looking at the confusion matrix shown in Figure 11, we see that the broken situations of "Elm 4115" (class "1") are always perfectly classified. However, there are 51 false positive classifications of the undamaged system as "Elm 4115" damaged (class "2") and 138 false negative classifications of "Elm 4115" damaged as an undamaged structure. By the way, these numbers are similar to the case of the binary classification reported in Figure 9, therefore showing a peculiarity on some time series related to this structural element when damaged. A totally different situation pertains to other elements with a poor classification accuracy, for instance, as illustrated in Figure 12 for element "Elm 3823". In this case, there is an even distribution of false positive classifications (i.e., 46/47 cases) between class "9" and "10", as well as an equal distribution of 139 false negative recognitions of an undamaged system instead of a broken or damaged element.

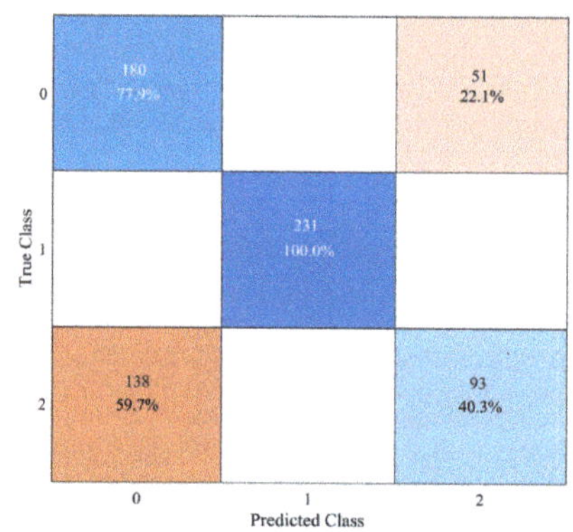

Figure 11. Confusion matrix for the 3-class recognition problem regarding the undamaged system (class "0") and the failures of element "Elm 4115" (class "1" and "2").

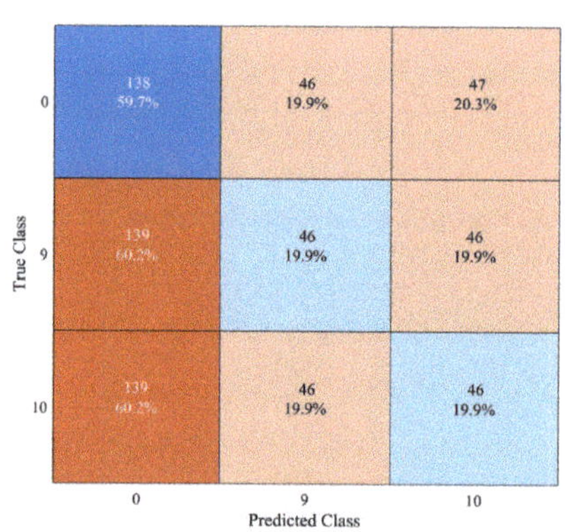

Figure 12. Confusion matrix for the 3-class recognition problem regarding the undamaged system (class "0") and the failures of element "Elm 3823" (class "9" and "10").

6. Discussion

The obtained outcome is comprehensively interpreted and discussed in Table 6. The analysed sensing architecture proved to be optimal to classify the damages for the elements near the antenna attachment area with the central satellite. It can be noticed, however, that the presented results clearly depend on the physical location of the damage with respect to the sensors network and, simultaneously, on the physical properties of the system itself. This opens the possibility to future investigations on the optimal positioning of sensing devices with respect to the most sensitive structural elements based on machine learning approaches, with particular regards to the problem of investigating and differentiating similar data originating in different attitude manoeuvres.

Table 6. Results interpretation.

Structural Element	Discussion
Elm 4110	The DNN architecture proved to well detect and identify the failure, both in the break and the partial damage cases. The reason for this outcome can be interpreted as three-fold. On one hand, Elm 4110 and Elm 4115, located at the antenna attachment point with the satellite, are the ones affecting the overall system behaviour the most—yet still limitedly, even from a structural point of view—when compared to the other considered elements. Indeed, as proved by the previous research discussed in [9], failures near a substructure attachment area to a satellite can be straightforwardly identified by using LSTM-based machine learning approaches. This is true in terms of structural stiffness modifications and consequently of natural frequencies changes and different antenna nodal elastic displacements. On the other hand, local partial damage—which could be more difficult to classify due to its minor impact on the satellite attitude dynamics—can be identified with high precision mainly due to its vicinity to Sensor "1", which is able to better "see" the local effects of the failure. Finally, the location of Elm 4110—as well as 4115—is peculiar with respect to the other elements: it is positioned in a way to not be impacted by the symmetry effects of the overall system structure and also in terms of the registered acceleration profiles (please refer to further discussions in this table for more details).
Elm 4115	The DNN architecture demonstrated to be able to well detect and classify the total break of Elm 4115. Similarly to Elm 4110, this behaviour can be assumed as due to the more relevant impact on the system dynamics than other structural elements and on its position. In detail, while a good classification performance can be observed for the three-class problem in Table 5, the partial damage cannot be identified with the same accuracy when identified with respect to the undamaged condition. This result is interpreted as related to the absence of any local sensors near the damaged element.
Elm 3972 Elm 3920 Elm 3872 Elm 3823	The DNN architecture showed not to be able to properly detect and classify both the total break and partial damage cases with respect to the undamaged condition nor in the three-class problem. This behaviour gives relevant information on the ability of detecting failures on distributed elements in relation to their position on the structure. It can be noticed, indeed, that the same classification performance can be observed for all the damaged elements, except Elm 4115 and Elm 4110. This can be explained by the fact that the structural parts are symmetrically positioned with respect to the plane, including the Z-axis, perpendicular to the Y-axis, which is indeed a symmetry plane of the satellite (please refer to Figure 1): Elm 3972 is symmetric to Elm 3823, and Elm 3920 to 3872. Likewise, Sensors "1" and "75" are the only sensors positioned in the symmetry plane, while the others are located symmetrically with respect to each other. Hence, they are inducing a comparable effect/change in the system dynamics which results in similar classification results. Indeed, according to the specific manoeuvre, they are producing either a response with the same magnitude and same sign, or the same response with opposite sign. In general, nevertheless, this proves that the local failures, both breaks and damages, have a limited impact on the system from a structural point of view. They are all indeed elements which are part of the antenna backbone supporting structure, as opposed to the attachment point. This, on the other hand, indicates that, even with local sensors near the damage/failure of the element, the damage on the supporting backbone loop does not change the properties of the system enough to be clearly identified by the DNN architecture. Because the DNN system relies purely on data, without knowledge of the physical system, which could help discriminate the area of the failure, the accuracy level is significantly worse than the one of both Elm 4115 and 4110.

7. Conclusions

This paper investigated the problem of detecting structural breaks and partial damages for a complex system, such as a large space antenna, while assessing the SHM framework/sensors sensitivity to local damages with regard to their geometric location and impact on the system dynamics. The adopted data-driven SHM approach was based on a state-of-the-art deep recurrent neural network (Bi-LSTM) with a tailored structure for damage classification. The proposed DL architecture was trained with data produced by using a 3D simulator of the attitude/flexible dynamics of a spacecraft equipped with structural appendages, including not only a mesh antenna but also a pair of symmetric solar panels. The obtained results proved the SHM system was able to accurately identify breaks in the antenna attachment area with the hosting spacecraft while showing a lower performance in the case of distributed failures in the backbone antenna supporting elements. Therefore, for the considered study case, a minimal set of two sensors is deemed sufficient to detect those failures which could produce detectable changes in the system dynamics.

Future research will deepen the obtained results by using the DL approach to investigate optimal sensors placement configurations for different classes of large space structures. Moreover, an improvement in the presented outcome will be explored by implementing upgraded DL strategies, such as methods augmented/coupled with a certain degree of knowledge of the physical system. This would aim at discriminating among the specific types of failures and/or similar acquired data and to knowingly address areas judged to have a minimal impact on the overall system dynamics. A possible approach could be the use of physics-informed machine learning techniques to better discern the types and location of damages, thus improving the system sensitivity.

Further work might also consider adopting automatic validation/testing methods to properly tune the complexity of the proposed network (based on the type and number of implemented sensors) and to use the DL approach to assess the best sensors to detect a specific structural failure or damage.

Author Contributions: Conceptualisation, F.A., P.I., P.G., M.P. and A.R.; data curation, F.A. and A.R.; formal analysis, F.A., P.I., M.P. and A.R.; methodology, F.A., P.I., P.G., M.P. and A.R.; project administration, F.A.; software, F.A., P.I., M.P. and A.R.; supervision, P.G. and M.P.; validation, M.P. and A.R.; visualisation, F.A. and P.I.; writing—original draft, F.A., P.I., P.G., M.P. and A.R.; writing—review and editing, F.A., P.I., P.G., M.P. and A.R. All authors have read and agreed to the published version of the manuscript.

Funding: This research received no external funding.

Institutional Review Board Statement: Not applicable.

Informed Consent Statement: Not applicable.

Data Availability Statement: Not applicable.

Conflicts of Interest: The authors declare no conflicts of interest.

Abbreviations

The following abbreviations are used in this manuscript:

ANN	Artificial Neural Network
DL	Deep Learning
DNN	Deep Neural Network
ELM	Element
FEM	Finite Element Method
LMRM	Large Mesh Reflector Model
LSS	Large Space Structure
LSTM	Long Short-Term Memory
ML	Machine Learning

NN	Neural Network
RNN	Recurrent Neural Network
SHM	Structural Health Monitoring

References

1. Ketner, G.L. *Survey of Historical Incidences with Controls-Structures Interaction and Recommended Technology Improvements Needed to Put Hardware in Space*; Technical Report; Pacific Northwest Lab: Richland, WA, USA, 1989. [CrossRef]
2. Chu, Z.; Deng, Z.; Qi, X.; Li, B. Modeling and analysis of a large deployable antenna structure. *Acta Astronaut.* **2014**, *95*, 51–60. [CrossRef]
3. Wang, Y.; Liu, R.; Yang, H.; Cong, Q.; Guo, H. Design and Deployment Analysis of Modular Deployable Structure for Large Antennas. *J. Spacecr. Rocket.* **2015**, *52*, 1101–1111. [CrossRef]
4. Angeletti, F.; Gasbarri, P.; Palmerini, G.; Sabatini, M. Learning-based control scheme to deploy modular space structures. In Proceedings of the 2018 IEEE Aerospace Conference, Big Sky, MT, USA, 3–10 March 2018; pp. 1–16. [CrossRef]
5. Kellogg, K.H.; Thurman, S.W.; Edelstein, W.N.; Spencer, M.W.; Chen, G.S.; Underwood, M.L.; Njoku, E.G.; Goodman, S.D.; Jai, B. NASA's Soil Moisture Active Passive (SMAP) observatory. In Proceedings of the 2013 IEEE Aerospace Conference, Big Sky, MT, USA, 2–9 March 2013; pp. 1–20.
6. Council, N.R. *Earth Science and Applications from Space: National Imperatives for the Next Decade and Beyond*; The National Academies Press: Washington, DC, USA, 2007.
7. Jiménez, C.; Tenerelli, J.; Prigent, C.; Kilic, L.; Lavergne, T.; Skarpalezos, S.; Høyer, J.L.; Reul, N.; Donlon, C. Ocean and Sea Ice Retrievals From an End-To-End Simulation of the Copernicus Imaging Microwave Radiometer (CIMR) 1.4–36.5 GHz Measurements. *J. Geophys. Res. Ocean.* **2021**, *126*, e2021JC017610. [CrossRef]
8. Iannelli, P.; Angeletti, F.; Gasbarri, P. A model predictive control for attitude stabilization and spin control of a spacecraft with a flexible rotating payload. *Acta Astronaut.* **2022**, *199*, 401–411. [CrossRef]
9. Iannelli, P.; Angeletti, F.; Gasbarri, P.; Panella, M.; Rosato, A. Deep learning-based Structural Health Monitoring for damage detection on a large space antenna. *Acta Astronaut.* **2022**, *193*, 635–643. [CrossRef]
10. Giurgiutiu, V.; Cuc, A. Embedded non-destructive evaluation for structural health monitoring, damage detection, and failure prevention. *Shock Vib. Dig.* **2005**, *37*, 83–105. [CrossRef]
11. Tessler, A. Structural Analysis Methods for Structural Health Management of Future Aerospace Vehicles. *Key Eng. Mater.* **2007**, *347*, 57–66. [CrossRef]
12. Giurgiutiu, V.; Zagrai, A.; Bao, J.J. Piezoelectric Wafer Embedded Active Sensors for Aging Aircraft Structural Health Monitoring. *Struct. Health Monit.* **2002**, *1*, 41–61. [CrossRef]
13. Kim, J.S.; Vinoy, K.J.; Varadan, V.K. Wireless health monitoring of cracks in structures with MEMS-IDT sensors. In Proceedings of the SPIE's 9th Annual International Symposium on Smart Structures and Materials, San Diego, CA, USA, 18–21 March 2002. [CrossRef]
14. Liu, Y.; Kim, S.B.; Chattopadhyay, A.; Doyle, D. Application of System-Identification Techniquest to Health Monitoring of On-Orbit Satellite Boom Structures. *J. Spacecr. Rocket.* **2011**, *48*, 589–598. [CrossRef]
15. Haridas, A.; Speckmann, H. *Structural Health Monitoring (SHM) of Space Structures*; Materials Research Forum LLC: Millersville, PA, USA, 2021. [CrossRef]
16. Dworakowski, Z.; Kohut, P.; Gallina, A.; Holak, K.; Uhl, T. Vision-based algorithms for damage detection and localization in structural health monitoring. *Struct. Control. Health Monit.* **2016**, *23*, 35–50. [CrossRef]
17. Ye, X.; Yi, T.H.; Dong, C.; Liu, T. Vision-based structural displacement measurement: System performance evaluation and influence factor analysis. *Measurement* **2016**, *88*, 372–384. [CrossRef]
18. Sony, S.; Laventure, S.; Sadhu, A. A literature review of next-generation smart sensing technology in structural health monitoring. *Struct. Control. Health Monit.* **2019**, *26*, e2321. [CrossRef]
19. Beitia, J.; Loisel, P.; Fell, C. Miniature accelerometer for High-Dynamic, Precision Guided Systems. In Proceedings of the 2017 IEEE International Symposium on Inertial Sensors and Systems (INERTIAL), Kauai, HI, USA, 27–30 March 2017; pp. 35–38. [CrossRef]
20. Bennani, S.; Ankersen, F.; Arcioni, M.; Casasco, M.; Silvestrin, P.; Massotti, L. Robust Attitude Control Design for the BIOMASS Satellite (Earth Explorer Core Mission Candidate). *IFAC Proc. Vol.* **2011**, *44*, 5130–5135. [CrossRef]
21. Angeletti, F.; Iannelli, P.; Gasbarri, P.; Gonzalez, J.A.P.; Ellero, N.; Wattrelot, T.; Ankersen, F.; Sabatini, M.; Celani, F.; Palmerini, G. Robust Collocated Control of Large Flexible Space Structures. *IFAC-PapersOnLine* **2022**, *55*, 85–90. [CrossRef]
22. Callipari, F.; Sabatini, M.; Angeletti, F.; Iannelli, P.; Gasbarri, P. Active vibration control of large space structures: Modelling and experimental testing of offset piezoelectric stack actuators. *Acta Astronaut.* **2022**, *198*, 733–745. [CrossRef]
23. Mancini, S.; Tumino, G.; Gaudenzi, P. Structural Health Monitoring for Future Space Vehicles. *J. Intell. Mater. Syst. Struct.* **2006**, *17*, 577–585. [CrossRef]
24. Palmerini, G.B.; Angeletti, F.; Iannelli, P. Multiple Model Filtering for Failure Identification in Large Space Structures. *Lect. Notes Civ. Eng.* **2021**, *128*, 171–181. [CrossRef]

25. Tansel, I.; Chen, P.; Wang, X.; Yenilmez, A.; Ozcelik, B. Structural health monitoring applications for space structures. In Proceedings of the 2nd International Conference on Recent Advances in Space Technologies, Istanbul, Turkey, 9–11 June 2005; pp. 288–292. [CrossRef]
26. Martins, B.; Kosmatka, J. In-Situ Health Monitoring of Aerospace Structures via Dynamic Strain Measurements. In Proceedings of the AIAA Scitech 2019 Forum, San Diego, CA, USA, 7–11 January 2019. [CrossRef]
27. Ferreira Gomes, G.; Souza Chaves, J.A.; de Almeida, F.A. An inverse damage location problem applied to AS-350 rotor blades using bat optimization algorithm and multiaxial vibration data. *Mech. Syst. Signal Process.* **2020**, *145*, 106932. [CrossRef]
28. Siebel, T.; Majer, D. Damage Detection on a Truss Structure Using Transmissibility Functions. In Proceedings of the EURODYN, Leuven, Belgium, 4–6 July 2011; p. 211.
29. Yu, L.; Cheng, L.; Yam, L.; Yan, Y. Application of eigenvalue perturbation theory for detecting small structural damage using dynamic responses. *Compos. Struct.* **2007**, *78*, 402–409. [CrossRef]
30. Li, J.H.; Yang, Q.W.; Shen, X.; Liang, C.F. Structural Damage Diagnosis Using an Improved Eigenvalue Perturbation Method. *Adv. Mater. Res.* **2013**, *753*, 2347–2350. [CrossRef]
31. Mucchielli, P.; Bhowmik, B.; Hazra, B.; Pakrashi, V. Higher-Order Stabilized Perturbation for Recursive Eigen-Decomposition Estimation. *J. Vib. Acoust.* **2020**, *142*, 061010. [CrossRef]
32. Bhowmik, B.; Krishnan, M.; Hazra, B.; Pakrashi, V. Real-time unified single- and multi-channel structural damage detection using recursive singular spectrum analysis. *Struct. Health Monit.* **2018**, *18*, 147592171876048. [CrossRef]
33. Kun, L.; Law, S.; Xia, Y.; Zhu, X. Singular spectrum analysis for enhancing the sensitivity in structural damage detection. *J. Sound Vib.* **2013**, *333*, 392–417. [CrossRef]
34. Liu, L.; Yan, Y.J.; Epureanu, B.; DSouza, K. Singular spectrum analysis based structural damage detection from nonlinear vibration measurements containing noise. *Noise Control. Eng. J.* **2015**, *63*, 402–414. [CrossRef]
35. Roemer, M.; Ge, J.; Liberson, A.; Tandon, G.; Kim, R. Autonomous impact damage detection and isolation prediction for aerospace structures. In Proceedings of the 2005 IEEE Aerospace Conference, Big Sky, MT, USA, 5–12 March 2005; pp. 3592–3600. [CrossRef]
36. Ghaffarzadeh, H.; Raeisi, F. Damage identification in truss structures using finite element model updating and imperialist competitive algorithm. *Jordan J. Civ. Eng.* **2016**, *10*, 266–277. [CrossRef]
37. Vitola, J.; Vejar, M.A.; Burgos, D.A.T.; Pozo, F. Data-Driven Methodologies for Structural Damage Detection Based on Machine Learning Applications. In *Pattern Recognition*; Ramakrishnan, S., Ed.; IntechOpen: Rijeka, Croatia, 2016; Chapter 6. [CrossRef]
38. Cross, E.J.; Gibson, S.J.; Jones, M.R.; Pitchforth, D.J.; Zhang, S.; Rogers, T.J. Physics-Informed Machine Learning for Structural Health Monitoring. In *Structural Integrity*; Springer International Publishing: Berlin/Heidelberg, Germany, 2021; pp. 347–367. [CrossRef]
39. Voss, S. *Application of Deep Learning for Spacecraft Fault Detection and Isolation*; TU Delft Aerospace Engineering: Delft, The Netherlands, 2019.
40. Rosato, A.; Panella, M.; Araneo, R.; Andreotti, A. A Neural Network Based Prediction System of Distributed Generation for the Management of Microgrids. *IEEE Trans. Ind. Appl.* **2019**, *55*, 7092–7102. [CrossRef]
41. Stolfi, A.; Angeletti, F.; Gasbarri, P.; Panella, M. A Deep Learning Strategy For On-Orbit Servicing Via Space Robotic Manipulator. *Aerotec. Missili Spaz.* **2019**, *98*, 273–282. [CrossRef]
42. Succetti, F.; Rosato, A.; Araneo, R.; Panella, M. Deep Neural Networks for Multivariate Prediction of Photovoltaic Power Time Series. *IEEE Access* **2020**, *8*, 211490–211505. [CrossRef]
43. Gregor, K.; Danihelka, I.; Graves, A.; Rezende, D.; Wierstra, D. DRAW: A Recurrent Neural Network For Image Generation. In Proceedings of the 32nd International Conference on Machine Learning, Lille, France, 6–11 July 2015; Bach, F., Blei, D., Eds.; PMLR: Lille, France, 2015; Volume 37, pp. 1462–1471.
44. Gasbarri, P.; Monti, R.; De Angelis, C.; Sabatini, M. Effects of uncertainties and flexible dynamic contributions on the control of a spacecraft full-coupled model. *Acta Astronaut.* **2014**, *94*, 515–526. [CrossRef]
45. Angeletti, F.; Gasbarri, P.; Sabatini, M. Optimal design and robust analysis of a net of active devices for micro-vibration control of an on-orbit large space antenna. *Acta Astronaut.* **2019**, *164*, 241–253. [CrossRef]
46. Mazzini, L. *Flexible Spacecraft Dynamics, Control and Guidance*, 1st ed.; Springer Aerospace Technology, Springer: Cham, Switzerland, 2016.
47. Angeletti, F.; Iannelli, P.; Gasbarri, P.; Sabatini, M. End-to-end design of a robust attitude control and vibration suppression system for large space smart structures. *Acta Astronaut.* **2021**, *187*, 416–428. [CrossRef]
48. Morterolle, S.; Maurin, B.; Dube, J.F.; Averseng, J.; Quirant, J. Modal behavior of a new large reflector conceptual design. *Aerosp. Sci. Technol.* **2015**, *42*, 74–79. [CrossRef]
49. Wie, B.; Barba, P.M. Quaternion feedback for spacecraft large angle maneuvers. *J. Guid. Control. Dyn.* **1985**, *8*, 360–365. [CrossRef]
50. Wie, B.; Weiss, H.; Arapostathis, A. Quarternion feedback regulator for spacecraft eigenaxis rotations. *J. Guid. Control. Dyn.* **1989**, *12*, 375–380. [CrossRef]
51. Ceschini, A.; Rosato, A.; Succetti, F.; Luzio, F.D.; Mitolo, M.; Araneo, R.; Panella, M. Deep Neural Networks for Electric Energy Theft and Anomaly Detection in the Distribution Grid. In Proceedings of the 2021 IEEE International Conference on Environment and Electrical Engineering and 2021 IEEE Industrial and Commercial Power Systems Europe (EEEIC / I&CPS Europe), Bari, Italy, 7–10 September 2021; pp. 1–5. [CrossRef]

52. Che, Z.; Purushotham, S.; Cho, K.; Sontag, D.; Liu, Y. Recurrent Neural Networks for Multivariate Time Series with Missing Values. *Sci. Rep.* **2018**, *8*, 6085. [CrossRef] [PubMed]
53. Siami-Namini, S.; Tavakoli, N.; Namin, A.S. The Performance of LSTM and BiLSTM in Forecasting Time Series. In Proceedings of the 2019 IEEE International Conference on Big Data (Big Data), Los Angeles, CA, USA , 9–12 December 2019; pp. 3285–3292. [CrossRef]
54. Mughees, N.; Mohsin, S.A.; Mughees, A.; Mughees, A. Deep sequence to sequence Bi-LSTM neural networks for day-ahead peak load forecasting. *Expert Syst. Appl.* **2021**, *175*, 114844. [CrossRef]
55. Kingma, D.; Ba, J. Adam: A Method for Stochastic Optimization. *arXiv* **2014**, arXiv:1412.6980.

sensors

MDPI

Article

Lock-In Thermal Test Simulation, Influence, and Optimum Cycle Period for Infrared Thermal Testing in Non-Destructive Testing

António Ramos Silva [1,2,*], Mário Vaz [1,2], Sofia Leite [3] and Joaquim Mendes [1,2]

1 Faculty of Engineering, University of Porto, 4200-465 Porto, Portugal
2 INEGI—Institute of Science and Innovation in Mechanical and Industrial Engineering, 4200-465 Porto, Portugal
3 CINTESIS—Center for Health Technology and Services Research, 4200-465 Porto, Portugal
* Correspondence: ars@fe.up.pt; Tel.: +351-22-041-3413

Abstract: Lock-in thermal tests (LTTs) are one of the best ways to detect defects in composite materials. The parameter that most affects their performance is the cycle period of the stimulation wave. Its influence on the amplitude-phase results was determined by performing various numeric simulations and laboratory tests. The laboratory tests were used to infer part of the simulation parameters, namely the input and output heat, corresponding to the stimulation and natural convection. The simulations and the analysis of their results focus on the heat flow inside the sample and the manner they change for different geometries. This was performed for poly(methyl methacrylate (PMMA) and carbon fiber-reinforced polymers (CFRPs). The simulation of these materials was also used to create prediction surfaces and equations. These predict the amplitude and phase for a sample with a thickness *l* and a cycle period. These new findings were validated with new laboratory tests and two new samples. These validated the prediction surfaces and equations and can now be used as a reference for future works and industrial applications.

Keywords: thermography; nondestructive testing; lock-in; simulation; laboratory tests

Citation: Ramos Silva, A.; Vaz, M.; Leite, S.; Mendes, J. Lock-In Thermal Test Simulation, Influence, and Optimum Cycle Period for Infrared Thermal Testing in Non-Destructive Testing. *Sensors* **2023**, *23*, 325. https://doi.org/10.3390/s23010325

Academic Editor: Zenghua Liu

Received: 18 November 2022
Revised: 19 December 2022
Accepted: 21 December 2022
Published: 28 December 2022

1. Introduction

1.1. Background

Non-Destructive Tests/Testing (NDT) is characterized by the ability to test components without diminishing their service period or degrading their performance. Field view techniques allow the examination of the area of a component, thus increasing the test speed. Most techniques use wave propagation like ultrasounds, visible, ultraviolet, X-rays, or infrared to test a component area. Infrared thermography measures the radiation naturally emitted by objects near the ambient temperature and creates temperature images using this information. External energy sources are used to amplify or create thermal patterns that provide insight into the component state external energy sources are used. These will provoke a temperature to unbalance. Studying and analyzing these patterns and how they evolve during the test enables the determination of some conditions and the identification of defects. Using optical stimulation (light) usually leads to good results and a setup that is easy to arrange with good portability, along with the good results provided. This type of stimulation is mainly used in two manners: (i) a simple uniform stimulation applied during a period (transient thermal tests) and (ii) a cyclic stimulation modulated as a sinusoidal wave. These last ones are called Lock-in Thermal Tests (LTTs) [1,2].

The temperature response is a combination of three factors: (i) heat received by radiation (optical stimulation), (ii) heat transfer by radiation and convection with the air, and (iii) conduction inside the component. The radiated heat from the sample is small and is usually discarded. With the increase in temperature, the heat loss by convection will also increase, and vice versa, meaning the heat will not travel from the surface to the interior of the sample but will travel from the inside to the outside of the sample [3]. Less energy (shorter stimulation) results in less thermal variations and less heat exchange by convection since the temperature difference is lower. Likewise, lower temperatures lead to less heat flow by conduction. With this type of test it is possible to analyze the temperature throughout its entire duration. However, determining the amplitude and phase response leads to precise and sensitive results.

The amplitude and phase images are determined by comparing (a) the stimulation and (b) individual pixel temperature. These are used to calculate the amplitude and phase response for each pixel. Two maps (images) of the object's thermal response are then created. These are sometimes called ampligrams and phasegrams Since each of these images is obtained using several temperature measures, the results contain less noise than the temperature images. However, special care is necessary when choosing and using these sinusoidal stimulations. If they are not perfect sinusoidal waves, significant errors can occur [4,5]. A higher number of measures (thermal images) is preferable when calculating the phase delay and amplitude response. The best situation or characteristic response has been found for stimulations with periods of a few seconds and some cycles. Longer cycles result in the application of more energy to the object under analysis. However, the longer stimulation disadvantage is the inherent duration of the global thermal test. In industrial applications, a longer stimulation means that the NDT and the entire maintenance or preventive operation will have a higher cost or might become a bottleneck.

Numeric simulations predict the behavior of structures or components under predetermined conditions [6]. These are particularly useful in situations where exact solutions are unknown or complex. Another alternative to study processes is to build prototypes and test them. Apart from the cost, this process can also be time-consuming. Simulations can show the inside of a component or structures difficult or impossible in experimental testing processes. Currently, numeric simulations are almost entirely run on computers, being implemented in several programming languages, such as Python, FORTRAN, C, and Matlab®, among others. Some companies commercialize specific programs to perform simulations; the most common are COMSOL®, SIMULIA®, ANSYS®, Nastran®, etc. Using a virtual environment to manipulate the results also facilitates the management, documentation, communication, and pre/post-processing of data.

The most common numeric simulation uses one of three methodologies: Finite Differences Method (FDM), Finite Elements Method (FEM), or Finite Volume Method (FVM) [7]. They are also called the classic choices to solve Partial Differential Equations (PDEs). FDM uses a local Taylor expansion to calculate the PDE. However, it has some difficulties for complex geometries [8]. FVM equations are integrated for each volume with linear variations inside, resulting in errors when using tetrahedral meshes or elements with unbalanced dimensions [9]. Usually, FEM uses the weighted Galerkin method, being well suited for complex geometries. The development of the FEM in the last years made them the most common choice in commercial simulation software [10]. Due to the robust formulation of the FEM and its flexibility for a given mesh, this was the selected method in this work. The analysis of the evolution and distribution of the temperature in the samples should use Fourier's first law of heat conduction. Here, a heat flux (q) that enters a surface has a thermal conductivity (K) of the material that travels for a certain distance (x) transversal to the surface with a temperature (T) [11]. The meshing process is also critical, with a constant spacing in the nodal points distribution leading to an equilateral and balanced element, preventing significant errors and singularities in the shape functions.

The selection of the materials is an essential part of the design of any component, even more in the case of high-performance elements. One of the most used materials is composite materials, particularly carbon fiber-reinforced polymers (CFRPs). These have a very high stress-to-weight ratio, among other characteristics. However, they present some disadvantages such as being anisotropic, having a costly and complex manufacturing process, and their failure is unpredictable, among other things [12]. For the latter, the importance of NDT in CFRP is even higher, especially in the amount of energy applied.

1.2. Recent and Relevant Works

Recently, artificial intelligence has gained much attention from the research community. It is also being used to improve NDT methods. Al-Athel et al. present a hybrid thermography, computational, and Artificial Neural Networks (ANN) approach to characterize sub-surface defects in composite materials. Their *"defects consisted of grill holes and the simulation data is fed to the ANN"*. Their results are not on par with other applications of ANN but are promising [13].

Stoynova et al. performed numerical simulations of a thermal 3D model with artificial defects under a lock-in test. This is one of several works that used LTT to perform subsurface defect detection and characterization [14]. Other similar works were performed by Shrestha [15], Chulkov [16], Pitarresi [17], An, Y.K, [18], Zoecke [19], and Peng [20]. The usage of lasers as an excitation source is not the most common. The delamination problem was researched by Swiderski using a numerical and experimental approach. With a laser stimulation with a wavelength of 808 nm, a maximum power of 32 W, and a FLIR 7600 SC camera, they showed it is possible to detect delaminations on CFRP [21].

Lock-in thermography is mainly used to analyze carbon-fiber reinforced polymers (CFRPs) or glass fiber-reinforced polymers (GFRPs). However, at a macroscale, electronic circuits, especially Printed Circuit Boards (PCBs), can be considered composite materials. Here, the most desired feature is not the improvement of the mechanical properties but rather the electrical connection between two or more points. If these connections are damaged, the PCB is useless. The usage of (a) flash, (b) transient, or (c) lock-in thermography has been used by Stoynova to study the shape and better understand their root cause [22,23]. Other similar and relevant works, even if not using simulations but focusing on laboratory tests, were presented by Breitenstein [24], Brand [25], Andersson [26], Leppänen [27], and Hovhannisyan [28].

Flash, Transient, and lock-in thermography are the classic Infrared Thermography NDT techniques. These require an intense heat source (transient test) or a well-controlled and linear power source [2,5]. To overcome these drawbacks, frequency-modulated stimulations have been studied by Mulaveesala and by Rani [29,30]. The highly damped behavior of the heat waves in LTT leads to dispersion in the surface thermal patterns. Hedayatrasa studied the viability and sensitivity of using thermal wave radar (TWR) to inspect CFRP [31]. Here, analog frequency-modulated (sweep), discrete phase-modulated (Barker binary coded) waveforms, and frequency-phase-modulated (FPM) were considered. The 3D Finite Element Analysis (FEA) showed *"outstanding performance of TWR at relatively high excitation frequencies is highlighted, particularly when approaching the so-called blind frequency of a defect"*.

Despite most research centering itself on CFRP or GFRP, Philipp et al. studied the application of thermal diffusivity of semitransparent polymer films. They presented a theoretical heat loss model by convection and radiation in semitransparent film surfaces. They claim that *"the slopes method is valid for any semitransparent film in the thermally thin regime"* [32].

Despite the several works studying LTT with experimental techniques and numerical simulations, several issues are persisting/missing in the literature and are addressed in this work. Most works are difficult to compare due to their differences in the test settings or do not present this information at all. This work aims to analyze the influence of the cycle period in LTT applied to CFRP samples for different thicknesses. An in-depth analysis of the sample's heat flow and temperature variations is presented. The analysis of the temperature profiles at the sample surface for different cycle periods and the temperature profiles' inversion pattern (related to blind frequencies) is exceptionally detailed. To summarize and facilitate the inclusion of the presented findings, prediction equations are presented. These predict the amplitude and phase results according to the component thickness and cycle period used. For this purpose, a numerical model was created using FEM and validated with poly(methyl methacrylate) (PMMA) and CFRP samples. This document and the work followed the structure presented in Figure 1.

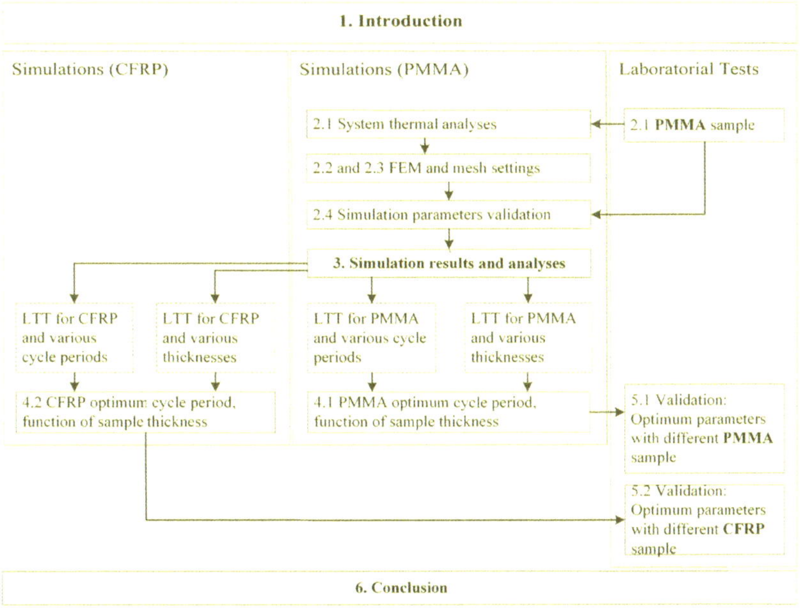

Figure 1. Workflow and article structure.

2. Materials and Methods (Simulating LTT)

The present chapter presents the analyses and simplification of a 3D sample into a 2D surface. A rapid description of the equations implemented for the FEM and the mesh evaluation is shown. A comparison between the temperature results from the laboratory tests and simulations is also present.

CFRP is anisotropic material built from overlapping sheets of carbon with resin as a bonding element. The initial LTT was done with a sample made from PMMA and used as a reference and had the geometry presented in Figure 2. In this 3D representation of the designed sample, the thickness and slot width were increased to improve visualization. The overall measures are 210 × 160 × 4.5 mm, with slots having 10 mm in width and 0.5, 1.5, 2.5, and 3.5 depth. A mate varnish (kameralack) from Tetenal® is used to paint the transparent PMMA. This provides a high and uniform emissivity (0.98), thus enabling the usage of optical stimulations. This procedure was also used in previous works [33].

The last analyses refer to a CFRP sample. Its size is 206 × 170 × 8 mm, built using 28 carbon fiber sheets, and has circular blind holes. There were twelve blind holes with a geometry that combined diameters of 10, 16, 20, and 25 mm, along with depths of 2, 4, and 6 mm.

The laboratory tests consisted in applying an optical stimulation modulated as a sinusoidal wave. The thermal camera is positioned horizontally and perpendicular to the sample at approximately 1 m. The halogen lamp was 0.8 m above the thermal camera. The stimulation and thermal measures were executed on the sample side without visible slots, using the configuration known as reflection mode [34]. In the initial test, a stimulation period of 20 s during 15 cycles was applied. The stimulations were applied using a halogen lamp from Hedler®, model H25S, with 2500 W. These were controlled using the power box model from AT—Automation Technology GmbH®. The sample temperature was measured using a thermal camera from FLIR® model 7500, with a measuring waveband ranging from

1.5 to 5.1 and less than 25 mK of Noise Equivalent Differential Temperature (NEDT). A computer synchronized the stimulation and the image acquisition through the application IRNDT V.1.7, also from AT—Automation Technology GmbH. These initial tests were used to optimize the simulation script and are mentioned further ahead.

It was intended to simulate the temperature variations in the sample with the slots having a uniform geometry, vertically oriented, for an isotropic material. Here, it is possible to discard the vertical dimension (Z dimension in Figure 2). This simplifies the geometry into a two-dimensional shape. This approximation was validated by analyzing the heat transfer during a sample LTT.

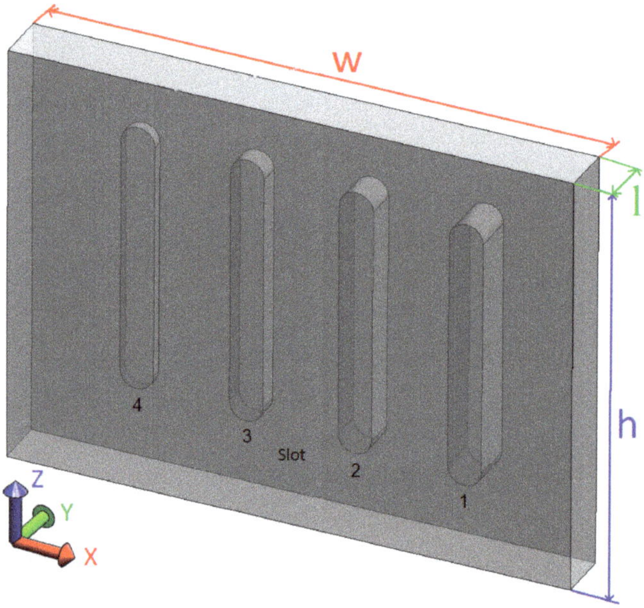

Figure 2. PMMA sample and its slots.

Depending on the temperature gradients, this simplification of the 3D sample into a 2D area, corresponding to the sample horizontal cross-section, might be incorrect. Thus, it was necessary to determine the Biot number. If this is less than 0.1, the simplification is valid. An accurate model to calculate the heat transfer coefficient was proposed by Churchill and Chu in 1975 [35]. This model requires the previous calculation of the Rayleigh number (Ra) and the Prandtl number (Pr) and can be calculated using Equation (4).

During the LTT, the sample is warmed, appearing as a vertical airflow due to the natural convection. By multiplying the Grashof number (Gr, Equation (1)) by the Prandtl number (Pr, Equation (2)), the Rayleigh number (Ra, Equation (3)) is obtained [36]. With temperature $T\infty$ equal to the laboratory tests and Ts twice the maximum amplitude recorded in the initial tests, Ra equals 3.84×10^6. It is a significant laminar flow (101 and 109), meaning the airflow due to convection is significant.

Equation (4), proposed by Churchill and Chu in 1975, determines the heat transfer coefficient [35]. With this coefficient h is calculated as a Biot of 0.08 (Equation (5)), thus being less than the established limit of 0.1. Indicating the vertical temperature gradients inside the sample due to the natural convection can be discarded, and any temperature patterns are a direct result of a defect (in this case, a machined slot). Therefore, simplifying the sample into a two-dimension horizontal surface is a good and correct approximation (1).

$$GR = \frac{g\beta(T_s - T_\infty)L_c^3}{\upsilon^2} \rightarrow GR = 5.39 \times 10^6 \quad (1)$$

$$PR = \frac{C_p\mu}{K} \rightarrow PR = 0.7 \quad (2)$$

$$Ra = GR \times Pr \rightarrow Ra = 3.84 \times 10^6 \quad (3)$$

$$h = \frac{K}{L_c}\left[0.68 + \frac{0.67 \times Ra^{1/4}}{[1 + (0.492/Pr)^{9/16}]^{4/9}}\right] \rightarrow h = 0.384\left(\frac{\mathrm{W}}{\mathrm{m^2K}}\right) \quad (4)$$

$$Bi = \frac{hl}{K_{PMMA}} \rightarrow Bi = 0.08 \quad (5)$$

where:

υ—Kinematic viscosity, $15.68 \times 10^{-6}\ \left(\frac{\mathrm{m^2}}{\mathrm{s}}\right)$
T_s—Temperature at the sample surface, 306.5 (K)
T_∞—Ambient temperature, 296.5 (K)
β—Volumetric thermal expansion coefficient, $3.3 \times 10^{-6}\ \left(\frac{\mathrm{m^2}}{\mathrm{s}}\right)$
L_c—Characteristic length, 0.16 (m)
Cp—specific heat, 1005 $\left(\frac{\mathrm{J}}{\mathrm{KgK}}\right)$
μ—Viscosity, $1.857 \times 10^{-5}\ \left(\frac{\mathrm{Kg}}{\mathrm{s \times m}}\right)$
k—Thermal conductivity, 0.0262 $\left(\frac{\mathrm{W}}{\mathrm{m \times K}}\right)$
l—Sample thickness, 4.5×10^{-3} (m)
K_{PMMA}—PMMA Thermal conductivity, 0.21 $\left(\frac{\mathrm{W}}{\mathrm{m \times K}}\right)$

The simulation of a transient response is well described in the literature. A sequence of transient stimulations can be used to discretize a sinusoidal one, like the one used in lock-in tests. Fourier's first law, in its basic form, must be expanded to match the tested sample geometries and the PDE. Since the Z axis is discarded, the geometry is defined in the X and Y coordinates and time (t). By expanding the PDE, commonly used in the finite element approximation (Equation (6)), we obtain the final and the system governing equation. The simulations were implemented in MATLAB® and followed the methodologies described by Erik Thompson [11].

$$\frac{\partial}{\partial x}\left[K_x\frac{\partial T}{\partial x}\right] + \frac{\partial}{\partial x}\left[K_y\frac{\partial T}{\partial y}\right] + Q_i = \rho \times Cp\frac{dT}{dt} \quad (6)$$

where:

Q_i—interior heat source per unitary area (stimulation);
K_x—conductivity coefficients in the **X** direction;
K_y—conductivity coefficients in the **Y** direction;
ρ—material relative density;
Cp—Heat capacity.

The simulated area required the definition of boundary conditions that mimicked the laboratory tests. It applied a natural convection coefficient to the external contour, the blue line in Figure 3. Along with the natural convection (energy output), the stimulus or energy input was simulated, the green line in Figure 3. In these nodes, the heat generation coefficient (Qi), contrary to all the other nodes, was not zero. Due to the uncertainty in determining the heat applied by the halogen lamps, the remaining parameters were determined by fitting the temperature curves to the laboratory data (the results are in the next section).

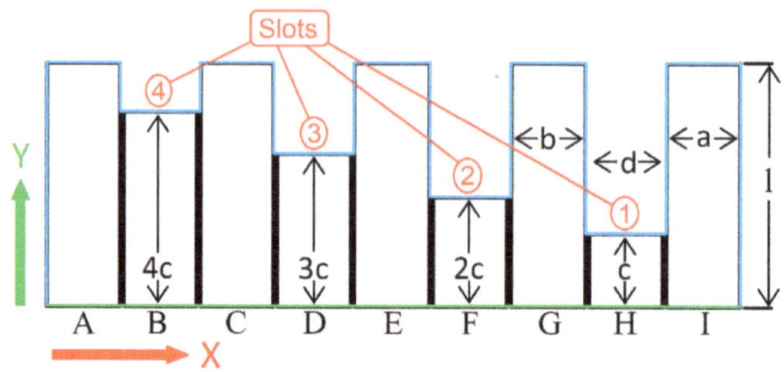

Figure 3. PMMA sample and its slots.

For the simulated section, X has the range [0–210], Y has the range [0–4.5], and t ranges from zero up to 15 times the cycle period (in seconds). The sample horizontal dimensions a, b, and d were kept equal to the samples, more precisely with 40, 30, and 10 mm. The mesh had a uniform spacing in the Y direction (l and c). The surface was divided into quadrilateral shapes with one nodal point at each corner, thus four nodes per element. Some recommendations for the mesh size and shape are available in the literature and provide some guidelines [37,38]. Fine-tuning of the mesh was performed through several simulations with different meshes. The results are presented in Table 1. The mesh's name indicates the number of elements in dimensions *d*, *a*/*b*, and *c* represented in Figure 3. The results in Table 1 show that there is a temperature conversion. In all the sets of meshes evaluated, the selected mesh is highlighted in bold, and the final mesh is in red.

Table 1. Reference temperatures from the center of the sound area and slot 1, obtained with various meshes.

Mesh	**Peak fig 5.4 and 5.5**	
Num. Elements in *d*, *a*/*b*, and *c*	**Sound Area**	**Slot 1**
20 × 20 × 2	301.97281	306.77600
20 × 20 × 3	301.89808	306.73894
20 × 20 × 4	**301.92363**	**306.72390**
20 × 20 × 5	301.95672	306.72745
20 × 20 × 6	301.91589	306.71402
20 × 20 × 8	301.92080	306.71607
20 × 4 × 4	301.92380	307.29634
20 × 6 × 4	301.92373	306.94711
20 × 8 × 4	301.92369	306.84510
20 × 12 × 4	301.92366	306.76780
20 × 16 × 4	**301.92364**	**306.73865**
20 × 20 × 4	301.92363	306.72390
20 × 40 × 4	301.92362	306.70019
12 × 16 × 4	301.92388	306.69013
16 × 16 × 4	301.92372	306.72122
20 × 16 × 4	**301.92364**	**306.73865**
30 × 16 × 4	301.92356	306.75944
40 × 16 × 4	301.92353	306.76843

The initial laboratory tests served to fine-tune the simulation parameters. Both correspond to a test with 15 cycles and 20 s of the cycle period. Figure 4 presents the temperatures inside the sample. These results are from a simulation at the end of the 15th cycle. Video S1 presents these same temperatures. In both, the Y scale was increased to improve visualization.

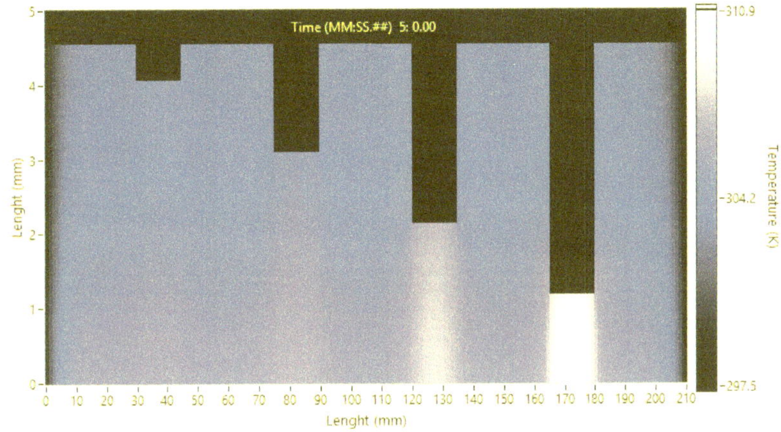

Figure 4. Temperature distribution after the initial simulation, 15 cycles of 20 s each.

The temperature profiles from the laboratory tests and simulation (Figure 5) show a significant difference. In the laboratory tests, the maximum temperatures are higher than in simulations. This is due to the reflected radiation from the stimulation. For the lower stimulation amplitude, the reflected radiation is almost null. This makes the temperatures in the laboratory tests and simulations very close. As a result, the stimulation amplitude value was set to approximate the temperatures in the lower part of the cycles (valleys).

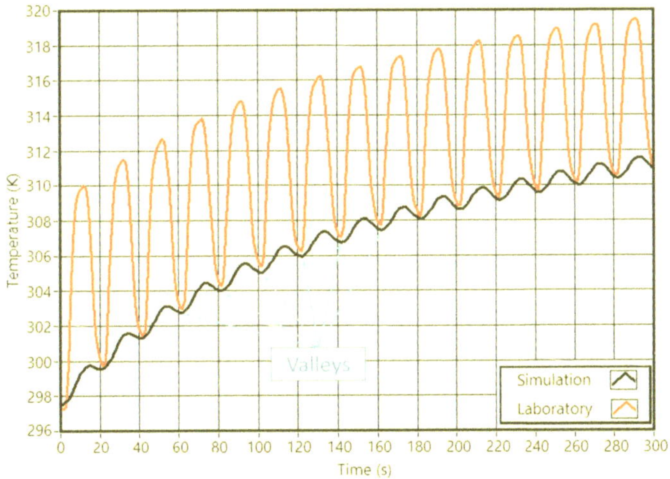

Figure 5. Obtained temperatures in laboratory and simulation at the center of slot 1.

Observing the temperature profile at the last frame of the tests (Figure 6) is the second validation of the simulation parameters (Figure 6). From the laboratory test, slot one thermal profile base was slightly wider than the simulated one, and the maximum temperature for slot 4 was lower. Apart from these two differences, all the other temperatures were very similar. Therefore, these differences are considered irrelevant.

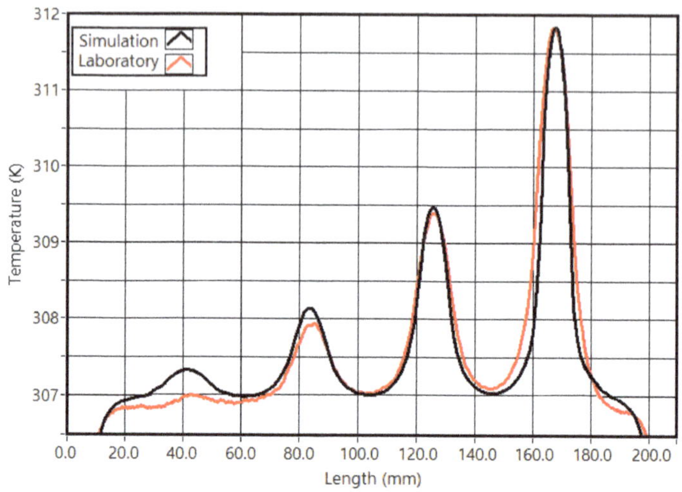

Figure 6. Temperature profiles at the end of the test in the stimulation surface (20 s).

3. Simulation Results

The sample simulations had a sinusoidal stimulation, with each simulation having 15 cycles and 65 different cycle periods, ranging from 0.02 to 80 s. The analyses were performed for all of these; however, to facilitate the representation and its observation, only the results of 15 are presented. Figure 7 presents the temperature evolution of the central node from slot one, through the entire LTT. For shorter cycle periods, it is visible that the temperature evolution is a linear oscillating curve. Longer cycle periods present a temperature response that combines sinusoidal stimulation and a negative exponential evolution. The temperature curves for cycles of 20 to 80 s present higher temperature variations. These higher variations in the slot temperature occur in the first three to four cycles. After these, the slot temperature evolution is mainly due to sinusoidal stimulation. The sinusoidal temperature variation in the entire profile is uniform. Among the several simulations, the temperature variations in the center of slot 4 were considerably small.

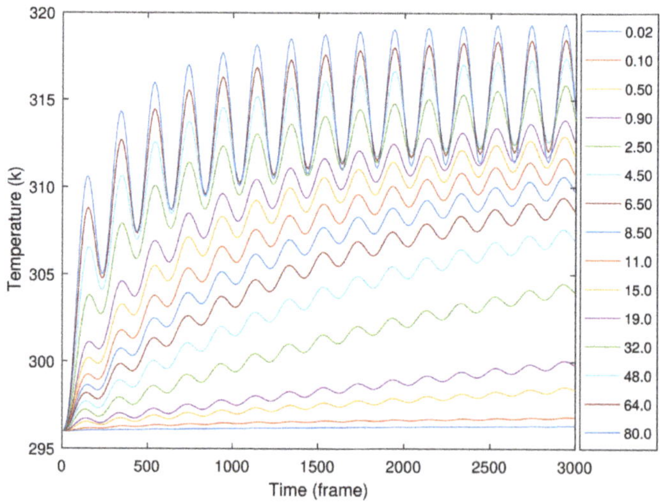

Figure 7. Temperature evolution at the center of slot 1 for cycle periods of 0.02, 0.10 up to 80 s.

Figure 8 represents the temperature profiles in the stimulus area after 15 cycles, also obtained with cyclic stimulations from 0.02 to 80 s per cycle. Slots 1, 2, 3, and 4 are visible for stimulations longer than 0.1, 0.2, 0.7, and 2 s, respectively. Increasing the stimulation period increases the overall temperatures. However, as the slots present higher temperatures, their profiles also become blurred, indicating an attenuation in the temperature distribution. For 15-s periods, the temperature response starts to invert the temperature response. The temperature difference between slot one and its boundaries starts to decrease. Increasing the cycle period makes the boundaries harder to identify. For stimulations longer than 32 s, the temperatures observed for the center of slots one and two are lower than those in its sound areas. Figure 9 presents the temperature profiles during the entire test with a cycle period of 40 s.

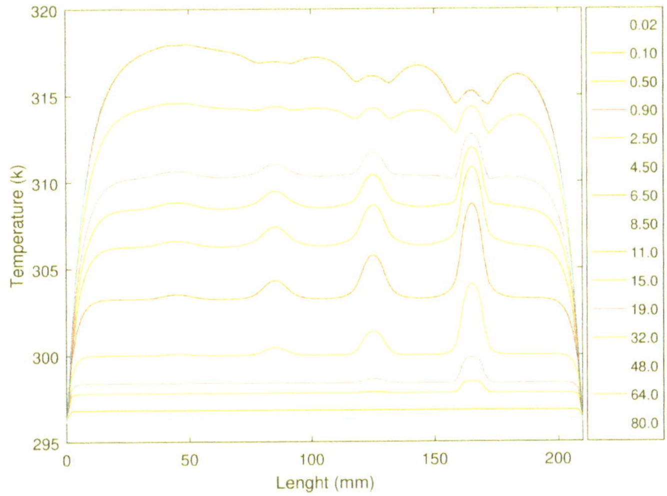

Figure 8. Temperature profiles in the stimulation surface after 15 cycles for cycle periods of 0.02, 0.10 up to 80 s.

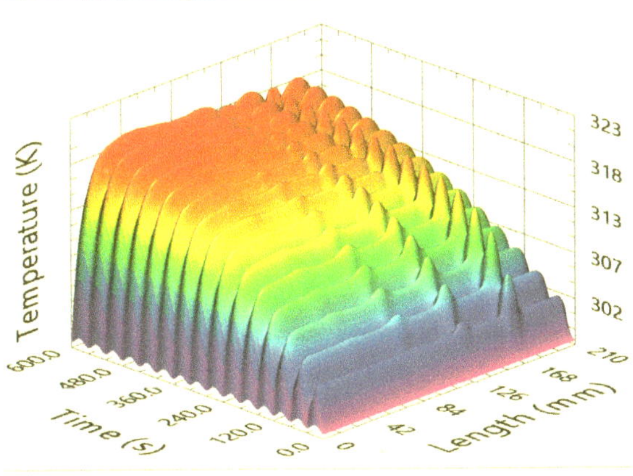

Figure 9. Temperature profiles throughout the entire test for a stimulation period of 40 s.

One of the main principles behind using the amplitude analyses is that any slight difference will be amplified (cumulative effect) at each cycle. Thus, observation of the temperature distribution after a certain number of cycles is a starting point to identify the defects in thermal patterns. The amplitude and phase delay calculus was executed at an application in LabVIEW®. It uses a Discrete Fourier Transform (DFT) to extract the relative gain and phase between the normalized sinusoidal reference and the temperature at each node. Figure 10a presents 15 of the 65 amplitude results to facilitate the visualization. The small temperature decays observed at the boundaries of the slot become visible for stimulations longer than 11 s (Figure 10a). A similar amplitude behavior is observed in slot 2. However, they are softer and tend to appear for slightly longer stimulations. The difference between slots 1 and 2 is the complete inversion of the amplitude profile at the slot 2 location, passing for a near "zero" identification for the stimulation of 14 s (Figure 10a). The thermal response of slot 4 is low, despite the cycle period used. We define the amplitude response as the temperature difference between a slot and its two sound areas. One can consider this to be technique sensitivity. Slot 1 amplitude response has a local maximum near the three-cycle period, a local minimum at the 11 cycle period, and starts to increase for stimulations longer than 15 s (Figure 10b). This behavior is observed for all the slots as visible in Figure 10b,c for different cycle periods. These local maximum and minimum result in a null response for the second and third slots. This can indicate the presence of blind frequencies. A stimulation frequency for which the slot presents zero thermal amplitude. This last aspect reveals the limitations of amplitude images. When comparing the amplitude response with the final temperature, the fourth slot was visible in the temperature profiles. This is a way to overcome the existence of blind frequencies.

Figure 11a presents the phase profiles for cycle periods from 0.02 to 80 s per cycle. With a cycle time of 0.02 s, the phase delay was considerably high and diminished for longer cycles. Slot 1 was once again the most visible, presenting the higher profile variation (increase in the phase delay) and appearing earlier (stimulation of 0.2 s). Slots 2, 3, and 4 were visible with stimulations of 0.4, 0.5, and 1 s, respectively. In Figure 11b and for slot 1, we observe a continuous increase in the phase difference. Since all the slot temperatures are a response to the stimulus, these present a time delay. Thus, all the phase differences presented negative values. Apart from this, the phase difference of slots 2, 3, and 4 present similar profiles to the amplitude difference. They have a local maximum followed by a local minimum and decrease with the increase of the cycle period. While not having this local maximum and minimum, slot 1 also presents a similar pattern.

Stimulations with a small amplitude response lead to phase images with high noise. These usually result from short stimulation periods and should be ignored. A slot can be observed as an area with a smaller thickness, thus, more heat per volume of material. However, the thicker area will have a higher damping factor, resulting in a slower dynamic. So, detecting and characterizing defects is accomplished by observing the different dynamics of the sound areas. The area corresponding to slot 1 is thinner than the sample thickness. Thus, slots 2 and 3 better represent the behavior of an "industrial" defect. Slot 2 and 3 behaviors are difficult to fully understand just by observing Figure 11. Thus, a detailed observation is presented in Figure 12. The evolution of the phase difference of the slots resembles the amplitude response. Slots 2 and 3 reached a local maximum with cycle periods of 4.5 and 6.5 s (Figure 12a,b). The phase delay decreases with the increase of the stimulus, being most visible in slot two and less visible in slot three. A local minimum is reached for slots two and three with cycle periods of 14 and 28 s, respectively. Slot 3 passes through an almost null response for a 36 s stimulation period. This is a blind frequency in the phase image. For slot two, the curve in Figure 12b does not cross the zero axis, being very close. With the existence of noise in laboratory tests, this slot might not be visible. However, slot 3 reaches a minimum of 0.01, leading to the intersection of the zero phase delay, thus the existence of two blind frequencies. These conditions that result in minimum phases should be avoided, and the thermal tests should be conducted with cycle periods that are shorter or longer. Longer stimulation will result in phase images with the contrast presented in Figure 11b.

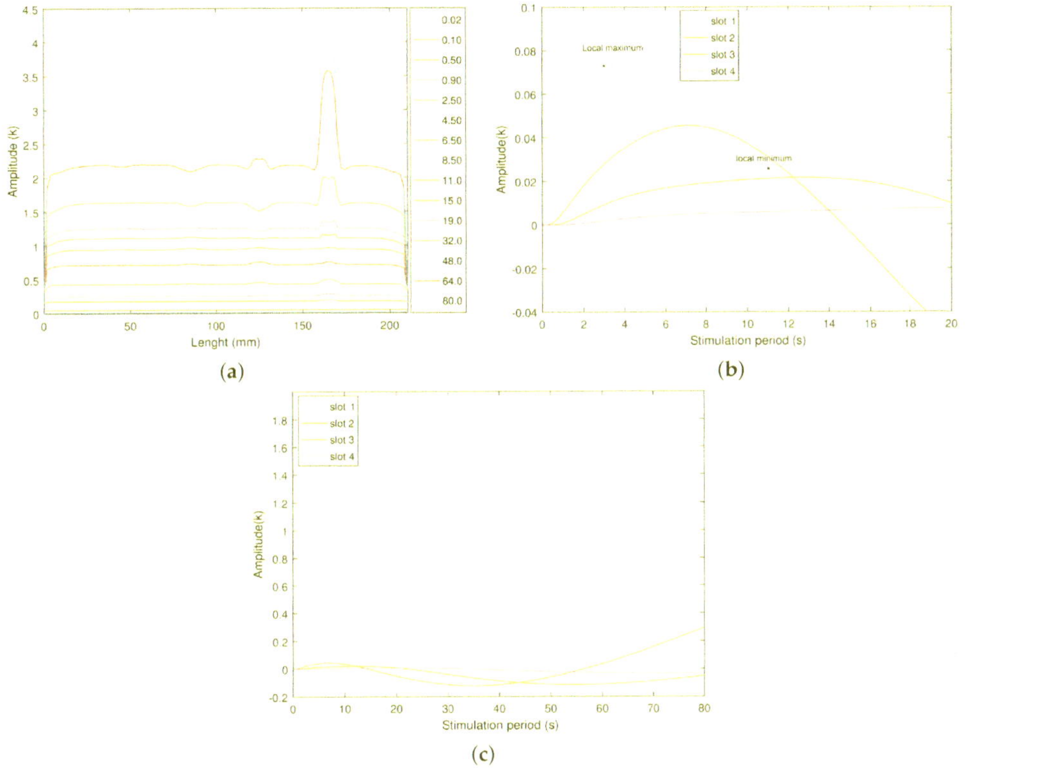

Figure 10. Amplitude response and Amplitude difference, function of the cycle period. (**a**) Amplitude responses from the stimulation surface after 15 cycles. (**b**) Amplitude difference between slots 1, 2, 3, and 4 and their sound areas, for cycle periods of 0.02 to 20 s. (**c**) Amplitude difference between the slot and its surroundings from 0.02 to 80 s per cycle.

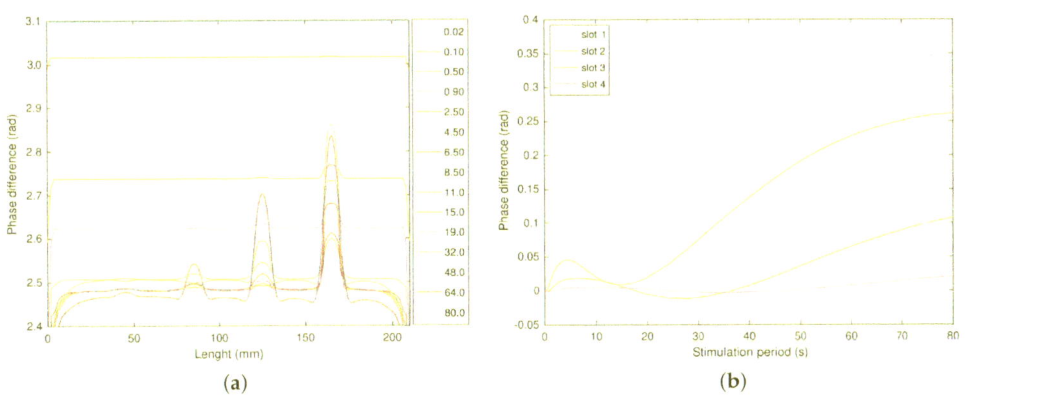

Figure 11. Phase delay and difference function of the cycle period. (**a**) Phase delay of the stimulation surface after 15 cycles. (**b**) Phase difference for periods from 0.02 to 80 s per cycle.

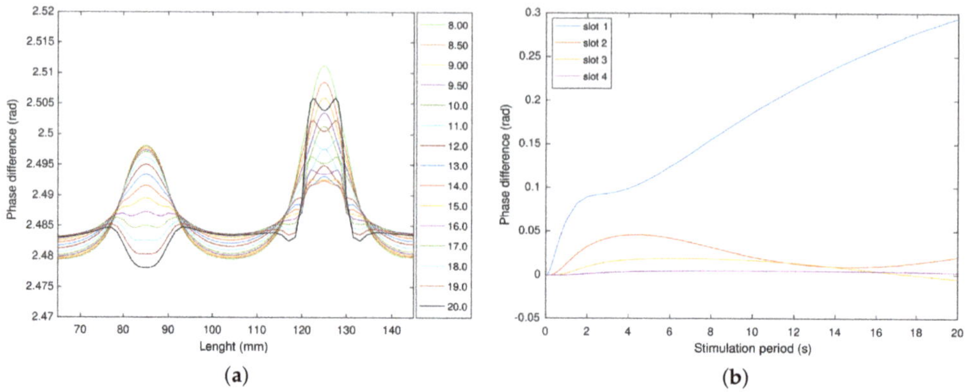

Figure 12. Detailed view of slots 2 and 3 for periods between 8 to 20 s. (**a**) Phase delay of slots 2 and 3 for stimulations ranging from 8 and 20 s. (**b**) Phase difference for periods from 0.02 to 20 s per cycle.

4. Analyses of the LTT Simulations

The first aspect to highlight from the results is the difference between the laboratory and simulated temperature curves. The reflected radiation prevented a direct comparison between the two sets of data. Comparing the cycle's lower temperatures (valleys) was the selected criterion to validate the FEA parameters. The comparison of the temperatures at the end of the test was used as a confirmation. Overall, the laboratory tests validated the settings used in the FEM simulations.

Using a thermal image during LTT is one manner to identify some defects. This analysis is relevant if the stimulation period is near a blind frequency. Using cyclic stimulation also has a cumulative effect and amplifies temperature profile patterns, showing different temperature patterns than the ones observed when using a constant and single stimulation. The samples used were made of PMMA. Its thermal conductivity is slightly lower than a CFRP but is fully isotropic. A material with low thermal conductivity is expected to increase its global temperature during cyclic stimulation. Thus, when a fast sinusoidal stimulation is applied, low-temperature variations are expected, and increasing the stimulation cycle period delivers more energy to the specimen, and higher temperatures will show. This type of behavior is observed in over-damped mechanical systems. Since all the tests used sinusoidal stimulations, it was impossible to separate the negative exponential from the sinusoidal response in the temperature data. When these present a constant sinusoid evolution, this indicates the test was long enough for the temperature to reach a steady-state regime. If longer sinusoidal stimulations are used, an exponential thermal response will appear, overlapping with the sinusoidal thermal response. Increasing the period of cyclic stimulation also increases the amount of energy received by the sample.

Thinner areas receive more energy per volume, resulting in higher temperatures and vice versa. For long cycle periods, areas have higher temperatures, and with the stimulation being in a lower instance of the cycle, these areas will start to warm their surroundings. This lateral temperature conduction (along the X-axis) will result in the dispersion of the temperature gradients, like in a transition area such as the slot boundaries. This is observed in the eight boundaries of the slots and at the profile extremities. As the cycle period increases, the time available for this lateral heat flux to become visible also increases. A higher difference between the slot temperature and the surrounding sound areas increases the lateral heat flux and consequentially softens the temperature profile. The resulting temperature profiles have blurred boundaries (difficult to identify), especially for long stimulations.

To optimize an LTT, a low number of cycles should always be used. Since most cameras can reach tens of frames per second, obtaining more than 100 images per cycle is an easy setup to accomplish. To calculate the amplitude image, it should use at least two cycles. Figure 13 presents the variations in the phase as a function of the number of cycles and cycle periods. For a short stimulation (cycle period of 5 s for example), the hardest slots to detect require more cycles, while the deepest slots reach the maximum phase difference with 5 to 9 cycles. With the stimulation of 30 s, this introduces blind frequencies and slots 2 and 3 detectability decreases with each cycle. This is particularly critical for slot 3, reaching a zero phase for nine cycles, while the third slot indicates that with a few more cycles this slot would not be detected. For longer stimulations (80 s for example), the smallest slots are mainly insensitive to the variation of the number of cycles. In contrast, the third and fourth slots present some big variations. The fourth slot presented a higher phase delay with stimulation of 15 s and one cycle when compared with the 15 cycles and stimulation of 80 s per cycle, which results in a considerably shorter test; however, this is not a usual situation and should not be used as a reference.

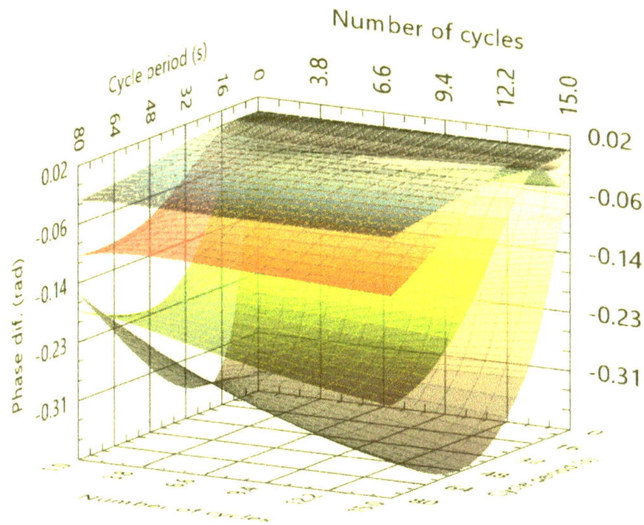

Figure 13. Phase difference function of cycle period, number of cycles, and for slots 1 to 4.

In LTT, it is impossible to identify a single image that produced the desired results. The best results (better detectability) occur when analyzing the amplitude and phase images. Thus, for these types of data/images, the optimal test parameters can be calculated. As such, we performed new simulations for PMMA and CFRP samples. These simulations focused on a sample with one slot (equivalent to slot 3) with various sample thicknesses and cycle periods. For each simulation, an amplitude and phase difference are calculated. Finally, the optimum cycle period was determined. For this, both phase and amplitude images were considered.

Figure 14 represents the amplitude variation as a function of the cycle periods (from 0.1 to 60 s) and for sample thickness ranging from 0.1 to 10 mm. Here, longer stimulations lead to higher temperature differences. When increasing the cycle period, a peak was not obtained, and the locations of the lower increasing rate were identified. These locations are represented by the green points in Figure 14. Due to the amplitude response stability (Equation (7)), this is considered the ideal location to conduct the LTT for PMMA samples. From the analyses, we concluded that the amplitude images do not provide the most accurate data. Despite a long cycle period leading to a higher amplitude result, the defect boundaries will become blurred. Therefore, the amplitude images provide a direct analysis and quality and accuracy control for the phase images. A higher amplitude response means that the temperature variations are significant and thus lead to less noise in the phase delay images. They can also act as a control preventing any false detection due to blind frequencies. Therefore, Equation (7) indicates the ideal stimulation for the amplitude analyses. However, it is less relevant than the ideal phase curve.

$$CyclePeriod_{PMMAamplitude}(s) = 4.1615 + 11.6358 \times l - 0.5648 \times l^2 \quad (7)$$

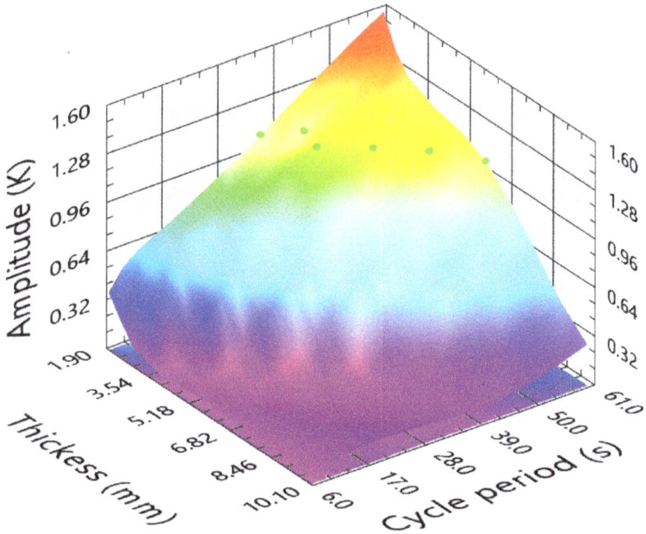

Figure 14. Amplitude difference function of PMMA sample thickness and cycle period.

In previous chapters, the phase images were considered one of the most sensitive ones. Apart from the blind frequencies, these should always be used to assess the sample results. Figure 15 presents the phase differences between the reference slot and its sound areas. The "negative" areas reveal the crossing of the zero axis. So, at least one blind frequency should occur, corresponding to the inversion of the phase profiles. The surface in Figure 15 can be divided into three main areas: (i) the peak for thicker samples and short cycle period; (ii) the valley on the diagonal of the surface; and (iii) the maximum area for the thinner samples and longer stimulations. In area (i), the amplitudes are very low; thus, the phase images will be very noisy. The area that presents a valley (area (ii) is a direct result of the existence of a blind frequency. From the analyses of the several slots, we conclude that a cycle period producing a minimum phase delay of a defect could correspond to the exact blind frequency of another. As a result, cycle periods and sample thicknesses that lead to these locations should be avoided. Finally, area (iii) corresponds to the maximum obtained for long stimulus periods. This maximum area indicates the presence of a higher difference in the phase profiles. These points occur for long stimulations and samples thinner than 5 mm. The ideal stimulations should be here. A regression of these green points is presented in Equation (8). This is the recommended cycle period function of the sample thickness. The LTT ideal parameters depend on the method used to identify and characterize the defect. The ideal parameters should consider where the best phase or amplitude images are obtained. While the phase images tend to present higher detectability and signal-to-noise ratio (SNR), the amplitude response should be high enough to enable the phase image to eliminate most of the existing noise.

$$CyclePeriod_{PMMAphase}(s) = 2.029 + 8.694 \times l - 0.309 \times l^2 \quad (8)$$

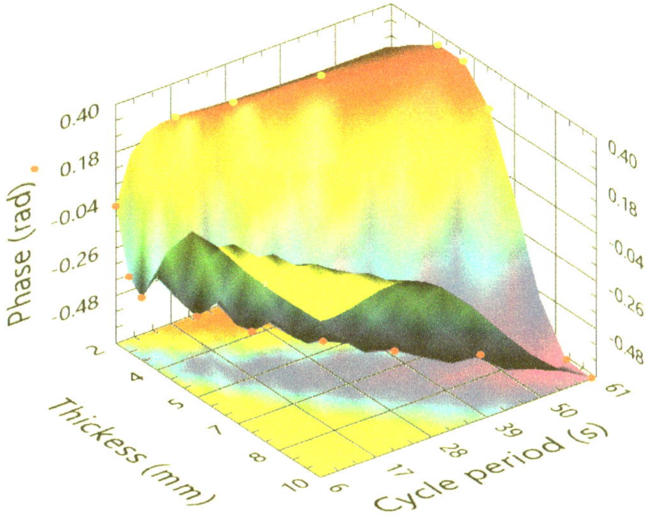

Figure 15. Phase difference function of the sample thickness and cycle period for PMMA samples.

The next step was to find the best parameters to conduct an LTT in CRFP. This is an anisotropy material, particularly for its thermal properties [39]. Different fiber orientations, layouts, and resign percentages also change its mechanical properties [40,41]. The simulations of the CFRP sample were the same as used previously, with thermal conductivities for the X and Y directions of 6.3 and 0.6, respectively [14]. It should be considered that the following results and analyses might be different for different properties such as different resin percentages or fiber alignment [42].

Figure 16 presents the amplitude of different functions of the sample thicknesses and cycle periods. The amplitude difference response presents a shape similar to the one observed in the PMMA simulations. Comparing both surfaces, the amplitude response for the PMMA sample is higher than for the CFRP. This may result from the different and higher thermal conductivity in the fiber direction and parallel to the surface, thus helping to blur and reduce the thermal patterns in the CFRP samples surface. Higher thermal conductivity decreases the temperature variations during the test. If only considering the amplitude images, using cycle periods indicated by the green dots is recommended. A fitting to these green points results in Equation (9).

$$CyclePeriod_{CFRPamplitude}(s) = 7.935 + 1.875 \times l - 0.0356 \times l^2 \quad (9)$$

One of the main tasks of this work is represented in Figure 17, the phase difference as a function of the cycle period and sample thickness for the phase response of CFRP samples. An obvious observation is the resemblance between the phase surface corresponding to the CFRP and the PMMA. Here, the surface also crosses the zero phase plane, revealing the existence of blind frequencies in some situations. Long cycle periods are also recommended, despite the higher values obtained with shorter cycle periods for thicker samples (surface valleys). In this manner, the low amplitude responses are avoided, and the noise in the phase images decreases. In the PMMA phase surface, when reaching the 5 mm thickness, the phase revealed the incapacity to maintain a good response. In the phase surface corresponding to the CFRP, the cycle period for which the maximum phase is almost proportional to the sample's thickness. This is obvious in the best phase response given with Equation (10). Thus, it was obtained by a curve fitting the green points in Figure 17.

$$CyclePeriod_{CFRPphase}(s) = 2.15 + 5.53 \times l - 0.25 \times l^2 \quad (10)$$

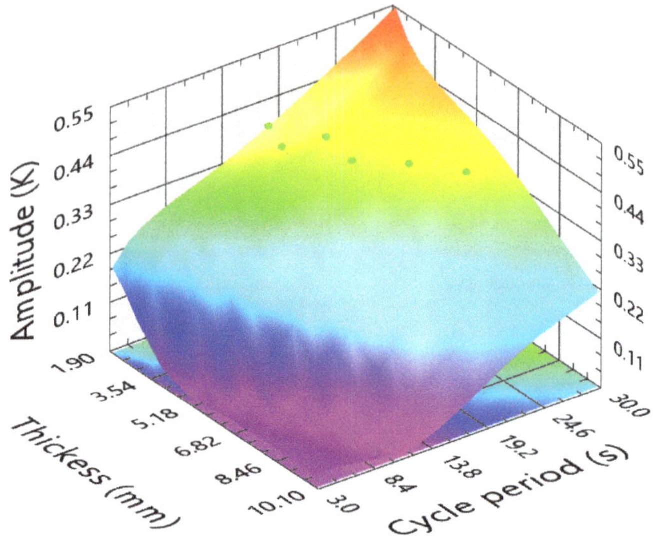

Figure 16. Amplitude difference function of the sample thickness and cycle period, for CFRP samples.

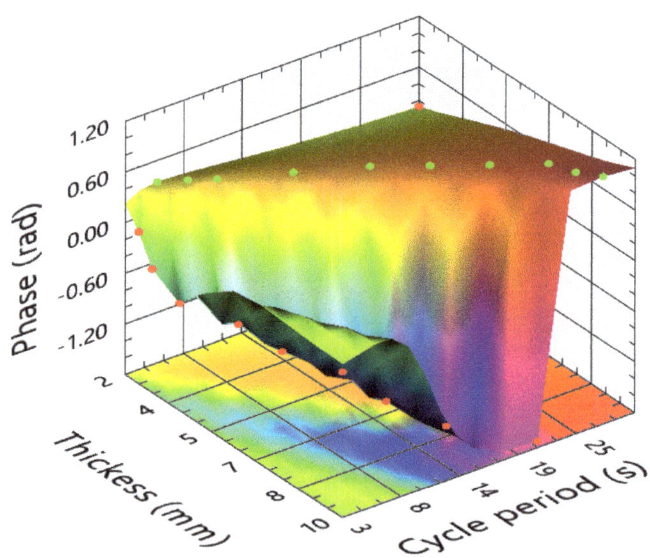

Figure 17. Phase difference function of the sample thickness and cycle period of CFRP samples.

5. Lock-In Thermal Test Validation

5.1. Validation of a Sample Made of PMMA

The previous chapter presented the best parameters for performing an LTT for PMMA and CFRP samples. The results show that certain conditions resulting in phase images can lead to inconclusive results. To prevent this, amplitude images should also be considered in the evaluation. The following validation tests used a different sample. It had four slots with a depth equal to slot three and widths of 5.0, 7.5, 10.0, and 12.5 mm. These validation tests consisted of laboratory tests LTT with various cycle periods.

Figure 18a,b illustrate the averaged amplitude and phase cross-section profiles obtained for the validation LTT. The amplitude response is proportional to the cycle period, as observed previously. Observing Figure 15, a small phase response was expected for shorter cycle periods as presented in Figure 18b. With the increase in the cycle period, the phase response in the slots' evidence started to increase. Similar to the simulation pattern and the evolution described previously, the phase difference at the slots becomes constant. The downside of longer stimulations (60 s) is the fading/blur of the boundaries. In the phase profiles, this is evident. Apart from the possible identification of defects, these long cycle periods do not provide better results and should not be considered. For the sample of PMMA and using Equation (8), the ideal cycle period has 35 s. Observing Figure 18a, one can say that the ideal cycle period is between 30 and 40 s.

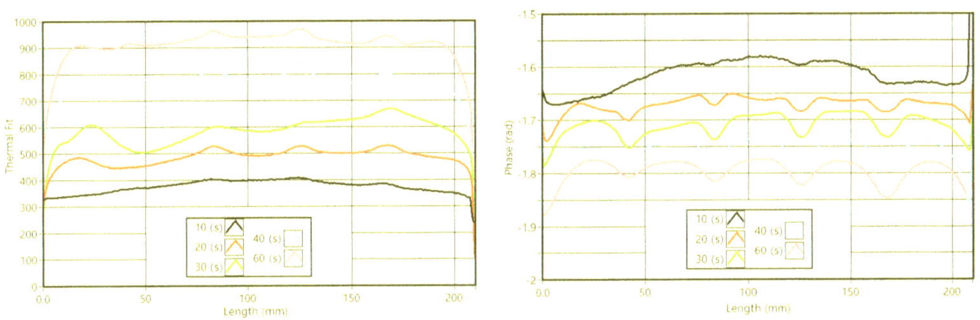

(a) (b)

Figure 18. Results from the PMMA validation tests. (**a**) Average amplitude profiles. (**b**) Average phase profiles.

5.2. Validation of a Sample Made of CFRP

Performing an analogy to the observations of the PMMA sample, the overall thermal response for the amplitude images should increase with the cycle period. Figure 19a,b present the vertical profiles for two groups of slots, corresponding to the amplitude profiles of 2 and 4 mm thick (slots areas). According to the pattern described, the amplitude response in the slots should increase for long cycle periods. Naturally, a deeper slot and a long cycle period lead to better slot visibility. An important aspect is a similarity in the amplitude differences between 20 and 30 s. This indicates that the amplitude variation resulting from a possible increase in the cycle period is becoming smaller, as predicted by the observed amplitude surface for CFRP (Figure 16).

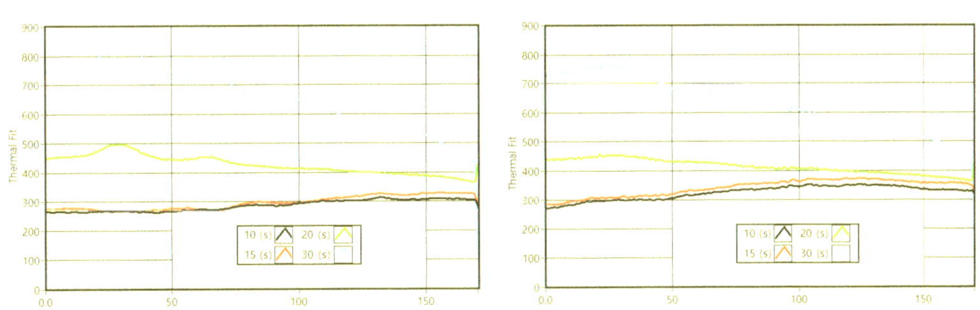

(a) (b)

Figure 19. Vertical amplitude profiles for two slots. (**a**) Thermal profiles for the slot 2 mm thick. (**b**) Thermal profiles for the slots 4 mm thick.

Figure 20 presents the phase profiles for the two sets of slots discussed. Similar to the PMMA samples and their phase, the CFRP profiles present a crescendo with the cycle period. The phase delay revealed low variations for the two shorter cycles, presenting a considerable variation for the 20 and 30 s profile. This pattern follows the behavior of the PMMA sample and the results of previous simulations. The phase delay for the cycle periods of 20 and 30 s are similar, thus validating the optimum cycle period of 26 s (Figure 20a,b).

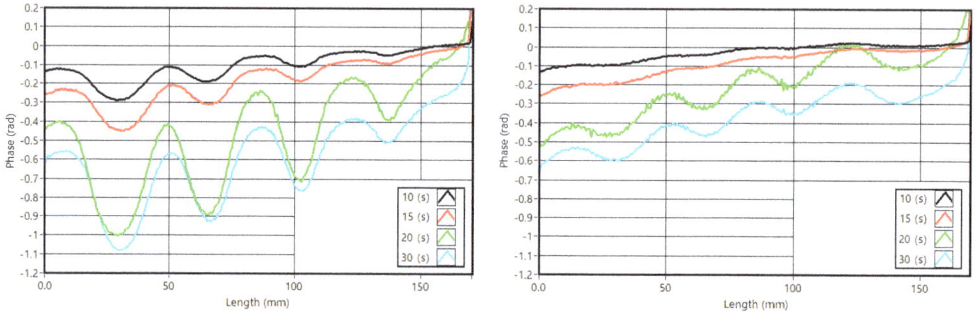

(a) (b)

Figure 20. Vertical phase profiles for two rows of slots of the CFRP samples and various cycle periods. (**a**) Phase profiles for the slots 2 mm thick. (**b**) Phase profiles for the slots 4 mm thick.

6. Conclusions

This work intended to study the influence of the cycle period in LTT for CFRP samples for different thicknesses. This was accomplished using simulations and validated experimentally using PMMA and CFRP samples. The analysis of laboratory conditions determined if the assumptions and simplifications were correct. Here, the Biot number was significant and low enough for the intended simplification to be possible. The mesh optimization diminished the computational cost and allowed accurate results. Fitting the temperature results through time and at the end of a cyclic simulation with 15 sinusoidal cycles, it was possible to determine the stimulation parameters.

The temperature differences resulted from the stimulation reflected radiation and were impossible to remove from the thermal measurements. Apart from this expected difference between the laboratory tests and the simulation, the similarity was very high.

Numerous LTT simulations were performed, producing the amplitude and phase results for different sample thicknesses and cycle periods. Despite the current results, some different behaviors occur when changing the number of cycles. Even if not compromising the evaluations, the number of cycles might decrease the test sensibility. The initial stimulations provided a comprehension of the temperature through the entire thermal cyclic test, thus providing good visualization of the heat flow inside the sample, inaccessible experimentally. An important aspect is the signal-to-noise ratio presented by the amplitude and phase. The amplitude images and the resulting profiles revealed a lower signal-to-noise ratio than the phase images. From all of the data, the phase images demonstrated a higher sensitivity to the stimulus and cycle period. The results also show that a shorter stimulation reveals a better boundary definition. In some situations, the blind frequencies can result in a null response in the phase images, only visible in the amplitude response. The rapid variation in the phase profiles was also predictable.

The recommended cycle period from the prediction surfaces was validated experimentally with two samples. These are different from the ones used to create the mentioned surfaces. After various laboratory experiments, the temperature, amplitude, and phase results validated the previous equations and prediction surfaces relating to the PMMA and CFRP sample. It is worth mentioning the CFRP validation sample was 8 mm thick and had circle blind holes.

To present this work's findings and facilitate their usage in future research, prediction surfaces and equations for the amplitude and phase difference were created. These consider PMMA and CFRP. The prediction surfaces can help in the identification of blind frequencies and prevent false negatives for a determined component thickness.

Supplementary Materials: The following supporting information can be downloaded at: https://www.mdpi.com/article/10.3390/s23010325/s1, Video S1: Simulation with 15 cycles and 20 s of the cycle period.

Author Contributions: Conceptualization, A.R.S., M.V., S.L. and J.M.; methodology, A.R.S., M.V. and J.M.; software, A.R.S.; validation, A.R.S., M.V. and J.M.; formal analysis, A.R.S., S.L. and J.M.; investigation, A.R.S.; resources, M.V. and J.M.; data curation, A.R.S. and S.L.; writing—original draft preparation, A.R.S.; writing—review and editing, A.R.S., M.V., S.L. and J.M.; visualization, A.R.S. and S.L.; supervision, M.V. and J.M.; funding acquisition, A.R.S., M.V. and J.M. All authors have read and agreed to the published version of the manuscript.

Funding: This research was funded by Projects LAETA UIDB/50022/2020 and UIDP/50022/2020, Fundação para a Ciência e a Tecnologia.

Institutional Review Board Statement: Not applicable.

Informed Consent Statement: Not applicable.

Data Availability Statement: Not applicable.

Conflicts of Interest: The authors declare no conflict of interest.

Abbreviations

The following abbreviations are used in this manuscript:

T_s	Temperature at the sample surface, 306.5 (K)
υ	Kinematic viscosity, $15.68 \times 10^{-6} \left(\frac{m^2}{s}\right)$
T_∞	Ambient temperature, 296.5 (K)
β	Volumetric thermal expansion coefficient, $3.3 \times 10^{-6} \left(\frac{m^2}{s}\right)$
L_c	Characteristic length, 0.16 (m)
Cp	specific heat, 1005 $\left(\frac{J}{KgK}\right)$
μ	Viscosity, $1.857 \times 10^{-5} \left(\frac{Kg}{s \times m}\right)$
k	Thermal conductivity, 0.0262 $\left(\frac{W}{m \times K}\right)$
l	Sample thickness, 4.5×10^{-3} (m)
K_{PMMA}	PMMA Thermal conductivity, 0.21 $\left(\frac{W}{m \times K}\right)$
Q_i	interior heat source per unitary area (stimulation)
K_x	conductivity coefficients in the **X** direction
K_y	conductivity coefficients in the **Y** direction
ρ	material relative density
Cp	Heat capacity
GR	Grashof number
PR	Prandtl number
Ra	Rayleigh number
Bi	Biot number

References

1. Ramos Silva, A.; Vaz, M.; Leite, S.R.; Mendes, J. Non-Destructive Infrared Lock-in Thermal Tests: Update on the Current Defect Detectability. *Russ. J. Nondestruct. Test.* **2019**, *55*, 772–784. [CrossRef]
2. Ramos Silva, A.; Vaz, M.; Leite, S.R.; Mendes, J. Analyzing the Influence of Thermal NDT Parameters on Test Performance. *Russ. J. Nondestruct. Test.* **2021**, *57*, 727–737. [CrossRef]
3. Ramos Silva, A.J.; Vaz, M.; Ribeirinho Leite, S.; Mendes, J. Analyzing the Influence of the Stimulation Duration in the Transient Thermal Test—Experimental and FEM Simulation. *Exp. Tech.* **2022**, 1–14. [CrossRef]

4. Ramos Silva, A.; Vaz, M.; Leite, S.R.; Mendes, J. Infrared Lock-in Thermography Tests with Optical Feedback. *U. Porto J. Eng.* **2022**, *8*, 1–12. [CrossRef]
5. Silva, A.R.; Vaz, M.; Leite, S.; Gabriel, J. Lock-in thermal test with corrected optical stimulation. *Quant. Infrared Thermogr. J.* **2021**, 1–22. [CrossRef]
6. Ferreira, A. *Problemas de Elementos Finitos em Matlab*; Fundacao Calouste Gulbenkian: Lisbon, Portugal, 2010.
7. Allaire, G. *Numerical Analysis and Optimization: An Introduction to Mathematical Modelling and Numerical Simulation*; Numerical Mathematics and Scientific Computation; Oxford University Press: Oxford, UK, 2007.
8. Ozisik, N. *Finite Difference Methods in Heat Transfer*; Heat Transfer; Taylor & Francis: London, UK, 1994.
9. Moukalled, F.; Mangani, L.; Darwish, M. *The Finite Volume Method in Computational Fluid Dynamics: An Advanced Introduction with OpenFOAM and Matlab*; Fluid Mechanics and Its Applications; Springer International Publishing: Zurich, Switwerland, 2015.
10. Fang, Q.; Tsuchiya, T.; Yamamoto, T. Finite difference, finite element and finite volume methods applied to two-point boundary value problems. *J. Comput. Appl. Math.* **2002**, *139*, 9–19. [CrossRef]
11. Thompson, E. *An Introduction to the Finite Element Method: Theory, Programming, and Applications*; John Wiley: New York, NY, USA, 2004.
12. Wang, J.; Gangarao, H.; Liang, R.; Liu, W. Durability and prediction models of fiber-reinforced polymer composites under various environmental conditions: A critical review. *J. Reinf. Plast. Compos.* **2016**, *35*, 179–211. [CrossRef]
13. Al-Athel, K.S.; Alhasan, M.M.; Alomari, A.S.; Arif, A.F.M. Damage characterization of embedded defects in composites using a hybrid thermography, computational, and artificial neural networks approach. *Heliyon* **2022**, *8*, e10063. [CrossRef]
14. Stoynova, A.; Bonev, B. Post-processing improvement of lock-in thermography study of MCM-L for better hidden defect localization. *Int. J. Circuits Syst. Signal Process.* **2022**, *16*, 941–947. [CrossRef]
15. Shrestha, R.; Choi, M.; Kim, W. Thermographic inspection of water ingress in composite honeycomb sandwich structure: A quantitative comparison among Lock-in thermography algorithms. *Quant. Infrared Thermogr. J.* **2021**, *18*, 92–107. [CrossRef]
16. Chulkov, A.O.; Gaverina, L.; Pradere, C.; Batsale, J.C.; Vavilov, V.P. Water detection in honeycomb composite structures using terahertz thermography. *Russ. J. Nondestruct. Test.* **2015**, *51*, 520–523. [CrossRef]
17. Pitarresi, G. Lock-In Signal Post-Processing Techniques in Infra-Red Thermography for Materials Structural Evaluation. *Exp. Mech.* **2015**, *55*, 667–680. [CrossRef]
18. An, Y.K.; Min Kim, J.; Sohn, H. Laser lock-in thermography for detection of surface-breaking fatigue cracks on uncoated steel structures. *Ndt Int.* **2014**, *65*, 54–63. [CrossRef]
19. Zoecke, C.; Langmeier, A.; Arnold, W. Size retrieval of defects in composite material with lockin thermography. In *Proceedings of the Journal of Physics: Conference Series*; IOP Publishing: Bristol, UK, 2010; Volume 214, p. 012093.
20. Peng, D.; Jones, R. Modelling of the lock-in thermography process through finite element method for estimating the rail squat defects. *Eng. Fail. Anal.* **2013**, *28*, 275–288. [CrossRef]
21. Swiderski, W. Non-destructive testing of CFRP by laser excited thermography. *Compos. Struct.* **2019**, *209*, 710–714. [CrossRef]
22. An, Y.K.; Yang, J.; Hwang, S.; Sohn, H. Line laser lock-in thermography for instantaneous imaging of cracks in semiconductor chips. *Opt. Lasers Eng.* **2015**, *73*, 128–136. [CrossRef]
23. Stoynova, A.; Bonev, B. Defects' Shape Influence on the Thermographic Control in Production of PCB. In Proceedings of the 2021 44th International Spring Seminar on Electronics Technology (ISSE), Bautzen, Germany, 5–9 May 2021. [CrossRef]
24. Breitenstein, O.; Warta, W.; Langenkamp, M. *Lock-in Thermography: Basics and Use for Evaluating Electronic Devices and Materials*; Springer Series in Advanced Microelectronics; Springer: Berlin/Heidelberg, Germany, 2013.
25. Brand, S.; Altmann, F. Lock-In-Thermography, Photoemission, and Time-Resolved GHz Acoustic Microscopy Techniques for Nondestructive Defect Localization in TSV. *IEEE Trans. Components Packag. Manuf. Technol.* **2018**, *8*, 735–744. [CrossRef]
26. Andersson, C.; Kristensen, O.; Miller, S.; Gloor, T.; Iannuzzo, F. Lock-in thermography failure detection on multilayer ceramic capacitors after flex cracking and temperature-humidity-bias stress. *IEEE J. Emerg. Sel. Top. Power Electron.* **2018**, *6*, 2254–2261. [CrossRef]
27. Leppänen, J.; Ross, G.; Vuorinen, V.; Ingman, J.; Jormanainen, J.; Paulasto-Kröckel, M. A humidity-induced novel failure mechanism in power semiconductor diodes. *Microelectron. Reliab.* **2021**, *123*, 114207. [CrossRef]
28. Hovhannisyan, R.A.; Kapran, O.M.; Golod, T.; Krasnov, V.M. Accurate determination of the josephson critical current by lock-in measurements. *Nanomaterials* **2021**, *11*, 2058. [CrossRef]
29. Mulaveesala, R.; Tuli, S. Theory of frequency modulated thermal wave imaging for nondestructive subsurface defect detection. *Appl. Phys. Lett.* **2006**, *89*, 191913. [CrossRef]
30. Rani, A.; Mulaveesala, R. Depth resolved pulse compression favourable frequency modulated thermal wave imaging for quantitative characterization of glass fibre reinforced polymer. *Infrared Phys. Technol.* **2020**, *110*, 103441. [CrossRef]
31. Hedayatrasa, S.; Poelman, G.; Segers, J.; Van Paepegem, W.; Kersemans, M. Performance of frequency and/or phase modulated excitation waveforms for optical infrared thermography of CFRPs through thermal wave radar: A simulation study. *Compos. Struct.* **2019**, *225*, 111177. [CrossRef]
32. Philipp, A.; Pech-May, N.W.; Kopera, B.A.; Lechner, A.M.; Rosenfeldt, S.; Retsch, M. Direct measurement of the in-plane thermal diffusivity of semitransparent thin films by lock-in thermography: An extension of the slopes method. *Anal. Chem.* **2019**, *91*, 8476–8483. [CrossRef] [PubMed]

33. Silva, A.J.R.; Moreira, P.M.G.; Vaz, M.A.P.; Gabriel, J. Temperature profiles obtained in thermoelastic stress test for different frequencies. *Int. J. Struct. Integr.* **2017**, *8*, 51–62. [CrossRef]
34. Maierhofer, C.; Myrach, P.; Reischel, M.; Steinfurth, H.; Röllig, M.; Kunert, M. Characterizing damage in CFRP structures using flash thermography in reflection and transmission configurations. *Compos. Part B Eng.* **2014**, *57*, 35–46. [CrossRef]
35. Churchill, S.W.; Chu, H.H.S. Correlating equations for laminar and turbulent free convection from a vertical plate. *Int. J. Heat Mass Transf.* **1975**, *18*, 1323–1329. [CrossRef]
36. Bergman, T.L.; Incropera, F.P.; Lavine, A.S. *Fundamentals of Heat and Mass Transfer*; John Wiley & Sons: New York, NY, USA, 2011.
37. Cignoni, P.; Montani, C.; Scopigno, R. A comparison of mesh simplification algorithms. *Comput. Graph.* **1998**, *22*, 37–54. [CrossRef]
38. Kim, S.J.; Kim, C.H.; Levin, D. Surface simplification using a discrete curvature norm. *Comput. Graph.* **2002**, *26*, 657–663. [CrossRef]
39. Tian, T.; Cole, K.D. Anisotropic thermal conductivity measurement of carbon-fiber/epoxy composite materials. *Int. J. Heat Mass Transf.* **2012**, *55*, 6530–6537. [CrossRef]
40. Siddiqui, N.A.; Woo, R.S.; Kim, J.K.; Leung, C.C.; Munir, A. Mode I interlaminar fracture behavior and mechanical properties of CFRPs with nanoclay-filled epoxy matrix. *Compos. Part A Appl. Sci. Manuf.* **2007**, *38*, 449–460. [CrossRef]
41. Yokozeki, T.; Iwahori, Y.; Ishiwata, S.; Enomoto, K. Mechanical properties of CFRP laminates manufactured from unidirectional prepregs using CSCNT-dispersed epoxy. *Compos. Part A Appl. Sci. Manuf.* **2007**, *38*, 2121–2130. [CrossRef]
42. Liang, J.; Saha, M.C.; Altan, M.C. Effect of carbon nanofibers on thermal conductivity of carbon fiber reinforced composites. *Procedia Eng.* **2013**, *56*, 814–820. [CrossRef]

sensors

MDPI

Article

Raspberry Pi Platform Wireless Sensor Node for Low-Frequency Impedance Responses of PZT Interface

Quang-Quang Pham [1], Quoc-Bao Ta [1], Jae-Hyung Park [2] and Jeong-Tae Kim [1,*]

[1] Department of Ocean Engineering, Pukyong National University, 45 Yongso-ro, Nam-gu, Busan 48513, Republic of Korea
[2] CNS Solution Co., Ltd., 21 Century City Office, 312 Suyeong-ro, Nam-gu, Busan 48513, Republic of Korea
* Correspondence: idis@pknu.ac.kr; Tel.: +82-51-629-6585

Abstract: A wireless impedance monitoring system, called SSeL-Pi, is designed to have cheap, mobile, and handy practical features as compared to wired commercial impedance analyzers. A Raspberry Pi platform impedance sensor node is designed to measure signals at a low-frequency range of up to 100 kHz. The low-frequency impedance measurement via the proposed node has been combined with a new PZT interface technique for measuring local responses sensitive to structural damage. The new PZT interface can work as a surface-mounted or embedded sensor, and its local dynamic characteristics are numerically analyzed to pre-determine an effective impedance resonant frequency range of less than 100 kHz. Next, a software scheme was designed to visualize the input/output parameters of the proposed SSeL-Pi system (i.e., Raspberry Pi platform and PZT interface) and automate signal acquisition procedures of the impedance sensor node. The calibration for impedance signals obtained from the proposed system was performed by a series of procedures, from acquiring real and imaginary impedance to adjusting them with respect to a commercial impedance analyzer (HIOKI-3532). The feasibility of the wireless impedance monitoring system was experimentally evaluated for PZT interfaces that were subjected to various compressive loadings. The consistent results analyzed from signals measured by the SSeL-Pi and HIOKI 3532 systems were observed. Additionally, the strong relationships between impedance features (frequency shift and RMSD index) and compressive stresses of the PZT interfaces showed the potential for axial force/stress variation monitoring in real structures using the Raspberry Pi platform impedance sensor node and developed PZT interface.

Keywords: Raspberry Pi; wireless sensor node; impedance response; PZT interface; stress variation

Citation: Pham, Q.-Q.; Ta, Q.-B.; Park, J.-H.; Kim, J.-T. Raspberry Pi Platform Wireless Sensor Node for Low-Frequency Impedance Responses of PZT Interface. *Sensors* **2022**, *22*, 9592. https://doi.org/10.3390/s22249592

Academic Editor: Zenghua Liu

Received: 27 October 2022
Accepted: 5 December 2022
Published: 7 December 2022

1. Introduction

Electromechanical impedance (EMI)-based techniques have been developed for structural health monitoring (SHM) in civil, mechanical, and aerospace engineering areas [1–3]. The fundamental theory of the EMI-based methods was developed by Liang et al. [4], and it was successively followed by many researchers [5,6]. However, the field implementation of the techniques is limited due to the lack of affordable hardware. As they are designed for the laboratory environment, most impedance analyzers (e.g., HIOKI 3532 or HP4149A) are bulky and expensive but unsuitable for use in the field. Therefore, there is a need to develop portable, handy, and inexpensive measurement units for impedance-based SHM of real structures in the field.

Wireless impedance monitoring methods have been developed for SHM [7–15]. Mascarenas et al. [7] proposed a prototype of a wireless impedance sensor node for structural health monitoring. The developed sensor node used an AD5933 impedance chip [16] for measuring the impedance signals of a piezoelectric (PZT) transducer, an ATmega128L microcontroller for computing, and a 2.4 GHz Xbee radio-frequency telemetry module for wireless communication. Many researchers have then followed and improved the initial prototype to obtain a low-cost multi-functional wireless impedance sensor node. Park S. et al. [10] improved the prototype by adding some functions such as a memory card slot, multi-channel measurement, and temperature recording. Min et al. [11] embedded the algorithms on the microcontroller for structural damage detection/sensor self-diagnosis. Kim et al. [13] developed an SSeL-I impedance sensor board operated on an Imote2 platform to enhance the wireless monitoring capacity. As a high-performance sensor platform, the Imote2 [17] consists of a low-power microcontroller of 416 MHz clock speed, 32 MB SDRAM (synchronous dynamic random-access memory), and a radio transceiver of 250 kb/s transmission rate with 16 channels in the 2.4 GHz band. The Imote2 platform impedance sensing system (called Imote2/SSeL-I) was applied for local damage detection in structural connections [13,15,18] and prestress-loss monitoring in tendon anchorage [19,20].

Recently, Raspberry Pi has been developed for a commercial microcomputer [21,22]. As compared to the Imote2 platform, Raspberry Pi has many advanced features, such as a powerful microprocessor of 1.5 GHz clock speed, 1~8 GB SDRAM, and a wireless LAN with 600 Mb/s transmission rate in the 2.4/5.0 GHz bands. The Raspberry platform simultaneously supports multiple sensors via a 40-pin GPIO (general-purpose input/output) header. For example, a unit of Raspberry Pi costs about US$80 for a model with 4 GB SDRAM. As a cheap, powerful, and multi-functional unit, Raspberry Pi is an effective platform for controlling distributed sensors to continuously monitor structural responses [23–26]. Mahmud et al. [24] used a Raspberry Pi to generate an excitation signal on a PZT sensor, detect damage in a metal plate, and send a structural health status to an Internet server. This study used a combination of pitch-catch and pulse-echo techniques for monitoring the structure's health. Ghosh et al. [25] employed a Raspberry Pi as a computing unit to code the piezoelectric sensors and receive the data from these sensors in structural health monitoring for concrete beams. Meng and Zhu [26] combined a Raspberry Pi 4 and a microelectromechanical systems accelerometer to form an Internet of Things sensing system for vibration monitoring.

So far, the Rasberry Pi platform has not been used for wireless impedance-based SHM. At least three techniques should be developed to make it feasible. Firstly, a data acquisition system should be integrated with a miniature impedance analyzer. A sensing module 'PmoldIA' for measuring impedance signals up to 100 kHz [27] (which costs about US$50) is a good choice for the Raspberry Pi platform. Secondly, the operating software of the Raspberry Pi platform should be programmed for autonomous measurement and feature extraction. Thirdly, the data acquisition protocol in the impedance sensor node should be calibrated based on the commercial impedance analyzer.

In this study, a new wireless impedance monitoring system, called SSeL-Pi, is designed to have cheap, mobile, and handy practical features as compared to wired commercial impedance analyzers (e.g., HIOKI 3532 or HP4149A). A Raspberry Pi platform sensor node is designed to acquire impedance signals at a low-frequency range of up to 100 kHz. The low-frequency impedance measurement was combined with a new PZT interface technique, which was developed following the study of Huynh et al. [28]. The new PZT interface helps pre-determine effective frequency ranges of less than 100 kHz, which is used for measuring local responses sensitive to structural damage. A software scheme was designed to operate the proposed SSeL-Pi system (i.e., Raspberry Pi platform and PZT interface). Next, a calibration procedure was designed for the impedance sensor node. The calibration results from the proposed system were compared to signals obtained from the commercial impedance analyzer system (HIOKI 3532). Then, the feasibility of the newly designed impedance sensing system was experimentally evaluated for PZT interfaces subjected to various compressive loadings.

This paper is organized as follows: (i) Section 2 presents the development of the hardware, the software, and the PZT interface technique for the proposed SSeL-Pi system; (ii) Section 3 presents the calibration procedures for the monitoring system; (iii) Section 4 presents the experiment performed to evaluate the feasibility of the SSeL-Pi system; and (iv) Section 5 draws the findings and conclusions. The major contributions of this study are described as follows: (i) the newly developed impedance sensor node was developed based on the Raspberry Pi platform; (ii) the new PZT interface model with a low predetermined frequency range of less than 100 kHz was integrated with the Raspberry Pi platform impedance sensor node; and (iii) the newly developed impedance sensor node was applied to monitor stress variation in concrete structures.

2. Design of Raspberry Pi Platform Impedance Monitoring System

According to Nagayama et al. [29], a smart sensor node has five main features: an on-board microprocessor, sensing capability, wireless radio, battery power, and low cost. The sensing capability is implemented by a data acquisition unit that enables a signal generator, amplifier, and anti-aliasing filter. The microcontroller operates on-board computation of signal process and feature extraction for the data acquisition unit. As proposed by Mascarenas et al. [7], the data acquisition procedure for an impedance-based sensor node has three steps. Firstly, an AD5933 impedance chip with a 12-bit analog-to-digital converter (ADC) measures impedance signals from a PZT sensor. Then, impedance signals are stored in data storage (e.g., secure digital card—SD card). And lastly, the stored data are transferred to a master device (e.g., laptop) via the wireless communication module.

2.1. Schematic of Impedance Monitoring System

In this study, a Raspberry Pi platform impedance monitoring system 'SSeL-Pi' was designed as schematized in Figure 1. The main components of the SSeL-Pi system include a power supply, a Raspberry Pi platform [22], an impedance sensing module PmodIA from Digilent [27], and a PZT interface device. The SSeL-Pi system requires a power supply of 5 V/3 A. The Raspberry Pi platform manages a micro-central processing unit (mCPU), a wireless radio (WiFi), and a few other parts described in the following Section 2.2. The Raspberry Pi platform runs on Raspbian, which is an operating software (OS) based on the Debian Linux distribution optimized for the Rasberry Pi family.

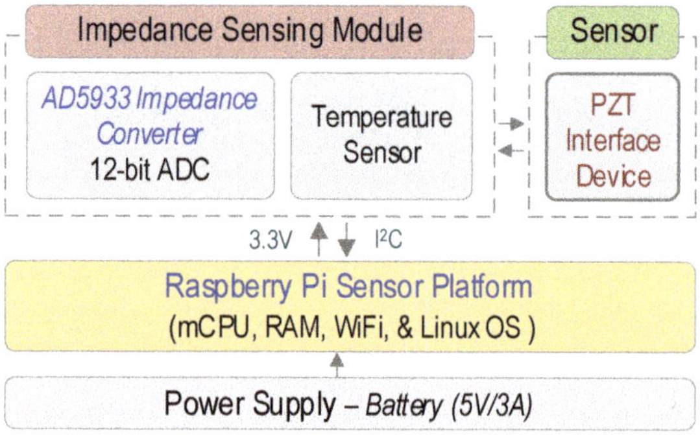

Figure 1. Raspberry Pi platform impedance monitoring system 'SSeL-Pi'.

The impedance sensing module manages a built-in AD5933 impedance chip with a 12-bit ADC converter. The board is also equipped with a temperature sensor. The built-in AD5933 impedance chip generates frequency-domain signals of less than 100 kHz to the PZT sensor (e.g., PZT 5 A patch). The impedance sensor node was designed to be combined with a new PZT interface device, which was designed following the study of Huynh et al., 2017 [28]. The proposed PZT interface helps ensure impedance signals (less than 100 kHz) that are sensitive to local damage and stress change. The inter-integrated circuit (I^2C) communication protocol and a power supply (3.3/5 V) are utilized for communicating between the Raspberry Pi platform and the impedance sensing module. The impedance sensing module is connected to the Raspberry Pi platform via GPIO pins [21] (see Figure 1). Additionally, the PZT sensor is connected to the impedance sensing module by electrical wires.

As outlined in Table 1, the specification of the proposed SSeL-Pi system is compared to those of the impedance measurement system HIOKI 3532 [30]. The proposed system can measure an impedance range of up to 10 MΩ with a frequency range of up to 100 kHz [27]. The proposed system is smaller in dimension and weight as compared to the commercial one. The voltage peak-to-peak ($V_{p\text{-}p}$) is 2 $V_{p\text{-}p}$ for the board PmodIA [27], while it is up to 14 $V_{p\text{-}p}$ for HIOKI 3532 [30]. Note that the Raspberry Pi platform sensing system costs around US$280 (in E-commerce, the price for Raspberry Pi 4 B with full options is around US$230 and for PmodIA about US$50), which is much lower than the HIOKI 3532 system.

Table 1. Raspberry Pi platform impedance monitoring system 'SSeL-Pi' versus commercial impedance analyzer 'HIOKI-3532'.

Parameter	SSeL-Pi System [22,27]	HIOKI 3532 [30]
Impedance range	100 Ω–10 MΩ	10 mΩ–200 MΩ
Frequency range	1 kHz–100 kHz	42 Hz–5 mHz
Voltage peak-to-peak	2 $V_{p\text{-}p}$	14 $V_{p\text{-}p}$
Cost	US$280	US$15,000
Dimension	88 × 58 × 19.5 mm	352 × 323 × 124 mm
Weight	80 g	6.5 kg

2.2. *Specification of Impedance Monitoring System*

2.2.1. Raspberry Pi Platform

A Raspberry Pi is a single-board computer with good features such as small size, low price, Linux operation, and flexible form factor. In this study, a Raspberry Pi 4 B platform was used for the impedance sensing system. Its main specifications are outlined in Table 2. It has a micro central processing unit (mCPU) of 64-bit quad-core Cortex-A72 (1.5 GHz), RAM memory of LPDDR4 4 GB, wireless radio 802.11 ac of WI-FI 2.4/5 GHz, and storage capacity of 16 GB SD card. The Raspberry Pi platform can be powered by a USB-C connection supplying at least 400 mA at 5 V. An on-board 40-pin GPIO header connects externally to the impedance sensing module. Others include USB and HDMI ports for external devices (e.g., keyboard and monitor), CSI and audio ports for video/audio input, and Gigabit Ethernet.

Table 2. Main specifications of Raspberry Pi 4 B.

No.	Name	Feature	No.	Name	Feature
1	mCPU	64-bit quad-core Cortex-A72 (1.5 GHz)	5	GPIO	40 pinouts
2	RAM	4 GB LPDDR4	6	Storage	MicroSD-16 GB
3	Wireless	802.11 ac (2.4/5 GHz)	7	USB ports	2 × USB 3.0, 2 × USB 2.0
4	Power supply	USB Type-C/5 V and 3 A	8	HDMI ports	2 × Micro-HDMI up to 1080 p60

2.2.2. Impedance Sensing Board

Figure 2 shows a portable impedance analyzer board (the so-called PmodIA module [27]) which consists of an AD5933 chip, temperature sensor, I^2C communication ports, and two SMA (subminiature version A) sensor connectors. The Analogue Devices AD5933 chip is a high-precision impedance converter that combines an on-board frequency generator with 12-bit ADC [16,27]. The I^2C communication has four pins, which are SCL (serial clock), SDA (serial data), GND (ground), and VCC (voltage common collector) (see Figure 2). There is a pair for each type of pin. The impedance measurement using the AD5933 is performed in the following three procedures. Firstly, a frequency generator excites complex dynamic signals at an assigned frequency that is commended externally via the I^2C communication. Secondly, the response signal is sampled by the on-board ADC. Thirdly, a discrete Fourier transform (DFT) is processed by an on-board DSP (digital signal processor) engine. The DFT algorithm returns a real (R) and imaginary (I) data word at each output frequency.

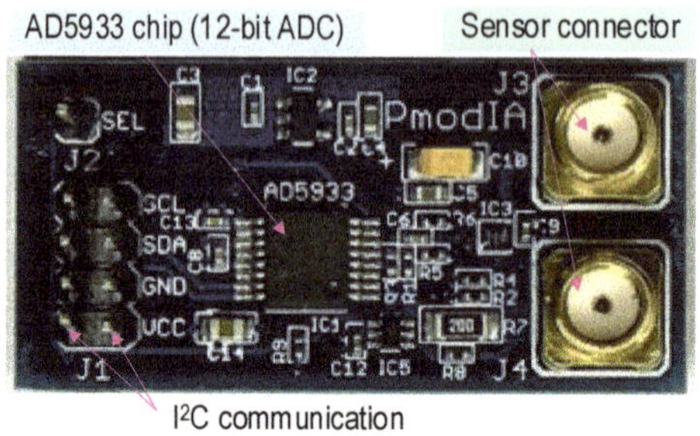

Figure 2. Impedance sensing module 'PmodIA'.

As outlined in Table 3, a few parameters of the PmodIA module [27,31,32] were orientated for the measurement of impedance signatures from the PZT interface device (which is described later). The internal oscillator frequency was set at 16.776 mHz to run the device [27]. The current-to-voltage amplifier gain resistor and peak-to-peak voltage were set at 20,000 Ohm and 2 $V_{p\text{-}p}$, respectively [31,32]. The programmable gain amplifier (PGA) was ×5 times, and the number of setting time cycles was 100. The supply voltage was 3.3 V.

Table 3. Initial setup of impedance sensing module 'PmodIA'.

Specification	Value	Unit
Internal oscillator frequency	16.776	mHz
Current-to-voltage amplifier gain resistor	20,000	Ohm
Programmable gain amplifier (PGA gain)	×5	-
Output excitation voltage	2 $V_{p\text{-}p}$	V
Number of settling time cycles	100	-
Supply voltage	3.3	V

2.2.3. Circuit Design of 'SSeL-Pi'

Figure 3 shows the connection method between Raspberry Pi 4, the PmodIA, and an example of the PZT interface. First, the PmodIA was linked with Raspberry Pi 4 via I^2C communication. The SDA and SCL pins of the PmodIA were connected with the GPIO SDA (pin #3) and GPIO SCL (pin #5) of Raspberry Pi 4, respectively. Additionally, the VCC and GND pins of the PmodIA were linked to pin #1 (3V3 PWR) and pin #9 (Ground) of the Raspberry Pi 4, correspondingly. Then, the example of the PZT interface was connected to the PmodIA via the SMA connectors on this board.

2.2.4. PZT Interface for Impedance Measurement

Impedance Concept of PZT Interface

Huynh et al. [28] developed a PZT interface technique to monitor the damage in a tendon anchorage induced by the variation of tendon forces. The developed interface, which worked as a beam-like structure, included a flexible part in the middle and two outside fixed parts. The proposed technique helped pre-determine the effective frequency band less than 100 kHz, which was suitable for wireless impedance measurement [7,12].

By following the above-mentioned study, a new type of aluminum PZT interface was proposed for structural health monitoring using a low pre-determined frequency range (less than 100 kHz) of impedance responses. As shown in Figure 4a, the newly designed PZT interface has a vibrating plate (i.e., flexural section). Then, the fabricated component is covered by two plates on its top and bottom, respectively. The vibrating plate, where the PZT patch is installed, is designed to help the interface fluctuate according to the inverse-piezoelectric behaviors of the PZT patch. With the support and protection of the cover walls and plates (i.e., top and bottom plates), the PZT interface works as a multi-functional sensor: (1) surface-bonded PZT sensor and (2) embedded PZT sensor.

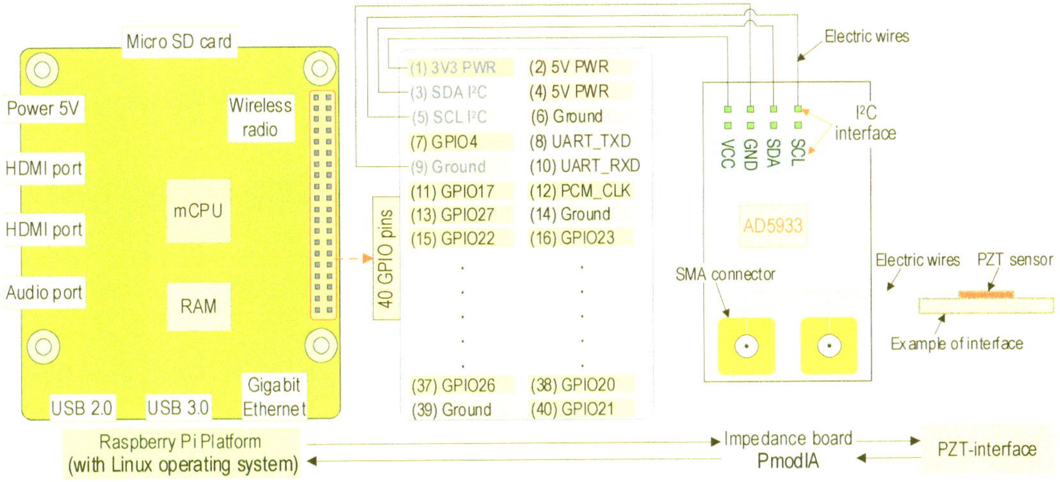

Figure 3. Circuit design of 'SSeL-Pi'.

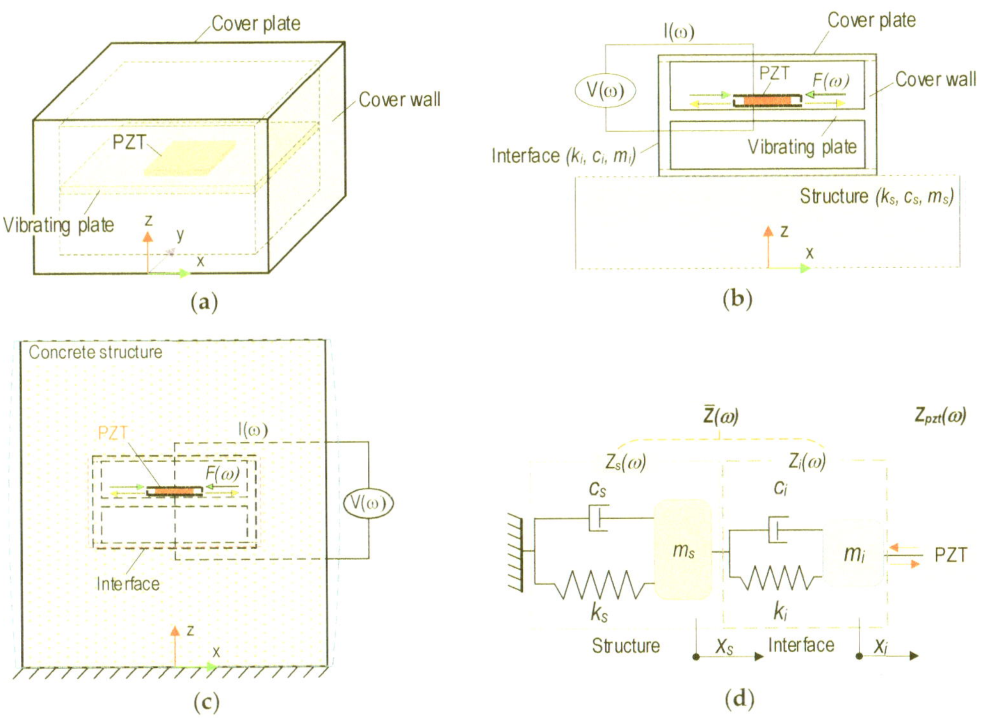

Figure 4. A PZT interface device for impedance monitoring. (**a**) PZT interface, (**b**) Surface-mounted PZT interface on structure, (**c**) Embedded PZT interface in concrete structure, (**d**) 2-DOF impedance model.

For the surface-bonded sensor (see Figure 4b), the PZT interface could be contacted with a host structure via a bonding layer between the bottom plate of the PZT interface and structural surface. With the protected cover walls and plates, the integrity of the PZT sensor could be ensured during the construction and operation of the host structure. Furthermore, the ambient effects [33,34] on the impedance responses could be reduced.

For the embedded PZT sensor (see Figure 4c), the PZT interface could be embedded in the concrete structures and work as a piezoelectric-based smart aggregate [35]. By embedding into the concrete structures, the PZT interface could directly observe the internal stress variation and concrete transformation induced by the externally applied forces [36]. Furthermore, with the pre-determined frequency range, the impedance measurement using the embedded PZT interface could be processed faster than that using the conventional smart aggregates, which used the trial-and-error method to find the effective monitoring frequency ranges [37]. Thus, the proposed PZT interface could be cost-saving and time-efficient for the impedance monitoring procedure.

As described in Sections 2.2.1 and 2.2.2, the Raspberry Pi platform impedance sensor node was designed to measure signals at the frequency range of up to 100 kHz. Thus, the developed measuring system can be combined with the new PZT interface device for low pre-determined frequency impedance monitoring.

As shown in Figure 4d, the interaction between the PZT interface and target structure could be simplified as a two-DOF impedance model [28]. In the two-DOF impedance model, one DOF stands for the motion of the PZT interface, and the other represents the motion of the inspected structure. The coupled structural-mechanical (SM) impedance $\overline{Z}(\omega)$ of the interface structure at the PZT-driven point can be written as [28]:

$$\overline{Z}(\omega) = \frac{f_i(\omega)}{\dot{x}_i(\omega)} = \frac{K_{11}(\omega)K_{22}(\omega) - K_{12}^2(\omega)}{i\omega K_{22}(\omega)} \quad (1)$$

where dynamic stiffness coefficients are derived as $K_{11} = -\omega^2 m_i + i\omega c_i + k_i$, $K_{12} = -i\omega c_i - k_i$, and $K_{22} = -\omega^2 m_s + i\omega(c_i + c_s) + (k_i + k_s)$. The terms m, c, and k are mass, damping coefficient, and spring stiffness.

The subscripts s and i stand for the inspected structure and interface, respectively. The stiffness coefficients depend on the structural parameters of the inspected structure and interface. The SM impedance of the PZT patch, $Z_p(\omega)$, and that of the interface-inspected structure, $Z_s(\omega)$, are coupled together (Liang et al. 1994 [4]):

$$Z(\omega) = \frac{V}{I} = \left\{ i\omega \frac{w_a l_a}{t_a} \left[\hat{\varepsilon}_{33}^T - \frac{1}{Z_a(\omega)/\overline{Z}(\omega) + 1} d_{31}^2 \hat{Y}_{11}^E \right] \right\}^{-1} \quad (2)$$

where w_a, l_a, t_a are geometric constants of the PZT patch; $\hat{Y}_{11}^E$, d_{31}, and $\hat{\varepsilon}_{33}^T$ are the complex Young's modulus patch at zero electric fields, the piezoelectric coupling constant, and the complex dielectric constant zero stress, respectively. Once the PZT patch has no changes in mechanical properties or electrical characteristics and no energy loss induced by contact conditions between the interface structure, the SM impedance $Z_p(\omega)$ remains constant. Thus, any changes in the monitored structure (e.g., stress change or structural damage) that cause shifts in impedance frequencies can be measured via the interface technique.

The 2-DOF impedance model contains resonant peaks in its impedance signatures that represent coupled vibration modes out of the PZT interface structure system. Therefore, the effective frequency bands of the impedance signatures can be predetermined by controlling the structural parameters of the PZT interface. Additionally, the impedance model represents the structural parameters of both the interface device and host structure. Any structural changes (e.g., damage or stress variation) result in a change in the impedance response of the model.

Numerical Impedance Response and Local Dynamic Characteristics of PZT Interface

As studied by [28,38], the effective frequency ranges for impedance measurement via the PZT interface have a relationship with the dynamic characteristics of this device. Therefore, the local dynamic characteristics of the PZT interface needed to be examined.

Figure 5a shows a prototype design of the PZT interface. Aluminum was selected for the interface's material. A PZT 5 A patch, which had a size of 10 × 10 × 0.51 mm, was surface-mounted on the middle of a component including the vibrating plate (21 × 21 × 1.5 mm) and the cover walls (height of 7 mm and thickness of 2 mm). Super glue was used for a bonding layer (0.1 mm thickness). After that, the fabricated combo was covered by two plates (25 × 25 × 2.0 mm) on the top and bottom to form the PZT interface (Figure 5a). The material properties of the aluminum, the PZT 5 A [38], and superglue are listed in Table 4. The epoxy's properties, which were used in the interface's fabrication in this study, are also presented in this table.

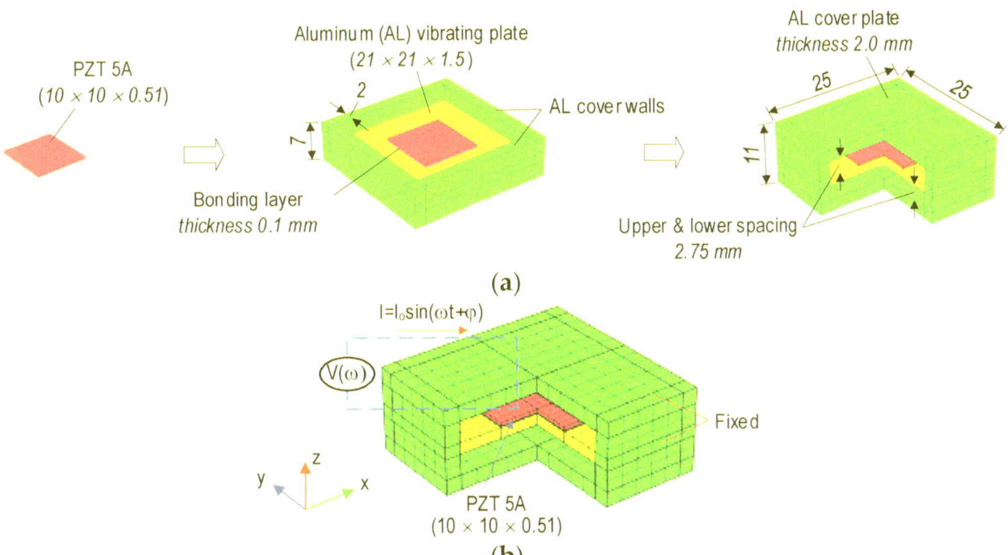

Figure 5. Geometric parameters of the PZT interface (dimension in mm). (**a**) Prototype design, (**b**) Meshing and boundary conditions.

Table 4. Material properties of aluminum, PZT patch, and bonding layers.

Properties	Aluminum 6061-T6	PZT 5 A	Bonding Layer Super Glue	Bonding Layer Epoxy
Young's modulus, E (GPa)	68.9	62.1	5	0.75
Poisson's ratio, ν	0.33	0.35	0.38	0.3
Mass density, ρ (kg/m^3)	2700	7750	1700	1090
Damping loss factor, η	0.02	0.0125		0.02
Yield strength, σ_y (MPa)	241			
Compressive strength, σ_c (MPa)				32.3
Dielectric constant, ε_{33}^T (F/m)		1.53×10^{-8}		
Coupling constant, d_{31} (m/V)		-1.71×10^{-10}		
Dielectric loss factor, δ		0.015		

Figure 5b shows three-quarters of a finite element (FE) model of the PZT interface, which was simulated using Comsol Multiphysics. The FE model had 968 elements, of which 100 were for the bonding layer, 100 for the PZT patch, 260 for the vibrating plate,

220 for the cover walls, and 288 for two cover plates. Quadratic hexahedron elements were used for the PZT interface. It was assumed that the PZT interface would be embedded in the concrete structures for impedance monitoring. Thus, all outer surfaces of the interface were assigned as fixed boundary conditions, and only the vibrating plate had the ability to oscillate in the structures.

To acquire the impedance responses of the PZT interface, a 1 V harmonic excitation was assigned on the top surface of the PZT, while the bottom one was simulated as the ground electrode. Figure 6 shows the impedance signals of the PZT interface in the frequency range of 5–100 kHz with 1901 sweeping points. The PZT interface shows the first and second resonant impedance peaks at 24.65 kHz (Peak 1) and 93.05 kHz (Peak 2). Since the vibrating plate (flexible section) of the PZT interface had a square shape, the transverse displacements along the x-axis were identical to the ones along the y-axis. The impedance responses of Peaks 1–2 were the 1st (at 24.14 kHz) and 6th (at 91.41 kHz) flexural motions of the vibrating plate in the PZT interface (see Figure 6). Note that the low-frequency range of the PZT interface's impedance signals can be pre-determined via the modal analysis, which is similar to previous studies [28,38].

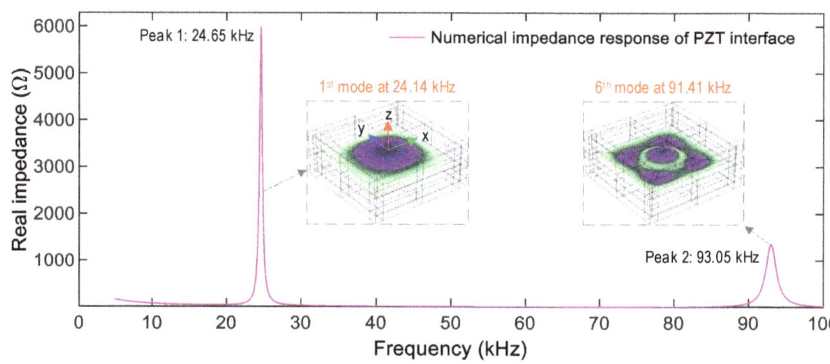

Figure 6. Numerical impedance responses of the PZT interface.

2.3. *Software Design for Raspberry Pi Platform Impedance Monitoring System*

2.3.1. Prototype of SSeL-Pi Impedance Monitoring System

A prototype of the SSeL-Pi impedance monitoring system was designed as shown in Figure 7. A prototype of the SSeL-Pi system was fabricated by a laptop, a Linux-operating Raspberry Pi platform, an impedance sensor board (PmodIA module), and a PZT interface device ensuring frequency responses less than 100 kHz. The way to connect the Raspberry Pi, PmodIA, and PZT interface is presented in Figure 3. The laptop controls the Raspberry Pi platform to command the PmodIA module for sensing impedance signals of the PZT sensor. The measurement command is transferred to the sensor node via WiFi communication.

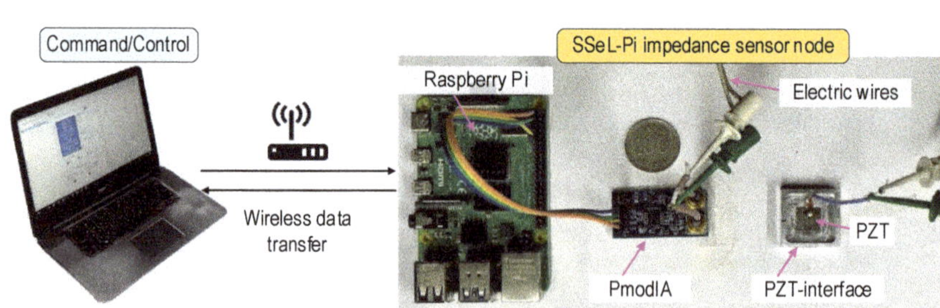

Figure 7. Prototype of impedance monitoring system 'SSeL-Pi'.

At the sensor node, Raspberry Pi automatically decodes the command to control the impedance sensor board. The PmodIA module introduces an excitation of 2 V into the PZT sensor at the designed frequency. At the same time, the corresponding impedance response at the exciting frequency is received backward from the sensor. The PZT interface device ensures the impedance signals (less than 100 kHz) that are sensitive to local damage and stress change. The decoding procedure of the input/output signals is carried out by the PmodIA module. Then, the impedance signatures are stored in the Raspberry Pi platform, and the saved data are wirelessly transferred to the laptop, where the post-process is conducted for the measured impedance features.

2.3.2. Operation Scheme for Impedance Sensor Node

An operation scheme was designed for the Raspberry Pi platform impedance monitoring system. The impedance sensor node is controlled by an embedded Flask server, which is an application for building a web-server using Python language. The Flask server is used for REST API (Representational State Transfer Application Programming Interface) to communicate with other units. Since REST API uses the standard HTTP protocol, a unit to control the impedance sensor node can be easily developed by any programming language.

As shown in Figure 8, five routines were designed for the control scheme of the impedance sensor node, which includes "setting", "operate", "data", "delete_data", and "measureone". The "setting" routine is used for setting impedance measurement parameters such as start frequency, frequency increment, and measuring interval or for getting stored parameters from the sensor node. The "operate" routine starts or stops scheduled monitoring by the pre-defined measuring interval. The "data" routine sends the measured data to the control unit. The "delete_data" routine deletes all measured data in the sensor node. The "measureone" routine measures impedance signals at once and sends the measured data. The setting parameters and measured data are stored in a database by SQLite DBMS (database management system). The sensor node returns JSON (JavaScript Object Notation) formatted data to the server.

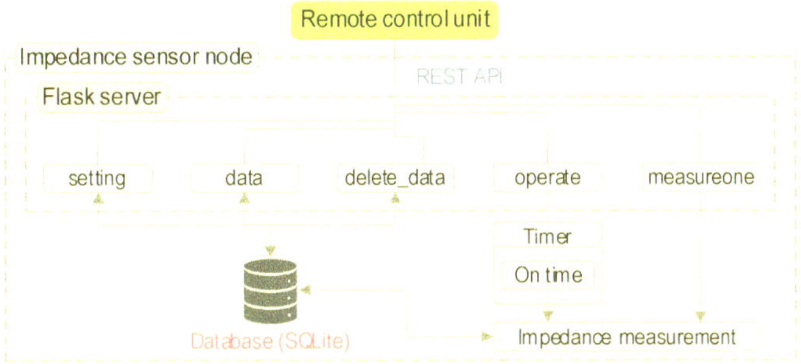

Figure 8. Control scheme of impedance monitoring system 'SSeL-Pi'.

2.3.3. GUI Software for Impedance Monitoring

As a general-purpose computer language, Delphi [39] was used to program the operating software of the SSeL-Pi impedance monitoring system. It is known that Delphi uses the Object Pascal (OP) programming language and provides an integrated development environment (IDE) for rapid application development of console software. The Delphi-based programming environment is free to academic users. A graphic user interface (GUI) software for the Raspberry Pi platform impedance monitoring was designed, as illustrated in Figure 9. As marked in the figure, the GUI software includes six main features to visualize the software interface, input command, and output results:

① Check and link to a 'SSeL' impedance sensor node. In this feature, the identity (ID) of the node is displayed. A previously set ID is a reference factor to distinguish a node from others. The tab "IP address" is also shown to visualize the dynamic IP of the node. A node's ID must be input to identify the selected node for post-usage.
② Set impedance measurement parameters for the selected node. Eleven parameters and corresponding options should be selected according to the test requirement. The parameter 'Measuring Interval' should be set for periodic monitoring tasks.
③ Control the selected node to acquire impedance signatures in periodic monitoring. The measurement interval should be set correctly to measure impedance signals.
④ Test the status of the impedance measurement of the selected node. There are four options: Calibration, Measurement (No Save), Measurement (Save), and Reference measurement. These options should be selected based on the user's purpose.
⑤ Display the real and phase parts of the impedance measurement results.
⑥ Save the measurement results into the master device (e.g., laptop).

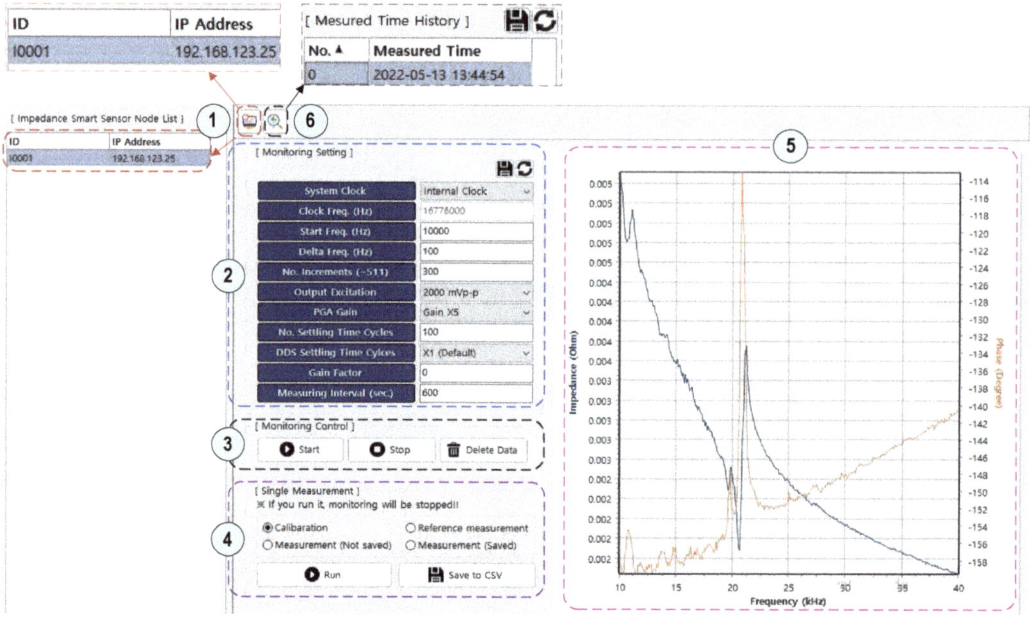

Figure 9. GUI software for Raspberry Pi system—'SSeL' impedance sensor node.

3. Calibration of Impedance Monitoring System 'SSeL-Pi'

The Raspberry Pi platform impedance sensing system (i.e., the prototype shown in Figure 7) was calibrated for accurate and stable performance. The calibration was carried out in the following four steps: (1) Acquisition of real and imaginary impedance, (2) Calculation of magnitude, gain factor, and phase angles, (3) Calibration of real and phase impedance, and (4) Extraction of correct real, and imaginary, and phase impedance. The calibration of impedance signals was conducted mainly on the impedance sensor board (Pmod IA).

Figure 10 shows the PZT interface device fabricated based on the conceptual design illustrated in Figure 4. A PZT 5 A patch [40] was surface-bonded on an aluminum vibrating plate (21 mm × 21 mm × 1.5 mm). The vibrating plate was supported by outer edges, which are aluminum walls (2 mm thickness and 7 mm height). Then, the fabricated combo was covered by two aluminum plates (25 × 25 × 2.0 mm). The cover plates had holes (diameter 2.0 mm) for passing electric wires.

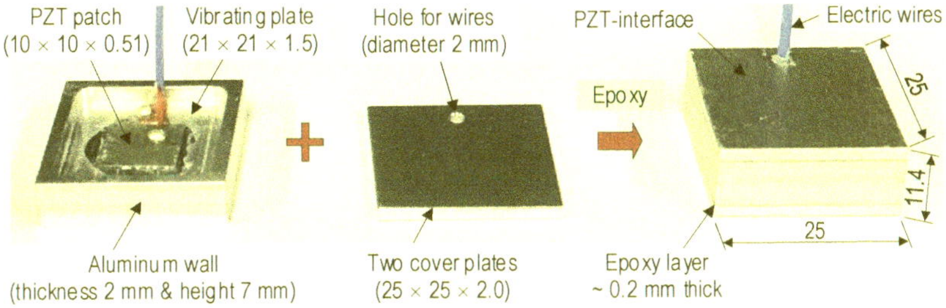

Figure 10. PZT interface device tested for calibration of impedance monitoring system 'SSeL-Pi'. (dimension in mm).

A series of calibration tests were performed on the PZT interface connected to the PmodIA board (see Figure 7). The PZT interface was freely placed on a table without any external loading. As the reference signature, the impedance signal of the PZT interface device was measured under the non-constraint boundary condition by using the HIOKI 3532 system. As shown in Figure 11, four impedance peaks (Peak 1 at 22.6 kHz, Peak 2 at 32.2 kHz, Peak 3 at 80.2 kHz, and Peak 4 at 88.2 kHz) were acquired in the frequency range of 5–95 kHz. As compared to the numerical impedance signal (see Figure 6), the experimental impedance signature of the PZT interface sample had a similar pattern at Peak 1 (highest peak). Furthermore, there was a difference between experimental and numerical impedance signals in other peaks. The difference could be induced by the boundary conditions of the PZT interface (non-constraint in the experimental measurement and fixed in the numerical analysis) and the effect of the fabrication process. It is known that the resonant impedance frequency bands represent meaningful structural characteristics [38,41]. The frequency range of 18–28 kHz (containing the highest peak—Peak 1) was selected for the calibration test. The calibration procedure was demonstrated by comparing it with the HIOKI 3532 analyzer.

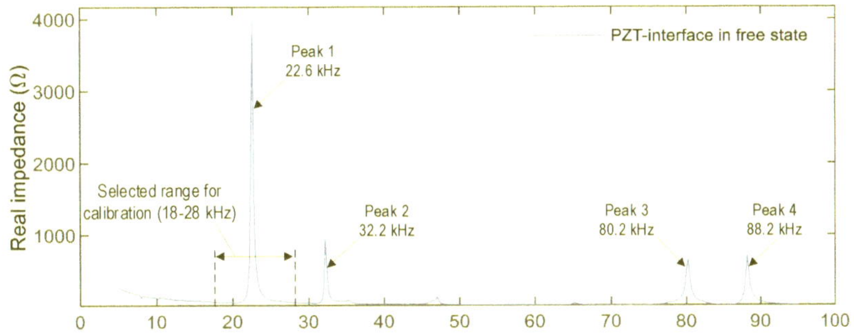

Figure 11. Impedance signal of PZT interface device measured by HIOKI-3532.

Step 1: Acquisition of Real (R_P) and Imaginary (I_P) Impedance

An impedance board (PmodIA) was connected to the PZT sensor interface, and a wide range of frequency-domain excitation was applied to the PZT sensor. Figure 12 shows real (R_P) and imaginary (I_P) parts of impedance responses measured by the SSeL-Pi system. Figure 13 shows the corresponding impedance signature (R_H) and phase (φ_H) measured by the HIOKI 3532 system. The frequency band of 18~28 kHz was selected for both the SSeL-Pi and HIOKI systems.

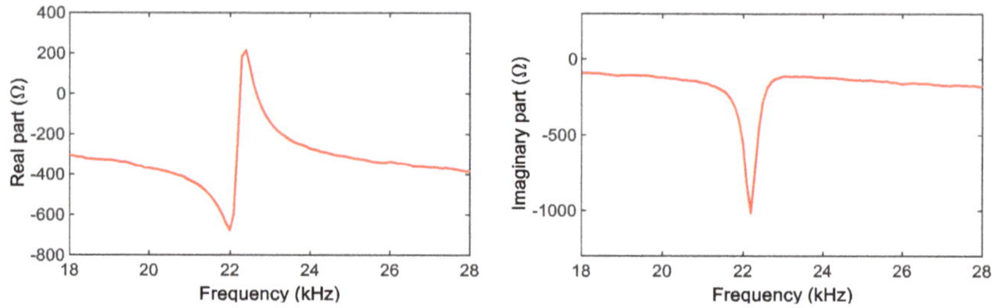

(**a**) (**b**)

Figure 12. Impedance responses of PZT interface measured by SSeL-Pi system. (**a**) Real part R_P, (**b**) Imaginary part I_P.

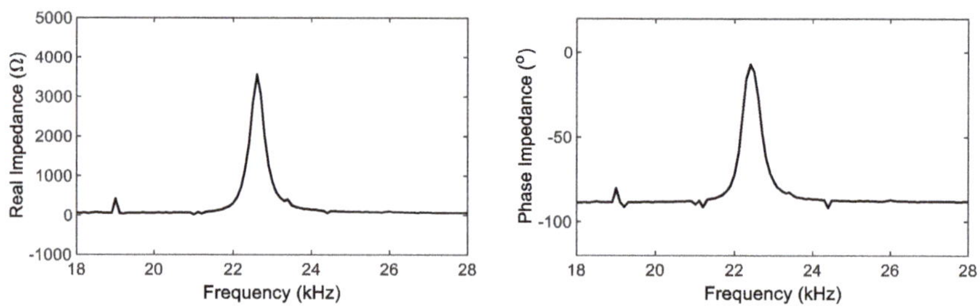

(**a**) (**b**)

Figure 13. Impedance responses of PZT interface measured by HIOKI-3532. (**a**) Real impedance R_H, (**b**) Phase impedance φ_H.

Step 2: Calculation of Impedance Magnitude M_P, Gain Factor G_1, and Phase φ_P, φ_1.

For the real and imaginary impedance data of the SSeL-Pi system (see Figure 12), the impedance magnitude was calculated as follows:

$$M_P = \sqrt{R_P^2 + I_P^2} \tag{3}$$

The gain factor G_1 was obtained for the calculated magnitude M_P by treating the real impedance data R_H of the HIOKI 3532 (see Figure 13a) as the reference:

$$G_1 = \frac{1}{M_P \times R_H} \tag{4}$$

At each sensor point of the SSeL-Pi system, the phase angle φ_P was calculated from the real and imaginary data, R_P and I_P:

$$\varphi_P = \tan^{-1}\left(\frac{I_P}{R_P}\right) \times \frac{180^o}{\pi} \tag{5}$$

in which the signs of R_P and I_P components were positive. In the case of the positive R_P and negative I_P, the phase angle should add 180° into Equation (5) [16]. For the calibration of the impedance sensing system, the gap of phase angles between φ_P (SSeL-Pi system) and φ_H (HIOKI-3532) was calculated as φ_1:

$$\varphi_1 = \varphi_P - \varphi_H \tag{6}$$

The impedance magnitude M_P and phase angle φ_P were calculated for the SSeL-Pi system, as shown in Figure 14. The impedance phase gap and gain factor were calculated for the SSeL-Pi system, as also shown in Figure 15.

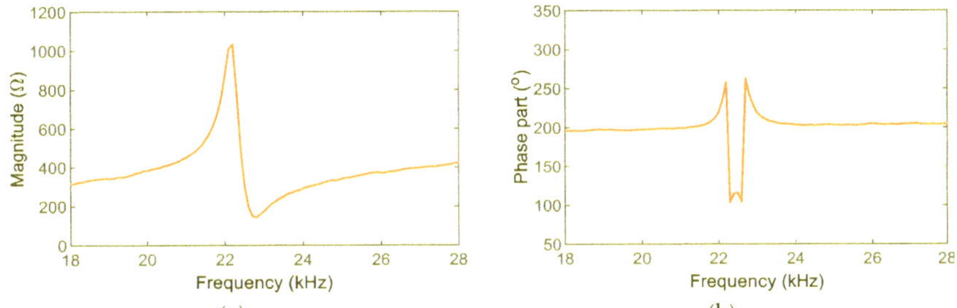

Figure 14. Calculation of impedance magnitude and phase angle for Raspberry Pi system. (**a**) Magnitude M_P, (**b**) Phase angle φ_P.

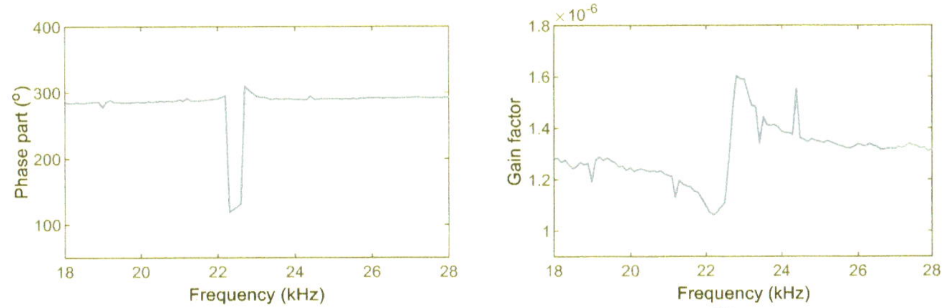

Figure 15. Calculation of impedance phase gap and gain factor for SSeL-Pi system. (**a**) Phase gap φ_1, (**b**) Gain factor G_1.

Step 3: Calibration of Real and Phase Impedance

As previously observed in Figure 15, the impedance phase gap φ_1 and gain factor G_1 were different along the monitored frequency (f). Linear regression models were established to manage these variations. The phase angle and gain factor values were standardized on the basis of the phase angle (φ_{1-i}, f_i) obtained in the *ith* frequency and the gain factor (G_{1-i}, f_i) obtained in the *ith* frequency (as described in Step 2).

Figure 16 shows two regression models for estimating the phase fit φ_2 (see Figure 16a), and the gain factor fit G_2 (see Figure 16b) at investigated frequencies. Then, the impedance magnitude and phase angle were corrected as follows:

$$Z = \frac{1}{G_2 \times M_P} \tag{7}$$

$$\theta = \varphi_P - \varphi_2 \tag{8}$$

in which Z and θ denote the corrected impedance magnitude and corrected phase angle.

Step 4: Extraction of Correct Real, Imaginary, and Phase Impedance

The real (R_C), imaginary (I_C), and phase $((\varphi_C)$ components of the impedance were calibrated based on the corrected magnitude Z and the corrected phase angle θ described in Step 3. The calibration was made by the vector projection of the impedance magnitude and phase angle onto the real and imaginary impedance components, as follows:

$$|R_C| = |Z| \times \cos(\theta) \tag{9}$$

$$|I_C| = |Z| \times \sin(\theta) \tag{10}$$

$$\varphi_C = -\theta \tag{11}$$

The accuracy of the calibrated impedance signals of the SSeL-Pi system was evaluated by comparing it to the HIOKI 3532 system. As shown in Figure 17, the imaginary impedance signal measured by the SSeL-Pi impedance sensing node had a good consistency with the one measured by the HIOKI 3532 system. The real impedance and phase impedance components were relatively similar, and the frequencies corresponding to the peaks were almost the same. Accordingly, the performance of the SSeL-Pi system could be successfully guaranteed after the calibration.

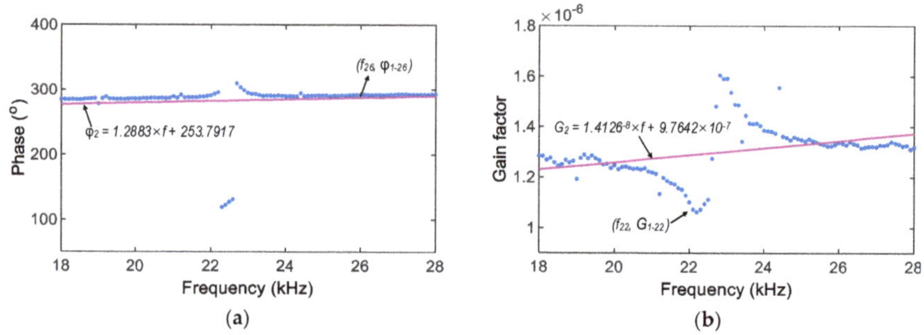

Figure 16. Regression models for impedance phase angle and gain factor. (**a**) Phase fit φ_2, (**b**) Gain factor fit G_2.

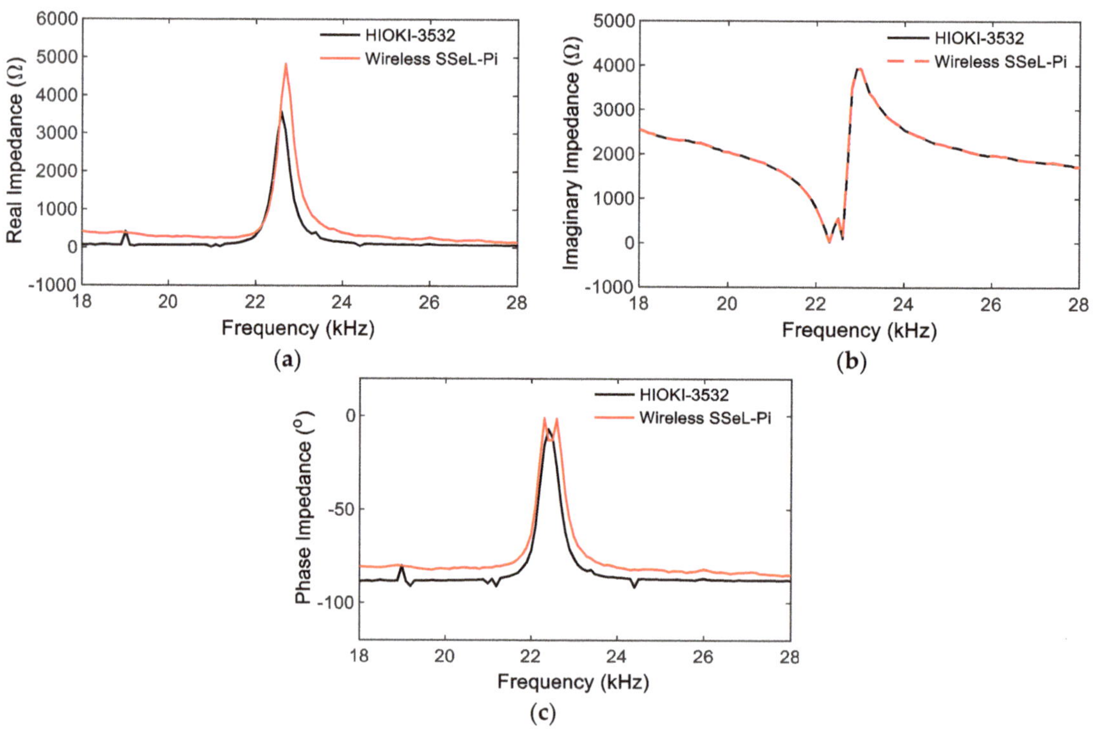

Figure 17. Performance of impedance calibration between SSeL-Pi and HIOKI-3532. (**a**) Real impedance R_C, (**b**) Imaginary impedance I_C, (**c**) Phase impedance φ_C.

4. Experimental Evaluation of Impedance Monitoring System 'SSeL-Pi'

4.1. Testing Setup and Test Scenarios

As shown in Figure 18, laboratory experiments were performed on PZT interfaces under compressive loadings to acquire impedance signals from the SSeL-Pi system. A PZT sensor interface (see Figure 10) was positioned in the compression testing machine, as shown in Figure 18a. Compression forces controlled by a load cell were applied to the PZT interface with a loading speed of 0.05 mm/min. Two PZT interfaces (i.e., two test samples), PZT 1 and PZT 2 were measured for impedance responses under the z-directional compressive loading. As the loading scenarios, the forces were incrementally applied from P1 (2.0 kN) to P5 (4.0 kN) with an interval of 0.5 kN. Note that P1 (2.0 kN) was treated as the baseline intact state to compare with P2~P5 loading cases.

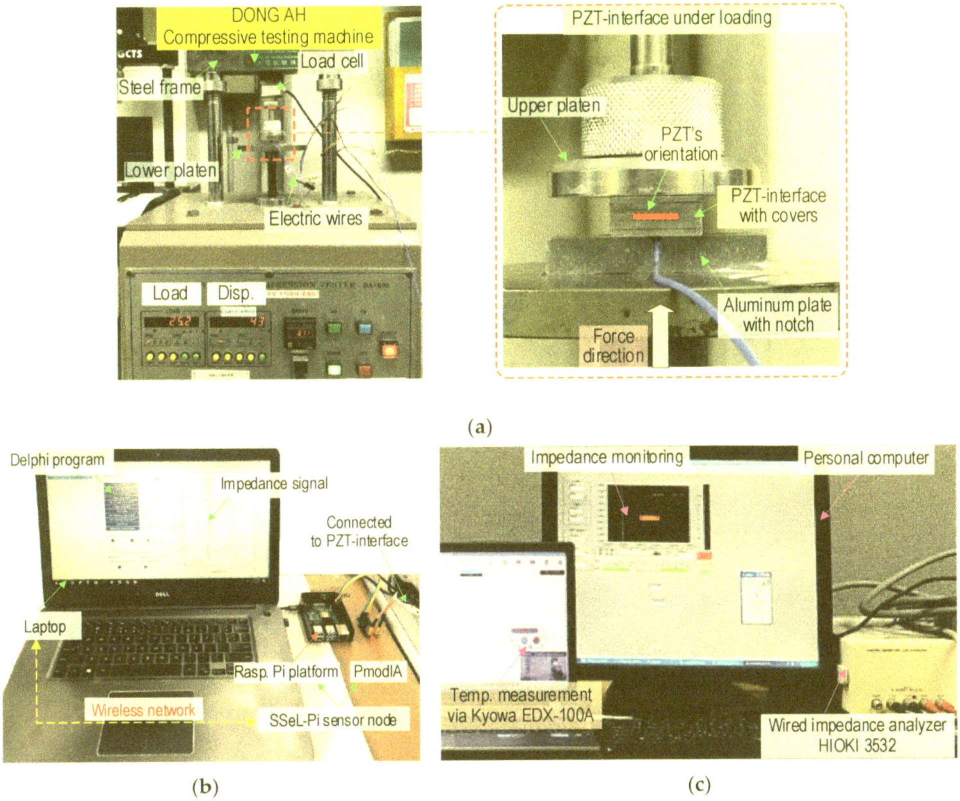

Figure 18. Impedance measurement of PZT interface device via SSeL-Pi and HIOKI-3532. (**a**) PZT interface device under compression test, (**b**) SSeL-Pi system, (**c**) HIOKI 3532 system.

As shown in Figure 18b, the SSeL-Pi system was utilized to measure impedance responses of the two PZT interfaces (PZT 1 and PZT 2) as the compressive loadings varied from P1 to P5. For the z-directional loadings, the impedance signals were measured at 31 points in a frequency range of 20.5–23.5 kHz by a constant interval of 0.1 kHz. At the same time, the impedance analyzer HIOKI 3532 (see Figure 18c) was set up to measure the impedance signals of the two PZT interfaces (PZT 1 and PZT 2) for the same loading scenarios. The impedance signatures measured from the HIOKI 3532 system were used as the reference to evaluate the confidence levels of impedance signatures measured from the SSeL-Pi sensor node. The laboratory temperature was kept near constant, around 20 °C (measured via Kyowa EDX-100A), to minimize the effect of temperature variation on impedance features. For each loading case, four ensembles of impedance responses were measured to determine an upper control limit (UCL) threshold.

4.2. Impedance Measurement of Raspberry Pi Impedance Monitoring System

4.2.1. Impedance Responses of PZT Interface under Compression

As shown in Figure 19, impedance responses of a PZT interface (PZT 1) were measured via the wireless SSeL-Pi and wired HIOKI-3532, respectively. For the five loading cases (i.e., P1–P5), the impedance signals of the two systems had similar patterns for the series of loadings. The impedance signals of both systems shifted to the right with the increment of compressive loadings. The impedance signals measured from both systems consistently responded to the series of loadings. Additionally, Figure 20 shows the impedance responses of the other interface (PZT 2) under five loading cases (i.e., P1–P5). Like the PZT 1, the impedance signals measured from the wireless SSeL-Pi and wired HIOKI 3532 were similar in their patterns. The variation tendency of the experimental impedance responses from the SSeL-Pi system was consistent with those from the HIOKI 3532 system. As observed in Figures 19 and 20, the alteration in impedance responses of PZT 2 was higher than those of PZT 1. The difference could be caused by the sensor fabrication, bonding condition, and compressive testing setup.

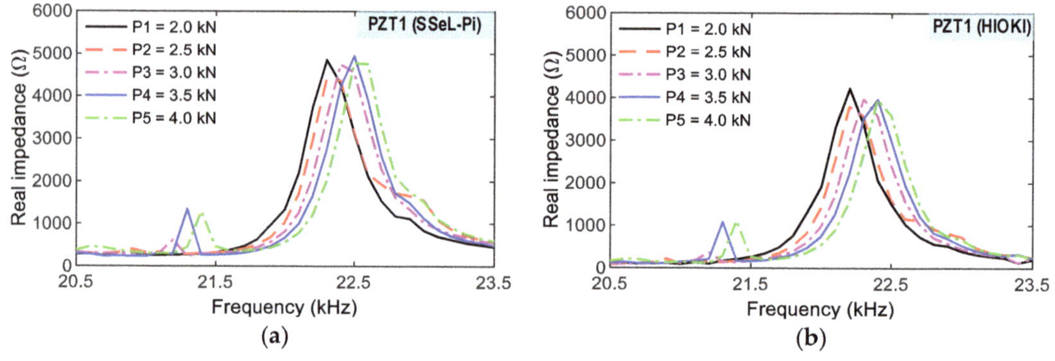

Figure 19. Impedance signals of PZT 1 for compressive loadings P1–P5. (**a**) SSeL-Pi, (**b**) HIOKI-3532.

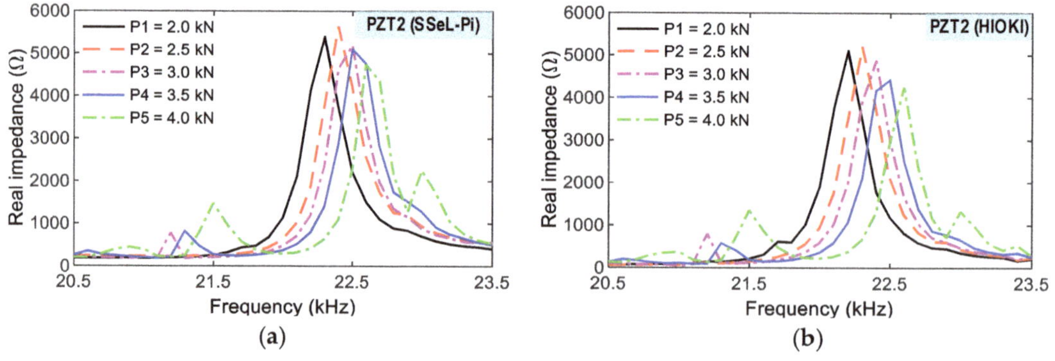

Figure 20. Impedance signals of PZT 2 for compressive loadings P1–P5. (**a**) SSeL-Pi, (**b**) HIOKI-3532.

4.2.2. Frequency Shifts in Impedance Responses of PZT Interfaces under Compression

Figures 21 and 22 show peak impedance frequencies of the two PZT interfaces (PZT 1 and PZT 2) that were acquired by the SSeL-Pi and HIOKI 3532 systems under applied forces P1–P5. There were slight differences in the peak frequencies of PZTs 1–2 measured by the SSeL-Pi and HIOKI 3532 under the reference case P1. By comparing to the reference P1, the peak frequencies increased linearly under the incremental loadings P2–P5. For PZT 1 (see Figure 21), the peaks rose from 0.05 kHz (0.25%) to 0.26 kHz (1.18%), which was measured via the SSeL-Pi, and from 0.05 kHz (0.25%) to 0.23 kHz (1.06%), which was measured via the HIOKI 3532. For PZT 2 (see Figure 22), the peaks increased from 0.09 kHz (0.40%) to 0.33 kHz (1.48%), which was measured via the SSeL-Pi, and from 0.09 kHz (0.42%) to 0.40 kHz (1.79%), which was measured via the HIOKI 3532.

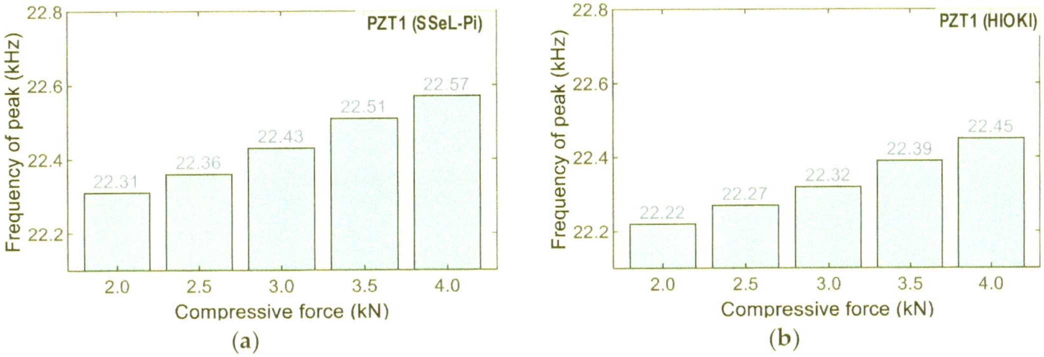

Figure 21. Frequency shifts of PZT 1 corresponding to compressive loadings P1–P5. (**a**) SSeL-Pi, (**b**) HIOKI-3532.

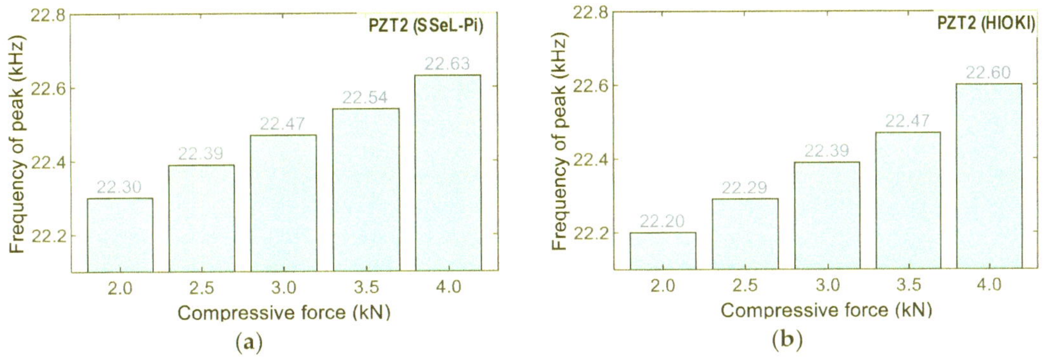

Figure 22. Frequency shifts of PZT 2 corresponding to compressive loadings P1–P5. (**a**) SSeL-Pi, (**b**) HIOKI-3532.

As analyzed, it can be observed that there was a similarity in the frequency shifts in the impedance responses measured by the SSeL-Pi and HIOKI 3532 for both PZTs 1–2, thus suggesting the potential of the SSeL-Pi system for impedance measurement. Furthermore, the linear variations of the peak frequencies in the impedance signals of the PZT interfaces (PZTs 1–2) showed the feasibility of the new PZT interface device for impedance monitoring.

4.2.3. RMSD Changes in Impedance Responses of PZT Interfaces under Compression

To quantify the changes in impedance signals, the RMSD (root mean square deviation) index is commonly used as a damage indicator for the characterization of structural damage.

The RMSD index is computed as follows [5]:

$$RMSD(Z, Z^*) = \sqrt{\left(\sum_{i=1}^{N}[Z^*(\omega_i) - Z(\omega_i)]^2\right) / \sum_{i=1}^{N}[Z(\omega_i)]^2} \quad (12)$$

where $Z(\omega_i)$ and $Z^*(\omega_i)$ are the impedance signals in the intact and damaged states of structure at frequency, respectively, and N denotes the number of frequency sampling points in the sweep.

Figures 23 and 24, respectively, show the RMSD indices of the impedance signatures of PZT 1 and PZT 2 computed for the five compressive loadings P1–P5. The upper control limit UCL [42] was calculated using the impedance signals at the applied force P1 (as the reference). The error bar was also computed for each loading case. Generally, an upward trend in the percentage of RMSD indices was found as the compressive force rose. The RMSD indices were negligible in the intact case (P1), but they were linearly increased and beyond the UCLs in other cases (P2–P5), thus suggesting that the variations of the compressive forces were successfully monitored. The small error bars in Figures 23 and 24 also pointed out that the impedance signals were relatively stable. Note that a similar tendency was observed in the SSeL-Pi and HIOKI-3532. Thus, the analyzed results demonstrated the feasibility of the SSeL-Pi system for impedance monitoring in the PZT interfaces.

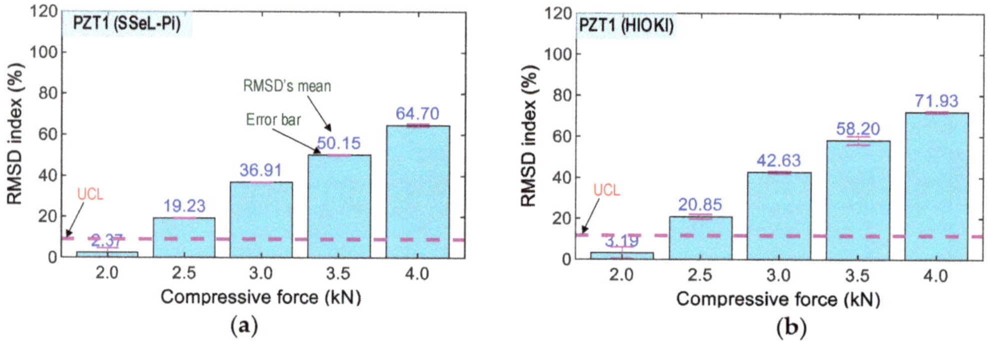

Figure 23. Impedance features of PZT 1 corresponding to compressive loadings P1–P5. (**a**) SSeL-Pi, (**b**) HIOKI-3532.

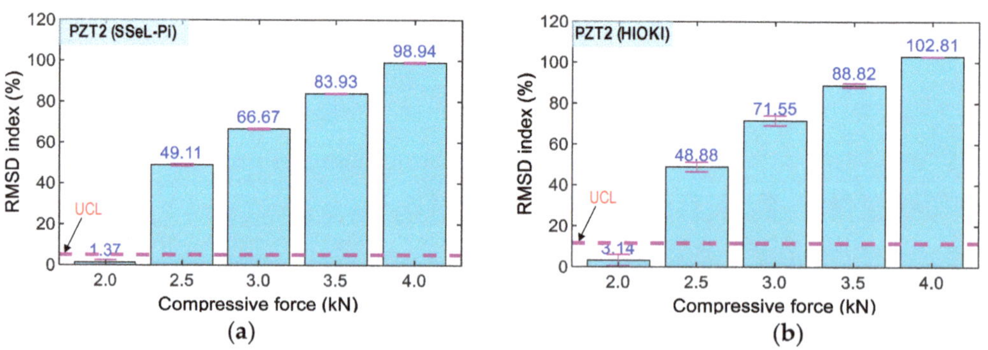

Figure 24. Impedance features of PZT 2 corresponding to compressive loadings P1–P5. (**a**) SSeL-Pi, (**b**) HIOKI-3532.

4.2.4. Empirical Relationship of Impedance Features and Stress

Figures 25 and 26 show the relationships between the impedance features of PZTs 1–2 and compressive stresses induced by loading cases P1–P5. The relationships were built based on impedance signals, which were obtained from the SSeL-Pi and HIOKI 3532 systems. These correlations were compared to each other to show the potential of the proposed measuring system for impedance measurement.

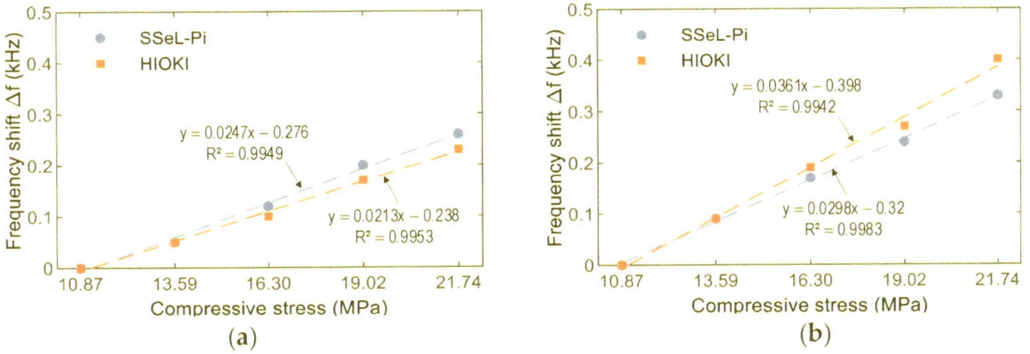

Figure 25. Relationship between frequency shift and corresponding compressive stress. (**a**) PZT1, (**b**) PZT2.

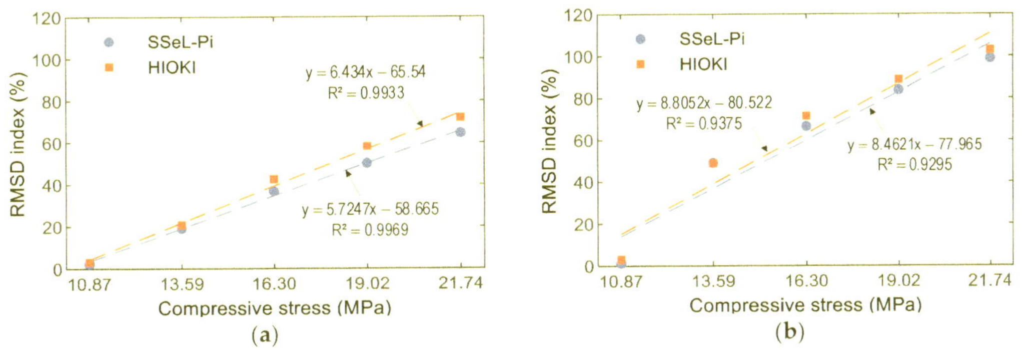

Figure 26. Relationship between RMSD index and corresponding compressive stress. (**a**) PZT1, (**b**) PZT2.

Figure 25 presents the correlation between the frequency shift Δf and the stress. As observed in this figure, strong positive linear relations could be found between the frequency shifts and stresses. The correlation coefficients R^2 were beyond 0.994 for both PZTs 1–2. The same tendency could be observed for the correlation between the RMSD index and the stress, as shown in Figure 26. It can be noted that significant correlations could be found between the RMSD indices and stresses. R^2 was beyond 0.993 for PZT 1 and 0.929 for PZT 2. Note that the strong relationships between the impedance features of PZT interfaces (i.e., PZTs 1–2) could be observed. Furthermore, the relationships built on impedance signals of the SSeL-Pi and HIOKI 3532 (see Figures 25 and 26) were quite consistent. Therefore, it can be concluded that the new wireless impedance sensing node SSeL-Pi has a high potential for impedance-based structural health monitoring.

From the experimental analyses, the following observations were made to verify the feasibility of the Raspberry Pi platform impedance monitoring system "SSeL-Pi": (1) the measurable impedance frequency range of less than 100 kHz was controlled using the new PZT interface device; (2) the impedance analyses using signals obtained from the SSeL-Pi were quite consistent with the ones using signatures measured from the commercial analyzer HIOKI 3532; (3) the impedance features measured by the PZT interfaces were relatively similar and sensitive to the variation in compressive forces; and (4) the relationship between impedance signatures and stresses could be used to quantitatively monitor stress change in structures using the new PZT interface.

It is also noted that more tests on the SSeL-Pi system should be conducted in the future to verify the practicability of the developed system and also to determine empirical formulas on impedance signatures and compressive forces, which would be reliable enough for impedance-based stress monitoring in concrete structures.

5. Concluding Remarks

The wireless impedance monitoring system was developed to have cheap, mobile, and handy practical features as compared to wired commercial impedance analyzers (e.g., HIOKI 3532 or HP4149A). The Raspberry Pi platform sensor node was designed to acquire impedance signals by collaborating with the PZT interface technique. The software scheme was designed to operate the Raspberry Pi platform and impedance sensor node. The calibration procedure was designed for the impedance sensor node. The feasibility of the proposed Raspberry Pi platform SSeL-Pi system was experimentally evaluated for PZT interfaces that were subjected to various compressive loadings.

From the experimental evaluation, at least four concluding remarks can be made as follows: Firstly, the prototype of the 'SSeL-Pi' impedance monitoring system was successfully fabricated by the remote-control laptop, Linux-operating Raspberry Pi platform, impedance sensor board (PmodIA module), and the PZT interface device ensuring frequency responses of less than 100 kHz. Secondly, the Delphi-based GUI software was programmed to visualize the input/output parameters for the Raspberry Pi platform and automate the signal acquisition procedure for the PmodIA module via the PZT interface. Thirdly, the calibration of the impedance signals was carried out on the PmodIA module, from the acquisition of real and imaginary impedance to the adjustment of them, with respect to a commercial impedance analyzer, HIOKI-3532. Finally, the impedance responses of the PZT interface that was subjected to a series of compressive loadings were accurately measured by the SSeL-Pi system compared to the HIOKI-3532. There were strong linear relationships between impedance features (frequency shift and RMSD index) and compressive stresses of the PZT interface, thus suggesting that the developed SSeL-Pi system has the potential for stress variation monitoring in concrete structures.

Author Contributions: Conceptualization, J.-T.K. and Q.-Q.P.; development of sensing devices and software, J.-H.P. and Q.-B.T.; monitoring method and evaluation, Q.-Q.P., Q.-B.T. and J.-T.K.; original manuscript preparation, Q.-Q.P. and J.-T.K.; supervision, J.-T.K. All authors have read and agreed to the published version of the manuscript.

Funding: This work was supported by a grant (21CTAP-C163708-01) from Technology Advancement Research Program funded by Korea Agency for Infrastructure Technology Advancement (KAIA).

Institutional Review Board Statement: Not applicable.

Informed Consent Statement: Not applicable.

Data Availability Statement: Data available on reasonable request from the corresponding author.

Conflicts of Interest: The authors declare no conflict of interest.

References

1. Annamdas, V.G.M.; Radhika, M.A. Electromechanical impedance of piezoelectric transducers for monitoring metallic and non-metallic structures: A review of wired, wireless and energy-harvesting methods. *J. Intell. Mater. Syst. Struct.* **2013**, *24*, 1021–1042. [CrossRef]
2. Na, W.S.; Baek, J. A Review of the Piezoelectric Electromechanical Impedance Based Structural Health Monitoring Technique for Engineering Structures. *Sensors* **2018**, *18*, 1307. [CrossRef] [PubMed]
3. Fan, X.; Li, J.; Hao, H. Review of piezoelectric impedance based structural health monitoring: Physics-based and data-driven methods. *Adv. Struct. Eng.* **2021**, *24*, 3609–3626. [CrossRef]
4. Liang, C.; Sun, F.P.; Rogers, C.A. Coupled Electro-Mechanical Analysis of Adaptive Material Systems-Determination of the Actuator Power Consumption and System Energy Transfer. *J. Intell. Mater. Syst. Struct.* **1994**, *8*, 335–343. [CrossRef]

5. Sun, F.P.; Chaudhry, Z.; Liang, C.; Rogers, C.A. Truss Structure Integrity Identification Using PZT Sensor-Actuator. *J. Intell. Mater. Syst. Struct.* **1995**, *6*, 134–139. [CrossRef]
6. Park, G.; Cudney, H.H.; Inman, D.J. Feasibility of using impedance-based damage assessment for pipeline structures. *Earthq. Eng. Struct. Dyn.* **2001**, *30*, 1463–1474. [CrossRef]
7. Mascarenas, D.L.; Todd, M.D.; Park, G.; Farrar, C.R. Development of an impedance-based wireless sensor node for structural health monitoring. *Smart Mater. Struct.* **2007**, *16*, 2137–2145. [CrossRef]
8. Park, S.; Lee, J.-J.; Yun, C.-B.; Inman, D.J. Electro-Mechanical Impedance-Based Wireless Structural Health Monitoring Using PCA-Data Compression and k-means Clustering Algorithms. *J. Intell. Mater. Syst. Struct.* **2007**, *19*, 509–520. [CrossRef]
9. Overly, T.G.S.; Park, G.; Farinholt, K.M.; Farrar, C.R. Development of an extremely compact impedance-based wireless sensing device. *Smart Mater. Struct.* **2008**, *17*, 065011. [CrossRef]
10. Park, S.; Shin, H.-H.; Yun, C.-B. Wireless impedance sensor nodes for functions of structural damage identification and sensor self-diagnosis. *Smart Mater. Struct.* **2009**, *18*, 055001. [CrossRef]
11. Min, J.; Park, S.; Yun, C.-B.; Song, B. Development of a low-cost multifunctional wireless impedance sensor node. *Smart Struct. Syst.* **2010**, *6*, 689–709. [CrossRef]
12. Park, J.-H.; Kim, J.-T.; Hong, D.-S.; Mascarenas, D.; Lynch, J.P. Autonomous smart sensor nodes for global and local damage detection of prestressed concrete bridges based on accelerations and impedance measurements. *Smart Struct. Syst.* **2010**, *6*, 711–730. [CrossRef]
13. Kim, J.-T.; Park, J.-H.; Hong, D.-S.; Ho, D.-D. Hybrid acceleration-impedance sensor nodes on Imote2-platform for damage monitoring in steel girder connections. *Smart Struct. Syst.* **2011**, *7*, 393–416. [CrossRef]
14. Park, S.; Kim, J.-W.; Lee, C.; Park, S.-K. Impedance-based wireless debonding condition monitoring of CFRP laminated concrete structures. *NDT E Int.* **2011**, *44*, 232–238. [CrossRef]
15. Kim, J.-T.; Nguyen, K.-D.; Park, J.-H. Wireless Impedance Sensor Node and Interface Washer for Damage Monitoring in Structural Connections. *Adv. Struct. Eng.* **2012**, *15*, 871–885. [CrossRef]
16. Analog Devices. 1 MSPS, 12-Bit Impedance Converter, Network Analyzer, AD5933 Data Sheet. 2017. Available online: https://www.Analog.Com/Media/En/Technical-Documentation/Data-Sheets/Ad5933.pdf (accessed on 6 May 2022).
17. Crossbow Technology, Inc. Imote2 Hardware Reference Manual, Report PN: 7430-0409-01. 2007. Available online: http://wsn.cse.wustl.edu/images/9/90/Imote2_hardware_ref (accessed on 5 July 2022).
18. Huynh, T.C.; Dang, N.L.; Kim, J.T. Preload Monitoring in Bolted Connection Using Piezoelectric-Based Smart Interface. *Sensors* **2018**, *18*, 2766. [CrossRef] [PubMed]
19. Nguyen, K.-D.; Kim, J.-T. Smart PZT-interface for wireless impedance-based prestress-loss monitoring in tendon-anchorage connection. *Smart Struct. Syst.* **2012**, *9*, 489–504. [CrossRef]
20. Park, J.-H.; Huynh, T.-C.; Kim, J.-T. Temperature effect on wireless impedance monitoring in tendon anchorage of prestressed concrete girder. *Smart Struct. Syst.* **2015**, *15*, 1159–1175. [CrossRef]
21. Halfacree, G. *The Official Raspberry Pi Beginner's Guide*; John Wiley & Sons: Hoboken, NJ, USA, 2020; Available online: https://magazines-attachments.raspberrypi.org/books/full_pdfs/000/000/038/original/BeginnersGuide-4thEd-Eng_v2.pdf (accessed on 7 July 2022).
22. Raspberry Pi Educational Services Private Limited, Raspberry Pi 4 Computer Model B. 2021. Available online: https://datasheets.raspberrypi.com/rpi4/raspberry-pi-4-product-brief.pdf (accessed on 12 March 2022).
23. Abdelgawad, A.; Yelamarthi, K. Structural health monitoring: Internet of things application. In Proceedings of the 2016 IEEE 59th International Midwest Symposium on Circuits and Systems (MWSCAS), Abu Dhabi, United Arab Emirates, 16–19 October 2016; pp. 1–4.
24. Mahmud, M.A.; Bates, K.; Wood, T.; Abdelgawad, A.; Yelamarthi, K. A complete Internet of Things (IoT) platform for Structural Health Monitoring (SHM). In Proceedings of the 2018 IEEE 4th World Forum on Internet of Things (WF-IoT), Singapore, 5–8 February 2018; pp. 275–279.
25. Ghosh, A.; Edwards, D.J.; Hosseini, M.R.; Al-Ameri, R.; Abawajy, J.; Thwala, W.D. Real-time structural health monitoring for concrete beams: A cost-effective 'Industry 4.0' solution using piezo sensors. *Int. J. Build. Pathol. Adapt.* **2020**, *39*, 283–311. [CrossRef]
26. Meng, Q.; Zhu, S. Developing IoT Sensing System for Construction-Induced Vibration Monitoring and Impact Assessment. *Sensors* **2020**, *20*, 6120. [CrossRef]
27. Digilent. PmodIATM Reference Manual. 2016. Available online: https://digilent.com/reference/_media/reference/pmod/pmodia/pmodia_rm.pdf (accessed on 7 May 2022).
28. Huynh, T.-C.; Kim, J.-T. Quantitative damage identification in tendon anchorage via PZT interface based impedance monitoring technique. *Smart Struct. Syst.* **2017**, *20*, 181–195.
29. Nagayama, T.; Spencer, B.F., Jr.; Structural Health Monitoring Using Smart Sensors. Urbana-Champaign, U. O. I. A., Report NSEL-001. 2007. Available online: https://www.ideals.illinois.edu/items/3728 (accessed on 2 March 2022).
30. HIOKI, E.E. Corporation. 3522/3532 LCR HiTESTER. 2011. Available online: https://www.hioki.cn/discon/pdf/component/3522_32E.pdf (accessed on 1 May 2022).
31. Digilent. PmodIA Schematic. 2011. Available online: https://digilent.com/reference/_media/reference/pmod/pmodia/pmodia_sch.pdf (accessed on 7 May 2022).

32. Digilent. Digilent Pmod Interface Specification. 2011. Available online: https://digilent.com/reference/_media/reference/pmod/digilent-pmod-interface-specification.pdf (accessed on 7 May 2022).
33. Huynh, T.-C.; Kim, J.-T. Quantification of temperature effect on impedance monitoring via PZT interface for prestressed tendon anchorage. *Smart Mater. Struct.* **2017**, *26*, 125004. [CrossRef]
34. Han, G.; Su, Y.-F.; Nantung, T.; Lu, N. Mechanism for using piezoelectric sensor to monitor strength gain process of cementitious materials with the temperature effect. *J. Intell. Mater. Syst. Struct.* **2020**, *32*, 1128–1139. [CrossRef]
35. Gu, H.; Song, G.; Dhonde, H.; Mo, Y.L.; Yan, S. Concrete early-age strength monitoring using embedded piezoelectric transducers. *Smart Mater. Struct.* **2006**, *15*, 1837–1845. [CrossRef]
36. Pham, Q.Q.; Dang, N.L.; Ta, Q.B.; Kim, J.T. Optimal Localization of Smart Aggregate Sensor for Concrete Damage Monitoring in PSC Anchorage Zone. *Sensors* **2021**, *21*, 6337. [CrossRef]
37. Xu, D.; Banerjee, S.; Wang, Y.; Huang, S.; Cheng, X. Temperature and loading effects of embedded smart piezoelectric sensor for health monitoring of concrete structures. *Constr. Build. Mater.* **2015**, *76*, 187–193. [CrossRef]
38. Dang, N.L.; Huynh, T.C.; Kim, J.T. Local Strand-Breakage Detection in Multi-Strand Anchorage System Using an Impedance-Based Stress Monitoring Method-Feasibility Study. *Sensors* **2019**, *19*, 1054. [CrossRef]
39. Embarcadero Technologies. RAD Studio 11.2—Delphi (Software). 2021. Available online: https://www.Embarcadero.Com/Products/Delphi (accessed on 10 July 2022).
40. Piezo.com. Materials Technical Data (Typical Values). 2022. Available online: https://info.piezo.com/hubfs/Data-Sheets/piezo-material-properties-data-sheet-20201112.pdf (accessed on 2 August 2022).
41. Zhao, S.; Fan, S.; Yang, J.; Kitipornchai, S. Numerical and experimental investigation of electro-mechanical impedance based concrete quantitative damage assessment. *Smart Mater. Struct.* **2020**, *29*, 055025. [CrossRef]
42. Huynh, T.-C.; Kim, J.-T. Impedance-Based Cable Force Monitoring in Tendon-Anchorage Using Portable PZT-Interface Technique. *Math. Probl. Eng.* **2014**, *2014*, 784731. [CrossRef]

sensors

MDPI

Article

Research on Inherent Frequency and Vibration Characteristics of Sandwich Piezoelectric Ceramic Transducer

Yuren Lu, Chunguang Xu, Qinxue Pan *, Quanpeng Yu and Dingguo Xiao

Key Laboratory of Fundamental Science for Advanced Machining, Beijing Institute of Technology, Beijing 100081, China
* Correspondence: panqx@bit.edu.cn; Tel.: +86-010-68912714

Abstract: Great progress has been made in the field of ultrasonic processing in recent years, and piezoelectric ceramic transducers have been widely used as drive sources. In this paper, a sandwich piezoelectric ceramic transducer is designed, and the vibration of each part of the transducer is analyzed by elastic mechanics and piezoelectric theory. According to its mechanical and electrical boundary conditions, the vibration model of the piezoelectric transducer was established. Based on the equivalent elastic modulus method for simplifying the pre-stressed bolts into a one-dimensional transducer vibration model, the relationship between the one-dimensional axial response frequency of the transducer and the length of each component was obtained. Based on the half wavelength theory, a transducer with the vibration node in the crystal stack and an inherent frequency of 15 kHz was designed and fabricated. In order to verify the natural frequency and vibration characteristics of the piezoelectric transducer, a laser vibration measurement system was built in this study. The vibration characteristics of the transducer under different parameters such as voltage and frequency were analyzed, and the accuracy of the vibration model was verified. The vibration states of the end surface of the transducer and the radial surface were evaluated at the first-order inherent frequency and second-order inherent frequency. The results show that the equivalent simplified model established in this study can effectively design the inherent frequency of the transducer, and the operation at the first-order inherent frequency meets the one-dimensional assumptions of this study. The transducer operating conditions measured in this study also provide a more detailed reference for ultrasonic processing applications.

Keywords: piezoelectric ceramics; vibration model; inherent frequency; transducer vibration characteristics

Citation: Lu, Y.; Xu, C.; Pan, Q.; Yu, Q.; Xiao, D. Research on Inherent Frequency and Vibration Characteristics of Sandwich Piezoelectric Ceramic Transducer. *Sensors* **2022**, *22*, 9434. https://doi.org/10.3390/s22239431

Academic Editor: Iren E. Kuznetsova

Received: 18 October 2022
Accepted: 22 November 2022
Published: 2 December 2022

1. Introduction

With the wide application of power ultrasound in advanced machining fields, sandwich piezoelectric transducers are valued and studied in ultrasonic milling [1,2], ultrasonic casting [3,4], and ultrasonic vibration for machining residual stress removal [5–9] because of their good electromechanical conversion characteristics. Although the structure and performance of piezoelectric ceramic transducers for different ultrasonic processing fields are not consistent, the inherent frequency of sandwich piezoelectric transducers is the most important performance parameter, which is the key to achieving high power output and high-efficiency energy conversion in practical applications. The excitation signal that deviates from the inherent frequency not only leads to the low efficiency of ultrasonic vibration but also to other vibration modes of the transducer, so it is very necessary to determine the inherent frequency before the application of the transducer.

Various methods have been used to study the relationship between the physical parameters of the PZT sandwich structure and its inherent frequency as well as other dynamic parameters. The most commonly used analytical method is the electromechanical equivalent model [10–12], whose transducer frequency characteristic equation can be obtained by drawing the mechanical-circuit 3-port or 4-port equivalent model according to mechanical boundary conditions. However, with the increasing demand for sound power of power transducers, the structure of multi-layer piezoelectric ceramic stacked pressure transducers is more popular, resulting in a very complex electromechanical equivalent model. In order to simplify the calculation model, the entire piezoelectric ceramic stack is regarded as an equivalent homogeneous body. The matrix transfer method [13–16] separates transducers into different independent elements to simulate multi-layer piezoelectric stack vibration, thus obtaining the vibration characteristics of piezoelectric ceramic stack transducers. Even more complex piezoelectric devices are still applicable under the assumption of one-dimensional wave propagation. Fu [17] proposed a hybrid transfer matrix approach for the modeling of the Rangzwan transducer, in which the prestressed bolt was modeled as a single quadrupole element, which was connected in parallel with other elements to establish a quadrupole element description of the piezoelectric Rangzwan transducer, thereby deriving the total transfer matrix and the inherent frequency of the transducer. Zhang [18] developed a simplified parametric analysis model of a piezoelectric transducer's distribution that can effectively analyze piezoelectric transducers with thickness vibration modes. Finite element simulation is also a commonly used method to study the inherent frequency and vibration characteristics of the transducer. Abdullah, A [19] analyzed the mechanical behavior and inherent frequency of a sandwich transducer using the finite element method. Then, the equivalent circuit of the transducer was simulated to study the inherent frequency of the transducer that had been calculated [20]. Pérez-Sánchez [21] designed an ultrasonic transducer for acoustic cavitation using finite element analysis and found that when the ultrasonic transducer was used with a load, the frequency response of the transducer needed to be adjusted in terms of inherent frequency and voltage. Jiang, X [22] proposed an equivalent length method based on half-wavelength theory to analyze the inherent frequency of a sandwich transducer and analyzed the effect of different thicknesses of piezoelectric ceramics on the inherent frequency. Wei, X. [23] proposed a PSpice loss model for a sandwiched piezoelectric ultrasonic transducer undergoing longitudinal vibration using one-dimensional wave and transmission line theory to obtain the transducer's resonance and anti-resonance frequencies. Chen [24] estimated the frequency of a sandwich transducer under various parameters by giant fuzzification-learning methods. Although the equivalent circuit method and matrix transfer method can model continuous systems with discrete components, which are valid around an inherent frequency, modeling the transducer's continuous system vibration [25] can be valid in a wider range of frequencies. It can also describe the effect of geometry and material properties on the frequency response of the transducer. Based on the plane stress–strain assumption [26], many scholars have analyzed the vibrational characteristics of transducers. Piao [27] analyzed the inherent frequency and vibration modes of a transducer by solving the differential equations of piezoelectric motion. Other studies consider the coupling effect of radial and axial vibrations. The coupling vibration of a metal hollow cylinder is simplified to two equivalent one-dimensional vibrations by the apparent elasticity method [28–31]: one is the axial vibration of the equivalent cylinder and the other is the equivalent plane vibration in the radial direction. The inherent frequency equation of the metal hollow cylinder in the coupled vibration was derived, and the longitudinal and radial coupled resonant frequencies were calculated.

Although there are many studies and methods for analyzing the inherent frequency of a transducer, most of the current studies only focus on the operating frequency of the transducer, and the study of the operating state of the transducer near the inherent frequency is often neglected. At present, the evaluation of transducers' performance is mainly described by the speed or amplitude of the front mass block—which is not only unintuitive but also incomplete with respect to the sandwich transducer—to evaluate the overall working state of the transducer. The operating conditions of the sandwich transducer lead to the observation of complex working conditions. Currently, the acoustic emission (AE) method [32,33], optical measurement method [34,35], mode degradation method [36], and other single-point methods are mainly used to measure the vibration amplitude of the transducer. Albareda [37] used a combination of optoelectronic methods, voltage and current measurements with an oscilloscope, and the vibration velocity of the transducer point with a laser vibrometer to evaluate a transducer's operating state.

In this paper, the sandwich transducer is simplified into three parts: the front mass block, backing, and piezoelectric ceramic crystal stack, and the vibration model of the sandwich piezoelectric ceramic transducer is derived by continuous dielectric mechanics and a planar stress–strain assumption. A sandwich piezoelectric ceramic transducer with an inherent frequency of about 15 kHz is designed and fabricated by considering the mass effect of the compression bolt using the equivalent elastic modulus method. A doppler laser vibration measurement system was built, the vibration signal of the transducer was collected by the single-point method, and the inherent response frequency of the transducer was derived by FFT (Fast Fourier Transform). The operating state of the transducer under different parameters is discussed, the overall operating state of the transducer is swept and imaged, and the vibration state of the transducer in the axial and radial directions at the first-order and second-order resonant frequencies is measured.

2. Transducer Structure and Equivalent Models

The sandwich transducer consists of front and back mass blocks, a piezoelectric ceramic stack consisting of piezoelectric ceramic circular pieces, pre-tensioning bolts, insulating sleeves, electrode sheets, etc. It is a continuous elastomer with a complex structure, as shown in Figure 1. The PZT-8 piezoelectric ceramic crystal stack is fixed between the front mass block and the metal backing by pre-stressed bolt connections.

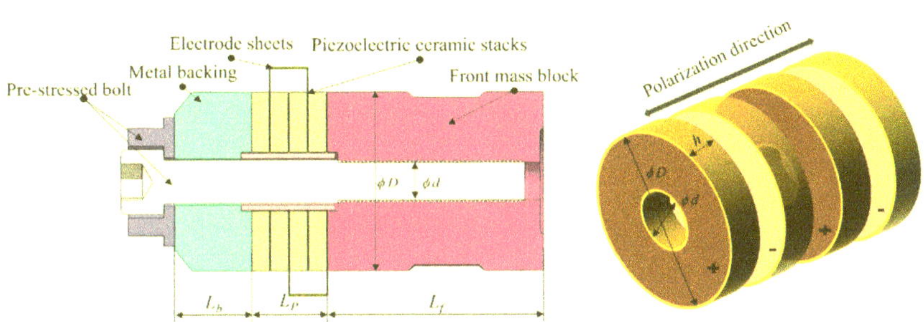

Figure 1. Schematic diagram of the structure of the sandwich piezoelectric transducer.

According to the wave propagation theory, the standing wave along the longitudinal axis of the transducer satisfies Equation (1) [22] during the resonant oscillation of the transducer:

$$L = N\frac{\lambda}{2}, \quad (1)$$

where L is the total length of the transducer; λ is the wavelength of the wave generated by the transducer's vibration; N is a positive integer—when $N = 1$, the corresponding sandwich transducer becomes a half-wavelength transducer; and the total phase length of each component wave is π. As shown in Figure 2, both the metal backing and the end of the front mass block are at the peak of the displacement wave, where the vibration is most intense. The interface along the longitudinal axis where the wave propagation phase occurs is $\pi/2$ and the interface with zero displacement amplitude is the displacement node plane. The length of the metal backing is L_b, the cross-sectional area is S_b, the vibration velocity is V_b, the length of the front mass block is L_f, the cross-sectional area is S_f, the vibration velocity is V_f, the length of the piezoelectric ceramic stack is L_p, and the cross-sectional area is S_p.

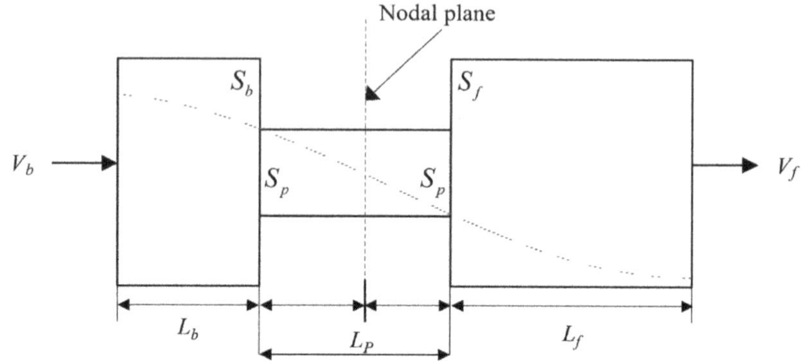

Figure 2. Schematic diagram of standing waves in a simplified sandwich transducer.

To satisfy the engineering requirements without affecting the working performance of the transducer, we need to make the following assumptions.

- The motion of the sandwich transducer along the axial direction is uniform, which means that the vibration caused by the piezoelectric ceramic is generated as a plane wave.
- It is assumed that the contact of the surface is perfect. Therefore, the reflection of the wave is neglected.
- Each ceramic piece in the piezoelectric ceramic stack is excited by a continuous sinusoidal wave, and the signal is in the same phase.
- The length of the transducer is more than three times its diameter, so the transverse Poisson effect is ignored and the transducer can be simplified to a one-dimensional model.
- The effect of the chamfer and thread of the transducer can be ignored; the transducer will be regarded as equal to the outside diameter.
- The acoustic impedance of air is much smaller than the transducer, so it is ignored. When the transducer is operating in the air, it is referred to as being under the "no-load" condition, and its stress is considered to be zero.
- The matched upper part is considered to be a solid cylinder with the center bolt integrated to simplify the boundary conditions.

3. Analysis Method of Transducer Inherent Frequency

3.1. Mechanical Vibration Equation of Piezoelectric Ceramic Stack

The thickness vibration of a thin, circular piezoelectric ceramic has an axial polarization direction and the thickness of the piezoelectric ceramics is h. The mechanical and electrical connections of the piezoelectric ceramics are shown in Figure 3, and the stress and electric field conditions of the piezoelectric equation are shown in Equation (2):

$$\begin{cases} T_3 \neq 0, T_1 = T_2 = T_{12} = T_{13} = T_{23} = 0 \\ E_3 \neq 0 \\ \frac{\partial D_3}{\partial z} = 0 \end{cases}, \quad (2)$$

where T is the stress on the piezoelectric ceramic and E is the electric field strength on the surface of the piezoelectric ceramic. Since this study simplifies the transducer into the one-dimensional axial continuous model, only the axial stress and electric field strength are considered. T_3 is the stress in the axial direction, D_3 is the electric displacement in the axial direction of the piezoelectric ceramic, and E_3 is the electric field strength applied to the piezoelectric ceramic in the axial direction.

When a voltage is applied on the surface of the piezoelectric ceramic, mechanical and electrical energy are converted into each other. Neglecting the transverse Poisson effect and assuming that only axial stresses exist, it can be concluded that $T_3 \neq 0$. Since the axially applied voltage has $E_3 \neq 0$, and considering that the ceramic is an insulating medium with no free charge in space, the potential shift vector is uniformly distributed and can be derived as $\partial D_3/\partial z = 0$. The piezoelectric ceramic plates are polarized along the thickness direction, and every two adjacent piezoelectric ceramic plates are polarized in opposite directions so that the vibration can be better transmitted and superimposed.

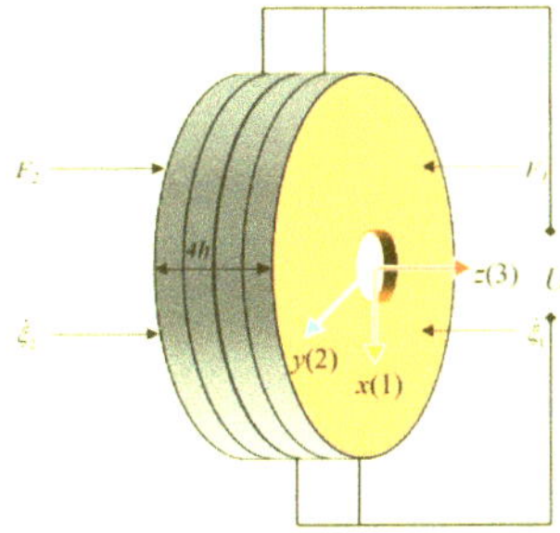

Figure 3. Electrical and mechanical boundary model of piezoelectric ceramic crystal stack.

According to the above boundary conditions, the piezoelectric matrix is shown in Equation (3):

$$\begin{bmatrix} S_3 \\ E_3 \end{bmatrix} = \begin{bmatrix} s_{33}^D & -g_{33} \\ -g_{33} & \beta^T{}_{33} \end{bmatrix} \begin{bmatrix} T_3 \\ D_3 \end{bmatrix}, \tag{3}$$

where S_3 is the strain in the axial direction, s_{33}^D denotes the elastic constants of the piezoelectric ceramic at the equipotential boundary condition, g_{33} is the piezoelectric constant, and $\beta^T{}_{33}$ is the dielectric isolation rate of the piezoelectric ceramic.

From the partial differential equation of motion, it follows that:

$$\rho_p \frac{\partial^2 \zeta(z,t)}{\partial t^2} = E_{p3} \frac{\partial \zeta(z,t)}{\partial z}, \tag{4}$$

where ρ_p is the density of the piezoelectric ceramic, E_{p3} is the equivalent modulus of elasticity in the axial direction of the piezoelectric ceramic, and $\zeta(z,t)$ is the longitudinal displacement of the cross-section.

Merging Equation (4) into the piezoelectric matrix yields:

$$\rho \frac{\partial^2 \zeta(z,t)}{\partial t^2} = \frac{\partial T_3(z,t)}{\partial z} - \frac{g_{33}}{s_{33}^D} \frac{\partial D_3(z,t)}{\partial z}, \tag{5}$$

The boundary condition, Equation (5), can be simplified as:

$$\frac{\partial^2 \zeta(z,t)}{\partial t^2} = v^2 \frac{\partial^2 \zeta(z,t)}{\partial z^2}, \tag{6}$$

where $v = \sqrt{\frac{1}{\rho s_{33}^D}}$ is the propagation velocity of the one-dimensional longitudinal vibration in the piezoelectric ceramic crystal stack.

Since a continuous sinusoidal signal is applied to the piezoelectric ceramic, the vibrational response of the transducer is also a simple harmonic:

$$\zeta(z,t) = \zeta(z) \cdot e^{jwt} = (m \sin kz + n \cos kz) \cdot e^{jwt}, \tag{7}$$

where m and n are coefficients to be determined, which can be determined by the boundary conditions of the piezoelectric ceramics.

The displacement distribution in the axial direction of the piezoelectric ceramic crystal stack can be derived as:

$$\begin{cases} \zeta(z) = \frac{\zeta_1 \sin k(l-z)+\zeta_2 \sin kz}{\sin kl} \\ \zeta_1 = \zeta|_{z=0} \\ \zeta_2 = \zeta|_{z=l} \end{cases}, \tag{8}$$

where ζ_1, ζ_2 denote the vibration displacement of the two ends of the piezoelectric ceramic, so according to the mechanical boundary conditions $F_1 = -ST_3|_{z=0}, F_2 = -ST_3|_{z=l}$, the mechanical vibration equation can be obtained as:

$$\begin{cases} F_1 = [j\rho v S \tan\frac{kh}{2} + (\frac{\rho v S}{j \sin kh} - \frac{n^2}{j\omega c_0})]\dot{\zeta}_1 + (\frac{\rho v S}{j \sin kh} - \frac{n^2}{j\omega c_0})]\dot{\zeta}_2 + nV \\ F_2 = (\frac{\rho v S}{j \sin kh} - \frac{n^2}{j\omega c_0})]\dot{\zeta}_1 + [j\rho v S \tan\frac{kh}{2} + (\frac{\rho v S}{j \sin kh} - \frac{n^2}{j\omega c_0})]\dot{\zeta}_2 + nV \end{cases}' \tag{9}$$

3.2. Vibration Model of Transducer

For this system, the parts of the transducer are all columns with the same cross-section, and assuming that the cross-sectional rods are composed of uniform, isotropic materials and neglecting their mechanical losses, the stress distribution on the rods' cross-section is considered to be uniform. Figure 4 represents the forces and displacements of the unit body at any interface of the equal diameter sandwich transducer, which can be expressed as:

$$dF = E\frac{\partial}{\partial z}(\frac{\pi D^2}{4}\frac{\partial \zeta(z,t)}{\partial z})dz = \rho\frac{\pi D^2}{4}dz\frac{\partial^2 \zeta(z,t)}{\partial t^2}, \tag{10}$$

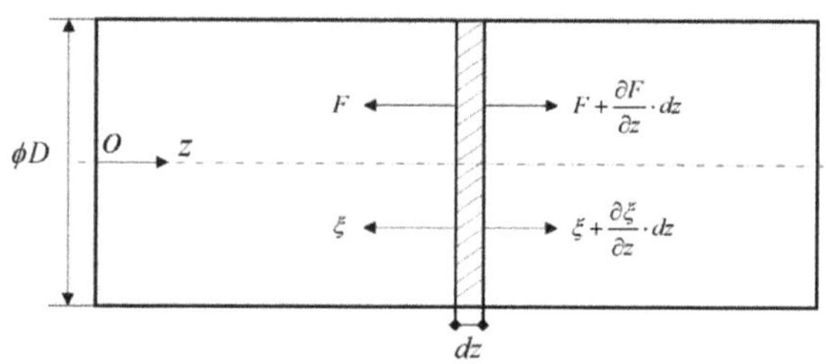

Figure 4. Simplified model of transducer vibration.

Combinning Equation (8) into Equation (10) can lead to:

$$\frac{\partial^2 \zeta(z,t)}{\partial z^2} = \frac{\rho}{E}\zeta(z,t), \tag{11}$$

Since the longitudinal vibration of the multilayer piezoelectric ceramic sandwich transducer generally exists at the nodal plane of the piezoelectric ceramic element part, it is divided into four parts, as shown in Figure 2, where the subscript n denotes each section of the sandwich transducer: n = 1, 2, 3, and 4. The vibration equation of each section is a simple harmonic vibration, so the vibration speed equation of each section can be expressed as:

$$\frac{\partial^2 v_n(z)}{\partial {z_n}^2} + {k_n}^2 v_n(z) = 0, \tag{12}$$

The vibration speed and force on the mass block part can be expressed as:

$$\begin{cases} v_n(z_n) = A_n \sin(k_n z_n) + B_n \cos(k_n z_n) \\ F_n(z_n) = \frac{E\pi D^2}{j4\omega} \cdot \frac{\partial v_n}{\partial z_n} = -jZ_n[A_n \cos(k_n z_n) - B_n \sin(k_n z_n)] \end{cases}' \tag{13}$$

where Z_n is the acoustic impedance of each part of the transducer and A_n, B_n are the coefficients to be determined by the boundary conditions.

The vibration node's cross-section is used to analyze both sides separately, and the boundary conditions are shown in Table 1.

Table 1. Boundary conditions of transducer vibration node's sections.

Cross-Sectional Position	Boundary Conditions
The right side of the cross-section	$v_1(0) = 0$ $v_1(L_1) = v_2(0)$ $v_1(L_2) = v_f$ $F_1(L_1) = F_2(0)$ $F_2(L_2) = -Z_f v_f$
The left side of the cross-section	$v_3(L_3) = 0$ $v_3(0) = v_4(L_4)$ $v_4(0) = v_b$ $F_3(0) = F_4(L_4)$ $F_4(0) = 0$

The frequency equation describing the nodal plane can be obtained by bringing the boundary conditions in Table 1 into Equation (13).

$$\begin{cases} \cot(k_1 L_1) = \frac{z_2}{z_1}\left(\tan(k_2 L_2) - j\frac{z_f}{z_2}\right) \\ \tan(k_3 L_3)\cdot \tan(k_4 L_4) = \frac{z_3}{z_4} \end{cases}, \tag{14}$$

If the input impedance z_f is neglected, the frequency equation for the left and right sides of the transducer vibration node can be simplified as:

$$\begin{cases} \tan(k_1 L_1)\cdot \tan(k_2 L_2) = \frac{z_1}{z_2} \\ \tan(k_3 L_3)\cdot \tan(k_4 L_4) = \frac{z_3}{z_4} \end{cases}, \tag{15}$$

3.3. Equivalent Elastic Modulus Method of Transducer

The piezoelectric ceramic group of the longitudinal vibration sandwich transducer consists of four PZT-8 piezoelectric ceramics and four copper electrode sheets bonded by epoxy resin. The front and rear mass blocks of the transducer are made of a Duraluminum alloy, and the nuts and preload bolts are made of C45E4 steel; the properties of the materials are shown in Table 2 [38].

Table 2. Parameters of each component material of the transducer.

Materials	Velocity of Sound m/s	Density kg/m^3	Elastic Modulus Gpa	Poisson's Ratio	Elastic Constants $S^D{}_{33}$ 10^{10} m^2/N	Piezoelectric Voltage Constants g_{33} 10^{-3} V·m/N
PZT-8	3100	7600	72	-	8.5	25.4
duralumin	5200	2790	72	0.34	-	-
C45E4 steel	5170	7850	200	0.28	-	-

As can be seen from Figure 1, the transducer is not composed of the same material within the same cross-section, so the equivalent elastic modulus method is used to correct the material parameters in this paper. Assuming that the velocity and displacement of each cross-section of the pre-stressed bolt are consistent with the external piezoelectric ceramic wafer, then the equivalent density ρ', equivalent sound velocity c', and equivalent cross-sectional area s' of the piezoelectric ceramic and pre-tensioned bolt can be chosen to replace the original ceramic acoustic parameters:

$$\begin{cases} \rho' = \frac{\rho_1 s_1 + \rho_2 s_2}{s'} \\ c' = \frac{c_1 s_1 + c_2 s_2}{s'} \\ Z' = \rho' c' s' \\ s' = s_1 + s_2 \\ k' = \frac{2\pi f}{c'} \end{cases}, \tag{16}$$

where ρ_1, s_1, c_1 are the density, cross-sectional area, and sound velocity of PZT-8, respectively. ρ_2, s_2, and c_2 are the density, cross-sectional area, and sound velocity of the pre-stressed bolt. Z' and k' and are the equivalent impedance and equivalent wave number. Structural equivalence means that the structure of different materials in the same section of the vibration direction is equal to the structure of a single material, and the material property parameters of the equivalent structure are shown in Table 3.

Table 3. Equivalent parameters of each component material of the transducer.

Structure of the Transducer	Velocity of Sound c (m/s)	Density ρ (Kg/m^3)	Elastic Modulus E (Gpa)
Piezoelectric Ceramics PZT-8	3367	7632	88.52
Aluminum front mass block	5019	3272	72
Aluminum back mass block	5019	3272	72

4. Experimental Methods

4.1. Design and Inherent Frequency Measurement of Piezoelectric Transducer

The propagation speeds of the acoustic waves in the three components of the transducer are different. Therefore, Equation (1) cannot be used to directly calculate the structural parameters of each part of the actuator, but the propagation frequencies of each part of the transducer are the same:

$$\begin{cases} \frac{\theta_b}{L_b} = \frac{2\pi}{\lambda_b}, \frac{\theta_f}{L_f} = \frac{2\pi}{\lambda_f}, \frac{\theta_p}{L_p} = \frac{2\pi}{\lambda_p} \\ L_b + L_f + L_p = L \\ \theta_b + \theta_f + \theta_p = \pi \\ \lambda = f \cdot c_{(b,f,p)} \end{cases}, \tag{17}$$

where c_b, c_p, and c_f represent the propagation velocities in the backing, the piezoelectric ceramic, and the front mass block, respectively. λ_b, λ_p, and λ_f represent the wavelengths in the corresponding components, respectively. As shown in Figure 2, the lengths of the backing, the piezoelectric ceramic stack, and the front mass block are L_b, L_P, and L_f, respectively, and the phase lengths of each element are θ_b, θ_p, and θ_f, respectively.

In order to verify the accuracy of calculating the inherent frequency of the one-dimensional simplified sandwich transducer, the test system is composed of a polytec doppler laser vibrometer (controller model OFV-5000; laser head model OFV-503), Power ultrasonic transducer, high-power supply (Yongke, Ningbo, China), signal-generating device (RIGOL-DG1022U, Suzhou, China), digital oscilloscope-MDO3102 (Tektronix, USA), and vibration isolation test bench, as shown in Figure 5. The pulse signal is generated by the signal-generating device to enable the transducer to produce free attenuation vibration, and the vibration velocity of the end face of the transducer is measured by the laser vibrometer. After the vibration signal is converted into a voltage signal, it is obtained on the digital oscilloscope and analyzed by FFT to obtain the inherent frequency of the transducer.

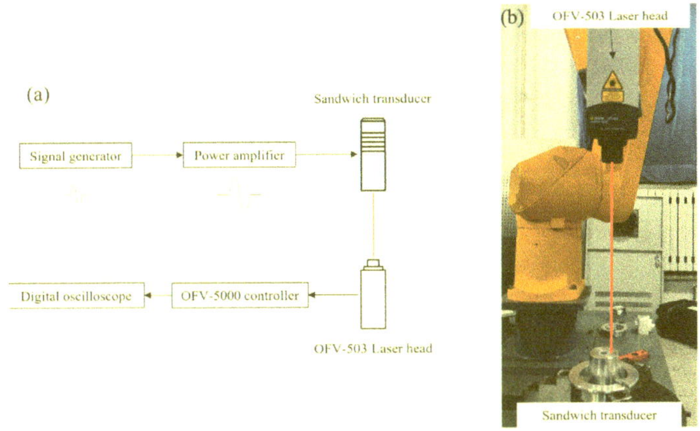

Figure 5. Single-point transducer vibration signal acquisition device: (**a**) Schematic diagram of transducer free vibration signal acquisition system; (**b**) Free vibration signal acquisition system of transducer.

4.2. Measurement of Transducer Working State

In order to accurately evaluate the working state of the transducer, the surface vibration energy of the transducer is used as the evaluation index in this paper [31].

$$I = \frac{1}{2}\rho c v^2, \tag{18}$$

Previous studies have only used laser vibrometry and acoustic emission methods to detect the vibration amplitude or vibration velocity of a single point of a piezoelectric sheet or sandwich transducer, which only enables the acquirement of the vibration state of a single point of the transducer and does not characterize the overall surface at the end of the transducer, so the characterization of the overall working state of the transducer is not complete. Therefore, this paper adopts a manipulator-based sandwich transducer sound intensity measurement system, as shown in Figure 6, which consists of a six-degree-of-freedom manipulator with a laser vibrometer that sweeps the working surface of the transducer along a preset trajectory, collecting the coordinates of the manipulator's position and surface vibrational data through a position acquisition card and a data acquisition card, wherein the position acquisition card triggers the data acquisition card to achieve synchronous acquisition and uploads the data to the host computer software via a PCI bus. The above computer software processes the data in real time, displays the vibration information in A-sweep, extracts its eigenvalues, matches them with the positional coordinates one by one, and displays the sound field distribution in C-sweep. The robot adopted was a six-degree-of-freedom Staubli RX160 robot with a repeatability of 0.05 mm. The selected laser vibrometer was Germany Polytec laser interferometer, including the controller OFV-5000 and the laser head OFV-503. The data acquisition card was M2i.3120 from Spectrum, Germany, with a frequency up to 10 MHz. The signal generator was DG1022U dual-channel arbitrary waveform generator from Puyuan. Power amplifier was the signal source small excitation signal employing the linear method and impedance matching to achieve high-power excitation of the power transducer.

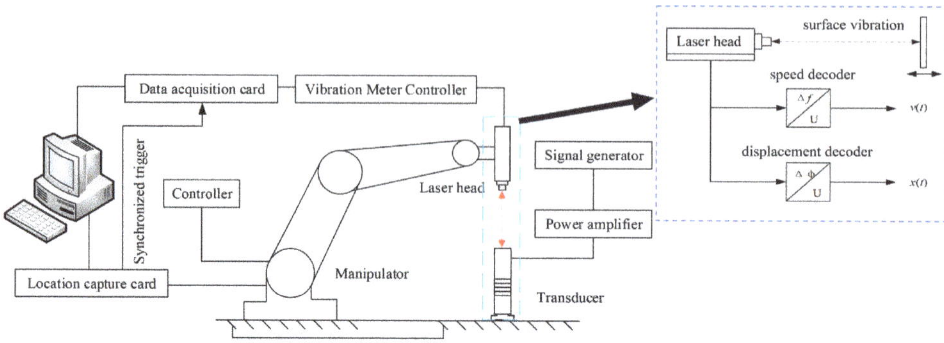

Figure 6. Working state measurement system of sandwich transducer.

5. Results and Discussion

5.1. Transducer Inherent Frequency Analysis

The length of each part of the transducer is set as $L_P = L_1 + L_3 = 41$ mm, $L_2 = 74$ mm, and $L_4 = 32$ mm; the outer diameter of the piezoelectric ceramic $\phi D_r = 60$ mm; the inner diameter $\phi d_r = 30$ mm; the thickness is $h = 10$ mm; and the thickness of the electrode sheet is 0.25 mm. Therefore, the piezoelectric ceramic's inherent frequency of 15.15 kHz can be obtained from Equations (15) and (17), where $L_1 = 3$ mm and $L_3 = 38$ mm, and the node's cross section is located in the piezoelectric ceramic stack and near the front mass block.

The free vibration signal of the transducer after being excited by the pulse is observed using a digital oscilloscope, as shown in Figure 7a. The obtained spectrum is shown in Figure 7b. There is only one peak in the analysis's frequency range, and the corresponding frequency value is 15.10 kHz, which is a 0.05 kHz difference from the design. The inherent frequency of the transducer is 15.44 kHz when ignoring the pressing bolt without an equivalent calculation of each part of the transducer.

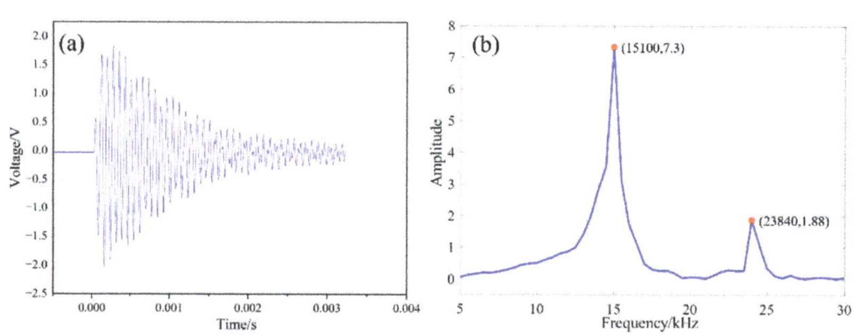

Figure 7. Free attenuation signal and spectrum analysis of high power piezoelectric transducer: (**a**) Free vibration signal of transducer; (**b**) FFT diagram of vibration signal.

5.2. Working Performance Measurement and Analysis of Sandwich Piezoelectric Transducer

The working performance of the sandwich transducer will directly affect the operational performance of the power ultrasonic system. In order to maintain the best performance of the transducer in its working state, it is necessary to analyze the factors affecting the working performance of the transducer. The influence of the excitation signal on the transducer is considered in this study. The characteristics of the excitation signal include voltage, frequency, and continuity.

Combined with the working range of the piezoelectrically powered transducer, the voltage range of the excitation signal is selected to be 80~400 V. The excitation signal is selected near the first-order inherent frequency, and the frequency range is 14.5~15.5 kHz.

The signal generator outputs a continuous sine wave, and the laser vibrometer is used to detect the vibration velocity of the end face of the front mass block. The measurement results are read on the digital oscilloscope and the performance cloud diagram of the piezoelectric transducer is obtained after fitting, as shown in Figure 8. It can be seen that as the excitation voltage increases, the vibration velocity of the transducer also increases, and the closer the excitation frequency is to the inherent frequency, the higher the vibration velocity at the end of the transducer. However, with the increase in the excitation voltage, the optimal working frequency of the transducer is not fixed, and the vibration velocity at the end of the transducer is not linear with respect to the excitation voltage.

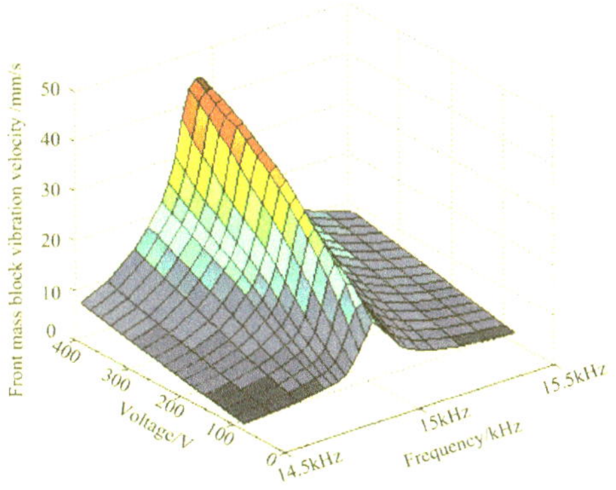

Figure 8. Performance cloud diagram of sandwich transducer.

In order to analyze the influence of the excitation frequency on the working characteristics of the sandwich transducer, six groups of constant voltage excitation were selected to obtain vibration velocity curves with different frequencies, as shown in Figure 9.

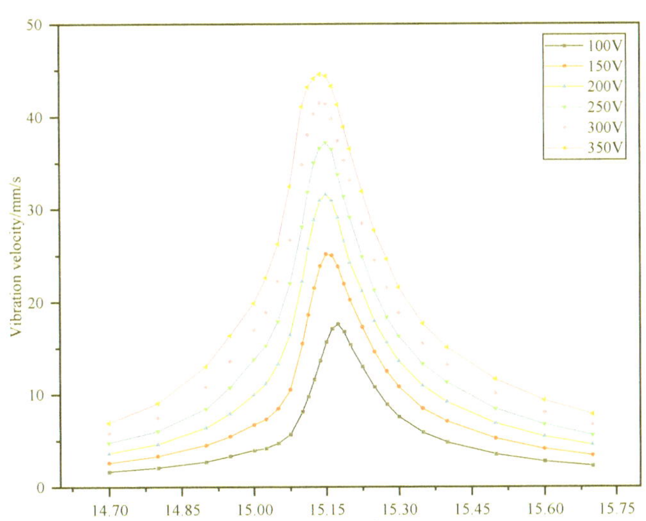

Figure 9. Power transducer frequency characteristic curve.

Corresponding to each voltage value, there is a peak in the frequency range, and the transducer deviates from the inherent frequency of the vibration, which decreases significantly. This shows that if the power transducer does not work at the optimal frequency, its working efficiency is greatly reduced. At the same time, the shape of the curve is similar to the peak at the resonant frequency shown in Ref [25], and the experimental results are consistent with the theory. In addition, it can be found that the resonant frequency of the transducer is different under different excitation voltages. For example, the resonant frequency of the transducer is 15.16 kHz when the excitation voltage is 100 V, and the resonant frequency is 15.15 kHz when the excitation voltage is 150 V and 200 V. As the excitation voltage continues to increase, the change in the resonant frequency is no longer obvious—specifically, it is 15.14 kHz. In general, the change in the resonant frequency is not obvious. In practical applications, the maximum working efficiency can be obtained by fine-tuning the frequency according to the excitation voltage.

Based on the continuous excitation method, the inherent frequency obtained by analyzing the above data is about 15.15 kHz. Based on the electromechanical equivalent circuit method proposed in Ref [25], the inherent frequency of the transducer is 15.25 kHz. The difference in the inherent frequency obtained after analysis is due to the machining assembly's accuracy and preload. The inherent frequency based on the pulse excitation method is 15.10 kHz. This can be explained by the different excitation modes of the pulse wave and the continuous wave of the transducer, which lead to the difference in the mechanical vibration state of the transducer.

Since the inherent frequency of the transducer is 15.15 kHz, several sets of frequencies near the two sides of the frequency were selected, and the voltage characteristic curve is shown in Figure 10.

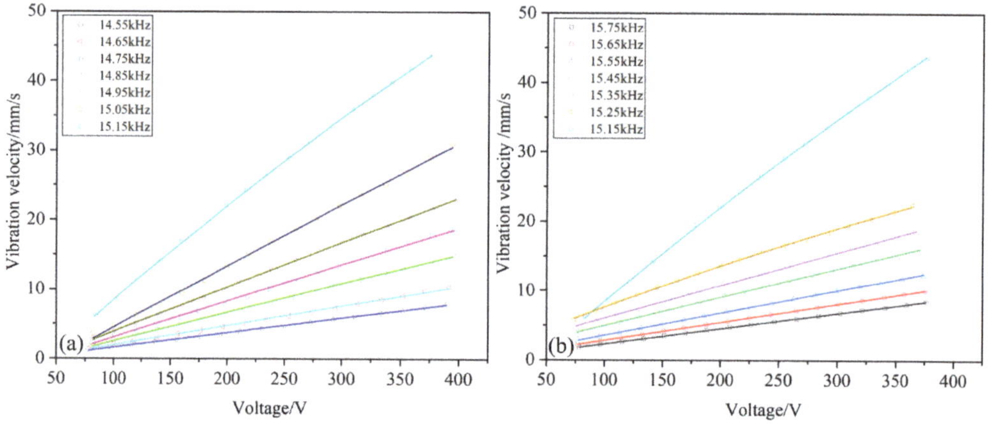

Figure 10. Voltage response curve of frequency for piezoelectric transducer near 15.15 kHz: (**a**) diagram of voltage-frequency relationship in 14.55–15.15 kHz frequency, (**b**) diagram of voltage-frequency relationship in 15.15–15.75 kHz frequency.

It can be found that at a frequency far from the inherent frequency, the change in the vibration velocity with the voltage is basically linear, but the closer to the inherent frequency, the weaker the linear relationship. When the vibration frequency is 15.15 kHz, a nonlinear relationship between the vibration velocity and the voltage can be seen, indicating that when the power transducer is close to the inherent frequency, the vibration amplitude changes dramatically; consequently, the bolt preload is not sufficient to offset the stress generated by the large displacement vibration, resulting in a gap between the transducer structures, thereby blocking the propagation of vibrations, and, finally, causing the measured velocity to differ from the theoretical value.

5.3. Measurement of Working State of Sandwich Transducer

The sandwich transducer uses the vibration of the piezoelectric ceramic along the thickness direction to realize its overall vibration along the axis. Under ideal conditions, the piezoelectric ceramic is in a uniform vibration state. However, the sandwich transducer relies on the pre-tightening of the bolts; the vibration amplitude and phase of the particles in the end face are different along the radial direction, which will affect the application of the power transducer system. Since the sandwich transducer is a vibration system, its vibration mode is related to the excitation frequency, and the function of the sandwich transducer is to use the longitudinal vibration to transfer energy; thus, it is necessary to find the frequency corresponding to the longitudinal vibration. It can be seen that the first-order and second-order inherent frequencies of the transducer are 15.10 kHz and 23.84 kHz. The vibrational intensity distribution of the transducer's surface obtained by continuous signal excitation is shown in Figure 11. Since the pre-stressed bolt of the transducer runs through the whole transducer, the working area of the face is annular.

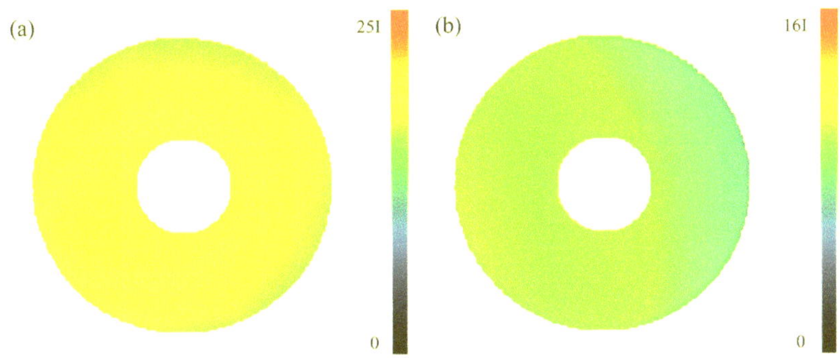

Figure 11. Radiation intensity distribution of sandwich transducer end face: (**a**) vibration intensity cloud diagram of transducer end under first order natural frequency; (**b**) Vibration intensity cloud diagram of transducer end under second order natural frequency.

In Figure 11, I is the sound intensity value of a unit, which is 7 W/m^2, and the corresponding vibration velocity is 1 mm/s. Comparing the first-order and second-order radiation intensity distribution maps, it is clear that the second-order radiation intensity of the end face of the transducer is almost reduced to one-quarter of the first-order. According to the sound intensity calculation formula, the particle vibration velocity of the end face is reduced to about one-half. Through Figure 7b, it can also be seen that during vibration, the composition of the first-order frequency is much higher than the second-order frequency; that is, the energy ratio is large, indicating that the sandwich transducer works under the first-order inherent frequency.

It can be seen from Figure 11a that the first-order radiation intensity distribution is relatively uniform, indicating that the vibration amplitude of the first-order end face particles is basically the same, indicating that there is no vibration mode other than longitudinal vibration, or that the proportion of other vibration modes is very small. Figure 11b shows that the second-order sound field distribution is not uniform. It can be seen that the surface of the transducer is not equal in terms of amplitude vibration, and the radiation intensity distribution is different from the center, indicating that the structural coupling between the piezoelectric ceramic, the front and rear cover plates, and the preload bolt is not uniform, resulting in the overall vibration of the transducer not only along the axial vibration but also in the radial vibration mode. The ideal vibration state of the transducer is that in which there is only longitudinal vibration, that is, there is no radial vibration of the transducer, which does not include the small side vibrations caused by material deformation. When there is radial vibration, that is, when there is vibration along the normal direction of the cylinder, it will have a super positional effect on the longitudinal vibration. Therefore, a radiation intensity scanning experiment was carried out on the cylinder of the transducer. The scanning area is a quarter of the circumferential area of the front mass block. The scanning results are shown in Figure 12.

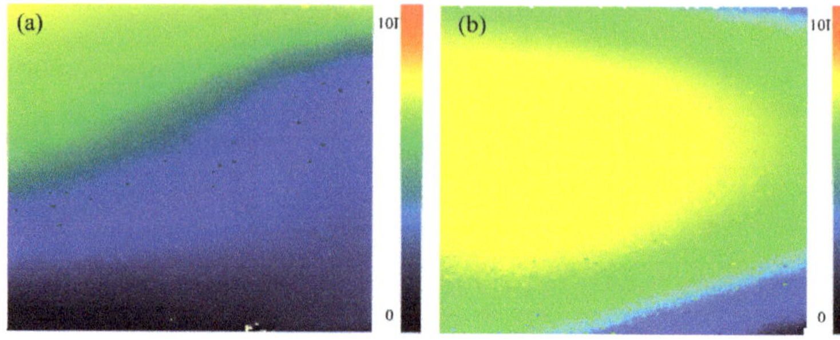

Figure 12. Radial radiation intensity distribution cloud diagram of sandwich transducer: (**a**) vibration intensity cloud diagram of transducer shaft side first order natural frequency; (**b**) Vibration intensity cloud diagram of transducer shaft side under second order natural frequency.

Compared with the end face's radiation intensity, the radial radiation intensity is very small, indicating that the radial vibration amplitude is relatively small. Figure 12a is the first-order side radiation intensity distribution. Although it is unevenly distributed along the circumferential direction, considering that its value is very low, the influence of bending vibration on the end face radiation intensity distribution can be ignored, which is consistent with the measurement results in Figure 11a. Figure 12b shows a uniform distribution along the circumferential direction and a uniform variation along the height direction, which is consistent with the radial vibration caused by radial deformation due to longitudinal tensile and compressive deformation.

6. Conclusions

In this paper, a natural frequency calculation model of a sandwich transducer is established based on continuum mechanics and a one-dimensional wave equation. Considering the influence of the preload bolt on the overall response frequency of the transducer, the equivalent elastic model-based quantitative method is proposed to calculate the equivalent sound velocity and elastic modulus of each part of the transducer. Based on the half-wavelength theory, the mathematical relationship between the natural frequency of the transducer and the length of each equivalent component is obtained. According to the one-dimensional simplified model proposed in this study, a sandwich transducer with a natural frequency of 15 KHz was designed and fabricated. The vibration signal of the end face of the transducer was collected by a laser vibrometer for spectrum analysis. The accuracy of the one-dimensional vibration model of the transducer was verified, and the effects of different voltages, operating frequencies, and signal continuity on the working state of the transducer were analyzed. Finally, a transducer end vibration-scanning system was built to evaluate the working state of the transducer, and the vibration state of the first-order and second-order end surfaces and the radial surface of the transducer were obtained. The results show that the axial vibration response of the transducer at the first-order resonant frequency is much larger than the radial vibration response. This study not only simplifies the calculation of the natural frequency of the transducer but also provides a reference for the application of the power transducer in the field of ultrasonic machining. This study plays a guiding role in quantifying the working state of the transducer and adjusting the process parameters of ultrasonic machining.

Author Contributions: Conceptualization, C.X., Q.P. and D.X.; methodology, Y.L.; investigation, Y.L. and Q.Y.; writing—original draft preparation, Y.L. and Q.Y.; writing—review and editing, C.X., Q.P. and D.X.; project administration, C.X. and Q.P. All authors have read and agreed to the published version of the manuscript.

Funding: This research and its publication were supported by the project of the National Defence Science and Industry Bureau of China: (Grant No. JSJL2019205B002); Basic Technology Research, which was funded by the National Natural Science Foundation of China (grant no. 51975050); Technology and Quality Division of the Ministry of Industry and Information Technology (grant no. JSZL2018602C001).

Institutional Review Board Statement: Not applicable.

Informed Consent Statement: Informed consent was obtained from all subjects involved in the study.

Data Availability Statement: Not applicable.

Conflicts of Interest: The authors declare no conflict of interest.

References

1. Pang, Y.; Feng, P.; Wang, J.; Zha, H.; Xu, J. Performance analysis of the longitudinal-torsional ultrasonic milling of Ti-6Al-4V. *Int. J. Adv. Manuf. Technol.* **2021**, *113*, 1255–1266. [CrossRef]
2. Ying, N.; Feng, J.; Bo, Z.; Guofu, G.; Jing-jing, N. Theoretical investigation of machining-induced residual stresses in longitudinal torsional ultrasonic–assisted milling. *Int. J. Adv. Manuf. Technol.* **2020**, *108*, 3689–3705. [CrossRef]
3. Lü, S.; Wu, S.; Dai, W.; Lin, C.; An, P. The indirect ultrasonic vibration process for rheo-squeeze casting of A356 aluminum alloy. *J. Mater. Process. Technol.* **2012**, *212*, 1281–1287. [CrossRef]
4. Haghayeghi, R.; Heydari, A.; Kapranos, P. The effect of ultrasonic vibrations prior to high pressure die-casting of AA7075. *Mater. Lett.* **2015**, *153*, 175–178. [CrossRef]
5. Zohrevand, M.; Aghaie-Khafri, M.; Forouzan, F.; Vuorinen, E. Internal stress relief and microstructural evolution by ultrasonic treatment of austeno-ferritic 2205 duplex stainless steel. *Mater. Sci. Eng. A* **2021**, *815*, 141290. [CrossRef]
6. Zhang, Q.; Yu, L.; Shang, X.; Zhao, S. Residual stress relief of welded aluminum alloy plate using ultrasonic vibration. *Ultrasonics* **2020**, *107*, 106164. [CrossRef]
7. Du, L.; Wang, Q.; Zhang, X. Reduction of internal stress in SU-8 photoresist layer by ultrasonic treatment. *Sci. China Technol. Sci.* **2010**, *53*, 3006–3013. [CrossRef]
8. Gu, B.; Jin, Z.; Kong, D.; Lai, J.; Yang, Z.; Pan, L. Reduction of pulsed-laser surface irradiation induced residual stress using ultrasonic vibration method. *Int. J. Adv. Manuf. Technol.* **2017**, *88*, 755–765. [CrossRef]
9. Song, W.-Y.; Xu, C.-G.; Pan, Q.-X.; Li, P.-L.; Wang, L.; Yu, T. The Residual Stress and Deformation Control of TC4 Thin-Walled Outer Ring Components by Ultrasonic Regulation. *Machines* **2022**, *10*, 598. [CrossRef]
10. Lin, S.; Xu, C. Analysis of the sandwich ultrasonic transducer with two sets of piezoelectric elements. *Smart Mater. Struct.* **2008**, *17*, 065008. [CrossRef]
11. Meng, X.; Lin, S. Analysis of a Cascaded Piezoelectric Ultrasonic Transducer with Three Sets of Piezoelectric Ceramic Stacks. *Sensors* **2019**, *19*, 580. [CrossRef]
12. Hu, J.; Lin, S.; Zhang, X.; Wang, Y. Radially Sandwiched Composite Transducers Composed of the Radially Polarized Piezoelectric Ceramic Circular Ring and Metal Rings. *Acta Acust. United Acust.* **2014**, *100*, 418–426. [CrossRef]
13. Bloomfield, P.E. Multilayer transducer transfer matrix formalism. *IEEE Trans. Ultrason. Ferroelectr. Freq. Control* **2002**, *49*, 1300–1311. [CrossRef] [PubMed]
14. Xue, Y.; Bolton, J.S.; Liu, Y. Modeling and coupling of acoustical layered systems that consist of elements having different transfer matrix dimensions. *J. Appl. Phys.* **2019**, *126*, 165102. [CrossRef]
15. Wang, L.; Hofmann, V.; Bai, F.; Jin, J.; Twiefel, J. Modeling of coupled longitudinal and bending vibrations in a sandwich type piezoelectric transducer utilizing the transfer matrix method. *Mech. Syst. Signal Process.* **2018**, *108*, 216–237. [CrossRef]
16. Zhang, Y.; Tu, Z.; Lu, T.-F.; Al-Sarawi, S. A simplified transfer matrix of multi-layer piezoelectric stack. *J. Intell. Mater. Syst. Struct.* **2017**, *28*, 595–603. [CrossRef]
17. Fu, B.; Li, C.; Zhang, J.; Huang, Z.; Hemsel, T.; Bo, F.; Chao, L.; Jianming, Z.; Zhenwei, H.; Tobias, H. Modeling of Piezoelectric Langevin Transducers by Using Mixed Transfer Matrix Methods. *J. Korean Phys. Soc.* **2010**, *57*, 929–932. [CrossRef]
18. Zhang, Y.; Lu, T.-F.; Al-Sarawi, S. Formulation of a simple distributed-parameter model of multilayer piezoelectric actuators. *J. Intell. Mater. Syst. Struct.* **2016**, *27*, 1485–1491. [CrossRef]
19. Abdullah, A.; Pak, A. Correct Prediction of the Vibration Behavior of a High Power Ultrasonic Transducer by FEM Simulation. *Int. J. Adv. Manuf. Technol.* **2008**, *39*, 21–28. [CrossRef]
20. Abdullah, A.; Pak, A.; Abdullah, M.M.; Shahidi, A.; Malaki, M. Study of the behavior of ultrasonic piezo-ceramic actuators by simulations. *Electron. Mater. Lett.* **2014**, *10*, 37–42. [CrossRef]
21. Pérez-Sánchez, A.; Segura, J.A.; Rubio-Gonzalez, C.; Baldenegro-Pérez, L.A.; Soto-Cajiga, J.A. Numerical design and analysis of a langevin power ultrasonic transducer for acoustic cavitation generation. *Sens. Actuators Phys.* **2020**, *311*, 112035. [CrossRef]
22. Jiang, X.; Zhu, X.; Li, S.; Geng, D.; Zhang, D. Design of Sandwich Transducer Based on the Equivalent Length Algorithm. *Comput. Mater. Contin.* **2019**, *61*, 23–28. [CrossRef]
23. Wei, X.; Yang, Y.; Yao, W.; Zhang, L. PSpice Modeling of a Sandwich Piezoelectric Ceramic Ultrasonic Transducer in Longitudinal Vibration. *Sensors* **2017**, *17*, 2253. [CrossRef] [PubMed]
24. Chen, Y.-C. A comparative assessment of classification methods for resonance frequency prediction of Langevin piezoelectric transducers. *Appl. Math. Model.* **2011**, *35*, 3334–3344. [CrossRef]

25. Karafi, M.; Kamali, S. A continuum electro-mechanical model of ultrasonic Langevin transducers to study its frequency response. *Appl. Math. Model.* **2021**, *92*, 44–62. [CrossRef]
26. Wang, J.; Qin, L.; Song, W.; Shi, Z.; Song, G. Electromechanical Characteristics of Radially Layered Piezoceramic/Epoxy Cylindrical Composite Transducers: Theoretical Solution, Numerical Simulation, and Experimental Verification. *IEEE Trans. Ultrason. Ferroelectr. Freq. Control* **2018**, *65*, 1643–1656. [CrossRef]
27. Piao, C.; Kim, J.O. Vibration characteristics of an ultrasonic transducer of two piezoelectric discs. *Ultrasonics* **2017**, *74*, 72–80. [CrossRef]
28. Lin, S. Coupled vibration of isotropic metal hollow cylinders with large geometrical dimensions. *J. Sound Vib.* **2007**, *305*, 308–316. [CrossRef]
29. Meng, X.; Lin, S. Analysis on coupled vibration of piezoelectric ceramic stack with two piezoelectric ceramic elements. *J. Acoust. Soc. Am.* **2019**, *146*, 2170–2178. [CrossRef]
30. Karafi, M.R.; Mirshabani, S.A. An Analytical Approach to Design of Ultrasonic Transducers Considering Lateral Vibrations. *J. Stress Anal.* **2019**, *3*, 47–58. [CrossRef]
31. Abdullah, A.; Shahini, M.; Pak, A. An approach to design a high power piezoelectric ultrasonic transducer. *J. Electroceramics* **2009**, *22*, 369–382. [CrossRef]
32. Bhuiyan, M.S.H.; Choudhury, I.A.; Dahari, M.; Nukman, Y.; Dawal, S.Z. Application of acoustic emission sensor to investigate the frequency of tool wear and plastic deformation in tool condition monitoring. *Measurement* **2016**, *92*, 208–217. [CrossRef]
33. Zhang, H.; Pei, Y.-C.; Liu, Q.-J. A measuring method for the resonant frequency of flexural horn-type transducer using non-contacting piezoelectric sensor. *Measurement* **2020**, *164*, 108016. [CrossRef]
34. Derusova, D.A.; Vavilov, V.P.; Druzhinin, N.V.; Kolomeets, N.P.; Chulkov, A.O.; Rubtsov, V.E.; Kolubaev, E.A. Investigating vibration characteristics of magnetostrictive transducers for air-coupled ultrasonic NDT of composites. *NDT E Int.* **2019**, *107*, 102151. [CrossRef]
35. Wu, J.; Mizuno, Y.; Nakamura, K. Vibration characteristics of polymer-based Langevin transducers. *Smart Mater. Struct.* **2018**, *27*, 095013. [CrossRef]
36. Al-Budairi, H.; Lucas, M.; Harkness, P. A design approach for longitudinal–torsional ultrasonic transducers. *Sens. Actuators A Phys.* **2013**, *198*, 99–106. [CrossRef]
37. Albareda, A.; Casals, J.A.; Pérez, R.; de Espinosa, F.M. Nonlinear Measurements of High Power 1–3 Piezo-Air-Transducers with Burst Excitation. *Ferroelectrics* **2002**, *273*, 47–52. [CrossRef]
38. DeAngelis, D.A.; Schulze, G.W. Performance of PIN-PMN-PT Single Crystal Piezoelectric versus PZT8 Piezoceramic Materials in Ultrasonic Transducers. *Phys. Procedia* **2015**, *63*, 21–27. [CrossRef]

sensors

MDPI

Article

The Smart Nervous System for Cracked Concrete Structures: Theory, Design, Research, and Field Proof of Monolithic DFOS-Based Sensors

Łukasz Bednarski [1], Rafał Sieńko [2], Tomasz Howiacki [2,3,*] and Katarzyna Zuziak [4]

1 Faculty of Mechanical Engineering and Robotics, AGH University of Science and Technology in Krakow, Mickiewicza 30, 30-059 Krakow, Poland
2 Faculty of Civil Engineering, Cracow University of Technology, Warszawska 24, 31-155 Krakow, Poland
3 SHM System Sp. z o.o., Sp. kom., Libertów, ul. Jana Pawła II 82A, 30-444 Kraków, Poland
4 Nerve-Sensors, Libertów, ul. Jana Pawła II 82A, 30-444 Kraków, Poland
* Correspondence: th@shmsystem.pl; Tel.: +48-505-870-669

Abstract: The article presents research on the performance of composite and monolithic sensors for distributed fibre optic sensing (DFOS). The introduction summarises the design of the sensors and the theoretical justification for such an approach. Lessons learned during monitoring cracked concrete are summarised to highlight what features of the DFOS tools are the most favourable from the crack analysis point of view. Later, the results from full-size laboratory concrete specimens working in a cracked state were presented and discussed in reference to conventional layered sensing cables. The research aimed to compare monolithic sensors and layered cables embedded in the same reinforced concrete elements, which is the main novelty. The performance of each DFOS nondestructive tool was investigated in the close vicinity of the cracks—both the new ones, opening within the tension zone, and the existing ones, closing within the compression zone. The qualitative (detection) and quantitative (widths estimation) crack analyses were performed and discussed. Finally, the examples of actual applications within concrete structures, including bridges, are presented with some examples of in situ results.

Keywords: distributed fibre optic sensing DFOS; composite sensors; monolithic sensors; strains; cracks; widths; detection; reinforced concrete; laboratory; civil engineering

Citation: Bednarski, Ł.; Sieńko, R.; Howiacki, T.; Zuziak, K. The Smart Nervous System for Cracked Concrete Structures: Theory, Design, Research, and Field Proof of Monolithic DFOS-Based Sensors. *Sensors* **2022**, *22*, 8713. https://doi.org/10.3390/s22228713

Academic Editor: Zenghua Liu

Received: 20 October 2022
Accepted: 9 November 2022
Published: 11 November 2022

1. Introduction

1.1. Structural Safety

An acceptable level of safety is an obvious and essential requirement for structural design, construction, and maintenance. During each phase, relevant control procedures should always be specified and applied to ensure optimal decision-making [1,2]. These decisions are mainly about finding the balance between financial savings and keeping the failure risk at an acceptable level. However, this task is particularly important and, at the same time, challenging for:

- **ageing infrastructure** [3], which is many years old and in a technical condition problematic for correct assessment,
- **new engineering structures** with unusual geometry, construction technologies or material solutions for which there is a lack of common experiences.

Because of the complexity of civil engineering projects, including material and geometrical imperfections, uncertainties, time-depended effects, construction technologies and many others, structural safety cannot be ensured only by theoretical considerations, even taking into account the newest achievements in advanced numerical simulations. The knowledge must be developed based on the analysis of real structures' behaviour, provided by in situ measurements. This task is currently being handled by structural health monitoring (SHM) systems [4,5] operating based on different sensing technologies selected according to a given project's specific requirements. Usually, sensors for measurements of various physical quantities, like strains, displacements, inclinations, accelerations, forces, and others, are installed within the structure or construction site to monitor changes in these values over time during changing loading and thermal conditions. A well-designed system should provide reliable data for structural assessment, validation of simplified models, and decrease the risk of failure [6]—meaning an increase in safety.

It must be emphasised that failure consequences in civil engineering and geotechnical facilities include not only financial, social, and environmental losses but also loss of human health or life. This is the reason for the constant development of new solutions improving structures' reliability and long-term high-quality performance.

1.2. Distributed Measurements of Concrete

Despite the advantages of conventional (discrete) structural health monitoring systems, a direct answer to the question about the state (e.g., the crack state) of the structure between the measurement points is still not possible. The solution to this problem is provided by distributed fibre optic sensing (DFOS), enabling the measurements with spatial resolution so high that they could be treated as geometrically continuous from an engineering point of view. Figure 1 shows the scheme comparing the capabilities of conventional spot strain gauges and distributed linear sensors installed within the example damaged concrete element. While performing distributed measurements, it is impossible to omit any extreme values, so one of the main goals of structural health monitoring—direct damage detection and its size estimation [7]—can finally be fulfilled.

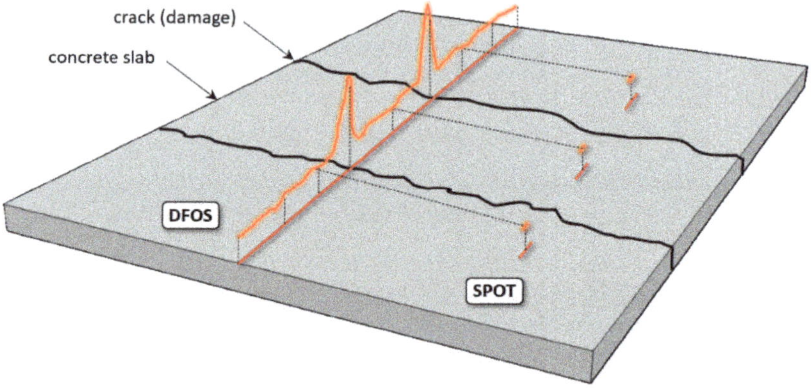

Figure 1. Comparison between capabilities of distributed fibre optic sensing DFOS and spot measurements in crack detection [8].

The concept of smart structures with integrated systems based on distributed fibre optic sensors has been intensively developed for several years. Instead of a single number obtained by standard spot gauge, these sensors measure the profiles of strains [9], displacements [10–12], temperatures [13] or vibrations [14] over the entire structural length. DFOS-based systems are an analogue of the human nervous system [15,16], as the linear sensors can be compared to the individual nerves informing of dangers at any point and the data logger to the brain processing the data.

The above feature is especially useful regarding concrete because of its heterogeneity, local imperfections, changes in stiffness caused by the random presence of aggregate and dependence on many other factors like mixture components, time (rheology), curing process or load procedure. The most crucial aspect that should be underlined, though, is concrete discontinuities in the form of multiple cracks. They are a common and natural feature of reinforced and prestressed concrete structures [17]. Cracking already starts during early-age concrete hydration, when thermal-shrinkage strains are blocked by internal (reinforcement) and external (other elements, formworks, supports, boundary conditions) constraints [18,19]. Later, under operating conditions, mechanical loads, external temperature impacts and time-dependent structural degradation, cracks develop and change their widths. Knowing the actual crack state of concrete structures is one of the essential requirements for technical assessments of their condition and safety.

Comprehensive data gathered by distributed fibre optic sensors, replacing thousands of conventional spot-type sensors, provide a new quality in structural measurements and become the reason why DFOS nowadays is finding a growing acceptance in the laboratory, and above all, in field applications. It is worth underlining that it is possible and favourable to install the DFOS sensors inside the structural components [20] during construction, e.g., embedding them inside the concrete [21,22], which brings many benefits, including:

- The possibility of analysis of the structural behaviour from a real zero stress-strain state (usually, conventional gauges are installed within existing structures with the unknown initial level of stress and deformation).
- Integration inside the structure (concrete) provides a more accurate transfer of the measured physical quantity from the structure to the sensor (no additional mounting brackets or installation methods are needed).
- Natural protection of the embedded sensors against mechanical damages, aggressive external environmental or direct sunlight. The expected operation lifetime of such a system should be comparable with the operation lifetime of the structure itself.

1.3. DFOS Techniques (Data Loggers)

The general idea of distributed sensing described above can be utilised using different techniques reflected in the data loggers available on the market. Three main optical phenomena used for that purpose are Rayleigh, Brillouin and Raman scatterings, named so to honour their discoverers (Figure 2). It is worth noting that both Rayleigh and Raman were awarded the Nobel Prize in the early twentieth century for their achievements in physics.

(a) (b) (c)

Figure 2. (**a**) John William Strutt (Lord Rayleigh); (**b**) Léon Brillouin; (**c**) Chandrasekhara Venkata Raman [6].

Each of these approaches is characterised by its own advantages and limitation related to measurement parameters. The final choice should always be preceded by a detailed analysis of the requirements of the project in question. However, general insights can be presented as follows:

1. **Rayleigh scattering** [23] is used to measure the strains in optical fibre, which are caused both by mechanical and thermal loads. It provides the best spatial resolution, starting from the mm range [24] (1000 measurement gauges per 1 m of the linear sensor), which is particularly useful when analysing localised events like cracks.

Moreover, dynamic readings with a frequency of up to 250 Hz are also possible [25]. However, the main limitation of that approach lies in a measurement distance range limited to 100 m (while keeping the high spatial resolution).

2. **Brillouin scattering** [26,27] is used to measure the strains in optical fibre, which are caused both by mechanical and thermal loads. It provides multiple lower spatial resolution (from 20 to 100 cm) compared to Rayleigh scattering. On the other hand, it allows the measurements to be performed over very long distances (e.g., 25 km or more). This makes it suitable for monitoring linear structures like tunnels, roads, embankments, dams or mining and landslide areas.
3. **Raman scattering** [28,29] is used to measure the temperature profile in optical fibre so that it could be used as one of the compensation solutions for Rayleigh and Brillouin measurements. Standard spatial resolution starts from 50 cm, while distance ranges from 25 km and more. This approach can be successfully used as a thermo-detection method [30] to localise fires or leakages.

It must be underlined that thanks to the dynamic development of commercially available optical solutions, their parameters are constantly improved. It is also possible to apply a few DFOS techniques to read the same sensors and thus, obtain extended benefits like automatic thermal compensation [31] or simultaneous measurements of different physical quantities.

1.4. DFOS Sensing Tools

The efficiency of the entire DFOS-based system depends not only on the optical data loggers' parameters but also on the quality and performance of the measuring tools. Optical devices are usually designed by physicists, optical and electronics engineers, and IT specialists. However, the design of the tools being installed within the concrete, very often in challenging construction conditions, requires the knowledge of civil engineers, materials scientists and mechanics.

There are three main groups of DFOS sensing tools [6], including optical fibres in their primary coating, layered sensing cables and monolithic, composite strain sensors (Figure 3). The first group is mainly used in laboratory conditions because of the dimensions of standard optical fibre (outer diameter of 250 µm—Figure 3a), which makes it susceptible to break or be disturbed by any local and transverse pressures. That is why installation conditions must be strictly controlled, which is impossible on construction sites or geotechnical fields.

A few aspects must be considered regarding the measurements of cracked concrete. The basic thing is the type of the primary coating, which influence the maximum strain range, strain transfer mechanism or the resistance of the fibre to an alkaline concrete environment. Two commonly used primary coating types are acrylate (softer) and polyimide (stiffer). The final choice should be made considering the following relationships:

- the softer the coating, the higher the maximum strain range [6,20] (during crack detection, very high and local strain values are expected, so the acrylate coating minimises the risk of the fibre's breakage);
- the stiffer the coating, the better the strain transfer mechanism and the shorter the length of strain mobilisation [32,33] (it means that acrylate coatings are not preferable for short measurement sections);
- alkaline concrete environment degrades the polyimide coating even after 14 days, while no influence is observed on the acrylate one [34] (that is why polyimide coatings are not advised for long-term measurements of concrete structures).

To enable safe handling of the measurement fibre on site, it must be somehow protected. There are two basic concepts utilised in practice. The earliest approach involves techniques known from telecom applications, where a set of intermediate layers are arranged around the fibre for its mechanical protection. In the worldwide literature, these solutions are known as layered sensing cables [35–37]. The layers are usually made of plastics, but steel strengthening inserts are sometimes applied (Figure 3b) to improve the strength parameters of the entire cable. However, it is important to note that the yielding point of steel is approximately equal to 2000 με, while the maximum range of the fibre itself could exceed even 50,000 με [6]. Moreover, crack-induced strain values are usually multiple times higher that yielding point of steel [6,8].

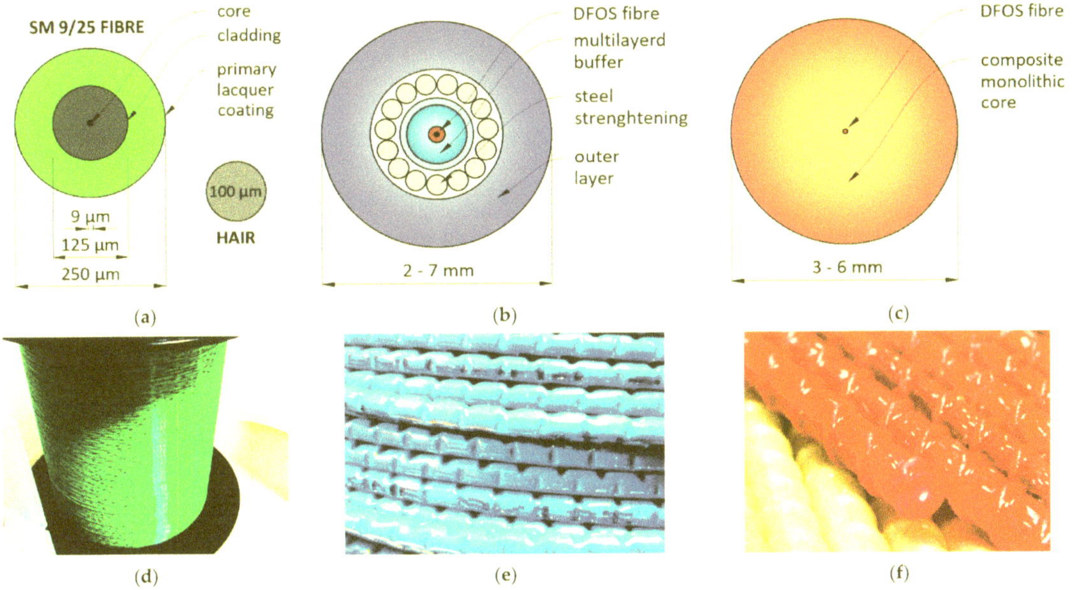

Figure 3. Example cross-sections of (**a**) single-mode optical fibre in its primary coating, (**b**) layered sensing cable with steel strengthening insert and (**c**) monolithic strain sensor with the corresponding views of their external surfaces (**d**–**f**).

The presence of intermediate layers implicates the risk of mutual slippage between them and thus, influences the correct strain transfer from the concrete to the measuring fibre inside the cable. The problem of slippage is especially noticeable during high strain values and high strain gradients close to the cracks. This phenomenon was observed and described in many works [6,38–41], confirming that the presence of a coating or layer results in differences between the structural strains and those sensed by the cable.

Another uncertainty is caused by thermal changes, which, in the case of different and unknown thermal expansion coefficients of subsequent layers, may produce unknown strains in the fibre. Thus, the mechanical and physical properties of the cable's components and their adhesion parameters must be carefully studied before application within the cracked concrete structures to enable the correct data interpretation. Moreover, the name 'cable' corresponds to the feature of these tools, i.e., their usability only when subjected to tension loads. If compression strains are expected, the pre-tensioning process should be performed during installation [42], which is difficult, and sometimes even impossible, in the case of site or geotechnical conditions.

Finally, the type of outer surface of the cable should be considered to provide its best possible bonding between the surrounding concrete. Figure 3e shows the example surface with tiny, perforated grooves. However, most conventional cables' outer surfaces are entirely smooth, which increases the unfavourable effect of internal slippage.

Another concept for fibre protection is proposed by monolithic sensors, which are produced as composite bars in the pultrusion process [43]. The glass measuring optical fibre is fully integrated within the monolithic core without any intermediate layers (Figure 3c). Thus, any potential disturbances caused by the internal slippage within the DFOS tool are excluded. External unidirectional (Figure 3e) or bi-directional braid should provide the best bonding properties with surrounding concrete, similarly to reinforcing bars (not only by adhesion but also by mechanical clampings).

Although the general idea and operation rule of monolithic sensors remain the same, their properties could be adjustable depending on the specific requirements of the project in question. Both geometrical (diameter, cross-section shape) and mechanical (strength, elastic modulus, limit strain) parameters can be considered. The design of a DFOS-based system involving monolithic sensors for crack detection in concrete is presented hereafter and discussed based on actual data from laboratory and field tests.

The research shows the performance of monolithic sensors when monitoring cracked concrete structures in comparison with selected layered cables. All of these tools are commercially available on the market and widely used, so the presented results would have important meaning for practical applications in the future.

2. Design of DFOS-Based System for Cracked Concrete

2.1. Main Objectives

The main goal of distributed fibre optic sensing for cracked concrete is to enable the advanced analysis of cracks. However, it should be clarified what this analysis means in practice. Two issues should be considered:

- **qualitative analysis**: cracks' detection and location (Figure 4a); a significant limitation of spot techniques for crack measurements is that they can be used only when the presence and the position of the crack are known. Knowledge about the location of the crack is not necessary during distributed sensing because its detection is one of the system's objectives. The possibility of crack detection was checked and confirmed in much research, e.g., [44–46]. However, this does not mean that the effectiveness of each DFOS system is unconditional and absolute.
- **cracks' width estimation** (Figure 4b); knowledge about the crack's presence and location is necessary but insufficient for assessing structural safety. The DFOS system should provide information about the actual width of the crack (mm), changing over time, which could be compared to the thresholds defined in relevant standards to answer the question about the crack's significance for load capacity or durability. The measured strain profile in the close vicinity of the crack should be converted into the crack width (in the literature called also crack opening displacement COD) with reasonable accuracy, useful from a practical point of view (not worse than 0.05 mm). A few procedures are presented in the literature [47–50], taking into account applied spatial resolution and the assumed physical model of the entire system (also internal design of the DFOS tool).

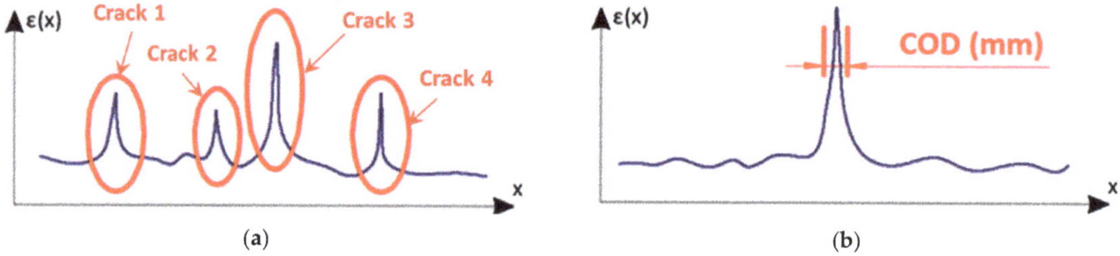

Figure 4. The scheme for crack analysis based on the DFOS system: (**a**) qualitative analysis (detection, location); (**b**) quantitative analysis (width estimation) [6].

2.2. Selection of DFOS Technique

Distributed sensing through a fibre optic sensor is supported by several optical phenomena already outlined in Section 1.3. The cracks formed within the concrete structures usually have widths much lower than 1 mm, so they should be treated as a local event. This is the basis for selecting the DFOS technique with appropriate high spatial resolution.

Despite the crack widths or their changes reaching only the tenths of a millimetre, the spatial resolution of mm or cm order is sufficient due to the deboning effect between the DFOS tool and the concrete. This effect is similar to that observed during the interaction of the concrete and reinforcing bar within the crack (Figure 5a), where debonding length occurs over a few to several centimetres, influencing the crack-induced strain profiles. A simplified graphical interpretation of high spatial resolution defined for the sensor installed in concrete along the reinforcing bars is presented in Figure 5b. The bases and spacings of virtual gauges, which are connected in series in a chain, allows for the identification and analysis of localised events.

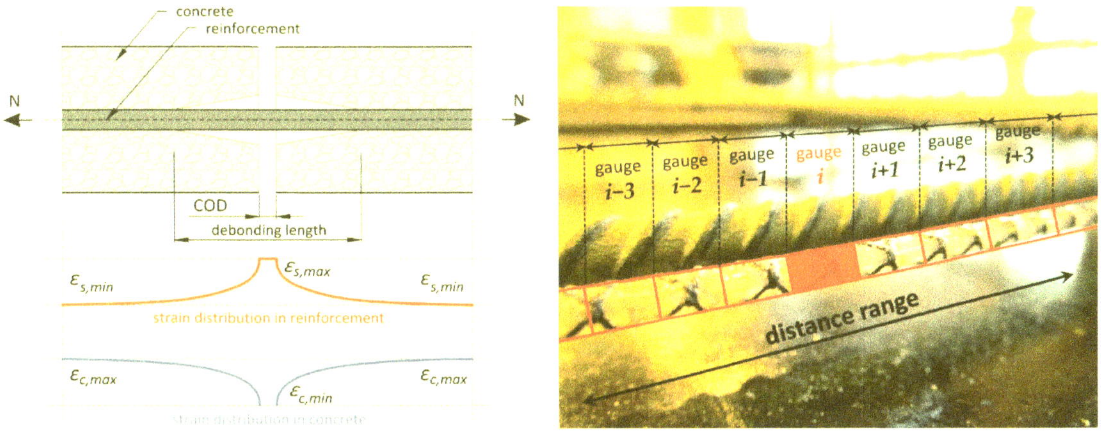

(**a**) (**b**)

Figure 5. (**a**) Simplified scheme for a bond-slip crack model for reinforced concrete [51]; (**b**) simplified graphical interpretation of spatial resolution in distributed sensing.

Today, the above requirements on high spatial resolution are usually met by Rayleigh scattering. For standard Brillouin-based systems, spatial resolutions range from 200 to 1000 mm, which is insufficient to solve the stated measurement problem (qualitative and quantitative crack analysis). However, it should be clearly emphasised that thanks to the dynamic development of DFOS techniques, the performance of optical data loggers (including spatial resolution) is constantly being improved. Other parameters should be counted when selecting the appropriate one, like accuracy, strain resolution, precision, or repeatability.

2.3. *Design of DFOS Tools*

DFOS tools are a critical component of the entire system. They are fully integrated with the monitored concrete structure (by mounting within the surface or embedding inside the structural member) to provide reliable measurements during the predicted period. Preferably, this time is equal to the operational lifetime of the structure itself. In most projects, it is difficult, and sometimes even impossible, to replace the DFOS tools, while the reflectometer can be easily replaced in case of failure.

The high-quality and long-term performance of DFOS tools dedicated to civil engineering and geotechnical applications must be ensured by meeting the set of the following requirements:

1. **High accuracy** ensured by unambiguous strain transfer from the structure to the measuring fibre inside the DFOS tool. This feature is characteristic of a monolithic cross-section of the sensor without any intermediate layers, which disturb measurements by extensive slippage. The lack of slippage within the sensor itself allows for

reducing uncertainties and simplifying mathematical models used for strain transfer analysis and crack width calculation;
2. **High strain range** allowing for undisturbed readings of crack-induced deformations without fear of damaging the sensor's components. Fast-yielding materials like steel or plastic tend to remember the localised historical strains rather than reflect the actual deformation state of the structure. This is especially dangerous during long-term monitoring while cyclic loads are expected;
3. **Rough outer surface** of the sensor must provide the best possible bonding with the surrounding concrete, not only through the adhesion but also mechanical clamping (similar to the reinforcing bars). For instance, it could be done by ribs, braids or perforated grooves [52];
4. **Resistance to harsh conditions**, which are expected during the construction and operation of civil engineering structures. The design of the sensor must provide appropriate protection against mechanical damages, local transverse forces (e.g., pressure of aggregate grains or mounting elements), alkaline concrete environment and other aggressive factors;
5. **High durability** provided by appropriate material. As the sensors are usually fully integrated within the concrete, the expected operation lifetime should be equal to the lifetime of the structure itself;
6. DFOS tools **cannot require pretension**, as this process in construction or geotechnical conditions is challenging or often impossible. Selected stiffness of the sensor's core must ensure correct positioning without extensive waving, as deviations from the designed position will result in additional errors during data interpretation. Reliable readings must be possible both in the tension and compression zone;
7. **Prove the sensors' high performance** in at least tens of engineering projects.

Debonding effect between the outer surface of the sensor and the surrounding concrete is natural and cannot be entirely eliminated. Perfect bonding means the full compliance of concrete and sensors strains, but in case of concrete discontinuity (crack), it would result in the sensor's breakage. Thus, debonding over the section called "effective length" appears, reducing the infinite theoretical strain to the level which will not cause the sensor's breakage. This is how the reinforcement bars work in the close vicinity of the crack (Figure 5a). Designing monolithic sensors for crack detection and analysis is thus about finding the balance between bonding quality, the core's elasticity and its maximum strain.

Example parameters of two different monolithic strain sensors are summarised in Table 1 for comparison.

Table 1. Comparison between the parameters of two different monolithic strain sensors [16].

Parameter	EpsilonSensor	EpsilonRebar
Strain resolution	±1 με	±1 με
Maximum strain	±40,000 με (±4%)	±20,000 με (±2%)
Standard diameter	Ø3 mm	Ø5 mm
Elastic modulus	3 GPa	50 GPa
Axial stiffness EA	21 kN	982 kN
Core material	PLFRP (polyester + epoxide)	GFRP (glass + epoxide)
Bending radius	50 mm	350 mm
Sensor weight	13 kg/km	45 kg/km
Light scattering [1]	Rayleigh, Brillouin, Raman	
Delivery method	coils or straight sections	
Length	any length made to order	

[1] Compatible with optical devices based on such phenomena.

Monolithic sensors can replace the composite reinforcing bars and be included in strength calculations of the structure itself. Figure 6a shows the footbridge example [15], where stiff monolithic sensors (called "EpsilonRebars", Ø5 mm, E = 50 GPa) were used to reinforce the hybrid deck over the entire length of 80 m.

(a)

(b)

Figure 6. (**a**) Footbridge reinforced with stiff monolithic sensors; (**b**) concrete slab with the flexible monolithic sensor not influencing the structural behaviour [8].

On the other hand, to increase the sensitivity in crack detection and to avoid strengthening the structural elements, it is favourable to reduce the axial stiffness as much as possible [6]. Figure 6b shows the example of a laboratory reinforced concrete slab, where the applied sensors (called "EpsilonSensors", Ø3 mm, E = 3 GPa) provided an almost 50 times reduction in stiffness, being invisible to the structure itself.

2.4. Installation Methods

After selecting the DFOS technique and type of monolithic sensor, the next important issue influencing the system's reliability and efficiency is the installation method. It depends on the type of structure (small-size laboratory specimens or full-scale engineering facilities) and the mounting time (existing, ageing infrastructure or new one).

The common approach for existing structures is to glue the sensors to the prepared (grinded, cleaned and degreased) concrete surface [6,52]. The monolithic sensors can be produced as flat bars (tapes) to facilitate that process and increase the bonding surface. However, thinking about long-term measurements, the more favourable solution is to install the sensors in pre-made near-to-surface grooves [52] filled with two-component epoxy or mortar injection (Figure 7a). This approach is preferable due to:

- the maximum bonding surface between the sensor and the surrounding concrete (from three sides instead of one in case of surface installation);
- natural protection against mechanical damages;
- significant reduction of thermal influence (direct sunlight) on strain results;
- the beast aesthetics without mounting elements visible on the surface.

The adhesive parameters, similar to the sensors themselves, should be selected carefully to ensure the best strain transfer mechanism. Too stiff adhesive can crack itself, while too soft can underestimate the actual structural strains.

Sensors can be installed not only within the concrete surface but also on the reinforcing bars along their longitudinal rib (Figure 7b). Based on that approach, crack detection will still be possible, but the main goal will be rather to assess crack-induced stress in the reinforcement instead of precise width estimation (because of the very high stiffness of reinforcing bars, the effective length of debonding will be higher in comparison to the effective length of the flexible sensor itself).

(a) (b) (c)

Figure 7. Possibilities of the sensors' installation methods: (**a**) installation in near-to-surface grooves; (**b**) surface gluing (on steel bar); (**c**) embedding inside the concrete.

The most convenient way of installation is applied in the new, reinforced structures. The sensors cut at the required length are usually delivered on site in coils, and the only task is to unroll them and stabilise them in designed positions by tying to the existing reinforcement using cable ties (Figure 7c). The significant advantage of the approach is that the measurements can be started at real zero stress-strain-crack state and being performed through all the constructions stages [53]: from early-age concrete, thermal-shrinkage strains, and corresponding micro-cracks, through prestressing and construction stages to the load tests and final operation. Structural assessment can be done more effectively based on such data because they refer to the absolute values of measured parameters (strains or crack widths) instead of their changes (increments).

2.5. Thermal Compensation

The measured strain profiles using Rayleigh or Brillouin scattering are affected both by mechanical loads and thermal changes. In most general terms, the following equation can be presented:

$$\Delta\varepsilon(x) = f(\Delta L(x), \Delta\, T(x)) \tag{1}$$

where $\Delta\varepsilon(x)$ is the strain measured over the distance x, $\Delta L(x)$ is the mechanical strain change and $\Delta T(x)$ is the temperature change.

Providing that temperature changes between the subsequent readings are constant, measured strains are directly equal to the mechanical strains (caused by external forces), i.e., generating the stress inside the structural component. This situation is possible only in laboratory conditions or during short-term field measurements, especially overnight or within underground installations.

However, during long-term monitoring, knowledge of thermal changes is unconditionally necessary to enable the correct engineering interpretation of structural behaviour. Appropriate correction can be applied based on the technical specifications of monolithic sensors and the guidelines of their producers. Usually, in the standard operating range (from −20 °C to +60 °C), it is enough to use a single (linear) coefficient, including the sensor core's thermal expansion and the refractive index's temperature-dependence of the optical fibre itself. This task became more complicated and uncertain with coated fibres [31] and layered cables, as the individual coatings have different (usually unknown) thermal expansion coefficients. With temperature changes, they interact with each other causing additional mechanical strains in the measuring fibre.

Nevertheless, the profiles of temperature changes between the subsequent readings over the entire monitored length are the basis for further compensation procedures. Today, there are four possible approaches to obtaining them:

1. Using the strain DFOS sensors with a Raman-based optical datalogger, which is insensitive to mechanical loads and thus, allows only for temperature measurements; DTS (distributed temperature sensing). This solution is primarily dedicated to long distances (km order).
2. Using the strain sensors with both Rayleigh-based and Brillouin-based optical datalogger. These two techniques are depended on mechanical strains and temperatures to varying degrees. Knowing the individual coefficients for each technique, it is possible to solve a system of equations in which the unknowns are mechanical strains and temperatures [31]. It is worth noticing that there are already hybrid data loggers available on the market that use different optical phenomena in their design.
3. Using special DFOS temperature sensors and one from DFOS techniques for strain measurements (Rayleigh or Brillouin). The idea behind this solution is to isolate the measuring optical fibre from mechanical strains, for instance, by placing it inside the tube. The fibre is then, apparently, subjected only to temperature changes. However, the free fibre does not exist because of the friction between the tube and the fibre. It can cause disturbances in the measured temperature profiles, especially considering longer distances. In addition, high mechanical strains expected while monitoring the cracks in concrete can consume excess fibre inside the tube.
4. Using conventional spot temperature gauges and approximating the temperature field between the measurement points. That approach is justified when no high gradients over length are expected (like in underground installations or other horizontal sections with similar conditions over the entire length).

2.6. System's Design—Summary

Multiple aspects must be considered when designing the DFOS system for crack monitoring in concrete structures. After selecting the DFOS technique, appropriate sensors, installation methods and thermal compensation approach, the last thing is to elaborate algorithms and procedures for data post-processing, allowing the system's main objectives to be achieved. First, raw strain data should be validated, and eventual distortions should be removed. For the inexperienced user, cross-correlation failed points can be misinterpreted as cracks [7]. Then, thermal compensation should be applied to distinguish between the strains causing the mechanical stress in the element (ε_σ) and those related to the change in length (ε_L)—actual shortenings or elongations consist of the stress-free part (unconstrained thermal strains) and part generating the stress. For correct crack width estimation, strains related to the real change in length should be considered. For example, for ultimate limit state analysis, only strains generating the stress are essential.

As described in Section 2.1, there are a few methods for crack width estimation and choosing the appropriate one should be preceded by the studies on uncertainties and possibilities of applied software (in case of automatic analysis). It is recommended to define thresholds in reference to the relevant standard and the decision-making procedure after possibly exceeding these values. One of the important steps is the intuitive representation of the data, both in length and time domain. The convenient way is to present the strains in relation to the geometry of the structure in question, which makes physical interpretation easier.

The proposed step-by-step procedure of designing a DFOS system dedicated to crack monitoring in concrete structures, with general descriptions, is summarised in Table 2.

Table 2. Design procedure of DFOS system for crack monitoring in concrete structures.

Step	Selection of	Details Considered
1	DFOS technique (Rayleigh, Brillouin, hybrid)	Measurement parameters (spatial resolution, accuracy, strain resolution, distance range, acquisition time, number of channels, etc.).
2	Monolithic strain sensor	Geometrical and mechanical properties (diameter, elastic modulus, strength, maximum strain, external braid, bending radius).
3	Installation method	Surface or near-to-surface installation for existing structures (with analysis of the adhesive's properties). Embedding inside the new structures.
4	Thermal compensation	Raman technique, hybrid measurements, special DFOS temperature sensors, conventional spot temperature gauges (depending on distance range or expected temperature distributions and changes).
5	Post-processing algorithms	Data validation, thermal compensation algorithms, strain presentation, crack detection, width estimation, assessment of uncertainties, results visualisation (in length and time domain).

3. Laboratory Tests

3.1. The Concept and Preparation of the Specimens

The laboratory tests described hereafter were one of the first in the world, allowing for direct comparison between different types of DFOS tools embedded in the same reinforced concrete beams. A total of six beams with rectangle cross-section 250 × 350 mm were made, with tools installed in the lower (tension) and upper (compression) zones. The spatial visualisation of the single specimen is presented in Figure 8a. Two types of layered cables and two types of monolithic sensors commercially available on the market were selected (Figure 8b), which makes the results particularly important in terms of practical applications:

- C1: sensing cable with three plastic layers (Ø2.8 mm, E = unknown).
- M1: monolithic, reinforcing sensor (Ø5.0 mm, E = 50 GPa)
- M2: monolithic, flexible sensor (Ø3.0 mm, E = 3 GPa)
- C2: sensing cable with two plastic layers and steel insert (Ø3.2 mm, E = unknown).

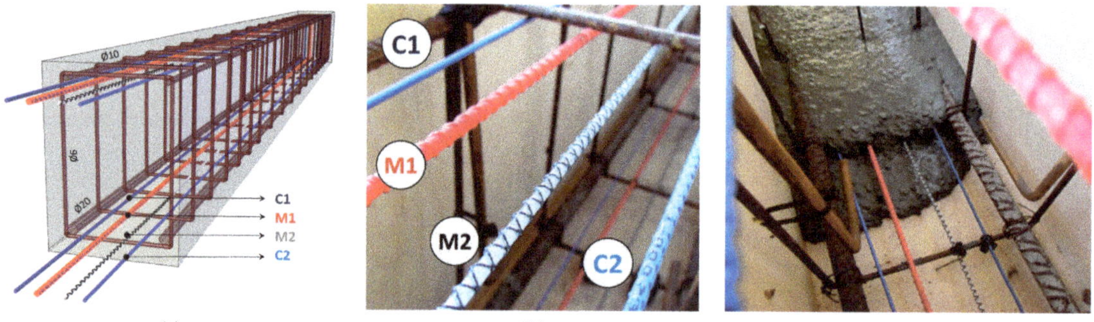

(a) (b) (c)

Figure 8. (**a**) Spatial visualisation of the reinforced concrete beam equipped with DFOS tools; (**b**) close-up to the selected tools; (**c**) the view of the beam during concreting.

To avoid damage or changing the position of the tools during concreting, self-compacting concrete was applied (Figure 8c). There were three beams with 10 mm main reinforcing bars (named "S") and three with 20 mm (named "L"). Thanks to that, different crack patterns (including the cracks' number, widths and spacing) were obtained.

Another implemented approach was to use a concrete mixture with a large amount of Portland cement and to leave the top surface of the beams after concreting without any curing (Figure 9a). The idea behind that procedure was intended to intensify the shrinkage process and thus, the formation of cracks within the upper part of the beams, what was finally succeeded (Figure 9b). Thanks to that, during mechanical tests, the effectiveness of DFOS tools was analysed in relation to the new cracks being opened in the tension zone (Figure 9c), simultaneously with the existing cracks being closed in the compression zone.

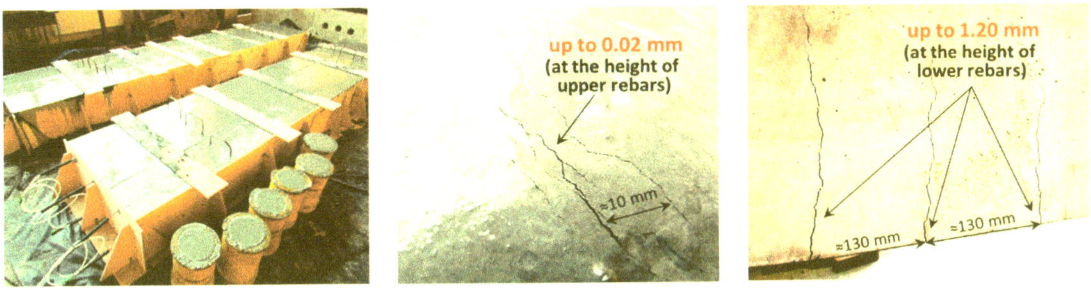

(a) (b) (c)

Figure 9. (**a**) The view of the beams after concreting; (**b**) thermal-shrinkage cracks within the upper part; (**c**) mechanical cracks within the lower part during bending tests.

3.2. Course of the Study and Measurements

The research was performed in the Building Materials and Structures Research Laboratory at Cracow University of Technology, Poland. Beams were loaded in four-point bending tests using the universal machine with the max. load cell capacity of 1000 kN, which is permanent equipment of the laboratory (Figure 10a). The machine's accuracy is ±0.5% of the reading down to 1/500 of the full scale, while the maximum force applied during research was equal to 210 kN. The static scheme of the simply supported beam has been provided by rolled supports, allowing for both rotations and horizontal movements.

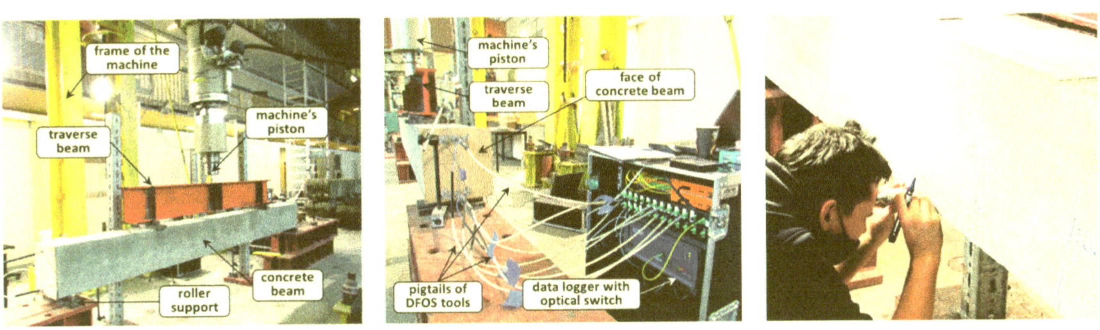

(a) (b) (c)

Figure 10. (**a**) Example beam during four-point bending test; (**b**) Rayleigh-based data logger with an optical switch connected to embedded DFOS tools; (**c**) reference readings.

The DFOS readings were done statically step-by-step (while increasing force) using a Rayleigh-based datalogger with an optical switch connected to all embedded tools inside the beam (Figure 10b). During selected load steps, the reference readings were performed by portable optical microscope with a resolution of 0.02 mm on the side surface of the beam (Figure 10c). An in-built scale above the focusing ring allowed for convenient readings. The results were systematically written on the beams using a marker. This method was applied for the crack widths up to 0.4 mm, while the plate crack-meters were used for the wider ones.

The test course has been established separately for S- and L-type beams due to their different structural response resulting from different reinforcement (4 times smaller cross-sectional area of main bars for beams type S). Due to the many beams, load steps and multiple cracks formed during research, in the following part of the article, the entire procedure and results will be presented and discussed for the example beam S3 for the sake of transparency. In the beginning, the experiment was steered (controlled) by the displacement of a hydraulic cylinder, which was stopped at selected force values. Figure 11 shows the measurement schedule graphically with designed load cycles. Blue dots represent the load steps when DFOS measurements were taken, while green arrows indicate the reference readings of crack widths. In S-type beams, after reaching the force of 55 kN (27th load step), it was impossible to keep it constant when the piston stopped. Thus, it was decided to move the piston 2 mm down at each load step, no matter the force value. It was because of the significant and uncontrolled crack development and reduction in stiffness.

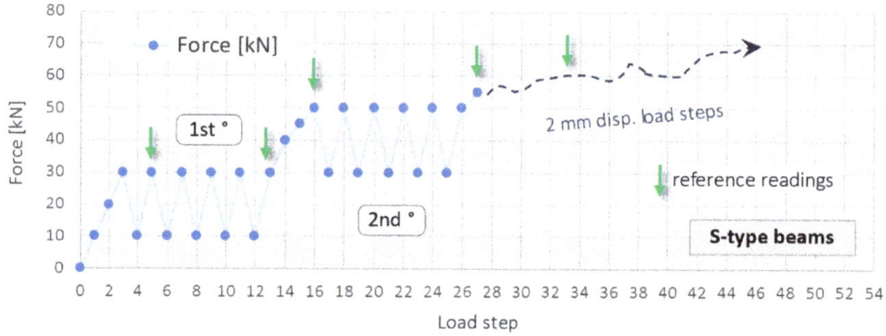

Figure 11. Designed research schedule for S-type beams.

All the beams were loaded up to the structural failure, which was achieved by losing the bending capacity in S-type beams and shearing capacity in L-type beams.

3.3. Compression—Example Results

A detailed comparative analysis of all DFOS tools in crack detection and width estimation will be the subject of another article [54]. This paper chose two fundamentally different tools for the abbreviated presentation: a monolithic sensor with reduced axial stiffness (M2) and a sensing cable (C2) with two plastic layers and a steel insert. That approach aims to underline the differences resulting from the tool's internal design. In the following part of the article, the results for the monolithic sensor will be represented by the red scale lines, and for the layered cable, by the blue scale lines.

The graphs presented in Figure 12 show the strain profiles obtained over the entire length of the beam (4 m) within the compression zone, where existing shrinkage cracks were closing under mechanical load. Presented data refer to the first twenty-seven measurement steps (up to the force of 55 kN), in which the beam behaviour was stable and the results, therefore, comparable. The zero reading was assumed just before the loading (the dead weight is not included). Negative strain values correspond to the increase in compression, while the positive ones to the tension. The closing cracks induced all the local negative strain peaks. Five of them were chosen for further analysis and marked on the plots. The maximum registered strain was approximately equal to only 550 με because the width changes of the existing cracks in the compression zone, in contrast to the opening cracks in tension, are limited by the initial crack width.

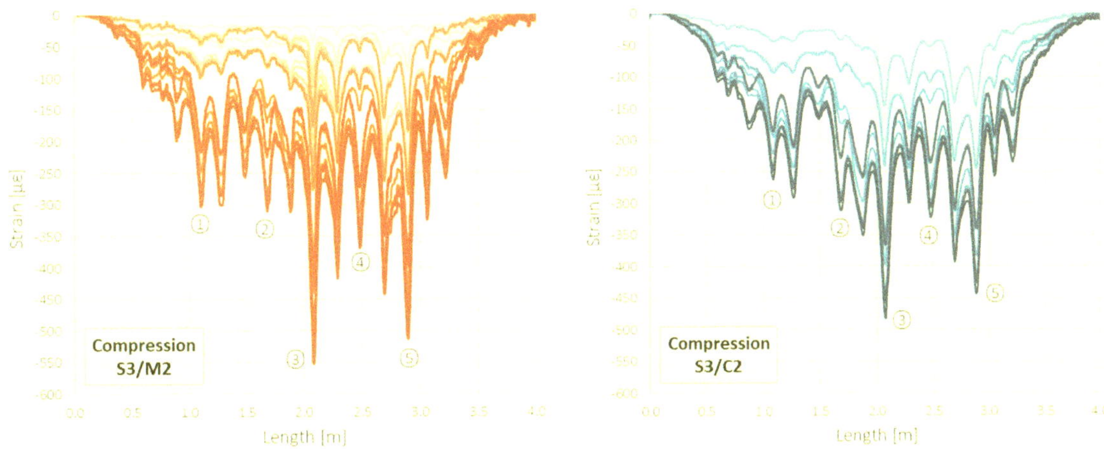

(**a**) (**b**)

Figure 12. Beam S3/compression zone: strain profiles registered by (**a**) monolithic sensor M2 and (**b**) layered sensing cable C2 with steel strengthening insert.

Despite the minor differences in the shape of strain profiles, which could be caused by the random course of the cracks and their varying size over the beam's width, it should be stated that analysed tools could detect all the existing cracks correctly. The locations of strain peaks were in line with the external observations. Moreover, the crack width changes were calculated by strain integration over the effective length, and the maximum difference between the tools during subsequent load steps did not exceed the value of 0.01 mm, while the mean difference was a few times smaller. Such a value, treated as an error in width estimation, allows for correct structural safety assessment and can be usually neglected from an engineering point of view.

3.4. Tension—Example Results

Although the results provided by all the analysed DFOS tools in the compression zone were satisfying, it should be underlined that the obtained strain range was minimal in comparison to the declared maximum strains (±40,000 με for sensor M2 and ±10,000 με for cable C2). Moreover, while analysing the behaviour of reinforced concrete structures, the tension zone is of the main interest, including cracks increasing their widths.

Figure 13 shows the strain profiles in the tension zone over the entire structural length during subsequent load steps, corresponding to those presented in Figure 12 for the compression zone. The scale of the horizontal and vertical axes is the same to facilitate visual comparison. For the load step no. 27 and the corresponding force of 55 kN, the maximum strain value within the crack-induced area registered by the monolithic sensor was equal to 16,780 με, while by the layered cable, it was almost four times smaller—4569 με. Given that both tools are theoretically dedicated to the same purpose and that the above maximum values are still well below the declared limits, such a difference in strain values is not acceptable.

The above observation about a significant decrease in the crack-induced strain peaks in the tension zone is valid for all the beams and all the cracks. Figure 14 presents the maximum strain values changes in subsequent load steps within the five selected cracks, which locations are marked in Figure 13.

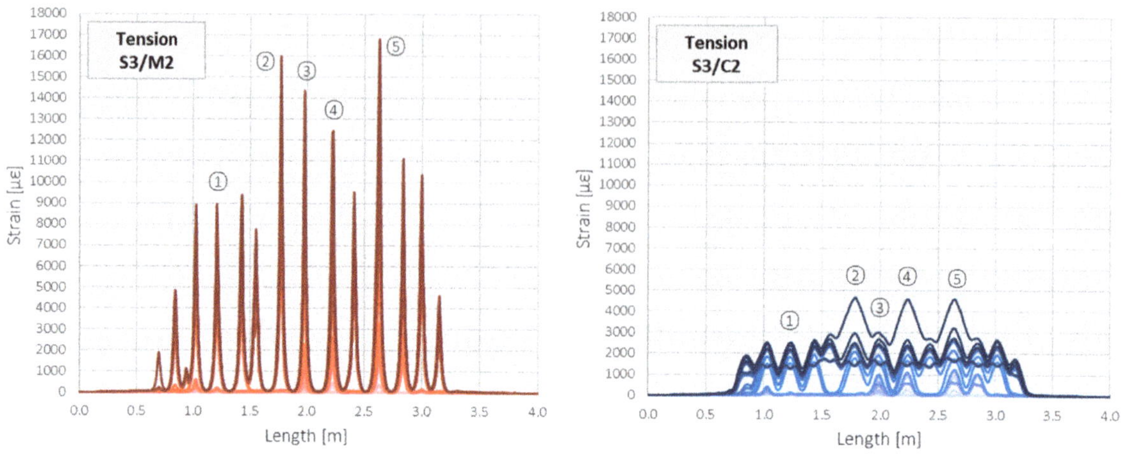

(a) (b)

Figure 13. Beam S3/tension zone: strain profiles registered by (**a**) monolithic sensor M2 and (**b**) layered sensing cable C2 with steel strengthening insert.

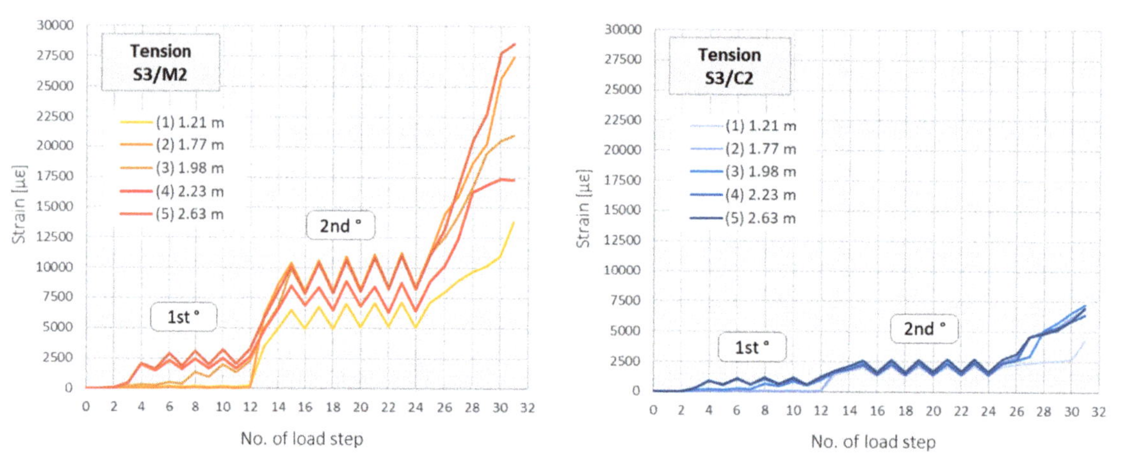

(a) (b)

Figure 14. Beam S3/tension zone: crack-induced strain peaks over subsequent load steps registered by (**a**) monolithic sensor M2 and (**b**) layered sensing cable C2 with steel strengthening insert.

Because the strain data gained through distributed sensing are expressed in two domains: structural length and time (load step), they could be conveniently presented in the form of spatial visualisations. The example plots corresponding to data in Figure 13 for monolithic sensor M2 and layered cable C2 are shown in Figures 15a and 15b, respectively.

The locations of all the cracks were identified unequivocally by a monolithic sensor with an accuracy equal to applied spatial resolution (10 mm), which was confirmed by reference readings using a metric tape on the side surface of the beam and photo-camera. On the other hand, the indications from the layered cable are not conclusive, especially given the negligible increments during cyclic loading. In real projects, very often, the zero reading is taken after the formation of the cracks and width changes are usually much smaller than in laboratory conditions (under the applied force). Thus, the risk of data misinterpretation is too high. Figure 16 summarises on the same plot the strain profiles measured by monolithic sensor (red line) and layered cable (blue line) in load step no. 26 (50 kN), in reference to the documented cracks' locations on a side surface.

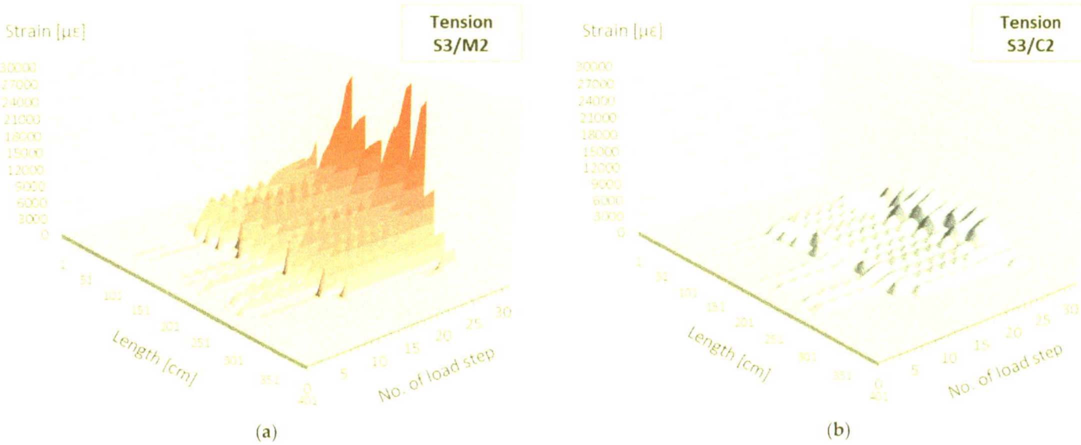

Figure 15. Beam S3/tension zone: strain profiles in length and load step domain registered by (**a**) monolithic sensor M2 and (**b**) layered sensing cable C2 with steel strengthening insert.

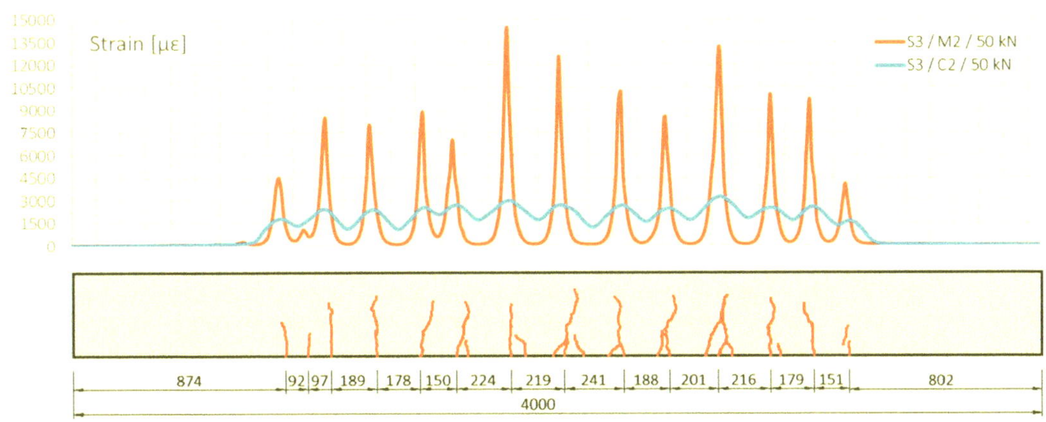

Figure 16. Strain profiles over length measured by monolithic sensor M2 and layered cable C2 (beam S3, load step no. 26, force = 50 kN).

The next step of the analysis was focused on the crack width estimation, which was done by strain integration over the effective length. This length was assumed to be equal to half of the mean spacing between the multiple cracks (the conservative assumption about overlapping the effective lengths). The local crack-induced strain peaks indicated the centre of this length. The crack widths calculated for the example load step no. 26 are summarised in Table 3, showing that the cable's results underestimate the sensor's results by almost 0.15 mm (≈30%). In addition, the results from the external microscope were provided to give the general overlook on the actual crack widths. However, these values cannot be compared directly, as the readings were taken in different place (on the side surface).

Table 3. Calculated crack widths based on strain profiles measured by monolithic sensor M2 and layered cable C2 (beam S3, load step no. 26, force = 50 kN).

Crack	Width (mm)			Diff. (mm)	Diff. (%)
	M2	C2	Ext. ref.	(M2 − C2)	(M2 − C2)/M2 × 100%
①	0.329	0.267	0.30	0.062	19.0
②	0.548	0.351	0.60	0.197	35.9
③	0.451	0.318	0.40	0.133	29.6
④	0.409	0.309	0.35	0.100	24.5
⑤	0.593	0.382	0.45	0.211	35.6
mean	**0.466**	**0.325**	**0.42**	**0.141**	**28.9**

3.5. Findings

Considering the above specific example, but also global statistics presented in [7], the most sensitive and reliable tool for crack detection and analysis is a monolithic sensor with reduced axial stiffness. Unambiguous measurements are further supported by theory: the internal design of the cross-section includes the full integration of measurement fibre with composite core during its pultrusion, eliminating the slippage phenomena. Moreover, minimal axial stiffness means that the sensor is no longer the reinforcement for the concrete and does not influence its structural response.

Another advantage is the highest strain range. The maximum value of tensile strain registered in the last load step, just before the structural failure, was equal to 2.86% (28,575 με), while the strain range declared by the producer is ±40,000 με. The corresponding crack width exceeded 1.3 mm at that time. The ability to measure extensive cracks without fear of damaging the sensor is an essential aspect while monitoring cracked reinforced concrete structures.

Although the layered cable with steel insert inside provided reasonable results in the compression zone in the very limited strain range (<550 με), its performance was unacceptable within the tension zone. The registered crack-induced strain peaks were almost four times lower compared to the monolithic sensor, and the shape of strain profiles did not allow for certain crack detection. The maximum declared value of correct strain measurements (equal to 1%) is not feasible. It is worth noticing that the steel layer inside the cable yields itself at the strain level of approx. 0.2%. The main reason for unreliable readings, though, is disturbed strain transfer from the concrete to the measuring fibre caused by the internal slippage between the intermediate layers in the cable.

Considering the above findings, the layered tools are not recommended for high-spatial-resolution DFOS measurements of reinforced concrete structures when crack detection and analysis are of the main interest.

4. Field Proofs

Field proofs are the last and most important step in verifying the effectiveness of DFOS-based monolithic sensors. This section describes three examples of actual concrete structures equipped with such sensors, giving the example results related to crack analysis. The first is a new bridge with sensors embedded inside the concrete slab. The second refers to the existing 10-years-old bridge, where the near-to-surface installation method was applied. Finally, the very old concrete sewer collector is described as an example of an ageing infrastructure requiring the appropriate control and maintenance.

4.1. Railway Bridge near Frankfurt, Germany

Three monolithic sensors were used to monitor the performance of the concrete slab in a railway bridge north of Frankfurt am Main, Germany. Since the decision to create the system was made before construction was completed, it was possible to embed the sensors inside the slab. The installation consisted of stabilising the sensors in their designed positions by tying them to the existing reinforcement along the main bars (Figure 17).

Figure 17. Installation of the monolithic sensor by tying to the existing reinforcement.

Sensors with appropriate lengths were delivered on site in coils, and the ease of installation allowed it to be completed within half a day. As the sensors were installed before concreting (Figure 18a), the structural performance analysis was possible from an actual zero state, i.e., zero readings were taken before the first cracks were formed. The blue lines on the strain plot in Figure 18b correspond to the measurements performed a few hours after concreting. There are still no characteristic crack-induced strain peaks. The data are presented over the 10 m long section.

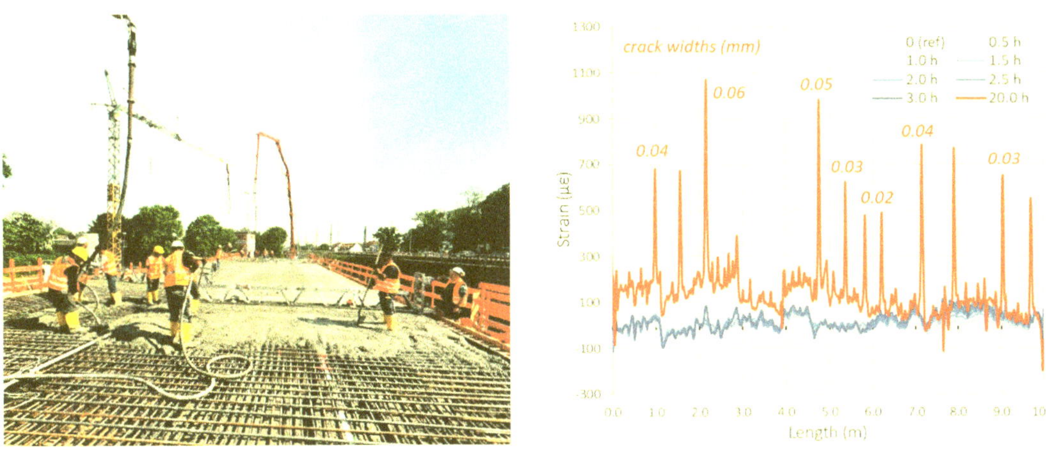

(a) (b)

Figure 18. (**a**) The view of the slab during concreting; (**b**) example strain profiles with crack analysis 20 h after concreting.

However, the measurement performed the next day, 20 h after concreting is completed, clearly indicates the presence of the first ten cracks (red line on the plot). Their spacings and locations depend on the geometry of the structure and boundary conditions. The estimated crack widths did not exceed the value of 0.06 mm.

Cracking in concrete structures is a normal process taking place from the first days, and it usually cannot be avoided. The structural engineer's main task is to monitor and control the crack positions and widths over time. Distributed fibre optic sensing allows all the cracks to be detected and analysed, even inside the structural elements, where visual inspection is impossible. However, a precondition for the system efficiency is the use of high-definition monolithic sensors, high-spatial-resolution DFOS technique and appropriate data processing algorithms.

The DFOS-based system in the Frankfurt bridge was used to detect the first cracks formed during early-age concrete behaviour (thermal shrinkage strains) but also to monitor structural response according to the planned schedule, including long-term static measurements as well as dynamic readings during the train passages.

4.2. Largest Concrete Cable-Stayed Bridge in Poland

Although embedding sensors inside the components of new structures is most beneficial from the quality and quantity of information obtained, it is also necessary to be aware of the existing, ageing infrastructure [55]. The following example is related to one of the largest concrete bridges in Poland—Rędziński Bridge [56,57] (Figure 19a). It was opened to traffic on August 31, 2011, and is the most significant bridge along the Wrocław motorway ring road. It is a four-span cable-stayed bridge over the Odra River with spans of 50 m + 2 × 256 m + 50 m long. The two separated concrete box girders are suspended to a single H-shaped pylon with a height of 122 m, which makes it the tallest in Poland. Its shape allows connecting 160 stays arranged in four planes.

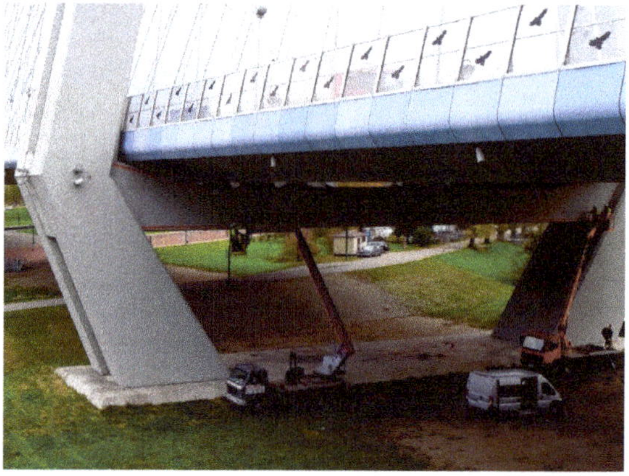

(a) (b)

Figure 19. (**a**) Rędziński Bridge in Wrocław, Poland—general view (photo: W. Kluczewski); (**b**) the close-up of the lower crossbeam connecting two inclined legs of the pylon.

Due to the limited width of the route below the lower crossbeam, the pylon's legs are inclined to minimise the size of the foundation (massive concrete slab with a base of 67.4 × 28.0 m and thickness from 2.5 to 6.5 m, placed on 160 reinforced concrete bored piles with a diameter of 1.5 m and 18 m in length). The pylon's legs have a rectangular cross-section with variable dimensions: 6.0 × 7.0 m at the foundation level, 4.0 × 4.0 m near the upper crossbeam and 4.0 × 6.0 at the top.

Although the decks are suspended by the set of cables, they are also (to a lesser extent) supported by the prestressed lower crossbeam of the pylon. However, the main task of this prestressed element, with the midspan rectangle cross-section of 2.5 m (height) and 4.0 m (width), is to connect two inclined legs of the pylon, which, due to their geometry, tend to move outwards (Figure 19b).

After ten years of the bridge's operation, multiple cracks (with widths up to 0.1 mm) were observed over the length of this crossbeam, which is a natural consequence of its static scheme and loading. However, due to the unfavourable environment above the river, controlling cracks widths changes over time (according to the standard's requirements) is very important.

The Rędziński Bridge was equipped with DFOS monolithic sensors in 2020, joining the honourable group of several bridges in Poland with this innovative monitoring solution. Four sensors were glued inside the pre-made near-to-surface grooves (Figure 20a) located in each corner of the lower crossbeam of the pylon (Figure 20b) over its entire length (Figure 20c). This installation approach is the best in terms of bonding properties and mechanical protection, but also in terms of reducing the direct impact of the sunlight and ensuring the appropriate aesthetics in the facility being operated.

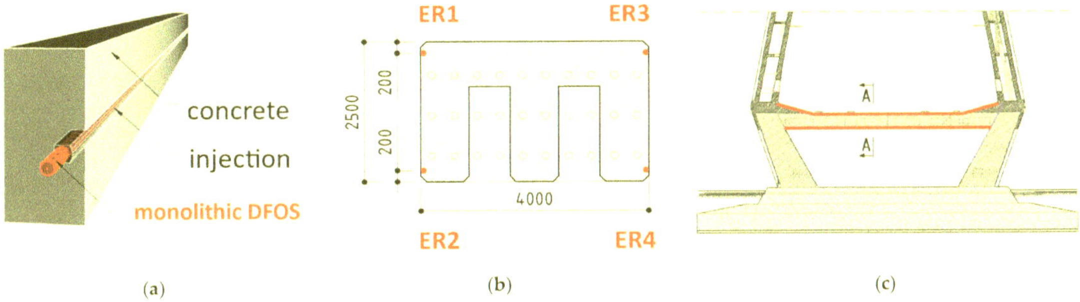

Figure 20. (**a**) Installation in near-to-surface grooves; (**b**) cross-section of the beam with locations of the sensors; (**c**) side view of the crossbeam with sensors marked over its length.

The cyclic measurement sessions are now performed according to a planned schedule, with reference taken shortly after the installation. In a single reading, the strain data from 7400 locations are recorded, providing direct crack detection and estimation of their width changes (both while closing and opening).

Next to the long-term analysis, the short-term readings (at an interval of a few seconds) were also done under the random traffic loads. The width changes identified then did not exceed thousandths of a millimetre. Example results from the lower and upper sensor are presented in Figure 21. Strain profiles with crack width analysis are shown on a few metre-long sections, selected in reference to the structure's geometry, to keep appropriate clarity.

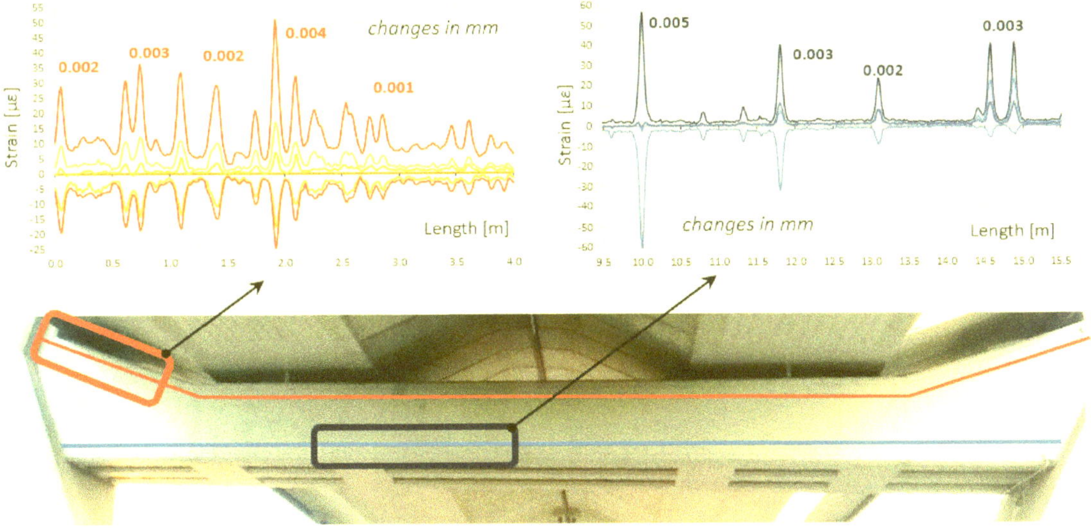

Figure 21. Example strain profiles over selected sections of the lower and upper monolithic sensor with crack width analysis.

Despite such small values of crack widths changes, the system was able for a reliable diagnosis thanks to the appropriate strain transfer from the concrete to the optical fibre fully integrated inside the monolithic core of the high-sensitivity sensors.

Four spot temperature gauges were installed within the crossbeam to compensate for the long-term thermal influences. It is also worth noticing that the applied strain sensors can be used simultaneously with other dataloggers (based on various optical phenomena) to measure different physical quantities like distributed temperatures or vibrations. Finally, thanks to the known arrangement of the sensors within the beam's cross-section, it is possible to calculate horizontal and vertical displacements (deflections, shape changes), as well as axial shortenings or elongations.

In 2022, the force adjustment within eight shortest stays was performed, relieving the pylon's lower crossbeam and bending it up slightly. Distributed fibre optic monolithic sensors were used for the advanced and simultaneous analysis of the beam's strains, cracks and displacements.

4.3. Renovated Sewage Collector in Poland

The last example refers to the Burakowski sewage collector, an essential element of the sewage system in Warsaw, Poland [58]. It was built in the 1960s using the mining method in a concrete casing, with segments of lengths from 2 to 3 m and an internal diameter of approximately 3 m, buried under the ground surface at a depth of 4.5 to 7.5 m. Its designed capacity is about 12 m^3/s, while the recorded capacity may be higher during heavy rainfalls.

The unfavourable external conditions in the close vicinity of the collector include the edge of the high bank of the Vistula River, as well as subway and tram lines and stations. In addition, several tall and deep-founded buildings have been built in this area over the last 15 years. In 2015, the construction of a new "Burakowski Bis" collector was completed passing along the existing one, which resulted in environmental changes (including ground-water conditions) compared to the times when it was constructed.

After analysing the collector's technical condition in 2019, a decision was made to retrofit a 4.8 km long section using non-circular 3 m-long GRP module panels and relining technology. The space between the panels placed inside and the existing inner walls of the concrete collector was filled with a cement injection. Thus, the possibility of a visual inspection of the collector was lost.

This was one of the reasons for equipping the old concrete part with a DFOS-based system. Three monolithic sensors were installed longitudinal over the length of the 146 m-long section between the maintenance holes. One top sensor and two side sensors (Figure 22a,b) in the known configuration were mounted in near-to-surface grooves filled with mortar injection (Figure 22c). The approach, like this applied in the above bridge example, aimed for simultaneous analysis of strains, cracks and vertical displacements during the retrofitting process, but also after putting the collector to the normal operation again.

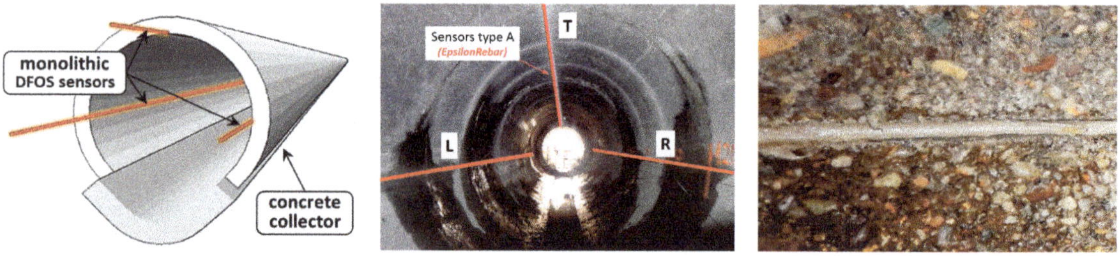

(a) (b) (c)

Figure 22. (**a**) Design of the collector with monolithic sensors; (**b**) the view of the collector with marked locations of monolithic sensors; (**c**) near-to-surface groove filled with injection.

What is more, one monolithic strain sensor was installed along the circumference of the collector's cross-section (Figure 23a), creating four loops located in the most safety-critical area (Figure 23b), where a few cracks were documented through visual inspection before the installation of the sensors (Figure 23c).

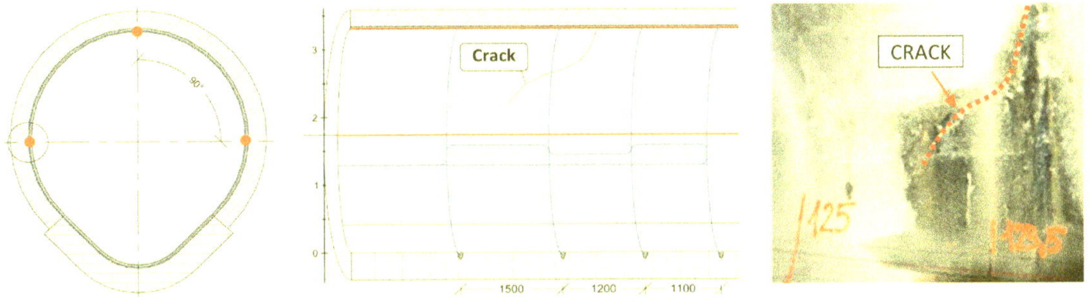

(**a**) (**b**) (**c**)

Figure 23. (**a**) Cross-section of the collector [58]; (**b**) location of the circumference monolithic sensor [58]; (**c**) the view of the crack documented before installation.

Strain measurements were taken during the retrofitting and strengthening works, including the following stages:

- before GRP modules were provided inside the collector (zero reading),
- after the GRP modules were placed (before grout injection),
- after grout (mortar) injection,
- and finally, after the collector was put back into service and filled with sewage.

DFOS readings allowed observing the collector's structural response under gradual load changes.

The dead weight of the new panels and mortar injection, as well as the dead weight of the launched sewage flow, caused the original structure of the concrete collector casing was reflected in DFOS readings. Increasing mechanical load during the renovation work revealed cracks caused by technological breaks, and changes in their widths were carefully analysed during this process. Figure 24 shows strain profiles registered by the top sensor over a 10 m-long section in the subsequent load steps. Crack-induced strain peaks are negative, which means they correspond to the compression (decreasing the crack width). The maximum changes in crack widths during retrofitting did not exceed the value of 0.02 mm. The crack spacing results from the original construction technology of the collector, where concrete casing sections had the length from 2.2 to 2.9 m.

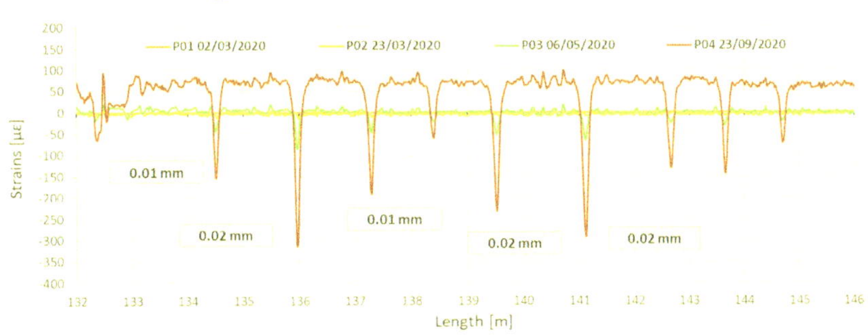

Figure 24. Strain profiles measured by the monolithic sensor (T) along the test section showing the primary discontinuities in the concrete collector closing during strengthening process. Example section from 132 to 146 m. Measurement P01—after installation of GRP panels, P02—during grout injection, P03—after completed grout injection, P04—after filling the collector with sewage [58].

The application of the DFOS system based on monolithic strain sensors allowed for detection of all discontinuities in the collector structure, most of which were undetectable by the naked eye during visual inspection. The system is now used to control the crack widths over time during periodical measurement sessions, supporting the expert in assessing the technical condition and structural safety of the collector.

5. Conclusions

There has been a noticeable increase in distributed fibre optic sensing deployment in recent years, not only in laboratory research but, above all, in real civil engineering and geotechnical projects. The main advantage is the possibility of continuously measuring selected physical quantities over length to fulfil one of the fundamental goals for structural health monitoring: direct damage detection and estimation of its size. However, to utilise the advantages of DFOS approach, a set of requirements must be understood and implemented to obtain high-quality and reliable data.

The versatility of this technique lies in the ability to customise system elements (including the type of sensor, datalogger parameters, light scattering, installation methods, software and post-processing algorithms) according to the individual needs of a given project. In reference to reinforced concrete structures, it is essential to be aware that they operate in a cracked state and that a reliable assessment of local discontinuities (cracks) of a very small size (usually < 1 mm) is one of the primary challenges for the DFOS system.

The lessons learned while monitoring cracked concrete structures were summarized in the article to highlight what features of the DFOS tools are the most favourable from the crack analysis point of view. During the selection, not only their internal design should be considered, but also the strain range and resolution, axial stiffness, cross-section area, elasticity modulus, bending radius, material and its resistance to corrosion, durability, no need for pre-tensioning, the tendency to wave and more.

Theoretical considerations were supported by laboratory tests on full-size components. The novelty of the research lies in direct comparison between specific monolithic sensors and layered cables commercially available on the market, and often used for the same purposes. The obtained results have, thus, practical meaning, showing significant differences in the performance of different DFOS tools.

The findings include that layer-free sensors are adequate solution for crack analysis. The measuring fibre does not slip inside the cross-section, as it does in layered cables, providing that the structural strains are transferred in an as undisturbed way as possible. Based on the performed measurements, a reliable interpretation of the deformation state of the structure was possible.

The DFOS-based monitoring of cracked concrete structures is a challenging process. In addition to the monolithic sensors, which are one of the key components of the entire system, the appropriate high-spatial-resolution technique should be selected (preferably, Rayleigh-based), as well as installation methods which directly influence the physical interpretation of the obtained data. This article is a part of a discussion about the best DFOS configuration, which could be recommended for application in cracked concrete, depending on whether we are dealing with new or existing structures. A set of good practices based on lessons learned while applying monolithic DFOS strain sensors were presented in this article and discussed through laboratory and field proofs.

Author Contributions: Conceptualization, R.S. and T.H.; methodology, Ł.B. and R.S.; software, T.H.; validation, R.S. and Ł.B.; formal analysis, T.H. and Ł.B; investigation, T.H. and K.Z.; data curation, T.H.; writing—original draft preparation, T.H.; writing—review and editing, K.Z.; visualization, K.Z.; supervision, R.S. and Ł.B.; project administration, R.S.; funding acquisition, R.S. and Ł.B. All authors have read and agreed to the published version of the manuscript.

Funding: This research received no external funding.

Institutional Review Board Statement: Not applicable.

Informed Consent Statement: Not applicable.

Data Availability Statement: The data presented in this study are available on request from the corresponding author.

Acknowledgments: The authors would like to show gratitude to the SHM System company (Kra-ków, Poland, www.shmsystem.pl (accessed on 7 November 2022), www.nerve-sensors.com (accessed on 7 November 2022)), working on the "Innovative fibre optic sensor for measuring strain and temperature" project (POIR.01.01.01-00-1154/19). It was funded by the National Centre for Research and Development within the Intelligent Development Operational Program 2014-2020. The authors would also like to thank the Marx Krontal Partner company (Hannover, Germany, www.marxkrontal.com (accessed on 7 November 2022)), with which they effectively collaborated to implement the DFOS-based monitoring system for the bridge near Frankfurt, described as an example in this article.

Conflicts of Interest: The authors declare no conflict of interest.

References

1. *EN 1990:2002;* Eurocode: Basis of Structural Design. European Committee for Standardisation: Brussels, Belgium, 2002.
2. Faber, M.H. *Statistics and Probability Theory in Pusuit of Engineering Decision Support;* Springer: Berlin/Heidelberg, Germany, 2012; ISBN 978-94-007-4055-6.
3. Frangopol, D.; Tsompanakis, Y. *Maintenance and Safety of Aging Infrastructure;* CRC Press: Boca Raton, FL, USA, 2014; ISBN 978-04-292-2058-6.
4. Balageas, D.; Fritzen, C.P.; Güemes, A. *Structural Health Monitoring;* Wiley-ISTE: New York, NY, USA, 2006; ISBN 978-1-905-20901-9.
5. Xu, Y.L.; Xia, Y. *Structural Health Monitoring of Long-Span Suspension Bridges;* Spon Press: London, UK, 2012; ISBN 9781138075634.
6. Kim, B.; Min, C.; Kim, H.; Cho, S.; Oh, J.; Ha, S.-H.; Yi, J.-H. Structural Health Monitoring with Sensor Data and Cosine Similarity for Multi-Damages. *Sensors* **2019**, *19*, 3047. [CrossRef] [PubMed]
7. Howiacki, T. Analysis of Cracks in Concrete Structures with the Use of Distributed Optical Fibre Measurements. Ph.D. Thesis, Faculty of Civil Engineering, Cracow University of Technology, Kraków, Poland, 2022.
8. Buda-Ożóg, L.; Zięba, J.; Sieńkowska, K.; Nykiel, D.; Zuziak, K.; Sieńko, R.; Bednarski, Ł. Distributed fibre optic sensing: Reinforcement yielding strains and crack detection in concrete slab during column failure simulation. *Measurement* **2022**, *195*, 111192. [CrossRef]
9. Di Sante, R. Fibre Optic Sensors for Structural Health Monitoring of Aircraft Composite Structures: Recent Advances and Applications. *Sensors* **2015**, *15*, 18666–18713. [CrossRef]
10. Bednarski, Ł.; Sieńko, R.; Grygierek, M.; Howiacki, T. New Distributed Fibre Optic 3DSensor with Thermal Self-Compensation System: Design, Research and Field Proof Application inside Geotechnical Structure. *Sensors* **2021**, *21*, 5089. [CrossRef]
11. Siwowski, T.; Rajchel, M.; Howiacki, T.; Sieńko, R.; Bednarski, Ł. Distributed fibre optic sensors in FRP composite bridge monitoring: Validation through proof load tests. *Eng. Struct.* **2021**, *246*, 113057. [CrossRef]
12. Amanzadehab, M.; Aminossadatia, S.M.; Kizil, M.S.; Rakićb, A.D. Recent developments in fibre optic shape sensing. *Measurement* **2018**, *128*, 119–137. [CrossRef]
13. Liu, Y.; Li, X.; Li, H.; Fan, X. Global Temperature Sensing for an Operating Transformer Based on Raman Scattering. *Sensors* **2020**, *20*, 4903. [CrossRef]
14. Hill, D. Distributed Acoustic Sensing (DAS): Theory and Applications. In *Frontiers in Optics;* Optica Publishing Group: San Jose, CA, USA, 2015. [CrossRef]
15. Sieńko, R.; Bednarski, Ł.; Howiacki, T. Smart Composite Rebars based on DFOS Technology as Nervous System of Hybrid Footbridge Deck: A Case Study. In *European Workshop on Structural Health Monitoring. EWSHM 2020. Lecture Notes in Civil Engineering;* Rizzo, P., Milazzo, A., Eds.; Springer: Berlin/Heidelberg, Germany, 2021; Volume 128. [CrossRef]
16. NERVE Composite DFOS Sensors. Available online: http://nerve-sensors.com (accessed on 22 September 2022).
17. Pise, N.; Meshram, T.; Doijad, Y.; Gathe, R.; Bobade, A.; Kutemate, A.; Patil, S.; Raut, P.; Gudadhe, A. A Brief Study on Causes of Cracks, Prevention and Pattern of Cracks on Concrete. *Int. J. Sci. Res. Sci. Eng. Technol.* **2021**, *8*, 439–443. [CrossRef]
18. Jędrzejewska, A.; Kanavaris, F.; Zych, M.; Schlicke, D.; Azenha, M. Experiences on early age cracking of wall-on-slab concrete structures. *Structures* **2020**, *27*, 2520–2549. [CrossRef]
19. Zych, M. Degree of external restraint of wall segments in semi-massive reinforced concrete tanks: Part II: Rectangular and cylindrical segments. *Struct. Concr.* **2018**, *19*, 3. [CrossRef]
20. Kulpa, M.; Howiacki, T.; Wiater, A.; Siwowski, T.; Sieńko, R. Strain and displacement measurement based on distributed fibre optic sensing (DFOS) system integrated with FRP composite sandwich panel. *Measurement* **2021**, *175*, 109099. [CrossRef]
21. Barrias, A.; Casas, J.R.; Villalba, S. Application study of embedded Rayleigh based Distributed Optical Fiber Sensors in concrete beams. *Procedia Eng.* **2017**, *199*, 2014–2019. [CrossRef]
22. Imai, M.; Igarashi, Y.; Shibata, M.; Miura, S. Experimental study on strain and deformation monitoring of asphalt structures using embedded fiber optic sensor. *J. Civ. Struct. Health Monit.* **2014**, *4*, 209–220. [CrossRef]

23. Coscetta, A.; Catalano, E.; Cerri, E.; Oliveira, R.; Bilro, L.; Zeni, L.; Cennamo, N.; Minardo, A. Distributed Static and Dynamic Strain Measurements in Polymer Optical Fibers by Rayleigh Scattering. *Sensors* **2021**, *21*, 5049. [CrossRef] [PubMed]
24. Sang, A.K.; Froggatt, M.E.; Kreger, S.T.; Gifford, D.K. Millimeter resolution distributed dynamic strain measurements using optical frequency domain reflectometry. *Proc. SPIE* **2011**, *7753*, 432–435. [CrossRef]
25. Luna—Optical Distributed Sensor Interrogator. Available online: https://lunainc.com/product/odisi-6000-series (accessed on 22 September 2022).
26. Bao, X.; Chen, L. Recent Progress in Brillouin Scattering Based Fiber Sensors. *Sensors* **2011**, *11*, 4152–4187. [CrossRef]
27. Hong, C.Y.; Zhang, Y.F.; Li, G.W.; Zhang, M.X.; Liu, Z.X. Recent progress of using Brillouin distributed fiber sensors for geotechnical health monitoring. *Sens. Actuators A Phys.* **2017**, *258*, 131–145. [CrossRef]
28. Wang, W.; Chang, J.; Lv, G.; Wang, Z.; Liu, Z.; Luo, S.; Jiang, S.; Liu, X.; Liu, X.; Liu, Y. Wavelength dispersion analysis on fiber-optic Raman distributed temperature sensor system. *Photonic Sens.* **2013**, *3*, 256–261. [CrossRef]
29. Muanenda, Y.; Oton, C.J.; Di Pasquale, F. Application of Raman and Brillouin Scattering Phenomena in Distributed Optical Fiber Sensing. *Front. Phys.* **2019**, *7*, 155. [CrossRef]
30. Yan, B.; Li, J.; Zhang, M.; Zhang, J.; Qiao, L.; Wang, T. Raman Distributed Temperature Sensor with Optical Dynamic Difference Compensation and Visual Localization Technology for Tunnel Fire Detection. *Sensors* **2019**, *19*, 2320. [CrossRef]
31. Kishida, K.; Yamauchi, Y.; Guzik, A. Study of Optical Fibers Strain-Temperature Sensitivities Using Hybrid Brillouin-Rayleigh System. *Photonic Sens.* **2014**, *4*, 1–11. [CrossRef]
32. Falcetelli, F.; Rossi, L.; Di Sante, R.; Bolognini, G. Strain Transfer in Surface-Bonded Optical Fiber Sensors. *Sensors* **2020**, *20*, 3100. [CrossRef] [PubMed]
33. Li, D.; Ren, L.; Li, H. Mechanical Property and Strain Transferring Mechanism in Optical Fiber Sensors. In *Fiber Optic Sensors*; Yasin, M., Harun, S.W., Arof, H., Eds.; InTechOpen: London, UK, 2012. [CrossRef]
34. Bremer, K.; Alwis, L.S.M.; Zheng, Y.; Weigand, F.; Kuhne, M.; Helbig, R.; Roth, B. Durability of Functionalized Carbon Structures with Optical Fiber Sensors in a Highly Alkaline Concrete Environment. *Appl. Sci.* **2019**, *9*, 2476. [CrossRef]
35. Barrias, A.; Casas, J.R.; Villalba, S. A Review of Distributed Optical Fiber Sensors for Civil Engineering Applications. *Sensors* **2016**, *16*, 748. [CrossRef] [PubMed]
36. Schenato, L. A Review of Distributed Fibre Optic Sensors for Geo-Hydrological Applications. *Appl. Sci.* **2017**, *7*, 896. [CrossRef]
37. Soga, K.; Luo, L. Distributed fiber optics sensors for civil engineering infrastructure sensing. *J. Struct. Integr. Maint.* **2018**, *3*, 1–21. [CrossRef]
38. Weisbrich, M.; Holschemacher, K. Comparison between different fiber coatings and adhesives on steel surfaces for distributed optical strain measurements based on Rayleigh backscattering. *J. Sens. Sens. Syst.* **2018**, *7*, 601–608. [CrossRef]
39. Sieńko, R.; Zych, M.; Bednarski, Ł.; Howiacki, T. Strain and crack analysis within concrete members using distributed fibre optic sensors. *Struct. Health Monit.* **2018**, *18*, 1510–1526. [CrossRef]
40. Bado, M.S.; Casas, J.R.; Kaklauskas, G. Distributed Sensing (DOFS) in Reinforced Concrete members for reinforcement strain monitoring, crack detection and bond-slip calculation. *Eng. Struct.* **2021**, *226*, 111385. [CrossRef]
41. Li, Q.; Li, G.; Wang, G. Effect of the plastic coating on strain measurement of concrete by fiber optic sensor. *Measurement* **2003**, *34*, 215–227. [CrossRef]
42. Kania, J.; Sørensen, K.K. A Static Pile Load Test on a Bored Pile Instrumented with Distributed Fibre Optic Sensors. In Proceedings of the International Symposium on Field Measurements in Geomechanics, Rio de Janeiro, Brazil, 16–20 July 2018.
43. Barkanov, E.; Akishin, P.; Namsone-Sile, E. Effectiveness and Productivity Improvement of Conventional Pultrusion Processes. *Polymers* **2022**, *14*, 841. [CrossRef] [PubMed]
44. Imai, M.; Feng, M. Sensing optical fiber installation study for crack identification using a stimulated Brillouin-based strain sensor. *Struct. Health Monit.* **2012**, *11*, 501–509. [CrossRef]
45. Barrias, A.; Casas, J.R.; Villalba, S. Embedded Distributed Optical Fiber Sensors in Reinforced Concrete Structures—A Case Study. *Sensors* **2018**, *18*, 980. [CrossRef] [PubMed]
46. Liu, T.; Huang, H.; Yang, Y. Crack Detection of Reinforced Concrete Member Using Rayleigh-Based Distributed Optic Fiber Strain Sensing System. *Adv. Civ. Eng.* **2020**, *2020*, 8312487. [CrossRef]
47. Fischer, O.; Thoma, S.; Crepaz, S. Distributed fiber optic sensing for crack detection in concrete structures. *Civ. Eng. Des.* **2019**, *1*, 97–105. [CrossRef]
48. Alj, I.; Quiertant, M.; Khadour, A.; Grando, Q.; Terrade, B.; Renaud, J.-C.; Benzarti, K. Experimental and Numerical Investigation on the Strain Response of Distributed Optical Fiber Sensors Bonded to Concrete: Influence of the Adhesive Stiffness on Crack Monitoring Performance. *Sensors* **2020**, *20*, 5144. [CrossRef]
49. Kwak, H.-G.; Song, J.-Y. Cracking analysis of RC members using polynomial strain distribution function. *Eng. Struct.* **2002**, *24*, 455–468. [CrossRef]
50. Rodríguez, G.; Casas, J.R.; Villaba, S. Cracking assessment in concrete structures by distributed optical fiber. *Smart Mater. Struct.* **2015**, *24*, 035005. [CrossRef]
51. Tang, Y.; Wang, Z.; Song, M. Self-sensing and strengthening effects of reinforced concrete structures with near-surfaced mounted smart basalt fibre-reinforced polymer bars. *Adv. Mech. Eng.* **2016**, *8*, 10. [CrossRef]
52. Bastianini, F.; Di Sante, R.; Falcetelli, F.; Marini, D.; Bolognini, G. Optical Fiber Sensing Cables for Brillouin-Based Distributed Measurements. *Sensors* **2019**, *19*, 5172. [CrossRef]

53. Sieńko, R.; Bednarski, Ł.; Howiacki, T. Smart prestressed concrete girders with integrated composite distributed fibre optic sensors (DFOS): Monitoring through all construction stages. In Proceedings of the 10th International Conference on Structural Health Monitoring of Intelligent Infrastructure, Porto, Portugal, 30 June–2 July 2021.
54. Sieńko, R.; Bednarski, Ł.; Howiacki, T. Crack shape coefficient: Comparison between DFOS tools used for crack monitoring in concrete. *Sensors*, 2022; *abstract accepted*.
55. Barrias, A.; Rodriguez, G.; Casa, J.R. Application of distributed optical fiber sensors for the health monitoring of two real structures in Barcelona. *Stucture Infrastruct. Eng.* **2018**, *14*, 967–985. [CrossRef]
56. Biliszczuk, J.; Barcik, W.; Onysyk, J.; Toczkiewicz, R.; Tukendorf, A.; Tukendorf, K. Rędziński Bridge in Wrocław—the largest concrete cable-stayed bridge in Poland. *Struct. Eng. Int.* **2014**, *24*, 285–292. [CrossRef]
57. Biliszczuk, J.; Hawryszków, P.; Teichgraeber, M. Structural Health Monitoring System of a Concrete Cable-Stayed Bridge. *Archit. Civ. Eng. Environ.* **2018**, *11*, 69–77. [CrossRef]
58. Popielski, P.; Bednarz, B.; Sieńko, R.; Howiacki, T.; Bednarski, Ł.; Zaborski, B. Monitoring of Large Diameter Sewage Collector Strengthened with Glass-Fiber Reinforced Plastic (GRP) Panels by Means of Distributed Fiber Optic Sensors (DFOS). *Sensors* **2021**, *21*, 6607. [CrossRef] [PubMed]

sensors

MDPI

Article

Evaluation of Axial Preload in Different-Frequency Smart Bolts by Laser Ultrasound

Guanpin Ren [1,2], Huan Zhan [2], Ziqian Liu [2,3], Wei Jiang [2], Ru Li [2] and Shuang Liu [1,*]

[1] Department of Applied Physics, College of Mathematics and Physics, Chengdu University of Technology, Chengdu 610059, China
[2] Chengdu Development Center of Science and Technology of CAEP, Chengdu 610299, China
[3] School of Big Data and Artificial Intelligence, Chengdu Technological University, Chengdu 611730, China
* Correspondence: liushuang19@cdut.edu.cn

Abstract: We report here on a laser ultrasonic system to indirectly evaluate the preload force of different-frequency piezoelectric bolts. This newly developed system enables us to achieve the goal of non-contact excitation and synchronously collects the laser-induced ultrasonic signal by the combination of a smart piezoelectric sensor and a magnetically mounted transducer connector. A numerical model based on the finite element method (FEM) was developed to simulate the propagation and displacement distribution of laser-generated ultrasonic waves along the axial direction. The measured A-scan waveform basically coincided with the counterpart obtained from a theoretical simulation, confirming the effectiveness of the proposed system to measure a bolt. By comparison, a laser spot diameter of 6 mm was the optimal beam diameter for the excitation of the ultrasonic wave in the bolt. The linear relationship between time of flight (TOF) of the ultrasonic longitudinal wave and bolt torque was almost independent from the center frequency of the smart bolt. By contrast, a piezoelectric patch centered at 5 MHz was more suitable as an ultrasonic sensor in terms of the nonlinear effects component suppression and linear fitting degree between TOF and torque. The results indicate that the proposed system based on a surface-mounted piezoelectric sensor is a promising system for evaluating the axial preload change of connector and fastener and is an additional potential laser ultrasonic system for nondestructive tests.

Keywords: laser inspection system; bolt preload; intelligent sensor; ultrasonic TOF; non-destructive testing

Citation: Ren, G.; Zhan, H.; Liu, Z.; Jiang, W.; Li, R.; Liu, S. Evaluation of Axial Preload in Different-Frequency Smart Bolts by Laser Ultrasound. *Sensors* **2022**, *22*, 8665. https://doi.org/10.3390/s22228665

Academic Editor: Zenghua Liu

Received: 11 October 2022
Accepted: 9 November 2022
Published: 10 November 2022

1. Introduction

Bolts are important fasteners indispensable in the manufacturing and national defense industries and are critical components of important infrastructure, such as aerospace, vehicle, ship, railway, bridge, and building facilities [1,2]. As the use of bolt increases, there is a greater need for the accurate control of bolt preload during tightening and in situ online monitoring or the evaluation of the residual axial force during the service life of the connection. Insufficient preload will cause vibration, slack, and slipping of structural connections, which can damage the integrity of the structure and affect the normal operation of any bolt-assembled equipment. Instead, an excessive preload will cause a severe plastic deformation of the bolt, which could lead to failure due to fatigue or fracture [3,4]. On this basis, researchers and engineers around the world are striving to develop on-line reliable methods to evaluate and measure the bolt preload force. These methods can be roughly divided into three categories: percussion detection, vision detection and sensor detection. It is difficult to achieve high precision using percussion and vision detection because of variations in the actual working conditions, low accuracy of the construction equipment and lack of experience among workers [5–7]. Other sensors, such as the strain sensor and the fiber Bragg grating (FBG) sensor, have also been developed in recent years and have been proven to evaluate the bolt preload force [8–11]. However, most of the sensors have different intrinsic shortcomings and are found unsuitable to evaluate or measure the bolt preload force [8–11]. For instance, conventional strain gauges cannot adhere to fully threaded bolts, and the FBG sensor embedded into the bolt body measures the preload force by the relative displacement of Bragg Wavelength. It is of no use for small bolts and causes damage to the bolt structure. In addition, its performance is strongly attenuated at high temperature, thus this sensor is not suitable for high-temperature environments.

In contrast to the above methods, benefiting from their advantages of high precision, excellent real−time performance, and strong sensing penetration, ultrasonic sensors have been widely used for the detection of axial stress in bolts since the 1980s [12–15]. A great deal of research has been conducted on ultrasonic theory and equipment to determine or evaluate the axial preload in bolts [16–24]. Although piezoelectric transducers are widely used, the accuracy of ultrasonic detection is relatively low due to poor bolt surface conditions and unstable coupling properties. During measurement, the ultrasonic wave passes through the coupling layer twice. As a result, unevenness and inconsistency of the coupling layer can produce errors of the same order of magnitude as the measured value. In order to improve the coupling performance, an electromagnetic acoustic transducer (EMAT) and a permanent mounted transducer system (PMTS) were introduced, gradually replacing the piezoelectric transducer in recent years. Nevertheless, EMAT has obvious drawbacks, such as low conversion efficiency and difficulty in exciting the longitudinal wave on the end faces of ferromagnetic bolts. Currently, the widespread application of PMAT is also limited by special handling processes of the bolt.

In addition to the previous methods, a few researchers [25–27] apply laser ultrasonic systems to monitor and estimate the looseness in bolted joints. For example, based on the Hertzian theory of contact, Yang Zhou et al. [27] applied laser-induced wave energy dissipation to monitor the looseness of a bolt. Ju−Won Kim et al. [26] reported the automatic estimation of looseness in a bolt joints by the combination of a non-contact laser scanning system and a deep convolutional neural network (DCNN). Studies in the literature applied a laser−induced guide wave rather than a single longitudinal wave (a longitudinal wave is more sensitive to preload changes) [12,13] to monitor and estimate the looseness of a bolt. Because there are few reports on the measurement of the bolt preload force by using a laser as a signal excitation source, all studies have used the Hertz contact method with this excitation mode, and the area of laser action was the surface of the connected parts; by analyzing the ultrasonic attenuation signal on the other side, the looseness of bolts was evaluated. In this work, a nanosecond pulsed laser used as a pump source directly irradiated the bottom of the bolt. The laser thermal effect area on the bolt itself, strictly in line with the thermoelastic effect, did not cause damage to the bolt, in line with non−destructive testing requirements, and the bolt head received the single longitudinal wave signals through a smart piezoelectric sensor and a magnetically mounted transducer connector. A numerical simulation of the longitudinal wave in the bolt induced by a nanosecond laser pulse was carried out. The influence of the laser spot diameter on a single longitudinal wave signal was investigated. By the TOF change, the axial preload change of the smart bolt could be evaluated effectively. The 5 MHz piezoelectric sensor showed advantages over the 2 MHz and 10 MHz piezoelectric sensors in terms of nonlinear effects and linear fitting degree between the TOF and the torque.

2. Research Methods

2.1. Measurement Principle of Bolt Preload

The preload force measured by an ultrasonic sensor is based on the acoustic−elastic effect. The mono−wave method, i.e., the use of a single transverse wave or longitudinal wave, is widely used to measure the axial stress in a bolt. The longitudinal wave is frequently used in engineering applications because it is more sensitive to changes in stress or preload force. Figure 1 shows the basic principle of the mono-wave method. The preload force can be calculated by measuring the TOFs of the pulse–echo before and after tightening. The relationship between TOF and preload force can be expressed as follows:

$$F = \frac{E \cdot S \cdot \Delta L}{L} \quad (1)$$

where F is the preload force, $\Delta L = \Delta T \cdot V$ is the change of bolt length, ΔT is the change in TOF from before to after tightening, and V is the propagation speed of the ultrasonic longitudinal wave in the material; E is the elastic modulus of the bolt material, S is the effective cross–sectional area, and L is the clamping length of the bolt. In this relation, the TOF of the ultrasonic wave is positively correlated with the preload force. It is well known that different batch bolts correspond to different elastic moduli E and therefore to the absolute preload values. Considering this, we only measured the TOF of the first longitudinal wave to evaluate the bolt axial preload in this work. The whole experiment was performed in a temperature–controlled laboratory. Thus, the influence of the temperature on the velocity of the longitudinal wave and the elastic modulus of the bolt material was not considered here to simplify the process.

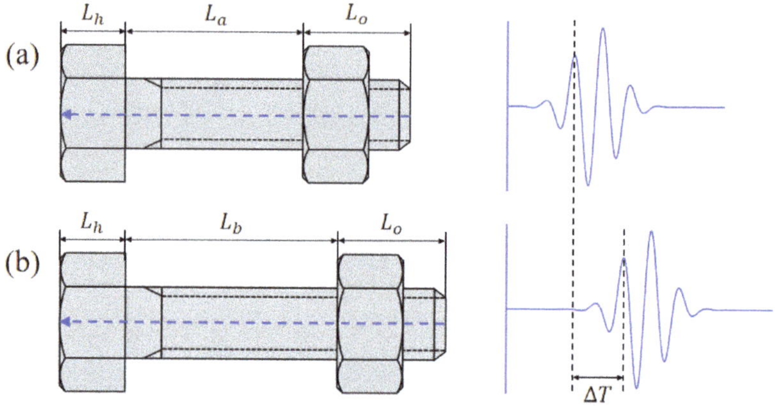

Figure 1. Fundamental principle to measure the bolt preload force by the mono-wave method: (**a**) TOF of the pulse–echo before tightening; (**b**) TOF of the pulse–echo after tightening.

2.2. Laser Ultrasonic Measurement System

The excitation source of the ultrasonic wave was a commercial nanosecond pulsed laser at 1064 nm in this work. The collimation laser beam had a spot diameter of 8 mm and a pulse width of 10 ns. The repetition rate could be adjusted from 1 Hz to 20 Hz. The laser output energy was 30 mJ, and the energy stability (RMS) was less than 1%. The independently developed instrument could collect ultrasonic wave signals at a frequency from 0.2 MHz to 25 MHz. The corresponding sampling frequency was 100 MHz, and the maximum gain value was as high as 89 dB. A schematic diagram of the laser ultrasonic measurement principle is presented in Figure 2. The laser beam pulse directly irradiated the bottom of the bolt, and then induced an ultrasonic longitudinal wave which propagated along the axial direction of the bolt through the thermo-elastic effect. Simultaneously, a smart piezoelectric sensor received an ultrasonic signal, and a magnetically mounted transducer connector converted the ultrasonic signal covert into a voltage signal above 100 mV. To ensure the effectiveness and reliability of the collected signal, the use of a BNC connector line achieved synchronization between the laser excitation unit and the signal receiver unit.

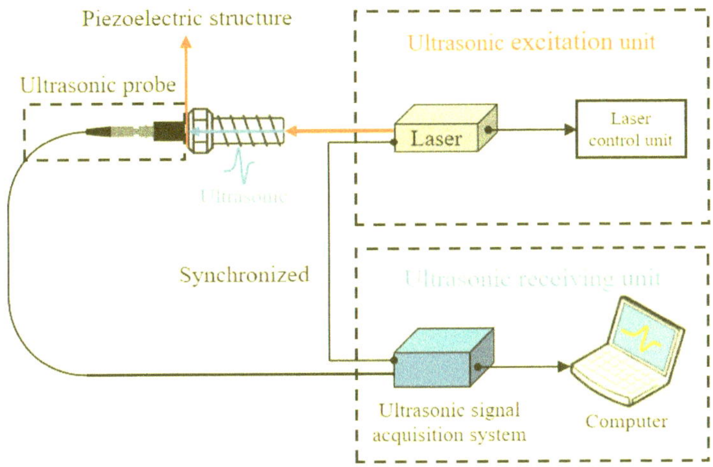

Figure 2. Schematic diagram of the laser ultrasonic system for bolt preload force evaluation.

The optical path of the laser excitation unit is shown in Figure 3. The ultrasonic probe was the developed magnetically mounted transducer connector. Through a commercial laser cleaning and polishing system, the surface roughness of the bolt head was controlled below 1 μm. Three different−frequency smart piezoelectric sensors were directly bonded on the surface of the bolt head by epoxies such as Loctite ABLESTIK 104, as a viscous coupling layer. After baking–curing in a low-temperature furnace for 2 h, a viscous couplant layer formed. The corresponding center frequencies were 2 MHz, 5 MHz, and 10 MHz, respectively. After passing through the aperture diaphragm, the collimated laser vertically irradiated the bottom of the bolt. Subsequently, an ultrasonic wave was induced through a thermo-elastic effect on the surface [28]. The induced signal component was mainly the ultrasonic longitudinal wave propagating from the bottom to the bolt head. The measured bolt in this work was an M8 carbon steel bolt with a nominal diameter of 8 mm and a length of 35.2 mm (Figure 4a). Considering that the laser irradiation area is the solid region free from thread, an aperture diaphragm was used to adjust the laser spot diameter from 2 mm to 6 mm. The influence of the laser spot diameter on the ultrasonic wave signal was investigated and is shown in Figure 4b. By comparison, the spot diameter of 6 mm corresponded to a higher signal amplitude and better signal quality. Therefore, the 6 mm diameter laser beam was the optimal ultrasonic wave excitation source in this work.

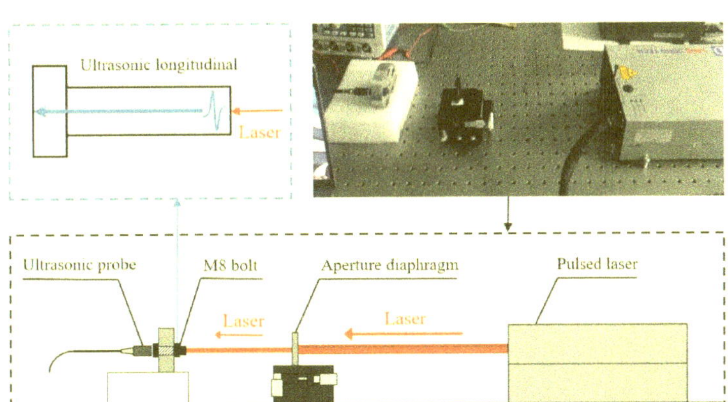

Figure 3. Schematic diagram and experimental setup of the laser ultrasonic excitation unit.

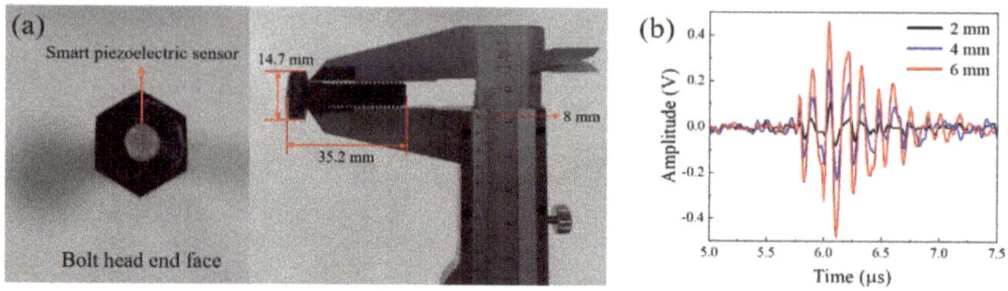

Figure 4. (**a**) M8 bolt with a smart piezoelectric sensor; (**b**) ultrasonic signal strength for different spot diameters.

3. Ultrasonic Measurement of the Bolt

3.1. Theoretical Model and Simulation

Laser energy absorption occurs on the surface and sub-surface of a bolt, regarded as the boundary heat source. The energy expression is

$$Q = I \cdot f(x) \cdot g(t) \tag{2}$$

where Q is the heat source, I is the energy absorbed by the material, and $f(x)$ and $g(t)$ are the laser spatial distribution function and the time distribution function, respectively. The laser source beam is the Gaussian beam, and its distribution function can be expressed as

$$\begin{cases} f(x) = exp\left(-\frac{x^2}{r_0^2}\right) \\ g(t) = \frac{t}{t_0} exp\left(-\frac{t}{t_0}\right) \end{cases} \tag{3}$$

where r_0 is the laser spot radius, and t_0 is the laser pulse width.

The finite element method [29,30] is used to simulate the propagation process of laser ultrasound inside the bolt. Figure 5 shows the displacement distribution of the ultrasonic wave from 4 μs to 8 μs. It is important to note that the main vibration energy was transmitted along the axial direction of the bolt and was reflected on both sides of the bolt to generate the trailing wave (Figure 5a). At 6 μs, the main vibration reached the boundary of the bolt head (Figure 5b) and was then reflected (Figure 5c). Part of the reflection was transmitted along the opposite axial direction, and the remaining wave secondary reflection occurred at the internal boundary of the bolt head (Figure 5d). Figure 6 shows the propagation process of the ultrasonic wave between 4 μs to 8 μs. The wave front propagated along the bolt axis (Figure 6a), reached the bolt head boundary at 6 μs (Figure 6b), and was reflected at the bolt head (Figure 6c). Some reflected waves appeared in the secondary reflection at the bolt head structural boundary (Figure 6d). Since the ultrasonic wave was generated by mechanical vibration displacement, the propagation process in Figures 5 and 6 is consistent.

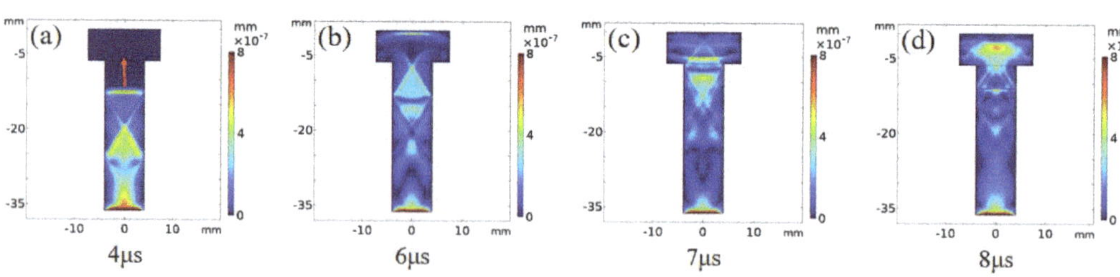

Figure 5. Displacement of the mechanical vibration caused by thermo−elastic effects in the bolt; dissemination process: (**a**)→(**b**)→(**c**)→(**d**).

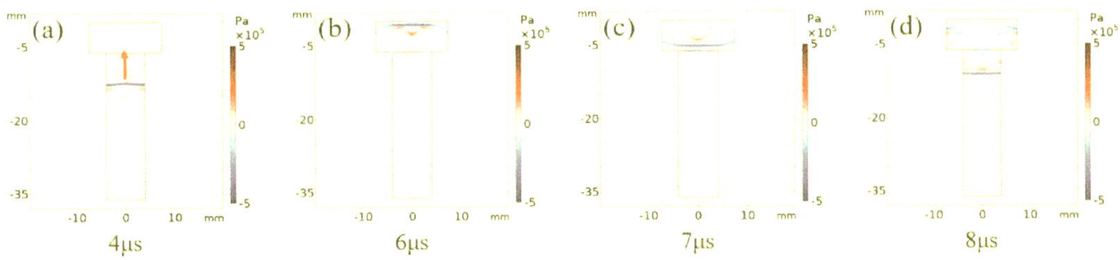

Figure 6. Ultrasonic longitudinal wave propagating along the axial direction in the bolt; dissemination process: (**a**)→(**b**)→(**c**)→(**d**).

3.2. Ultrasonic Signal Acquisition

To examine the effectiveness of the laser ultrasonic system for bolt preload evaluation, the simulated A−scan waveform and measured A−scan waveform from the ultrasonic acquisition unit were compared, as shown in Figure 7a. We noted that the profile curve of the simulated A−scan waveform was almost in accordance with that of the measured one. The first longitudinal wave signal envelope can be easily observed in Figure 7 and originated from the interaction between the laser-induced ultrasonic wave and the head boundary. After propagating three times along the axial direction, the bolt head could detect the second longitudinal wave. The multiple small peaks between the first and the second longitudinal wave correspond to trail waves and reflection waves from the bolt head region. The amplitude intensity of the second longitudinal wave presented a reduction compared with its counterpart in the simulation results. This was due to the bolt intrinsic loss caused by dissipation, absorption and scattering and the slight deviation of the wave transmission direction caused by the roughness and uncleanness of the bolt surface. Thus, the first longitudinal wave was more suitable for the measurement of the bolt preload force in our work.

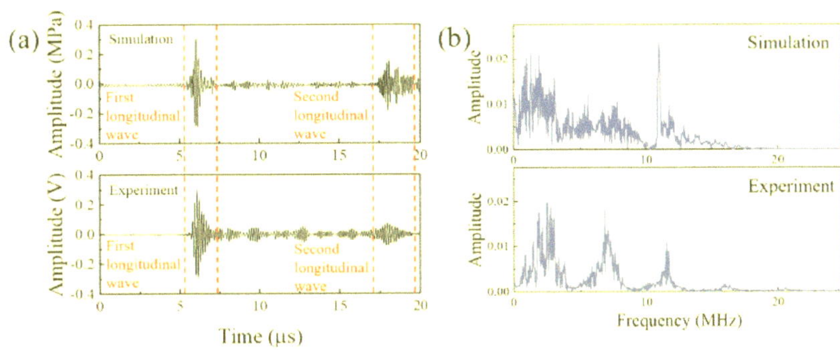

Figure 7. (**a**) Ultrasonic signals obtained from the simulation and the experiment; (**b**) FFT spectrum of the simulation and experiment's ultrasonic signal.

Limited by the 100 MHz of the available data acquisition frequency, the time accuracy was relatively low. In our experiment, the TOFs of the pulse−echo presented no change when the axial preload force was below 1 kN, as shown in Figure 8a,b; although the signal waveform showed a slight phase shift, the TOF of the peak point was still the same, because of the original sampling frequency. To address this issue, we applied the spline function interpolation [31] to increase the data points and improve the resolution of the propagation time difference. We inserted 100 interpolation points between the two original time data points. The time interval was improved from 10 ns to 100 ps. The original temporal signal A and B correspond to the longitudinal wave signal before and after a slight tightening, as shown in Figure 8a. We found no change of TOF between signal A and signal B. The interpolation algorithm made the signal curves A and B convert into the corresponding curves A′ and B′. The curve profiles became smooth, as indicated in Figure 8b. With the introduction of the interpolation algorithm, a small TOF change was observed at the signal peak position, and the corresponding change value was 0.7 ns (see Figure 8b).

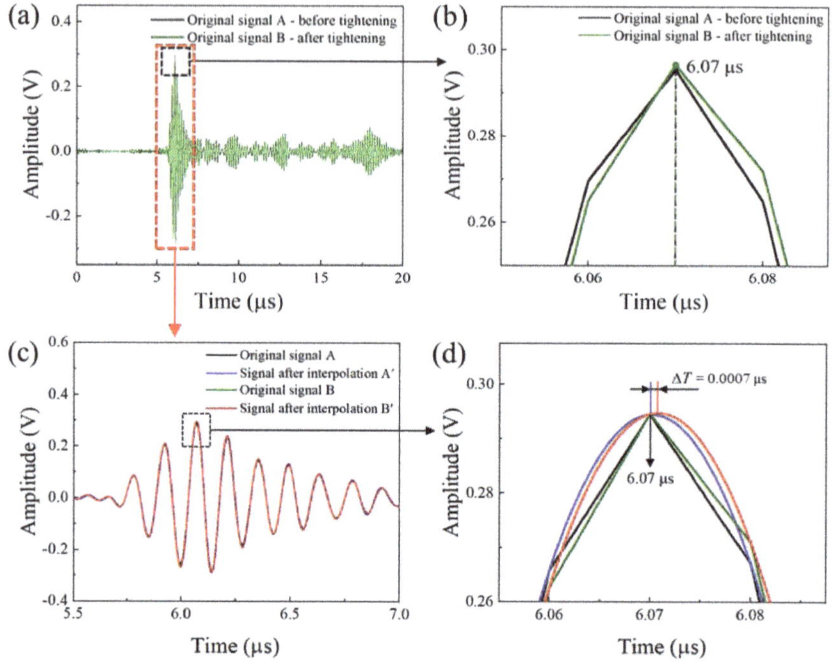

Figure 8. (**a**) Original signal before and after applying a small preload; (**b**) the waveform underwent a slight phase shift, but the data of the sampling points at the peak remained the same; (**c**) original sample data (temporal signals A, B) and interpolation waveforms (interpolated signals A′, B′) before and after loading; (**d**) rectangular window for comparing the TOF at the peak position between pre- and post−interpolation.

The preload force of the bolt of this work was inconvenient for a commercial tensile machine. A manual torque wrench was used here to indirectly control the bolt preload through a tightening torque, although only about 10% of the torque was converted into axial preload. A bolt was gradually installed by a torque wrench with a torque from 0 N·m to 15 N·m, and the preload increase interval was designed to be 2.5 N·m. For each torque, the laser ultrasonic system improved the measurement precision by averaging the testing ultrasonic signal of measurements repeated 20 times. In addition, time−domain averaging and an FIR filter were adopted to suppress noise interference. In order to investigate the relationship between the torque and the TOF change, the ultrasonic signals during different torque tightening processes were received by a smart sensor, as shown in Figure 9a. With the increase of the torque, the highest peak position of the first longitudinal wave presented the time−decay phenomenon. The ultrasonic signal amplitude had a small range of fluctuations, as indicated in Figure 9b. This could be attributed to a slight variation of the laser beam irradiation position and the small noise inference of the acquisition unit.

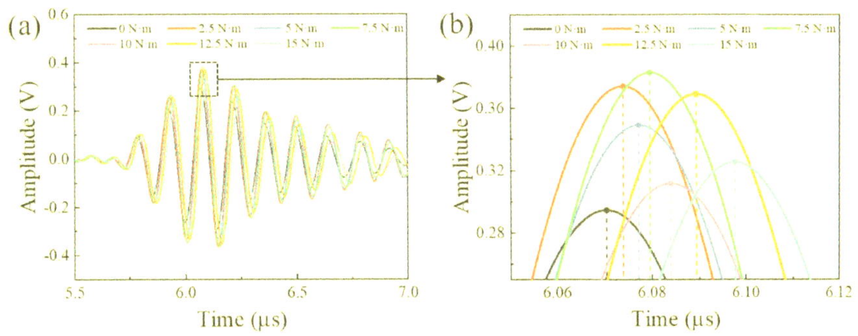

Figure 9. (**a**) Received signals from the ultrasonic sensor during different torque−tightening processes; (**b**) rectangular window demonstrating the TOF change before and after different torque tightening processes.

For different torques, the bolt was subjected to different forces, which made the bolt elongation change. The time variation of signal T (TOF) is the key data we use to evaluate a preload. It directly reflects the elongation change of the bolt force, but the amplitude of the signal is not the value we pay attention to. In the experiment, we also found that due to the small interference of the measuring equipment and external signals, the amplitude of the signal constantly fluctuated up and down, but this vibration did not affect our acquisition of the TOF and the overall measurement results. Figure 9 shows one set of data from all repeated experiments. For each set of different data, the amplitude of the signal was also different. However, the phase shift (TOF) on the time axis was strictly in line with physical laws and expectations.

For the increase in the amount of phase shift from 12.5 N·m to 15 N·m, as shown in Figure 10a, the relationship was not strictly linear. This was true only for bolts with 2 MHz sensors, as the linearity of 5 MHz and 10 MHz bolts was obviously better. However, this did not affect the validity of the preload measurement, which, we think, was affected by the sensor itself.

The TOF of the laser-induced longitudinal wave as a function of the torque for three different-frequency smart bolts was measured and is shown in Figure 10a,c,e. With the increase of the preload applied to the bolts, the TOF of the longitudinal wave presented a nearly linear increase, demonstrating an extension of the axial length of the bolts. Nonlinear ultrasonic effects were easily induced due to the special bolt shape, heterogeneous extension of the length, and intense ultrasonic amplitude induced by the laser. On this basis, frequency spectra of the three different sensor bolts were obtained by the Fastest Fourier Transform in the West (FFTW) algorithm. The corresponding results are presented in Figure 10b,d,f. It is important to note that all the three different sensor bolts presented broad frequency spectra. This was due to a wide−band ultrasonic signal induced by the pulsed laser. In addition to this, obvious harmonic waves induced by nonlinear effects can be easily observed. As shown in Figure 10b, the center frequency of the 2 MHz smart bolt was at 2.6 MHz, and the components of the third and four harmonic generations can be clearly seen. The deviation of the frequency center value could be caused by the piezoelectric material and frequency shift induced by the special bolt structure. In addition, no second harmonic generation component was found. This was probably related to the intrinsic property of the piezoelectric material. The frequency spectrum of the 5 MHz smart bolt also presented no second harmonic generation component. The intensity of the higher harmonic generations was weaker than the center frequency intensity and counterpart of the 2 MHz smart bolt. For the 10 MHz smart bolt, the center frequency peaked at 11.2 MHz. No higher harmonic generation component was observed. However, a strong low−frequency response centered at 5.2 MHz could be observed. The reason is still unclear and will be further explored in the near future. According to the nonlinear effects component proportion and the linearly fitting degree between the TOF and the torque, the 5 MHz piezoelectric sensor appeared more suitable as an ultrasonic sensor for the evaluation of the axial preload.

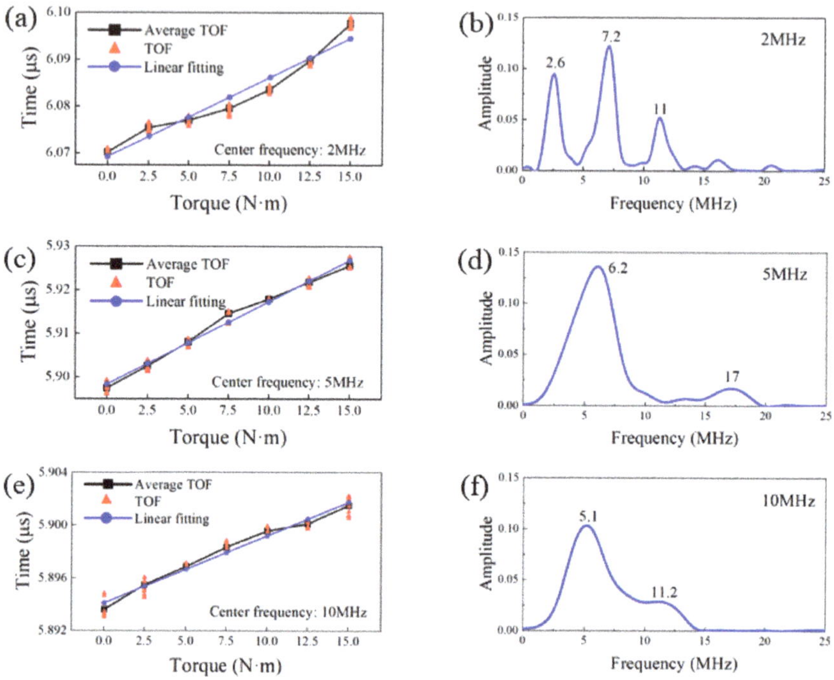

Figure 10. Relationship between the TOF and the preload torque for different center frequency sensors: (**a**) 2 MHz, (**c**) 5 MHz, (**e**) 10 MHz; spectrum of the first longitudinal wave signal for (**b**) 2 MHz, (**d**) 5 MHz, and (**f**) 10 MHz PZT sensors.

To further analyze the deviation between the actual measurement and the linearly fitted result, a photograph of the bottom end face of the smart bolts was taken, as shown in Figure 11a. The bottom of the bolt presented an obvious, uneven distribution. A two dimensional (2D) cross−sectional snapshot of the bolt is presented in Figure 11b. The deviation was mainly induced by the uneven bottom of the bolt, the length error Δl caused by the offset of the laser incidence center position during multiple tests, and the angular error $\Delta\theta$ of the incident position originated from the incomplete collimation of optical path and the tilted bottom surface of the bolt. The slight loosening of the bolts during testing caused by friction failure of the bolt die also affected the stress stability of the bolts. Additionally, signal interference of the laser excitation source and electronic noise of the ultrasonic acquisition unit contributed to the deviation.

The pure laser−based ultrasonic technique, i.e., the laser generation−laser receiving technique, has been intensely investigated recently in various industrial fields. Nevertheless, for strongly scattering surfaces, high−powered detection lasers are needed as well as complex interferometer set−ups, which may be costly and difficult to miniaturize. The laser ultrasonic system proposed here combines the advantages of laser non−contact excitation and good−stability ultrasonic sensing free from a couplant. In addition to this, the combination of a smart piezoelectric sensor and a magnetically mounted transducer connector allows for much higher sensitivity with a lower cost in ultrasound detection than laser interferometers. In this work, we applied this proposed system to successfully measure the TOF of a laser ultrasonic longitudinal wave and indirectly evaluate the relationship between the bolt preload and the TOF of the ultrasonic wave. This work indicates that this system can be used as another effective laser ultrasonic method for non−destructive testing (NDT) applications in the near future.

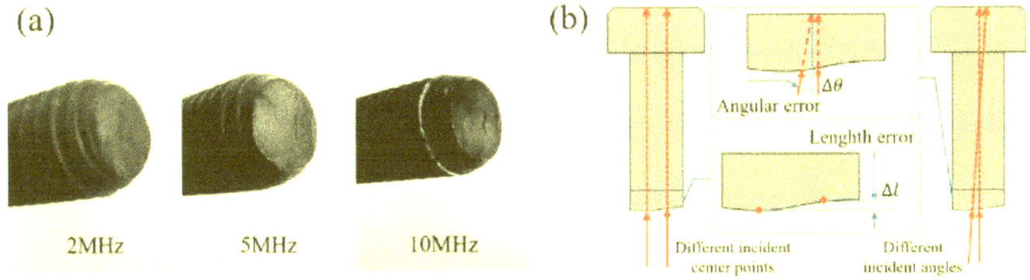

Figure 11. (**a**) Photograph of the bottom−end faces of three intelligent bolts; (**b**) errors caused by incident center and incident angle offset.

4. Conclusions

In this paper, we successfully measured the TOF of an ultrasonic longitudinal wave by applying a pulsed laser as the ultrasonic excitation in combination with a newly developed ultrasonic sensor (i.e., a smart piezoelectric sensor and a magnetically mounted transducer connector). The propagation process and displacement distribution of the laser-induced ultrasonic wave were fully demonstrated along the axial direction of the bolt. The experimental testing results agreed very well with the phenomena observed in the simulation results. The introduction of interpolation made the TOF change measure of the ultrasonic longitudinal wave accurate and effective. The TOF change of the first longitudinal wave presented an almost linear increase with the increase of the torque. Compared with the piezoelectric patches centered at 2 MHz and 10 MHz, the piezoelectric patch centered at 5 MHz showed a better suppression ability of more intense nonlinear effects and a better degree of linear fitting between the TOF and the torque. Future research work will focus on the optimization of the laser ultrasonic measurement system to reduce external errors, the analysis and verification of nonlinear ultrasonic effects of the piezoelectric sensor at a microscopic level, and the combination with deep learning algorithms for axial preload evaluation.

Author Contributions: Conceptualization, G.R., H.Z., R.L. and S.L.; methodology, G.R., Z.L., W.J. and S.L.; software, G.R. and Z.L.; validation, G.R., R.L. and W.J.; formal analysis, Z.L., H.Z. and S.L.; investigation, R.L. and W.J.; writing—original draft preparation, G.R.; writing—review and editing, H.Z. and S.L.; visualization, G.R. and S.L.; supervision, H.Z. and S.L.; project administration, S.L. All authors have read and agreed to the published version of the manuscript.

Funding: This research received no external funding.

Institutional Review Board Statement: Not applicable.

Informed Consent Statement: Not applicable.

Acknowledgments: This work was supported by grants from the public service platform of innovation achievements industrialization, Chengdu Development Center of Science and Technology, China Academy of Engineering Physics (CAEP). Great thanks are given to Xia Ding (TECLAB (CHINA) LIMITED) for experimental instruction and results discussion.

Conflicts of Interest: The authors declare no conflict of interest.

References

1. Basava, S.; Hess, D. Bolted joint clamping force variation due to axial vibration. *J. Sound Vib.* **1998**, *210*, 255–265. [CrossRef]
2. Izumi, S.; Yokoyama, T.; Iwasaki, A.; Sakai, S. Three-dimensional finite element analysis of tightening and loosening mechanism of threaded fastener. *Eng. Fail. Anal.* **2005**, *12*, 604–615. [CrossRef]
3. Dong, Z.Y.; Zhang, H.L. Bolt static finite element numerical simulation and experimental research. In *Applied Mechanics and Materials*; Trans Tech Publications Ltd.: Bäch, Switzerland, 2014; pp. 373–376.

4. Liu, M.M.; Zhai, Q.G.; Shi, H. Analysis of Vibration Testing for the Structure with Bolted Flange Joints. In *Applied Mechanics and Materials*; Trans Tech Publications Ltd.: Bäch, Switzerland, 2015; pp. 205–210.
5. Yella, S.; Gupta, N.K.; Dougherty, M.S. Comparison of pattern recognition techniques for the classification of impact acoustic emissions. *Transp. Res. Part C Emerg. Technol.* **2007**, *15*, 345–360. [CrossRef]
6. Wang, F.; Song, G. Bolt-looseness detection by a new percussion-based method using multifractal analysis and gradient boosting decision tree. *Struct. Health Monit.* **2020**, *19*, 2023–2032. [CrossRef]
7. Zhang, Y.; Sun, X.; Loh, K.J.; Su, W.; Xue, Z.; Zhao, X. Autonomous bolt loosening detection using deep learning. *Struct. Health Monit.* **2020**, *19*, 105–122. [CrossRef]
8. Nikravesh, S.M.Y.; Goudarzi, M. A review paper on looseness detection methods in bolted structures. *Lat. Am. J. Solids Struct.* **2017**, *14*, 2153–2176. [CrossRef]
9. Khomenko, A.; Koricho, E.G.; Haq, M.; Cloud, G.L. Bolt tension monitoring with reusable fiber Bragg-grating sensors. *J. Strain Anal. Eng. Des.* **2016**, *51*, 101–108. [CrossRef]
10. Yeager, M.; Whitaker, A.; Todd, M. A method for monitoring bolt torque in a composite connection using an embedded fiber Bragg grating sensor. *J. Intell. Mater. Syst. Struct.* **2018**, *29*, 335–344. [CrossRef]
11. Wang, Z.; Liu, M.; Zhu, Z.; Qu, Y.; Wei, Q.; Zhou, Z.; Tan, Y.; Yu, Z.; Yang, F. Clamp looseness detection using modal strain estimated from FBG based operational modal analysis. *Measurement* **2019**, *137*, 82–97. [CrossRef]
12. Sun, Q.; Yuan, B.; Mu, X.; Sun, W. Bolt preload measurement based on the acoustoelastic effect using smart piezoelectric bolt. *Smart Mater. Struct.* **2019**, *28*, 055005. [CrossRef]
13. Liu, Y.; Liu, E.; Chen, Y.; Wang, X.; Sun, C.; Tan, J. Measurement of fastening force using dry-coupled ultrasonic waves. *Ultrasonics* **2020**, *108*, 106178. [CrossRef] [PubMed]
14. Pan, Q.; Pan, R.; Chang, M.; Xu, X. A shape factor based ultrasonic measurement method for determination of bolt preload. *NDT E Int.* **2020**, *111*, 102210. [CrossRef]
15. Chen, P.; He, X.; Song, W. Parameter Recognition of Mode-Converted Wave in Single-Source Ultrasound Using Gabor Transform for Bolt Axial Stress Evaluation. *J. Sens.* **2020**, *2020*, 8883845. [CrossRef]
16. Parvasi, S.M.; Ho, S.C.M.; Kong, Q.; Mousavi, R.; Song, G. Real time bolt preload monitoring using piezoceramic transducers and time reversal technique—A numerical study with experimental verification. *Smart Mater. Struct.* **2016**, *25*, 085015. [CrossRef]
17. Tao, W.; Shaopeng, L.; Junhua, S.; Yourong, L. Health monitoring of bolted joints using the time reversal method and piezoelectric transducers. *Smart Mater. Struct.* **2016**, *25*, 025010. [CrossRef]
18. Zhao, Z.; Chen, P.; Zhang, E.; Lu, G. Health monitoring of bolt looseness in timber structures using PZT-enabled time-reversal method. *J. Sens.* **2019**, *2019*, 2801638. [CrossRef]
19. Kędra, R.; Rucka, M. Preload monitoring in a bolted joint using Lamb wave energy. *Bull. Pol. Acad. Sci.-Tech. Sci.* **2019**, *67*, 1161–1169.
20. Jiang, T.; Wu, Q.; Wang, L.; Huo, L.; Song, G. Monitoring of bolt looseness-induced damage in steel truss arch structure using piezoceramic transducers. *IEEE Sens. J.* **2018**, *18*, 6677–6685. [CrossRef]
21. Hei, C.; Luo, M.; Gong, P.; Song, G. Quantitative evaluation of bolt connection using a single piezoceramic transducer and ultrasonic coda wave energy with the consideration of the piezoceramic aging effect. *Smart Mater. Struct.* **2020**, *29*, 027001. [CrossRef]
22. Zhang, Z.; Liu, M.; Su, Z.; Xiao, Y. Quantitative evaluation of residual torque of a loose bolt based on wave energy dissipation and vibro-acoustic modulation: A comparative study. *J. Sound Vib.* **2016**, *383*, 156–170. [CrossRef]
23. Zhang, Z.; Xu, H.; Liao, Y.; Su, Z.; Xiao, Y. Vibro-acoustic modulation (VAM)-inspired structural integrity monitoring and its applications to bolted composite joints. *Compos. Struct.* **2017**, *176*, 505–515. [CrossRef]
24. Zhang, Z.; Liu, M.; Liao, Y.; Su, Z.; Xiao, Y. Contact acoustic nonlinearity (CAN)-based continuous monitoring of bolt loosening: Hybrid use of high-order harmonics and spectral sidebands. *Mech. Syst. Signal Process.* **2018**, *103*, 280–294. [CrossRef]
25. Haynes, C.; Yeager, M.; Todd, M.; Lee, J.-R. Monitoring bolt torque levels through signal processing of full-field ultrasonic data. In *Health Monitoring of Structural and Biological Systems*; SPIE: Bellingham, WA, USA, 2014; pp. 650–658.
26. Tran, D.Q.; Kim, J.-W.; Tola, K.D.; Kim, W.; Park, S. Artificial intelligence-based bolt loosening diagnosis using deep learning algorithms for laser ultrasonic wave propagation data. *Sensors* **2020**, *20*, 5329. [CrossRef]
27. Zhou, Y.; Yuan, C.; Sun, X.; Yang, Y.; Wang, C.; Li, D. Monitoring the looseness of a bolt through laser ultrasonic. *Smart Mater. Struct.* **2020**, *29*, 115022. [CrossRef]
28. Zhou, Z.; Zhang, K.; Zhou, J.; Sun, G.; Wang, J. Application of laser ultrasonic technique for non-contact detection of structural surface-breaking cracks. *Opt. Laser Technol.* **2015**, *73*, 173–178. [CrossRef]
29. Ren, G.; Xu, L.; Zhan, H.; Liu, S.; Jiang, W.; Li, R. Quantifying the shape effect of plasmonic gold nanoparticles on photoacoustic conversion efficiency. *Appl. Opt.* **2022**, *61*, 4567–4570. [CrossRef]
30. Ren, G.; Sun, Z.; Dai, X.; Liu, S.; Zhang, X.; Chen, X.; Yan, M. Laser ultrasonic nondestructive evaluation of sub-millimeter-level crack growth in the rail foot weld. *Appl. Opt.* **2022**, *61*, 6414–6419. [CrossRef] [PubMed]
31. Sun, S.; Li, S.; Lin, L.; Yuan, Y.; Li, M. A novel signal processing method based on cross-correlation and interpolation for ToF measurement. In Proceedings of the 2019 IEEE 4th International Conference on Signal and Image Processing (ICSIP), Wuxi, China, 19–21 July 2019; pp. 664–668.

sensors

MDPI

Article

Extending the Incidence Angle of Shear Vertical Wave Electromagnetic Acoustic Transducer with Horizontal Magnetization

Zhengyang Qu, Zhichao Li *, Runjie Yang, Songtao Hu and Shujuan Wang

School of Electrical Engineering and Automation, Harbin Institute of Technology, Harbin 150001, China
* Correspondence: zhichao.li@hit.edu.cn

Abstract: Angled shear vertical (SV) waves have been successfully employed in the non-destructive testing of welds, pipes, and railways. Non-contact meander-line coil electromagnetic acoustic transducers (EMAT) have many benefits in generating angled SV waves. The most important benefit is that the incidence angle of an SV wave can be controlled by the excitation frequency. However, the incidence angle of a traditional SV-wave EMAT is reported to be under 45 degrees in many cases. In this work, such cases are tested, and the problems of the received signal at large incidence angles are found to be due to wave interference and small signal amplitudes. An equivalent finite element (FE) model is established to analyze the problem, and the main reason is found to be the head wave. An alternative configuration of angled SV-wave EMAT with horizontal magnetization is proposed to reduce the influence of the head wave. Finally, the results from simulations and experiments show that the proposed EMAT has a larger signal amplitude and significantly reduced interference in large-incidence angle scenarios. Moreover, an incidence angle of an SV wave of up to 60 degrees can be achieved, which will help improve the performance and capability of nondestructive testing.

Keywords: SV wave; EMAT; incidence angle; horizontal magnetization; head wave

Citation: Qu, Z.; Li, Z.; Yang, R.; Hu, S.; Wang, S. Extending the Incidence Angle of Shear Vertical Wave Electromagnetic Acoustic Transducer with Horizontal Magnetization. *Sensors* **2022**, *22*, 8589. https://doi.org/10.3390/s22228589

Academic Editor: Zenghua Liu

Received: 19 October 2022
Accepted: 5 November 2022
Published: 8 November 2022

1. Introduction

Ultrasonic testing is a popular tool in nondestructive testing and evaluation. Guided ultrasonic waves can be used for distant flaw detection and structural health monitoring [1–3], while bulk ultrasonic waves can be used for local flaw sizing and imaging [4]. Angled shear vertical waves are one of the common bulk ultrasonic wave types for ultrasonic testing. An angled SV wave propagates along a path that deflects from the normal direction of the testing plane at a specified angle, which is usually called the incidence angle. With an incidence angle, SV waves could be applied in the detection of oriented flaws and situations where space is restricted such that normal probes cannot be employed. Therefore, angled SV waves have had remarkable success in the nondestructive testing of welds [5,6], pipes [7,8], railways [9,10], etc.

Angled SV waves are normally generated by a piezoelectric transducer assembled on a plastic wedge [6,11]. The wedge alters the incidence angle of longitudinal waves generated by the piezoelectric transducer and transforms them into shear waves through mode conversion and refraction. The angle of the wedge should be predesigned knowing the acoustic velocities or impedance of the wedge and the tested object. However, the incidence angle of SV waves cannot be changed without replacing the wedge once the angle of the wedge is determined. A meander-line coil (MC) electromagnetic acoustic transducer (EMAT) can also be used for angled SV wave generation [12]. The incidence angle of an SV wave generated by an MC-EMAT is controlled by the spacing of the coil, the excitation frequency, and the SV wave velocity in the specimen. As a result, the incidence angle of an SV wave can be easily manipulated by changing the excitation frequency of an MC-EMAT without redesigning the transducer. An EMAT is a type of non-contact transducer, making it suitable for some applications where wedged piezoelectric transducers cannot be applied, such as in elevated temperatures, online testing etc. [13–18]. Moreover, an EMAT has no requirement for a couplant, making it suitable for some applications where accurate time-of-flight (ToF) measurement is needed by getting rid of the influence of the inconsistent coupling condition [19,20].

The main drawback of an EMAT is its low transduction efficiency such that the signal-to-noise ratio (SNR) of EMAT signal is small and prone to noise interference [21]. Therefore, improving the efficiency of SV-wave EMAT has drawn a lot of attention. Line-focusing [22] or point-focusing [23] SV-wave EMATs employ MC with variable coil spacing, focusing the SV-wave beam on a line or a point, respectively, inside the specimen with designed frequency. The beam profile and intensity of the SV waves generated by these transducers are improved. However, the coil structure and the excitation frequency are bonded together to achieve beam focusing for a focusing SV-wave EMAT, which means the focusing line or point is fixed and cannot be simply controlled through switching the excitation frequency. Accordingly, the focusing-type SV-wave EMAT is less flexible than typical MC SV-wave EMAT, especially in some circumstances such as scanning tests of large areas or angles.

Much previous research has shown that the signal amplitude of SV-wave EMAT drops sharply when the incidence angle is beyond a specific angle (typically below 45 degrees) [21], which limits the scanning range. Even with a phased array of SV EMATs, large incidence angles still seem unavailable [24]. In our experiments with a pitch-catch configuration on an aluminum specimen, the receiving transducer picked up some interference signals when operating at large incidence angle. The interval between the interference signal and the SV-wave signal varies when the incidence angle changes. This phenomenon is not conducive to applications such as flaw detection and precise ToF measurement as the interference signal may cause false alarms or partially overlap the actual SV wave signal. Therefore, solving the two problems is critical to enabling the large-angle operation of SV-wave EMAT. With the aid of the FE model, the main cause is analyzed and found to be the head wave. An alternative configuration of SV-wave EMAT is designed to solve the problem based on this analysis.

This article first provides a quick overview of the basic working principles of angled SV-wave EMAT. Then, the problems are illustrated with experimental results, and an equivalent simulation model is established to show the internal acoustic field. In the next section, the acoustic field generated by a point source is analyzed to find differences from the assumptions made in the working principles. Finally, a new configuration of SV-wave EMAT is proposed, and its performances are assessed with both simulations and experiments. The performances of the traditional angled SV-wave EMAT and the proposed one are also compared.

2. Working Principles of Angled SV-Wave EMAT

The typical SV-wave EMAT configuration is made up with an MC with equal spacing and a permanent magnet with vertical magnetization, as shown in Figure 1.

When the MC-EMAT is placed on an aluminum material, the Lorentz force mechanism is the only consideration since it dominates the transduction process on paramagnetic materials. In the MC coil, current flows in adjacent wires have opposite direction. Consequently, a spatial alternating eddy current $\boldsymbol{J}$ is induced in the specimen. Under the bias magnetic field $\boldsymbol{B}$ provided by the permanent magnet, the Lorentz force $\boldsymbol{f}$ can be calculated by

$$\boldsymbol{f} = \boldsymbol{J} \times \boldsymbol{B}. \tag{1}$$

Inside the specimen, $\boldsymbol{f}$ has a similar distribution to the eddy current. This force makes ultrasonic waves form and propagate. In practice, the eddy current decays exponentially along the depth at a rate related to skin depth δ, which is given by

$$\delta = \sqrt{2/(\omega\mu\sigma)}, \tag{2}$$

where ω is the angular frequency of the excitation current, μ is the magnetic permeability, and σ is the electrical conductivity. Skin depth is at the sub-millimeter scale in the common EMAT operating frequency range. As a result, the Lorentz force mainly acts on the surface of the specimen.

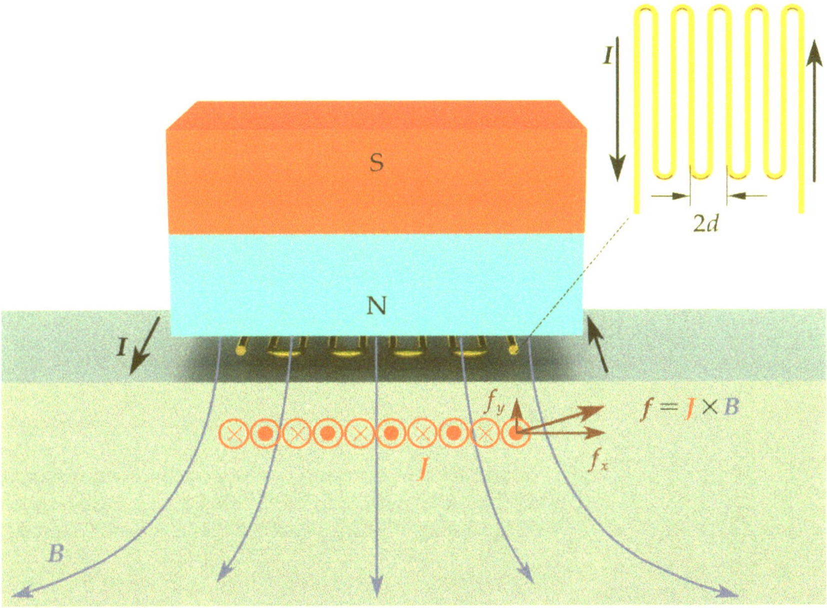

Figure 1. Illustration for the working principle of MC-EMAT on an aluminum specimen.

An ultrasonic wave, when propagating inside the solid media, obeys the frequency-wavelength relationship, given by

$$c = \lambda f \,, \tag{3}$$

where c is the wave propagation velocity, λ is the wavelength and f is the frequency. Figure 2 employs a linearly distributed point force on the top boundary to illustrate the criteria for angled incidence. To make the incident waves interfere constructively, the wave generated by each source should arrive at an arbitrary plane perpendicular to the propagation direction with an identical phase. These planes are the wavefronts of SV waves. In Figure 2, the formation of one wavefront (black dot-dash line) is illustrated, and those of the other wavefronts (grey dot-dash lines) are omitted for simplicity. This means the equivalent wavelength λ_{eq} is $2d \sin\theta$ by the geometric relationship, assuming each wave initiates at the same phase and has a cylindrical wavefront. Substituting λ_{eq} into Equation (3), the equation for incidence angle calculation can then be written as

$$\theta = \sin^{-1}\frac{c_S}{2df}, \tag{4}$$

where c_s is the shear wave velocity and $2d$ is the distance between two point sources with the same initial phase. In the MC-EMAT, d is the distance between two adjacent wires.

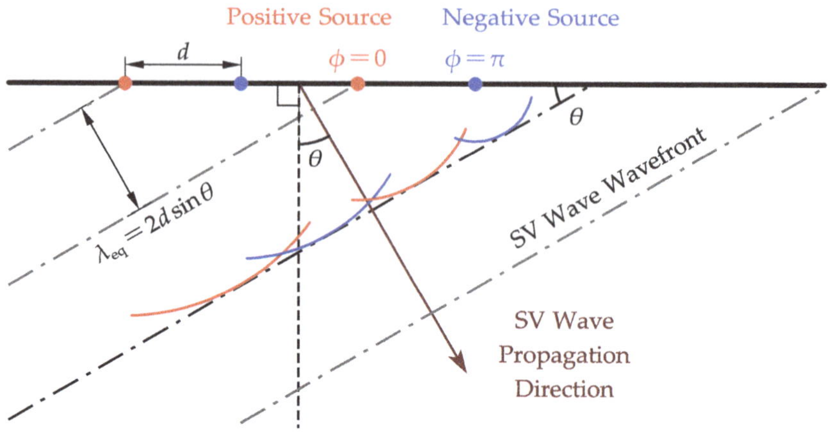

Figure 2. Illustration for geometric relations in SV wave angled incidence.

3. The Problems of Common Angled SV-Wave EMAT

The present experiment was set up to provide a more intuitive explanation of the problems mentioned in the introduction. The configuration of the experiment is shown in Figure 3, where the picture in the top-right corner shows the coil and the magnet of the EMAT used in this experiment. In this configuration, the high-power pulser (RITEC RAM-5000 [25]) generates high voltage tone-burst signals and drives the transmitting EMAT connected by a coaxial cable. The parameters (cycles, frequency, power, etc.) of the tone-burst signal can be controlled by the computer. The transmitting EMAT generates an angled SV wave in the specimen, and when the wave arrives, the receiving EMAT transduces the ultrasonic waves to the electrical signals. The voltage amplitude of the received signal is typically within tens of microvolts, so the receiving EMAT is wired to a homemade variable gain amplifier (40–80 dB) to make the signal measurable in the following data acquisition (DAQ) process. The amplified signal is fed to an oscilloscope (Agilent MSO-X 3024A) under the control of a computer for acquisition, and the data are saved on the computer for further processing. The transmitting and DAQ processes are synchronized by the trigger output of the pulser. Two identical EMATs are arranged on both sides of the specimen in a pitch-catch configuration. Fixing the transmitting EMAT on one side of the specimen, the incidence angle of the SV wave is expected to be controlled via changing the excitation frequency according to Equation (4). The receiving EMAT can then be moved along the other side to a corresponding position to receive the angled SV-wave signal.

The specimen was made of 6061-T6 aluminum alloy. The acoustic velocities of the specimen from the experiment were different from those calculated from the standard material properties (density, Young's modulus, and Poisson's ratio), so the acoustic velocities were measured firstly with an electromagnetic acoustic thickness gauge (Orisonic ETGmini-X1 [26]) with a thickness measurement resolution of 0.01 mm and a Vernier caliper with a resolution of 0.01 mm.The pieces of equipment with calibration functions were calibrated before the experiment. The corresponding properties of the specimens and the EMAT in this experiment are detailed in Table 1.

The excitation signal used in the experiment was an 8-cycle sinusoidal tone burst with a voltage amplitude of 1000 V. The frequencies for 30 degrees and 45 degrees of incidence are calculated to be 1.210 MHz and 0.856 MHz, respectively. The environmental temperature was measured to be 26 degrees Celsius, and the temperature variation was within ± 1 degree Celsius, so the acoustic velocities had almost no changes during the experiments.

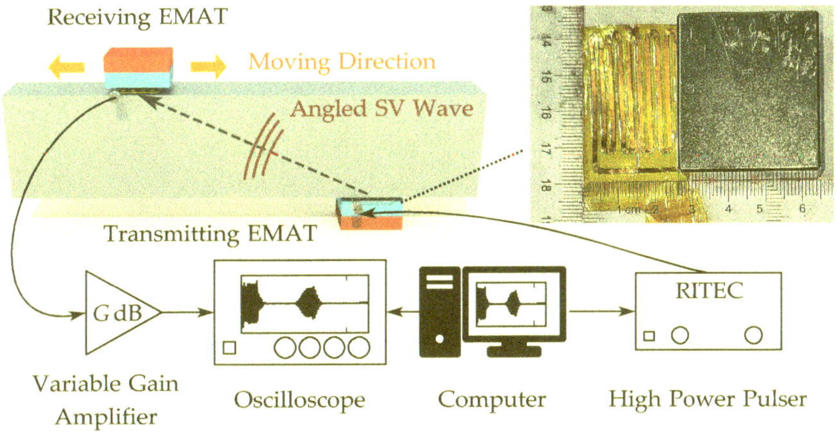

Figure 3. Configuration of angled SV-wave EMAT testing experiment.

Table 1. Properties of specimens and EMAT used in experiment.

Parameter	Value
Specimen dimension	35 mm × 70 mm × 400 mm
Specimen longitudinal wave velocity c_l	6436 m/s
Specimen shear wave velocity c_s	3146 m/s
Magnet dimension	25 mm × 35 mm × 40 mm
Magnetic flux density on surface of magnet	0.55 T (Teslas)
Coil turns	7
Coil space interval d	2.6 mm
Coil wire diameter	0.12 mm
Coil element width	1.2 mm
Coil element length	30 mm
Number of coil elements	10

Figure 4 shows the received SV-wave signals. The signal at 30 degrees looks promising, while the signal at 45 degrees suffers from severe discontinuity and interference. Moreover, the amplifier gain settings for the two signals are different. The real amplitude of the SV-wave signal at 30 degrees is 16 dB larger than that at 45 degrees at the same amplifier gain. The small signal amplitude is still resolvable through increasing the gain and the averaging level. However, the discontinuity and interference in Figure 4b will cause some problems such as false alarms or incorrect ToF measurements. Therefore, a two-dimensional FE model is established to further investigate the problem, which contributes to the analysis of wave motions inside the specimen. The model is built with COMSOL Multiphysics [27] using parameters consistent with the above experiment. Figure 5 shows the schematic diagram of the FE model. There are two things that need special care: one is that the left and right edge should be a low-reflection boundary to lower disturbances from the reflections on those edges, while the other is that the mesh size of the subsurface layer should be small enough to accurately analyze the eddy current as well as the Lorentz force.

This full coupled model employs a magnetic field and solid mechanics module to model the electromagnetic field and the mechanic field inside the specimen, respectively. The two modules interact through Lorentz force coupling. The detailed parameters used in this model are shown in Table 2 in addition to those in Table 1 above, where δ refers to the skin depth under the corresponding frequency and can be calculated from Equation (2).

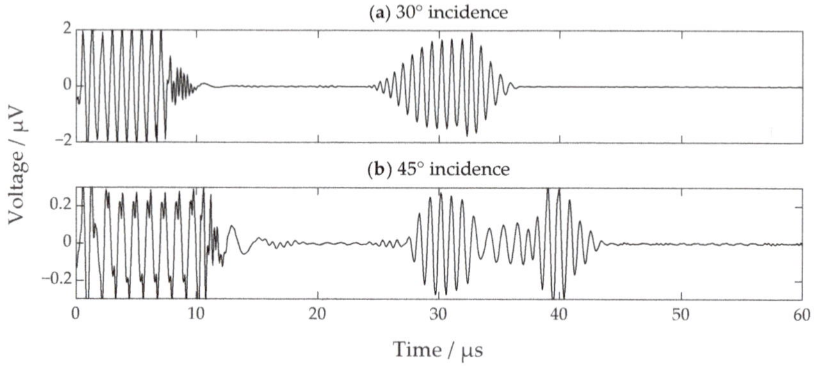

Figure 4. The received SV-wave signals when the incidence angles are (**a**) 30 degrees and (**b**) 45 degrees.

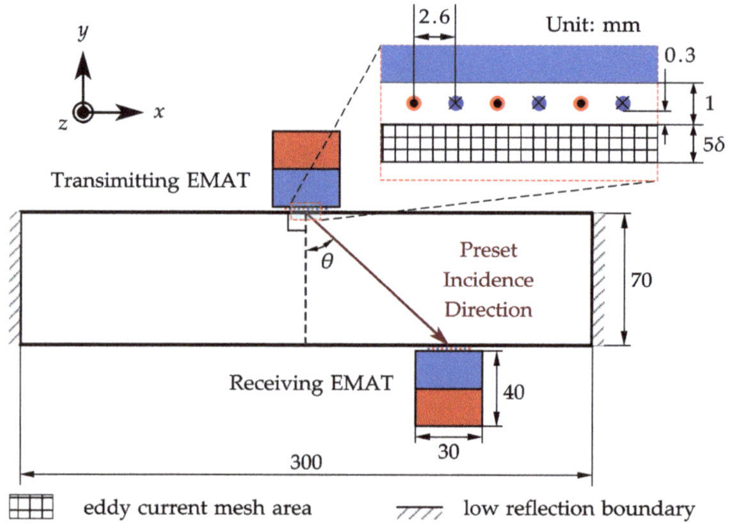

Figure 5. The schematic diagram of the angled SV wave EMAT FE model.

Table 2. Extra parameters used in FE model.

Parameter	Value
Model thickness	35 mm
Specimen dimension	70 mm × 300 mm
Specimen density ρ	2700 kg/m^3
Specimen relative permeability μ_{rc}	1
Specimen conductivity σ_s	3.774×10^7 S/m
Coil relative permeability μ_{rs}	1
Coil conductivity σ_c	5.998×10^7 S/m
Lift-off of coil	0.3 mm
Magnet residual flux density B_r	1.4 T
Lift-off of magnet	1 mm
Eddy current area meshing thickness	5δ
Eddy current area meshing size	$\delta/3$

The excitation signal is loaded through voltage inside the transmitting coil. The receiving signal can be extracted from the receiving coil. The simulated signals for incidence angles of 30 degrees and 45 degrees are shown in Figure 6, which are in good agreement with the experimental results. There are minor differences between the simulation and experimental results. The reasons for these differences are mainly due to the following: (1) slight anisotropy of the specimen, (2) misalignment of the transducers and (3) inaccurate parameters in the FE model. During the ultrasound transmitting, propagating, and receiving processes, both mechanical and electromagnetic fields are involved. Therefore, the good agreement between the simulated and experimental results can verify the validity of the FE model. This FE model will be used to simulate and analyze the acoustic field inside the specimen.

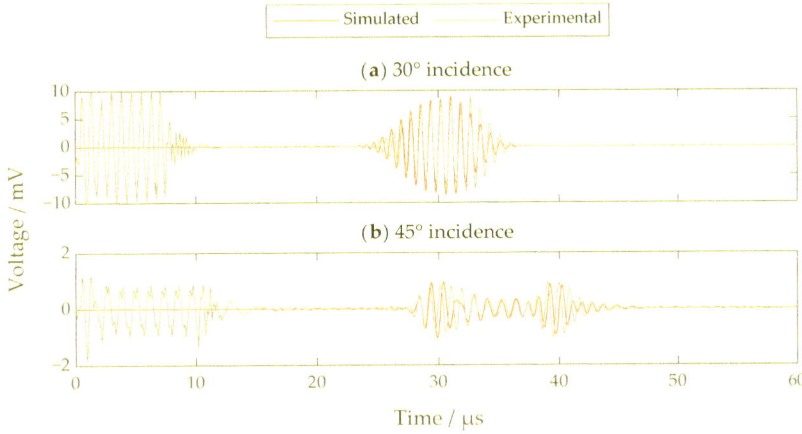

Figure 6. The simulated and experimental SV wave signals when the incidence angles are (**a**) 30 degrees and (**b**) 45 degrees. Here, the amplitudes of experimental signals are scaled to those of the simulated ones for comparison.

There are many quantities that can be used for visualizing acoustic fields, including displacement, velocity, acceleration, stress, etc. However, most of them simply reflect local motions of particles and are not intuitive for analyzing wave components when the acoustic field is relatively complex. Here, we use the second principal stress invariant [28] to visualize the acoustic field inside the specimen, which is given by

$$I_2 = \mathrm{tr}(\sigma)^2 - \mathrm{tr}(\sigma^2) = \sigma_{ii}\sigma_{jj} - \sigma_{ij}\sigma_{ji}, \tag{5}$$

where σ is the Cauchy stress tensor, $\mathrm{tr}(\cdot)$ is the trace of a matrix, and the component form uses the Einstein summation convention. The sign of I_2 divides the whole acoustic field into two types—one with a minus sign representing a shear wave, and the other with a plus sign representing a longitudinal wave. The magnitude of I_2 corresponds to the power of the acoustic field. Thus, the signed squared root of the second principal stress invariant can be used to visualize the bulk shear and longitudinal wave in a much clearer way, which is given by

$$\mathrm{A} \equiv \sqrt{|I_2|}\mathrm{sgn}(I_2), \tag{6}$$

where $\mathrm{sgn}(\cdot)$ is the sign of a number. A symmetric linear color map with blue and red for negative and positive values, respectively, is used in the following visualizations. Meanwhile, the acoustic field is highly symmetric, so only half of the field is shown in the following figures unless otherwise specified.

The acoustic field with an incidence angle of 30 degrees at 20 μs is shown in Figure 7a. Blue parts represent the shear wave, or more specifically the SV wave, and it propagates at the shear wave velocity. Red parts represent the longitudinal wave, and it propagates at the longitudinal wave velocity. The surface wave is not considered here. The wave front angle is defined as the angle between the wave front and the horizontal direction, which should be equal to the incidence angle θ by a geometry relationship. The shear wave front angle is measured as roughly 30.3 degrees, which is basically the same as the incidence angle.

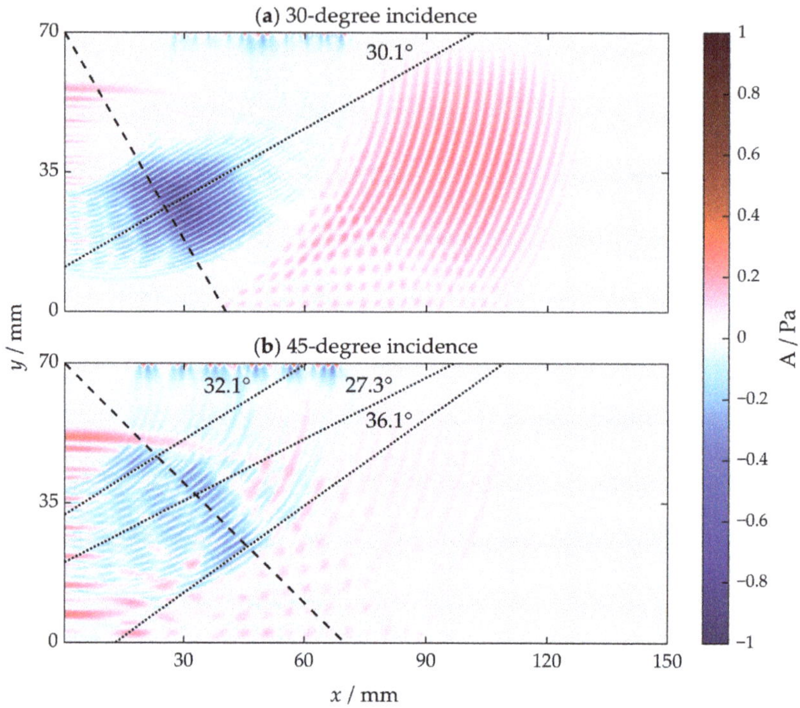

Figure 7. The simulated acoustic field from SV-wave EMAT at 20 μs when the incidence angles are (**a**) 30 degrees and (**b**) 45 degrees.

The acoustic field with an incidence angle of 45 degrees at 20 μs is shown in Figure 7b. Ideally, there should be an SV wave propagating along the direction identical to the preset incidence direction. However, in the configuration with an incidence angle of 45 degrees, the shear wave fronts have multiple angles, measured roughly as 32.1 degrees, 27.3 degrees and 36.1 degrees from top to bottom, respectively. None of them are 45 degrees, which is unexpected. Moreover, the wave front is not continuous, resulting in the discontinuity and interference of the time domain signal.

The acoustic field generated by the SV-wave EMAT is too complex to analyze. Therefore, the model was simplified to an elemental source to investigate the acoustic field and analyze the working principle of SV waves.

4. Analysis

Based on Equation (4), the working principle of SV waves requires that the waves generated by each source arrive at a plane with the same phase. The related calculation also assumes that the bulk wave by the same source has a cylinder wavefront and the same phase at every angle of an instant circle. The Lorentz force mainly acts within the skin depth (around 0.08 mm at 1MHz in the used specimen) of the specimen, so a boundary point source is a suitable approximation in a 2D space. Therefore, another model with a point load on the center of the top boundary is established to analyze the basic acoustic field, as shown in Figure 8. This model uses the solid mechanics module only. The geometry of the specimen is changed to 100 × 50 mm. The point load is set to be a 3-cycle 1MHz sinusoidal tone-burst with a direction parallel to the top boundary.

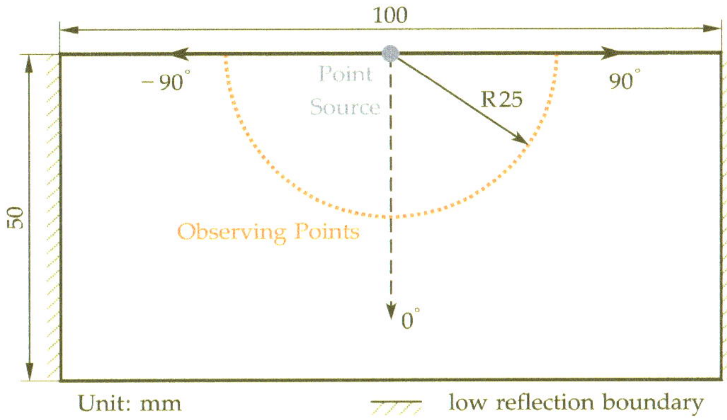

Figure 8. The schematic diagram of the point source model.

The acoustic field at 7.5 μs is shown in Figure 9. There are four different types of waves that can be distinguished from the results: a circular-front shear wave, a longitudinal wave, a surface wave, and a planar-front shear wave. The planar-front shear wave intersects with the circular-front shear wave, and the superposition of these two shear waves results in the non-uniform distribution of phase and amplitude on a circle.

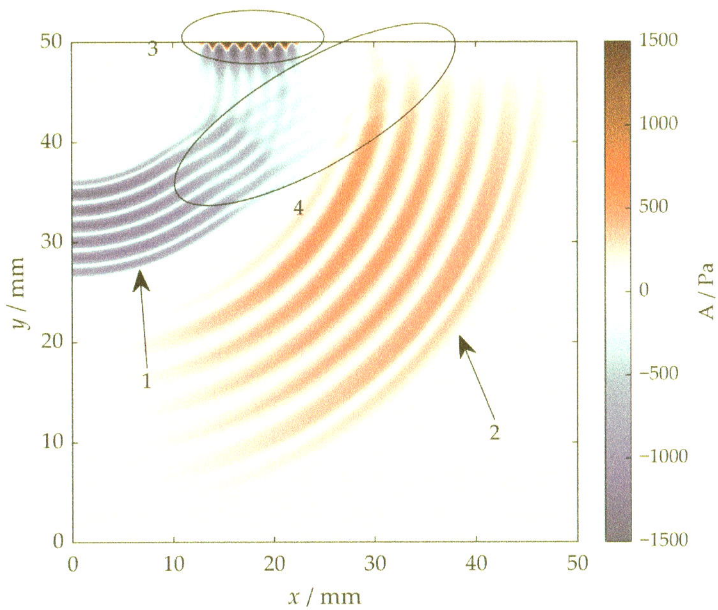

Figure 9. The acoustic field from a horizontal point source at 7.5 μs. There are 4 types of waves: circular-front (1) SV wave; (2) longitudinal wave; (3) surface wave; (4) planar-front SV wave.

A set of observing points is arranged in a 25 mm-radius semicircle with an even angle interval to further evaluate the amplitude and phase of the circular shear wave front versus the angle, as shown in Figure 8. A time-domain signal of the quantity A can be calculated for each observing point. Then a B-scan-like image is obtained via combining the time domain signal of every observing point, as shown in Figure 10. The *X*-axis represents the angle of the observing points when defining the direction normal to the surface as 0 degrees, as indicated in Figure 8. The Y-axis represents time, and the color represents the quantity A.

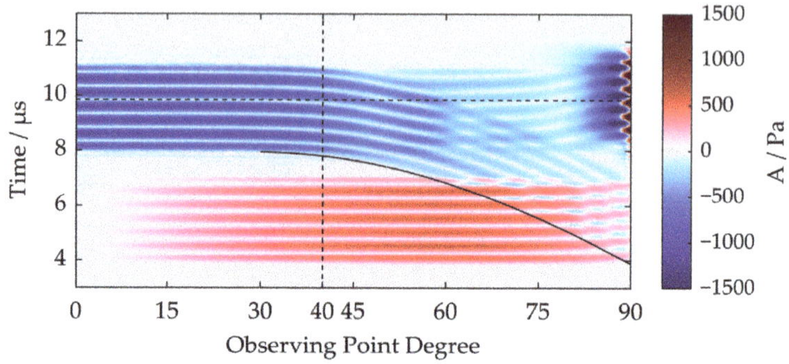

Figure 10. Time-angle distribution from a horizontal point source with observers on a 25 mm semicircle; the solid line refers to the calculated head wave ToF.

As indicated in Figure 10, the shear wave is uniform within 30 degrees, while abrupt changes in the wave front happen after 40 degrees, which is troublesome for beam-forming in those angles. Therefore, weakening the planar front shear wave should be a solution. In fact, this type of wave has been extensively studied in the past and has names such as head wave, von Schmidt wave, or refraction wave. In the half-space medium above, the head wave forms due to the interface and the surface-skimming longitudinal (SL) wave according to research originated by Goodier and Bishop [29]. Here, the SL wave refers to the part of the circular front longitudinal wave near the surface, whose propagation direction and polarization direction are both parallel to the surface. The SL wave, if considered as a plane wave with a horizontal incidence angle, will emit a planar front shear wave at the surface, which is a head wave and propagates at an angle equals to the critical refraction angle given by

$$\theta_h = \sin^{-1}\frac{c_s}{c_l}, \tag{7}$$

where the subscription h refers to the head wave. θ_h is 29.3 degrees for the material studied in this paper.

The time-of-flight (ToF) of the head wave from the point source to the observing point can be calculated through a geometric relationship, as illustrated in Figure 11, given by

$$t = \frac{r\cos(\theta - \theta_h)}{c_s}, \tag{8}$$

where r is the radius of the circle the observing point lies on, and θ is the observing point angle. The predicted ToF of the head wave is drawn in Figure 10 in a solid line to confirm that the interference wave is a head wave. Evidently, the head wave should be considered for reducing the interference between the circular shear wave and the planar shear wave. In addition, the amplitude of the head wave is proportional to that of the SL wave, so lowering the SL wave amplitude will help reduce the interference.

The horizontal vibration source creates a strong shear wave in the normal direction. However, it also creates a strong longitudinal wave in the horizontal direction, part of which is the unwanted SL wave. To reduce the SL wave and head wave, a vertical vibration source can be used. Switching the point load direction in the previous model to the vertical direction, the corresponding acoustic field at 7.5 µs is shown in Figure 12a. The amplitudes of both the SL wave and head wave decrease as expected. The circular shear wave is affected when the incidence is close to the normal direction, since there is no shear vibration in the normal direction. Fortunately, this is not a problem for large incident angle operation, indicated by the B-scan-like image obtained with the same observing points shown in Figure 12b.

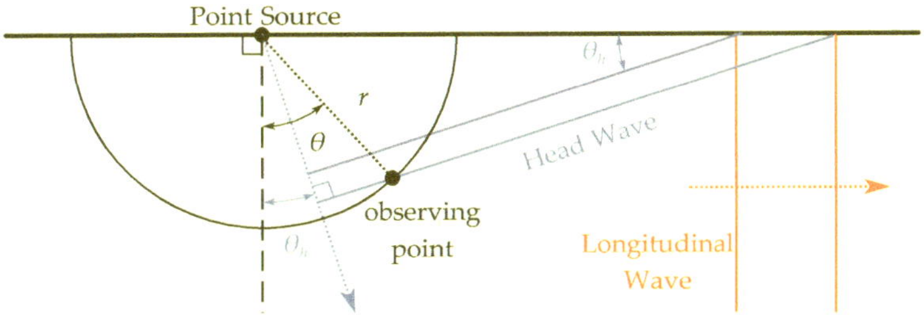

Figure 11. Illustration for head wave ToF calculation at the observing point.

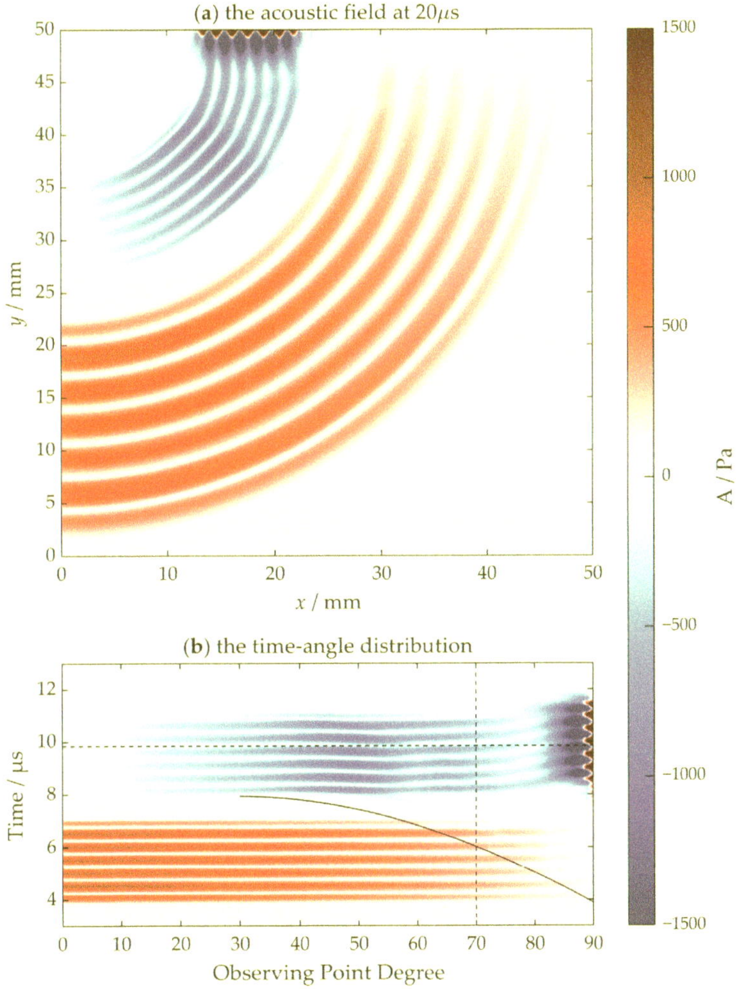

Figure 12. The simulation results of a vertical point source: (**a**) the acoustic field at 7.5 μs and (**b**) the time-angle distribution from observing points.

The influence of the head wave is not eliminated but has been significantly lowered—there is only a small change in the phase of the circular shear wave for a larger angle range, especially at large incidence angles. To be more precise, the incidence angle range of SV waves has been enlarged to be between 15 and 70 degrees. The head wave still needs to be considered at large angles, since the head wave arrives earlier than the circular shear wave.

5. New Configuration of SV-Wave EMAT for Large Incidence Angles

From the analysis, vertical force should outperform the horizontal force when generating SV waves at large incidence angles. According to Equation (1), the Lorentz force direction is related to the eddy current direction and the magnetic field direction, so using horizontal magnetization instead of vertical magnetization will generate the vertical Lorentz force. Figure 13 shows both the common and the proposed EMAT configurations.

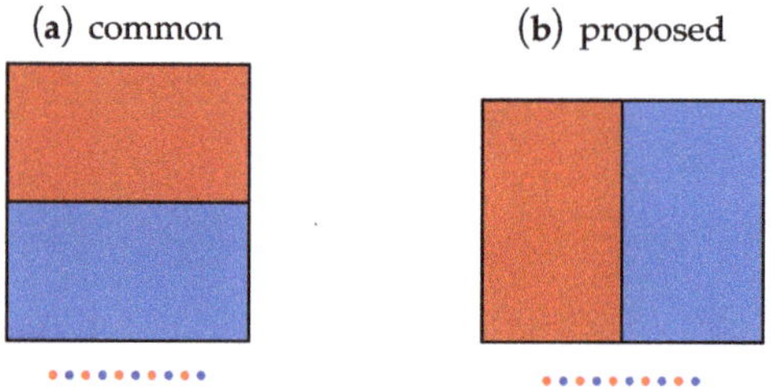

Figure 13. The structure diagram of (**a**) common and (**b**) proposed angled SV-wave EMAT.

The magnet direction in the full coupled FE model is set to be horizontal for both transmitting and receiving EMATs. Firstly, the Lorentz force distribution at $\delta/3$ under the upper surface with x between −15 mm and 15 mm is extracted and visualized in Figure 14 at an instant when it comes to the maximum. The corresponding force components of interest for the related configuration are drawn in red. With horizontal magnetization, the vertical Lorentz force predominates as intended.

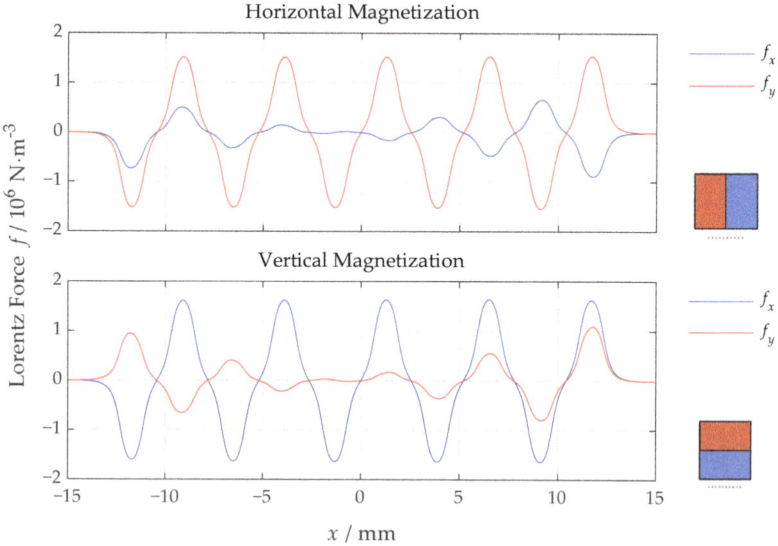

Figure 14. The subsurface Lorentz force distributions along x of both configurations.

Then, the acoustic fields in incidence configurations of 30 degrees and 45 degrees at 20 µs were examined separately. The result is shown in Figure 15. The corresponding received signals in the SV-wave EMAT are shown in Figure 16. From Figures 15 and 16, the SV wave with an incidence of 30 degrees with horizontal magnetization makes no improvement compared to that with vertical magnetization, while the situation with an incidence of 45 degrees makes a notable difference. The wave no longer has discontinuity and interference but has a uniform propagation direction, which precisely matches the preset incidence angle. In addition, the intensity of the wave is enforced as well.

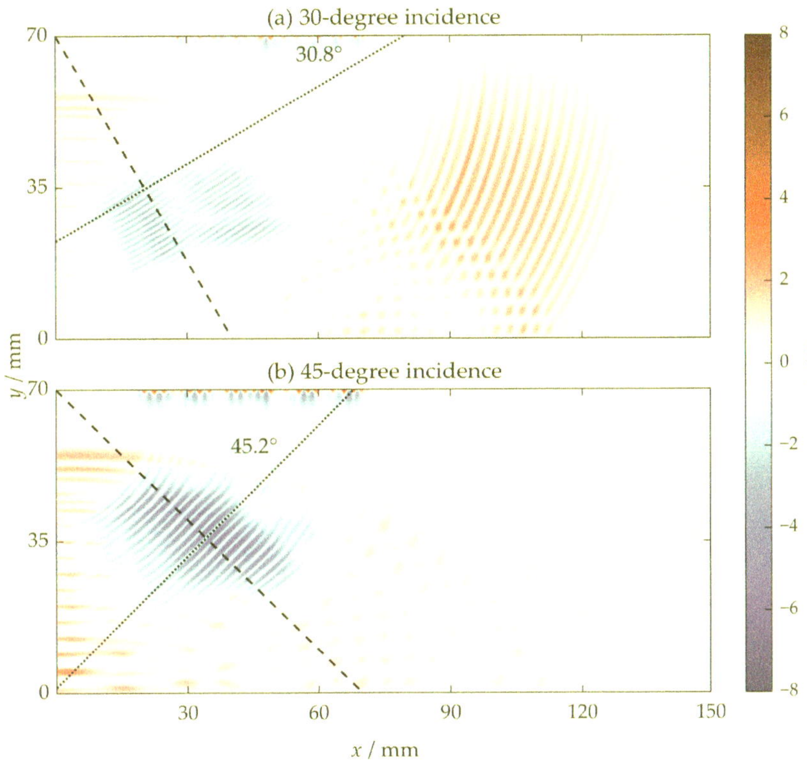

Figure 15. The acoustic field from the proposed horizontal magnetization EMAT at 20 μs when the incidence angles are (**a**) 30 degrees and (**b**) 45 degrees.

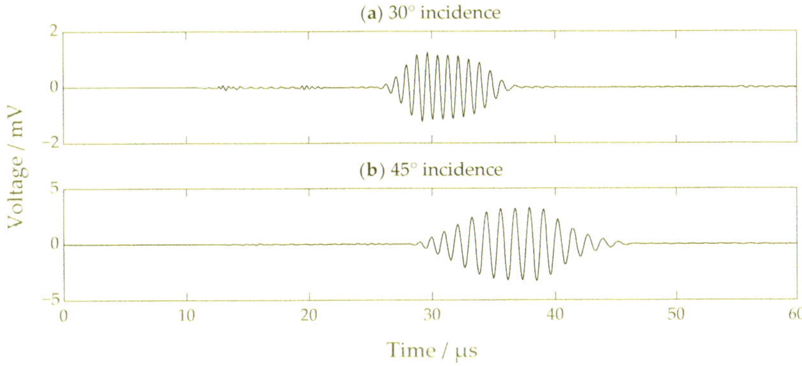

Figure 16. The simulated wave signals with proposed horizontal magnetization EMAT when the incidence angles are (**a**) 30 degrees and (**b**) 45 degrees.

6. Comparison between SV-Wave EMATs with Vertical and Horizontal Magnetization

Simulations and experiments were carried out for SV-wave EMATs with vertical and horizontal magnetic field configurations, respectively. The incidence angles varied from 30 degrees to 60 degrees with a 5 degree step, and the received SV wave signals were captured and compared for both configurations. The signals were normalized through dividing the maximum amplitude of the signals among all angles with the same configuration, and the experimental signals were compensated from corresponding amplifier gains. The simulation and experimental signals are shown in Figures 17 and 18, respectively. The signals with the horizontal magnetization configuration show great uniformity of the time-domain signal in the angle range of interest. As for the vertical magnetization configuration, the overall problems are the interfering head wave and weak signals at large incidence angles.

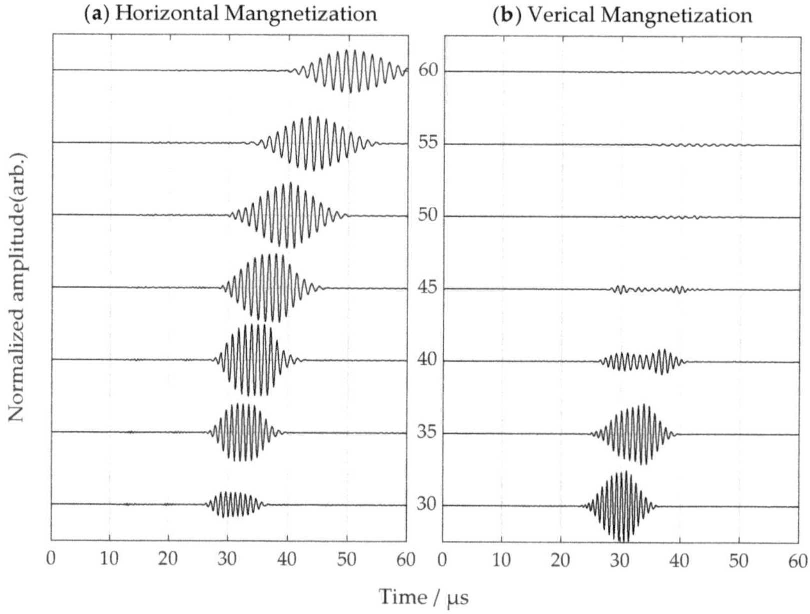

Figure 17. The simulated signals from (**a**) proposed horizontal magnetization EMAT and (**b**) common vertical magnetization EMAT; incidence angle ranges from 30 to 60 degrees. The amplitudes for normalization are 3.49 and 8.87 for (**a**,**b**), respectively.

It should be noted that the vertical magnetization configuration has a larger signal amplitude and better SNR than the horizontal magnetization configuration when the incidence angle of the SV wave is small (e.g., <40 degrees). Therefore, the vertical magnetization configuration should be considered when improving SNR is the priority. When the incidence angle of SV wave is the same or close to the angle of the head wave (e.g., the 30 degree scenario in the paper), the two shear waves interfere constructively. The signal is much stronger with the superposition of these two shear waves. The horizontal magnetization configuration is suitable for a wide range of incidence angles, making it suitable for wide-angle scanning tests. The horizontal magnetization configuration should be used if the incidence angle steerability is the concern. The incidence angle near the angle of the head wave should be avoided because the head wave may introduce errors in the incidence angle. Part of the interference signal can be filtered out through the differences in the ToF of the SV wave and the head wave.

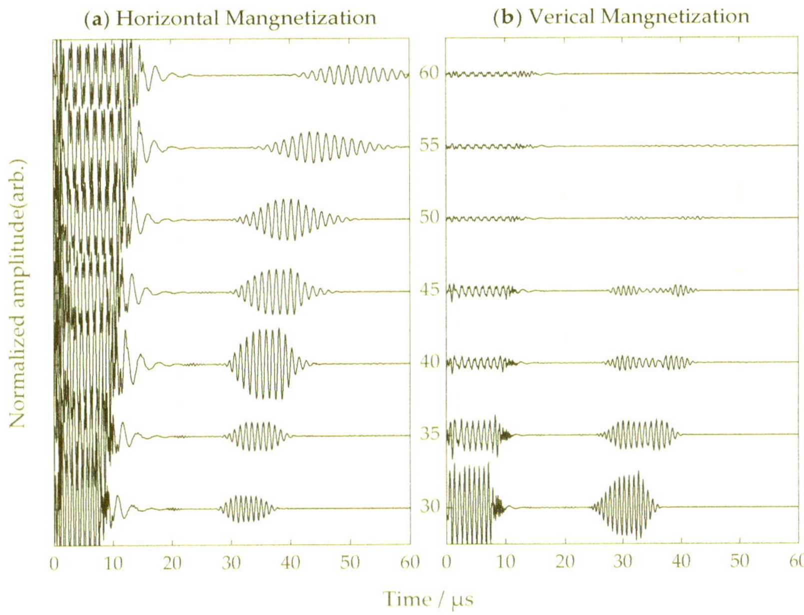

Figure 18. The experimental signals from (**a**) the proposed horizontal magnetization EMAT and (**b**) the common vertical magnetization EMAT. The incidence angle ranges from 30 to 60 degrees. The amplitudes for normalization are 0.33 and 1.92 for (**a**,**b**), respectively.

7. Conclusions

The typical angled SV-wave EMAT suffers from interference and weak signals when used for large-incidence-angle applications. In this paper, a fully coupled FE model was established and verified. The acoustic fields of an angled SV-wave were visualized with the aid of the FE model to analyze the interference in the SV-wave signal. Through 2D space point source analysis, the main reason was found to be the head wave, which is directly related to the surface skimming longitudinal wave. Accordingly, an EMAT with a horizontal magnetization configuration was proposed to suppress the impact of the head wave. The simulation and experimental results showed that the proposed configuration makes a great improvement in large incidence angles. Finally, the performances of both configurations for 30- to 60-degree incidence angles were compared with simulations and experiments. The proposed horizontal magnetization configuration outperforms the vertical configuration on both signal amplitude and interference when the incidence angle is larger than 40 degrees. Moreover, the horizontal magnetization configuration is suitable for wide-angle scanning tests.

The proposed EMAT adopts horizontal magnetization to solve the problem mentioned in this article. Further studies will be carried out to optimize the horizontal magnetic field of the EMAT, such as using a magnet array or a specially designed magnetic flux concentrator. Moreover, the parameters of the EMAT can be optimized with suitable optimization methods, and the directivity profile and steerability of the EMAT can be assessed and further improved.

Author Contributions: Conceptualization, Z.Q. and Z.L.; Formal analysis, Z.Q. and Z.L.; Funding acquisition, Z.L. and S.W.; Investigation, Z.Q., R.Y. and S.H.; Methodology, Z.Q. and R.Y.; Supervision, S.W.; Visualization, Z.Q.; Writing—original draft, Z.Q.; Writing—review & editing, Z.L. and S.W. All authors have read and agreed to the published version of the manuscript.

Funding: This research was funded by a Start-up Research Grant at the Harbin Institute of Technology under grant number AUGA5710012022 and by the National Natural Science Foundation of China under grant number 51977044.

Data Availability Statement: The data used in visualizations in this paper are available at https://github.com/joyqat/data-svemat.

Acknowledgments: The authors would like to thank Xu Zhang from the Hubei University of Technology for help with conceptions.

Conflicts of Interest: The authors declare no conflict of interest.

References

1. Hu, S.; Shi, W.; Lu, C.; Chen, Y.; Chen, G.; Shen, G. Rapid detection of cracks in the rail foot by ultrasonic B-scan imaging using a shear horizontal guided wave electromagnetic acoustic transducer. *NDT E Int.* **2021**, *120*, 102437. [CrossRef]
2. Wang, S.; Li, C.; Ma, X.; Liang, Q.; Zhai, G. Broadband torsional guided wave magnetostrictive patch transducer with circumferential alternating permanent magnet array for structural health monitoring. *Ultrasonics* **2022**, *125*, 106805. [CrossRef] [PubMed]
3. Dackermann, U.; Yu, Y.; Niederleithinger, E.; Li, J.; Wiggenhauser, H. Condition Assessment of Foundation Piles and Utility Poles Based on Guided Wave Propagation Using a Network of Tactile Transducers and Support Vector Machines. *Sensors* **2017**, *17*, 2938. [CrossRef] [PubMed]
4. Felice, M.V.; Fan, Z. Sizing of flaws using ultrasonic bulk wave testing: A review. *Ultrasonics* **2018**, *88*, 26–42. [CrossRef] [PubMed]
5. Feng, Q.; Li, R.; Nie, B.; Liu, S.; Zhao, L.; Zhang, H. Literature Review: Theory and Application of In-Line Inspection Technologies for Oil and Gas Pipeline Girth Weld Defection. *Sensors* **2017**, *17*, 50. [CrossRef] [PubMed]
6. Zhang, Y.; Qin, Z.; Luo, S.; Hyunjo, J.; Zhang, S. Design and Application of Partial Immersion Focused Ultrasonic Transducers for Austenitic Weld Inspection. *Sensors* **2022**, *22*, 2671. [CrossRef] [PubMed]
7. Nagamizo, H.; Kawashima, K.; Miyauchi, J. NDE of Pipe Inner Corrosion with Delayed Echoes of SV Wave Propagated Circumferentially in Liquid-Filled Pipes. In Proceedings of the Proceedings of the ASME 2003 Pressure Vessels and Piping Conference, Cleveland , OH, USA, 20–24 July 2003; pp. 7–12. [CrossRef]
8. Nakamura, N.; Ashida, K.; Takishita, T.; Ogi, H.; Hirao, M. Inspection of stress corrosion cracking in welded stainless steel pipe using point-focusing electromagnetic-acoustic transducer. *NDT E Int.* **2016**, *83*, 88–93. [CrossRef]
9. Clark, R. Rail flaw detection: Overview and needs for future developments. *NDT E Int.* **2004**, *37*, 111–118. [CrossRef]
10. Wang, W.; Zheng, Y.; Sun, Q.; Liu, S.; Yan, S.; Zhao, Z.; Liao, X.; Sun, X.; Phua, S.Y.; Kelvin, S.Y.P.; et al. Investigation and Study for Rail Internal-Flaw Inspection Technique. In Proceedings of the 2018 International Conference on Intelligent Rail Transportation (ICIRT), Singapore, 12–14 December 2018; pp. 1–4. [CrossRef]
11. Jiang, X.; Han, Q.; Zheng, H.; Tian, G.; Wang, P. Study on Sectorial Scan Angle Range for Phased Array Ultrasonic Testing in standard setting. In Proceedings of the 19th World Conference on Non-Destructive Testing (WCNDT 2016), Munich, Germany, 13–17 June 2016; NDT.net: Munich, Germany, 2016; p. 6.
12. Thompson, R.B. 4-Physical Principles of Measurements with EMAT Transducers. In *Physical Acoustics*; Thurston, R.N., Pierce, A.D., Eds.; Ultrasonic Measurement Methods; Academic Press: Cambridge, MA, USA, 1990; Volume 19, pp. 157–200. [CrossRef]
13. Zhai, G.; Liang, B.; Li, X.; Ge, Y.; Wang, S. High-temperature EMAT with double-coil configuration generates shear and longitudinal wave modes in paramagnetic steel. *NDT E Int.* **2022**, *125*, 102572. [CrossRef]
14. Ge, Y.; Wang, S.; Liang, B.; Li, X. Development of High Temperature Rayleigh Wave Electromagnetic Acoustic Transducer with Double Coil Structure. In Proceedings of the 2021 IEEE Far East NDT New Technology & Application Forum (FENDT), Kunming, China, 14–17 December 2021; pp. 119–123. [CrossRef]
15. Lunn, N.; Dixon, S.; Potter, M. High temperature EMAT design for scanning or fixed point operation on magnetite coated steel. *NDT E Int.* **2017**, *89*, 74–80. [CrossRef]
16. Lunn, N.; Potter, M.; Dixon, S. Shear wave EMAT thickness measurements of low carbon steel at 450 °C without cooling. In *AIP Conference Proceedings*; American Institute of Physics: College Park, MD, USA, 2017; Volume 1826, p. 050009. [CrossRef]
17. Burrows, S.; Fan, Y.; Dixon, S. High temperature thickness measurements of stainless steel and low carbon steel using electromagnetic acoustic transducers. *NDT E Int.* **2014**, *68*, 73–77. [CrossRef]
18. Petcher, P.; Potter, M.; Dixon, S. A new electromagnetic acoustic transducer (EMAT) design for operation on rail. *NDT E Int.* **2014**, *65*, 1–7. [CrossRef]
19. Qian, Z.; Li, X.; Qu, Z.; Wang, S. A Uniform Transversely Polarized Shear Horizontal Electromagnetic Acoustic Transducer for Stress Measurement. In Proceedings of the 2021 IEEE Far East NDT New Technology & Application Forum (FENDT), Kunming, China, 14–17 December 2021; pp. 129–133. [CrossRef]
20. Murav'ev, V.V.; Volkova, L.V.; Platunov, A.V.; Kulikov, V.A. An electromagnetic-acoustic method for studying stress-strain states of rails. *Russ. J. Nondestruct. Test.* **2016**, *52*, 370–376. [CrossRef]
21. Hirao, M.; Ogi, H. *Electromagnetic Acoustic Transducers*, 2nd ed.; Springer Series in Measurement Science and Technology; Springer: Tokyo, Japan, 2017. [CrossRef]

22. Ogi, H.; Hirao, M.; Ohtani, T. Line-focusing of ultrasonic SV wave by electromagnetic acoustic transducer. *J. Acoust. Soc. Am.* **1998**, *103*, 2411–2415. [CrossRef]
23. Takishita, T.; Ashida, K.; Nakamura, N.; Ogi, H.; Hirao, M. Development of shear-vertical-wave point-focusing electromagnetic acoustic transducer. *Jpn. J. Appl. Phys.* **2015**, *54*, 07HC04. [CrossRef]
24. Parra-Raad, J.; Cegla, F. On the steerability of phased array EMATs: The dipole element. *NDT E Int.* **2022**, *125*, 102563. [CrossRef]
25. RITEC Inc. Available online: https://www.ritecinc.com/ (accessed on 3 November 2022).
26. Orisonic Techonology. Available online: http://www.orisonic.com/ (accessed on 3 November 2022).
27. COMSOL Multiphysics ® v. 6.0. COMSOL AB, Stockholm, Sweden. Available online: https://www.comsol.com (accessed on 3 November 2022).
28. Ground Motion After Seismic Event: Scattering off a Small Mountain. Available online: https://www.comsol.com/model/ground-motion-after-seismic-event-scattering-off-a-small-mountain-78241 (accessed on 15 October 2022).
29. Goodier, J.N.; Bishop, R.E.D. A Note on Critical Reflections of Elastic Waves at Free Surfaces. *J. Appl. Phys.* **1952**, *23*, 124–126. [CrossRef]

sensors

MDPI

Article

Small Ultrasound-Based Corrosion Sensor for Intraday Corrosion Rate Estimation

Upeksha Chathurani Thibbotuwa [1,*], Ainhoa Cortés [1,2,*] and Andoni Irizar [1,2]

1 CEIT-Basque Research and Technology Alliance (BRTA), Manuel Lardizabal 15, 20018 Donostia-San Sebastián, Spain; airizar@ceit.es
2 Department of Electronics and Communications, Universidad de Navarra, Tecnun, Manuel Lardizabal 13, 20018 Donostia-San Sebastián, Spain
* Correspondence: uthibbotuwa@ceit.es (U.C.T.); acortes@ceit.es (A.C.); Tel.: +34-943212800 (A.C.)

Abstract: The conventional way of studying corrosion in marine environments is by installing corrosion coupons. Instead, this paper presents an experimental field study using an unattended corrosion sensor developed on the basis of ultrasound (US) technology to assess the thickness loss caused by general atmospheric corrosion on land close to the sea (coastal region). The system described here uses FPGA, low-power microcontroller, analog front-end devices in the sensor node, and a Beaglebone black wireless board for posting data to a server. The overall system is small, operates at low power, and was deployed at Gran Canaria to detect the thickness loss of an S355 steel sample and consequently estimate the corrosion rate. This experiment aims to demonstrate the system's viability in marine environments and its potential to monitor corrosion in offshore wind turbines. In a day, the system takes four sets of measurements in 6 hour intervals, and each set consists of 5 consecutive measurements. Over the course of 5 months, the proposed experiment allowed for us to continuously monitor the corrosion rate in an equivalent corrosion process to an average thickness loss rate of 0.134 mm/year.

Keywords: corrosion monitoring; corrosion rate; FPGA; offshore wind turbines; ultrasound; thickness loss

Citation: Thibbotuwa, U.C.; Cortés, A.; Irizar, A. Small Ultrasound-Based Corrosion Sensor for Intra-Day Corrosion Rate Estimation. *Sensors* **2022**, *22*, 8451. https://doi.org/10.3390/s22218451

Academic Editor: Zenghua Liu

Received: 16 September 2022
Accepted: 26 October 2022
Published: 3 November 2022

1. Introduction

As the offshore wind-energy industry is prepared to increase its capacity in the following years [1] it is also facing important challenges. The need for wind farms farther away from land (10–50) km and at increasing depths (>200 m) [2] has led to new wind tower designs and deployment strategies [3] to reduce capital expenditure (CAPEX). However, in the medium and long term, wind-farm operators face increasing costs due to the operation of the wind farms in extremely harsh environments. Maintenance tasks in offshore wind farms need to be thoroughly prepared in advance because access to the wind towers is expensive in terms of both trained staff and trip costs [4].

Therefore, structural health monitoring (SHM) has become one of the major disciplines for offshore wind turbines (OWTs) with the necessity of improving operation and maintenance strategies [5]. SHM provides a strategy for the damage detection of structures in the mean of changes in material or geometric properties that affect current or future performance [6]. Hence, the continuous structural health monitoring of wind turbines would be beneficial in improving structural reliability and optimizing maintenance tasks at minimal associated costs [7]. Corrosion is one of the main roots for degradation of offshore structure materials which could eventually lead to damage and weaken the structure [8]. The deterioration of the materials and their properties due to corrosion is a fact that reduces the useful service life of the material and leads to fail the structures, equipment and other objects from their intended functioning. Studies show that the estimated cost of corrosion is about 3–4%, of each nation's gross domestic product (GDP). It was estimated that between 15% and 35% of the cost of corrosion could be saved through improved corrosion management [9].

Corrosion Monitoring in Offshore Wind Turbines

Corrosion in the tower and other equipment within the wind tower is one of the main consequences of an offshore environment [10]. In particular, corrosion in the tower and its foundations is critical because it could potentially lead to a structural failure of catastrophic consequences. The form of corrosion on offshore steel wind turbine towers varies for each tower zone depending on the access to the levels of oxygen, humidity, and water, including water water temperature, salinity, and depth [11].

Corrosion coupons are the most conventional technique for calculating the corrosion rate in an offshore environment. Corrosion rates for construction steel below sea level are 0.2 mm/year on average; in tidal and splash zones, the rate may fluctuate from around 0.4 to 1.2 mm/year [12]. Reported corrosion rates of steel from different marine locations are available in [13]. Moreover, a 0.83 mm/year of corrosion was experimentally estimated in the North Sea, which is quite a higher value compared to other reported values [14].

Hence, the early detection of the possible risks of structural failure allows for avoiding severe damage to the structure and saving on associated costs. For that, maintenance operators are very keen on proper and efficient solutions that can continuously monitor corrosion unattended while using wireless connectivity to obtain real-time data, with the possibility to move around the tower to inspect new areas as required.

Corrosion monitoring covers a broad range of techniques that involve measuring and converting a measured parameter into corrosion loss or rate [15–17]. On the basis of the sensing principle, different types of corrosion detection sensors are available [18]. The change in the material could be sensed via the loss of weight, the alteration of physical, chemical, electrical, magnetic, or mechanical properties, and the loss of component integrity (e.g., cracking). Nondestructive testing (NDT)-based corrosion detection allows for inspecting the material without disturbing the material properties. Hence, NDT techniques are more suitable for corrosion monitoring solutions with real-time data acquisition systems and remote data logging.

Ultrasound techniques for corrosion monitoring are based on the continuous reduction in the thickness of a material immersed in a corrosive environment. As such, they are normally classified as a direct, nondestructive, and nonintrusive technique. Nonintrusiveness means that the sensor itself does not influence the corrosion mechanism taking place. Furthermore, the ultrasound sensor can be located on the "clean" side of the material being corroded. There are many commercial ultrasound devices that provide valuable information about various forms of degradation (cracks, deformations, thinning, corrosion, etc.). However, in general, they are classified as inspection devices rather than monitoring devices. Inspection means measurements carried out by trained staff over a predetermined time frame in accordance with maintenance schedules. Instead, corrosion monitoring devices perform very frequent measurements with the aim of detecting small fluctuations, in this case related to corrosion phenomena. Thus, the data coming from the corrosion monitoring devices can be used as input for a general assessment of corrosion. The precision obtained with ultrasound signals allows for very frequent thickness measurements that can be used to track corrosion.

In this paper, we present a miniaturized corrosion monitoring solution based on ultrasound technology for offshore platforms. The ultrasound probe was glued onto an S355 bare steel sample of 5 mm in thickness. The developed system was deployed at a corrosive environment in Gran Canaria with the aim of analyzing the thickness loss evolution and estimating the corrosion rate in a real and harsh scenario.

This paper is organized as follows. The theory of the ultrasound-based sensor is presented in Section 2, including the theory of ultrasound with ToF estimation in Section 2.1, the description of the corrosion sensor in Section 2.2, the discussion of the ToF calculation algorithm with the temperature compensation in Section 2.3, and how to calculate the relative thickness loss is discussed in Section 2.4. Next, the introduction to the experiment of measuring thickness loss due to corrosion in real conditions is provided in Section 3, followed by the experiment's results and discussion in Section 4.

2. Theory

2.1. Ultrasound Theory

In the ultrasound technique, the corrosion loss rate is determined on the basis of wall thickness loss. During the ultrasound test, a short duration of ultrasound energy is introduced into the test object at periodic intervals of time. The sound energy propagates through the material in the form of mechanical vibrations. This propagation attenuates and weakens the sound wave, with the travel distance mainly as a result of scattering, absorption, and reflection [19]. When a sound wave strikes a medium interface, part of the energy is transmitted into the next medium across the boundary, while some is reflected into the first medium as a result of different acoustic impedances in different media or materials. The amount of reflected or transmitted energy gives information about the size of the reflector.

On that basis, there are two approaches that can be used to estimate thickness during ultrasound testing: through-transmission and pulse-echo techniques. In the through-transmission technique, the transmitted part of the signal through the test specimen is monitored, whereas in the pulse-echo technique, the reflected signal at different boundaries is considered [20]. Hence, in through transmission, two transducers are necessary for transmission and reception, whereas in the pulse-echo technique, a single transducer is enough for both tasks.

Generally, ultrasound nondestructive-testing-based measurements involve wave propagation (transit) time across a given distance and a degree of attenuation that takes place at that time [21]. Measurement modes used to find the transit time of a sound wave in the process of thickness calculation can be classified on the basis of the choice of echoes [22]:

- Mode 1: excitation signal—first back-wall echo.
- Mode 2: near surface—first back-wall echo.
- Mode 3: two successive back-wall echoes.

The probes used to generate and receive ultrasonic energy are known as ultrasonic transducers. Typically, these transducers are produced from piezoelectric ceramic or composite material [23]. The performance efficiency of a piezoelectric transducer depends on the proper matching of electrical and acoustic impedance. Electrical impedance matching needs the design of an electrical matching circuit between the driving circuit and the transducer; the acoustic impedance between the piezoelectric generators/detectors and the propagating media likewise needs to be matched. A mismatch could lead to a loss in signal transmission and a low signal-to-noise ratio (SNR) [24]. The electrical impedance can be matched in different ways, such as using an impedance matching electrical network, and matching the characteristics and properties of the cable connected to the transducer [25]. A significant acoustic impedance mismatch results in most of the signal's energy being reflected. Hence, a couplant material (often a liquid, such as a thin film of oil, glycerin, or water) is typically required in nondestructive material testing to reduce the acoustic impedance mismatch between air and a test specimen.

In our solution, we used the pulse-echo technique with a single direct-contact transducer, and Mode 3 was selected to estimate the thickness. Moreover, relative measurements were performed to observe the thickness loss in a certain location where the sensor was placed, which we discuss further in Section 2.4.

The capacity of the ultrasound technique was presented in [26], measuring the wall thickness loss of a mild steel sample under accelerated laboratory conditions. Our approach presents a novel solution, applying the ultrasound technique to a smart continuous monitoring system deployed on a marine environment over the course of several months.

Estimation of Time of Flight

Using ultrasonic waves, corrosion condition is evaluated on the basis of the wall thickness loss caused by corrosion. Generally, the two common ways to determine the geometric distance measurement on the basis of an ultrasound response are phase shift and time of flight (ToF) [27,28].

ToF is the time taken by ultrasound signal to arrive from transmitter to receiver corresponding to the thickness d of the test specimen; when a single transducer is used, this relation can be stated as in Equation (1).

$$d = c \times \text{ToF}/2 \quad (1)$$

where c is the speed of sound in the material. This value depends on the physical properties of the material/medium (elastic moduli and the density of the material) [29]. The most common technique employed in many ultrasonic applications to estimate a signal's transit time through a test material is the ToF technique, which was also applied to our approach.

Applying a threshold with an analog comparator at a certain amplitude level is the most fundamental and straightforward method of calculating the ToF value. A more accurate and reliable way is to apply appropriate digital signal processing techniques. In that case, the ultrasound response is processed on the basis of its echo parameters. One of the most common methods is to consider the received signal as a delayed and attenuated (amplitude-scaled) version of a reference signal with noncorrelated additive noise [30]. The objective is to estimate a ToF measurement by matching the received signal with a reference signal. For that, the cross-correlation function is widely used to mathematically quantify the highest similarity between the reference and delayed version of signals/echoes in the time domain. An estimation of the ToF value is the time at which the cross-correlation result reaches its maximum.

However, the precision of a ToF measurement depends on the quality of the received signal. ToF value estimation can become more complex in the presence of noise, interference, scattering, and attenuation of the signals [31]. Thus, it may require a preprocessing step of filtering before the cross-correlation process. Even afterwards, ambiguities detecting the actual ToF value in the cross-correlated result are possible; in such situations, the precision of the ToF estimation could be improved with additional computational steps [31–34].

Our solution uses a single transducer based on the pulse-echo method. As the same transducer is employed for transmission and reception, it is important to consider the dead zone of the sensor, which is the interval of the transducer ringing due to the initial pulse and not yet completely ready to receive/detect reflected echoes [35]. Figure 1 shows a raw signal obtained for a bare steel sample of 5 mm thickness. There is an interval of approximately 440 samples (3.5 μs) produced by the piezoelectric sensor's dead zone, in which the output of the sensor presented transient behavior due to the characteristics of the signal's generation circuitry and the electrical adaptation inside the piezoelectric sensor. Since the first back-wall echo falls into the dead zone of the piezoelectric sensor, the next two consecutive back-wall echoes were used to calculate the ToF value (using Mode 3 as mentioned in Section 2.1).

2.2. Description of the Corrosion Sensor

Our miniaturized and low-cost monitoring solution (see the electronics in Figure 2) is expected to operate unattended under the harsh conditions of offshore platforms for a long period, estimating the thickness loss due to corrosion. A small size (110 × 60 × 15 mm), wireless connectivity, and low power (5.4 μW in standby mode, 850 mW average power for measurement events with a duration of 200 μs) are some important features to facilitate deployment and to provide higher autonomy. On top of that, the low weight of the system (around 100 g) is important to obtain a feasible solution integrated into a mobile platform with the aim of covering large structures by using the same sensor node.

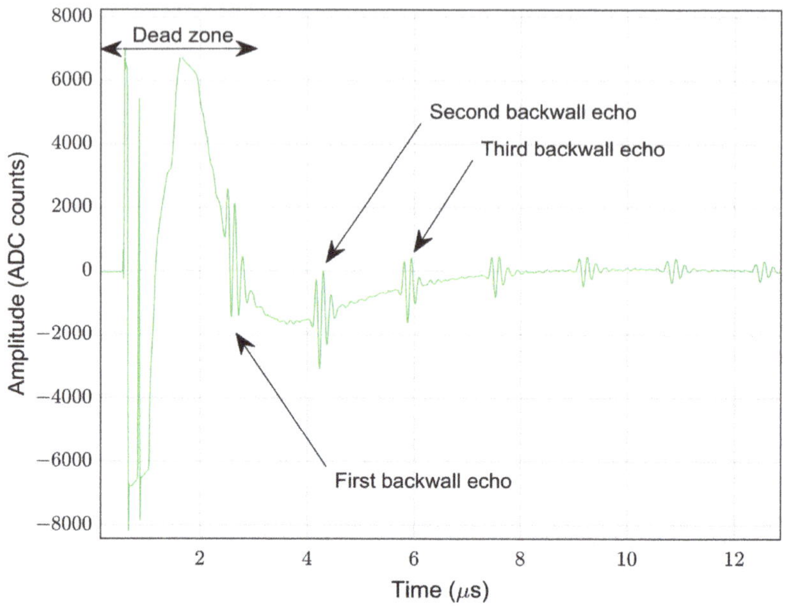

Figure 1. Raw piezoelectric signal with echo representation.

Figure 2. Electronics of the ultrasound corrosion sensor node.

The ultrasound probe/transducer used in the proposed solution is V111 from Olympus [36] with a peak frequency of 8.44 MHz and 15 mm of diameter of contact. The sensor was chosen after analyzing the performance specifications prioritizing its lower waveform duration (higher bandwidth) and sound pressure power.

Architecture of the Sensor Node

Typically, an ultrasound testing system consists of a pulser, a transmitter/receiver transducer, an amplifier, and data collection and display devices/platform. The hardware architecture of our corrosion sensor comprised a low-power Cortex-M4-based microcontroller interfacing with an Intel MAX 10 FPGA by using a serial peripheral interface (SPI) link and analog front end (AFE) devices as shown in Figure 3. More details related to the microcontroller and the wireless connectivity of the system are provided in [37].

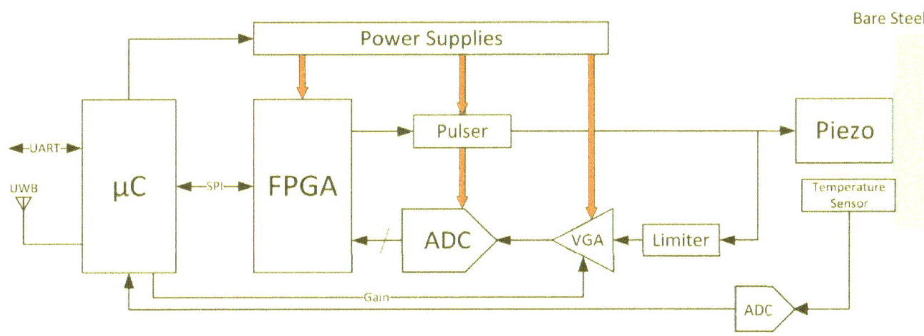

Figure 3. Hardware architecture of the ultrasound corrosion sensor node.

When the microcontroller requests a ToF measurement, the field-programmable gate array (FPGA) takes control of the devices that form the data acquisition circuit [36]. Thus, the FPGA is responsible for generating enabling signals for the ultrasound pulser and the trigger to start the data acquisition. The ultrasound pulser circuit used in this work can generate high-voltage, high-frequency, unipolar, or bipolar pulses. In the this paper's experiment, the excitation pulse is a bipolar square wave with 50% duty cycle. The pulse frequency in MHz was set according to Equation (2), where 125 MHz is the sampling frequency and FREQC was set as an even positive integer to maintain a duty cycle of 50%.

$$f = 125/\mathrm{FREQC} \tag{2}$$

Once a measurement is performed, the FPGA processes the acquired data and provides the calculated ToF estimation to the microcontroller. This signal processing is controlled by the microcontroller through the design parameters explained in Table 1 and shown in Figure 4 over the filtered ultrasound response. The LOC and PEAK values shown in Figure 4 are calculated in the FPGA and are used internally to extract the different echoes. They are also accessible via SPI as read-only registers.

The ultrasound response processed by the FPGA was previously amplified by a low-noise variable gain amplifier (VGA) that was composed of two cascaded amplifiers, with each one providing a maximal amplification of up 38 dB. Both amplifiers were controlled with analog voltage V_{ctrl} from 0.2 to 2.3 V. The relation between the control voltage V_{ctrl} and the total amplification G is given by Equation (3).

$$G(\mathrm{dB}) = 44 \times V_{ctrl} - 23.6 \tag{3}$$

The system is capable of generating the V_{ctrl} for the amplifier using the microcontroller and allowing for further degrees of flexibility of its operation. The corresponding signal level of the ultrasound response (root mean square (RMS) value of the first detected echo) is measured by the FPGA and sent to the microcontroller via SPI with the aim of implementing an automatic gain controller. The signal processing algorithm for ultrasound response analysis takes place in the FPGA and is explained next in Section 2.3.

Table 1. Parameters to configure the digital processing from the microcontroller.

Parameter Name	Description	Value
SIZEUS	Number of data samples received from the ADC	2048
NT	Number of samples that must be trimmed from the envelop signal to remove of the piezo's dead zone	0
FREQC	Number of cycles of the sampling period to obtain the piezo sensor's resonance frequency	16
NCYC	Number of periods of the excitation pulse	1
TOFREF	Approximate expected value of the ToF in number of samples	210
SIZEW	Typical echo duration in number of samples	192
ECHON	First echo to be detected and processed	2
SIZEWL	Number of cross-correlation samples taken on the left-hand side of the correlation index 0	32
SIZEXC	Total number of processed cross-correlation samples	64

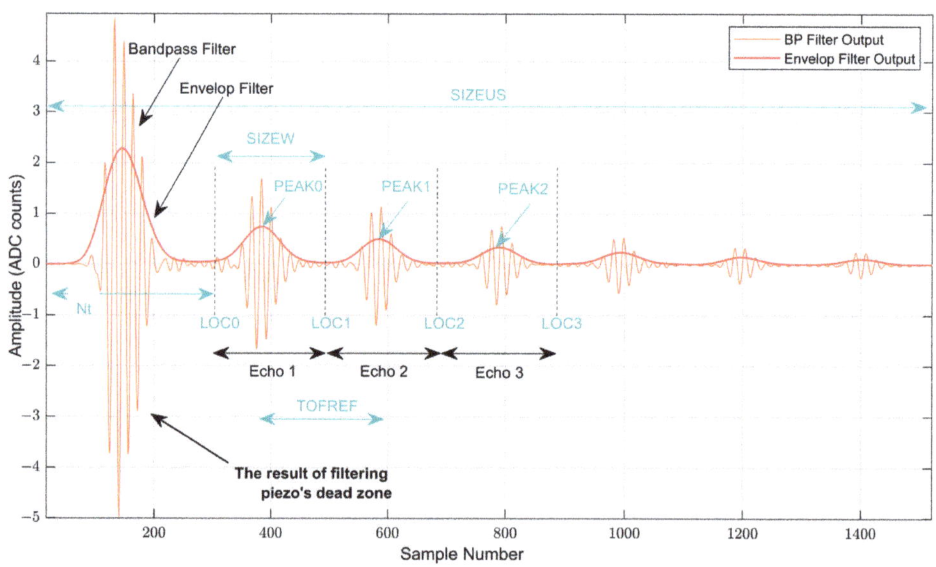

Figure 4. Design parameters used for output response signal analysis.

Furthermore, the sensor node reads from a 16-bit analog-to-digital converter (ADC) (LT2451) [38] through an interintegrated circuit (I2C) to link temperature values to the corrosion measurements using the NTC10K3A1I thermistor [39] as the temperature sensor to acquire the temperature of the steel sample. The variation of ToF with temperature is discussed in Section 2.4.

2.3. Algorithm Description of ToF Calculation

The algorithm extracts two consecutive back-wall echoes from the received ultrasound response with an echo windowing process that separates the echoes for the ToF estimation.

The thickness at a certain time and position is determined by using the ToF technique as given in Equation (1). The size of the S355 bare steel sample to test the algorithm was $75 \times 150 \times 5$ mm. Therefore, considering the initial thickness of the sample to be 5mm, the expected delay between echoes was $2 \times 5/5.9 \times 10^6 = 1.69$ µs (the speed of sound in steel is approximately 5.9×10^6 mm/s).

A block diagram of the digital processing stages performed inside the FPGA to calculate the ToF value between two consecutive echoes is given in Figure 5. The signal from the piezoelectric sensor was first bandpass (BP)-filtered; then, the envelop of that signal was obtained. Both signals, the outputs of the bandpass filter, and the envelop filter are shown in Figure 4.

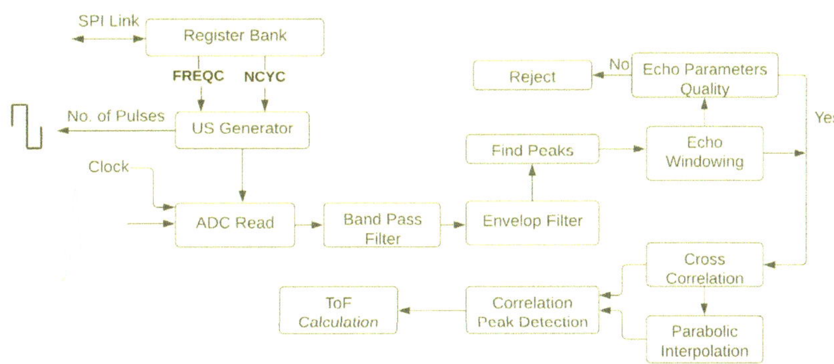

Figure 5. Block diagram of digital signal processing in FPGA.

The two extracted consecutive echoes (see Figure 4), Echos 2 and 3 from the bandpass signal, were cross-correlated to detect the point of maximal matching to determine the ToF value. In the cross-correlation approach, the obtained time delay at the maximal cross-correlation peak was used for estimating this ToF value. Generally, time delays in cross-correlation results are not integral multiples of the sampling period. Therefore, the largest cross-correlation peak position could be between the indices of the time-delay vector. Thus, to improve the precision of the ToF value, a cross-correlation function (ccf) can be interpolated between the samples [40]. In our solution, parabolic interpolation was used to estimate the location of the ccf peak with the aim of reducing the computational complexity. The final cross-correlation output from the second and third consecutive echoes is shown in Figure 6a. To obtain the final ccf peak, curve fitting was performed for the maximal cross-correlation peak, and its two nearest neighbors were as shown in Figure 6b. The peak offset (δToF) was estimated using parabolic interpolation given in Equation (4), where $r[0]$ is the largest peak of the cross-correlation result, and $r[-1]$ and $r[+1]$ are its two nearest neighbors.

$$\delta \text{ToF} = \frac{0.5 \times (r[-1] - r[+1])}{r[-1] + r[+1] - 2 \cdot r[0]} \tag{4}$$

Accordingly, the final cross-correlation peak after parabolic interpolation was obtained at δ ToF. In the example shown in Figure 6b, the maximal correlation peak was obtained at correlation index -10. Therefore, the estimated delay between the two echoes was $-10 + \delta\text{ToF}$.

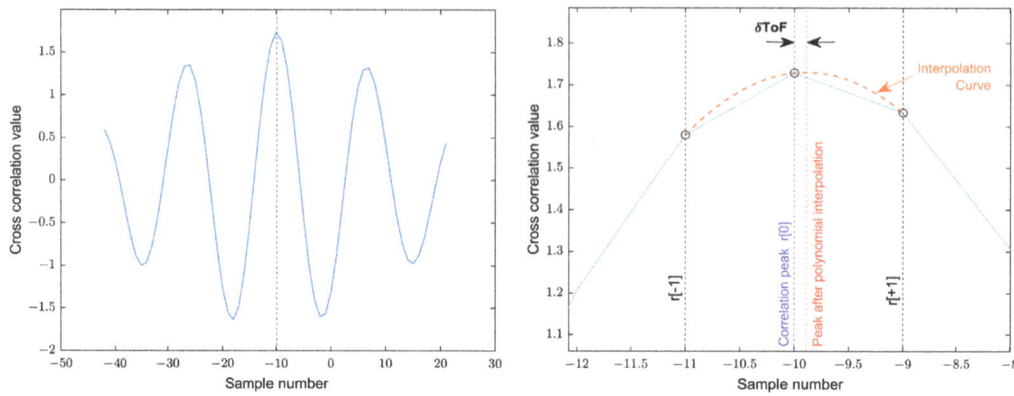

(a) **(b)**

Figure 6. (**a**) Final cross-correlation output; (**b**) parabolic interpolation around the maximum of cross correlation.

The methodological design presented in [35] was followed to validate this digital processing. The first step was to use our own testbed on the basis of Red Pitaya [41] to acquire ultrasound signals. Then, a model was developed in MATLAB to process the ultrasound responses. Afterwards, the digital processing algorithm was implemented in Red Pitaya and compared against the MATLAB model. Lastly, this signal processing was implemented into the FPGA and verified using the MATLAB model.

2.4. Calculation of Relative Thickness Loss Using ToF

The measurement of ToF in ultrasound signals propagating through steel is directly related to variation in the speed of sound with temperature and, to a lesser extent, to thermal expansion. Therefore, we needed to measure the temperature of the steel to compensate for this temperature effect in the measurement of ToF. Let us assume that, in the absence of corrosion, we were measuring the ToF of a bare steel sample of thickness L_{s0} at a room temperature T_0 and the speed of sound v_{s0} also at T_0. For the relatively small temperature ranges that we considered, the propagation speed of the ultrasonic signal in steel at a given temperature T follows this expression:

$$v_s(\Delta T) = v_{s0} \cdot (1 + \xi_1 \Delta T + \xi_2 \Delta T^2 + \cdots), \tag{5}$$

where ξ_i are constants of the material (°C^{-i}), $\Delta T = T - T_0$ is the temperature difference with respect to T_0. For the range of working temperatures, 15 °C$< T <$ 25 °C, we can safely assume that a first-order approximation is more than enough, and $\xi = \xi_1$. At a given temperature T, the measured ToF is:

$$\text{ToF} = \frac{2L_s}{v_s} = \frac{2L_{s0}(1 + c_x \Delta T)}{v_{s0}(1 + \xi \Delta T)} = \text{ToF}_0 \frac{1 + c_x \Delta T}{1 + \xi \Delta T}, \tag{6}$$

where c_x is the thermal expansion coefficient of steel ($c_x \approxeq 12 \times 10^{-6}$ °C^{-1}). The value of ξ for steel is small and negative, so the speed of sound decreases as temperature increases.

$$\begin{aligned} \text{ToF} &\approx \text{ToF}_0(1 + c_x \Delta T)(1 - \xi \Delta T) = \text{ToF}_0[1 + (c_x - \xi)\Delta T] \\ \text{ToF}/\text{ToF}_0 &\approx 1 + (c_x - \xi)\Delta T \\ \frac{\Delta \text{ToF}}{\text{ToF}_0} &\approx (c_x - \xi)\Delta T \end{aligned} \tag{7}$$

The value of $(c_x - \xi)$ can be experimentally determined by running an experiment that measures the ToF in a bare steel sample in a climatic chamber (see Section 2.4.1).

In the initial state, the bare sample presented no corrosion, and the first ToF measure was taken at temperature T_0. We call that value ToF_0, which corresponds to an initial thickness L_{s0}:

$$\mathrm{ToF}_0 = \frac{2L_{s0}}{v_{s0}} \tag{8}$$

If we now consider the situation of a corroded sample at temperature T, the thickness of that sample is related to ToF as follows:

$$\begin{aligned} L_n &= \frac{\mathrm{ToF} \times v_s}{2} = \frac{\mathrm{ToF}}{2} \times v_{s0}(1+\xi\Delta T) = \frac{\mathrm{ToF}}{2} \times \frac{2L_{s0}}{\mathrm{ToF}_0}(1+\xi\Delta T) \\ &= \frac{\mathrm{ToF}}{\mathrm{ToF}_0} \times L_{s0}(1+\xi\Delta T) \end{aligned} \tag{9}$$

However, L_n is the thickness after corrosion at temperature T. We must reference all measurements to the same temperature T_0. Thickness L_n is related to the thickness after corrosion at temperature T_0, L_{n0}, as follows:

$$L_n = L_{n0}(1 + c_x\Delta T) \tag{10}$$

Therefore, the loss of thickness due to corrosion at temperature T_0 substituting Equation (9) into the above equation is:

$$\begin{aligned} \frac{\Delta L_c}{L_{s0}} &= \frac{\mathrm{ToF}}{\mathrm{ToF}_0} \times (1 + (\xi - c_x)\Delta T) - 1 \\ &= \frac{\Delta\mathrm{ToF}}{\mathrm{ToF}_0} + \frac{\mathrm{ToF}(\xi - c_x)\Delta T}{\mathrm{ToF}_0} \end{aligned} \tag{11}$$

2.4.1. Temperature Experiment

To experimentally determine the value of ξ, we conducted an experiment that consisted of placing the sample in a climatic chamber and linearly varying the temperature from 20 to 40 °C. The process has to be as slow as possible to avoid transient behavior in the time-of-flight and temperature measurements. In this way, the process is almost stationary. Because corrosion is an even slower process, from the above equation, we can write that, for this experiment,

$$\frac{\Delta L_c}{L_{s0}} = 0 \tag{12}$$

Substituting the value of $\Delta L_c / L_{s0}$ into Equation (11), we obtain:

$$\frac{\Delta\mathrm{ToF}}{\mathrm{ToF}} = -(\xi - c_x) \cdot \Delta T = r \cdot \Delta T \tag{13}$$

Therefore, if we plot Equation (13) in the $(\Delta T, \Delta\mathrm{ToF}/\mathrm{ToF})$ axis using the results of the experiment and perform a linear fitting, the calculated slope is $r = -(\xi - c_x)$. With the obtained data (see Figure 7), we estimated the value of $\xi = -1.054 \cdot 10^{-4}\ °\mathrm{C}^{-1}$.

To understand the needed amount of correction given the known values for ξ and c_x, this results in a correction in thickness of around −0.6 µm/°C for a 5 mm thick steel sample.

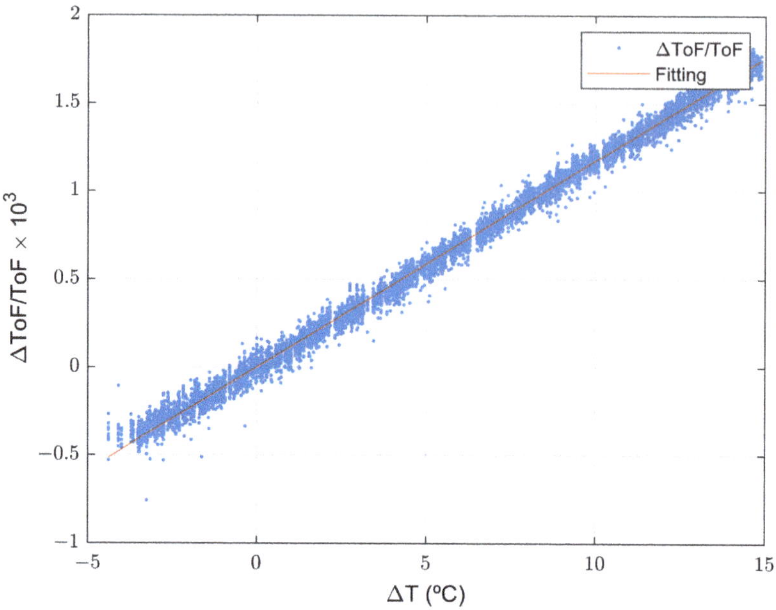

Figure 7. Experimental results to estimate the value of parameter ξ.

3. Experiment: Evolution of Thickness Loss Due to Corrosion in Real Conditions

With the aim of analyzing the evolution of thickness loss due to corrosion in a real environment, an experiment was carried out deploying our ultrasound sensor node in Gran Canaria, near the coast (see Figure 8).

(**a**) (**b**)

Figure 8. Location of the experimental setup. (**a**) Photo of the final location of the system (provided by PLOCAN); (**b**) experimental setup location on the map.

The experimental setup was arranged by permanently adhering the ultrasound probe onto a certain spot on the (75 × 150 × 5) mm bare steel sample (as shown in Figure 9) using bicomponent epoxy adhesive Structalit 1028 R from Panacol [42]. After fixing the sensor to the sample, a bicomponent layer of epoxy resin PX900D [43] was applied over the sensor-attached side of the sample as a corrosion protection layer. As is shown in Figure 10b, the sensor was attached from the inner (clean) side of the sample, and the other side was fully exposed to the marine environment. Since the sensor was fixed to one particular location of the sample, this experiment allowed for monitoring the thickness loss due to corrosion, avoiding larger uncertainties due to thickness variations on different spots of the same sample. The thickness variations underneath the whole sensor (contact area of 15 mm) were even smaller than those when measuring different spots. The solution is averaging a shot of quick measurements. As was said in [35], thickness at different locations of the same sample varied even for the unexposed bare steel samples. These small thickness variations (at the μm level) were due to common existing imperfections during the sample production process. Therefore, in order to achieve more precise ToF measurements, the positioning of the sensor on the same location is very important.

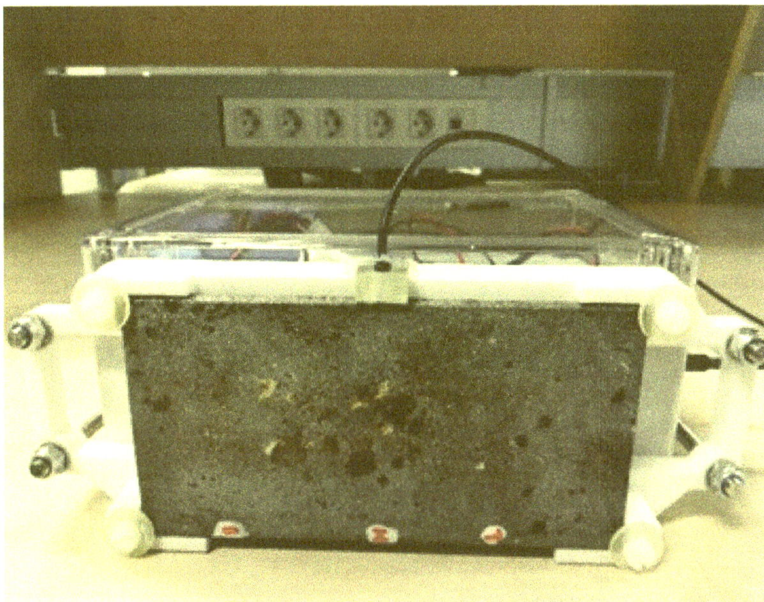

Figure 9. Bare steel sample attached to the monitoring solution.

To easily collect data from this experiment, the ultrasound sensor node was connected to a Beaglebone black wireless board (see Figure 10b). The Beaglebone was used to receive the ultrasound signals and to post the data to a data server in json format. Both the sensor node and Beaglebone are battery-powered (rechargeable, 3.6 V/7000 mAh), and the amplifier gain was set to 17.7 dB in this experiment.

Although our sensor node is capable of autonomously operating without a Beaglebone, in this case, the microcontroller from the sensor node communicated with the Beaglebone via the Universal Asynchronous Receiver and Transmitter (UART) protocol. The Beaglebone was basically in charge of sending the updated TOFREF value (see Table 1), estimated from the previous measurement event, to the sensor node at the beginning of an ultrasound measurement event, and then receiving and posting the data to the cloud after each measurement event. Thus, in a day, the system performed four measurement events (every 6 h). Each event consisted of five successive ToF measurements.

The corrosion monitoring setup was deployed at the PLOCAN premises in Gran Canaria (near the coast, see Figure 10) from the end of February to the beginning of August to receive thickness loss data, and to estimate the intraday corrosion rate thanks to the fast continuous measurements and high-precision estimates. Figure 11 shows how the bare steel sample looked like after five months of exposure in this marine environment.

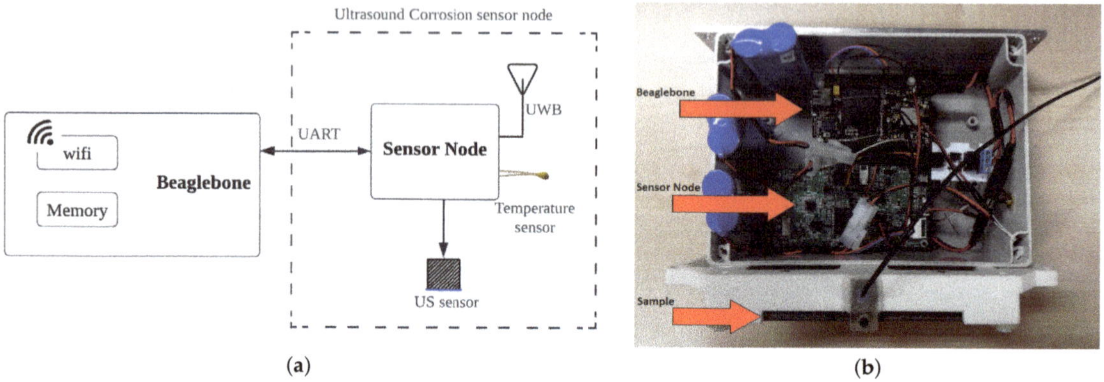

Figure 10. Final electronics setup used in the experiment. (**a**) Block digram of the setup; (**b**) photograph of the actual setup.

Figure 11. Photo of the bare steel sample after five months of exposure in the Gran Canaria (provided by PLOCAN).

The system gathered data for approximately 5 months, performing four measurement shots/events per day, and each shot was composed of five successive measures. Table 2 presents the main features of the experimental setup explained throughout this paper.

Table 2. Main features of the experiment.

Experimental Setup	Value
Sample material	S355
Sample thickness	5 mm
Ultrasound probe	V111 (Olympus)
Probe diameter of contact	15 mm
Probe nominal element size	13 mm
Adhesive/Couplant	Structalit 1028 R
US signal frequency	7.8 MHz
US signal amplitude	30 V (±15 V)
Speed of sound in S355 (long. waves)	5950 m/s
Speed of sound thermal coefficient ξ	-1.054×10^{-4} °C^{-1}
Thermal expansion coefficient c_x	12×10^{-6} °C^{-1}
Experimental location in latitude–longitude (decimal degrees)	27.9920, −15.3686
No. of measurement events per day	4 (one every 6 h)

4. Results of the Experiment and Discussion

The corrosion rate is the speed of deterioration of a material due to corrosion. One of the most important aspects of the proposed corrosion monitoring solution is being able to perform very frequent measurements in order to calculate the corrosion rate. The proposed experiment was carried out, analyzing the thickness loss evolution due to uniform corrosion in a real marine environment, and estimating the corrosion rate in real time (for practical purposes, taking into account that corrosion is an extremely slow process).

After computing the average of the five successive measurements taken every 6 h, we processed a linear fitting of the last 28 measurements, which represent 1 week of data. Figure 12 shows the thickness estimations in mm obtained during the experiment.

During the first month, we chose the location and prepared the system setup. Once the experiment had begun, we could see that the measured temperature was higher than what was expected due to the setup being exposed to direct sunlight. We corrected that during May by shadowing the setup. This means that we measured higher temperatures from March to May (particularly during sunlight periods), but taking into account that the error for a 5 mm thick steel sample is −0.6 µm/°C , the measurements could still be considered to be acceptable. On the other hand, we lost data at the beginning of July because of the accumulation of dirt and dust on the cable of the ultrasound probe. Once the cable had been cleaned, the system resumed working as usual.

The thickness estimations were corrected on the basis of the reference temperature; in this case, we chose them to be the initial temperature that we had measured at the beginning of the experiment (21.3 °C). The correction to temperature variations was applied using the second term of Equation (11) and considering that $(\xi - c_x)$ was -1.174×10^{-4} °C^{-1}.

As commented in Section 2.3, we followed the methodological design presented in [35] to validate the system processing. Figure 13 compares the ToF estimations performed by the deployed system and the MATLAB model using the raw data acquired from the experiment and configured with the same design parameters as the deployed system. The differences between ToF estimations were so small that it was impossible to visualize them, and most of the data points seemed to be overlapped. Then, Figure 14 presents the error among the ToF estimations shown in Figure 13. In the worst case, we had a maximal error of around 4.3×10^{-3} %, and the error increased with exposure time. Figure 15 shows how the signal level (SL) of the US response decreased during the experiment, explaining why the error increased. The SL refers to the root mean square (RMS) value of the first detected echo, measured after the amplifier [35]. This is an indication of the SNR of the digitized signal after bandpass filtering.

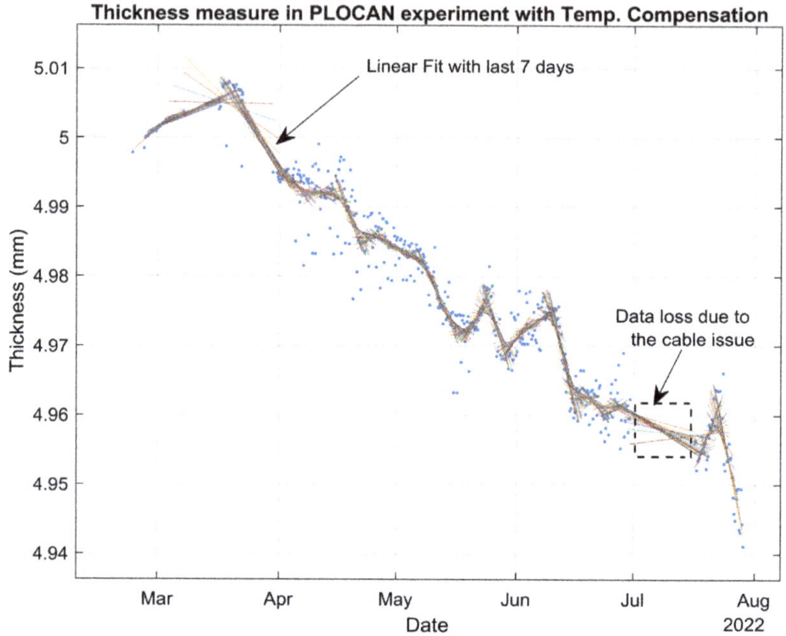

Figure 12. Thickness estimations from the experiment's data with temperature compensation.

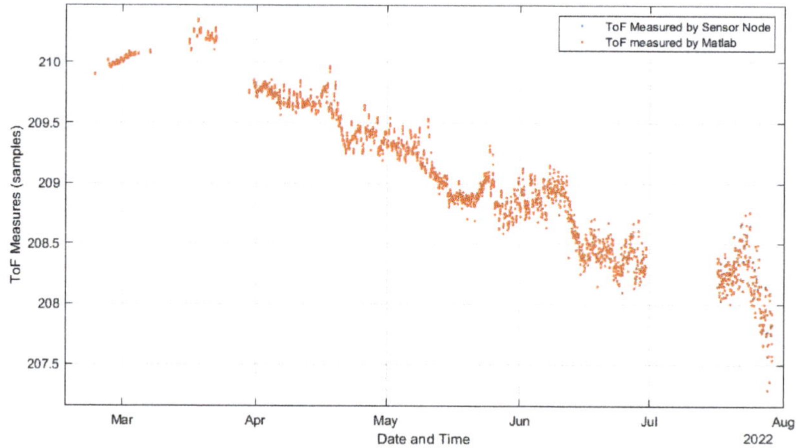

Figure 13. Comparison of ToF estimations with the MATLAB model.

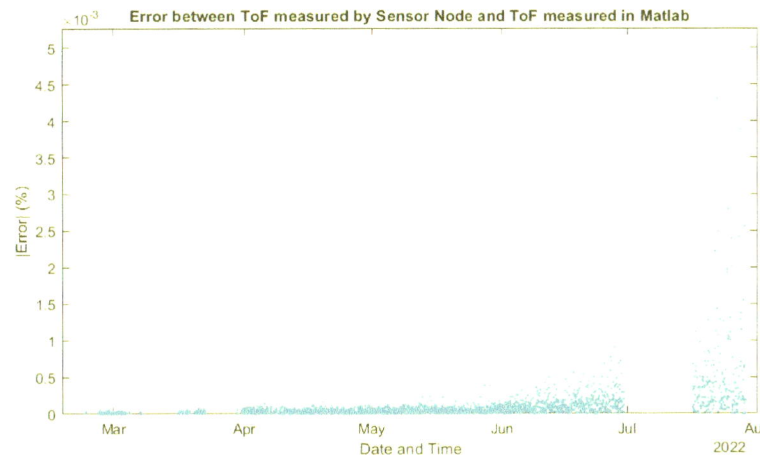

Figure 14. Error between ToFs estimated by the sensor node and by the MATLAB model.

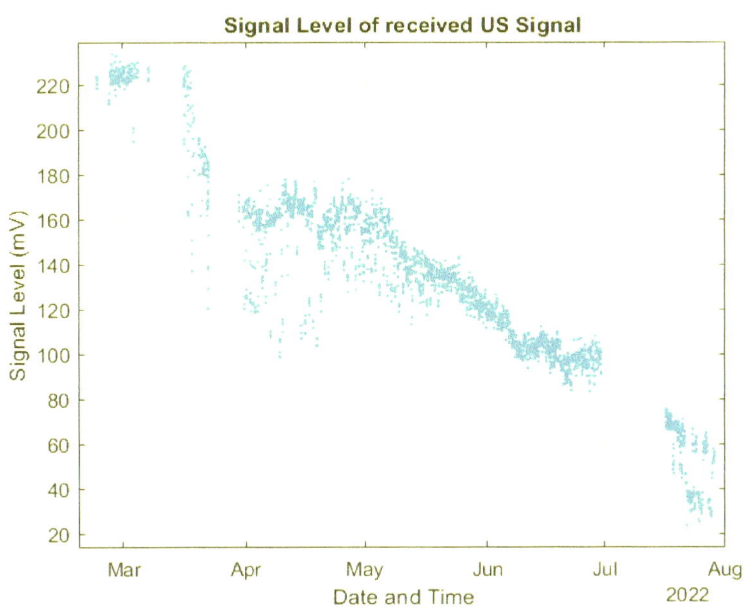

Figure 15. Signal level of the received US signal during the experiment.

Next, we took the slope of the linear fittings shown in Figure 12 as an estimation of the corrosion rate at each measurement point. Afterwards, a filtered corrosion rate was obtained by running the average of the last four slopes as is shown in Figure 16. Thus, the intraday corrosion rate could be estimated from the ToF measurements taken every 6 h and had an approximate delay of 4 days.

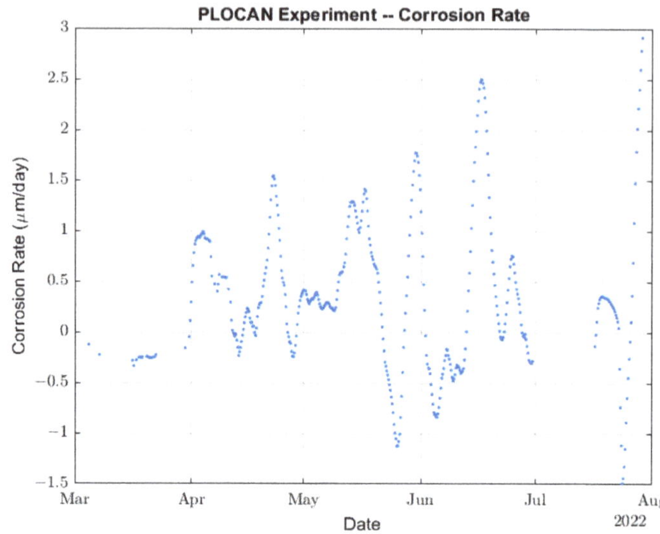

Figure 16. Estimation of the intraday corrosion rate.

As we can see from the thickness measurements, Figure 12 shows increments of thickness followed by a rapid and greater decrease in thickness. A similar behavior was observed even in a laboratory experiment with permanently installed ultrasound transducer: an increase in thickness followed by rapid thickness loss [26,44,45]. This was a consequence of the ToF measuring method that relies on the speed of sound being constant at all points of the steel sample. However, corrosion provokes a gradual degradation of a small layer of steel that reduces the speed of sound and acoustic impedance in that layer. At the beginning of the corrosion process, acoustic impedance is still large enough to allow for the ultrasound signal to bounce off the initial wall, while slower speed of sound produces an apparent increase in thickness. When the acoustic impedance decreases so much that the ultrasound signal bounces off the new wall, the result is a rapid reduction in total thickness. The net result is, of course, thickness loss (see Figure 12). Therefore, although the average corrosion rate calculated below is correct (considering positive and negative slopes), the positive instantaneous corrosion rates were overestimated using the ToF method.

The proposed experiment allowed for us to measure an average corrosion rate equivalent to 0.134 mm/year, which is in very good agreement with typically reported corrosion rates for uniform corrosion (0.1–0.2 mm/year). It is also possible to reduce the sample interval from the 6 h of the current experiment. The experimental setup could be modified to drastically reduce its power consumption by removing the Beaglebone board, resulting in a more frequent waking up of the sensor node using the same batteries. This means that ToF measurements can be taken, for example, every hour instead of 6 h, reducing the time response of the corrosion rate estimation algorithm.

Author Contributions: Conceptualization, A.C. and A.I.; formal analysis, A.I. and U.C.T.; funding acquisition, A.C. and A.I.; investigation, A.C., A.I. and U.C.T.; methodology, A.I. and U.C.T.; project administration, A.C.; software, A.I. and U.C.T.; supervision, A.C. and A.I.; validation, A.I. and U.C.T.; writing—original draft, U.C.T.; writing—review and editing, A.C. and A.I. All authors have read and agreed to the published version of the manuscript.

Funding: This work was supported by the WATEREYE project, which has received funding from the European Union's Horizon 2020 research and innovation programme under grant agreement no. 851207.

Institutional Review Board Statement: Not applicable.

Informed Consent Statement: Not applicable.

Data Availability Statement: Not applicable.

Acknowledgments: This work was possible thanks to the cooperation of CEIT with all the WATEREYE partners, especially SINTEF Industry, who are responsible for producing the samples, and PLOCAN, who have provided us with the necessary support to run the experiment at their premises.

Conflicts of Interest: The authors declare no conflict of interest.

Abbreviations

The following abbreviations are used in this manuscript:

ADC	Analog-to-digital converter
AFE	Analog front end
BP	Bandpass
CAPEX	Capital expenditure
CCF	Cross-correlation function
FPGA	Field-programmable gate array
GDP	Gross domestic product
I2C	Interintegrated circuit
NDT	Nondestructive testing
OWTs	Offshore wind turbines
RMS	Root mean square
SNR	Signal-to-noise ratio
SHM	Structural health monitoring
SL	Signal level
SPI	Serial peripheral interface
ToF	Time-of-flight
UART	Universal asynchronous receiver and transmitter
US	Ultrasound
VGA	Variable gain amplifier

References

1. IRENA. Future of Wind: Deployment, Investment, Technology, Grid Integration and Socio-Economic Aspects by International Renewable Energy Association. 2019. Available online: https://www.irena.org/-/media/files/irena/agency/publication/2019/oct/irena_future_of_wind_2019.pdf (accessed on 7 July 2022).
2. The European Offshore Wind Industry: Key Trends and Statistics 2017, Wind Europe. 2018. Available online: https://windeurope.org/about-wind/statistics/offshore/european-offshore-wind-industry-key-trends-statistics-2017/ (accessed on 9 July 2022).
3. Zheng, D.; Bose, S. Offshore wind turbine design. In *Wind Power Generation and Wind Turbine Design*; WIT Press: Southampton, UK, 2010; p. 363.
4. Márquez, F.P.G.; Jimenez, A.A.; Muñoz, C.Q.G. Non-destructive testing of wind turbines using ultrasonic waves. In *Non-Destructive Testing and Condition Monitoring Techniques for Renewable Energy Industrial Assets*; Elsevier: Boston, MA, USA, 2020; pp. 91–101.
5. Ciang, C.C.; Lee, J.R.; Bang, H.J. Structural health monitoring for a wind turbine system: A review of damage detection methods. *Meas. Sci. Technol.* **2008**, *19*, 122001. [CrossRef]
6. Martinez-Luengo, M.; Kolios, A.; Wang, L. Structural health monitoring of offshore wind turbines: A review through the Statistical Pattern Recognition Paradigm. *Renew. Sustain. Energy Rev.* **2016**, *64*, 91–105. [CrossRef]
7. Smarsly, K.; Hartmann, D.; Law, K.H. An integrated monitoring system for life-cycle management of wind turbines. *Int. J. Smart Struct. Syst.* **2013**, *12*, 209–233. [CrossRef]
8. Igwemezie, V.; Mehmanparast, A.; Kolios, A. Materials selection for XL wind turbine support structures: A corrosion-fatigue perspective. *Mar. Struct.* **2018**, *61*, 381–397. [CrossRef]
9. Koch, G.; Varney, J.; Thompson, N.; Moghissi, O.; Gould, M.; Payer, J. *NACE International Impact Report: International Measures of Prevention, Application, and Economics of Corrosion Technologies Study*; Technical Report; NACE International: Houston, TX, USA, 2016.
10. Hossain, M.L.; Abu-Siada, A.; Muyeen, S. Methods for advanced wind turbine condition monitoring and early diagnosis: A literature review. *Energies* **2018**, *11*, 1309. [CrossRef]
11. Price, S.; Figueira, R. Corrosion Protection Systems and Fatigue Corrosion in Offshore Wind Structures: Current Status and Future Perspectives. *Coatings* **2017**, *7*, 2017. [CrossRef]

12. Corrosion Classifications for Offshore Wind. Available online: https://www.onropes.co.uk/corrosion-classifications#:~:text=In%20the%20tidal%20and%20splash,yearly%20depending%20on%20the%20location (accessed on 19 August 2022).
13. Nassar, N.E.A. Corrosion in marine and offshore steel structures: Classification and overview. *Int. J. Adv. Eng. Sci. Appl.* **2022**, *3*, 7–11. [CrossRef]
14. Khodabux, W.; Causon, P.; Brennan, F. Profiling corrosion rates for offshore wind turbines with depth in the North Sea. *Energies* **2020**, *13*, 2518. [CrossRef]
15. International, N. *Techniques for Monitoring Corrosion and Related Parameters in Field Applications*; NACE International: Houston, TX, USA, 1999.
16. Bardal, E.; Drugli, J. Corrosion detection and diagnosis. *Mater. Sci. Eng.* **2004**, *3* .
17. Forsyth, D.S. Non-destructive testing for corrosion. In *Corrosion Fatigue and Environmentally Assisted Cracking in Aging Military Vehicles (RTO-AG-AVT-140)*; NATO: Washington, DC, USA, 2011.
18. Wright, R.F.; Lu, P.; Devkota, J.; Lu, F.; Ziomek-Moroz, M.; Ohodnicki, P.R. Corrosion sensors for structural health monitoring of oil and natural gas infrastructure: A review. *Sensors* **2019**, *19*, 3964. [CrossRef]
19. Hendee, W.R.; Ritenour, E.R. *Medical Imaging Physics*; Wiley-Liss, Inc.: New York, NY, USA, 2002; pp. 303–353.
20. Wróbel, G.; Pawlak, S. A comparison study of the pulseecho and through-transmission ultrasonics in glass/epoxy composites. *J. Achiev. Mater. Manuf. Eng.* **2007**, *22*, 51–54.
21. Blitz, J.; Simpson, G. *Ultrasonic Methods of Non-Destructive Testing*; Springer Science & Business Media: New York, NY, USA, 1995; Volume 2.
22. Fowler, K.A.; Elfbaum, G.M.; Smith, K.A.; Nelligan, T.J. Theory and application of precision ultrasonic thickness gauging. *Insight* **1996**, *38*, 582–587.
23. An Introduction to Ultrasonic Transducers for Nondestructive Testing, OLYMPUS. Available online: https://www.olympus-ims.com/en/resources/white-papers/intro-ultrasonic-transducers-ndt-testing/ (accessed on 18 August 2022).
24. Rathod, V.T. A review of acoustic impedance matching techniques for piezoelectric sensors and transducers. *Sensors* **2020**, *20*, 4051. [CrossRef]
25. Rathod, V.T. A review of electric impedance matching techniques for piezoelectric sensors, actuators and transducers. *Electronics* **2019**, *8*, 169. [CrossRef]
26. Zou, F.; Cegla, F. High accuracy ultrasonic corrosion monitoring. In *CORROSION 2017*; OnePetro: New Orleans, LA, USA, 2017.
27. Queirós, R.; Martins, R.C.; Girao, P.S.; Serra, A.C. A new method for high resolution ultrasonic ranging in air. In Proceedings of the XVIII IMEKO World Congress, Rio de Janeiro, Brazil, 17–22 September 2006; pp. 17–22.
28. Huang, Y.; Wang, J.S.; Huang, K.; Ho, C.; Huang, J.; Young, M.S. Envelope pulsed ultrasonic distance measurement system based upon amplitude modulation and phase modulation. *Rev. Sci. Instruments* **2007**, *78*, 065103. [CrossRef]
29. Franco, E.E.; Meza, J.M.; Buiochi, F. Measurement of elastic properties of materials by the ultrasonic through-transmission technique. *Dyna* **2011**, *78*, 58–64.
30. Svilainis, L. Review of high resolution time of flight estimation techniques for ultrasonic signals. In Proceedings of the 2013 International Conference NDT, Telford, UK, 10–12 September 2013; pp. 1–12.
31. Herter, S.; Youssef, S.; Becker, M.M.; Fischer, S.C. Machine Learning Based Preprocessing to Ensure Validity of Cross-Correlated Ultrasound Signals for Time-of-Flight Measurements. *J. Nondestruct. Eval.* **2021**, *40*, 1–9. [CrossRef]
32. Schäfer, M.; Theado, H.; Becker, M.M.; Fischer, S.C. Optimization of the Unambiguity of Cross-Correlated Ultrasonic Signals through Coded Excitation Sequences for Robust Time-of-Flight Measurements. *Signals* **2021**, *2*, 366–377. [CrossRef]
33. Lu, Z. Estimating time-of-flight of multi-superimposed ultrasonic echo signal through envelope. In Proceedings of the 2014 International Conference on Computational Intelligence and Communication Networks, Bhopal, India, 14–16 November 2014; pp. 300–303.
34. Lu, Z.; Yang, C.; Qin, D.; Luo, Y.; Momayez, M. Estimating ultrasonic time-of-flight through echo signal envelope and modified Gauss Newton method. *Measurement* **2016**, *94*, 355–363. [CrossRef]
35. Thibbotuwa, U.C.; Cortés, A.; Irizar, A. Ultrasound-Based Smart Corrosion Monitoring System for Offshore Wind Turbines. *Appl. Sci.* **2022**, *12*, 808. [CrossRef]
36. Contact Transducers—Olympus-IMS.com. Available online: https://www.olympus-ims.com/en/ultrasonic-transducers/contact-transducers/#!cms[focus]=cmsContent10861 (accessed on 15 November 2021).
37. Guisasola, A.; Cortés, A.; Cejudo, J.; da Silva, A.; Losada, M.; Bustamante, P. Reliable and Low-Power Communications System Based on IR-UWB for Offshore Wind Turbines. *Electronics* **2022**, *11*, 570. [CrossRef]
38. Ultra-Tiny, 16-Bit Delta-Sigma ADC with I2C Interface, LTC2451. Available online: https://www.analog.com/media/en/technical-documentation/data-sheets/2451fg.pdf (accessed on 25 June 2022).
39. GA10K3A1i SERIES II THERMISTORS. Available online: https://il.farnell.com/te-connectivity/ga10k3a1ib/thermistor-ntc-10k/dp/3397780 (accessed on 20 June 2022).
40. Cespedes, I.; Huang, Y.; Ophir, J.; Spratt, S. Methods for estimation of subsample time delays of digitized echo signals. *Ultrason. Imaging* **1995**, *17*, 142–171. [CrossRef]
41. Red Pitaya STEMlab 125-14. Available online: https://redpitaya.com/stemlab-125-14/ (accessed on 1 June 2022).
42. Structalit® 1028 R. Available online: https://www.panacol.com/panacol/datasheets/structalit/structalit-1028r-english-tds-panacol-adhesive.pdf (accessed on 24 June 2022).

43. PX900D A Low Viscosity Unfilled Epoxy Resin System. Available online: https://www.blelektronik.com.pl/wp-content/uploads/2019/07/PX900D.pdf (accessed on 24 June 2022).
44. Zou, F.; Cegla, F.B. On quantitative corrosion rate monitoring with ultrasound. *J. Electroanal. Chem.* **2018**, *812*, 115–121. [CrossRef]
45. Technical Discussion: Can Corrosion Deposits on Hidden Face Affect Thickness Reading? Available online: https://www.ndt.net/forum/thread.php?rootID=52247 (accessed on 1 September 2021).

sensors

MDPI

Article

An Improved RAPID Imaging Method of Defects in Composite Plate Based on Feature Identification by Machine Learning

Fei Deng *, Xiran Zhang, Ning Yu and Lin Zhao

School of Electrical and Electronic Engineering, Shanghai Institute of Technology, Shanghai 200235, China
* Correspondence: dengfei@sit.edu.cn

Abstract: The RAPID (reconstruction algorithm for probabilistic inspection of defect) method based on Lamb wave detection is an effective method to give the position information of a defect in composite plate. In this paper, an improved RAPID imaging method based on machine learning (ML) is proposed to precisely visualize the location and features of defects in composite plate. First, the specific feature information of the defect, such as type, size and direction, can be identified by analyzing the detection signals through multiple machine learning models. Then, according to the obtained defect features, the scaling parameter β of the RAPID method which controls the size of the elliptical area is revised, and weights are set to the important detection paths which are related to defect features to realize precise defect imaging. The simulation results show that the proposed method can intuitively characterize the location and related feature information of the defect, and effectively improve the accuracy of defect imaging.

Keywords: lamb wave; composite plate; nondestructive testing; defect imaging; machine learning

Citation: Deng, F.; Zhang, X.; Yu, N.; Zhao, L. An Improved RAPID Imaging Method of Defects in Composite Plate Based on Feature Identification by Machine Learning. *Sensors* **2022**, *22*, 8413. https://doi.org/10.3390/s22218413

Academic Editor: Zenghua Liu

Received: 26 September 2022
Accepted: 30 October 2022
Published: 1 November 2022

1. Introduction

Composite materials with fatigue resistance, light weight, and high strength are widely utilized in aerospace, construction, and other fields [1,2]. Surface or internal defects such as delamination, crack, debonding, pore, inclusion, looseness, and rich resin may occur in the manufacturing and service process of composite materials [3]. Lamb wave-based detection has been widely used in the detection of composite materials due to its long propagation distance, small attenuation, and sensitivity to small defects [4,5]. The anisotropy of the composite plate results in the change of the propagation behavior of guided waves in the plate structure [6], and the complex transformation of waveform aggravates the complexity of interpreting the detection signal [7,8]. At the same time, the complexity of the detection signal also increases the difficulty of precise defect imaging in the composite plate.

Many damage imaging methods can be used for defect detection in composite plates, such as the ellipse imaging method [9,10], delay-and-sum imaging method [11,12], time-reversal imaging method [13,14], RAPID imaging method [15–18], etc. The RAPID can intuitively present the location of defects by analyzing the statistical characteristics of the detection signal, and has been widely studied by researchers. Wang et al. [19] studied the mode of directional scattering of cracks in aluminum plate and proposed an improved probabilistic damage imaging algorithm, which achieved defect imaging by finding the path with the smallest signal difference coefficient to predict the crack direction. Wu et al. [20] studied the influence mechanism of several parameters in the probabilistic damage imaging method, and proposed a method to optimize the sensor network and determine these parameters, which was verified on the composite stiffened panel. In the above studies of probabilistic damage imaging methods, the specific characteristics of defects (e.g., length and direction of cracks, size of delaminated areas) cannot be precisely known, which reduces the accuracy of defect imaging and affects the application of imaging methods in practice.

The combination of Lamb wave detection and ML can be used to identify defects. In general, according to prior knowledge in the field, damage-sensitive features are analyzed and extracted as input training model through a variety of signal processing methods to realize the identification of damage types and degrees [21]. Zhou et al. [22] extracted the relevant characteristics of the time domain, the frequency domain, and the energy of the Lamb wave signal as input to the classifier, and studied the degree of corrosion damage at the bend of the pipeline. Experimental results show that machine learning can be used to identify the damage category of defects, and support vector machine (SVM) has a better identification effect under the condition of small samples. Peng et al. [23] estimated the size and location of the defect by obtaining the damage-sensitive features of the guided wave signal and incorporating them into a Bayesian update framework. Mardanshahia et al. [24] extracted different features of Lamb wave signal to train three supervised ML methods to classify the density of cracks. However, for anisotropic composite plates, the propagation law of Lamb waves is more complicated, and it relies too much on existing prior knowledge to extract features and cannot perform damage detection highly efficiently.

In this paper, a defect imaging method based on an improved RAPID algorithm is proposed for defects in composite plates. Before the defect detection, the guided wave detection baseline signal is obtained by a circular sensor array arranged in the defect-free composite plate through simulation or experiment. Then defects with different types and parameters are pre-constructed in the center of the array to obtain the corresponding damage detection signal for constructing the defect feature identification sample library. And based on these defect detection signals, the ML models for defect feature identification are trained, respectively. For the composite plate to be inspected, the above-mentioned circular sensor array is first used to obtain the detection signals of the inspected area. Then, based on these detection signals and the reference signals, the DI values of each path are calculated, and path selection is performed based on these DI values. For the selected paths, pre-location is performed using RAPID to obtain the center coordinates of the defect. After that, the center of the sensor array is moved to the predicted defect location to obtain the detection signals for feature recognition. On this basis, the corresponding models are used to identify the defect types and main geometric parameters (such as the length and direction of cracks, the size of the delamination area, etc.). Finally, after obtaining the characteristic parameters of the defects, the RAPID algorithm is improved by modifying the β values and setting the weights of the relevant detection paths to refine the imaging of different types of defects. Based on theoretical analysis, this paper takes two types of defects, notably crack and delamination, in composite plates as examples to carry out simulation experiments and research to verify the effectiveness of the above methods. The simulation results show that the proposed method can intuitively characterize the location and related feature information of the defect, and effectively improve the accuracy of defect imaging.

The abbreviations involved in the proposed method and their meanings are shown in Table 1.

Table 1. The acronyms in the defect imaging method.

Symbol	Meaning	Symbol	Meaning
RAPID	The Reconstruction Algorithm for Probabilistic Inspection of Defect	SVM	Support Vector Machine
ML	Machine Learning	PCA	The Principal Component Analysis
DI	Damage Index	C	The Penalty Factor

2. Methods

The RAPID is based on the correlation coefficient of the reference and damage signals, and the location relationship among the transmitter, the receiver, and the defect. It avoids the complex process analysis of Lamb wave propagation to achieve defect localization. The proposed method fuses the defect feature predicted by the ML method into the RAPID algorithm to achieve precise defect imaging.

The mode conversion, reflection, and transmission process of the Lamb wave is affected by the characteristics of the defect which include type, size, location, and other factors. Such influence causes the detection signal to change. The closer the defect is to the sensing path, the more influence it has on Lamb wave signal. Conversely, the farther the relative distance is, the smaller the influence on the Lamb wave signal. Therefore, RAPID imaging methods can roughly estimate the location of defects.

2.1. Path Selection and Defect Pre-Locating

In the process of using the RAPID, the composite plate is divided into uniform grids, and each grid corresponds to a pixel. For each pixel, its value is the probability value of existence of damage. The probability of the presence of the defect at position (x, y) can be defined as:

$$P(x,y) = \sum_{i=1}^{N} DI_i \cdot \frac{\beta - R_i(x,y)}{\beta - 1} \tag{1}$$

where N is the number of participating sensing paths, $[\beta - R_i(x,y)]/(\beta - 1)$ is the spatial distribution function of the ith path, and its outline is an ellipse. The parameter β is a scaling parameter that controls the size of the ellipse distribution area. It is usually set around 1.0, which is an empirical value determined on a case-by-case basis, but such treatment is not always suitable for all situations in practice. In this paper, the pre-locating process sets $\beta = 1.05$ [25]. The term $R_i(x,y)$ is defined as follows:

$$R_i(x,y) = \begin{cases} RD_i(x,y), RD_i(x,y) < \beta \\ \beta, RD_i(x,y) \geq \beta \end{cases} \tag{2}$$

where $RD_i(x,y) = [D_{a,i}(x,y) + D_{s,i}(x,y)]/D_i$,$D_{a,i}(x,y)$,$D_{s,i}(x,y)$ is the distance from the point (x,y) to the excitation point and the receiving point, respectively. D_i is the distance from the excitation point to the receiving point, as shown in Figure 1a.

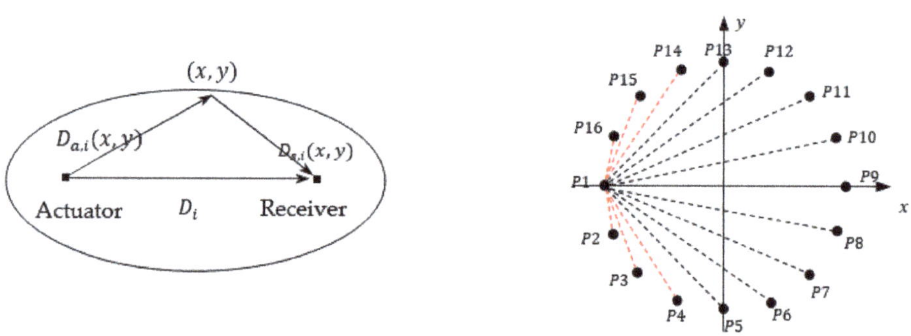

(a) (b)

Figure 1. Schematic diagram of RDI and path selection: (**a**) Schematic diagram of RD_i; (**b**) Schematic diagram of the path selection.

DI_i is the damage factor of the ith path, defined as follows:

$$\begin{aligned} DI = 1 - \rho &= 1 - \frac{\sum_{k=1}^{n}(X_k - \overline{X_k})(Y_k - \overline{Y_k})}{\sqrt{\sum_{k=1}^{n}(X_k - \overline{X_k})^2 (Y_k - \overline{Y_k})^2}} \\ &= 1 - \frac{n\sum_{k=1}^{n} X_k Y_k - \sum_{k=1}^{n} X_k \cdot \sum_{k=1}^{n} Y_k}{\sqrt{[n\sum_{k=1}^{n} X_k^2 - (\sum_{k=1}^{n} X_k)^2][n\sum_{k=1}^{n} Y_k^2 - (\sum_{k=1}^{n} Y_k)^2]}} \end{aligned} \tag{3}$$

where ρ is the correlation coefficient between the baseline data and the detection data. X_K and Y_k are the baseline signal and inspection signal, respectively. $\overline{X}_k$ and $\overline{Y}_k$ are the mean of X_K and Y_K, respectively. k is the number of recorded points in the data.

The magnitude of the DI value reflects the degree of correlation between the sensing path and the defect. The rule for path selection is as follows: using P1 as the excitation sensor and the rest of the sensors as the receiving sensors as an example, the circular array is shown in Figure 1b. A coordinate system with path P1–P9 as the x-axis and path P5-P13 as the vertical y-axis is created. The sensing paths in the second and third quadrants are ignored, and the remaining paths are the filtered paths. The DI values of these remaining paths are calculated according to Equation (3), then the larger two are selected as valid *DI* values. On this basis, the final *DI* values is obtained by filtering again through considering the reciprocity of the paths.

2.2. Defect Classification and Feature Identification

Under the same detection conditions, the detection signals generated by the same type of defect have some common characteristics, and this lays a foundation for the use of ML methods to realize defect feature recognition.

SVM is a ML method based on statistical theory. It maximizes the interval between classes by finding a "hyperplane". Additionally, the kernel function is utilized to map the data from a low-dimensional space to a high-dimensional space to make the data linearly separable. SVM has great advantages for data with high dimensions, small sample size, and nonlinearity [26,27]. It can be used for the identification of defect types and defect sizes.

Light GBM [28] is a distributed gradient lifting framework with the decision tree as a weak learner. It innovatively uses the gradient-based unilateral sampling algorithm and mutually exclusive feature binding algorithm to conduct data sampling and feature sampling, respectively. In addition, it does not need to traverse all the data in each iteration and can improve the training speed with the same accuracy. Although the model has high complexity, it is very suitable for classification tasks with high dimension. The model is used to identify crack direction in this study.

2.3. RAPID Algorithm for Fusing Defect Features

Sheen et al. [29] proposed that the β value can be further adjusted to achieve precise imaging based on knowing the size of the defect. Therefore, according to the main characteristics (type, size, orientation) of the defect obtained by the ML method, the parameter β is precisely corrected to improve the quality of the defect imaging. Figure 2 is a flow chart of the improved imaging algorithm.

The calculation of β is shown in Figure 3, where the defect range is described as the maximum vertical distance between the defect boundary and the detection path. In this paper, two kinds of defects (delamination and cracks) are studied as follows.

2.3.1. Delamination

In the detection path, the sum of the distances from the farthest point of the delamination defect to the excitation and receiver is calculated first, then the ratio of this sum value and the distance between excitation and receiver can be used to revise the β value as follows:

$$\beta = R_{Max\_defect\_range} = \frac{\sqrt{(x_a - x)^2 + (y_a - y)^2} + \sqrt{(x_s - x)^2 + (y_s - y)^2}}{\sqrt{(x_a - x_s)^2 + (y_a - y_s)^2}} \tag{4}$$

Here, (x_a, y_a), (x_s, y_s), and (x, y) are the coordinates of the excitation, receiver, and of the point which is farthest from the sensing path on the delamination boundary, respectively.

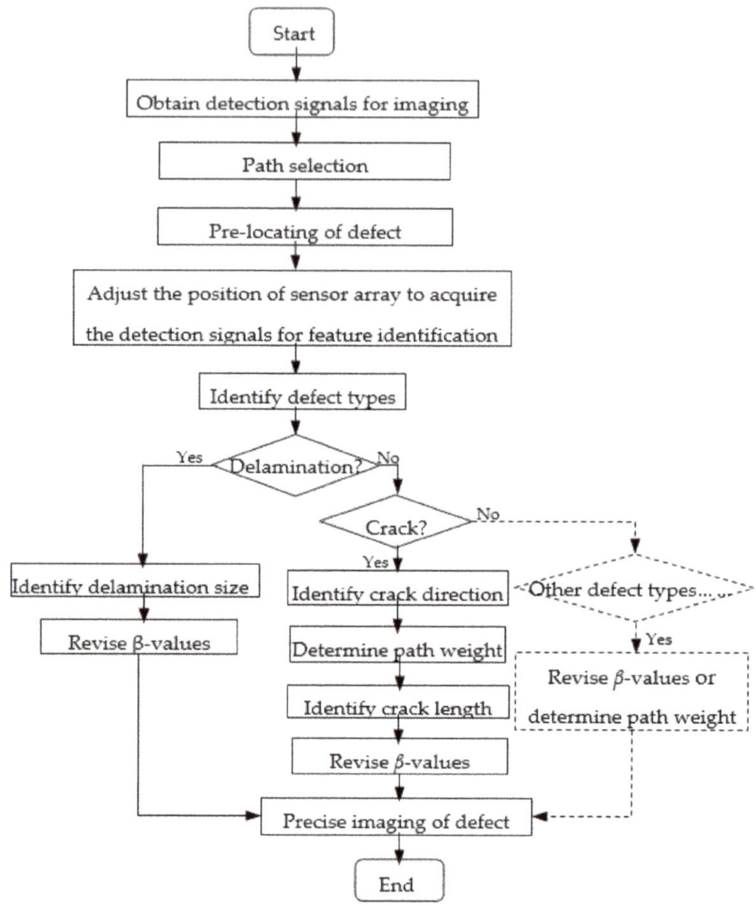

Figure 2. The flow diagram of the proposed method.

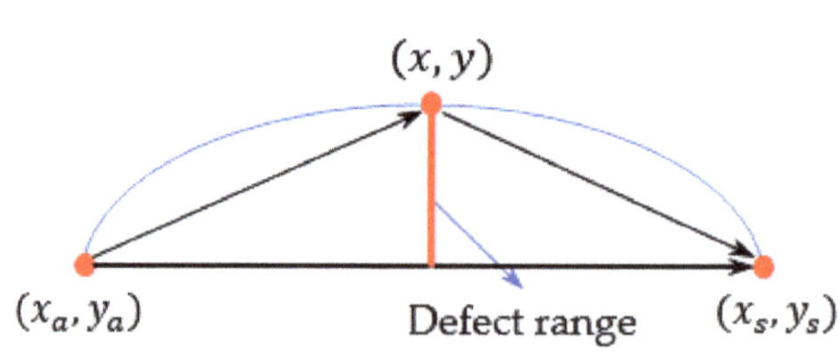

Figure 3. Schematic diagram of the calculation of β.

2.3.2. Crack

The width of the crack is extremely narrow. Therefore, if the direction of path i is consistent with the direction of crack (defined as the angle between the crack and the positive x-axis in the coordinate system shown in Figure 4a), it can be empirically set to $\beta_0 = 1.0001$. After the length value of the crack is predicted by the ML model, half of the length value is considered to be the defect range. On the basis of obtaining the defect range, and knowing (x_a, y_a) and (x_s, y_s), the β can be adjusted according to the following formula:

$$\beta = \begin{cases} RMax\_defect\_range, & \text{the direction of path } i \text{ is not consistent with the crack direction} \\ \beta_0, & \text{the direction of path } i \text{ is consistent with the crack direction} \end{cases} \quad (5)$$

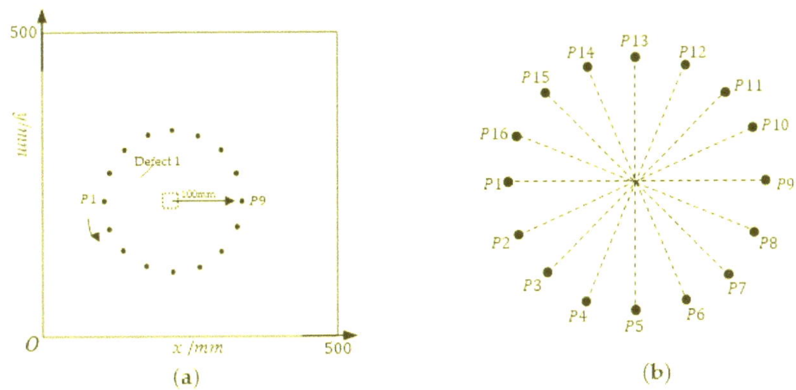

Figure 4. Schematic diagram of the experimental setup: (**a**) sensor arrangement and defect setting; (**b**) sensing path.

In addition, the path closest to the crack direction and the path perpendicular to the crack direction form the "cross path". The weight W_i of the above two paths should be set much larger than the other paths, and its weight can be set as:

$$W_i = \begin{cases} \frac{L}{l}, & \text{cross paths} \\ 1, & \text{other paths} \end{cases} \quad (6)$$

where $L(mm)$ is the diameter of the circular sensing array; and $l(mm)$ is the crack length. The smaller the crack length is, the greater the weight applied.

In summary, for the crack, the modified damage imaging algorithm is defined as:

$$P(x,y) = \sum_{i=1}^{N} DI_i \cdot \frac{\beta - R_i(x,y)}{\beta - 1} \cdot W_i \quad (7)$$

In order to illustrate the proposed method more specifically, the pseudocode of the improved methods is given in Algorithms 1–3. Algorithms 1 and 2: the pseudocode for feature recognition. Algorithm 3: the main pseudocode of crack in the proposed method.

Algorithm 1: Train the SVM model

```
Input: D: training set
D1←StandardScaler (D)
D2←Dimension reduction: PCA (D1)
Params←[{C}, {kernel function}]
C, kernel function←GridSearchCV(SVM, params)
return the SVM model
```

Algorithm 2: Feature identification by ML models

Input: *D*1: detection signal acquired at paths P1–P9 after adjusting the position of the sensing array
*D*2: detection signals acquired in paths P1–P9, P2–P10, P3–P11, P4–P12, P5–P13, P6–P14, P7–P15, P8–P16 after adjusting the position of the sensing array
Input: I_1: the model for identifying the defect type
I_2: the model for identifying the delamination size
I_3: the model for identifying the crack direction
I_4: the model for identifying the crack size
preds←I_1. predict (*D*1)
if preds==delamination **then**
| the size of delamination←I_2. predict (*D*1)
else if preds==crack **then**
| *f* (the direction of crack)←I_3. predict (*D*2), *l* (the size of crack)←I_4. predict (*D*1)
else
| ... ...
end
return the features of the defect

Algorithm 3: Precise imaging of crack

Input: *sig_d*: damage signals, *sig_b*: baseline signals,
Input: (x_{ai}, y_{ai}): the coordinates of the excitation sensor for path *i*, (x_{si}, y_{si}) the coordinates of the receiver for path *i*, *d:* the distance between excitation point and reception point, *N*: the number of all sensing paths, *n*: the number of the final paths after the path selection
Input: *k*: number of sensors, *L*: the diameter of the circular sensing array,
Input: *l*: the size of crack, *f*: the direction of crack,β_0: the scaling parameters of the path which is in the same direction as the crack
Input: *DI*: damage index,*m*: the number of grids per row/column
Calculate ρ ←for *i* in 1…*N*
Calculate *DI* = $1 - \rho$ ←for *i* in 1…*N*
| **for**$i = 1$ tok**do**
| | **for**$j = 1$ $tok - 1$**do**
| | $DI_{1nd} \leftarrow sorted[DI_1 : DI_j]$
| | $DI_{1nd} \leftarrow sorted[DI_1 : DI_j]$
Multi-path image fusion←path selection
return the center coordinates←the image thresholding

Algorithm 3 *Cont.*

$R_{Max\_defect\_range} \leftarrow \sqrt{d^2 + l^2}/2d$
if the direction of path i == the direction of crack **then**
| $\beta = \beta_0$
else
| $\beta = R_{Max\_defect\_range}$
end
$W \leftarrow L/l$
if the path i == cross paths **then**
| $W_i \leftarrow W$
else
| $W_i \leftarrow 1$
end
for $x = 1 to m$ **do**
for $y = 1$ *to m* **do**
$RD_1(x,y) \leftarrow (\sqrt{(x - x_{a1})^2 + (y - y_{a1})^2} + \sqrt{(x - x_{s1})^2 + (y - y_{s1})^2})/\sqrt{(x_{s1} - x_{a1})^2 + (y_{s1} - y_{a1})^2}$
⋮
$RD_n(x,y) \leftarrow (\sqrt{(x - x_{an})^2 + (y - y_{an})^2} + \sqrt{(x - x_{sn})^2 + (y - y_{sn})^2})/\sqrt{(x_{sn} - x_{an})^2 + (y_{sn} - y_{an})^2}$
if $RD_1(x,y) \geq \beta$ **then**
$R_1(x,y) \leftarrow \beta$
else
$R_1(x,y) \leftarrow RD_1(x,y)$
end
⋮
if $RD_n(x,y) \geq \beta$ **then**
$R_n(x,y) \leftarrow \beta$
else
$R_n(x,y) \leftarrow RD_n(x,y)$
end
$P(x,y) \leftarrow \sum_{i=1}^{n} DI_i \cdot \frac{\beta - R_i(x,y)}{\beta - 1} \cdot W_i$
F←imagesc(P)
return F

3. Experimental Setup

A three-dimensional finite element analysis is performed with ABAQUS software. A [45°/−45°/0°/90°] quasi-isotropic composite laminate with a dimension of 500 mm × 500 mm × 1.28 mm is considered in the coordinate system shown in Figure 4a. The material properties of the laminate are listed in Table 2. A circular sensor array (radius r = 100 mm) composed of 16 evenly distributed sensors is arranged on the surface of the laminate. The excitation signal used in the numerical study is a 160-kHz narrow-band five-cycle sinusoidal tone burst modulated by a Hanning window. Loading is performed at each sensor position in turn, and the received signal is obtained at the remaining sensor positions. A total duration of 300 μs time length is solved using ABAQUS/Explicit dynamical analysis with a fixed step size of 10^{-8} s for all the numerical cases.

Table 2. Elastic properties of composite plate in the numerical study.

E_1/GPa	$E_2 = E_3$/GPa	$G_{12} = G_{13}$/GPa	G_{23}/GPa	$\upsilon_{12} = \upsilon_{13}$	υ_{23}	$\rho/(\text{kg/m}^3)$
128	82	4.7	3.44	0.27	0.2	1560

3.1. Building the Sample Library

This study only discusses two typical types of defects in the composite plate: delamination and cracks. On the defect-free simulation model mentioned above, the delamination or crack defect with changing geometric parameters is set at the center of the circular sensor array which is shown in Figure 4b, and the detection signals are acquired by one-excitation, one-receive pattern to construct a sample library.

As shown in Figure 4a, a rectangle of variable length and width corresponding to the dashed box is used to simulate the delamination defects. Due to computational limitations, the obtained samples do not cover all delamination sizes. In the sample library for this simulation experiment, only the following nine classes are considered: 5 mm × 5 mm, 10 mm × 10 mm, 15 mm × 15 mm, 5 mm × 10 mm, 5 mm × 15 mm, 10 mm × 5 mm, 10 mm × 15 mm, 15 mm × 5 mm, and 15 mm × 10 mm, and the corresponding sample labels are set as 1, 2, 3, 4, 5, 6, 7, 8, and 9, respectively, as shown in Table 3. For the composite plate studied in this paper, the delamination defects are set between the top layer and the adjacent second layer, and between the second layer and the adjacent third layer, respectively. The variations of the length and width of the defects are shown in Table 3. Classes 1, 2 and 3 contain 84 samples, respectively, and classes 4, 5, 6, 7, 8 and 9 contain 44 samples, each. In addition, in order to increase the robustness of the model, samples that are between classes are also supplemented in the experiments, such as adding samples of 5 mm × 8 mm, 5 mm × 14 mm, 10 mm × 6 mm, 10 mm × 16 mm, 15 mm × 6 mm, and 15 mm × 11 mm. Although the length and width directions of the rectangles can be changed, this paper does not consider such changes for the time being and only assumes that the length and width directions of the rectangles are the same as the length and width directions of the plates. Since in the method proposed in this paper, the detection signal is closely related to the variation of defects in the direction perpendicular to the direction in which the path is located, only the detection signal acquired on the path P1-P9 as shown in Figure 4b is selected here for the parametric classification identification of layered defects.

Table 3. The variation of length and width of delamination between top layer and the adjacent second layer (mm).

Label	Length (mm)	Width (mm)	Number of Samples	Label	Length (mm)	Width (mm)	Number of Samples
1	5, 5.1	5, 5.01, ······ 5.2	42	6	10, 10.1	5, 5.01, ······ 5.1	22
2	10, 10.1	10, 10.01, ······ 10.2	42	7	10, 10.1	15, 15.01, ······ 15.1	22
3	15, 15.1	15, 15.01, ······ 15.2	42	8	15, 15.1	5, 5.01, ······ 5.1	22
4	5, 5.1	10, 10.01, ······ 10.1	22	9	15, 15.1	10, 10.01, ······ 10.1	22
5	5, 5.1	15, 15.01, ······ 15.1	22				

The main geometrical features of the crack include length and direction. For the direction of the crack, there are 7 categories: 0°, 30°, 45°, 60°, 90°, 120°, 135°. The ply direction of composite plate and the crack direction itself will comprehensively affect the detection signal, so the detection signals on the detection paths which is in different directions need to be acquired for direction identification. The detection paths include path P1–P9, P2–P10, P3–P11, P4–P12, P5–P13, P6–P14, P7–P15, and P8–P16 as shown in Figure 4b. For the same crack direction, the crack length is set to 3 mm, 4 mm, ..., 12 mm, respectively. In the same crack length category, the length is varied in 0.01mm steps to obtain 88 samples of defects with length differences. For example, 2.68 mm, 2.69 mm, ..., 3 mm, 3.01 mm, ..., 3.55 mm is considered to be a variation of the length of the 3 mm class. Thus, 880 samples are included under each direction class, and a total of 6160 samples are obtained.

3.2. Building the Testing Defects

To illustrate the implementation process of the optimized RAPID imaging algorithm, three representative defects are selected as test samples: defect 1, an oblique crack with a length of 12.05 mm and an angle of 45° is set on the surface of the composite plate, and its coordinates are $(x, y) = (186, 264)$; defect 2, a vertical crack with a length of 4.1 mm and an angle of 90° is set on the surface of the composite plate, and its coordinates are $(x, y) = (263, 225)$; defect 3, a rectangle shape representative delamination with the size of 5 mm × 5.2 mm is modeled at the coordinates of the composite plate $(x, y) = (225, 263)$ between the first layer (top surface) and the adjacent second layer of the composite plate. The sensor array and the setting of defect 1 are shown in Figure 4a.

4. Identification Model

4.1. Identification Model of Defect Type

The effect of delamination and cracks on the propagation behavior of guided waves is different. On the detection path passing through the center of the circle, the signal is obtained by one-excitation and one-receive pattern for building a sample library. According to exciting frequency and sampling frequency, each detection signal contains 30,000 data points, and the signal is taken as a 30,000-dimensional sample. The sample library of defect type is constituted by 522 detection signals of delamination and 6160 detection signals of the crack. It is randomly divided into training set and test set with a ratio of 8:2. Because the dimension of the samples is too high, the Principal Component Analysis (PCA) method [30,31] is utilized for pre-processing. Samples of the training set are used as input of the SVM classifier, and the label set is formed by the defect classes of the corresponding samples (0 and 1 represent crack and delamination, respectively). The grid optimization method and cross-validation are used to adjust the kernel function type and the main model parameters. The kernel function is set as the linear kernel function, and the penalty factor (C) is 3. The average accuracy rate, the recall rate, and F_1 score obtained from 10 experiments are about 96.74%, 97.81%, and 96.04%, respectively. It can be seen from the scatter plot in Figure 5 that most of the samples are correctly classified and only one sample from the crack class is misclassified into the delamination class. Crack and delamination have different effects on the propagation behavior of Lamb waves, which enables the model to better identify the two types of defects.

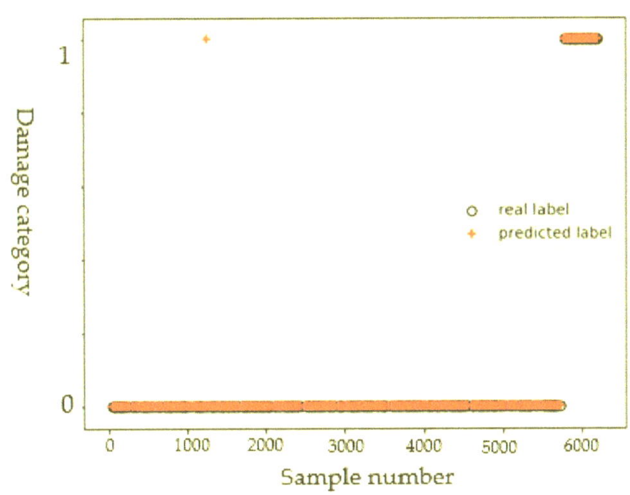

Figure 5. Scatter diagram of identification results of the defect type.

4.2. Parameter Identification Model for Delamination

The delamination defects mainly include 9 categories, with a total of 516 samples (6 supplemented samples between different classes are taken as test samples), which are randomly divided into training set and test set in a ratio of 8:2. Referring to the above method, the model kernel function is set as a linear kernel function, the penalty parameter

$C = 2$, and the average accuracy rate, recall rate, and F_1 score are 97.78%, 96.67%, and 97.57%, respectively. Figure 6 is the scatter diagram of the results of identification of delamination size. As can be seen from the figure, only one sample with a size of 5 mm × 8 mm is misclassified into the 5 mm × 5 mm class. The variation in the width of the rectangle has a certain influence on the detection signal. When the sample size is not sufficient to fully cover this variation, it will lead to misclassification of the model.

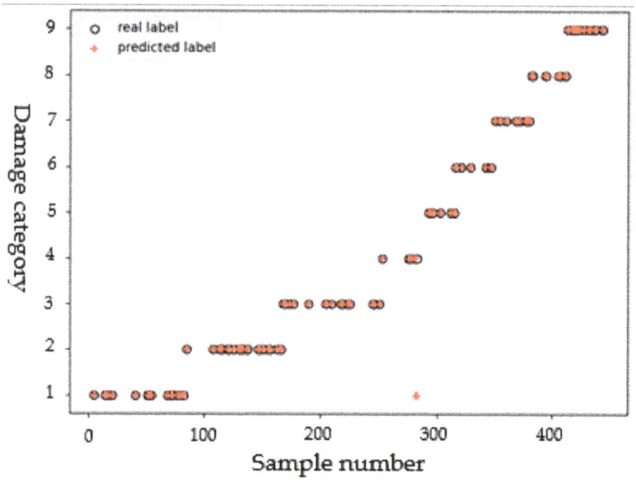

Figure 6. Scatter diagram of identification results of delamination size.

4.3. Parameter Identification Model for Crack

4.3.1. Direction

The scattered field formed by the crack has a certain influence on the guided wave. However, it is difficult to clearly know its law of propagation. In this paper, the LightGBM algorithm is used to identify the crack direction, because it is suitable for high-dimensional data, and its calculation speed is fast. The Hyperparameter Optimization (Hyperopt) method is used to automatically optimize the model parameters.

First, the LightGBM model is used to identify the crack directions on each single path separately, but the effect is not good after parameter adjustment. Taking the path P3–P11 as an example, the average accuracy rate is only 54.29%. This fully shows that in the single detection path, the influence of the ply direction of the composite board and the crack direction cannot be peeled off, which is not conducive to the correct identification of the crack direction by the model.

Comparing the detection signals obtained from different paths for the same defect, as shown in Figure 7a, the distance between the defect and the excitation and receiving point is relatively consistent, resulting in the wave packets of all samples appearing in the time period [100, 300] μs. Therefore, for all defects, the detection samples obtained on the paths P1–P9, P2–P10, P3–P11, P4–P12, P5–P13, P6–P14, P7–P15, and P8–P16 are intercepted and then connected to form new samples with a dimension of 160,000. Dataset with high-dimensional features is randomly divided into training set and test set according to a 9:1 ratio. During parameter adjustment, it is found that the parameters learning_rate and max_depth have a great influence on the classification results. The optimization range of learning_rate is [0.01, 1], the optimization range of max_depth is [1, 15]. Therefore, the optimization result is learning_rate = 0.05, max_depth = 7, and other parameters are set to default. The average accuracy rate, recall rate, and F_1 score are 97.44%, 98.43%, and 97.71%, respectively.

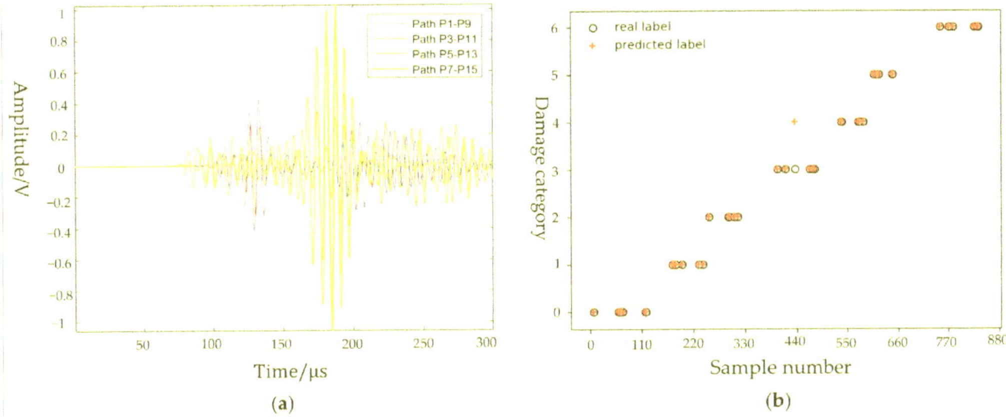

Figure 7. Identification of crack direction: (**a**) comparison of detection signals obtained from different paths for the same defect; and (**b**) scatter diagram of identification results of crack direction.

Figure 7b shows the scatter plot of the classification results after updating the sample size. As can be seen from the figure, only 1 sample in the 60° class is misclassified in the 90° class. This shows that after considering the relationship among the crack direction, the ply direction, and the direction of detection paths to update the dimension of samples, the classification accuracy of the model has been significantly improved.

4.3.2. Length

In each crack direction, when using the detection signals to identify the crack length, these detection signals should be taken from the path perpendicular to the crack direction. For example, the detection signal for a direction of 45° is taken from the path P7–P15; and the detection signal for a direction of 90° is taken from the path P1–P9. The 880 samples obtained in each direction are trained separately to identify the crack length in the corresponding direction. Taking the dataset of crack length in the 45° as an example, the 880 crack samples are randomly divided into a training set and a test set according to an 8:2 ratio. Referring to the method of identifying delamination parameters, the kernel function is set as a linear kernel function, and the penalty parameter $C = 2$. It can be seen from the scatter diagram in Figure 8 that two samples are misclassified to other classes. Since some paths for acquiring detection signals cannot be strictly perpendicular to the crack direction, the deviation in directions sometimes leads to misclassification of some samples (so, a sample with a length of 6.31 mm is misclassified into the 7 mm class; and a sample with a length of 7.07 mm is misclassified into the 9 mm class). However, most samples are classified correctly, indicating that the model has good performance. Based on the identification results of crack length in all directions, the mean of the classification accuracy, recall, and F1 scores of length samples in each crack direction are 96.84%, 97.12%, and 96.52%, respectively.

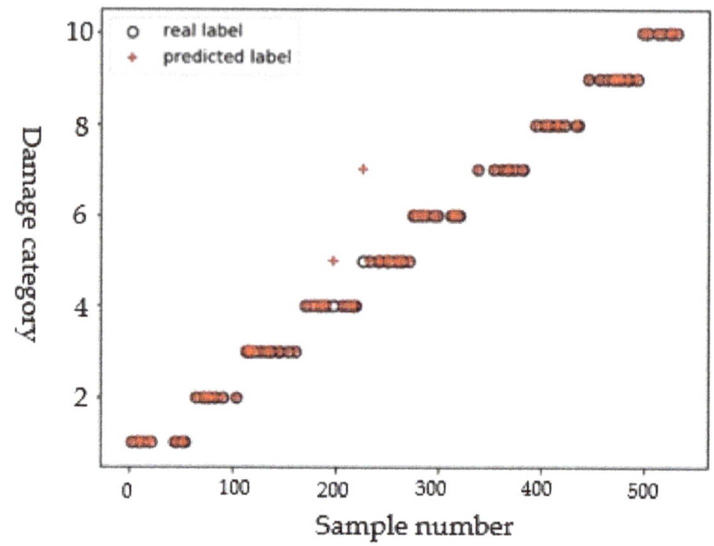

Figure 8. Scatter diagram of crack length.

5. Imaging for Testing Defects

5.1. Pre-Locating of Defects

Taking defect 1 as an example, 18 valid *DI* values are obtained by the above method for reconstructing the image. The image of the defect is reconstructed by RAPID, and the pre-locating imaging result is shown in Figure 9. The defect center is located at (178,266), which is close to the actual position of the defect shown by "+" in the figure. This shows the effectiveness of RAPID for defect location, yet at the same time that the algorithm is not precise enough and cannot give the type and geometric parameters of defects.

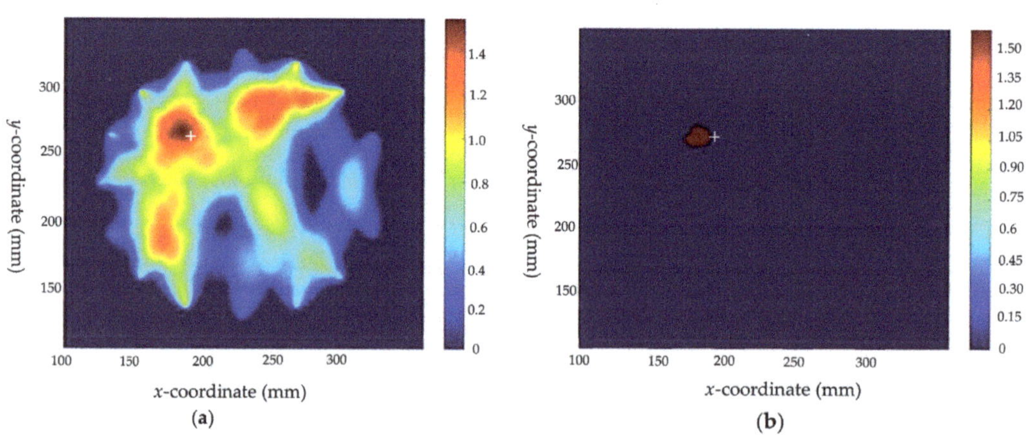

Figure 9. Imaging results before optimization of defect 1: (**a**) before optimization; and (**b**) threshold processing result before optimization.

5.2. Identification of Defect Features

The circular sensor array is repositioned so that its center is located at the predetermined position as described in Section 5.1. Obtaining detection signals for defect feature parameter identification was performed as follows: First, the defect type identification model is used to determine the type of defect, and the identification result shows that defect 1 is a crack. Then, the direction of the crack is determined by the crack direction identification model, and the identification result shows that the direction of defect 1 is 45°. Finally, the length of the crack under the 45° direction is identified by the crack length model, and the identification result shows that the length of defect 1 is 12 mm.

5.3. Precise Imaging of Defects

Based on the known crack size, the scaling parameter β is calculated according to Equation (4) for paths that are inconsistent with crack direction. Taking paths P7–P15 as an example, the scaling parameter β calculated for this path is 1.004. According to the identification results of the crack direction model, the weight of the cross paths can be calculated by Equation (6) as $W_i = 17$, where the cross paths are P1–P13 (in line with the crack direction) and P7-P15 (perpendicular to the crack direction). The optimized results are shown in Figure 10, where "+" is marked as the real position. Comparing the results of Figures 9 and 10, the optimized results not only reflect the location of the defect, but also intuitively indicate the type, direction, and size of the defects. However, because the crack is extremely narrow in the width direction, and the scaling parameters in this direction are not considered, the obtained localization results still have a certain deviation from the real position.

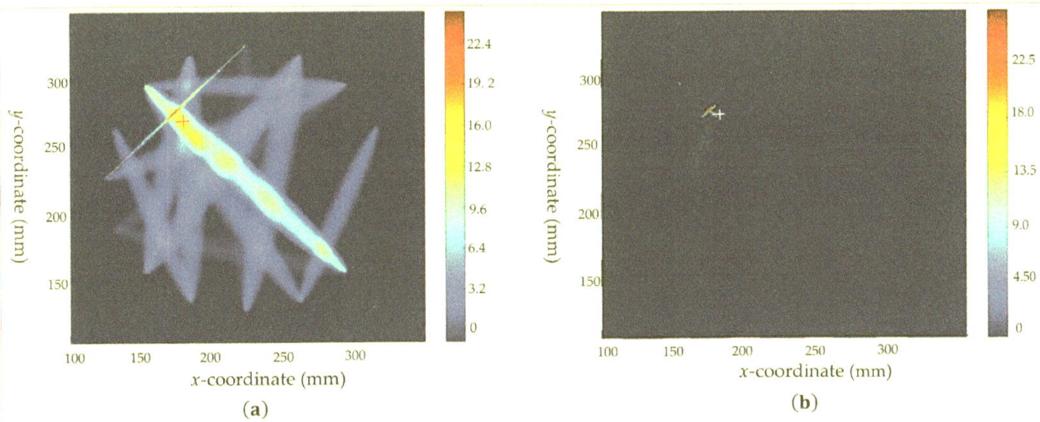

Figure 10. The imaging results of defect 1: (**a**) after optimization; (**b**) threshold processing result after optimization.

Defects 2 and 3 are also studied according to the analysis process of defect 1. Firstly, the RAPID algorithm is used for pre-locating and the results are obtained as shown in Figures 11a and 12a, respectively. After the sensor array position adjustment to regain the required detection signals, the results of crack and delamination are derived from the identification model of defect type, respectively. Next, different feature parameter identification models are applied to predict the crack tilt angle, crack length, and the size of the delamination defect, respectively. The results are shown in Table 4. The scaling parameter β is calculated as mentioned before (taking the paths P1–P9 of defect 2 and P10–P16 of defect 3 as examples, the scaling parameters β are calculated as 1.00053 and 1.0022, respectively) and to set the weights for the "cross paths" (P1–P9 and P6–P12) of defect 2. The coordinate positions of defect centers before and after optimization are shown in Table 4. The comparison plots of the results are shown in Figures 11 and 12, which indicate that the features of the defects are visualized and the imaging positions are closer to the real positions.

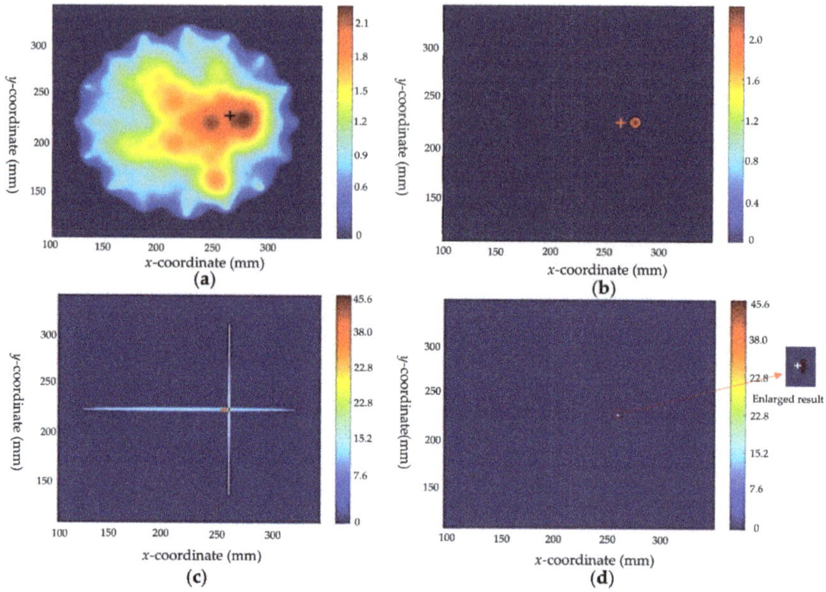

Figure 11. Comparison of imaging results before and after optimization of defect 2: (**a**) imaging results before optimization; (**b**) the threshold processing result before optimization; (**c**) image result after optimization; and (**d**) the threshold processing result after optimization.

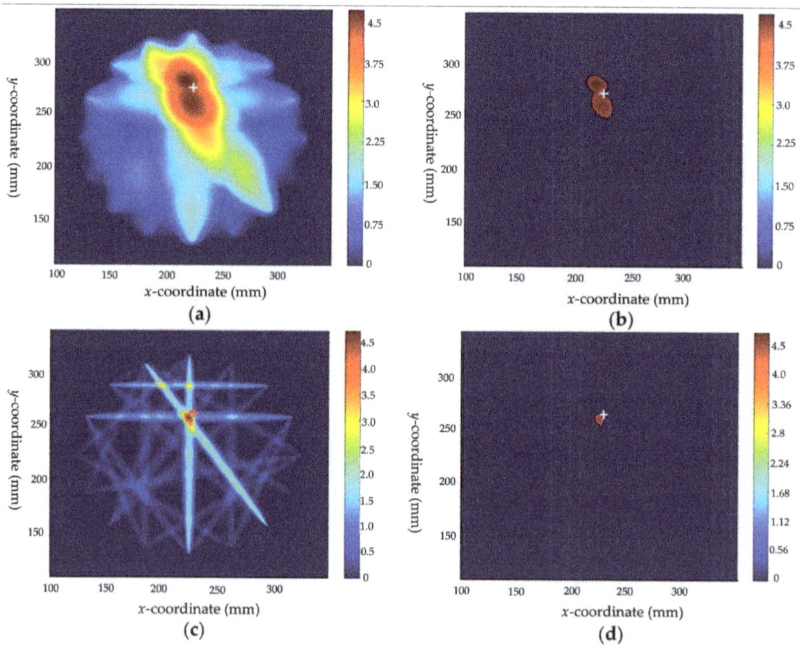

Figure 12. Comparison of imaging results before and after optimization of defect 3: (**a**) imaging results before optimization; (**b**) the threshold processing result before optimization; (**c**) image result after optimization; and (**d**) the threshold processing result after optimization.

Table 4. Comparison of parameters and center coordinates of three defects before and after optimization with the actual situation.

	Actual Defect Parameters and Center Coordinates				Predicted Defect Parameters and Center Coordinates				
	Type	Size (mm)	Direction	Center Coordinates	Type	Size (mm)	Direction	Before Optimization	After Optimization
Defect 1	Crack	12.05	45°	(186,264)	Crack	12	45°	(178,266)	(180,265)
Defect 2	Crack	4.1	90°	(263,225)	Crack	4	90°	(277,227)	(264,225)
Defect 3	Delamination	5 × 5.2	×	(225,263)	Delamination	5 × 5	×	(222,268)	(224,261)

6. Discussion

In this paper, the proposed method can visually characterize the defects and improve the accuracy of the imaging results. However, there are still some shortcomings in the method.

First, we need to further supplement and improve the sample library; the special case where the size is exactly in the middle of the two classes should be especially considered. For the study of delamination defects, our sample library currently only considers the case when the length and width directions are fixed. In subsequent studies, it will also be necessary to further investigate how to identify the size and direction of delamination when both the length and width directions vary.

Secondly, in relation to the algorithm, the effect of the ellipse factor of the width direction on the imaging results is not considered at present, resulting in the optimized imaging results still having a certain degree of offset compared with the actual position. At the same time, the influence of the neglected path on the imaging results during the path selection process, and whether there is a correspondence between the geometric parameters of the defect and the path selection scheme, need to be further explored in subsequent studies.

Finally, the limited number of sensors in the sensor array can degrade defect imaging quality. The lower the density of the sensor array, the lower the image quality. For oblique cracks, the fewer the sensors arranged in the detection area, the fewer sensing paths there will be. An insufficient number of sensors will result in the inability to select a suitable path to characterize the crack. As described above, the detection signal of the path perpendicular to the crack direction is selected to identify the crack length. When the direction of the selected path deviates from the desired direction, the recognition result of the crack length will be affected. Even when the number of sensors is severely insufficient, this will also lead to the inability to select a suitable path to characterize the crack. Determining how many cells should be contained in the sensor array to be most suitable for the final imaging for defects of different sizes will also need to be further discussed in further research.

7. Conclusions

In this paper, an improved RAPID imaging method for composite plates based on machine learning is proposed. The improved method consists of two main parts: (1) Using multiple machine learning models, the main features of defects, including type, size and direction, are identified by analyzing the detection signals. (2) Based on these defect features, the ellipse area scaling parameter β of the RAPID method is modified and weights are set for the detection paths associated with the features to visualize the defect features and effectively improve the accuracy of defect imaging.

(1) Appropriate ML models can be used to predict characteristics of defects such as the type of defect, the size of the delamination, and the length and the direction of the crack.
(2) The RAPID can be used for imaging of composite plates. After the type of defects and main parameter are predicted by the ML models, the β value in the algorithm can be further corrected and weights applied to the relevant paths, so that the predicted position is closer to the real position and the geometric characteristics of the defect can be visually displayed in the imaging results.

(3) The relationship between the crack direction, the ply direction, and the angle of detection paths comprehensively affects the detection signal and cannot be easily peeled off from the waveform curve. Therefore, the detection samples obtained under a single detection path are not sufficient to accurately predict the crack direction. Combining the information obtained under multiple detection paths leads to a larger sample dimension, and the combination of different information facilitates the model's ability to accurately identify the crack direction.

Author Contributions: Conceptualization, F.D. and X.Z.; Data curation, X.Z.; Formal analysis, F.D., X.Z., N.Y. and L.Z.; Funding acquisition, F.D.; Investigation, X.Z.; Methodology, F.D. and X.Z.; Project administration, F.D.; Software, X.Z.; Supervision, F.D.; Validation, X.Z., N.Y. and L.Z.; Visualization, X.Z.; Writing—original draft, X.Z.; Writing—review & editing, F.D. All authors have read and agreed to the published version of the manuscript.

Funding: This research was funded by the National Natural Science Foundation of China (11202137), Shanghai Alliance Program (2019025), Collaborative innovation fund of Shanghai Institute of Technology (XTCX2018-11).

Institutional Review Board Statement: Not applicable.

Informed Consent Statement: Not applicable.

Data Availability Statement: Not applicable.

Conflicts of Interest: The authors declare no conflict of interest.

References

1. Soutis, C. Fibre reinforced composites in aircraft construction. *Prog. Aerosp. Sci.* **2005**, *41*, 143–151. [CrossRef]
2. Ye, L.; Lu, Y.; Su, Z.Q.; Meng, G. Functionalized composite structures for new generation airframes: A review. *Compos. Sci. Technol.* **2005**, *65*, 1436–1446. [CrossRef]
3. Khan, A.; Kim, N.; Shin, J.K.; Kim, H.S.; Youn, B.D. Damage assessment of smart composite structures via machine learning: A review. *JMST Adv.* **2019**, *1*, 107–124. [CrossRef]
4. Qiu, L.; Liu, B.; Yuan, S.F.; Bao, Q. A spatial filter and two linear PZT arrays based composite structure imaging method. *J. Vibroeng.* **2015**, *17*, 1218–1231.
5. Purekar, A.S.; Pines, D.J. Damage Detection in Thin Composite Laminates Using Piezoelectric Phased Sensor Arrays and Guided Lamb Wave Interrogation. *J. Intell. Mater. Syst. Struct.* **2010**, *21*, 995–1010. [CrossRef]
6. Moll, J.; Golub, M.V.; Glushkov, E.; Glushkova, N.; Fritzen, C.P. Non-axisymmetric Lamb wave excitation by piezoelectric wafer active sensors. *Sens. Actuators A Phys.* **2012**, *174*, 173–180. [CrossRef]
7. Prasad, S.M.; Balasubramaniam, K.; Krishnamurthy, C.V. Structural health monitoring of composite structures using Lamb wave tomography. *Smart Mater. Struct.* **2004**, *13*, N73. [CrossRef]
8. Rhee, S.H.; Lee, J.K.; Lee, J.J. The group velocity variation of Lamb wave in fiber reinforced composite plate. *Ultrasonics* **2007**, *47*, 55–63. [CrossRef]
9. Moll, J.; Schulte, R.T.; Hartmann, B.; Fritzen, C.P.; Nelles, O. Multi-site damage localization in anisotropic plate-like structures using an active guided wave structural health monitoring system. *Smart Mater. Struct.* **2010**, *19*, 126–134. [CrossRef]
10. Deng, F.; Chen, H.L. A Defects localization Algorithm Based on the Lamb Wave of Plate Structure. *J. Nondestruct. Eval. Diagn. Progn. Eng. Syst.* **2021**, *4*, 021001. [CrossRef]
11. Nokhbatolfoghahai, A.; Navazi, H.M.; Groves, R.M. Evaluation of the sparse reconstruction and the delay-and-sum damage imaging methods for structural health monitoring under different environmental and operational conditions. *Measurement* **2021**, *169*, 108495. [CrossRef]
12. Nokhbatolfoghahai, A.; Navazi, H.M.; Groves, R.M. Using the hybrid DAS-SR method for damage localization in composite plates. *Compos. Struct.* **2020**, *247*, 112420. [CrossRef]
13. Wang, C.H.; Rose, J.T.; Chang, F.K. A synthetic time-reversal imaging method for structural health monitoring. *Smart Mater. Struct.* **2004**, *13*, 415. [CrossRef]
14. Eremin, A.; Glushkov, E.; Glushkova, N.; Lammering, R. Guided wave time-reversal imaging of macroscopic localized inhomogeneities in anisotropic composites. *Struct. Health Monit.* **2019**, *18*, 1803–1819. [CrossRef]
15. Guo, J.; Zeng, X.; Liu, Q.; Qing, X. Lamb Wave-Based Damage Localization and Quantification in Composites Using Probabilistic Imaging Algorithm and Statistical Method. *Sensors* **2022**, *22*, 4810. [CrossRef]
16. Liu, G.; Wang, B.; Wang, L.; Yang, Y.; Wang, X. Probability-based diagnostic imaging with corrected weight distribution for damage detection of stiffened composite panel. *Struct. Health Monit.* **2022**, *21*, 1432–1446. [CrossRef]

17. Liu, Z.; Zhong, X.; Dong, T.; He, C.; Wu, B. Delamination detection in composite plates by synthesizing time-reversed Lamb waves and a modified damage imaging algorithm based on RAPID. *Struct. Control Health Monit.* **2017**, *24*, e1919. [CrossRef]
18. Su, C.; Bian, H.; Jiang, M.; Zhang, F.; Sui, Q. Research on composites damage identification based on power spectral density and lamb wave tomography technology in strong noise environment. *Compos. Struct.* **2022**, *289*, 115466. [CrossRef]
19. Wang, S.; Wu, W.; Shen, Y.; Liu, Y.; Jiang, S. Influence of the PZT sensor array configuration on Lamb wave tomography imaging with the RAPID algorithm for hole and crack detection. *Sensors* **2020**, *20*, 860. [CrossRef]
20. Wu, Z.; Liu, K.; Wang, Y.S.; Zheng, Y.B. Validation and evaluation of damage identification using probability-based diagnostic imaging on a stiffened composite panel. *J. Intell. Mater. Syst. Struct.* **2015**, *26*, 2181–2195. [CrossRef]
21. Zhang, Z.; Pan, H.; Wang, X.; Lin, Z. Machine learning-enriched lamb wave approaches for automated damage detection. *Sensors* **2020**, *20*, 1790. [CrossRef] [PubMed]
22. Zhou, C.; Deng, F.; Liu, Y.; Liu, X.C.; Chen, H.L.; Liu, Z.H. Identification of corrosion damage degree of guided wave bend pipe based on neural network and support vector machine. *J. Mech. Eng.* **2021**, *57*, 136–144.
23. Peng, T.; Saxena, A.; Goebel, K.; Xiang, Y.; Sankararaman, S.; Liu, Y. A novel Bayesian imaging method for probabilistic delamination detection of composite materials. *Smart Mater. Struct.* **2013**, *22*, 125019. [CrossRef]
24. Mardanshahi, A.; Nasir, V.; Kazemirad, S.; Shokrieh, M. Detection and classification of matrix cracking in laminated composites using guided wave propagation and artificial neural networks. *Compos. Struct.* **2020**, *246*, 112403. [CrossRef]
25. Huo, H.; He, J.; Guan, X. A Bayesian fusion method for composite damage identification using Lamb wave. *Struct. Health Monit.* **2020**. [CrossRef]
26. Cortes, C.; Vapnik, V. Support-vector networks. *Mach. Learn.* **1995**, *20*, 273–297. [CrossRef]
27. Bahonar, M.; Safizadeh, M.S. Investigation of real delamination detection in composite structure using air-coupled ultrasonic testing. *Compos. Struct.* **2022**, *280*, 114939. [CrossRef]
28. Ke, G.L.; Meng, Q.; Finley, T.; Wang, T.F.; Chen, W.; Ma, W.D.; Ye, Q.W.; Liu, T.Y. Lightgbm: A highly efficient gradient boosting decision tree. In Proceedings of the 31st Conference on Neural Information Processing Systems (NIPS 2017), Long Beach, CA, USA, 4–9 December 2017; Volume 30.
29. Sheen, B.; Cho, Y. A study on quantitative lamb wave tomogram via modified RAPID algorithm with shape factor optimization. *Int. J. Precis. Eng. Manuf.* **2012**, *13*, 671–677. [CrossRef]
30. Zhou, Z.H. *Machine Learning*; Tsinghua University Press: Beijing, China, 2016.
31. Li, R.; Gu, H.; Hu, B.; She, Z. Multi-feature fusion and damage identification of large generator stator insulation based on Lamb wave detection and SVM method. *Sensors* **2019**, *19*, 3733. [CrossRef]

sensors

MDPI

Article

Characterization of Tensile Stress-Dependent Directional Magnetic Incremental Permeability in Iron-Cobalt Magnetic Sheet: Towards Internal Stress Estimation through Non-Destructive Testing

Borel Toutsop [1,2,3], Benjamin Ducharne [3,4,*], Mickael Lallart [3], Laurent Morel [2] and Pierre Tsafack [1]

[1] Faculty of Engineering and Technology, University of Buea, Buea P.O. Box 63, Cameroon
[2] Univ Lyon, INSA Lyon, Ecole Centrale de Lyon, Université Claude Bernard Lyon 1, CNRS, Ampère, UMR5005, 69622 Villeurbanne, France
[3] Univ Lyon, INSA Lyon, LGEF EA682, 69621 Villeurbanne, France
[4] ELyTMaX IRL3757, Univ Lyon, INSA Lyon, Centrale Lyon, Université Claude Bernard Lyon 1, Tohoku University, Sendai 980-8577, Japan
* Correspondence: benjamin.ducharne@insa-lyon.fr

Abstract: Iron-Cobalt ferromagnetic alloys are promoted for electrical energy conversion in aeronautic applications, but their high magnetostrictive coefficients may result in undesired behaviors. Internal stresses can be tuned to limit magnetostriction but must be adequately assessed in a non-destructive way during production. For this, directional magnetic incremental permeability is proposed in this work. For academic purposes, internal stresses have been replaced by homogenous external stress, which is easier to control using traction/compression testbench and results in similar effects. Tests have been limited to tensile stress stimuli, the worst-case scenario for magnetic stress observation on positive magnetostriction coefficient materials. Hysteresis cycles have been reconstructed from the incremental permeability measurement for stability and reproducibility of the measured quantities. The directionality of the sensor provides an additional degree of freedom in the magnetic response observation. The study reveals that an angle of $\pi/2$ between the DC ($H_{surf\,DC}$) and the AC ($H_{surf\,AC}$) magnetic excitations with a flux density B_a at $H_{surf\,DC}$ = 10 kA·m^{-1} constitute the ideal experimental situation and the highest correlated parameter to a homogeneous imposed tensile stress. Magnetic incremental permeability is linked to the magnetic domain wall bulging magnetization mechanism; this study thus provides insights for understanding such a mechanism.

Keywords: magnetic control; internal stress; local characterization; domain wall bulging; multi-axis magnetization

Citation: Toutsop, B.; Ducharne, B.; Lallart, M.; Morel, L.; Tsafack, P. Characterization of Tensile Stress-Dependent Directional Magnetic Incremental Permeability in Iron-Cobalt Magnetic Sheet: Towards Internal Stress Estimation through Non-Destructive Testing. *Sensors* **2022**, 22, 6296. https://doi.org/10.3390/s22166296

Academic Editor: Zenghua Liu

Received: 3 July 2022
Accepted: 16 August 2022
Published: 21 August 2022

1. Introduction

The aeronautic domain is undergoing profound changes, visible in the proliferation of electrical comfort equipment, electronic controllers, and navigation aid systems. As the level of required electrical power increases, so do the mass of electrical appliances since the weight/power ratio remains challenging to reduce [1].

In the case of onboard transformers [2], an industrial solution for the increase in power density consists of developing new magnetic materials such as iron-cobalt alloys. High power density gains (up to 10% and more) are, for instance, expected with Fe-27%Co compound. This material exhibits the highest saturation magnetization of all known soft ferromagnetic materials (>2.4 T) [3].

Unfortunately, FeCo alloys' astonishing saturation magnetization is associated with substantial magnetostrictive effects. These effects result in large deformation and significant undesired acoustic noise [4,5]. There are many ways to reduce magnetostriction, including the development of appropriate textures [4,6]. In [7], B. Nabi et al. claimed that magnetostriction in Fe-27%Co alloy is a consequence of magnetocrystalline anisotropy partly linked to the sheet crystallographic texture and, by developing a Goss texture, much better magnetic properties will be observed including a lower magnetostriction. Tuning internal stresses is another promising solution [8].

Mechanical Internal Stress (MIS) is crucial in iron alloys' magnetic and mechanical performance. MIS originates from various mechanisms, including plastic deformations, temperature gradients, or microstructural changes. Manufacturing processes including machining, welding, shot peening, heat treatment, and grinding are sources of MIS.

All MIS estimation methods are indirect. They start with the measurement of a coupled physical quantity and finish with a calculation stage [9]. The hole-drilling method [10] (see Figure 1 for illustration), the contour method [11], the crack compliance method [12], and the stripping method [9] are the most conventional mechanical methods. Together with the chemical processes [9], they constitute the so-called destructive techniques.

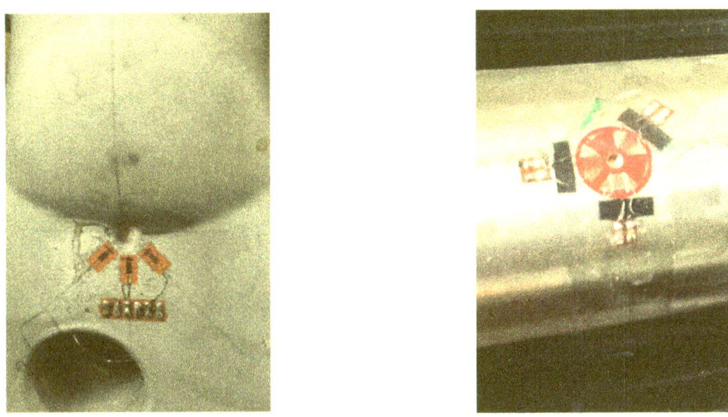

Figure 1. Illustration of internal stress evaluation by destructive hole-drilling tests (courtesy of CETIM, Senlis, France).

Non-destructive Testing (NDT) methods have also been described [13–15]. These methods include diffraction (X-ray, Neutron), ultrasounds, or acoustic emission [16]. In the case of conductive and/or ferromagnetic parts, electromagnetic NDT is often indicated. Eddy currents testing [17], magneto-acoustic emission [18,19], electromagnetic acoustic transducer [6], and Magnetic Barkhausen Noise (MBN) [20–22] are the most popular ones [23]. These methods offer fast response, low cost, small size, and easy maintenance, but their current industrial developments are limited, and their efficiency is questionable. Very few of them have crossed the threshold between academia and industry.

Based on MBN measurements, the Stresstech controller (Jyväskylä, Finland) is, however, one of them [9]. A significant problem for this device comes from a quasi-impossibility to distinguish the effect of MIS from other dependent properties (dislocations, grain size, texture, plastic strain, precipitates, phase changes, impurities, etc.). The micromagnetic, multi-parametric, microstructure and stress analysis (3MA) developed by IZFP Fraunhofer institute (Saarbrücken, Germany) is an interesting alternative. 3MA accumulates and combines data from different magnetization signatures and identifies the ultimate magnetic combination of indicators to a given targeted property (hardness, internal stress, yield strength, etc.) [24]. 3MA is pragmatic and efficient but needs time-consuming experimental campaigns and provides non-transposable results.

All magnetic MIS controllers have been designed on the same principles that use a unidirectional magnetization induced by a powerful electromagnet combined with local surface sensors to observe the magnetic answers [25]. A current research trend of alternative and/or self-made magnetic sensors for magnetic NDT measurements is perceptible in the literature [26,27]. While attractive in terms of sensor reproducibility and stability, these innovations fail to accurately showcase the claim for directional selectivity of the magnetic sensor as they are applied in angular-dependent measurements. In Figure 14 of [21], however, simulation predictions on FeCo alloys show that stress effects on the magnetic response can be more visible when the magnetic excitation and the stress direction are not colinear and that a directional sensor would probably bring significant improvement in the stress observation.

Among the magnetic methods, just a few can enforce directional measurements. Point probes [28,29] and MBN are one of them [30] but not tested in this study. Instead, we opted for Magnetic Incremental Permeability (MIP, μ_{MIP}), as the first results over stress dependency were auspicious ([31], Figure 13 in [32]). MIP is based on the domain walls bulging magnetization mechanism (reversible magnetization variations), as observed under low amplitude magnetic excitation variations. This mechanism appears to be very sensitive to MIS [31].

So far, MIP has almost always been observed with pancake (Figure 2, left-hand side) or wound coils preventing access to directional information. Still, this limitation can be overcome using double coils Transmitter/Receptor adjacent TR-probes as depicted on the right-hand side of Figure 2.

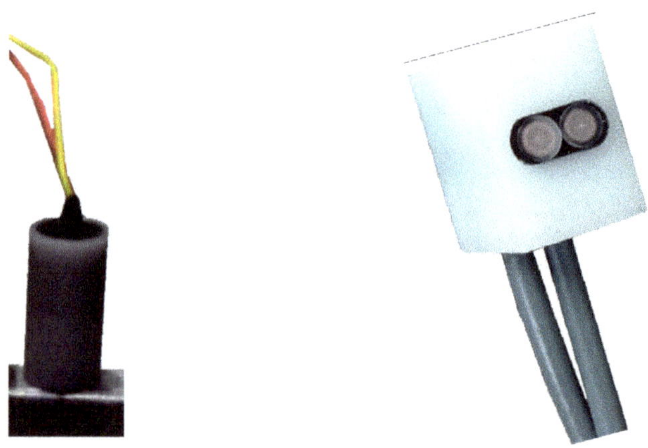

Figure 2. (**Left**): single pancake coil illustration; (**Right**): adjacent Transmitter/Receptor, TR-probes illustration.

We propose to overpass the directional limitation in the present study by using a miniature U-shape ferrite magnetic core. MIS is impossible to be precisely controlled; thus, it has been replaced by external stress. The effect on the magnetization process is considered independent of the stress origin. Tests were limited to tensile stress, constituting the worst-case scenario for stress observation. Iron-Cobalt alloys have positive magnetostriction coefficients, and tensile stress softens the magnetic behavior. As an already soft material, limited tensile stress effects can be observed in the $B_a(H_{surf})$ classic hysteresis cycles (where B_a is the flux density, and H_{surf} is the tangential magnetic excitation field). Hysteresis cycles were reconstructed for all MIP measurements providing a stable and reproducible signature. Then, indicators read on these cycles were defined and plotted vs. the homogeneous tensile stress. Pearson factors were calculated and analyzed to establish their correlation level with the external tensile stress. Conclusions were drawn regarding the non-destructive testing objective and the physical properties of the domain wall bulging.

This manuscript is organized as follows: Section 2 describes the experimental conditions, including sensors and tested specimens. Section 3 gives the experimental results. Section 4 provides analysis, discussions, and conclusions. Among them, the optimal conditions arise for a mechanical tensile stress estimation based on directional MIP.

2. Experimental Setup

2.1. Description of the Specimens

FeCo Iron-Cobalt laminations were tested in this study. This material has a high yield strength (1000 MPa). All specimens were extracted from the same batch. Strips were cut by Electrical Discharge Machining (EDM) to get tensile-shape type specimens (Figure 3). Two specimens were tested for consistency.

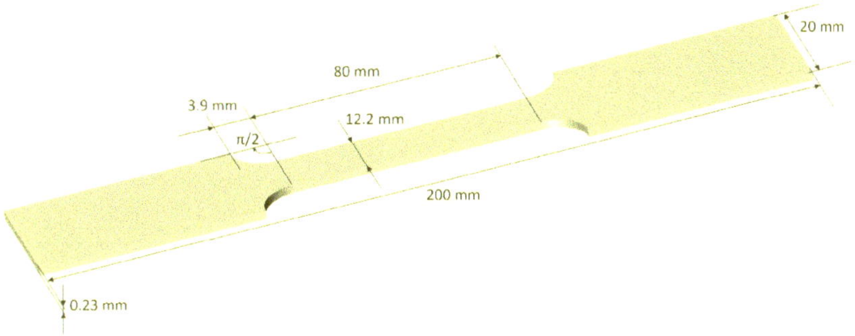

Figure 3. FeCo tensile stress specimen dimensions.

Table 1 gives the detailed composition, physical and mechanical properties of the tested specimens.

Table 1. Composition, physical and mechanical properties of the tested specimens.

• Composition:					
C (Mass %)	**Si**	**Mn**	**Co**	**V**	**Fe**
<0.015	<0.1	<0.15	49	2	Bal.
• Physical properties:					
Density ($g{\cdot}cm^3$)	**Elect. Res. ($\mu\Omega{\cdot}cm$)**	**Exp. Coef. ($\cdot K^{-1}$)**	**Therm. Cond. ($W{\cdot}cm^{-1}\ K^{-1}$)**	**Curie Temp. (°C)**	
8.12	40	9×10^{-6}	0.3	950	
• Mechanical properties:					
Yield Strength (MPa)	**Tens. Strength (MPa)**	**Young Mod. (GPa)**	**Hardness (HV10)**		
1000	1345	250	300		

2.2. Description of the Experimental Setup

The IEC 60404-3 standard details using a single sheet tester for the magnetic characterization of a ferromagnetic lamination [33]. For geometrical reasons, it is impossible to combine a traction bench while respecting all the recommendations. Still, the experimental setup we developed was inspired by this standard. Figure 4 gives pictures and a 2D overview of the test bench.

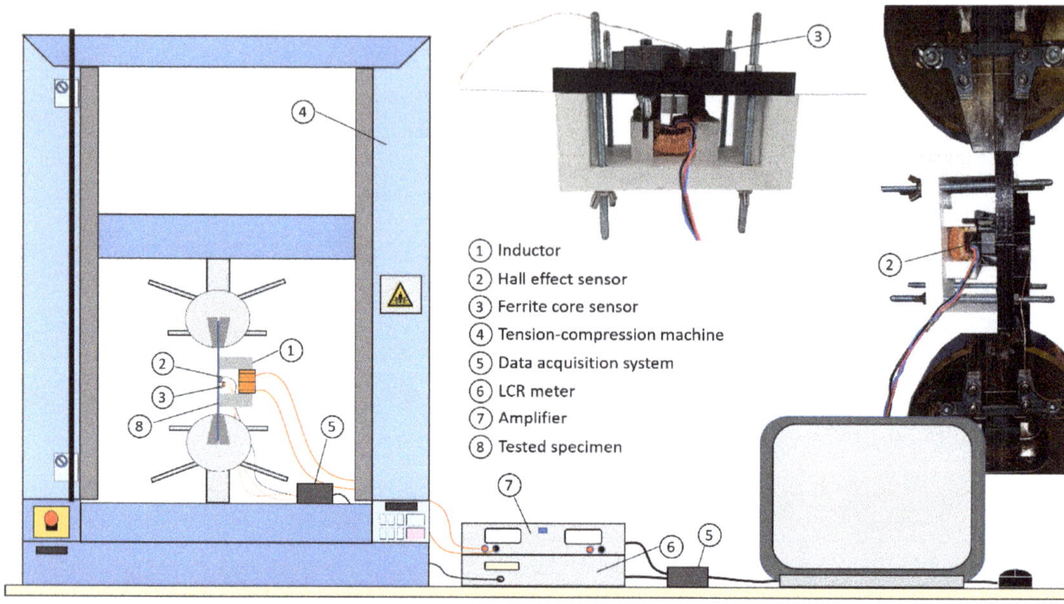

Figure 4. 2D overview of the experimental setup.

A tension–compression machine Shimadzu AGS-X series (Kyoto, Japan) was used for the tensile stress application. All tests were done at room temperature and under constant imposed stress conditions without initial load. The distance between the grips was 100 mm. Once the constant stress was imposed, a minimum of 60 s was waited before starting the acquisition process to avoid any drift issues. The magnetic inductor was made of a U-shaped FeSi 3% electrical steel yoke. The leg size of the yoke was 12 mm × 12 mm, and the inner distance between the legs was 30 mm. A 500 Turns excitation coil was wound around the yoke and supplied by a Kepco BOP 100-4 M (New York, NY, USA) power amplifier for the $B_a(H_{surf})$ characterization and by an RS pro 180 W bench power supply (Corby, UK) for the MIP. The data acquisition and analog signal generation were ensured by the DEWESoft X2 (Trbovlje, Slovenia) data acquisition software associated with a SIRIUS 8 × CAN data acquisition. A noise-shielded radiometric linear Hall probe SS94A from Honeywell (Charlotte, NC, USA) was positioned tangentially to the surface of the tested sample for the surface magnetic field (H_{surf}) measurement.

2.3. Magnetic Sensors

Two sensors dedicated to evaluating the magnetic state of the tested specimens have been used in this study. Figure 5 depicts a 3D overview of both sensing solutions.

The first sensor was an n = 50 turns wound coil. The magnetic flux density B_a was obtained by numerical integration (Equation (1)) of the voltage drop $e(t)$ during the magnetization cycle:

$$B_a(t) = -\frac{1}{n \cdot S}\int_0^t e(t)dt \tag{1}$$

where S is the specimen cross-section. A numerical correction was done to cancel the undesired drift due to the integration process. The second sensor (Figure 6) combines half a toroidal ferrite core (WE-TOF EMI from Würth Elektronik, Künzelsau, Germany), and a 54 turns wound coil. Specific 3D-printed support (see Figures 5 and 6) has been designed to hold the sensor and authorize plane measurements at every $\Delta q = \pi/18$ rad angle step. A

Flashforge Dreamer (Jinhua, China) and PLA polymer were used. This support does not influence the magnetic measurements.

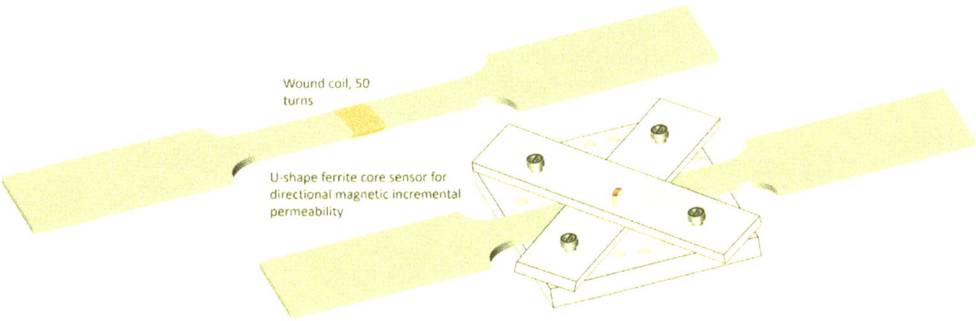

Figure 5. Illustration of the magnetic sensors used in this study.

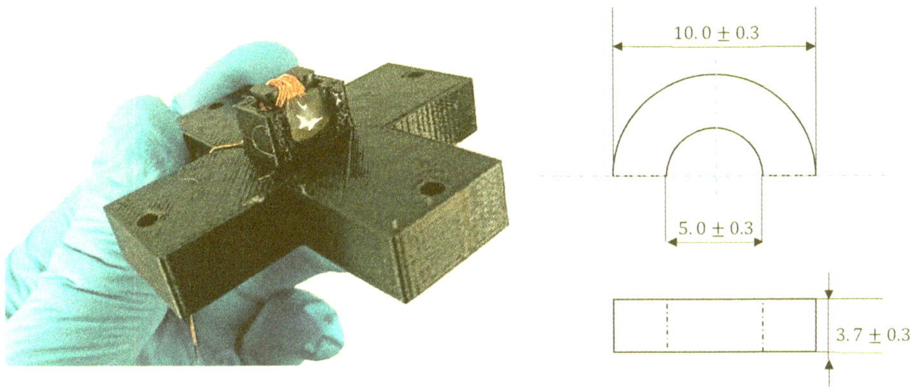

Figure 6. Photo and dimensions of the directional MIP sensor.

A precision Keysight (Santa Rosa, CA, USA) LCR meter was used to record the impedance Z of the second sensor during the experimental phases.

2.4. Experimental Process

The experimental campaign was divided into two phases:

- The first phase was dedicated to evaluating the tensile stress effect on the evolution of the classic $B_a(H_{surf})$ hysteresis cycles. This effect has already been thoroughly described in the literature [28,34], and our objective was to validate the specimens' conformity;
- A similar tensile stress sequence was run in the second phase but combined with directional magnetic incremental permeability measurements. For each tensile stress level, a set of ten $Z(H_{surf})$ curves were plotted (for different values of angle q from 0 to $\pi/2$ rad with a $\Delta q = \pi/18$ rad step).

3. Experimental Results

3.1. $B_a(H_{surf})$ Hysteresis Cycles

Tensile stress σ on Iron-Cobalt ferromagnetic materials is expected to soften the magnetic behavior. In the low-frequency range, softer magnetic behavior means higher permeability but lower coercivity. This behavior is well-known and has already been displayed in the scientific literature (Figure 6 in [21] and Figure 2.34 in [34]).

Experimental results depicted in Figure 7 confirm our expectations. They validate the conformity of our test bench and the tested specimens.

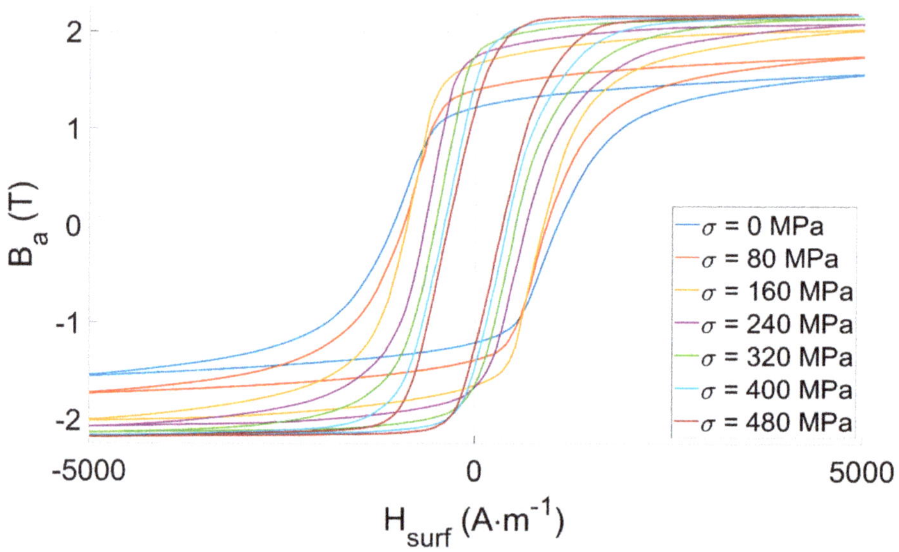

Figure 7. FeCo: experimental measurements for the tensile stress-dependent $B_a(H_{surf})$ hysteresis cycles.

$B_a(H_{surf})$ hysteresis cycles are firmly tensile stress-dependent and could be considered for σ observation. Still, the impossibility of using wound coils in an industrial context where large surfaces must be controlled is worth mentioning. This observation encourages users toward other methods, including the directional MIP non-destructive method described below.

3.2. *Directional Incremental Permeability Z(H_{surf})*

3.2.1. Magnetic Incremental Permeability

In general, magnetic permeability μ can be described as a measure of the material's response to an applied magnetic field.

More precisely and according to the German standard DIN1324 (part II: magnetic field for material quantities [35,36]), the Magnetic Incremental Permeability (MIP, μ_{MIP}) is defined as the slope of inner asymmetric loops (Equation (2), Figure 8). These loops, also called minor cycles, are obtained when the tested material is exposed to the superimposition of two magnetic contributions:

1. A low-frequency (quasi-static), high amplitude magnetic excitation, that provides a bias magnetization;
2. A high-frequency, low amplitude magnetic excitation, allowing the measurement of the relative magnetic incremental permeability μ_{MIP_r} as:

$$B_a(t) = -\frac{1}{n \cdot S}\int_0^t e(t)dt \quad (2)$$

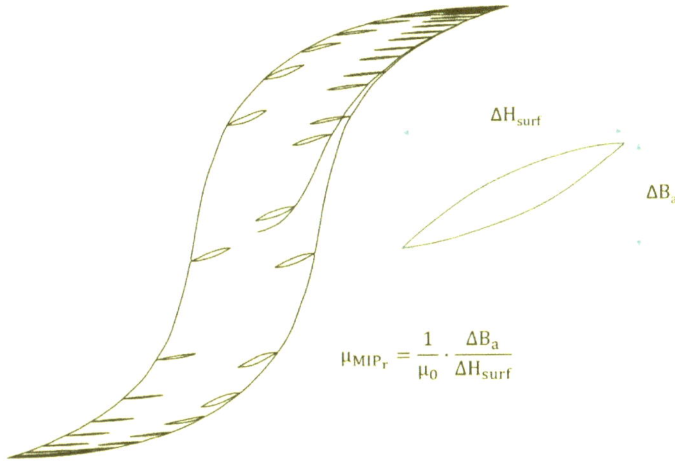

Figure 8. MIP illustration and equation [37].

3.2.2. $Z(H_{surf})$ Butterfly Loops

The experimental setup described in Section 2.2 measures the impedance of the MIP sensor shown in Figures 5 and 6. Figure 9 consists of the evolution of *IZI*, the sensor impedance modulus vs. the quasi-static magnetic excitation for four levels of tensile stress σ. For this test, the sensor is aligned with the tested specimen's length and the magnetic excitation (see Figure 9 top left-hand corner).

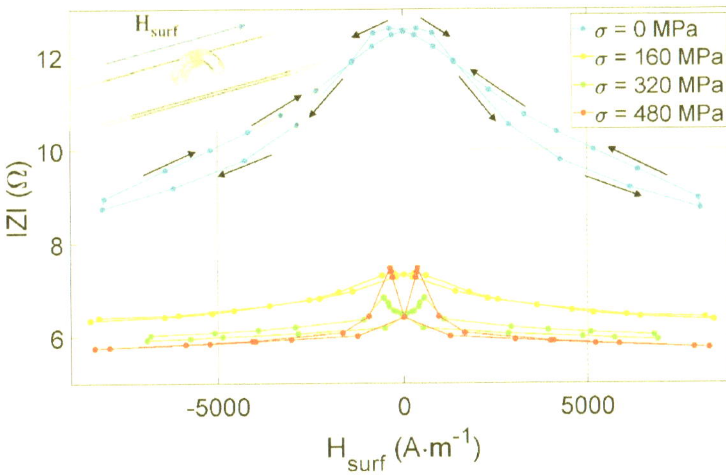

Figure 9. $IZI(H_{surf})$ for different tensile stress levels and with the MIP sensor aligned with the length of the tested specimen as illustrated in the top left-hand side incrustation.

3.2.3. From Z to the FeCo μ_{MIP}

The interpretation of measurements displayed in Section 3.2.2 is facilitated by returning the physical quantities. For this, the relation between *IZI* and μ_{MIP} for the FeCo sheet MIP needs to be established.

The solution opted out in this study is based on a magnetic reluctance scheme. Even inexact as relying on simplifying assumptions, it allows quickly estimating the permeability with satisfactory accuracy. Additional post-processing checking the validity of the resulting MIP was also performed: we ran comparisons between incremental μ_{MIP} and differential

μ_{Diff} permeabilities (from Figure 9 results) in the saturated zone where they are supposed to be similar. These comparisons are available in Table 1. Figure 10 describes the method: reluctances $\mathcal{R}_1$ and $\mathcal{R}_2$ are associated with the ferrite core and the FeCo sheet, respectively. The magnetic contact between the ferrite core and the sheet is supposed ideal; thus, with a total absence of airgap. Then, by assuming $\mathcal{R}_2 >> \mathcal{R}_1$, this simplified approach leads to a linear relationship between the imaginary part of Z'' and the relative permeability μ_r:

$$\mu_r = \frac{Z'' \cdot l_2}{\mu_0 \cdot N^2 \cdot A_2 \cdot \omega} \tag{3}$$

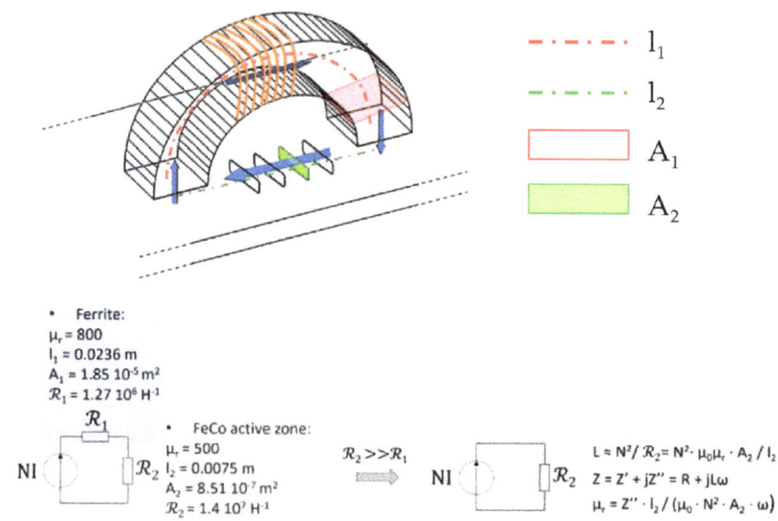

Figure 10. Simplified reluctance scheme for a linear relation between Z'' and μ_r.

Figure 11 gives the resulting $\mu_r(H_{surf})$ for all Figure 9 tests. It can be noted that the curve shape is conserved compared to the impedance modulus |Z|, denoting negligible electrical losses in the device.

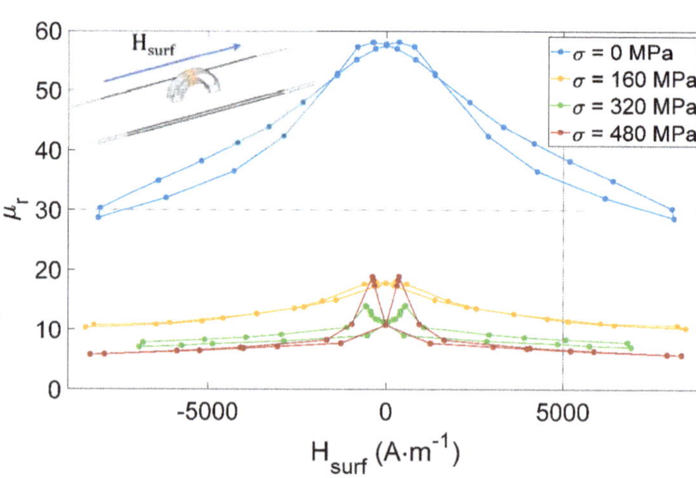

Figure 11. $\mu_r(H_{surf})$ for different tensile stress levels and with the MIP sensor aligned with the length of the tested specimen as illustrated in Figure 9.

Table 2 comparisons confirm our expectations and the relatively similar permeabilities in the saturated range.

Table 2. Comparisons between $\mu_{r\ MIP}$ and $\mu_{r\ Diff}$ in the high H_{surf} amplitude range (saturated zone).

σ (MPa)	H_{surf} (A·m^{-1})	$\mu_{r\ MIP}$	$\mu_{r\ Diff}$
–	4500	37.5	40
160	4500	12	17
320	4500	9	10
480	4500	7	9

3.2.4. From μ_{MIP} to $B_{a\ MIP}(H_{surf})$ Hysteresis Cycles

The butterfly loop (Figure 11) is the MIP method conventional signature. In NDT, defining indicators read directly on this signature and plot them vs. the targeted properties (i.e., the property to be assessed, MIS, microstructural information, etc.) is classical. For MIP tests, those indicators include the maximum amplitude, the amplitude at the remanence point, the curve width at 50% of max(μ_{MIP}), to name but a few [38]. In this study and for comparison purposes, the indicator definition has been postponed to a final step consisting of reconstructing a MIP hysteresis cycle.

Same indicators can be used for the classic and the MIP hysteresis cycles and comparing the stress influence on both these magnetic signatures is facilitated. The resulting induction field $B_{a\ MIP}$ is obtained from Equation (4):

$$B_{a\ MIP} = \int \mu_0 \mu_{r\ MIP}.dH_{surf} \tag{4}$$

The integration process brings stability and eases the determination of the indicators. Figure 12 depicts the $B_{a\ MIP}(H_{surf})$ for all tests performed in Figure 9.

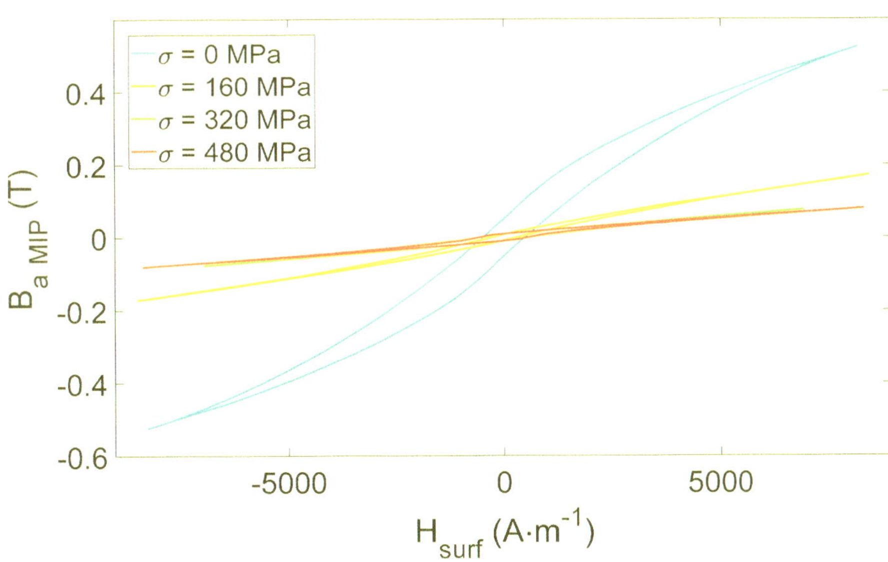

Figure 12. $B_{a\ MIP}(H_{surf})$ hysteresis cycles for all Figure 9 tests.

3.2.5. Directional $B_{a\ MIP}(H_{surf})$ Hysteresis Loop

In the following series of tests, five tensile stress levels and ten angle positions were implemented. The applied stress was limited to 250 MPa. For saturation reasons, the stress effect is especially weak beyond the [0–250] MPa range.

4. Discussion

$B_a(H_{surf})$ standard cycles are straightened by external tensile stress (Figure 7). Quite interestingly, $B_{a\,MIP}(H_{surf})$ loops show opposite behaviors (i.e., the resulting hysteresis loop is laying down, see Figure 13). Iron-Cobalt alloys are characterized by positive magnetostriction, resulting in a magnetic behavior softened by tensile stress. As an already soft material, tensile stress effects on $B_a(H_{surf})$ are limited. Oppositely, a much stronger influence can be detected on the $B_{a\,MIP}(H_{surf})$ cycles. This observation opens exciting perspectives and confirms MIP as an excellent alternative for estimating tensile stress stimuli.

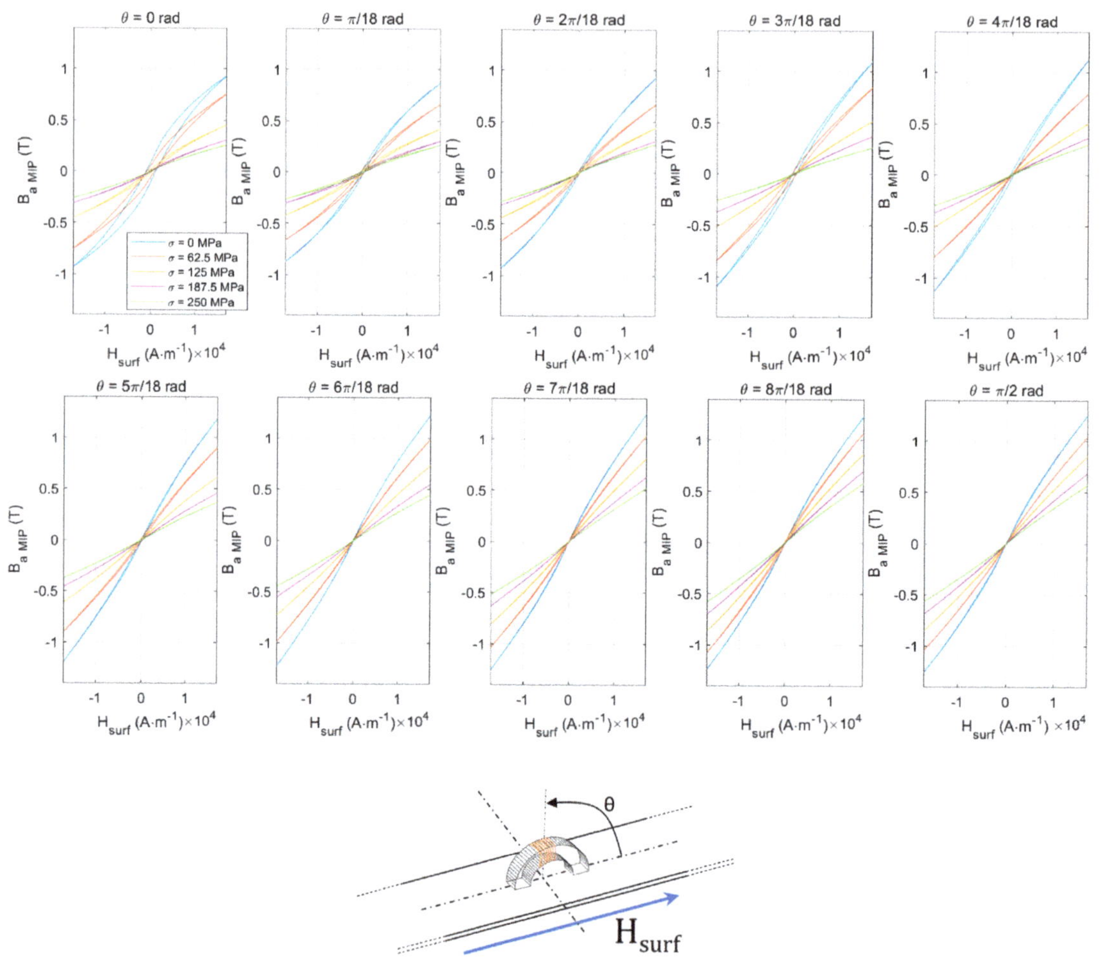

Figure 13. $B_{a\,MIP}(H_{surf})$ hysteresis cycles in the [0–π/2] angle range.

Coercivity is supposed to be independent of the nature of the magnetic signature tested ($B_a(H_{surf})$, Barkhausen noise, MIP, etc.). Therefore, comparing $H_c(\sigma)$ and $H_{c\,MIP}(\sigma)$ (at θ = 0 rad) can be considered a way to check the reliability of the MIP measurements. Figure 14 depicts this comparison in the σ = 0–250 MPa range (Figure 13), and as expected, coercivities of both charaterizations ($B_a(H_{surf})$ and $B_{a\,MIP}(H_{surf})$) remain close as σ varies.

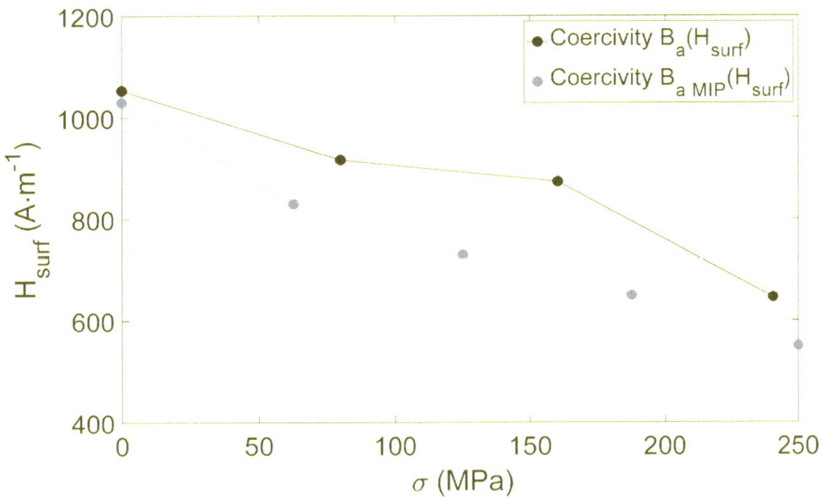

Figure 14. Coercivity vs. stress for $B_a(H_{surf})$ and $B_{a\ MIP}(H_{surf})$ hysteresis cycles.

From its atomic origin to human-scale observation, magnetization in ferromagnetic steels depends on distinct mechanisms. Each mechanism is characterized by its geometrical scale, time constant, and sensitivity to the physical environment (magnetic, mechanical, thermal, etc.). These mechanisms overlap; thus, isolating and listing them is complex. Still, three categories can be established:

1. Structure and kinetics of the magnetic domains (10^{-4}–10^{-6} m):
    - Domain walls bulging (reversible, in the low excitation range);
    - Irreversible domain wall motions (middle excitation range);
    - Nucleation and annihilation (high range);
2. Orientation and amplitude of atomic magnetic moments (10^{-11}–10^{-9} m):
    - Magnetization rotation (high and very high magnetic excitation).
3. Human scale mechanisms:
    - Macroscopic eddy currents

Such a combination makes impossible to isolate the effect of stress on a given mechanism by a unique observation of standard $B_a(H_{surf})$ hysteresis cycles. Oppositely, MIP experimental observation provides privileged access to domain walls bulging. In the recommendation of the literature [39–41], MIP is suggested to be run under an alternating magnetic field of amplitude half the coercivity, limiting the domain wall motions to reversible ones. Domain walls bulging is sensitive to the low amplitude of magnetic excitation. Scaling approaches and analogies encourage us to check its sensitivity in a low range of stress.

The directionality of domain wall bulging is an open question. Even if according to the manufacturer, FeCo laminations exhibit a not very pronounced crystallographic texture, every material is somehow anisotropic in its magnetic answer, and so it should be for MIP.

Figure 13 confirms this statement by showing substantial variations over the tested angles.

Five indicators directly read on the q-dependent $B_{a\ MIP}(H_{surf})$ cycles have been tested to identify the most adapted to a tensile stress estimation:

1. Coercivity $H_{c\ MIP;}$
2. Remanence $B_{r\ MIP;}$
3. $B_{a\ MIP}(H_{surf})$ Hysteresis area;
4. $B_{a\ MIP}$ at $H_{surf} = 2$ kA·m$^{-1;}$
5. $B_{a\ MIP}$ at $H_{surf} = 10$ kA·m$^{-1.}$

A refined analysis based on Pearson correlation factors ρ reveals $B_{a\ MIP}$ at $H_{surf} = 10$ kA·m^{-1} as the best indicator. Figure 15a depicts its variations vs. σ and for the ten sensor positions that have been tested. An astonishing 0.99 linear correlation was obtained for $\theta = \pi/2$.

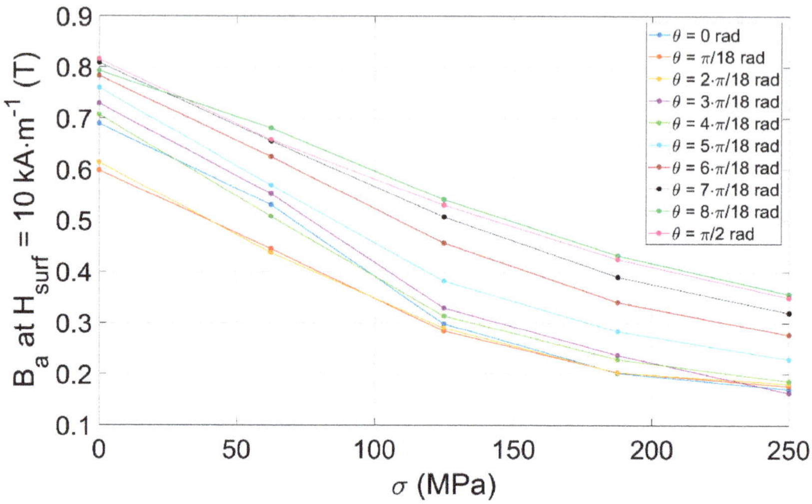

(a)

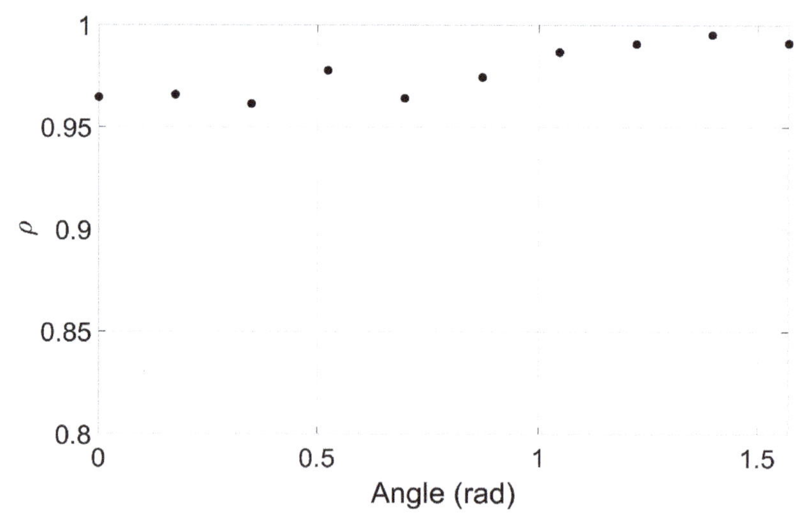

(b)

Figure 15. (**a**) B_a at $H_{surf} = 10$ kA·m^{-1} for θ in the [0, π/2] range. (**b**) Related Pearson correlation factors.

Even if read on the $B_{a\ MIP}(H_{surf})$ cycles, $H_{c\ MIP}$ depends on every magnetization mechanism. Oppositely, $B_{a\ MIP}$ at $H_{surf} = 10$ kA·m^{-1} provides a cumulative reflection all along the magnetization cycle of the domain wall bulging mechanism.

Concerning the sensor orientation, there is no energetical reason for the domain wall to bulge differently in one direction or the other in a supposedly isotropic material. σ and H_{surf} being parallel, on the first hand, if $\theta = 0$ rad, the static and dynamic contributions are superimposed, creating an even stronger softening behavior, and accentuating the effect on the material and the behavior differences between the stress levels. Oppositely when

$\theta = \pi/2$ rad, the static and the dynamic contribution operate in quadratic directions, and the sensor measurement is less affected by the softening effect, so that the measure becomes quasi-linearly dependent on σ.

5. Conclusions

The directionality influence of MIP sensors has barely been studied before. In this work, we focus on observing the effect of a tensile stress σ on the magnetic response for several angles. We found that ideal conditions are reached when $H_{surf\,DC}$ and $H_{surf\,AC}$ exhibit quadratic directions.

$B_a(H_{surf})$ standard cycles are straightened by external tensile stress and show little changes. $B_{a\,MIP}(H_{surf})$ loops show opposite behaviors and significant variations (Figure 13), confirming MIP as an excellent way to estimate tensile stress stimuli.

Among $B_{a\,MIP}(H_{surf})$ indicators, $H_{c\,MIP}$ depends on all the magnetization mechanisms. Its σ dependency is evident but not the most linear one. Other $B_{a\,MIP}(H_{surf})$ indicators are more specific to the domain wall bulging mechanism and can be used to interpret the influence of σ on this mechanism. The best result in terms of linear correlation was obtained with $B_{a\,MIP}$ at $H_{surf} = 10$ kA.m^{-1} and for $q = \pi/2$ rad (Pearson coefficient $\rho \approx 0.995$).

Many possibilities can be listed in the follow-up associated with this study; this includes, in the short term, the influence of homogeneous compressive stress. Also, the proposed results should be confirmed on MIS of pre-characterized specimens (by X-ray diffraction, for instance).

Eventually, a new type of materials (martensitic stainless steel, low carbon steel, etc.) could be tested. In an even longer-term vision, it would be interesting to check the capability of the directional MIP on other usual magnetic NDT targeted properties, including microstructural properties, aging, etc.

Author Contributions: Conceptualization, B.D., M.L. and L.M.; methodology, B.D., M.L. and L.M.; software, B.T. and B.D.; validation, B.D., M.L. and L.M.; investigation, B.T., B.D., M.L. and L.M.; writing—original draft preparation, B.D.; writing—review and editing, B.T., B.D., M.L. and L.M.; supervision, B.D., M.L., L.M. and P.T. All authors have read and agreed to the published version of the manuscript.

Funding: This research received no external funding.

Institutional Review Board Statement: Not applicable.

Informed Consent Statement: Not applicable.

Data Availability Statement: The study did not report any data.

Acknowledgments: This work was supported by the Service de Coopération et d'Action Culturelle (SCAC) of the French Embassy in Cameroon.

Conflicts of Interest: The authors declare no conflict of interest.

References

1. Dorn-Gomba, L.; Ramoul, J.; Reimers, J.; Emadi, A. Power electronic converters in electric aircraft: Current status, challenges, and emerging technologies. *IEEE Trans. Transp. Electrif.* **2020**, *6*, 1648–1664. [CrossRef]
2. Ram, B.S.; Paul, A.K.; Kulkarni, S.V. Soft magnetic materials and their applications in transformers. *J. Magn. Magn. Mater.* **2021**, *537*, 168210. [CrossRef]
3. Turgut, Z.; Huang, M.Q.; Horwath, J.C.; Hinde, R.; Kubicki, J.; Fingers, R.T. Effect of tensile stress and texture on magnetic properties of FeCo laminates. *IEEE Trans. Magn.* **2004**, *40*, 2742–2744. [CrossRef]
4. Savary, M.; Hubert, O.; Helbert, A.L.; Baudin, T.; Waeckerlé, T. Magnetostrictive and magnetic effects in Fe-27% Co laminations. *AIP Adv.* **2017**, *8*, 047711. [CrossRef]
5. Hubert, O.; Chaabane, M.; Jumel, J.; Maurel, V.; Alves, F.; Bensalah, A.; Besbes, M.; Azoum, K.; Bouillault, F. A new experimental setup for the characterisation of magneto-mechanical behaviour of materials submitted to biaxial stresses. Application to Fe-Co alloys. *Prz. Elektrotechniczny* **2005**, *81*, 19–23.
6. Szpunar, J.A.; Atherton, D.L. Magneto-striction and the effect of stress and texture. In *Nondestructive Characterization of Materials II*; Springer: Boston, MA, USA, 1978; pp. 577–584.

7. Nabi, B.; Helbert, A.L.; Brisset, F.; Waeckerle, T.; Batonnet, R.; Baudin, T. Effect of the hot rolling on Goss development and magnetic induction in an advanced soft magnetic Fe–27% Co alloy. *J. Alloys Compd.* **2020**, *834*, 155149. [CrossRef]
8. Ke, Y.; Wu, H.H.; Lan, S.; Jiang, H.; Ren, Y.; Liu, S.; Jiang, C. Tuning magnetostriction of Fe–Ga alloys via stress engineering. *J. Alloys Compd.* **2020**, *822*, 153687. [CrossRef]
9. Jiang, G.U.O.; Haiyang, F.U.; Bo, P.A.N.; Renke, K.A.N.G. Recent progress of residual stress measurement methods: A review. *Chin. J. Aeronaut.* **2021**, *34*, 54–78.
10. *ASTM E837*; Standard Test Method for Determining Residual Stresses by the Hole-Drilling Strain-Gauge Method. ASTM International: West Conshohocken, PA, USA, 2008.
11. Pagliaro, P.; Prime, M.B.; Swenson, H.; Zuccarello, B. Measuring multiple residual-stress components using the contour method and multiple cuts. *Exp. Mech.* **2010**, *50*, 187–194. [CrossRef]
12. Prime, M.B. Residual stress measurement by successive extension of a slot: The crack compliance method. *Appl. Mech. Rev.* **1999**, *52*, 75–96. [CrossRef]
13. Hellier, C. *Handbook of Non-Destructive Evaluation*; McGraw-Hill: New York, NY, USA, 2003.
14. Höller, P.; Hauk, V.; Dobmann, G.; Ruud, C.O.; Green, R.E. (Eds.) *Non-Destructive Characterization of Materials, Proceedings of the 3rd International Symposium, Saarbrücken, Germany, 3–6 October 1988*; Springer: Berlin/Heidelberg, Germany, 1988.
15. Du, W.; Zhao, Y.; Roy, R.; Addepalli, S.; Tinsley, L. A review of miniaturised Non-destructive Testing technologies for in-situ inspections. *Procedia Manuf.* **2018**, *16*, 16–23. [CrossRef]
16. Withers, P.J.; Bhadeshia, H.K.D.H. Residual stress. Part 1–measurement techniques. *Mater. Sci. Technol.* **2001**, *17*, 355–365. [CrossRef]
17. Blodgett, M.P.; Nagy, P.B. Eddy current assessment of near-surface residual stress in shot-peened nickel-base superalloys. *J. Non-Destr. Eval.* **2004**, *23*, 107–123. [CrossRef]
18. Su, F. Methodology for the Stress Measurement of Ferromagnetic Materials by Using Magneto Acoustic Emission. *Exp. Mech.* **2014**, *54*, 1431–1439. [CrossRef]
19. Shibata, M.; Ono, K. Magnetomechanical acoustic emission—A new method for non-destructive stress measurement. *NDT E Int.* **1981**, *14*, 227–234. [CrossRef]
20. Gauthier, J.; Krause, T.W.; Atherton, D.L. Measurement of residual stress in steel using the magnetic Barkhausen noise technique. *NDT E Int.* **1998**, *31*, 23–31. [CrossRef]
21. Fagan, P.; Ducharne, B.; Daniel, L.; Skarlatos, A.; Domenjoud, M.; Reboud, C. Effect of stress on the magnetic Barkhausen noise energy cycles: A route for stress evaluation in ferromagnetic materials. *Mater. Sci. Eng. B* **2022**, *278*, 115650. [CrossRef]
22. Stewart, D.M.; Stevens, K.J.; Kaiser, A.B. Magnetic Barkhausen noise analysis of stress in steel. *Curr. Appl. Phys.* **2004**, *4*, 308–311. [CrossRef]
23. Wilson, J.W.; Tian, G.Y.; Moorthy, V.; Shaw, B.A. Magneto-acoustic emission and magnetic Barkhausen emission for case depth measurement in En36 gear steel. *IEEE Trans. Magn.* **2009**, *45*, 177–183. [CrossRef]
24. Dobmann, G.; Altpeter, I.; Wolter, B.; Kern, R. Industrial applications of 3MA–micromagnetic multiparameter microstructure and stress analysis. *Electromagn. Nondestr. Eval.* **2008**, *31*, 18–25.
25. Stupakov, O.; Pal'a, J.; Takagi, T.; Uchimoto, T. Governing conditions of repeatable Barkhausen noise response. *J. Magn. Magn. Mater.* **2009**, *321*, 2956–2962. [CrossRef]
26. Santa-aho, S.; Laitinen, A.; Sorsa, A.; Vippola, M. Barkhausen noise probes and modelling: A review. *J. Nondestruct. Eval.* **2019**, *38*, 94. [CrossRef]
27. Capó Sánchez, J.; De Campos, M.F.; Padovese, L.R. Comparison between different experimental setups for measuring the magnetic Barkhausen noise in a deformed 1050 steel. *J. Nondestruct. Eval.* **2017**, *36*, 66. [CrossRef]
28. Kouakeuo, S.N.; Deffo, Y.T.; Ducharne, B.; Morel, L.; Raulet, M.A.; Tsafack, P.; Garcia-Bravo, J.M.; Newell, B. Embedded printed magnetic needle probes sensor for the real-time control of the local induction state through a laminated magnetic core. *J. Magn. Magn. Mater.* **2020**, *505*, 166767. [CrossRef]
29. Kouakeuo, S.N.; Ducharne, B.; Solignac, A.; Morel, L.; Raulet, M.A.; Toutsop, B.; Deffo, Y.T.; Tsafack, P. Non-invasive local magnetic hysteresis characterization of a ferromagnetic laminated core. *J. Magn. Magn. Mater.* **2021**, *527*, 167783. [CrossRef]
30. Ducharne, B.; Deffo, Y.T.; Tsafack, P.; Kouakeuo, S.N. Directional magnetic Barkhausen noise measurement using the magnetic needle probe method. *J. Magn. Magn. Mater.* **2021**, *519*, 167453. [CrossRef]
31. Li, K.; Li, L.; Wang, P.; Liu, J.; Shi, Y.; Zhen, Y.; Dong, S. A fast and non-destructive method to evaluate yield strength of cold-rolled steel via incremental permeability. *J. Magn. Magn. Mater.* **2020**, *498*, 166087. [CrossRef]
32. Stevens, K.J. Stress dependence of ferromagnetic hysteresis loops for two grades of steel. *NDT E Int.* **2000**, *33*, 111–121. [CrossRef]
33. *IEC 60404-3*; Magnetic Materials—Part 3: Methods of Measurement of the Magnetic Properties of Electrical Strip and Sheet by Means of a Single Sheet Tested. International Electrotechnical Commission: Geneva, Switzerland, 2010.
34. Rekik, M. Mesure et modélisation du comportement magnéto-mécanique dissipatif des matériaux ferromagnétiques à haute limite élastique sous chargement multiaxial: Application aux génératrices à grandes vitesses pour l'aéronautique. Ph.D. Thesis, LMT ENS Cachan—Paris-Saclay Normal School, Gif-sur-Yvette, France, 2014. (In French).
35. Gabi, Y.; Jacob, K.; Wolter, B.; Conrad, C.; Strass, B.; Grimm, J. Analysis of incremental and differential permeability in NDT via 3D-simulation and experiment. *J. Magn. Magn. Mater.* **2020**, *505*, 166695. [CrossRef]
36. *DIN 1324 Teil*; Deutsches Institut für Normung e.V.: Elektromagnetisches Feld. DIN: Berlin, Germany, 2017.

37. Gupta, B.; Uchimoto, T.; Ducharne, B.; Sebald, G.; Miyazaki, T.; Takagi, T. Magnetic incremental permeability non-destructive evaluation of 12 Cr-Mo-WV Steel creep test samples with varied ageing levels and thermal treatments. *NDT E Int.* **2019**, *104*, 42–50. [CrossRef]
38. Wolter, B.; Gabi, Y.; Conrad, C. Non-destructive testing with 3MA—An overview of principles and applications. *Appl. Sci.* **2019**, *9*, 1068. [CrossRef]
39. Yashan, A.; Dobmann, G. using eddy current coil in the presence of magnetic hysteresis. *Electromagn. Nondestruct. Eval.* **2002**, *6*, 150.
40. Matsumoto, T.; Uchimoto, T.; Takagi, T.; Dobmann, G.; Ducharne, B.; Oozono, S.; Yuya, H. Investigation of electromagnetic non-destructive evaluation of residual strain in low carbon steels using the eddy current magnetic signature (EC-MS) method. *J. Magn. Magn. Mater.* **2019**, *479*, 212–221. [CrossRef]
41. Gupta, B.; Ducharne, B.; Uchimoto, T.; Sebald, G.; Miyazaki, T.; Takagi, T. Comparison of electromagnetic inspection methods for creep-degraded high chromium ferritic steels. *NDT E Int.* **2021**, *118*, 102399. [CrossRef]

sensors

MDPI

Article

3D Measurement of Large Deformations on a Tensile Structure during Wind Tunnel Tests Using Microsoft Kinect V2

Daniele Marchisotti, Paolo Schito and Emanuele Zappa *

Department of Mechanical Engineering, Politecnico di Milano, 20156 Milano, Italy
* Correspondence: emanuele.zappa@polimi.it

Abstract: Wind tunnel tests often require deformation and displacement measures to determine the behavior of structures to evaluate their response to wind excitation. However, common measurement techniques make it possible to measure these quantities only at a few specific points. Moreover, these kinds of measurements, such as Linear Variable Differential Transformer LVDTs or fiber optics, usually influence the downstream and upstream air fluxes and the structure under test. In order to characterize the displacement of the structure not just at a few points, but for the entire structure, in this article, the application of 3D cameras during a wind tunnel test is presented. In order to validate this measurement technique in this application field, a wind tunnel test was executed. Three Kinect V2 depth sensors were used for a 3D displacement measurement of a test structure that did not present any optical marker or feature. The results highlighted that by using a low-cost and user-friendly measurement system, it is possible to obtain 3D measurements in a volume of several cubic meters (4 m $\times$ 4 m $\times$ 4 m wind tunnel chamber), without significant disturbance of wind flux and by means of a simple calibration of sensors, executed directly inside the wind tunnel. The obtained results highlighted a displacement directed to the internal part of the structure for the side most exposed to wind, while the sides, parallel to the wind flux, were more subjected to vibrations and with an outwards average displacement. These results are compliant with the expected behavior of the structure.

Keywords: wind tunnel; Kinect V2; 3D measurements; 3D reconstruction; point cloud registration

Citation: Marchisotti, D.; Schito, P.; Zappa, E. 3D Measurement of Large Deformations on a Tensile Structure during Wind Tunnel Tests Using Microsoft Kinect V2. *Sensors* **2022**, *22*, 6149. https://doi.org/10.3390/s22166149

Academic Editor: Zenghua Liu

Received: 18 July 2022
Accepted: 12 August 2022
Published: 17 August 2022

Publisher's Note: MDPI stays neutral with regard to jurisdictional claims in published maps and institutional affiliations.

1. Introduction

Wind tunnel tests usually investigate the response of structures to wind in terms of deformation and displacement [1]. For these tests, it is essential to measure deformations and displacements on a large number of points on a structure, for a comparison with mathematical models [2]. In addition, measuring the 3D shape and deformations of an entire structure could be more suitable for simulation model comparison, since usually mathematical models refer to the entire structure [3], while traditional measurement techniques can often return data referred to a few points (e.g., Linear Variable Differential Transformers LVDTs, fiber optics, and strain gauges).

Displacement measures could be performed using common and traditional approaches and instruments. For example, the application of LVDT sensors or laser doppler vibrometers [4] can be helpful also for determining natural frequencies, as well as for time domain analysis. In addition, the application of fiber optics [5] can return small deformation measurements at specific points and they are also commonly used for structural monitoring [6]. However, these techniques can be invasive for wind tunnel tests [3]. For example, LVDT sensors have to be placed in direct contact with the surface for measuring displacement, changing the wind flow, and affecting the test. Fiber optics, accelerometers, and pressure sensors are less intrusive, but, if it is necessary to determine the shape of an object, it will be difficult to acquire many different sensor signals, leading to a quite complex setup and acquisition system. Moreover, they should be applied to the surface to be measured leading to possible changes in its properties.

With these considerations, non-contact measurement techniques can be addressed as less invasive measurement methods. In this sense, laser doppler vibrometers could be used for wind tunnel tests [7], but they could interfere with the air flux and they could have a limited Field of View (FOV) and measurement range. In addition, the application of a vibrometer could disturb the wind flux. Therefore, to obtain the shape of an object during wind tunnel tests, a sensor with a large FOV, to be placed far away from the structure, and can return a point cloud at a specific time, can be of great advantage in the case of measurements of entire object displacements.

Within this framework, in this article, the application of 3D Time-of-Flight (ToF) cameras, such as Kinect V2, is investigated in wind tunnel tests, for 3D shape measurement, with the objective of measuring the structure displacement for different wind velocities. For this purpose, the application of 3D cameras returning a point cloud for a specific time instant can help to obtain the entire 3D shape under different testing conditions. Traditional optical measurement techniques can be used as well [8,9], but they would involve the application of markers or spraying patterns [10], leading to a modification of the model surface. In the case of reflective markers consisting of solid spheres, object surfaces could be modified, and the response of the structure to wind would be different. Three-dimensional laser scanners, such as the ones used in [11,12], could be used as well, but they usually require a scanning time of a few seconds to complete a scan and if the object to measure is subjected to vibration, the final result will be affected by noise. In addition, these techniques have a limited measurement range and a limited FOV, while the proposed method could perform measurements in a volume of a few meters, it is possible to place the sensors close to the walls of the wind tunnel chamber, reducing the effect on the downstream and upstream wind fluxes. The proposed method is one of the less expensive wind tunnel tests for 3D measurements. For these reasons, it would be difficult and not very meaningful to perform a comparison with the cited techniques, which are more expensive and have very different characteristics. In this paper, the feasibility of the application of 3D ToF cameras, such as Kinect V2, is studied, since they represent a valid option for their compactness, lightness, and low cost and, to the best of our knowledge, are not present in the literature. At the same time, they do not require markers and they can be used for 3D shape measurement to obtain the displacement of the entire structure. In addition, the greater measurement range, compared to the devices mentioned in this paragraph, permits the sensors to be placed close to the walls of the wind tunnel chamber, reducing the effect on the downstream and upstream wind fluxes. The measurement method is based on:

1. Calibration of the 3 Kinect V2 placed in the wind tunnel to align all the three sensors to a unique reference system (Section 3.3);
2. Extraction of sections of the structure measured to identify the displacement on those sections for different wind velocities (Section 3.5);
3. Subdivision of each section into different sides of the object (Section 3.5);
4. Definition of the undeformed object sections given by fitting 3D reconstructions obtained from undeformed object acquisitions (Section 3.5);
5. Calculation of the displacement for each side provided as the average and standard deviation displacement, in relation to the undeformed condition, along each side coordinate (Section 3.5);
6. Evaluation of the average displacement of each side to obtain more compact data (Section 4).

The paper is structured as follows: in Section 2, a literature review is presented to introduce the scientific context of the article; in Section 3, an experimental setup is described; in Section 3, the sensors point cloud registration and the data analysis method are explained; test results are presented and discussed in Section 4; and in Section 5, conclusions about the test method and results are drawn.

2. Related Works

In the literature, vision-based measurements may require the presence of markers or at least features to track in order to obtain the displacement of an object on the image plane [13]. On the other hand, 3D scanners can project infrared or visible light patterns, whose reflection can be used to obtain depth information. Using this method, 3D scanners do not require the presence of markers to obtain a set of points in a 3D environment and to perform displacement measurements for monitoring, e.g., monitoring of human breathing [14]. In particular, Time-Of-Flight (ToF) devices, such as Kinect V2, project IR light on the entire scene and measure the elapsed time between the light source emission and reception after its reflection on a target. This time delay is determined for Kinect V2 as the reflected energy and sampled at every pixel, using two windows with a phase shift of 180° (pulsed modulation) [15]. In this way, it is possible to perform wind tunnel tests without interfering with the aerodynamical properties of the structure under test. Moreover, it was also demonstrated that the application of depth cameras, in wind tunnel tests, is suitable to estimate the pose of an object subjected to wind, as addressed in [16], or to determine the position and orientation of an ultra-small airplane [17]. Indeed, pose estimation is one of the main applications of low-cost depth cameras, such as Kinect V2, using also multiple cameras [18], and also for cameras covered by a protective glass [19].

Regarding the 3D measurement applications, the main limitation of depth cameras is their uncertainty of a few millimeters [20], when it is required to measure objects with high stiffness and subjected to submillimeter displacement. For these conditions, it could be more appropriate to use one single camera or stereo cameras with a high-quality sensor and high framerate to perform Digital Image Correlation (DIC) [21]. This technique can be used for measuring strains of mechanical components, also using 3D-DIC [22], also in a wind tunnel environment [23]. However, using these techniques, it is possible to measure only a reduced part of an object to perform DIC effectively, with speckle reproduced on the object. On the other hand, further studies presented in the literature involve the application of photogrammetry for the 3D shape measurement of large offwind yacht sails, which were also carried out in wind tunnels [24,25]. As explained in [24], the precision of this measurement system is of few centimeters, which is not significant for measuring the apparent wind angles (AWA) and spinnaker shapes, but it can be important for other objects' wind tunnel tests. Moreover, for 3D reconstructions in a wind tunnel, more than one camera or stereo camera (four cameras in [24]) can be required, leading to a quite expensive and cumbersome setup. On the contrary, the accuracy can significantly decrease in the case of the application of Light Detection and Ranging (LIDAR) sensors, as in [26]. However, in this case, the measurement system is based on a custom sensor, which can be quite expensive in terms of instrumentation and involves considerable work to set up the system. As for the system presented in this article, if the displacement of a structure, such as the one used for the test presented in this article, is much higher than the uncertainty of the Kinect V2 sensor, the application of our measurement method is a valid option for measuring 3D displacements without advance equipment and with a more user-friendly acquisition. In particular, the uncertainty of Kinect V2 is 1.2 mm at about 1500 mm to 3.3 mm at the maximum reliable distance (4200 mm) [20], (much lower than photogrammetry measures), while the displacement of the structure under test was estimated to be of few centimeters. At the same time, the wind tunnel chamber used for the test is 4 m $\times$ 4 m $\times$ 4 m; thus, Kinect V2 can measure inside its measurement range, if the object to measure is placed at the center of the chamber. Thus, in this case, the application of depth cameras is suitable for this test.

In the literature, 3D reconstruction of a civil structure could be performed with different state-of-the-art techniques. These algorithms for obtaining a 3D shape reconstruction usually involve the movement of a camera around the structure to capture a set of 2D images to be processed using Structure from Motion (SfM) algorithms [27] or a set of RGBD images to be processed using Simultaneous Localization And Mapping (SLAM) algorithms [28]. However, during wind tunnel tests, it is not possible to change the position of a 3D scanner and the structure is not still during tests; therefore, these algorithms would not be able to obtain reliable results. For these reasons, in this research, the application of multiple Kinect V2 sensors for 3D reconstruction is presented with the specific application of 3D displacement measurement, for wind tunnel tests, without interference from the upstream and downstream wind and with calibration of the system executed directly in the wind tunnel.

3. Materials and Methods

The experimental setup is composed of two parts. The first one is defined by the structures to be tested and the second is the optical measurement system.

3.1. Tensile Structures

For the wind tunnel test, the structures under test were based on a squared and a decagonal base prism, respectively (Figure 1). The two structures are based on a tissue (cover) made of polyester micro-pierced tissue, which is the part being evaluated and the one more subjected to greater displacement. The cover is linked to the steel structure only at the top and bottom extremities (referred to as steel to extremity and steel bottom extremity in Figure 2) and it can be easily deformed by a few centimeters by hand. This link between the cover and the steel extremities is obtained with glue and small screws to prevent them from fluttering, creating a disturbance to the flow and leading to potential local damage of the cover tissue. The tissue displacement under the effect of the wind is much higher than the state-of-the-art uncertainty of ToF depth cameras. The height of both structures is 3500 mm and the cover starts from the bottom steel extremity at 580 mm from the wind tunnel floor. For the squared structure, the side length is 1200 mm, and for the decagonal structure, the diameter of the circle containing the prism base is 1230 mm. The tensile structure is made out of micro-pierced polyester tissue. This cover is connected at the extremities to a steel border base and the same border is replicated at the top of the structure. The steel structure is connected to a central steel column. A scheme of both structures is visible in Figure 2. To reinforce the central part of the structure, elastic belts connect the two borders, constraining the cover to a smaller displacement. Elastic belts are highlighted in Figure 1 using red lines in the top view of the schematic in Figure 2. In Figure 2, the belts are represented as short lines to schematically represent that they are attached to the steel extremities and do not cross the top and bottom base of the structures. The cross section of the elastic belts is 60×5 mm and the length of the elastic belts is equal to the height of the cover. The cover thickness is 1 mm. This tissue could be broken by wind or it could be plastically deformed if it is not reinforced by elastic belts; this is the reason for the presence of elastic belts. A side is equipped with a small entrance to access the internal part of the structure. This side is considered as the door side. The entire structure was fixed on a rotating plate to make it possible to perform tests for different wind directions. The two described structures are shown in Figures 1 and 2. The two structures, squared and decagonal, are simple structures whose displacements can be measured using the proposed technique, since the displacement of the cover is much higher than the Kinect V2 measurement uncertainty. This measurement method can also be extended to more complex structures with known geometry. There is no particular reason why these two structures were selected. They simply represent a case study of the measurement method.

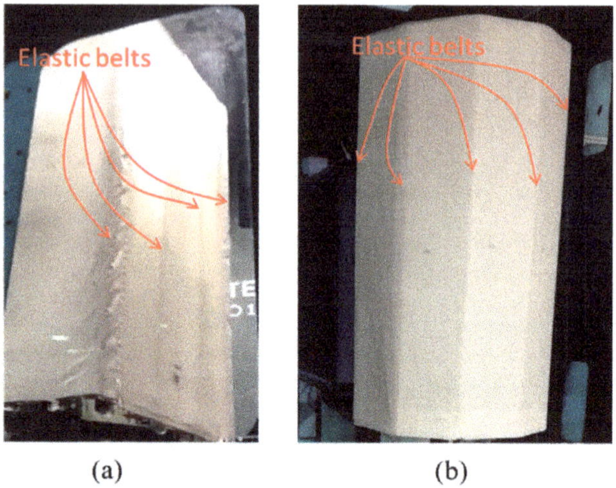

Figure 1. Squared (**a**) and decagonal (**b**) structures under test.

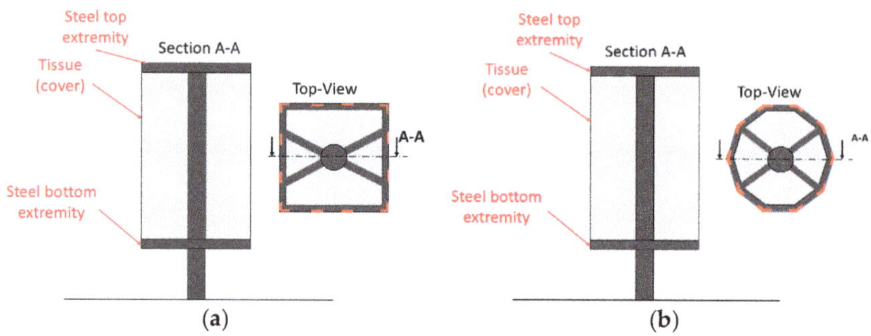

Figure 2. A schematic of the squared (**a**) and decagonal (**b**) structures under test. The positions of elastic belts are highlighted by using red lines, in the top view.

3.2. Measurement System

The measurement system is based on 3 Kinect V2 3D cameras used to reconstruct the 3D geometry of the two structures with the main goal to determine the average displacement of the object subjected to different wind velocities.

For our test, among the different technologies and sensors available on the market, Kinect V2 was chosen, since its characteristics (Table 1) are compatible with the requirements of the test. In particular:

1. Total measurement range: 0.7–4.2 m [20];
2. Displacement of the tensile structure, which is a few centimeters and it is much higher than the random error of the Kinect V2 sensor, which is 1.2 mm at about 1500 mm to 3.3 mm at the maximum reliable distance (4200 mm) [20];
3. Low latency time of 20 ms to acquire a depth image [20];
4. Absence of markers, which could affect or damage the structure during the test;
5. The structure is convex; thus, the multiple reflection errors that are typical of Time-Of-Flight devices are not present [29];
6. Low cost: during wind tunnel tests many cameras could be required and their usage could be limited to a few tests. Thus, large investments for very low uncertainty devices could be not justified.

Table 1. Kinect V2 characteristics [20].

Characteristic	Kinect V2
Working principle	Pulsed Time-of-Flight (ToF)
Depth Range	0.7–4.2 m
Max Depth Resolution	512 × 424
Max Color Resolution	1920 × 1080
Field Of View (FOV)	H: 70°, V: 60°
Max Acquisition Frequency	30 Hz
Latency time	20 ms
Software, SDK	Libfreenect2

"Max Depth Resolution" and "Max Color Resolution" in Table 1 are the resolutions of the depth sensor and RGB camera of Kinect V2. They are referred to as max since it is possible to acquire depth and color images with a lower resolution by changing the acquisition mode of the sensor. In the current work, the maximum resolution is used.

Since it is not possible to move Kinect V2 around the structure during the test to perform the 3D reconstruction in a conventional way, 3 Kinect V2 sensors were placed according to the scheme in Figure 3. The number of sensors was limited to the minimum, since a higher number of sensors would not significantly reduce the uncertainty, unless a very high number of sensors was applied with a high overlap of point clouds, to have a large number of measurements at the same point.

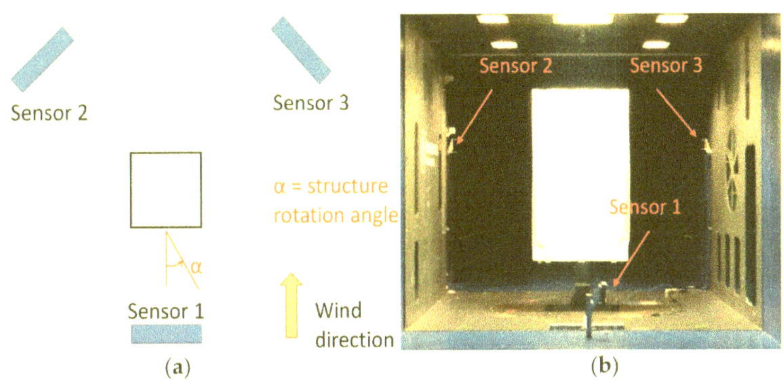

Figure 3. Scheme of the sensor positions (**a**) and actual positions (**b**).

Sensor 1 was placed at a height of about 0.2 m from the ground and it was inclined with respect to the vertical direction at about 20° (Figure 4). On the other side, sensors 2 and 3 are connected to the walls of the wind tunnel at a height of about 2.5 m and rotated toward the structure and downwards. All sensors were installed with their longer dimension close to the vertical direction to exploit the higher FOV of Kinect V2 on the horizontal direction (Table 1). In this way, the structure is centered on the depth image of each sensor, limiting the errors since it was demonstrated that close to the borders of the depth image, the Kinect V2 sensor is less precise, due to the projection IR cone that is less intense going to the corners [20].

The acquisition of data from sensors was performed using *libfreenect2* API [30]. This API permitted us to acquire up to 5 sensors simultaneously by using the same PC, instead of the *Kinect for Windows SDK2.0,* for which only one single device acquisition is permitted. The synchronization of devices was not required for our case, since the average and standard deviation of displacement do not require particular synchronization of devices. The average and standard deviation of displacement were computed for specific sections of the structure, with reference to the sides of the structure, as described in Section 3.5.

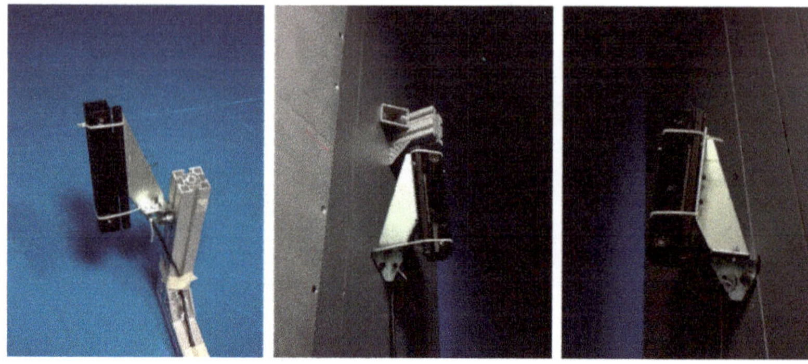

Figure 4. Pictures of the Kinect V2 1, 2, and 3, respectively, for wind tunnel test (cable ties to fix the sensors do not affect the point cloud acquisition).

3.3. Measurement System Calibration: Point Cloud Registration

Before starting the test, it is necessary to calibrate the vision system. This procedure is performed to find those transformation matrices to align the 3 Kinect V2 sensors with respect to the same reference system. Then, point clouds from the 3 sensors are registered in post-processing to obtain the full 3D geometry of the structure for each frame acquired, assuming the 3D cameras are not moving during the test. This assumption is based on the consideration that the wind force acting on a single sensor, in the worst case, was estimated to be 20–30 N, while sensors were fixed using cable ties able to resist a force of about 10 times higher.

Calibration was performed to ensure that the transformations aligned with the point clouds from sensors 1 and 2 to sensor 3. To obtain this result, a thin wooden plane was used in order to acquire the point cloud from all 3 sensors at the same time in static conditions, and the point clouds obtained from the sensors were segmented to extract the plane. To better identify the plane, high-reflective tape was put on the corners of the plane since it can lead to out-of-range measures for ToF cameras, such as Kinect V2 [29]. Using this method, it was easier to find the corners of the plane and to extract the part of the point cloud related to the plane. All the 3 point clouds representing a single plane were then aligned with the Iterative Closest Point (ICP) algorithm [31], and the transformation obtained matrices were referred to the one of sensor 3. To consider the 10 mm thickness of the wooden plane of the point clouds from sensor 1, the extracted point cloud was modeled by using a plane, and then the point cloud was translated 10 mm along the normal direction of the plane.

To investigate the level of uncertainty in the registration process, this procedure was executed on 20 different acquisitions with the plane rotated with different quantities and directions. The uncertainty was estimated by measuring the distance between the corners of the plane obtained from the 3 sensors after the registration process. There were 3 estimations of each corner of the plane, one for each sensor. The distance between these points, which referred to the same corner, was used to determine the uncertainty of the calibration procedure. This distance was estimated to be on average 16 mm, and the standard deviation 5.5 mm. These values are definitely below the range of displacements to be measured during the described wind tunnel test. At the same time, it should be noted that the measured quantity is the displacement of the structure, which does not depend directly on the registration procedure, since a large part of the structure displacement is not related to overlapping parts of the point clouds (Figure 5). A similar registration procedure is presented in [28], with an accuracy ranging a few centimeters.

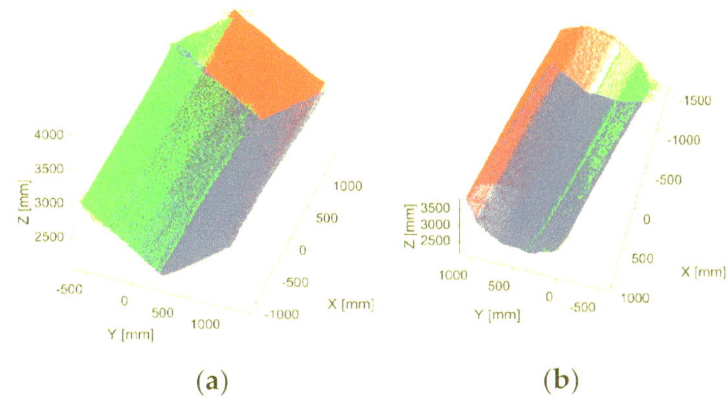

Figure 5. Three-dimensional reconstruction of the rectangular (**a**) and decagonal (**b**) tensile structures example (different colors refer to point clouds acquired from the three sensors).

A reconstruction of the tensile structure for both geometries can be seen in Figure 5. Only the points related to the object under inspection are visible since the background was removed through segmentation by considering the Euclidean distance of points (outliers' removal [32]).

3.4. Tests Summary

To study the cover deformation due to wind exposure, the test was performed at different wind speeds and for different angles of incidence of the wind. The variation in the angle of incidence was obtained by rotating the structure that was linked to a rotating platform. A summary of the tests carried out is shown in Table 2. The wind velocities were related to the power percentage of the engines of the wind tunnel.

Table 2. Tests summary.

Structure Type	Wind Velocities (km/h)	Wind Exposure Angles (°)
Squared base	41.4, 66.6, 90, 101.5, 112, 124, 135, 144, 158.5	0, 45, 90, 135, 180
Decagonal base	41.4, 66.6, 90, 101.5, 112, 124, 135, 144, 158.5, 176.5	0, 45, 180

For each wind velocity and exposure angle, 30 depth frames for each sensor were acquired. Having set the sensor acquisition at 30 fps, the acquisition time was 1 s, for each wind velocity. Acquisitions were performed also for each angle with no wind.

3.5. Data Analysis Method

After the point cloud registration process, to obtain a measure of the displacement of the structure, it is necessary to find a relation between the point clouds acquired during the test and the ones before the tests, with no wind.

To find this relation, 3 sections of the object were considered. These sections were defined by the planes shown in Figure 6. The points within 10 mm of distance from the sections were projected on the planes and a section of the object could be extracted. The position of each section compared to the height of the structure is shown in Figure 7. The planes of the sections were found by obtaining the sides of structure prisms and by computing the height from the basis of the structures on each side.

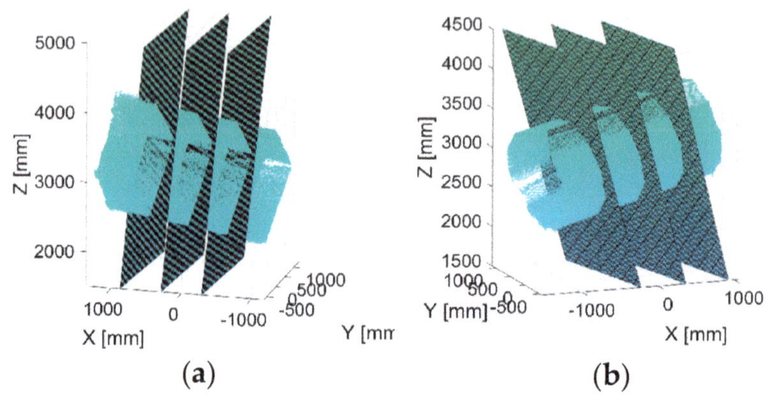

Figure 6. Planes to extract the sections of the structures of rectangular (**a**) and decagonal (**b**) structures.

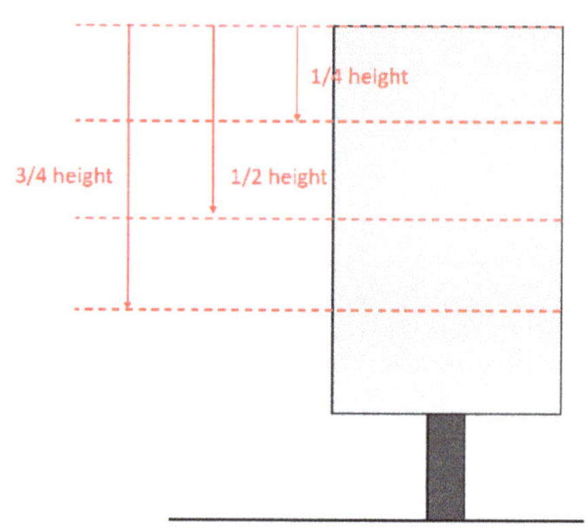

Figure 7. Schematic diagram showing the height of the sections.

Point clouds of the 3 sensors after registration were defined with respect to the reference system of sensor 3. In particular, the edges of the squared and decagonal structures in the undeformed conditions were extracted from acquisitions of the structures not subjected to wind. After extracting the sections of the structures by projecting points within 10 mm distance from the sections, as described before, the sides of the structures for each section are given by fitting a straight line for each side. Using this procedure, the displacement can be computed as the point-line distance of the points of the deformed condition in relation to the straight lines of the undeformed condition, for each side.

The final result of this procedure is the displacement calculated and displayed in graphs in Figures 8 and 9 (similar graphs, but with different values, were calculated for each velocity and angle). The error source, which influences the displacement measures, is the uncertainty of the sensor, since the bias is deleted when the displacement is computed as the difference between deformed and undeformed object measurements. The second source of measurement error can be given by the point cloud registration to align the 3 sensors to the same reference system. The influence of the point cloud registration on the results has already been discussed in Section 3.3. Displacement is computed for each frame acquired and mean and standard deviation are extracted for each wind speed. For each side of the structure, the deformation trend is visible. In addition, the dashed line curves are defined, as the mean displacement $\pm$ standard deviation (Std. Dev.). Displacement is computed as the mean displacement for each millimeter along the length of the side. In order to avoid outliers, in particular at the corners of each side, in the case of a low number of points per millimeter along the length of the side, the average and standard deviation were not considered. This was done to reduce the noise related to error measurement and to discard unreliable data.

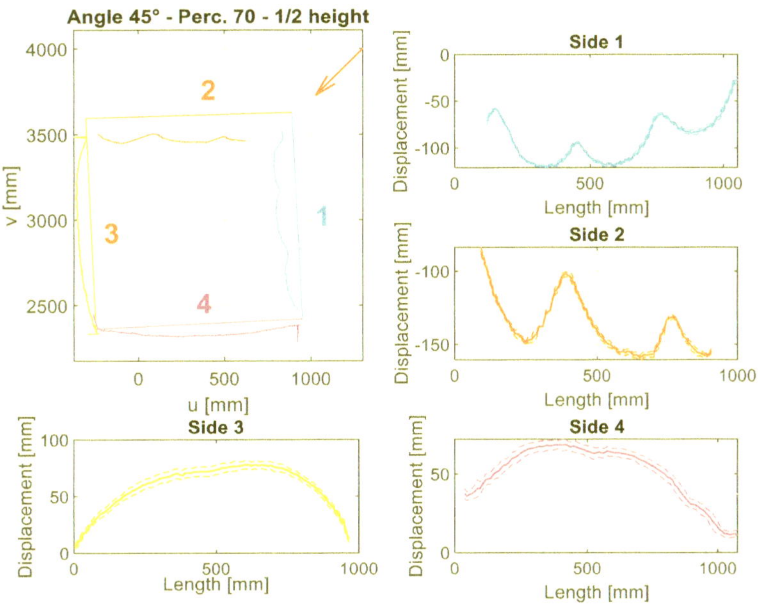

Figure 8. Example of the analysis performed for the squared structure of section at 1/2 of total height at wind speed of 158.5 km/h. The wind direction is indicated by the red arrow.

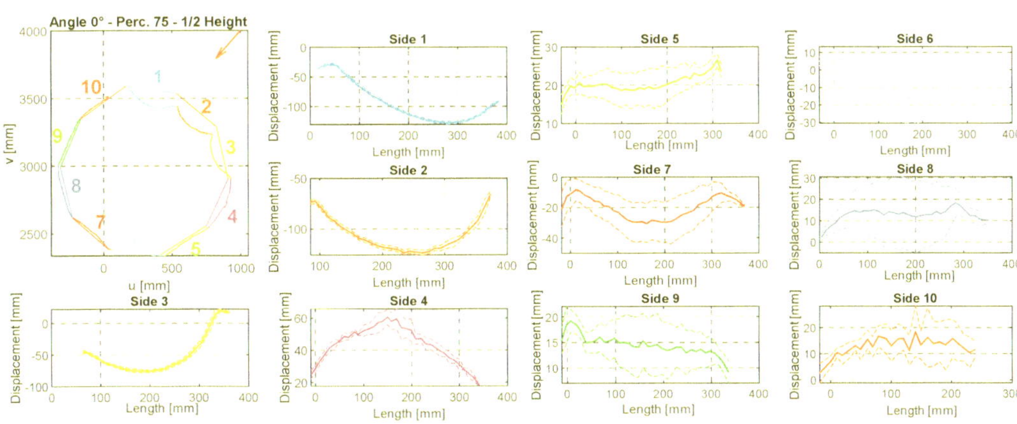

Figure 9. Example of the analysis performed for the decagonal structure of section at 1/2 of total height at wind speed of 176.5 km/h. The wind direction is indicated by the red arrow.

At the corners of the structure, the displacement can be subjected to mistakes due to possible errors of Kinect V2 sensors, described as mixed pixels errors, in [29]. On the other hand, the computed standard deviation is caused by both vibration and by sensor uncertainty. However, the sensor uncertainty can be quantified, since the sensor noise is given by the distance at which the object is framed. Since the object was placed between

2 and 3 m from the sensors, the uncertainty given by the sensors is expected to be about 2–3 mm, according to the sensors' uncertainty [20]. Moreover, point cloud registration can have an impact, as described in Section 3.3. This level of uncertainty can be confirmed by lower standard deviation levels for the sides of the structures that are the most exposed to wind (sides 1-2 Figure 8 and sides 1-2-3 Figure 9). On these sides, the vibrations are almost negligible and the standard deviation is mainly caused by the sensor uncertainty. Moreover, it is difficult to split the points of the deformed structure section and relate them to the correct side of the structure, when they are close to corners. For these reasons, from displacement graphs, the corners of the structure are partially excluded. The 2D graph in Figure 8 that represents the squared structure refers to a single frame, while graphs are related to the entire acquisition for one single wind speed (30 frames). The 2D graph scales are related to the sensor 3 origin and they are the coordinates of the plane at 1/2 of the total height section.

The displacement of the graphs in Figures 8 and 9 are referred to in the reference systems in Figures 10 and 11. By convention, the reference exposure angle compared to the wind is defined as shown in Figures 10 and 11. Positive displacement is always directed outwards and the reference systems are named in a clockwise direction. Reference systems are shown for each angle of the tests performed (Table 2).

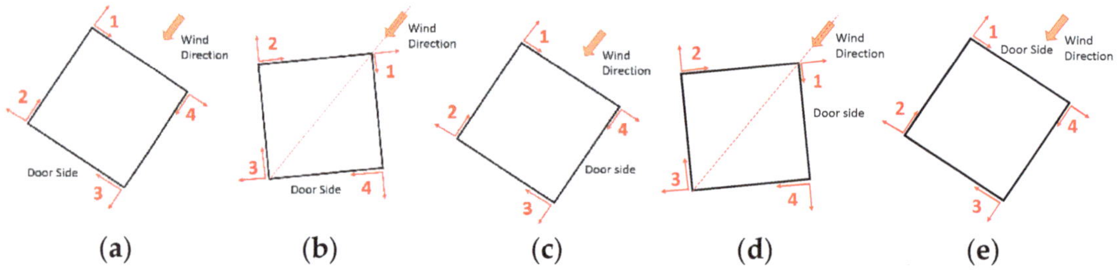

Figure 10. Reference systems for squared structure test for (**a**) 0°, (**b**) 45°, (**c**) 90°, (**d**) 135°, and (**e**) 180°.

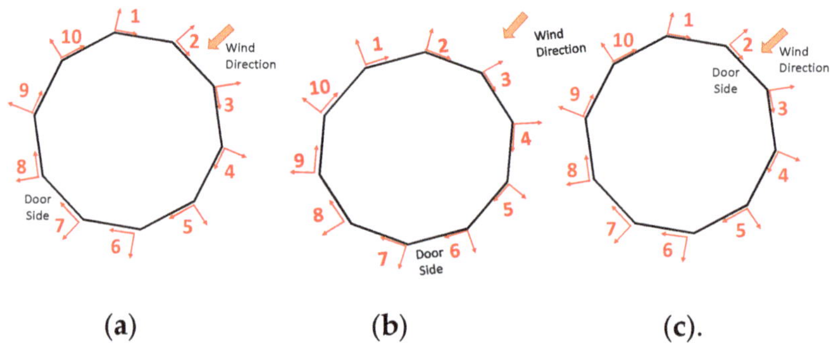

Figure 11. Reference systems for squared structure test for (**a**) 0°, (**b**) 45°, and (**c**) 180°.

By looking at the results in Figures 7 and 8, it is possible to observe the symmetry between sides 1 and 2 in Figure 8, by noting the reference systems used to compute displacements for sides 1 and 2 in Figure 9. To observe this symmetry, one of the two graphs of the two sides should be flipped with respect to the center of the horizontal axis (length of the side). Regarding Figure 8, a symmetry can be observed between sides 1 and 3 with a reversed horizontal axis of side 1 or of side 3, since the reference systems are placed as in the scheme of Figure 10a.

To make a comparison between tests performed at different angles and wind speed, the average displacement of an entire side was computed.

4. Results

Using the procedure described before, the average displacement was calculated for each side, wind velocity, and structure section.

In general, for both structures, tests evidenced the following results (Figures 12–14):

1. The average displacement of the sides of the structure more exposed to wind changes approximately linearly compared to the increase in the wind velocity;
2. A higher average displacement and a larger vibration level were observed for the middle section with respect to the other sections, which are closer to the sides of the structure. This is reasonable since the central section is the farthest from steel borders that link the cover to the extremities of the object (Figure 6);
3. Greater displacements were measured at 1/4 of the total length section with respect to 3/4 of the total length section, since the 1/4 section is more distant from the connection of the tensile structure to the ground (Figure 13);
4. The most exposed to wind sides are the ones that have the highest displacement and their adjacent sides are the ones subjected to the highest vibrations (standard deviation of displacement). This is likely because the wind first impacts against the sides more exposed and then it slides to the closest ones generating a separated unsteady flow, causing vibrations. This happens especially for the decagonal structure;
5. The displacements of the most exposed to the wind side and of the opposite side are directed inwards, suggesting an internal pressure lower than the external one, while the displacement of the other sides is usually directed outwards;
6. The most exposed to wind sides are the ones with the lowest vibration level;
7. Elastic belts connecting the two steel borders of the structure (Figure 1) can reduce significantly the average displacement when it is directed to the center of the structure, while they do not change the shape of the expected deformation when the displacement is outwards;
8. A higher displacement of the side most exposed to wind has a corresponding lower displacement of the opposite side.

The application of consumer 3D depth cameras, such as Kinect V2, is not a common practice for wind tunnel tests. However, the Kinect V2's advantage is its low-cost, its wide FOV (suitable for large dimension objects 3D reconstruction), and its capability to acquire the point cloud at a frame rate of 30 fps or higher with a random error well below 10 mm [20] in the entire operative range. Thus, 3D reconstruction can be applied to measure the displacement of the entire structure, rather than perform measures of just a few points.

In this article, a method to measure the displacement of a structure in harsh environments, such as a wind tunnel, is presented. This method is based on the acquisition of three sensors' point clouds. Since all the sensors remained in the same positions during the entire test, point cloud registration was performed only once before the test. The calculation of displacements was obtained by extracting three sections of the object. This procedure can be applied to the measurement of displacement for any structure, with a similar shape, for which a section can be extracted.

One of the main advantages of this kind of measurement system is the application of low-cost and user-friendly sensors for 3D displacement measurements, in cases of a large deformation tensile structure. At the same time, this method does not require the installation of markers or speckles for optical measurements. Another advantage is the possibility to have the displacement of the entire structure under evaluation, given by the 3D shape measurements obtained for each point of the structure (with traditional measurement systems, it would be difficult to obtain such resolution in terms of displacement measures). On the other hand, a limitation of this procedure can be determined by the uncertainty of Kinect V2, which would be not suitable in cases of a low displacement of a structure. Indeed, these kinds of measurements require a sensor that has an uncertainty much lower than the expected deformation. The presence of edges makes it possible to locate the section to extract; in the case of irregular or very complicated geometry, the section to extract might be hard to detect.

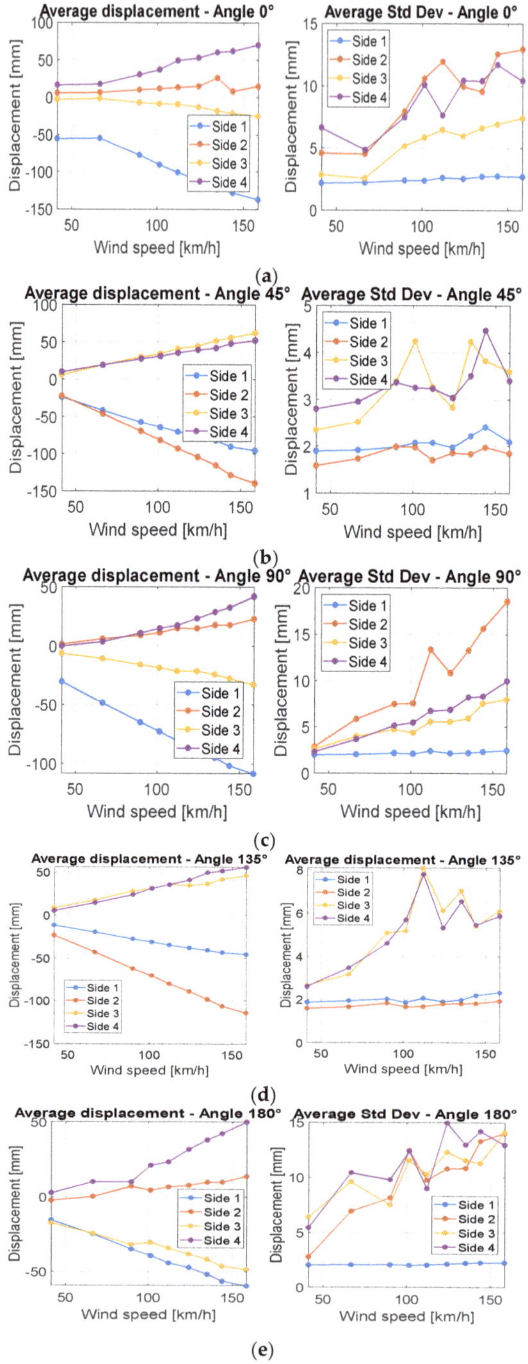

Figure 12. Average displacements and the standard deviation for each side of the squared base structure at 1/2 of the total height for different wind velocities for angles 0° (**a**), 45° (**b**), 90° (**c**), 135° (**d**), 180° (**e**).

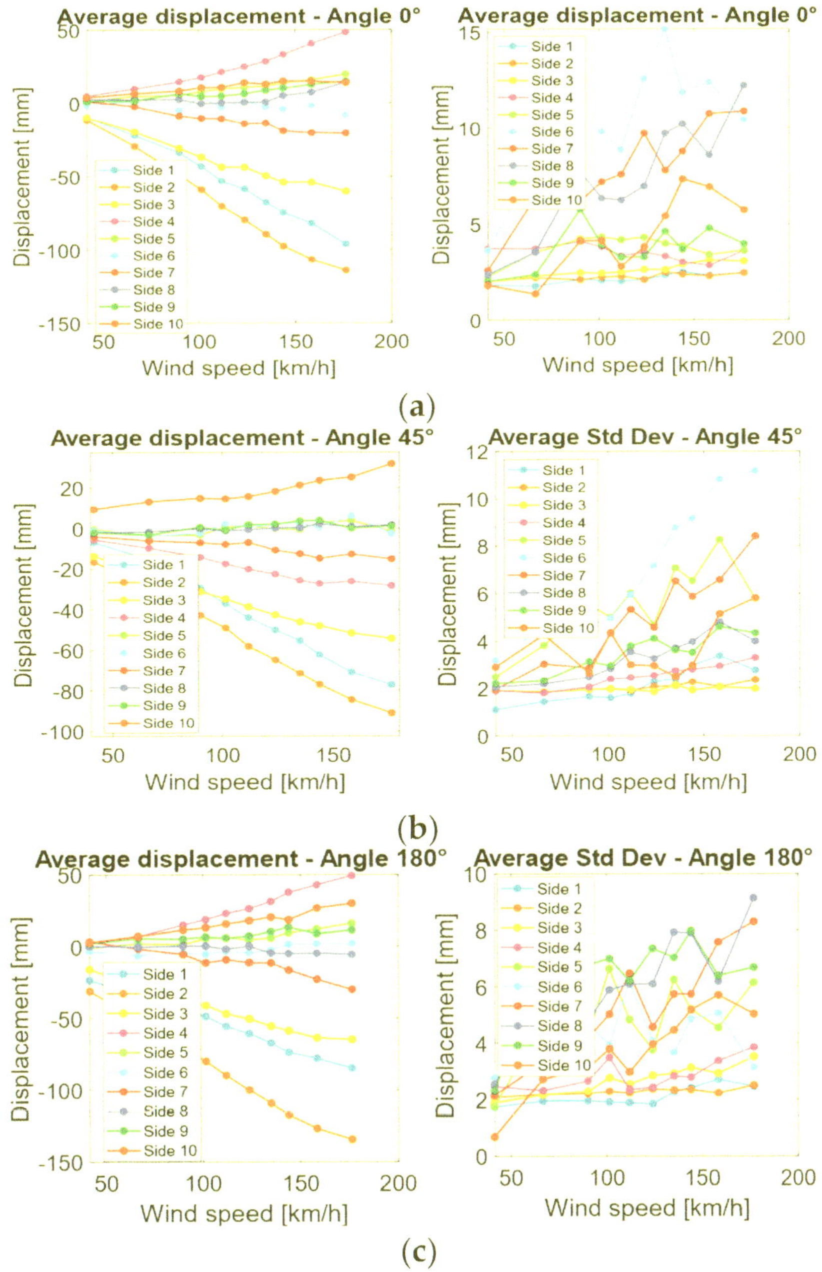

Figure 13. Average displacements and standard deviation for each side of the decagonal base structure at 1/2 of the total height for different wind velocities for angles 0° (**a**), 45° (**b**), 180° (**c**).

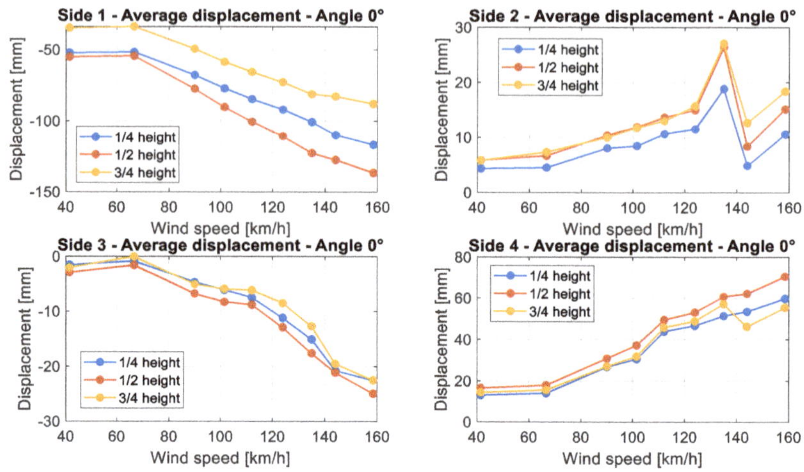

Figure 14. Comparison between average displacements at different heights of sections for the squared structure at 0° angle, for each side of the structure going from side 1 to 4.

The results allowed the behavior of the two structures starting from their deformed shape to be studied. Results at points 1 and 2 of Section 4 are expected and they allow the goodness of the measurement system to be checked.

It was possible to observe the shape of the structures for different speed velocities and angles. The average displacement and the standard deviation were computed since the results for these kinds of tests are related to a quantification of the displacement and 3D shape measurement of a structure when it is exposed to wind, in order to compare experimental results with numerical ones, for stationary wind conditions.

5. Conclusions

In this article, a measurement system involving the application of 3D cameras for a wind tunnel test is presented. The specific application of displacement measurement from a 3D shape is treated in this article and the analysis of acquired data allowed us to obtain the displacement of each side of the structures, for different sections. With reference to traditional measurement systems, this measurement system permits us to obtain the entire shape of the object being evaluated, and it is not measured only for specific points. As matter of fact, any section of the object could be extracted to determine the displacement. In comparison with other non-contact measurement systems, it does not require the application of markers on the object to measure. For these reasons, its application is suitable for the test described in this article. Other measurement systems, based on different 3D cameras, can be created with the same procedure of data analysis.

As a consequence, this measurement system can be seen as a complete system for measuring displacement, higher than the uncertainty of 3D sensors, and as a system for measuring the 3D geometry of structures in harsh environments. The frequency of vibration was not accurately measured in this experiment, since the purpose of the proposed technique was to measure the average displacement distribution of the 3D structure under test and even possible aliasing of the vibration would not affect the mean displacement. The measurement of the frequency of vibration can be part of future work, by acquiring sensors using an external trigger and by acquiring at an fps that is large enough to describe properly the vibrations of the structure (twice that of the vibration frequencies of the structure). Further improvements could follow these considerations involving the study of structural properties of structures with depth cameras, e.g., modal parameters or natural frequencies.

Author Contributions: Conceptualization, D.M. and P.S.; methodology, D.M.; software, D.M.; validation, D.M., P.S. and E.Z.; formal analysis, D.M.; investigation, D.M.; resources, D.M.; data curation, D.M. and E.Z.; writing—original draft preparation, D.M.; writing—review and editing, D.M., P.S. and E.Z.; visualization, D.M.; supervision, P.S. and E.Z.; project administration, P.S.; funding acquisition, P.S. All authors have read and agreed to the published version of the manuscript.

Funding: This research received no external funding.

Institutional Review Board Statement: Not applicable.

Informed Consent Statement: Not applicable.

Data Availability Statement: The data presented in this study are available upon request from the corresponding author.

Conflicts of Interest: The authors declare no conflict of interest.

References

1. Cermak, J.E.; Jack, E. Wind-tunnel development and trends in applications to civil engineering. *J. Wind Eng. Ind. Aerodyn.* **2003**, *91*, 355–370. [CrossRef]
2. Kalvig, S.; Manger, E.; Hjertager, B. Comparing different CFD wind turbine modelling approaches with wind tunnel measurements. *J. Phys. Conf. Ser.* **2014**, *555*, 12056. [CrossRef]
3. Min, Z. *Prediction and Validation Technologies of Aerodynamic Force and Heat for Hypersonic Vehicle Design*; Springer: Berlin/Heidelberg, Germany, 2021.
4. Nassif, H.H.; Gindy, M.; Davis, J. Comparison of laser doppler vibrometer with contact sensors for monitoring bridge deflection and vibration. *NDT E Int.* **2005**, *38*, 213–218. [CrossRef]
5. Guo, H.; Xiao, G.; Mrad, N.; Yao, J. Fiber optic sensors for structural health monitoring of air platforms. *Sensors* **2011**, *11*, 3687–3705. [CrossRef] [PubMed]
6. Lawal, O.; Najafi, A.; Hoang, T.; Shajihan, S.A.V.; Mechitov, K.; Spencer, B.F. Development and validation of a framework for smart wireless strain and acceleration sensing. *Sensors* **2022**, *22*, 1998. [CrossRef] [PubMed]
7. Cheng, L.; Zhou, Y.; Zhang, M.M. Perturbed interaction between vortex shedding and induced vibration. *J. Fluids Struct.* **2003**, *17*, 887–901. [CrossRef]
8. Najafi, N.; Paulsen, U.S. Operational modal analysis on a VAWT in a large wind tunnel using stereo vision technique. *Energy* **2017**, *125*, 405–416. [CrossRef]
9. Soinne, E.; Rosnell, T. Aerodynamic laser scanning and photogrammetric measurements of cold soaked fuel frost. In Proceedings of the Atmospheric and Space Environments Conference (ASEC 2018), Atlanta, GA, USA, 25–29 June 2018.
10. Gramola, M.; Brucel, P.; Santer, M. Photogrammetry for accurate model deformation measurement in a supersonic wind tunnel. *Exp. Fluids* **2019**, *60*, 8. [CrossRef]
11. Bayati, I.; Muggiasca, S.; Vandone, A. Experimental and numerical wind tunnel investigation of the aerodynamics of upwind soft sails. *J. Waterw. Port Coastal Ocean Eng.* **2019**, *182*, 395–411. [CrossRef]
12. Vandone, A. Measurement and Analysis of Free-Form Objects. Development of a Solution for Flyng Sail Shape Reconstruction. Ph.D. Thesis, Politecnico di Milano, Milan, Italy, March 2016.
13. Xu, Y.; Brownjohn, J. Review of machine-vision based methodologies for displacement measurement in civil structures. *J. Civ. Struct. Health Monit.* **2018**, *8*, 91–110. [CrossRef]
14. Tahavori, F.; Alnowami, M.; Wells, K. Marker-less respiratory motion modeling using the Microsoft Kinect for Windows. In Proceedings of the SPIE Medical Imaging, San Diego, CA, USA, 15–20 February 2014.
15. Li, L. *Time-of-Flight Camera—An Introduction*; Tech. Rep. SLOA190B; Texas Instruments: Dallas, TX, USA, May 2014.
16. Liu, S.; Feng, Y.; Shen, K.; Wang, Y.; Chen, S. An RGB-D-based cross-field of view pose estimation system for a free flight target in a wind tunnel. *Complexity* **2018**, *2018*, 7358491. [CrossRef]
17. Paces, P.; Yu, W.K.; Klesa, J. Optical measurement methods for attitude determination of unmanned aerial systems. In Proceedings of the 37th Digital Avionics Systems Conference(DASC 2018), London, UK, 23–27 September 2018.
18. Hong, S.; Kim, Y. Dynamic pose estimation using multiple RGB-D cameras. *Sensors* **2018**, *18*, 3865. [CrossRef] [PubMed]
19. Liu, W.; Ma, X.; Li, X.; Pan, Y.; Wang, F.; Jia, Z. A novel vision-based pose measurement method considering the refraction of light. *Sensors* **2018**, *18*, 4348. [CrossRef] [PubMed]
20. Corti, A.; Giancola, S.; Mainetti, G.; Sala, R. A metrological characterization of the Kinect V2 time-of-flight camera. *Rob. Autom. Syst.* **2016**, *75*, 584–594. [CrossRef]
21. Schreier, H.; Orteu, J.; Sutton, M. *Image Correlation for Shape, Motion and Deformation Measurements: Basic Concepts, Theory and Applications*; Springer: London, UK, 2009; pp. 1–321.
22. Helfrick, M.; Niezrecki, C.; Avitabile, P.; Schmidt, T. 3D digital image correlation methods for full-field vibration measurement. *Mech. Syst. Sig. Process.* **2011**, *25*, 917–927. [CrossRef]

23. Beberniss, T.J.; Ehrhardt, D.A. High-speed 3D digital image correlation vibration measurement: Recent advancements and noted limitations. *Mech. Syst. Sig. Process.* **2017**, *86*, 35–48.
24. Deparday, J.; Bot, P.; Hauville, F.; Augier, B.; Rabaud, M. Full-scale flying shape measurement of offwind yacht sails with photogrammetry. *J. Waterw. Port Coastal Ocean Eng.* **2016**, *127*, 135–143. [CrossRef]
25. Mausolf, J.; Deparday, J.; Graf, K.; Renzsch, H.; Böhm, C. Photogrammetry based flying shape investigation of downwind sails in the wind tunnel and at full scale on a sailing yacht. In Proceedings of the 20th Chesapeake Sailing Yacht Symposium (CSYS 2011), Annapolis, MD, USA, 18–19 March 2011; pp. 33–43.
26. Fossati, F.; Bayati, I.; Orlandini, F.; Muggiasca, S.; Vandone, A.; Mainetti, G.; Sala, R.; Bertorello, C.; Begovic, E. A novel full scale laboratory for yacht engineering research. *J. Waterw. Port Coastal Ocean Eng.* **2015**, *104*, 219–237. [CrossRef]
27. Ma, Z.; Liu, S. A review of 3D reconstruction techniques in civil engineering and their applications. *Adv. Eng. Inf.* **2018**, *37*, 163–174. [CrossRef]
28. Berger, K. A state of the art report on multiple RGB-D sensor research and on publicly available RGB-D datasets. In *Advances in Computer Vision and Pattern Recognition*; Computer Vision and Machine Learning with RGB-D Sensors; Springer: Cham, Switzerland, 2014; pp. 27–44.
29. Giancola, S.; Valenti, M.; Sala, R. *A Survey on 3D Cameras: Metrological Comparison of Time-of-Flight, Structured-Light and Active Stereoscopy Technologies*; Springer: Cham, Switzerland, 2018.
30. Libfreenect2. Available online: https://github.com/OpenKinect/libfreenect2 (accessed on 12 August 2022).
31. Chen, Y.; Medioni, G. Object modelling by registration of multiple range images. *Image Vision Comput.* **1992**, *10*, 145–155. [CrossRef]
32. Rusu, R.B.; Marton, Z.C.; Blodow, N.; Dolha, M.; Beetz, M. Towards 3D point cloud based object maps for household environments. *Robot Auton Syst.* **2008**, *56*, 927–941. [CrossRef]

sensors

MDPI

Article

Modeling Full-Field Transient Flexural Waves on Damaged Plates with Arbitrary Excitations Using Temporal Vibration Characteristics

Dan-Feng Wang [1], Kuo-Chih Chuang [1,*], Jun-Jie Liu [2] and Chan-Yi Liao [3,*]

1 Key Laboratory of Soft Machines and Smart Devices of Zhejiang Province, School of Aeronautics and Astronautics, Institute of Applied Mechanics, Zhejiang University, Hangzhou 310027, China
2 Applied Mechanics and Structure Safety Key Laboratory of Sichuan Province, School of Mechanics and Aerospace Engineering, Southwest Jiaotong University, Chengdu 610031, China
3 Department of Mechanical Engineering, National Central University, Taoyuan 32001, Taiwan
* Correspondence: chuangkc@zju.edu.cn (K.-C.C.); chanyi@ncu.edu.tw (C.-Y.L.)

Abstract: We propose an efficient semi-analytical method capable of modeling the propagation of flexural waves on cracked plate structures with any forms of excitations, based on the same group of vibration characteristics and validated by a non-contact scanning Laser Doppler Vibrometer (LDV) system. The proposed modeling method is based on the superposition of the vibrational normal modes of the detected structure, which can be applied to analyze long-time and full-field transient wave propagations. By connecting the vibration-based transient model to a power flow analysis technique, we further analyze the transient waves on a cracked plate subjected to different excitation sources and show the influence of the damage event on the path of the propagating waves. The experimental results indicate that the proposed semi-analytical method can model the flexural waves, and through that, the crack information can be revealed.

Keywords: flexural waves; vibrations; structural health monitoring; semi-analytical method

Citation: Wang, D.-F.; Chuang, K.-C.; Liu, J.-J.; Liao, C.-Y. Modeling Full-Field Transient Flexural Waves on Damaged Plates with Arbitrary Excitations Using Temporal Vibration Characteristics. *Sensors* **2022**, *22*, 5958. https://doi.org/10.3390/s22165958

Academic Editor: Zenghua Liu

Received: 8 July 2022
Accepted: 4 August 2022
Published: 9 August 2022

1. Introduction

Structural health monitoring (SHM) technologies are crucial in maintaining the safety and integrity of engineering structures and thus have attracted much attention in the research community [1–3]. Generally, non-destructive damage detection techniques can be characterized by the vibration-based [4,5] and guided wave-based methods [6,7]. The guided wave-based detection techniques are suitable for sensitively locating small or local damages. However, detected signals in guided waves are liable to be contaminated by environmental noises, such as temperature effects [8,9], dispersion properties [10], multiple modes [11], and material uncertainties [12]. The vibration-based detection techniques, on the other hand, can detect global damage events and have the advantage of requiring no a priori information of the vicinity of damage. The existence or onset of damage will alter the vibration characteristics, such as shifts of natural frequencies, changes of modal damping, or derivatives of mode shapes, of the detected structures, and thus, the vibration characteristics contain the information of the damage events [13–16]. Generally, the vibration-based and guided wave-based methods are regarded as two distinct approaches despite the fact that vibration and waves are essentially closely related.

Waves propagating in plate structures, known as Lamb waves, are sensitive to various types of damage and their propagation characteristics and interpretation are crucial in developing related monitoring strategies [17–20]. Numerically, Lamb waves are generally analyzed by the finite element method (FEM) [21,22], finite difference method (FDM) [23,24], the spectral element method (SEM) [25], the boundary element method (BEM) [26], or the semi-analytical finite element method (SAFE) [27,28], where the SAFE has recently been developed to model guided wave on anisotropic composite laminates and in situations with thermal effect [29,30]. Although vibrations- and waves-based damage detection techniques are characterized as two distinct approaches, the vibration can be regarded as standing waves while waves can be regarded as the superposition of vibrational modes. Thus, the damage-induced changes of the modal parameters in the time-domain should contribute to the behavior of the spatial-domain propagating waves. In addition to capturing the wave propagation characteristics with plate structures with cracks, analysis of spatial-domain propagating Lamb waves based on time-domain vibration characteristics enables us to calculate the instantaneous wave responses at any given time under different arrangements or forms of excitations with the same database of vibration information and time-consuming iterative time-stepping process from the initial condition required in general numerical methods can be avoided.

Aiming to provide a modeling method for cracked plate structures by linking the vibration and wave, in this work, we propose a semi-analytical method for modeling the full-field transient propagation of flexural waves (i.e., the approximation of the low-frequency A0 Lamb wave mode) on a cracked plate based on temporal vibration characteristics and the superposition of the deviated vibrational normal modes. The proposed semi-analytical method is further connected to a power flow analysis technique to analyze the transient waves excited by different excitation sources. Finally, the proposed semi-analytical method is validated by a non-contact scanning Laser Doppler Vibrometer (LDV) system.

2. Semi-Analytical Model Linking the Vibrations and Waves

Classical plate theory claims that if the coordinate in the thickness direction of a three-dimensional solid can be eliminated, the three-dimensional plate problem can be reduced to a two-dimensional problem and physical quantities, such as displacements, strains, and stresses, at all points in the mid-plane of the plate can be used to determine the associated quantities at any point in the three-dimensional plate structure [31]. When the thin plate has cracks symmetrical in the direction of thickness, where the mid-plane of the plate remains continuous, the classical plate theory can still be applied. Based on the classical plate theory, we now propose a semi-analytical method to analyze the transient behaviors of a cracked plate.

Consider a homogeneous thin plate, also known as the Kirchhoff plate, with an external loading $p(x_1, x_2, t)$ applied on the upper surface of the plate. The governing equation for the out-of-plane deformation u_3 is [32]

$$D\left(\frac{\partial^4 u_3}{\partial x_1{}^4} + 2\frac{\partial^4 u_3}{\partial x_1{}^2 \partial x_2{}^2} + \frac{\partial^4 u_3}{\partial x_2{}^4}\right) + \rho_s h \frac{\partial^2 u_3}{\partial t^2} = p(x_1, x_2, t), \tag{1}$$

where $D = \frac{Eh^3}{12(1-\nu^2)}$, E is Young's modulus, ρ_s is the density, ν is Poisson's ratio, and $h(x_1, x_2)$ is the thickness at position (x_1, x_2) of a plate. The out-of-plane transient displacement of the plate can be written as a superposition of the vibration normal modes as follows

$$u_3(x_1, x_2, t) = \sum_{i=1}^{\infty} T_i(t) W_i(x_1, x_2), \tag{2}$$

where $W_i(x_1, x_2)$ is the ith mode shape, $T_i(t)$ is the time function representing the weight coefficients of the contribution of the ith mode shape at a specific time to the transient wave field. By the orthogonal properties between the mode shapes in the modal space, the transient displacement solution of the flexural wave (i.e., the approximation of the low-frequency A0 Lamb wave mode) can be derived from Equation (1) as [32]

$$u_3(x, y, t) = \sum_{i=1}^{\infty} \frac{W_i(x_1, x_2)}{\rho_s \omega_i} \left\{ \frac{\int_0^t \left[\int_0^b \int_0^a p(x_1, x_2, \tau) W_i dx_1 dx_2\right] \sin \omega_i (t - \tau) d\tau}{\int_0^b \int_0^a \left(h \cdot W_i{}^2\right) dx_1 dx_2} \right\}, \tag{3}$$

where ω_i is the resonant frequency of the ith resonant mode of the plate and a and b are the length and width of the plate, respectively. If the external loading with a loading history $F(t)$ is applied within a specific range $R(x_1, x_2)$, Equation (3) can be written as the following Duhamel's integral

$$u_3(x, y, t) = \sum_{i=1}^{\infty} \left[\frac{W_i(x_1, x_2) \cdot \int_0^b \int_0^a R(x_1, x_2) \cdot W_i dx_1 dx_2}{\rho_s \omega_i \cdot \int_0^b \int_0^a \left(h \cdot W_i^2 \right) dx_1 dx_2} \right] \cdot \int_0^t F(\tau) \sin \omega_i (t - \tau) d\tau. \quad (4)$$

From the term $\int_0^b \int_0^a R(x_1, x_2) \cdot W_i dx_1 dx_2$ shown in the temporal solution in Equation (4), we can see that the applying range of the external loading will influence the contribution of the individual mode. Note that $R(x_1, x_2)$ is expressed as a Dirac-delta function $\delta(x_1 - x_{1p})\delta(x_2 - x_{2p})$ when a pointwise excitation is applied at position (x_{1p}, x_{2p}). Although the out-of-plane displacement is dependent on space and time, the vibration characteristics ω_i and $W_i(x_1, x_2)$ in Equation (4) (i.e., resonant frequencies and mode shapes) are determined by the boundary conditions, material constants, and geometry of the plate and are independent of time, magnitude of the external loadings, and the excitation locations. Thus, even for complicated transient responses, if the vibration characteristics can be obtained beforehand through theoretical calculations or finite element simulations, transient behaviors at any time and any location can be obtained by Equation (4) by the fact that the vibration characteristics and wave source are decoupled and thus can be analyzed separately. We note that the frequencies and mode shapes of the analyzed plate can also be experimentally measured using techniques, such as experimental modal analysis (EMA) or electronic speckle pattern interferometry (ESPI) [33]. However, it should be noted that limited by the bandwidth of the excitations as well as the resolution of the sensing systems, high-frequency resonant frequencies and mode shapes are usually difficult to be experimentally identified. In this work, resonant frequencies and mode shapes of the intact plate are theoretically derived [32]. However, for the damaged plate, vibration characteristics are obtained using finite element analysis.

The analysis of the instantaneous wave propagations on the damaged plate is based on the temporal contributions of each vibrational mode to the overall transient wave field. Although each vibrational mode contributes to the overall transient waves, only a finite number of vibrational modes are required to efficiently converge the calculation. However, when the excitation frequency is high with a short wavelength, more modes are required, and the convergence speed of the calculation would be lower. In addition, for the calculation of the overall transient responses, only a specific range of dominated resonant frequencies centered around the excitation frequency are required. Furthermore, according to the Nyquist sampling theory, when the excitation frequency is at N Hz, the considered highest mode in the modal contribution should have resonant frequency higher than $2N$ Hz for the convergence of the transient solution [34].

This semi-analytical approach based on the vibration characteristics enjoys the following advantages:

(1) When the loading range $R(x_1, x_2)$ or the loading history $F(t)$ is changed, the same vibration characteristics in the database of the vibration information can be substituted by Equation (4) to obtain the transient solutions.

(2) There is no time-consuming iterative time-stepping process in calculating the temporal integral term $\int_0^t F(\tau) \sin \omega_i (t - \tau) d\tau$. Thus, we can efficiently obtain on-demand full-field wave propagations at a specific time t.

(3) The form of the transient solution in Equation (4) can be further transformed to other physical quantities, such as strain or velocity fields, without having to perform any additional numerical post-processing operations, which minimizes the possible sacrifice of the resolution (spatial or temporal) or distortion of the transient solutions coming from numerical processing. Based on this advantage, in addition to the relation between the vibration modes and the instantaneous wave responses, in this

work, a power flow analysis (PFA) is also considered to reveal the transmission paths of the propagating waves on the plate. The PFA uses the instantaneous power flow density vector defined as the dot product of the velocity vector and stress tensor in the form of $q_j = -\sigma_{ji}v_i$, i, $j = 1,2,3$, to trace the instantaneous energy transmission paths from one location to another within the plate [35–37].

3. Simulation of the Instantaneous Flexural Waves Using the Semi-Analytical Model

To demonstrate the proposed semi-analytical model for analyzing instantaneous wave propagation, here, we investigate a damaged cantilever rectangular aluminum plate ($400 \times 150 \times 2$ mm) with a rectangular slot ($60 \times 5 \times 2$ mm) in the center, which has been studied in the literature, as shown in Figure 1a [4,38]. The isotropic plate is made of aluminum material with a Young's modulus of 69 GPa, a Poisson's ratio of 0.33, and a density of 2705 kg/m^3. The dispersion curves, obtained by the FEM analysis using the COMSOL software for the analyzed plate, are shown in Figure 1b. For the Lamb waves on plates, each frequency corresponds to both the symmetric and asymmetric modes on the dispersion curves. However, by using a windowed excitation signal around a low-dispersion frequency with an asymmetric arrangement of the actuator, a dominated mode (here, the low-frequency A0 mode) can be generated. The transient behaviors of the excited flexural wave (i.e., the low-frequency approximation of the A0 mode) can be analytically captured based on the Kirchhoff plate theory given in Section 2. Later, to model and study the influence of the crack, we considered low-frequency one-sided excitations, and thus, practically, the flexural waves with a high signal-to-noise ratio are mainly excited [39].

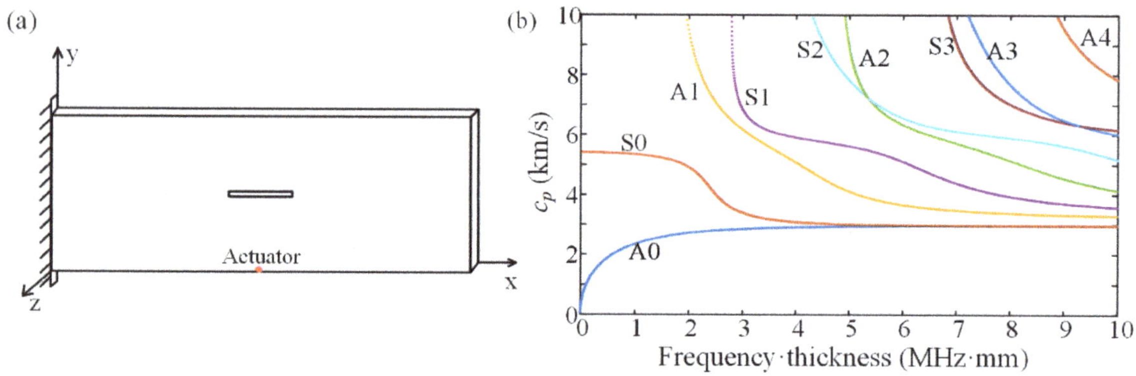

Figure 1. (**a**) Dimension of the damaged plate and the location of the actuator. (**b**) The dispersion curves.

Before demonstrating the disturbed waves on the damaged plate, we first consider an intact plate with a point source in the form of a five-cycle tone burst at an operation frequency f_0 located at coordinate (200 mm, 0 mm, 2 mm) as marked in Figure 1a. The expression of the five-cycle tone burst is as follows:

$$A_1(t) = \frac{1}{2}[H(t) - H(t - 5/f_0)][1 - \cos(2\pi f_0 t/5)]\sin(2\pi f_0 t), \quad (5)$$

where $H(t)$ is the Heaviside step function. To avoid numerical errors, Equation (5) will be substituted by the temporal solution in Equation (4) and a closed form of the integral can be obtained.

Then, we show the number of modes required to determine the transient displacement field. We compute the root-mean-square (RMS) of the displacement field ${u_3}^{rms}$ at the coordinate (200 mm, 150 mm, 2 mm), with a 1 ms duration and a 1 MHz sampling rate, where

$$u_3^{rms} = \sqrt{\frac{\sum_{i=1}^{M} [u_3^2(t_1) + u_3^2(t_2) + u_3^2(t_3) + \ldots u_3^2(t_M)]}{M}}. \quad (6)$$

The number of the composed vibration modes with respect to the RMS displacement field at excitation frequencies of 5, 10, and 20 kHz is considered, as shown in Figure 2a. Although only a finite number of vibrational modes is required for converging the solution, the convergent speed of the RMS displacement field is related to the excitation frequency. When the excitation frequency is high, more modes are required for the convergence of the RMS displacement field. Then, we show the contribution of each mode (i.e., $u_{3,i}^{rms}$) to the overall transient responses with respect to the resonant frequency in Figure 2b. We can see that as the excitation frequency gets higher, the bandwidth of the contributing modes gets wider, where the dominating modes of the transient responses are centered around the excitation frequency.

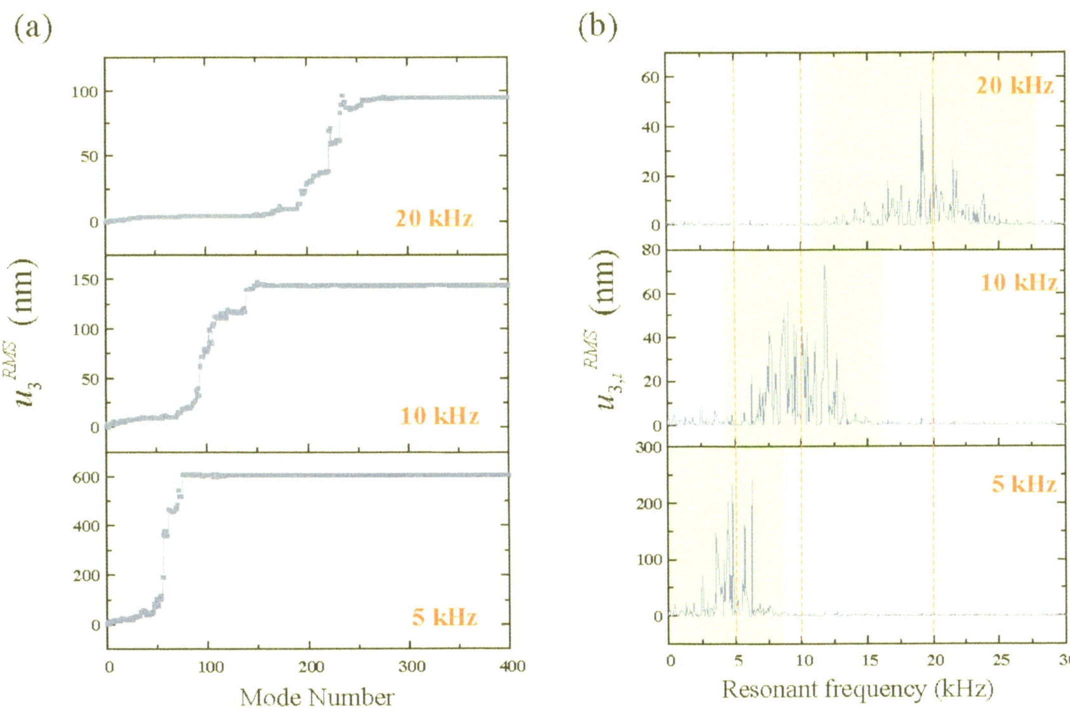

Figure 2. (**a**) The relation between the root-mean-square (RMS) transient displacement and the required number of vibrational modes. (**b**) The relation between the RMS transient displacement due to the *i*th mode and the resonant frequency.

The normalized full-field transient wave fields of the intact plate when the operation frequency is 10 kHz are plotted in Figure 3, where the Solid Mechanics Module with the time-dependent solver is utilized. Good correspondences can be obtained between the semi-analytical method and the FEM results, where the maximum displacements are 4.51×10^{-7} m and 4.57×10^{-7} m, respectively. Here, in calculating the semi-analytical method, there are a total of 400 vibrational modes considered, with a span from 0 to 35 kHz. The PFA results that give the energy transport at every location and at every specific time are also listed in Figure 3a, where the arrows demonstrate the instantaneous transmission paths of the propagating waves, and their lengths refer to the intensity of the vectors. We can see that, as time increases, the waves expand from the point source and are finally reflected from the boundary to interfere with the coming waves. Since the boundary condition is clamped at the left end, the wave pattern is not symmetric.

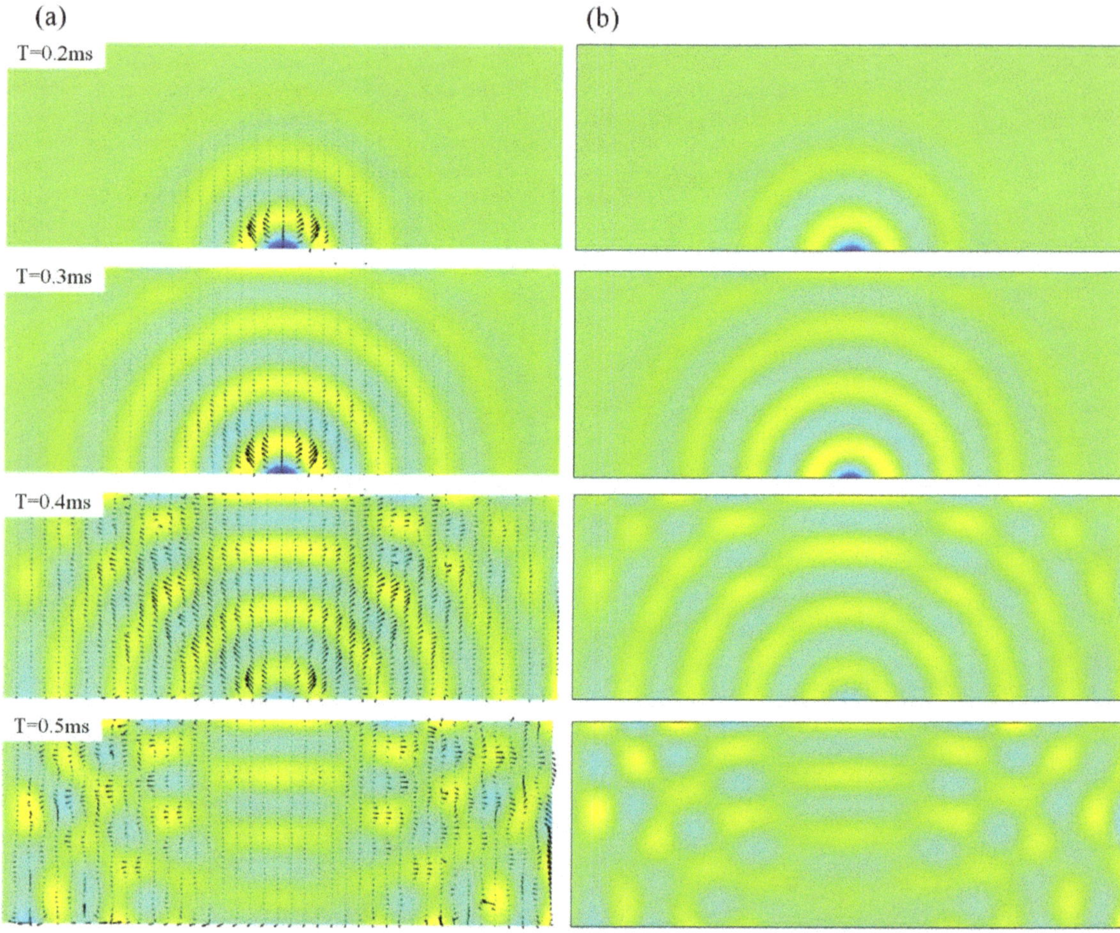

Figure 3. The full-field displacement response within 0.2 to 0.5 ms of the intact plate (**a**) the semi-analytical method with the power transmission paths. (**b**) The FEM methods.

Benefiting from avoiding the time-consuming iterative time-stepping process, the proposed semi-analytical method needs less calculation time compared to the FEM. To compare the simulation efficiency, a computer with the configuration of an AMD Ryzen 7 5800X 8-Core Processor 3.8 GHz and 32 GB RAM was used to run the simulations using the semi-analytical method (i.e., a combination of the FEM for Eigen Frequency Module and the theoretical Duhamel's integral solution) and the FEM (i.e., time-dependent solver). The mesh of the model consisted of 27,229 elements and the duration of the observation time is 1 ms. The comparisons of the calculation time between the FEM and the proposed semi-analytical method with respect to three different sampling times are shown in Table 1. As the sampling time is decreased, the calculation time for the FEM is increased. However, mainly spent on the obtainment of the vibration modes, the additional calculation time for the semi-analytical method is all very short. For the case of a 1 μs sampling time, the calculation time for the semi-analytical method is only 53% of that used in the FEM analysis.

Table 1. Comparison of the computational time between the FEM and semi-analysis method.

Sampling Time Δt	FEM (Time-Dependent)	Semi-Analytical Method
3 μs	159 s	241.4 s (235 + 6.4)
2 μs	242 s	242.4 s (235 + 7.4)
1 μs	462 s	244.1 s (235 + 9.1)
0.5 μs	855 s	247.0 s (235 + 12.0)

The relationship of the amplitude of the Lamb waves with respect to the time and each vibration mode contribution at its resonant frequency is given as a three-dimensional plot shown in Figure 4. We can see that, as time passes by, more and more modes are involved in the contribution of the instantaneous wave field. However, although each vibration mode contributes to the overall wave field, only the vibration modes centered around the excitation frequency will dominate the transient wave field, where the amplitude of a specific vibration mode is the largest. Similar to the Nyquist sampling theory, well known in signal analysis, for the convergence of the transient solution, vibrational modes higher than $2N$ Hz are required when the excitation frequency is at N Hz. However, from Figure 4, we can see that the dominated vibration modes are within $2N$ Hz [34]. We also note that, even for a specific mode, its contribution to the full-field wave field is not uniform and instead will oscillate as time increases, and the oscillation frequency is $2N$ Hz when the excitation frequency is N Hz.

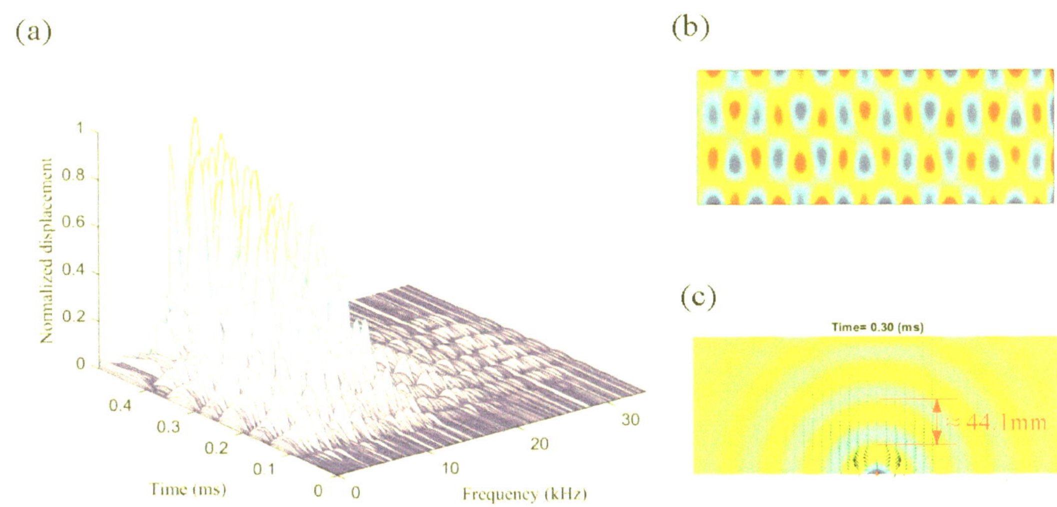

Figure 4. (**a**) The relationship of the displacement of the flexural waves with respect to the time and each vibration mode at its resonant frequency. (**b**) The 129th mode shape around 10 kHz with a resonant frequency of 10,077 Hz (**c**). The full-field displacement response at 0.3 ms, where the flexural wavelength is 44.1 mm.

It is interesting to point out that the wavelength of the dominated mode is related to the excitation frequency of the transient wave as

$$\lambda_f = \frac{c_f}{f} = \sqrt{\frac{2\pi}{f}}\left[\frac{Eh^2}{12\,\rho_s(1-\nu^2)}\right]^{1/4}, \tag{7}$$

where c_f is the phase velocity of the flexural wave, which has been given in Figure 1b. By substituting the 10 kHz excitation frequency by Equation (8), we can get a wavelength of 44.1 mm, which is close to the distance between the nodal lines of the dominating mode, the 129th mode, plotted in Figure 4. Thus, through the excitation frequency, one can know the resonant frequency and wavelength of the dominating modes of the instantaneous displacement field.

3.1. Simulation of Instantaneous Wave Field on the Damaged Plate

Figure 5 shows the simulated instantaneous wave field on the damaged plate. Good correspondence can be seen between the proposed semi-analytical method and the FEM. We can see the energy transport vectors are reduced when the instantaneous waves pass behind the slot. To separate the damaged information for clear observation, we then substrate the wave fields on the intact plate from those on the damaged plate, and the result is shown in Figure 6. Diffraction of the scattered elastic waves from the slot can be clearly seen due to the fact that the wavelength of the excited wave is larger than the corresponding dimension of the slot, and thus, the baseline subtraction method amplifies the influence of the presence of the damage on the propagating waves [40]. The intensity of the incoming wave distributes to the reflection waves and the scattered waves. From Figure 6, we can see that the presence of the damage reduces the intensity of the incoming transient waves, and that of the reflection wave is increased after performing the baseline subtraction method, and an antisymmetric displacement pattern can be observed. The pattern of the scattered waves follows the principle of diffraction, where a small obstacle would span the wave fields.

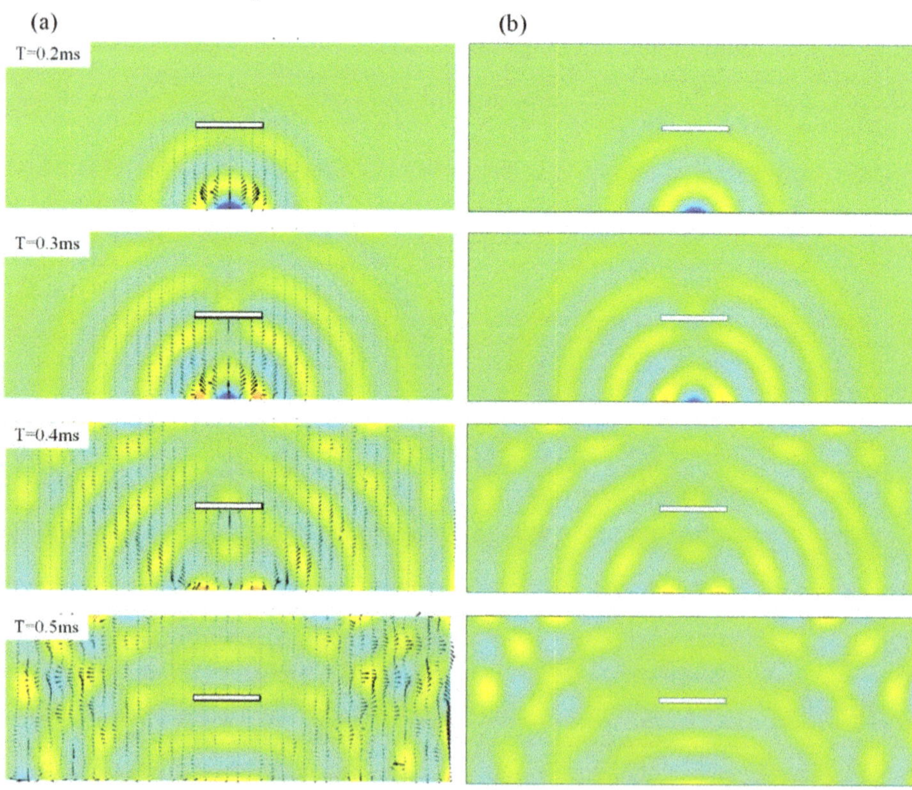

Figure 5. The full-field displacement response within 0.2 to 0.5 ms of the damaged plate (**a**) the semi-analytical method with the power transmission paths. (**b**) The FEM methods.

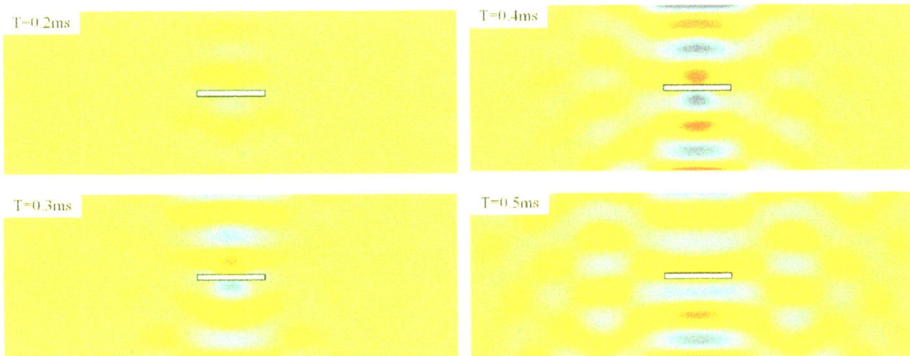

Figure 6. The subtracted full-field displacement response within 0.2 to 0.5 ms by the baseline subtraction method.

3.2. Simulation for the Pointwise Excitations

One advantage of the proposed semi-analytical method is that, when calculating the full-field instantaneous wave field on the damaged plate with different arrangements of the excitations, we only need to substitute different forms of excitations without the requirement of recalculating the vibration modes. Next, using the proposed semi-analytical method, we respectively model full-field transient Lamb waves on damaged plates with any spatial or temporal arrangement of excitations, as illustrated in Figure 7.

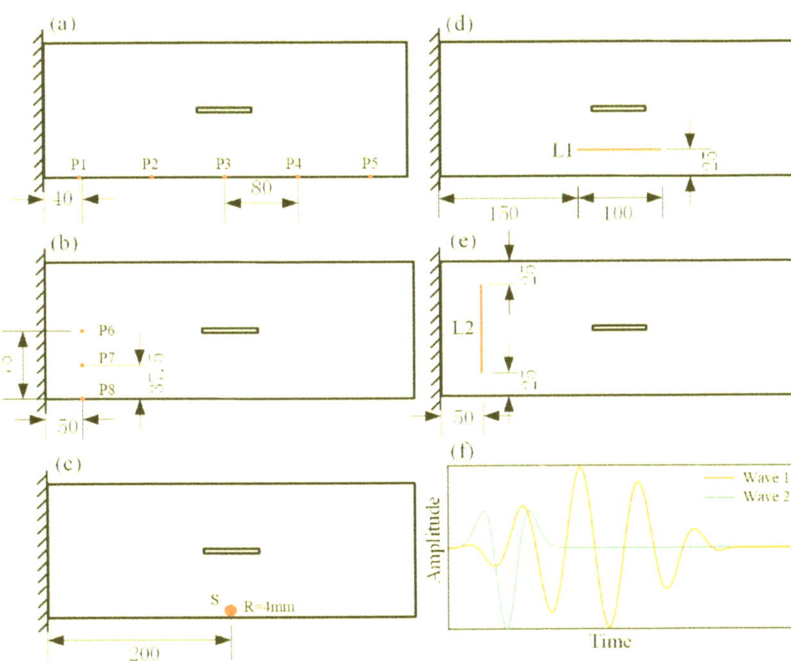

Figure 7. (**a**) Pointwise excitations at different horizontal locations. (**b**) Pointwise excitations at different vertical locations. (**c**) Surface excitation with a radius of R = 4 mm. (**d**) Horizonal line excitation. (**e**) Vertical line excitation. (**f**) Different temporal waveforms of excitations.

First, we consider pointwise excitations at different locations, marked as P1∼P5 in Figure 7a. Since different excitation locations lead to different arrival times and different incident angles of the Lamb waves, excitation locations will affect the instantaneous Lamb waves and the corresponding differential wave responses from the baseline subtraction method. Figure 8 shows the differential instantaneous wave fields when the excitations are at P1, P2, P4, and P5, where the results at P3 have been shown in Figure 6. When the Lamb waves are excited directly facing the long side of the slot (i.e., at P3), the diffraction of the Lamb waves is more significant. In addition, when exciting at P3, the differential instantaneous waves reflect the damage information much earlier than the excitations at other locations. For the instantaneous waves, the influence of the boundary condition is not significant in affecting the wave propagation, and we can see that the wave patterns are symmetric to the center axis in the y direction of the plate when the waves are excited at P1 and P5 or P2 and P4.

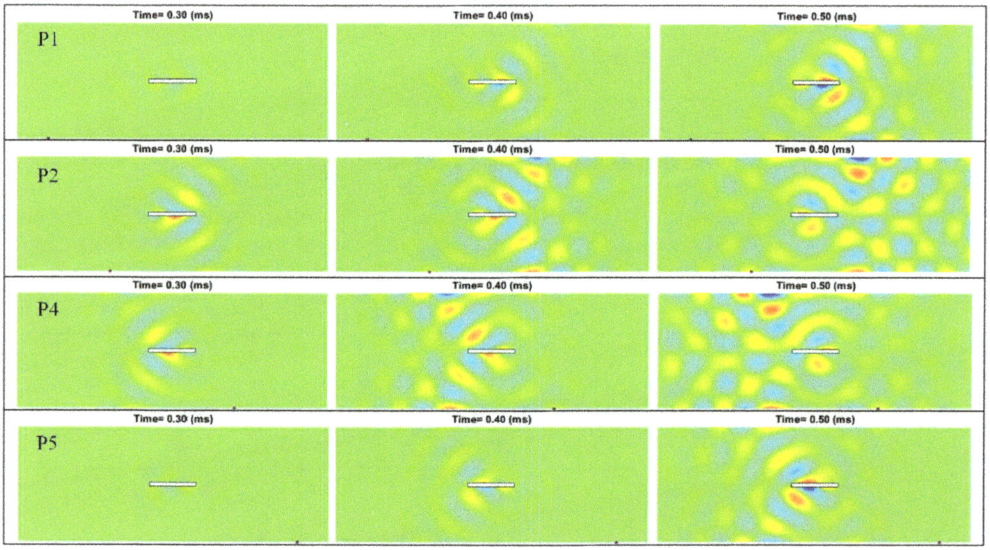

Figure 8. Differential instantaneous wave field when the pointwise excitations are at different locations.

The simulation results when the pointwise excitations are at P6, P7, and P8 at the left side of the plate are shown in Figure 9. Since the damage "seen" by the incoming waves is smaller when the excitations are located normal to the short side of the slot, the intensity of the scattered waves is smaller compared to those in Figure 8. Note that unlike the results obtained when the excitation is at P6, the differential instantaneous wave field obtained when the excitation is at P8 is asymmetric to the slot due to the fact that the waves are asymmetrically reflected by the slot. These simulation results support the concept of identifying the damage information through the full-field instantaneous wave patterns by using pointwise excitations at different locations.

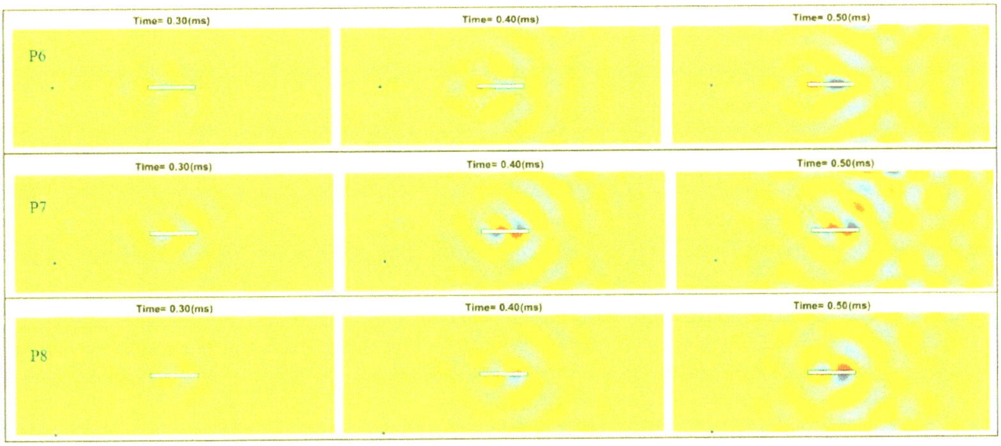

Figure 9. Differential instantaneous wave field when the pointwise excitations are at different locations from the left side of the damaged plate.

3.3. Simulation for the Surface or Linear Excitations

Now, we consider more practical linear or surface excitations. We first replace the pointwise excitation located at P3 (as shown in Figure 7a and studied in Section 3.1) as a round surface excitation (as shown in Figure 7c) with a radius of 4 mm, a dimension identical to that of a commercial piezoelectric actuator. Under the surface excitation, the reflected A0 mode Lamb waves in front of the plate edges interfere with the incoming waves and distort the wave pattern from a circular one to that excited by the pointwise excitation [41]. Even generating a distorted circular wave pattern due to the presence of the boundary conditions, the differential instantaneous wave field shown in Figure 10a is similar to that when the pointwise excitations are at location P3 (i.e., Figure 6). Linear excitations are also practical, and the results of the linear excitations, as arranged in Figure 7d (parallel to the long edge of the plate) and Figure 6e (parallel to the short edge of the plate), are shown in Figure 10b,c. Similar to the pointwise excitations, the differential wave field in the damaged plate is also dependent on the relative locations between the excitation and the damage event as well as its geometry [4,12,38]. These simulations are efficiently carried out in a sense that the same group of temporal vibration modes is utilized.

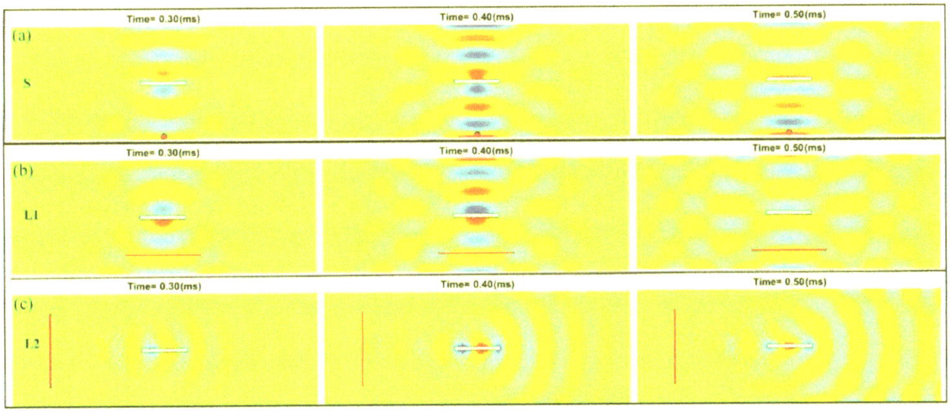

Figure 10. Differential instantaneous wave field when the excitations are (**a**) surface, (**b**) linear parallel to the long edge of the damaged plate, and (**c**) linear parallel to the short edge of the damaged plate.

3.4. Simulation for the Excitations with Specific Temporal Waveforms

After considering the simulation of different spatial excitations, we next consider excitations with different temporal waveforms. We consider the general five-cycle tone bursts, marked as Wave 1, at different frequencies and a Ricker wavelet, marked as Wave 2, as shown in Figure 7f. The expression of the Ricker wavelet is as follows:

$$A_2(t) = (2(\pi f_0(t - 1/f_0))^2 - 1)\exp(-(\pi f_0(t - 1/f_0))^2). \quad (8)$$

The expression of the Ricker wavelet is substituted by Equation (4) to obtain the transient displacement solution of the flexural wave. However, unlike the wave source of the tone burst where a simpler closed form can be obtained, here, Simpson's 1/3 rule is used to perform Duhamel's integral.

Figure 11a,b shows the differential wave field when the excitations are the 5 kHz and 20 kHz five-cycle tone bursts (i.e., Wave1), respectively. Obviously, dependent on the relation of the excitation frequency and the excited wavelength, the advantage of high-frequency excitations can be seen in damage detection. Without having to recalculate the vibration modes, we can efficiently simulate the full differential wave field, as shown in Figure 11c, by simply replacing the tone bursts as the Ricker wavelet (i.e., Wave2) with an operation frequency of 10 kHz.

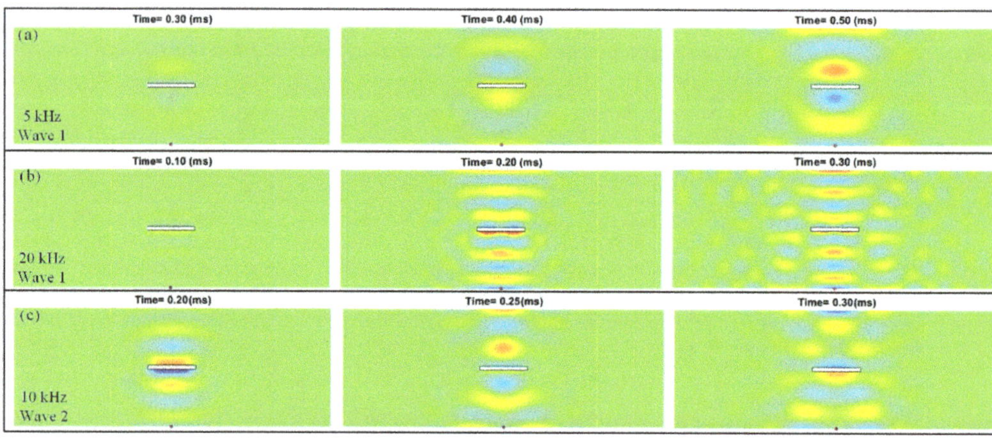

Figure 11. Differential instantaneous wave field when the excitations are (**a**) 5 kHz five-cycle tone burst, (**b**) 20 kHz five-cycle tone burst, and (**c**) the 10 kHz Ricker wavelet.

4. Experimental Results and Discussions

To validate the proposed semi-analytical method for modeling the full-field transient Lamb waves, we performed experiments on an intact and a damaged aluminum plate (150 × 500 × 2 mm) utilizing the surface and the linear excitations. For the convenience of clamping one side of the plate, the long edge of the plate is longer than that of the plate considered in the previous modeling. For the surface excitation, the A0 mode Lamb waves were generated by a circular piezoelectric patch (Haiying Group, Wuxi, China), bonded on the location S (as shown in Figure 7c) of the plate. For the linear excitation, an array of the piezoelectric patches was bonded along the location L2 (as shown in Figure 7e). A 10 kHz five-cycle tone burst was excited by the piezoelectric patches using a function generator (DG4102, Rigol Technologies, Suzhou, China) and a voltage amplifier (Trek Model PZD350). A laser scanning vibrometer system (Polytec PSV-500) was used to measure the out-of-plane component of the propagating displacement wave fields of the intact plate and damaged plate over a grid of points.

The measured displacement fields at 10 kHz frequencies are given in Figure 12a,c and Figure 13a,c, which agree well with the results obtained from the semi-analytical method in Figure 12b,d and Figure 13a,c. From the wave patterns in Figure 12a,b, we can see the interference between the incoming and reflected waves due to the presence of the boundary when using a non-pointwise excitation source. In addition, from the experimental results in Figure 12, we can see that the slot disturbs the instantaneous incident wave fields. However, agreed with the analytical prediction, when the Lamb waves generated from the linear excitation meet the short edge of the slot, the incident wave is almost not disturbed due to the fact that the dimension of the damage is smaller than the wavelength.

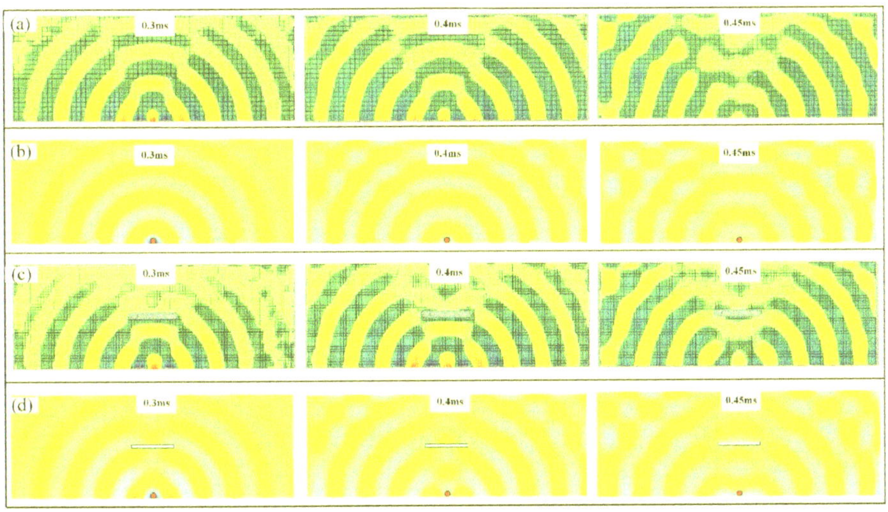

Figure 12. Instantaneous wave field under the surface excitations: (**a**) Experimental results of the intact plate; (**b**) semi-analytical results of the intact plate; (**c**) experimental results of the damaged plate; (**d**) semi-analytical results of the damaged plate.

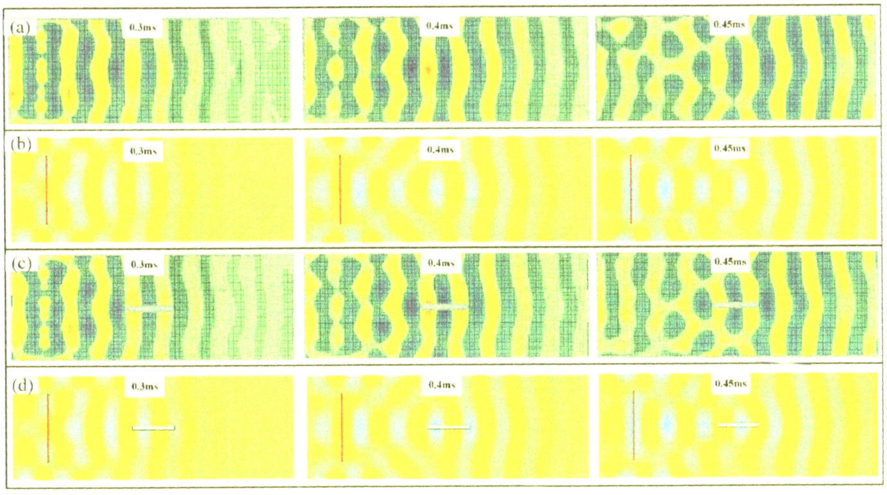

Figure 13. Instantaneous wave field under the linear excitations: (**a**) Experimental results of the intact plate; (**b**) semi-analytical results of the intact plate; (**c**) experimental results of the damaged plate; (**d**) semi-analytical results of the damaged plate.

5. Conclusions

In this paper, a semi-analytical method for modeling the propagation of Lamb waves based on the vibration characteristics of the intact or damaged plate is proposed. Using the same group of the temporal vibrational modes, the proposed method can model the full-field transient Lamb waves efficiently with any forms of excitations, where the temporal resolution relies on the number of the vibration modes. As an additional advantage, the wave fields expressed as the superposition of the vibration modes can be transformed into other physical quantities without having to perform any additional numerical post-processing operations. As an application, a power flow analysis is employed in this work to reveal the transmission paths of the propagating waves on the damaged plate. In addition to demonstrating the modeling capability of the semi-analytical method, we also examine its role in structure health monitoring on a classic damage model by combing the baseline subtraction method. After comparing the analytical results with the finite element method simulation, the proposed semi-analytical method is validated by a non-contact scanning Laser Doppler Vibrometer (LDV) system.

Author Contributions: Conceptualization, K.-C.C.; Formal analysis, D.-F.W. and C.-Y.L.; Funding acquisition, K.-C.C., J.-J.L. and C.-Y.L.; Investigation, D.-F.W. and C.-Y.L.; Methodology, K.-C.C. and C.-Y.L.; Resources, D.-F.W.; Software, D.-F.W. and C.-Y.L.; Supervision, K.-C.C.; Validation, K.-C.C. and J.-J.L.; Writing—original draft, D.-F.W., K.-C.C. and C.-Y.L.; Writing—review & editing, K.-C.C. and C.-Y.L. All authors have read and agreed to the published version of the manuscript.

Funding: This research was funded by the National Natural Science Foundation of China (Nos. 11972318 and 12102370), the Opening Project of Applied Mechanics and Structure Safety Key Laboratory of Sichuan Province (No. SZDKF-202001), the Fundamental Research Funds for the Central Universities (No. 2682021CX082), and the Ministry of Science and Technology of the Republic of China (Grant No. MOST110-2222-E-008-001-MY2).

Institutional Review Board Statement: Not applicable.

Informed Consent Statement: Not applicable.

Data Availability Statement: The data that support the findings of this study are available from the corresponding author upon reasonable request.

Conflicts of Interest: The authors declare no conflict of interest.

References

1. Golub, M.V.; Shpak, A.N.; Mueller, I.; Fomenko, S.I.; Fritzen, C. Lamb wave scattering analysis for interface damage detection between a surface-mounted block and elastic plate. *Sensors* **2021**, *21*, 860. [CrossRef] [PubMed]
2. Poddar, B.; Kumar, A.; Mitra, M.; Mujumdar, P.M. Time reversibility of a Lamb wave for damage detection in a metallic plate. *Smart Mater. Struct.* **2011**, *20*, 025001. [CrossRef]
3. Yeung, C.; Ng, C.T. Time-domain spectral finite element method for analysis of torsional guided waves scattering and mode conversion by cracks in pipes. *Mech. Syst. Signal. Process.* **2019**, *128*, 305–317. [CrossRef]
4. Lee, B.C.; Staszewski, W.J. Sensor location studies for damage detection with Lamb waves. *Smart Mater. Struct.* **2007**, *16*, 399. [CrossRef]
5. Yu, T. Plate waves scattering analysis and active damage detection. *Sensors* **2021**, *21*, 5458. [CrossRef]
6. Bagheri, A.; Amiri, G.G.; Razzaghi, S.S. Vibration-based damage identification of plate structures via curvelet transform. *J. Sound Vib.* **2009**, *327*, 593–603. [CrossRef]
7. Lele, S.P.; Maiti, S.K. Modelling of transverse vibration of short beams for crack detection and measurement of crack extension. *J. Sound Vib.* **2002**, *257*, 559–583. [CrossRef]
8. Fan, W.; Qiao, P. A 2-D continuous wavelet transform of mode shape data for damage detection of plate structures. *Int. J. Solids Struct.* **2009**, *46*, 4379–4395. [CrossRef]
9. Mariani, S.; Heinlein, S.; Cawley, P. Location specific temperature compensation of guided wave signals in structural health monitoring. *IEEE Trans. Ultrason. Ferroelectr. Freq. Control* **2019**, *67*, 146–157. [CrossRef]
10. Alleyne, D.N.; Pialucha, T.P.; Cawley, P. A signal regeneration technique for long-range propagation of dispersive Lamb waves. *Ultrasonics* **1993**, *31*, 201–204. [CrossRef]
11. Ghosh, T.; Kundu, T.; Karpur, P. Efficient use of Lamb modes for detecting defects in large plates. *Ultrasonics* **1998**, *36*, 791–801. [CrossRef]

12. Ng, C.T.; Veidt, M.; Lam, H.F. Guided wave damage characterisation in beams utilising probabilistic optimisation. *Eng. Struct.* **2009**, *31*, 2842–2850. [CrossRef]
13. Chinchalkar, S. Determination of crack location in beams using natural frequencies. *J. Sound Vib.* **2001**, *247*, 417–429. [CrossRef]
14. Qian, G.; Gu, S.; Jiang, J. The dynamic behaviour and crack detection of a beam with a crack. *J. Sound Vib.* **1990**, *138*, 233–243. [CrossRef]
15. Jena, P.K.; Thatoi, D.N.; Nanda, J.; Parhi, D. Effect of damage parameters on vibration signatures of a cantilever beam. *Procedia Eng.* **2012**, *38*, 3318–3330. [CrossRef]
16. Moore, E.Z.; Murphy, K.D.; Nichols, J.M. Crack identification in a freely vibrating plate using Bayesian parameter estimation. *Mech. Syst. Signal Process.* **2011**, *25*, 2125–2134. [CrossRef]
17. Peng, H.; Meng, G.; Li, F. Modeling of wave propagation in plate structures using three-dimensional spectral element method for damage detection. *J. Sound Vib.* **2009**, *320*, 942–954. [CrossRef]
18. Lu, G.; Li, Y.; Zhou, M.; Feng, Q.; Song, G. Detecting damage size and shape in a plate structure using PZT transducer array. *J. Aerosp. Eng.* **2018**, *31*, 04018075. [CrossRef]
19. Han, S.; Palazotto, A.N.; Leakeas, C.L. Finite-element analysis of Lamb wave propagation in a thin aluminum plate. *J. Aerosp. Eng.* **2009**, *22*, 185–197. [CrossRef]
20. Lu, G.; Feng, Q.; Li, Y.; Wang, H.; Song, G. Characterization of ultrasound energy diffusion due to small-size damage on an aluminum plate using piezoceramic transducers. *Sensors* **2017**, *17*, 2796. [CrossRef]
21. Palacz, M.; Żak, A.; Krawczuk, M. FEM-based wave propagation modelling for SHM: Certain numerical issues in 1D structures. *Materials* **2020**, *13*, 2051. [CrossRef] [PubMed]
22. Shen, Y.; Giurgiutiu, V. Combined analytical FEM approach for efficient simulation of Lamb wave damage detection. *Ultrasonics* **2016**, *69*, 116–128. [CrossRef] [PubMed]
23. Carrino, S.; Nicassio, F.; Scarselli, G.; Vitolo, R. Finite difference model of wave motion for structural health monitoring of single lap joints. *Int. J. Solids Struct.* **2019**, *161*, 219–227. [CrossRef]
24. Lee, B.C.; Staszewski, W.J. Lamb wave propagation modelling for damage detection: I. Two-dimensional analysis. *Smart Mater. Struct.* **2007**, *16*, 249. [CrossRef]
25. Krawczuk, M.; Palacz, M.; Ostachowicz, W. The dynamic analysis of a cracked Timoshenko beam by the spectral element method. *J. Sound Vib.* **2003**, *264*, 1139–1153. [CrossRef]
26. Galan, J.M.; Abascal, R. Boundary element solution for the bidimensional scattering of guided waves in laminated plates. *Comput Struct.* **2005**, *83*, 740–757. [CrossRef]
27. Ahmad, Z.; Gabbert, U. Simulation of Lamb wave reflections at plate edges using the semi-analytical finite element method. *Ultrasonics* **2012**, *52*, 815–820. [CrossRef]
28. Ahmad, Z.; Vivar-Perez, J.M.; Gabbert, U. Semi-analytical finite element method for modeling of lamb wave propagation. *CEAS Aeronaut. J.* **2013**, *4*, 21–33. [CrossRef]
29. Duan, W.; Gan, T. Investigation of guided wave properties of anisotropic composite laminates using a semi-analytical finite element method. *Compos. Part B Eng.* **2019**, *173*, 106898. [CrossRef]
30. Yang, Z.; Liu, K.; Zhou, K.; Liang, Y.; Zhang, J.; Zheng, Y.; Gao, D.; Ma, S.; Wu, Z. Investigation of thermo-acoustoelastic guided waves by semi-analytical finite element method. *Ultrasonics* **2020**, *106*, 106141. [CrossRef]
31. Bhaskar, K.; Varadan, T.K. *Plates: Theories and Applications*; Springer Nature: Cham, Switzerland, 2021.
32. Liao, C.; Ma, C. Transient behavior of a cantilever plate subjected to impact loading: Theoretical analysis and experimental measurement. *Int. J. Mech. Sci.* **2020**, *166*, 105217. [CrossRef]
33. Chuang, K.; Lin, S.; Ma, C. Efficient excitation of transverse vibrational modes using improved configurations of PFCs connected to an isotropic plate. *Compos. Struct.* **2021**, *265*, 113718. [CrossRef]
34. Chuang, K.; Wang, D.; Liu, J.; Liao, C. Linking Time-Domain Vibration Behaviors to Spatial-Domain Propagating Waves in a Leaf-like Gradient-Index Phononic Crystal Lens. *Crystals* **2021**, *11*, 1490. [CrossRef]
35. Xing, J.; Price, W.G. A power–flow analysis based on continuum dynamics. *Proc. R. Soc. Lond. Ser. A Math. Phys. Eng. Sci.* **1999**, *455*, 401–436. [CrossRef]
36. Huang, W.; Ji, H.; Qiu, J.; Cheng, L. Wave energy focalization in a plate with imperfect two-dimensional acoustic black hole indentation. *J. Vib. Acoust.* **2016**, *138*, 6. [CrossRef]
37. Zhu, C.; Yang, J.; Rudd, C. Vibration transmission and power flow of laminated composite plates with inerter-based suppression configurations. *Int. J. Mech. Sci.* **2021**, *190*, 106012. [CrossRef]
38. Lee, B.C.; Staszewski, W.J. Modelling of Lamb waves for damage detection in metallic structures: Part II. Wave interactions with damage. *Smart Mater. Struct.* **2003**, *12*, 815. [CrossRef]
39. Castaings, M.; Cawley, P. The generation, propagation, and detection of Lamb waves in plates using air-coupled ultrasonic transducers. *J. Acoust. Soc. Am.* **1996**, *100*, 3070–3077. [CrossRef]
40. Melville, J.; Alguri, K.S.; Deemer, C.; Harley, J.B. Structural damage detection using deep learning of ultrasonic guided waves. *AIP Conf. Proc.* **2018**, *1949*, 230004.
41. Li, J.; Khodaei, Z.S.; Aliabadi, M.H. Boundary element modelling of ultrasonic Lamb waves for structural health monitoring. *Smart Mater. Struct.* **2020**, *29*, 105030. [CrossRef]

sensors

MDPI

Article

Concrete Curing Performance Assessment Based on Gas Permeability Testing in the Lab and on Site

Lisa Ptacek [1,*], Alfred Strauss [1], Clémence Bos [2], Martin Peyerl [2] and Roberto Torrent [3,4]

[1] Department of Civil Engineering and Natural Hazards, University of Natural Resources and Life Sciences, 1180 Vienna, Austria; alfred.strauss@boku.ac.at
[2] Smart Minerals GmbH, 1030 Vienna, Austria; bos@smartminerals.at (C.B.); peyerl@smartminerals.at (M.P.)
[3] Materials Advanced Services, Buenos Aires 1425, Argentina; torrent.concrete@gmail.com
[4] Materials Advanced Services, 6877 Coldrerio, Switzerland
* Correspondence: lisa.ptacek@boku.ac.at

Abstract: Durability is an essential aspect of the lifetime performance of concrete components. The adequate surface quality and thus the service life of concrete can be achieved, among other things, by appropriate curing during hydration. To measure and control the curing quality, appropriate procedures are required. Gas permeability allows conclusions to be drawn about the porosity of concrete, which has a significant impact on durability. In this contribution, the effect of different curing methods on gas permeability is presented with the help of laboratory and on-site tests, showing that inadequate curing leads to increased permeability in the near-surface area of concrete. The measurement results of concrete samples and components with the same composition but varying curing treatment are compared and evaluated. Influences such as concrete composition and environmental factors on the quality of concrete are observed, and recommendations are made for a reliable assessment of the surface quality as a result of the investigated curing measures.

Keywords: concrete curing; non-destructive testing (NDT); gas permeability testing; durability; quality control

Citation: Ptacek, L.; Strauss, A.; Bos, C.; Peyerl, M.; Torrent, R. Concrete Curing Performance Assessment Based on Gas Permeability Testing in the Lab and on Site. *Sensors* **2022**, *22*, 4672. https://doi.org/10.3390/s22134672

Academic Editor: Zenghua Liu

Received: 16 May 2022
Accepted: 15 June 2022
Published: 21 June 2022

1. Introduction

1.1. Motivation

The durability of concrete has received increasing attention in recent years. In the past, it was assumed that concrete structures are maintenance-free if certain basic rules of concrete technology are followed. The experience of the past decades, however, showed that only minor deviations from these rules as well as incorrectly assessed or harsh environmental conditions can lead to considerable damage [1]. Due to the addition of ground granulated blast furnace slag and fly ash to reduce CO_2 emissions, modern cement in particular tends toward slower reacting, which is why modern concrete requires a more extensive curing [2].

The curing of concrete is essential for the durability of a structure [3]. Various guidelines and standards as [4–6] give information on measures and durations for the adequate curing of concrete. However, during the construction phase, in many cases, little importance is attached to curing due to a lack of control options, tight schedules, and cost reasons. Excerpting from Prof. Neville [2]: '*Curing concrete is the lowest of low-tech operations ... it is seen by many as a silly operation, a non-job ... and bad curing does not show .... If I emphasize ensuring curing, it is because curing can make all the difference between having good concrete and having good concrete ruined by the lack of a small effort*'.

Different to bad compaction that 'shows', lack of curing does not show; therefore, the finding of test methods for proving sufficient curing is of great interest.

Numerous methods are already cited in the literature to evaluate the curing efficiency. In addition to standardized methods such as the determination of the abrasion resistance [7], the compressive strength [8] or the rate of absorption [9,10], alternative methods were investigated that provide information about the concrete quality as a result of curing. The procedures range from monitoring during the hydration phase to determine the appropriate time for formwork removal [11] to destructive procedures in the laboratory [12] to evaluate the curing effects afterwards. The authors of [13] investigate the effect of curing water availability on cement hydration using isothermal calorimetry. Consequential damage or damage processes occurring due to inadequate curing such as freeze–thaw attack or microcracks have been determined in studies with mercury intrusion porosimetry testing [14] or acoustic emission [15].

Various non-destructive methods are already being used to assess the durability of concrete structures, such as computer tomography, radar techniques and ultrasonic testing [16]. There is also the approach to combine non-destructive tests in order to obtain results for the evaluation of the curing performance [17].

Many of the methods mentioned are associated with great effort or represent the curing effectiveness inadequately. Therefore, the objective of this contribution is to present a non-destructive, simple, and low-cost test method for curing quality control.

1.2. Durability

According to [18], durability is defined as the capability of structures, products or materials to fulfill the requirements defined, which are determined after a specified period of time and usage. The durability and performance of concrete is reduced by indirect and direct damage processes during its entire service life. Direct damage processes on concrete are freeze–thaw attack, acid action and leaching processes. The indirect damage processes include chloride penetration and carbonation, which can lead to the corrosion of reinforcing steel [18]. The prevention of the mentioned damage processes and a sufficient durability is achieved amongst other things by:

- Durability-oriented design of the structure, e.g., keep away from attacking substances and water;
- Correct selection of the concrete raw materials and a concrete composition suitable for the job;
- Proper production and processing techniques, especially curing of the concrete;
- Passive protection of the concrete, e.g., by impregnation, coating [1].

1.3. Concrete Curing

The concrete cover needs to be as impermeable as possible to achieve a proper resistance against chemical and physical attack of the reinforcement steel, such as chloride and carbon dioxide [3]. Therefore, a high hydration rate of the concrete surface must be achieved. Cement can hydrate only when there is enough moisture available. Curing measures are carried out to ensure optimal humidity and temperature conditions in the first days of hydration and subsequently the achievement of the expected concrete quality [3]. Under a lack of proper curing, the capillary pores of the outer concrete surface dry out. This leads to a standstill of the hydration and eventually to plastic shrinkage, a reduced surface strength and higher surface permeability [18]. As a result of this, the concrete in the areas close to the surface, which are highly exposed to the weather and environmental conditions, are less durable [4,19,20].

The following methods are suitable for proper curing of the concrete surface, and they can also be combined:

- Keeping the concrete in the formwork;
- Covering the concrete surface with foils;
- Wetting the concrete surface;
- Application of an effective curing compound [18,21–24].

1.4. Permeability Testing

The durability of concrete depends heavily on the tightness of the concrete pore system, which, according to [1], can be demonstrated, among other things, by gas permeability.

Permeable concrete favors transport processes in the concrete and therefore damage processes such as carbonation, chloride ingress and freeze–thaw deterioration [25]. It has been shown that there is a close relationship between the time it takes to reach a certain carbonation depth and the gas permeability. The authors of [26–28] developed an analytical model for the correlation of gas permeability with resistance to carbonation, while [29] found that the chloride input in concrete correlates with its gas permeability. Moreover, a linear relationship between freeze–thaw deterioration and gas permeability has been demonstrated [27]. The parameter gas permeability can cover the permeability of concrete for various corrosion and aging mechanisms and represents a suitable test method for characterizing the transport properties of concrete [30]. The gas permeability of a concrete surface depends on many key factors, including:

- Concrete mixture composition;
- The composition and physical properties of the cement and aggregates;
- Curing quality [31–33];
- Degree of hydration or age;
- Presence of microcracks;
- Presence of surface treatments;
- Moisture content at the time of the test [30,34,35].

Gas permeability is, therefore, a relevant durability indicator for the tested concrete [36].

The procedure for collecting and evaluating gas permeability data for determining the durability of the concrete surface is described in the Swiss standard SIA 262/1: 2019 [24].

As shown in Figure 1, the equipment for gas permeability testing basically consists of:

- A two-chamber vacuum cell composed of two concentric chambers (inner and external chambers);
- A control system, consisting of valves, pressure sensors and a pressure regulator that keeps the pressure of both vacuum chambers balanced during the measurement (Pe = Pi) [37].

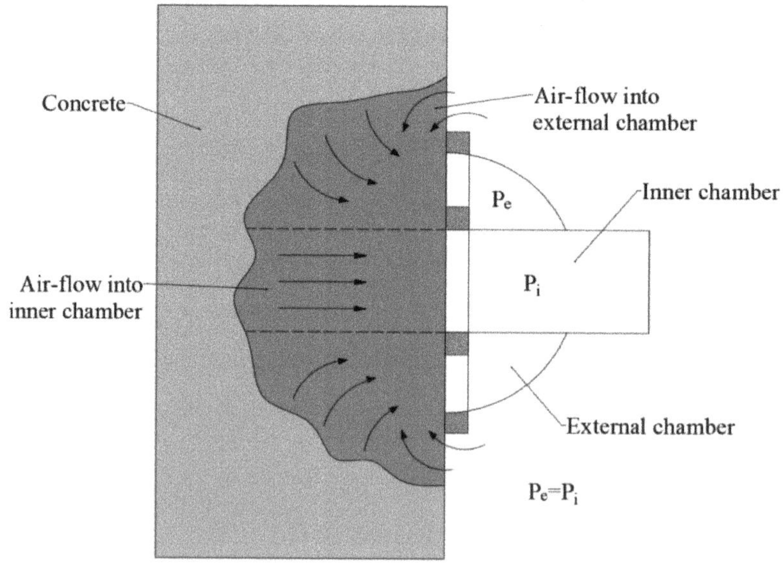

Figure 1. Vacuum cell, pressure regulation and air flow into its two concentric chambers [38].

In the inner test chamber (Ø = 50 mm) and the protective external ring surrounding this chamber, both of which are open toward the concrete surface, a vacuum is generated using a vacuum pump. Depending on the concrete and the measuring device, the vacuum is approx. 10–40 mbar. At 60 s, the inner chamber is isolated from the vacuum pump, which is the moment at which its pressure begins to build up due to the gas flowing through the concrete. The gas permeability is calculated from the change in pressure over time and other parameters. The purpose of the protective ring surrounding the test chamber (the pressure of which is permanently balanced with that of the inner chamber) is to ensure that the gas only flows into the test chamber in one dimension, which means that it only comes from inside the component (see Figure 1). For laboratory tests, cubes of at least 150 mm side length are the most suitable specimen for measuring the gas permeability coefficient kT [39]. The moisture of the concrete has a very strong influence on the gas permeability, as the water in the pores hinders the flow of air [30,33,40,41]. The concrete moisture has to be measured with an impedance-based moisture meter [42] and should be a maximum of 5.5%. This can, e.g., be achieved through drying at 50 °C for a few days prior to measurement.

In the case of in situ tests, it should be noted that according to [24], the test should be carried out when the concrete is aged between 28 and 120 days, so that the requirements for moisture content can be met. The moisture content of the concrete component to be examined must be determined in accordance with the laboratory measurements before the gas permeability measurement. In addition, it must be ensured that the air and component temperature should be at least 5 °C. Ref. [24] stipulates that 6–12 measurements need to be made per test area, which are separated by horizontal and vertical distances of at least 0.2 m. The coefficient of permeability to gas kT of hardened concrete is measured in m^2, corresponding to the intrinsic coefficient of permeability of the material, and it is calculated with Equation (1) according to [29].

$$kT = \left(\frac{V_c}{A}\right)^2 \cdot \frac{\mu}{2 \cdot \varepsilon \cdot P_a} \cdot \left(\frac{\ln\left(\frac{P_a+\Delta P}{P_a-\Delta P}\right)}{\sqrt{t_f} - \sqrt{t_0}}\right)^2, \tag{1}$$

where

V_c = volume of the inner test chamber (m^3);
A = area of the inner test chamber (m^2);
μ = dynamic viscosity of air (Ns/m^2);
ε = open porosity of the concrete (-) which, by default is taken as 0.15;
P_a = atmospheric pressure (N/m^2);
ΔP = increase in effective pressure in the inner chamber between time t_0 and t_f (N/m^2);
t_0 = time at which the increase in pressure is measured (s);
t_f = time at which the test is finished (s).

The kT values recorded can be assessed using the permeability classes from [29]. Table 1 shows the limit values for "very good" to "very poor" concrete cover quality for the kT value measured after 28 days.

Table 1. Classification of the quality of the concrete cover according to kT [29].

Classification of the Quality of the Concrete Cover		kT Measured at 28 Days [10^{-16} m^2]
1	very good	kT < 0.01
2	good	0.01 ≤ kT < 0.1
3	normal	0.1 ≤ kT < 1.0
4	bad	1.0 ≤ kT < 10
5	very bad	kT ≥ 10

1.5. Objectives

With the following studies, the authors want to present and evaluate an innovative possibility to test the curing quality in a non-destructive and user-friendly way. As outlined in previous sections, the gas permeability allows making specific statements about the surface condition of concrete. Since the surface condition of concrete heavily depends on the quality of curing, the following objectives will be addressed:

1. Can curing quality be classified and evaluated using gas permeability measurements?
2. How can the curing quality on a structure be assessed? What kind of requirements must be met for reliable results with onsite measurements, considering environmental influences such as moisture, solar radiation, ambient temperature, wind, etc.?
3. Which recommendations for the implementation and analysis of gas permeability measurements for curing quality evaluation can be given on the basis of the laboratory and construction site investigations for different concrete types?

2. Laboratory Test Campaign

In the laboratory, it was possible to exclude influences such as wind, temperature fluctuations and solar radiation and only examine the effect of various curing conditions on concrete. By evaluating concrete samples that were treated in the laboratory under precisely defined curing conditions, connections between gas permeability and curing quality could be established.

2.1. Concrete and Curing Characteristics

With the involvement of a group of experts, three common types of concrete used in Austrian civil engineering were selected for the laboratory tests. Since the area of interest was particularly in road and railway construction, the types of concrete should cover as many different areas of application as possible. The names and composition of the concrete types and their fresh concrete properties according to the Austrian standard [5] are summarized in Table 2. Possible areas of application for concrete type B3 are traffic areas without de-icing agents. B5 is used for components that are exposed to spray mist containing a de-icing agent, such as bridge structures. The third type, BS1C, is used for waterproof structures directly exposed to de-icing agents.

Table 2. Material properties of the examined concrete types in the course of the laboratory tests.

(Austrian) Concrete Name	**B3**	**B5**	**BS1C Plus**
Strength class	C25/30	C30/37	C25/30
Cement (kg/m^3):	290 (CEM II/A-M (S-L) 42.5 N)	320 (CEM II/A-M (S-L) 42.5 N)	258 (CEM I 42.5N SR0 WT27 C_3A-free)
Additions (kg/m^3): additions type 2 according to EN 206	40	40	103
Total water (kg/m^3)	171	170	165
w/c-ratio	0.53	0.48	0.48
Aggregates (kg/m^3): Gravel 16/32 round grain Gravel 8/16 round grain Gravel 4/8	 457 292 311	 451 289 307	 333 333 298
Sand 0/4 round grain	770	761	792
Air-entraining agent (m%-Cement): Mapei Mapeair LP 100 Mapei Mapeair LP 200 Sika Addiment LPS A NEU	 0.30–0.97	 0.00–0.50 0.00–0.50	 0.00–1.20 0.00–0.70 0.00–0.15
Superplasticizer (m%-Cement):			
Mapei Dynamon LZF Sika VC4030Ultra	0.00–0.31 -	0.20–0.39 -	0.28–0.64 0.00–0.15
Air content (fresh concrete) (%)	2.6–3.6	2.5–4.2	5.8–8.0
Bulk density (fresh concrete) (kg/m^3)	2379–2431	2362–2442	2224–2313
Fresh concrete temperature (°C)	21.5–22.9	21.8–24.1	19.8–23.3

In order to analyze various curing effects on those three concrete types, nine concrete cubes with a side length of 0.15 m were produced for each type of concrete. The cubes were subjected to three different types of curing, with a distinction being made between "good", "no" and "bad" treatment. The applied curing conditions were discussed in advance in a panel of experts. Based on empirical values, it was assumed that with surface protection and a moderate temperature of 20 °C and a relative humidity of 60%, satisfactory concrete properties can be achieved ("good curing"). On the other extreme, being unprotected under an ambient temperature of 30 °C and a lower relative humidity of 40% should simulate more extreme conditions for "bad" curing. As can be seen in Table 3, variations for the curing types were made in the storage conditions with regard to storage with or without foil protection and also in terms of temperature and relative humidity. Since the duration of the curing treatments also influences hydration, the curing was carried out with different durations of one, three or seven days. All samples were then dried at 50 °C for three days to ensure dry samples for the measurements.

Table 3. Curing types for the laboratory tests, see also [43].

Curing Type	Storage	Temperature	Relative Humidity
	Curing conditions until day 1, 3 or 7		
good	in foil	20 °C	60%
no	without foil	20 °C	60%
bad	without foil	30 °C	40%
	Curing conditions from day 1, 3 or 7		
all types	Storage at 50 °C for 3 days		

2.2. Measurements and Monitoring

Since there is the possibility of measurement deviations due to the application of different gas permeability devices, two devices were used for the measurements in order to compare their results, including the PermeaTORR device from Materials Advanced Services [44] and the Torrent Permeability Tester from Proceq SA [45].

Before the start of the measurements, the instruments were calibrated at least two times, until the recorded maximum pressure did not exceed 5 mbar nor did it differ by more than 0.5 mbar from the previous calibration measurement, according to [24].

A minimum of three faces per sample were measured, including the top surface that was not covered by the formwork during concreting and two lateral faces that were protected by the formwork. The measurement of two "formwork faces" was considered sufficient, since the scatter of the results between two faces is very small, and a further measurement would change the standard deviation only slightly.

As recommended in [24], the moisture content of each sample was measured with a impedance-based moisture meter CMEXII from Tramex [46] before measuring kT. To ensure the creation of a sufficient vacuum and protect the instrument, the existing loose dust on the surface was removed with a brush. In case a repetition of a measurement might be needed, due to abnormal results or to find if the permeability changes with time, the outer perimeter of the vacuum cell was marked on the surfaces. In addition to the permeability and moisture measurements, the compressive strength at the concrete age of 28 days was measured for all samples to determine whether the minimum requirements for the concrete types examined were met and to see to what extent the strength of the samples changes because of the varying curing treatments.

2.3. Findings

According to [29], the results of the permeability measurements are presented in a diagram which refers to the permeability classes. For a clear visualization of the results, the geometric mean values of the coefficient of gas permeability (kT) values in 10^{-16} m^2 are entered as points on a logarithmic scale also showing segments representing ± the standard deviation of the logarithms of the results. A high gas permeability corresponds to a poor surface quality and vice versa. All results of a series are summarized in a separate graph for further analysis. Series B3, B5 and BS1C Plus data are shown in Figures 2–4, respectively. Each graph includes the measured kT values, divided into "good", "no" and "bad" cured samples depending on the duration of the curing treatment. When presenting the results, the focus was not only placed on the classification into surface classes but also on differentiating the results of the two devices used and the distinction of the values from measurements at the side and the top faces of the samples, which are displayed with different symbols.

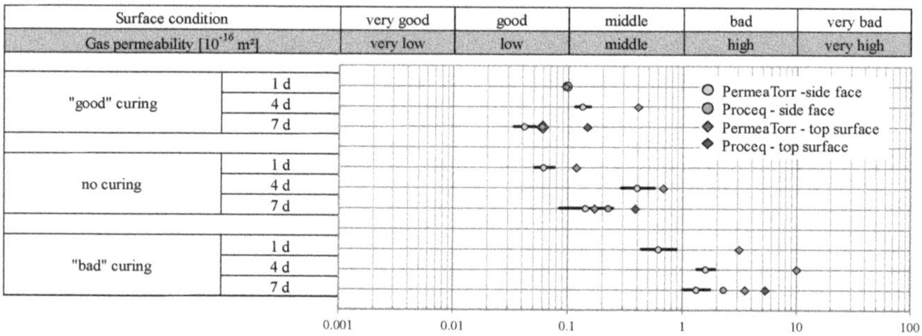

Figure 2. Gas permeability kT [10^{-16} m^2] of the B3 test samples depending on the curing quality.

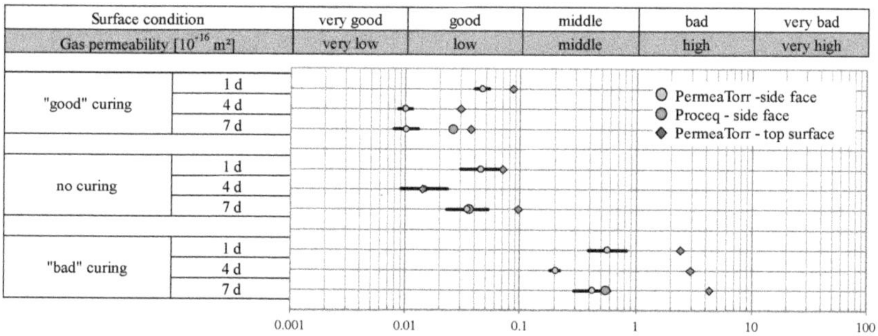

Figure 3. Gas permeability kT (10^{-16} m^2) of the B5 test samples depending on the curing quality.

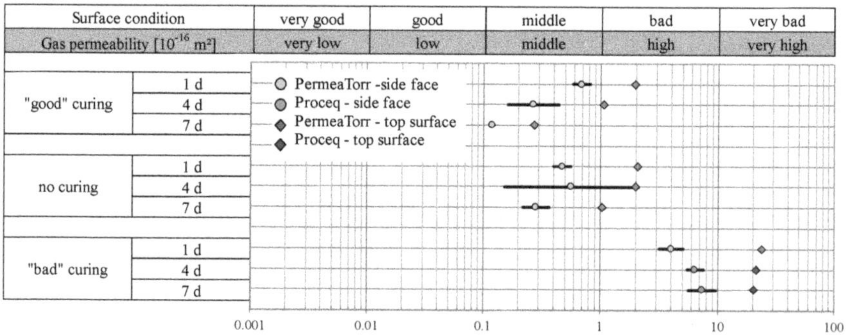

Figure 4. Gas permeability kT (10^{-16} m^2) of the BS1C Plus test samples depending on the curing quality.

2.3.1. Permeability Properties of Concrete B3

Clear differences in the results of the three curing types of the samples of concrete type B3 are visualized in Figure 2. As expected, the kT values of the samples with "good" curing, shown in the first row of Figure 2, are in a good to middle surface condition class, while the values of the samples with "bad" curing, represented in the last row, are in the bad surface condition range, i.e., have a one to two orders of magnitude higher gas permeability. The samples that were not subjected to any curing treatment (middle row in Figure 2) show a middle to good surface condition, but on average, they have a slightly higher permeability than the samples with "good" curing. It can be confirmed that the duration of the curing treatment has an influence on the gas permeability. The sample with the longest treatment (7 days) of the "well"-treated samples had the lowest kT values. It follows that the protection of a sample with an adequate duration has a positive effect on the quality of the concrete. The top surfaces, marked as squares in Figure 2, tend to have a higher gas permeability, as the surface dries out faster than the side faces (round symbols) protected with the formwork and is therefore more porous. A settlement effect can also contribute to the higher permeability of the top surface as cast.

The Proceq permeability tester (blue symbols) was only used on samples with 7 days of curing. By comparing the results of the two gas permeability devices used, it can be seen that the kT values of the Proceq device are generally higher, which was a difference that was detected also by other researchers [47].

In parallel to the gas permeability measurements, the moisture content (%) of the samples was recorded according to Table 4. As listed in the third column, the average moisture decreases with increasing age of the concrete and with increasing poor curing. While the moisture content is almost the same for all samples after one day of curing, it differs more and more with increasing age in the three types of curing. Samples with "good" curing still have a moisture level of 3.6% after 7 days of curing and 3 days of drying treatment, while poorly cured samples with same curing duration only have a moisture level of 3.2%.

Table 4. Average moisture and compressive strength after 28d tested on 15cm cubes for series B3.

Curing Type	Curing Duration (d)	Average Moisture on Measurement Day (%)	Compressive Strength after 28 d (N/mm^2)
good	1	4.0	41.1
	4	3.8	
	7	3.6	
no	1	4.0	39.3
	4	3.4	
	7	3.3	
bad	1	3.9	36.5
	4	3.1	
	7	3.2	

An influence of the curing can also be demonstrated by determining the compressive strength after 28 days. The last column of Table 4 shows the compressive strengths for all three curing types. "Good" curing samples show the highest compressive strength with 41.1 N/mm^2, while the strength of the "bad" cured samples is the lowest with only 36.5 N/mm^2.

2.3.2. Permeability Properties of Concrete B5

As shown in Figure 3, the measured kT values of the samples of concrete type B5 (w/c = 0.48) are lower for all curing types compared to series B3 (w/c = 0.53). On average, they have a better surface quality, which suggests that the concrete composition has a significant influence on the gas permeability.

With series B5, it can be demonstrated that a "good" curing has a positive effect on the surface quality over the long term, as already could be concluded with series B3. The lowest kT values were measured in samples with "good" curing and a curing period of 4 and 7 days with about 0.01×10^{-16} m^2. Samples with no curing, such as the samples with "good" curing, are in the good surface quality range with slightly higher kT values between 0.01 and 0.1×10^{-16} m^2. The results of the "badly" cured samples are in the middle to bad quality range reaching from 0.2 to 5×10^{-16} m^2 depending on the curing duration and the used device. Especially, the top surfaces (orange squares in Figure 3) of the samples with "bad" curing show that the quality decreases with the duration of the poor treatment.

In this series, too, it can be confirmed that the permeability tester from Proceq SA delivers higher kT values than the PermeaTORR device.

The moisture contents of the concrete type B5 are listed in Table 5 for the three curing types with different curing durations. They partly show a different effect than the samples of the B3 series. Samples with "good" or no curing show higher moisture content with increasing treatment time. "Bad" curing leads to drier samples with longer lasting curing duration. The highest humidity with 4.3% could be measured on "well" cured samples with 7 days of curing treatment followed by 3 days of drying. The lowest moisture content was found after 7 days of "bad" curing and 3 days of drying, with 3.4%.

Table 5. Average moisture and compressive strength after 28 d tested on 15 cm cubes for series B5.

Curing Type	Curing Duration (d)	Average Moisture on Measurement Day (%)	Compressive Strength after 28 d (N/mm^2)
good	1	3.6	49.7
	4	4.1	
	7	4.3	
no	1	3.7	46.7
	4	4.0	
	7	4.1	
bad	1	3.7	38.9
	4	3.4	
	7	3.4	

As with series B3, "good" curing has a positive effect on the 28 days compressive strength. In Table 5, "good" curing samples from series B5 show a compressive strength of 49.7 N/mm^2, while "badly" cured samples only achieve a strength of 38.9 N/mm^2.

2.3.3. Permeability Properties of Concrete BS1C Plus

The results of the gas permeability coefficient measurements for series BS1C Plus are presented in Figure 4. The measured values of this series also clearly confirm the influence of the curing on the kT value and the surface quality.

While the measured values of the samples with "good" and no curing are in the range of a middle to bad surface quality, reaching from 0.1 to 2×10^{-16} m^2, the condition of the "bad" cured samples is correspondingly worse and is in the bad to very bad range between 2 and 12×10^{-16} m^2. As it was already evident in the case of series B3 and B5, the extended "good" curing duration up to day 7 also has a positive effect on the surface quality. As with the other two series, the kT values of the top surfaces of the BS1C Plus samples are higher than those of the side faces.

The samples in this series were examined with the PermeaTorr device. Only the "bad" cured samples with a curing period of 4 and 7 days were measured with the Proceq SA permeability tester. These measurements have the highest permeability values, as shown with the blue symbols.

The average moisture in the BS1C Plus samples is given in Table 6. The values are similar to those of the B5 series, where the 7-day lasting "good" curing results in a moisture level of 4.1%. The moisture of the samples without curing treatment hardly differs due to the curing time, while the "bad" cured samples show the lowest moisture with 2.7% after 7 days of treatment.

Table 6. Average moisture and compressive strength after 28 d tested on 15cm cubes for series BS1C Plus.

Curing Type	Curing Duration (d)	Average Moisture on Measurement Day (%)	Compressive Strength after 28 d (N/mm^2)
good	1	3.5	33.7
	4	3.9	
	7	4.1	
no	1	3.6	31.5
	4	3.8	
	7	3.6	
bad	1	3.2	29.1
	4	2.9	
	7	2.7	

The samples of the concrete type BS1C Plus achieve a compressive strength after 28 days of 33.7 N/mm^2 at "good" curing conditions and 29.1 N/mm^2 at "bad" curing conditions (Table 6). It can be concluded that also with this series, the curing type affects the compressive strength of the samples.

2.3.4. Summary of the Laboratory Findings

In all three series, an effect of the curing quality on the gas permeability could be demonstrated. Samples with "good" curing generally had a lower kT value than samples with "no" or "poor" curing. As shown in Figure 5, there is a correlation between the permeability coefficient kT (geometric mean) and the compressive strength (arithmetic mean). Considering the kT values of the three concrete series as a function of the curing conditions after 7 days of curing, samples with higher gas permeability have a lower compressive strength. Therefore, better curing results in a low permeability coefficient and increases the compressive strength of concrete. On average, the BS1C Plus series achieved the highest gas permeability and therefore the worst surface condition and at the same time had the lowest compressive strength. This may be due to the higher content of additions, the hydraulic contribution of which is not very significant in the first 7 days, even under "good" curing conditions. The samples of series B5 had the highest compressive strength compared to B3 and BS1C, had lower kT values and thus achieved a better surface condition on average. The B3 series results were in the middle with regard to compressive strength and gas permeability. Overall, the laboratory tests confirm that the quality of the curing influences both the gas permeability and the compressive strength of concrete and that the type of concrete plays an important role in the evaluation of the curing quality.

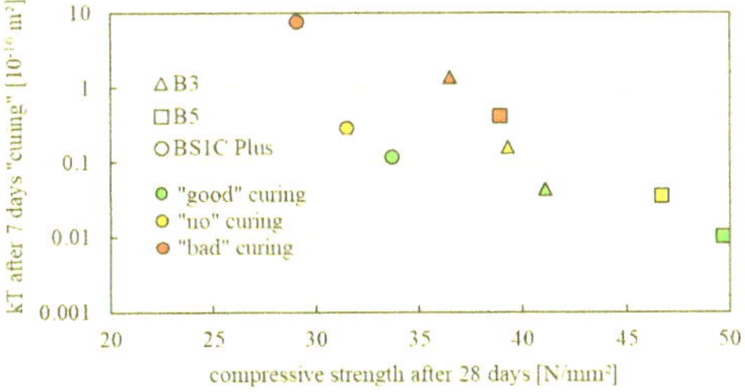

Figure 5. Relation between kT (10^{-16} m^2) after 7 days "good", "no", "bad" curing and the compressive strength after 28 days of B3, B5 and BS1C Plus test samples.

3. On Site Test Campaign

Under controlled laboratory conditions, clear differences in the gas permeability of samples with varying curing treatments could be determined. In construction site conditions, components are exposed to a wide variety of environmental factors, such as solar radiation, temperature fluctuations, rain, wind, etc., which also strongly influence the development of the pore structure of concrete during the early hydration phase and therefore the permeability. The suitability of the gas permeability measurement for on-site evaluation of the curing quality was investigated by further tests directly on concrete structures. Concrete walls of two objects were examined at two different construction sites at different times of the year (summer and late autumn) in order to check the influence of the environmental conditions and the concrete composition on the measurement results.

3.1. Object 1—Measured in Summer

The first on-site investigations were carried out during dry and warm weather conditions on a concrete wall at a construction site of a railway passage of the Austrian Federal Railways in Obersiebenbrunn (48°14′43.6″ N 16°42′14.1″ E). The wall was executed with the Austrian type of concrete BS1C, which was composed according to Table 7.

Table 7. Material properties of the examined concrete type of object 1.

(Austrian) Concrete Name	BS1C
Strength class	C25/30
Cement (kg/m^3):	275 (CEM I 42.5N SR0 WT27 C_3A-free)
Additions (kg/m^3): additions type 2 according to EN 206	80
Total water (kg/m^3)	165
w/c-ratio	0.49
Aggregates (kg/m^3): Gravel 16/32 round grain Gravel 4/16 round grain Sand 0/4 round grain	 442 619 705
Air-entraining agent (kg/m^3): WTB Air 102	1.53
Superplasticizer (kg/m^3): WTB Plast 100/3	2.17
Air content (fresh concrete) (%)	5.8
Bulk density (fresh concrete) (kg/m^3)	2290
Fresh concrete temperature (°C)	20

Different curing areas were prepared by dividing the west-facing side of the 4 × 11 m large concrete wall into three parts, each of which was given a different curing treatment by the construction company. A distinction was made between "good" (curing A), "moderate" (curing B) and "bad" curing (curing C). The area with "bad" curing was left in the formwork for one day with no further treatment afterwards. The formwork of the "moderately" cured area was removed after four days, and the "good" area was left in formwork for seven days. Both the "good" and the "moderate" areas were encased in fleece for three days after removing the formwork and were sprinkled with water in the mornings. The outside temperature during the curing period was on average 22.5 °C, and the air humidity was 67.3%. The gas permeability tests on the concrete wall were carried out 13 days after concreting.

As pictured in Figure 6a, the side of the wall to be examined was close to a sheet pile wall, which is why the lower area of the concrete wall was better protected from environmental influences such as solar radiation and wind than the upper area. In order to take the more humid conditions in the lower area of the wall into account, the measurements were carried out both in the upper and in the lower part and were evaluated separately. A sketch of the division of the areas can be seen in Figure 6b.

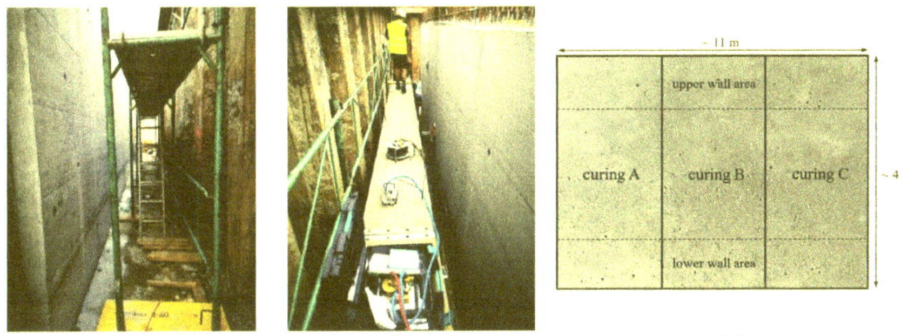

(a) (b) (c)

Figure 6. (**a**) Lower and (**b**) upper area of the concrete wall on the construction site, (**c**) division of the measured wall areas according to curing type.

As performed in the laboratory, the measurements were carried out with both the PermeaTorr and the Proceq devices. The results of the two devices are shown separately in Figure 7. The kT values in Figure 7 were given separately for curing A ("good" curing area), curing B ("moderate" curing area) and curing C ("bad" curing area) and for the lower and upper wall area. Each given kT value represents the geometric mean value of the results of three measurements taken per area. The graphic shows that the surface condition with curing A ("good") is generally in the good to middle range, while curing B has a medium quality and curing C is in the middle to bad surface condition class. It is noticeable that the Proceq device measures a significantly higher gas permeability than the PermaTorr device, especially in the upper wall area. This effect could already be observed with the laboratory samples. The significantly poorer values in the upper wall area can be explained by the more intensive exposure of this area to wind and solar radiation and therefore more severe drying of the concrete surface. Since there was a more humid and shadier environment in the lower wall area, lower kT values were measured. This effect is confirmed by the moisture meter results. The humidity in the upper area averaged 3.6%, while an average of 4.1% was measured in the lower area.

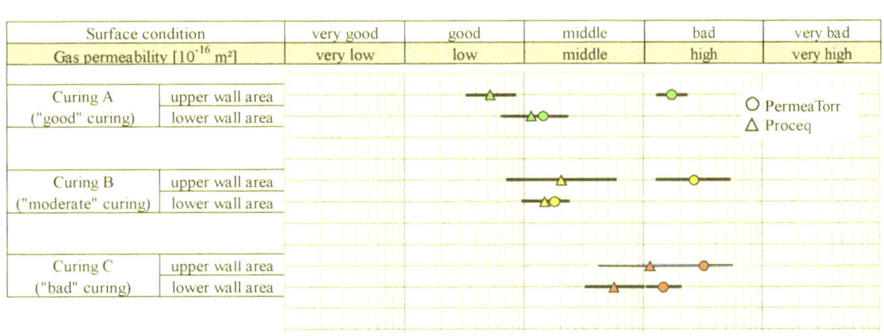

Figure 7. Gas permeability kT (10^{-16} m^2) of the BS1C concrete wall depending on the curing quality.

The results of the construction site tests show that the quality of the curing can be measured under natural environmental influences by means of gas permeability. It can also be seen that exposure and environmental conditions have an impact on the development of concrete quality.

3.2. Object 2—Measured in Late Autumn

Another series of tests was carried out during cooler weather on the concrete walls of a sedimentation basin at an Asfinag (Austrian highway operator) construction site near Göttlesbrunn (48°03′55.3″ N 16°41′46.0″ E). The concrete walls were cast with Austrian concrete type B7 as described in Table 8.

Table 8. Material properties of the examined concrete type of object 2.

(Austrian) Concrete Name	**B7**
Strength class	C25/30
Cement (kg/m^3):	330
Additions (kg/m^3): additions type 2 according to EN 206	80
Total water (kg/m^3)	178
w/c-ratio	0.43
Aggregates (kg/m^3):	1669
Air content (fresh concrete) (%)	6.0
Bulk density (fresh concrete) (kg/m^3)	2251
Fresh concrete temperature (°C)	30.4

Same as for construction site object 1, three areas with varying curing treatment were prepared. The curing for these tests was carried out on differently oriented areas. According to Figure 8, curing A ("good" curing) was carried out on the southeast side of a wall. Curing A means that the concrete was left in the formwork for 8 days. Curing B was on the opposite side, facing northeast. This area B was left in the formwork for 5 days and is referred to as "moderate" curing. Finally, a poor curing treatment (curing C) also took place on a wall facing northeast. This area was left in formwork for just three days. No further curing treatment was carried out in any of the three areas after removing the formwork.

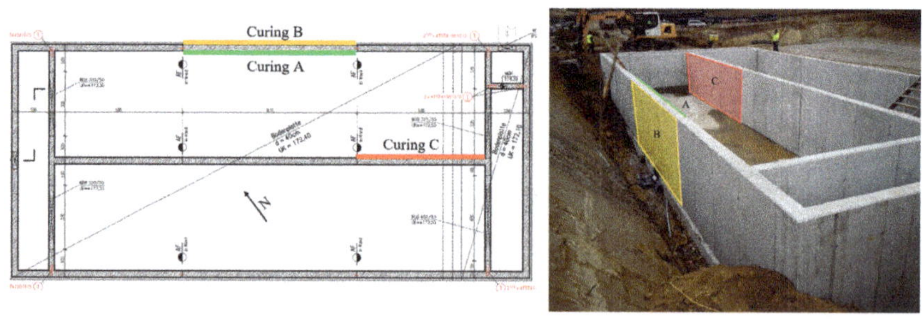

Figure 8. Orientation of the measured concrete walls depending on their curing treatments A, B and C.

The measurements took place when the concrete was 15 days old. The average outside temperature from concreting to the day of measurement was 6.5 °C. The humidity was relatively high at 84%.

Due to the prevailing damp weather, the moisture content of the structure was also measured. With 2.5%, the moisture content was below the maximum recommended value for gas permeability measurements. The low moisture content in the concrete was attributed to the strong wind conditions.

Three measurements were made per curing area with the Proceq permeability tester. The results in Figure 9 show the geometric mean kT values in 10^{-16} m^2 with $\pm$ the standard deviations of the logarithms for each area. The tendency that the surface quality is impaired with increasingly poor curing is also confirmed in this series of tests. The values in the "good" curing area A show a very good to good surface condition, while the poorly treated wall area C only achieves middle quality. A good surface condition could be demonstrated through "moderate" curing. The results of "good" and "moderate" curing are very close to each other. It can be assumed that the favorable location of the area of curing B had a positive effect on hydration due to the adjacent moist soil, while curing A is more exposed and is therefore less protected from wind and consequently less protected from drying out. The stronger solar radiation on the southwest side in area A could be considered as a further aspect, which would also promote dehydration. However, due to the cloudy circumstances, solar radiation plays only a minor role in the period from concreting to the day of measurement. The variability of the kT data, for the three curings (especially for curings A and C) was extremely high, which was possibly due to the young age and exposure conditions prone to the development of microcracks; this experience indicates that taking at least six readings on each test area, as defined in the Swiss standard SIA 262/1: 2019 [24], is recommendable.

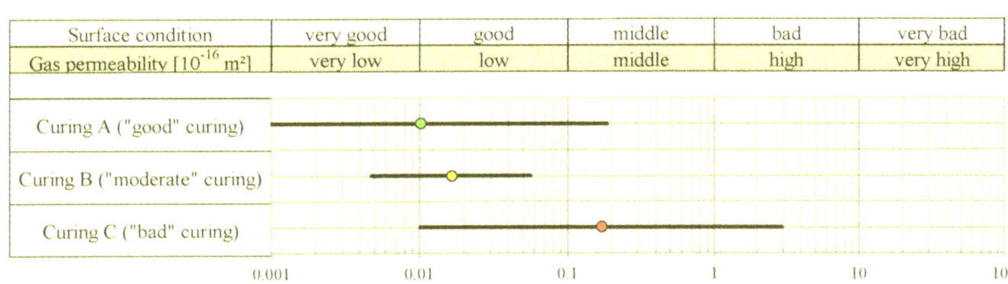

Figure 9. Gas permeability kT (10^{-16} m^2) of the B7 concrete wall depending on the curing quality.

4. Discussion of the Main Findings

The laboratory tests show that the influence of the curing quality on concrete can be detected by determining the gas permeability. Concrete samples of the same composition have different properties due to different curing treatments. Sufficiently cured concrete has a lower gas permeability than untreated or insufficiently cured concrete. Furthermore, the compressive strength is reduced for laboratory samples with increasingly poor curing treatment. In addition to the curing quality, other factors, such as the composition of the concrete, are also decisive for the quality of the concrete surface. Therefore, the results of the examined types of concrete differ from each other when the curing treatment was carried out in the same way. While the concrete type BS1C Plus generally had higher kT values in the laboratory tests, the results for type B5 were significantly lower. Nevertheless, the trend of decreasing gas permeability with higher curing quality is reflected within all examined series.

It is not possible to set a general kT value as a threshold value for adequate curing. Rather, test results already collected for different types of concrete represent guide values for further measurements.

Different curing quality could also be clearly identified on the construction site. The same effect was found there as with the laboratory samples. Longer or better curing treatment had a positive effect on the concrete quality and thus lowered the gas permeability. However, the results of the construction site tests cannot be directly compared with the laboratory results, as the climatic conditions, the type of curing and the concrete composition differed.

When evaluating the quality of curing on a structure, numerous other factors are responsible for the surface quality of concrete in addition to the composition of the concrete. The gas permeability of structures on site was not only influenced by the concrete composition and curing quality (achieved in the laboratory due to variations in ambient temperature, humidity, and protection against drying out) but also by exposure to wind, solar radiation and precipitation. Concrete surfaces exposed to sunlight and wind showed a higher gas permeability than surfaces that were protected against these influences. Environmental influences can have both positive and negative effects on the hydration of the concrete. Their effect is reflected in the results of the gas permeability.

In addition, when determining the gas permeability, the age of the concrete must be taken into account, which according to the Swiss standard [24] has at least to be 28 days. The measurements of this study were already carried out approx. 14 days after concreting. A difference in the quality of the concrete due to the curing treatment was nevertheless measurable. However, the measured values should not be used as a reference for older or newer structures.

Since a single gas permeability measurement only captures a small sub-area, several measurements need to be carried out for the overall assessment of a component in order to achieve reliable results. The measurements should not be carried out before the concrete is 14 days old, as young concrete still has a very high level of moisture, which may falsify the results. In addition, the humidity of the investigated component should be determined before each permeability measurement so that the recommended limit humidity of 5.5% is not exceeded. If the limit of 5.5% is exceeded, it is advisable to repeat the measurements later once the concrete has dried sufficiently.

As already mentioned, no general recommendation for a limit gas permeability value for sufficiently cured concrete can be given. A possibility for evaluation of the results is the use of the permeability classes approach from [29] as guide values. A possible solution could be to establish reference areas where good curing conditions have been consistently applied, taking the resulting kT values as control values, not to be exceeded.

5. Conclusions

In the research work presented, different curing treatments were compared under otherwise almost identical conditions. It was found that the type and duration of curing has a strong influence on the gas permeability and, subsequently, on the durability of the structure. Thus, the quality of the curing of concrete structures can be measured indirectly by determining the gas permeability.

This method is characterized by the simple and non-destructive implementation of measurements both in the laboratory and on the construction site. The evaluation of the results is quick and uncomplicated. Recommendations for the representation of the permeability values and the assessment of the surface quality are already available.

In summary, the following main statements could be derived from the investigations carried out:

- Investigations in the laboratory have shown that the gas permeability measurement method is a promising and reliable method for evaluating the curing quality of concrete.
- The comparison of a "good" with "no" or a "poor" curing treatment is reflected in the permeability coefficient. In addition to the curing conditions, the gas permeability also depends on the composition of the concrete, which is why no general limit values can be specified for the curing quality evaluation.
- The higher permeability due to inadequate curing is accompanied by lower compressive strength than for optimal curing.
- The construction site tests show that the gas permeability measured directly on the structure also provides reliable information on the curing quality of concrete.

- Environmental influences play a major role in on-site measurements, since the concrete quality can be influenced both positively and negatively by factors such as wind, temperature, and humidity. For this reason, these factors must be considered when selecting the measurement points, the time of measurement and when evaluating the curing quality.
- For site testing, it is recommended to perform at least six individual measurements of kT per testing area, as set in the Swiss standard SIA 262/1:2019 [24].

As the measured permeability coefficient not only depends on curing conditions, but also on the type of concrete and other influencing factors, a direct measurement of the curing quality is not possible. One possibility would be to set a limit value in the planning phase as a minimum requirement for achieving the desired concrete quality. This limit value could be determined on the basis of reference measurements on similar concrete components with the same compositions. To implement this approach, however, further comprehensive investigations would be necessary in order to obtain a sufficient sample size for a reliable determination of these limit values.

In a follow-up research project, further investigations are currently being carried out on concrete structures with different properties and under varying climatic conditions with gas permeability measurements to evaluate the curing quality. At the same time, reference values are collected from test samples with the same formulation and optimal curing conditions for evaluating the construction site results. This promising study still requires extensive research to determine the reliability of this type of assessment.

Author Contributions: Writing—original draft, L.P.; Writing—review & editing, A.S., C.B., M.P. and R.T. All authors have read and agreed to the published version of the manuscript.

Funding: This research was supported by Bundesminsterium für Klimaschutz, Umwelt, Energie, Mobiliät, Innovation und Technologie, ÖBB, ASFINAG and the Austrian Research Promotion Agency (FFG).

Institutional Review Board Statement: Not applicable.

Informed Consent Statement: Not applicable.

Acknowledgments: This work was carried out during VIF OptiNB activities. The authors would like to acknowledge Bundesminsterium für Klimaschutz, Umwelt, Energie, Mobiliät, Innovation und Technologie, ÖBB and ASFINAG als well as the Austrian Research Promotion Agency (FFG) for supporting this project.

Conflicts of Interest: The authors declare no conflict of interest.

References

1. Stark, J.; Wicht, B. *Dauerhaftigkeit von Beton*; Springer Vieweg: Berlin/Heidelberg, Germany, 2013.
2. Neville, A.M. *Neville on Concrete—An Examination of Issues in Concrete Practice*; American Concrete Institute: Farmington Hills, MI, USA, 2003.
3. Alexander, M.; Bentur, A.; Mindess, S. *Durability of Concrete: Design and Constructio*; CRC Press: Boca Raton, FL, USA, 2017.
4. Verein Deutscher Zementwerke. *Zement-Taschenbuch*, 51st ed.; Verlag Bau + Technik: Düsseldorf, Germany, 2008.
5. *ÖNORM B4710-1*; Beton—Festlegung, Eigenschaften, Herstellung, Verwendung und Konformität Teil 1. Austrian Standards Institute: Vienna, Austria, 2018.
6. *ACI PRC-308-16*; Guide to External Curing of Concrete. American Concrete Institute: Farmington Hills, MI, USA, 2016.
7. Senbetta, E.; Scholer, C.F. A New Approach for Testing Concrete Curing Efficiency. *ACI J. Proc.* **1984**, *81*, 82–86.
8. *ASTM C803/C803M-18*; Standard Test Method for Penetration Resistance of Hardened Concrete. ASTM International: West Conshohocken, PA, USA, 2019.
9. *ASTM C1585-13*; Standard Test Method for Measurement of Rate of Absorption of Water by Hydraulic-Cement Concretes. ASTM International: West Conshohocken, PA, USA, 2020.
10. Wang, L.C.; Cheng, B.J.; Bao, J.W. Experimental Study on the Influence of Curing Conditions on Capillary Absorption of Concrete. In Proceedings of the 5th International Conference on the Durability of Concrete Structure, Shenzhen, China, 30 June–1 July 2016; Purdue University Press: Shenzhen, China, 2016; pp. 51–56. [CrossRef]
11. Lim, Y.Y.; Smith, S.T.; Padilla, R.V.; Soh, C.K. Monitoring of concrete curing using the electromechanical impedance technique: Review and path forward. *Struct. Health Monit.* **2021**, *20*, 604–636. [CrossRef]

12. Wedatalla, A.M.O.; Jia, Y.; Ahmed, A.A.M. Curing Effects on High-Strength Concrete Properties. *Adv. Civ. Eng.* **2019**, *2019*, 1683292. [CrossRef]
13. Siddiqui, M.S.; Nyberg, W.; Smith, W.; Blackwell, B.; Riding, K.A. Effect of Curing Water Availability and Composition on Cement Hydration. *ACI Mater. J.* **2013**, *110*, 315–322. [CrossRef]
14. Al-Assadi, G.; Casati, M.J.; Fernández, J.; Galvez, J.C. Effect of the curing conditions of concrete on the behavior under freeze-thaw cycles. *Fatigue Fract. Eng. Mater. Struct.* **2010**, *34*, 461–469. [CrossRef]
15. Trąmpczyński, W.; Goszczyńska, B.; Bacharz, M. Acoustic Emission for Determining Early Age Concrete Damage as an Important Indicator of Concrete Quality/Condition before Loading. *Materials* **2020**, *13*, 3523. [CrossRef] [PubMed]
16. Bień, J.; Sadowski, Ł.; Hoła, J.; Schabowicz, K. Non-destructive and semi-destructive diagnostics of concrete structures in assessment of their durability. *Bull. Pol. Acad. Sci. Tech. Sci.* **2015**, *63*, 87–96. [CrossRef]
17. Barluenga, G.; Puentes, J.; Palomar, I.; Guardia, C. Methodology for monitoring Cement Based Materials at Early Age combining NDT techniques. *Constr. Build. Mater.* **2018**, *193*, 373–383. [CrossRef]
18. Fib. *Fib Model Code for Concrete Structures 2010*; Wilhelm Ernst & Sohn: Berlin, Germany, 2013.
19. Springenschmid, R. *Betontechnologie für die Praxis*; Beuth Verlag GmbH: Berlin, Germany, 2018.
20. Wasserman, R.; Bentur, A. Efficiency of curing technologies: Strength and durability. *Mater. Struct.* **2013**, *46*, 1833–1842. [CrossRef]
21. Pickhardt, R.; Schäfer, W. *Nachbehandlung und Schutz des jungen Betons*; Verein Deutscher Zementwerke e.V.: Düsseldorf, Germany, 2014.
22. Taylor, P.C. *Curing Concrete*; CRC Press: Boca Raton, FL, USA, 2013.
23. Beitzel, H.; Beitzel, M. Herstellung, Verarbeitung und Nachbehandlung von Beton. In *Beton Kalender 2017*; John Wiley & Sons, Ltd.: Hoboken, NJ, USA, 2016; pp. 163–233.
24. *Schweizer Norm SIA 262/1:2019*; Betonbau—Ergänzende Festlegungen, Anhang E. SIA Zurich: Zürich, Switzerland, 2019.
25. Basheer, L.; Kropp, J.; Cleland, D.J. Assessment of the durability of concrete from its permeation properties: A review. *Constr. Build. Mater.* **2001**, *15*, 93–103. [CrossRef]
26. Imamoto, K.; Neves, R.; Torrent, R. Carbonation rate in old structures assessed with air-permeability site NDT. In *IABMAS 2016*; Paper 426; CRC Press: Boca Raton, FL, USA, 2016.
27. Hilsdorf, H.K.; Schönlin, K.F.; Burieke, F. *Dauerhaftigkeit von Betonen.*; IRB-Verlag: Darmstadt, Germany, 1995.
28. Neves, R.; da Fonseca, B.S.; Branco, F.; de Brito, J.; Castela, A.; Montemor, M.F. Assessing concrete carbonation resistance through air permeability measurements. *Constr. Build. Mater.* **2015**, *82*, 304–309. [CrossRef]
29. Torrent, R.; Frenzer, G. A method for the rapid determination of the coefficient of permeability of the "covercrete". In Proceedings of the International Symposium Non-Destructive Testing in Civil Engineering (NDT-CE), Berlin, Germany, 26–28 September 1995; pp. 985–992.
30. Hilsdorf, H.; Kropp, J. Performance Criteria for Concrete Durability. In *RILEM Report 12*; E & FN SPON: London, UK, 1995.
31. Mu, S.; Wu, Y.; Jiang, Q.; Shi, L. Study on Transient Air Permeability of Concrete under Different Curing Conditions. *Key Eng. Mater.* **2015**, *629–630*, 223–228. [CrossRef]
32. Kawaai, K.; Ujike, I.; Kunikata, S.; Okazaki, S. Effect of curing condition on air permeability coefficient measured by in-situ test method. In *Life-Cycle of Structural Systems: Design, Assessment, Maintenance and Management, Proceedings of the 4th International Symposium on Life-Cycle Civil Engineering, Tokyo, Japan, 16–19 November 2014*; Taylor & Francis Group: London, UK; Tokyo, Japan, 2015.
33. Jacobs, F.; Hunkler, F.; Mühlan, B. *Wirksamkeit und Prüfung der Nachbehandlungsmethoden von Beton (Efficiency and Testing of Curing Methods of Concrete)*; Eidgenössisches Departement für Umwelt, Verkehr, Energie und Kommunikation UVEK: Bern, Switzerland, 2014.
34. Ngo, V.T.; Hosoda, A.; Komatsu, S.; Ikawa, N. Effect of moisture content on surface water absorption test and air permeability test. In Proceedings of the JCI Annual Convention, Sapporo, Japan, 10–12 July 2019.
35. Neville, A.M. *Properties of Concrete*, 5th ed.; Pearson Education Limited: London, UK, 2012.
36. Torrent, R.; Szychowski, J. Innovation in Air-Permeability NDT: Concept and Performance. In Proceedings of the XIV DBMC, Ghent, Belgium, 29–31 May 2017.
37. Torrent, R. A two-chamber vacuum cell for measuring the coefficient of permeability to air of the concrete cover on site. *Mater. Struct.* **1992**, *25*, 358–365. [CrossRef]
38. Ptacek, L.; Strauss, A.; Bos, C.; Peyerl, M. Optimal Curing of Concrete. In Proceedings of the Fib Symposium 2021, Lisbon, Portugal, 14–16 June 2021.
39. Materials Advanced Services SRL. *User Manual of PermeaTORR AC (Active Cell), v.1.5 (19.11.17)*; Materials Advanced Services SRL: Buenos Aires, Argentina, 2017.
40. Jacobs, F. Permeabilität und Porengefüge Zementgebundener Werkstoffe. Ph.D. Thesis, ETH Zurich, Zürich, Switzerland, 1994.
41. Jacobs, F. Permeability to gas of partially saturated concrete. *Mag. Concr. Res.* **1998**, *50*, 115–121. [CrossRef]
42. *ASTM F2659-10*; Standard Guide for Preliminary Evaluation of Comparative Moisture Condition of Concrete, Gypsum Cement and Other Floor Slabs and Screeds Using a Non-Destructive Electronic Moisture Meter. ASTM International: West Conshohocken, PA, USA, 2015.
43. Bos, C.; Peyerl, M.; Strauss, A.; Ptacek, L.; Boisits, P. *Optimale NachBehandlung für verbesserte Qualität bei der Bauausführung—OptiNB*; Final Report FFG Project nr. 866978; Bundesministerium für Klimaschutz: Vienna, Austria, 2020.
44. MAS—Advanced Materials Services. Product Permea-TORR. Available online: http://www.m-a-s.com.ar/eng/product.php (accessed on 8 January 2022).

45. Proceq, S.A. Corrosion Potential, Concrete Resistivity and Permeability Testers. Available online: https://www.proceq.com/compare/corrosion-potential-concrete-resistivity-and-permeability/ (accessed on 8 January 2022).
46. Tramex. TRAMEX CMEXII. Available online: http://tramexmeters.com/cmex-concrete-moisture-encounter-expert (accessed on 8 January 2022).
47. Torrent, R. Non-destructive air-permeability measurement: From gas-flow modelling to improved testing. In Proceedings of the 2nd International Conference on Microstructural-Related Durability of Cementitious Composites, Nanjing, China, 13–15 October 2008; RILEM Publications SARL: Marne la Vallée, France, 2013; pp. 467–476.

sensors

MDPI

Article

Lamb Waves Propagation Characteristics in Functionally Graded Sandwich Plates

Jie Gao [1], Jianbo Zhang [2], Yan Lyu [1,*], Guorong Song [1] and Cunfu He [1]

[1] Faculty of Materials and Manufacturing, Beijing University of Technology, Beijing 100124, China; jiegao@emails.bjut.edu.cn (J.G.); grsong@bjut.edu.cn (G.S.); hecunfu@bjut.edu.cn (C.H.)
[2] The 3th Research Institute of China Electronics Technology Group Corporation, Beijing 100015, China; zhangjianbo@emails.bjut.edu.cn
* Correspondence: lvyan@bjut.edu.cn

Abstract: Functionally graded materials (FGM) have received extensive attention in recent years due to their excellent mechanical properties. In this research, the theoretical process of calculating the propagation characteristics of Lamb waves in FGM sandwich plates is deduced by combining the FGM volume fraction curve and Legendre polynomial series expansion method. In this proposed method, the FGM plate does not have to be sliced into multiple layers. Numerical results are given in detail, and the Lamb wave dispersion curves are extracted. For comparison, the Lamb wave dispersion curve of the sliced layer model for the FGM sandwich plate is obtained by the global matrix method. Meanwhile, the FGM sandwich plate was subjected to finite element simulation, also based on the layered-plate model. The acoustic characteristics detection experiment was performed by simulation through a defocusing measurement. Thus, the Lamb wave dispersion curves were obtained by $V(f, z)$ analysis. Finally, the influence of the change in the gradient function on the Lamb wave dispersion curves will be discussed.

Keywords: functionally graded materials; Legendre polynomial series expansion method; Lamb wave dispersion curve; volume fraction curve; finite element simulation

Citation: Gao, J.; Zhang, J.; Lyu, Y.; Song, G.; He, C. Lamb Waves Propagation Characteristics in Functionally Graded Sandwich Plates. *Sensors* **2022**, *22*, 4052. https://doi.org/10.3390/s22114052

Academic Editor: Luca De Marchi

Received: 6 May 2022
Accepted: 22 May 2022
Published: 27 May 2022

1. Introduction

Functionally graded materials (FGM) are based on computer-aided material design, using advanced material compounding technology to make the elements (composition, structure, etc.) of the constituent materials continuously change from one side to the other along the thickness direction. Thus, the properties and functions of the material also vary in gradient. Functional gradient materials of metal-ceramics were proposed and prepared in 1984 [1]. Since the volume content of the FGM components is continuously changed in the spatial position, and there is no sudden change in physical properties, the interlayer stress problem can be avoided and the stress concentration phenomenon can be reduced. At the same time, FGM is a good devisable material, in which one can change the spatial distribution of composition and content of the material by a target function, so as to achieve the purpose of optimizing the internal stress distribution of the structure [2].

The FGM sandwich plate consists of three layers: the top layer, the middle layer and the bottom layer. Generally, FGM sandwich plates are divided into two categories. One is FGM as the top and bottom layers of the sandwich plate, and the homogeneous isotropic materials as the intermediate layer. The other type is FGM as the middle layer of the sandwich plate, and homogeneous isotropic materials as the top and bottom layers. FGM sandwich plates have excellent overall performance, and have been used in optical, biomedical, electromagnetic and mechanical engineering, etc. [3].

The elastic waves in the FGM sandwich plate contain ultrasound guided waves and body waves. Ultrasonic guided waves cover Lamb waves, surface waves, Love waves, etc. Ultrasonic guided waves provide unique capabilities for the structural health monitoring of plate-like structures [4]. However, the guided waves have multi-mode and dispersion characteristics during propagation, and the dispersion appears to be a unique physical property. It mainly indicates that the propagation characteristics of the guided waves are affected by frequency. That is to say, the propagation velocity of a guided wave will change by frequency, which is called dispersion [5]. In addition, most guided wave modes have strong dispersion characteristics. Therefore, studying the relationship between the dispersion curve of FGM sandwich plates and material property parameters is an important part of theoretical research. Zhu et al. [6] used the matrix recursion method to establish the characteristic equations of Lamb waves of multi-layer free plates, and analyzed the dispersion characteristics of double-layer plates and sandwich plates. Wu et al. [7] studied the propagation dispersion characteristics of Lamb waves from single-layer plates to multi-layer FGM plates, and obtained the relationship between the continuous change in material properties and the Lamb wave velocity and displacement. Bruck [8] analyzed the propagation of stress waves in FGM by establishing a one-dimensional FGM model, and transitioned the FGM layered model to a continuously changing gradient model. Chen et al. [9] used a layered plate model to analyze the dispersion characteristics of FGM plates under large frequencies and thick product conditions. In all the above research, the FGMs were divided into many homogeneous or inhomogeneous layers, in order to solve the wave propagation problem. However, the layer number of FGMs plays a vital role in the numerical accuracy of the calculations.

In addition, Lefebvre et al. [10] proposed the Legendre orthogonal polynomial series expansion (LOPSE) method to study the propagation properties of waves in layered-plate structures. Yu et al. [11] further introduced the Legendre series expansion method into the dispersion curve calculation of an anisotropic multilayer piezoelectric material plate with a greater difference in mechanical parameters. Compared with the rotation matrix method, a good calculation result is obtained. Dong et al. [12] studied the SH surface wave in the piezoelectric gradient half space, considering the horizontal shear direction displacement by using Laguerre orthogonal polynomials. Salah et al. [13] proposed a layered model to analyze the Love wave over a half space of an elastic substrate covered by a functionally gradient piezoelectric material plate. As mentioned above, the studies treated the FGM structures as a continuously gradient medium, and they effectively calculated the propagation characteristics of acoustic waves in FGMs without separating them into multilayer plates. However, there are few reports on the numerical simulation of Lamb wave propagation in FGM sandwich structures.

Likewise, the finite element method is a numerical method with both a theoretical basis and practical significance. It was originally used by Zienkiewicz [14] to simulate wave propagation and scattering, but then Finnveden [15] successively used the spectral finite element method to study the periodic waveguide structure and the guided wave in the viscoelastic damped waveguide structure. Cheng et al. [16] studied the propagation of surface acoustic waves excited by lasers in functionally graded materials, and simulated the gradients of various mechanical and thermal parameters in functionally graded materials. Kim and Paulino [17] proposed an isoparametric gradient element model, and applied the shape function of the model to obtain the material properties of the attribute of the element node to the inner difference. Zhang and Xiao [18] applied this method to prove that the finite element model based on isoparametric gradient elements can better reflect the gradient variation in material properties. Wang and Gross [19] proposed a layered model of FGM. The material parameters of each layer change according to a continuous function and are continuous at the interface. Such a layered model achieved good results in the crack analysis of FGM structures. Nevertheless, little research paid attention to the complex multi-mode dispersion characteristics of functionally graded materials, which can provide more abundant information for non-destructive testing and the evaluation of the characteristics of FGM plates.

In this research, we use the Legendre polynomial series expansion method to study the propagation of Lamb waves in functionally graded material sandwich plates, and discuss their convergence problem. The influence of gradient layer parameter changes on Lamb wave dispersion curves will also be given. In addition, the finite element model of FGM sandwich plates was established by PZFlex (Division of Applied Science, Mountain View, United States), and the experimental process of defocusing the measurement of line-focused ultrasonic transducers based on acoustic microscopy, also known as the $V(f, z)$ measurement, was simulated.

2. Theoretical Derivation and Numerical Results

2.1. Modeling

For a functionally graded sandwich plate, as shown in Figure 1, the propagation direction of the Lamb wave is along the x_1 axis. The thickness of the sandwich plate is $h_1 + h_2 + h_3$, in which h_2 is the thickness of the FGM layer, and h_1 and h_3 are the thicknesses of steel and copper, respectively. The material parameters of the FGM layer vary continuously in the thickness direction. Here, we are referring to the density and the elastic constants, which are functions of x_3.

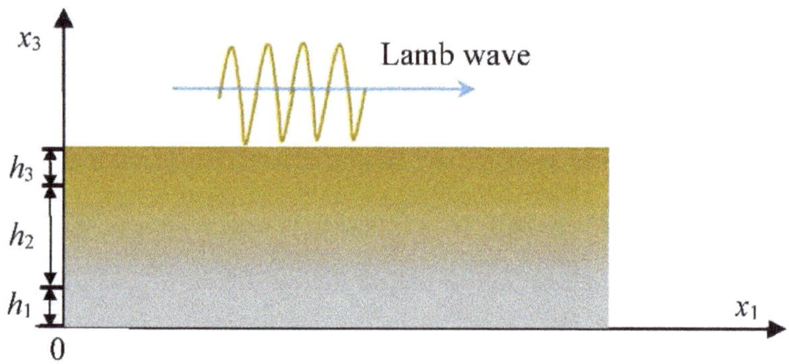

Figure 1. Schematic diagram of Lamb wave propagation and spatial coordinate system in a functionally graded material sandwich panel.

Assuming that the displacement components of the Lamb wave are the following:

$$\begin{aligned} u_1 &= U(x_1, x_3, t) \\ u_2 &= 0 \\ u_3 &= W(x_1, x_3, t) \end{aligned} \tag{1}$$

then the equations of motion will be given as follows:

$$\begin{aligned} \frac{\partial \sigma_{11}}{\partial x_1} + \frac{\partial \sigma_{13}}{\partial x_3} &= \rho \frac{\partial^2 u_1}{\partial t^2} \\ \frac{\partial \sigma_{31}}{\partial x_1} + \frac{\partial \sigma_{33}}{\partial x_3} &= \rho \frac{\partial^2 u_3}{\partial t^2} \end{aligned} \tag{2}$$

Geometric relationship under the assumption of small deformation is as follows:

$$\varepsilon_{ij} = \frac{1}{2}\left(\frac{\partial u_i}{\partial x_j} + \frac{\partial u_j}{\partial x_i}\right) \quad (i, j = 1, 2, 3) \tag{3}$$

Free harmonics of the particle displacement can be written as follows:

$$\begin{aligned} u_1 &= U(x_3) \cdot e^{[i(kx_1 - \omega t)]} \\ u_3 &= W(x_3) \cdot e^{[i(kx_1 - \omega t)]} \end{aligned} \tag{4}$$

where σ_{ij} and ε_{ij} represent stress and strain, respectively, $U(x_3)$ and $W(x_3)$ are the amplitudes of particle vibrations on the x_1 and x_3 direction, k is the wave number, and ω is the angular frequency.

Considering the boundary problem of isotropic plates, the rectangular window function can be introduced by the following:

$$\pi_{h_n}(x_3) = \begin{cases} 1, \ 0 \le x_3 \le h_1 + h_2 + h_3 \\ 0, \qquad\qquad\quad elsewhere \end{cases} \tag{5}$$

The elastic constant and density of the material are expressed as a function of position, as follows:

$$\begin{aligned} C_{ij} &= \sum_{n=1}^{N} C_{ij}^{n} \pi_{h_n}(x_3) \\ \rho &= \sum_{n=1}^{N} \rho^{n} \pi_{h_n}(x_3) \end{aligned} \tag{6}$$

where N is the total number of layers, and here, $N = 3$. Therefore, the elastic constants and density in the sandwich plate can be expressed as follows:

$$\begin{aligned} C_{ij}(x_3) &= C_{ij}^{1} \pi_{h_1}(x_3) + C_{ij}^{2}(x_3) \pi_{h_2}(x_3) + C_{ij}^{3} \pi_{h_3}(x_3) \\ \rho(x_3) &= \rho^{1} \pi_{h_1}(x_3) + \rho^{2}(x_3) \pi_{h_2}(x_3) + \rho^{3} \pi_{h_3}(x_3) \end{aligned} \tag{7}$$

The middle layer of the sandwich plate is the FGM layer; the volume fraction of copper of this layer is represented as V_{Cu}, which can be written by a power function, as follows:

$$V_{\mathrm{Cu}} = \left(1 - \frac{x_3 - h_1}{h_2}\right)^{n} \ (h_1 \le x_3 \le h_1 + h_2, \ 0 \le n \le \infty) \tag{8}$$

where n is the exponent of the power function. The propagation characteristics of Lamb waves in the FGM layer under different gradient distributions can be obtained by changing the power exponent n. Then, in the FGM layer, the relationships between elastic constants/density and volume fraction are as follows:

$$\begin{aligned} C_{IJ}^{\mathrm{steel}} &= C_{IJ}^{\mathrm{steel}} + \left(C_{IJ}^{\mathrm{Cu}} - C_{IJ}^{\mathrm{steel}}\right) V_{\mathrm{Cu}} \\ \rho^{\mathrm{steel}} &= \rho^{\mathrm{steel}} + \left(\rho^{\mathrm{Cu}} - \rho^{\mathrm{steel}}\right) V_{\mathrm{Cu}} \end{aligned} \tag{9}$$

Substituting Equation (8) into Equation (9) yields the functions of the elastic constant and density in the FGM layer, with respect to x_3:

$$\begin{aligned} C_{IJ}^{\mathrm{steel}}(x_3) &= C_{IJ}^{\mathrm{steel}} + \left(C_{IJ}^{\mathrm{Cu}} - C_{IJ}^{\mathrm{steel}}\right)\left(1 - \tfrac{x_3 - h_1}{h_2}\right)^{n} \\ \rho^{\mathrm{steel}}(x_3) &= \rho^{\mathrm{steel}} + \left(\rho^{\mathrm{Cu}} - \rho^{\mathrm{steel}}\right)\left(1 - \tfrac{x_3 - h_1}{h_2}\right)^{n} \end{aligned} \tag{10}$$

Thus, the constitutive relationship is given as follows:

$$\begin{aligned} \sigma_{11} &= \left[C_{11}(x_3)\varepsilon_{11} + C_{13}(x_3)\varepsilon_{33}\right] \cdot \pi(x_3) \\ \sigma_{33} &= \left[C_{13}(x_3)\varepsilon_{11} + C_{33}(x_3)\varepsilon_{33}\right] \cdot \pi(x_3) \\ \sigma_{13} &= 2C_{55}(x_3)\varepsilon_{13} \cdot \pi(x_3) \end{aligned} \tag{11}$$

2.2. Legendre Orthogonal Polynomial Expansion

Substituting Equations (3), (4), (7), (10) and (11) into Equation (2) yields the wave control equation in the x_1–x_3 plane. Then, the wave control equation in the x_1 direction is as follows:

$$
\begin{aligned}
&[C_{11}^{1}\pi_{h_1}(x_3)+C_{11}^{2}(x_3)\pi_{h_2}(x_3)+C_{11}^{3}\pi_{h_3}(x_3)]\cdot i^2k^2\cdot U\\
&+[C_{13}^{1}\pi_{h_1}(x_3)+C_{13}^{2}(x_3)\pi_{h_2}(x_3)+C_{13}^{3}\pi_{h_3}(x_3)]\cdot ik\cdot W'\\
&+C_{55}^{2}{}'(x_3)\pi_{h_2}(x_3)\cdot U'+[C_{55}^{1}\pi_{h_1}(x_3)+C_{55}^{2}(x_3)\pi_{h_2}(x_3)+C_{55}^{3}\pi_{h_3}(x_3)]\cdot U''\\
&+C_{55}^{2}{}'(x_3)\pi_{h_2}(x_3)\cdot ik\cdot W+[C_{55}^{1}\pi_{h_1}(x_3)+C_{55}^{2}(x_3)\pi_{h_2}(x_3)+C_{55}^{3}\pi_{h_3}(x_3)]\cdot ik\cdot W'\\
&+\begin{bmatrix} C_{55}^{1}[\delta(x_3-0)-\delta(x_3-h_1)]+C_{55}^{2}(x_3)[\delta(x_3-h_1)-\delta(x_3-h_1-h_2)]\\ +C_{55}^{3}[\delta(x_3-h_1-h_2)-\delta(x_3-h_1-h_2-h_3)]\end{bmatrix}\cdot(U'+ik\cdot W)\\
&=-[\rho^1\pi_{h_1}(x_3)+\rho^2(x_3)\pi_{h_2}(x_3)+\rho^3\pi_{h_3}(x_3)]\cdot\omega^2\cdot U
\end{aligned}
\tag{12}
$$

The wave control equation in the x_3 direction is as follows:

$$
\begin{aligned}
&[C_{55}^{1}\pi_{h_1}(x_3)+C_{55}^{2}(x_3)\pi_{h_2}(x_3)+C_{55}^{3}\pi_{h_3}(x_3)]\cdot ik\cdot U'\\
&+[C_{55}^{1}\pi_{h_1}(x_3)+C_{55}^{2}(x_3)\pi_{h_2}(x_3)+C_{55}^{3}\pi_{h_3}(x_3)]\cdot i^2k^2\cdot W\\
&+C_{13}^{2}{}'(x_3)\pi_{h_2}(x_3)\cdot ik\cdot U+[C_{13}^{1}\pi_{h_1}(x_3)+C_{13}^{2}(x_3)\pi_{h_2}(x_3)+C_{13}^{3}\pi_{h_3}(x_3)]\cdot ik\cdot U'\\
&+C_{33}^{2}{}'(x_3)\pi_{h_2}(x_3)\cdot W'+[C_{33}^{1}\pi_{h_1}(x_3)+C_{33}^{2}(x_3)\pi_{h_2}(x_3)+C_{33}^{3}\pi_{h_3}(x_3)]\cdot W''\\
&+\begin{bmatrix} C_{13}^{1}[\delta(x_3-0)-\delta(x_3-h_1)]+C_{13}^{2}(x_3)[\delta(x_3-h_1)-\delta(x_3-h_1-h_2)]\\ +C_{13}^{3}[\delta(x_3-h_1-h_2)-\delta(x_3-h_1-h_2-h_3)]\end{bmatrix}\cdot ik\cdot U\\
&+\begin{bmatrix} C_{33}^{1}[\delta(x_3-0)-\delta(x_3-h_1)]+C_{33}^{2}(x_3)[\delta(x_3-h_1)-\delta(x_3-h_1-h_2)]\\ +C_{33}^{3}[\delta(x_3-h_1-h_2)-\delta(x_3-h_1-h_2-h_3)]\end{bmatrix}\cdot W'\\
&=-[\rho^1\pi_{h_1}(x_3)+\rho^2(x_3)\pi_{h_2}(x_3)+\rho^3\pi_{h_3}(x_3)]\cdot\omega^2\cdot W
\end{aligned}
\tag{13}
$$

The amplitudes of $U(x_3)$ and $W(x_3)$ of the displacements are expanded into the form of a summation of the Legendre orthogonal polynomials, which can be written as follows:

$$
\begin{aligned}
U(x_3)&=\sum_{m=0}^{\infty}p_m^1\cdot Q_m(x_3)\\
W(x_3)&=\sum_{m=0}^{\infty}p_m^3\cdot Q_m(x_3)
\end{aligned}
\tag{14}
$$

where (i = 1, 3; m = 1, 2, . . . , M) are the expansion coefficients of $Qm(x_3)$. Theoretically, m takes from zero to infinity, but, in fact, m takes a finite value of M. Higher-order terms can be considered as infinitesimal quantities, and M is the cutoff order of Legendre orthogonal polynomial series. It should be noted that $Qm(x_3)$ is an orthogonally normalized polynomial group, as follows:

$$
Q_m(x_3)=\sqrt{\frac{2m+1}{h_1+h_2+h_3}}P_m\left(\frac{2x_3-h_1-h_2-h_3}{h_1+h_2+h_3}\right)
\tag{15}
$$

Substituting the displacement amplitude Equation (14) into the wave control Equations (12) and (13) will derive the final form of Legendre polynomial expansion equations. Multiply both sides of the expanded equation by $Q_j(x_3)$ and integrate x_3 from zero to $h_1 + h_2 + h_3$. Using the orthogonal properties of the Legendre polynomial, a matrix form of the equations can be given, as follows:

$$
\begin{bmatrix} A_{11}^{j,m} & A_{12}^{j,m}\\ A_{21}^{j,m} & A_{22}^{j,m}\end{bmatrix}\begin{Bmatrix} p_m^1\\ p_m^3\end{Bmatrix}=-\omega^2\begin{bmatrix} M_m^j & 0\\ 0 & M_m^j\end{bmatrix}\begin{Bmatrix} p_m^1\\ p_m^3\end{Bmatrix}
\tag{16}
$$

where $A_{ij}^{j,m}$ and M_m^j can be obtained from the wave control equations after expansion, which are shown in Appendix A. According to the matrix Equation (16), the relationship between the wave number k and the angular frequency ω can be obtained by solving the eigenvalues. That is how the dispersion curves of the Lamb wave in the FGM sandwich plate can be extracted.

2.3. Numerical Results and Discussion

2.3.1. Convergence Analysis of Cutoff Order M

The material selected was a copper–FGM–steel sandwich plate, and the mechanical performance parameters of copper and steel are shown in Table 1. In total, the thickness of the plate is 0.4 mm, in which both the thicknesses of copper and steel are 0.1 mm, and the thickness of the FGM layer is 0.2 mm. According to Equation (8), the volume fraction of copper in the FGM layer along the thickness direction will take the indices $n = 0.2, 0.5, 1, 2, 10$, respectively, as illustrated in Figure 2.

Table 1. Parameters of Cu and steel [20].

Material	C_{11} (GPa)	C_{13} (GPa)	C_{33} (GPa)	C_{55} (GPa)	ρ (kg/m^3)
Cu	154.8	81.5	154.8	36.7	8292
Steel	275.0	113.2	275.0	80.9	7900

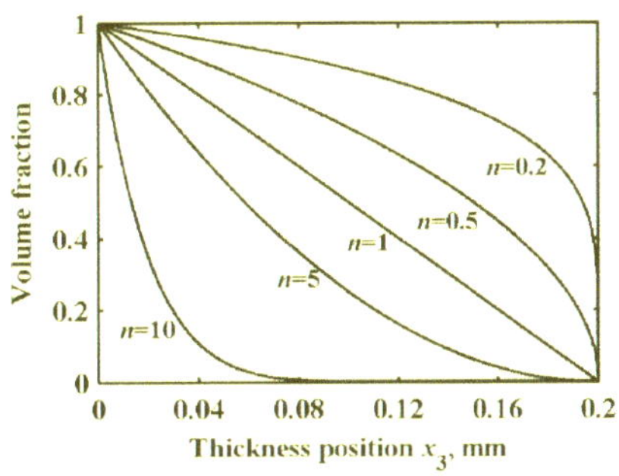

Figure 2. Volume fraction distribution of copper.

From the LOPSE method, it can be concluded that in the process of solving the Lamb wave dispersion curves, once the number of polynomials exceeds a certain threshold, the phase velocity will infinitely approach the eigenvalue. Calculations of the Lamb wave dispersion curves in the frequency range of 0–10 MHz under seven cutoff orders ($M = 3, 4, 5, 6, 7, 8, 9$) are shown in Figure 3a–g, where the volume fraction index is $n = 0.2$. It can be observed that as the cutoff order M increases, the Lamb wave dispersion curve shows a convergence trend, which is consistent with the characteristics of the LOPSE method. This verifies the feasibility of the theoretical method. As can be observed from Figure 3h, when $M = 8$ and $M = 9$, the curves are substantially coincident, so the Lamb wave dispersion curves obtained at $M = 8$ are taken as a convergence solution.

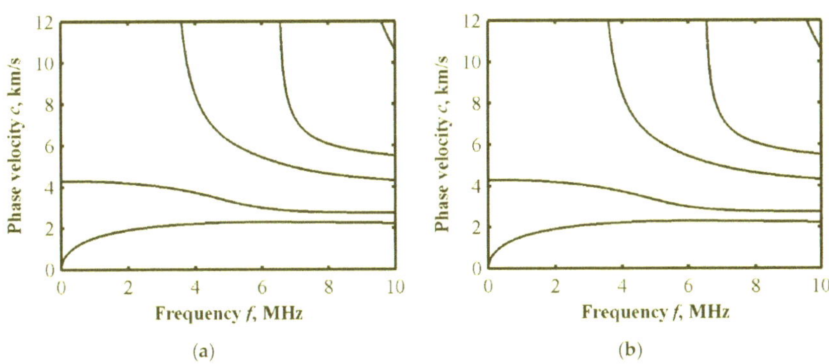

Figure 3. *Cont.*

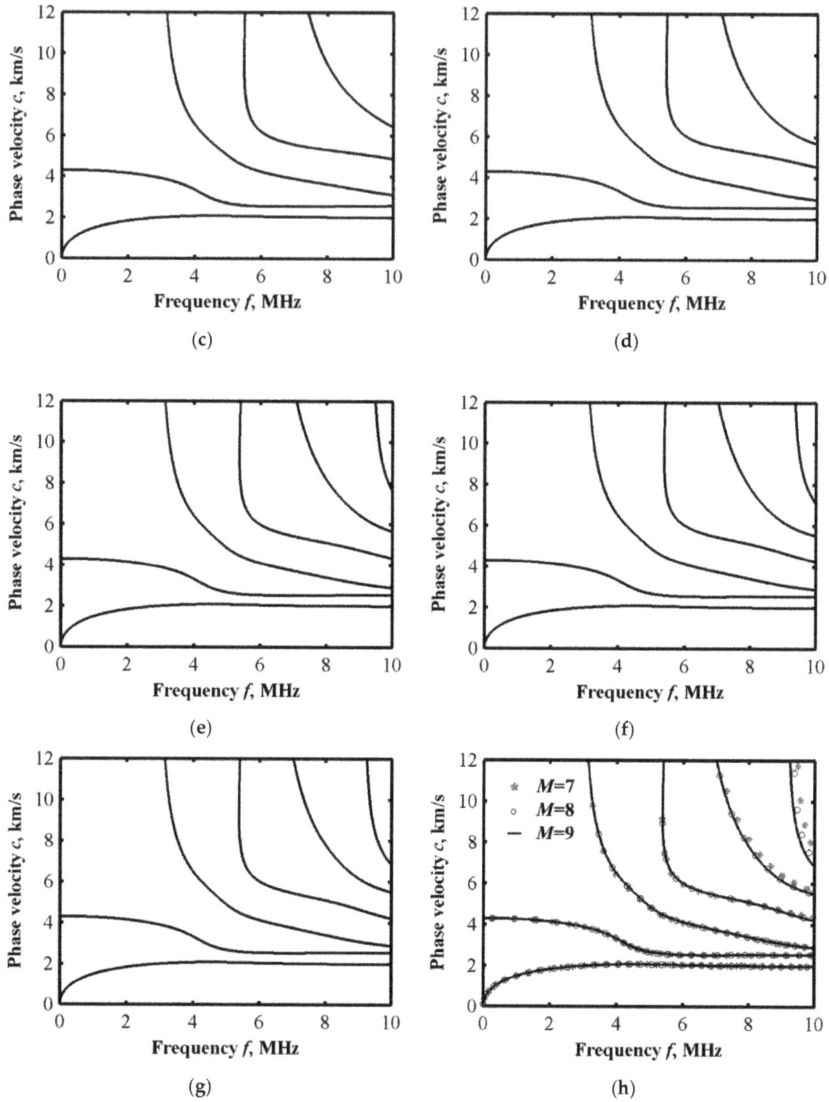

Figure 3. Lamb wave dispersion curves under different cutoff orders: (**a**) $M = 3$; (**b**) $M = 4$; (**c**) $M = 5$; (**d**) $M = 6$; (**e**) $M = 7$; (**f**) $M = 8$; (**g**) $M = 9$; (**h**) $M = 7, 8, 9$ comparison.

Taking the volume fraction curve at $n = 0.2$ as an example, according to Equation (10), the elastic constant C_{IJ} and density ρ of the FGM layer can be sliced into 10 equal minor sub layers. Meanwhile, when $N = 1$, the corresponding material layer is Cu; and when $N = 10$, the corresponding material layer is steel, and the material parameters can be obtained from Table 1. The thickness of each layer is 0.02 mm, and the material parameters vary in the same step. The corresponding parameters of all layers can be obtained from Equation (10), which are shown in Table 2.

Table 2. Parameters of sliced FGM layer.

Layer	C_{11} (GPa)	C_{13} (GPa)	C_{33} (GPa)	C_{55} (GPa)	ρ (kg/m^3)
$N = 1$	154.8	81.5	154.8	36.7	8292.0
$N = 2$	157.6	82.2	157.6	37.7	8282.9
$N = 3$	160.7	83.0	160.7	38.8	8272.8
$N = 4$	164.2	84.0	164.2	40.1	8261.5
$N = 5$	168.1	85.0	168.1	41.6	8248.5
$N = 6$	172.8	86.2	172.8	43.3	8233.3
$N = 7$	178.5	87.7	178.5	45.4	8214.7
$N = 8$	186.0	89.7	186.0	48.2	8190.2
$N = 9$	197.6	92.8	197.6	52.4	8152.6
$N = 10$	275.0	113.2	275.0	80.9	7900.0

The parameters from Table 2 are used to obtain a comparison result from Disperse (Imperial College NDT Laboratory, London, UK), also under $n = 0.2$. Compared with the convergent solution from the Legendre orthogonal polynomial expansion method, the results are shown in Figure 4. It can be observed that the curve is basically consistent in the frequency range of 0–10 MHz, which means that the theoretical solutions of the non-sliced model by Legendre expansion are consistent with the solutions of the sliced model by Disperse. In this case, $M = 8$ will approach closely enough to the results of the global matrix method [21].

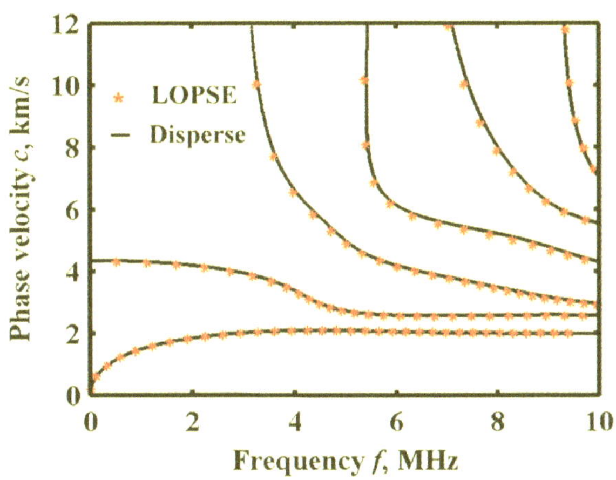

Figure 4. Lamb wave dispersion curve with cutoff order $M = 8$.

2.3.2. Effect of Volume Fraction n on Dispersion Curves

Under different power exponents, the gradient distribution of material parameters in the FGM layer is different, which has a certain influence on the Lamb wave dispersion curves. The dispersion curves of Lamb waves in the sandwich plate under different gradient distributions are calculated, as shown in Figure 5. The cutoff order of the Legendre orthogonal polynomial is also $M = 8$. Figure 5a–e show Lamb wave dispersion curves in five different sandwich plates, with $n = 0.5, 1, 5, 10, 20$, respectively. It can be observed that as the power exponent increases, the phase velocity of $S0$ mode at a low frequency range gradually increases. Meanwhile, the same phenomenon shows up in the higher-order modes. According to Figure 5, when the power index is gradually increased to infinity, the copper content in the FGM layer almost reduces to zero, and the sandwich plates can be regarded as double-layered plates with a top layer of 0.1 mm copper and a bottom layer of 0.3 mm steel. The Lamb wave dispersion curve in the copper–steel double-layered plate calculated by the Legendre orthogonal polynomial method is shown in Figure 5f.

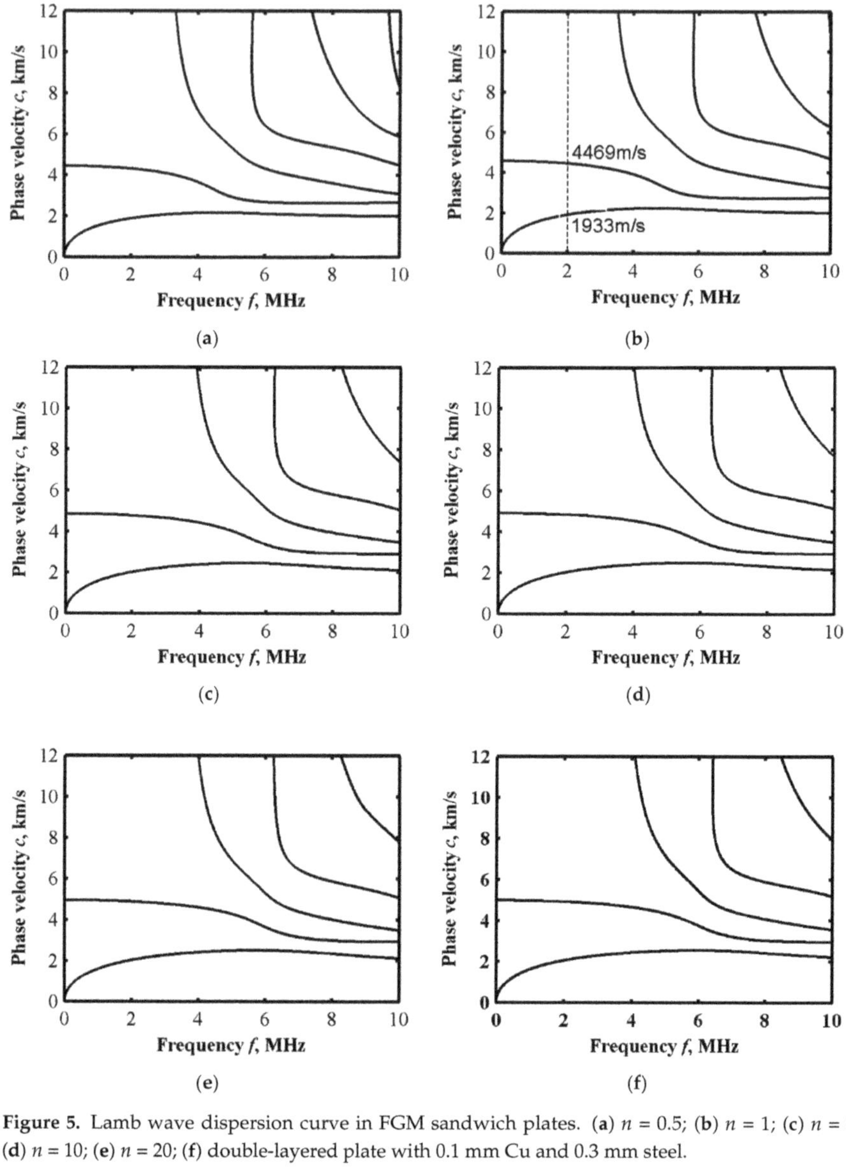

Figure 5. Lamb wave dispersion curve in FGM sandwich plates. (**a**) n = 0.5; (**b**) n = 1; (**c**) n = 5; (**d**) n = 10; (**e**) n = 20; (**f**) double-layered plate with 0.1 mm Cu and 0.3 mm steel.

2.3.3. Displacement and Stress Distribution

The amplitude distribution of displacements and stress components along the thickness direction is the wave structure. According to the calculation result of the dispersion curve at n = 1 in Figure 5b, the eigenvector and its corresponding eigenvalue are calculated. Then, the displacement distribution of the different Lamb wave modes at different frequencies can be obtained. An arbitrary frequency f = 2 MHz is selected, and the Lamb wave velocities corresponding to the A0 (anti-symmetric zero-order mode) and S0 (symmetrical zero-order mode) modes at this frequency are 1933 m/s and 4469 m/s, respectively. The matrix eigenvectors p_m^1 and p_m^3 are inversely obtained by using the angular frequency ω corresponding to the two wave velocities as the eigenvalues. Substituting p_m^1 and p_m^3 into the Equation (16), the displacement and stress distribution in the FGM sandwich plate can be obtained, as shown in Figures 6 and 7. So, the displacement and stress distribution curves corresponding to the arbitrary modes of the Lamb wave at any frequency can be obtained.

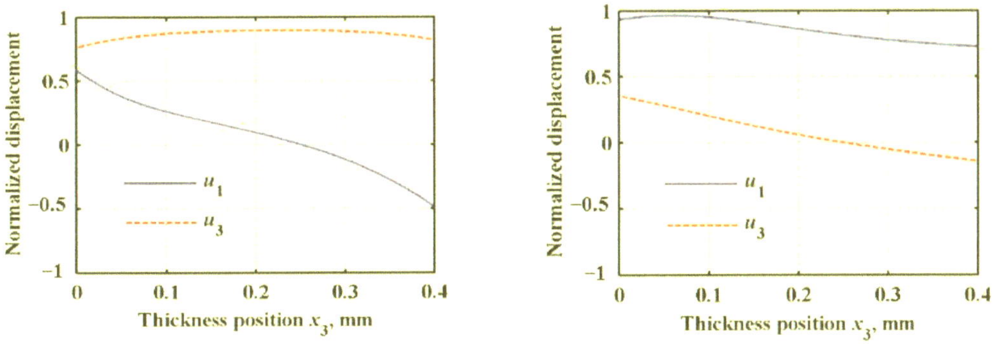

(a) (b)

Figure 6. Lamb wave displacement distribution curve in FGM sandwich plate. (**a**) A_0 mode; (**b**) S_0 mode.

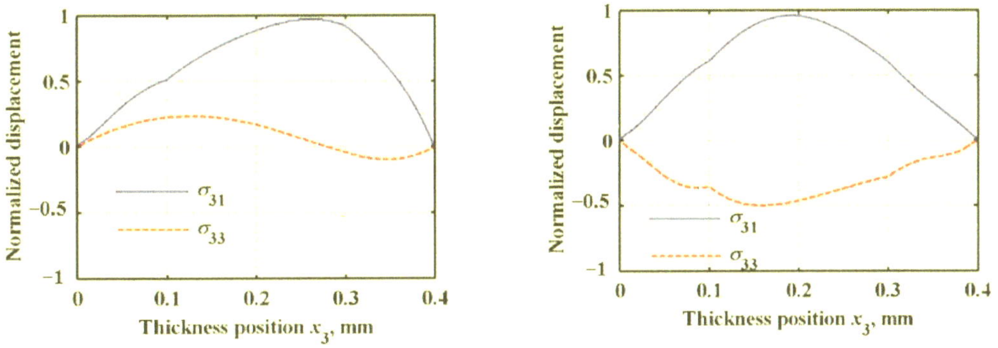

(a) (b)

Figure 7. Lamb wave stress distribution curve in FGM sandwich plate. (**a**) A_0 mode; (**b**) S_0 mode.

It can be observed from Figure 6 that, with the gradual change in the material composition in the FGM sandwich plate along the thickness direction, the LOPSE method can ensure that the displacement variation in the plate is continuous. Additionally, due to the gradual change in the material composition, its displacement distribution no longer has a strict "symmetric" or "asymmetric" distribution, with respect to the center position of the plate. The advantage of the LOPSE method is that the sandwich plate can be calculated globally without delamination, thus solving the problem of stress discontinuity at the boundary. In the calculation, the stress distribution of the Lamb wave can be obtained by simply substituting the obtained displacement solution into the constitutive equation and the geometric equation. As can be observed from Figure 7, the stress components $\sigma 31$ and $\sigma 33$ are continuously distributed in the FGM sandwich plate, and the stress components at the top and bottom boundaries are zero.

3. Finite Element Analysis

3.1. Simulation Model

Based on ultrasonic microscope technology, an acoustic measurement simulation model with an FGM sandwich plate was established, and the corresponding Lamb wave dispersion curve was extracted. For the functionally graded material sandwich panel, the thickness of the sandwich plate is $h_1 + h_2 + h_3$, in which h_2 is the thickness of the FGM layer, and h_1 and h_3 are the thicknesses of steel and copper, respectively. In order to simulate the structural characteristics of nonhomogeneous materials (FGM layer), the corresponding material properties should vary between homogeneous steel and copper. Meanwhile, it is assumed that the material properties of each element layer are constant, and the material properties mesh uniformly along the thickness direction [22,23]. A number of subdivisions can approximate the continuous property variation; the corresponding propagation characteristics of acoustic waves are close to the graded type at this time [24]. On the other hand, when using the commercial finite element package PZFlex to simulate the distribution of sound field in materials, it is very important to assign mechanical property parameters to the corresponding layer of the FGM sandwich plate. In this problem, the uniform element with a thickness of 0.02 mm can solve the numerical simulation of sound field distribution for functionally graded material layers with a thickness of 0.2 mm.

In this section, a two-dimensional finite element model for a line-focusing ultrasound transducer was built in PZFlex. The dimensional parameters and material properties of the finite element model of the line-focusing ultrasonic transducer were referred to with the ultrasonic transducer used in the experiment. In the model, a piezoelectric polymer of polyvinylidene fluoride (PVDF) film was selected as the excitation/receiving element, and the polarization direction is directed to the center of the circle. The upper surface of the film is the positive electrode and the lower surface is the negative electrode. Back10 (tungsten-loaded epoxy, 10% VF, 5.8 Mray1) was used as the backing. Water was selected as the coupling medium for detection, and a copper–FGM–steel sandwich plate was used as the specimen.

Table 3 shows the material parameters of the model. The top layer of the specimen is copper, the middle is layered FGM, and the bottom is steel. The transverse/longitudinal wave velocity and density of copper and steel are known. The material parameters of the FGM layered model are obtained from the volume fraction curve ($n = 0.2$). The parameters of each layer are shown in Table 3. The thickness, focus radius, and full opening angle of PVDF film were set to 40 μm, 20 mm, and 80°, respectively. Then, this finite element model can be simplified to a two-dimensional model, as shown in Figure 8. The signal excited by the line-focusing ultrasonic transducer is a transient wide-band signal. Therefore, the excitation signal in the simulation selects the Sine-Impulse broadband signal with a central frequency of 7 MHz.

Table 3. The material property parameters of the model.

Material	Density ρ (Kg/m^3)	Longitudinal Wave Velocity C_L (m/s)	Transverse Wave Velocity C_T (m/s)
Back10	2975	1960	1047
PVDF	1780	—	—
Water	1000	1496	—
Cu	8292	4321	2103
FGM layer	8282.9	4357.7	2128.5
	8272.8	4398.4	2156.8
	8261.5	4444.0	2188.5
	8248.5	4496.1	2224.7
	8233.3	4557.4	2267.2
	8214.7	4632.5	2319.4
	8190.2	4731.2	2388.0
	8152.6	4882.5	2493.1
Steel	7900	5900	3200

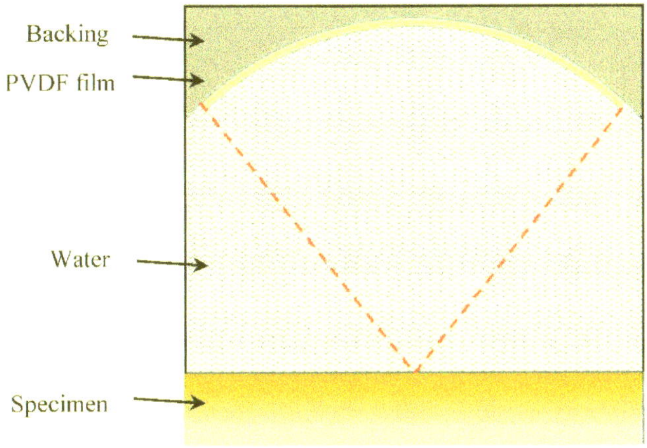

Figure 8. Finite element simulation model.

The simulation started at the focusing plane. Generally, at around 28 μs, the PVDF film receives the reflected echo from the bottom surface of the specimen for the first time. Thus, in this simulation, the propagating times of the acoustic waves were set to 35 μs. The finite element model is discretized by a rectangular grid, and a unit wavelength is divided by 20 grid nodes in water. It should be noted that the bottom surface of the model is set as a free boundary. In order to prevent reflection, the other boundaries of the model are set as absorbing boundaries.

3.2. Simulation Results

By changing the relative position of the ultrasonic transducer to achieve equal interval defocusing, a defocusing measurement simulation based on an ultrasonic microscopy technique was simulated, which is called $V(f, z)$ analysis [25]. The defocus distance was 15 mm and the step was 0.025 mm. The finite element simulation was performed on each defocus position, and, in total, 600 sets of simulation data were obtained. The Lamb wave dispersion curve can be extracted by performing 2D Fourier transform of the time and space domains, as shown in Figure 9.

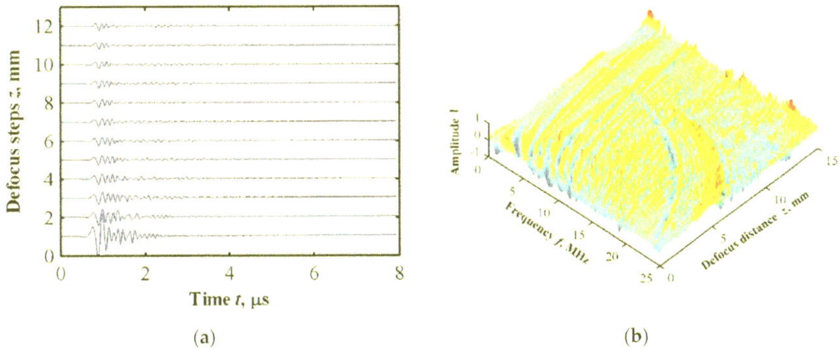

Figure 9. *Cont.*

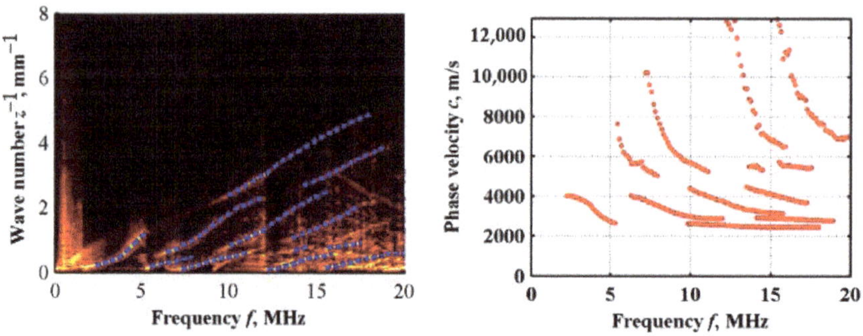

(c) (d)

Figure 9. Defocusing experiment simulation results. (**a**) Time domain signal waveform at different defocusing positions; (**b**) frequency domain diagram after time domain Fourier transform; (**c**) frequency peak tracing after spatial Fourier transform; (**d**) Lamb wave dispersion curves.

The Lamb wave dispersion curves from the simulation were superimposed with the dispersion curves from the LOPSE method, as shown in Figure 10. It can be observed from the figure that the theoretical results solved by the LOPSE method using the volume fraction index are consistent with the finite element simulation results using the layered model. Therefore, this result lays the theoretical foundation for FGM characterization by acoustic microscopy.

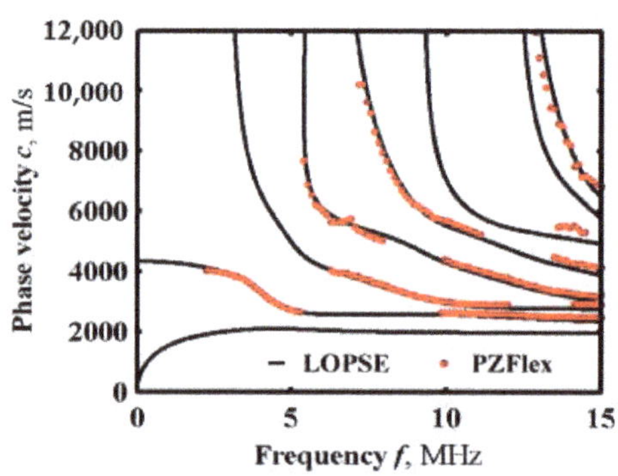

Figure 10. Comparison of simulation results with theoretical results.

4. Conclusions

In this research, the problem of Lamb wave propagation in the FGM sandwich plate without discretizing the gradient structure into a homogeneous multilayered model is solved numerically.

(1) The LOPSE method is employed for solving the Lamb wave dispersion curves and their displacement and stress distributions, even when the material parameters vary continuously along the thickness direction. The convergence of the results by a polynomial method is analyzed, and the convergence solution is also obtained. Moreover, the convergence solution is basically consistent with the results calculated using the global matrix method.

(2) The middle layer of the sandwich plate is FGM, in which the material parameter changes gradiently along the thickness direction. By solving the Lamb wave dispersion curve of the sandwich plate under different gradient distributions, it is obvious that the volume fraction of the top layer material in the FGM layer decreases and the volume fraction of the underneath layer material increases when the power exponent increases, then the dispersion relation of the Lamb wave gradually approaches a double-layer plate.

(3) The finite element model of the FGM sandwich plate is established by slicing the FGM into layers, and the defocus measurement simulation by a line-focusing ultrasonic transducer was carried out based on an acoustic microscopy technique. The extracted Lamb wave dispersion curves are basically consistent with the theoretical calculation results, which further verifies the LOPSE method. Then, this research provides an approach for the FGM characterization method based on acoustic microscopy.

Author Contributions: Conceptualization, Y.L.; methodology, J.G.; software, J.Z.; validation, G.S.; data curation, J.Z.; writing—original draft preparation, J.G.; writing—review and editing, C.H.; visualization, Y.L. and C.H. All authors have read and agreed to the published version of the manuscript.

Funding: This research was funded by National Natural Science Foundation of China (Nos. 12072004, 11872082); Beijing Municipal Education Commission and Beijing Natural Science Foundation (No. KZ202110005005).

Institutional Review Board Statement: Not applicable.

Informed Consent Statement: Not applicable.

Data Availability Statement: Not applicable.

Acknowledgments: The authors thank the support from National Natural Science Foundation of China and Beijing Municipal Education Commission and Beijing Natural Science Foundation.

Conflicts of Interest: The authors declare no conflict of interest.

Appendix A

In the integration process, the common factor term is extracted for the convenience of calculation, as follows:

The common factor of the first layer:

$$\begin{aligned} u_1(j,m,l) &= \int_0^{h_1} Q_j^*(x_3)\cdot\frac{\partial^l Q_m(x_3)}{\partial x_3{}^l}\cdot[H(x_3-0)-H(x_3-h_1)]dx_3 \\ K_1(j,m,l) &= \int_{-\infty}^{\infty} Q_j^*(x_3)\cdot\frac{\partial^l Q_m(x_3)}{\partial x_3{}^l}\cdot[\delta(x_3-0)-\delta(x_3-h_1)]dx_3 \end{aligned} \quad (A1)$$

The common factor of the second layer:

$$\begin{aligned} u_2(j,m,l) &= \int_{h_1}^{h_1+h_2} Q_j^*(x_3)\cdot\frac{\partial^l Q_m(x_3)}{\partial x_3{}^l}\cdot[H(x_3-h_1)-H(x_3-h_1-h_2)]dx_3 \\ K_2(j,m,l) &= \int_{-\infty}^{\infty} Q_j^*(x_3)\cdot\frac{\partial^l Q_m(x_3)}{\partial x_3{}^l}\cdot[\delta(x_3-h_1)-\delta(x_3-h_1-h_2)]dx_3 \end{aligned} \quad (A2)$$

The common factor of the third layer:

$$\begin{aligned} u_3(j,m,l) &= \int_{h_1+h_2}^{h_1+h_2+h_3} Q_j^*(x_3)\cdot\frac{\partial^l Q_m(x_3)}{\partial x_3{}^l}\cdot[H(x_3-h_1-h_2)-H(x_3-h_1-h_2-h_3)]dx_3 \\ K_3(j,m,l) &= \int_{-\infty}^{\infty} Q_j^*(x_3)\cdot\frac{\partial^l Q_m(x_3)}{\partial x_3{}^l}\cdot[\delta(x_3-h_1-h_2)-\delta(x_3-h_1-h_2-h_3)]dx_3 \end{aligned} \quad (A3)$$

Then the matrix items $A_{ij}^{j,m}$ and M_m^j can be obtained by the following:

$$\begin{aligned} A_{11}^{j,m} &= i^2k^2\cdot\left[C_{11}^1\cdot u_1(j,m,0)+C_{11}^2(x_3)\cdot u_2(j,m,0)+C_{11}^3\cdot u_3(j,m,0)\right] \\ &+C_{55}^2{}'(x_3)\cdot u_2(j,m,1)+C_{55}^1\cdot u_1(j,m,2)+C_{55}^2(x_3)\cdot u_2(j,m,2) \\ &+C_{55}^3\cdot u_3(j,m,2)+C_{55}^1\cdot K_1(j,m,1)+C_{55}^2(x_3)\cdot K_2(j,m,1)+C_{55}^3\cdot K_3(j,m,1) \end{aligned} \quad (A4)$$

$$A_{12}^{j,m} = ik\cdot\begin{bmatrix} C_{13}^1\cdot u_1(j,m,1)+C_{13}^2(x_3)\cdot u_2(j,m,1)+C_{13}^3\cdot u_3(j,m,1) \\ +C_{55}^2{}'(x_3)\cdot u_2(j,m,0)+C_{55}^1\cdot u_1(j,m,1)+C_{55}^2(x_3)\cdot u_2(j,m,1) \\ +C_{55}^3\cdot u_3(j,m,1)+C_{55}^1\cdot K_1(j,m,0)+C_{55}^2(x_3)\cdot K_2(j,m,0) \\ +C_{55}^3\cdot K_3(j,m,0) \end{bmatrix} \quad (A5)$$

$$A_{21}^{j,m} = ik \cdot \begin{bmatrix} C_{55}^1 \cdot u_1(j,m,1) + C_{55}^2(x_3) \cdot u_2(j,m,1) + C_{55}^3 \cdot u_3(j,m,1) \\ +C_{13}^{2}{}'(x_3) \cdot u_2(j,m,0) + C_{13}^1 \cdot u_1(j,m,1) + C_{13}^2(x_3) \cdot u_2(j,m,1) \\ +C_{13}^3 \cdot u_3(j,m,1) + C_{13}^1 \cdot K_1(j,m,0) + C_{13}^2(x_3) \cdot K_2(j,m,0) \\ +C_{13}^3 \cdot K_3(j,m,0) \end{bmatrix} \quad \text{(A6)}$$

$$\begin{aligned} A_{22}^{j,m} &= i^2k^2 \cdot \left[C_{55}^1 \cdot u_1(j,m,0) + C_{55}^2(x_3) \cdot u_2(j,m,0) + C_{55}^3 \cdot u_3(j,m,0)\right] \\ &+C_{33}^{2}{}'(x_3) \cdot u_2(j,m,1) + C_{33}^1 \cdot u_1(j,m,2) + C_{33}^2(x_3) \cdot u_2(j,m,2) \\ &+C_{33}^3 \cdot u_3(j,m,2) + C_{33}^1 \cdot K_1(j,m,1) + C_{33}^2(x_3) \cdot K_2(j,m,1) + C_{33}^3 \cdot K_3(j,m,1) \end{aligned} \quad \text{(A7)}$$

$$M_{11}^{j,m} = \rho^1 \cdot u_1(j,m,0) + \rho^2 \cdot u_2(j,m,0) + \rho^3 \cdot u_3(j,m,0) \quad \text{(A8)}$$

$$M_{22}^{j,m} = \rho^1 \cdot u_1(j,m,0) + \rho^2 \cdot u_2(j,m,0) + \rho^3 \cdot u_3(j,m,0) \quad \text{(A9)}$$

References

1. Kumakawa, A.; Niino, M.; Kiyoto, S.; Nagataet, S. Proceedings of the international symposium on functionally gradient materials. *J. Am. Ceram. Soc.* **1992**, *34*, 213–219.
2. Guo, J.; Chen, J.; Pan, E. Size-dependent behavior of functionally graded anisotropic composite pate. *Int. J. Eng. Sci.* **2016**, *106*, 110–124. [CrossRef]
3. Ankit Gupta, M. Talha, Recent development in modeling and analysis of functionally graded materials and structures. *Prog. Aerosp. Sci.* **2015**, *79*, 1–14. [CrossRef]
4. Sha, G.; Lissenden, C.J. Modeling Magnetostrictive Transducers for Structural Health Monitoring: Ultrasonic Guided Wave Generation and Reception. *Sensors* **2021**, *21*, 7971. [CrossRef]
5. Jie, G.; Yan, L.; Zheng, M.F.; Liu, M.K.; Liu, H.Y.; Wu, B.; He, C.F. Modeling guided wave propagation in functionally graded plates by state-vector formalism and the Legendre polynomial method. *Ultrasonics* **2019**, *99*, 105953.
6. Zhu, J.Y.; Chen, L.Z.; Wu, S.M. Dispersion of Lamb Waves in Layered Plates. *J. Vib. Eng.* **1998**, *3*, 366–372.
7. Wu, R.X.; Yu, L.Z.; Li, X.D.; Qiu, Y. Propagation of Lamb Wave in a Functionally Graded Plate with a Layered Model. *Appl. Acoust.* **2016**, *35*, 199–205.
8. Bruck, H.A. One-dimensional model for designing functionally graded materials to manage stress waves. *Int. J. Solids. Struct.* **2000**, *37*, 6383–6395. [CrossRef]
9. Chen, W.Q.; Wang, H.M.; Bao, R.H. On calculating dispersion curves of waves in a functionally graded elastic plate. *Compos. Struct.* **2007**, *81*, 233–242. [CrossRef]
10. Lefebvre, J.E.; Zhang, V.; Gazalet, J.; Gryba, T. Legendre polynomial approach for modeling free-ultrasonic waves in multilayered plates. *J. Appl. Phys.* **1999**, *85*, 3419–3427. [CrossRef]
11. Yu, J.G.; Lefebvre, J.E.; Guo, Y.; Elmaimouni, L. Wave propagation in the circumferential direction of general multilayered piezoelectric cylindrical plates. *IEEE Trans. Ultrason. Ferroelectr. Freq. Control* **2012**, *59*, 2498–2508.
12. Salah, I.B.; Wali, Y.; Ghozlen, M.H. Love waves in functionally graded piezoelectric materials by stiffness matrix method. *Ultrasonics* **2011**, *51*, 310–316. [CrossRef] [PubMed]
13. Gao, L.M.; Wang, J.; Zhong, Z.; Du, J. An analysis of surface acoustic wave propagation in functionally graded plates with homotopy analysis method. *Acta. Mech.* **2009**, *208*, 249–258. [CrossRef]
14. Taylor, R.L.; Zienkiewicz, O.C. *The Finite Element Method*, 6th ed.; McGraw-Hill: London, UK, 2005.
15. Finnveden, S. Evaluation of modal density and group velocity by a finite element method. *J. Sound Vib.* **2004**, *273*, 51–75. [CrossRef]
16. Cheng, X.; Xu, X.D.; Liu, X.J. Studying the Propagating of Surface Wave Traveling in Functional Gradient by Laser Ultrasonic. *Acta Acustica.* **2011**, *2*, 145–149.
17. Kim, J.H.; Paulino, G.H. Isoparametric Graded Finite Elements for Nonhomogeneous Isotropic and Orthotropic Materials. *J. Appl. Mech.* **2002**, *69*, 502–514. [CrossRef]
18. Zhang, C.; Xiao, J.Y. Research on Finite Element Methods for Functional Graded Material. *Aircr. Design* **2007**, *27*, 31–33.
19. Wang, Y.S.; Gross, D. Analysis of a Crack in a Functionally Gradient Interface Layer under Static and Dynamic Loading. *Key Eng. Mater.* **2000**, *183–187*, 331–336.
20. Gao, J.; Lyu, Y.; Zheng, M.F.; Liu, M.K.; Liu, H.Y.; Wu, B.; He, C.F. Modeling guided wave propagation in multi-layered anisotropic composite laminates by state-vector formalism and the Legendre polynomials. *Compos. Struct.* **2019**, *228*, 111319. [CrossRef]
21. Lowe, M.J.S. Matrix techniques for modeling ultrasonic waves in multilayered media. *IEEE Trans. Ultrason. Ferroelectr. Freq. Control* **1995**, *42*, 525–542. [CrossRef]
22. Martínez-Pañeda, E. On the finite element implementation of functionally graded materials. *Materials* **2019**, *12*, 287. [CrossRef] [PubMed]
23. Santare, M.H.; Lambros, J. Use of graded finite elements to model the behavior of nonhomogeneous materials. *J. Appl. Mech.* **2000**, *67*, 819–822. [CrossRef]

24. Gao, J.; Lyu, Y.; Song, G.R.; Liu, M.K.; Zheng, M.F.; He, C.F.; Lee, Y.C. Legendre orthogonal polynomial method in calculating reflection and transmission coefficients of fluid-loaded functionally gradient plates. *Wave Motion* **2021**, *104*, 102754. [CrossRef]
25. Lyu, Y.; He, C.F.; Song, G.R.; Wu, B.; Lee, Y.C. Elastic Properties Inversion of an Isotropic Plate by Hybrid Particle Swarm-Based-Simulated Annealing Optimization Technique from Leaky Lamb Wave Measurements Using Acoustic Microscopy. *J. Nondestr. Eval.* **2014**, *33*, 651–662.

sensors

MDPI

Article

Pulse-Modulation Eddy Current Evaluation of Interlaminar Corrosion in Stratified Conductors: Semi-Analytical Modeling and Experiments

Zhengshuai Liu [1], Yong Li [1,*], Shuting Ren [1], Yanzhao Ren [1], Ilham Mukriz Zainal Abidin [2] and Zhenmao Chen [1]

[1] State Key Laboratory for Strength and Vibration of Mechanical Structures, Shaanxi Engineering Research Centre of NDT and Structural Integrity Evaluation, School of Aerospace Engineering, Xi'an Jiaotong University, Xi'an 710049, China; liuzhengshuai@stu.xjtu.edu.cn (Z.L.); renshuting1@stu.xjtu.edu.cn (S.R.); renyanzhao@stu.xjtu.edu.cn (Y.R.); chenzm@mail.xjtu.edu.cn (Z.C.)

[2] Leading Edge NDT Technology (LENDT) Group, Malaysian Nuclear Agency, Bangi 43000, Malaysia; mukriz@nuclearmalaysia.gov.my

* Correspondence: yong.li@mail.xjtu.edu.cn; Tel.: +86-029-8266-5721

Abstract: Interlaminar corrosion (ILC) poses a severe threat to stratified conductors which are broadly employed in engineering fields including aerospace, energy, etc. Therefore, for the pressing concern regarding the safety and integrity of stratified conductors, it is imperative to non-intrusively and quantitatively interrogate ILC via non-destructive evaluation techniques. In this paper, pulse-modulation eddy current (PMEC) for imaging and assessment of ILC is intensively investigated through theoretical simulations and experiments. A semi-analytical model of PMEC evaluation of ILC occurring at the interlayer of two conductor layers is established based on the extended truncated region eigenfunction expansion (ETREE) along with the efficient algorithm for the numerical computation of eigenvalues for reflection coefficients of the stratified conductor under inspection. Based on theoretical investigation, PMEC evaluation of ILC in testing samples are further scrutinized by using the PMEC imaging system built up for the experimental study. The theoretical and experimental results have revealed the feasibility of PMEC for imaging and evaluation of ILC in stratified conductors.

Keywords: electromagnetic non-destructive evaluation; pulse-modulation eddy current technique; semi-analytical model; interlaminar corrosion; defect imaging and evaluation

Citation: Liu, Z.; Li, Y.; Ren, S.; Ren, Y.; Abidin, I.M.Z.; Chen, Z. Pulse-Modulation Eddy Current Evaluation of Interlaminar Corrosion in Stratified Conductors: Semi-Analytical Modeling and Experiments. *Sensors* **2022**, *22*, 3458. https://doi.org/10.3390/s22093458

Academic Editor: Zenghua Liu

Received: 2 April 2022
Accepted: 28 April 2022
Published: 1 May 2022

1. Introduction

The stratified conductor which consists of a train of thin metallic layers is widely utilized in engineering fields including aerospace, energy, chemical, etc. However, the penetration of moisture and corrosive substances and abrasion of interlaminar surfaces may result in the interlaminar corrosion (ILC) at the interface between each two layers [1]. The fact that ILC is essentially buried within the structure body severely leaves the layered conductor vulnerable to structural failure, since the non-destructive testing methods such as visual testing [2], ultrasonic testing [3], etc., are inapplicable for the inspection and assessment of ILC. Therefore, advanced non-destructive evaluation (NDE) techniques are required to detect and evaluate ILC in order to guarantee the integrity and safety of layered structures in service for an extended period.

In view of the conductive characteristics of stratified conductors, eddy current (EC) testing as well as pulsed eddy current (PEC) testing, which barely requires any contact between the test piece and probe, is one of the preferred NDE techniques for the efficient inspection of conductive structures [4]. In recent years, EC and PEC have been successfully applied to detect the anomalies of railways [5] and delamination in carbon-fiber-reinforced plastic [6], and to estimate the yield strength of ferromagnetic materials [7]. To further improve the testing performance of EC and PEC, the related research is focused on the probe design, signal processing and feature extraction, etc. Wang et al. proposed a non-destructive testing technique integrating EC with PEC to localize the micro-crack in metals and quantitatively characterize its depth separately [8]. Yu et al. designed transverse probes to render the eddy currents perpendicular to circumferential cracks and improved the detection sensitivity to the defects [9]. Bernieri et al. designed a double-coil-based differential probe to detect buried thin cracks with a giant magneto resistance (GMR) sensor [10]. Ge et al. proposed a bobbin probe with two excitation coils to induce more uniform eddy current and pick up signals with array sensors [11]. Besides the novel design of probes, signal features immune to lift-off variations have been investigated. Wang et al. used the dynamic apparent time constant of the PEC-induced coil voltage decay signal to measure wall thinning under the large lift-off variations [12]. Song et al. applied the last peak point of differential PEC signals to measure thickness for non-ferrimagnetic metal under large lift-offs [13]. Zhang et al. proposed Euclidean distances as a signal feature by decoupling interferences of insulations, claddings and the lift-off [14].

As one of the extended NDE methods from EC and PEC, pulse-modulation eddy current (PMEC) testing has been found to be superior to EC and PEC in terms of higher sensitivity and accuracy in detection, imaging and assessment of subsurface defects in planar and tubular conductors [15,16]. The technical advantage of PMEC lies in the fact that the majority of the excitation energy can be effectively allocated to the eddy currents induced right within a conductor body, and thus the dedicated interrogation of the conductor is realized [15]. In a bid to further improve the testing sensitivity, a funnel-shaped PMEC probe has been proposed, and found to be advantageous over the traditional pancake probe in an evaluation of back-surface flaws [17]. Based on this, Yan et al. investigated the methods for imaging of subsurface corrosion in the planar conductor [18], and subsequently established an inversion scheme for the 3D profile reconstruction of detected corrosions [19]. However, to the authors' knowledge, little research on PMEC inspection of ILC in layered conductors has been conducted.

It is also noticeable that in the aforementioned research, despite the analytical models established based on the extended truncated region eigenfunction expansion (ETREE) modeling [20] for theoretical investigation of PMEC, the corrosion involved in the related model is deemed to be the wall-thinning defect under the assumption that the defect size is appreciably larger than the probe dimension. The scenario regarding the localized corrosion is barely taken into account, leaving a research blank in regard to the theory of PMEC inspection of localized defects and particularly ILC in layered structures. This issue could be tackled in reference to the relevant analytical modeling of EC. Theodoulidis et al. constructed the analytical modeling for calculating eddy currents in a plate with a long slot flaw by the average of an odd and even parity solution [21]. Jiang et al. established an analytical model for eddy current testing of an angled slot with different upper and bottom length in 2D system [22]. Yu et al. proposed an analytical expression for the magnetic field for a cylindrical defect in metal in eddy current testing system by solving the partial differential equations [23]. Tytko et al. presented an axially symmetric mathematical model of an I-cored coil placed over a two-layered conductive material with a cylindrical surface hole [24]. However, more and more complex geometries in analytical models with defect flaws make the numerical implementation more difficult [25]. To solve this problem, some researchers tried to develop more efficient calculation approach to solve the analytical modeling. The main obstacle for implementation of the analytical model is solving discrete complex eigenvalues of the partial differential equations in a defect domain. To find eigenvalues in a defect domain, Theodoulidis et al. used the Newton–Raphson algorithm [21] and a derived Cauchy principle [26] developed by Delves and Lyness [27] for the computation of eigenvalues. Darko et al. further improved Delves and Lyness' work and designed an iteration algorithm for dividing regions in complex planes for guaranteeing the efficient solution of eigenvalues [28]. Strakova adapted Cauchy's residue theorem and presented a contour integral method for the localization of eigenvalues of a matrix pencil in a bounded domain in the complex plane [29]. Tytko proposed a method for the multilevel computation of complex eigenvalues by combining both the Newton method and Cauchy principle and applying them in different regions of the complex plane [25]. The related analytical modeling of EC with localized defects could be supportive of the establishment of theoretical models for PMEC evaluation of ILC in stratified conductors.

In a bid to scrutinize the feasibility of PMEC for the evaluation of ILC in stratified conductors, a series of theoretical simulations of PMEC testing of localized ILC with different radii and depths are carried out with a semi-analytical model established based on the ETREE modeling. In parallel, a PMEC system has been built up for experimental investigation in a bid to further confirm the applicability of PMEC for the detection, imaging and assessment of ILC. The rest of the paper is organized as follows: Section 2 elaborates the semi-analytical modeling of PMEC for ILC evaluation. The theoretical and experimental investigations involving characteristics and features of PMEC responses to ILC, etc., are intensively presented in Sections 3 and 4, respectively. The investigation results are summarized and concluded in Section 5.

2. Semi-Analytical Modeling Regarding PMEC Evaluation of ILC

2.1. Field Formulation

In an effort to investigate PMEC evaluation of ILC in stratified conductors, a semi-analytical model based on ETREE modeling is established. A 2D axisymmetric model of PMEC evaluation of a two-layer conductor subject to an ILC is portrayed in Figure 1. The model comprises a funnel-shaped PMEC probe and a two-layer conductor. The funnel-shaped probe consists of: (1) an excitation coil with the parallelogram cross-section for generation of the incident magnetic field; and (2) a solid-state magnetic-field sensor which is deployed at the bottom center of the excitation coil and used for acquiring the testing signal of the net magnetic field (superposition of the incident and eddy-current-induced fields). It is noted that the ILC residing at the interface between the upper and bottom layers is in the shape of a cylinder with the radius of c and depth/thickness of D, $D = d_2 - d_1$. The region of interest (ROI) truncated with the radial distance of h in Figure 1 is divided into a series of sub-regions numbered from Region II to Region VI in reference to the continuity of electromagnetic field over the boundaries of the coil, conductor and ILC with the z coordinates of z_1, 0, $-d_1$, $-d_2$ and $-d_3$.

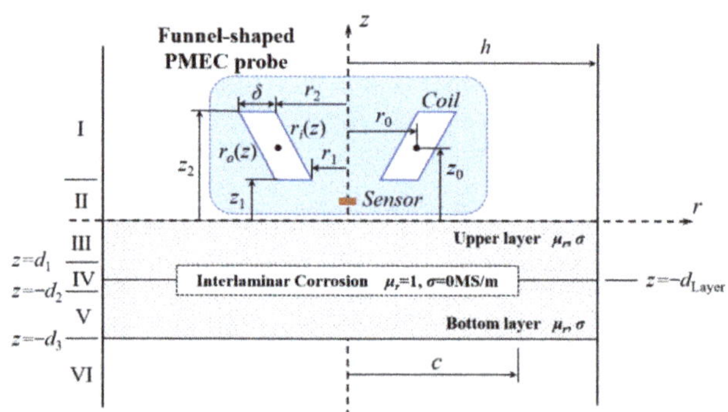

Figure 1. A 2D axisymmetric model of a funnel-shaped PMEC probe placed over a two-layer stratified conductor with an ILC at the interface between the upper and bottom layers.

Based on ETREE modeling for transient eddy current testing [15], the closed-form expression of z-component of the transient net magnetic field which is sensed by the sensor at an arbitrary coordinate of (r, z) in Region II can be written as:

$$B_z(r,z,t) = I(t) \otimes \mathcal{F}^{-1}[B_z(r,z,\omega)], \quad (1)$$

where $I(t)$ denotes the excitation signal of the electric current driving the coil. $\otimes$ stands for the circular convolution. F^{-1} denotes the inverse Fourier transform. B_z (r, z, ω) is the spectral response with respect to every harmonic at an angular frequency of ω within the excitation signal. For each harmonic, B_z (r, z, ω) can be written in a form of a double integral of the z-component of the total magnetic field resulted from a filament excitation coil (with the radius of r_0 and liftoff of z_0), i.e., B_z^f (r, z, ω) as:

$$B_z(r,z,\omega) = \frac{N_{coil}}{\delta(z_2 - z_1)} \int_{z_1}^{z_2} \int_{r_i(z)}^{r_o(z)} B_z^f(r,z,\omega) dr dz, \tag{2}$$

where N_{coil} is the number of turns of the excitation coil. δ denotes the coil radial thickness. z_1 and z_2 are the z-integral limits corresponding to the lower and upper boundaries of the coil. r_i (z) and r_o (z), which stand for the inner and outer boundaries of the coil, respectively, are taken as the lower and upper limits for the r-integral. They are written as:

$$\begin{cases} r_i(z) = r_1 + \frac{\delta z}{z_2 - z_1} \\ r_o(z) = r_1 + \delta + \frac{\delta z}{z_2 - z_1} \end{cases}. \tag{3}$$

It is noteworthy that since the localized ILC is taken into account in the model, in Equation (2) $B_z^f(r, z, \omega)$ can hardly be expressed in the form of series expansion, but is formulated in the matrix notation. Since the magnetic field can be derived from the identity, i.e., the curl of the magnetic vector potential, in the cylindrical coordinate system it is thus written as:

$$\vec{B}^f(r,z,\omega) = -\frac{\partial A^f(r,z,\omega)}{\partial z}\vec{r} + \frac{1}{r}\frac{\partial [r A^f(r,z,\omega)]}{\partial r}\vec{z}, \tag{4}$$

where $\vec{r}$ and $\vec{z}$ are the unit vectors. A^f denotes the magnetic vector potential resulting from the filament excitation coil. Referring to [23], it is expressed as:

$$\begin{cases} A^f(r,z,\omega) = J_1(\boldsymbol{\kappa}^T r)(e^{\boldsymbol{\kappa} z}\mathbf{C} + e^{-\boldsymbol{\kappa} z}\mathbf{D}) \\ \mathbf{C} = \frac{\mu_0 r_0}{2} e^{-\boldsymbol{\kappa} z_0} \boldsymbol{\kappa}^{-1} \mathbf{E}^{-1} J_1(\boldsymbol{\kappa} r_0) \\ \mathbf{D} = \boldsymbol{\Gamma}\mathbf{C} \end{cases}, \tag{5}$$

where J_1 $(\boldsymbol{\kappa}^T r)$ and J_1 $(\boldsymbol{\kappa} r_0)$ represent the $1 \times N_s$ and $N_s \times 1$ matrices with the corresponding elements of J_1 $(\kappa_i r)$ and J_1 $(\kappa_i r_0)$, respectively. J_n is the first-kind Bessel function of the order of n. $e^{\boldsymbol{\kappa} z}$ and $e^{-\boldsymbol{\kappa} z}$ denote the $N_s \times N_s$ diagonal matrices with the individual elements of $e^{\kappa_i z}$ and $e^{-\kappa_i z}$, respectively. μ_0 is the permeability of vacuum. $\mathbf{C}$ and $\mathbf{D}$ are $N_s \times 1$ matrices. Based on Equations (4) and (5), the magnetic field at the sensor position is formulated as:

$$\vec{B}^f(r,z,\omega) = -J_1(\boldsymbol{\kappa}^T r)\boldsymbol{\kappa}(e^{\boldsymbol{\kappa} z}\mathbf{C} - e^{-\boldsymbol{\kappa} z}\mathbf{D})\vec{r} + J_0(\boldsymbol{\kappa}^T r)\boldsymbol{\kappa}(e^{\boldsymbol{\kappa} z}\mathbf{C} + e^{-\boldsymbol{\kappa} z}\mathbf{D})\vec{z}. \tag{6}$$

In Equations (5) and (6), $\boldsymbol{\kappa}$ is a diagonal matrix with the diagonal elements of κ_i, $i = 1, 2, 3 \ldots N_s$. κ_i are eigenvalues for the air domains, which are the positive roots of the equation:

$$J_1(\kappa_i h) = 0. \tag{7}$$

$\mathbf{E}$ is a diagonal matrix with the elements expressed as:

$$E_{ii} = \frac{h^2 J_0^2(\kappa_i h)}{2}. \tag{8}$$

$\boldsymbol{\Gamma}$ is a $N_s \times N_s$ full matrix representing the conductor reflection coefficient. It can be derived from the boundary conditions implying the continuity of electromagnetic field over each interface in the domain under the excitation coil (i.e., Regions II, III, ... and VI), and thus written as:

$$\boldsymbol{\Gamma} = \mathbf{K}_2^{-1}\mathbf{K}_1, \tag{9}$$

where $\mathbf{K_1}$ and $\mathbf{K_2}$ are formulated as:

$$\begin{cases} \mathbf{K}_1 = (\mathbf{L}_1\mathbf{M}_1 - \mathbf{L}_2\mathbf{M}_2)\mathbf{E}e^{-\boldsymbol{\lambda}d_1}\mathbf{N}_1 - (\mathbf{L}_2\mathbf{M}_1 - \mathbf{L}_1\mathbf{M}_2)\mathbf{E}e^{\boldsymbol{\lambda}d_1}\mathbf{N}_2 \\ \mathbf{K}_2 = (\mathbf{L}_2\mathbf{M}_1 - \mathbf{L}_1\mathbf{M}_2)\mathbf{E}e^{\boldsymbol{\lambda}d_1}\mathbf{N}_1 - (\mathbf{L}_1\mathbf{M}_1 - \mathbf{L}_2\mathbf{M}_2)\mathbf{E}e^{-\boldsymbol{\lambda}d_1}\mathbf{N}_2 \end{cases}. \tag{10}$$

It is noted that $\mathbf{L_1}$, $\mathbf{L_2}$, $\mathbf{M_1}$, $\mathbf{M_2}$, $\mathbf{M_3}$, $\mathbf{M_4}$, $\mathbf{N_1}$ and $\mathbf{N_2}$ in Equation (10) are the coefficient matrices which are expressed as:

$$\begin{cases} \mathbf{L}_1 = \left[\mathbf{N}_1^{-1}e^{\boldsymbol{\lambda}(d_2-d_3)}\mathbf{E}^{-1}\mathbf{M}_3 - \mathbf{N}_2^{-1}e^{\boldsymbol{\lambda}(d_3-d_2)}\mathbf{E}^{-1}\mathbf{M}_4\right]e^{\mathbf{p}(d_1-d_2)} \\ \mathbf{L}_2 = \left[\mathbf{N}_2^{-1}e^{\boldsymbol{\lambda}(d_3-d_2)}\mathbf{E}^{-1}\mathbf{M}_3 - \mathbf{N}_1^{-1}e^{\boldsymbol{\lambda}(d_2-d_3)}\mathbf{E}^{-1}\mathbf{M}_4\right]e^{\mathbf{p}(d_2-d_1)} \end{cases}, \tag{11}$$

$$\begin{cases} \mathbf{M}_1 = \mathbf{U}^{-1} + (\mathbf{Up})^{-1}\boldsymbol{\lambda},\ \mathbf{M}_3 = \mathbf{U} + \boldsymbol{\lambda}^{-1}\mathbf{Up} \\ \mathbf{M}_2 = \mathbf{U}^{-1} - (\mathbf{Up})^{-1}\boldsymbol{\lambda},\ \mathbf{M}_4 = \mathbf{U} - \boldsymbol{\lambda}^{-1}\mathbf{Up} \end{cases}, \tag{12}$$

$$\begin{cases} \mathbf{N}_1 = \mathbf{I} + \boldsymbol{\lambda}^{-1}\boldsymbol{\kappa} \\ \mathbf{N}_2 = \mathbf{I} - \boldsymbol{\lambda}^{-1}\boldsymbol{\kappa} \end{cases}. \tag{13}$$

In these expressions, $\mathbf{I}$ denotes the $N_s \times N_s$ identity matrix. $\boldsymbol{\lambda}$ is the $N_s \times N_s$ diagonal matrix with the diagonal elements of λ_i which are eigenvalues for the layered conductor free of ILC, and can be expressed as:

$$\lambda_i = \sqrt{\kappa_i^2 + j\omega\mu_0\mu_r\sigma}, \tag{14}$$

where σ and μ_r denote the apparent conductivity and relative permeability of the conductive layers, respectively. The elements U_{ij} in the $N_s \times N_s$ full matrix $\mathbf{U}$ can be computed out by using the equation written as:

$$U_{ij} = \frac{-ck^2}{(p_j^2 - \kappa_i^2)(q_j^2 - \kappa_i^2)}\left[\kappa_i J_0(\kappa_i c)J_1(p_j c) - p_j J_1(\kappa_i c)J_0(p_j c)\right]R_1(q_j c), \tag{15}$$

where $k^2 = j\omega\mu_0\mu_r\sigma$. R_n $(q_i c)$ is formulated as:

$$R_n(q_i c) = Y_1(q_i h)J_n(q_i c) - J_1(q_i h)Y_n(q_i c), \tag{16}$$

where Y_n denotes the second-kind Bessel function of the order of n. The diagonal elements p_i in the $N_s \times N_s$ diagonal matrix $\mathbf{p}$ are eigenvalues for the ILC region. They can be sought by finding the roots of the equation:

$$p_i J_0(p_i c)R_1(q_i c) - q_i J_1(p_i c)R_0(q_i c) = 0. \tag{17}$$

In Equations (15)–(17), q_i is also the eigenvalues for the ILC domain, and can be derived by using the equation:

$$q_i = \sqrt{p_i^2 - j\omega\mu_0\mu_r\sigma}. \tag{18}$$

By using Equations (1) and (2), the temporal responses of PMEC to localized ILC at the interface of the layered conductor can be readily predicted.

2.2. Numerical Calculation of Eigenvalues

The calculation of eigenvalues particularly p_i is a fundamental step in predicting the PMEC signals via the semi-analytical model. In this paper, eigenvalues in the ILC domain are acquired based on the Delves and Lyness algorithm calculating the eigenvalues and further improving the precision by Newton–Raphson iteration. When applying the Delves and Lyness algorithm to seek eigenvalues, the roots of Equation (17) are equivalent to zeros of the function written as:

$$f(x) = xJ_0(xc)R_1(qc) - qJ_1(xc)R_0(qc). \tag{19}$$

According to the Cauchy argument principle developed by Delves and Lyness, the summation of nth power of zeros of an analytic function enclosed by the contour C in the complex plane can be expressed as:

$$s_n = \frac{1}{2\pi j}\oint_C x^n \frac{f'(x)}{f(x)}dx = \sum_{i=1}^{N} \xi_i^n \; n = 1,2,\ldots N, \tag{20}$$

$$s_0 = \frac{1}{2\pi j}\oint_C x^0 \frac{f'(x)}{f(x)}dx = \sum_{i=1}^{N} \xi_i^0 = N, \tag{21}$$

where $f'(x)$ is the derivative function of $f(x)$. ξ_i are zeros of the $f(x)$ enclosed by the contour C, giving the solutions to p_i in Equation (17). By dividing the complex plane into a series of discretized subareas, information of eigenvalues in each subarea is derived from Equations (20) and (21). In each subarea, the number of zeros N is equal to s_0. To obtain to solutions to p_i in defect domain, it is required to solve the Nth degree system of equations which is written as:

$$\begin{cases} s_1 = p_1 + p_2 + \ldots + p_N \\ s_2 = p_1^2 + p_2^2 + \ldots + p_N^2 \\ \vdots \\ s_N = {p_1}^N + p_2^N + \ldots + p_N^N \end{cases}. \tag{22}$$

In a bid to mitigate the potential overflow and rounding error in computation of Bessel functions, Equation (19) is further modified and formulated in an exponentially scaled form as:

$$\begin{cases} f(x) = e^{|\Im(xc)|+|\Im(qh)|+|\Im(qc)|}\hat{f}(x) \\ \hat{f}(x) = x\hat{J}_0(xc)\hat{R}_1(qc) - q\hat{J}_1(xc)\hat{R}_0(qc) \end{cases}, \tag{23}$$

where $\Im$ denotes the imaginary part of a complex value. This scaling method is implemented based on the embedded functions "*besselj* (n, x, 1)" and "*bessely* (n, x, 1)" in MATLAB, and gives solutions to $e^{-|\Im(x)|}J_n(x)$ and $e^{-|\Im(x)|}Y_n(x)$ which are efficient in mitigation regarding overflow and reduction in computational accuracy of Bessel functions. The scaled forms of $J_n(x)$ and $Y_n(x)$ are $\hat{J}_n(x)$ and $\hat{Y}_n(x)$ whilst $\hat{R}_n(qc)$ is the scaled form of $R_n(qc)$. $\hat{R}_n(qc)$ can thus be expressed as:

$$\hat{R}_n(qc) = \left[\hat{Y}_1(qh)\hat{J}_n(qc) - \hat{J}_1(qh)\hat{Y}_n(qc)\right]. \tag{24}$$

In addition, $\hat{f}'(x)$, which is the scaled form of $f'(x)$ can also be derived:

$$f'(x) = e^{|\Im(xc)|+|\Im(qh)|+|\Im(qc)|}\hat{f}'(x). \tag{25}$$

In the virtue of the uniform coefficient in Equations (23) and (25), the ratio term, i.e., $f'(x)/f(x)$ in Equations (20) and (21) can be replaced by $\hat{f}'(x)/\hat{f}(x)$. It is noteworthy that: (1) the function ratio in Equations (20) and (21) is essentially the integrand of the loop-integration for finding the eigenvalue in each discretized subarea when applying the Delves and Lyness algorithm in Reference [27]; and (2) by substituting $\hat{f}'(x)/\hat{f}(x)$ for $f'(x)/f(x)$ and taking it as the integrand, the overflow and rounding error in computation can be suppressed. In a bid to predict PMEC responses to the ILC, the efficient computation of the complex roots of $f(x)$ is realized via the modified Delves and Lyness formula which is written in the scaled form as:

$$s_n = \frac{1}{2\pi j}\oint_C x^n \frac{\hat{f}'(x)}{\hat{f}(x)}dx = \sum_{i=1}^{N} p_i^n \; n = 1,2,\ldots N, \tag{26}$$

$$s_0 = \frac{1}{2\pi j}\oint_C \frac{\hat{f}'(x)}{\hat{f}(x)}dx = \sum_{i=1}^{N} p_i^0 = N. \tag{27}$$

The loop-integration can readily be computed with the MALTAB routine *"integral"*. It is further revealed that in Newton–Raphson algorithm for improving the computation accuracy regarding the complex eigenvalues, the ratio term, i.e., $f(x)/f'(x)$ can also be replaced with the exponentially scaled form to avoid computational overflow or rounding errors.

It is noted that the accuracy regarding the computation of complex eigenvalues depends also on the discretized subareas in the complex plane. In the event that the number of eigenvalues in each subarea is increased, it becomes more difficult to solve the system of equations shown in Equation (22). Previous investigation has revealed that the computational accuracy for p_i is undermined when the discretized subarea in the complex plane encloses too many eigenvalues [25]. The relatively large number of N also brings about the tedious computation of s_n with n varying from 1 to N. Therefore, it is vital to properly divide the complex plane into the discretized subareas and limit the maximum number of eigenvalues in each subarea. Different from the proposed iteration algorithms to make the subareas denser in the complex plane to reach the limited number of eigenvalues in Reference [28], in this paper characteristics of eigenvalues' distribution are firstly investigated and utilized to optimize the shape and size of the subarea in the complex plane.

When the ILC radius varies from 0 (the defect-free case) to h (the wall-thinning case where the ILC radius is considerably larger than the probe size), in the complex plane, the corresponding eigenvalues change from λ_i to κ_i. Due to the continuity in numerical values, the distribution of p_i is thus bounded by λ_i and κ_i. An example regarding the distribution characteristics of eigenvalues p_i in the complex plane for the harmonic case of $\omega = 2000\pi$ rad/s and the subareas are portrayed in Figure 2. It can be observed from Figure 2a that p_i (red dots) sought in the plane is essentially enclosed by a region with its lower and upper boundaries defined by κ_i and λ_i, respectively. Interestingly, after discretizing the region into a train of rectangular subareas (when $\kappa_i < \Re(\lambda_1)$, where $\Re$ denotes the real part of a complex value) and trapezoidal subareas (when $\kappa_i > \Re(\lambda_1)$) with their vertices set as κ_i and λ_i, it can be found that within each resultant subarea the total number of p_i is up to 2. Further investigation indicates that such a finding still holds for the other harmonic cases. The distribution characteristics of eigenvalues provide an advantageous reference to the division of discretized subareas. In light of this, a new approach for discretizing the complex plane is thus proposed. Prior to the computation of the loop integrals in Equations (26) and (27), the complex plane is discretized into a train of subareas in the shapes of rectangles and trapezoids. The coordinates of the vertices defining the contour of each subarea are set as κ_i and λ_i. For the rectangular-shaped and trapezoidal-shaped subareas, the coordinates regarding the vertices of the subarea contour are exhibited in Figure 2b.

It should be pointed out that the proposed approach ensures that there are up to 2 eigenvalues of p_i in each subarea whilst the number of discretized subareas is highly relied on the number of κ_i and λ_i. This makes the subarea discretization free of multiple iteration to obtain denser subareas in the localized complex region where eigenvalues are concentrated and are beneficial to the efficient seeking of all p_i with high computation speed and accuracy. Further investigation also reveals that by taking the p_i sought with the proposed approach as the initial estimates, the number of iterations for refinement of eigenvalues via the Newton–Raphson algorithm for further precision improvement is significantly decreased, which benefits the fast solution to predicted PMEC responses to ILC.

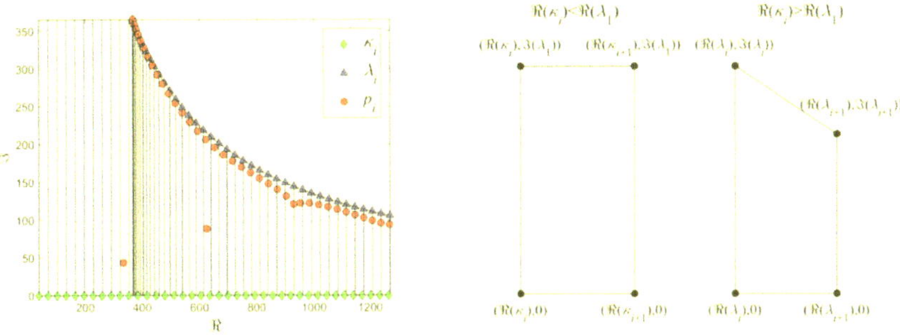

(a) (b)

Figure 2. Characteristics regarding the distribution of eigenvalues in the complex plane and defined subareas: (**a**) distribution of eigenvalues p_i, κ_i and λ_i along with the discretized subareas; (**b**) coordinates of the subareas.

2.3. Verification with the Finite Element Modeling

In conjunction with the proposed efficient algorithm for computation of eigenvalues, the established semi-analytical model and formulated closed-form expression of the PMEC signal could be adopted for efficient prediction of testing signals. Prior to theoretical simulations, the semi-analytical model is corroborated by the finite element modeling (FEM). The specimen under PMEC inspection is chosen as a double-layered conductor with cylindrical ILC. The material of each conductor layer is the aluminum alloy. The parameters of the PMEC probe and specimen are listed in Tables 1 and 2, respectively.

Table 1. Parameters of the PMEC probe.

Symbol	Quantity	Value
r_1	Inner radius of the coil bottom	8.0 mm
r_2	Inner radius of the coil top	14.8 mm
δ	Coil radial thickness	0.8 mm
z_1	Lift off	2.0 mm
z_2	Position of the coil top	12.1 mm
N_{coil}	The number of coil turns	205
(r, z)	Sensor location	(0, 1) mm

Table 2. Parameters of the specimen.

Symbol	Quantity	Value
h	Specimen length	100 mm
$d_3 = 2d_{Layer}$	Specimen thickness	8.0 mm
σ	Specimen conductivity	34.0 MS/m
μ_r	Specimen relative permeability	1
d_1	Upper boundary of ILC	2.0 mm
D	ILC thickness	4.0 mm
c	ILC radius	15.0 mm

The excitation current driving the funnel-shaped PMEC probe and testing signals are exhibited in Figure 3. The maximum amplitude of the excitation current is 1A. It is noted that regarding the excitation-current signal in the pulse-modulation waveform, as per the frequency-selection strategy in Reference [15] the frequencies of modulation wave and carrier wave are set at 24 Hz and 204 Hz, respectively. Different from the excitation signal presented in Reference [15], the carrier wave is truncated by the modulation wave when its amplitude reaches maximum. The comparison of the PMEC signals predicted by the semi-analytical model and FEM is presented in Figure 3b.

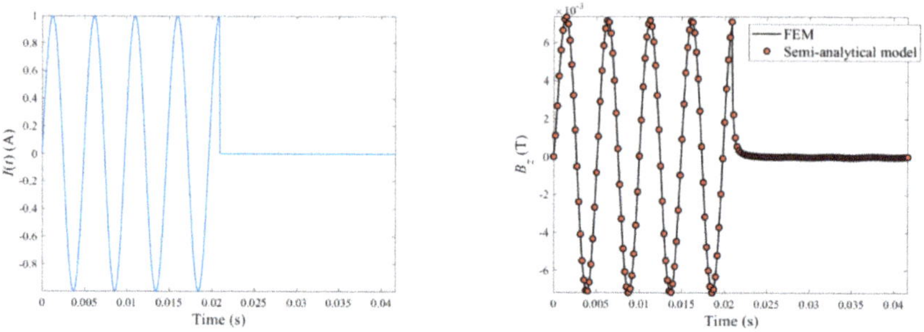

(a) (b)

Figure 3. Excitation current and PMEC signals: (**a**) excitation current driving the probe; (**b**) signals predicted by the semi-analytical model and FEM.

It can be observed from Figure 3b that the predicted PMEC signal from the semi-analytical model agrees well with that from FEM. Further analysis implies that compared with the FEM results, the relative error of the computed signal from the semi-analytical model is less than 1%, whilst the simulation time of the semi-analytical model is 10~15 times faster than FEM, indicating the superiority in terms of efficiency regarding the established semi-analytical model to FEM.

3. Theoretical Simulations and Discussion

Following verification, the PMEC responses to ILCs with different radii and depths are scrutinized through theoretical simulations based on the established semi-analytical model. For the ILC scenarios of various radii (c = 1~15 mm with D fixed at 4 mm) and depths (D = 1~5 mm, while c = 15 mm), the predicted PMEC signals are exhibited in Figure 4.

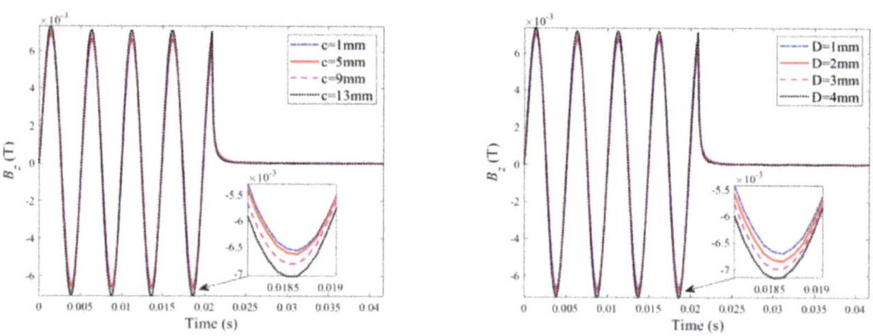

(a) (b)

Figure 4. Testing signals predicted by the semi-analytical model: (**a**) ILCs with various radii; (**b**) ILCs with various depths.

It can be seen from Figure 4 that the signal amplitude increases as the ILC size in terms of the radius (shown in Figure 4a) and depth (shown in Figure 4b) rises. The reasoning lies in the fact that the increase in the ILC volume enhances the perturbation of eddy currents induced within the specimen, causing the drop in the eddy-current-induced field and thus the increase in the net magnetic field. Further signal processing is conducted by subtracting the reference signal for the defect-free case from the defect signals for different ILC scenarios, giving the so-called difference signals. The peak value (Pv) is subsequently extracted from the difference signal. The correlations of Pv with the radius and depth of ILC are presented in Figure 5.

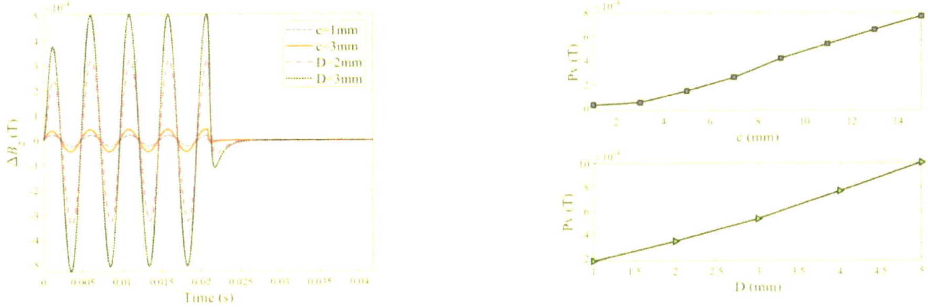

(a) (b)

Figure 5. Difference signal and correlations of Pv: (**a**) difference signals with various radii and depths of ILC; (**b**) correlations of Pv with the radius and depth of ILC.

As can be observed from Figure 5b, amplitude of difference signals will rise with the increase of radius and depth of ILC. In Figure 5b, Pv which is the extreme value extracted from the PMEC difference signal is closely associated with the sizing parameters of the ILC with a positive correlation. It is directly proportional to either radius or depth of the ILC, indicating that more eddy currents are perturbed in the presence of the ILC with the increasing size. Therefore, Pv is utilized as the signal feature for detection, imaging and evaluation of ILC in layered conductors.

4. Experiments

In parallel to the theoretical investigation, a PMEC inspection system has been built up for detection, imaging and assessment of ILC in stratified conductors. The schematic illustration of the PMEC system along with the picture of the fabricated funnel-shaped probe and testing specimen is portrayed in Figure 6.

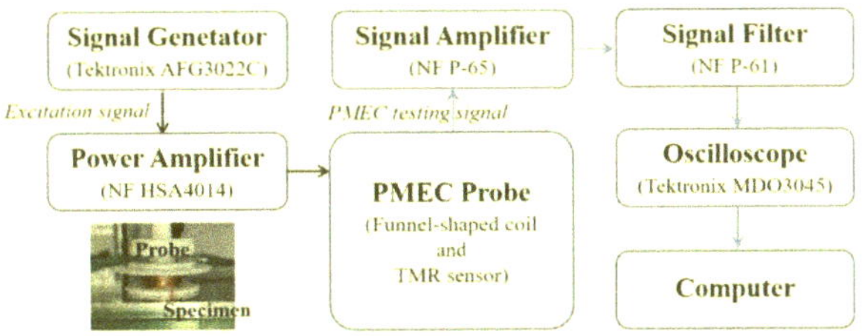

Figure 6. Schematic illustration of the PMEC inspection system (inset: the funnel-shaped probe and testing specimen).

The parameters of the probe and specimen are same as those tabulated in Tables 1 and 2. The in-house PMEC probe generates the incident field with the funnel-shaped coil whilst a Tunnel Magneto-Resistance (TMR) sensor (MultiDimension TMR 2505) deployed at the bottom center of the coil is used to pick up the z-component of the net magnetic field. The excitation signal driving the probe is in pulse-modulation waveform with the frequencies of modulation wave and carrier wave set at 24 Hz and 204 Hz, respectively, whilst its maximum amplitude is 0.3 A.

The layered specimens adopted in experiments consist of two aluminum-alloy layers whose conductivity and relative permeability are 34 MS/m and 1, respectively. Flat bottom holes imitating ILCs with different radii (from 12.5 mm to 17.5 mm) and depths (from 1 mm to 4 mm) are fabricated at the interface of the layers and reside at: (1) the back surface of the upper layer for ILC Scenario #1; and (2) the surface of the bottom layer for ILC Scenario #2. During experiments, the probe is scanned over the specimen by a scanning table with the spatial resolution of 1 mm whilst the lift-off of the probe is fixed at 2 mm. The reference signal is obtained when the probe is right over the defect-free area of the specimen. Difference signals are acquired by subtracting the testing signal at each scanning position from the reference signal. The experimental signals and derived difference signals when the PMEC probe is placed right over the center of each ILC are portrayed in Figure 7.

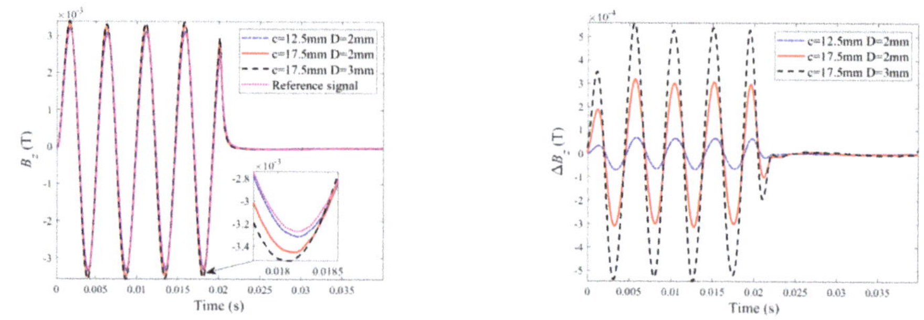

(a) (b)

Figure 7. Experimental signals for different ILC parameters: (**a**) testing signals; (**b**) difference signals.

It can be observed from Figure 7a that the amplitude of the testing signal increases as the size of ILC is increased due to more perturbation of eddy currents by ILC whilst the magnitude of the difference signal presented in Figure 7b rises when the radius and depth of ILC are increased. This is consistent with findings from the theoretical investigation. It is also noticeable that the amplitudes of the testing signals are lower than those for the back-surface corrosion in a conductive plate (analogous to a single-layer conductor). This is due to the fact that more eddy currents are induced within the multilayered conductor, and thus the eddy-current-induced field opposing the incident field becomes stronger, resulting in the decrease in strength of the net magnetic field.

It is also noticed from Figure 7b that Pv of the difference signal is directly proportional to the ILC size. Therefore, Pv extracted from the difference signal is subsequently used for ILC imaging. With all Pvs obtained at scanning positions, the Pv-based images of ILCs with different radii and depths are produced. The imaging results for various ILC scenarios are exhibited in Figures 8 and 9 against different sizes of ILCs in either circle or square shape for ILC Scenarios #1 and ILC Scenarios #2. It is observed from Figures 8 and 9 that the position of each ILC can be readily localized by using the imaging results along with the indication regarding the ILC size.

Based on the acquired ILC images, further investigation is carried out to analyze the feasibility of PMEC for evaluation of ILCs in terms of the ILC opening size and depth. A Canny-algorithm-based edge recognition method [16] is applied for identification of the ILC edge with Pv-based images. The processed image results are shown in Figures 10 and 11. It can be seen from Figures 10 and 11 that the identified ILC edge agrees well with the real ILC opening profile. Following this, the ILC opening size is estimated based on the identification results. The estimation results of the ILC opening size for ILC Scenarios #1 and ILC Scenarios #2 are listed in Tables 3 and 4. It is noticeable from Tables 3 and 4 that the approximated area of ILC opening has good agreement with the actual size. The maximum of relative errors is less than 10%.

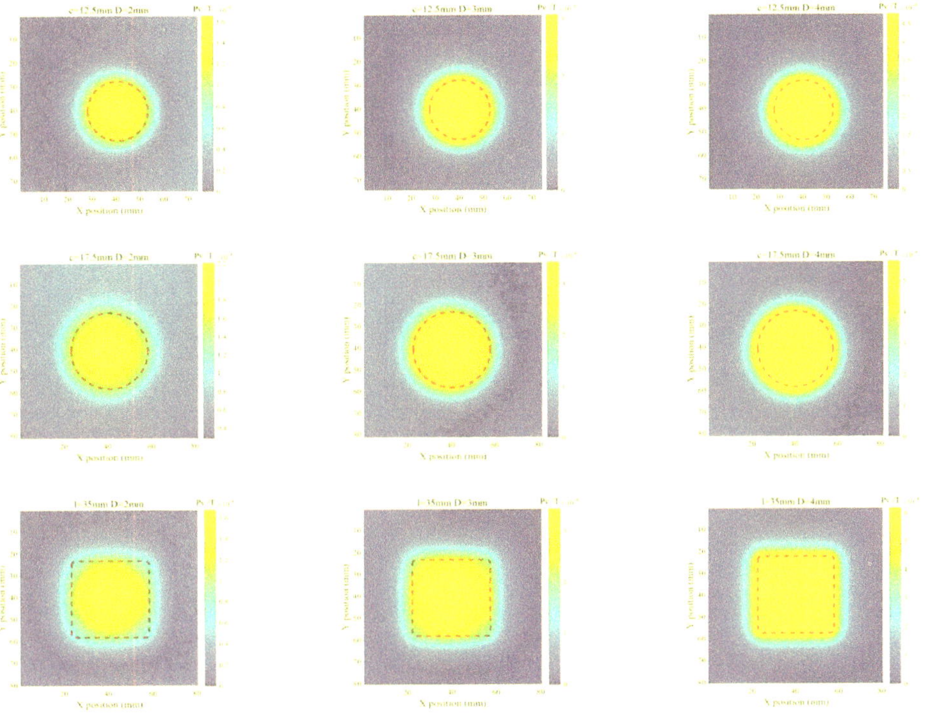

Figure 8. Imaging results for ILC Scenario #1 (where *l* denotes the length of the square-shaped ILC).

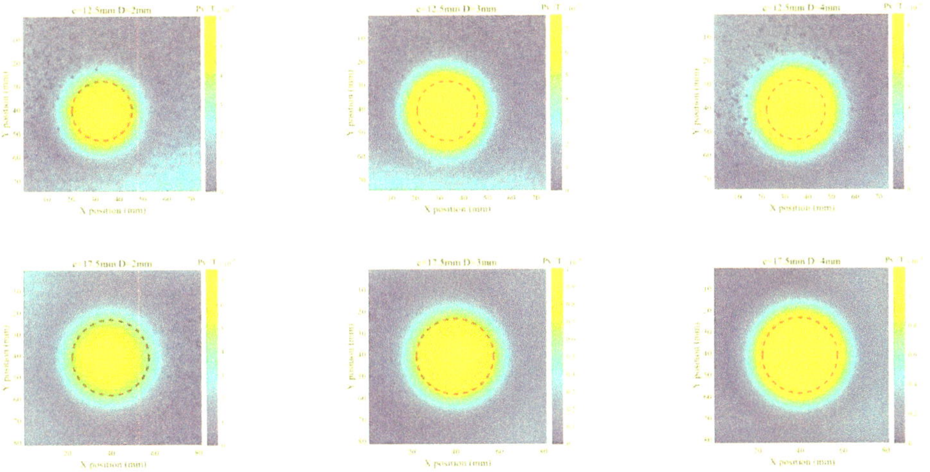

Figure 9. *Cont.*

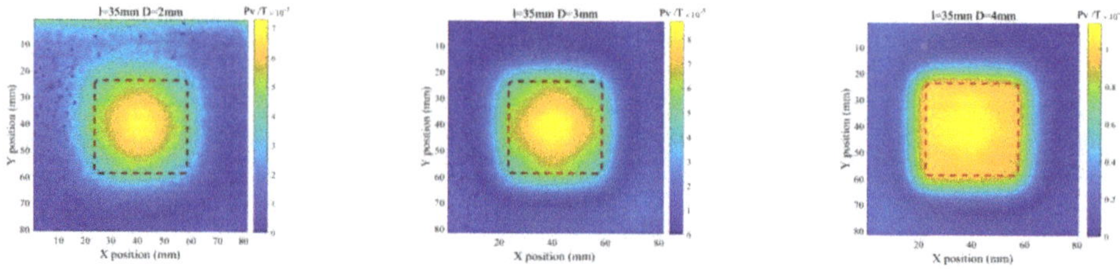

Figure 9. Imaging results for ILC Scenario #2.

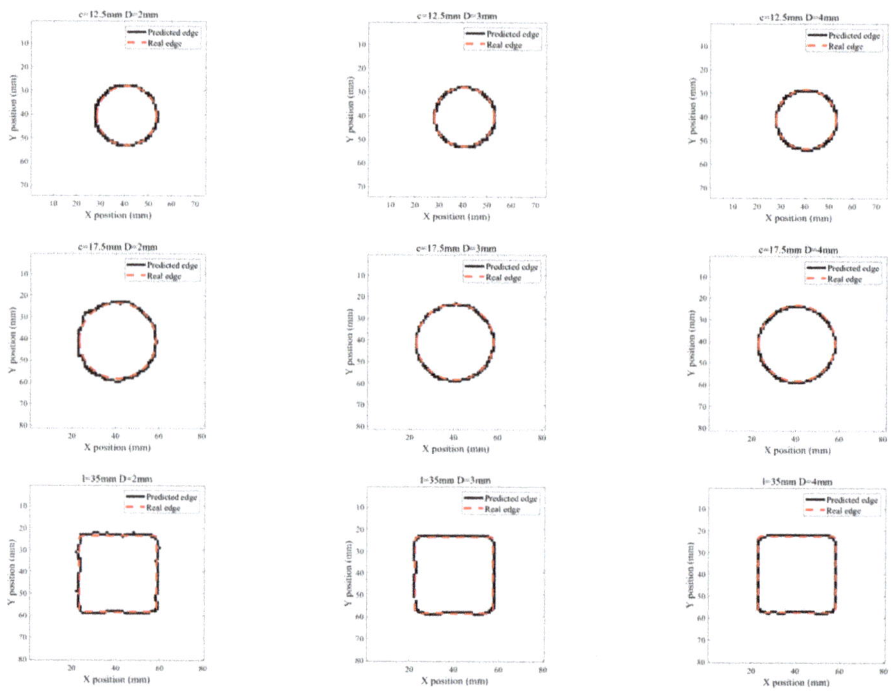

Figure 10. Edge-identification results for ILC Scenario #1.

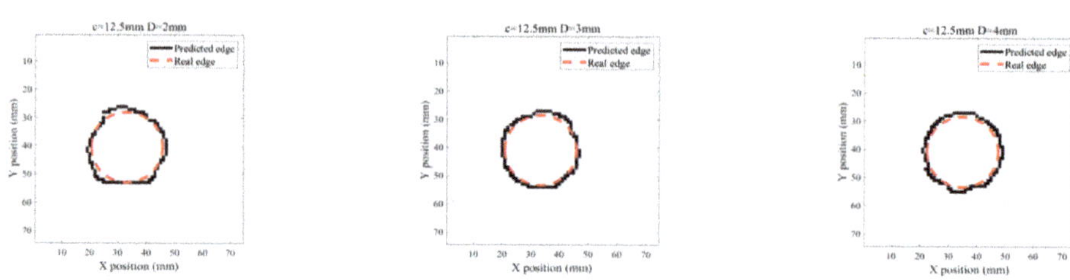

Figure 11. *Cont.*

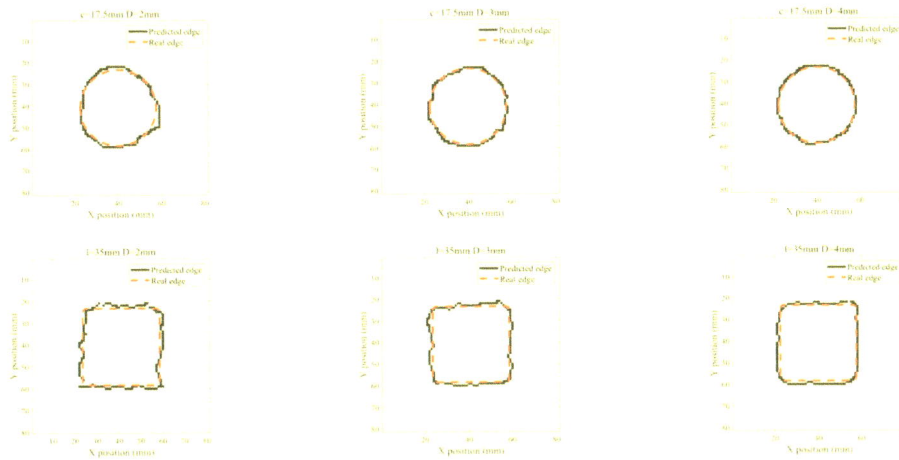

Figure 11. Edge-identification results for ILC Scenario #2.

Table 3. Assessment results regarding ILC opening areas for ILC Scenario #1.

ILC Radius/Length	ILC Depth	Estimated Area	Actual Area	Relative Error
c = 12.5 mm	2 mm	471 mm^2	490.87 mm^2	4.05%
	3 mm	449 mm^2	490.87 mm^2	8.53%
	4 mm	456 mm^2	490.87 mm^2	7.10%
c = 17.5 mm	2 mm	957 mm^2	962.11 mm^2	0.53%
	3 mm	910 mm^2	962.11 mm^2	5.42%
	4 mm	903 mm^2	962.11 mm^2	6.14%
l = 35 mm	2 mm	1196 mm^2	1225 mm^2	2.37%
	3 mm	1181 mm^2	1225 mm^2	3.59%
	4 mm	1157 mm^2	1225 mm^2	5.55%

Table 4. Assessment results regarding ILC opening areas for ILC Scenario #2.

ILC Radius/Length	ILC Depth	Estimated Area	Actual Area	Relative Error
c = 12.5 mm	2 mm	526 mm^2	490.87 mm^2	7.16%
	3 mm	519 mm^2	490.87 mm^2	5.73%
	4 mm	539 mm^2	490.87 mm^2	9.81%
c = 17.5 mm	2 mm	964 mm^2	962.11 mm^2	0.20%
	3 mm	954 mm^2	962.11 mm^2	0.84%
	4 mm	950 mm^2	962.11 mm^2	1.26%
l = 35 mm	2 mm	1224 mm^2	1225 mm^2	0.08%
	3 mm	1244 mm^2	1225 mm^2	1.55%
	4 mm	1260 mm^2	1225 mm^2	2.86%

Based on the identification results regarding the opening of each ILC, the center of every ILC is determined, and the Pv corresponding to the ILC center is extracted from the Pv-based image in order to establish the correlation between the Pv and ILC depth. The obtained correlation curves are shown in Figure 12. As can be observed from Figure 12 that for every ILC scenario, the Pv has a monotonic correlation with the ILC depth whilst Pv rises with the ILC depth increased. This agrees with the finding from the simulations. Both theoretical simulation and experiment indicate that the ILC depth can readily be evaluated by using the correlation curve. The ILC radius can be assessed by using the processed ILC image with the Canny-algorithm-based edge recognition method. The feasibility of PMEC for the detection, imaging and evaluation of ILC in stratified conductors is confirmed.

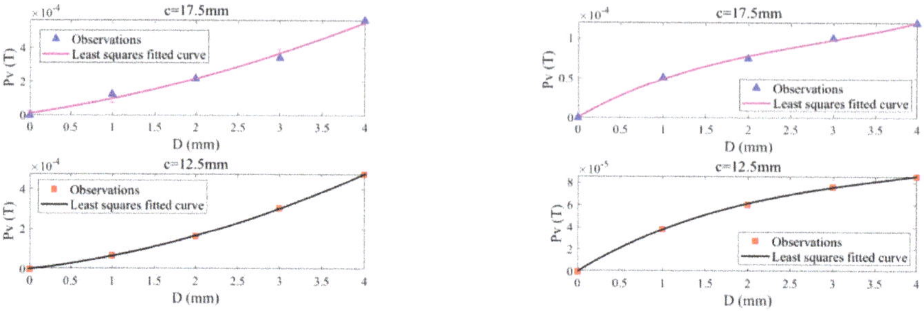

(**a**) (**b**)

Figure 12. Pv against the ILC depth for: (**a**) ILC Scenario #1; (**b**) ILC Scenario #2.

5. Conclusions

In an effort to intensively investigate the feasibility of PMEC for the detection, imaging and evaluation of ILCs in layered conductors, in this paper a semi-analytical model for the efficient prediction of PMEC responses to ILC has been established along with the resulting formulation of the closed-form expression of the PMEC testing signal. More efforts have been given to the reliable computation of eigenvalues for the ILC region. The exponentially scaled form of the integrand in the loop integral is deduced in a bid to mitigate the overflow and rounding errors in computation. An approach for discretizing the complex plane for an efficient solution to eigenvalues has been proposed. Following the establishment of the semi-analytical model, the characteristics of PMEC responses to ILC and correlation of the PMEC signal feature with the sizing parameter are investigated through theoretical simulations. It has been found that the Pv extracted from the difference signal has a monotonic relation with the ILC size. Therefore, it is used for ILC imaging.

In conjunction with the finding from simulations, the feasibility and applicability of PMEC for imaging and evaluation of ILC in a stratified conductor are further investigated via experiments. It can be noticed that the Pv-based ILC image implies the position and size of the ILC. The processed image with the Canny-algorithm-based edge recognition method can be adopted for approximation regarding the opening area of the detected ILC. Complying with the finding from simulations, the correlation of Pv with the ILC depth is monotonic, and the resultant correlation curve is applicable for the estimation of ILC depths.

Following the affirmation regarding the feasibility of PMEC for the imaging and evaluation of ILCs, further work involves: (1) investigation regarding the advantages of PMEC over other NDE methods such as PEC in particular; (2) imaging of ILCs in irregular shapes/profiles; and (3) 3D reconstruction of ILCs in layered conductors.

Author Contributions: Methodology, Y.L. and Z.C.; establishment of the semi-analytical model, Z.L. and Y.L.; design and realization of the probe, Z.L. and S.R.; model validation, I.M.Z.A. and Y.R.; experiments, I.M.Z.A. and Z.L.; signal analysis, Z.L. and Y.L.; writing—original draft preparation, Z.L. and Y.L.; writing—review and editing, Y.L. and Z.C.; supervision, Y.L. and Z.C.; funding acquisition, Y.L. and Z.C. All authors have read and agreed to the published version of the manuscript.

Funding: This research was funded by: the National Natural Science Foundation of China, grant number 52177007 and 11927801; the National Magnetic Confinement Fusion Program of China, grant number 2019YFE03130003.

Institutional Review Board Statement: Not applicable.

Informed Consent Statement: Not applicable.

Data Availability Statement: Not applicable.

Conflicts of Interest: The authors declare no conflict of interest.

References

1. Geng, Q.; North, W. The detection of inter laminar corrosion in rivetted thin aluminum skins. In Proceedings of the Nondestructive Evaluation of Aging Aircraft, Airports, and Aerospace Hardware II, San Antonio, TX, USA, 31 March 1998; Volume 3397, pp. 256–262.
2. Yang, S.M.; Sun, Z.Y.; Sun, Q.L.; Wang, Y.; An, S.L. Study on Detection System of Precision Casting Cracks Based on Image Processing. *Proc. Spie.* **2017**, *10420*, 104201A.
3. Kolkoori, S.; Hoehne, C.; Prager, J.; Rethmeier, M.; Kreutzbruck, M. Quantitative evaluation of ultrasonic C-scan image in acoustically homogeneous and layered anisotropic materials using three dimensional ray tracing method. *Ultrasonics* **2014**, *54*, 551–562. [CrossRef] [PubMed]
4. Garcia-Martin, J.; Gomez-Gil, J.; Vazquez-Sanchez, E. Non-Destructive Techniques Based on Eddy Current Testing. *Sensors* **2011**, *11*, 2525–2565. [CrossRef] [PubMed]
5. Alvarenga, T.A.; Carvalho, A.L.; Honorio, L.M.; Cerqueira, A.S.; Luciano, M.A.; Nobrega, R.A. Detection and Classification System for Rail Surface Defects Based on Eddy Current. *Sensors* **2021**, *21*, 7937. [CrossRef] [PubMed]
6. Kokurov, A.M.; Malushin, D.S.; Chichigin, B.A.; Subbotin, D.E.; Kusnetsov, A.O. Defect Characterization in Layered Composites Using a Pulsed Eddy-Current Technique. *Tech. Phys. Lett.* **2020**, *46*, 1116–1119. [CrossRef]
7. Li, K.Y.; Qiu, P.C.; Wang, P.; Lu, Z.X.; Zhang, Z.D. Estimation method of yield strength of ferromagnetic materials based on pulsed eddy current testing. *J. Magn. Magn. Mater.* **2021**, *523*, 167647. [CrossRef]
8. Wang, Z.W.; Yu, Y.T. Traditional Eddy Current-Pulsed Eddy Current Fusion Diagnostic Technique for Multiple Micro-Cracks in Metals. *Sensors* **2018**, *18*, 2909. [CrossRef]
9. Yu, Z.H.; Fu, Y.W.; Jiang, L.F.; Yang, F. Detection of circumferential cracks in heat exchanger tubes using pulsed eddy current testing. *NDT E Int.* **2021**, *121*, 102444. [CrossRef]
10. Bernieri, A.; Ferrigno, L.; Laracca, M.; Rasile, A. Eddy Current Testing Probe Based on Double-Coil Excitation and GMR Sensor. *IEEE Trans. Instrum. Meas.* **2019**, *68*, 1533–1542. [CrossRef]
11. Ge, J.H.; Hu, B.W.; Yang, C.K. Bobbin pulsed eddy current array probe for detection and classification of defects in nonferromagnetic tubes. *Sens. Actuat. A Phys.* **2021**, *317*, 112450. [CrossRef]
12. Wang, H.W.; Huang, J.B.; Liu, L.H.; Qin, S.Q.; Fu, Z.H. A Novel Pulsed Eddy Current Criterion for Non-Ferromagnetic Metal Thickness Quantifications under Large Liftoff. *Sensors* **2022**, *22*, 614. [CrossRef] [PubMed]
13. Song, Y.; Wu, X.J. Thickness measurement for non-ferromagnetic metal under the large liftoff based on the last peak point of differential pulsed eddy current signals. *Int. J. Appl. Electromagn. Mech.* **2020**, *64*, 1119–1126. [CrossRef]
14. Zhang, Q.; Wu, X.J. Wall Thinning Assessment for Ferromagnetic Plate with Pulsed Eddy Current Testing Using Analytical Solution Decoupling Method. *Appl. Sci.* **2021**, *11*, 4356. [CrossRef]
15. Li, Y.; Yan, B.; Li, D.; Jing, H.; Li, Y.; Chen, Z. Pulse-modulation eddy current inspection of subsurface corrosion in conductive structures. *NDT E Int.* **2016**, *79*, 142–149. [CrossRef]
16. Li, Y.; Yan, B.; Li, W.J.; Jing, H.Q.; Chen, Z.M.; Li, D. Pulse-modulation eddy current probes for imaging of external corrosion in nonmagnetic pipes. *NDT E Int.* **2017**, *88*, 51–58. [CrossRef]
17. Li, Y.; Liu, Z.S.; Yan, B.; Wang, Y.; Abidin, I.M.Z.; Chen, Z.M. A funnel-shaped probe for sensitivity enhancement in pulse-modulation eddy current inspection of subsurface flaws in conductors. *Sens. Actuat. A Phys.* **2020**, *307*, 111991. [CrossRef]
18. Yan, B.; Li, Y.; Liu, Z.S.; Ren, S.T.; Abidin, I.M.Z.; Chen, Z.M. Pulse-modulation eddy current imaging and evaluation of subsurface corrosion via the improved small sub-domain filtering method. *NDT E Int.* **2021**, *119*, 102404. [CrossRef]
19. Yan, B.; Li, Y.; Liu, Z.S.; Ren, S.T.; Chen, Z.M.; Lu, X.Z.; Abidin, I.M.Z. Pulse-Modulation Eddy Current Imaging for 3D Profile Reconstruction of Subsurface Corrosion in Metallic Structures of Aviation. *IEEE Sens. J.* **2021**, *21*, 28087–28096. [CrossRef]
20. Li, Y.; Tian, G.Y.; Simm, A. Fast analytical modelling for pulsed eddy current evaluation. *NDT E Int.* **2008**, *41*, 477–483. [CrossRef]
21. Theodoulidis, T.; Bowler, J. Eddy-current interaction of a long coil with a slot in a conductive plate. *IEEE T Magn.* **2005**, *41*, 1238–1247. [CrossRef]
22. Jiang, F.; Liu, S.L. Evaluation of cracks with different hidden depths and shapes using surface magnetic field measurements based on semi-analytical modelling. *J. Phys. D Appl. Phys.* **2018**, *51*, 125002. [CrossRef]
23. Yu, Y.T.; Gao, K.H.; Theodoulidis, T.; Yuan, F. Analytical solution for magnetic field of cylindrical defect in eddy current nondestructive testing. *Phys. Scr.* **2020**, *95*, 015501. [CrossRef]
24. Tytko, G.; Dziczkowski, L. I-Cored Coil Probe Located above a Conductive Plate with a Surface Hole. *Meas. Sci. Rev.* **2018**, *18*, 7–12. [CrossRef]
25. Tytko, G.; Dawidowski, L. Locating complex eigenvalues for analytical eddy-current models used to detect flaws. *Compel* **2019**, *38*, 1800–1809. [CrossRef]
26. Theodoulidis, T.; Bowler, J.R. Interaction of an Eddy-Current Coil with a Right-Angled Conductive Wedge. *IEEE T Magn.* **2010**, *46*, 1034–1042. [CrossRef]

27. Delves, L.M.; Lyness, J.N. A numerical method for locating the zeros of an analytic function. *Math. Comput.* **1967**, *21*, 543–560. [CrossRef]
28. Vasic, D.; Ambrus, D.; Bilas, V. Computation of the Eigenvalues for Bounded Domain Eddy-Current Models with Coupled Regions. *IEEE T Magn.* **2016**, *52*, 7004310. [CrossRef]
29. Strakova, E.; Lukas, D.; Vodstrcil, P. Finding Zeros of Analytic Functions and Local Eigenvalue Analysis Using Contour Integral Method in Examples. *Adv. Electr. Electron.* **2017**, *15*, 286–295. [CrossRef]

sensors

MDPI

Article

Acoustic Emission Source Location Using Finite Element Generated Delta-T Mapping

Han Yang [1,2,*], Bin Wang [2], Stephen Grigg [1], Ling Zhu [3,4], Dandan Liu [1,2] and Ryan Marks [1]

1 TWI Ltd., Granta Park, Great Abington, Cambridge CB21 6AL, UK; stephen.grigg@twi.co.uk (S.G.); dandan.liu@brunel.ac.uk (D.L.); ryan.marks@twi.co.uk (R.M.)
2 Department of Mechanical and Aerospace Engineering, Brunel University London, Uxbridge UB8 3PH, UK; bin.wang@brunel.ac.uk
3 Key Laboratory of High Performance Ship Technology (Wuhan University of Technology), Ministry of Education, Wuhan 430063, China; zl79111@hotmail.com
4 School of Naval Architecture, Ocean and Energy Power Engineering, Wuhan University of Technology, Wuhan 430063, China
* Correspondence: han.yang9407@gmail.com; Tel.: +44-(0)-7529-148173

Abstract: One of the most significant benefits of Acoustic Emission (AE) testing over other Non-Destructive Evaluation (NDE) techniques lies in its damage location capability over a wide area. The delta-T mapping technique developed by researchers has been shown to enable AE source location to a high level of accuracy in complex structures. However, the time-consuming and laborious data training process of the delta-T mapping technique has prevented this technique from large-scale application on large complex structures. In order to solve this problem, a Finite Element (FE) method was applied to model training data for localization of experimental AE events on a complex plate. Firstly, the FE model was validated through demonstrating consistency between simulated data and the experimental data in the study of Hsu-Nielsen (H-N) sources on a simple plate. Then, the FE model with the same parameters was applied to a planar location problem on a complex plate. It has been demonstrated that FE generated delta-T mapping data can achieve a reasonable degree of source location accuracy with an average error of 3.88 mm whilst decreasing the time and effort required for manually collecting and processing the training data.

Keywords: Acoustic Emission; Non-Destructive Evaluation; Structural Health Monitoring; delta-T mapping; finite element; Hsu-Nielsen sources; source location; complex plate

Citation: Yang, H.; Wang, B.; Grigg, S.; Zhu, L.; Liu, D.; Marks, R. Acoustic Emission Source Location Using Finite Element Generated Delta-T Mapping. *Sensors* **2022**, *22*, 2493. https://doi.org/10.3390/s22072493

Academic Editor: Zenghua Liu

Received: 16 February 2022
Accepted: 22 March 2022
Published: 24 March 2022

1. Introduction

Ageing-induced cracks, human-caused hazards and natural disasters such as earthquakes or hurricanes can result in the deterioration of in-service metallic structures and loss of structural integrity in the oil and gas industry [1]. Without appropriate remedial measures, structural integrity is likely to reduce further with potential to lead to catastrophic damage with risks of loss of lives and capital. In order to eliminate unplanned shutdowns and reduce inspection and maintenance costs, there is a need for a proactive monitoring approach to assess the existing structural condition and facilitate safe operation. As a method of monitoring the structural behaviour, evaluating structural performance and identifying damage at an early stage, Structural Health Monitoring (SHM) systems such as systems which include strain and vibration monitoring are gaining significant interest and increased application. Recent industrial studies suggest that detecting damage at an early stage can reduce costs and increase the life expectancy of a structure [2]. Therefore, a practical and cost-efficient SHM method that can identify micro and nano defects at an early stage and continually assess the structural health is desirable to ensure structures operate safely and efficiently throughout the service life. Adaptation of Non-Destructive Evaluation (NDE) techniques for SHM is gathering significant interest because of the ability to provide quantitative information on defects based on measured physical parameters [3]. Inspection of large-scale complex structures using NDE techniques such as X-ray and ultrasonic techniques can be laborious and expensive. However, if the damage location can be identified accurately in advance by one of the NDE techniques, i.e. the Acoustic Emission (AE) measurement technique, only the area of concern needs to be inspected. AE is defined as the transient elastic energy released by a material when it experiences a change of deformation in its structure [4]. As a passive technique, only active defects can be detected by AE testing.

Difficulties in applying the most commonly used source location method within the field of AE, Time of Arrival (TOA) technique, for locating active defects in complex structures were reported by Baxter et al. [5] and Pearson et al. [6]. The main assumptions of constant wave speed and unobstructed wave path between the AE source and sensor in TOA technique were not met due to the inhomogeneity of certain materials such as composites and the complexity of the real structures that contain features such as welds, holes and thickness. In order to overcome these difficulties, the delta-T mapping technique was developed by Baxter et al. [5]. This has been shown to improve AE source location accuracy in complex structures. In this technique, a training grid is placed on an area of special interest. The generation of a well-established and widely used artificial source of AE, Hsu-Nielsen (H-N) source [7], at each node of the grid can provide arrival time data of waves for each sensor. The difference in arrival time of waves (or delta-T) for each sensor pair can be found at each point; these data can be employed to generate maps. Once real test data has been collected, the actual delta-Ts can be used to identity a line with same and equal delta-T in corresponding map. By overlaying all the resulting maps, the intersections of all lines should converge into one point, which is the most likely location of a real AE event. Due to the good performance of this technique, it was adopted in further lab-based studies by Eaton et al. [8], Pearson et al. [6] and Grigg et al. [9]. Researchers have also improved localisation by utilising the minimum difference method [10] and generalized regression neural network [11] for identifying the source location from the delta-T map. The details of delta-T mapping technique can be found in Baxter et al. [5]. However, there is a disadvantage to this technique in that the manual collection and processing of training data requires significant time and human input. This lengthy and time-consuming process has prevented this technique from broad applications on large complex structures.

To overcome these problems, Marks [12] used Local Interaction Simulation Approach (LISA) to train the delta-T maps and locate real damage events on a plane of a complex structure. Results of fatigue crack event locations using simulated training data were compared to those from experimentally acquired training data. Similarity in the accuracy of the results of the two approaches demonstrated great potential for the use of modelling data to locate real AE events. The commercially available package based on LISA, cuLISA3D, bears high computational efficiency and thus has a short runtime (for example, running a 370 mm $\times$ 200 mm $\times$ 3 mm model takes approximately 3 min [12]). However, the application of cuLISA3D was very limited. The reported disadvantages of cuLISA3D include lacking the ability to define irregular mesh sizes and to model anisotropic materials such as composites [12]. As the Finite Element (FE) method allows for easier modelling of complex geometrical structures with anisotropic materials properties, it was employed in this study to produce simulated data required for delta-T mapping. Some studies [13–18] for simulation of AE signals from H-N sources using FE method can be found. A 2-D (two-dimensional) Dynamic Finite Element Method (DFEM) was firstly proposed by Gary and Hamstad [13] and extended to a 3-D (three-dimensional) DFEM [14] to simulate AE waveforms generated by an H-N source in thin plates. A step force was used in 2-D and 3-D models to excite the AE waves. Later, the lead break in an H-N source was also simulated using an abrupt change of boundary conditions by Sause [15]. The simulated AE waveforms were compared with those obtained by analytical source functions and the best agreement was found for the 'cosine bell' function [16]. Le Gall et al. [17] also proposed a monopole point source in FE model to simulate AE signals from H-N sources.

The FE model was validated and the influence of different AE sources, specimen geometry and piezoelectric sensor on the AE wave were investigated. Recently, AE waves induced from H-N sources were simulated by Cheng et al. [18] using excitation signals modulated as tone burst with a central frequency of 150 kHz at the FE model. The artificial neural network models were trained using the data collected in FE modelling and shown to predict AE source locations with satisfactory accuracy on a steel–concrete composite girder. Although FE modelling of AE signals from H-N sources has been widely studied over the past 20 years, the authors were unable to source any studies where FE modelled training data were used for delta-T mapping in the literature.

In this study, a numerical method for simulation of an H-N source was validated and applied on a complex plate to train a delta-T map for localization of experimental H-N sources. Results of the present investigation indicated that the FE generated delta-T map decreases the time and effort required for manually collecting and processing the training data, whilst maintaining a reasonable degree of source location accuracy. To the best of our knowledge, this is the first paper in which FE method was used to collect data to train delta-T maps. The FE generated delta-T mapping technique can be applied on monitoring active defects in complex structures as an accurate, low-cost and time-efficient method. The paper is arranged as follows: The numerical method for simulation of an H-N source is described and validated on a simple structure. Then, the delta-t map generated by the FE method is compared with the map from the experiment and traditional TOA techniques in a complex plate. Finally, findings are discussed and concluded in the last section.

2. Numerical Method for Simulating H-N Sources and Its Validation

A numerical method in which a monopole point force pulse acts in the out-of-plane direction was used to model an H-N source. Before applying this approach on a complex plate model, it was used to simulate an H-N source on a simple plate and numerical results were compared with the experimental results to validate the FE model.

2.1. Numerical Method for Simulating H-N Sources

Using the break of a pencil lead to generate acoustic waves, the H-N source [19–21] creates a step-function from a maximum compressive contact force and to 0 value in a very short time period [22]. A broadband range of frequencies is excited by the step-function force, which has been used for the primary sensor calibration in codes and standards [23]. Therefore, a commonly used step-function, 'cosine bell' function, which was proposed by Hamstad [16], was selected to simulate the time-dependent surface deflection on the plate caused by H-N source. The 'cosine bell' force pulse is described in Equation (1). The characteristics of 'cosine bell' function sources are plotted and shown in Figure 1.

$$F(t) = \begin{cases} 0\,N & ,t \leq 0 \\ 0.5 - 0.5 \times \cos\left(\frac{\pi \times t}{T}\right)\,N & ,0 < t < T \\ 1\,N & ,t > T\ (T = 1.5\ \mu\text{s}) \end{cases} \quad (1)$$

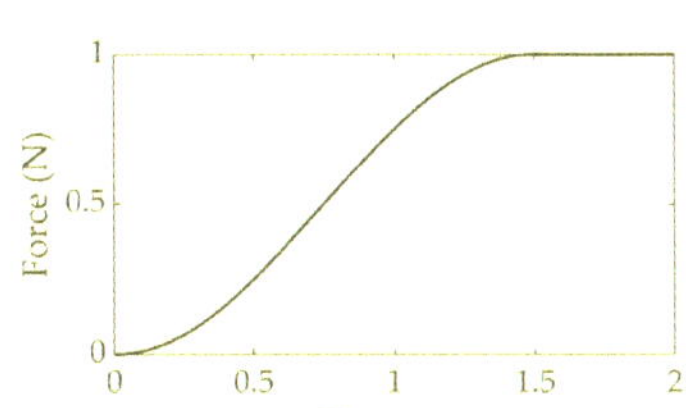

Figure 1. 'Cosine bell' source characteristic.

The pulse was applied to a point on the top surface of the FE model in the out-of-plane direction. As the actual load during a H-N source is unknown, the magnitude of the force was assumed to be 1 N [24]. As shown in Figure 1, the force pulse ramps up from 0 N to

1 N over a time duration of 1.5 μs, with a temporal resolution of 0.1 μs and then remains constant.

2.2. FE Model for Simulating an H-N Source on a Simple Plate

To test the effectiveness of the numerical method described in Section 2.1, a simple plate model was created using a commercial code ABAQUS (Dassault Systèmes Simulia Corp., Providence, RI, USA). This model has a width of 300 mm in the X direction, length 625 mm in the Y direction and thickness 3 mm in the Z direction. Material properties of the steel plate are a density value of 7850 kg/m^3, Young's modulus of 210 GPa, and Poisson's ratio of 0.3. A schematic layout of the sensors and the simulated H-N source is shown in Figure 2. The point force pulse described in Section 2.1 was excited at the location of the simulated H-N source. For the sake of simplicity and saving computing time [18], it was decided not to model an actual sensor and to only consider the out-of-plane displacement since the focus of the work was to model the delta-T mapping process rather than to model a sensor. A single node acted as the basis of the output signal; displacements in Z direction (out-of-plane) over time of all nodes at locations of sensors were used for the analysis. The central 500 mm section of the specimen was considered as test area in the investigation for the purpose of minimizing the influence of the edge reflections on the recorded waves. A coordinate system was introduced, and the origin was placed on the bottom left corner of the test area. The coordinates relative to the origin of four sensors and the H-N source are summarized in Table 1. The four outermost nodes at the bottom surface of the plate (Z = 0 mm) were fixed in X, Y, Z direction (Ux = Uy = Uz = 0) to prevent rigid body motion.

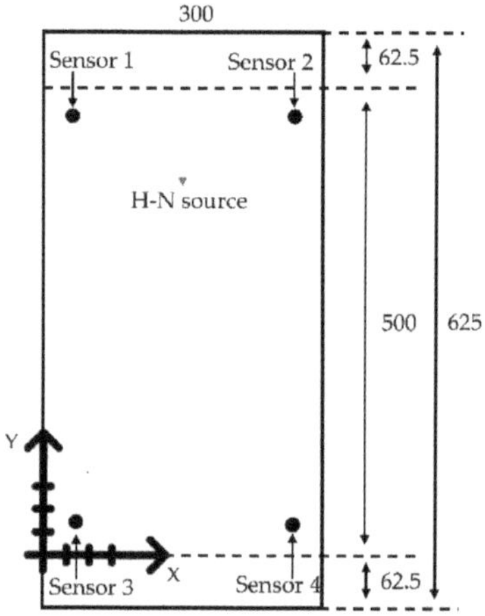

Figure 2. Schematic layout of the sensors and H-N source (units: mm).

Table 1. Coordinates of sensors and H-N source (units: mm).

	X Coordinate	Y Coordinate
Sensor 1	35	465
Sensor 2	265	465
Sensor 3	35	35
Sensor 4	265	35
H-N source	150	400

The convergence of numerical results is strictly dependent upon the temporal and the spatial resolution used in the FE model. To avoid numerical instability and enhance wave propagation accuracy, the shortest wavelength of the simulated elastic waves needs to be resolved by the mesh resolution [25–27].

According to the wavelength formula, the relationship between wavelength λ and frequency f is described by the following simple equation:

$$\lambda = v/f \quad (2)$$

where v is the wave velocity.

The maximum frequency of the wave was set as 500 kHz to align with the frequency range (100–450 kHz) of VS150-RIC (Vallen Systeme GmbH, Wolfratshausen, Germany) sensors used in the following experiments. Assuming the longitudinal and shear wave velocities in steel are 5940 m/s and 3220 m/s, respectively [28], the smallest wavelength, thereby the maximum mesh size for 500 kHz frequency was calculated to be 6.44 mm. As reported by Le Gall et al. [17], at least 5 nodes per wavelength are sufficient to simulate AE waves accurately with frequencies in the range of 500 kHz. Therefore, the mesh resolution of 1 mm, which gives 6 nodes per wavelength, was used for the simulations for both the accuracy and computational efficiency. The mesh convergence test also showed that 1 mm was sufficient mesh resolution for accurate simulation of elastic wave propagation. The general purpose 8-node linear brick element, C3D8R, provided by ABAQUS was used for the FE model. The FE model and mesh are shown in Figure 3.

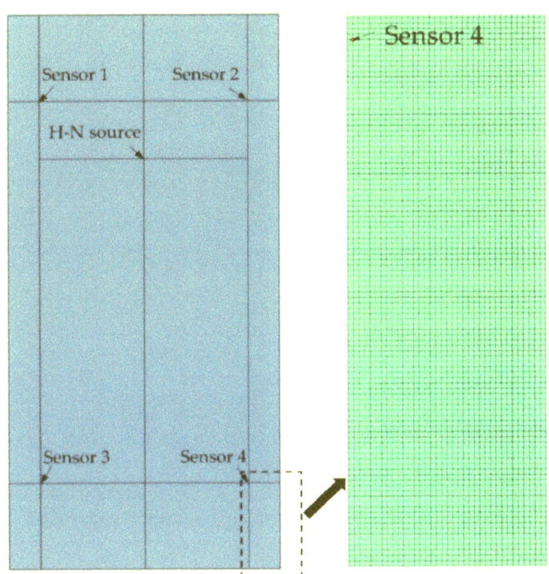

Figure 3. FE model and mesh in ABAQUS.

Choosing an appropriate time integration step is also critical to achieve numerical stability and resolve the high frequency components accurately. An equation shows the relationship between time increment Δt and maximum of frequency f_{max} was recommended by Moser et al. [29] and expressed as:

$$\Delta t < 1/(20 f_{max}) \quad (3)$$

Given that the highest frequency of interest is 500 kHz, a maximum time integration step of 0.1 μs was calculated. In order to save computational time while maintain a sufficient a temporal resolution, a time integration step was set as 0.01 μs in accordance to the reference [15]. The time increment equalled to a sampling rate of 100 MHz, which did not correspond to the sample rate of the experimental tests (5 MHz). However, it should be noted that both the sampling rates used in the modelling and experiments are sufficient to resolve the signals' highest frequency content of interest (500 kHz). Therefore, the error on time estimation between experiments and modelling caused by different sampling rates was marginal.

2.3. Experimental Verification of FE Model for Simulating an H-N Source on a Simple Plate

In parallel, experimental testing of H-N sources on a simple mild steel plate was performed to validate the FE model. The dimensions of this plate were the same as those of the FE model in Section 2.2, 625 mm × 300 mm × 3 mm. The setup of the experiment is shown in Figure 4. In order to easily specify the locations of AE sensors and H-N sources on the plate, a grid with 10 mm spacing was drawn on the central 500 mm section of the specimen. A coordinate system identical to that of FE model was introduced. The coordinates relative to the origin of four sensors and the H-N source are the same as those presented in Table 1.

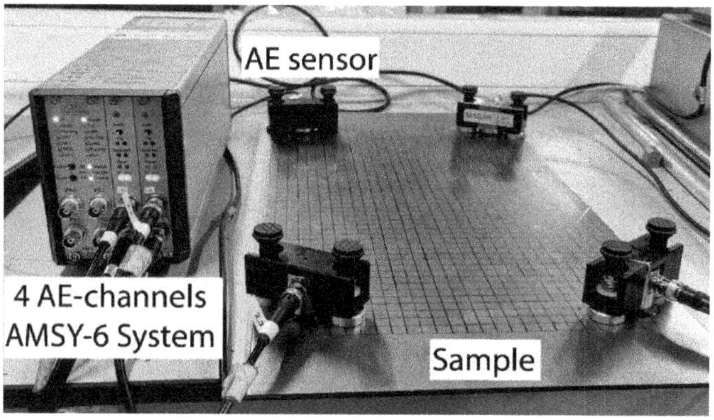

Figure 4. AE sensors mounted on 10 × 10 mm spaced grid specimen and connected to the AE systems.

The AE waveforms were measured by four VS150-RIC sensors, which were acoustically coupled using ultrasound gel and pressure was applied using magnetic clamps. The integral pre-amplifiers with 34 dB gain amplified the signals recorded by sensors and outputs of the transducers were linked to a Vallen 4-channel AMSY-6 System. The signals were acquired using the AE system with the acquisition settings shown in Table 2.

Table 2. AE acquisition settings.

Threshold (dB)	Sample Length (ms)	Sample Rate (MHz)	Pre-Trigger (ms)	Rearm Time (ms)	Duration Discrimination Time (ms)
45	1.6	5	0.1	0.8	0.8

The experiment was conducted in the lab at TWI Ltd., Cambridge, UK. Bubble wrap was used to acoustically de-couple the sample from the workbench surface. Before starting the experiment, an H-N source was generated close to each AE sensor to assess the sensor sensitivity and the coupling between the sensors and testing plate. Hits with amplitude above 98 dB at each sensor were recorded, which verified the coupling. A total of three H-N sources with 0.5 mm 2H type pencils were performed on the location of H-N source given in Table 1 to reduce erroneous data and provide a reliable average result. The amplitude remained constant with discrepancies within a couple of dBs for each sensor.

2.3.1. Modal Analysis

When two surfaces are introduced in a medium (i.e., a plate-like structure), the longitudinal and transverse waves couple at the surfaces and form new surface waves known as Lamb waves. Lamb wave propagation is very complex. Velocities of Lamb wave modes have a strong dependency on frequency content and plate thickness. For a specific geometry, the relationship between wave velocity and frequency of each mode can be described by a set of curves known as dispersion curves. Two major Lamb waves are symmetric zero-order (S_0) and antisymmetric zero-order (A_0) modes.

Given the dispersive properties of elastic waves in flat plates, each wave mode with different frequencies has varied phase and group velocities, which results in the overlapping of some modes. Wavelet Transform (WT) is one of the most useful analysis tools for transient signals. It can be used to plot a time-domain graph showing the magnitude of the frequency spectrum of the signal. Hence, the amplitude of the AE waveform in both time and frequency domains can be viewed simultaneously. By overlapping the calculated dispersion curves with WT, the oncoming wave modes can be clearly distinguished.

Numerical and experimental signals at sensor 1, which were induced by an H-N source whose location was given in Table 1, are presented with WT diagrams processed using a Gabor wavelet in Figures 5 and 6 respectively. Dispersion curves for Lamb wave propagation in a 3 mm steel plate were superimposed onto the WT diagrams. These have been shifted manually to achieve a best fit between the arrival of the fastest S_0 mode in dispersion curve with the arrival of S_0 mode on the WT colour plot.

The movements of particles are mainly parallel to the plate for S_0 mode and perpendicular to the plate for the A_0 mode [30]. When the H-N source is conducted in plane to the sensor face, a large portion of the resulting displacement is normal to the plane of the plate giving rise to A_0 mode, while there is still a small portion of displacement tangential to the plate due to the Poisson effect, giving rise to the S_0 mode. Consequently, A_0 modes have a substantially larger out-of-plane amplitude than S_0 modes [31]. WT for each signal were analysed under two time periods (i.e., before A_0 arrival and after A_0 arrival).

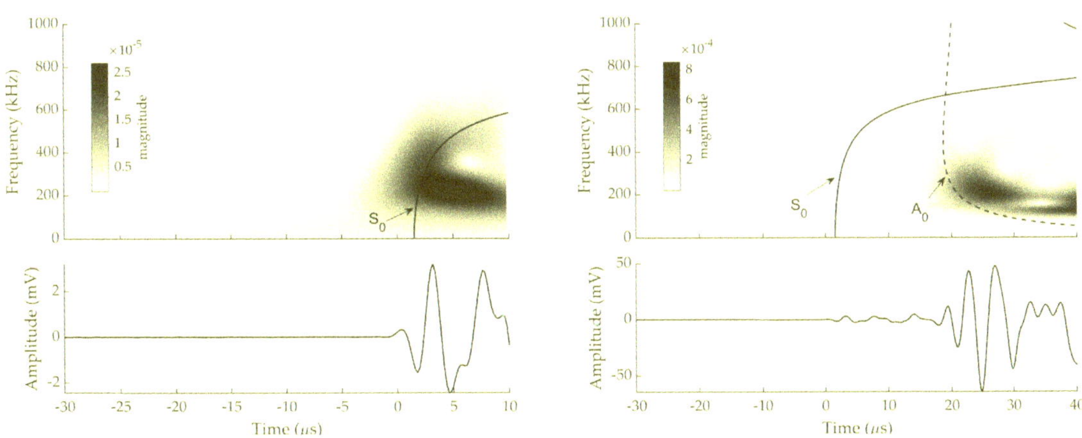

Figure 5. Experimental signal at sensor 1 and corresponding WT diagram with superimposed dispersion curves from: (**a**) −30 to 10 μs (scale was altered to make the S_0 visible); (**b**) −30 to 40 μs.

As can be seen in Figures 5 and 6, S_0 and A_0 are discernible within both experimental and numerical signals even though S_0 and A_0 components were not well separated in the responses of the sensor 1. The S_0 components have a faster velocity than the A_0 components [32]. It is shown that the arrival of low amplitude S_0 components are followed by high amplitude A_0 component. The first-arrival waves were S_0 modes with low amplitude, which were followed by the A_0 modes with very high amplitude as expected.

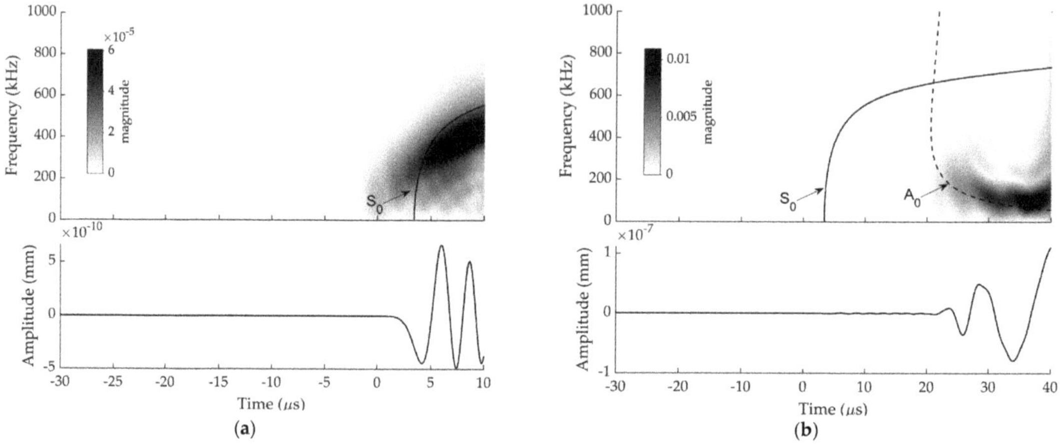

Figure 6. Numerical signal at sensor 1 and corresponding WT diagram with superimposed dispersion curves from: (**a**) −30 to 10 μs (scale was altered to make the S_0 visible); (**b**) −30 to 40 μs.

It can be observed that the experimental and numerical WT diagrams show a reasonably good agreement. The discrepancy between the experimental and numerical WT magnitude at sensor 1 can be explained by the frequency response of the actual sensors not being simulated in the FE modelling.

2.3.2. Arrival Time Estimation

Determination of arrival time of signals is crucial for localization of the AE source. Three arrival time estimation methods, which are threshold crossing, the Akaike Information Criterion (AIC) and WT analysis are discussed in the following.

The default time arrival estimation method in commercially available AE systems is threshold crossing. In this method, the arrival timing will be triggered, and an AE hit will be recorded only when a signal crosses a pre-set threshold level. Because of factors including the assumptions on the magnitude of the force, the time duration during an H-N source and a simplified sensor model, accurate prediction of AE signals' amplitudes using the FE model were not achieved and not within the scope of this work. Moreover, displacement was measured in FE model while voltage was measured in the experiment. Consequently, a same threshold level cannot be found for both experimental and numerical signals in the threshold-crossing method. Therefore, this method was only used to estimate arrival times of experimental signals.

For the threshold-crossing method, it is difficult to choose the appropriate threshold level. Triggering errors might come from missing the start of the wave as shown in Figure 7b. A reduced threshold can improve the arrival time measurement accuracy [6]. However, background noise bearing a lower amplitude than the pre-set threshold can falsely trigger AE acquisition, resulting in erroneous arrival time. Therefore, in addition to threshold-crossing method, AIC was also used in this study to estimate the arrival times of AE signals. The AIC function [33] compares the difference in classic variance before and after each point in the wave, typically when it is at its minimum represents the point where the transition from noise to waveform is occurring, and therefore the arrival time of the wave. It can be observed that the threshold-crossing error was around 2 μs on a high-amplitude signal from an H-N source in Figure 7a, while it was around 65 μs on a low-amplitude signal from a corrosion test in Figure 7b which is presented for demonstration. However, the AIC method detected the accurate onset time of signal regardless of changes to the amplitude.

Moreover, the robust performance of AIC function to pick the arrival times of signals with varied signal-to-noise ratios was demonstrated by Kurz et al. [33] and Pearson et al. [6]. Therefore, the AIC method has been demonstrated to be a reliable onset time determination tool.

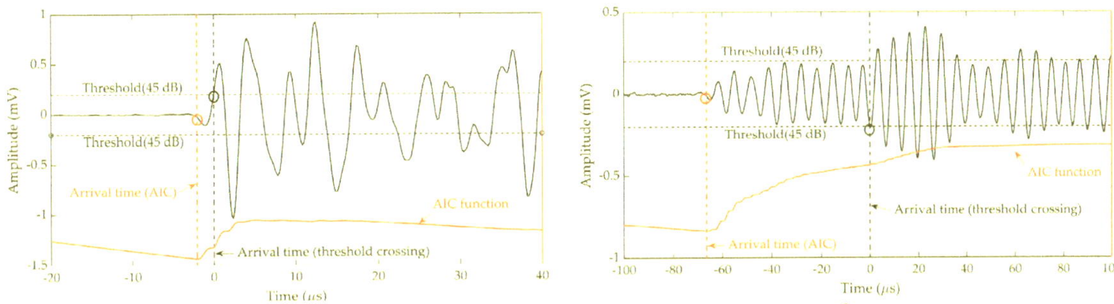

Figure 7. An example of arrival time estimation using AIC and threshold crossing on (**a**) high amplitude and (**b**) low amplitude AE signals.

The arrival times can also be predicted by considering the peak WT magnitudes at a mode known frequency, as shown by Yamada et al. [34], Hamstad et al. [35] and Jeong and Jang [36]. It was demonstrated that the maximum magnitude of WT in the time-frequency domain corresponds to the arrival time of mode wave traveling with group velocity. Hence, the arrival times of specific frequency component can be determined. In this study, different dominant Lamb wave modes over time were shown using WT in Section 2.3.1. The frequency component of 200 kHz was used in the WT analysis to determine the arrival time of mode waves. This frequency was chosen because S_0 modes can be observed at this frequency according to the spectral analysis of captured signals and the peak wavelet coefficients of S_0 modes are conspicuous at this frequency. The arrival time of S_0 modes at 200 kHz was determined based on the peaks of the WT coefficients at 200 kHz.

In general, threshold crossing was only used to estimate the arrival time of experimental signals induced by an H-N source on a simple plate and AIC and WT analysis were used on both experimental and numerical signals. For simplicity, delta-T for sensor 1 and sensor 3 was calculated. Results are summarized in Table 3.

Table 3. Delta-T for sensor 1 and sensor 3 in the experiment and FE modelling (units: μs).

	Threshold Crossing	AIC	WT Analysis
Experiment	48.4	46.8	47.2
FE modelling	-	46.74	46.57
Difference	-	0.06	0.63

The estimated delta-T produced by threshold crossing, AIC and WT analysis in the experiment showed very little difference. Compared with delta-T determined by AIC, threshold crossing and WT analysis showed a discrepancy of 1.6 μs and 0.4 μs, respectively. The simulation predictions appeared to be well in agreement with experimental observations. For the delta-T estimated by AIC, a 0.06 μs discrepancy was given between the experiment and FE modelling. Good agreement was also demonstrated between delta-T in the experiment predicted by WT analysis and that of modelling, with a discrepancy of 0.63 μs. Moreover, for the simulation results, delta-T given by WT analysis was very close to that given by AIC.

Overall, experimental and numerical WT diagrams show a reasonably good agreement. A high level of accuracy of the delta-T from FE modelling can be observed. It can thus be concluded that a numerical method for simulation of an H-N source on a simple plate was validated. The validation of this numerical method demonstrated great potential on making it a useful tool for predicting arrival times of AE waveforms induced by H-N sources on a complex FE plate model. For arrival time estimation, results show that AIC represents a viable option for estimating the onset of signals. As the accuracy of threshold crossing depends on the pre-set threshold level, and the arrival time determined by WT analysis is the time at a single frequency, AIC was used to estimate arrival time of signals in numerical and experimental delta-T techniques.

3. Experimental and Numerical Delta-T Mapping Training on Complex Plate

Using the validated numerical method described in Section 2.1, H-N sources were simulated on a complex plate model. Information including arrival times of signals at each sensor was collected to build the numerical delta-T map. In parallel, experimental delta-T mapping was also calculated on a complex plate for the localization of experimental AE events. Localization results produced by numerical and experimental delta-T mapping were compared.

3.1. Experimental Delta-T Mapping Training on Complex Plate

The delta-T mapping technique requires the structure to be mapped whereby test data are compared to the training maps which are created by performing H-N sources and evaluating the difference of arrival times for each sensor pair to estimate the source location. The procedural steps of implementing the delta-t mapping technique [5] are presented in Figure 8. By performing H-N sources on each node of the grid, factors including complex geometric features can be considered within mapping area and detailed accurate wave speed data are not required. Therefore, this technique shows its advantage on source location in complex structures.

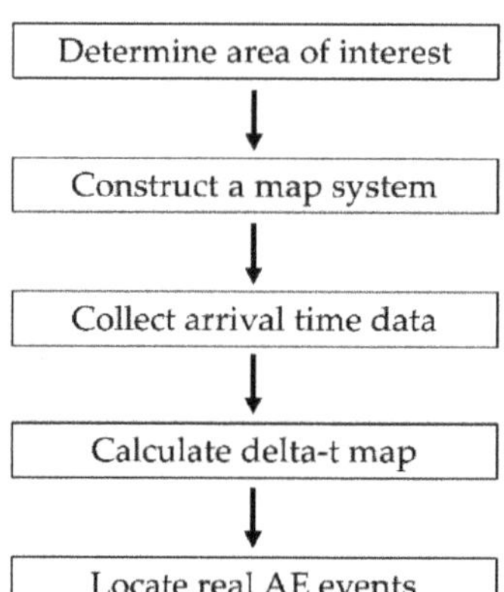

Figure 8. Procedural steps of implementing delta-T mapping technique [5].

A complex geometry mild steel plate with four holes was used for experimental delta-T mapping training. The dimensions of the specimen with coordinates of the centres of four holes and radiuses are shown in Figure 9. The locations of all the sensors were the same as those in the experiment on simple plate described in Section 2.3. Following procedural steps shown in Figure 8, the central 500 mm section was determined as the area of interest and a coordinate system was introduced. After that, a training grid was placed on this area. The spacing of the grid was determined to be 20 mm as prior research [6] has shown that this is the point at which further reduction in size has minimal influence on accuracy. The arrival time data, which were determined by AIC, were collected after the generation of an H-N source at each node of the grids. For each sensor pair, a delta-T map was produced. The number of maps produced will depend on the number of sensors given by the equation

$N * (N-1)/2$, where N is the number of sensors in the network. By comparing the delta-T for an unknown source with those of the maps, the location can be determined. It should be noted that there is a significant interaction of waves with holes and plate edges. However, the basis of the delta-T mapping is the first arrival time of signals, and this interaction is beyond the scope of this study.

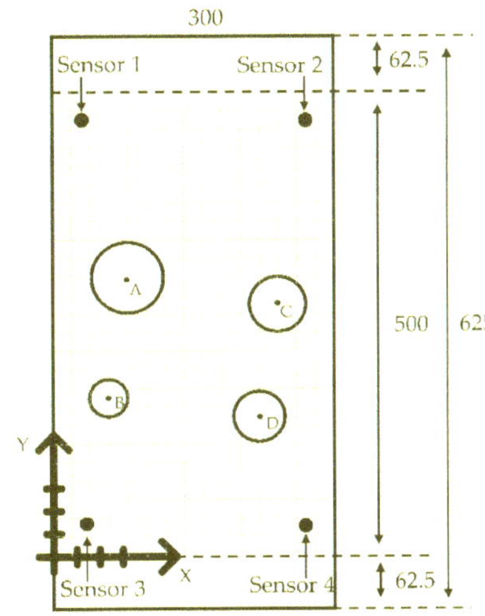

	X coordinate	Y coordinate	Radius
A	80	300	40
B	60	170	20
C	240	270	30
D	220	150	30

Figure 9. Schematic layout of the sensors on a complex plate model with 20 mm spacing grid (unit: mm).

3.2. FE Generated Delta-T Mapping Training on Complex Plate

In parallel, FE modelling was performed to produce a FE generated delta-T map or numerical delta-T map. As shown in Figure 10, a complex plate model same as the plate in Section 3.1 was created in ABAQUS. As with the FE modelling of H-N sources on simple plate in Section 2.2, steel material properties, mesh resolution of 1 mm, time integration step of 0.01 µs, C3D8R element and same boundary conditions were used for simulation. The locations of sensors and the grid were chosen to be the same as those of the experiment on complex plate in Section 3.1. The validated numerical method described in Section 2.1 was used to simulate an H-N source on each node of the grid. Displacements in Z direction (out-of-plane) over time of all nodes at locations of sensors were recorded in FE modelling to estimate the arrival time. AIC was used to determine the arrival times of simulated signals. The FE generated delta-T map was built in the same way as experimental delta-T map in Section 3.1.

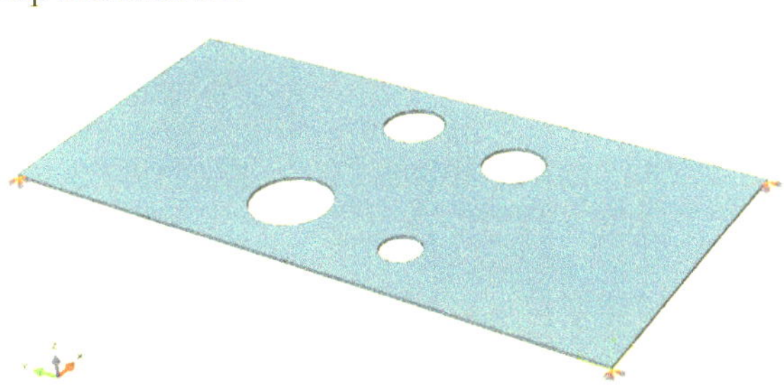

Figure 10. Complex geometry plate modelled in ABAQUS.

3.3. Experimental Test Data on Complex Plate

In order to test the performance of experimental and numerical delta-T maps, six extra H-N sources (H-N source 1, 2, 3, 4, 5 and 6) were conducted at six off grid locations on complex plate in the experiment. The coordinates of the locations of six extra H-N sources relative to the origin are given in Table 4.

Table 4. Actual locations of H-N sources and results calculated by TOA-TC, TOA-AIC, numerical delta-T and experimental delta-T (units: mm).

	H-N Source 1			H-N Source 2			H-N Source 3			H-N Source 4			H-N Source 5			H-N Source 6		
	X	Y	Error	X	Y	Error	X	Y	Error	X	Y	Error	X	Y	Error	X	Y	Error
Actual	70	70	-	70	210	-	130	90	-	170	390	-	230	230	-	250	430	-
TOA-TC	74.42	59.19	11.68	83.04	207.21	13.34	134.54	79.42	11.51	168.99	397.51	7.58	213.60	223.92	17.49	244.54	438.59	10.18
TOA-AIC	77.13	61.63	11.00	81.29	210.88	11.32	133.89	83.19	7.84	167.79	393.18	3.87	222.44	224.29	9.47	246.58	434.70	5.81
Numerical delta-T	74.82	67.80	5.30	73.99	211.46	4.25	132.26	91.41	2.66	169.33	386.67	3.40	225.27	229.86	4.73	251.17	432.7	2.94
Experimental delta-T	71.54	66.80	3.55	67.39	210.13	2.61	132.89	90.03	2.89	169.93	390.35	0.36	231.13	230.45	1.22	248.96	433.33	3.49

3.4. Results

The AE test data collected from H-N source 1, 2, 3, 4, 5 and 6 were assessed with the traditional TOA location approach based on threshold crossing (or TOA-TC for short), TOA location approach based on AIC method (or TOA-AIC), experimental delta-T mapping (or experimental delta-T) and numerical delta-T mapping (or numerical delta-T) techniques. Table 4, Figures 11 and 12 documented source location results and the associated Euclidian distance errors.

For H-N source 1 and 2, it can be observed that a small improvement in accuracy was observed after the AIC method was used in the TOA localization technique. The errors were reduced significantly using the numerical delta-T and experimental delta-T respectively compared with that of the TOA-TC. For H-N source 3 and 6, the TOA-AIC produced location errors of 7.84 mm and 5.81 mm compared with 11.51 mm and 10.18 mm calculated by TOA-TC, respectively. Results at H-N source 3 and 6 showed that the experimental and numerical training approaches produced similar levels of error, with a slightly higher accuracy given by numerical training approach. For H-N source 4, TOA-TC had a location error of 7.58 mm compared with 3.87 mm using TOA-AIC, a 49% reduction using TOA-AIC. The results showed little difference on location error when TOA-AIC and numerical delta-T were used, with numerical delta-T having slightly increased accuracy. However, the accuracy of source location was significantly better using experimental delta-T with a Euclidean distance error of 0.36 mm. For H-N source 5, the TOA-TC, TOA-AIC and numerical delta-T located the H-N source slightly to the left of the actual position with a Euclidean error distance of around 17.49 mm, 9.47 mm and 4.73 mm, respectively, while experimental delta-T located the source to the right of the actual location with an error of 1.22 mm. Experimental delta-T and numerical delta-T showed an obvious improvement over the TOA-TC and TOA-AIC. The average errors of H-N source locations calculated by experimental and numerical delta-T are 2.35 mm and 3.88 mm, respectively.

In general, the TOA-TC technique produced highest location errors because of indirect wave path and unreliable arrival time estimation method. Since the AIC method is considered to be more accurate in arrival time determination than threshold crossing [6], TOA-AIC led to a marked improvement in accuracy. As had been discovered in previous works [1,6,8,9,37,38], the delta-T mapping technique improved the source location accuracy on the complex plate. The presented results show the noticeable improvement in source location accuracy using experimental delta-T and numerical delta-T methods over TOA-TC and TOA-AIC. Because a coarse training grid was used in experimental delta-T and numerical delta-T, the accuracy of the location results may be further improved with a higher spatial resolution of the training grid. Other than H-N source 3 and 6, experimental delta-T shows a higher degree of accuracy over numerical delta-T, which can be explained by errors of FE analysis in numerical delta-T including the modelling errors associated with the simplification and mesh discretization error. It is also worth noting that the locations of the artificial damage were calculated with sufficient accuracy by both experimental delta-T and numerical delta-T.

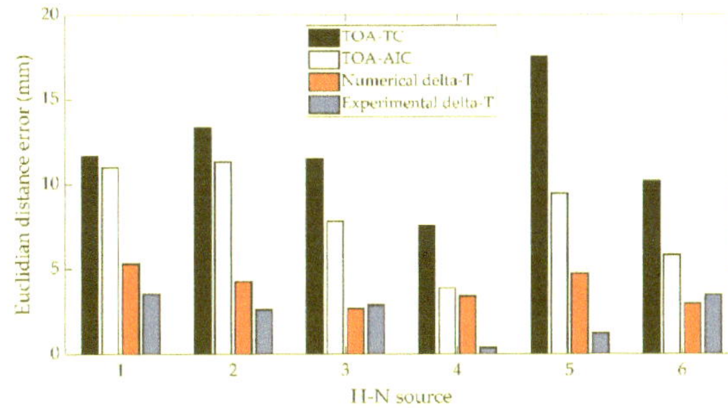

Figure 11. Euclidian distance errors of source location results.

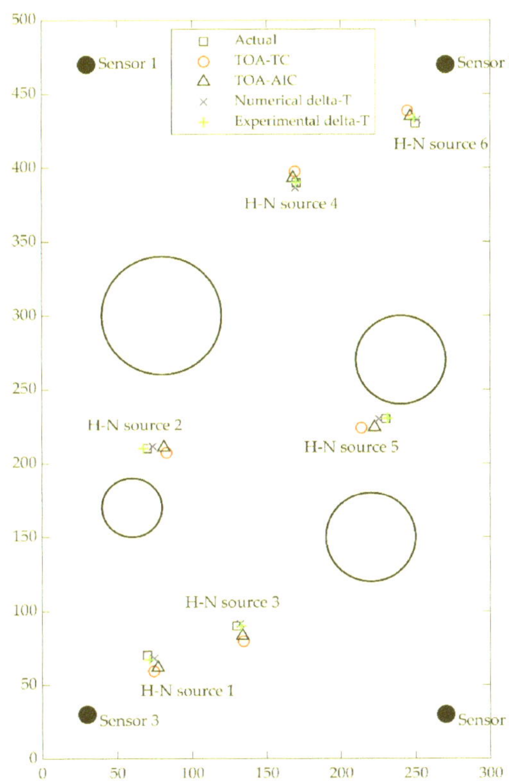

Figure 12. Actual locations of H-N sources and results calculated by TOA-TC, TOA-AIC, numerical delta-T and experimental delta-T (units: mm).

4. Discussions

The FE generated or numerical delta-T mapping techniques offer certain advantages over experimental delta-T. Firstly, for experimental delta-T, it requires several days to set up the experiment, perform H-N sources on all the grid points, select and prepare the AE data to build the delta-T maps. For the numerical delta-T, it takes 0.5 days to build an FE model. The computation time for simulation of an H-N source on each node of the grid is 5 min on a personal laptop. There is a total of 285 nodes in the present model, which results in a computation time of 1 day. The total time required for the whole process is 1.5 days. This reveals that the numerical delta-T decreases the time required for constructing a delta-T map. It is worth noting that processes of experimental and numerical delta-T mapping could be optimised to reduce these times. To ensure maximum accuracy, an operator with an AE background is required to manually sort and discard useless arrival time data being captured experimentally [38]. With numerical delta-T, data are captured autonomously, which means that collection and processing of training data do not rely on experience and human errors are removed. Moreover, the expenses associated with the equipment and labour required for experiments can be eliminated. Furthermore, the numerical delta-T mapping method scales better compared with experimental delta-T. With the FE method, large-scale models with numerical sensors can be easily created for collection of training data, whilst experimental collection of delta-T data may be limited by several factors including large size or complex structures. Moreover, instead of artificial AE sources which can only be performed on the surface in experiments, the FE method allows multiple training sources such as fatigue [39] and fibre breaks [40] and internal damage to be modelled. This might provide a solution to a 3-D delta-T location problem. A final benefit of numerical delta-T is that the risk of foreign object debris from broken pencil leads is eliminated, which in some sectors is a major consideration.

However, numerical delta-T also bears some disadvantages. One of the main difficulties lies in the modelling of wave propagation in highly complex structures, such as composites which are being used abundantly in safety critical structures in recent years. Furthermore, fine mesh is required to model AE waves with high accuracy. The number of degrees of freedom in a large complex FE model will be huge, leading to a significant demand for computational resources such as memory space and processor time. The solution for this could be the use of fine meshes over the area of interest and coarse meshes over other areas.

Although the FE generated delta-T was carried out successfully, there are some limitations to this study. Firstly, one such limitation is that TOA-TC, TOA-AIC, numerical delta-T and experimental delta-T have not been trialled to locate real AE data. As real damage sources [12] usually exhibit a smaller amplitude and different frequency content compared with H-N sources, it will be of interest to examine the performance of four methods with AE signals generated from real damage mechanisms. Secondly, only holes were considered in the complex plate in the present study; however, other complexities of real structures including multiple thickness changes, stiffeners, holes, nozzles, welds, etc., need to be considered in future study. Thirdly, the defined boundary conditions in the model were not the same as real boundary conditions of the plate. The wave propagation will be slightly influenced by the edge of the hole on the complex plate and thus might result in errors in arrival time. By applying realistic boundary conditions in the FE model, the influence of the edges of holes on the first arrival of AE waves can be eliminated. The accuracy of arrival time of signals recorded in complex FE model can be improved. Moreover, another source of error is that the AE sensor face is not considered, as the sensor is simplified as a point in the modelling. It could be stated that the signal in experimental training is captured at the sensors' circumference rather than sensors' centre point where simulated sensors are located, thereby resulting in different arrival times. A numerical study conducted by Tsangouri and Aggelis [41] showed that the size of the simulated AE sensor bears a direct effect on wave content in time domains. The source location accuracy produced by finite element generated delta-T maps may be improved if the AE sensor with the sensor face can be modelled. Furthermore, the material defined in the FE model was an isotropic homogenous metallic material which was considered in this investigation. As composites are being increasingly used in the oil and gas industry, it would be beneficial to examine the performance of a modelled delta-T training dataset on composite materials. Moreover, it should be noted that the delta-T mapping technique and other localization techniques do not account for factors such as temperature, loading and sensor coupling changes. It will be of interest to investigate how these factors will influence the delta-T maps and identifying the methods to update delta-T maps accordingly. Finally, current commercial AE systems remain incapable of the live application of experimental and numerical delta-T mapping techniques. However, it will be beneficial to integrate these methods with commercial AE systems to observe online AE events.

5. Conclusions

In this investigation, a numerical method from the literature was used to simulate an H-N source on a simple plate and was validated by the experiment results. The validation of the FE model exhibited great promise for application of the FE method to a planar location problem on a complex plate. Using the same numerical method, H-N sources were simulated on a complex plate and a numerical delta-T map was generated. The location results of the numerical delta-t map technique were compared with those of traditional TOA techniques and an experimental delta-t map technique. The viability of using the FE method was demonstrated to decrease the time and labour required for manually collecting and processing training data whilst maintaining a reasonable degree of source location accuracy with an average error of 3.88 mm. With such a high source location accuracy, the specific area of concern requiring inspection using other NDE techniques can be greatly reduced. Moreover, if the specific location of an AE event is identified on an area with certain geometric features or loading conditions, the number of potential damage source mechanisms can be decreased.

Although the FE generated delta-T mapping was conducted successfully, it would be preferable to examine the performance of present FE generated delta-T maps with AE signals from real damage mechanisms. Development of the FE model such as improved boundary conditions, realistic AE sensors, additional complexity and anisotropic materials requires further investigation.

Author Contributions: Conceptualization, H.Y., S.G. and R.M.; methodology, H.Y., R.M., S.G. and L.Z.; software, H.Y. and L.Z.; validation, R.M., S.G. and B.W.; formal analysis, H.Y.; investigation, H.Y. and D.L.; resources, R.M., S.G. and B.W.; data curation, H.Y.; writing—original draft preparation, H.Y.; writing—review and editing, R.M., S.G. and B.W.; visualization, H.Y., R.M., S.G. and B.W.; supervision, R.M., S.G. and B.W. All authors have read and agreed to the published version of the manuscript.

Funding: This research is funded by Lloyd's Register Foundation.

Institutional Review Board Statement: Not applicable.

Informed Consent Statement: Not applicable.

Data Availability Statement: The data presented in this study are available on request from the corresponding author.

Acknowledgments: This publication was made possible by the sponsorship and support of Lloyd's Register Foundation and Brunel University London. The work was enabled through and undertaken at the National Structural Integrity Research Centre (NSIRC), a postgraduate engineering facility for industry-led research into structural integrity established and managed by TWI through a network of both national and international Universities. Lloyd's Register Foundation helps to protect life and property by supporting engineering-related education, public engagement and the application of research.

Conflicts of Interest: The authors declare no conflict of interest.

References

1. Biondini, F.; Frangopol, D.M. Life-cycle performance of deteriorating structural systems under uncertainty. *J. Struct. Eng.* **2016**, *142*, F4016001. [CrossRef]
2. Jinachandran, S.; Rajan, G. Fibre Bragg Grating Based Acoustic Emission Measurement System for Structural Health Monitoring Applications. *Materials* **2021**, *14*, 897. [CrossRef]
3. Raj, B.; Jayakumar, T.; Rao, B.P.C. Non-destructive testing and evaluation for structural integrity. *Sadhana* **1995**, *20*, 5–38. [CrossRef]
4. Miller, R.K.; Hill, E.V.K.; Moore, P.O. *Nondestructive Testing Handbook, Acoustic Emission Testing*; The American Society For Nondestructive Testing Inc.: Columbus, OH, USA, 2005; Volume 5, p. 541.
5. Baxter, M.G.; Pullin, R.; Holford, K.M.; Evans, S.L. Delta T source location for acoustic emission. *Mech. Syst. Signal Process.* **2007**, *21*, 1512–1520. [CrossRef]
6. Pearson, M.R.; Eaton, M.; Featherston, C.; Pullin, R.; Holford, K. Improved acoustic emission source location during fatigue and impact events in metallic and composite structures. *Struct. Health Monit.* **2017**, *16*, 382–399. [CrossRef]
7. *ASTM E650*; Standard Guide for Mounting Piezoelectric Acoustic Emission Sensors. ASTM International: West Conshohocken, PA, USA, 2017; Volume 97, pp. 1–4.
8. Eaton, M.J.; Pullin, R.; Holford, K.M. Acoustic emission source location in composite materials using Delta T Mapping. *Compos. Part A Appl. Sci. Manuf.* **2012**, *43*, 856–863. [CrossRef]
9. Grigg, S.; Featherston, C.A.; Pearson, M.; Pullin, R. Advanced Acoustic Emission Source Location in Aircraft Structural Testing. *IOP Conf. Ser. Mater. Sci. Eng.* **2021**, *1024*, 12029. [CrossRef]
10. Liu, Z.; Peng, Q.; He, C.; Wu, B. Time difference mapping method for locating the acoustic emission source in a composite plate. *Chin. J. Acoust.* **2019**, *38*, 490–503.
11. Liu, Z.; Peng, Q.; Li, X.; He, C.; Wu, B. Acoustic emission source localization with generalized regression neural network based on time difference mapping method. *Exp. Mech.* **2020**, *60*, 679–694. [CrossRef]
12. Marks, R. Methodology Platform for Prediction of Damage Events for Self-Sensing Aerospace Panels Subjected to Real Loading Conditions. Ph.D. Thesis, Cardiff University, Cardiff, UK, 2016.
13. Gary, J.; Hamstad, M.A. On the far-field structure of waves generated by a pencil lead break on a thin plate. *J. Acoust. Emiss.* **1994**, *12*, 157–170.
14. Hamstad, M.A.; Gary, J.; O'Gallagher, A. Far-field acoustic emission waves by three-dimensional finite element modeling of pencil-lead breaks on a thick plate. *J. Acoust. Emiss.* **1996**, *14*, 103–114.
15. Sause, M. Investigation of pencil-lead breaks as acoustic emission sources. *J. Acoust. Emiss.* **2011**, *29*, 184–196.
16. Hamstad, M.A. Acoustic emission signals generated by monopole (pencil lead break) versus dipole sources: Finite element modeling and experiments. *J. Acoust. Emiss.* **2007**, *25*, 92–106.
17. Le Gall, T.; Monnier, T.; Fusco, C.; Godin, N.; Hebaz, S.-E. Towards quantitative acoustic emission by finite element modelling: Contribution of modal analysis and identification of pertinent descriptors. *Appl. Sci.* **2018**, *8*, 2557. [CrossRef]
18. Cheng, L.; Xin, H.; Groves, R.M.; Veljkovic, M. Acoustic emission source location using Lamb wave propagation simulation and artificial neural network for I-shaped steel girder. *Constr. Build. Mater.* **2021**, *273*, 121706. [CrossRef]
19. Nielsen, A. *Acoustic Emission Source Based on Pencil Lead Breaking*; The Danish Welding Institute Publication: Copenhagen, Denmark, 1980; pp. 15–18.
20. Hsu, N.N.; Breckenridge, F.R. Characterization and calibration of acoustic emission sensors. *Mater. Eval.* **1981**, *39*, 60–68.
21. *ASTM E976-15*; Standard Guide for Determining the Reproducibility of Acoustic Emission Sensor Response. ASTM International: West Conshohocken, PA, USA, 2021; pp. 1–7.
22. Grosse, C.U.; Ohtsu, M. *Acoustic Emission Testing*; Springer Science & Business Media: Berlin/Heidelberg, Germany, 2008; ISBN 3540699724.
23. *ASTM E1106-12*; Standard Test Method for Primary Calibration of Acoustic Emission Sensors. ASTM International: West Conshohocken, PA, USA, 2021; pp. 1–13.
24. Prosser, W.H.; Hamstad, M.A.; Gary, J.; O'Gallagher, A. Finite Element and Plate Theory Modeling of Acoustic Emission Waveforms. *J. Nondestruct. Eval.* **1999**, *18*, 83–90. [CrossRef]
25. Castaings, M.; Bacon, C.; Hosten, B.; Predoi, M.V. Finite element predictions for the dynamic response of thermo-viscoelastic material structures. *J. Acoust. Soc. Am.* **2004**, *115*, 1125–1133. [CrossRef]
26. Greve, D.W.; Neumann, J.J.; Nieuwenhuis, J.H.; Oppenheim, I.J.; Tyson, N.L. Use of Lamb waves to monitor plates: Experiments and simulations. In Proceedings of the Smart Structures and Materials 2005: Sensors and Smart Structures Technologies for Civil, Mechanical, and Aerospace Systems, San Diego, CA, USA, 7–10 March 2005; Volume 5765, pp. 281–292.
27. Nienwenhui, J.H.; Neumann, J.J.; Greve, D.W.; Oppenheim, I.J. Generation and detection of guided waves using PZT wafer transducers. *IEEE Trans. Ultrason. Ferroelectr. Freq. Control* **2005**, *52*, 2103–2111. [CrossRef]
28. Engineering ToolBox. Solids and Metals—Speed of Sound. Available online: https://www.engineeringtoolbox.com/sound-speed-solids-d_713.html (accessed on 4 February 2022).
29. Moser, F.; Jacobs, L.J.; Qu, J. Modeling elastic wave propagation in waveguides with the finite element method. *NDT E Int.* **1999**, *32*, 225–234. [CrossRef]
30. Gorman, M.R. Plate wave acoustic emission. *J. Acoust. Soc. Am.* **1991**, *90*, 358–364. [CrossRef]

31. Pullin, R.; Holford, K.M.; Baxter, M.G. Modal analysis of acoustic emission signals from artificial and fatigue crack sources in aerospace grade steel. *Key Eng. Mater.* **2005**, *293–294*, 217–224. [CrossRef]
32. Mohd, S.; Holford, K.M.; Pullin, R. Continuous wavelet transform analysis and modal location analysis acoustic emission source location for nuclear piping crack growth monitoring. *AIP Conf. Proc.* **2014**, *1584*, 61–68.
33. Kurz, J.H.; Grosse, C.U.; Reinhardt, H.W. Strategies for reliable automatic onset time picking of acoustic emissions and of ultrasound signals in concrete. *Ultrasonics* **2005**, *43*, 538–546. [CrossRef]
34. Yamada, H.; Mizutani, Y.; Nishino, H.; Takemoto, M.; Ono, K. Lamb wave source location of impact on anisotropic plates. *J. Acoust. Emiss.* **2000**, *18*, 51.
35. Hamstad, M.A.; Gary, J. A wavelet transform applied to acoustic emission signals: Part 2: Source location. *J. Acoust. Emiss.* **2002**, *20*, 62–82.
36. Jeong, H.; Jang, Y.-S. Wavelet analysis of plate wave propagation in composite laminates. *Compos. Struct.* **2000**, *49*, 443–450. [CrossRef]
37. Pullin, R.; Baxter, M.; Eaton, M.; Holford, K.M.; Evans, S.L. Novel acoustic emission source location. *J. Acoust. Emiss.* **2007**, *25*, 215–223.
38. Al-Jumaili, S.K.; Pearson, M.R.; Holford, K.M.; Eaton, M.J.; Pullin, R. Acoustic emission source location in complex structures using full automatic delta T mapping technique. *Mech. Syst. Signal Process.* **2016**, *72*, 513–524. [CrossRef]
39. Cuadra, J.; Vanniamparambil, P.A.; Servansky, D.; Bartoli, I.; Kontsos, A. Acoustic emission source modeling using a data-driven approach. *J. Sound Vib.* **2015**, *341*, 222–236. [CrossRef]
40. Hamam, Z.; Godin, N.; Fusco, C.; Doitrand, A.; Monnier, T. Acoustic Emission Signal Due to Fiber Break and Fiber Matrix Debonding in Model Composite: A Computational Study. *Appl. Sci.* **2021**, *11*, 8406. [CrossRef]
41. Tsangouri, E.; Aggelis, D.G. The influence of sensor size on acoustic emission waveforms—A numerical study. *Appl. Sci.* **2018**, *8*, 168. [CrossRef]

sensors

Article

A New Probabilistic Ellipse Imaging Method Based on Adaptive Signal Truncation for Ultrasonic Guided Wave Defect Localization on Pressure Vessels

Qinfei Li, Zhi Luo, Gangyi Hu and Shaoping Zhou *

School of Mechanical and Power Engineering, East China University of Science and Technology, Shanghai 200237, China; liqinfei@mail.ecust.edu.cn (Q.L.); zhiluo@ecust.edu.cn (Z.L.); gy_hu0202@163.com (G.H.)
* Correspondence: shpzhou@ecust.edu.cn

Abstract: Pressure vessels are prone to defects due to environmental conditions, which may cause serious safety hazards to industrial production. The probabilistic ellipse imaging method, based on ultrasonic guided wave, is a common method for locating defects on plate-like structures. In this paper, the research showed that the accuracy of the traditional probabilistic ellipse imaging method was severely affected by the truncation length of the signal. In order to improve the defect location accuracy of the probabilistic elliptic imaging algorithm, an adaptive signal truncation method based on signal difference analysis was proposed, and a novel probabilistic elliptic imaging method was developed. Firstly, the relationship model between the signal difference coefficient (SDC) and the distance coefficient was constructed. Through this model, the distance coefficient of each group signal can be calculated, so that the adaptive truncation length for each group of signals can be determined and the truncated signals used for defect imaging. Secondly, in order to improve the robustness of the new imaging method, the relationship between the defect location accuracy and SDC thresholds were investigated and the optimal threshold was determined. The experimental results showed that the probabilistic ellipse imaging algorithm, based on the new adaptive signal truncation method, can effectively locate a single defect on a pressure vessel.

Keywords: pressure vessel; ultrasonic guided wave; defect location; signal analysis; probabilistic elliptic algorithm; adaptive signal truncation

Citation: Li, Q.; Luo, Z.; Hu, G.; Zhou, S. A New Probabilistic Ellipse Imaging Method Based on Adaptive Signal Truncation for Ultrasonic Guided Wave Defect Localization on Pressure Vessels. *Sensors* **2022**, 22, 1540. https://doi.org/10.3390/s22041540

Academic Editor: Zenghua Liu

Received: 19 January 2022
Accepted: 12 February 2022
Published: 17 February 2022

1. Introduction

Pressure vessels are devices that can hold gas or liquid under a certain pressure, which are widely used in industry production, civil industries, etc. Under the influence of actual working conditions and environment, pressure vessels are prone to cracks, corrosion, holes, and other defects, which may cause safety hazards to industrial production [1]. Therefore, in order to improve the working safety of pressure vessels, it is necessary to detect and locate the defects on pressure vessels. Ultrasonic guided wave testing technology is one of the more popular non-destructive testing technologies used in recent years. Compared with traditional non-destructive testing technologies, ultrasonic guided wave testing is widely used with the advantages of simpler detection equipment, larger detection range, faster detection speed, and online monitoring. Ultrasonic guided wave testing has also been applied in the structural health monitoring of pressure vessels and has achieved certain results [2–4]. At present, the research of ultrasonic guided wave defect detection mainly focuses on plate and pipe structure, with few studies reporting on pressure vessels. In terms of the propagation characteristics of guided waves, Li et al. [5] studied the dispersion characteristics of different guided wave modes in pressure vessels, and optimized the excitation center frequency and waveform parameters of guided waves according to the simulation and experimental results. Yang et al. [6] proposed an elliptic positioning algorithm based on coordinate transformation, based on the propagation characteristics of guided waves in hollow spheres and cylinders. Zhai et al. [7] studied the direct wave difference values of different distance coefficients in pressure vessels, and Chen et al. [8] proposed a fuzzy C-means clustering algorithm based on direct wave. Although some scholars have done some work around the defect detection of pressure vessels based on guided wave technology, the imaging algorithm and positioning accuracy are still worthy of further research.

The probabilistic ellipse algorithm is a commonly used imaging algorithm based on ultrasonic guided wave to detect defects in plate-like structures. For defect detection and localization in flat-panel components, the probabilistic ellipse imaging algorithm has relatively mature applications. Many scholars have studied the improvement and application of the probabilistic ellipse algorithm. Liu et al. [9] used the probabilistic elliptic positioning algorithm to perform imaging of artificial defects on board members, and the experimental results showed that the direction and shape of the artificial defects could be effectively identified. Tua et al. [10] estimated defect-reflected wave time-of-flight (ToF) through the energy spectrum of the Hilbert–Huang transform, and used the probabilistic elliptic positioning algorithm to locate defects in plates. This provides a basis for the signal difference analysis of ultrasonic guided wave. Li et al. [11] combined wavelet transform and the probabilistic elliptic imaging algorithm to realize the location of artificial defects in a curved, spherical, thin shell structure. Chen et al. [12] proposed a signal screening method based on the distance coefficient on the basis of the traditional probabilistic elliptic imaging algorithm, and the experiments showed that the specific data filtered by the distance coefficient could improve the defect positioning accuracy of plate structures. These studies have improved the imaging algorithm through signal analysis and signal preprocessing. This idea inspired the following research in this paper. Hu et al. [13] studied the propagation characteristics of guided waves in 30 CrMo steel bent plates with different radii and depths using experimental and numerical methods, and used the probabilistic elliptic imaging algorithm to locate defects in bent plates. In these studies, the probabilistic elliptic imaging algorithm was widely applied in the defect detection of pressure vessels based on guided waves, but the feasibility of the probabilistic elliptic imaging algorithm for different arrays has not been discussed. In particular, the signal truncation strategy, which plays a decisive role in imaging accuracy, has also not been discussed. There are few studies on the imaging accuracy of the probabilistic elliptic imaging algorithm applied to the defect detection of pressure vessels. Therefore, the probabilistic ellipse imaging algorithm for defect detection in pressure vessels is worth studying.

The feasibility of the traditional probabilistic elliptic imaging algorithm for defect detection and location in pressure vessels will be studied in this paper. In order to improve the accuracy of the imaging positioning, the factors affecting the imaging accuracy should be discussed. In this study, a new adaptive signal truncation method based on signal difference analysis is proposed. The structure of this paper is as follows. The factors affecting the imaging accuracy of the traditional probabilistic elliptic imaging algorithm for defect detection in pressure vessels are discussed in Section 2. In Section 2, the concepts of the signal difference coefficient (SDC) and the distance coefficient are briefly introduced, and a new idea of signal truncation based on signal difference analysis is constructed. The experimental objects and parameters of the experimental operations are described in detail in Section 3. In Section 4, the relationship model between the SDC and the distance coefficient is constructed based on experimental data, and the new adaptive signal truncation method is verified. Then, the imaging accuracy of the probabilistic ellipse imaging algorithm based on the new method is discussed and further improved. Finally, the conclusion is given in Section 5.

2. Theory

2.1. Traditional Probabilistic Elliptic Imaging Algorithm and Signal Processing

The probabilistic ellipse imaging algorithm is an algorithm developed based on the ellipse imaging algorithm. Both are localization algorithms based on the time-of-flight of scattered waves. By calculating the difference in signal between the healthy signal and the damaged signal, the peak wave packet of the defect scattered wave can be obtained. The arrival time t_1 of the scattered wave is generally taken as the time when the peak of the scattered wave appears. t_1 represents the time taken for the guided wave to propagate from the excitation source to the defect to the receive sensor. Thus, the sum of the distances of the defect to the two sensors can be calculated. According to the sum of the distances from the defect to the two sensors, the possible trace of the defect can be obtained, like an ellipse, as shown in Figure 1. The location of the defect can be determined from the intersection of the elliptical trajectories of multiple sensor pairs. Based on the ellipse imaging algorithm, the probabilistic ellipse imaging algorithm combines the probability functions, and then the entire area of the detected object is meshed. The distance from each cell to the elliptical trajectory is calculated, and a probability function is used to express the probability of the defect location in that cell. After all groups of signals are processed, the position of the maximum probability value in the imaging result is considered as the position of the defect.

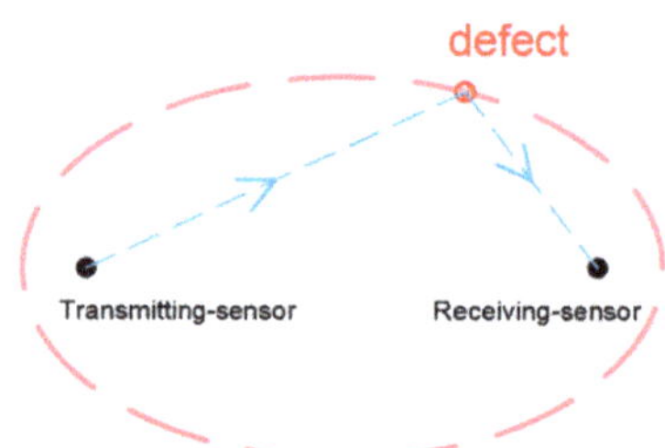

Figure 1. Ellipse trajectory of defect location.

In traditional probabilistic elliptic imaging, the guided wave signal is analyzed by using baseline subtraction to extract the scattered wave packet. Firstly, the ultrasonic guided wave signal acquired by the experiment is preliminarily filtered and denoised. Then, the damaged signal is subtracted from the healthy signal to obtain the residual signal containing the defect scattered wave. Finally, the envelope of the residual signal is obtained by Hilbert transformation. A case of signal processing from the defect detection experiment on pressure vessel is shown in Figure 2.

By analyzing the envelope signal (Figure 2c), it can be found that in addition to the scattered signal caused by the defect, there are also interference components. These interference components may be caused by factors such as multiple reflected wave and electromagnetic noise. In Figure 2c, it can be seen that the amplitude of the interference component peak is even greater than the scattered wave. In this case, the arrival time of the interference component packet (t_2), rather than the arrival time of the scattered wave packet (t_1), is easy to be mistaken for the index representing the defect. In order to avoid the influence of these interference signals on the imaging accuracy, the signal should be truncated after the abovementioned signal processing steps.

2.2. Signal Difference Coefficient and Signal Truncation

In order to avoid the influence of unimportant signals on defect detection, the difference analysis of the signal is carried out. The signal difference coefficient (SDC) was first used in rapid algorithms in ultrasonic guided wave imaging technology to evaluate the correlation between a healthy signal and a damaged signal [11], and the equation is:

$$\mathrm{SDC} = 1 - \frac{C_{XY}}{\sigma_X \sigma_Y} \tag{1}$$

where C_{XY} is the covariance of signal X and Y, σ_X and σ_Y are the variance of signal X and signal Y.

$$C_{XY} = \sum_{k=1}^{K} (X_k - \mu_X)(Y_k - \mu_Y) \tag{2}$$

$$\sigma_X \sigma_Y = \sqrt{\sum_{k=1}^{K} (X_k - \mu_X)^2} \sqrt{\sum_{k=1}^{K} (Y_k - \mu_Y)^2} \tag{3}$$

where μ is the mean value of the signal amplitude and K is the signal length.

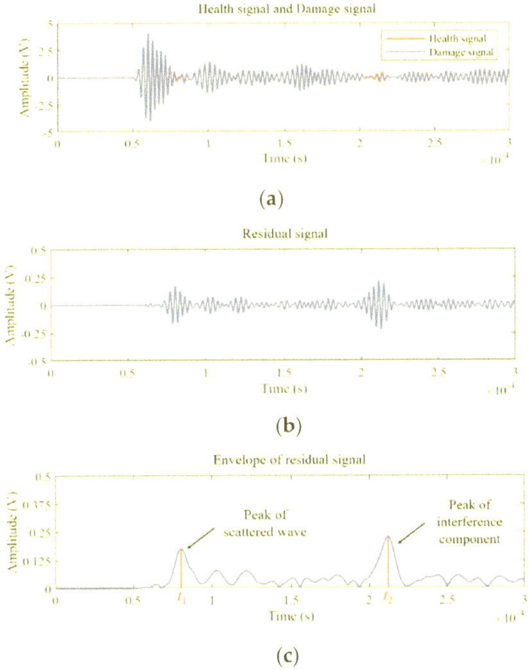

Figure 2. Signal processing of the traditional elliptic imaging method: (**a**) healthy signal and damaged signal; (**b**) residual signal obtained by baseline subtraction. (**c**) Envelope of residual signal.

In the traditional probabilistic ellipse imaging algorithm, the signals of all groups are usually truncated at a fixed length of time. The arrival time of boundary reflection/diffraction wave (t_s) is usually used as the fixed length. Theoretically, the amplitude of the scattered wave packets caused by defect reflection in the truncated signal should be more prominent. However, in practice, some interference components may be larger than the residual signals, like the case in Figure 2c. Therefore, different truncation lengths should be discussed. Rather than residual signal analysis, signal difference analysis is conducted. Figure 3 shows the signal difference analysis between the healthy signal and the damaged signal between the sensor pairs of different paths after truncation. Before t_s, 100 equidistant signal interception is performed. The time of each curve's peak in Figure 3 corresponds to the arrival time of each group of damaged reflection wave signals (t_1). In the curves of sensor pairs 1, 2, 4, and 5, the damaged reflection waves obviously correspond with the curve's peak. Figure 3b shows the SDC curve of sensor pair 3. According to the experimental parameters, the SDC peak value of sensor pair 3 should appear where it is circled in Figure 3b. However, with the extension of the truncation time, a peak that is not related to the damaged reflection signal was shown in the curve.

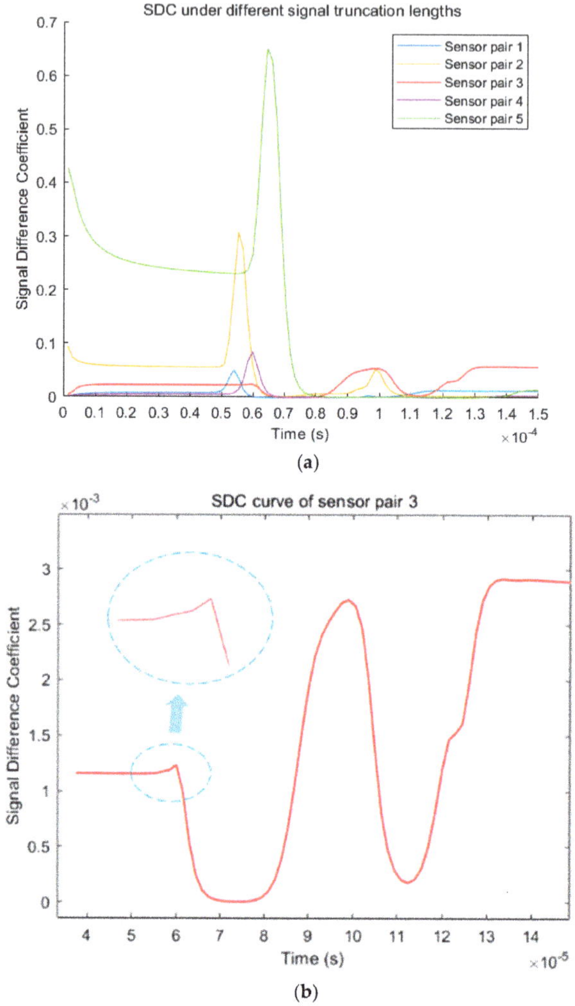

Figure 3. Signal difference analysis of healthy signal and damaged signal under different truncated length: (**a**) SDC curves of different sensor pairs; (**b**) SDC curves of sensor pair 3.

From Figure 3a, the fixed truncation length used in the whole group is not suitable for the signal of sensor pair 3. If each group of signals is truncated at a fixed length, some interference components that are more prominent than the scattered wave packages from defects may be included in the truncated signals. The actual ellipse track will be missed due to these truncated signals in the defect location. When the processed signal contains multiple interference components, the pixel value at the defect position in the imaging result will be sharply reduced, which will eventually lead to the failure of defect localization. Therefore, in order to accurately locate defects, appropriate truncated lengths must be selected for each group of signals to overcome the influence of interference components.

2.3. Signal Difference Coefficient and Signal Truncation

In actual ultrasonic guided wave defect detection and positioning experiments, the large deviation of the imaging results is caused by the inaccurate truncated length. Due to the uncertainty of the defect location information and the difference in the relative position of the sensor pairs, the optimal truncated length is inconsistent. In order to obtain more accurate imaging results, the relationship between the optimal truncated length and the distance coefficient of the signals will be discussed in this section.

Figure 3a shows that the peak values of different curves have different amplitudes and delay times. According to Chen's studies [12], the peak value of the SDC varies with the distance between defects and the path of a transmitter/receiver pair. Figure 3 shows that the damaged reflected signal in each group of signals will cause SDC mutations, and this mutation can be observed even in sensor pair 3. However, the severity of the mutation is different in these signals. In order to adaptively determine the signal's truncated length, it is necessary to analyze the relationship between the arrival time of the scattered waves and the peak value of the SDC mutation. The arrival time of the damaged reflection wave is related to the defect position and the distance of the sensor pair. Therefore, for the convenience of description, the concept of distance coefficient c_d is defined, and it is defined as in Equation (4):

$$c_d = \frac{d_{TD} + d_{RD}}{d_{TR}} = \frac{d_{TRD}}{d_{TR}} \tag{4}$$

where d_{TR} is the distance from the excitation sensor to the receiving sensor; d_{TD} is the distance from the excitation sensor to the defect; d_{RD} is the distance from the defect to the receiving sensor; and d_{TRD} is the sum of the distances from the excitation sensor to the defect and from the defect to the receiving sensor. Then, the ideal truncation time t_1 can be expressed by Equation (5) [12]:

$$t_1 = c_d d_{TR} / v_g \tag{5}$$

where v_g is the group velocity of the S0 mode of the ultrasonic guided wave.

If the signal truncated time is set to t_1, the wave package from the defect is included in the truncated signal and the interference components are excluded. However, the defect location is unknown in the actual detection, and the distance coefficient of each sensor pair cannot be determined directly. Although the arrival time of the scattered waves cannot be obtained directly, Equation (5) provides a method to estimate the t_1 of the scattered waves using a distance coefficient. Relevant studies have shown that with the increase of the distance between a defect and sensor pairs, the difference between the healthy signal and damaged signal will decrease [14,15]. According to this characteristic, the distance coefficient can be judged by the difference between the healthy signal and damaged signal, and then the t_1 of the scattered wave can be estimated. In order to determine the optimal truncated length, the relationship between the peak value of the SDC mutation and the distance coefficient should be studied. For the convenience of presentation, the equational relationship between the SDC and the distance coefficient is expressed as the exponential equation form of Fourier series:

$$S(c_d) = F_0 + \sum_{n=1}^{n} \left(F_n \cdot e^{jn\omega c_d} + F_{-n} \cdot e^{-jn\omega c_d}\right) \tag{6}$$

where n is the number of series to expand, ω is the angular frequency, F_0 is the constant term to be solved, and F_n and F_{-n} are the coefficients of the two terms of the nth level in the expansion.

Obviously, the relationship between SDC and the distance coefficient monotonically decreases, so the above equation is simplified to Equation (7). Through the experimental data in Section 3 and the discussion in Section 4, the specific parameters in Equation (7) will be determined and the accuracy of the formula will be verified.

$$S(c_d) = a_0 + b_0 \cdot e^{f(c_d)} \tag{7}$$

where a_0 is the constant term to be solved, and b_0 is the coefficient of the exponential term to be solved.

In order to solve the relationship equation between the SDC and the distance coefficient, the signals of each sensor pair need to be tested at different defect positions. The experimental setup is described in detail in Section 3. The technical route of the new signal difference analysis method is shown in Figure 4. Firstly, the SDC mutation of the healthy signal and the damaged signal will be calculated according to Equation (1). Then, according to Equation (7), the distance coefficient can be estimated. Finally, the optimal truncated lengths for different group signals can be determined.

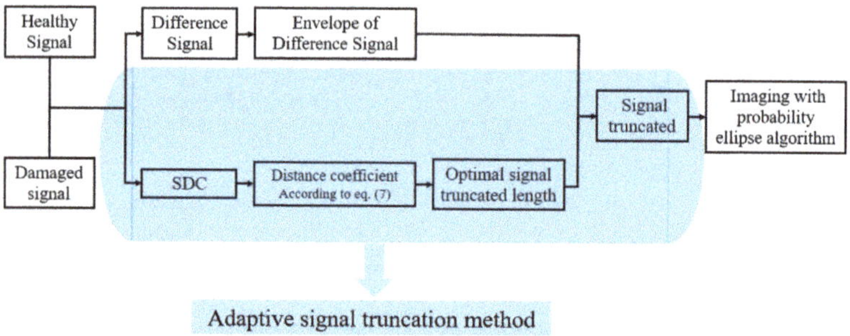

Figure 4. Imaging process of the probabilistic elliptical algorithm based on the adaptive signal truncation method.

3. Experiment

3.1. Experiment Platform

In this study, the whole experimental platform of multi-channel ultrasonic guided wave detection and the connection of the system are shown in Figure 5. The experimental detection steps are as follows. Firstly, five-cycle sinusoidal wave signals modulated by the Hanning window are generated by a waveform generator (Tektronix AFG 3021C, Tektronix, Beaverton, OR, USA), and then the signals are amplified using a power amplifier (AG 1006, TC Power Conversion, Rochester, NY, USA) and transmitted to one piezoelectric sensor. The piezoelectric sensor converts the electrical signal into a vibration signal through the inverse piezoelectric effect, and then the ultrasonic guided wave signal is excited in the pressure vessel. After the ultrasonic guided wave vibration signal propagates in the pressure vessel, the corresponding vibration signal is received by the piezoelectric sensors in other positions. The guided wave vibration signals are then converted into electrical signals by the piezoelectric effect and displayed on an oscilloscope (Tektronix DPO 2012B, Tektronix, Beaverton, OR, USA), and finally, data are automatically collected and stored in the computer through the data acquisition system (NI PXIe 5105, Pickering PXI 40-533B-032, Apex Waves, Cary, NC, USA).

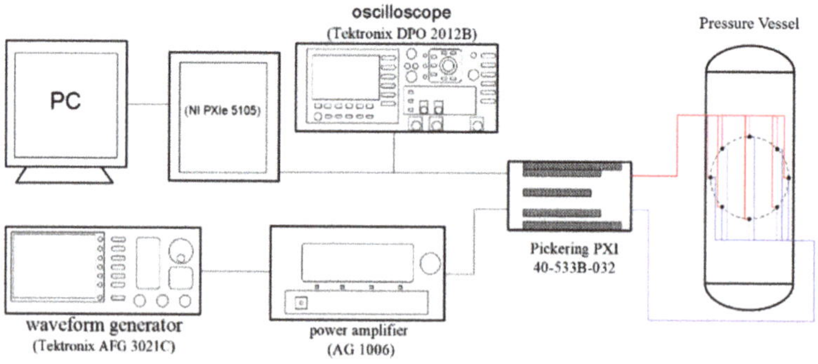

Figure 5. Experimental system.

3.2. Experimental Parameters

The size and material composition of the pressure vessel in the experiment are shown in Table 1. The size of the ceramic piezoelectric wafers used to excite and receive the ultrasonic guided waves are 10 mm (diameter) × 1 mm (thickness). The signal sampling frequency involved in the experiment in this paper is 50 MHz, the trigger voltage is set at 200 mV, and the single sampling time is 0.5 ms. According to the dispersion curve and amplitude–frequency characteristic (Figures 6 and 7), the L (0,2) mode guided wave signal under 275 kHz is selected as excited signal for the defect detection experiment.

Table 1. Parameter of pressure vessel.

Material	Outside Diameter (mm)	Inside Diameter (mm)	Height (mm)	Density (kg/mm^3)	MOE (GPa)	Poisson's Ratio
30 CrMo	350	340	1000	7850	210	0.30

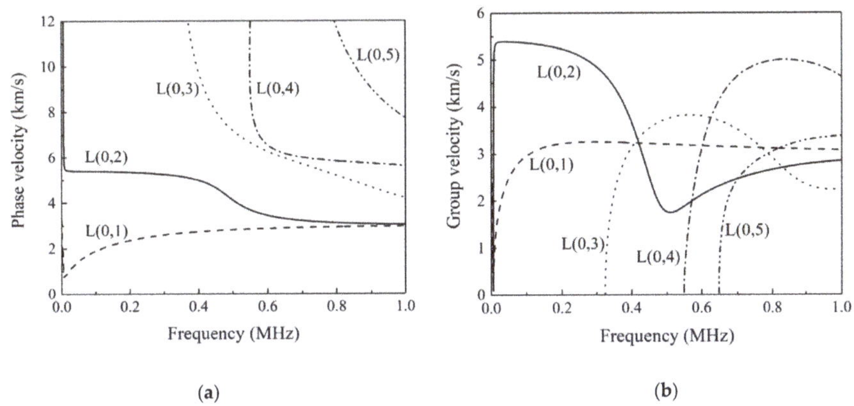

Figure 6. Dispersion curves in 5 mm thick pressure vessel: (**a**) dispersion curve of phase velocity; (**b**) dispersion curve of group velocity.

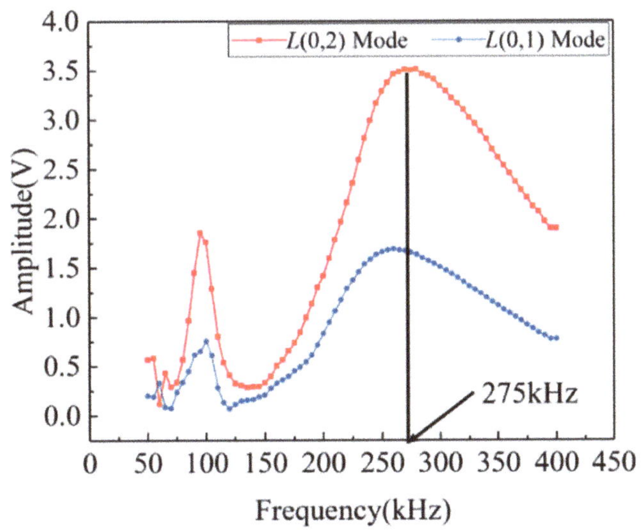

Figure 7. Amplitude–frequency characteristics of L (0,1) and L (0,2) modes.

Defects are simulated by bonding a steel ingot (diameter 10 mm) to the surface of the pressure vessel. To visually represent the relative positions of sensors and defects in the four groups, the images of the curved cylinder are converted into 2D flat images, and then the positions of the sensors and defects are quantitatively described on the plane. In order to avoid errors, four groups of experiments with different defect locations and arrays are set up. The locations of the sensors and defects are shown in Figure 8. The coordinates of the sensors and defects in each array are shown in Table 2.

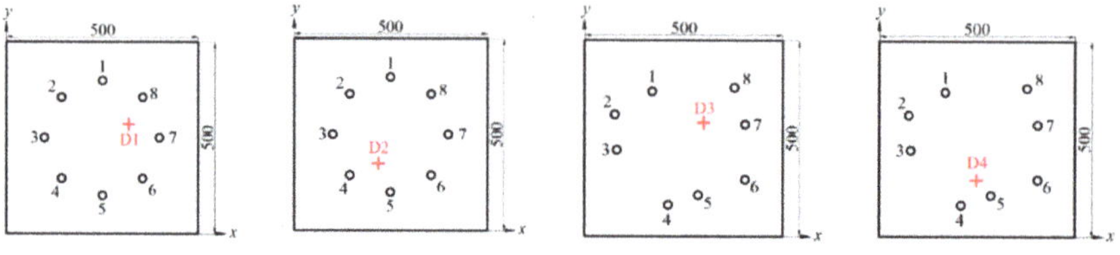

(a) Case 1 **(b)** Case 2 **(c)** Case 3 **(d)** Case 4

Figure 8. Position of sensors and defects. ("O" represents the positions of the sensors and "+" represents the actual positions of the defects.)

Table 2. Coordinates of sensors and defects.

Sensor/Defect Number	Circular Array	Irregular Array
S1	(250,400)	(170,385)
S2	(144,356)	(75,325)
S3	(100,250)	(80,235)
S4	(144,144)	(210,95)
S5	(250,100)	(285,120)
S6	(356,144)	(405,160)
S7	(400,250)	(405,300)
S8	(356,356)	(380,395)
D1	(320,275)	-
D2	(220,175)	-
D3	-	(300,305)
D4	-	(250,160)

4. Results and Discussion

4.1. Adaptive Signal Truncation Method Based on Signal Difference Analysis

Through the experiment in Section 3, the data of each case are collected and recorded. The data are shown in Figure 9. It can be seen that the SDC decreases as the distance coefficient increases. Based on these data points, the relationship models of the SDC and the distance coefficient in each case are first established, and then exponential fitting is performed according to the relationship equation of the four cases. The fitting curve is shown as the black curve in the figure. The analytical formula of the curve is as follows:

$$S(c_d) = 0.0013 \cdot e^{\frac{4.2922}{c_d}} \tag{8}$$

After obtaining the approximate relationship between the SDC and the distance coefficient, the distance coefficient of the sensor pair can be estimated to be t_1. It should be noted that the distance coefficient–SDC curve that fits all data is located in the "middle" of the data. Therefore, when calculating the distance coefficient according to the SDC, the distance coefficient of the "outside" part of the curve is smaller than the actual distance coefficient, resulting in an estimated t_1 that is too small, which may cause the reflected wave packages from defects to be missed. In order to avoid this situation, the edge data points are used to fit the curve, and a new fitting curve is obtained (the red curve in Figure 9). The analytical formula of the new curve is as follows:

$$S(c_d) = 0.0039 \cdot e^{\frac{2.4293}{c_d}} \quad (9)$$

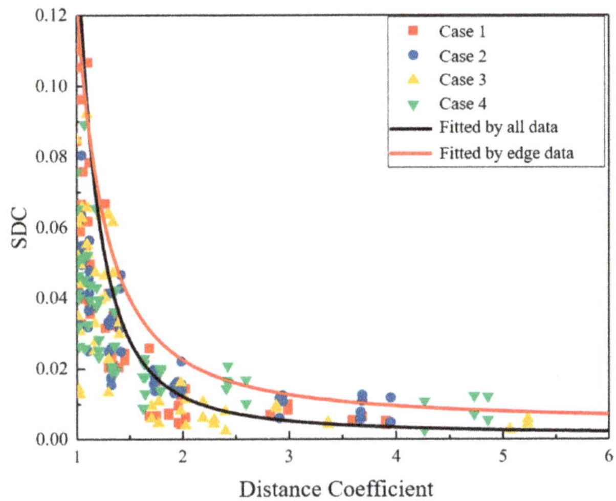

Figure 9. Relationship between distance coefficient and signal difference coefficient.

Thus, the specific parameters of Equation (7) in Section 2.3 are determined through experiments. Therefore, an adaptive signal truncation method based on signal difference analysis is proposed in this study. The technical route of this method, described in Figure 3 in Section 2.3, is supplemented as shown in Figure 10. The imaging effect based on this new signal truncation method will be discussed in the next section.

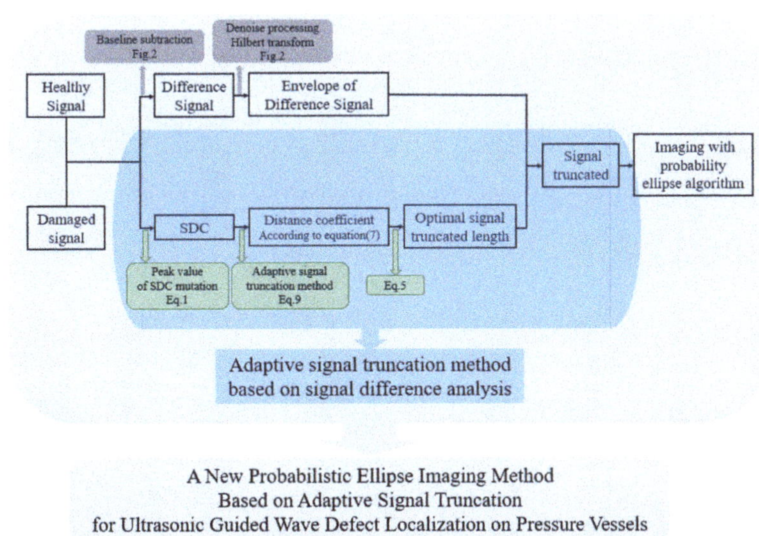

Figure 10. Adaptive signal truncation method based on signal difference analysis.

4.2. Analysis of Imaging Effect under New Adaptive Signal Truncation Method

In order to explore the effect of the probabilistic ellipse imaging algorithm based on the new adaptive signal truncation method, the improved imaging algorithm is used to image the experimental data. The traditional probabilistic elliptic imaging results are shown in Figure 11. The imaging results are shown in Figure 12. From the imaging results, it can be seen that the image obtained by the improved probabilistic ellipse imaging algorithm can effectively characterize the defect location. The defect location errors of case 1 to case 4 are 5.30 mm, 2.80 mm, 6.25 mm, and 4.51 mm, respectively. In addition, compared with the imaging results in Figure 11, the number and range of artifacts in Figure 12 are significantly reduced. The defect location accuracy is significantly improved. Obviously, the new signal truncation method can be effectively applied to defect detection in pressure vessels.

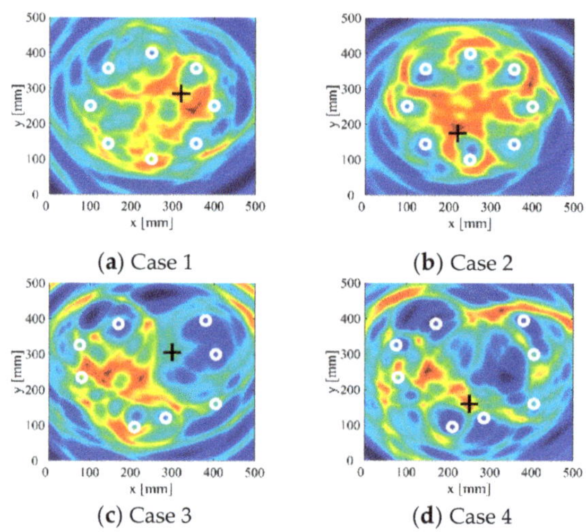

Figure 11. Imaging results of traditional probabilistic ellipse imaging algorithm. (The white circles are the sensor locations, and the black crosses are the actual locations of the defect.)

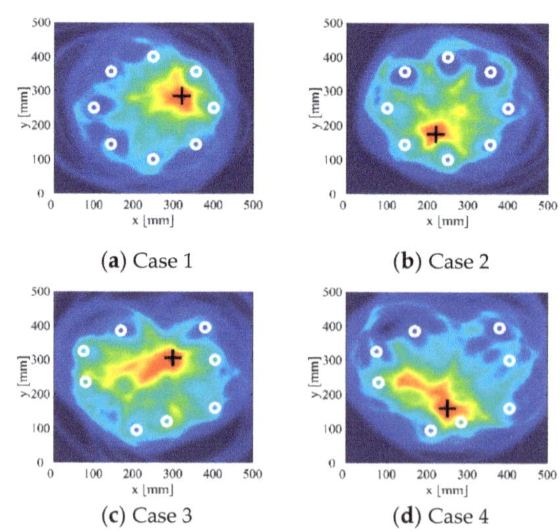

Figure 12. Imaging results of probabilistic ellipse based on adaptive signal truncation method.

In Figure 12c,d, the smear outside the defect area can be seen. This phenomenon can be explained as follows: when calculating the distance coefficient according to the SDC–distance coefficient fitting Equation (9), if the SDC is very small, the deviation of the calculated distance coefficient can be large. Moreover, when all signals are directly used for imaging, the program can easily report errors. Therefore, it may be beneficial to set a minimum SDC threshold Q during the imaging process to perform a preliminary screening of signals, and to exclude signals with an SDC less than the threshold Q from the final imaging result.

4.3. Threshold Optimization of Adaptive Signal Truncation Method

In order to determine an appropriate threshold Q, the imaging results under different thresholds are compared. In this study, the threshold Q value is set from 0 to 0.03 at 0.002 intervals. The imaging results of Case 1 under 15 different thresholds are shown in Figure 13. Among them, sub-picture 1 corresponds to the imaging result with threshold Q = 0, sub-picture 2 corresponds to the imaging result with threshold Q = 0.002, and so on, and sub-picture 15 corresponds to the imaging result with threshold Q = 0.028. It can be observed that as the threshold Q increases from 0 to about 0.02, the high pixel value area close to the defect position in the imaging result gradually shrinks. That is, the pixel value contrast between the defect position and the non-defect position is improved, and the accuracy of the imaging result is also improved. As the threshold Q increases from 0.02, the high pixel value area of the defect tends to expand, and the imaging quality gradually deteriorates.

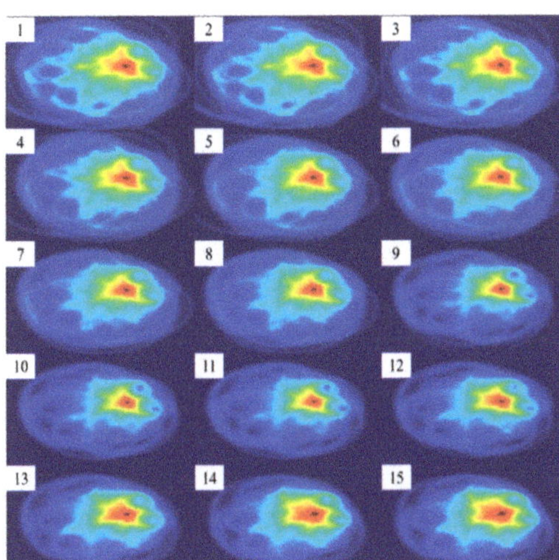

Figure 13. Case 1 imaging results of probabilistic ellipse imaging algorithm at different SDC thresholds.

In order to quantitatively evaluate the influence of threshold Q on the positioning results, a positioning error is used to quantify the deviation between the defect position in the imaging result and the actual defect position. The aggregation coefficient *c* [16] is used to quantify the aggregation degree of pixels in the imaging results, that is, the color contrast between the defect area and the non-defect area in the image. The definition of positioning error *d* is given by Equation (10), and the definition formula of aggregation coefficient *c* is given by Equation (11):

$$d = \sqrt{(x_p - x_r)^2 + (y_p - y_r)^2} \tag{10}$$

$$\hat{c} = \underset{c}{\operatorname{argmin}} \sum \left(\ln\left(\frac{P(x)}{P(0)} \right) + cx \right)^2 \quad (11)$$

where x_p and y_p represent the horizontal and vertical coordinates of the defect obtained from the imaging results; x_r and y_r represent the actual horizontal and vertical coordinates of the defect;$P(x)$ represents the pixel value of a discrete point whose distance is x from the maximum pixel value point, and in particular, $P(0)$ is the pixel value of the maximum pixel value point.

According to the definition of clustering coefficient c, when the pixel value contrast increases, the aggregation coefficient also increases. In order to compare the imaging effects under different SDC thresholds Q, the defect location errors and aggregation coefficients of different sensors and defects at each threshold Q are calculated, as shown in Figure 14a,b. It can be observed that in terms of the positioning error, when the threshold Q is in the range of 0–0.02, the positioning error slightly decreases with the increase of the threshold value, but the decrease range is very small; when the threshold Q exceeds 0.02, the positioning error begins to increase with the increase of the threshold value. In terms of aggregation coefficient c, when the threshold Q is in the range of 0–0.02, the aggregation coefficient c increases with the increase of the threshold value; when the threshold value exceeds 0.02, the aggregation coefficient begins to decrease. Considering the comprehensive influence of the threshold on the positioning error and aggregation coefficient, it can be found that the imaging results with a threshold between 0.018 and 0.022 are better. The imaging results of four cases when the SDC threshold Q is set to 0.02 are shown in Figure 15.

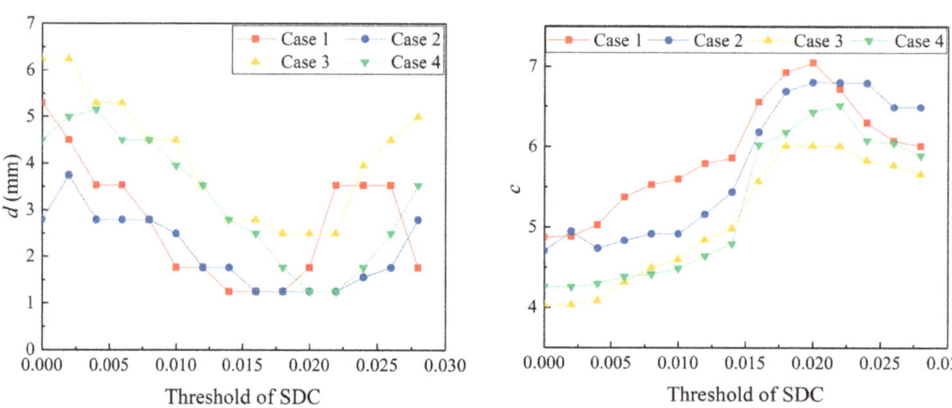

(a) SDC threshold – d **(b)** SDC threshold – c

Figure 14. Imaging location deviation and degree of aggregation under different SDC thresholds.

In Table 3, the defect location error and aggregation coefficient of the traditional probabilistic ellipse imaging algorithm, the probabilistic ellipse imaging algorithm based on the fixed time signal truncation method, and the new adaptive signal truncation method are listed. (The fixed truncated length is set as 0.10 ms.) It can be seen that the traditional probabilistic ellipse imaging algorithm has the worst localization quality, and its defect location error is much greater than that of the improved algorithm; the fixed time signal truncation method effectively reduces the positioning error of the traditional probabilistic ellipse imaging algorithm, but the improvement of the aggregation coefficient of the positioning result is limited; the results show that the new image algorithm based on the adaptive signal truncation method has the best positioning quality when the threshold set to 0.02, and the positioning accuracy and aggregation coefficient of the positioning results are significantly improved. The minimum positioning error is 1.25 mm, and the maximum aggregation coefficient is 14.10.

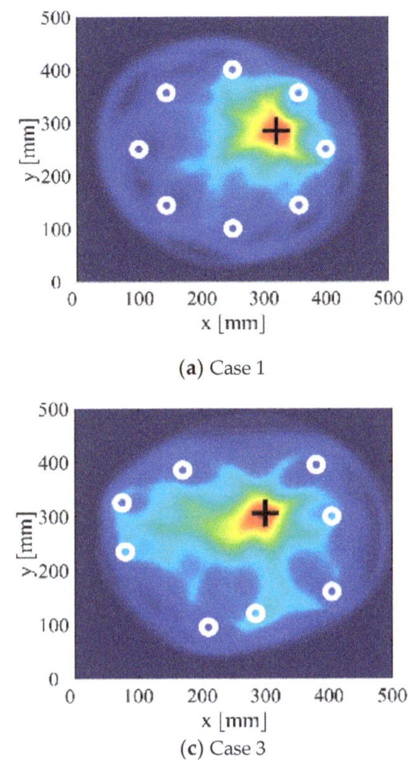

Figure 15. Imaging results with SDC threshold set to 0.02.

Table 3. Deviations and aggregation coefficients of localization results of different probabilistic ellipse imaging algorithms.

	Imaging Algorithms		Case 1	Case 2	Case 3	Case 4
(1)	Traditional probabilistic ellipse imaging algorithm without signal truncation	Positioning error *d*	48.54	17.15	106.09	137.39
		Aggregation coefficient *c*	0.89	0.89	0.82	0.92
(2)	Probabilistic ellipse imaging algorithm with fixed signal truncation length	Positioning error *d*	13.51	15.61	22.21	25.61
		Aggregation coefficient *c*	1.74	1.67	1.60	1.71
(3)	Probabilistic ellipse imaging algorithm based on adaptive signal truncation method (no threshold)	Positioning error *d*	5.30	2.80	6.25	4.51
		Aggregation coefficient *c*	9.75	9.40	8.07	8.52
(4)	Probabilistic ellipse imaging algorithm based on adaptive signal truncation method (set threshold = 0.02)	Positioning error *d*	1.77	1.25	2.50	1.25
		Aggregation coefficient *c*	14.10	13.61	12.03	12.86

It can be seen intuitively from Table 3 and Figure 15 that the signal truncation method based on signal difference analysis proposed in this study can effectively improve the imaging accuracy of the probabilistic ellipse imaging algorithm for single defect detection in pressure vessels.

5. Conclusions

The feasibility of the traditional probabilistic ellipse imaging algorithm for defect detection in pressure vessels were theoretically and experimentally studied in this paper. The experimental results show that the imaging results are affected by the signal truncation length. There are large deviations in the imaging results based on the fixed-length truncated signal. On this basis, in view of the confusion of imaging results, the relationship between signal truncation length and imaging accuracy was discussed. In order to improve the positioning accuracy of the probabilistic ellipse imaging algorithm applied to the defect detection on the pressure vessel, experiments, theoretical analysis, and discussion have been conducted in this paper. The main conclusions are as follows:

(i). In this study, it was found that the positioning accuracy of the probabilistic ellipse imaging algorithm is affected by the signal truncation length. By analyzing the healthy signal and the damaged signal, it was found that the truncation length can be determined according to the SDC and the distance coefficient. Then, the functional relationship between the SDC and the distance coefficient was constructed through experiments in this study.

(ii). In this paper, an adaptive signal truncation method based on signal difference analysis is proposed, and based on this, a new probabilistic ellipse imaging method was developed in this study for ultrasonic guided wave defect localization on pressure vessels. The new probabilistic ellipse imaging method can effectively improve the accuracy in detecting and positioning defects on pressure vessels.

(iii). Further study is recommended to analyze the effect of different thresholds of the SDC on the imaging results. The location error and aggregation coefficient are useful for measuring the location accuracy of different imaging methods. The experimental results show that when the SDC threshold is 0.02, the defect location accuracy can be further improved.

Author Contributions: Conceptualization, Q.L. and G.H.; data curation, Q.L. and G.H.; methodology, Q.L. and Z.L.; visualization Q.L. and Z.L.; writing—original draft preparation, Q.L.; writing—review and editing, Z.L. and S.Z.; supervision, S.Z. All authors have read and agreed to the published version of the manuscript.

Funding: This research was funded by Natural Science Foundation of China, Grant No. 52175137.

Institutional Review Board Statement: Not applicable.

Informed Consent Statement: Not applicable.

Data Availability Statement: Not applicable.

Conflicts of Interest: The authors declare no conflict of interest.

References

1. Ennaceur, C.; Laksimi, A.; Hervé, C.; Cherfaoui, M. Monitoring crack growth in pressure vessel steels by the acoustic emission technique and the method of potential difference. *Int. J. Press. Vessel. Pip.* **2006**, *83*, 197–204. [CrossRef]
2. Ihn, J.B.; Chang, F.K. Pitch-catch active sensing methods in structural health monitoring for aircraft structures. *Struct. Health Monit.* **2008**, *7*, 5–19. [CrossRef]
3. Giurgiutiu, V. Tuned lamb wave excitation and detection with piezoelectric wafer active sensors for structural health monitoring. *J. Intell. Mater. Syst. Struct.* **2005**, *16*, 291–305. [CrossRef]
4. Lin, X.; Yuan, F.G. Damage detection of a plate using migration technique. *J. Intell. Mater. Syst. Struct.* **2001**, *12*, 469–482. [CrossRef]
5. Li, F.; Liu, Z.; Sun, X.; Li, H.; Meng, G. Propagation of guided waves in pressure vessel. *Wave Motion* **2015**, *52*, 216–228. [CrossRef]
6. Yang, B.; Xiang, Y.; Xuan, F.Z.; Hu, C.; Xiao, B.; Zhou, S.; Luo, C. Damage localization in hydrogen storage vessel by guided waves based on a real-time monitoring system. *Int. J. Hydrogen Energy* **2019**, *44*, 22740–22751. [CrossRef]
7. Chen, S.; Zhou, S.; Yong, L.; Zhang, L. A Novel Defect Location Method for Plate-Like Structure by Using Forward-Scattering Wave and Fuzzy C-Means Clustering. In Proceedings of the ASME 2017 Pressure Vessels and Piping Conference, Waikoloa, HI, USA, 16–20 July 2017.
8. Zhai, S.; Zhou, S.; Chen, S.; Yang, B.; Li, Y. Novel defect location method for pressure vessel by using L (0, 2) mode guided wave. *J. Press. Vessel Technol.* **2019**, *141*, 010910. [CrossRef]
9. Liu, Z.; Chun-Guang, X.U.; Meng, F.W.; Zhou, S.Y.; Xi-Peng, L.I. Ultrasonic lamb wave array technique for damage defection in plates. *Instrum. Tech. Sens.* **2013**, *9*, 86–88.
10. Tua, P.S.; Quek, S.T.; Wang, Q. Detection of cracks in plates using piezo-actuated lamb waves. *Smart Mater. Struct.* **2004**, *13*, 643. [CrossRef]
11. Li, X.; Xu, C.; Yang, L.; Wang, L.; Yan, H. Defect detection of curved thin shell based on ultrasonic transducer array. *Int. J. Microstruct. Mater. Prop.* **2014**, *9*, 224–239. [CrossRef]
12. Chen, S.-J.; Zhou, S.-P.; Li, Y.; Xiang, Y.-X.; Qi, M.-X. Distance-Coefficient-Based Imaging Accuracy Improving Method Based on the Lamb Wave. *Chin. Phys. Lett.* **2017**, *34*, 55–59. [CrossRef]
13. Hu, C.; Yang, B.; Xuan, F.-Z.; Yan, J.; Xiang, Y. Damage orientation and depth effect on the guided wave propagation behavior in 30crmo steel curved plates. *Sensors* **2020**, *20*, 849. [CrossRef] [PubMed]

14. Zhao, X.; Gao, H.; Zhang, G.; Ayhan, B.; Rose, J.L. Active health monitoring of an aircraft wing with embedded piezoelectric sensor/actuator network: I. defect detection, localization and growth monitoring. *Smart Mater. Struct.* **2007**, *16*, 1208. [CrossRef]
15. Chen, S.; Zhou, S.; Chen, C.; Li, Y.; Zhai, S. Detection of double defects for plate-like structures based on a fuzzy c-means clustering algorithm. *Struct. Health Monit.* **2018**, *18*, 147592171877204. [CrossRef]
16. Hall, J.S.; Michaels, J.E. Minimum variance ultrasonic imaging applied to an in situ sparse guided wave array. *IEEE Trans. Ultrason. Ferroelectr. Freq. Control* **2010**, *57*, 2311. [CrossRef] [PubMed]

sensors

MDPI

Article

Modeling and Analysis of Acoustic Emission Generated by Fatigue Cracking

Weilei Mu [1,2], Yuqing Gao [1], Yuxue Wang [1], Guijie Liu [1] and Hao Hu [3,*]

[1] Engineering College, Ocean University of China, Qingdao 266100, China; muweilei@ouc.edu.cn (W.M.); gaoyuqing@stu.ouc.edu.cn (Y.G.); wangyuxue@stu.ouc.edu.cn (Y.W.); liuguijie@ouc.edu.cn (G.L.)
[2] Suzhou Academy, Xi'an Jiaotong University, Suzhou 215123, China
[3] Qingdao Research Institute & School of Marine Science and Technology, Northwestern Polytechnical University, Xi'an 710072, China
* Correspondence: huhao@nwpu.edu.cn

Abstract: The acoustic emission (AE) method is a popular and well-developed method for passive structural health monitoring of metallic and composite structures. The current study focuses on the analysis of one of its processes, sound source or signal propagation. This paper discusses the principle of plate wave signal sensing using piezoelectric transducers, and derives an analytical expression for the response of piezoelectric transducers under the action of stress waves, to obtain an overall mathematical model of the acoustic emission signal from generation to reception. The acoustic emission caused by fatigue crack extension is simulated by a finite element method, and the actual acoustic emission signal is simulated by a pencil lead break experiment. The results predicted by the mathematical model are compared with the experimental results and the simulation results, respectively, and show good agreement. In addition, the presence of obvious S0 mode Lamb waves is observed in the simulation results and experimental results, which further verifies the correctness of the analytical model prediction.

Keywords: acoustic emission (AE); fatigue crack growth; predictive modeling

Citation: Mu, W.; Gao, Y.; Wang, Y.; Liu, G.; Hu, H. Modeling and Analysis of Acoustic Emission Generated by Fatigue Cracking. *Sensors* **2022**, *22*, 1208. https://doi.org/10.3390/s22031208

Academic Editor: Zenghua Liu

Received: 8 January 2022
Accepted: 3 February 2022
Published: 5 February 2022

1. Introduction

Today, many kinds of materials are used for construction infrastructure, aviation and ocean sailing. Applying the technical conditions of continuous monitoring to the security of these infrastructures is a challenge [1]. AE technology has shown great advantages in monitoring large structures, and it allows effective health monitoring and life prediction of materials. It is important to detect damage in the early stages to prevent catastrophes from occurring [2]. Mu et al. used visible graph modelling [3] and acoustic emission beamforming [4] to achieve AE source localization. In terms of material life prediction, Roberts et al. linked the AE count rate with crack growth [5] in an effort to predict the material's remaining life. Similar methods have been used to link observed AE data trends with fracture [6] and crack growth [7] in metallic materials.

However, the AE electrical signal output by the AE transducer is often far from the real AE source signal [8]. In contrast with other nondestructive testing and evaluation methods for acoustics, AE lacks precise information related to the source, which complicates efforts to relate AE activity to, e.g., fracture location and size [9]. Therefore, determining how to obtain more information about the AE source, based on the electrical signal of the sensor, has become a challenge for the relevant practitioners.

Previous research on AE mostly uses statistical methods, such as duration, rise time, count and frequency, which reflect the physical characteristics of AE. W. Kaewwaewno et al. used AE detection technology to analyze the characteristics of valve leakage at low pressure, and the effect of differences in valve size and pressure on the effective voltage value of AE characteristic parameters [10]. Guo-Yang Ye et al. used Gaussian process regression (GPR) to develop a multivariate mathematical model to characterize the relationship between the AE signal and the pressure and leakage rate [11]. On the other hand, many researchers have proved that time–frequency conversion technology, as a very useful tool for AE signal analysis, can provide more reliable and comprehensive characteristic information about AE phenomena [12]. E. Mland et al. used kernel partial least squares regression (KPLSR) to establish the relationship between spectral components and leakage rate for both gas valves and liquid valves [13].

Another research method, analytical modeling of the AE signal, can be performed to predict the AE signal through the mathematical model. These studies help to understand the AE signal signature and source mechanism from both theoretical and experimental perspectives. Research on wave propagation and other basic theories is one of the development trends of AE technology [14].

Based on the theory of elastic dynamics and the dislocation model, Ohtsu and Ono developed the generalized theory of AE and the representation of the AE source mechanism in half-space [15], and further continued their work by performing a simulation of tensile cracks and shear cracks in half-space [16]. Researchers are also analyzing and studying the AE of the plate structure. Weaver and Pao [17] studied the numerical results of the surface response of a plate at different locations from the source for different modes of the Rayleigh–Lamb spectrum. Gorman and Prosser [18] studied the normal-mode solution to the classical plate bending equation to understand its applicability to AE. Aiming at the AE phenomenon caused by crack propagation, Lysak [19] studied the AE of growing cracks from the point of view of fracture mechanics, and described the models of crack initiation and subcritical growth in quasi-brittle materials. Giurgiutiu et al. [20] used a piezoelectric wafer active sensor (PWAS) to study the AE signal characteristics of fatigue crack growth in metal sheets.

For metallic or composite materials, the in-service conditions and failure modes are generally complex and may not be accurately predicted [21]. Guo-Yang Ye et al. [22] combined the time–frequency domain characteristics of acoustic emission signals and the random forest method to develop a multi-variable classification model that relates the internal leakage acoustic emission signal and the leakage rate under varying pressure. Ajay and Carlos [23] made efforts to accurately characterize guided wave excitation. They derived equations for the output voltage response of surface-bonded piezo-sensors in guided fields based on 3D linear elasticity equations. The above study focuses on the modeling of acoustic emission sources. Maillet and Morscher [24] proposed a new waveform-based procedure for localization of the AE source based on the Akaike information criterion (AIC). The energy-based approach also offers a potential for damage monitoring that could be used to improve the description of AE sources. Sause et al. [25] modeled the acoustic emission signal of hybrid multi-layer plates, focusing on the acoustic emission source and the signal propagation process. This model can solve the anisotropy problem in the plate, and it is more widely used. Sengupta et al. [26] proposed a refined 2D plate theory more applicable to plate structures, which satisfies the transverse shear stress continuity at the layer interface. This method has better computational efficiency compared to the 3D plate theory. These studies are mainly devoted to the modeling of acoustic emission signal propagation processes. Victor [27] described a model of the Lamb waves' tuning mechanism with transducers. In this model, the piezoelectric wafer also used a thin adhesive layer to achieve structural coupling. Subsequently, he innovatively developed a theoretical model for the analysis of Lamb wave time reversal, and used it to analyze the Lamb wave propagation process [28]. His research focuses on the corresponding modeling of sensors. Zelenyak et al. [29] changed the geometry of the wave guide by varying the radius and height to investigate the influence on the detected signals. Similarly, this paper explores the effects of changes in the source, plate thickness and receiving sensor parameters on the detected signal. Unlike previous studies, the paper establishes a mathematical model of the overall process of the acoustic emission signal from generation and propagation to reception, and the accepted signal can be predicted based on the parameters of the source, propagation and reception.

In this paper, the AE induced by fatigue crack growth in a plate is analytically modeled by using a piezoelectric transducer. Sections 3 and 4 verify the correctness of the analytical modeling using a simulation and an experiment, respectively. In Section 5, the factors influencing the spectrum of the received AE signal are discussed. The final part presents the summary and conclusions, and makes suggestions for further work.

2. Modeling Method of AE

The excitation in the fatigue crack growth process can be represented by the M11 dipole moment, and the dipole moment is used to deduce the complete solution of the wave field caused by AE.

As shown in Figure 1, suppose the torque is generated due to the force vector F_i, applied at the position $\xi(\alpha, \beta)$, and the reverse force vector F_i applied at the position $\xi'(\alpha, \beta + \Delta X_j)$, ΔX_j is a minute distance in the X_j direction.

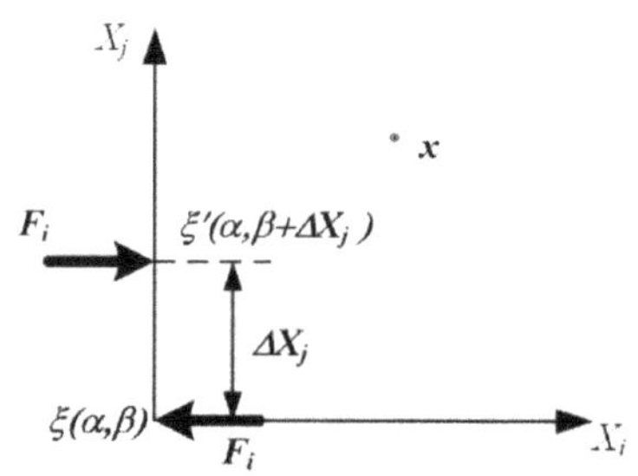

Figure 1. Schematic diagram of coupling force generating torque at small distances.

The displacement field at an arbitrary position x, due to the force vector F_i, is u_i^{ξ} and $u_i^{\xi'}$, respectively. The displacement field in the force couple M_{ij} can be written as follows:

$$u_i^{M_{ij}} = M_{ij} \frac{\partial}{\partial X_j} G(x; \alpha, \beta) \tag{1}$$

where G is Green's function of point forces and the force couple $M_{ij} = F_i \Delta X_j$.

In order to obtain the excitation field of the M11 dipole moment through the thickness of the plate, the dynamic reciprocity theorem was used to solve the elastic wave field generated by the force excitation through the thickness of the plate [30]. The M11 dipole moment excitation of the inertial plate thickness is shown in Figure 2, and the minute distance of the moment M11 is in the X direction.

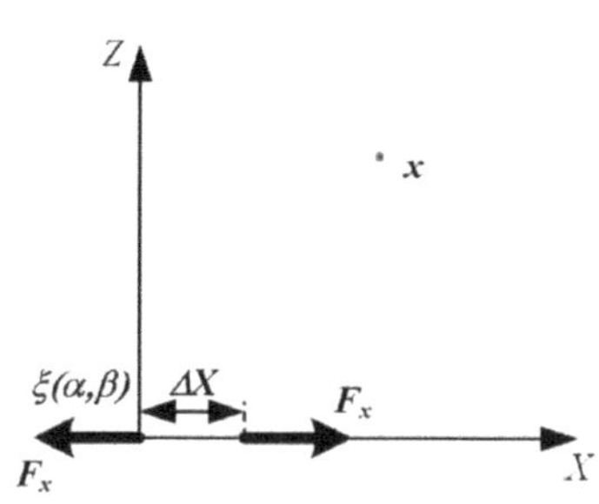

Figure 2. M11 dipole moment excitation.

The displacement of the wave field can be obtained as follows:

$$\begin{aligned} u_{M_{11}} &= -ik_n M_{11} \sum_{n=0}^{\infty} \frac{U_n(h)}{2I_{nn}} U_n(z) \exp(-ik_n x) \\ w_{M_{11}} &= k_n M_{11} \sum_{n=0}^{\infty} \frac{U_n(h)}{2I_{nn}} W_n(z) \exp(-ik_n x) \end{aligned} \tag{2}$$

where the harmonic factor $\exp(i\omega t)$ is omitted; parameter n represents the nth mode; k_n represents the beam of the nth mode at a given frequency; h is half of the plate thickness; $[U_n(z), W_n(z)]$ is the mode shape, and its expression is $U_n^S = s_1 \cos pz + s_2 \cos qz$ and $W_n^S = s_3 \sin pz + s_4 \sin qz$; T_m is the stress mode shape of the normal mode, and $I_{nn} = \int_{-h}^{h} [T_{xx}^n U_n(z) - T_{xz}^n W_n(z)] dz$.

The strain can be obtained by the first derivative of Equation (2), as follows:

$$\begin{aligned} \varepsilon_{M_{11}}^x &= \frac{\partial}{\partial x} u_{M_{11}} = -k_n^2 M_{11} \sum_{n=0}^{\infty} \frac{U_n(h)}{2I_{nn}} U_n(z) \exp(-ik_n x) \\ \varepsilon_{M_{11}}^z &= \frac{\partial}{\partial z} w_{M_{11}} = k_n M_{11} \sum_{n=0}^{\infty} \frac{U_n(h)}{2I_{nn}} W_n{}'(z) \exp(-ik_n x) \\ &= k_n M_{11} \sum_{n=0}^{\infty} \frac{U_n(h)}{2I_{nn}} (s_3 p \cos pz + s_4 q \cos qz) \exp(-ik_n x) \end{aligned} \tag{3}$$

The PZT sensor is located on the plate. The PZT sensor detects the surface strain caused by the AE waveform and converts it to an equivalent voltage. Due to the physical characteristics of the PZT sensor, it does not significantly change the strain field of the incident wave. The voltage at both ends of the sensor electrode can be written as follows:

$$V_s = -\int_{b/2}^{b/2+h_s} E_z \quad dz = h_s \frac{e_{31}^p \varepsilon_y - D_z}{\epsilon_{33}^p} \tag{4}$$

where b is the thickness of the plate and h_s is the thickness of the sensor. D_z is defined as the amount of charge per unit area.

The electrical boundary condition of the piezoelectric sensor is open circuit and the total charge on the electrode area is zero [31]. In the frequency domain, Equation (4) can be written as follows:

$$\overline{V}_s = C_{vs} \frac{1}{A} \int_A \bar{\varepsilon} dA \tag{5}$$

where A in Equation (5) represents the area where the PZT can generate charge. The electromechanical conversion coefficient can be defined as follows:

$$C_{vs} = \frac{h_s e_{31}^p}{\epsilon_{33}^p} = h_s \frac{e_{31} - e_{33} c_{13}/c_{33}}{\epsilon_{33} + e_{33}^2/c_{33}} \tag{6}$$

The M11 dipole moment excitation through the plate thickness, as shown in Figure 3, is used to represent the dipole moment excitation generated by the type I crack. Therefore, the complete solution of the response of the M11 dipole moment excitation at the AE sensor is derived for the prediction modeling of the type I crack.

The distance between the PZT sensor and the AE source is r_c; a_s is the radius of the PZT sensor. The traditional AE sensor mainly inducts the vibration of the structure to be measured by off-plane displacement. Therefore, the following equation can be obtained by substituting the strain of the excitation wave field of the dipole into the induction formula of the piezoelectric sensor:

$$\begin{aligned}\overline{V}_s(x,\omega) &= C_{vs}\frac{1}{A}\int_A k_n M_{11}\sum_{n=0}^{\infty}\frac{U_n(h)}{2I_{nn}}W_n{}'(z)\exp(-ik_n x)dA \\ &= k_n C_{vs} M_{11}\sum_{n=0}^{\infty}\frac{U_n(h)}{2I_{nn}}\frac{\sin k_m a_s}{k_m a_s}W_n{}'(z)\exp(-ik_m x_0) \\ &= k_n C_{vs} M_{11}\sum_{n=0}^{\infty}\frac{U_n(h)}{2I_{nn}}\frac{\sin k_m a_s}{k_m a_s}(s_3 p\cos pz + s_4 q\cos qz)\exp(-ik_m x_0)\end{aligned} \quad (7)$$

Translated to the time domain by the inverse Fourier transform, the following is obtained:

$$\begin{aligned}V_s(x,t) &= \frac{1}{2\pi}\sum_{n=0}^{\infty}k_n C_{vs} M_{11}\sum_{n=0}^{\infty}\frac{U_n(h)}{2I_{nn}}\frac{\sin k_m a_s}{k_m a_s}W_n{}'(z)\exp(i\omega t)d\omega \\ &= \frac{1}{2\pi}\sum_{n=0}^{\infty}k_n C_{vs} M_{11}\sum_{n=0}^{\infty}\frac{U_n(h)}{2I_{nn}}\frac{\sin k_m a_s}{k_m a_s}(s_3 p\cos pz + s_4 q\cos qz)\exp(i\omega t)d\omega\end{aligned} \quad (8)$$

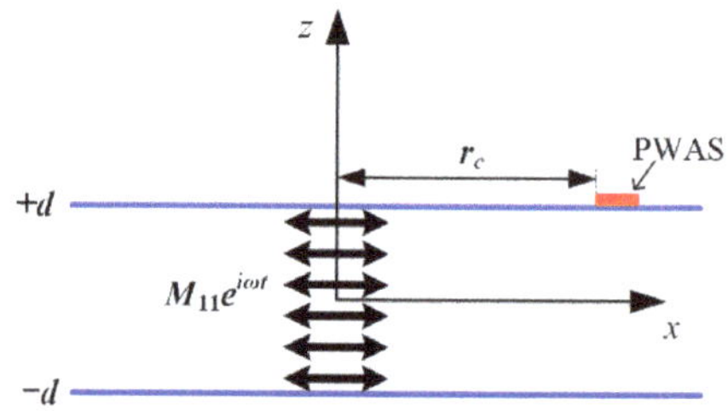

Figure 3. M11 excitation through plate thickness in thin plate.

3. AE Simulation of Type I Fatigue Crack

3.1. Finite Element Simulation Settings

For validating the effectiveness of the analytical modeling, ANSYS software was used to conduct finite numerical analysis of the response precision of the piezoelectric sensor in the finite element model. The plate is set as 600 mm × 1.6 mm aluminum material, and is modeled by the two-dimensional unit Plan-182. The piezoelectric sensor is made of APC-850 material, and the two-dimensional multi-physical field unit Plan-13 is used for modeling. The size of the piezoelectric sensor is set as 20 mm × 0.42 mm. The characteristics of aluminum and piezoelectric materials are detailed in Table 1.

Table 1. Material parameters.

Material	Density (kg/m^3)	Elastic Modulus (GPa)	Poisson's Ratio
Aluminum plate	2700	69	0.33
APC-850	7700	84.3	0.31

For the time-domain excitation of the dipole moment source, the wide-band cosine response function is used, and its mathematical representation is as follows [32]:

$$E(t) = \begin{cases} 0 & t < 0 \\ 0.5(1-\cos(\pi t/\tau)) & 0 \le t \le \tau \\ 1 & t > \tau \end{cases} \quad (9)$$

where $\tau = 1.5$ μs is the rise time of the signal. The distance between dipoles is 0.2 mm.

The excitation of the M11 dipole moment is located in the central position of the plate, and piezoelectric sensors are set on the upper and lower surfaces of the thin plate, 100 mm away from the AE source. The size of the grids is set as 0.1 mm × 0.5 mm. The simulation model is shown in Figure 4.

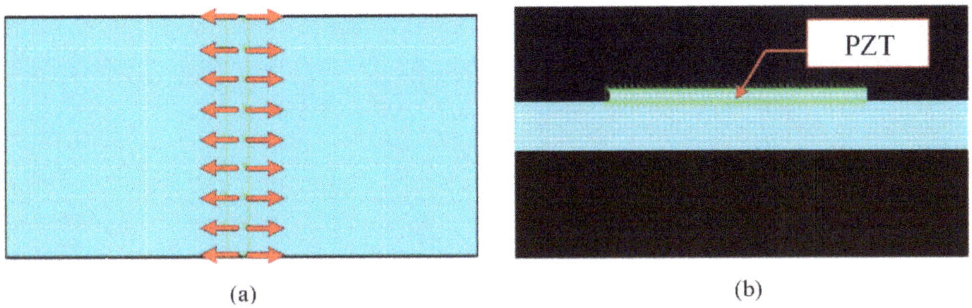

Figure 4. The simulation model: (**a**) dipole moment diagram; (**b**) overall model.

3.2. Verification and Analysis of Models

The in-plane and off-plane displacements of the nodes on the upper and lower surfaces of the thin plate were extracted at a distance of 100 mm from the excitation point, and the displacement waveform is shown in Figure 5. It can be observed that the in-plane displacements of the upper and lower surface nodes are in the same direction, while the off-plane displacements are in the opposite direction, which is in line with the particle motion characteristics of S0 mode Lamb waves.

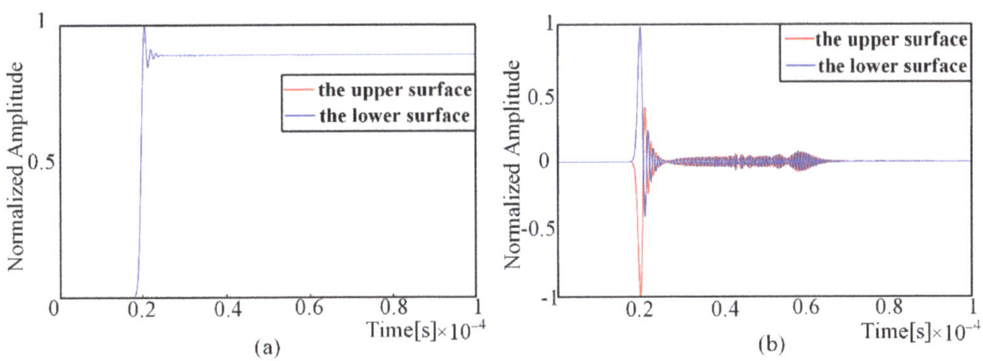

Figure 5. Displacement waveform of finite element simulation: (**a**) in-plane displacement; (**b**) off-plane displacement.

The results of the signal comparison between the finite element simulation and the prediction model are shown in Figure 6. As can be observed from the analysis results, Figure 6a,b show that the finite element simulation and analytical prediction results have a superb matching effect in the time domain. As the finite element model does not take into account issues such as wave absorption setting and high-frequency filtering, there are slight differences in the waveforms, which have little impact on the description of the AE phenomenon. Figure 6c,d show the results of the time–frequency analysis after the short-time Fourier transform of the time-domain waveform. It can be observed that the time–frequency analysis results have a high overlap with the time–frequency curve of the S0 mode Lamb wave. Type I crack excitation through the plate thickness only produces the S0 mode in the sheet pattern, and there is no A0 mode. The comparison of finite element simulations and theoretical simulations effectively proved the correctness of the analytical predictions.

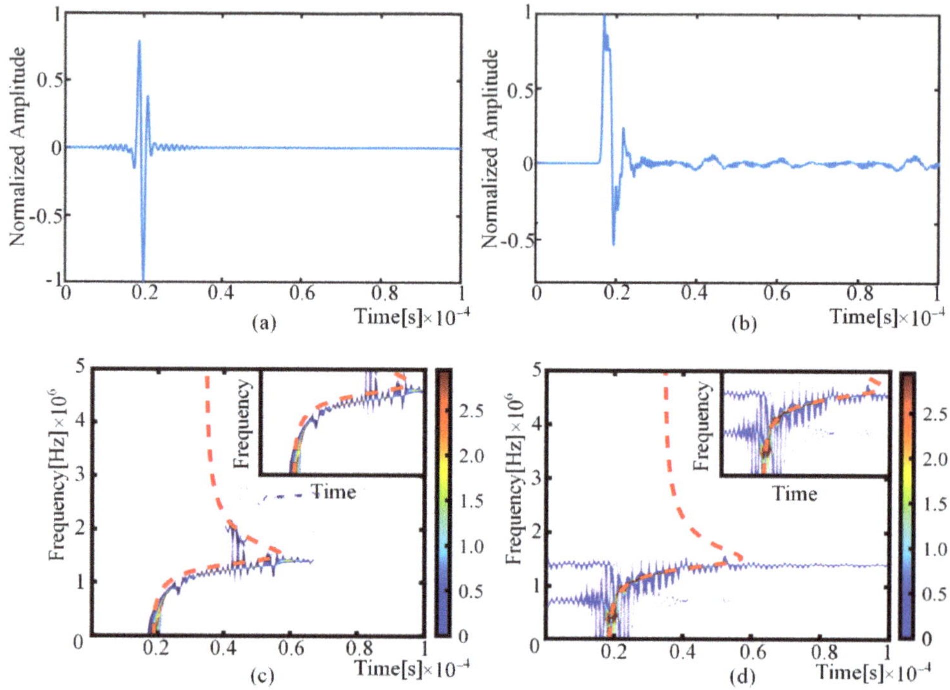

Figure 6. Sensor waveforms of finite element simulation and analytical prediction: (**a**) the analytical modeling result (time domain); (**b**) the finite element simulation results (time domain); (**c**) the analytical modeling result (short-time Fourier transform); (**d**) the finite element simulation results (short-time Fourier transform).

4. Experimental Verification

4.1. Experimental Setup

A square plate, with the dimensions 600 mm × 600 mm × 1.6 mm (length × width × thickness), was adopted as the experimental sample. The sensor was placed 10 mm away from the edge of the plate. The experimental setup consisted of a DS2−8B data acquisition instrument, a smart AE charge amplifier and a computer. The amplifier magnification was 10 times; the resonant frequency of the sensor was 150 kHz; a silicone coupling agent was used between the sensor and the plate. The experimental equipment and excitation details are shown in Figure 7.

It is impossible to accurately monitor the AE signal of simple type I crack growth in the laboratory environment. Based on the analysis of the crack and the load form, it can be concluded that the AE of type I crack growth can be simplified to the assumption of the load excitation in the inner surface of the thin plate. Therefore, a pencil lead break on the edge of the plate is used to simulate the AE phenomenon of the actual type I crack growth. It should be noted that the location of the broken lead should be, as far as possible, in the middle of the plate thickness to ensure the experimental effect.

4.2. Experimental Results

The experimental results are shown in Figure 8. It can be observed from Figure 8a,b that the experimental results and the analytical waveform have a good matching effect in the time domain. The time–frequency analysis results can be obtained through the synchronous compression wavelet transform of the time–domain waveform, as shown in Figure 8c,d. Because the frequency band of the pencil lead break is narrow, the analytical results and the experimental results are set at the same frequency band. The time–frequency diagram showed the waveform component as vertical linear distributed, which completely accords with the time–frequency performance of the S0 modal. Although the two spectra showed differences in details, this does not affect the verification of the correctness of the model.

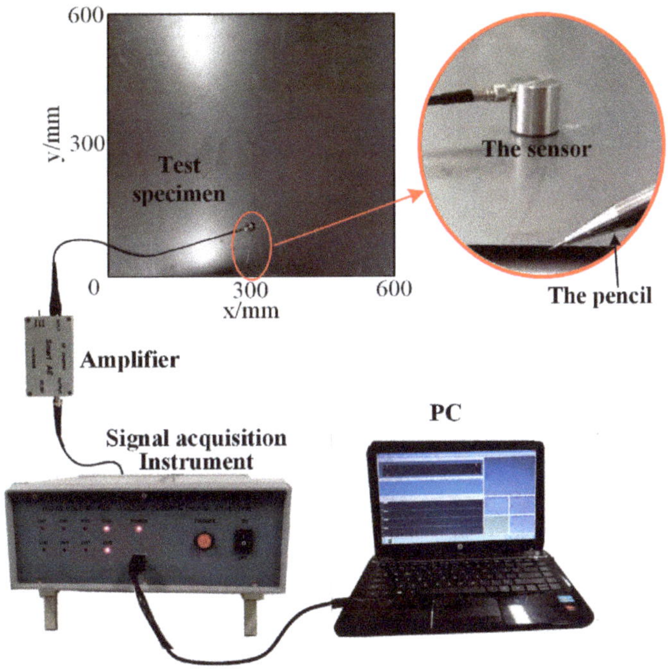

Figure 7. Experimental setup.

In the AE experiments, in addition to the obvious presence of S0 wave packets, the presence of slow-moving wave packets and a mass of interference noise can be observed. Since their frequency domain components differ significantly from the wave packets generated by AE, the frequency spectrum of the signal is affected. From Equation (8), it can be deduced that, for the same acoustic emission signal, the main frequency of the signal received by sensors with different resonant frequencies should also be different. As shown in Figure 9a, the sensors with resonant frequencies of 150 kHz and 80 kHz are arranged at the same distance from the excitation point. The frequency spectrum envelopes of the signals received by the two sensors are shown in Figure 9b,c. The experimental results are consistent with the predictions of analytical modeling.

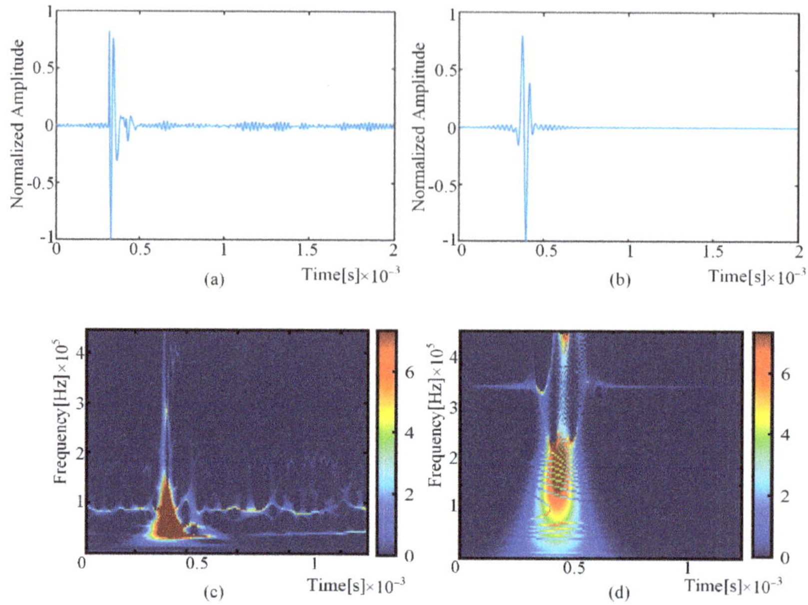

Figure 8. Comparison between the experimental result and the analytical modeling prediction result: (**a**) the waveform of experimental detection; (**b**) the waveform of analytical prediction; (**c**) experimental results (synchronous compression wavelet transform); (**d**) analytical prediction results (synchronous compression wavelet transform).

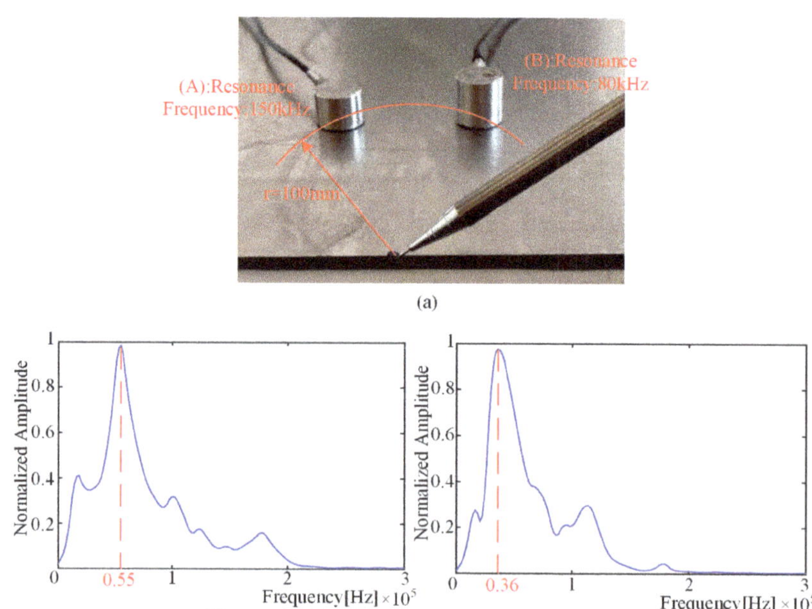

Figure 9. The experiment with sensors of different resonant frequencies: (**a**) experiment setup; (**b**) the frequency spectrum of the signal received by sensor A; (**c**) the frequency spectrum of the signal received by sensor B.

5. Influencing Factors of AE Frequency Spectrum

The modeling of the AE monitoring process is divided into the following three components: the AE source response, the response of the plate structure, and the response of the sensor. These aspects will be discussed in this section.

5.1. Effect of AE Sources

The AE source is modeled with reference to the dipole model, and the excitation function is a unified cosine clock function. Therefore, adjusting the rise time parameter τ of the function can effectively change the frequency of the excitation function, and, thus, obtain the best similarity between the experiment and the simulation. However, the actual AE signal cannot be measured directly, and the material properties, stress level and other factors may affect the AE signal frequency. By varying the rise time parameter τ of the cosine clock function in the prediction model, differences in the perceived AE signals under different rise time conditions are obtained, as shown in Figure 10. It can be observed that, with the increase in rise time τ, the main frequency of the AE sensing signal shows a decreasing trend, but the corresponding amplitude of the main frequency rises significantly. Although the main frequency and the corresponding amplitude basically show a linear change, the rise time of the function has a more significant effect on the main frequency amplitude, compared with the small change in the main frequency.

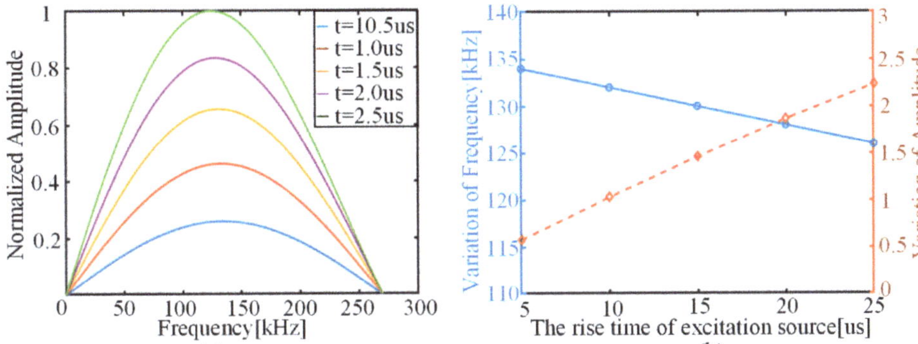

Figure 10. Variation in predicted AE results for different rise times: (**a**) change in model-predicted AE frequency spectrum; (**b**) trend of model-predicted main frequency and corresponding amplitude.

5.2. Effect of Structural Response

The role of the structural response in the AE prediction model is mainly reflected in the structural factor $U_n(h)W_n{}'(z)/I_{nn}$. Analysis of the detailed expressions for the structural factor shows that the wave number k of the Lamb wave in the structure varies with the frequency component, in addition to the parameters related to the material properties of the structure. For an isotropic metal sheet with known material parameters, the frequency thickness product is directly related to the phase velocity of the Lamb wave. Therefore, the effect of wave number k on the frequency spectrum of the signal can be reflected by varying the thickness of the thin plate. The main peaks of the signal spectrum perceived by the piezoelectric sensor under different plate thickness conditions are shown in Figure 11. As the thickness of the thin plate increases, both the main frequency and the corresponding amplitude of the sensed signal show a decreasing trend, and the changes are relatively weak. Because the AE signal is mostly band-pass filtered at 30–700 kHz, and the S0 mode of the Lamb wave has a small phase velocity change in this frequency range, the corresponding wave number change is not obvious, and the change range is relatively small when reflected in the final spectrum.

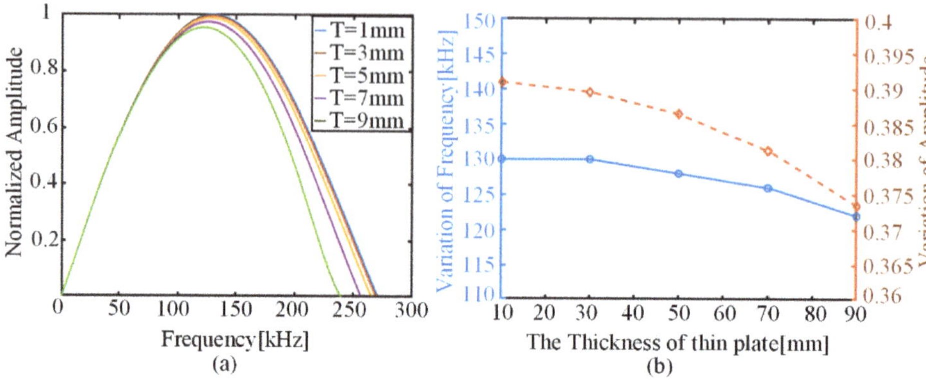

Figure 11. Variation in predicted AE results for different plate thicknesses: (**a**) variation in model-predicted AE spectrum; (**b**) trend of model-predicted main frequencies and corresponding amplitudes.

5.3. *Effect of Sensor Parameters*

The electromechanical conversion coefficient C_{vs} is related to the material properties of the piezoelectric material, and affects the amplitude of the signal. The parameter $\sin(k_n a_s)/a_s$ of the sensor plays the role of signal frequency modulation. Considering that the wave number k does not vary much in this frequency range, the effect of the radius of the piezoelectric sensor is discussed. The main peaks of the signal spectrum received by piezoelectric sensors with different radiuses are shown in Figure 12. As the radius of the piezoelectric sensor becomes larger, the main frequency of the sensed signal decreases rapidly and the corresponding amplitude shows an inverse trend. It should also be noted that, unlike the effect of rise time, the sensor radius mostly affects the final signal main frequency, and the corresponding amplitude change is much smaller. This is related to the fact that the parameter $\sin(k_n a_s)/a_s$ mainly plays the role of signal modulation.

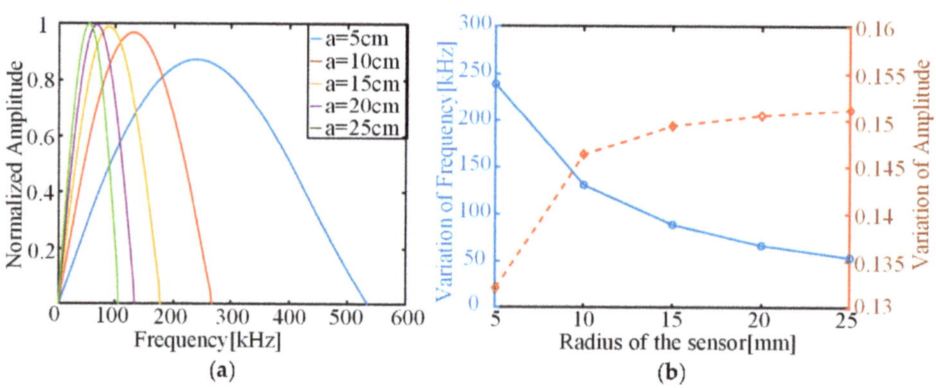

Figure 12. Variation in predicted AE results for different sensor radiuses: (**a**) variation in model-predicted AE spectrum; (**b**) trend of model-predicted main frequencies and corresponding amplitudes.

6. Conclusions

This paper presents an analytical modeling method for AE caused by fatigue crack growth in a thin plate using a piezoelectric sensor. Assuming the existence of type I fatigue cracks in an isotropic thin metal plate, the dynamic analytical expression of Lamb wave propagation in the plate is derived by using the reciprocity theorem. On this basis, the concept of the dipole moment is used to model the type I fatigue crack growth AE source, and the constitutive equation of piezoelectric materials is combined. Finally, a complete analytical prediction model of the piezoelectric sensor AE signal caused by normal fatigue crack growth is obtained. By comparing the finite element simulations, experimental validation and analytical modeling, it is found that they are almost consistent in the time and frequency domains, and the presence of the Lamb wave S0 mode is clearly observed. These validate the accuracy of the analytical modeling predictions. The model is useful as a guide for the selection of detection equipment. The developed model allows the AE signal to be predicted in advance and prepared for the upcoming monitoring. For the same AE signal, different resonant frequencies of the sensor produce different main frequencies of the electrical signal. Therefore, when the ambient noise frequency is high, a sensor with a lower resonant frequency can be selected to distinguish the received AE signal from the noise.

This paper considers an ideal two-dimensional case and does not consider the effect of the third dimensional crack length and the difference in the three-dimensional propagation of guided waves in thin plates. Additionally, it only considers a type I open crack in the modeling process; the modeling of a slip-open crack and tear-open crack will be addressed in the future, as an extension of the current research. In the current study, the models of structural response and sensor response are relatively clear, while the crack expansion AE source is mostly simulated by the cosine clock function. In fact, there are differences in the signal spectrum at different stages of cracking. Therefore, it is important to investigate the relationship between the actual AE source and the stress level and material properties, and to construct a reasonable AE source simulation function. The above results obtained by AE analytical modeling still have certain simplifying conditions and sub-optimal considerations. For example, the signal modulation effect of the parameters is affected not only by the radius of the piezoelectric transducer, but also by the wave number k, which cannot be ignored. However, our conclusions are still important for the selection of piezoelectric transducers and the improvement in AE monitoring sensitivity.

Author Contributions: The first and second author contributed equally to this paper. W.M. and Y.G. prepared the original draft. Y.G. and Y.W. carried out the simulation experiments. G.L. and H.H. provided review and editing help. All authors have read and agreed to the published version of the manuscript.

Funding: This research was funded by the National Natural Science Foundation of China, grant number 52171283, Key R&D Projects of Shandong Province (No. 2019GHY112083), Natural Science Foundation of Shandong Province (No. ZR2020ME268), and Natural Science Foundation of Jiangsu Province (No. BK20201189). The authors are grateful to these foundation committees for their support.

Institutional Review Board Statement: Not applicable.

Informed Consent Statement: Not applicable.

Data Availability Statement: The supplementary data and simulation programs involved in this paper will be uploaded by the first author on the website of https://www.researchgate.net/profile/Weilei_Mu.

Acknowledgments: The valuable comments from the anonymous reviewers are highly appreciated.

Conflicts of Interest: The authors declare no conflict of interest with respect to the research, authorship, and publication of this article.

References

1. Khon, H.; Bashkov, O.V.; Zolotareva, S.V.; Solovev, D.B. Modeling the Propagation of Elastic Ultrasonic Waves in Isotropic and Anisotropic Materials When Excited by Various Sources. *Mater. Sci. Forum* **2018**, *945*, 926–931. [CrossRef]
2. Shen, G.; Gong, R.; Liu, S. AE source localization technology. *Nondestruct. Test.* **2002**, *125*, 114–117.
3. Mu, W.; Zou, Z.; Sun, H. Research on the time difference of arrival location method of an acoustic emission source based on visible graph modelling. *Insight* **2018**, *60*, 575–579. [CrossRef]
4. Mu, W.; Qu, W.; Liu, G.; Zou, Z. Acoustic emission beamforming localisation approach based on particle swarm optimisation. *Insight* **2018**, *60*, 575–581. [CrossRef]

5. Roberts, T.; Talebzadeh, M. Fatigue life prediction based on crack propagation and acoustic emission count rates. *J. Constr. Steel Res.* **2003**, *59*, 679–694. [CrossRef]
6. McBride, S.; MacLachlan, J.; Paradis, B. Acoustic emission and inclusion fracture in 7075 aluminum alloys. *J. Nondestruct. Eval.* **1981**, *2*, 35–41. [CrossRef]
7. Vanniamparambil, P.; Guclu, U.; Kontsos, A. Identification of crack initiation in aluminum alloys using acoustic emission. *Exp. Mech.* **2015**, *55*, 837–850. [CrossRef]
8. Shen, G.; Gong, R.; Liu, S. Parametric analysis method of AE signal. *Nondestr. Test.* **2002**, *24*, 72–77.
9. Wisner, B.; Mazur, K.; Perumal, V. Acoustic Emission Signal Processing Framework to Identify Fracture in Aluminum Alloys. *Eng. Fract. Mech.* **2018**, *210*, 367–380. [CrossRef]
10. Kaewwaewnoi, W.; Prateepasen, A.; Kaewtrakulponng, P. Investigation of the relationship between internal fluid leakage through a valve and the acoustic emission generated from the leakage. *Measurement* **2010**, *43*, 274–282. [CrossRef]
11. Ye, G.; Xu, K.; Wu, W. Multivariable modeling of valve inner leakage acoustic emission signal based on Gaussian process. *Mech. Syst. Signal Process.* **2020**, *140*, 106675. [CrossRef]
12. Shen, G.; Wu, Z.; Zhang, J. Advances in AE Technology. In *Proceedings of the World Conference on AE–2017*; Springer: Berlin/Heidelberg, Germany, 2017.
13. Meland, E.; Thornhill, N.F.; Lunde, E.; Rasmussen, M. Quantification of valve leakage rates. *Am. Inst. Chem. Eng.* **2012**, *58*, 1181–1193. [CrossRef]
14. Shen, G.; Dai, G.; Liu, S. Advances in AE Detection Technology in China. *Nondestruct. Test.* **2003**, *25*, 302–307.
15. Ono, K.; Ohtsu, M. A generalized theory of AE and Green's functions in a half space. *J. AE* **1984**, *3*, 27–40.
16. Ohtsu, M.; Ono, K. The Generalized Theory and Source Representation of AE. *J. AE* **1986**, *5*, 124–133.
17. Weaver, R.L.; Pao, Y.H. Axisymmetric Elastic Waves Excited by a Point Source in a Plate. *J. Appl. Mech.* **1982**, *49*, 84. [CrossRef]
18. Gorman, M.R.; Prosser, W.H. Application of Normal Mode Expansion to Accoustic Emission Waves in Finite Plates. *J. Appl. Mech.* **1996**, *63*, 73–80. [CrossRef]
19. Lysak, M.V. Development of the theory of AE by propagating cracks in terms of fracture mechanics. *Eng. Fract. Mech.* **1996**, *55*, 443–452. [CrossRef]
20. Bhuiyan, M.Y.; Giurgiutiu, V. The Signatures of AE Waveforms from Fatigue Crack Advancing in Thin Metallic Plates. *Smart Mater. Struct.* **2017**, *27*, 015019. [CrossRef]
21. Qing, X.; Li, W.; Wang, Y.; Sun, H. Piezoelectric transducer-based structural health monitoring for aircraft applications. *Sensors* **2019**, *19*, 545. [CrossRef]
22. Ye, G.; Xu, K.; Wu, W. Multi-variable classification model for valve internal leakage based on acoustic emission time–frequency domain characteristics and random forest. *Rev. Sci. Instrum.* **2021**, *92*, 025108. [CrossRef]
23. Raghavan, A.; Cesnik, C.E.S. Finite-dimensional piezoelectric transducer modeling for guided wave based structural health monitoring. *Smart Mater. Struct.* **2005**, *14*, 1448–1461. [CrossRef]
24. Maillet, E.; Morscher, G.N. Waveform-based selection of acoustic emission events generated by damage in composite materials. *Mech. Syst. Signal Process.* **2015**, *52*, 217–227. [CrossRef]
25. Sause, M.G.R.; Hamstad, M.A.; Horn, S. Finite element modeling of lamb wave propagation in anisotropic hybrid materials. *Compos. Part B Eng.* **2013**, *53*, 249. [CrossRef]
26. Sengupta, S.; Roy, P.; Topdar, P. Investigation of layered composite plates under acoustic emission using an appropriate FE model. *Can. J. Civ. Eng.* **2021**, *48*, 1639–1651. [CrossRef]
27. Giuruiutiu, V. Tuned Lamb wave excitation and detection with piezoelectric wafer active sensors for structural health monitoring. *J. Intell. Mater. Syst. Struct.* **2005**, *16*, 291–305. [CrossRef]
28. Xu, B.; Giuruiutiu, V. Single mode tuning effects on Lamb wave time reversal with piezoelectric wafer active sensors for structural health monitoring. *J. Nondestruct. Eval.* **2007**, *26*, 123–134. [CrossRef]
29. Zelenyak, A.M.; Hamstad, M.A.; Sause, M.G.R. Modeling of Acoustic Emission Signal Propagation in Waveguides. *Sensors* **2015**, *15*, 11805–11822. [CrossRef]
30. Achenbach, J.D.; Xu, Y. Wave motion in an isotropic elastic layer generated by a time-harmonic point load of arbitrary direction. *J. Acoust. Soc. Am.* **1999**, *106*, 83–90. [CrossRef]
31. Tzou, H.S.; Zhong, J.P. Electromechanics and Vibrations of Piezoelectric Shell Distributed Systems. *J. Dyn. Syst. Meas. Control* **1993**, *115*, 506–517. [CrossRef]
32. Hamst, A.M.A. AE signals generated by monopole (pencil-lead break) versus dipole sources: Finite element modeling and experiments. *J. AE* **2007**, *25*, 92–106.

sensors

Article

TR Self-Adaptive Cancellation Based Pipeline Leakage Localization Method Using Piezoceramic Transducers

Yanbin Mo [1] and Lvqing Bi [1,2,*]

1 School of Physics and Telecommunications Engineering, Guangxi Colleges and Universities Key Laboratory of Complex System Optimization and Big Data Processing, Yulin Research Institute of Big Data, Yulin Normal University, Yulin 537000, China; moyanbin9559@163.com
2 Department of Electronic Engineering, Xiamen University, Xiamen 361005, China
* Correspondence: bilvqing108@163.com

Abstract: In this paper, we propose a novel time reversal-based localization method for pipeline leakage. In the proposed method, a so-called TR self-adaptive cancellation is developed to improve the leak localization resolution. First of all, the proposed approach time reverses and back-propagates the captured signals. Secondly, the time reversed signals with the various coefficients are superposed. Due to the synchronous temporal and spatial focusing characteristic of time reversal, those time reversed signals will cancel each other out. Finally, the leakage location is distinguished by observing the energy distribution of the superposed signal. In this investigation, the proposed method was employed to monitor a 58 m PVC pipeline. Three manually controllable valves were utilized to simulate the leakages. Six piezoceramic sensors equipped on the pipeline, recorded the NWP signals generated by the three valves. The experimental results show that the leak positions can accurately revealed by using the proposed approach. Furthermore, the resolution of the proposed approach can be ten times that of the conventional TR localization method.

Keywords: time reversal; localization; PZT transducer; negative pressure wave; pipeline leakage

Citation: Mo, Y.; Bi, L. TR Self-Adaptive Cancellation Based Pipeline Leakage Localization Method Using Piezoceramic Transducers. *Sensors* **2022**, *22*, 696. https://doi.org/10.3390/s22020696

Academic Editor: Zenghua Liu

Received: 28 November 2021
Accepted: 13 January 2022
Published: 17 January 2022

1. Introduction

Structural health monitoring (SHM) is a newly developed research field devoted to the monitoring and assessment of structural health and durability. SHM is especially useful for remotely monitoring the health of high-profile mechanical systems such as spacecraft, ships, offshore structures, and pipelines where onsite monitoring is difficult or even impossible [1]. Optic fiber bragg grating is employed for the applications for SHM [2–4], such as monitoring the load level of the bolts [5,6] and the detection of hoop-strain in the pipeline [7–9]. Another SHM systems are based on lead zirconate titanate (PZT) transducer. For the PZT based passive sensing, the PZT transducers are employed to acquire the stress signals from impact or leakage [10–13]. Furthermore, the active sensing based on PZT is applied more widely [14,15]. For example, the monitoring of debonding in the structures [16–18], the pipeline damage detection [19], the grouting compactness monitoring of post-tensioning tendon duct [20,21], the pre-stress monitoring of rock bolts [22], the evaluation of the fatigue damage in the modular bridge expansion joints [23], structure damage imaging [24,25], and real-time monitoring of soil compaction [26].

As an important investigation field of SHM, pipeline leakage monitoring is attracting extensive attention [27,28] since the pipeline leakage caused numbers of catastrophic accidents every year around the world. The detection methods of the pipeline leakage have been explored in numerous studies in the available literature. The representative pipeline leak localization methods include the transient model method [29–31], the magnetic flux detection method [32,33], the acoustics method [34–37], the image analysis method [38], the hybrid method [39,40], the negative pressure wave (NPW) method [41,42], etc. For example, Ni et al. [31] compared the leak localization based on the particle swarm optimization theory with that based on the transient model. Yan et al. presented the magnetic flux leakage detection technique which uses magnetic sensitive sensors to detect the magnetic leakage field of defects on both the internal and external surfaces of pipelines [32]. Bian et al. developed a leakage localization method which analyzes the space-time correlation of the ultrasonic signals of the leakage hole [37]. Su et al., detected leaking defects by using an image segmentation method [38]. Zhang et al. [40] proposed a hybrid leak detection technology which combined the negative-pressure wave method and the transient modeling method. Liu et al. [41] developed a novel leakage detection approach based on the amplitude attenuation model of dynamic pressure waves. Although the aforementioned technologies are capable of detecting the existence of leakages, they may suffer from high computational cost [31], expensive equipment [32,33], or large location error [41]. For example, the location error of the hybrid leak detection technology is about 10%. Although the transient model method has a high localization accuracy, its complex models mean a lot of training and expensive cost which prevent its wide applications. Compared with other methods, the NPW method is widely utilized due to its simple operation and high accuracy.

When a leak occurs, a negative pressure wave propagates toward the two ends of a pipeline from the leak position. After recording the NPW signals via the sensors mounted on the outside layer of the pipeline, the signal processing algorithms are utilized to analyze the NPW signals, in order to localize the leakage points. For instance, Jia et al. [42] developed a leakage localization approach which detected the NPW by measuring the hoop strain. Zhu et al. [43] localized the leakages by calculating the arriving time of the NPW and analyzed the localization accuracy. Hou et al. [8] developed a compound Simpson formula and dichotomy searching-based leak localization formula. Zhou et al. [44] developed an improved spline-local mean decomposition to process the negative pressure wave signal, and revealed the leak by using the convolutional neural network to analyze the processed signal. Li et al., developed a novel localization algorithm based on the attenuation of negative pressure wave [45]. Those methods aimed to improve the location error via suppressing the noise [44,45]. No matter that the localization methods are based on the time delay estimation [42–44] or the attenuation of negative pressure wave [45], most of them merely revealed the leakage positions. Other acoustic signatures such as −3 dB width which sets a boundary limit between leakage points and non-leakage points, cannot be observed from the results obtained by using the existing methods. However, Ing et al. proved that −3 dB width and the maximum peak which are the two main acoustic characteristics can be identified by using the time reversal (TR) localization method at the same time [46].

In the computational time-reversal localization [47], the signals are computationally re-radiated into the area of interest [48–50] in a computer [48]. In the computational process, the TR methods generate grids in the monitoring area, and obtain the signals at the grids by convoluting the time reversed signal with channel impulse response in time domain. Due to the spatial reciprocity, TR techniques can force the signals to focus at their sources. The sources' locations can be revealed by plotting the distribution of the TR signals' energy [51,52]. Recently, time reversal technology is studied for non-destructive testing and SHM [53,54]. For example, Amitt et al. performed a computational TR run, and obtained an identification solution at the reference time [55]. In concrete model, the peak of the TR signal is utilized to identify the source's position [56,57]. Zhao et al., built the defect image map by using the maximum of the time reversal field [58].

However, for the passive detection, such as the pipeline leakage detection, the time reversal localization suffers several issues. Due to the long signal duration which is the result of the superposition of incident and reflected NPW, although the conventional TR localization method can reveal the positions of the pipeline leakages, it can merely offer a poor resolution which is dozens of meters in size (i.e., the maximum −3 dB width is dozens of meters in size) [46,47]. For improving the localization resolution, a new TR localization method is developed in this paper. In this approach, the time reversed signals with specific coefficients are superposed. Since time reversed signals are multiplied by the coefficients, the localization functional value based on the proposed method decrease significantly with the movement of the observed point. Therefore, the resolution gets improved. For investigating the performance of the proposed method in passive detection application, a pipeline leakage localization experiment was conducted. For a 58 m pressurized pipeline, three manually controllable valves were utilized to simulate the leakages. Six piezoceramic transducers equipped on the pipeline, recorded the NWP signals generated by the three valves. The experimental results show that the proposed approach accurately reveals the leak locations with a maximum −3 dB width of 2 m. Meanwhile, the maximum −3 dB width of the conventional TR localization method is 20 m. That means the proposed approach can improve the resolution, compared to the conventional one.

2. Theoretical Approach of the Proposed Localization Method

For corresponding to the pipeline leakage experiment, we describe the proposed method by using the model of pipeline. For a pressurized gas pipeline, a leakage will generate a negative pressure wave (NPW) which will propagate toward the two ends of a pipeline. The NPW can be detected by lead zirconate titanite sensors equipped on the pipeline wall. We assume that N transducers are used along a pipeline, and the nth transducer is located at $\mathbf{r_n}$, as shown in Figure 1. We further assume that the leakage occurs at $\mathbf{r_L}$. In this paper, plain symbols denote scalar quantities, whereas vectors and matrices are denoted by bold symbols.

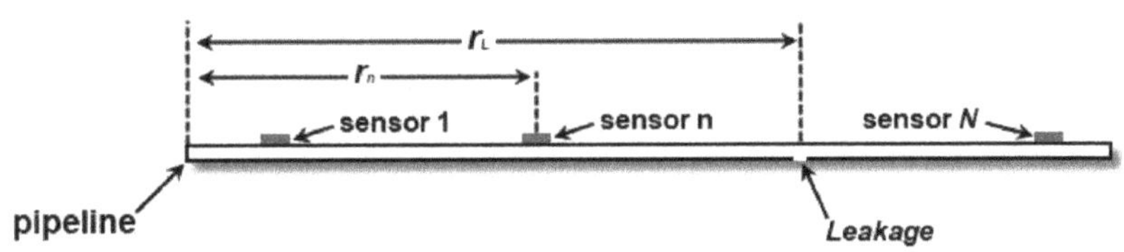

Figure 1. Model of the proposed method.

For the purpose of convenience, we describe the proposed method in the time domain. Assume a NPW signal generated by the leakage is represented as $x(t)$ and the leakage occurrence time is $t = T$. All transducers are synchronous. The leakage signal recorded by the nth transducer can be modeled as,

$$y(t, \mathbf{r_n}, \mathbf{r_L}) = \mathrm{g}_r(\mathbf{r_n}, \mathbf{r_L}, \mathrm{t}) \otimes x(t - T) \tag{1}$$

where "$\otimes$" represents the convolution operation. The function $\mathrm{g}_r(\mathbf{r_n}, \mathbf{r_L}, \mathrm{t})$ is the actual channel impulse response between the nth transducer and the leakage at $\mathbf{r_L}$, obtained via measurement.

The time reverse signal $y(t, \mathbf{r_n}, \mathbf{r_L})$, and the time-reversal version of Equation (1) can be represented as,

$$y(-t, \mathbf{r_n}, \mathbf{r_L}) = \mathrm{g}_r(\mathbf{r_n}, \mathbf{r_L}, -\mathrm{t}) \otimes x(-t - T) \tag{2}$$

A time reversal process is indeed equivalent to a correlation of impulse responses. Therefore, accounting for back-propagation from the nth transducer to the observation point $\mathbf{r_k}$ in the monitoring domain, the time reversed signal at a point $\mathbf{r_k}$ can be illustrated as,

$$f(t, \mathbf{r_n}, \mathbf{r_k}) = \mathrm{g}_r(\mathbf{r_n}, \mathbf{r_L}, -\mathrm{t}) \otimes x(-t - T) \otimes \mathrm{g}_c(\mathbf{r_n}, \mathbf{r_k}, \mathrm{t}) \tag{3}$$

where $\mathrm{g}_c(\mathbf{r_n}, \mathbf{r_k}, \mathrm{t})$ is the computational channel impulse response from the nth transducer to the point $\mathbf{r_k}$.

Then, the so-called self-adaptive cancellation is used to process the time reversed signals: firstly, multiplying $f(t, \mathbf{r_n}, \mathbf{r_k})$ by a coefficient $(-1)^{\mathrm{n}}$. Secondly, summing over all time reversed signals multiplied by the coefficients. On the other hand, due to the various attenuation coefficients of the actual channels, the various sensors' TR signals cannot cancel each other out via the superposition at the leak position. For eliminating that influence, the TR signals are normalized. The corresponding result can be written as,

$$q(\mathbf{r_k}, t) = \sum_{n=1}^{N} (-1)^n f(t, \mathbf{r_n}, \mathbf{r_k}) / \max(f(t, \mathbf{r_n}, \mathbf{r_k})) \tag{4}$$

Due to reciprocity principle, and assuming that the computational channel response function matches the measured data perfectly, namely $g_c(\mathbf{r_n}, \mathbf{r_L}, t) = g_r(\mathbf{r_n}, \mathbf{r_L}, t)$. Then, all time reversed signals $f(t, \mathbf{r_n}, \mathbf{r_k})$ $(n = 1 \ldots\ldots N)$ will focus on the leakage location $(\mathbf{r_k} = \mathbf{r_s})$ at the time $t = -T$, with similar waveform. Therefore, all time reversed signals multiplied by coefficients will cancel each other out at $\mathbf{r_s}$ as the amplitude of $q(\mathbf{r_s}, t)$ approaches zero.

Since the time reversed signals cancel each other out at the leakage's location, the signal energy in the leakage area is low than those in other areas. Therefore, the localization functional is designed as following,

$$I_p(\mathbf{r_k}) = \left(\int |q(\mathbf{r_k}, t)| dt \right)^{-1} \tag{5}$$

In the localization map, the localization functional value at leakage locations will be larger than those at other points.

For the convenience of explanation, the diagram of signal processing based on the proposed method is shown in Figure 2. After time reversal and back-propagation, the time reversed signals focus on the leakage positions at the reference time $t = -T$. Due to the temporal focusing, those time reversed signals are the same to each other. Via multiplying the coefficient $(-1)^n$, part of the time reversed signals is upside down. Then, when the time reversed signals multiplied by coefficients are superposed, they cancel each other out. Therefore, via the self-adaptive cancellation, the amplitude of the final output signal at the leakage position approaches to minimum.

Generally speaking, in the proposed method, the following steps are taken to localize the leakages, as shown in Figure 3. First of all, the sensors capture the NPW signals. Secondly, time reverse and back-propagate the captured signals (convolve the time reversed signal with channel impulse response in time domain). The time reversed signals at a generic observation point are normalized and multiplied with the coefficients (i.e., $(-1)^n$). Then, the sum of all time reversed signals is multiplied by the coefficients, as Equation (4) shows. Finally, plot the localization map is plotted by calculating the localization functional (i.e., Equation (5)).

3. Experiment

As one kind of passive detection, a PVC pipeline leak experiment was executed. The pipeline was composed by a series of straight sections connected by twenty-four 90°-elbow connectors and twelve 0.05 m sections. The pipeline was fixed to a steel frame and connected to another pipeline with the pressure of 0.8 MPa, as shown in Figure 4. Six lead zirconate titanite (PZT) sensors whose size is 15 mm × 10 mm × 2 mm were directly mounted on the pipe wall and utilized to measure the local hoop strain change. Their special locations are listed at Table 1. The locations of three manually controllable valves equipped in the pipeline are shown in Table 2. The inner diameter of the valves is 12.7 mm. A leakage was simulated by opening any one of the valves. The data acquisition system is NI USB-6366 with the sampling rate of 2 MS/s. Since the distance between the first sensor and the last one was 58 m, the proposed method was utilized to monitor a pipeline of length = 58 m in this experiment. The host computer, which was utilized for localization computation, was equipped with i7-6700 CPU, 8 GB DDR3 memory.

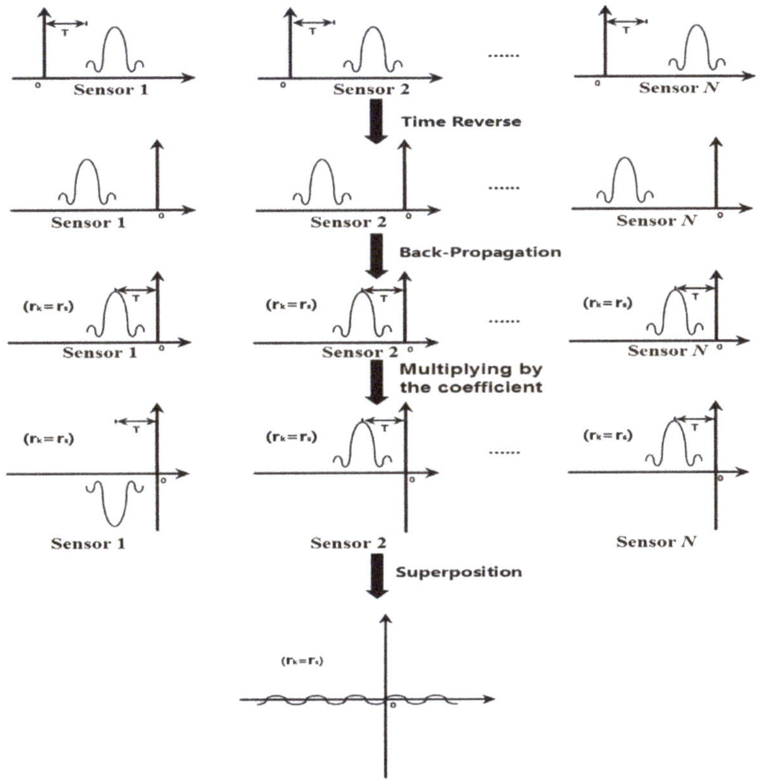

Figure 2. The diagram of signal processing based on the proposed method.

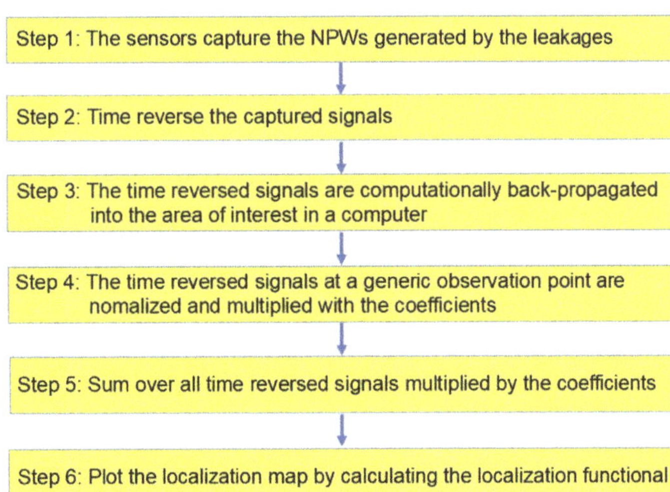

Figure 3. The procedure of the localization method.

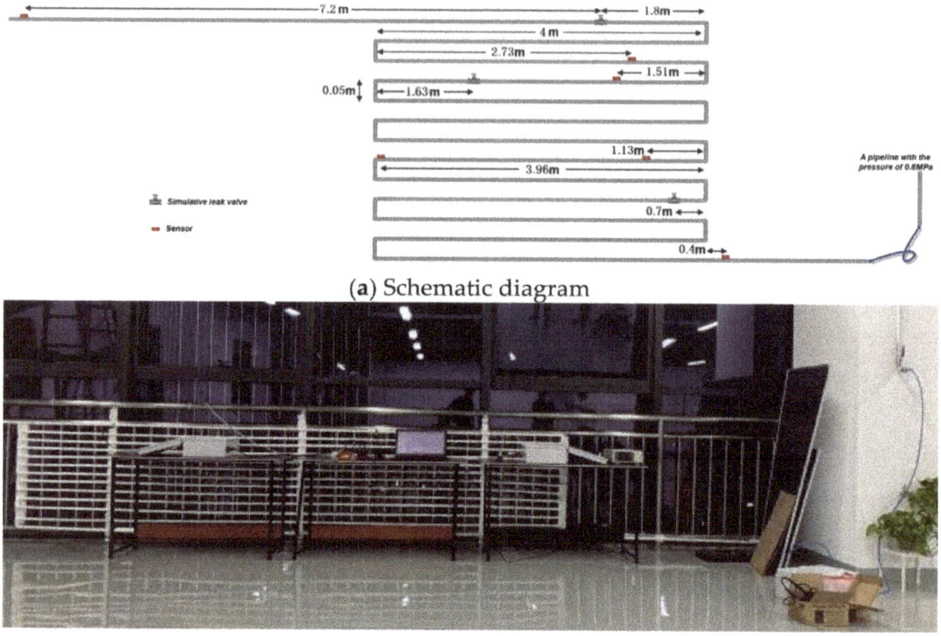

(**a**) Schematic diagram

(**b**) Photo

Figure 4. Schematic diagram and photo of the measured experiment.

Table 1. Specific positions of the sensors.

Sensors	Distance from Sensor 1 (Unit: m)
Sensor 2	15.83
Sensor 3	18.66
Sensor 4	34.48
Sensor 5	37.31
Sensor 6	58

Table 2. Specific locations of the leakages.

Leakages	Distance from Sensor 1 (Unit: m)
Leakage 1 (L1)	7.2
Leakage 2 (L2)	23.52
Leakage 3 (L3)	42.15

The experiment was performed by taking the following steps: (1) The three valves were initially closed; (2) for each test, the pipeline was pressurized up to 0.8 MPa via the pipeline with the pressure of 0.8 MPa; (3) an event of leakage was created through opening one of the valves in the pipeline; (4) the data acquisition system recorded the NPW signals captured by PZT sensors; (5) by using the proposed method, the leakage position could be distinguished by analyzing the recorded signals.

Under the usual operating states of pressurized pipelines, the internal pressure of the pipeline is invariable and much higher than the external pressure [43]. When a leakage happens (i.e., a crack develops), the pipeline content escapes through the crack, thus, internal pressure drops significantly at the leakage point. The content in the pipeline shifts from both upstream and downstream towards the crack. Following the decrease of the internal pipeline pressure, the contraction of pipe wall decreases hoop strain variation on the surface of the pipe. The six piezoelectric sensors will detect the hoop strain variation.

4. Results and Discussion

The typical waveform of the NPW captured by PZT sensors are shown in Figure 5. As aforementioned [43], for an undamaged pipeline, the amplitude is held at zeros as a result of the internal pressure keeping constant. Then, the pulse-like waveform is generated since the NPW reaches the sensors. The initial upward edge of the pulse indicates the drop of internal pressure associated with the NPW, while the downward edge means the internal pressure settling at a new baseline pressure. The peak of the pulse is the result of the NPW being arriving at the sensor's position.

In the signal processing phase, all NPW signals generated the leakages are processed by using the conventional TR localization algorithms and the proposed algorithm, respectively. During re-constructing the leakage area, we made the channel impulse response $g_c(\mathbf{r_n}, \mathbf{r_k}, t)$ in Equation (3), by using the NPW propagation time of the corresponding channel. $g_c(\mathbf{r_n}, \mathbf{r_k}, t)$ can be written as the following [51]:

$$g_c(\mathbf{r_n}, \mathbf{r_k}, t) = a_{\mathbf{n,k}}\delta(t - \frac{distanc\, e_{\mathbf{n,k}}}{v_g}) \tag{6}$$

where $\delta(t)$ is the Dirac function, $distance_{\mathbf{n,k}}$ is the distance between $\mathbf{r_n}$ and $\mathbf{r_k}$, $a_{\mathbf{n,k}}$ is the attenuation coefficient of the computational channel from point $\mathbf{r_n}$ and the localization point $\mathbf{r_k}$. During the computational back-propagation, it can reduce computation cost for obtaining the attenuation coefficients and the waveform deterioration caused by the back-propagation to set the attenuation coefficients to 1 (i.e., $a_{\mathbf{n,k}} = 1$). Furthermore, as mentioned above, the normalization eliminates the influence of the attenuation coefficients of the channels (including the actual channels and the computational channels). Therefore, by using the parameter setting ($a_{\mathbf{n,k}} = 1$), the proposed method can obtain the results easily and accurately. The NPW velocity is considered as 300 m/s, which is also the theoretical NPW velocity reported in [42,43]. Furthermore, this value was chosen as it is situated in the upper margin of velocity estimation and for convenience of calculations [59,60].

Since the system recorded the NPW signals under voltage trigger mode, the leakage happened before the data acquisition system began to obtain the signals. Therefore, there is a time delay between the leakage occurrence time and the starting time of the received leakage signal, namely the parameter "T" of Section 2. After time reversing and re-transmitting the received NPW signals, the time reversed signals focus at the time $t = -T$. The conventional TR localization algorithm which used the signal values at the time $t = 0$ to re-construct the image of the leakage location is shown in following equation:

$$I_c(\mathbf{r_k}) = \sum_{n=1}^{N} f(t, \mathbf{r_n}, \mathbf{r_k})\bigg|_{t=0} \tag{7}$$

In order to focus the time reversed signals at the time $t = 0$, the time delay needed to be compensated. However, since the pipeline leakage occurrence time is unknown, the compensation cannot be done. In this case, the conventional TR localization based on the focal time will result in a wrong localization result.

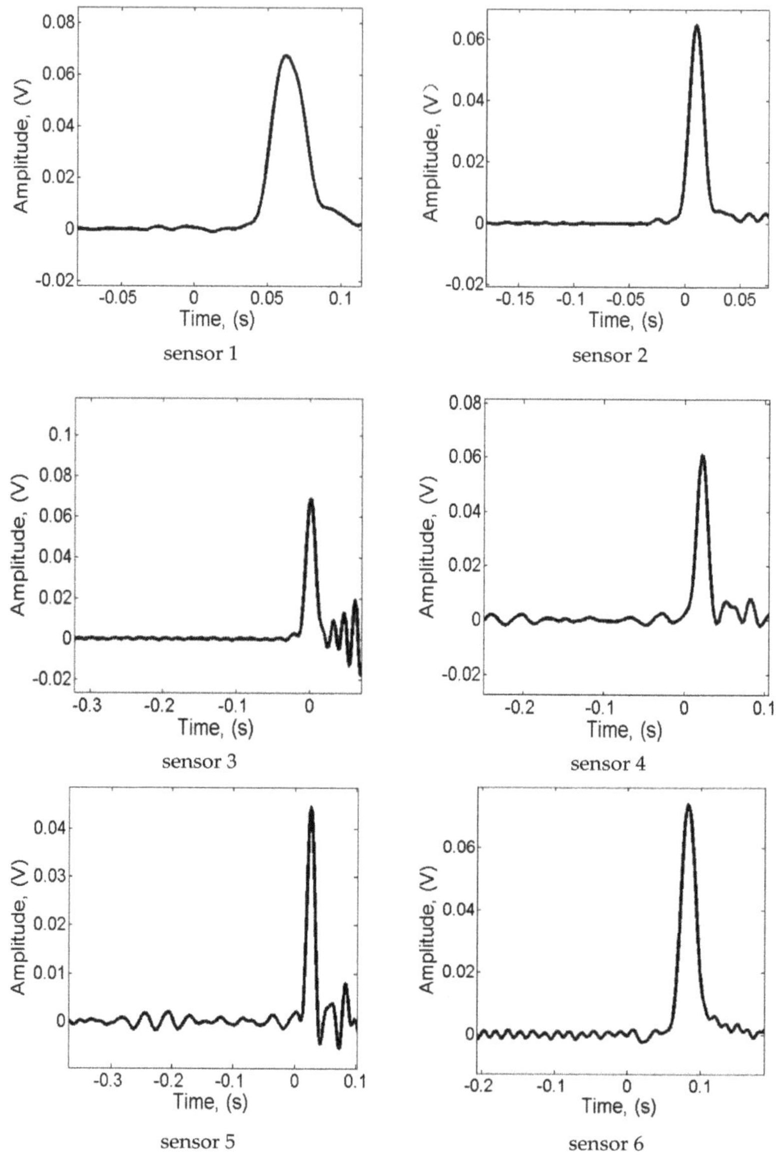

Figure 5. The NPW signals from leakage L2.

Therefore, the conventional TR localization method based on the maximum of the time reversed signal is utilized to process the data for the purpose of comparison. The localization functional of the conventional TR localization algorithm based on the maximum value is represented as follows:

$$I_c(\mathbf{r_k}) = \max \sum_{n=1}^{N} f(t, \mathbf{r_n}, \mathbf{r_k}) \tag{8}$$

The results base on the proposed method and the conventional TR localization method are shown in Figure 6. The localization functional values are normalized, and the position of sensor 1 is considered as the starting point of the results. Since the distance between the first sensor and the last sensors is 58 m, the length of the pipe in a computer is set as 60 m for convenience, and the length of the grid is set as 0.2 m. As shown in Figure 6, the conventional TR localization method based on the peak of the time reversed signal can reveal the positions of the leakages. However, due to the long signal duration which is the result of the superposition of incident and reflected NPW, the time reversed signals still superposed with each other at the points beyond the leakage point. Therefore, the maximum value of the output signal attenuated slowly with the movement of the observed point, the localization functional values at a lot of points are very close to that at the leakage position. The leakage area revealed by the conventional TR localization method based on the maximum of the time reversed signal covers a large range. By using the conventional TR localization method, the −3 dB area of the L1 cover from length = 0 m to length = 19 m, the −3 dB area of the L2 is from length = 23 m to length = 27 m and the −3 dB area of the L3 ranges from length = 37 m to length = 60 m, as shown in Figure 6.

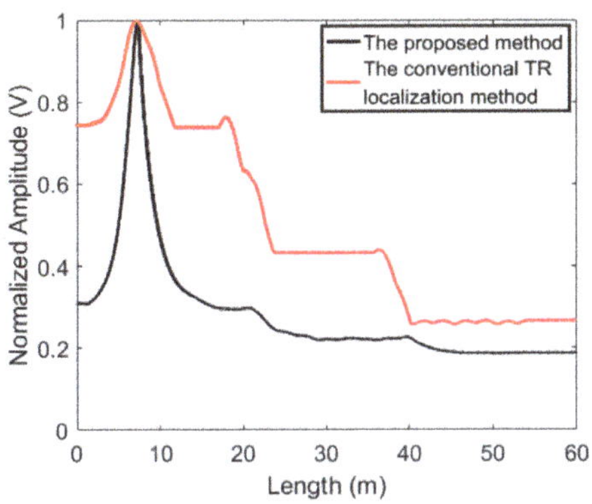

L1

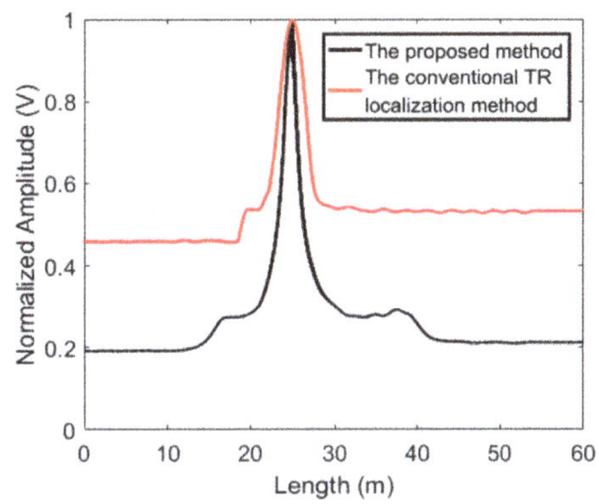

L2

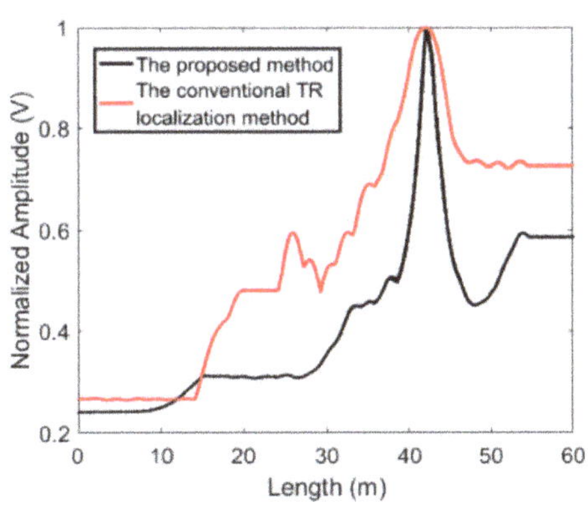

L3

Figure 6. The localization results obtained by using the proposed method and the conventional TR localization method based on the maximum of the time reversed signal.

All time reversed signals will focus at the leakage positions with a similar waveform. That means that the time reversed signals can cancel each other out completely via Equation (4) at the leakage position. After integrating the superposed signals via Equation (5), the localization functional values at the leakage position will be larger than those at other positions. Due to the adoption of the temporal integration, we do not need to calculate the time delay which cannot be obtained. As shown in Figure 6, the plot based on the proposed method gives a good estimate of the leakage positions. By using the proposed method, the three estimated leakage positions are 7.4 m, 25 m, and 42.2 m, respectively. The maximum deviation is about 2.7% over a 58-m monitoring area. Furthermore, the resolution of the proposed method is superior to that of the conventional TR localization algorithm based on the peak of the TR signal. As shown in Figure 6, the lengths of the three leakages' −3 dB areas are about 2 m, 2 m, and 4 m respectively. Obviously, the −3 dB area of the results based on the proposed method is much smaller than that of the conventional one [46].

At various observation points, the L2's time reversed signals from the all sensors, the output signals of the proposed method and the output signals of the TR localization method based on the maximum signal are shown in Figures 7–9. As shown in Figure 7, the time reversed signals completely overlap with each other at the leakage position. Furthermore, due to the long duration of the signals, most of the time reversed signals still superpose with each other at the points beyond the leakage. Therefore, the output signal of the TR localization method based on the maximum signal slowly attenuates at the points beyond the leakage position, as shown in Figure 8. On the other hand, by using the proposed method, most of the signals are canceled after the superposition. The output signal based on the proposed localization method rapidly enlarges at the points away from the leakage, as shown in Figure 9. The output signal of the proposed method at 24.68 m is twice larger than that at 23.52 m. However, the signal of the conventional TR localization method based on the maximum value only drops to 92%. At 25.68 m, the same phenomenon can also be found; the output signal of the proposed method at the leakage position is one third of that at 25.68 m, but the output signal of the conventional TR localization method based on the maximum value drops to 69%. Since the change of the output signal based on the proposed method is larger than that based on the conventional TR localization method, the −3 dB leakage area revealed by the proposed method contains less points, compared to that of the conventional TR localization method. Therefore, the resolution of the proposed method is better than that of the conventional TR localization method based on the maximum.

Generally speaking, by using the proposed method, the three estimated leakage positions are 7.4 m, 25 m, and 42.2 m respectively. That means that the maximum deviation is about 2.7% over a 58 m monitoring area. Furthermore, the lengths of the three leakages' −3 dB areas are about 2 m, 2 m, and 4 m, which are superior to those of the conventional localization method. Therefore, the plot based on the proposed method gives a better estimate of the leakage positions, compared to the result based on the conventional one. That improvement is a result of the localization functional value based on the proposed method decreasing significantly with the movement of the observed point. Besides, the estimated localization errors of the proposed method are the result of the NPW reflecting off the ends of the pipeline. The superposition of incident and reflected NPW postpones the peak of the NPW signal. Consequently, the estimated leak locations shift from the actual locations slightly. However, the estimated localization error of 2.7% is less and comparative.

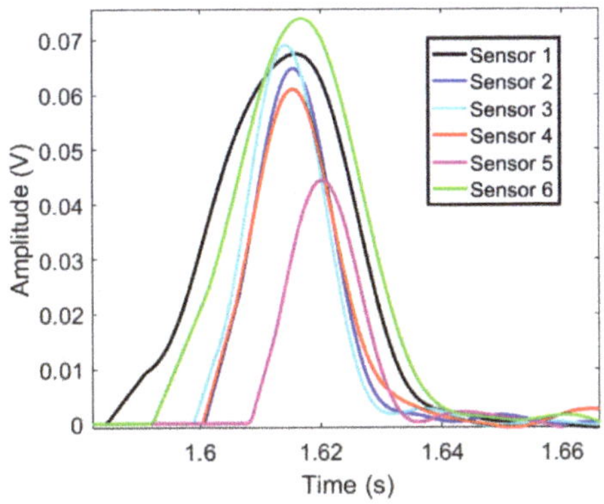

At 23.52 m

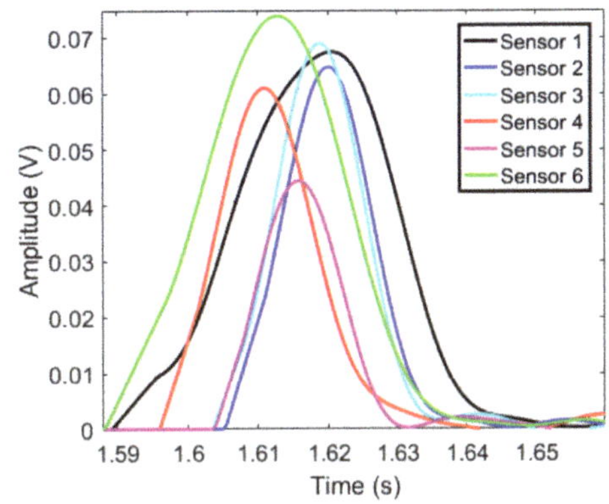

At 24.68 m

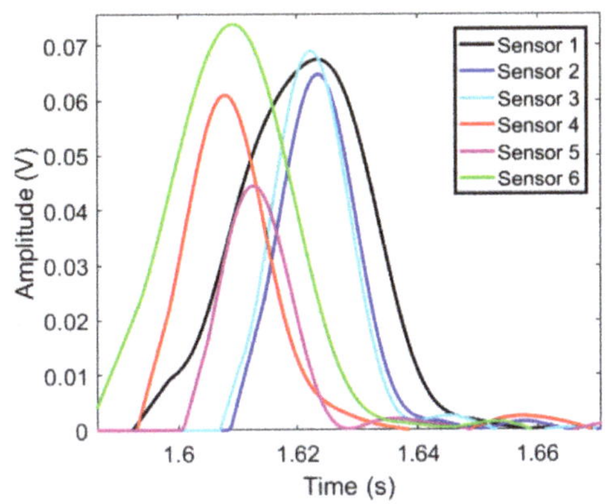

At 25.68 m

Figure 7. The L2's time reversed signals.

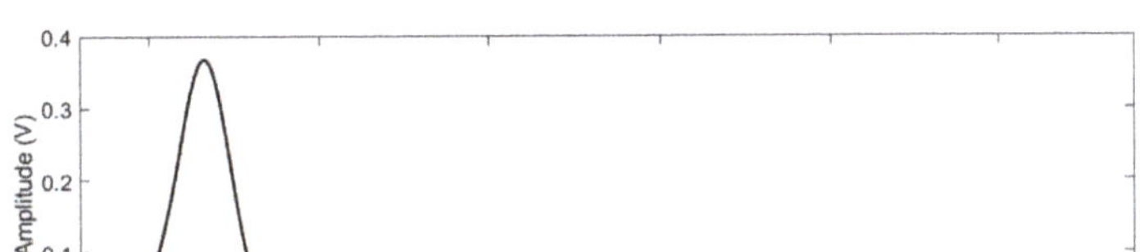

At 23.52 m

Figure 8. *Cont.*

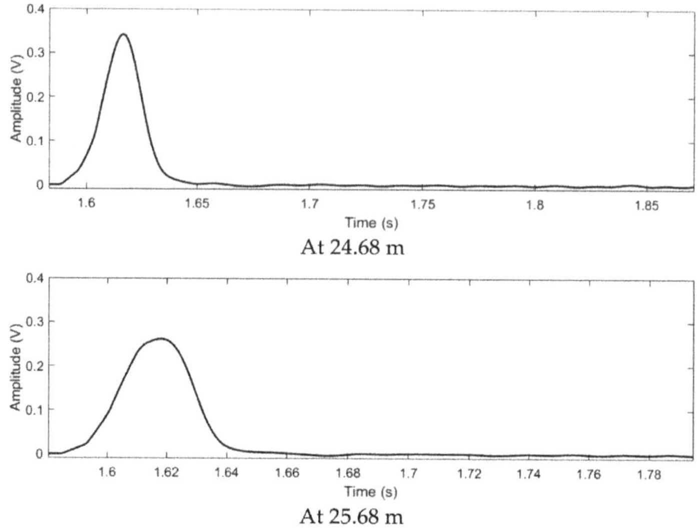

Figure 8. After the superposition, the L2's signals of the TR localization method based on the maximum value.

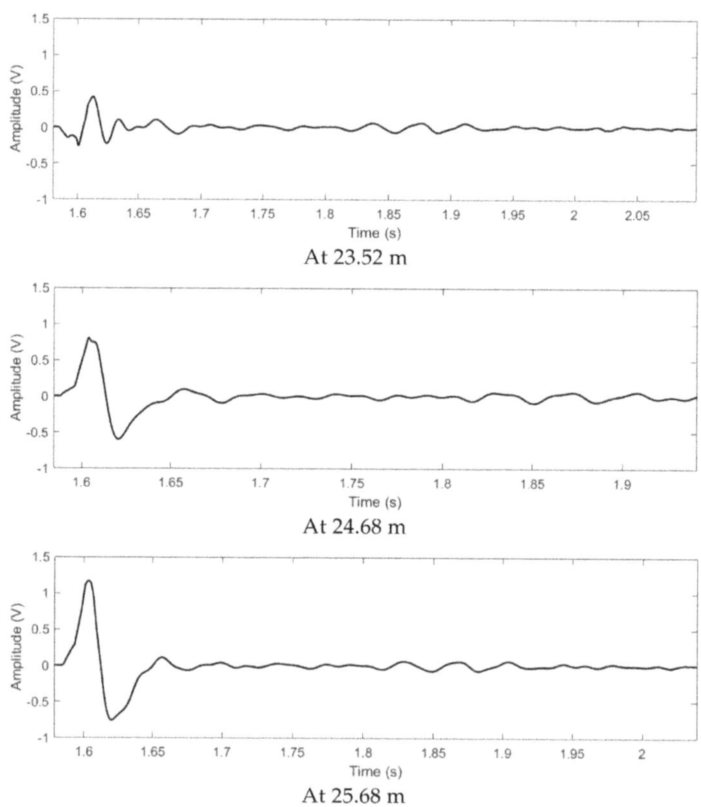

Figure 9. After the superposition, the L2's signals of the proposed method.

5. Conclusions

When the conventional TR localization methods are employed for pipeline leakage localization, they suffer from some issues, such as a low resolution. In this paper, a novel time reversal localization method is designed for localizing pipeline leakages. Via the self-adaptive cancellation and the temporal integration, the proposed localization method can reveal NPW sources with an accurate and high-resolution result. We applied the proposed method for leakage localization in a pressurized pipeline with PZT sensors. The results indicate the proposed method can localize the leakages with a high accuracy. Furthermore, the resolution of the proposed approach can be ten times that based on the conventional TR localization method.

Author Contributions: Conceptualization, Y.M. and L.B.; methodology, Y.M. and L.B.; writing—original draft preparation, Y.M.; writing—review and editing, L.B.; supervision, L.B. All authors have read and agreed to the published version of the manuscript.

Funding: This research was funded by the Opening Foundation of Yulin Research Institute of Big Data grant number [2020YJKY04].

Institutional Review Board Statement: Not applicable.

Informed Consent Statement: Not applicable.

Data Availability Statement: Not applicable.

Conflicts of Interest: The authors declare no conflict of interest.

References

1. Song, G.; Wang, C.; Wang, B. Structural Health Monitoring (SHM) of Civil Structures. *Appl. Sci.* **2017**, *7*, 789. [CrossRef]
2. Chen, D.; Huo, L.; Li, H.; Song, G. A Fiber Bragg Grating (FBG)-Enabled Smart Washer for Bolt Pre-Load Measurement: Design, Analysis, Calibration, and Experimental Validation. *Sensors* **2018**, *18*, 2586. [CrossRef]
3. Ho, S.C.M.; Li, W.; Wang, B.; Song, G. A load measuring anchor plate for rock bolt using fiber optic sensor. *Smart Mater. Struct.* **2017**, *26*, 057003. [CrossRef]
4. Ho, S.C.M.; Ren, L.; Li, H.N.; Song, G. A fiber Bragg grating sensor for detection of liquid water in concrete structures. *Smart Mater. Struct.* **2013**, *22*, 055012. [CrossRef]
5. Song, G.; Li, W.; Wang, B.; Ho, S.C. A Review of Rock Bolt Monitoring Using Smart Sensors. *Sensors* **2017**, *17*, 776. [CrossRef] [PubMed]
6. Ren, L.; Feng, T.; Ho, M.; Jiang, T.; Song, G. A smart "shear sensing" bolt based on FBG sensors. *Measurement* **2018**, *122*, 240–246. [CrossRef]
7. Ren, L.; Jia, Z.G.; Li, H.N.; Song, G. Design and experimental study on FBG hoop-strain sensor in pipeline monitoring. *Opt. Fiber Technol.* **2014**, *20*, 15–23. [CrossRef]
8. Hou, Q.; Ren, L.; Jiao, W.; Zou, P.; Song, G. An Improved Negative Pressure Wave Method for Natural Gas Pipeline Leak Location Using FBG Based Strain Sensor and Wavelet Transform. *Math. Probl. Eng.* **2013**, *2013*, 278794. [CrossRef]
9. Li, W.; Ho, S.C.M.; Song, G. Corrosion detection of steel reinforced concrete using combined carbon fiber and fiber Bragg grating active thermal probe. *Smart Mater. Struct.* **2016**, *25*, 045017. [CrossRef]
10. Du, G.; Zhang, J.; Zhang, J.; Song, G. Experimental Study on Stress Monitoring of Sand-Filled Steel Tube during Impact Using Piezoceramic Smart Aggregates. *Sensors* **2017**, *17*, 1930. [CrossRef]
11. Zhu, J.; Wang, N.; Ho, S.C.; Song, G. Method for Rapid Impact Localization for Subsea Structures. *IEEE Sens. J.* **2018**, *18*, 3554–3563. [CrossRef]
12. Huo, L.; Li, X.; Chen, D.; Li, H.; Song, G. Identification of the impact direction using the beat signals detected by piezoceramic sensors. *Smart Mater. Struct.* **2017**, *26*, 085020. [CrossRef]
13. Zhu, J.; Ho, S.C.M.; Patil, D.; Wang, N.; Hirsch, R.; Song, G. Underwater pipeline impact localization using piezoceramic transducers. *Smart Mater. Struct.* **2017**, *26*, 107002. [CrossRef]
14. Zhang, J.; Li, Y.; Du, G.; Song, G. Damage Detection of L-Shaped Concrete Filled Steel Tube (L-CFST) Columns under Cyclic Loading Using Embedded Piezoceramic Transducers. *Sensors* **2018**, *18*, 2171. [CrossRef]
15. Xu, K.; Deng, Q.; Cai, L.; Ho, S.; Song, G. Damage Detection of a Concrete Column Subject to Blast Loads Using Embedded Piezoceramic Transducers. *Sensors* **2018**, *18*, 1377. [CrossRef] [PubMed]
16. Feng, Q.; Kong, Q.; Jiang, J.; Liang, Y.; Song, G. Detection of Interfacial Debonding in a Rubber–Steel-Layered Structure Using Active Sensing Enabled by Embedded Piezoceramic Transducers. *Sensors* **2017**, *17*, 2001. [CrossRef] [PubMed]
17. Di, B.; Wang, J.; Li, H.; Zheng, J.; Zheng, Y.; Song, G. Investigation of Bonding Behavior of FRP and Steel Bars in Self-Compacting Concrete Structures Using Acoustic Emission Method. *Sensors* **2019**, *19*, 159. [CrossRef]

18. Zeng, L.; Parvasi, S.M.; Kong, Q.; Huo, L.; Lim, I.; Li, M.; Song, G. Bond slip detection of concrete-encased composite structure using shear wave based active sensing approach. *Smart Mater. Struct.* **2015**, *24*, 125026. [CrossRef]
19. Yan, S.; Li, Y.; Zhang, S.; Song, G.; Zhao, P. Pipeline Damage Detection Using Piezoceramic Transducers: Numerical Analyses with Experimental Validation. *Sensors* **2018**, *18*, 2106. [CrossRef]
20. Jiang, T.; Zheng, J.; Huo, L.; Song, G. Finite Element Analysis of Grouting Compactness Monitoring in a Post-Tensioning Tendon Duct Using Piezoceramic Transducers. *Sensors* **2017**, *17*, 2239. [CrossRef]
21. Jiang, T.; Kong, Q.; Wang, W.; Huo, L.; Song, G. Monitoring of Grouting Compactness in a Post-Tensioning Tendon Duct Using Piezoceramic Transducers. *Sensors* **2016**, *16*, 1343. [CrossRef]
22. Wang, B.; Huo, L.; Chen, D.; Li, W.; Song, G. Impedance-Based Pre-Stress Monitoring of Rock Bolts Using a Piezoceramic-Based Smart Washer—A Feasibility Study. *Sensors* **2017**, *17*, 250. [CrossRef]
23. Jiang, T.; Zhang, Y.; Wang, L.; Zhang, L.; Song, G. Monitoring Fatigue Damage of Modular Bridge Expansion Joints Using Piezoceramic Transducers. *Sensors* **2018**, *18*, 3973. [CrossRef] [PubMed]
24. Gao, W.; Huo, L.; Li, H.; Song, G. An Embedded Tubular PZT Transducer Based Damage Imaging Method for Two-Dimensional Concrete Structures. *IEEE Access* **2018**, *6*, 30100–30109. [CrossRef]
25. Lu, G.; Li, Y.; Wang, T.; Xiao, H.; Huo, L.; Song, G. A multi-delay-and-sum imaging algorithm for damage detection using piezoceramic transducers. *J. Intell. Mater. Syst. Struct.* **2016**, *10*, 2545. [CrossRef]
26. Yang, W.; Kong, Q.; Ho, S.C.M.; Mo, Y.L.; Song, G. Real-Time Monitoring of Soil Compaction Using Piezoceramic-Based Embeddable Transducers and Wavelet Packet Analysis. *IEEE Access* **2018**, *6*, 5208–5214. [CrossRef]
27. Papadakis, G.A. Assessment of requirements on safety management systems in EU regulations for the control of major hazard pipelines. *J. Hazard. Mater.* **2000**, *78*, 63–89. [CrossRef]
28. Hu, J.; Zhang, L.; Liang, W. Detection of small leakage from long transportation pipeline with complex noise. *J. Loss Prev. Process Ind.* **2011**, *24*, 449–457. [CrossRef]
29. Verde, C.; Molina, L.; Torres, L. Parameterized transient model of a pipeline for multiple leaks location. *J. Loss Prev. Process Ind.* **2000**, *29*, 177–185. [CrossRef]
30. Duan, H.-F.; Lee, P.J.; Ghidaoui, M.S.; Tung, Y.-K. Essential system response information for transient-based leak detection methods. *J. Hydraul. Res.* **2010**, *48*, 650–657. [CrossRef]
31. Ni, L.; Jiang, J.; Pan, Y. Leak location of pipelines based on transient model and PSO-SVM. *J. Loss Prev. Process Ind.* **2013**, *26*, 1085–1093. [CrossRef]
32. Shi, Y.; Zhang, C.; Li, R.; Cai, M.; Jia, G. Theory and Application of Magnetic Flux Leakage Pipeline Detection. *Sensors* **2015**, *15*, 31036–31055. [CrossRef]
33. Wu, J.; Fang, H.; Huang, X.; Xia, H.; Kang, Y.; Tang, C. An Online MFL Sensing Method for Steel Pipe Based on the Magnetic Guiding Effect. *Sensors* **2017**, *17*, 2911. [CrossRef] [PubMed]
34. Yan, Y.; Shen, Y.; Cui, X.; Hu, Y. Localization of multiple leak sources using acoustic emission sensors based on MUSIC algorithm and wavelet packet analysis. *IEEE Sens. J.* **2018**, *18*, 9812–9820. [CrossRef]
35. Oh, W.; Yoon, D.-B.; Kim, G.J.; Bae, J.-H.; Kim, H.S. Acoustic data condensation to enhance pipeline leak detection. *Nucl. Eng. Des.* **2018**, *327*, 198–211. [CrossRef]
36. Liu, C.; Li, Y.; Fang, L.; Xu, M. New leak-localization approaches for gas pipelines using acoustic waves. *Measurement* **2019**, *134*, 54–65. [CrossRef]
37. Bian, X.; Li, Y.; Feng, H.; Wang, J.; Qi, L.; Jin, S. A Location Method Using Sensor Arrays for Continuous Gas Leakage in Integrally Stiffened Plates Based on the Acoustic Characteristics of the Stiffener. *Sensors* **2015**, *15*, 24644–24661. [CrossRef] [PubMed]
38. Su, T.-C.; Yang, M.-D. Application of Morphological Segmentation to Leaking Defect Detection in Sewer Pipelines. *Sensors* **2014**, *14*, 8686–8704. [CrossRef] [PubMed]
39. Ni, L.; Jiang, J.C.; Pan, Y.; Wang, Z. Leak location of pipelines based on characteristic entropy. *J. Loss Prev. Process Ind.* **2014**, *30*, 24–36. [CrossRef]
40. Zhang, T.T.; Tan, Y.F.; Zhang, X.D.; Zhao, J. A novel hybrid technique for leak detection and location in straight pipelines. *J. Loss Prev. Process Ind.* **2015**, *35*, 157–168. [CrossRef]
41. Liu, C.W.; Li, Y.X.; Fang, L.P.; Han, J.K.; Xu, M.H. Leakage monitoring research and design for natural gas pipelines based on dynamic pressure waves. *J. Process Control.* **2017**, *50*, 66–76. [CrossRef]
42. Jia, Z.G.; Ren, L.; Li, H.N.; Ho, S.C.; Song, G. Experimental study of pipeline leak detection based on hoop strain measurement. *Struct. Control. Health Monit.* **2015**, *22*, 799–812. [CrossRef]
43. Zhu, J.; Ren, L.; Ho, S.C.; Jia, Z.G.; Song, G. Gas pipeline leakage detection based on PZT sensors. *Smart Mater. Struct.* **2017**, *26*, 025022. [CrossRef]
44. Zhou, M.; Pan, Z.; Liu, Y.; Zhang, Q.; Cai, Y.; Pan, H. Leak detection and location based on ISLMD and CNN in a pipeline. *IEEE Access* **2019**, *7*, 30457–30464. [CrossRef]
45. Li, J.; Zheng, Q.; Qian, Z.; Yang, X. A novel location algorithm for pipeline leakage based on the attenuation of negative pressure wave. *Process Saf. Environ. Prot.* **2019**, *123*, 309–316. [CrossRef]
46. Ing, R.K.; Quieffin, N.; Catheline, S.; Fink, M. In solid localization of finger impacts using acoustic time-reversal process. *Appl. Phys. Lett.* **2005**, *87*, 204104. [CrossRef]

47. Fink, M. Time-reversal of ultrasonic fields-part I: Basic principles. *IEEE Trans. Ultrason. Ferroelectr. Freq. Control* **1992**, *39*, 555–566. [CrossRef]
48. Liu, D.; Kang, G.; Li, L.; Chen, Y. Electromagnetic time-reversal imaging of a target in a cluttered environment. *IEEE Trans. Antennas Propag.* **2005**, *53*, 3058–3066.
49. Zhao, A.; Zeng, C.; Hui, J.; Ma, L.; Bi, X. An Underwater Time Reversal Communication Method Using Symbol-Based Doppler Compensation with a Single Sound Pressure Sensor. *Sensors* **2018**, *18*, 3279. [CrossRef]
50. He, C.; Jing, L.; Xi, R.; Li, Q.; Zhang, Q. Improving Passive Time Reversal Underwater Acoustic Communications Using Subarray Processing. *Sensors* **2017**, *17*, 937. [CrossRef]
51. Cai, J.; Shi, L.; Yuan, S. High spatial resolution localization for structural health monitoring based on virtual reversal. *Smart Mater. Struct.* **2011**, *20*, 901–904. [CrossRef]
52. Wang, C.H.; Rose, J.T.; Chang, F.K. A synthetic time-reversal imaging method for structural health monitoring. *Smart Mater. Struct.* **2004**, *13*, 415–423. [CrossRef]
53. Huo, L.; Wang, B.; Chen, D.; Song, G. Monitoring of Pre-Load on Rock Bolt Using Piezoceramic-Transducer Enabled Time Reversal Method. *Sensors* **2017**, *17*, 2467. [CrossRef] [PubMed]
54. Liang, Y.; Feng, Q.; Li, D. Loosening Monitoring of the Threaded Pipe Connection Using Time Reversal Technique and Piezoceramic Transducers. *Sensors* **2018**, *18*, 2280. [CrossRef]
55. Amitt, E.; Dan, G.; Turkel, E. Time reversal for crack identification. *Comput. Mech.* **2014**, *54*, 443–459. [CrossRef]
56. Kocur, G.K.; Vogel, T.; Saenger, E.H. Crack localization in a double-punched concrete cuboid with time reverse modeling of acoustic emissions. *Int. J. Fract.* **2011**, *171*, 110. [CrossRef]
57. Saenger, E.H.; Kocur, G.K.; Jud, R.; Torrilhon, M. Application of time reverse modeling on ultrasonic non-destructive testing of concrete. *Appl. Math. Model.* **2011**, *35*, 807–816. [CrossRef]
58. Zhao, G.; Zhang, D.; Zhang, L.; Wang, B. Detection of Defects in Reinforced Concrete Structures Using Ultrasonic Nondestructive Evaluation with Piezoceramic Transducers and the Time Reversal Method. *Sensors* **2018**, *18*, 4176. [CrossRef] [PubMed]
59. Benkherouf, A.; Allidina, A.Y. Leak detection and location in gas pipelines. *IEE Proc.-Control. Theory Appl.* **1988**, *135*, 142–148. [CrossRef]
60. Zhao, Y.; Xiong, Z.; Shao, M. A new method of leak location for the natural gas pipeline based on wavelet analysis. *Energy* **2010**, *35*, 3814–3820.

sensors

MDPI

Article

Strength Development Monitoring of Cemented Paste Backfill Using Guided Waves

Wen He [1,2,3,*], Changsong Zheng [1,2], Shenhai Li [1,2], Wenfang Shi [1,2] and Kui Zhao [1,2]

1 School of Resources and Environmental Engineering, Jiangxi University of Science and Technology, Ganzhou 341000, China; zhengcs666@163.com (C.Z.); jxust-lsh1995@163.com (S.L.); wenfangshi@126.com (W.S.); yglmf_zk@163.com (K.Z.)
2 Jiangxi Provincial Key Laboratory of Mining Engineering, Jiangxi University of Science and Technology, Ganzhou 341000, China
3 Engineering Research Center for High-Efficiency Development and Application Technology of Tungsten Resources, Jiangxi University of Science and Technology, Ministry of Education, Ganzhou 341000, China
* Correspondence: hewen@jxust.edu.cn

Abstract: The strength of cemented paste backfill (CPB) directly affects mining safety and progress. At present, in-situ backfill strength is obtained by conducting uniaxial compression tests on backfill core samples. At the same time, it is time-consuming, and the integrity of samples cannot be guaranteed. Therefore guided wave technique as a nondestructive inspection method is proposed for the strength development monitoring of cemented paste backfill. In this paper, the acoustic parameters of guided wave propagation in the different cement-tailings ratios (1:4, 1:8) and different curing times (within 42 d) of CPBs were measured. Combined with the uniaxial compression strength of CPB, relationships between CPB strength and the guided wave acoustic parameters were established. Results indicate that with the increase of backfill curing time, the guided wave velocity decreases sharply at first; on the contrary, attenuation of guided waves increases dramatically. Finally, both velocity and attenuation tend to be stable. When the CPB strength increases with curing time, guided wave velocity shows an exponentially decreasing trend, while the guided wave attenuation shows an exponentially increasing trend with the increase of the CPB strength. Based on the relationship curves between CPB strength and guided wave velocity and attenuation, the guided wave technique in monitoring the strength development of CPB proves feasible.

Keywords: cemented paste backfill; guided wave; guided wave velocity; attenuation of guided wave

Citation: He, W.; Zheng, C.; Li, S.; Shi, W.; Zhao, K. Strength Development Monitoring of Cemented Paste Backfill Using Guided Waves. *Sensors* **2021**, *21*, 8499. https://doi.org/10.3390/s21248499

Academic Editor: Zenghua Liu

Received: 30 November 2021
Accepted: 16 December 2021
Published: 20 December 2021

1. Introduction

The utilization of mineral resources has extensively promoted the development and progress of human society, but it has also brought many environmental and safety problems. For example, many environmental or geotechnical problems in the waste rock and tailings are produced in ore mining and processing and the void left by mining [1]. One way to solve this problem is to backfill the void with waste rocks and tailings, supporting the ground and decreasing the management on the mine-site surface [2–4]. The backfilling method can directly transport waste rocks and tailings to the void or transport the waste rock and tailings to the void after adding water or cementitious materials [5]. It is called cemented paste backfill (CPB) mining method to backfill the void after mixing the tailings with cementitious material. Cemented paste backfill mainly consists of 70–85% (by solid weight) tailings discharged by the mineral processing plant, 3–7 wt% binder, and a certain amount of water [6]. Compared to other mining methods, it performs better in preventing surface subsidence and controlling ground pressure activities. It has a higher ore recovery rate, which reduces the accumulation of mine wastes on the surface, releases a large amount of ground space, and reduces the mining industry's impact on the environment [7]. Since the first underground CPB was implemented at the Bad Grund Mine in Germany in

1979, which employed the fine mill rejects aggregate and silty filtered tailings as its main components [8], it is still widely used worldwide. The strength of backfilling body is an important index to evaluate the quality of CPB. Moreover, it is a crucial factor affecting the safety production of adjacent stope [9]. If the backfill strength is insufficient, safety accidents will quickly occur, and the environment will be polluted. At present, in-situ backfill strength is obtained by conducting uniaxial compression tests on backfill core samples, while it is time-consuming and the integrity of samples cannot be guaranteed [10].

The ultrasonic wave propagates as a bulk wave in an infinite medium, while the guided wave is confined to propagate in a medium due to the medium boundary, and the propagation direction is parallel to the medium boundary. The most significant difference between bulk and guided wave propagation is that guided wave propagation needs the boundary's 'guidance'. A large number of reflections, refractions, and modal transformations occur when acoustic waves propagate to the boundary of the waveguide structure, resulting in mutual interference between acoustic waves and continuous superposition, forming stable guided waves in a waveguide structure. The formation process of guided waves determines that the guided wave propagation has high sensitivity to defect characteristics and mechanical boundary changes. Based on this feature, guided wave detection technology had been formed [11]. Guided wave technology is a popular and relatively novel technology in the field of nondestructive testing. Compared with the traditional ultrasonic wave test, it has the advantages of long propagation distance, high detection efficiency, and wide detection range. However, the application of guided waves has only gradually matured in the past 20 years. Cawley et al. [12–17] conducted in-depth research on the propagation of guided waves in plates and pipelines and first applied it to the field of nondestructive testing. Simonetti et al. [18] studied the corrosion of pipelines by guided ultrasonic wave tomography. He C, Liu Z, et al. [19–22] proposed a series of methods for the nondestructive testing of steel strands and wind blades by ultrasonic guided waves [23–25]; meanwhile, they have developed a pipeline guided wave testing system, which is based on a time-reversal defect-identifying method and time-space focus theory [26–29]. D. Zou and Wang C et al. [30–35] conducted an in-depth study on the guided wave characteristics of the rock bolts. Li J and Zima B [36,37] explored the application of guided wave nondestructive testing technology to the integrity testing of reinforced concrete materials through theories and experiments. Literature review shows that guided waves have rarely been reported in the strength detection of mine backfill.

As the curing time goes on, the types and speciation of hydration reaction products inside the cemented paste backfill changes, which will cause the strength of CPB to differ [38] and affect the propagation of guided waves in cemented paste backfill mass. Then a variety of acoustic parameters such as wave velocity and attenuation during age means the change of backfilling mechanical properties. Therefore, the corresponding acoustic parameters of the guided wave to CPB strength, the guided wave monitoring method of strength development of cemented paste backfill mass can be realized.

In this paper, we try to find out the relationship between acoustic characteristics of guided waves propagating in CPB and the mechanical properties of CPB and propose a new method for detecting the strength of CPB using guided waves. The frequency and period of the excitation longitudinal guided waves were first optimized, and then the propagation velocity and attenuation changes of guided waves in CPBs at different cement-tailings ratios and different curing ages were tested. Combined with the uniaxial compression strength of CPB, a relation between CPB strength and the guided wave acoustic parameters was established.

2. Materials and Methods

2.1. Materials

The unclassified tailings come from the tailings pond of a copper mine. The chemical composition of the tailings is shown in Figure 1. The main chemical compositions are SiO_2, CaO, Fe, and S, which account for 71.43% of the tailings' weight. The particle size distribution of tailings is listed in Table 1. The tailings have a specific gravity of 2.75 and an average moisture content of 1.17%. It can be seen from Table 1 that the tailings with a particle size larger than 0.074 mm account for 42.16%. Therefore, according to the commonly used classification standard [39], the whole tailings in the test are classified as coarse tailings.

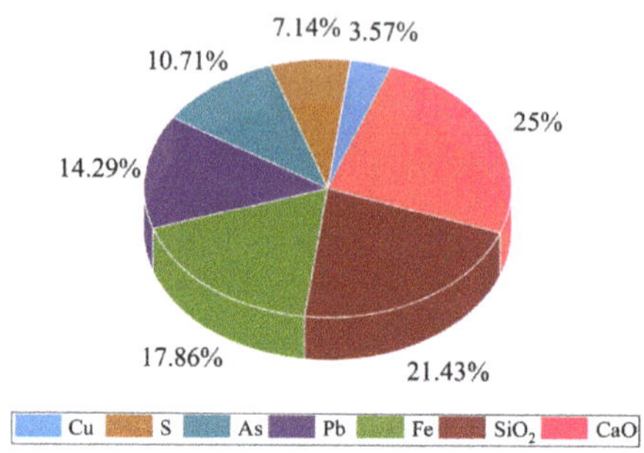

Figure 1. Main chemical properties of the tailings.

Table 1. Particle size distribution of the tailings samples.

Particle Size/um	0~33	33~45	45~74	>74	
Content/%	38.82	4.31	14.71	42.16	100

No. 32.5 ordinary Portland cement was used as a binder agent, commonly used in the copper mine. Ordinary tap water was used to mix binder and tailings in this study.

2.2. Specimens for Test

2.2.1. Background

Cemented paste backfill mining method is adopted in a copper mine, and the void is backfilled in two phases. The concentration of the first stage backfilling slurry is 70%, and the cement-tailings ratio is 1:4. When backfilling to 1/2 of an inlet section, the first stage backfilling is terminated, the second stage backfilling slurry concentration is 70%, but the cement-tailings ratio is 1:8. To be combined with the field, the concentration of test specimens prepared in this test is 70%, and cement-tailings ratios are 1:4 and 1:8, respectively, representing the first-stage and second-stage backfill.

2.2.2. Specimens for Guided Wave Test

Generally, the waveguide structure, including a waveguide rod and several sensors, is used to excite and receive signals. Therefore, a waveguide structure was buried in the CPB specimen, with part of the waveguide structure exposed to the air.

The size of guided wave test pieces is 400 mm × 400 mm × 300 mm, the concentration of test specimens prepared in this test is 70%, and cement-tailings ratios are 1:4 and 1:8. With a diameter of 20 mm and length of 460 mm, a waveguide was buried in the longitudinal center of the piece, letting 30 mm at each end of the waveguide rod fix the sensor. The dimension diagram of specimens is shown in Figure 2. For the convenience of explanation, the 1:4 CPB sample was numbered 'A', and the 1:8 CPB sample was numbered 'B'.

Before preparing test pieces, we placed the waveguide rod in the prefabricated hole. The weighed materials were mixed and homogenized for about 10 min. Then, the stirred slurry was poured into the test mold to be vibrated evenly. Demoulding after curing for three days, and placed the test piece into an environmental chamber with the controlled temperature at 20 ± 2 °C and 90% ± 5% relative humidity.

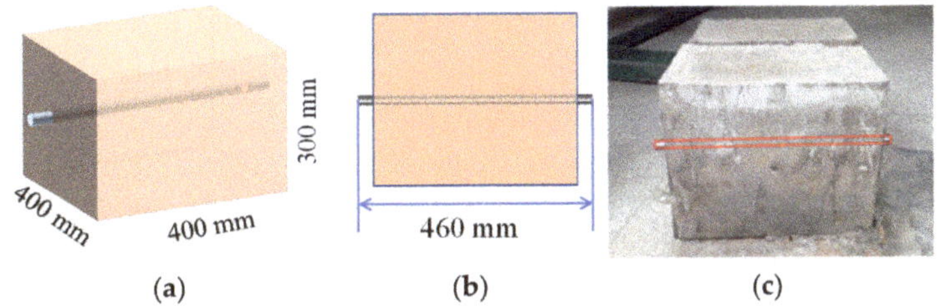

Figure 2. The dimension of specimens for guided wave testing. (**a**) Side view; (**b**) Top view; (**c**) Photograph.

2.2.3. Specimens for Uniaxial Compression Strength (UCS) Test

In this section, specimens for UCS tests are prepared with a concentration of 70%, and the cement-tailings ratio is 1:4 and 1:8, respectively, and the curing time is 3, 7, 14, and 28 days. The tailings, binder, and water with different proportions were prepared by an electronic scale with an accuracy of 0.01 g and thoroughly homogenized for about 5 min to produce the desired CPB mixtures. Then the CPB slurry was molded to a 50 mm × 100 mm cylinder. Next, put the prepared test specimens into an environmental chamber with the controlled temperature at 20 ± 2 °C and 90% ± 5% relative humidity.

2.3. Guided Wave Test

During the consolidation process, the phase state and microstructure of CPB constantly change, making the acoustic characteristics of guided wave propagation change. In this experiment, the parameters of the guided wave are optimized. Firstly, the same frequency with different cycles (within eight cycles) of excitation waves was adopted to determine the optimal cycle, and then the same cycle with different frequencies (20–70 kHz) of excitation waves was used to find the best frequency. These tests were carried out on CPB with the ratio of 1:4 and 1:8 and the concentration of 70%. The monitoring time was from 1 d to 42 d. The solidification process of the specimen is changing from a fluid to an elastoplastic body, and the change of this process will definitely lead to the change of the mechanical parameters of the specimen. The wave velocity and attenuation characteristics are obtained through the guided wave test to reflect the solidification process of the filling. In this way, the relationship between the strength of the filled body and the wave velocity and attenuation is established.

2.3.1. Test Set-Up and Procedure

Guided wave testing system includes waveform generator, oscilloscope, signal power amplifier, sensors, et al. Connect as shown in Figure 3. Firstly, sensors were connected with the waveguide rod. Then, a thin layer of coupling agent was coated on the contact surfaces of the sensor and waveguide rod to ensure complete contact. Only one channel was used for the waveform generator after turning on the signal waveform generator, preheated for 10 min. The waveform generator generated excitation waves by reading the waveform file. After amplifying by the power amplifier, excitation wave entered the waveguide rod through the sensor. Finally, the received signal was denoised by the signal acquisition system and then stored in a U disk.

The excitation wave in this test comes from the following signal function Equation (1):

$$A \sin(2\pi f t) \times (1 - \cos(2\pi f t/n)) \tag{1}$$

where A is amplitude, f is the frequency, t is the duration, and n is the number of input cycles. The larger n is, the higher the energy of the excitation wave and the easier the resulting reflected wave is to identify. However, increasing n also increases the width of the wave packet, resulting in the overlap of the wave packet of the excitation wave with the wave packet of the reflection wave. Therefore, n cannot be too large or too small.

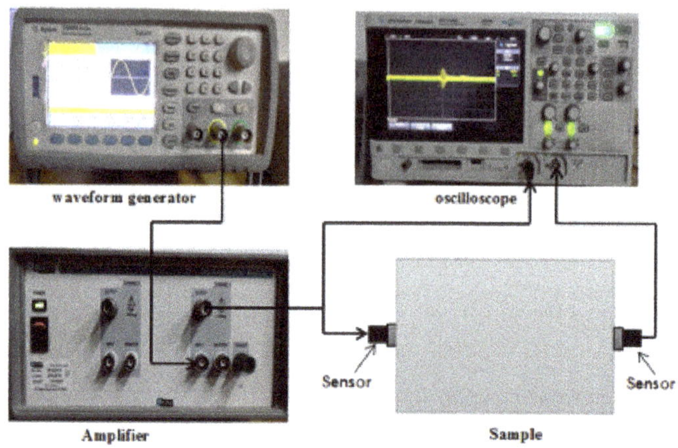

Figure 3. Diagram of guided wave testing system.

2.3.2. Guided Wave Velocity and Attenuation Data Processing

The collected data need to be processed to obtain the desired guided wave acoustic parameters, wherein the wave velocity of the guided wave is calculated by the following Equation (2):

$$V = \frac{L}{\Delta t} \tag{2}$$

where V represents wave velocity, L represents the waveguide length, Δt represents the time difference between the original signal and the received signal.

To reduce the error, the time difference of data processing takes a value between the starting point of the original signal and the first received wave, as shown in Figure 4. The first vibration change of the signal waveform is the starting point, and the starting point of the excitation and reception waves is specifically the point where the amplitude of the excitation and reception waves cross the threshold for the first time.

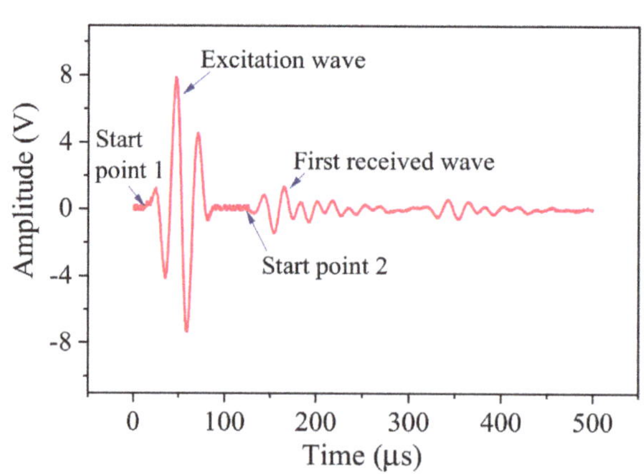

Figure 4. Pickup at the starting point and the time difference between the excitation and reception waves.

In addition to the velocity of guided waves, the attenuation of guided waves can also reflect the change of cemented paste backfill strength. Therefore, the attenuation value A_t is calculated by the following Equation (3):

$$A_t = -\frac{20}{L}\lg(\frac{P}{P_{ref}}) \tag{3}$$

where P_{ref} represents the peak-to-peak value of the excitation wave, P represents the peak-to-peak value of the wave captured from the other end of the waveguide, L is the length of the waveguide.

2.4. UCS Test

Uniaxial compressive strength is one of the important indexes to evaluate the quality of backfill. The High-pressure triaxial testing system produced by British GDS Instruments was used in this test, as shown in Figure 5. In the test progressing, specimens were loaded under a constant force ratio of 0.5 kN/s. To avoid random errors in the tests, referring to the test standard ASTMC 39, four specimens were taken for each group of tests.

Figure 5. The High-pressure triaxial testing system.

3. Results and Discussion

3.1. Optimization of Excitation Wave

To obtain the effective acoustic parameters of guided waves, it is necessary to select the optimal parameters of the excitation wave. The main parameters of the excitation wave are cycle and frequency. In this section, the appropriate cycle and frequency of the excitation wave were selected by analyzing the following two parts.

3.1.1. Optimization of the Excitation Wave Cycle

The signal waveforms of the same frequency (within 100 kHz) in different cycles were used to test. Due to space constraints, following Figure 6 shows four representative signal waveforms under different cycles (n = 3, 5, 6, 8) at 40 kHz.

As shown in Figure 6, when the excitation waves with the same frequency but different periods were generated, the reflection formed by the guided wave propagating in the rod was different. The more period the excitation wave, the more incredible energy it will generate and the more pronounced the received wave can be formed, which is convenient to analyze. As shown in Figure 6a,b, it can be seen that the first end-reflected wave magnitude of 5 cycles has less attenuation than three cycles. The amplitude of 3 cycles was reduced from 15.09 v to 2.73 v, a decrease of 81.90%; the amplitude of 5 cycles was reduced from 10.00 v to 2.33 v, a decrease of 76.74%, indicating that the reflected wave with five cycles has more energy than that with three cycles, and the peaks of the excitation wave and the reflection wave with five cycles are more straightforward to distinguish than the waves with three cycles. Hence, it is more convenient to calculate the time difference between the peaks. However, increasing the number of cycles also increases the width of the wave packet, as shown in Figure 6c,d, which will cause the wave packet of the excitation wave to overlap with the reflected wave of the waveguide and cannot be identified. The amplitude of guided wave with six cycles reduced from 30.12 v to 1.32 v, with a decrease of 95.61%, and the amplitude of guided wave with eight cycles decreased from 15.11 v to 0.97 v, with a decrease of 93.60%. In conclusion, the excitation wave with five periods is the most suitable for this test.

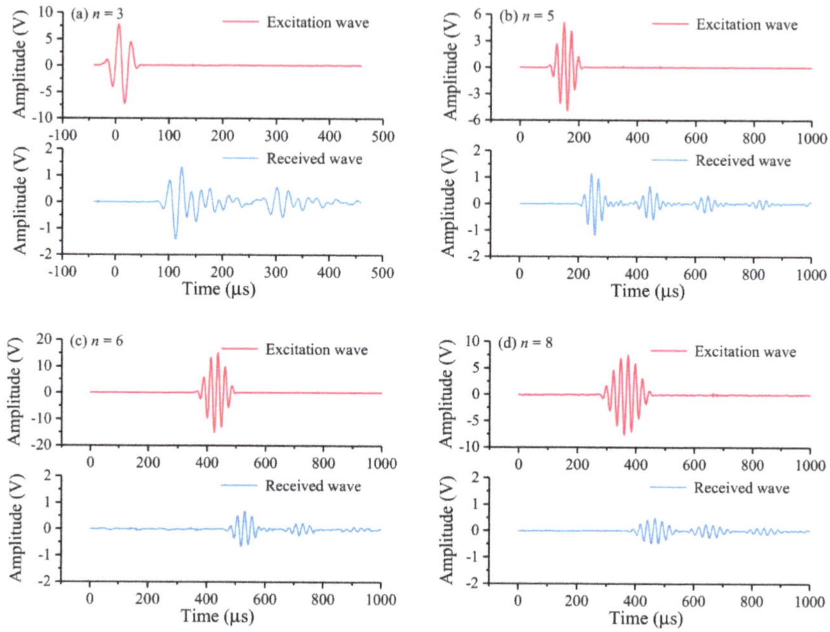

Figure 6. Effect of excitation wave parameter n on the receiving wave. The characteristics of the received wave are observed by changing the excitation wave parameter n. The amplitude decay, wave packet shape and overlap of the received wave are compared to optimize the parameter n.

3.1.2. Optimization of the Excitation Wave Frequency

The same cycle with different frequencies (20–70 kHz) of the excitation waves was used to detect guided waves' velocity and attenuation in CPBs. The results were illustrated in the following figures. Figure 7 shows the change of velocity with curing age at different guided wave frequencies. The result shows that the sensitivity of different frequency excitation wave to the detected structure is different. The overall trend of wave velocity changes at various frequencies decreases with the curing time, especially the initial wave velocity changes most apparent, and the wave velocity changes at the middle and late curing periods are small and tend to be stable. As shown in Figure 7, sample A has noticeable wave velocity variation at the excitation wave frequencies of 30 kHz, 50 kHz, and 60 kHz. The differences between the first day and the second day are all about 1000 m/s, among them, the wave velocity changes the most when the excitation frequency is 30 kHz, and the difference reaches 1364.33 m/s; For sample B, the wave velocity changes significantly when the excitation wave frequencies were 20 kHz, 30 kHz, and 50 kHz. At the frequency of 30 kHz, the maximum difference between the first day and the second day reaches 1300 m/s.

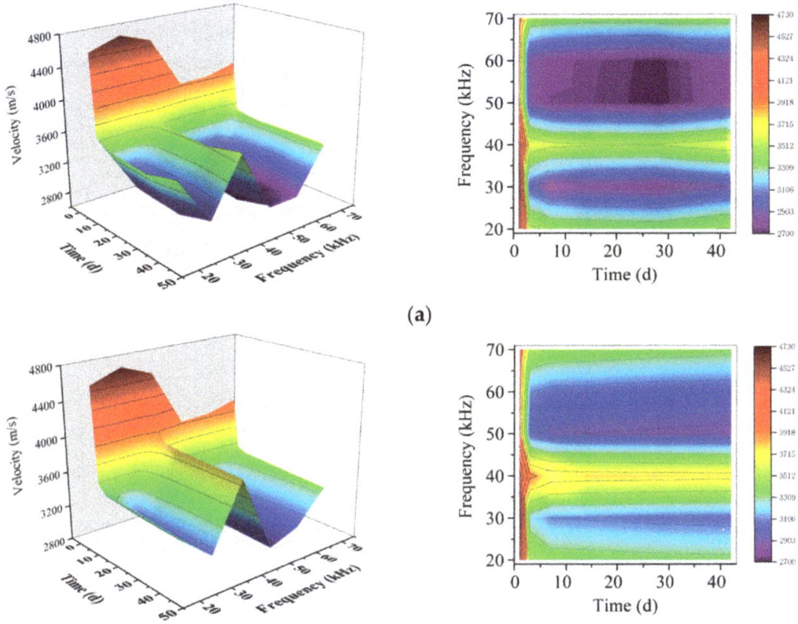

Figure 7. Velocities at different guided wave frequencies; (**a**) sample A; (**b**) sample B.

It can also be demonstrated from Figure 7 that the influence of CPB with a different cement-tailings ratio on wave velocity is different. From the stable value of wave velocity in the later curing period, velocity in sample A is lower than that in sample B.

Figure 8 shows the change of attenuation with curing age at different guided wave frequencies. It can be seen from the figure that according to the change of attenuation value with age, as the curing age increases, the attenuation change trend of guided waves in samples A and B is the same, which increases first and then decreases. From the initial curing to the age of 21 d, the attenuation shows an increasing trend, reaching the maximum value at 21 d and decreasing. When the excitation frequencies are 20 kHz, 30 kHz, and 70 kHz, the attenuation is more prominent, the maximum attenuation value of all three exceed 120 dB/m, and the guided wave attenuation is the smallest at 50 kHz. Among them, the attenuation value at 20 kHz and 70 kHz are relatively close, but the attenuation value of guided waves reaches the maximum at 30 kHz. The attenuation of guided waves in CPB with different cement-tailings ratios is also different. In the frequency range 20 kHz to 70 kHz, the attenuation value of specimens with a cement-tailings ratio of 1:8 is more significant than that with a cement-tailings ratio of 1:4. From the above results, it can be seen that the excitation wave with a frequency of 30 kHz is more sensitive to the detection of CPB, so the best frequency of the guided wave is 30 kHz. Therefore, combined with the conclusion of Section 3.1.1, a cycle of 5 and a frequency of 30 kHz are the most suitable parameters for monitoring CPB.

3.2. Relationship between UCS and Velocity of Guided Wave

When specimens were tested using guided waves, specimens with curing times of 3 d, 7 d, 14 d, and 28 d were conducted uniaxial compressive testing simultaneously. The test results are shown in Table 2.

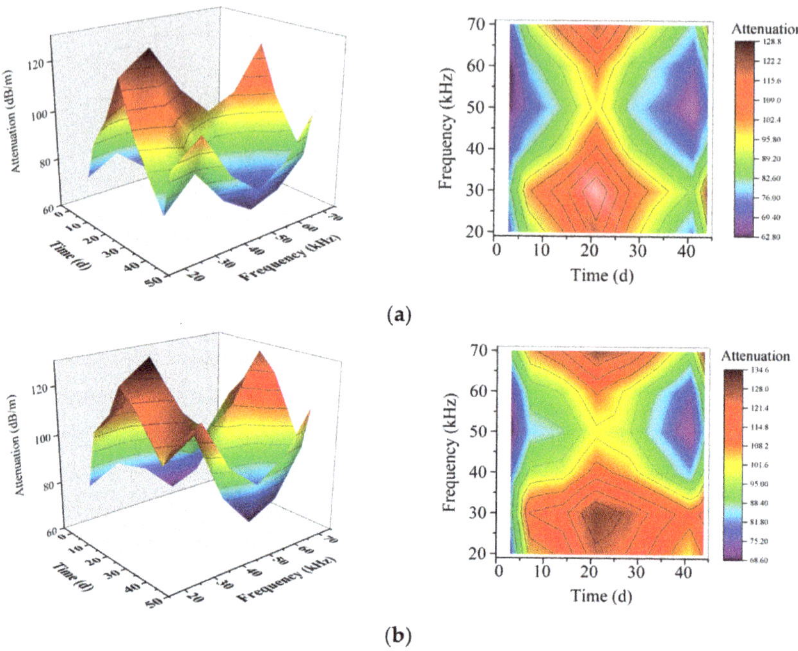

Figure 8. Attenuation at different guided wave frequencies; (**a**) sample A; (**b**) sample B.

Table 2. Uniaxial compressive strength of CPBs.

	UCS/MPa			
	3 d	**7 d**	**14 d**	**28 d**
Cement-tailings Ratio (1:4)	1.25	1.73	2.12	3.42
	1.23	2.01	2.15	3.33
	1.08	1.82	2.38	3.15
	1.32	1.64	2.34	3.65
Cement-tailings Ratio (1:8)	0.69	1.14	1.67	2.12
	0.73	1.12	1.47	2.08
	0.71	1.15	1.23	1.99
	0.75	1.08	1.51	2.29

As demonstrated in Table 2, the UCS of CPB regularly increases with curing time. From the results obtained in Section 3.1.2, the guided wave propagation in CPB also changes regularly with curing time. Based on this, the uniaxial compressive strength can be correlated with velocity or attenuation of guided waves in CPB under the corresponding curing time, and relationships between UCS and velocity or attenuation can be found. In engineering practice, the waveguide rod structure can be applied to monitor the strength of CPB, thus realizing the on-site monitoring of the strength of cemented paste backfill. The curve between the UCS and guided wave velocity of CPB was fitted in this section.

Compared with various fitting schemes, an exponential function was finally selected for fitting, and the results are shown in Figure 9. The fitting of UCS and wave velocity of CPB with cement-tailings ratio 1:4 is illustrated in Figure 9a, and the fitting formula is shown in the following Equation (4), with a correlation coefficient (R2) of 0.9868. The fitting of UCS and wave velocity of CPB with cement-tailings ratio 1:8 is shown in Figure 9b, and the fitting formula is shown in the following Equation (5), with a correlation coefficient (R2) of 0.9874.

$$y = e^{(313.9545-0.21249x+3.5965\times10^{-5}x^2)} \tag{4}$$

$$y = e^{(626.11828 - 0.39199x + 6.1315 \times 10^{-5}x^2)} \tag{5}$$

where y represents the UCS, and x represents the wave velocity of guided waves in CPB.

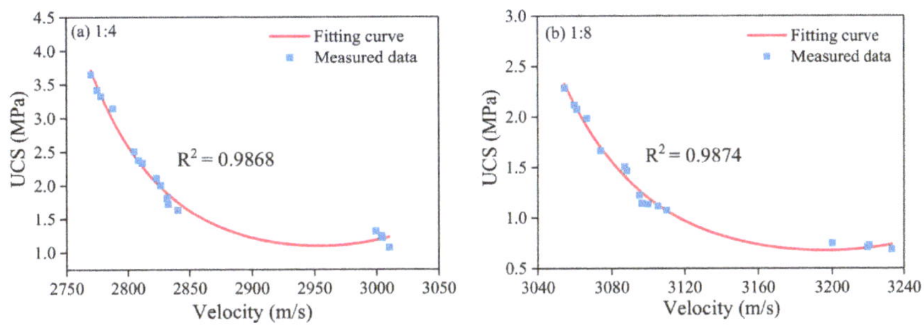

Figure 9. Relationship between UCS and velocity data for different cement-tailings ratio. Quantification of UCS by wave velocity.

The correlation coefficients of fitting are all greater than 0.98, so the fitting is reliable. The strength changes of the two kinds of cement-tailings ratio CPB are the same with the change of wave velocity, and both show an exponential downward trend with the increase of wave velocity. The higher the wave velocity of guided waves propagating in cemented paste backfill mass, the lower the strength of CPB and the smaller change value of the strength.

3.3. Relationship between UCS and Guided Wave Attenuation

This part fits the curve between UCS and the attenuation of guided waves in cemented paste backfill mass. Compared with various fitting schemes, the exponential function is still selected for fitting, and the results are shown in Figure 10.

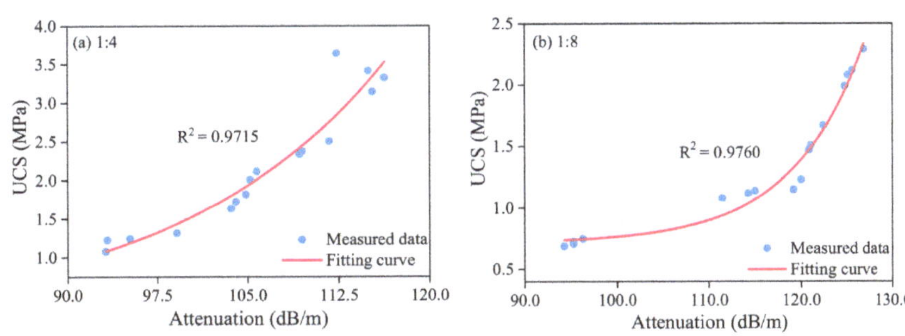

Figure 10. Relationship between UCS and attenuation data for different cement-tailings ratio. Quantification of UCS by wave attenuation.

The fitting of UCS and attenuation of guided wave in CPB with cement-tailings ratio 1:4 is shown in Figure 10a, and the fitting formula is shown in the following Equation (6), with a correlation coefficient (R2) of 0.9715. The fitting of UCS and attenuation of guided wave in CPB with cement-tailings ratio 1:8 is shown in Figure 10b, and the fitting formula is shown in the following Equation (7), with a correlation coefficient (R2) of 0.9760.

$$y = 0.22848 + 0.00372e^{0.05837x} \tag{6}$$

$$y = 0.71553 + 1.1902 \times 10^{-7}e^{0.1296x} \tag{7}$$

where y represents the UCS, and x represents the attenuation of guided wave in CPB.

As is shown from the fitting function, the correlation coefficients of the fitting are both more than 0.97, and it is considered that the fitting is ideal. The strength variations of the two kinds of cement-tailings ratio are the same with the change of attenuation. UCS has the trend of the exponential increase with the increase of the wave velocity. The higher the attenuation of guided wave in CPB, the higher the UCS, and the greater the intensity variation value.

4. Conclusions

This section is not mandatory but can be added to the manuscript if the discussion is unusually long or complex.

(1) Guided waves with a frequency of 40 kHz and cycle numbers of 3, 5, 6, and 8 were used to detect cemented paste backfill samples. The results show that the larger cycle of the excitation wave, the easier the excitation wave and received wave overlap; they are difficult to distinguish. Combined with the attenuation of guided waves, the optimal detection cycle of the guided wave was determined to be 5.
(2) Guided waves with frequencies of 20–70 kHz were used to test samples A and B, respectively. The general trend of wave velocity at each frequency decreases with curing time and will finally be stable. However, with the increase of the curing period, the attenuation of guided waves in CPB increases first and then decreases. The test results indicate that both the velocity and attenuation of guided wave change most obviously at the frequency of 30 kHz. Therefore, 30 kHz is the optimal guided wave frequency for cemented paste backfill monitoring.
(3) Through the uniaxial compression test and guided wave test, the Mechanical and acoustic parameters of CPB were studied. The relationship between strength and wave velocity, strength and attenuation, and the corresponding exponential function fitting equations were found.

This study implicated that the guided wave technique as a nondestructive inspection method can also be used to evaluate CPB's quality and properties besides defect detection. While some aspects still need to investigate the performance of CPB samples using the guided wave technique. In this paper, the guided wave testing uses the through-transmission method; whether the pulse-echo method still works remains determined. Furthermore, the influence of cement-tailings ratio on acoustic parameters has been studied, but the influence of slurry concentration and aggregate particle size on guided wave acoustic parameters still needs further study.

Author Contributions: Conceptualization, W.H. and C.Z.; methodology, S.L.; software, W.S.; validation, W.H., C.Z. and K.Z.; formal analysis, C.Z.; investigation, C.Z.; resources, W.H., S.L. and W.S.; data curation, C.Z.; writing—original draft preparation, C.Z.; writing—review and editing, W.H.; visualization, W.H.; supervision, K.Z.; project administration, W.H.; funding acquisition, W.H. All authors have read and agreed to the published version of the manuscript.

Funding: The research is funded by the National Natural Science Foundation of China (No.51604127 and No.51874268), China Postdoctoral Science Foundation (No.2019M650156), Jiangxi Postdoctoral Science Foundation (No.2018KY41), Key R&D Program of Jiangxi Province (No. 20202BBG73001) and the Science and Technology Innovation Talent Project of Ganzhou City.

Institutional Review Board Statement: Not applicable.

Informed Consent Statement: Not applicable.

Data Availability Statement: The data presented in this study are available on request from the corresponding author.

Conflicts of Interest: The authors declare no conflict of interest.

References

1. Nasir, O.; Fall, M. Modeling the heat development in hydrating CPB structures. *Comput. Geotech.* **2009**, *36*, 1207–1218. [CrossRef]
2. Benzaazoua, M.; Fall, M.; Belem, T. A contribution to understanding the hardening process of cemented pastefill. *Miner. Eng.* **2004**, *17*, 141–152. [CrossRef]
3. Fall, M.; Benzaazoua, M.; Saa, E.G. Mix proportioning of underground cemented tailings backfill. *Tunn. Undergr. Space Technol.* **2008**, *23*, 80–90. [CrossRef]
4. Yilmaz, E.; Kesimal, A.; Ercidi, B. Strength Development of Paste Backfill Simples at Long-Term Using Different Binders. In Proceedings of the 8th Symposium MineFill04, Beijing, China, 20–21 September 2004; pp. 281–285.
5. Sivakugan, N.; Rankine, R.M.; Rankine, K.J.; Rankine, K.S. Geotechnical considerations in mine backfilling in Australia. *J. Clean. Prod.* **2006**, *14*, 1168–1175. [CrossRef]
6. Fall, M.; Pokharel, M. Coupled effects of sulphate and temperature on the strength development of cemented tailings backfills: Portland cement-paste backfill. *Cem. Concr. Compos.* **2010**, *32*, 819–828. [CrossRef]
7. Pagé, P.; Li, L.; Yang, P.; Simon, R. Numerical investigation of the stability of a base-exposed sill mat made of cemented backfill. *Int. J. Rock Mech. Min. Sci.* **2019**, *114*, 195–207. [CrossRef]
8. Qi, C.C.; Fourie, A. Cemented paste backfill for mineral tailings management: Review and future perspectives. *Miner. Eng.* **2019**, *144*, 106025. [CrossRef]
9. Fall, M.; Adrien, D.; Célestin, J.C.; Pokharel, M.; Touré, M. Saturated hydraulic conductivity of cemented paste backfill. *Miner. Eng.* **2009**, *22*, 1307–1317. [CrossRef]
10. Jiang, H.; Yi, H.; Yilmaz, E.; Liu, S.; Qiu, J. Ultrasonic evaluation of strength properties of cemented paste backfill: Effects of mineral admixture and curing temperature. *Ultrasonics* **2020**, *100*, 105983. [CrossRef] [PubMed]
11. He, C.F.; Zheng, M.F.; Lv, Y. Development applications and challenges in ultrasonic guided waves testing technology. *Chin. J. Sci. Instrum.* **2016**, *37*, 1713–1735. [CrossRef]
12. Alleyne, D.N.; Lowe, M.J.S.; Cawley, P. The reflection of guided waves from circumferential notches in pipes. *J. Appl. Mech.* **1998**, *65*, 635–641. [CrossRef]
13. Fan, Z.; Lowe, M.J.S. Elastic waves guided by a welded joint in a plate. *Proc. R. Soc. A Math. Phys. Eng. Sci.* **2009**, *465*, 2053–2068. [CrossRef]
14. Pettit, J.R.; Walker, A.; Cawley, P.; Lowe, M.J.S. A stiffness reduction method for efficient absorption of waves at boundaries for use in commercial finite element codes. *Ultrasonics* **2014**, *54*, 1868–1879. [CrossRef] [PubMed]
15. Leinov, E.; Lowe, M.J.S.; Cawley, P. Investigation of guided wave propagation and attenuation in pipe buried in sand. *J. Sound Vib.* **2015**, *347*, 96–114. [CrossRef]
16. Leinov, E.; Lowe, M.J.S.; Cawley, P. Ultrasonic isolation of buried pipes. *J. Sound Vib.* **2016**, *363*, 225–239. [CrossRef]
17. Huthwaite, P.; Ribichini, R.; Cawley, P.; Lowe, M.J. Mode selection for corrosion detection in pipes and vessels via guided wave tomography. *IEEE Trans. Ultrason. Ferroelectr. Freq. Control.* **2013**, *60*, 1165–1177. [CrossRef] [PubMed]
18. Simonetti, F.; Alqaradawi, M.Y. Guided ultrasonic wave tomography of a pipe bend exposed to environmental conditions: A long-term monitoring experiment. *NDT E Int.* **2019**, *105*, 1–10. [CrossRef]
19. Liu, Z.H.; Liu, S.; Wu, B.; Zhang, Y.N.; He, C.F. Experimental research on acoustoelastic effect of ultrasonic guided waves in prestressing steel strand. *Chin. J. Mech. Eng.* **2010**, *46*, 22–28. (In Chinese) [CrossRef]
20. Liu, Z.; Liu, S.; Wu, B.; Zhang, Y.; He, C. Propagation characteristics of high order longitudinal modes in steel strands and their applications. *Acta Mech. Solida Sin.* **2008**, *21*, 573–579. [CrossRef]
21. Wang, W.B.; Feng, Z.J.; Ai, M.Y.; Wang, K. Image reconstruction of phased ultrasonic guided wave testing for storage tank floor defect. *Nondestruct. Test.* **2010**, *32*, 847–849. (In Chinese)
22. Liu, Z.H.; Zhang, Y.N.; He, C.F.; Cao, C.; Wu, B. Development of magnetostrictive transmitter and receiver for exciting ultrasonic longitudinal guided waves in steel strands. *Chin. J. Appl. Mech.* **2009**, *26*, 507–512. (In Chinese)
23. Liu, Z.H.; Yu, F.X.; Yu, H.T.; He, C.F.; Wu, B. Ultrasonic guided wave technology based on group velocity calibration and its application for defect detection in composite plates. *J. Mech. Eng.* **2012**, *48*, 8–15. (In Chinese) [CrossRef]
24. Wu, B.; Li, Y.; Zheng, Y.; He, C.F. Thickness measurement of surface attachment on plate with SH wave. *J. Mech. Eng.* **2012**, *48*, 78–84. (In Chinese) [CrossRef]
25. Liu, Z.H.; Xu, Y.Z.; He, C.F.; Wu, B. Experimental study on defect imaging based on single Lamb wave mode in plate-like structures. *Eng. Mech.* **2014**, *31*, 232–238. (In Chinese) [CrossRef]
26. Liu, Z.H.; Wu, B.; He, C.F.; Wang, X.Y. Torsional mode of ultrasonic guided wave propagation characteristics in pipes with viscoelastic coatings. *J. Basic Sci. Eng.* **2005**, *13*, 291–299. (In Chinese) [CrossRef]
27. Deng, F.; Wu, B.; He, C.F. A time-reversal defect-identifying method for guided wave inspection in pipes. *J. Press. Vessel. Technol.* **2008**, *130*, 021503. [CrossRef]
28. Yu, J.G.; Wu, B.; Chen, G.Q. Wave characteristics in functionally graded piezoelectric hollow cylinders. *Arch. Appl. Mech.* **2009**, *79*, 807–824. [CrossRef]
29. Liu, Z.H.; He, C.F.; Wu, B.; Wang, X.Y. Experimental research on longitudinal mode selection in pipes using angle beam probes. *Eng. Mech.* **2009**, *26*, 246–250. (In Chinese)
30. Madenga, V.; Zou, D.H.; Zhang, C. Effects of curing time and frequency on ultrasonic wave velocity in grouted rock bolts. *J. Appl. Geophys.* **2006**, *59*, 79–87. [CrossRef]

31. Cui, Y.; Zou, D.H. Numerical simulation of attenuation and group velocity of guided ultrasonic wave in grouted rock bolts. *J. Appl. Geophys.* **2006**, *59*, 337–344. [CrossRef]
32. Zou, D.H.; Cui, Y.; Madenga, V. Effects of frequency and grouted length on the behavior of guided ultrasonic waves in rock bolts. *Int. J. Rock Mech. Min. Sci.* **2007**, *44*, 813–819. [CrossRef]
33. Zou, D.H.; Cui, Y. A new approach for field instrumentation in grouted rock bolt monitoring using guided ultrasonic waves. *J. Appl. Geophys.* **2011**, *75*, 506–512. [CrossRef]
34. Cui, Y.; Zou, D.H. Assessing the effects of insufficient rebar and missing grout in grouted rock bolts using guided ultrasonic waves. *J. Appl. Geophys.* **2012**, *79*, 64–70. [CrossRef]
35. Wang, C.; He, W.; Ning, J.G.; Zhang, C.S. Propagation properties of guided wave in the anchorage structure of rock bolts. *J. Appl. Geophys.* **2009**, *69*, 131–139. [CrossRef]
36. Zima, B.; Kedra, R. Reference-free determination of debonding length in reinforced concrete beams using guided wave propagation. *Constr. Build. Mater.* **2019**, *207*, 291–303. [CrossRef]
37. Li, J.S.; Lu, Y.; Guan, R.Q.; Qu, W.Z. Guided waves for debonding identification in CFRP-reinforced concrete beams. *Constr. Build. Mater.* **2017**, *131*, 388–399. [CrossRef]
38. Xu, W.B.; Pan, W.D.; Ding, M.L. Experiment on evolution of microstructures and long-term strength model of cemented backfill mass. *J. Cent. South Univ.* **2015**, *46*, 2333–2341. (In Chinese) [CrossRef]
39. Torzaghi, K.; Peck, R.B. *Soil Mechanics in Engineering Practice*; John Wiley & Sons: New York, NY, USA, 1948; p. 22.

sensors

MDPI

Article

Ultrasound Defect Localization in Shell Structures with Lamb Waves Using Spare Sensor Array and Orthogonal Matching Pursuit Decomposition

Weilei Mu [1,2], Yuqing Gao [1] and Guijie Liu [1,*]

1 Engineering College, Ocean University of China, Qingdao 266100, China; muweilei@ouc.edu.cn (W.M.); gaoyuqing@stu.ouc.edu.cn (Y.G.)
2 Suzhou Academy, Xi'an Jiaotong University, Suzhou 215123, China
* Correspondence: liuguijie@ouc.edu.cn

Abstract: Lamb waves have multimodal and dispersion effects, which reduces their performance in damage localization with respect to resolution. To detect damage with fewest sensors and high resolution, a method, using only two piezoelectric transducers and based on orthogonal matching pursuit (OMP) decomposition, was proposed. First, an OMP-based decomposition and dispersion removal algorithm is introduced, which is capable of separating wave packets of different propagation paths and removing the dispersion part successively. Then, two simulation signals, with nonoverlapped and overlapped wave packets, are employed to verify the proposed method. Thereafter, with the proposed algorithm, the wave packets reflected from the defect and edge are all separated. Finally, a sparse sensor array with only two transducers succeeds in localizing the defect. The experimental results show that the OMP-based algorithm is beneficial for resolution improvement and transducer usage reduction.

Keywords: Lamb wave; modal separation; damage imaging

Citation: Mu, W.; Gao, Y.; Liu, G. Ultrasound Defect Localization in Shell Structures with Lamb Waves Using Spare Sensor Array and Orthogonal Matching Pursuit Decomposition. *Sensors* **2021**, *21*, 8127. https://doi.org/10.3390/s21238127

Academic Editor: Zenghua Liu

Received: 11 November 2021
Accepted: 2 December 2021
Published: 4 December 2021

1. Introduction

Ultrasonic guided waves have become the preferred tool in nondestructive testing and structural health monitoring (SHM). Lamb waved are one of the most commonly used ultrasonic guided waves in damage detection, due to their intrinsic advantages [1,2]. Lamb waves can propagate in a thin shell structure for a long distance with low energy loss. At the same time, they are sensitive to different types of damage such as cracks and corrosion [3]. Usually, a change of sensing signal amplitude or energy can be considered to be a change of wave-guided material, which is generally caused by damage.

In most mechanical structures, such as airplanes, ships, and deep submersibles, there are a lot of covering skins fixed by rivets. (e.g., Figure 1). Although the skins do not bear the structural gravity and workload, they play a great role in structure safety. There have been serious airline accidents caused by damage to skins. Usually, the damage is in the form of fatigue cracks in the rivet area. Therefore, the defects are mostly found at locations close to the edge. In this case, generated Lamb waves propagate to the defect and edge zone, and are reflected by the crack and edge. Consequently, the sensing signals contain the direct arrival wave, and the reflection waves from crack and edge. The wave signals may overlap each over, which makes signal explanation rather difficult. However, if these individual waves can be decomposed, it is beneficial to localize the damage, and moreover, it helps to reduce the number of deployed sensors. In this paper, we try to decompose the overlapped signals from the collected signal of two sensors. Ideally, it is best to deploy only one self-sensing sensor, which can generate and sense simultaneously.

Due to the dispersion nature of Lamb waves, waves of different frequencies will propagate with different speeds, which causes the wave packet to widen during propagation. Therefore, the dispersed waves of each path are overlap with each other more easily than without dispersion. Moreover, there are at least two modes of Lamb waves at any frequency, due to their multimodal nature, which makes signal processing more difficult [4]. Therefore, signal processing is essential for a Lamb wave-based SHM system. Wang et al. proposed a damage imaging algorithm based on time-of-arrival (TOA), in which they innovatively defined the possible location of the defect to be an ellipse with excitation and reception sensors as the focal points [5]. Following from that, Hall and Michaels improved the TOA algorithm using adaptive weighting coefficients, making the location imaging more accurate [6]. Another commonly used imaging method is based on time difference of arrival (TDOA) to different sensors; mathematically, the possible location of the defect seems to be a hyperbolic line. When using the TDOA algorithm, the number of sensors increases, but the imaging quality is not significantly enhanced [7]. Perelli et al. proposed a damage localization based on warped frequency transform (WFT) [8], which decomposes each wave packet from the spectrum of WFT and localizes the damage using these decomposed wave packets. However, the decomposed wave, not the original excitation wave, is dispersed, which may reduce the localization resolution. If damage occurs near the edge, there are multiple simultaneous propagation paths. Therefore, Ebrahimkhanlou et al. initially used a priori information of scattered wave fields to confirm each wave path. Then, the start and end points are taken as the focal points, and the TOA method is used to localize the damage. However, they used the original signal to confirm the possible localization line [9]. To improve the localization resolution, in this paper, it is similarly necessary to clarify the path of scattered waves. Moreover, accurately separating individual wave packets is also important.

(a) (b)

Figure 1. Skins of structures. (**a**) hull of ships and (**b**) aircraft skins.

In the history of scattered wave decomposition, decomposition methods based on dictionaries have gained great popularity. Alguri et al. constructed a dictionary containing wave signals of different travel distance using experiments, and then reconstructed the scattered waves from the online monitoring data [10]. Golato et al. built a dictionary consisting of multi-mode scattered Lamb waves and solved the sparse reconstruction problem of Lamb wave signals [11]. However, the above-mentioned methods above did not accomplish modal separation and elimination of the dispersion. Caibin Xu et al. built a dispersion signal dictionary using the dispersion curves to sparsely decompose the recorded dispersive guided waves, and innovatively obtained dispersion-compensated guided waves by using a non-dispersion signal dictionary [12]. Furthermore, they furthered an interesting method to obtain the dispersion curve of an unknown material, in which they acquired the unit pulse response of the measured structure, and interpolated the transfer function from the frequency domain into the wavenumber domain [13]. After this, they systematically proposed the construction method of over-completed dictionaries, which contained dispersive atoms and nondispersive atoms [14].

Recently, Caibin Xu et al. have proposed a focusing classical multiple-signal classification (MUSIC) algorithm based on the virtual time reversal technique, which can localize multi-defects without baseline subtraction. The proposed method is valuable for the practical use of Lamb wave-based structural health monitoring [4]. The matching pursuit (MP) algorithm, proposed by Mallat and Zhang, can decompose the signal into a combination of waveforms selected from the dictionary [15]. However, the vertical projection of the signal onto selected atoms is non-orthogonal, which makes the result of each iteration not optimal but suboptimal, requiring more iterations to achieve convergence. In this paper, we propose a guided-wave imaging method based on sparse reconstruction, which uses an analytical guided-wave propagation model to generate scattered signals for different propagation distances and modes, known as atoms of the dictionary. We constructed a dictionary containing wave signals of different travel distances by equations, and built a nondispersive dictionary by equations. In addition, then, the overcomplete dictionary is used to match the collected Lamb wave signals under the assumption of signal sparsity, which means there are only a few damage-related signals in the collected signal. Indeed, the damage occurs at only a few discrete locations in the structure. The orthogonal matching pursuit (OMP) algorithm is used to separate each wave component. The propagation distance is related to the labels of the selected atoms. The OMP algorithm orthogonalizes all the atoms needed at each step of the decomposition, and converges faster [16].

The rest of the article is organized as follows. Section 2 describes in detail how to construct the dictionary using the Lamb wave scattering model and the procedures of the OMP algorithm, and how to remove dispersion simultaneously. In Section 3, numerical simulations are carried out, nonoverlapped and overlapped signals are decomposed, and dispersion is removed by the OMP algorithm to verify its correctness. In Section 4, experimental studies are implemented to validate the proposed method. Finally, conclusions are drawn in Section 5.

2. Methodology

2.1. Introduction of OMP

If the signal is sparse on some basis functions during decomposition, it means that the energy of the signal gathers on a few basis functions. Therefore, the signal can be represented by these basis functions and their coefficients [17].

Suppose that given a dictionary $D = \{a_1, a_2, a_3 \ldots\ldots a_L\}$, $L \gg N$. For arbitrary signals $y \in R^N$, select K atoms in the dictionary D and make a K-term approximation to y.

$$y_K = \sum_{i \in I_K, |I_K| = K} \langle y, a_i \rangle a_i \tag{1}$$

where I_K is the subscripts set of the K atoms, corresponding to the first K maximum coefficient of $|y, a_i|$.

In other words, it prefers selecting the set of atoms with the sparsest decomposition coefficients from various possible combinations of dictionary atoms, which also reflects the idea of sparse signal decomposition.

2.2. Overcomplete Dictionary Construction

In an infinitely large thin plate, the structure response at a distance of d from the excitation can be expressed as [18]

$$u_n^{\mathrm{AorS}}(t) = \sqrt{\frac{1}{d}}\frac{1}{2\pi}\int_{-\infty}^{+\infty} S(\omega)e^{j\omega t}e^{-jk_n^{\mathrm{A\ or\ S}}(\omega)d}d\omega \tag{2}$$

where $u(t)$ is the time domain response signal of n-th symmetric mode (S) or anti-symmetric mode (A), t is the travel time, ω is the angular frequency, $S(\omega)$ is the frequency domain representation of the excitation signal $f(t)$, $k(\omega)$ is the wave number of the corresponding Lamb wave mode, and j is the imaginary unit. The frequency representation of Equation (2) can be written as

$$U(\omega) = \sqrt{\frac{1}{d}}S(\omega)e^{-jk(\omega)d} \quad (3)$$

The same as above, if there is a through-thickness damage in the plate, then the propagation distance from the excitation source to the damage then to the receiver is ds. Therefore, the received scattered signal $Y(\omega)$ in the frequency domain can be expressed as

$$Y(\omega) = \alpha(\omega)\beta(\omega)\sqrt{\frac{1}{d_s}}S(\omega)e^{-jk(\omega)d_s} \quad (4)$$

When there are multiple damages in the plate, the corresponding scattered signal can be approximated as a linear superposition of each individual case, without considering the scatter phenomenon between any two damages [18,19]

$$Y(\omega) = \sum_i \alpha_i(\omega)\beta_i(\omega)\sqrt{\frac{1}{{d_s}^i}}S(\omega)e^{-jk(\omega){d_s}^i} \quad (5)$$

In the above equation, $\alpha_i(\omega)$ is the scattered coefficient from the excitation source to the damage of the i-th damage; $\beta_i(\omega)$ is the scattered coefficient from the damage to the receiver of the i-th damage, and ${d_s}^i$ is the propagation distance of the i-th propagation path. From Equation (2), it can be seen that the signal is obtained by multiplying the atoms in the dictionary with their coefficients, so the value of the scattered coefficient $\alpha_i(\omega)$ will be reflected in the coefficient matrix θ. From this, the i-th column of the dictionary D of A or S mode can be expressed as

$$a_i = F^{-1}\left\{\sqrt{\frac{1}{d^i}}S(\omega)e^{-jk_n^{AorS}(\omega)d^i}\right\} \quad (6)$$

where d^i is the propagation distance of the i-th atom, $F^{-1}\{\cdot\}$ represents the inverse Fourier transform. Thus, the single-mode dictionary can be obtained in the form [12]

$$D^{\text{A or S}} = \begin{bmatrix} {a_1}^1{a_2}^1\cdots{a_i}^1\cdots{a_L}^1 \\ {a_1}^2{a_2}^2\cdots{a_i}^2\cdots{a_L}^2 \\ \vdots \quad \vdots \quad \ddots \quad \vdots \quad \ddots \quad \vdots \\ {a_1}^N\ {a_2}^N\ldots{a_i}^N\cdots{a_L}^N \end{bmatrix}$$

where the aim is the value of the m-th sampling point in the i-th atom.

2.3. Nondispersive Dictionary

The expressions of wave number, phase velocity, and group velocity are respectively [14]

$$\begin{aligned} k_a(\omega) &= \tfrac{1}{C_p(2\pi f)}\omega \\ C_p(\omega) &= \tfrac{\omega}{k_a} = C_p(2\pi f) \\ C_g(\omega) &= \tfrac{d\omega}{dk_a} = C_g(2\pi f) \end{aligned} \quad (7)$$

It is proved that if the wave number of the Lamb wave is linearly related to the frequency, the envelope of the response signal will be the same as the envelope of the excitation. In other words, if the wave number of Lamb wave is linear with respect to frequency, it will not be dispersive. Therefore, the propagation distance can be calculated by the group velocity of central frequency.

$$x_i = C_g(2\pi f_c)t_i \tag{8}$$

Bringing the above equation into Equation (3) and neglecting the amplitude change due to signal propagation, we can obtain the nondispersive signal $y_i(t)$.

$$y_i(t) = \frac{1}{2\pi}\int_{-\infty}^{+\infty} S(\omega)e^{j(\omega t - k_a x_i)}d\omega = \frac{1}{2\pi}\int_{-\infty}^{+\infty} S(\omega)e^{j\omega(t-t_i)}d\omega = f(t-t_i) \tag{9}$$

Equation (9) shows that the nondispersive signal is equal to the excitation signal with a phase shift, so the nondispersive dictionary D_n can be built.

$$D_n^{\text{A or S}} = \begin{bmatrix} f_1(t-t_1) & f_1(t-t_2) \cdots & f_1(t-t_i) \cdots & f_1(t-t_L) \\ f_2(t-t_1) & f_2(t-t_2) \cdots & f_2(t-t_i) \cdots & f_2(t-t_L) \\ \vdots & \vdots \ \ddots & \vdots \ \ddots & \vdots \\ f_m(t-t_1) & f_m(t-t_2) \cdots & f_m(t-t_i) \cdots & f_m(t-t_L) \\ \vdots & \vdots \ \ddots & \vdots \ \ddots & \vdots \\ f_N(t-t_1) & f_N(t-t_2) \cdots & f_N(t-t_i) \cdots & f_N(t-t_L) \end{bmatrix}$$

where $f_m(t\text{-}t_i)$ is the value of the *m-th* sampling point of the nondispersive signal with time shift t_i.

2.4. OMP-Based Decomposition and Dispersion Removal Algorithm

Assuming that y_s contains r wave packets, the received signal y_s can be represented by a linear combination of its atoms [18]

$$y_s = D\theta + n \tag{10}$$

where $D \in R^{N\times M}$ $(M \geq r)$ is the overcomplete dictionary, $\theta \in R^M$ is the coefficient column vector, and $n \in R^N$ is the residual noise term.

The above equation is an underdetermined equation. If the number of atoms M is large enough and the travel distance $d_s{}^i$ $(1 \leq i \leq r,\ i \in N^+)$ of the scattered signal is completely covered by the distance d^i $(1 \leq i \leq M,\ i \in N^+)$ corresponding to the given atoms, then, the scattered signal can be sparsely decomposed with the dictionary D by the OMP algorithm.

OMP is an iterative algorithm capable of selecting the best fitness atoms from the overcomplete dictionary for signal reconstruction. The OMP algorithm is described as [20]

(1) Initialization process. Determine sparsity degree K which means the number of potential wave packets. Build an overcomplete dictionary D as:

$$D = [a_1, a_2, a_3 \ldots a_i \ldots a_L]$$

where ai indicates the *i-th* atom in the dictionary.

(2) Orthogonal matching. Find the column a_λ in the dictionary according to the product value θ_λ of acquisition signal y and a_λ. Then, record the product value θ_λ, known as matching coefficients. $a_\lambda = \text{argmax}_{i=1,2\ldots N}|\langle y, a_i\rangle|$, λ indicates that the atom is the λ*-th* column in the dictionary, and $\theta_\lambda = \max_{i=1,2\ldots N}|\langle y, a_i\rangle|$.
(3) Update and iteration. Update the solution set $\theta = \theta \cup \{\theta_\lambda\}$, and update the residual signal by subtracting the selected atoms from the signal of last iteration.

$$y = \min\|y - a_\lambda\theta_\lambda\|_2^2$$

(4) Termination judgment. Determine whether the number of iterations is greater than K. If it is not satisfied, execute the matching and update procedures again.

It should be noted that the solved θ is a column vector with mostly zero elements, when using the overcomplete dictionary, and the non-zero element θ_λ of θ represents the scattering coefficient of the λ-*th* atom in the dictionary with travel distance L_λ. Each scattering wave packet with a unique travel distance in the collected signal can be recovered using the above equation and the overcomplete dictionary D. The propagation distance of the scattering wave packet can be visually expressed by the travel distance of the λ-*th* atom. With the same column label, the collected signal can be represented as the combination of nondispersive wave packets using the nondispersive dictionary. Sparsity degree K determines the number of atoms selected from the dictionary. When the value of K is small, some wave packets may not be matched. When the value of K is large, matching performance will be better but the amount of calculation is greatly increased. Therefore, the total number of wave packets m in the signal needs to be predicted, and in general, the value range of K is $2m \geq K \geq m$. The procedure of the proposed method is shown in Figure 2.

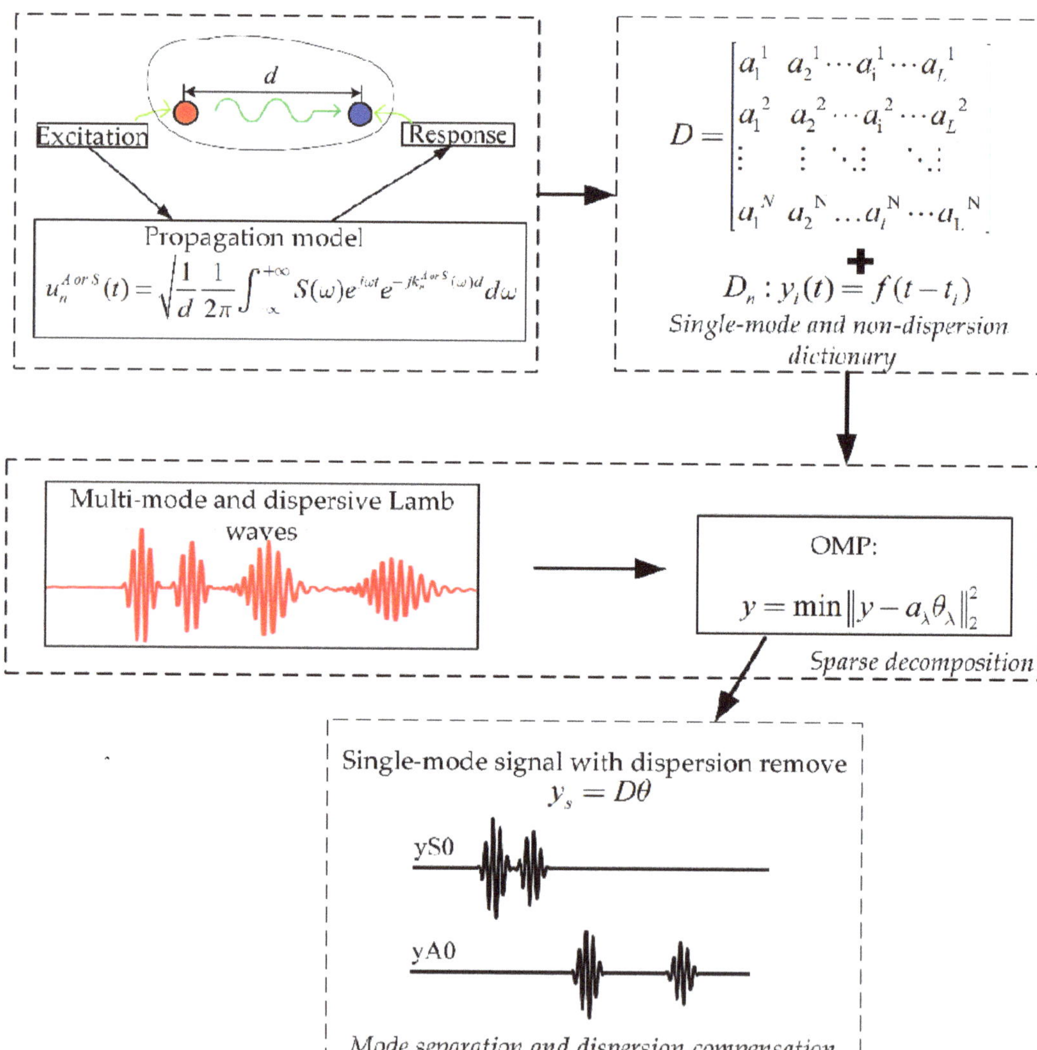

Figure 2. Schematic of the procedure of the proposed method.

3. Methodology Verification

Numerical Simulation

To verify the performance of the OMP-based dispersion removal approach, numerical simulations are first provided, including a nonoverlapped case and an overlapped case.

Abuqus CAE was adopted to simulate Lamb wave propagation, and a simulation model with the dimensions 1250 mm × 1250 mm × 2 mm (length × width × thickness) was conducted. At the edge regions, the wave absorption layer was built to eliminate the reflected wave. Moreover, two different actuators with the locations *A* (100,400) and *B* (300,400), and one receiver with location *C* (800,400) were set. Therefore, two different wave-guided paths were simulated in this case, and the distances between the actuator and receiver were different. Material parameters of the simulation model are listed in Table 1.

Table 1. Material parameters.

Material	Density (kg/m^3)	Elastic Modulus (Pa)	Poisson's Ratio
Q235	7800	2.1×10^{11}	0.33

In the numerical simulation, the larger the grid size, the less computational consumption. However, with the increase of the grid size, simulation error decreases. On the contrary, a smaller size might increase the computational consumption exponentially. Therefore, element size is usually limited to one-tenth of the wavelength. In this case, the central frequency of the excitation signal was 100 kHz, and the grid size in this study is set as 1 mm. According to Nyquist's theory, the acquisition frequency must be at least twice the maximum frequency component. In the simulation and experiment, the acquisition frequency was 2.5 MHz for full frequency range acquisition [21].

A five-cycle sinusoidal signal modulated with Hanning window was selected as the excitation signal, it can be expressed as follows:

$$y = A\sin(2\pi f_c t) \times (1 - \cos(2\pi f_c t/N)) \tag{11}$$

where A is the amplitude of the sinusoidal signal, N denotes the cycle number of the modulated signal, f_c represents the central frequency, and t is time serial ranging from zero to N/f_c.

Since the center frequency used is below the cutoff frequency of S1 and A1 modes, only two fundamental Lamb wave modes A0 and S0 are considered [22]. As shown in Figure 3a, the dispersive signal contains a total of four wave packets, which are two mode components for two propagation paths. By OMP algorithm, the original signal can be separated to four wave packets. The two separated S0 and A0 mode wave packets are as shown in Figure 3b,c respectively.

The A0 and S0 modes can be easily separated by the previously built single-mode dictionary, and it can be seen that the A0 mode has undergone more serious dispersion, as shown in Figure 3c. Reconstruction of the separated modal signals using a nondispersive dictionary can better eliminate the dispersion, as in Figure 4.

In the modal separation of Lamb waves, the sparse decomposition of the dispersive signal is based on the corresponding modal dictionary, which must contain all the corresponding atoms to fully compensate all the dispersive modes. Otherwise, the dispersive signal will be incompletely compensated [23]. In signal reconstruction using only the S0 modal dispersion dictionary, the dispersive signal is recovered for the S0 mode component, while the A0 mode component is not recovered, as shown in Figure 5. This is because there is no atom corresponding to the wave packet of A0 mode in the S0 mode dictionary. In conclusion, the proposed approach based on sparse decomposition cannot handle the case with an incomplete dictionary. In other words, the maximum propagation distance of the atom signal in the single-mode dictionary must be larger than that of the signal to be processed, and the same requirement is suitable for the atom signal in the nondispersive dictionary. In practice, the time information depending on the length of the dispersive signal acquisition can be known in advance, and then the dispersive and nondispersive dictionaries can be built to cover all possible wave packets.

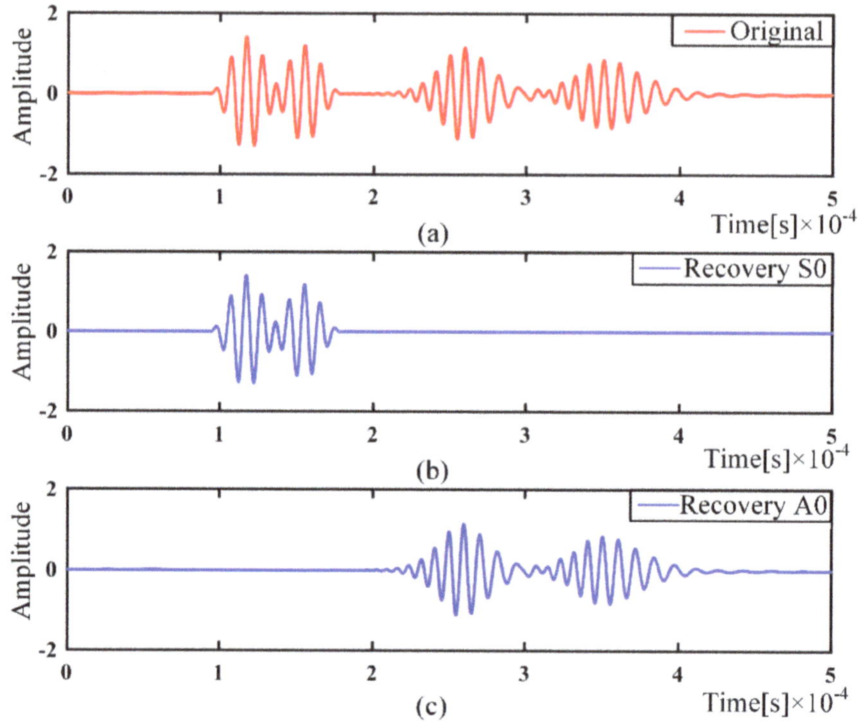

Figure 3. Mode separation: (**a**) Original signal; (**b**) Separated S0 mode dispersion signal; (**c**) Separated A0 mode dispersion signal.

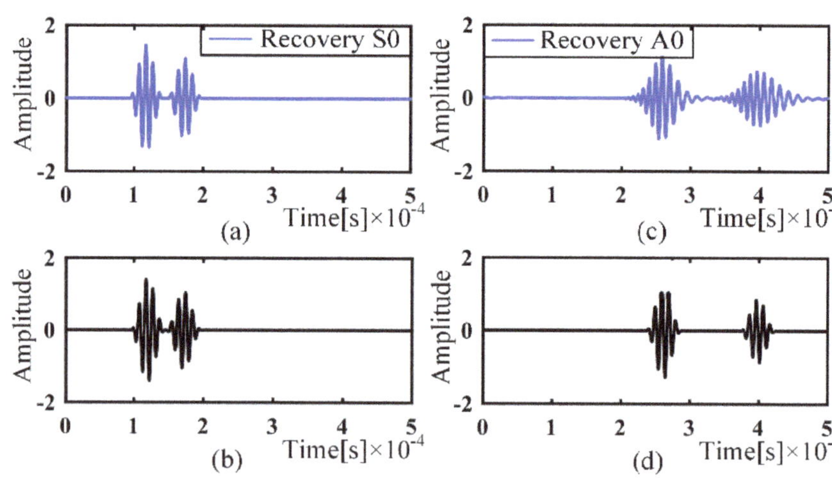

Figure 4. Dispersion compensation: (**a**) S0 mode signal; (**b**) S0 dispersion removal signal; (**c**) A0 mode signal; (**d**) A0 dispersion removal signal.

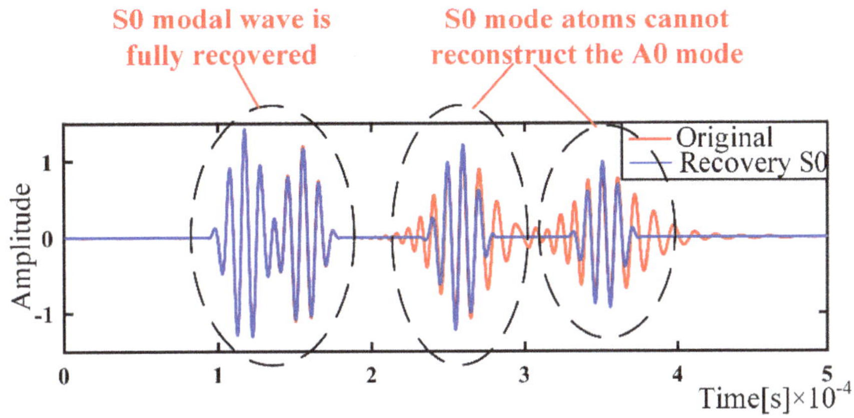

Figure 5. S0 modal recovery.

Usually, there are many wave packets from different propagation paths collected by the acquisition system simultaneously. Apparently, there are wave packets that overlap each other. For further verification of the proposed algorithm, the exciter positions are set to *A* (250,400) and *B* (300,400), and the receiver position remains unchanged. In contrast to Figure 5, there seems to be only two wave packets of two mode components, as shown in Figure 6, due to the overlap phenomenon.

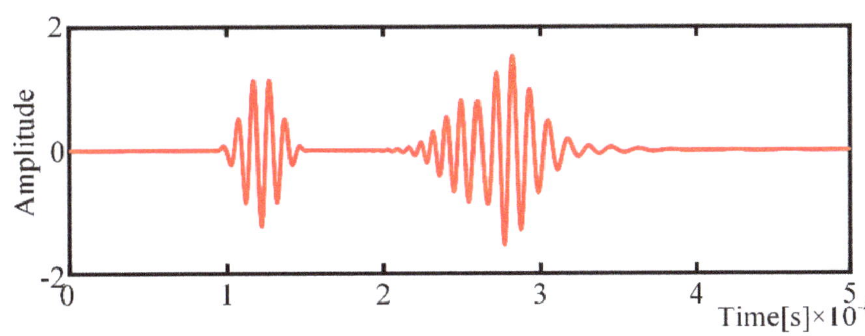

Figure 6. Overlapping wave packets.

However, when sparse decomposition of the signal is performed, it still requires two atoms from the A0 dictionary and two atoms from the S0 dictionary to match it. In the same way as above, the matched atoms are used to represent the wave packets of the signal, and finally the dispersion is removed using the nondispersive dictionary. The procedure is shown in Figure 7. It turns out that this approach is still able to separate the wave packets, even if they are overlapped.

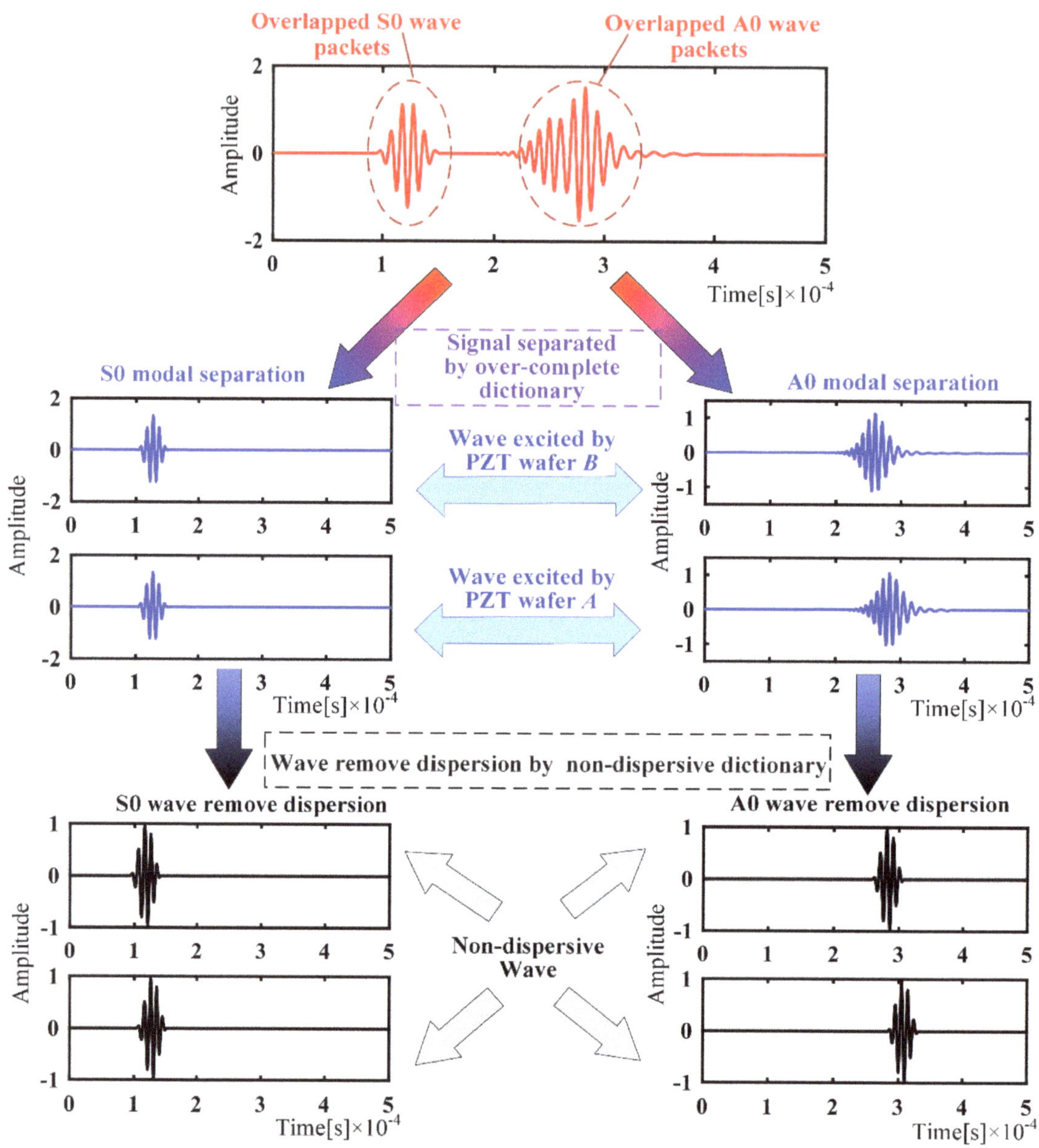

Figure 7. Separation of overlapped wave packets.

4. Experimental Verification

4.1. Sparse Sensor Array-Based Localization Method

The PZT wafer A and B with dimensions 13 mm × 0.5 mm (diameter × thickness) are mounted at (x_A, y_A) and (x_B, y_B), respectively. Therefore, the propagation distance of the primary scattered wave (as shown in Figure 8) that is reflected only by the defect can be described as

$$d = \sqrt{(x - x_A)^2 + (y - y_A)^2} + \sqrt{(x - x_B)^2 + (y - y_B)^2} \tag{12}$$

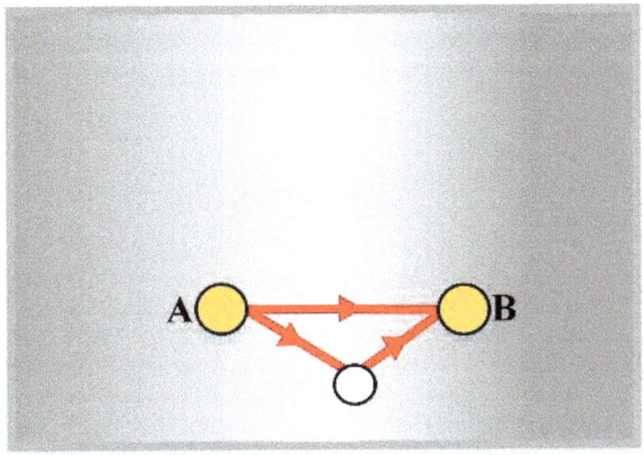

Figure 8. Experiment sample.

Although the defect location is unknown in practical detection, a through-thickness hole is fabricated to make an analogy to the defect and drilled at (x_c, y_c) for this experiment. In the process of defect localization, at least three different propagation distances need to be determined for a more accurate localization [24]. Traditionally, each propagation distance is confirmed by a pair of PZT wafers. Hence, at least three PZT wafers are needed. In addition to the direct arrival wave and the primary scattered waves, there are also secondary scattered waves reflected from the edges and defects. Figure 9a indicates the signal via path *a-b*, excited by the exciter A and reflected by the defect. Moreover, sensor B will also receive signal from paths *c-d*, *e-f*, and *e-d*, with two reflections by the defect and the edge. The latter two scattered waves are relatively weak compared with the first one, and are not considered here.

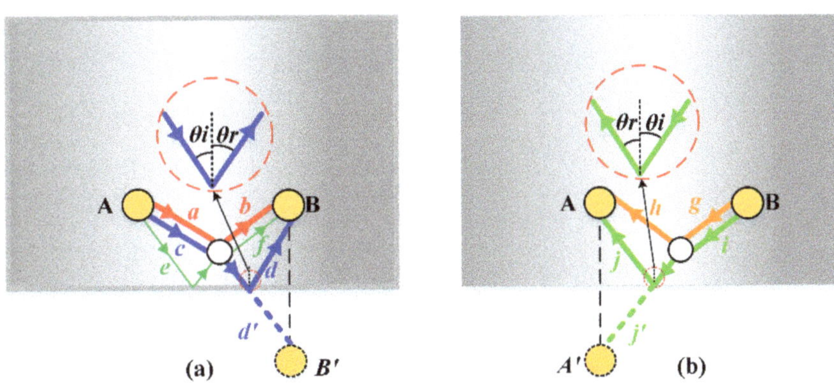

Figure 9. Guided-wave scattering path: (**a**) PZT A to PZT B; (**b**) PZT B to PZT A.

The incident angle into the edge is equal to the reflection angle that is $\theta_i = \theta_r$, so the scattered signal path *c-d* can be easily deduced out by the mirror point of sensor B, and its propagation distance is

$$D_{cd} = d_c + d'_d = \sqrt{(x - x_A)^2 + (y - y_A)^2} + \sqrt{(x - x'_B)^2 + (y - y'_B)^2} \quad (13)$$

where d_c denotes the distance of path c that is the distance from the excitation sensor $A(x_A, y_A)$ to the defect; d'_d denotes the distance from the defect to the mirror point $B'(x'_B, y'_B)$

With sensor B for excitation and sensor A for reception, as shown in Figure 9b, the propagation distance of the primary scattered wave is the same D_{ab}. The propagation distance of the secondary scattered wave is

$$D_{ij} = d_i + d'_j = \sqrt{(x - x_B)^2 + (y - y_B)^2} + \sqrt{(x - x'_A)^2 + (y - y'_A)^2} \tag{14}$$

where d_i denotes the distance of path i; d'_j denotes the distance from the defect to the mirror point $A'(x'_A, y'_A)$ of sensor A(x_A,y_A).

According to analyses above, three wave paths are obtained with only two PZT wafers. The propagation distance determined by each pair of excitation–reception or excitation-to-mirror sensors can determine an ellipse with the sensor as the focus, as shown in Figure 10. The red ellipse in Figure 10a confirmed by PZT A to PZT B is the same as that in Figure 10b, due to the same propagation distance of A–B and B–A. Additionally, the blue ellipse and green ellipse confirmed by the primary scattered waves are combined with the red ellipse in Figure 10c to access the damage location.

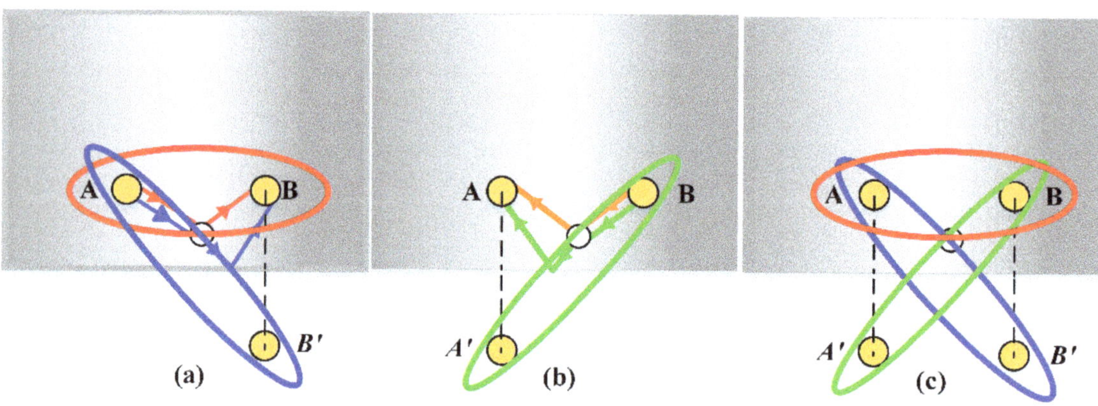

Figure 10. Elliptical trajectories: (**a**) PZT A to PZT B; (**b**) PZT B to PZT A; (**c**) combination.

When three ellipses are identified, their envelopes, or intersections, are the defect locations. As long as the measured wave packet has enough energy to be detected, the damage can theoretically be located with only one excitation [25]. However, in practice, the signal resolution and intensity are not high enough, so it is necessary to change the excitation to improve the imaging effect. PZT A and B will be taken as the actuator and sensor successively. When the plate is intact, the recorded signals are considered to be the reference signals. Ignoring the changes of wave guide material due to environment and damage, the wave signal varies only with respect to the scatter phenomenon of the damage [26]. The sparsity degree K can be set to be 8 at least, due to 4 propagation paths for S0 and A0 mode.

4.2. Experimental Setup

A square plate with dimensions 1250 mm × 1250 mm × 2 mm (length × width × thickness) was adopted as the experimental sample. Two PZT wafers A and B were mounted at (400, 150) and (700, 150), respectively. A prefabricated through-hole was located at (500, 50). The experimental setup consisted of a Tektronix AFG1022 arbitrary waveform generator, a DS2−8B data acquisition instrument and a smart AE charge amplifier. To coincide with the simulation case, the experimental parameters, such as the acquisition frequency and the excitation signal, were all the same as the simulation model.

As PZT A and B will be taken as the actuator and sensor successively, the actuator will be connected with Tektronix AFG1022, and the sensor will be connected with the charge amplifier which is connected to the data-acquisition device. At the same time, the waveform generator will be connected to the data acquisition device with an isolator to record the excitation and sensing wave packets simultaneously. The diagram of the experimental setup is shown in Figure 11.

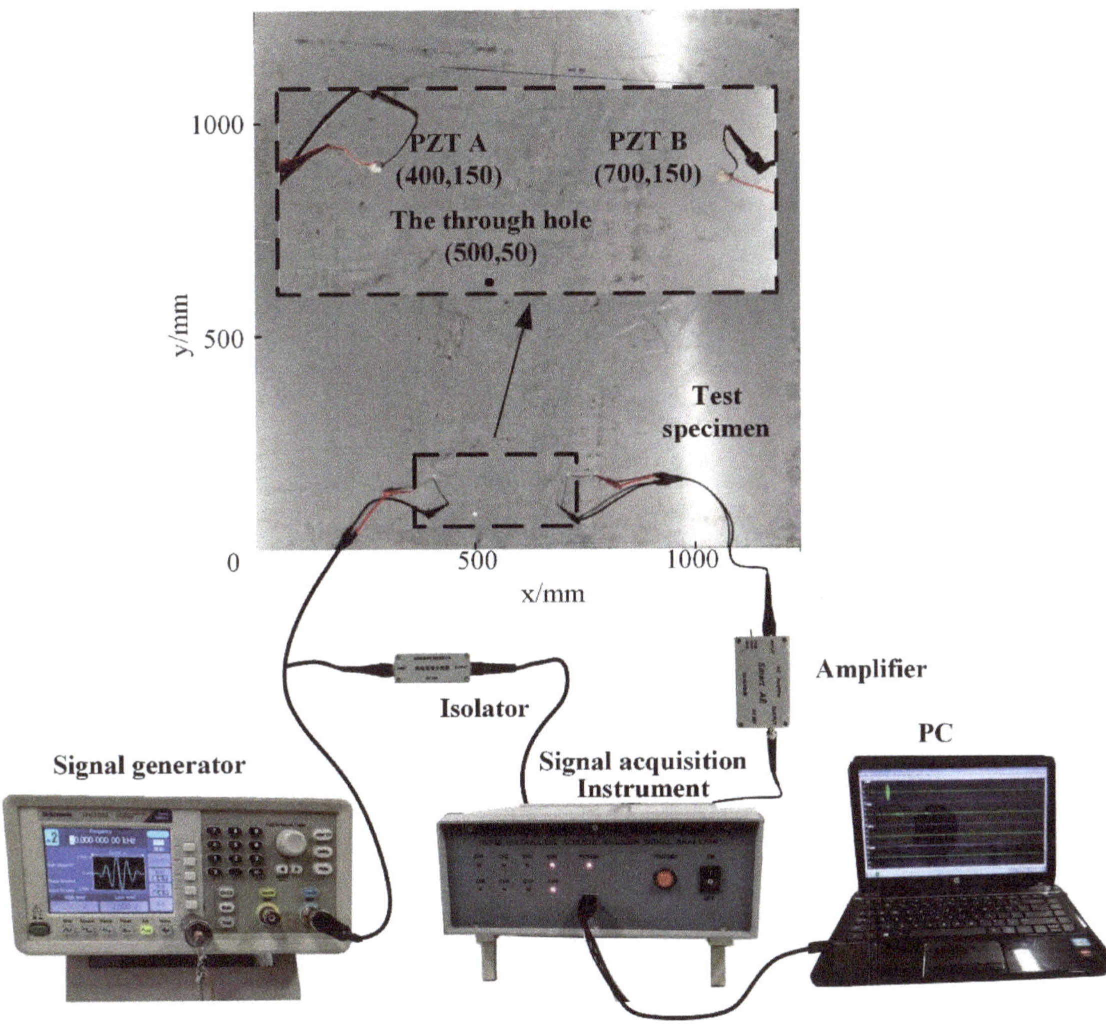

Figure 11. Experimental setup.

Therefore, the reference-based monitoring method can be carried out as follows.

(1) under the intact condition, excite a five-peak sinusoidal wave modulated by Hanning window with a center frequency of 100 kHz from PZT A, collect the signals from PZT B. Then, excite modulated wave from PZT B, and collect the signals from PZT B. These signals are considered to be the reference signals.
(2) in the practical monitoring, repeat the acquisition step above.
(3) subtract the intact signal from the acquisition signal with defects, known as the state-relative signals.

(4) decompose the state-relative signals into wave packets of a single mode by the OMP-based decomposition and dispersion compensation method.
(5) localize the defect by sparse sensor array-based localization method.

4.3. Experimental Results

Under the intact situation, PZT A and B are taken as actuator and sensor, respectively. The collected signal was taken from PZT B for the sample, as shown in Figure 12a, which contains disturbed noise using the excitation signal, direct wave of S0 and A0 mode, and reflected wave using the edge. If the sparsity degree K is 8, the wave packets with relatively high amplitude are extracted. The recovery signal is shown in Figure 12a. Theoretically, there is also a reflected wave of S0 mode, although it is not determined visually. With the largest two matching coefficients θ_λ, only the A0 mode waves are extracted, as shown in (b), because the amplitude of S0 mode wave is relatively low. Therefore, the OMP algorithm can remove the system noise well, because the predefined atomic functions are independent of the noise. Moreover, using the columns of the chosen atoms, it is easy to get the travel distance of the corresponding wave packets, which verifies the correctness of the method. Figure 12c shows the recomposed signal using the nondispersive dictionary to remove the dispersion.

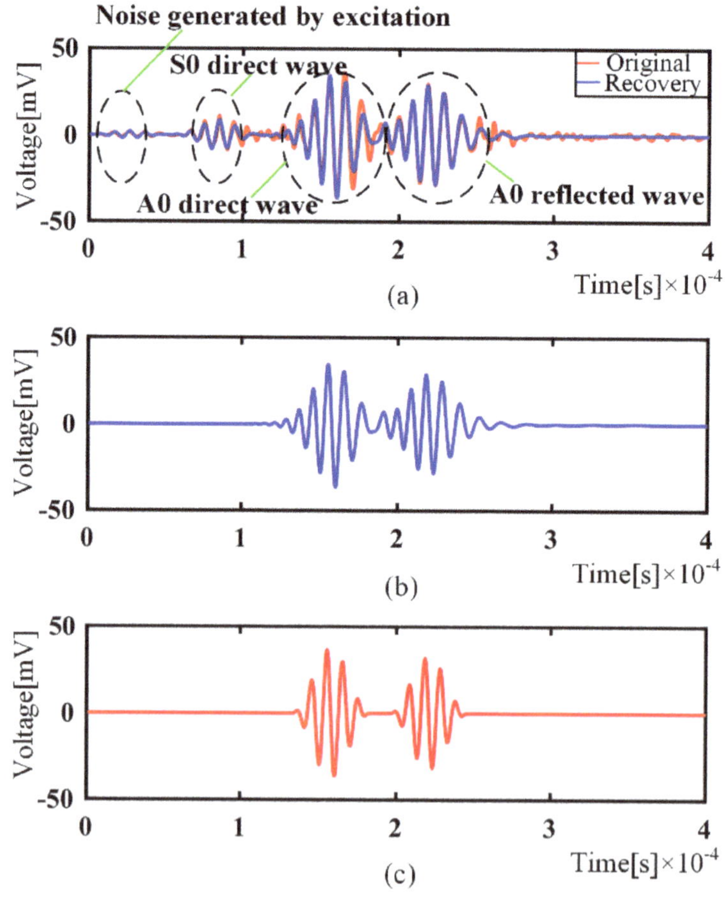

Figure 12. Defect-free signal processing: (**a**) Defect-free signal and recovery signal; (**b**) Recovery A0 with largest matching coefficients; (**c**) The result after dispersion removal.

After fabricating a through-thickness hole in the plate, we collect the signals from PZT A and B successively. Then, the received signals are subtracted from the corresponding signal when there is no defect, the residual signal is processed with the same process procedures, and the results are shown in Figure 13.

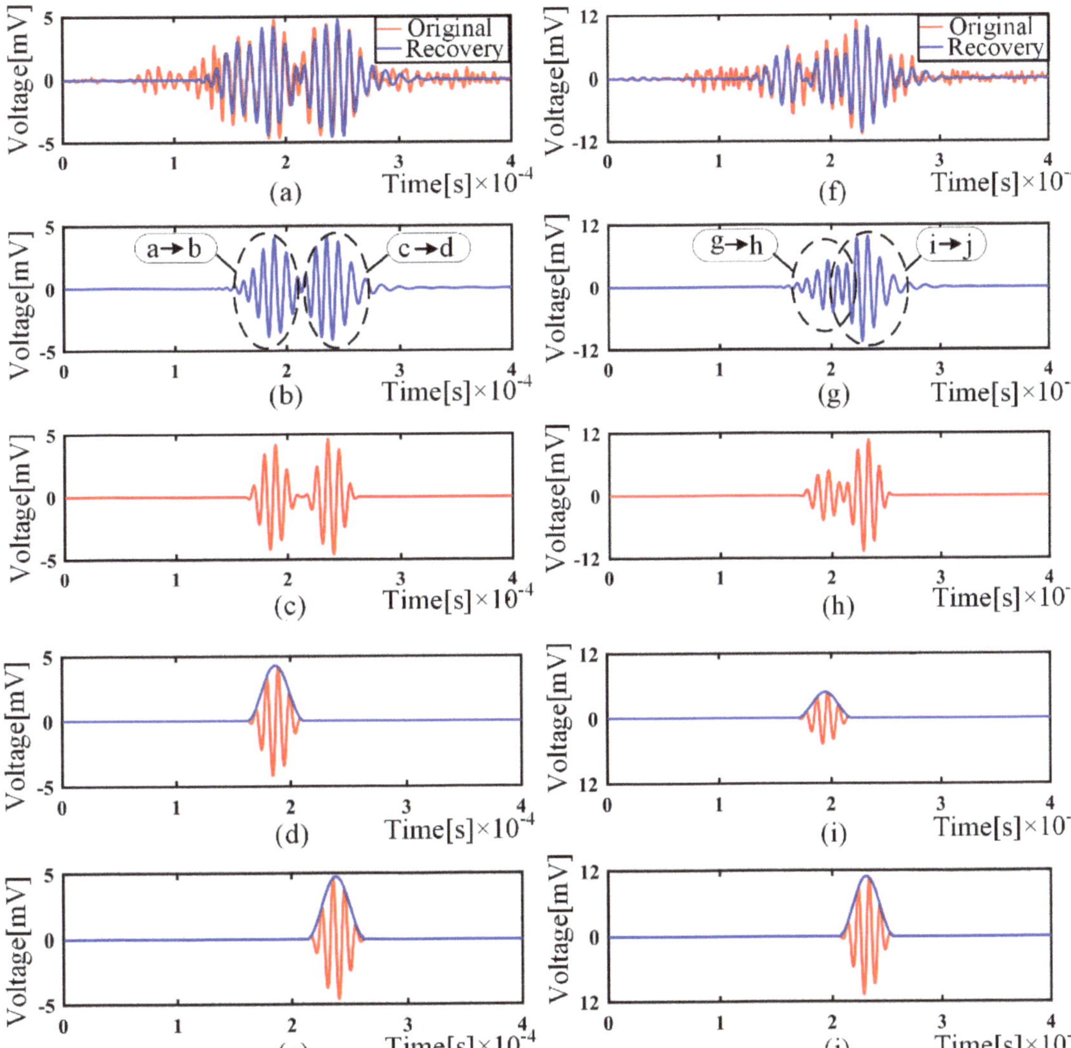

Figure 13. Residual signal processing: (**a**) Original residual signal and reconstructed signal; (**b**) Reconstructed signal with largest matching coefficients; (**c**) Reconstructed signal with dispersion removed; (**d**) Separated wave packet and envelope of path *a–b*; (**e**) Separated wave packet and envelope of path *c–d*; (**f**) Original residual signal and reconstructed signal; (**g**) Reconstructed signal with largest matching coefficients; (**h**)Reconstructed signal with dispersion removed; (**i**) Separated wave packet and envelope of *g–h*; (**j**) Separated wave packet and envelope of *i–j*;.

We decompose with the overcomplete dictionary and extract with the largest two matching coefficients. Only two scattered wave packets from two paths are obtained, as shown in Figure 13b. Therefore, wave packets on multiple paths are precisely separated, and the effect of measurement noise is significantly reduced. Then, we remove the wave dispersion by the non-dispersion dictionary, as shown in Figure 13c, which makes the energy more concentrated for damage imaging. To determine the case of the third propagation path, the signal is excited by PZT B and received by PZT A. The signal is processed using the same process and the results, as shown in Figure 13f–j. Meaningfully, Figure 13h indicates that the proposed decomposition approach can separate the overlapped wave packets, due to similar travel distance or the occurrence of dispersion. To improve the resolution of defect localization, the enveloped signals of the reconstructed waveform, as shown in Figure 13d,e,i,j, are taken to generate the virtual wavefield for each exciter–sensor pair. Each wavefield characterizes the possible trajectory of the defect as an ellipse, and the energies of multiple wavefields are superimposed at the actual defect location [26,27], as shown in Figure 14.

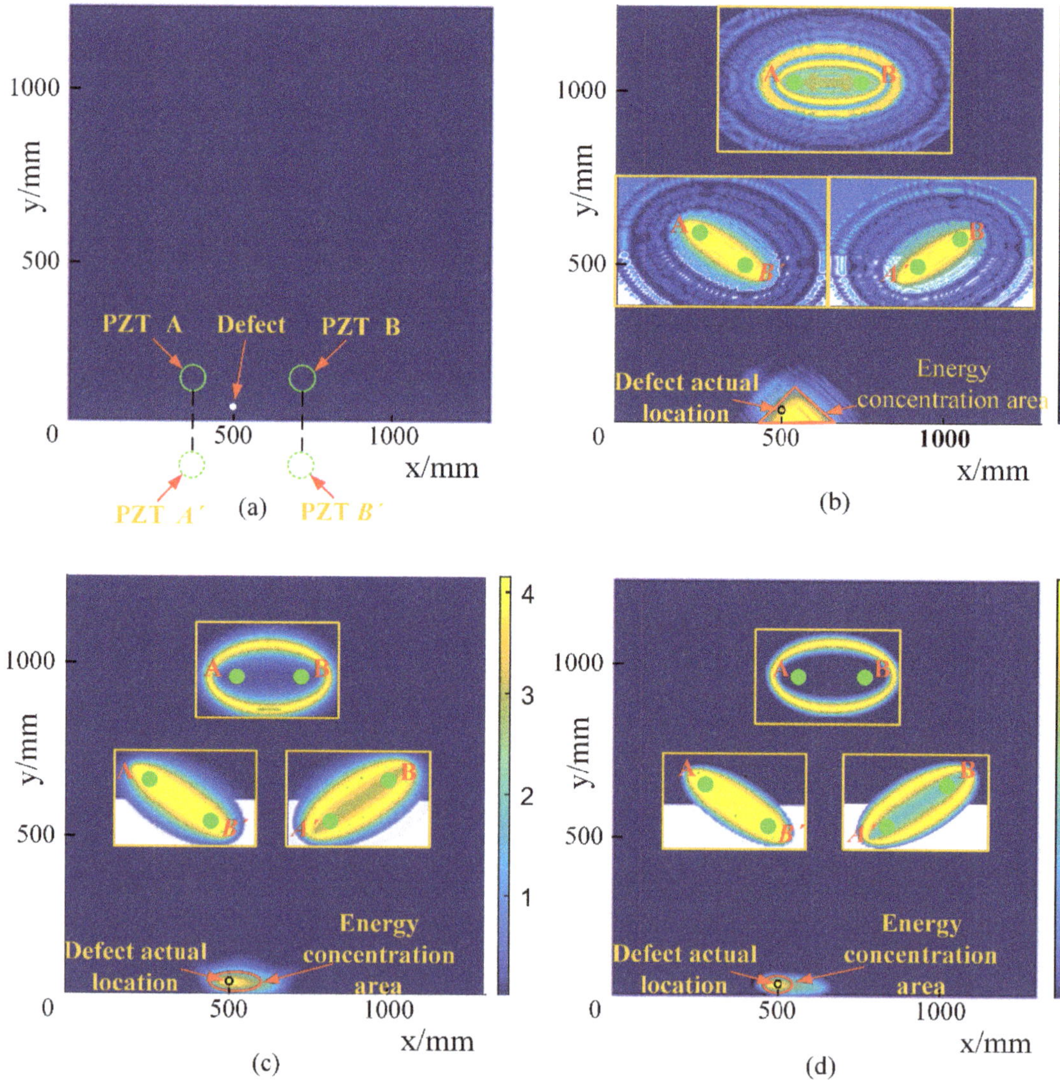

Figure 14. Damage imaging: (**a**) The location schematic of PZT; (**b**) The imaging result with original signal; (**c**) The imaging result with dispersion signal; (**d**) The imaging result with non-dispersion signal.

Figure 14a shows the actual location of the PZT wafer A and B, and the mirror wafer of PZT A and B. With the state-relative signal, the imaging result, as shown in Figure 14b, presents a large energy concentration area, which is as about 15 times the size of the actual defect area. With the reconstructed signals, a relatively clear image can be obtained by filtering out unwanted noise signals, as shown in Figure 14c. With the final non-dispersion signal, a much higher-resolution image, as shown in Figure 14d, succeeds in localizing the defect with a similar energy-concentrated area. Therefore, with the OMP-based sparse reconstruction, the signals from the sparse sensor array can be separated.

5. Conclusions

To detect damage with as few sensors and as high resolution as possible, a method, using only two piezoelectric transducers and based on orthogonal matching pursuit (OMP) decomposition, was proposed. With the proposed algorithm, the wave packets reflected from the defect and edge are all separated. The experimental results show that a sparse sensor array with only two transducers succeeds in localizing the defect, and the OMP-based algorithm is beneficial for resolution improvement and transducer usage reduction. Some conclusions can be drawn as follows.

1. with the over-completed dictionaries of A0 and S0 mode, the OMP-based algorithm can separate wave packets from collected signals, even if the wave packets are overlapped. Thereafter, with the nondispersive dictionaries, the dispersion part is removed, which transforms the deformed wave packets to the original excitation signal.
2. the wave packets reflected by the defect and edge are innovatively used for defect localization, which is the equivalent of mounting a virtual sensor at the mirroring position. The use of these multipath wave packets is beneficial for reducing the use of transducers.
3. the dispersion-removed wave packets of multipath can localize the defect position and improve the resolution of defect localization.

A little further discussion is required. The method in this article is mainly used to locate a defect. When there are two defects, the energy is concentrated at two or more potential points. However, when there are too many defects, the signals overlap and separation is difficult, and this method is no longer suitable.

The defect is located by first-order mode signals in the experiment. When other mode signals (e.g., A1, S1) are excited, it is possibly beneficial for high-resolution localization, if there are sufficient atoms of the corresponding mode in the over-completed dictionary. Although the method can separate overlapping signals, in practice, the OMP algorithm is no longer effective when the envelope peaks of the two signals completely overlap. When the travel distance of the wave packet is very short, the waveforms of A0 mode and S0 mode are similar to the excitation waveform, which may also produce matching errors. Therefore, the travel distance of the wave packet is suggested to be longer than 150 mm, when the A0 mode signal has been significantly scattered.

Author Contributions: The first and second author contributed equally to this paper. W.M. and Y.G. prepared the original draft. Y.G. carried out the simulation experiments. G.L. provided review and editing help. All authors have read and agreed to the published version of the manuscript.

Funding: This research was funded by the National Natural Science Foundation of China, grant number 52171283, Key R&D Projects of Shandong Province (No. 2019GHY112083), Natural Science Foundation of Shandong Province (No. ZR2020ME268), and Natural Science Foundation of Jiangsu Province (No. BK20201189). The authors are grateful to these foundation committees for their support.

Institutional Review Board Statement: Not applicable.

Informed Consent Statement: Not applicable.

Data Availability Statement: The supplementary data and simulation programs involved in this paper will be uploaded by the first author on the website of https://www.researchgate.net/profile/Weilei_Mu.

Acknowledgments: The valuable comments from the anonymous reviewers are highly appreciated.

Conflicts of Interest: The authors declare no conflict of interest with respect to the research, authorship, and publication of this article.

References

1. Zhongqing, S.; Lin, Y.; Ye, L. Guided lamb waves for identification of damage in composite structures: A review. *J. Sound Vib.* **2006**, *295*, 753–780.
2. Rose, J.L. A baseline and vision of ultrasonic guided wave inspection potential. *J. Press. Ves. Technol.* **2002**, *124*, 273–282. [CrossRef]
3. Wang, B.; Lin, X. Study of scattering characteristics of Lamb wave-defect interaction. *Intern. Combust. Engine Parts* **2020**, *18*, 3–4.
4. Caibin, X.; Jishuo, W.; Shenxin, Y. A focusing MUSIC algorithm for baseline-free Lamb wave damage localization. *Mech. Syst. Signal. Process.* **2021**, *164*, 108242.
5. Wang, C.H.; Rose, J.T.; Change, F. A synthetic time-reversal imaging method for structural health monitoring. *Smart Mater. Struct.* **2004**, *13*, 415–423. [CrossRef]
6. Hall, J.S.; Michaels, J.E. Multipath ultrasonic guided wave imaging in complex structures. *Struct. Health Monit.* **2015**, *14*, 345–358. [CrossRef]
7. Croxford, A.J.; Wilcox, P.D.; Drinkwater, B.W. Guided wave SHM with adistributed sensor network. *Health Monit. Struct. Biol. Syst.* **2008**, *6935*, 69350E.
8. Perelli, A.; de Marchi, L.; Flamigni, L. Best basis compressive sensing of guided waves in structural health monitoring. *Digit. Signal Process.* **2015**, *42*, 35–42. [CrossRef]
9. Ebrahimkhanlou, A.; Dubuc, B.; Salamone, S. Damage localization in metallic plate structures using edge-reflected lamb waves. *Smart Mater. Struct.* **2016**, *25*, 085035. [CrossRef]
10. Supreet, A.K.; Joseph, M.; Harley, J.B. Baseline-free guided wave damage detection with surrogate data and dictionary learning. *J. Acoust. Soc. Am.* **2018**, *143*, 3807–3818.
11. Golato, A.; Ahmad, F.; Santhanam, S.; Amin, M.G. Defect classification in sparsity-based structural health monitoring. In Proceedings of the Conference on Compressive Sensing VI—From Diverse Modalities to Big Data Analytics, Anaheim, CA, USA, 13–14 April 2017.
12. Xu, C.B.; Yang, Z.B.; Chen, X.F.; Tian, S.H.; Xie, Y. A guided wave dispersion compensation method based on compressed sensing. *Mech. Syst. Signal. Process.* **2018**, *103*, 89–104. [CrossRef]
13. Xu, C.; Yang, Z.; Qiao, B.; Chen, X. Sparse estimation of propagation distances in Lamb wave inspection. *Meas. Sci. Technol.* **2019**, *30*, 055601. [CrossRef]
14. Xu, C.; Yang, Z.; Qiao, B.; Chen, X. A parameter estimation based sparse representation approach for mode separation and dispersion compensation of Lamb waves in isotropic plate. *Smart Mater. Struct.* **2020**, *29*, 035020. [CrossRef]
15. Mallat, S.; Zhang, Z. Matching pursuit with time-frequency dictionaries. *IEEE Trans. Signal. Process.* **1993**, *41*, 3397–3415. [CrossRef]
16. Hao, L.; Zhongke, Y.; Jianying, W. Convergence research of orthogonal matching pursuit algorithm. *Microcomput. Inf.* **2008**, *3*, 209–210.
17. Hua, J.; Gao, F.; Zeng, L.; Lin, J. Modified sparse reconstruction imaging of lamb waves for damage quantitative evaluation. *NDT E Int.* **2019**, *107*, 102143. [CrossRef]
18. Xu, C.; Yang, Z.; Deng, M. Weighted Structured Sparse Reconstruction-Based Lamb Wave Imaging Exploiting Multipath Edge Reflections in an Isotropic Plate. *Sensors* **2020**, *20*, 3502. [CrossRef]
19. Hua, J.; Gao, F.; Zeng, L.; Lin, J. Sparse reconstruction imaging of damage for Lamb wave simultaneous excitation system in composite laminates. *Measurement* **2018**, *136*, 102143. [CrossRef]
20. Pati, Y.; Rezaiifar, R.; Krishnaprasad, P. Orthogonal matching pursuit: Recursive function approximation with applications to wavelet decomposition. In Proceedings of the IEEE Asilomar Conference on Signals, Systems, and Computers, Los Alamitos, CA, USA, 1–3 November 1993; Volume 1, pp. 40–44.
21. Mu, W.; Sun, J.; Liu, G.; Wang, S. High-resolution crack localization approach based on diffraction wave. *Sensors* **2019**, *19*, 1951. [CrossRef]
22. Viktorov, I. *Rayleigh and Lamb Waves (Physical Theory and Applications)*; PlenumPress: New York, NY, USA, 1967; pp. 123–135.
23. Harley, J.B.; Moura, J.M.F. Sparse recovery of the multimodal and dispersive characteristics of Lamb waves. *J. Acoust. Soc. Am.* **2013**, *133*, 2732–2745. [CrossRef]
24. Zhao, X.; Gao, H.; Zhang, G. Active health monitoring of an aircraft wing with embedded piezoelectric sensor/actuator network: I. Defect detection, localization and growth monitoring. *Smart Mater. Struct.* **2007**, *16*, 1208–1217. [CrossRef]
25. Kim, H.; Yuan, F. Enhanced damage imaging of a metallic plate using matching pursuit algorithm with multiple wavepaths. *Ultrasonics* **2018**, *89*, 84–101. [CrossRef] [PubMed]

26. Michaels, J.E.; Michaels, T.E. Detection of structural damage from the local temporal coherence of diffuse ultrasonic signals. *IEEE Trans. Ultrason. Ferroelectr. Freq. Control.* **2005**, *52*, 1769–1782. [CrossRef] [PubMed]
27. Mu, W.; Sun, J.; Xin, R.; Liu, G.; Wang, S. Time reversal damage localization method of ocean platform based on particle swarm optimization algorithm. *Mar. Struct.* **2020**, *69*, 102672. [CrossRef]

sensors

MDPI

Article

An Intelligence Method for Recognizing Multiple Defects in Rail

Fei Deng *, Shu-Qing Li, Xi-Ran Zhang, Lin Zhao, Ji-Bing Huang and Cheng Zhou

School of Electrical and Electronic Engineering, Shang Hai Institute of Technology, Shanghai 201418, China; 196131105@mail.sit.edu.cn (S.-Q.L.); 206102111@mail.sit.edu.cn (X.-R.Z.); 206101139@mail.sit.edu.cn (L.Z.); 196102104@mail.sit.edu.cn (J.-B.H.); 18656017843@163.com (C.Z.)

* Correspondence: dengfei@sit.edu.cn

Abstract: Ultrasonic guided waves are sensitive to many different types of defects and have been studied for defect recognition in rail. However, most fault recognition algorithms need to extract features from the time domain, frequency domain, or time-frequency domain based on experience or professional knowledge. This paper proposes a new method for identifying many different types of rail defects. The segment principal components analysis (S-PCA) is developed to extract characteristics from signals collected by sensors located at different positions. Then, the Support Vector Machine (SVM) model is used to identify different defects depending on the features extracted. Combining simulations and experiments of the rails with different kinds of defects are established to verify the effectiveness of the proposed defect identification techniques, such as crack, corrosion, and transverse crack under the shelling. There are nine channels of the excitation-reception to acquire guided wave detection signals. The results show that the defect classification accuracy rates are 96.29% and 96.15% for combining multiple signals, such as the method of single-point excitation and multi-point reception, or the method of multi-point excitation and reception at a single point.

Keywords: ultrasonic guided waves; principal component analysis; defect recognition; nondestructive testing; multi-signal combination

Citation: Deng, F.; Li, S.-Q.; Zhang, X.-R.; Zhao, L.; Huang, J.-B.; Zhou, C. An Intelligence Method for Recognizing Multiple Defects in Rail. *Sensors* **2021**, *21*, 8108. https://doi.org/10.3390/s21238108

Academic Editor: Zenghua Liu

Received: 28 October 2021
Accepted: 29 November 2021
Published: 3 December 2021

1. Introduction

As an infrastructure, the structural health of the rails has attracted much attention in the fields of engineering and NDT. Because of the influence of the manufacturing process, the operating situation, and the geographic conditions, rails are prone to various defects. Based on the analysis of the operating situation, rolling contact is the main reason for the rail surface crack. In the manufacturing process, the inclusion in the railhead can lead to an area-shaped section within the rail, which will lead to the formation of transverse cracks under the shelling. Moreover, the natural status, such as air pollution, natural rainfall, and temperature change, provides favorable conditions for the generation of corrosion. In the case of in-service rail, the expansion speed of defects will increase as the size of the defect increases, and the expansion speed is different for different types of defects [1–3]. Therefore, further research determining the type and size of rail defects is necessary to ensure proper and effective maintenance and replacement. Attracted by multi-mode and low attenuation, ultrasonic guided waves can perform nondestructive testing of multiple types of defects in long-range rails [4–8]. For instance, the vertical vibration mode is sensitive to cracks at the bottom of rail [9]; the SH mode can detect a transverse defect in rail [8]; the flexural mode can measure axial stress to monitor rail breakage [10]; Evans et al. [11] used ultrasonic guided waves to detect defects in rail level crossings. Lee et al. [12] presented a hybrid analytical-FEM technique based on the dispersion characteristics of the guided wave to design the sensor which can excite specific modes and frequencies for identifying transverse cracking under the shelling. Xing et al. [13] constructed a mathematical model composed of a modal vibration factor and a modal orthogonal factor to select the guided wave modal with the highest sensitivity to detect rail cracks. According to the variation regularity of dispersion characteristics along the longitudinal direction, a three-dimensional dispersion surface with cross-section position information was proposed by Chen et al. [14] for damage detection in turnout straight switch rail. Based on the Lamb wave, Deng et al. [15] proposed the method of automatically searching defects to achieve defect location in the plate structure.

Influenced by the mode of vibration and wavelength, ultrasonic guided waves display different sensitivity to different rail defects. These will induce changes in signal features [16–18]. Various signal processing techniques are used to extract signal features from the time, frequency, or time-frequency domain in the field of NDT. Power Spectral Density (PSD) is used to retrieve frequency domain information. The short-term Fourier transformation (STFT) can reflect the characteristics of the time-frequency domain. Researchers extracted various features, including maximum, peak power frequency (PPF), median power frequency (MPF), and STFT coefficients from the original signals, and used an SVM classifier to realize 4 kinds of corrosion defects recognition in the rail foot [19]. Jiang et al. [20] used laser ultrasound to detect defects of different depths on the rail surface. The information generated by the wave-packet transformation, including time-frequency coefficients, local entropy, and energy from the collected detection signals, are used for identifying different defects. Zhou et al. [21] proposed that extracting features by data-driven was used for recognizing defects. Furthermore, it is based on empirical analysis that six features from different domains are extracted from the detection signal. According to the experience, Li et al. [22] extracted six features from the time domain and frequency domain of the guided wave detection signal to identify rail cracks. Torkamani et al. [23] introduced an innovative time-domain damage index named Normalized Correlation Moment (NCM) for identifying laminated composites based on the guided wave. The result was shown to have significant advantages over the signal difference coefficient (SDC), including sensitivity to attenuation of the signal and decreased sensitivity to signal noise. Luca et al. [24] calculated damage indexes (DI's) to quantify the variations of the signal amplitude caused by the induced damage and chose the DI's for the Probability Ellipse (PE) method to estimate the probability of the presence of the damage. All the above methods can identify defects. However, these all require professional knowledge or experience to extract the damage index reflecting the defect from the guided wave detection signal to identify the defects.

This paper proposes a new method for identifying rail defects to get away from the limitation of professional knowledge and experience. It can distinguish the defects by combining multiple signals and extracting the features by the S-PCA. The process of identifying rail weaknesses is shown in Figure 1. First, guided wave detection signals are obtained through experiments and simulations of several single-excitation and single-reception methods, as shown in the area enclosed by the round dash line. Then, the S-PCA is used to extract features from the obtained signals, as shown in the area enclosed by the round dotted line. Finally, the SVM classification model enables one to qualitatively and quantitatively identify rail defects based on the extracted features. In signal processing, the restriction of professional knowledge is avoided by dividing the guided wave detection signal and extracting features from each segment. Compared with PCA, S-PCA can dig more weak damage information from guided wave detection signals, which is more conducive to defect identification.

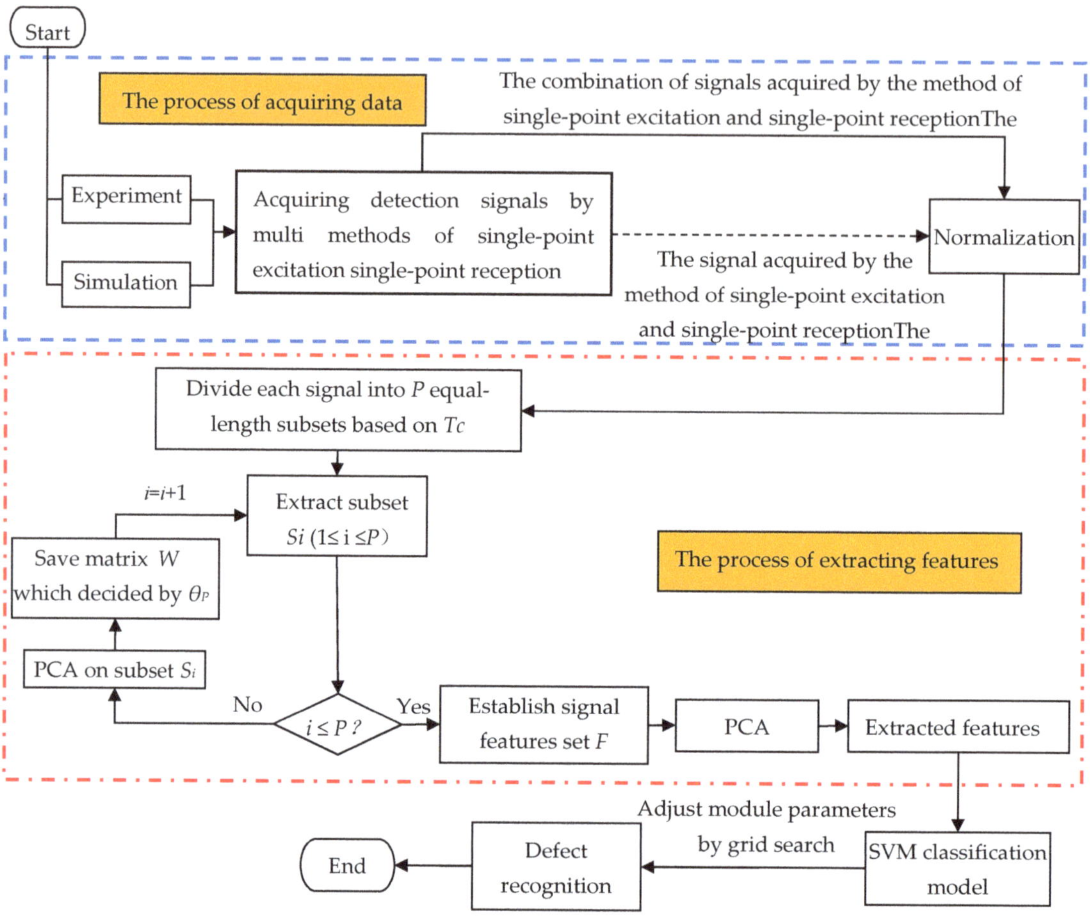

Figure 1. Flow chart of the segmented principal components analysis (S-PCA) for rail defect identification. The enclosed area with the dashed line represents the process of acquiring data; the enclosed area with the dotted line represents the process of extracting features by S-PCA, and the rest area represents the process of classification. S_i is the ith sub-signal of the origin signal. And W is the matrix of correlation coefficient obtained via PCA with the threshold θ_P.

2. Materials and Methods

2.1. Signal Dividing

The guided wave detection signal is a typical nonstationary time-series signal. The existent defects may cause modal conversion and dispersion of the guided wave signal, and the most intuitive effect is the change of the wave shape. The diverse defect information may be retained in the local zone of the guided wave detection signal. Therefore, we consider that extracting features from each local area realizes the feature extraction of the detection signal. The dividing is an effective means that the feature extraction of the high-dimensional signal is converted into the feature extraction of P low-dimensional signals.

$$\begin{cases} Tc = \frac{L}{V_p} = \frac{K}{f} \\ P = floor(\frac{T}{T_c}) \end{cases} \quad (1)$$

L is a parament in signal segmentation and is determined by both L_{min} and N. The special expression is $L = N \times L_{min}$, in which L_{min} is the smallest size of the defect detected by the guided wave selected. N is a constant that needs to be determined experimentally. The choice of N or L will be discussed in the following section. Tc, as shown in Equation (1), is equal to the time interval between two adjacent dividing points. V_P is the guided wave group velocity. Figure 2 is a schematic diagram of dividing signals. The guided wave detection signal of crack defect is divided by Tc (12.5 μs) or L (5 mm).

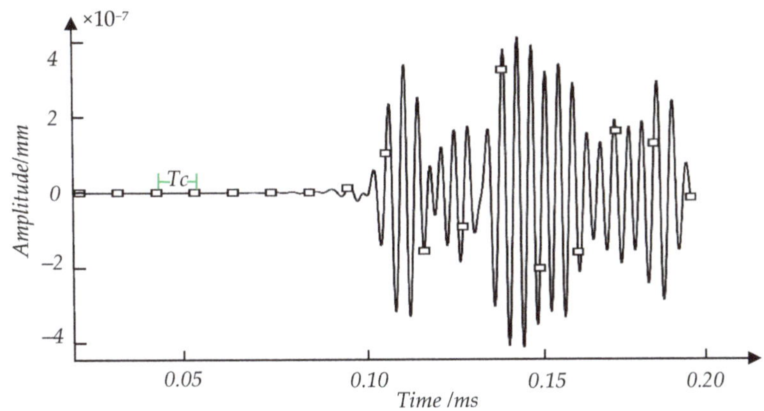

Figure 2. Diagram of the signal divided by Tc (12.5 μs) or L (5 mm), in which the black boxes represent the dividing points. Tc is the time interval between two adjacent dividing points. The vertical axis is the signal amplitude. The horizontal axis is the sampling time of the signal.

2.2. Principal Component Analysis

PCA is a common method to reduce data dimensions. Its theme is to use low-dimensional data to reflect the valuable information contained in high-dimensional data based on the mapping. Gottumukka et al. [25] used a modular PCA method to improve face recognition technology. Senneville B et al. [26] applied PCA to the motion estimation of abdominal organs. Mazzeo et al. [27] combined wavelet transformed with PCA for the preprocessing of bolts image.

The role of PCA for rail defect recognition is manifested in two aspects: removing redundant information and extracting features. The specific implementation steps are as follows:

- Construct a sample set X of the rail damage characteristics. Matrix A is composed of guided wave detection signals of different defect types or the same type of defects with different levels of damage. As shown in Equation (2), m is the number of detection signals, and n is the number of features in each sample; a_m represents the mth detected signal sample with n data, and a_{m1} represents the first data in the mth sample.

$$A = \begin{bmatrix} a_1 \\ a_2 \\ \vdots \\ a_m \end{bmatrix}_{m\times 1} = \begin{bmatrix} a_{11} & a_{12} & \cdots & a_{1n} \\ a_{21} & a_{22} & \cdots & a_{2n} \\ \vdots & \vdots & \ddots & \vdots \\ a_{m1} & a_{m2} & \cdots & a_{\mathrm{mn}} \end{bmatrix}_{m\times n} \tag{2}$$

The set of samples X of the rail damage characteristic is obtained after centralized processing A. The centralization is expressed in Equation (3), which $\overline{\mathrm{A}}$ represents the average value of each column in A.

$$\mathrm{X} = \mathrm{A} - \overline{\mathrm{A}} \tag{3}$$

- Construct the damage covariance matrix C. The correlation coefficient in guided wave detection signals can be used to characterize different defects. A covariance matrix is an efficient tool for characterizing the correlation coefficient. The covariance matrix C is described by Equation (4).

$$C = \frac{1}{m-1} X^T X \tag{4}$$

- Calculate the eigenvalues and eigenvectors of the damage covariance matrix C. According to the matrix decomposition method, the eigenvalues and corresponding eigenvectors of matrix C are solved. Then the eigenvalues are ranked in descending order, $\lambda_1 \geq \lambda_2 \geq \ldots \geq \lambda_n$, and the corresponding eigenvectors in sequential order, $x_1\ x_2 \ldots\ x_n$. Among them, each eigenvector represents a principal component.
- Determine the number of principal components K. The information percentage of a principal component is an important reference when the number of principal components is determined. The ratio of one eigenvalue to the sum of all eigenvalues is the information percentage of this principal component, as shown in Equation (5).

$$\theta_i = \frac{\lambda_i}{\sum_{i=1}^{n} \lambda_i} \tag{5}$$

Set a threshold θ ($0 < \theta < 1$) and accumulate the information percentages of the principal components sorted in sequence. When the cumulative sum of the information percentages of the Kth principal component is greater than or equal to θ, there is selected for the K principal components.

$$\theta \geq \frac{\sum_{i=1}^{K} \lambda_i}{\sum_{i=1}^{n} \lambda_i} \tag{6}$$

- Extract features. The K principal components extracted from the covariance matrix C are gathered to form a local weight matrix W. The process of feature extraction is shown as Equation (7), where X' is a feature set extracted.

$$X' = XW \tag{7}$$

2.3. Build S-PCA

Figure 3 is a schematic diagram of the S-PCA. There are two steps based on the S-PCA to extract features. The first step is to divide. According to the time interval Tc, the detection signal S is divided into P segments of equal length. S_i represents the ith signal segment ($1 \leq i \leq P$). The second step is feature extraction in two sub-steps. The first sub-step is extracting features from each segment by PCA with a threshold θ_P ($0 < \theta_P < 1$) to form a feature set named F. The second sub-step is to get the final features from F by PCA with a threshold θ ($0 < \theta < 1$).

2.4. SVM Classification Model

SVM realizes classification by searching for the optimal hyperplane determined by a certain number of support vectors. It is often used for solving linearly separable problems. For nonlinear separable problems, the introduction of the kernel function maps the sample data to a high-dimensional space, making it a linearly separable problem, which is then solved by linear classification. The Gaussian radial basis function (RBF) has excellent characteristics of nonlinearity and continuity, so it is often used as the kernel function of SVM. Equation (8) is the expression of RBF, in which g is the kernel size of the RBF, and X and X_i are feature vectors with the same dimension.

$$\mathbf{Kernel}(X, X_i) = \mathbf{exp}\left(\frac{\|\mathbf{X} - \mathbf{X}_i\|^2}{2g^2}\right) \tag{8}$$

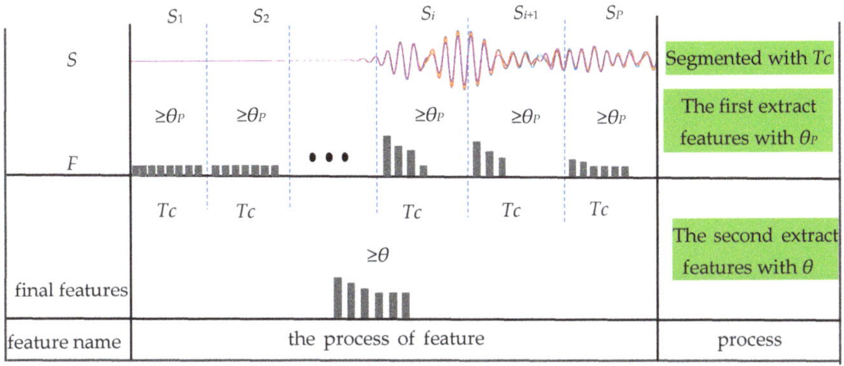

Figure 3. Diagram of segmented principal component analysis of ultrasonic guided waves. S is the original signal; F is a feature set that combines features from each segment. S_i represents the ith subsection; Tc is the time interval used for dividing S; θ_P is a threshold used for extracting features from each segment; θ is a threshold used for extracting features from F. The grey chip represents a feature.

2.5. Classification Model Evaluation

Evaluating the performance of the model provides a basis for the credibility of the model classification results. These are called respectively the precision rate and the recall rate, which are usually used for evaluating the model performance. The precision rate reflects the ability of the model to detect positive samples from the sample set, its expression is shown in Equation (9). The recall rate reflects the ability of the model to detect the number of correctly classified positive samples from all positive samples and is expressed as Equation (10). This article also introduces the F1-score to balance the precision rate and the recall rate to make the model evaluation more accurate, as shown in Equation (11). All the equations above, True Positive (TP) is the number of samples whose labels and the predictions are positive samples. False Positive (FP) is the number of samples where the object labels are negative and the predicted labels inverted. True Negative (TN) is the number of samples whose object labels are positive and the predicted labels inverse. False Negative (FN) is the number of samples whose labels of the object and predicted are negative samples.

$$P = \frac{TP}{TP + FP} \tag{9}$$

$$R = \frac{TP}{TP + FN} \tag{10}$$

$$F1 = \frac{2 \times P \times R}{P + R} \tag{11}$$

3. Experimental Setup

There are 7 different degrees of crack, 5 different degrees of transverse crack under the shelling, and 4 different degrees of corrosion within the railhead. The detect signals of the above defects are acquired by combining the experiments and the simulations.

3.1. Experimental Setup

The experimental system is shown in Figure 4. The experimental object is a 60 kg/m rail with a length of 250 mm. Ceramic piezoelectric sheets (PZT-5H, diameter 14 mm, thickness 1 mm, center frequency 200 kHz) are arranged symmetrically on the railhead and marked in sequence as 1, 2, 3, A, B, C. The distance between the center of the piezoelectric films and the rail end is 7 mm. The vertical distance between the rail tread and the center of the piezoelectric films 1, 3, A, or C is 23.57 mm. The piezoelectric plates at positions 1, 2, and 3 are used to excite guided wave signals, and the piezoelectric plates at positions A, B, and

C are used to receive detection signals. Any exciting sensors are paired with any receiving sensors to form a signal acquisition channel. There are 9 channels for acquiring signals, like 1TA, 1TB, 1TC, 2TA, 2TB, 2TC, 3TA, 3TB, 3TC. Figure 5 is a connection diagram of the detection equipment. The arbitrary function generator (TektronixAFG3021B) can modulate a sinusoid exciting signal of 200 kHz and 5 circles loaded alternately on piezoelectric slices 1, 2, or 3. The guided wave signal will propagate in the rail. The detection signals are collected using piezoelectric slices A, B, or C. Both the excitation signal and the reception signal are input into the oscilloscope (Tektronix DPO4054) with a specific sampling frequency.

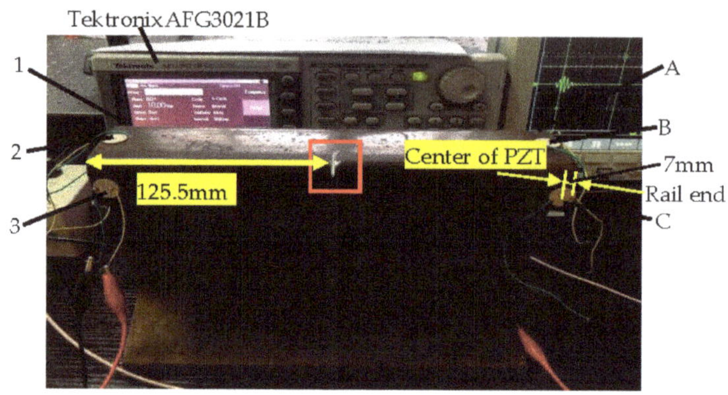

Figure 4. The experimental system for using ultrasonic guided waves detection, in which the position of the red box represents the location of the crack. The PZTs are located at the position of 1, 2, 3, A, B, C, respectively. The distance between the center of PZT and the rail end is 7 mm.

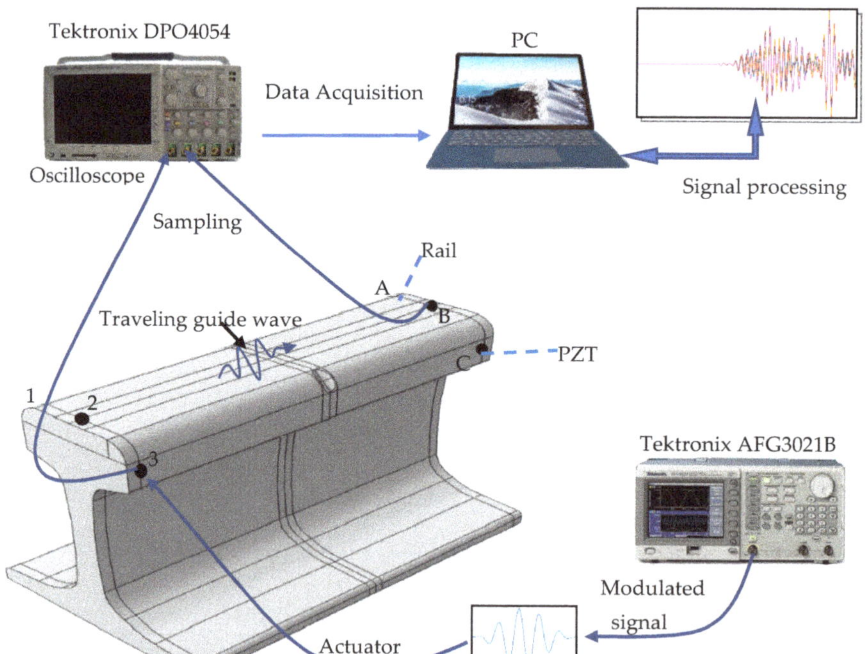

Figure 5. The connection diagram of the detection equipment, in which the Tektronix AFG3021B is the arbitrary function generator and the oscilloscope is used to sample the signals. The PZTs are coupled via ultrasonic gel with the rail surface.

As shown in Figure 4, the location of the crack is indicated by the red square. The crack is artificially cut in the fillet radius on one side of the railhead, located 125.5 mm from the rail end. The crack depth gradually increases at an interval of 1 mm to reflect the varying degrees of crack damage. In this study, 8 kinds of crack defects with different damage levels, 0 mm, 1 mm, 2 mm, 3 mm, 4 mm, 5 mm, 6 mm, 7 mm, are considered, and the corresponding sample labels are 1, 2, 3, 4, 5, 6, 7, 8, respectively.

In the experiment, uncertain physical factors are inevitable, such as the tightness of coupling between the piezoelectric sheet and the surface of the rail. Hence the experiments are repeated 10 times under each excitation-reception mode for each degree of crack to reduce the effect of tightness. The detection signal is picked up at a sampling frequency of 100 MHz, and 10 samples are taken accordingly for each crack degree. 80 samples may be obtained in each signal acquisition mode, and a total of 720 may be obtained from 9 signal acquisition channels. Table 1 shows the types of cracks and the number of samples taken through the experiments of 1TA.

Table 1. The types of cracks and the quantities of samples taken from the experiments of exciting at position 1 and receiving the signal at position A.

Damage Degree	Number of Samples	Damage Degree	Number of Samples
intact	10	4 mm	10
1 mm	10	5 mm	10
2 mm	10	6 mm	10
3 mm	10	7 mm	10

3.2. Numerical Simulation

The finite element simulation software ABAQUS is used to build a model consistent with the experimental object. Table 2 shows the material parameters of the rail model. ρ represents density, E represents the Elastic Modules, and v represents Poisson's ratio. The rail model is divided by grid cells C3D8, and the grid size is 1.5 mm. The arrangement of the piezoelectric sheets on the railhead is consistent with the experiment, as shown in Figure 6. The total time to analyze the model is 2 ms, and the time step is 0.01 µs.

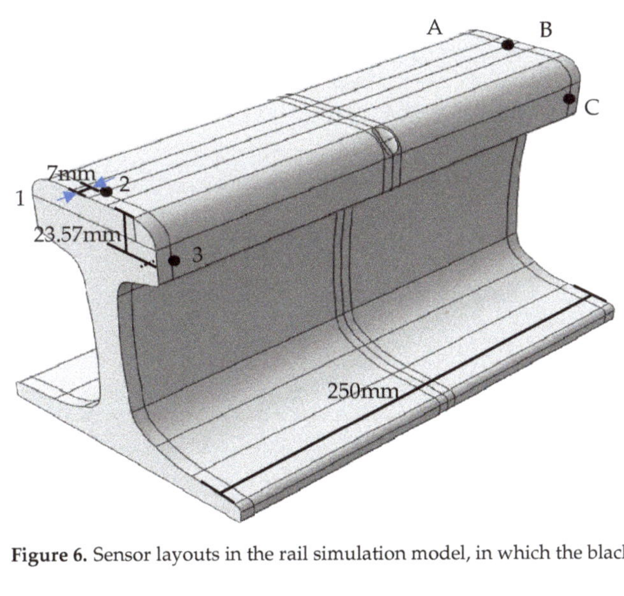

Figure 6. Sensor layouts in the rail simulation model, in which the black dots represent the PZTs.

Table 2. Configuring the rail model built by *ABAQUS*.

Model	Length	ρ	E	v
60 kg/m	250 mm	7.85×10^{-9}	2×10^5	0.29

Figure 7a is the model of rail defects made by ABAQUS, such as crack, corrosion. The position of the black box corresponds to the location of the failure. The method of fine-tuning the defect parameters is adopted to make the numerical simulation and the experimental effect more consistent. Moreover, more detect signals are collected. For crack, the depth of the defect is continuously adjusted with a deviation of 0.001 mm. Taking a 1 mm sample as an example, the crack depth of 0.999 mm, 0.998 mm, 1.001 mm, and 1.002 mm, etc., and these defects are classed as 1 mm cracks. According to the above method, 10 samples are expanded under each excitation-reception mode, there are a total of 80 samples expanded for 8 damage categories. In practice, the depth of crack defect is always irregular. The author uses rounding to make the mark of defects in integers. For instance, the depth of crack less than 1.5 mm carries a 1 mm tag, and the depth of crack which equals 1.5 mm or more carries a 2 mm tag. If the difference between the marked value and the defect depth is within 0.2 mm, the defect depth is indicated with the marked value. For example, the crack depth of 1.2 mm, 1.1 mm, 0.9 mm, and 0.8 mm are all classified as 1 mm cracks. The type of sample expansion is only in the crack depth of 1 mm and 7 mm in this study, and the total number of expanded samples is 58, as shown in Table 3. With respect to corrosion, four defect models of different damage levels, 6 mm^2, 12 mm^2, 25 mm^2 and 30 mm^2, are constructed by ABAQUS. And the corresponding samples are marked as 9, 10, 11, and 12, respectively. For each type of corrosion defect, the corrosion size is modified by 0.004 mm^2 to produce more samples. 80 samples are obtained per the excitation-reception method. The transverse crack under shelling is also an area defect. One rectangle is used to approximate the shape of the internal transverse crack. Figure 7b is a rail model with a 6 mm $\times$ 7 mm transverse crack under the shelling defect constructed by ABAQUS. 6 mm is the defect's width, and 7 mm is the length of the defect. The defect is located at a rail length of 125.5 mm and a depth of 12.5 mm from the rail tread. Five transverse cracks under shelling models with different damage levels, 2 mm $\times$ 3 mm, 3 mm $\times$ 4 mm, 4 mm $\times$ 5 mm, 5 mm $\times$ 6 mm, 6 mm $\times$ 7 mm, are constructed through numerical simulation. Moreover, sample labels corresponding to defect sizes are marked as 13, 14, 15, 16, 17. Under each excitation-reception channel, 21 samples are collected under each type of transverse crack under shelling by adjusting the length or width with the deviation of 0.001 mm.

Table 3. The type of crack and the quantity of samples collected in a simulation of exciting at position 1 and receiving the signal at position A.

Defection Type	Defection Category (mm)	Total Number of Samples	Actual Damage (mm)	Number of Samples
intact	intact	10	0	10
crack	1	30	1.0 1.1 1.2	10 10 10
	2	30	1.8 1.9 2.0	10 10 10
	3 4 5 6	10 10 10 10	3.0 4.0 5 6	10 10 10 10
	7	28	6.9 7 7.1 7.2	6 10 6 6

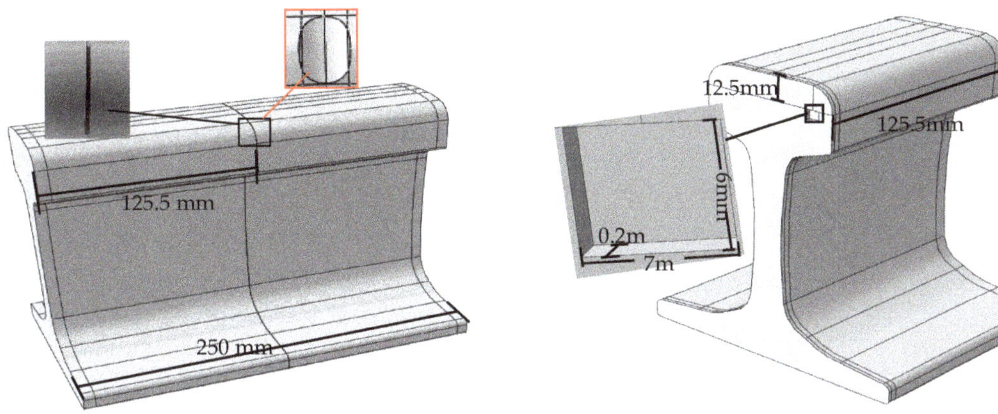

Figure 7. Rail defect model diagram. (**a**) describes the position of the crack and corrosion defect model where the black sequence is the position of the defect, such as corrosion and crack; (**b**) describes the position of the transverse crack under the shelling model.

Table 4 shows that the total number of samples obtained by experiments and simulations under the channel of 1TA, and the sample labels corresponding to defects. Subsequently, all samples obtained by 1TA are made into a dataset including samples and the sample labels. Moreover, the dataset is divided into two parts, which randomly take 80% of the dataset as the training set, the remaining 20% as the testing set. The classify model is trained and tested by training set and testing set, respectively. The data obtained from the remaining 8 signal acquisition methods are processed by the same method with 1TA for subsequent rail defect identification.

Table 4. Sample type and the total number of samples collected by exciting at the position of 1 and receiving the signal at position A.

Defection Type	Sample Label	Defection Category	Number of Samples	Total Number of Samples
intact	1	-	20	403
Crack defect	2	1 mm	40	
	3	2 mm	40	
	4	3 mm	20	
	5	4 mm	20	
	6	5 mm	20	
	7	6 mm	20	
	8	7 mm	38	
Corrosion defect	9	6 mm^2	20	
	10	12 mm^2	20	
	11	25 mm^2	20	
	12	30 mm^2	20	
Internal nuclear defect	13	2 mm × 3 mm	21	
	14	3 mm × 4 mm	21	
	15	4 mm × 5 mm	21	
	16	5 mm × 6 mm	21	
	17	6 mm × 7 mm	21	

4. Feature Extraction

4.1. Signal Analysis

Figure 8 shows the waveform diagram of the crack detection signal obtained at the time range of 0–0.002 s by 2TA in the experiment, and the cracks' depth is 2 mm, 4 mm, 6 mm, 7 mm, respectively. Based on the guided wave theory, the guided wave group velocity is about 3845 m/s in the experiment. And the speed in the simulation is 3850 m/s, approximately like the result of the experiment. In addition, the experimental signal waveform is like the simulated signal waveform, under different degrees of defect damage. Thus, the simulated signals are used for the expansion of defect samples.

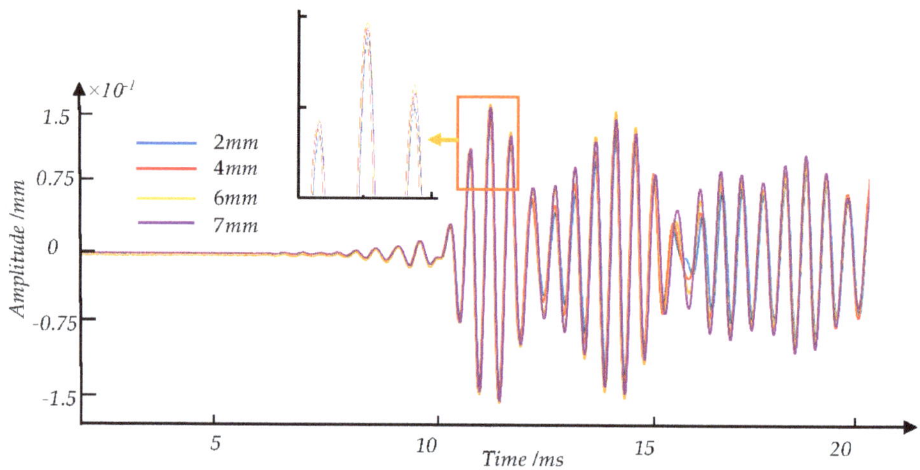

Figure 8. Comparison of experimental signals waveforms of four types of different crack defects with different degrees of damage.

Figures 8–10 show the signal waveforms of crack defects, transverse cracks under the shelling, and corrosion defects at different damage levels, respectively. We find that the waveforms of detection signals caused by different damage degrees of defects at the same position are generally similar. According to the elastic wave theory, the corresponding wave packet will cause amplitude changes and overlap due to some factors such as defects. However, the existing method is difficult to distinguish the specific defects by the time domain waveform. In this paper, the amplitude corresponding to each time point on the detection signal is regarded as a feature value in each sample. The statistical analysis and machine learning algorithms are used to find the features between the sample data of defects and correspondence defect types and eliminate redundant features.

4.2. Feature Extraction

Feature extraction is a significant part of structural health monitoring (SHM). Redundant features are removed from the detection signal through feature extraction, and several important features are retained to achieve accurate and rapid identification of defects.

Tc is an important parameter to achieve feature extraction by S-PCA. And the selection of *Tc* depends on parament *L* and the group velocity V_P of the guided wave signal. In this study, the guided wave group velocity V_P is 3850 m/s. The *L* participating in the discussion is shown in Table 5. K_P is the number of data points in each segment. The influence of different *Tc* on defect identification is stated in Section 6.

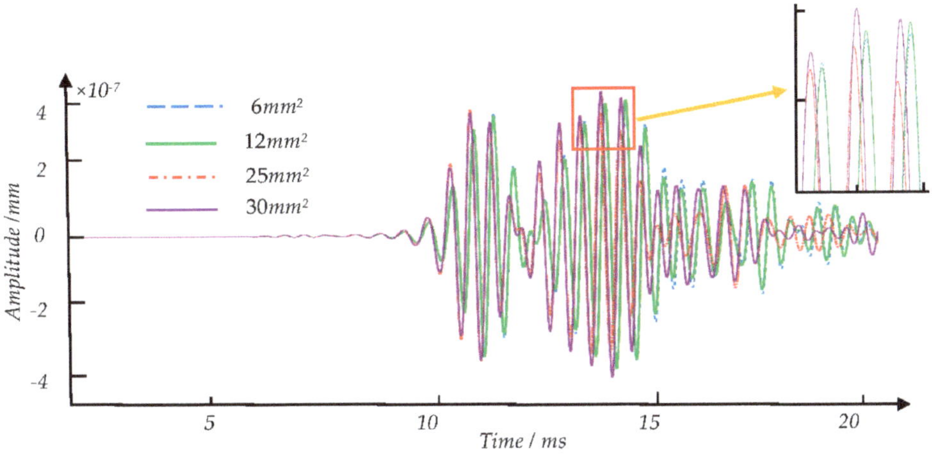

Figure 9. Comparison of the detection signal waveforms of four different corrosion defects with different degrees of damage.

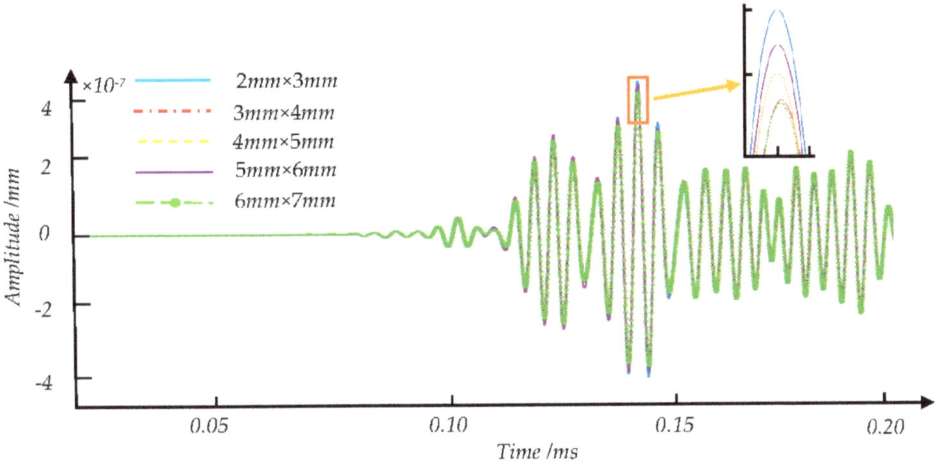

Figure 10. Comparison of five transverse cracks under shelling detection signal waveforms with different levels of damage.

Table 5. Selection of the parameter *L* and time interval *Tc*.

N	L(mm)	*Tc* (μs)	K_p	P
2	0.2	0.5	50	400
10	1.0	2.5	250	80
20	2.0	5.0	500	40
30	3.0	7.5	750	27
40	4.0	10.0	1000	20
50	5.0	12.5	1250	16

PCA is a means of globally reducing data dimensions. Selecting an adequate number of principal components is critical when the PCA algorithm filters the feature. In the process of extracting features from detection signals with S-PCA, θ_P is a threshold to extract principal components from sub-signal segments using PCA. θ is a threshold for extracting the principal components from feature set *F*. The influences of these two parameters on rail defect identification are outlined in Section 6, respectively.

5. Model Parameter Adjustment

The number of support vectors in the SVM classification model is an essential factor influencing the model performance [28]. The number of support vectors for nonlinear SVM is modified by adjusting the penalty coefficient C and the kernel radius g. There are always optimal C and g values for different datasets to achieve accurate classification of the samples. It is considered that the method of grid search automatically finds suitable C and g within the range of [−3, 20], and [−25, 0], for nine datasets.

6. Experiment Results and Discuss

6.1. Influence of *Tc* on the Defect Recognition

The guided wave detection signals are divided into P sub-signal segments with a time interval Tc. To select a suitable Tc, we assume that both θ_P and θ are equal to 99%. When Tc is equal to 0.5 μs, 2.5 μs, 5 μs, 7.5 μs, 10 μs, and 12.5 μs, the accuracy of SVM varies in shown in Table 6.

Table 6. The influence of *Tc* on the accuracy of defect classifications, in which the red number is the best-classified result in this dataset.

	0.5 μs	**2.5 μs**	**5.0 μs**	**7.5 μs**	**10.0 μs**	**12.5 μs**
1TA	86.21	86.21	87.36	86.21	86.21	86.21
1TB	94.25	95.40	95.40	90.80	91.95	90.80
1TC	77.01	77.01	79.31	78.16	79.31	78.16
2TA	94.25	95.40	95.40	95.40	95.40	93.10
2TB	90.80	90.80	93.10	91.95	89.66	88.51
2TC	88.51	88.51	88.51	89.66	89.66	88.51
3TA	72.41	74.71	75.86	73.56	72.41	72.41
3TB	89.66	89.66	90.80	89.66	89.66	88.51
3TC	90.81	91.95	91.95	93.10	90.81	91.95

In Table 6, the red number represents the maximum classification accuracy attained by the classifier based on the selected parameters in each dataset. When Tc is 5 μs, seven datasets, such as 1TA, 1TB, 1TC, 2TA, 2TB, 3TA, and 3TB, reach the maximum classification accuracy rate. When Tc is equal to 7.5 μs, both 2TC and 3TC reach the maximum classification accuracy rate. But when Tc is 5 μs, the classification results of 2TC and 3TC are not much different from the optimal result maintaining a relatively high classification level. According to the content above, most of the nine datasets may achieve the best classification effect when Tc is equal to 5 μs. So 5 μs is selected as the optimal parament of Tc for subsequent analysis.

6.2. Influence of θ_P on The Rail Defects Identification

θ_P controls the number of principal components extracted from each sub-signal. Different θ_P can extract different numbers of principal components from the sub-signal segments and affects the classification results of the classification model. It is the initial step that Tc is equal to 5 μs and θ is equal to 99%. Table 7 shows the variation of the classification results in the nine datasets with θ_P selected in turn from 10%, 20%, 30%, 40%, 50%, 60%, 70%, 80%, 90%, and 99%.

Table 7. The effect of θ_P on classification accuracy, in which the red number is the best classification result in this dataset.

	10%	**20%**	**30%**	**40%**	**50%**	**60%**	**70%**	**80%**	**90%**	**99%**
1TA	89.66	89.66	89.66	89.66	90.80	91.95	90.80	89.66	89.66	88.51
1TB	88.51	88.51	88.51	88.51	94.25	94.25	89.66	90.80	88.51	87.36
1TC	79.31	79.31	79.31	79.31	79.31	77.01	70.11	78.16	75.86	78.16
2TA	95.40	95.40	95.40	95.40	96.55	93.10	94.25	93.10	94.25	90.80
2TB	94.25	94.25	94.25	94.25	96.55	96.55	95.40	95.40	95.40	94.25

Table 7. *Cont.*

	10%	20%	30%	40%	50%	60%	70%	80%	90%	99%
2TC	90.80	90.80	90.80	90.80	90.80	88.51	88.51	88.51	88.51	83.91
3TA	72.41	72.41	72.41	72.41	73.56	68.97	68.97	67.82	68.97	67.82
3TB	81.61	81.61	81.61	81.61	89.66	94.25	91.95	87.36	83.91	87.36
3TC	86.21	86.21	86.21	86.21	90.80	89.66	87.36	88.51	89.66	86.21

As shown in Table 7, when θ_P equals 50%, any one of 1TB, 1TC, 2TA, 2TB, 2TC, 3TA, and 3TC reaches the maximum classification accuracy. For 1TB, when θ_P is equal to 60%, the classification accuracy is the same as the optimal classification result. However, when θ_P equals 50%, the number of the features used to classify defects is 5. When θ_P equals 60%, the number is 7. Thus 50% is taken at 1TB to reduce the loss of computation. When θ_P equals 60%, 1TA and 3TB reach the best classification outcomes at 91.95% and 94.25%, respectively. For 1TC and 2TC, when θ_P equals 10%, they attain the maximum classification accuracy. Moreover, when θ_P is equal to 20%, 30%, 40%, and 50%, in turn, the classification accuracy keeps the same classification result equaled to the optimal result. However, when θ_P equals 10%, 20%, 30%, and 40%, in turn, the 13th category cannot be identified in 3TB, as shown in Table 8. The reason is that the value of θ_P is too low to extract enough principal components. When θ_P equals 50%, all categories can be recognized, and when θ_P equals 60%, the classification accuracy rate is 94.25% that is the best-classified outcome for 3TB. According to the analysis above, most of the nine datasets can maintain a relatively high classification accuracy when θ_P equals 50%. Thus 50% is selected as the optimal parament of θ_P for subsequent research.

Table 8. 3TB's confusion matrix when θ_P equals 10%, Tc equals 5 μs, and θ equals 99%.

	1	2	3	4	5	6	7	8	9	10	11	12	13	14	15	16	17
precision	1.00	0.75	0.75	1.00	1.00	1.00	0.75	0.90	0.75	1.00	0.75	1.00	0	1.00	0.80	0.60	1.00
recall	1.00	0.67	0.75	1.00	1.00	1.00	1.00	0.90	0.75	0.80	1.00	1.00	0	1.00	0.80	0.50	0.71
F1-score	1.00	0.71	0.75	1.00	1.00	1.00	0.86	0.90	0.75	0.89	0.86	1.00	-	1.00	0.80	0.55	0.83

6.3. Influence of θ on the Rail Defect Identification

θ is an important parameter to extract features from F which are final features. Table 9 shows that when Tc equals 5 µs and θ_P equals 50%, the classification results of the nine datasets vary with θ changed from 10%, 20%, 30%, 40%, 50%, 60%, 70%, 80%, 90%, and 99%. According to Table 9, the classification accuracy rates of the nine datasets show an increasing trend as the θ value increases. Therefore, 99% is chosen as a fixed parameter of θ for follow-up research.

Table 9. The influence of θ on classification accuracy in which the red circle is the best classification result in this dataset.

	10%	20%	30%	40%	50%	60%	70%	80%	90%	99%
1TA	49.43	49.43	49.43	49.43	74.71	74.71	74.71	77.01	80.46	90.80
1TB	42.53	42.53	42.53	63.22	63.22	63.22	73.56	73.56	85.06	94.25
1TC	50.57	50.57	50.57	66.67	66.67	66.67	66.67	66.67	71.26	78.16
2TA	54.02	54.02	54.02	75.86	75.86	75.86	83.91	89.66	89.66	96.55
2TB	58.62	58.62	58.62	58.62	82.76	82.76	90.80	90.80	93.10	95.40
2TC	48.28	48.28	48.28	48.28	73.56	73.56	73.56	83.91	89.66	88.51
3TA	42.53	42.53	42.53	42.53	55.17	55.17	55.17	59.77	64.37	73.56
3TB	58.62	58.62	58.62	58.62	81.61	81.61	81.61	85.06	87.36	94.25
3TC	50.57	50.57	50.57	70.11	70.11	74.71	74.71	80.46	78.16	90.80

In summary, 5 µs, 50%, and 99% are selected, respectively, as the optimal parameters, which are the fixed values of Tc, θ_P, θ.

6.4. Classification Result of Single-Channel Acquisition Signal

The signals collected by different signal acquisition channels are turned into corresponding datasets. Once each dataset is normalized, the S-PCA model with selected paraments is utilized to extract features. The dataset is split into the training set and the testing set. The outcome of the classification is the average of the 10 results of the classification. The variance *S* represents the standard deviation of 10 classification results, which is used for evaluating the stability of the classification. And the results of nine datasets are presented in Table 10.

Table 10. The SVM's classification results of single-channel acquisition signal.

	Accuracy	S	Precision	Recall	F1-Score
1TA	90.99%	0.03	91.20%	90.13%	89.83%
1TB	90.22%	0.02	91.95%	90.07%	90.19%
1TC	76.15%	0.04	82.01%	73.76%	74.43%
2TA	93.63%	0.02	94.12%	93.54%	93.32%
2TB	92.75%	0.02	94.68%	92.36%	92.05%
2TC	90.11%	0.03	90.79%	89.26%	89.06%
3TA	76.53%	0.03	83.10%	79.71%	76.89%
3TB	90.77%	0.02	91.80%	90.26%	90.29%
3TC	88.02%	0.03	89.90%	87.88%	87.43%

Table 10 shows that among the 9 signal acquisition methods, the accuracy rates of the SVM classification model exceed 76% that is a medium score. In 1TA, 1TB, 2TA, 2TB, 2TC, and 3TB, the SVM classification accuracy rates reach nearly or more than 90%, which is a good score. Both 2TA and 2TB have excellent classification because the sensor at position 2 can be smoothly attached to the track surface, making the defect information in the collected detection signal more obvious. For 3TB, the distance between the sensor and the defect is relatively close, which makes the difference of the defect signals more prominent, which is conducive to the identification of the defect. Both 1TA and 1TB have higher classification accuracy because the signal reflected by the defect can be received by the A and B sensors, which is conducive to the recognition defect. In addition, the S-PCA is an effective method of extracting features and can extract the local differences in the detection signal, which is useful for the classifier to make the correct distinction. Similarly, the distance between the exciting sensor and the receiving sensor is relatively long, so the classification results of 1TC and 3TA are 76.15% and 76.53%, respectively, lower than the signals collected by other signal acquisition methods.

6.5. Classification Result of Multi-Channel Signal Combination

The above research shows that a single channel collecting signal can identify different types and degrees of defects damage on the railhead. Based on theoretical analysis, a single-channel acquisition signal can only reflect part of the characteristics of the defect. Thus, 1TC and 3TA classification accuracy rates are lower. The two types of combination methods are comprehensively analyzed to explore the influence of multi-channel signals on rail defect recognition, such as multi-point excitation and single-point reception or single-point excitation and multi-point reception. The types of combinations and the corresponding classification results are indicated in Tables 11 and 12.

Table 11 shows the results of the classification of the combined methods of single-point excitation and multi-points reception signals. With the number of combined signals increasing, it is found that the classification accuracy rate shows an increasing trend, such as 1TAC, 1TBC, and 1TABC; 2TAB, 2TAC, 2TBC, and 2TABC; 3TAB, 3TAC, 3TBC, and 3TABC. Moreover, compared to the classification of signals collected by one single channel, the standard deviation *S* of the classification is slightly reduced after multiple signal combinations. The above results show that the accuracy of the classification can be effectively enhanced by the combination method. Since the propagation path of each signal is different, the effective combination can more widely reflect the health status of the detected object. The same results can also be found in Table 12.

Table 11. The SVM's classification results of feature combinations method of single-point excitation and multi-point reception.

Combination Type	Accuracy	S	Precision	Recall	F1-Score
1TAB	96.15%	0.02	96.29%	95.59%	95.51%
1TAC	93.08%	0.01	93.67%	92.37%	92.08%
1TBC	91.21%	0.02	92.56%	90.295	90.50%
1TABC	94.73%	0.02	95.32%	94.00%	94.20%
2TAB	93.74%	0.02	95.42%	93.51%	93.20%
2TAC	94.07%	0.01	94.99%	93.66%	93.58%
2TBC	94.95%	0.02	96.34%	95.10%	94.87%
2TABC	96.15%	0.01	96.63%	96.13%	95.93%
3TAB	89.01%	0.03	90.25%	88.36%	88.51%
3TAC	88.79%	0.02	90.57%	88.49%	87.77%
3TBC	88.90%	0.02	90.62%	88.38%	88.34%
3TABC	90.55%	0.03	91.64%	90.31%	89.99%

Table 12. The SVM's classification results of the combination method of multi-point excitation and single-point reception.

Combination Type	Accuracy	S	Precision	Recall	F1-Score
1TA + 2TA	96.29%	0.02	96.34%	96.06%	95.70%
1TA + 3TA	89.78%	0.01	90.39%	87.82%	87.58%
2TA + 3TA	92.75%	0.02	93.23%	92.72%	92.43%
1TA + 2TA + 3TA	95.49%	0.02	96.12%	95.20%	95.03%
1TB + 2TB	95.38%	0.01	96.32%	95.35%	94.94%
1TB + 3TB	93.08%	0.02	94.54%	92.35%	92.59%
2TB + 3TB	94.03%	0.02	95.31%	94.71%	94.10%
1TB + 2TB + 3TB	93.41%	0.02	95.23%	93.79%	93.45%
1TC + 2TC	92.09%	0.02	93.02%	92.05%	92.00%
1TC + 3TC	90.22%	0.03	91.97%	89.74%	89.06%
2TC + 3TC	91.43%	0.03	93.35%	90.98%	90.83%
1TC + 2TC + 3TC	92.09%	0.03	93.44%	92.22%	91.95%

7. Conclusions

Aiming to identify multiple defects in rail, experiments and numerical simulations are used to focus on the ultrasonic guided wave detection signals of crack defect, transverse crack under shelling, and corrosion defect. By modifying and combining the sensor positions, a sample library of defect detection signals is obtained by 9 different signal excitation-reception methods. The S-PCA algorithm is proposed to extract the features of the signal to eliminate the dependency on professional knowledge. Furthermore, the extracting features are input into the SVM classifier to identify the type and extent of defects. At the end of the research, the following conclusions are drawn:

- The method of extracting features from the segments of detection signal by PCA can effectively eliminate redundant information in the signal and retain adequate information, which improves the accuracy of quantitative and qualitative identification of rail defects.
- The detection signals collected from different excitation-reception positions describe the overall health of the different parts of the rail. Obtaining the combined detection signal through single-point excitation and multi-point reception or multi-point excitation and single-point reception can more comprehensively describe the health status of the detection object, which is good to improve the accuracy of defect recognition.

- The S-PCA algorithm is an efficient method of extracting features based on statistical theory. It does not rely too much on the professional knowledge of guided wave detection, which reduces the difficulty of rail defect identification. Furthermore, the method could be more easily implemented in practical engineering in the future.

Author Contributions: Conceptualization, F.D. and S.-Q.L.; methodology, F.D. and S.-Q.L.; software, S.-Q.L.; validation, X.-R.Z., L.Z. and J.-B.H.; formal analysis, F.D., S.-Q.L., X.-R.Z., L.Z. and J.-B.H.; investigation, S.-Q.L.; resources, S.-Q.L. and C.Z.; data curation, S.-Q.L.; writing—original draft preparation, S.-Q.L.; writing—review and editing, X.-R.Z., L.Z. and J.-B.H.; visualization, L.Z.; supervision, J.-B.H.; project administration, F.D.; funding acquisition, F.D. All authors have read and agreed to the published version of the manuscript.

Funding: This research was funded by the National Natural Science Foundation of China (11202137), Shanghai Alliance Program (2019025), Collaborative innovation fund of the Shanghai Institute of Technology (XTCX2018-11).

Institutional Review Board Statement: Not applicable.

Informed Consent Statement: Not applicable.

Data Availability Statement: Not applicable.

Acknowledgments: The authors would like to express profound gratitude to Beijing University of Technology, Department of Mechanical and Electrical Engineering.

Conflicts of Interest: The authors declare no conflict of interest.

References

1. Tian, G.Y.; Gao, B.; Gao, Y.L.; Wang, P.; Wang, H.T.; Shi, Y.S. Summary of inspection and monitoring technology for railway rail defects and damage. *Chin. J. Sci. Instrum.* **2016**, *37*, 1763–1780.
2. Hernandez, F.C.; Plascencia, G.; Koch, K. Rail base corrosion problem for North American transit systems. *Eng. Fail. Anal.* **2009**, *16*, 281–294. [CrossRef]
3. Zerbst, U.; Lundén, R.; Edel, K.O.; Smith, R.A. Introduction to the damage tolerance behaviour of railway rails—A review. *Eng. Fract. Mech.* **2009**, *76*, 2563–2601. [CrossRef]
4. Karahaliou, A. Evaluation of Railway Rails with Non-Destructive Techniques. *Key Eng. Mater.* **2014**, *605*, 641–644. [CrossRef]
5. Zhang, H.; Song, Y.N.; Wang, Y.N.; Liang, Z.C.; Zhao, M. Summary of non-destructive testing and evaluation technology for rail defects. *Chin. J. Sci. Instrum.* **2019**, *40*, 11–25.
6. He, C.F.; Zheng, M.F.; Lv, Y.; Deng, P.; Zhao, H.M.; Liu, X.C.; Song, G.R.; Liu, Z.H.; Jiao, J.P.; Wu, B. Development, Application and Challenge of ultrasonic guided wave detection technology. *Chin. J. Sci. Instrum.* **2016**, *37*, 1713–1735.
7. He, C.F.; Wu, B.; Fan, J.W. Research progress and application of ultrasonic cylindrical guided wave technology. *Adv. Mech.* **2001**, *2*, 203–214.
8. Rose, J.L.; Avioli, M.J.; Mudge, P.; Sanderson, R. Guided wave inspection potential of defects in rail. *NDT E Int.* **2003**, *37*, 153–161. [CrossRef]
9. Lu, C.; Sheng, H.J.; Song, K.; Lin, J.M.; He, F.C. Scattering characteristics of ultrasonic guided waves of oblique cracks at the bottom of rails. *Non-Destr. Test.* **2016**, *38*, 18–22, 25.
10. Loveday, P.W. Guided Wave Inspection and Monitoring of Railway Track. *J. Nondestruct. Eval.* **2012**, *31*, 303–309. [CrossRef]
11. Evans, M.; Lucas, A.; Ingram, I. The inspection of level crossing rails using guided waves. *Constr. Build. Mater.* **2018**, *179*, 614–618. [CrossRef]
12. Lee, C.; Joseph, L.; Younho, C. A guided wave approach to defect detection under shelling in rail. *NDT E Int.* **2009**, *42*, 174–180. [CrossRef]
13. Xing, B.; Yu, Z.; Xu, X.; Zhu, L.; Shi, H. Mode Selection Model for Rail Crack Detection Based on Ultrasonic Guided Waves. *Shock Vib.* **2020**, *2020*, 19. [CrossRef]
14. Chen, R.; Hu, C.; Xu, J.; Gong, Z.; Liu, L.; Wang, P.; Chen, X. Research on guided wave propagation characteristics in turnout rails with variable cross-section. *J. Sound Vib.* **2021**, *494*, 115853. [CrossRef]
15. Deng, F.; Chen, H.L. A Defects localization Algorithm Based on the Lamb Wave of Plate Structure. *J. Nondestruct. Eval. Diagn. Progn. Eng. Syst.* **2021**, *4*, 021001. [CrossRef]
16. Liu, X.; Liu, Z.Q. Research on the inverse problem of ultrasonic nondestructive testing. *Non-Destr. Test.* **2001**, *10*, 441–443.
17. Wu, B.; Deng, F.; He, C.F. The research progress of signal processing in ultrasonic guided wave nondestructive testing. *J. Beijing Univ. Technol.* **2007**, *4*, 342–348.
18. Zhou, J.M.; Xu, Q.Y.; Li, P.; Wan, Q.; Liao, X.S. Ultrasonic guided wave technology in nondestructive testing of rails. *Instrum. Technol. Sens.* **2015**, *6*, 99–102, 106.

19. Moustakidis, S.; Kappatos, V.; Karlsson, P.; Selcuk, C.; Gan, T.H.; Hrissagis, K. An Intelligent Methodology for Railways Monitoring Using Ultrasonic Guided Waves. *J. Nondestruct. Eval.* **2014**, *33*, 694–710. [CrossRef]
20. Jiang, Y.; Wang, H.; Tian, G.Y.; Yi, Q.J.; Zhao, J.Y.; Zhen, K. Fast classification for rail defect depths using a hybrid intelligent method. *Optik* **2019**, *180*, 455–468. [CrossRef]
21. Zhou, C.; Deng, F.; Liu, Y.; Liu, X.C.; Cheng, H.L. Research on Identification of Corrosion Damage Degree of Guided Wave Bend Pipe Based on Neural Network and Support Vector Machine. *J. Mech. Eng.* **2021**, *57*, 136–144.
22. Li, S.Q.; Deng, F.; Zhou, C.; Zhang, X.R. Identification of rail crack defects based on support vector machine and artificial neural network. In Proceedings of the 15th Symposium on Piezoelectricity, Acoustic Waves and Device Applications, SPAWDA 2020, Zhengzhou, China, 16–19 April 2021.
23. Torkamani, S.; Roy, S.; Barkey, M.; Sazonov, E.; Burkett, S.; Kotru, S. A novel damage index for damage identification using guided waves with application in laminated composites. *Smart Mater. Struct.* **2014**, *23*, 095015. [CrossRef]
24. Luca, D.; Fenza, A.; Petrone, G.; Caputo, F. Guided wave SHM system for damage detection in complex composite. *Struct. Theor. Appl. Fract. Mech.* **2020**, *105*, 102408. [CrossRef]
25. Gottumukkal, R.; Asari, V.K. An improved face recognition technique based on modular PCA approach. *Pattern Recognit. Lett.* **2004**, *25*, 429–436. [CrossRef]
26. Senneville, D.B.; El Hamidi, A.; Moonen, C. A direct PCA-based approach for real-time description of physiological organ deformations. *IEEE Trans. Med. Imaging* **2015**, *34*, 974–982. [CrossRef]
27. Mazzeo, P.L.; Nitti, M.; Stella, E.; Distante, A. Visual recognition of fastening bolts for railroad maintenance. *Pattern Recognit. Lett.* **2004**, *25*, 669–677. [CrossRef]
28. Wang, Z.Y.; Zhu, D.H. An accurate detection method for surface defects of complex components based on support vector machine and spreading algorithm. *Measurement* **2019**, *147*, 106886. [CrossRef]

sensors

MDPI

Article

Nonlinear Ultrasound Crack Detection with Multi-Frequency Excitation—A Comparison

Frank Mevissen and Michele Meo *

Research Centre, Department of Mechanical Engineering, University of Bath, Bath BA2 7AY, UK; f.mevissen@bath.ac.uk
* Correspondence: m.meo@bath.ac.uk

Abstract: Nonlinear ultrasound crack detection methods are used as modern, non-destructive testing tools for inspecting early damages in various materials. Nonlinear ultrasonic wave modulation, where typically two or more frequencies are excited, was demonstrated to be a robust method for failure indicators when using measured harmonics and modulated response frequencies. The aim of this study is to address the capability of multi-frequency wave excitation, where more than two excitation frequencies are used, for better damage identification when compared to single and double excitation frequencies without the calculation of dispersion curves. The excitation frequencies were chosen in such a way that harmonic and modulated response frequencies meet at a specific frequency to amplify signal energy. A new concept of nonlinearity parameter grouping with multi-frequency excitation was developed as an early failure parameter. An analytical solution of the one-dimensional wave equation was derived with four fundamental frequencies, and a total of 64 individual and 30 group nonlinearity parameters. Experimental validation of the approach was conducted on metal plates with different types of cracks and on turbine blades where cracks originated under service conditions. The results showed that the use of multi-frequency excitation offers advantages in detecting cracks.

Keywords: nonlinear ultrasound; modulation; crack detection; gas turbines

Citation: Mevissen, F.; Meo, M. Nonlinear Ultrasound Crack Detection with Multi-Frequency Excitation—A Comparison. *Sensors* **2021**, *21*, 5368. https://doi.org/10.3390/s21165368

Academic Editors: Zenghua Liu and Dipen N. Sinha

Received: 23 June 2021
Accepted: 4 August 2021
Published: 9 August 2021

1. Introduction

Linear ultrasound techniques are proven testing methods for inspecting components. The failure detection entails measuring changes in elastic properties such as the speed of sound, attenuation, transmission coefficient, and reflection coefficient [1–3].

In 1964, Hikata et al. expanded this research and found that a sinusoidal ultrasonic wave distorts the fundamental frequency when it propagates in the presence of nonlinearities [4]. If an ultrasonic wave propagates into a solid body with the fundamental frequency and is disturbed during the propagation, harmonic frequencies are generated. When the amplitudes of these frequencies are measured and compared with the fundamental frequencies, these values represent an accurate indicator for detecting material changes [4]. When linear ultrasonic failure detection techniques reach their limits, nonlinear ultrasonic techniques revealed higher detection rates of incipient failures in materials [4,5].

The comparison of the amplitudes is conducted using derived nonlinearity parameters. Jhang et al. studied the second harmonic parameter β to explain its dependence on the wave shift, which enabled drawing conclusions about the material to be examined [6]. This work was extended by Frouin et al. [7], Rothenfusser et al. [8], and Yost et al. [9], who showed the direct dependence of the amplitudes of the fundamental frequency as well as the frequency of the second harmonic. The analytical derivations were conducted by Jeong et al. [10], while Ostrovky et al. investigated the nonlinearity parameters for geomaterials [11]. To determine the residual fatigue life of a component, Malfense Fierro and Meo developed the nonlinearity parameters for two superimposed frequencies [12–14], which was further developed by Jinpin et al. [15]. The third-order nonlinearity parameter with one driving frequency was derived from Amura and Meo [16]. The nonlinearity parameters of the second and third harmonic frequencies are shown in Equation (1) for one excitation frequency, where A is the amplitude of the fundamental frequency, A_2 is the amplitude of the second harmonic frequency, and A_3 is the amplitude of the third harmonic frequency:

$$\begin{aligned} \beta_{2f} &\propto \frac{A_2}{A^2}, \\ \gamma_{3f} &\propto \frac{A_3}{A^3}. \end{aligned} \tag{1}$$

The excitation with two different frequencies up to third-order nonlinearity was analytically derived and investigated [17]. A total of 12 derived nonlinearity parameters were successfully used to detect cracks in metal components and turbine blades (Equation (2)). The excitation with two different frequencies was defined with $u^{(1)} = A_1 sin\left[k_{f1}(x-ct)\right] + A_2 cos\left[k_{f2}(x-ct)\right]$, where A_1 and A_2 are the amplitudes and k_{f1} and k_{f2} are the wavenumbers of the fundamental frequencies f_1 and f_2. A_{f1+f2} are the amplitudes of the harmonic or modulated response frequencies:

$$\begin{aligned} \gamma_{f1} &\propto \frac{A_{f1+f2}}{{A_1}^3 + A_1 {A_2}^2}, \\ \gamma_{f2} &\propto \frac{A_{f1+f2}}{{A_2}^3 + {A_1}^2 A_2}, \\ \beta_{2f1} &\propto \frac{A_{f1+f2}}{{A_1}^2}, \\ \beta_{2f2} &\propto \frac{A_{f1+f2}}{{A_2}^2}, \\ \gamma_{3f1} &\propto \frac{A_{f1+f2}}{{A_1}^3}, \\ \gamma_{3f2} &\propto \frac{A_{f1+f2}}{{A_2}^3}, \\ \beta_{f2\pm f1} &\propto \frac{A_{f1+f2}}{A_1 A_2}, \\ \gamma_{2f1\pm f2} &\propto \frac{A_{f1+f2}}{{A_1}^2 A_2}, \\ \gamma_{2f2\pm f1} &\propto \frac{A_{f1+f2}}{A_1 {A_2}^2}. \end{aligned} \tag{2}$$

An overview of current research developments relating to nonlinear vibroacoustic modulation techniques is shown in [18]. Van Den Abeele et al. used two superimposed frequencies for excitation and examined the harmonics of both waves and their sideband frequencies to detect cracks [19]. They also developed a technique to investigate the influence of damaged materials on the amplitude-dependent resonance frequency shifts [20]. These methods are based on the nonlinear interactions of low-frequency and high-frequency waves. Straka et al. and Greenhall et al. worked on a nonlinear elastic wave modulation spectroscopy (NEWMS), where the effect of two superimposed waves was investigated, and low-frequency and high-frequency waves were combined to detect damages in components [21,22]. Using time-reversal-based imaging methods, an application for non-destructive material testing (NDT) with two multi-frequency signals was developed [23]. A technique was presented where two signals are correlated to estimate the Time-of-Flight (ToF) for determining distance information [24]. Novak et al. developed a method to extract system reactions for the second and third harmonic frequencies with a nonlinear convolutional signal analysis [25]. Pfleiderer et al. investigated the nonlinear response of subharmonics, ultra-subharmonics, and ultra-frequency pairs for detecting defects in components [26]. Abraham et al. used nonlinear multi-frequency ultrasound measurement to characterise grain size fluctuations and distributions in metal samples [27]. Gao et al. developed a method for estimating corrosion size, position, and depth in aircraft structures based on a local multi-frequency wavenumber estimation [28]. Malfense Fierro and Meo developed a phased array technique where three frequencies were used to identify cracks. The initiate signal f_2 is sent out to affect the dynamics of the crack topology. The subsequent transmitted pump signals f_1 and f_3 generate a modulation that can be used for crack detection, and it was demonstrated that using multiple frequency excitation improves crack detection capabilities [29].

Deng investigated shear horizontal (SH) mode propagation on solid plates, where two shear waves propagate vertically and parallel in the plate and intersect. If the phase velocity of the SH mode is equal to the longitudinal velocity of the plate material, this leads to a cumulative effect of the second harmonic excitation frequency [30,31]. Pruell et al. showed that phase and group velocity matching can be used to generate acoustic nonlinearities to determine plasticity-driven fatigue damage [32,33]. Approaches were developed where correction factors were integrated into the solution of the wave equation for other wave types [34,35].

To show possibly occurring Lamb waves, Figure 1 reveals the dispersion diagram of the phase velocity for the material Inconel 718, which is examined in Section 4. Due to comparable behaviour, the group velocity was not shown in this context. The fundamental frequencies from 6 MHz to 10 MHz were used in parallel and represented by the S_1 mode according to [32]. The S_2 mode demonstrates the range of the second harmonic frequency (12 MHz to 20 MHz) and S_3 demonstrates the third harmonic frequency range (18 MHz to 30 MHz). It becomes clear that phase synchronisation with several excitation frequencies in this frequency range is not possible. The difference in phase velocities decreases only at higher frequencies, such as in the third harmonic frequency range. The integration of the modulated response frequencies would make synchronisation highly complex.

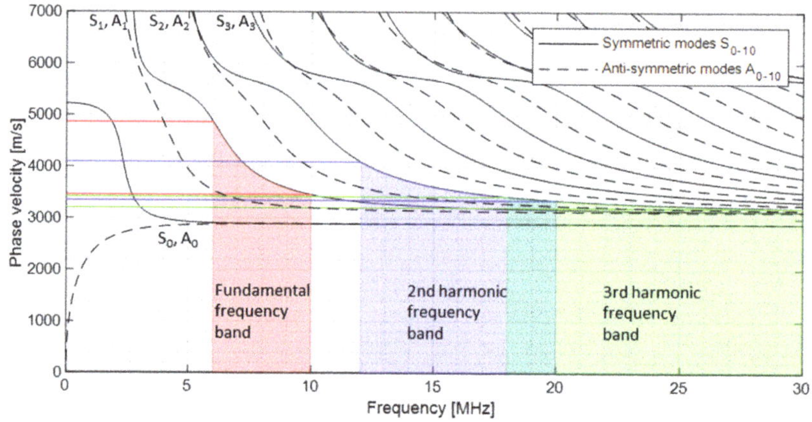

Figure 1. Dispersion curve—phase velocity.

In this paper, the strength of the quasi-chaotic generation of harmonic and modulated response frequencies was clearly demonstrated, while the dispersion curves were not considered.

Different excitation concepts were compared to detect cracks in metal plates and on turbine blades. For this purpose, excitations with up to four frequencies were examined. In the case of quadruple excitation, a new concept of nonlinearity grouping was developed analytically. The fundamental frequencies were defined in such a way that several harmonics and sideband frequencies meet at certain frequencies to improve crack detection and classification.

2. Multiple Nonlinearity Parameter

Several studies demonstrated that damages in materials could be detected with nonlinear ultrasound techniques using two superimposed fundamental frequencies. The

Table 2. Nonlinearity parameter group 1, δ_{G1}.

Group	A_x	Frequency [MHz]	$\Gamma_{f/h/s_x}$	Nonlinearity Parameter Group δ_G
1	1	2	$2f_1 - f_4$	$\delta^1_{G1} = \frac{A_{f1+f2+f3+f4}}{A_1{}^2 A_4}$
	1	4	$f_4 - f_1$	$\delta^2_{G1} = \frac{A_{f1+f2+f3+f4}}{A_4 A_1}$
	1	12	$2f_1$	$\delta^3_{G1} = \frac{A_{f1+f2+f3+f4}}{A_1{}^2}$
	1	13.3	$f_2 + f_1$	$\delta^4_{G1} = \frac{A_{f1+f2+f3+f4}}{A_2 A_1}$
	1	14	$2f_4 - f_1$	$\delta^5_{G1} = \frac{A_{f1+f2+f3+f4}}{A_4{}^2 A_1}$
	1	18	$3f_1$	$\delta^6_{G1} = \frac{A_{f1+f2+f3+f4}}{A_1{}^3}$
	1	18.7	$f_4 + f_3$	$\delta^7_{G1} = \frac{A_{f1+f2+f3+f4}}{A_4 A_3}$
	1	19.3	$2f_1 + f_2$	$\delta^8_{G1} = \frac{A_{f1+f2+f3+f4}}{A_1{}^2 A_2}$
	1	20	$2f_4$	$\delta^9_{G1} = \frac{A_{f1+f2+f3+f4}}{A_4{}^2}$
	1	28.7	$2f_4 + f_3$	$\delta^{10}_{G1} = \frac{A_{f1+f2+f3+f4}}{A_4{}^2 A_3}$
	1	30	$3f_4$	$\delta^{11}_{G1} = \frac{A_{f1+f2+f3+f4}}{A_4{}^3}$

Table 3. Nonlinearity parameter group 2, δ_{G2}.

Group	A_x	Frequency [MHz]	$\Gamma_{f/h/s_x}$	Nonlinearity Parameter Group δ_G
2	1 2	2.6	$f_3 - f_1$ $f_4 - f_2$	$\delta^1_{G2} = \frac{A_{f1+f2+f3+f4}}{A_3 A_1 + A_4 A_2}$
	1 2	3.3	$f_1 + f_2 - f_4$ $2f_1 - f_3$	$\delta^2_{G2} = \frac{A_{f1+f2+f3+f4}}{A_1 A_2 A_4 + A_1{}^2 A_3}$
	1 2	12.6	$f_3 + f_4 - f_1$ $2f_4 - f_2$	$\delta^3_{G2} = \frac{A_{f1+f2+f3+f4}}{A_3 A_4 A_1 + A_4{}^2 A_2}$
	1 2	14.7	$2f_2$ $f_3 + f_1$	$\delta^4_{G2} = \frac{A_{f1+f2+f3+f4}}{A_2{}^2 + A_3 A_1}$
	1 2	16	$f_4 + f_1$ $f_3 + f_2$	$\delta^5_{G2} = \frac{A_{f1+f2+f3+f4}}{A_4 A_1 + A_3 A_2}$
	1 2	17.3	$2f_3$ $f_4 + f_2$	$\delta^6_{G2} = \frac{A_{f1+f2+f3+f4}}{A_3{}^2 + A_4 A_2}$
	1 2	20.7	$2f_1 + f_3$ $2f_2 + f_1$	$\delta^7_{G2} = \frac{A_{f1+f2+f3+f4}}{A_1{}^2 A_3 + A_2{}^2 A_1}$
	1 2	27.3	$2f_3 + f_4$ $2f_4 + f_2$	$\delta^8_{G2} = \frac{A_{f1+f2+f3+f4}}{A_3{}^2 A_4 + A_4{}^2 A_2}$

Table 4. Nonlinearity parameter group 3, δ_{G3}.

Group	A_x	Frequency [MHz]	$\Gamma_{f/h/s_x}$	Nonlinearity Parameter Group δ_G
3	1 2 3	1.3	$f_2 - f_1$ $f_3 - f_2$ $f_4 - f_3$	$\delta^1_{G3} = \frac{A_{f1+f2+f3+f4}}{A_2 A_1 + A_3 A_2 + A_4 A_3}$
	1 2 3	6	f_1 $2f_2 - f_3$ $f_2 + f_3 - f_4$	$\delta^2_{G3} = \frac{A_{f1+f2+f3+f4}}{A_1{}^3 + A_1 A_2{}^2 + A_1 A_3{}^2 + A_1 A_4{}^2 + A_2{}^2 A_3 + A_2 A_3 A_4}$
	1 2 3	10	f_4 $2f_3 - f_2$ $f_2 + f_3 - f_1$	$\delta^3_{G3} = \frac{A_{f1+f2+f3+f4}}{A_4{}^3 + A_1{}^2 A_4 + A_2{}^2 A_4 + A_3{}^2 A_4 + A_3{}^2 A_2 + A_2 A_3 A_1}$
	1 2 3	22	$3f_2$ $2f_1 + f_4$ $f_1 + f_2 + f_3$	$\delta^4_{G3} = \frac{A_{f1+f2+f3+f4}}{A_2{}^3 + A_1{}^2 A_4 + A_1 A_2 A_3}$
	1 2 3	23.3	$2f_2 + f_3$ $2f_3 + f_1$ $f_1 + f_2 + f_4$	$\delta^5_{G3} = \frac{A_{f1+f2+f3+f4}}{A_2{}^2 A_3 + A_3{}^2 A_1 + A_1 A_2 A_4}$
	1 2 3	24.7	$2f_2 + f_4$ $2f_3 + f_2$ $f_1 + f_3 + f_4$	$\delta^6_{G3} = \frac{A_{f1+f2+f3+f4}}{A_2{}^2 A_4 + A_3{}^2 A_2 + A_1 A_3 A_4}$
	1 2 3	26	$3f_3$ $2f_4 + f_1$ $f_2 + f_3 + f_4$	$\delta^7_{G3} = \frac{A_{f1+f2+f3+f4}}{A_3{}^3 + A_4{}^2 A_1 + A_2 A_3 A_4}$

Table 5. Nonlinearity parameter group 4, δ_{G4}.

Group	A_x	Frequency [MHz]	$\Gamma_{f/h/s_x}$	Nonlinearity Parameter Group δ_G
4	1 2 3 4	4.6	$2f_2 - f_4$ $f_1 + f_2 - f_3$ $f_1 + f_3 - f_4$ $2f_1 - f_2$	$\delta^1_{G4} = \frac{A_{f1+f2+f3+f4}}{A_2{}^2 A_4 + A_1 A_2 A_3 + A_1 A_3 A_4 + A_1{}^2 A_2}$
	1 2 3 4	7.3	$2f_3 - f_4$ f_2 $f_1 + f_3 - f_2$ $f_1 + f_4 - f_3$	$\delta^2_{G4} = \frac{A_{f1+f2+f3+f4}}{A_3{}^2 A_4 + A_2{}^3 + A_1{}^2 A_2 + A_2 A_3{}^2 + A_2 A_4{}^2 + A_1 A_3 A_2 + A_1 A_4 A_3}$
	1 2 3 4	8.7	f_3 $2f_2 - f_1$ $f_2 + f_4 - f_3$ $f_1 + f_4 - f_2$	$\delta^3_{G4} = \frac{A_{f1+f2+f3+f4}}{A_3{}^3 + A_1{}^2 A_3 + A_2{}^2 A_3 + A_3 A_4{}^2 + A_2{}^2 A_1 + A_2 A_4 A_3 + A_1 A_4 A_2}$
	1 2 3 4	11.3	$2f_3 - f_1$ $f_2 + f_4 - f_1$ $f_3 + f_4 - f_2$ $2f_4 - f_3$	$\delta^4_{G4} = \frac{A_{f1+f2+f3+f4}}{A_3{}^2 A_1 + A_2 A_4 A_1 + A_3 A_4 A_2 + A_4{}^2 A_3}$

The derivation of this parameter is illustrated by Equation (8). For group 1, where only one harmonic or sideband per frequency were used, the parameters result from Equations (A2) and (A5). In group 2, two harmonics and sidebands always meet one frequency, which is described by the generalised shift of group 2 with u_{G2}. The amplitudes A_1 and A_2 reveal the summary of the solutions from Equations (A2) and (A5) which should be solved. The placeholder $\Gamma_{f/h/s}$ represents the combination of fundamental, harmonic, and sideband frequencies:

$$
\begin{aligned}
u_{G2} &= A_1 \genfrac{}{}{0pt}{}{sin}{cos} \left[\left(\Gamma_{f/h/s_1}\right)(x-ct)\right]x + A_2 \genfrac{}{}{0pt}{}{sin}{cos} \left[\left(\Gamma_{f/h/s_2}\right)(x-ct)\right]x, \\
u_{G3} &= A_1 \genfrac{}{}{0pt}{}{sin}{cos} \left[\left(\Gamma_{f/h/s_1}\right)(x-ct)\right]x \\
&+A_2 \genfrac{}{}{0pt}{}{sin}{cos} \left[\left(\Gamma_{f/h/s_2}\right)(x-ct)\right]x + A_3 \genfrac{}{}{0pt}{}{sin}{cos} \left[\left(\Gamma_{f/h/s_3}\right)(x-ct)\right]x, \\
u_{G4} &= A_1 \genfrac{}{}{0pt}{}{sin}{cos} \left[\left(\Gamma_{f/h/s_1}\right)(x-ct)\right]x + A_2 \genfrac{}{}{0pt}{}{sin}{cos} \left[\left(\Gamma_{f/h/s_2}\right)(x-ct)\right]x \\
&+A_3 \genfrac{}{}{0pt}{}{sin}{cos} \left[\left(\Gamma_{f/h/s_3}\right)(x-ct)\right]x + A_4 \genfrac{}{}{0pt}{}{sin}{cos} \left[\left(\Gamma_{f/h/s_4}\right)(x-ct)\right]x.
\end{aligned}
\tag{8}
$$

This procedure was also used for the further displacement terms u_{G3} and u_{G4}.

If several harmonics or sidebands meet on one frequency, it was expected that the signal energy increases at this point. This assumption was proved experimentally, and the groupings were conducted with two frequencies for comparison. In this case, the previously defined criteria for frequency combination selection do not apply. The frequencies f_1 = 5 MHz and f_2 = 10 MHz are selected for the experiments (Figure 3).

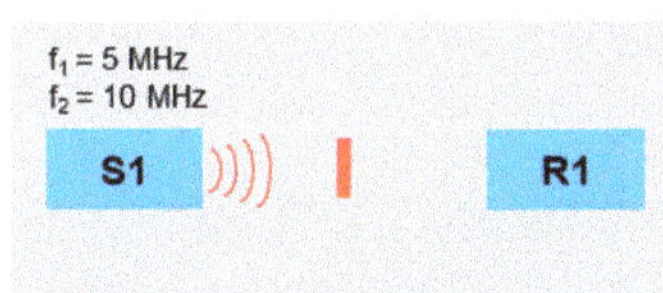

Figure 3. Sensor arrangement—evidence signal grouping.

Using the pulse generator Rigol DG1022Z, the sinus signals were sent to the sensor Olympus A5013. The same sensor was used on the receiving side. Table 6 illustrates the fundamentals, harmonics, and sidebands with the corresponding frequencies and groupings. With this frequency combination, there are several harmonics and sidebands at the frequencies 5 MHz, 10 MHz, 15 MHz, and 20 MHz.

Table 6. Signal grouping.

Group	$\Gamma_{f/h/s}$	Frequency [MHz]
-	$2f_1 - f_2$	0
1	f_1 $f_2 - f_1$	5 5
2	f_2 $2f_1$	10 10
3	$3f_1$ $f_2 + f_1$ $2f_2 - f_1$	15 15 15
4	$2f_2$ $2f_1 + f_2$	20 20
-	$2f_2 + f_1$	25
-	$3f_2$	30

The first measurement was separately executed with one frequency, f_1 and f_2. In the frequency spectrum, the amplitudes were measured at positions $f_{1/2}$, $2f_{1/2}$, and $3f_{1/2}$ (Figure 4a,b). The same experiments were repeated using two superimposing frequencies

(f_1 and f_2) transmitted in parallel (Figure 4c). For this purpose, 100 signals were sent out, an average frequency spectrum was formed, and the corresponding amplitudes were evaluated.

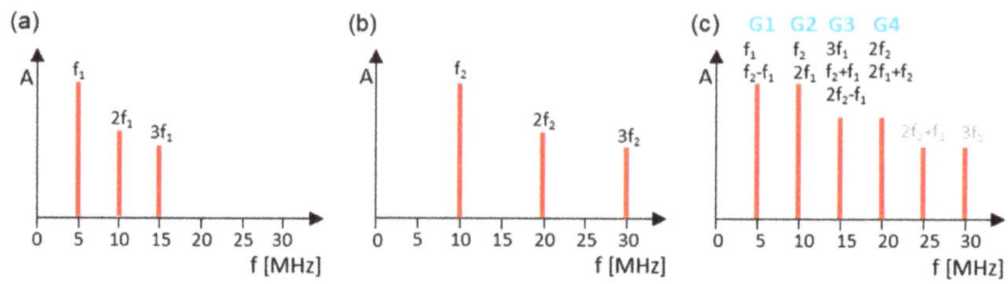

Figure 4. Grouping spectra: (**a**) f_1 = 5 MHz; (**b**) f_2 = 10 MHz; (**c**) f_1 = 5 MHz and f_2 = 10 MHz.

The signal energies were calculated from the measured amplitudes and shown as a comparison in Figure 5. The signal energy resulting from the excitation with one frequency is on the left side of the graphs (1f), while the energy from the two-frequency excitation is demonstrated on the right (2f). It is evident that more energy was transmitted in the multi-harmonic and sideband grouped signals (2f).

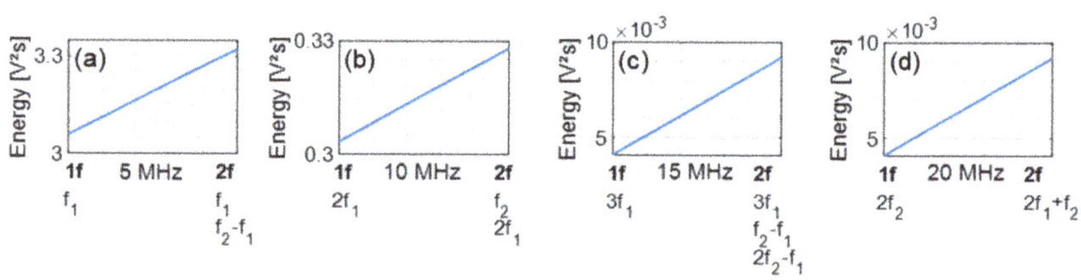

Figure 5. Energy comparison: (**a**) Group 1 (5 MHz); (**b**) Group 2 (10 MHz); (**c**) Group 3 (15 MHz); (**d**) Group 4 (20 MHz).

2.3. Evaluation Parameter

The aim of this work is to demonstrate a practicable application of the multi-frequency excitation and compare the results with one, two, and four excitation frequencies.

In other research, the total sums of the fundamental frequencies and those of the harmonic and sideband frequencies were compared for evaluation (Equation (9)) [17,36]:

$$E = \int_{-\infty}^{\infty} |\hat{x}(f_n)|^2 df_n \propto \sum|A_F| + \sum|A_H| \propto \sum|E_F| + \sum|E_H|, \quad (9)$$

where E is the energy, $\hat{x}$ is the signal, A_F are the fundamental amplitudes, A_H are the harmonic amplitudes, E_F is the signal energy of the fundamental frequencies, and E_H is the energy of the harmonic signal. This approach may not be applicable with the four-frequency excitation, however, because the grouping of the fundamental frequencies as well as the harmonic and sideband frequencies may overlap. Another possibility of evaluation with the same disadvantages is to use the acoustic moment concept, which is the integral of the power spectral density (PSD) function of a signal [37,38].

In addition to the nonlinearity parameters, the comparison value N_{index} was introduced for this purpose. This is a summation of the parameter values δ (Equation (10)):

$$N_{index}^{(1f,2f,2f-swept,\,4f)} = \sum_{n=1}^{m} |\delta_n|, \quad (10)$$

where m is the total number of nonlinearity parameters of the individual measurements.

3. Proof of the Use of the One-Dimensional Wave Equation

The derived nonlinearity parameters are based on one-dimensional wave propagation, and it was examined as to whether this model can also be applied to multi-dimensional applications. A round bar made of the material 1.4571 with a diameter of 5 mm and length of 100 mm was used to simulate a one-dimensional model. To generate artificial contact nonlinearities, the sample was cut in the middle and glued back together. At an area of 1 mm wide, no glue was used so that the two halves had metallic contact. For the measurements, the Olympus sensor A109S was used to send the longitudinal waves into the sample via the front side and to receive them on the other side (Figure 6).

One frequency (6 MHz), two frequencies (6 and 10 MHz), and 4 frequencies (6, 7.3, 8.7, and 10 MHz) were sent according to the group derivation from Section 2.2. The signals were generated with Rigol DG1022Z pulse generators over two channels each and an input voltage of 150 V. The output signal was amplified with the Phoenix ISL 40 dB amplifier.

Figure 6. Experimental setup—longitudinal waves.

These tests were also conducted with a reference sample without damage for comparison. The nonlinearity parameters were calculated from these measurements, and the determined N_{index} values are shown in Figure 7a. To clarify the comparison of the excitation methods, the normalised view was also selected (Figure 7b). It becomes clear that the difference with the 4f excitation is stronger than with the 1f excitation.

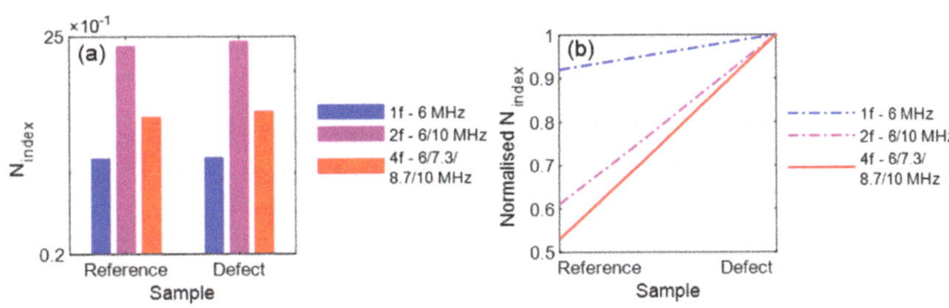

Figure 7. Evaluation: (**a**) N_{index}; (**b**) Normalised N_{index}.

4. Experimental Investigation—Plate Samples

In this section, the analytical derivations were proven experimentally and compared with the different excitation concepts.

4.1. Experimental Setup

Identical test setups were installed for all experiments (Figure 8a), where two Rigol DG1022Z pulse generators were used for signal generation. Each device had two output channels to send the single or superimposed frequencies to the sending sensor. For synchronisation, both devices were connected to each other via the trigger inputs and outputs. The combined signals were amplified with the amplifier Falco WMA-300. The S1 sensor was the Olympus A5013. The same sensor was used to receive the signal (R1), which was amplified with the Phoenix ISL 40 dB amplifier and sent to the Picoscope 4424 oscilloscope. The post-processing was conducted from this position.

Figure 8b reveals the experimental setup. The sensors were held in an additively manufactured holder and always ensured precise alignment. Two clamps were used for a certain contact pressure, and standard contact gel was applied under the sensors for better contact.

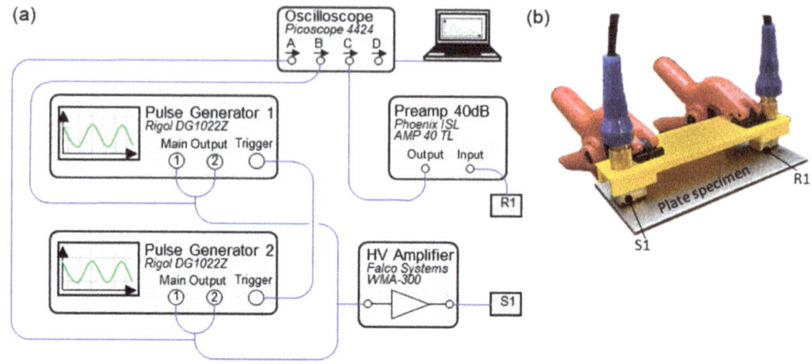

Figure 8. Experimental setup: (**a**) Scheme; (**b**) Sensor arrangement.

The samples used were Inconel 718 plates with the dimensions 2 mm × 70 mm × 30 mm, which consisted of two plates connected using a special micro laser welding process. Samples with a defect width of 5, 2, and 1 mm were used and arranged centrally (Figure 9). Due to these artificial defects, contact nonlinearities were generated and evaluated, and reference samples without defects were available for comparison. Six variants of each sample were examined, and the arithmetic mean values were processed.

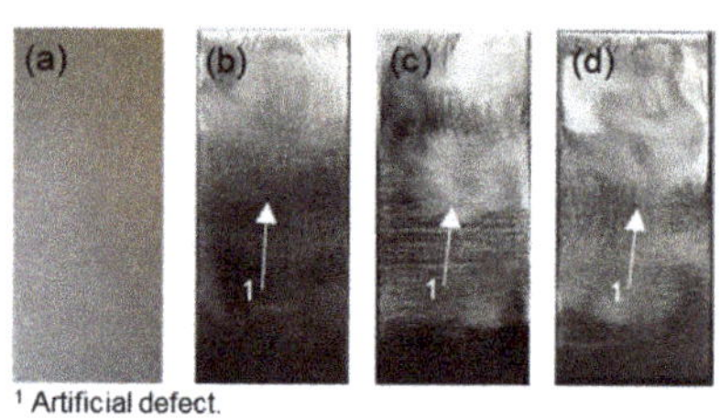

Figure 9. Specimen: (**a**) X1—reference sample; (**b**) 1M—failure size: 1 mm; (**c**) 2M—failure size: 2 mm; (**d**) 5M—failure size: 5 mm.

4.2. One-Frequency Excitation

In this section, the excitation via single frequencies was examined. A 6 MHz frequency signal and a 10 MHz signal were separately sent to the samples (Figure 10).

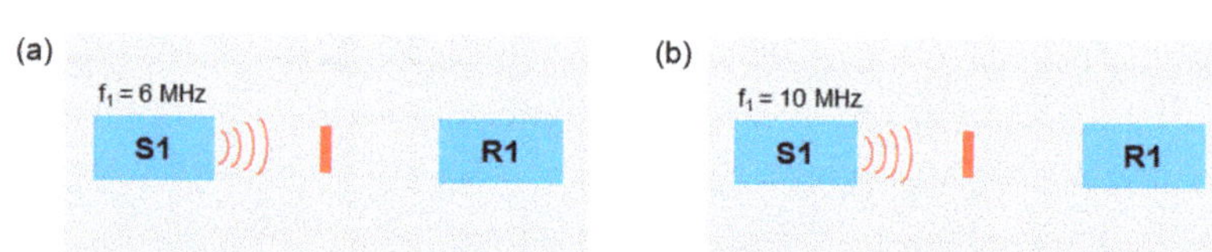

Figure 10. Excitation frequencies: (**a**) f_1 = 6 MHz; (**b**) f_1 = 10 MHz.

The second and third harmonic frequencies were considered accordingly. When 6 MHz is excited, the nonlinearity parameters of samples 1M and 2M increase significantly (Figure 11a). A form of saturation was observed as the values drop in the 5M sample. With the nonlinearity parameters of the third harmonic frequency, the value for the 1M sample decreases and then continuously increases for the other samples (2M and 5M). If the samples are sonicated with 10 MHz, a downward trend for both the second and third harmonic frequencies was given. Only the 5M sample demonstrated a significant increase in nonlinearities compared to the reference sample (Figure 11b).

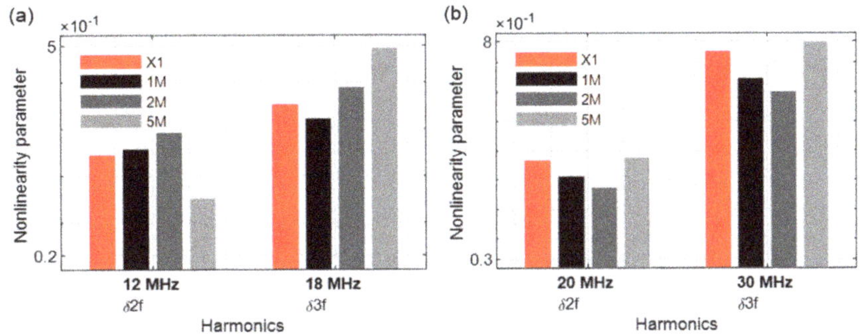

Figure 11. Nonlinearity parameter: (**a**) f_1 = 6 MHz; (**b**) f_1 = 10 MHz.

4.3. Two-Frequency Excitation

4.3.1. Constant Frequencies

If two frequencies emitted in parallel are used, the evaluation becomes more complex since a total of 12 harmonic and sideband frequencies are formed which can be evaluated. This technique was successfully used and proven in [17]. A 6 MHz and 10 MHz signal were generated, superimposed, and sent via sensor S1 (Figure 12). If the waves overlap, harmonic frequencies and more complex modulated response frequencies were generated, which are evaluated in the following. To compensate the overlapping effects of alternating compressive and tensile stresses on the various sine waves, 100 signals were transmitted per experiment and the mean values of the amplitudes were evaluated.

Figure 12. Excitation frequencies: f_1 = 6 MHz, f_2 = 10 MHz.

Figure 13 reveals the evaluation of the nonlinearity parameters, and all demonstrate a change compared to the reference sample. The parameter value $\delta 2f_2$ showed a clear pattern depending on the crack size. This phenomenon was moreover observed in [17,36], where the largest nonlinearities were not generated for the largest crack. These are strongly dependent on the fundamental frequencies and illustrate significantly more influence with the 1M and 2M samples.

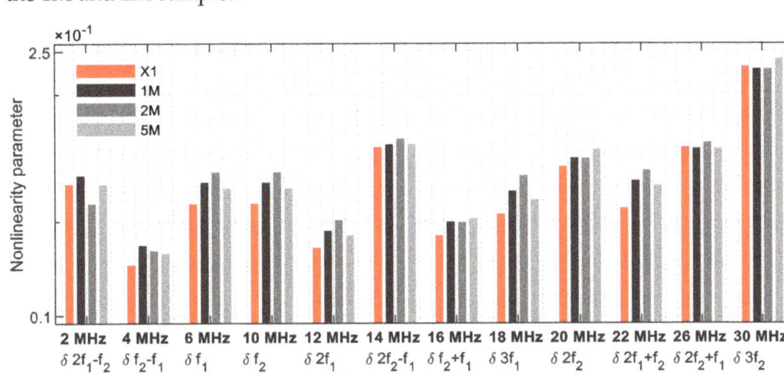

Figure 13. Nonlinearity parameter.

4.3.2. Linear Increasing Frequency

To investigate as many frequency combinations as possible, the concept of linearly increasing frequencies was also explored. A constant frequency of 10 MHz was transmitted via the sensor S1, and a 6 MHz signal was sent in parallel, which increases linearly to 10 MHz over a period of 60 s. The signals were received by sensor R1 (Figure 14), and 60 measurements per second were recorded.

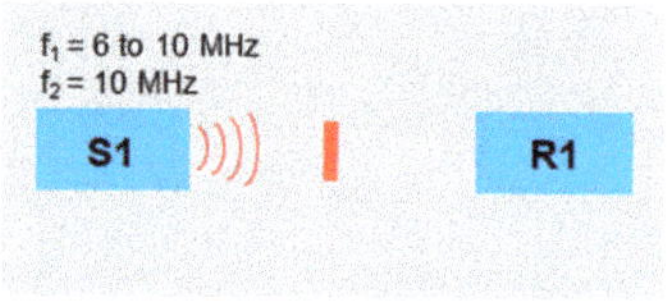

Figure 14. Excitation frequencies: f_1 = 6 to 10 MHz, f_2 = 10 MHz.

In Figures 15–19, the nonlinearity parameters are shown for the individual harmonics and sidebands.

In Figure 15a, the x-axis illustrates the actual increasing frequency combination starting with 6/10 MHz up to 10/10 MHz. At the excitation frequencies of 8.4/10 MHz, the values of the 5M sample increase drastically. The values of the 1M sample are consistently higher than those of the 2M sample, while the reference values are always lower, except for a peak at 7.4/10 MHz. With parameter $\delta 2f_2$ (Figure 15b), there were increased nonlinearities in the 5M sample at the frequency combinations 6.4/10 MHz, 9.25/10 MHz, and 9.6/10 MHz; therefore, the 1M sample reveals smaller peaks at 6.4/10 MHz and 7.8/10 MHz.

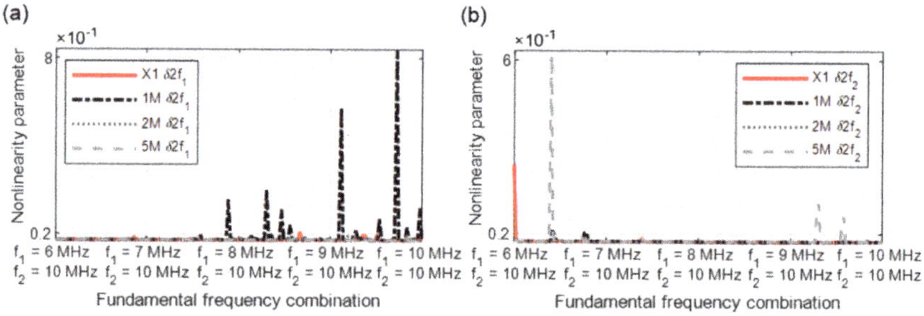

Figure 15. Nonlinearity parameter: (**a**) $\delta 2f_1$; (**b**) $\delta 2f_2$.

The third harmonic frequency of the sample 1M demonstrates several significant points at $\delta 3f1$ (Figure 16a). With $\delta 3f2$, increased material nonlinearities were detected in the reference sample at the start of the measurement. In the further course, several peaks of the 5M sample arise again (Figure 16b).

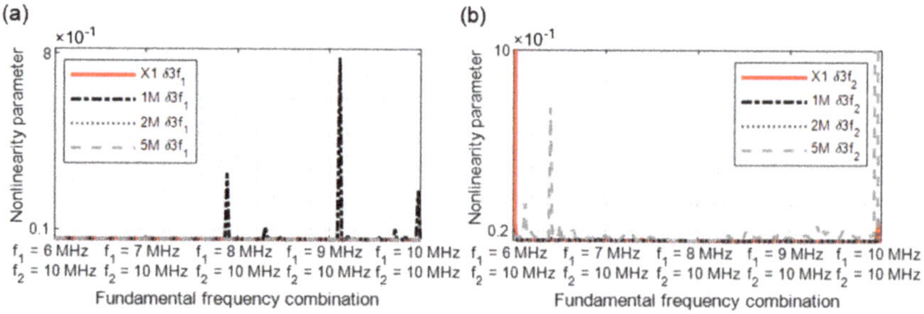

Figure 16. Nonlinearity parameter: (**a**) $\delta 3f_1$; (**b**) $\delta 3f_2$.

In Figure 17a, the strength of the dual frequency excitation with a modulated response parameter becomes clear. A comparison demonstrates that the $\delta f_2 - f_1$ values for the 1M sample as well as for the 2M and 5M samples reveal significant increases in nonlinearity. For the parameter $\delta f_2 + f_1$, all samples reveal peaks from different frequency excitation combinations (Figure 17b).

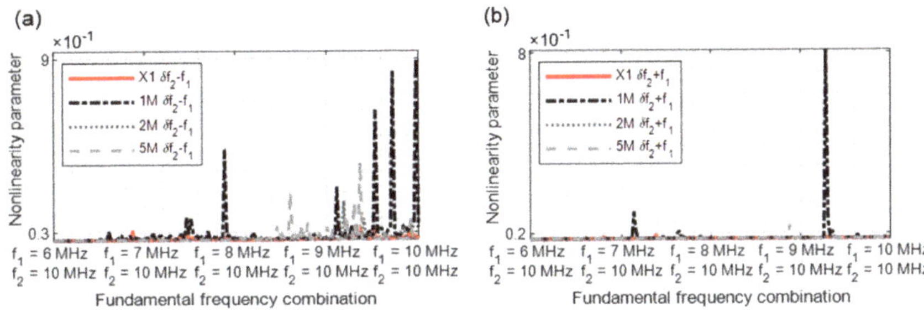

Figure 17. Nonlinearity parameter: (**a**) $\delta f_2 - f_1$; (**b**) $\delta f_2 + f_1$.

The parameter $\delta 2f_1 - f_2$ (Figure 18a) demonstrates a clear crack-predicting behaviour. All damaged samples reveal nonlinear features. The previously shown nonlinearity behaviour is continued in Figure 18b, with clear nonlinear signature particularly in the frequency range 9/10 MHz to 10/10 MHz.

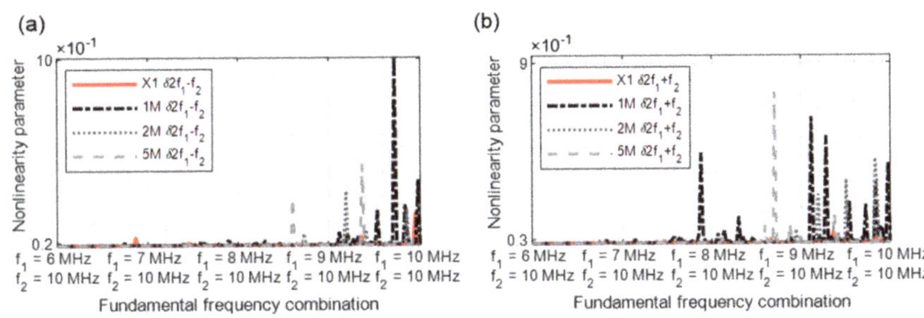

Figure 18. Nonlinearity parameter: (**a**) $\delta 2f_1 - f_2$; (**b**) $\delta 2f_1 + f_2$.

In Figure 19a, the sample 1M stands out again due to increased parameter values. In the higher frequency combination range, the damages of the samples 2M and 5M could be identified. Of all parameters, $2f_2 + f_1$ (Figure 19b) best demonstrates the nonlinearity increases as a function of the crack size. Although higher values were required from the frequency combination 7/10 MHz, the clearest results are from 8.6/10 MHz to 10/10 MHz.

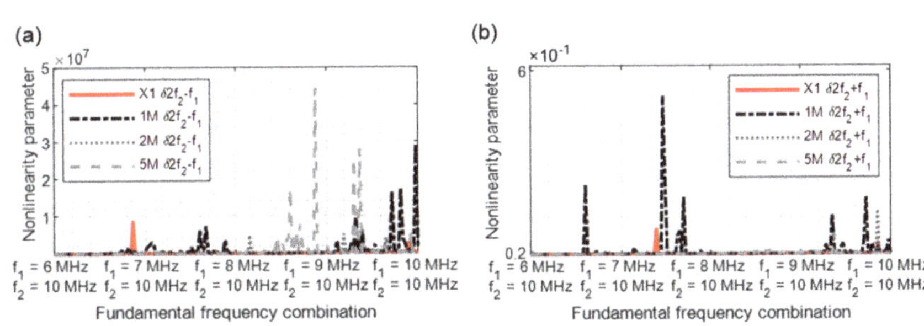

Figure 19. Nonlinearity parameter: (**a**) $\delta 2f_2 - f_1$; (**b**) $\delta 2f_2 + f_1$.

4.4. Four-Frequency Excitation

Four frequencies were sent into the sample with sensor S1 (Figure 20). The frequency selection should relate to the concept of frequency grouping (Section 2.2).

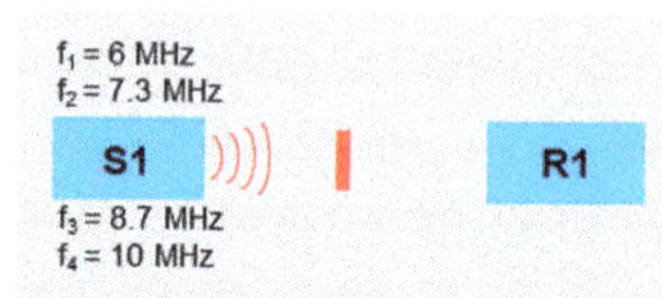

Figure 20. Excitation frequencies: f_1 = 6 MHz, f_2 = 7.3 MHz, f_3 = 8.7 MHz, f_4 = 10 MHz.

In Figures 21–24, the nonlinearity parameters were shown separately with the different groupings. Figure 21 demonstrates all nonlinearity parameters for one frequency (Table 2). The parameter values of the damaged samples usually differ significantly from those of the reference sample. With the 5M sample, insufficient nonlinearities are generated to observe a clear change to the reference sample. For the parameters $\delta 2f_1$, $\delta f_2 + f_1$, $\delta 2f_4 - f_1$, $\delta 3f_1$, $\delta 2f_1 + f_2$, $\delta 2f_4 + f_3$, and $\delta 3f_4$, the values of the reference sample and the damaged 5M sample are on the same level. With this nonlinearity group, frequencies with a large bandwidth from 2 MHz to 30 MHz were considered.

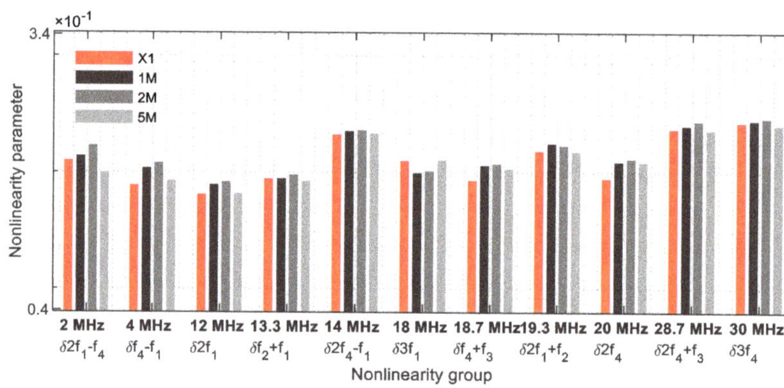

Figure 21. Nonlinearity parameter—δ_{G1}.

In the nonlinearity grouping with 2 frequencies (Table 3), eight parameters were evaluated (Figure 22), with a clear change to the reference sample. At 12.6 MHz ($\delta f_3 + f_4 - f_1/\delta 2f_4 - f_2$), 14.7 MHz ($\delta 2f_2/\delta f_3 + f_1$), 16 MHz ($\delta f_4 + f_1/\delta f_3 + f_2$), 17.3 MHz ($2f_3/\delta f_4 + f_2$) and 20.7 MHz ($\delta 2f_1 + f_3/\delta 2f_2 + f_1$), there is an increase in the nonlinearity parameters as a function of the crack size. The bandwidth ranges from 2.6 MHz to 27.3 MHz.

Figure 23 demonstrates the frequencies where three nonlinearity parameters meet (Table 4). This group consists of seven frequencies covering a bandwidth from 1.3 MHz to 26 MHz. These results revealed the strength of the multiple harmonic and modulated response frequencies on one frequency. They reflect the size of the cracks in the samples, with exception of the higher frequencies (23.3 MHz, 24.7 MHz, and 26 MHz). Contrary to group 1, an increase in the parameter values of the 5M sample is given; however, the values remain below the 1M and 2M samples.

The group with four nonlinearity parameters per frequency is shown Figure 24 (Table 5), which was revealed in a narrower frequency range from 4.6 MHz to 11.3 MHz. This illustrated that the energy of four fundamental, harmonic, and sideband frequencies are combined at one frequency, which is why this group has the greatest influence in the evaluation. At 7.3 MHz and 8.7 MHz, two fundamental frequencies are included.

At 4.6 MHz, the values of the 5M sample decrease. The other frequencies reveal the nonlinearity parameters as a function of the crack size.

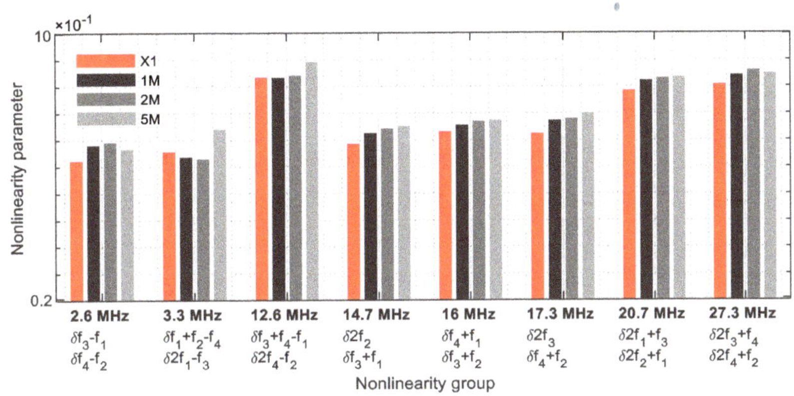

Figure 22. Nonlinearity parameter—δ_{G2}.

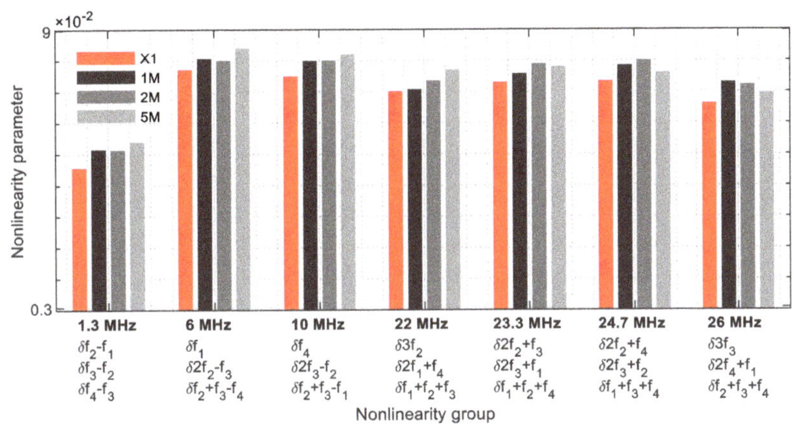

Figure 23. Nonlinearity parameter—δ_{G3}.

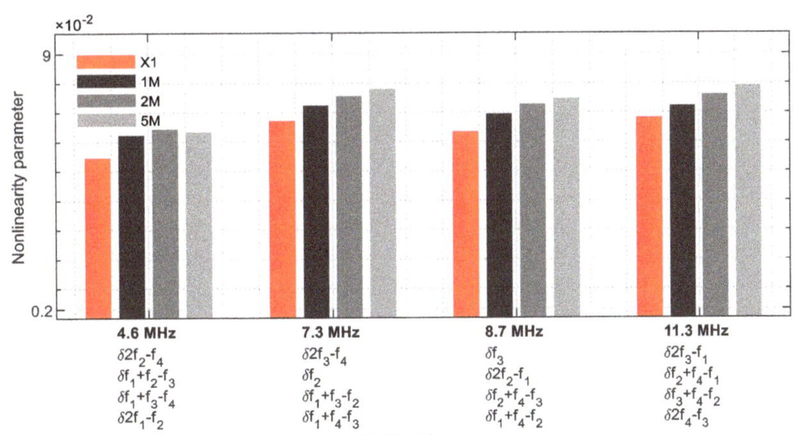

Figure 24. Nonlinearity parameter—δ_{G4}.

4.5. Results and Discussion

The various excitation techniques are compared in this section. Therefore, Figure 25 illustrates all N_{index} values. Since the reference sample X1 represents the basis of comparison (N_{index} = 0), only the difference to these values were shown. If the values are higher than zero, the nonlinearities increase; if they are smaller, the nonlinearities decrease. The values drop sharply with single-frequency excitation at 10 MHz. Only the values of the 5M sample are in the positive range. The 6 MHz excitation provided better values, even if the values of the 1M sample are minimally below the value of the reference sample. The constant excitation with two frequencies reveals a clear increase in the values, which, however, decrease in the still-positive range for the 5M sample. The arithmetic mean of all measurement data was used to evaluate the swept values. The N_{index} of the 2M sample is on the same level as the reference sample. In this graphic, it was only possible to deliver a course as a function of the crack size with the quadruple-excitation.

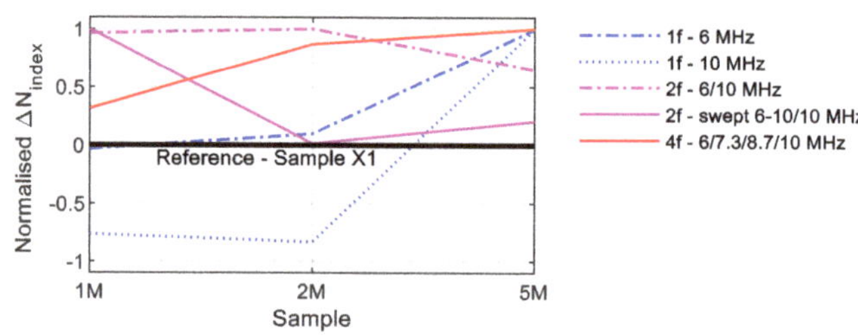

Figure 25. Comparison N_{index} values.

5. Experimental Investigation—Turbine Blades

In the previous section, the various excitation philosophies were tested on metal plates with artificial defects. In this section, turbine blades with cracks on the trailing edges were examined using the different excitation concepts.

5.1. Experimental Setup

Four Inconel 628 turbine blades from an industrial gas turbine were used for this study. Three blades had cracks on the trailing edges on different positions, while one blade was unused and undamaged and serves as a reference for comparison. These blades are cooled internally, with the cooling air directed into the blade via the blade root and exiting via various outlet slots at the trailing edge following the serpentine cooling. Figure 26 illustrates the blades used with the crack positions and Table 7 illustrates the corresponding crack lengths.

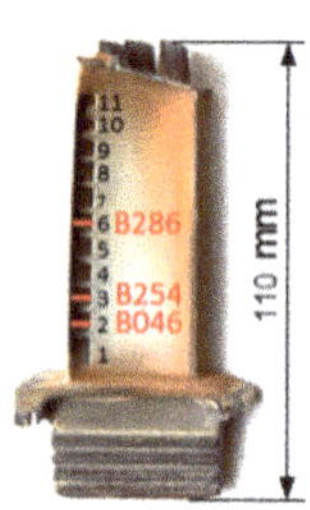

Figure 26. Damaged turbine blades.

Table 7. Overview—turbine blades.

Blade	Crack Position [Cooling Air Outlet Chamber (Figure 26)]	Crack Length [mm]
B292	-	-
B046	2	6.0
B254	3	6.8
B286	6	6.1

The experimental setup was implemented as described in Section 4. The emitting sensor S1 was positioned in the area of the inner shroud on a smooth area on the blade airfoil (Figure 27a). The receiving sensor R1 was located opposite, near the outer shroud. Using an additively manufactured sensor holder, the sensors are held in the same position during all tests, and an identical measuring distance was ensured (Figure 27b).

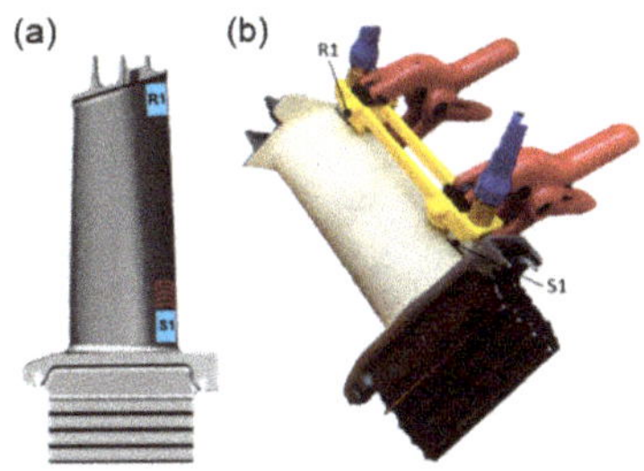

Figure 27. Measurements of turbine blades: (**a**) Sensor orientation; (**b**) Experimental setup.

5.2. Measurements with Results and Discussion

Figure 28 revealed the excitation with a frequency of 6 MHz (Figure 28a) and 10 MHz (Figure 28b). When excited with 6 MHz, the second harmonic parameter δf_2 demonstrates a clear increase in nonlinearities; the 10 MHz excitation, however, hardly allows conclusions to be drawn about the damage.

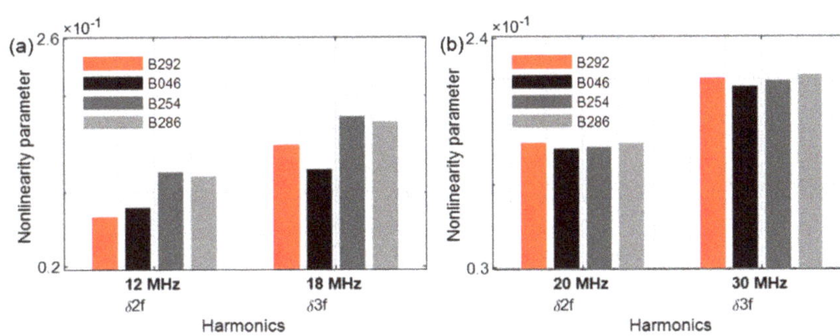

Figure 28. Nonlinearity parameter: (**a**) f_1 = 6 MHz; (**b**) f_1 = 10 MHz.

The dual frequency excitation with 6 MHz and 10 MHz clearly reveals increases in nonlinearity (Figure 29).

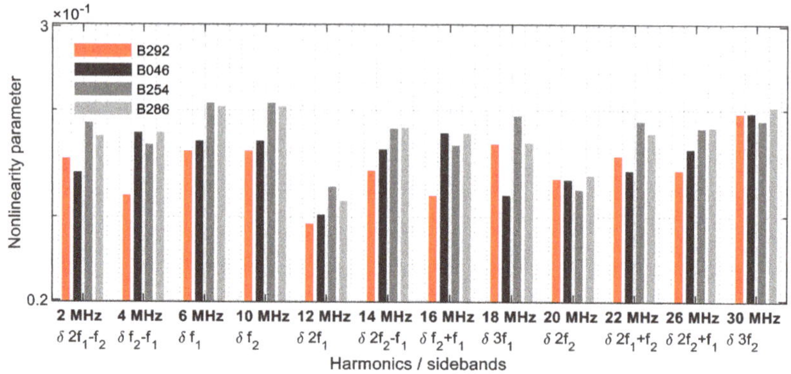

Figure 29. Nonlinearity parameter.

In Figures 30 and 31, the most relevant results are demonstrated with swept signals. It is noticeable that the reference blades (B292) also have a higher degree of nonlinearity peaks (Figure 30a). Using an excitation frequency combination starting with 8.2/10 MHz, both the reference blade B292 and blade B286 show harmonics. Further significant increases in the nonlinearity parameters were measured for all damaged blades. As with the metal plate samples, the excitation frequency combinations from 8/10 MHz lead to considerable increases in the nonlinearity parameters. Blade B254 achieved the highest values for both $\delta 2f_1$ and parameter $\delta 3f_1$ (Figure 30b).

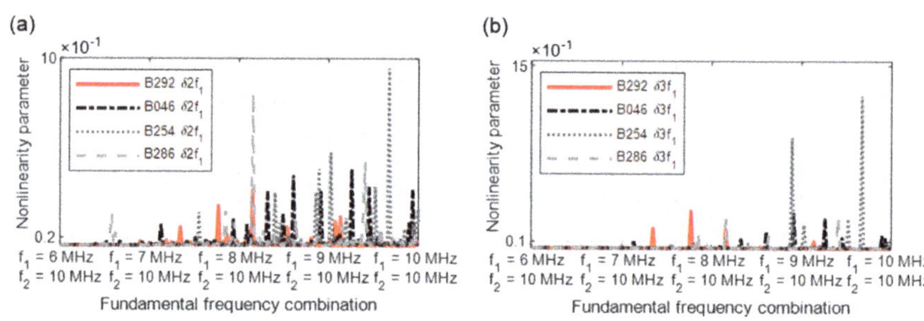

Figure 30. Nonlinearity parameter: (**a**) $\delta 2f_1$; (**b**) $\delta 3f_1$.

The results of two modulated response parameters ($\delta 2f_2 \pm f_1$) are illustrated in Figure 31. The blade B254 reveals the highest nonlinear values. The crack can also be diagnosed for blades B046 and B286 by the increased nonlinearity values.

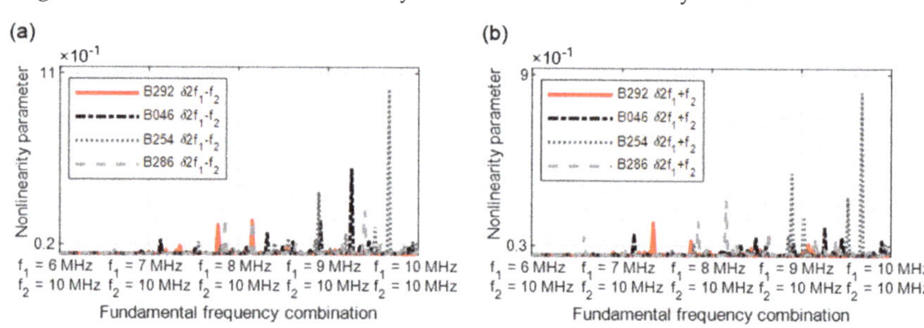

Figure 31. Nonlinearity parameter: (**a**) $\delta 2f_1 - f_2$; (**b**) $\delta 2f_1 + f_2$.

When evaluating the excitation with four frequencies, the procedure described in Section 2.2 was followed. The grouping of the nonlinearity parameters was used for crack detection. Figure 32 illustrates group 1, with one nonlinearity parameter per frequency. The parameters $\delta f_4 - f_1$, $\delta 2f_4 - f_1$, $\delta 3f_1$, $\delta f_4 + f_3$, $\delta 2f_4$, $\delta 2f_4 + f_3$, and $\delta 3f_4$ reveal a clear increase in the nonlinearity parameters and allow a positive crack prognosis.

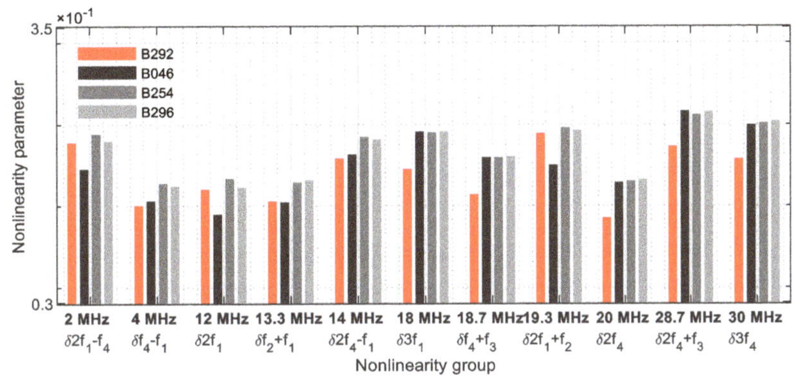

Figure 32. Nonlinearity parameter—δ_{G1}.

If two parameters decrease on one frequency (Figure 33), the results are strong, showing that detecting cracks was possible with this group evaluation alone.

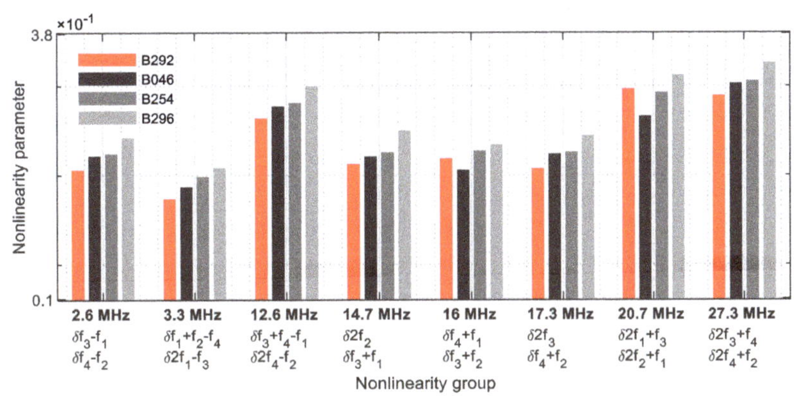

Figure 33. Nonlinearity parameter—δ_{G2}.

This behaviour continues and is also shown in the nonlinearity group 3 (Figure 34). Only $\delta 3f_2$ failed to clearly detect the crack on the blade B046.

As in Section 4, the differences between the various parameter values have become smaller (Figure 35). Because the signal energies of four fundamental, harmonic, and modulated response parameters each coincide on one frequency, this group approach provides the strongest informative value.

In Figure 36, the normalised N_{index} values for the different processes are used for evaluation again. If only one frequency is used for excitation, a crack forecast with these values is hardly possible; if two frequencies are used, both constant excitation and excitation by means of swept signals reveal a comparable, increasing behaviour; with four frequency excitations, these values increase continuously.

In most of the measurements, the B254 blade revealed the highest values. Although this blade had the longest crack with a crack length of 6.8 mm, the high nonlinearities can have various causes. Compared to the plate metal samples, the position of the cracks on the blades is variable. The crack of the blades B046 is only 5 mm behind the sensors, and the distance increases accordingly up to blade B286. In addition, real cracks behave differently, and many factors can lead to an increase of nonlinearities. The decisive factor is whether the crack is closed or open, and it is furthermore crucial how the asperities of the crack topography are in contact and how much of them are in interaction [39,40].

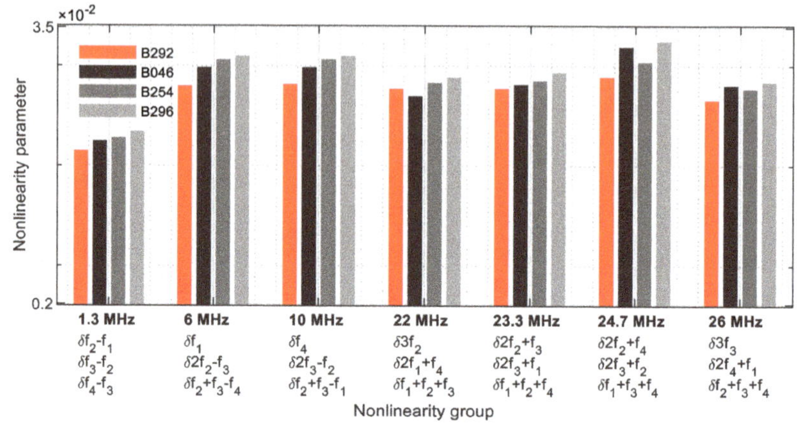

Figure 34. Nonlinearity parameter—δ_{G3}.

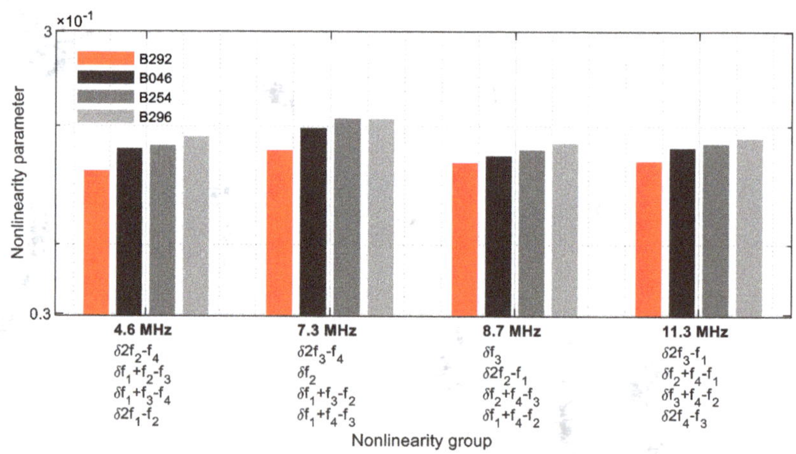

Figure 35. Nonlinearity parameter—δ_{G4}.

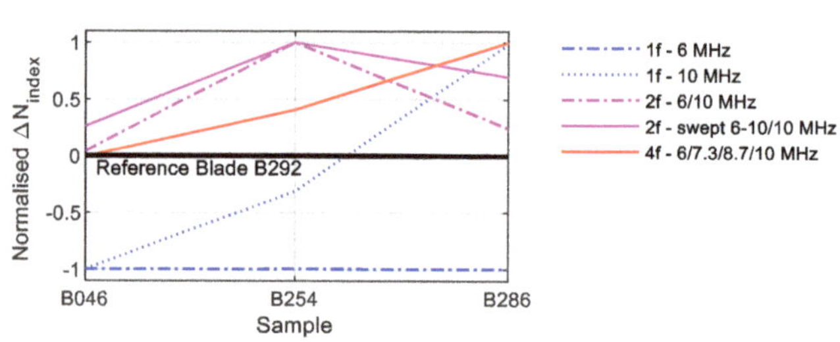

Figure 36. Comparison N_{index} values.

6. Conclusions

A comparison of various nonlinear ultrasonic wave modulation techniques was presented. The tests were conducted on metal plates, with artificially generated cracks, and on turbine blades with real cracks.

When using one excitation frequency, the results for both the plate samples and turbine blades were not convincing and showed that selecting the appropriate fundamental frequency is crucial to the success of the measurement.

The parallel excitation with two frequencies provides a total of 12 nonlinearity parameters to evaluate the components, and the new N_{index} values were used as a comparison with the other excitation techniques. For the plate samples, some nonlinearity parameters revealed a clear increase depending on the crack size. With the two-frequency excitation with a linearly increasing signal, the informative value of the failures had not improved in comparison.

The four-frequency excitation was analytically solved using the extended one-dimensional wave equation. A total of 64 individual nonlinearity parameters were derived. Furthermore, the four fundamental frequencies were chosen so that as many different fundamental frequencies, harmonic frequencies, and sidebands as possible meet at one frequency. The concept of grouping nonlinear parameters simplifies the evaluation with a total of 30 nonlinearity parameters, which facilitates identifying a defect in the material. This behaviour and advantages were proven experimentally. The N_{index} values of these measurements demonstrated behaviour as a function of the crack size. It was shown that, especially with the quasi-chaotic frequency evaluation, detecting the failure is possible without calculating dispersion curves.

The use of four fundamental frequencies offers clear advantages in detecting defects in materials by pairing with the frequency grouping method.

Author Contributions: Writing—original draft preparation, F.M.; writing—review and editing, M.M. All authors have read and agreed to the published version of the manuscript.

Funding: This research received no external funding.

Acknowledgments: Thanks to the MAN Energy Solutions SE for their support.

Conflicts of Interest: The authors declare no conflict of interest.

Appendix A

This section includes the further derivation for the solution of the one-dimensional wave equation with four excitation frequencies up to the third-order of nonlinearity (Section 2.1).

Substituting Equation (7) in the right side of Equation (5) with the second-order nonlinearity parameter β and making further transformations with trigonometric formulas lead to:

$$\frac{\partial^2 u}{\partial t^2} - c^2 \frac{\partial^2 u}{\partial x^2} = c^2\beta \left(\begin{array}{c} -\frac{1}{2}A_1^2 k_{f1}^3 sin\left[k_{f1}(x-ct)\right] \\ +\frac{1}{2}A_2^2 k_{f2}^3 sin\left[k_{f2}(x-ct)\right] \\ -\frac{1}{2}A_3^2 k_{f3}^3 sin\left[k_{f3}(x-ct)\right] \\ +\frac{1}{2}A_4^2 k_{f4}^3 sin\left[k_{f4}(x-ct)\right] \\ -\frac{1}{2}A_1 A_2 k_{f1} k_{f2}\left(k_{f2}-k_{f1}\right) cos\left[\left(k_{f2}-k_{f1}\right)(x-ct)\right] \\ -\frac{1}{2}A_1 A_2 k_{f1} k_{f2}\left(k_{f2}+k_{f1}\right) cos\left[\left(k_{f2}+k_{f1}\right)(x-ct)\right] \\ -\frac{1}{2}A_1 A_3 k_{f1} k_{f3}\left(k_{f3}-k_{f1}\right) sin\left[\left(k_{f3}-k_{f1}\right)(x-ct)\right] \\ -\frac{1}{2}A_1 A_3 k_{f1} k_{f3}\left(k_{f3}+k_{f1}\right) sin\left[\left(k_{f3}+k_{f1}\right)(x-ct)\right] \\ -\frac{1}{2}A_1 A_4 k_{f1} k_{f4}\left(k_{f4}-k_{f1}\right) cos\left[\left(k_{f4}-k_{f1}\right)(x-ct)\right] \\ -\frac{1}{2}A_1 A_4 k_{f1} k_{f4}\left(k_{f4}+k_{f1}\right) cos\left[\left(k_{f4}-k_{f1}\right)(x+ct)\right] \\ +\frac{1}{2}A_2 A_3 k_{f2} k_{f3}\left(k_{f3}-k_{f2}\right) cos\left[\left(k_{f3}-k_{f2}\right)(x-ct)\right] \\ -\frac{1}{2}A_2 A_3 k_{f2} k_{f3}\left(k_{f3}+k_{f2}\right) cos\left[\left(k_{f3}+k_{f2}\right)(x-ct)\right] \\ +\frac{1}{2}A_2 A_4 k_{f2} k_{f4}\left(k_{f4}-k_{f2}\right) sin\left[\left(k_{f4}-k_{f2}\right)(x-ct)\right] \\ +\frac{1}{2}A_2 A_4 k_{f2} k_{f4}\left(k_{f4}+k_{f2}\right) sin\left[\left(k_{f4}+k_{f2}\right)(x-ct)\right] \\ -\frac{1}{2}A_3 A_4 k_{f3} k_{f4}\left(k_{f4}-k_{f3}\right) cos\left[\left(k_{f4}-k_{f3}\right)(x-ct)\right] \\ -\frac{1}{2}A_3 A_4 k_{f3} k_{f4}\left(k_{f4}+k_{f3}\right) cos\left[\left(k_{f4}+k_{f3}\right)(x-ct)\right] \end{array} \right). \tag{A1}$$

Based on this expression and according to [17], the following solution results for $u^{(2)}$:

$$\begin{aligned} u^{(2)} = & -\frac{\beta A_1^2 k_{f1}^2}{8}\cos\left[2k_{f1}(x-ct)\right]x + \frac{\beta A_2^2 k_{f2}^2}{8}\cos\left[2k_{f2}(x-ct)\right]x \\ & -\frac{\beta A_3^2 k_{f3}^2}{8}\cos\left[2k_{f3}(x-ct)\right]x + \frac{\beta A_4^2 k_{f4}^2}{8}\cos\left[2k_{f4}(x-ct)\right]x \\ & +\frac{\beta A_1 A_2 k_{f1} k_{f2}}{4}\sin\left[\left(k_{f2}-k_{f1}\right)(x-ct)\right]x + \frac{\beta A_1 A_2 k_{f1} k_{f2}}{4}\sin\left[\left(k_{f2}+k_{f1}\right)(x-ct)\right]x \\ & +\frac{\beta A_1 A_3 k_{f1} k_{f3}}{4}\cos\left[\left(k_{f3}-k_{f1}\right)(x-ct)\right]x - \frac{\beta A_1 A_3 k_{f1} k_{f3}}{4}\cos\left[\left(k_{f3}+k_{f1}\right)(x-ct)\right]x \\ & -\frac{\beta A_1 A_4 k_{f1} k_{f4}}{4}\sin\left[\left(k_{f4}-k_{f1}\right)(x-ct)\right]x - \frac{\beta A_1 A_4 k_{f1} k_{f4}}{4}\sin\left[\left(k_{f4}+k_{f1}\right)(x-ct)\right]x \\ & -\frac{\beta A_2 A_3 k_{f2} k_{f3}}{4}\sin\left[\left(k_{f3}-k_{f2}\right)(x-ct)\right]x + \frac{\beta A_2 A_3 k_{f2} k_{f3}}{4}\sin\left[\left(k_{f3}+k_{f2}\right)(x-ct)\right]x \\ & -\frac{\beta A_2 A_4 k_{f2} k_{f4}}{4}\cos\left[\left(k_{f4}-k_{f2}\right)(x-ct)\right]x + \frac{\beta A_2 A_4 k_{f2} k_{f4}}{4}\cos\left[\left(k_{f4}+k_{f2}\right)(x-ct)\right]x \\ & +\frac{\beta A_3 A_4 k_{f3} k_{f4}}{4}\sin\left[\left(k_{f4}-k_{f3}\right)(x-ct)\right]x + \frac{\beta A_3 A_4 k_{f3} k_{f4}}{4}\sin\left[\left(k_{f4}+k_{f3}\right)(x-ct)\right]x. \end{aligned} \tag{A2}$$

The solution of Equation (5) should now be extended by the cubic nonlinearity parameter γ to obtain an encompassing description of the nonlinearity parameters. Substituting Equation (7) into the right side of Equation (5) in the third-order nonlinearity parameter γ gives:

$$\frac{\partial^2 u}{\partial t^2} - c^2 \frac{\partial^2 u}{\partial x^2} = \frac{c^2\gamma}{2}\left[\left(\begin{array}{c} A_1 sin\left[k_{f1}(x-ct)\right] + A_2 cos\left[k_{f2}(x-ct)\right] + \\ A_3 sin\left[k_{f3}(x-ct)\right] + A_4 cos\left[k_{f4}(x-ct)\right] \end{array} \right)^2 \left(\begin{array}{c} -A_1 {k_{f1}}^2 sin\left[k_{f1}(x-ct)\right] - A_2 {k_{f2}}^2 cos[k_2(x-ct)] \\ -A_3 {k_{f3}}^2 sin\left[k_{f3}(x-ct)\right] - A_4 {k_{f4}}^2 cos[k_4(x-ct)] \end{array} \right) \right]. \tag{A3}$$

Transformation with binominal and trigonometric formulas leads to:

$$
\frac{\partial^2 u}{\partial t^2} - c^2\frac{\partial^2 u}{\partial x^2} = \frac{c^2\gamma}{2}\left(\begin{array}{c}
-\frac{1}{2}\left(A_1A_2{}^2k_{f1}{}^2k_{f2}{}^2 + A_1A_4{}^2k_{f1}{}^2k_{f4}{}^2 + A_1A_3{}^2k_{f1}{}^2k_{f3}{}^2 + \frac{1}{2}A_1{}^3k_{f1}{}^4\right)sin\left[k_{f1}(x-ct)\right] \\
-\frac{1}{2}\left(A_1{}^2A_2k_{f1}{}^2k_{f2}{}^2 + A_2A_4{}^2k_{f2}{}^2k_{f4}{}^2 + A_2A_3{}^2k_{f2}{}^2k_{f3}{}^2 + \frac{1}{2}A_2{}^3k_{f2}{}^4\right)sin\left[k_{f2}(x-ct)\right] \\
-\frac{1}{2}\left(A_1{}^2A_3k_{f1}{}^2k_{f3}{}^2 + A_2{}^2A_3k_{f2}{}^2k_{f3}{}^2 + A_3A_4{}^2k_{f3}{}^2k_{f4}{}^2 + \frac{1}{2}A_3{}^3k_{f3}{}^4\right)sin\left[k_{f3}(x-ct)\right] \\
-\frac{1}{2}\left(A_1{}^2A_4k_{f1}{}^2k_{f4}{}^2 + A_2{}^2A_4k_{f2}{}^2k_{f4}{}^2 + A_3{}^2A_4k_{f3}{}^2k_{f4}{}^2 + \frac{1}{2}A_4{}^3k_{f4}{}^4\right)sin\left[k_{f4}(x-ct)\right] \\
-\frac{1}{4}A_1{}^3k_{f1}{}^4sin\left[3k_{f1}(x-ct)\right] + \frac{1}{4}A_2{}^3k_{f2}{}^4cos\left[3k_{f2}(x-ct)\right] \\
-\frac{1}{4}A_3{}^3k_{f3}{}^4sin\left[3k_{f3}(x-ct)\right] + \frac{1}{4}A_4{}^3k_{f4}{}^4cos\left[3k_{f4}(x-ct)\right] \\
-\frac{1}{2}A_1{}^2A_2\left(k_{f1}{}^3k_{f2} + \frac{1}{2}k_{f1}{}^2k_{f2}{}^2\right)cos\left[(2k_{f1}+k_{f2})(x-ct)\right] \\
+\frac{1}{2}A_1{}^2A_2\left(k_{f1}{}^3k_{f2} - \frac{1}{2}k_{f1}{}^2k_{f2}{}^2\right)cos\left[(2k_{f1}-k_{f2})(x-ct)\right] \\
-\frac{1}{2}A_1{}^2A_3\left(\frac{1}{2}k_{f1}{}^2k_{f3}{}^2 + k_{f1}{}^3k_{f3}\right)sin\left[(2k_{f1}+k_{f3})(x-ct)\right] \\
+\frac{1}{2}A_1{}^2A_3\left(\frac{1}{2}k_{f1}{}^2k_{f3}{}^2 - k_{f1}{}^3k_{f3}\right)sin\left[(2k_{f1}-k_{f3})(x-ct)\right] \\
-\frac{1}{2}A_1{}^2A_4\left(k_{f1}{}^3k_{f4} + \frac{1}{2}k_{f1}{}^2k_{f4}{}^2\right)cos\left[(2k_{f1}+k_{f4})(x-ct)\right] \\
+\frac{1}{2}A_1{}^2A_4\left(k_{f1}{}^3k_{f4} - \frac{1}{2}k_{f1}{}^2k_{f4}{}^2\right)cos\left[(2k_{f1}-k_{f4})(x-ct)\right] \\
+\frac{1}{2}A_1A_2{}^2\left(\frac{1}{2}k_{f1}{}^2k_{f2}{}^2 + k_{f1}k_{f2}{}^3\right)sin\left[(2k_{f2}+k_{f1})(x-ct)\right] \\
+\frac{1}{2}A_1A_2{}^2\left(k_{f1}k_{f2}{}^3 - \frac{1}{2}k_{f1}{}^2k_{f2}{}^2\right)sin\left[(2k_{f2}-k_{f1})(x-ct)\right] \\
+\frac{1}{2}A_2{}^2A_3\left(k_{f2}{}^3k_{f3} + \frac{1}{2}k_{f2}{}^2k_{f3}{}^2\right)sin\left[(2k_{f2}+k_{f3})(x-ct)\right] \\
+\frac{1}{2}A_2{}^2A_3\left(k_{f2}{}^3k_{f3} - \frac{1}{2}k_{f2}{}^2k_{f3}{}^2\right)sin\left[(2k_{f2}-k_{f3})(x-ct)\right] \\
+\frac{1}{4}A_2{}^2A_4k_{f2}{}^2k_{f4}{}^2cos\left[(2k_{f2}+k_{f4})(x-ct)\right] \\
+\frac{1}{4}A_2{}^2A_4k_{f2}{}^2k_{f4}{}^2cos\left[(2k_{f2}-k_{f4})(x-ct)\right] \\
-\frac{1}{2}A_1A_3{}^2\left(\frac{1}{2}k_{f1}{}^2k_{f3}{}^2 + k_{f1}k_{f3}{}^3\right)sin\left[(2k_{f3}+k_{f1})(x-ct)\right] \\
+\frac{1}{2}A_1A_3{}^2\left(\frac{1}{2}k_{f1}{}^2k_{f3}{}^2 - k_{f1}k_{f3}{}^3\right)sin\left[(2k_{f3}-k_{f1})(x-ct)\right] \\
-\frac{1}{2}A_1A_3{}^2\left(\frac{1}{2}k_{f2}{}^2k_{f3}{}^2 + k_{f2}k_{f3}{}^3\right)cos\left[(2k_{f3}+k_{f2})(x-ct)\right] \\
+\frac{1}{2}A_2A_3{}^2\left(k_{f2}k_{f3}{}^3 - \frac{1}{2}k_{f2}{}^2k_{f3}{}^2\right)cos\left[(2k_{f3}-k_{f2})(x-ct)\right] \\
-\frac{1}{2}A_3{}^2A_4\left(k_{f3}{}^3k_{f4} + \frac{1}{2}k_{f3}{}^2k_{f4}{}^2\right)cos\left[(2k_{f3}+k_{f4})(x-ct)\right] \\
+\frac{1}{2}A_3{}^2A_4\left(k_{f3}{}^3k_{f4} - \frac{1}{2}k_{f3}{}^2k_{f4}{}^2\right)cos\left[(2k_{f3}-k_{f4})(x-ct)\right] \\
+\frac{1}{2}A_1A_4{}^2\left(\frac{1}{2}k_{f1}{}^2k_{f4}{}^2 + k_{f1}k_{f4}{}^3\right)sin\left[(2k_{f4}+k_{f1})(x-ct)\right] \\
+\frac{1}{2}A_1A_4{}^2\left(k_{f1}k_{f4}{}^3 - \frac{1}{2}k_{f1}{}^2k_{f4}{}^2\right)sin\left[(2k_{f4}-k_{f1})(x-ct)\right] \\
+\frac{1}{4}A_2A_4{}^2k_{f2}{}^2k_{f4}{}^2cos\left[(2k_{f4}+k_{f2})(x-ct)\right] \\
+\frac{1}{4}A_2A_4{}^2k_{f2}{}^2k_{f4}{}^2cos\left[(2k_{f4}-k_{f2})(x-ct)\right] \\
+\frac{1}{2}A_3A_4{}^2\left(\frac{1}{2}k_{f3}{}^2k_{f4}{}^2 + k_{f3}k_{f4}{}^3\right)sin\left[(2k_{f4}+k_{f3})(x-ct)\right] \\
+\frac{1}{2}A_3A_4{}^2\left(k_{f3}k_{f4}{}^3 - \frac{1}{2}k_{f3}{}^2k_{f4}{}^2\right)sin\left[(2k_{f4}-k_{f3})(x-ct)\right] \\
-\frac{1}{2}A_1A_2A_3\left(k_{f1}{}^2k_{f2}k_{f3} + k_{f1}k_{f2}{}^2k_{f3} + k_{f1}k_{f2}k_{f3}{}^2\right)cos\left[(k_{f1}+k_{f2}+k_{f3})(x-ct)\right] \\
-\frac{1}{2}A_1A_2A_3\left(k_{f1}{}^2k_{f2}k_{f3} + k_{f1}k_{f2}{}^2k_{f3} - k_{f1}k_{f2}k_{f3}{}^2\right)cos\left[(k_{f1}+k_{f2}-k_{f3})(x-ct)\right] \\
-\frac{1}{2}A_1A_3A_4\left(k_{f1}{}^2k_{f3}k_{f4} + k_{f1}k_{f3}{}^2k_{f4} + k_{f1}k_{f3}k_{f4}{}^2\right)cos\left[(k_{f1}+k_{f3}+k_{f4})(x-ct)\right] \\
+\frac{1}{2}A_1A_3A_4\left(k_{f1}{}^2k_{f3}k_{f4} + k_{f1}k_{f3}{}^2k_{f4} - k_{f1}k_{f3}k_{f4}{}^2\right)cos\left[(k_{f1}+k_{f3}-k_{f4})(x-ct)\right] \\
+\frac{1}{2}A_1A_2A_4\left(k_{f1}{}^2k_{f2}k_{f4} + k_{f1}k_{f2}k_{f4}{}^2\right)sin\left[(k_{f1}+k_{f2}+k_{f4})(x-ct)\right] \\
+\frac{1}{2}A_1A_2A_4\left(k_{f1}{}^2k_{f2}k_{f4} - k_{f1}k_{f2}k_{f4}{}^2\right)sin\left[(k_{f1}+k_{f2}-k_{f4})(x-ct)\right] \\
+\frac{1}{2}A_1A_2A_3\left(k_{f1}{}^2k_{f2}k_{f3} - k_{f1}k_{f2}{}^2k_{f3} - k_{f1}k_{f2}k_{f3}{}^2\right)cos\left[(k_{f2}+k_{f3}-k_{f1})(x-ct)\right] \\
+\frac{1}{2}A_1A_3A_4\left(k_{f1}k_{f3}{}^2k_{f4} - k_{f1}{}^2k_{f3}k_{f4} - k_{f1}k_{f2}k_{f4}{}^2\right)cos\left[(k_{f1}+k_{f4}-k_{f3})(x-ct)\right] \\
+\frac{1}{2}A_1A_2A_4\left(k_{f1}{}^2k_{f3}k_{f4} - k_{f1}k_{f3}{}^2k_{f4} - k_{f1}k_{f3}k_{f4}{}^2\right)cos\left[(k_{f3}+k_{f4}-k_{f1})(x-ct)\right] \\
+\frac{1}{2}A_1A_2A_4\left(k_{f1}k_{f2}{}^2k_{f4} + k_{f1}k_{f2}k_{f4}{}^2\right)sin\left[(k_{f2}+k_{f4}-k_{f1})(x-ct)\right] \\
+\frac{1}{2}A_1A_2A_3\left(k_{f1}k_{f2}k_{f3}{}^2 - k_{f1}k_{f2}{}^2k_{f3} + k_{f1}{}^2k_{f2}k_{f3}\right)cos\left[(k_{f1}+k_{f3}-k_{f2})(x-ct)\right] \\
+\frac{1}{2}A_2A_3A_4\left(k_{f2}{}^2k_{f3}k_{f4} - k_{f2}k_{f3}k_{f4}{}^2\right)sin\left[(k_{f2}+k_{f3}+k_{f4})(x-ct)\right] \\
+\frac{1}{2}A_2A_3A_4\left(k_{f2}{}^2k_{f3}k_{f4} - k_{f2}k_{f3}k_{f4}{}^2\right)sin\left[(k_{f2}+k_{f4}-k_{f3})(x-ct)\right] \\
-\frac{1}{2}A_2A_3A_4\left(k_{f2}{}^2k_{f3}k_{f4} + k_{f2}k_{f3}k_{f4}{}^2\right)sin\left[(k_{f2}+k_{f3}-k_{f4})(x-ct)\right] \\
+\frac{1}{2}A_2A_3A_4\left(k_{f2}{}^2k_{f3}k_{f4} + k_{f2}k_{f3}k_{f4}{}^2\right)sin\left[(k_{f3}+k_{f4}-k_{f2})(x-ct)\right] \\
-\frac{1}{2}A_1A_2A_4\left(k_{f1}k_{f2}k_{f4}{}^2\right)sin\left[(k_{f1}+k_{f4}-k_{f2})(x-ct)\right]
\end{array}\right). \tag{A4}
$$

Based on [17], this results in the following solution:

$$
\begin{aligned}
u^{(3)} = &+\left(\frac{\gamma(A_1A_2{}^2k_{f1}k_{f2}{}^2+A_1A_4{}^2k_{f1}k_{f4}{}^2+A_1A_3{}^2k_{f1}k_{f3}{}^2+\frac{1}{2}A_1{}^3k_{f1}{}^4)}{8}\right)cos\left[k_{f1}(x-ct)\right]x\\
&-\left(\frac{\gamma(A_1{}^2A_2k_{f1}{}^2k_{f2}+A_2A_4{}^2k_{f1}k_{f4}{}^2+A_2A_3{}^2k_{f2}k_{f3}{}^2+\frac{1}{2}A_2{}^3k_{f2}{}^4)}{8}\right)sin\left[k_{f2}(x-ct)\right]x\\
&+\left(\frac{\gamma(A_1{}^2A_3k_{f1}{}^2k_{f3}+A_2{}^2A_3k_{f2}{}^2k_{f3}+A_3A_4{}^2k_{f3}k_{f4}{}^2+\frac{1}{2}A_3{}^3k_{f3}{}^4)}{8}\right)cos\left[k_{f3}(x-ct)\right]x\\
&-\left(\frac{\gamma(A_1{}^2A_4k_{f1}{}^2k_{f4}+A_2{}^2A_4k_{f2}{}^2k_{f4}+A_3{}^2A_4k_{f3}{}^2k_{f4}+\frac{1}{2}A_4{}^3k_{f4}{}^4)}{8}\right)sin\left[k_{f4}(x-ct)\right]x\\
&+\frac{\gamma A_1{}^3k_{f1}{}^3}{48}cos\left[3k_{f1}(x-ct)\right]x+\frac{\gamma A_2{}^3k_{f2}{}^3}{48}sin\left[3k_{f2}(x-ct)\right]x\\
&+\frac{\gamma A_3{}^3k_{f3}{}^3}{48}cos\left[3k_{f3}(x-ct)\right]x+\frac{\gamma A_4{}^3k_{f4}{}^3}{48}sin\left[3k_{f4}(x-ct)\right]x\\
&-\left(\frac{\gamma A_1{}^2A_2(k_{f1}{}^3k_{f2}+\frac{1}{2}k_{f1}{}^2k_{f2}{}^2)}{8(2k_{f1}+k_{f2})}\right)sin\left[\left(2k_{f1}+k_{f2}\right)(x-ct)\right]x+\left(\frac{\gamma A_1{}^2A_2(k_{f1}{}^3k_{f2}-\frac{1}{2}k_{f1}{}^2k_{f2}{}^2)}{8(2k_{f1}-k_{f2})}\right)sin\left[\left(2k_{f1}-k_{f2}\right)(x-ct)\right]x\\
&+\left(\frac{\gamma A_1{}^2A_3(\frac{1}{2}k_{f1}{}^2k_{f3}{}^2+k_{f1}{}^3k_{f3})}{8(2k_{f1}+k_{f3})}\right)cos\left[\left(2k_{f1}+k_{f3}\right)(x-ct)\right]x+\left(\frac{\gamma A_1{}^2A_3(\frac{1}{2}k_{f1}{}^2k_{f3}{}^2-k_{f1}{}^3k_{f3})}{8(2k_{f1}-k_{f3})}\right)cos\left[\left(2k_{f1}-k_{f3}\right)(x-ct)\right]x\\
&-\left(\frac{\gamma A_1{}^2A_4(k_{f1}{}^3k_{f4}+\frac{1}{2}k_{f1}{}^2k_{f4}{}^2)}{8(2k_{f1}+k_{f4})}\right)sin\left[\left(2k_{f1}+k_{f4}\right)(x-ct)\right]x+\left(\frac{\gamma A_1{}^2A_4(k_{f1}{}^3k_{f4}-\frac{1}{2}k_{f1}{}^2k_{f4}{}^2)}{8(2k_{f1}-k_{f4})}\right)sin\left[\left(2k_{f1}-k_{f4}\right)(x-ct)\right]x\\
&-\left(\frac{\gamma A_1A_2{}^2(\frac{1}{2}k_{f1}{}^2k_{f2}{}^2+k_{f1}k_{f2}{}^3)}{8(2k_{f2}+k_{f1})}\right)cos\left[\left(2k_{f2}+k_{f1}\right)(x-ct)\right]x-\left(\frac{\gamma A_1A_2{}^2(\frac{1}{2}k_{f1}{}^2k_{f2}{}^2-k_{f1}k_{f2}{}^3)}{8(2k_{f2}-k_{f1})}\right)cos\left[\left(2k_{f2}-k_{f1}\right)(x-ct)\right]x\\
&-\left(\frac{\gamma A_2{}^2A_3(k_{f2}{}^3k_{f3}+\frac{1}{2}k_{f2}{}^2k_{f3}{}^2)}{8(2k_{f2}+k_{f3})}\right)cos\left[\left(2k_{f2}+k_{f3}\right)(x-ct)\right]x-\left(\frac{\gamma A_2{}^2A_3(k_{f2}{}^3k_{f3}-\frac{1}{2}k_{f2}{}^2k_{f3}{}^2)}{8(2k_{f2}-k_{f3})}\right)cos\left[\left(2k_{f2}-k_{f3}\right)(x-ct)\right]x\\
&+\frac{\gamma A_2{}^2A_4k_{f2}{}^2k_{f4}{}^2}{16(2k_{f2}+k_{f4})}sin\left[\left(2k_{f2}+k_{f4}\right)(x-ct)\right]x+\frac{\gamma A_2{}^2A_4k_{f2}{}^2k_{f4}{}^2}{16(2k_{f2}-k_{f4})}sin\left[\left(2k_{f2}-k_{f4}\right)(x-ct)\right]x\\
&+\left(\frac{\gamma A_1A_3{}^2(\frac{1}{2}k_{f1}{}^2k_{f3}{}^2+k_{f1}k_{f3}{}^3)}{8(2k_{f3}+k_{f1})}\right)cos\left[\left(2k_{f3}+k_{f1}\right)(x-ct)\right]x+\left(\frac{\gamma A_1A_3{}^2(\frac{1}{2}k_{f1}{}^2k_{f3}{}^2-k_{f1}k_{f3}{}^3)}{8(2k_{f3}-k_{f1})}\right)cos\left[\left(2k_{f3}-k_{f1}\right)(x-ct)\right]x\\
&-\left(\frac{\gamma A_2A_3{}^2(\frac{1}{2}k_{f2}{}^2k_{f3}{}^2+k_{f2}k_{f3}{}^3)}{8(2k_{f3}+k_{f2})}\right)sin\left[\left(2k_{f3}+k_{f2}\right)(x-ct)\right]x+\left(\frac{\gamma A_2A_3{}^2(\frac{1}{2}k_{f2}{}^2k_{f3}{}^2-k_{f2}k_{f3}{}^3)}{8(2k_{f3}-k_{f2})}\right)sin\left[\left(2k_{f3}-k_{f2}\right)(x-ct)\right]x\\
&-\left(\frac{\gamma A_3{}^2A_4(k_{f3}{}^3k_{f4}+\frac{1}{2}k_{f3}{}^2k_{f4}{}^2)}{8(2k_{f3}+k_{f4})}\right)sin\left[\left(2k_{f3}+k_{f4}\right)(x-ct)\right]x+\left(\frac{\gamma A_3{}^2A_4(k_{f3}{}^3k_{f4}-\frac{1}{2}k_{f3}{}^2k_{f4}{}^2)}{8(2k_{f3}-k_{f4})}\right)sin\left[\left(2k_{f3}-k_{f4}\right)(x-ct)\right]x\\
&-\left(\frac{\gamma A_1A_4{}^2(\frac{1}{2}k_{f1}{}^2k_{f4}{}^2+k_{f1}k_{f4}{}^3)}{8(2k_{f4}+k_{f1})}\right)cos\left[\left(2k_{f4}+k_{f1}\right)(x-ct)\right]x-\left(\frac{\gamma A_1A_4{}^2(\frac{1}{2}k_{f1}{}^2k_{f4}{}^2-k_{f1}k_{f4}{}^3)}{8(2k_{f4}-k_{f1})}\right)cos\left[\left(2k_{f4}-k_{f1}\right)(x-ct)\right]x\\
&+\frac{\gamma A_2A_4{}^2k_{f2}{}^2k_{f4}{}^2}{16(2k_{f4}+k_{f2})}sin\left[\left(2k_{f4}+k_{f2}\right)(x-ct)\right]x+\frac{\gamma A_2A_4{}^2k_{f2}{}^2k_{f4}{}^2}{16(2k_{f4}-k_{f2})}sin\left[\left(2k_{f4}-k_{f2}\right)(x-ct)\right]x\\
&-\left(\frac{\gamma A_3A_4{}^2(\frac{1}{2}k_{f3}{}^2k_{f4}{}^2+k_{f3}k_{f4}{}^3)}{8(2k_{f4}+k_{f3})}\right)cos\left[\left(2k_{f4}+k_{f3}\right)(x-ct)\right]x-\left(\frac{\gamma A_3A_4{}^2(k_{f3}k_{f4}{}^3-\frac{1}{2}k_{f3}{}^2k_{f4}{}^2)}{8(2k_{f4}-k_{f3})}\right)cos\left[\left(2k_{f4}-k_{f3}\right)(x-ct)\right]x\\
&-\left(\frac{\gamma A_1A_2A_3(k_{f1}{}^2k_{f2}k_{f3}+k_{f1}k_{f2}{}^2k_{f3}+k_{f1}k_{f2}k_{f3}{}^2)}{8(k_{f1}+k_{f2}+k_{f3})}\right)sin\left[\left(k_{f1}+k_{f2}+k_{f3}\right)(x-ct)\right]x\\
&-\left(\frac{\gamma A_1A_2A_3(k_{f1}{}^2k_{f2}k_{f3}+k_{f1}k_{f2}{}^2k_{f3}-k_{f1}k_{f2}k_{f3}{}^2)}{8(k_{f1}+k_{f2}-k_{f3})}\right)sin\left[\left(k_{f1}+k_{f2}-k_{f3}\right)(x-ct)\right]x\\
&-\left(\frac{\gamma A_1A_3A_4(k_{f1}{}^2k_{f3}k_{f4}+k_{f1}k_{f3}{}^2k_{f4}+k_{f1}k_{f3}k_{f4}{}^2)}{8(k_{f1}+k_{f3}+k_{f4})}\right)sin\left[\left(k_{f1}+k_{f3}+k_{f4}\right)(x-ct)\right]x\\
&+\left(\frac{\gamma A_1A_24(k_{f1}{}^2k_{f3}k_{f4}+k_{f1}k_{f3}{}^2k_{f4}-k_{f1}k_{f3}k_{f4}{}^2)}{8(k_{f1}+k_{f3}-k_{f4})}\right)sin\left[\left(k_{f1}+k_{f3}-k_{f4}\right)(x-ct)\right]x\\
&-\left(\frac{\gamma A_1A_2A_4(k_{f1}k_{f2}{}^2k_{f4}+k_{f1}k_{f2}k_{f4}{}^2)}{8(k_{f1}+k_{f2}+k_{f4})}\right)cos\left[\left(k_{f1}+k_{f2}+k_{f4}\right)(x-ct)\right]x\\
&-\left(\frac{\gamma A_1A_2A_4(k_{f1}k_{f2}{}^2k_{f4}-k_{f1}k_{f2}k_{f4}{}^2)}{8(k_{f1}+k_{f2}-k_{f4})}\right)cos\left[\left(k_{f1}+k_{f2}-k_{f4}\right)(x-ct)\right]x\\
&+\left(\frac{\gamma A_1A_2A_3(k_{f1}{}^2k_{f2}k_{f3}-k_{f1}k_{f2}{}^2k_{f3}-k_{f1}k_{f2}k_{f3}{}^2)}{8(k_{f2}+k_{f3}-k_{f1})}\right)sin\left[\left(k_{f2}+k_{f3}-k_{f1}\right)(x-ct)\right]x\\
&+\left(\frac{\gamma A_1A_2A_4(k_{f1}k_{f3}{}^2k_{f4}-k_{f1}{}^2k_{f3}k_{f4}-k_{f1}k_{f2}k_{f3}{}^2)}{8(k_{f1}+k_{f4}-k_{f3})}\right)sin\left[\left(k_{f1}+k_{f4}-k_{f3}\right)(x-ct)\right]x\\
&+\left(\frac{\gamma A_3A_4A_1(k_{f1}{}^2k_{f3}k_{f4}-k_{f1}k_{f3}{}^2k_{f4}-k_{f1}k_{f2}k_{f4}{}^2)}{8(k_{f3}+k_{f4}-k_{f1})}\right)sin\left[\left(k_{f3}+k_{f4}-k_{f1}\right)(x-ct)\right]x\\
&-\left(\frac{\gamma A_1A_2A_4(k_{f1}k_{f2}{}^2k_{f4}+k_{f1}k_{f2}k_{f4}{}^2)}{8(k_{f2}+k_{f4}-k_{f1})}\right)cos\left[\left(k_{f2}+k_{f4}-k_{f1}\right)(x-ct)\right]x\\
&+\left(\frac{\gamma A_1A_2A_3(k_{f1}k_{f3}k_{f3}{}^2-k_{f1}k_{f2}{}^2k_{f3}+k_{f1}{}^2k_{f2}k_{f3})}{8(k_{f1}+k_{f3}-k_{f2})}\right)sin\left[\left(k_{f1}+k_{f3}-k_{f2}\right)(x-ct)\right]x\\
&-\left(\frac{\gamma A_2A_3A_4(k_{f2}{}^2k_{f3}k_{f4}-k_{f2}k_{f3}k_{f4}{}^2)}{8(k_{f2}+k_{f3}+k_{f4})}\right)cos\left[\left(k_{f2}+k_{f3}+k_{f4}\right)(x-ct)\right]x\\
&-\left(\frac{\gamma A_2A_3A_4(k_{f2}{}^2k_{f3}k_{f4}-k_{f2}k_{f3}k_{f4}{}^2)}{8(k_{f2}+k_{f4}-k_{f3})}\right)cos\left[\left(k_{f2}+k_{f4}-k_{f3}\right)(x-ct)\right]x\\
&+\left(\frac{\gamma A_2A_3A_4(k_{f2}{}^2k_{f3}k_{f4}+k_{f2}k_{f3}k_{f4}{}^2)}{8(k_{f2}+k_{f3}-k_{f4})}\right)cos\left[\left(k_{f2}+k_{f3}-k_{f4}\right)(x-ct)\right]x\\
&-\left(\frac{\gamma A_2A_3A_4(k_{f2}{}^2k_{f3}k_{f4}+k_{f2}k_{f3}k_{f4}{}^2)}{8(k_{f3}+k_{f4}-k_{f2})}\right)cos\left[\left(k_{f3}+k_{f4}-k_{f2}\right)(x-ct)\right]x\\
&-\left(\frac{\gamma A_1A_2A_4k_{f1}k_{f2}k_{f4}{}^2)}{8(k_{f1}+k_{f4}-k_{f2})}\right)cos\left[\left(k_{f1}+k_{f4}-k_{f2}\right)(x-ct)\right]x.
\end{aligned}
\tag{A5}
$$

Appendix B

Table A1 summarises the derived nonlinearity parameters with an increasing number of fundamental frequencies. If only one frequency is used, the second and third harmonic frequencies are available for evaluation. With two fundamental frequencies, 12 nonlinearity parameters were derived, including the linear-dependent parameter δf_n, the harmonic parameters $\delta 2f_n$, $\delta 3f_n$, and the modulated response parameters $\delta f_n \pm f_m$ and $\delta 2f_n \pm f_m$. If four frequencies are transmitted in parallel, the number of harmonics and sidebands increases to a maximum of 64. New parameters of the type $\delta f_n + f_m \pm f_p$ were added.

The second harmonic parameters $\delta 2f_n$ and sideband $\delta f_n \pm f_m$ were derived from the second-order nonlinearity parameter β. All others have their origin in the third-order nonlinearity parameter, γ.

Table A1. Comparison—nonlinearity parameters.

One Excitation Frequency		Two Excitation Frequencies		Four Excitation Frequencies	
Frequencies	Number	Frequencies	Number	Frequencies	Number
-	-	f_n	2	f_n	4
$2f_n$	1	$2f_n$	2	$2f_n$	4
$3f_n$	1	$3f_n$	2	$3f_n$	4
-	-	$f_n \pm f_m$	2	$f_n \pm f_m$	12
-	-	$2f_n \pm f_m$	4	$2f_n \pm f_m$	24
-	-	-	-	$f_n + f_m \pm f_p$	16
	$\sum 2$		$\sum 12$		$\sum 64$

References

1. Li, Z.L.; Achenbach, J.D. Reflection and Transmission of Rayleigh Surface Waves by a Material Interphase. *J. Appl. Mech.* **1991**, *58*, 688–694. [CrossRef]
2. Achenbach, J.D.; Lin, W.; Keer, L.M. Mathematical modelling of ultrasonic wave scattering by sub-surface cracks. *Ultrasonics* **1986**, *24*, 207–215. [CrossRef]
3. Angel, Y.C.; Achenbach, J.D. Reflection and transmission of obliquely incident Rayleigh waves by a surface-breaking crack. *Acoust. Soc. Am.* **1983**, *74*, S87. [CrossRef]
4. Hikata, A.; Chick, B.B.; Elbaum, C. Dislocation Contribution to the Second Harmonic Generation of Ultrasonic Waves. *J. Appl. Phys.* **1965**, *36*, 229–236. [CrossRef]
5. Cantrell, J.H.; Yost, W.T. Nonlinear ultrasonic characterization of fatigue microstructures. *Int. J. Fatigue* **2001**, *23*, 487–490. [CrossRef]
6. Jhang, K.-Y.; Kim, K.-C. Evaluation of material degradation using nonlinear acoustic effect. *Ultrasonics* **1999**, *37*, 39–44. [CrossRef]
7. Frouin, J.; Sathish, S.; Matikas, T.E.; Na, J.K. Ultrasonic Linear and Nonlinear Behavior of Fatigued Ti–6Al–4V. *J. Mater. Res.* **1999**, *14*, 1295–1298. [CrossRef]
8. Rothenfusser, M.; Mayr, M.; Baumann, J. Acoustic nonlinearities in adhesive joints. *Ultrasonics* **2000**, *38*, 322–326. [CrossRef]
9. Yost, W.T. Nonlinear ultrasonic pulsed measurements and applications to metal processing and fatigue. In Proceedings of the 27th Annual Review of Progress in Quantitative Nondestructive Evaluation, Ames, IA, USA, 16–20 July 2000; pp. 1268–1275.
10. Jeong, H.; Nahm, S.-H.; Jhang, K.-Y.; Nam, Y.-H. A nondestructive method for estimation of the fracture toughness of CrMoV rotor steels based on ultrasonic nonlinearity. *Ultrasonics* **2003**, *41*, 543–549. [CrossRef]
11. Ostrovsky, L.A.; Johnson, P.A. Dynamic nonlinear elasticity in geomaterials. *Riv. Nuovo Cim.* **2001**, *24*, 1–46. [CrossRef]
12. Malfense Fierro, G.P.; Meo, M. Residual fatigue life estimation using a nonlinear ultrasound modulation method. *Smart Mater. Struct.* **2015**, *24*, 25040. [CrossRef]
13. Malfense Fierro, G.P.; Meo, M. Nonlinear imaging (NIM) of flaws in a complex composite stiffened panel using a constructive nonlinear array (CNA) technique. *Ultrasonics* **2017**, *74*, 30–47. [CrossRef]
14. Malfense Fierro, G.P. Development of Nonlinear Ultrasound Techniques for Multidisciplinary Engineering Applications. Ph.D. Thesis, University of Bath, Bath, UK, 2014.
15. Jingpin, J.; Xiangji, M.; Cunfu, H.; Bin, W. Nonlinear Lamb wave-mixing technique for micro-crack detection in plates. *NDT E Int.* **2017**, *85*, 63–71. [CrossRef]
16. Amura, M.; Meo, M.; Amerini, F. Baseline-free estimation of residual fatigue life using a third order acoustic nonlinear parameter. *Acoust. Soc. Am.* **2011**, *130*, 1829–1837. [CrossRef]
17. Mevissen, F.; Meo, M. A Nonlinear Ultrasonic Modulation Method for Crack Detection in Turbine Blades. *Aerospace* **2020**, *7*, 72. [CrossRef]

18. Pieczonka, L.; Klepka, A.; Martowicz, A.; Staszewski, W.J. Nonlinear vibroacoustic wave modulations for structural damage detection: An overview. *Opt. Eng.* **2015**, *55*, 11005. [CrossRef]
19. Van den Abeele, K.E.-A.; Johnson, P.A.; Sutin, A. Nonlinear Elastic Wave Spectroscopy (NEWS) Techniques to Discern Material Damage, Part I: Nonlinear Wave Modulation Spectroscopy (NWMS). *Res. Nondestruct. Eval.* **2000**, *12*, 17–30. [CrossRef]
20. Van den Abeele, K.E.-A.; Carmeliet, J.; Ten Cate, J.A.; Johnson, P.A. Nonlinear Elastic Wave Spectroscopy (NEWS) Techniques to Discern Material Damage, Part II: Single-Mode Nonlinear Resonance Acoustic Spectroscopy. *Res. Nondestruct. Eval.* **2000**, *12*, 31–42. [CrossRef]
21. Straka, L.; Yagodzinskyy, Y.; Landa, M.; Hänninen, H. Detection of structural damage of aluminum alloy 6082 using elastic wave modulation spectroscopy. *NDT E Int.* **2008**, *41*, 554–563. [CrossRef]
22. Greenhall, J.; Grutzik, S.; Graham, A.; Sinha, D.N.; Pantea, C. Nonlinear acoustic crack detection in thermoelectric wafers. *Mech. Syst. Signal Process.* **2020**, *139*, 106598. [CrossRef]
23. Fan, C.; Yang, L.; Zhao, Y. Ultrasonic multi-frequency time-reversal-based imaging of extended targets. *NDT E Int.* **2020**, *113*, 102276. [CrossRef]
24. Queirós, R.; Corrêa Alegria, F.; Silva Girão, P.; Cruz Serra, A. A multi-frequency method for ultrasonic ranging. *Ultrasonics* **2015**, *63*, 86–93. [CrossRef] [PubMed]
25. Novak, A.; Bentahar, M.; Tournat, V.; El Guerjouma, R.; Simon, L. Nonlinear acoustic characterization of micro-damaged materials through higher harmonic resonance analysis. *NDT E Int.* **2012**, *45*, 1–8. [CrossRef]
26. Pfleiderer, K.; Aufrecht, J.; Solodov, I.; Busse, G. Multi-frequency ultrasonic NDE for early defect recognition and imaging. *IEEE Ultrason. Symp.* **2004**, *1*, 693–696.
27. Abraham, S.T.; Shivaprasad, S.; Sreevidya, N.; Das, C.R.; Albert, S.K.; Venkatraman, B.; Balasubramaniam, K. A Novel Multi-frequency Nonlinear Ultrasonic Approach for the Characterization of Annealed Polycrystalline Microstructure. *Metall. Mater. Trans. A* **2019**, *50*, 5567–5573. [CrossRef]
28. Gao, T.; Sun, H.; Hong, Y.; Qing, X. Hidden corrosion detection using laser ultrasonic guided waves with multi-frequency local wavenumber estimation. *Ultrasonics* **2020**, *108*, 106182. [CrossRef]
29. Malfense Fierro, G.P.; Meo, M. Non-linear Phased Array Imaging of Flaws Using a Dual and Tri Frequency Modulation Technique. *Front. Built Environ.* **2020**, *6*, 68. [CrossRef]
30. Deng, M. Cumulative second-harmonic generation accompanying nonlinear shear horizontal mode propagation in a solid plate. *J. Appl. Phys.* **1998**, *84*, 3500–3505.
31. Deng, M. Cumulative second-harmonic generation of Lamb-mode propagation in a solid plate. *J. Appl. Phys.* **1999**, *85*, 3051–3058. [CrossRef]
32. Pruell, C.; Kim, J.-Y.; Qu, J.; Jacobs, L.J. Evaluation of fatigue damage using nonlinear guided waves. *Smart Mater. Struct.* **2009**, *18*, 35003. [CrossRef]
33. Pruell, C.; Kim, J.-Y.; Qu, J.; Jacobs, L.J. Evaluation of plasticity driven material damage using Lamb waves. *Appl. Phys. Lett.* **2007**, *91*, 231911. [CrossRef]
34. Niethammer, M.; Jacobs, L.J.; Qu, J.; Jarzynski, J. Time-frequency representations of Lamb waves. *J. Acoust. Soc. Am.* **2001**, *109*, 1841–1847. [CrossRef]
35. Herrmann, J.; Kim, J.-Y.; Jacobs, L.J.; Qu, J.; Littles, J.W.; Savage, M.F. Assessment of material damage in a nickel-base superalloy using nonlinear Rayleigh surface waves. *J. Appl. Phys.* **2006**, *99*, 124913. [CrossRef]
36. Mevissen, F.; Meo, M. Detection of Cracks in Turbine Blades with Nonlinear Ultrasonic Frequency Modulation. In Proceedings of the 12th International Workshop on Structural Health Monitoring: Enabling Intelligent Life-Cycle Health Management for Industry Internet of Things (IIOT), IWSHM 2019, Stanford, CL, USA, 10–12 September 2019; pp. 824–834.
37. Malfense Fierro, G.P.; Meo, M. IWSHM 2017: Structural health monitoring of the loosening in a multi-bolt structure using linear and modulated nonlinear ultrasound acoustic moments approach. *Struct. Health Monit.* **2018**, *17*, 1349–1364. [CrossRef]
38. Amerini, F.; Meo, M. Structural health monitoring of bolted joints using linear and nonlinear acoustic/ultrasound methods. *Struct. Health Monit.* **2011**, *10*, 659–672. [CrossRef]
39. Blanloeuil, P.; Rose, L.; Guinto, J.A.; Veidt, M.; Wang, C.H. Closed crack imaging using time reversal method based on fundamental and second harmonic scattering. *Wave Motion* **2016**, *66*, 156–176. [CrossRef]
40. Ohara, Y.; Mihara, T.; Sasaki, R.; Ogata, T.; Yamamoto, S.; Kishimoto, Y.; Yamanaka, K. Imaging of closed cracks using nonlinear response of elastic waves at subharmonic frequency. *Appl. Phys. Lett.* **2007**, *90*, 11902. [CrossRef]

sensors MDPI

Article

Negative Index Metamaterial Lens for Subwavelength Microwave Detection

Srijan Datta [1,*], Saptarshi Mukherjee [2], Xiaodong Shi [1], Mahmood Haq [1], Yiming Deng [1], Lalita Udpa [1] and Edward Rothwell [1]

[1] Department of Electrical and Computer Engineering, Michigan State University, East Lansing, MI 48824, USA; shixiaod@msu.edu (X.S.); haqmahmo@egr.msu.edu (M.H.); dengyimi@egr.msu.edu (Y.D.); udpal@egr.msu.edu (L.U.); rothwell@egr.msu.edu (E.R.)

[2] Lawrence Livermore National Laboratory, Livermore, CA 94550, USA; mukherjee5@llnl.gov

* Correspondence: dattasri@egr.msu.edu

Abstract: Metamaterials are engineered periodic structures designed to have unique properties not encountered in naturally occurring materials. One such unusual property of metamaterials is the ability to exhibit negative refractive index over a prescribed range of frequencies. A lens made of negative refractive index metamaterials can achieve resolution beyond the diffraction limit. This paper presents the design of a metamaterial lens and its use in far-field microwave imaging for subwavelength defect detection in nondestructive evaluation (NDE). Theoretical formulation and numerical studies of the metamaterial lens design are presented followed by experimental demonstration and characterization of metamaterial behavior. Finally, a microwave homodyne receiver-based system is used in conjunction with the metamaterial lens to develop a far-field microwave NDE sensor system. A subwavelength focal spot of size 0.82λ was obtained. The system is shown to be sensitive to a defect of size 0.17λ × 0.06λ in a Teflon sample. Consecutive positions of the defect with a separation of 0.23λ was resolvable using the proposed system.

Keywords: metamaterial; lenses; refractive index; microwave sensors; nondestructive testing

Citation: Datta, S.; Mukherjee, S.; Shi, X.; Haq, M.; Deng, Y.; Udpa, L.; Rothwell, E. Negative Index Metamaterial Lens for Subwavelength Microwave Detection. *Sensors* **2021**, *21*, 4782. https://doi.org/10.3390/s21144782

Academic Editor: Anthony N. Sinclair

Received: 1 June 2021
Accepted: 10 July 2021
Published: 13 July 2021

1. Introduction

In 1968, V. Veselago theoretically introduced the electrodynamics of materials having simultaneous negative values of permittivity ϵ and permeability μ [1]. He showed that such materials will exhibit unusual properties such as negative refraction, reversal of Doppler shift and backward Cherenkov radiation. The electric field, magnetic field, and wave vector of a plane wave form a left-handed triplet in such a medium, instead of the conventional right-handed one. The word "metamaterial" was coined for such materials, alluding to their unusual properties, not generally encountered in nature. The first left-handed metamaterial (LHM) structure was realized by Smith et al., in their seminal paper of 2000, where they showed that an alternating periodic array of split-ring resonators (SRRs), and thin wires can produce an effective medium having a negative refractive index in the microwave regime [2]. Extensive research demonstrating and characterizing the left-handed behavior of such structures followed [3–5]. Early on, negative refractive index structures were a controversial topic, and their existence was disputed by researchers [6,7]. However, over the past two decades, there has been significant evidence that certain periodic structures can indeed have an effective negative refractive index over a limited range of frequencies [8–10]. Such periodic structures, even though inhomogeneous, can behave as a homogeneous medium in response to electromagnetic (EM) waves with appropriately long wavelength. The homogenized negative index behavior of inhomogeneous metamaterial structures has been described by an effective negative ϵ and μ of the periodic arrays [11].

Metamaterials have inspired many novel applications based on their negative refractive index. One of the most ingenious applications of LHM structures was put forward by J.B. Pendry, where he showed that a negative refractive index material can act as a "super lens", capable of achieving subwavelength focusing in the far field by restoring the amplitude of evanescent wave components [12]. The highest resolution that can be obtained using a conventional lens in the far field is limited by the operating wavelength, due to the physics of diffraction. The breaking of this diffraction limit using point source focusing (Figure 1a) and evanescent wave amplification of a LHM lens has been one of the significant driving factors for metamaterial research. Various metamaterial designs, operating from radio to optical frequencies, have been developed and shown to achieve subwavelength focusing [13–17].

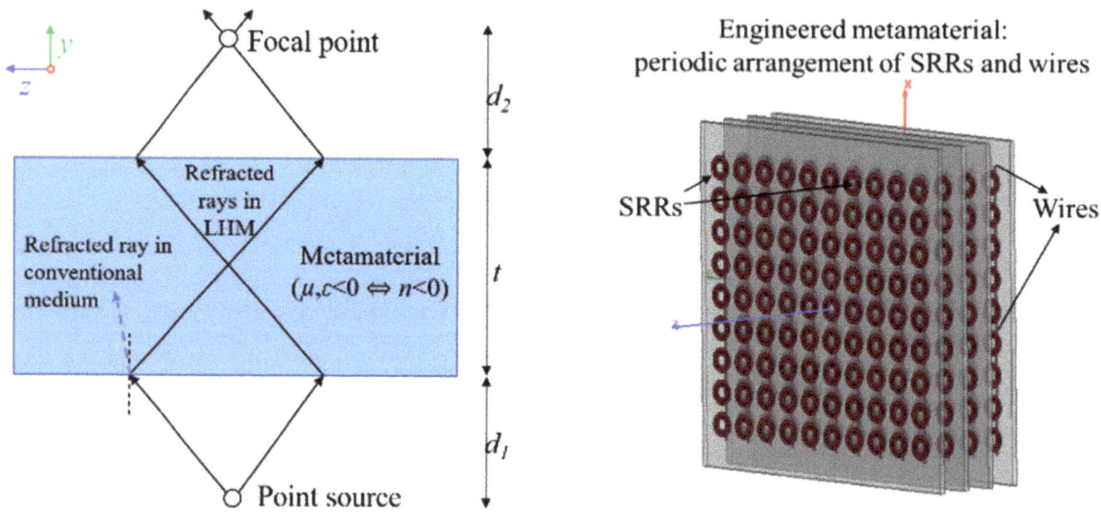

Figure 1. (**a**) Ray diagram showing reversal of Snell's law in a metamaterial medium. For a conventional medium, the diverging beams from a point source will not come into focus. (**b**) Printed circuit board (PCB) implementation of a metamaterial consisting of alternating periodic arrangement of SRRs and wires. The structure will exhibit an effective negative refractive index over a range of frequencies under specific incident wave polarization.

This paper reports the design of a metamaterial lens and its experimental implementation for far-field microwave detection of subwavelength defects. Far-field microwave NDE offers the advantage of rapid scan times, but is constrained by the diffraction limit from detecting smaller subwavelength defects [18]. A single SRR coupled with a transmission line behaves as a LC tank circuit, whose resonant frequency can be changed in the presence of a load. Although extensive research on such metamaterial-inspired near-field sensors have been described in literature, they do not offer the advantages of far-field systems [19–22]. While numerical studies of LHMs as lenses in the far field have been undertaken [23–26], the practical feasibility of their use has not been widely demonstrated. One study of far-field microwave imaging is reported by Shrieber et al., who present subwavelength defect detection in fiberglass composites [27]. An LHM lens-based microwave hyperthermia scheme for treatment of tumors is proposed in [28]. The metamaterial lens concept has been extended to ultrasonics as well [29], with various studies demonstrating subwavelength imaging using acoustic LHM lenses being reported [30–32]. The authors of the present paper recently reported a numerical study on enhancement of far-field microwave time-reversal imaging resolution using a homogenized model of a metamaterial lens [33]. Preliminary studies on the physical design of a metamaterial lens were presented by the authors in [34]. The present contribution focuses on detailed numerical and experimental characterization of the lens design and its feasibility for far-field subwavelength defect detection. Existing literature on using LHMs in the far field employs electromagnetic windowing techniques to realize subwavelength focusing or defect detection. In this paper, a homodyne receiver-based architecture is proposed to be used in conjunction with the metamaterial lens for far-field measurements. The high SNR associated with such synchronous detection allows the lens to be characterized in free space and, hence, provides a system that can be used in the field under practical conditions.

The theory governing the working principles of metamaterials is briefly discussed in the Section 2. The Section 3 provides a numerical study and EM parameter retrieval of the lens design using the commercial software HFSS. The Section 4 presents the experimental characterization of the metamaterial lens and its implementation for detection of subwavelength defects using the homodyne system.

2. Theory

A periodic array of conducting elements can act as an effective medium for EM scattering when the wavelength is much longer than the element dimensions, i.e.,

$$a \ll \lambda \tag{1}$$

where a is the dimension of a unit cell of the periodic array, and λ is the operating wavelength in the effective medium. The EM response of an effective medium is determined by the configuration of the unit cell and can be characterized by an effective relative permeability μ_{eff} and effective relative permittivity ϵ_{eff}. Metamaterials with simultaneous negative μ_{eff} and ϵ_{eff} over a specific range of frequencies are termed as "double negative" metamaterials. One typical example of a unit cell comprises two distinct structures: an SRR element with dominant magnetic response and a thin wire element with dominant electric response.

A periodic array of the SRR elements can exhibit an effective magnetic behavior, similar to that of magnetic plasmas, in the microwave regime [35]. Under excitation by an external magnetic field parallel to the axis of the SRRs, the array behaves as a bulk medium having an effective relative permeability given by

$$\mu_{eff}(\omega) = 1 - \frac{F\omega^2}{\omega^2 - \omega_0^2 + j\omega\Gamma} \tag{2}$$

where F is the fractional volume of the unit cell occupied by the rings, Γ is the dissipation factor, ω_0 is the resonant frequency of the SRRs, and ω is the frequency of the excitation field. Equation (2) shows that the real part of μ_{eff} is negative for frequencies ω greater than resonant frequency ω_0 and less than magnetic plasma frequency ω_{mp} given by $\omega_{mp} = \omega_0/\sqrt{1-F}$. Propagating wave modes are prohibited in this frequency band due to negative μ_{eff} of the SRR medium.

A periodic array of the thin metallic wire elements, under the influence of a time-varying electric field, can mimic an electric plasma at microwave frequencies [36]. The effective relative permittivity of such an array in the presence of an external electric field parallel to the wires is given by

$$\epsilon_{eff}(\omega) = 1 - \frac{\omega_{ep}^2}{\omega^2} \tag{3}$$

where ω_{ep} is the electric plasma frequency, and ω is the frequency of the excitation field. Equation (3) shows that the real part of ϵ_{eff} has negative values for frequencies $\omega < \omega_{ep}$. This causes the wire medium to prohibit propagating modes in that frequency regime.

Combining both the SRR and wire elements in a periodic array gives rise to a metamaterial medium having simultaneous negative μ_{eff} and ϵ_{eff} over a certain range of frequencies. Assuming that there is no direct interaction between the SRR and wire media, the refractive index n of the resulting metamaterial is given through

$$n^2 = \mu_{eff}(\omega)\epsilon_{eff}(\omega) \tag{4}$$

The negative square root in the calculation of n is chosen in (4) when both $\mu_{eff}(\omega)$ and $\epsilon_{eff}(\omega)$ are negative to account for propagation of left-handed waves in a metamaterial [37]. A left-handed transmission band occurs within the previously overlapping forbidden bands of negative μ_{eff} and negative ϵ_{eff}. The combined array behaves as a medium having an effective negative refractive index in this transmission band, and the transmission peak is referred to as a left-handed peak. Such a negative index medium not only focuses propagating waves but also enhances the evanescent wave component of the angular spectrum of the incident field, which contains high-resolution information. Subwavelength focusing beyond the diffraction limit is, thus, made possible by using a negative index metamaterial lens. As shown in Figure 1a, for an ideal lossless LHM lens of thickness t and refractive index $n = -1$, a diverging beam from a point source at a distance d_1 from the lens focuses first inside the lens and, then, outside the lens at a distance d_2 given by [12]

$$d_2 = t - d_1 \tag{5}$$

Although the presence of losses associated with fabricated LHMs causes them to deviate from perfect focusing capabilities of an ideal metamaterial lens, subwavelength focusing is still achievable using a lossy negative index lens [38].

3. Simulation

A metamaterial can be designed by simulating an infinite array of metamaterial unit cells using periodic boundary conditions. Figure 2a shows an HFSS model of the proposed unit cell along with incident field polarization and direction of propagation. The *Perfect E* and *Perfect H* boundary conditions of HFSS were applied on the *y-z* and *x-y* boundaries, respectively, to mimic an infinite array of unit cells and ensure correct polarization of the incident wave. FR4 ($\epsilon_r = 4.4$, $\tan\delta = 0.02$) of thickness 1.6 mm was used as the substrate for the PCB. Copper of thickness 35 micron was used as the conducting material. Wave ports were assigned on the *z-x* boundary to excite the model with a plane wave and obtain the S-parameters of the metamaterial medium. Figure 2b shows the dimensions of the unit cell. The dimensional parameters for the proposed design at 3.45 GHz are as follows: r = 1.5 mm, c = g = 0.2 mm, t = w = 0.9 mm, and a = 9.3 mm. The distance between consecutive PCB layers (length of unit cell model along z direction) is 6.5 mm. The dimensions were adapted from previous works by Aydin et al., where they demonstrated negative refraction and left handed focusing by a metamaterial lens in the 3–4 GHz regime [39]. Although the principal objective of designing a negative index LHM lens is high spatial resolution, which can be achieved by improving the losses in the design, the primary purpose of this work is to demonstrate the viability of using an LHM lens for subwavelength defect detection. Hence, optimization of the performance of the unit cell by parameterizing its dimensions was left for future work.

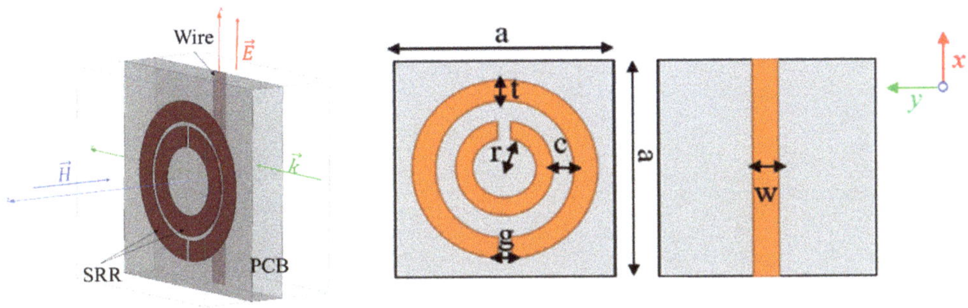

(a) (b)

Figure 2. (**a**) HFSS unit cell model. (**b**) Schematic of the metamaterial unit cell showing both sides of the PCB.

3.1. Scattering Parameters

Figure 3 shows the simulated S-parameters for three cases—an SRR-only medium, a wire-only medium, and a medium that combines both wires and SRRs. For the SRR-only medium, a dip in the transmission parameter S_{21} is observed around the resonant frequency of 3.45 GHz of the SRR (Figure 3a). This is due to the prohibition of propagating waves by the negative μ_{eff} of the medium. Figure 3b shows that the wire-only medium allows transmission (with less than 10 dB of insertion loss) above 5.5 GHz, which is the electric plasma frequency ω_{ep}. Propagating waves below this frequency are prohibited due to the negative ϵ_{eff} of the wire medium. Figure 3c shows that after combining both the SRR and wire, a transmission band is observed around 3.45 GHz. Left-handed waves are allowed to propagate in the frequency region where both μ_{eff} and ϵ_{eff} are simultaneously negative.

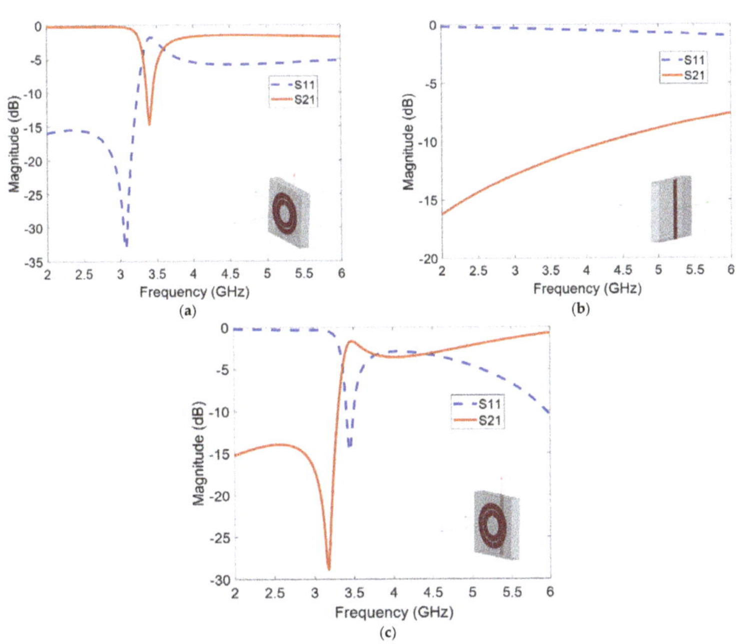

Figure 3. HFSS S-parameter results for (**a**) SRR-only medium, (**b**) wire-only medium, and (**c**) SRR and wire combined metamaterial medium. The respective HFSS models are shown in the insets of the figures.

3.2. Electromagnetic Parameter Retrieval

Extraction of the EM parameters from S-parameter data of the metamaterial design was done to verify left-handed nature of the transmission band. The procedure to determine the EM properties of the metamaterials is presented in Appendix A.

The extracted material properties for the combined SRR and wire medium simulation results are shown in Figure 4. From the normalized impedance curve in Figure 4a, a resonance near the two plasma frequencies (3.45 GHz and 5.5 GHz) of the metamaterial medium is observed as expected. Figure 4b shows that the real part of the extracted refractive index is negative, thus verifying the left-handed transmission band in this frequency region. The value of real part of n at the resonant frequency of 3.45 GHz is -2.18. The real parts of the extracted μ_{eff} and ϵ_{eff} are also simultaneously negative in the frequency region, as expected (Figure 4c,d). It should be noted that above 5.5 GHz, both μ_{eff} and ϵ_{eff} are simultaneously positive, rendering the refractive index to be positive above this frequency.

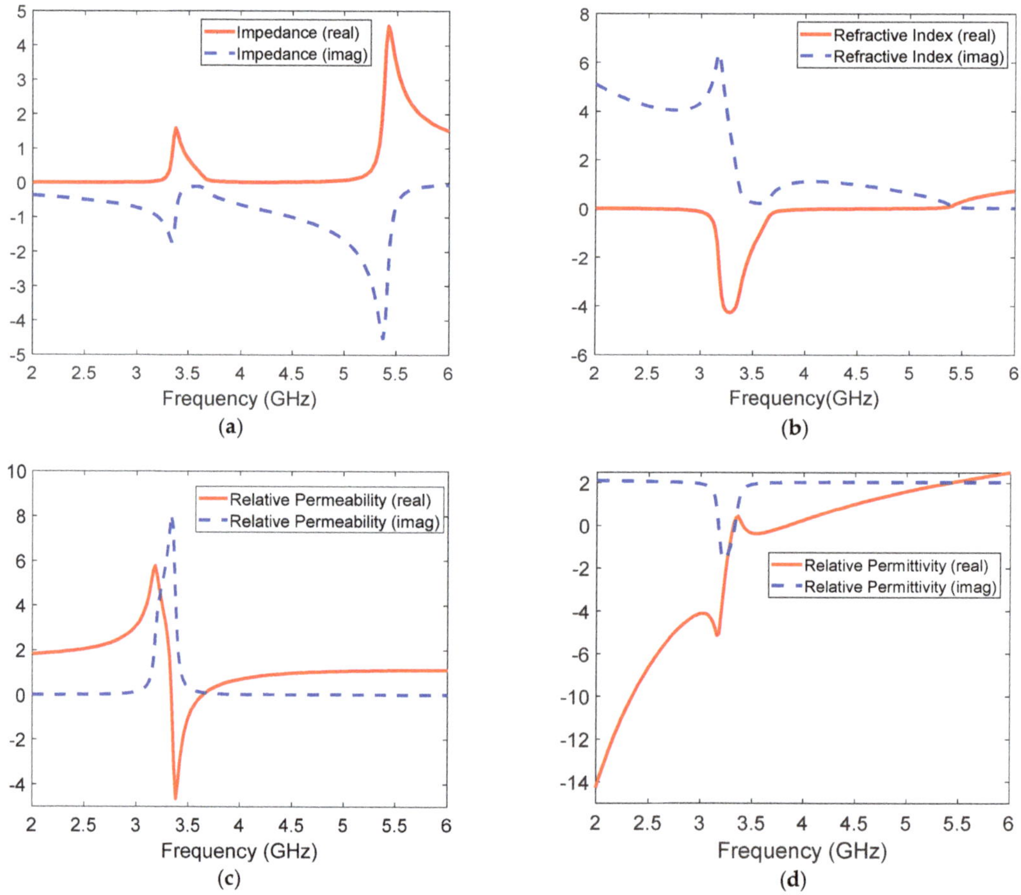

Figure 4. Extracted simulated EM parameters of the metamaterial design: (**a**) impedance, (**b**) refractive index, (**c**) permeability, and (**d**) permittivity.

4. Experiment

A metamaterial lens, consisting of $N_x = 20$, $N_y = 10$ and $N_z = 31$ unit cells, was fabricated for experimental validation. Figure 5 presents the fabricated metamaterial lens. An amount of 20 $\times$ 10 unit cells in the x-y plane were printed in a single FR4 PCB of thickness 1.6 mm, and 31 such boards were stacked in the z direction at an interval of 6.5 mm. The magnetic and electric field vectors are polarized along the z and x axes, respectively, while the wave propagation vector is along the y axis. The thickness of the lens, t, in direction of propagation is 100 mm. Frequency sweep measurements were done to confirm the presence of left-handed transmission peak. A homodyne detection-based scheme was used to experimentally validate the negative refractive index and determine the subwavelength focal spot at the left-handed transmission peak frequency. NDE results for a set of dielectric test samples are presented in Section 4.2.3 to show the feasibility of using an LHM lens for the detection of subwavelength defects.

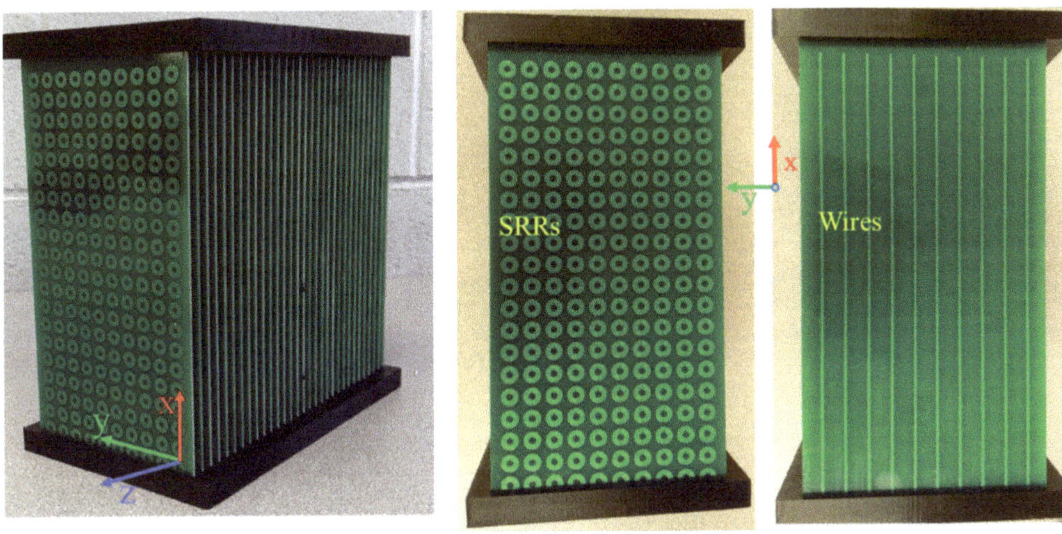

Figure 5. Fabricated metamaterial lens.

4.1. Transmission Characteristics

A large metallic screen (~10λ) with an aperture was used to obtain the transmission characteristics of the fabricated metamaterial lens. The metallic screen was implemented by attaching aluminum sheets to a Styrofoam board. An aperture of the size of the lens was cut in the middle of the board to allow for waves to pass through the lens only [40]. Wideband (675 MHz to 12 GHz) Vivaldi antennas were used as transmitter and receiver to illuminate the lens with a uniform plane wave. The antennas were placed 40 cm apart to ensure far-field measurements. The frequency sweep measurements were done using an Agilent EB070B vector network analyzer (VNA). Figure 6a shows the schematic of experimental set up. The measurements of S_{21} clearly indicate the presence of a left-handed transmission band with a peak transmission of -16 dB around 3.5 GHz (Figure 6b). The slight shift in frequency between the simulated and experimental results can be attributed to fabrication tolerances. Above 5.5 GHz, the metamaterial acts as a conventional medium having positive μ_{eff} and ϵ_{eff}. Conventional right handed waves are allowed to propagate in this frequency regime, and hence, the transmission band is observed.

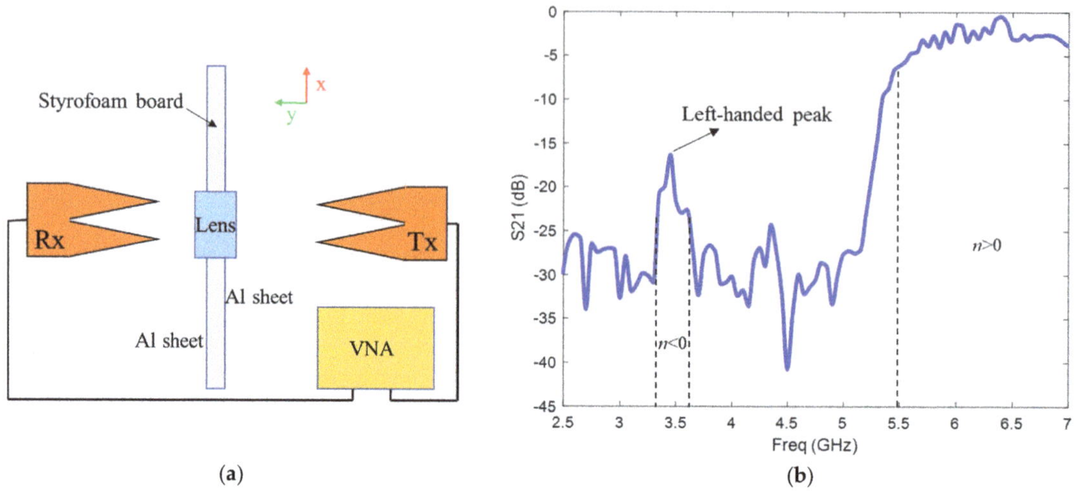

Figure 6. (**a**) Free space transmission experiment schematic. (**b**) Experimental transmission response of the metamaterial lens. The measurements were calibrated with respect to transmission in free space.

4.2. Left-Handed Characteristics

After experimentally confirming the left-handed transmission peak at 3.5 GHz, single frequency measurements using a homodyne receiver architecture were used to facilitate fast characterization of the metamaterial lens at 3.5 GHz. The schematic of the homodyne architecture is shown in Figure 7. The RF signal generator produces a continuous sinusoidal wave of frequency 3.5 GHz. The generated signal is passed through a splitter with one channel to the transmitting antenna and the other channel to the LO port of the mixer. The RF port of the mixer is connected to the receiver antenna. The DC signal produced at the IF of the mixer is read by a digital multimeter (DMM) and is proportional to the strength of the received signal. The high SNR associated with such synchronous detection allows the lens to be characterized in free space without the need for the windowing used in the frequency sweep measurements. Moreover, using homodyne detection circumnavigates the use of expensive RF instruments such as a VNA.

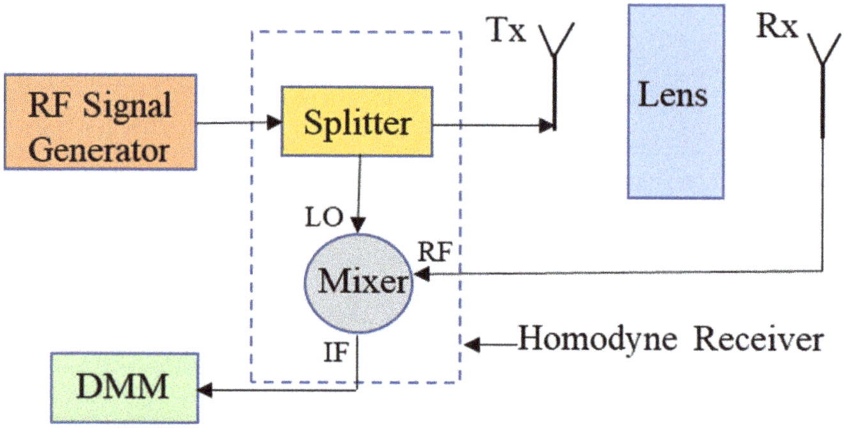

Figure 7. Schematic of homodyne architecture-based setup for single frequency measurements.

4.2.1. Negative Refraction

An imaging experiment was carried out to demonstrate negative refraction by the metamaterial lens and calculate its effective refractive index [33]. A standard gain horn antenna at 3.5 GHz was used as the transmitter, while a quarter wavelength monopole was used as the receiving probe. Outgoing spherical waves from the horn are incident at an angle θ_i at the first air-LHM interface. After undergoing negative refraction through the LHM of thickness t, the waves are shifted towards the side of the transmitter by a distance d at the second LHM–air interface. The angle of refraction can be determined by scanning the received signal amplitude can be calculated as $\theta_r = \tan^{-1}(d/t)$. The top view of the experimental setup is shown schematically in Figure 8.

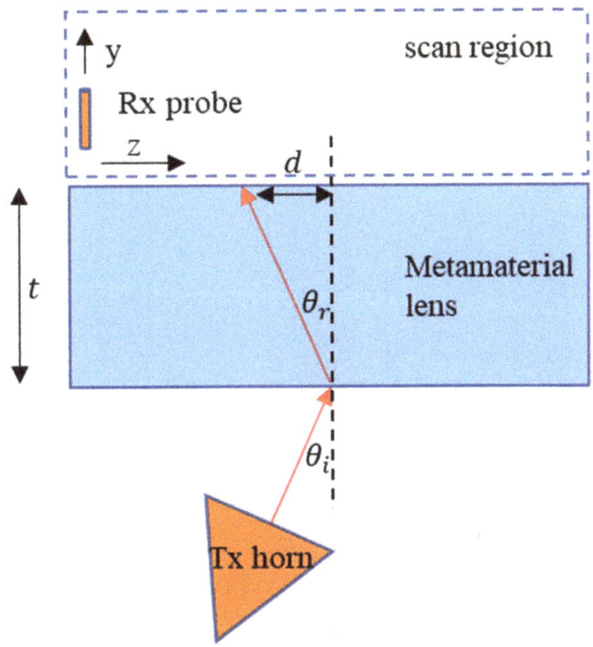

Figure 8. Negative refraction experiment schematic top view.

The receiving probe was mounted on a 2D scanner, moved in the y-z plane, and the received amplitude distribution was measured. A step size of 5 mm was used in both the z and y directions. The transmitting horn was placed at a distance of 12 cm (1.4λ) from the first air–LHM interface with an angle of incidence θ_i = 10°. The normalized measured amplitude distribution is shown in Figure 9a. The outgoing wave from the metamaterial lens has a beam profile centered towards the transmitting antenna. The angle of refraction, θ_r, through the fabricated lens of thickness t = 100 mm and beam shift d = 15 mm is calculated to be 8.53°. The real part of effective refractive index is thereby computed using Snell's law and is equal to −1.17 at 3.5 GHz.

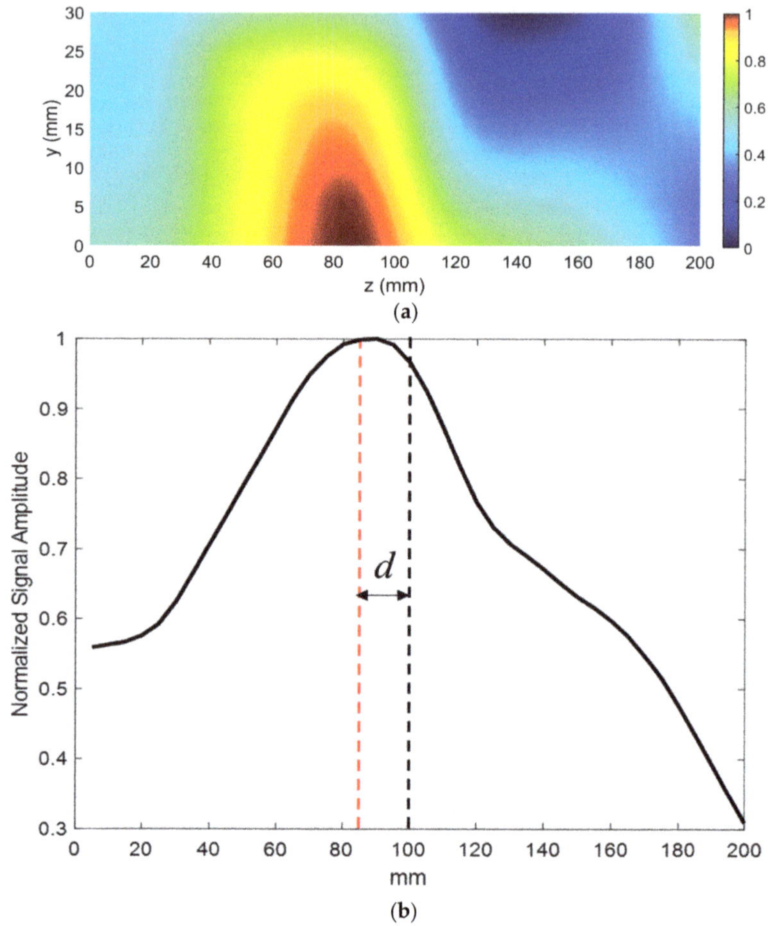

Figure 9. (**a**) Normalized received signal amplitude scan. The outgoing wave from the LHM has a beam profile centered at 85 mm. (**b**) Normalized line scan of the received signal amplitude at $y = 0$. The beam shift d is measured to be equal to 15 mm.

4.2.2. Subwavelength Focusing

The presence of negative refraction allows the possibility of using the fabricated metamaterial structure as a lens for subwavelength focusing. A monopole antenna produces an azimuthally symmetric field pattern, as does an ideal isotropic radiator. Therefore, a monopole with a resonant frequency of 3.5 GHz was used as the transmitter to demonstrate subwavelength focusing. Due to the negative refractive index of the metamaterial lens, diverging beams from the monopole antenna, placed at an appropriate distance, will be brought to focus outside the lens according to (5). Figure 10 presents the schematic top view of the experimental setup.

A quarter wavelength monopole was used as the probe for measuring the received signal. The receiving probe was mounted on a 2D scanner and moved on the y-z plane and the received amplitude distribution was measured. A step size of 5 mm was used in both the z and y directions. The transmitting monopole was placed at 60 mm (d_1) from the air–LHM interface. Figure 11a shows normalized received signal amplitude. A focal point is observed at a distance 30 mm (d_2) from the second LHM–air interface. The measured focal spot distance satisfies the relation in (5) for the fabricated lens of thickness 100 mm (t).

Figure 11b shows the normalized line scan at the focal plane (y = 30 mm). The full width at half maxima (FWHM) for the focal spot is found to be 70 mm (0.82λ).

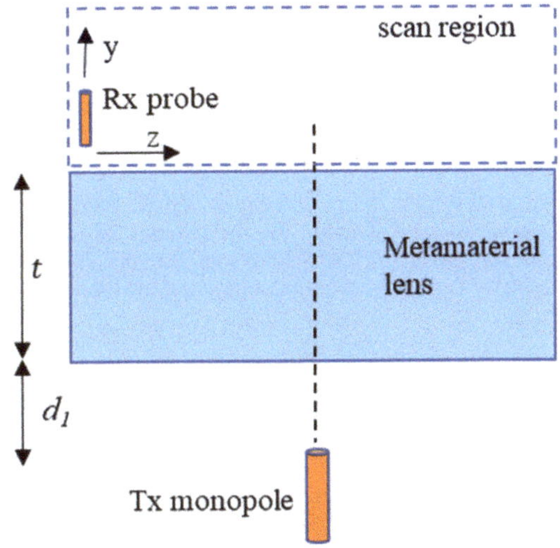

Figure 10. Subwavelength focusing experiment schematic top view.

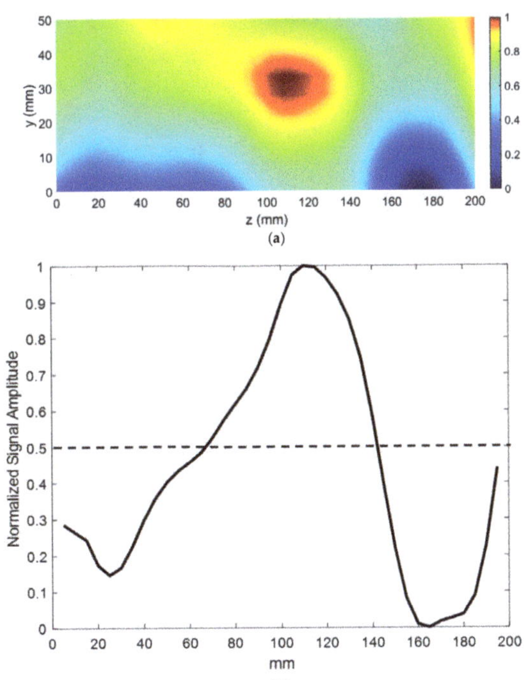

Figure 11. (**a**) Normalized received signal amplitude. A focal spot is observed at a distance of d_2 = 30 mm from the lens. (**b**) Normalized line scan at y = 30 mm. The dashed lines indicate −3 dB point.

4.2.3. Microwave NDE

Next, the capability of the LHM lens for detection of subwavelength defects with far-field microwave NDE data is demonstrated. The experimental setup is shown in Figure 12, and the schematic top view is shown in Figure 13a. Figure 13b shows the schematic of the sample under test. Teflon samples are used as the dielectrics for testing. A groove of size 15 mm (0.17λ) × 5 mm (0.06λ) is machined along the length of the sample. A similar Teflon sample with no machined groove is treated as the healthy sample. The samples are placed at the focal spot of the lens, and line scans are performed to obtain the scattered data. The contribution due to the defect is measured by the changes in the test signal relative to the baseline signal (found by measuring the healthy sample). The calibration and detection procedure is shown in Figure 14. The position of the defect can be determined by the minima in the line scans [26]. Three sets of measurements were taken at three positions of the sample. The distance p between consecutive defect positions was set to be 20 mm (0.23λ).

Figure 12. Proposed microwave NDE sensor. The sample is kept at the focal point of the LHM lens to allow for subwavelength defect detection.

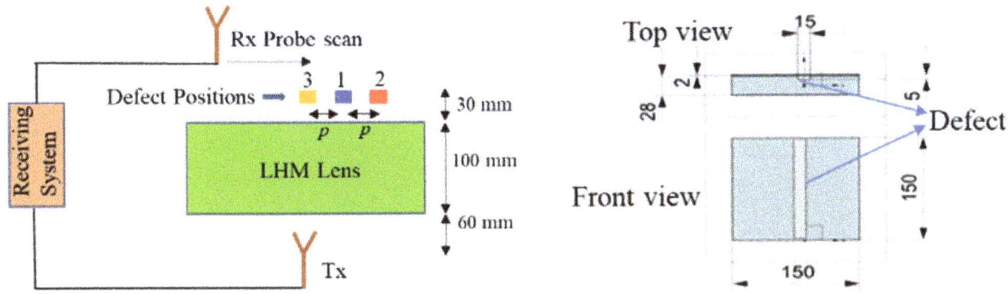

(a) (b)

Figure 13. (**a**) NDE experiment schematic top view. (**b**) Sample under test schematic. All dimensions are given in mm.

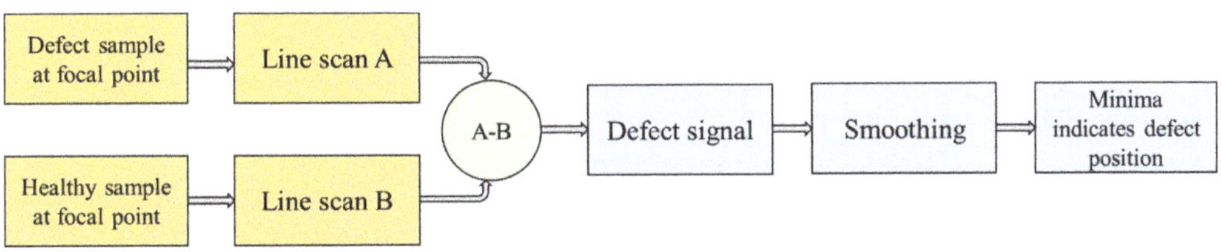

Figure 14. Calibration and defect detection flowchart.

Figure 15a shows the defect signal for the three measurements. The received line scans were fitted with smooth curves to obtain the minima. The minima in the three line scans are shifted with repositioning of the defect location, thus indicating the position of the subwavelength defect. This demonstrates the proposed system is sensitive to a $0.17\lambda \times 0.06\lambda$ defect and can determine consecutive defect positions with a separation of 0.23λ. The NDE measurements were repeated without the LHM lens to illustrate that the subwavelength defects are not detectable without the lens. Figure 15b shows the resulting received line scan signals without the lens. Since waves from the monopole transmitter are not focused, the received signal strengths are dominated by edge effects and scattering from the background. Therefore, no minima, as in the case of the LHM lens, are observed. Hence, these initial results clearly demonstrate that subwavelength defects, which are undetectable in free space in the far field, can be detected using a properly designed metamaterial lens.

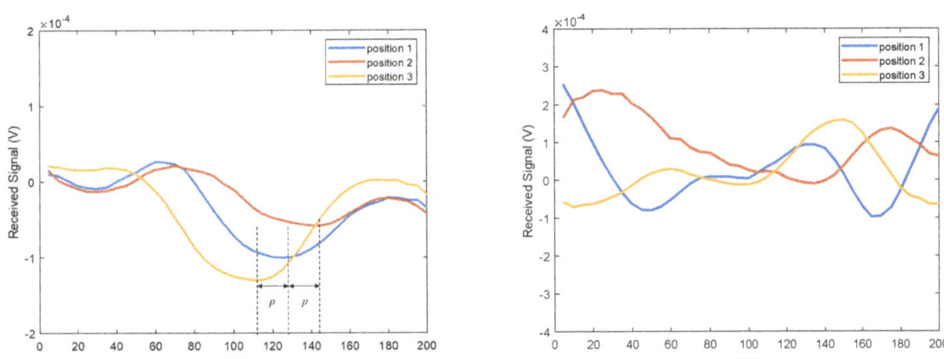

(a) (b)

Figure 15. NDE line scan results. (**a**) Measurements using the LHM lens. The minimum of the plot gives the defect position. (**b**) Measurements without the LHM lens. Waves are not focused in free space. Subwavelength defects position cannot be determined.

5. Discussion

This paper presents the design of a metamaterial lens and its application as a far-field microwave sensor. Numerical studies of the metamaterial unit cell design are presented. A fabricated metamaterial lens was used for experimental verification of the left-handed propagation characteristics of the lens. A homodyne detection setup was used with the LHM lens for NDE of subwavelength defects. Initial results demonstrate that a negative refractive index metamaterial lens can achieve resolution beyond the diffraction limit for far-field microwave NDE. The subwavelength resolution capability of a metamaterial lens is limited by the inherent losses associated with a fabricated LHM. Therefore, future work will involve the design of low loss LHMs to mitigate this issue. Moreover, work is in progress to design active metamaterials that can provide tunability that is lacking in passive metamaterial lens designs such as the one reported in this study. Finally, more extensive testing of experimental imaging of subwavelength defects is also underway to demonstrate the full potential of LHMs.

Author Contributions: Conceptualization, S.M. and L.U.; design, S.D.; fabrication, X.S. and S.D.; experiment, S.D., X.S., and E.R.; resources, Y.D., L.U., and M.H.; writing—original draft preparation, S.D.; writing—review and editing, E.R. and L.U.; supervision, L.U., E.R., and Y.D.; funding acquisition, M.H., S.M., Y.D., and L.U. All authors have read and agreed to the published version of the manuscript.

Funding: This work was sponsored by the National Science Foundation under the Manufacturing USA Award number 1762331.

Institutional Review Board Statement: Not applicable.

Informed Consent Statement: Not applicable.

Data Availability Statement: Not applicable.

Acknowledgments: The authors would like to thank Deepak Kumar of Michigan State University for his valuable contributions with the homodyne receiver system.

Conflicts of Interest: The authors declare no conflict of interest.

Appendix A

Estimation of permittivity and permeability of an engineered material from its S-parameters is a well-established method, first proposed in the original work of Nicolson and Ross [41] and Weir [42]. Using this approach, an inhomogeneous metamaterial structure, assumed to be a homogenous medium under the effective medium theory, can be characterized by a refractive index n and normalized impedance z [43,44].

The normalized impedance z of a metamaterial unit cell is related to its S parameters by the equation

$$z = \pm\sqrt{\frac{(1+S_{11})^2 - S_{21}^2}{(1+S_{11})^2 + S_{21}^2}} \tag{A1}$$

Since the metamaterial is a passive device, the reflected power cannot exceed the incident power. Therefore, the real part of z is positive, which in turn resolves the sign ambiguity in (A1).

The refractive index of a metamaterial unit cell of dimension d is related to the S-parameters by the following equation

$$e^{jnk_0d} = \frac{S_{21}}{1 - S_{11}\frac{z-1}{z+1}} \tag{A2}$$

The value of refractive index n can be evaluated from (A2) as

$$n = \frac{1}{k_0d}\left[\left\{Re\left(\ln\left(e^{jnk_0d}\right)\right) + 2m\pi\right\} - j\left\{Im\left(\ln\left(e^{jnk_0d}\right)\right)\right\}\right] \tag{A3}$$

where $Re(.)$ and $Im(.)$ are the real and imaginary operators, respectively, k_0 is the free space wavenumber, and m is an integer. The ambiguity in the branch selection of the multi-valued complex logarithmic function in (A3) can be resolved by choosing correct integer value m, which is dependent on the electrical length of the unit cell. Due to the small electrical length of the proposed unit cell design (9.3 mm) compared to the homogenized wavelength (38 mm), the fundamental branch ($m = 0$) is chosen for calculation of material parameters from the simulation model [45].

The effective permittivity and permeability of the metamaterial are computed from z and n using the following relations

$$\mu_{eff} = nz; \qquad \epsilon_{eff} = n/z \tag{A4}$$

References

1. Veselago, V.G. The electrodynamics of substances with simultaneously negative values of \epsilon and μ. *Sov. Phys. Uspekhi* **1968**, *10*, 509–514. [CrossRef]
2. Smith, D.R.; Padilla, W.; Vier, D.C.; Nemat-Nasser, S.C.; Schultz, S. Composite Medium with Simultaneously Negative Permeability and Permittivity. *Phys. Rev. Lett.* **2000**, *84*, 4184–4187. [CrossRef]
3. Shelby, R.A.; Smith, D.R.; Schultz, S. Experimental Verification of a Negative Index of Refraction. *Science* **2001**, *292*, 77–79. [CrossRef] [PubMed]
4. Parazzoli, C.G.; Greegor, R.B.; Li, K.; Koltenbah, B.E.C.; Tanielian, M. Experimental Verification and Simulation of Negative Index of Refraction Using Snell's Law. *Phys. Rev. Lett.* **2003**, *90*, 107401. [CrossRef]
5. Aydin, K.; Guven, K.; Soukoulis, C.M.; Ozbay, E. Observation of negative refraction and negative phase velocity in left-handed metamaterials. *Appl. Phys. Lett.* **2005**, *86*, 124102. [CrossRef]
6. Valanju, P.M.; Walser, R.M.; Valanju, A.P. Wave Refraction in Negative-Index Media: Always Positive and Very Inhomogeneous. *Phys. Rev. Lett.* **2002**, *88*, 187401. [CrossRef]
7. Munk, B.A. *Metamaterials: Critique and Alternatives*; Wiley: Hoboken, NJ, USA, 2009.
8. Houck, A.A.; Brock, J.B.; Chuang, I.L. Experimental Observations of a Left-Handed Material That Obeys Snell's Law. *Phys. Rev. Lett.* **2003**, *90*, 137401. [CrossRef]
9. Ziolkowski, R. Design, fabrication, and testing of double negative metamaterials. *IEEE Trans. Antennas Propag.* **2003**, *51*, 1516–1529. [CrossRef]
10. Aydin, K.; Guven, K.; Kafesaki, M.; Zhang, L.; Soukoulis, C.M.; Ozbay, E. Experimental observation of true left-handed transmission peaks in metamaterials. *Opt. Lett.* **2004**, *29*, 2623–2625. [CrossRef]
11. Alù, A. First-principles homogenization theory for periodic metamaterials. *Phys. Rev. B Condens. Matter Mater. Phys.* **2011**, *84*, 075153. [CrossRef]
12. Pendry, J. Negative Refraction Makes a Perfect Lens. *Phys. Rev. Lett.* **2000**, *85*, 3966–3969. [CrossRef]
13. Grbic, A.; Eleftheriades, G.V. Overcoming the Diffraction Limit with a Planar Left-Handed Transmission-Line Lens. *Phys. Rev. Lett.* **2004**, *92*, 117403. [CrossRef] [PubMed]
14. Aydin, K.; Bulu, I.; Ozbay, E. Subwavelength resolution with a negative-index metamaterial superlens. *Appl. Phys. Lett.* **2007**, *90*, 254102. [CrossRef]
15. Ozbay, E.; Li, Z.; Aydin, K. Super-resolution imaging by one-dimensional, microwave left-handed metamaterials with an effective negative index. *J. Phys. Condens. Matter* **2008**, *20*, 304216. [CrossRef]
16. Roy, T.; Rogers, E.; Zheludev, N. Sub-wavelength focusing meta-lens. *Opt. Express* **2013**, *21*, 7577–7582. [CrossRef]
17. Haxha, S.; Abdelmalek, F.; Ouerghi, F.; Charlton, M.D.B.; Aggoun, A.; Fang, X. Metamaterial Superlenses Operating at Visible Wavelength for Imaging Applications. *Sci. Rep.* **2018**, *8*, 16119. [CrossRef]
18. Mukherjee, S.; Tamburrino, A.; Haq, M.; Udpa, S.; Udpa, L. Far field microwave NDE of composite structures using time reversal mirror. *NDT E Int.* **2018**, *93*, 7–17. [CrossRef]
19. Savin, A.; Bruma, A.; Steigmann, R.; Iftimie, N.; Faktorova, D. Enhancement of Spatial Resolution Using a Metamaterial Sensor in Nondestructive Evaluation. *Appl. Sci.* **2015**, *5*, 1412–1430. [CrossRef]
20. Mukherjee, S.; Shi, X.; Udpa, L.; Udpa, S.; Deng, Y.; Chahal, P.; Shi, X. Design of a Split-Ring Resonator Sensor for Near-Field Microwave Imaging. *IEEE Sens. J.* **2018**, *18*, 7066–7076. [CrossRef]
21. Zhang, Y.; Zhao, J.; Cao, J.; Mao, B. Microwave Metamaterial Absorber for Non-Destructive Sensing Applications of Grain. *Sensors* **2018**, *18*, 1912. [CrossRef]
22. O'Hara, J.F.; Singh, R.; Brener, I.; Smirnova, E.; Han, J.; Taylor, A.J.; Zhang, W. Thin-film sensing with planar terahertz metamaterials: Sensitivity and limitations. *Opt. Express* **2008**, *16*, 1786–1795. [CrossRef]
23. Leggio, L.; Dadrasnia, E.; de Varona, O. Microwave Focusing within Arbitrary Refractive Index Media Using Left-Handed Metamaterial Lenses. *Prog. Electromagn. Res. M* **2016**, *45*, 51–58. [CrossRef]

24. Chen, J.J.; Grzegorczyk, T.M.; Wu, B.-I.; Kong, J.A. Limitation of FDTD in simulation of a perfect lens imaging system. *Opt. Express* **2005**, *13*, 10840–10845. [CrossRef]
25. Lu, J.; Grzegorczyk, T.M.; Wu, B.-I.; Pacheco, J.; Chen, M.; Kong, J.A. Effect of poles on subwavelength focusing by an LHM slab. *Microw. Opt. Technol. Lett.* **2005**, *45*, 49–53. [CrossRef]
26. Tassin, P.; Veretennicoff, I.; Van Der Sande, G. Veselago's lens consisting of left-handed materials with arbitrary index of refraction. *Opt. Commun.* **2006**, *264*, 130–134. [CrossRef]
27. Shreiber, D.; Gupta, M.; Cravey, R. Microwave nondestructive evaluation of dielectric materials with a metamaterial lens. *Sens. Actuators A Phys.* **2008**, *144*, 48–55. [CrossRef]
28. Tao, Y.; Yang, E.; Wang, G. Left-handed metamaterial lens applicator with built-in cooling feature for superficial tumor hyperthermia. *Appl. Comput. Electromagn. Soc. J.* **2017**, *32*, 1029–1034.
29. Zhang, S.; Yin, L.; Fang, N. Focusing Ultrasound with an Acoustic Metamaterial Network. *Phys. Rev. Lett.* **2009**, *102*, 194301. [CrossRef]
30. Amireddy, K.K.; Rajagopal, P.; Balasubramaniam, K.; Nadu, T. Holey-structured metalens for deep sub-wavelength resolution of delamination in layered materials using ultrasound. In Proceedings of the 15th Asia Pacific Conference for Non-Destructive Testing (APCNDT2017), Singapore, 13–17 November 2017; pp. 1–5.
31. Walker, E.L.; Jin, Y.; Reyes, D.; Neogi, A. Sub-wavelength lateral detection of tissue-approximating masses using an ultrasonic metamaterial lens. *Nat. Commun.* **2020**, *11*, 5967. [CrossRef]
32. Zaman, A.-U.-Z.; Song, K.; Lee, D.-G.; Hur, S. A novel approach to Fabry–Pérot-resonance-based lens and demonstrating deep-subwavelength imaging. *Sci. Rep.* **2020**, *10*, 10769. [CrossRef]
33. Mukherjee, S.; Su, Z.; Udpa, L.; Udpa, S.; Tamburrino, A. Enhancement of Microwave Imaging Using a Metamaterial Lens. *IEEE Sens. J.* **2019**, *19*, 4962–4971. [CrossRef]
34. Datta, S.; Shi, X.; Mukherjee, S.; Deng, Y.; Udpa, L. Model-Based Study of a Metamaterial Lens for Nondestructive Evaluation of Composites. *J. Nondestruct. Eval. Diagn. Progn. Eng. Syst.* **2020**, *3*, 041001. [CrossRef]
35. Pendry, J.; Holden, A.; Robbins, D.; Stewart, W. Magnetism from conductors and enhanced nonlinear phenomena. *IEEE Trans. Microw. Theory Tech.* **1999**, *47*, 2075–2084. [CrossRef]
36. Pendry, J.B.; Holden, A.J.; Stewart, W.J.; Youngs, I. Extremely Low Frequency Plasmons in Metallic Mesostructures. *Phys. Rev. Lett.* **1996**, *76*, 4773–4776. [CrossRef]
37. Smith, D.R.; Kroll, N. Negative Refractive Index in Left-Handed Materials. *Phys. Rev. Lett.* **2000**, *85*, 2933–2936. [CrossRef]
38. Smith, D.R.; Schurig, D.; Rosenbluth, M.; Schultz, S.; Ramakrishna, S.A.; Pendry, J.B. Limitations on subdiffraction imaging with a negative refractive index slab. *Appl. Phys. Lett.* **2003**, *82*, 1506–1508. [CrossRef]
39. Aydin, K.; Bulu, I.; Ozbay, E. Focusing of electromagnetic waves by a left-handed metamaterial flat lens. *Opt. Express* **2005**, *13*, 8753–8759. [CrossRef] [PubMed]
40. Wilson, P.; Ma, M.; Adams, J. Techniques for measuring the electromagnetic shielding effectiveness of materials. I. Far-field source simulation. *IEEE Trans. Electromagn. Compat.* **1988**, *30*, 239–250. [CrossRef]
41. Nicolson, A.M.; Ross, G.F. Measurement of the Intrinsic Properties of Materials by Time-Domain Techniques. *IEEE Trans. Instrum. Meas.* **1970**, *19*, 377–382. [CrossRef]
42. Weir, W. Automatic measurement of complex dielectric constant and permeability at microwave frequencies. *Proc. IEEE* **1974**, *62*, 33–36. [CrossRef]
43. Chen, X.; Grzegorczyk, T.M.; Wu, B.-I.; Pacheco, J.J.; Kong, J.A. Robust method to retrieve the constitutive effective parameters of metamaterials. *Phys. Rev. E Stat. Phys. Plasmas Fluids Relat. Interdiscip. Top.* **2004**, *70*, 016608. [CrossRef]
44. Arslanagic, S.; Hansen, T.V.; Mortensen, N.A.; Gregersen, A.H.; Sigmund, O.; Ziolkowski, R.; Breinbjerg, O. A Review of the Scattering-Parameter Extraction Method with Clarification of Ambiguity Issues in Relation to Metamaterial Homogenization. *IEEE Antennas Propag. Mag.* **2013**, *55*, 91–106. [CrossRef]
45. Rothwell, E.J.; Frasch, J.L.; Ellison, S.M.; Chahal, P.; Ouedraogo, R.O. Analysis of the Nicolson-Ross-Weir Method for Characterizing the Electromagnetic Properties of Engineered Materials. *Prog. Electromagn. Res.* **2016**, *157*, 31–47. [CrossRef]

sensors

MDPI

Article

Coherent Fiber-Optic Sensor for Ultra-Acoustic Crack Emissions

Ilaria Di Luch [1,*], Maddalena Ferrario [2], Davide Fumagalli [3], Michele Carboni [3] and Mario Martinelli [1]

[1] Dipartimento di Elettronica, Informazione e Biomedica, Politecnico di Milano, 20133 Milano, Italy; mario.martinelli@polimi.it
[2] Cohaerentia s.r.l., 20133 Milano, Italy; maddalena.ferrario@cohaerentia.com
[3] Dipartimento di Meccanica, Politecnico di Milano, 20156 Milano, Italy; davide11.fumagalli@mail.polimi.it (D.F.); michele.carboni@polimi.it (M.C.)
* Correspondence: ilaria.diluch@polimi.it; Tel.: +39-02-2399-8949

Abstract: A coherent optical fiber sensor with adequate sensitivity for detecting the acoustic emission (AE) during the propagation of a crack in a ferrous material is presented. The proposed fiber optic sensor is successfully compared in terms of the SNR (Signal to Noise Ratio) and detectable AE energy levels to commercially available AE piezo-transducers sensors and is proven to be an effective and advantageous alternative for sensing and monitoring fatigue damage in structural applications.

Keywords: fiber optic sensor; guided waves; acoustic emission; fatigue crack propagation

Citation: Di Luch, I.; Ferrario, M.; Fumagalli, D.; Carboni, M.; Martinelli, M. Coherent Fiber-Optic Sensor for Ultra-Acoustic Crack Emissions. *Sensors* **2021**, *21*, 4674. https://doi.org/10.3390/s21144674

Academic Editor: Zenghua Liu

Received: 20 May 2021
Accepted: 4 July 2021
Published: 8 July 2021

1. Introduction

Acoustic Emissions (AE) are an established real-time structural health monitoring approach for detecting, localizing, and monitoring damage initiation and development in many structural applications, for example in the mechanical, civil, automotive and aerospace fields made of different technical materials, such as metals, polymers, ceramics and composites. The onset and development of damaging events, such as cracks, delamination, corrosion, and debonding, release energy in terms of transient high-frequency elastic (ultrasonic) waves, which travel into the involved materials assuming the shape of either spherical volumetric waves or guided waves depending on the given boundary conditions.

By acquiring and suitably analyzing such sound waves, it is possible to detect and locate faults in complex structures, exactly as it happens in seismology from which the industrial application of AE was directly derived, even if adapted to a different scale. AE sensors based on piezoelectric transducers (PZT) are the most widespread sensor types currently adopted, featuring high-sensitivity and suitable bandwidth [1,2]. However, PZT transducers work with electric charges, which are low in intensity, and thus an electronic circuitry is needed close to the sensor transducer (e.g., a couple of meters) in order to avoid severe propagation losses as well as bandwidth limitations.

Moreover, the connecting cable must be electrically shielded in order to avoid electromagnetic noise (e.g., a coaxial cable), and the results are very sensitive to environmental perturbation (e.g., vibrations and acoustic noise) [3]. These "interface" limitations have shrunk the deployment of PZT sensors in applications that require a remote measurement, such as field applications.

In such a context, fiber-optic sensors, FOS, have been investigated through the years as a possible alternative. The greatest advantages of FOS are the total electromagnetic immunity, the possibility to easily have the interrogation unit remote from the transducer unit, the capability of withstanding harsh environments, and the availability of many multiplexing options (time, space, as well as wavelength). In front to these recognized advantages, it is not yet well established if FOS offer the same high-sensitiviy of PZT transducers in practical applications. The aim of this paper is to prove this, i.e., to prove in real experimental conditions that a FOS system achieves the same sensitivity to the acoustic emission as the standard PZT system.

Hence, the paper is organized in order to compare the PZT and the FOS sensor under the same experimental conditions. The AE events are generated by considering the case of fatigue crack propagation in a HEM220 carbon steel beam submitted to fatigue cycles. After a description of the FOS schemes, details of the FOS transducer are given. In the second half of the paper, the conventional and novel sensors are applied to the same experimental conditions, and the results are finally discussed.

2. The Coherent Fiber Optic Sensor

The detection of acoustic waves in solids emitted during fatigue and crack propagation phenomena dates back to the dawn of the invention of fiber optic sensors. An important paper appeared at the beginning of the 1990s [4], where the authors attempted to develop a fiber optic sensor dedicated not only to the detection but also to the localization of acoustic events. Starting from these years, a large number of contributions appeared in the literature for all the following decades. Over the first decade of the 2000s, the detection and localization of AE events has been assessed in different materials. In [5], an embedded optical-fiber sensor was proven to monitor crack propagation in concrete beams.

The sensing of AEs at a number of selected points on an optical fiber is achieved by the several fiber-optic transducer placed along the sensing fiber. The proposed sensor proved to be effective in localization of AE events, even though the collected signals are of high energy content and amplitude. The sensor sensitivity is not assessed in the paper, which limits the comparison with other systems. In [6], a fiber-optic interferometric sensor is proposed to detect and localize AE events on a steel plate, while a similar approach was adopted in [7] where AE events were studied on an aluminum plate.

In particular, AE have been simulated via steel ball-dropping on a plate, which produces high intensity signals. In [8], a fiber optic sensor for the detection of precursory acoustic emissions in rockfall events was proposed and experimentally characterized. The sensor, based on an interferometric scheme, was compared to a commercial PZT sensor. Another interferometric scheme was adopted in [9], where a fiber-optic based acoustic emission sensor was proposed to detect the vibrations produced by ultrasonic waves propagating in a solid body.

Similarly, in [10], a fiber-optic interferometric sensor was addressed to detect ultracoustic guided waves in pipes. In [11], a novel fiber-optic sensor based on a polarization modulation was presented for AE detection in large structures and compared to PZT transducers.

In more recent years, Fiber Bragg Gratings (FBG)-based techniques have also been investigated: for example, in [12], a FBG-based AE measurement technique was employed to monitor crack activity in civil engineering structures and sites. In [13–15], FBG sensors with high sensitivity over broad bandwidths were proposed for AE detection in composite plates. AE generated by crack formation in an aluminum panel was investigated through a FBG-based sensor in [16], while in [17], the performances of a FBG AE sensor on the surface of a thin polymer-bonded explosive (PBX) material were analyzed.

The localization of AE events, in addition to other acoustic events due to fretting within joints, was also demonstrated in [18] with a multiplexed array of three sensors. Even though these results are promising and, for the first time, a fiber-optic spatial resolution was investigated in AE sensing, the layout presented comprises a distributed-fiber laser sensor, which contributes to creating a complex and expensive solution, which suffers from directional sensitivity of the measurand. This angle-dependency is one of the limiting intrinsic factors correlated to FBG-based solutions.

One of the most recent contributions, presented by the authors in a previous work [19], analyzed a novel fiber-optic based solution for impact detection on an aluminum plate. In this case, the sensor showed promising resulta when compared with commercial PZT sensors. In the following year, in [20], another fiber-optic sensor wAs proposed for ultrasonic stress wave detection: the optical sensor was used to detect guided wave signals on an aluminum plate, and subsequently it was tested on a reinforced concrete beam.

In this case, as in most of the previous work here presented, the sensitivity of the proposed sensors was not fully assessed. Moreover, the analyzed signals were of high-intensity content. The topical interest of a fiber-optic based alternative for AE sensing was not, heretofore, fully addressed. The aim of the proposed work is to meet the need of a fiber-optic based solution for AE sensing, which can ensure comparable performance with respect to the current state-of-the art PZT sensors in detecting AE events.

When a light beam propagates along an optical fiber, some of its features, such as the propagating wave vector, are modulated according to the external influences, such as mechanical and thermal deformation, acting on the fiber itself [21]. Fiber optic interferometric sensors are designed to measure these external perturbations by looking at light phase changes induced by the propagation of wave vector modulation.

Surface-bonded sensors, such as the one proposed in this research work, read the strain transferred from the structure to the fiber: the strain field produces a fiber-optic perturbation, which, in turn, produces an optical response. The amount of phase changes in phase-based sensors is linearly proportional to the strain field transferred to the fiber, the fiber-optic material properties, and the configuration by which the optical fiber is bonded to the specimen. The optical phase delay ϕ (in radians) of light passing through a fiber is given by [21]:

$$\Delta\phi(t) = n\beta_0(1 - \frac{n^2}{2}[p_{12} - \mu(p_{11} + p_{12})]) \int_0^l \varepsilon_{xx}(x,t)\, dx, \quad (1)$$

where n is the effective optical fiber refractive index, β_0 is the vacuum propagaction wave-vector, p_{ij} are the photo-elastic constant, μ is the Poisson's ratio of silica, ε_{xx} is the strain in the optical fiber axis direction, and l is the optical fiber sensing length. Equation (1) shows that the optical phase change $\Delta\phi(t)$ is a function of integral deformation.

In particular, the overall phase modulation $\Delta\phi(t)$ of the light beam propagating into the sensing fiber is obtained by summing all the infinitesimal phase-change contributions $d\phi(x,t)$ along the length dl of the sensor; for this reason, it can be seen as an integral phase sensor.

Without loss of generality, it can be assumed that $p_{11} = +0.121$, $p_{12} = +0.270$, $\mu = 0.17$, $n = 1.45$ [21], and, if the sensor is operating at 1550 nm, Equation (1) becomes simply:

$$\Delta\phi(t) = 4.57 \cdot 10^6 \int_0^l \varepsilon_{xx}(x,t)\, dx, \quad (2)$$

which shows the extreme sensitivity that the coherent optical fiber sensor can reach.

The interferometric fiber optic sensor proposed in this work relies on a coherent detection scheme [22]. The coherent receiver comprises a 3 × 3 coupler in which two ports were chosen as the reference and sensing arm of the interferometer. These ports are terminated with Faraday-Rotator-Mirrors (FRMs), which guarantee the retracing of signal polarizations and, thus, fixed polarizations at the splitter in order to avoid signal fading in interferometric detection schemes [23]. The 3 × 3 coupler performs a passive stable homodyne demodulation where the output signals are 120° phase shifted and the corresponding receiver photo-currents I_1 and I_2 as depicted in the schematic of Figure 1.

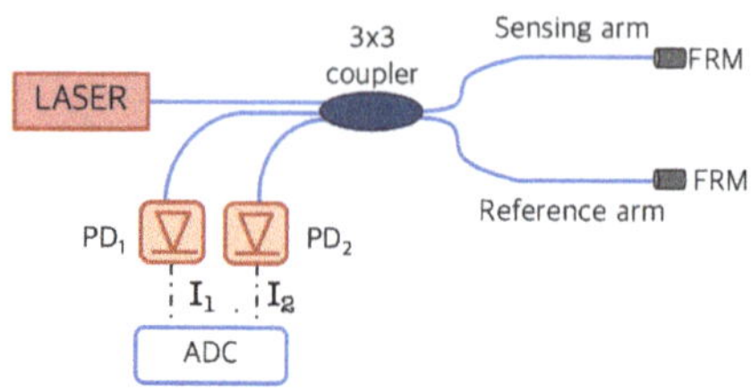

Figure 1. Michelson interferometer layout.

As the output of a 3 × 3 coupler, I_1 and I_2 can be written as [24]:

$$I_1 = C + B\cos(\Delta\phi(t) - 2/3\pi),$$
$$I_2 = C + B\cos(\Delta\phi(t) + 2/3\pi), \tag{3}$$

where $\Delta\phi(t)$ is the phase difference induced by the dynamic event between the sensing and reference arms of the interferometer, and C and B are constants depending on the coupling coefficient of the 3 × 3 coupler.

I_1 and I_2 can be digitally sampled by an opportune Analog to Digital Converter (ADC), properly normalized, and processed to retrieved the desired $\Delta\phi(t)$ through the *atan* of $(I_1 + I_2)/(I_1 - I_2)$ [25]. Due to the adopted coherent demodulation scheme, based on a 3 × 3 coupler, the phase signal $\Delta\phi(t)$ can be retrieved in a completely passive way without the need for an active feedback for quadrature point stabilization as in conventional fiber optic interferometry.

The minimum phase detection is established by the minimum achievable signals. Considering the spectral content of Acoustic Emission, we decided to adopt low-noise PIN photo-receivers with $B_{el} = 200$ kHz electrical bandwidth and Noise Equivalent Power $NEP = 0.19\ \mathrm{pW}/\sqrt{\mathrm{Hz}}$. Since the minimum detectable signal power is :

$$P_s^{min} = NEP\sqrt{B_{el}}, \tag{4}$$

this resulted in an expected $P_s^{min} = 0.12$ nW, corresponding to a minimum detectable phase of:

$$\Delta\phi^{min} = \frac{\Delta P_s^{min}}{P_s^{\pi}}, \tag{5}$$

where P_s^{π} is the power corresponding to a half-fringe excursion, which is equal to 250 µW in our set-up. By introducing this value in Equation (5), the expected $\Delta\phi_{min} \approx 0.5$ µrad can be obtained.

3. Fiber Optic Transducer

In acoustic sensing, the fiber optic transducer design is important because it has the task of transferring the ultra-acoustic perturbation of the material in a strain of the optical fiber. The evaluated $\Delta\phi_{min}$ of the previous chapter sets the minimum resolution of the fiber optic interferometer. However, this does not take into account the sensitivity of the fiber-optic transducer, which describes the smallest absolute amount of strain-induced phase change that can be detected by the sensor. In order to increase the sensitivity and to design a fiber-optic transducer able to read small perturbations, one has to increase the fiber sensing length. Intuitively, from Equation (2), it is clear that the longer the fiber length, the higher the integral strain value.

There are two main approaches to lengthen the sensing fiber: by wrapping the fiber in a coil around a solid cylinder (solid cylinder transducer, SCT) or by arranging it in multiple loops (stand alone spiral, SAS). The first approach is the one adopted in [26], in which the fiber was wrapped around an acrylic skeleton and tested for AE sensing with an aluminum plate. This solution offers the possibility to wrap meters of fiber and, thus, to easily increase the sensing fiber length; however, since this type of transducer does read the strain induced by the acoustic wave traveling in the cylinder, it suffers from energy dissipation due to the interface between the cylinder and the metallic plate under test.

In the second configuration instead, the fiber optic is wound like a spiral, and this latter is set at 90° with respect to the metal surface in order to be adhesively bonded directly onto the structure. This solution was investigated by the authors in a previous work [19], in which this layout showed promising results in impact detection on an aluminum plate. Regarding this configuration, the advantages include the direct detection of waves traveling in the metal and the possibility to increase the sensitivity by adding more loops to the ring, which, however, could compromise the sensor punctuality in the case of a high number of loops.

In order to properly design the fiber-optic transducer, two layouts were compared: the SCT and the SAS were directly compared by varying only the geometry while maintaining the same length of sensing fiber. By using a cylinder of 10 mm in height and a 10 mm radius, it is possible to wrap as much as 30 fiber loops in total for both configurations. However, while, in the first configuration, a solid quartz cylinder was adopted to better match the acoustic wave impedance at the interface between metal and cylinder, in the latter, a 3D printed polymeric ring was chosen to lighten the supporting structure.

The detailed layout for both configurations is depicted in Figure 2. Since both configurations present a sensitivity that is a function also of the bonded fiber section, 5 mm was selected as the sensing fiber length, being a good compromise between sensitivity and resolution, as will be discussed later in this section.

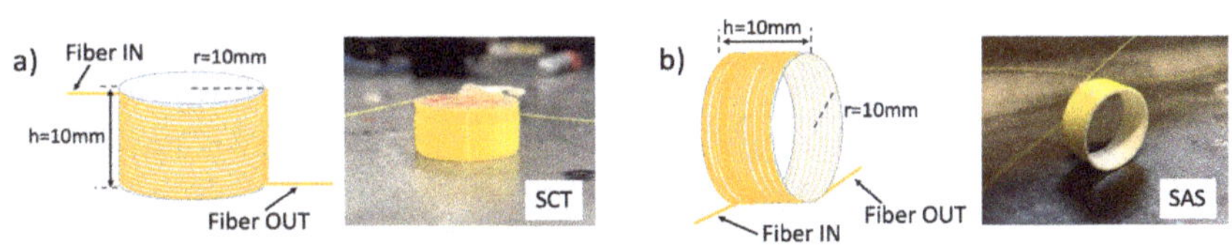

Figure 2. Fiber optic ultrasonic transducer (**a**) quartz cylinder (SCT) and (**b**) 3D-printed ring (SAS).

The sensitivity comparison was investigated by generating an AE signal and comparing the amplitude of the collected signals of the two configurations. The cylinder and the spiral were placed on a metal plate at a 100-mm distance from a PZT actuator. The AE signals were obtained by driving a PZT actuator with a five-cycle toneburst, shaped with a Hamming window, at 100 kHz. The results are shown in Figure 3. In particular, in Figure 3a,b, an example of the collected signal is reported, in the case of 180 V applied to the PZT actuator for the SCT and the SAS configurations. Both detected signals in Figure 3a,b refer to the same anti-symmetric A_0 wave mode propagating in the metal plate as the respective Time-of-Arrivals of the five-cycle toneburst.

Several experiments were performed by sweeping the voltage applied to the PZT actuator, and the results are reported in Figure 3c. The spiral configuration overdoes the cylinder performance: it is evident that the sensitivity obtained by the SCT was significantly lower than with the SAS. In the authors' opinions, this difference can be ascribed to the transducing process: in the SCT, the optical fiber wrapped read the cylinder deformations, which means that the acoustic wave energy had to be sufficient to strain the significant cylinder mass; on the contrary, in the SAS, the optical fiber read the acoustic wave strain directly.

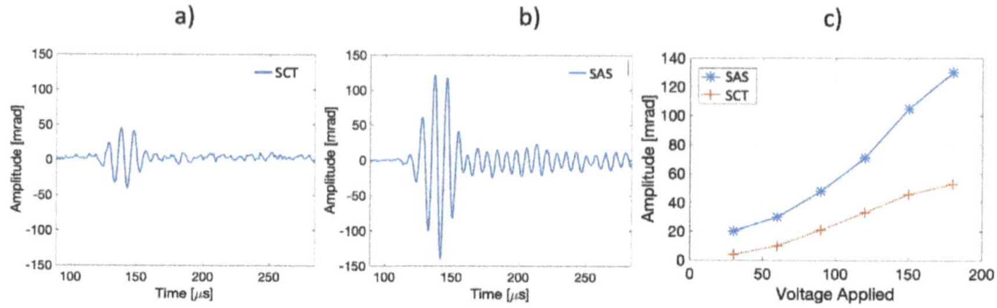

Figure 3. The recovered signal in the case of the (**a**) SCT configuration and (**b**) SAS configuration, (**c**) comparison between the detected amplitude as a function of different five-cycle toneburst amplitudes.

To conclude, the SAS configuration appeared as the most promising in detecting AE signals.

As stated in [19], a further key parameter affecting the FOS response was the ratio between the ultrasonic wavelength and the actual fiber length bonded to the structure. In fact, the sensor performed a spatial integration of the ultrasonic signals over the entire sensing length: if an excessive sensing length was chosen, the overall integration drastically degraded the sensor sensitivity.

On the other hand, if the number of loops is increased, the signal bending losses can become too high, and the sensor results may not be punctual, leading to an inaccurate time of arrival evaluation. An optimal choice in terms of the number of loops and bonded sensing fiber length, thus, resulted in a coil of 30 loops, each with a 5-mm bonded section, which guaranteed a good sensitivity in the whole acoustic wavelength range of interest of [18–110] mm.

4. Preliminary Tests: Pencil Lead Break

The experimental set up is shown in Figure 4. Only the fiber-optic sensor transducer is bonded on the metallic structure under test, which is the sensing arm of the MI, while the reference arm is kept close to the sensor interrogator. The signals coming from the photoreceivers are sampled by a 10 MS/s real-time oscilloscope board from Picoscope model 5442D, with 16 resolution bits and a dynamic range of ±0.2 V. These characteristics guarantee an effective overall noise of about 20 μrad on the reconstructed and filtered phase. This noise contribution, thus, represented the actual noise level of the overall FOS detection system.

To recover the phase information $\Delta\phi(t)$, signal I_1 and I_2 from the photodetectors were normalized and demodulated in real-time with the algorithm described in [25] developed in LabVIEW. For the AE signals in the band of [30–150] kHz, a digital band-pass filter was also applied on the recovered phase to reduce the environmental noise contribution. The acquisition algorithm implemented in LabVIEW is schematized in Figure 5.

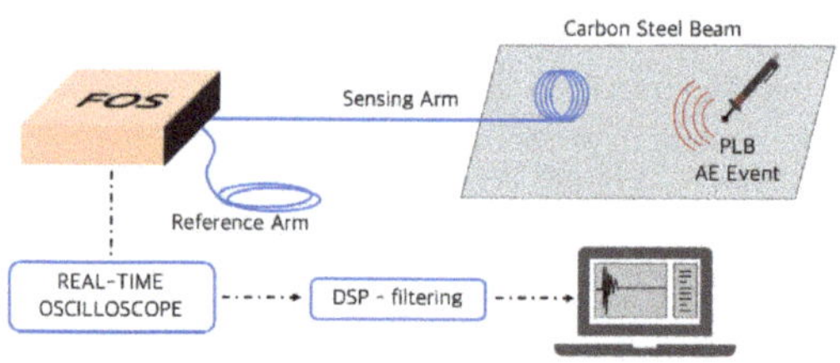

Figure 4. Experimental layout of the FOS sensor.

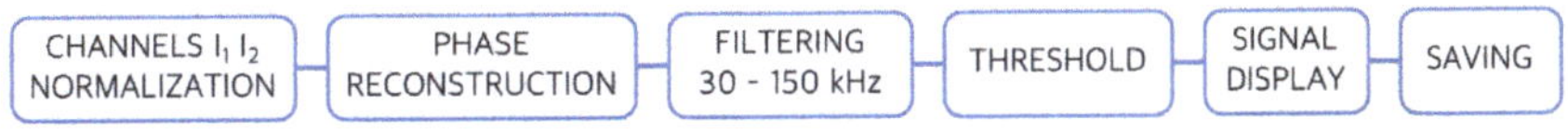

Figure 5. LabVIEW SW steps schematic implemented for the FOS demodulation algorithm.

The metallic structure under test is a 15-mm thick, 2-m long, and zinc-coated HEM220 beam (EN 10025:1990/A1:1993) made of *S*355 carbon steel. Its sketch and picture are reported in Figure 6. In this kind of slender and plate-like structures, AE events generate elastic waves propagating in terms of guided waves, i.e., multimodal (dispersive) waves whose number of excited modes depends on the excitation frequency. For most practical purposes, only the fundamental symmetric (compressive) *S*0 and anti-symmetric (flexural) *A*0 wave modes are used.

These fundamental wave modes are independent of each other and propagate at different velocities, depending on the thickness-frequency product, based on their relative dispersion curves [27]. Group and phase velocities values were calculated by Vallen AE-Suite Software R2017.0504.1, according to the parameters of the metallic structure. The chosen features for the simulations were the Young modulus $E = 210{,}000$ MPa, Poisson ratio $\nu = 0.30$, and density $\rho = 7.85\ \text{kg/dm}^3$.

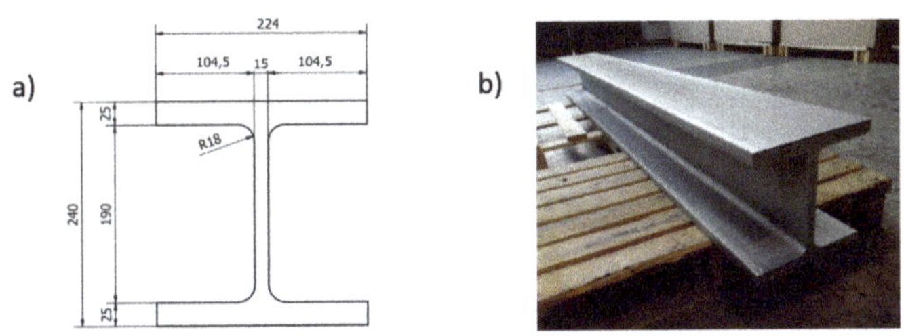

Figure 6. H-Shaped HEM220 beam: (**a**) sketch, (**b**) picture.

Before performing AE detection during a fatigue crack growth test, the proposed FOS was tested to detect guided waves generated by Pencil Lead Break (PLB) tests, also known as the Hsu–Nielsen source. The Hsu–Nielsen source is an artificial method of generating AE signals, which can roughly represent an acoustic emission damage source [28]. PLB tests have been conducted by increasing the distance between the spiral fiber and the PLB breaking point. As reported in Figure 7, PLB tests were performed at 100, 200, and 300 mm on the carbon steel beam.

Both the *S*0 and the *A*0 times of arrival could be identified, even at the greatest distance. The FOS was able to detect signals up to 300 mm distance with a good Signal-to-Noise Ratio (SNR), which also allowed the detection of the *S*0 wave packets, which had a lower energy content with respect to the *A*0 waves. The Δt between the *S*0 and the *A*0 times of arrival are in agreement with the theoretically evaluated wave group velocities for the carbon steel of Table 1. In particular, the theoretical $c_g(S0) = 5011$ m/s and $c_g(A0) = 3090$ m/s were measured at $f \approx 70$ kHz as the main frequency of the recorded waves.

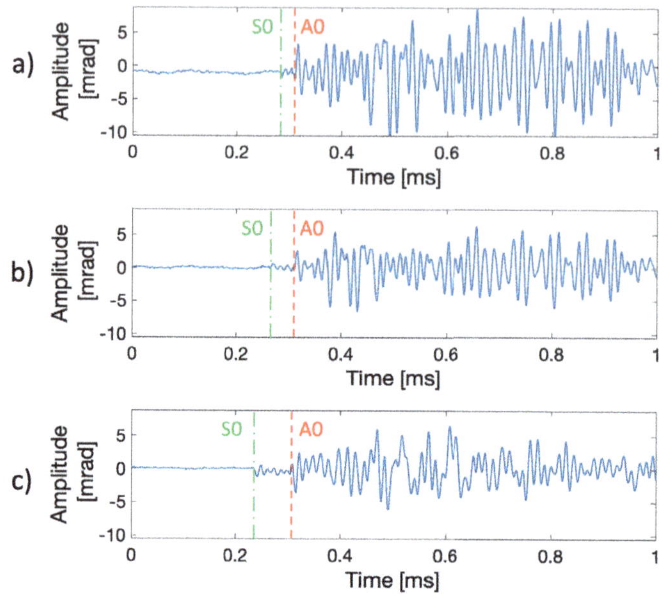

Figure 7. Detail of the recovered signals of FOS during PLB tests on the HEM220 beam: (**a**) 100 mm distance, (**b**) 200 mm distance, and (**c**) 300 mm distance.

Table 1. The theoretical and measured Δt between the $S0$ and $A0$ times of arrival, for the 100, 200, and 300 mm distances.

L [mm]	$c_g(S0)$ [m/s]	$c_g(A0)$ [m/s]	Δt_{TH} [μs]	Δt_{EVAL} [μs]
100	5011	3090	12.4	14.6
200	5011	3090	24.8	28.0
300	5011	3090	37.3	40.2

5. Fatigue Crack Propagation Test

A fatigue crack propagation test was run on a 500 mm long chunk cut from the HEM220 beam described in Section 4. A general view of the experimental set up for the fatigue crack propagation test and of the final result are summarized in Figure 8. In particular, an MTS servo-hydraulic actuator model 244.31, equipped with an MTS load-cell model 661.22D-01 with a maximum nominal load equal to 250 kN and a specifically designed and manufactured structural frame were adopted (Figure 8a) for carrying out the test. More details are described in the following.

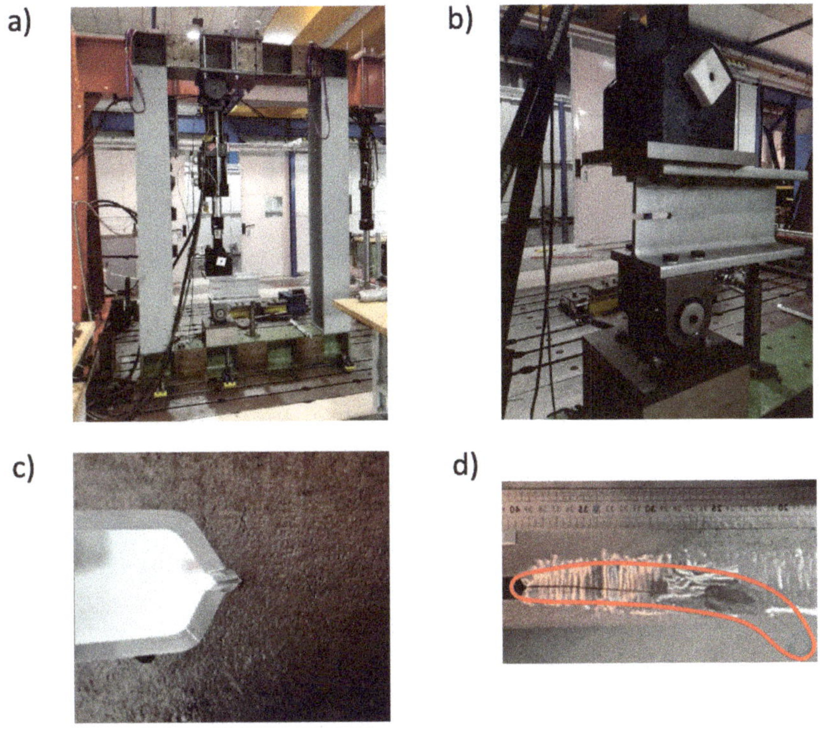

Figure 8. The experimental set up for the fatigue crack propagation test. The experimental set up for the fatigue crack propagation test. (**a**) MTS servo-hydraulic actuator, (**b**) rotational hinges, (**c**) sharp notch at the crack tip and (**d**) final path and morphology of the fatigue crack.

In order to induce crack initiation within the desired region of the beam, i.e., its web, and to decrease the crack initiation time, a sharp notch (Figure 8b,c) was manufactured, in the HEM220 beam by mechanical machining (milling) and sharpening through a razor blade. Then, the fatigue crack was initiated and propagated by subjecting the HEM beam to mode-I (opening) loading with respect to the prospective crack plane.

This loading configuration was achieved reproducing the same boundary conditions (Figure 8b, where the presence of the two required rotational hinges is highlighted) characterizing the standardized small-scale fracture mechanics specimen known as "compact tension" [29]. The test was carried out under load-controlled conditions applying a constant amplitude sinusoidal fatigue load with a frequency of 1 Hz, a maximum load level equal to 100 kN, and a minimum one equal to 10 kN, i.e., with a stress ratio $R = 0.1$. Finally, Figure 8d reports the final path and morphology of the fatigue crack at the end of the test.

Figure 9 shows the PZT sensors and FOS applied to acquire AE events during the fatigue crack propagation. As can be seen, the former were secured on the structures by means of magnetic holders, while the latter was protected by a 3D-printed case together with a protective tube for the fibers. The complete PZT measurement chain was composed of four Vallen VS150-M piezoelectric sensors, each equipped with a 34-dB Vallen AEP5 pre-amplifier and connected to an eight-channel Vallen ASMY-6 acoustic emission acquisition unit. Moreover, the rules for the acquisition of AE events and for the management of the acquired data were defined using Vallen AE-Suite Software R2017.0504.1.

In particular, the acquisition threshold was set at 50 dB (with respect to a reference voltage amplitude of 1 μV) based on the analysis of the background noise (≈40 μV) observed during the fatigue test, and a digital passband filter [25–300] kHz was adopted to further skim meaningless signals. The AE events fulfilling the requirements defined by the acquisition threshold and the digital filter were recorded in terms of the traditional parametric features of the AE transient hits (amplitude, energy, duration, counts, and rise time) and in terms of the full waveform sampled at 5 MS/s.

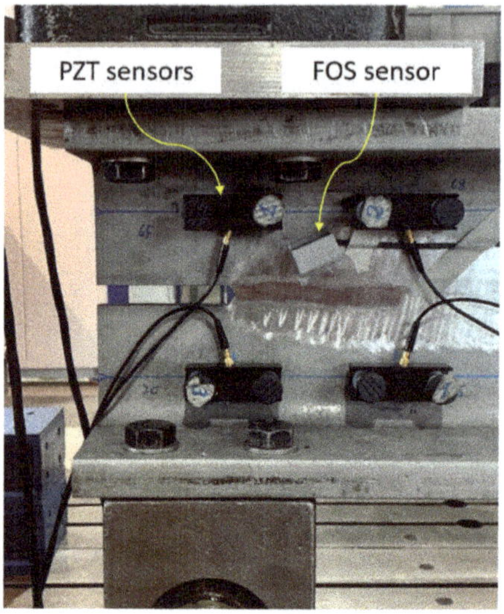

Figure 9. Layout for acoustic emission sensing by PZT sensors and the proposed FOS.

Figure 10 reports the signals recovered by the FOS sensor when the crack was approximately 100 mm long. AEs cannot be assumed as repeatable sources both in terms of the energy content and time duration, as noticeable from Figure 10. In fact, AE events typically show up as a random burst of pulses produced after a rapid release of energy from localized sources within the material. Therefore, the acquisition SW developed in LabVIEW requires the implementation of a specific threshold on the recovered phase only after filtering to select only the useful signals.

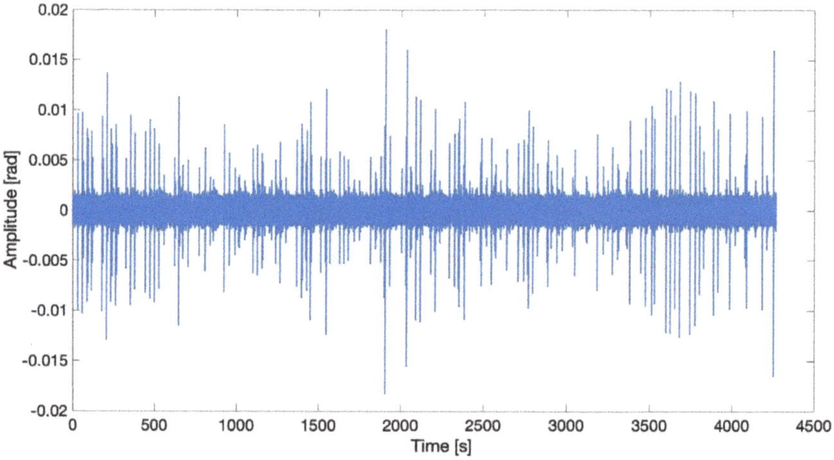

Figure 10. Example of a burst of AE signals detected by the FOS sensor during fatigue crack growth testing.

In Figure 11, the detailed AE signals recovered by the FOS and the PZT sensors are depicted, showing the typical temporal behavior of an AE event registered during a fatigue growth crack test. AE durations can be comparable, while the frequency content is different mainly due to the resonant frequency response of PZT that enhances the higher AE frequencies. To assess and compare the performance of the proposed FOS sensor with respect to PZT, the SNR of the two detected AE signals was evaluated, assumed as the ratio between the average background noise level and the maximum peak amplitude of the AE signal. In particular, the resulting SNRs were $SNR_{FOS} = 26.19$ dBV and $SNR_{PZT} = 27.88$ dBV, respectively, for the FOS sensor and the PZT sensor.

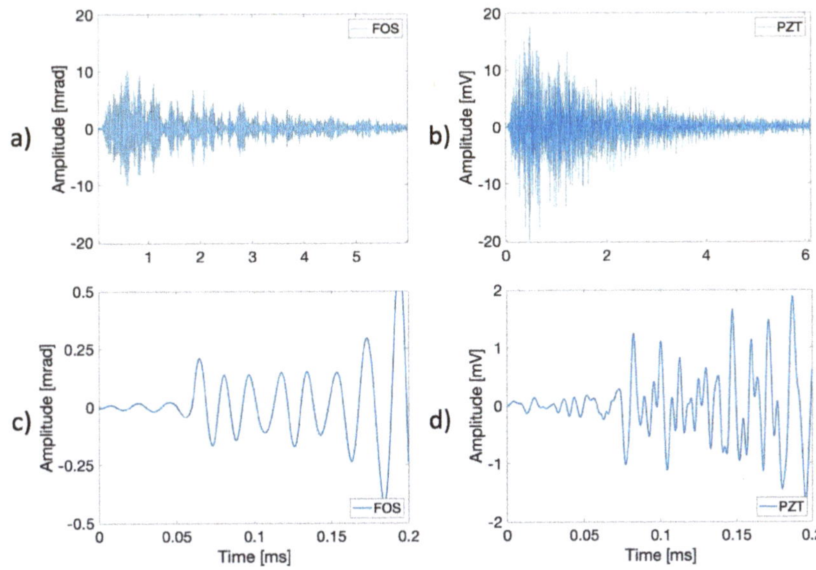

Figure 11. Comparison of the recovered signal during the fatigue growth crack test: (**a**,**c**) FOS-registered AE signal and (**b**,**d**) PZT-registered AE signal.

To further confirm this result, a z-test was run over 1000 recovered SNR samples, which were collected during a portion of the fatigue growth crack test. This type of test is suitable to determine whether two population means, that is the two evaluated SNR samples, are different when the variance is known and the number of sample is large. The results reported in Table 2 confirm that the two sensors guaranteed an almost equal SNR mean value and similar confidence interval of 95%, thus, proving comparable performance in AE monitoring. The large confidence interval calculated for both type of sensors is due to the random nature of AE events, which cannot be assumed as repeatable sources.

Table 2. The mean value and critical value at 95% for FOS- and PZT-collected SNRs.

Sensor	Mean Value μ [dB]	Confidence Interval at 95% [dB]
FOS	24.68	8.13
PZT	25.95	7.40

In addition to SNR assessment, a comparison of the detected energy levels was carried out, showing that PZT and FOS detected signals featuring similar energy peaks in time. The energy was calculated in the same time span of 6 ms as the integral of the modulus square of the signal. This statement holds if the compared signals are of the same AE event.

The energy values were evaluated in dB with respect to the reference voltage amplitude of 1 μV and phase of 1 μrad for PZT and FOS, respectively.

Part of the results from the comparison of the two sensors are reported in Figure 12, where, for the sake of clarity, only 1 min of registered high-energy signal values is detailed. Indeed, similar behavior to that depicted in Figure 12 was observed during the whole fatigue test, though with different energy levels depending on the position of the crack tip along its propagation path. As an example, Figure 13 shows the energy signals recovered when the crack tip was only 15-mm long.

In this case, the peak energy values were much lower with respect to those in Figure 12. Indeed, during the initial part of its propagation, the crack was typically not very energetic, while, after about 100 mm, in correspondence with the area where the crack began to deviate from its straight path, the emissions started to be highly energetic.

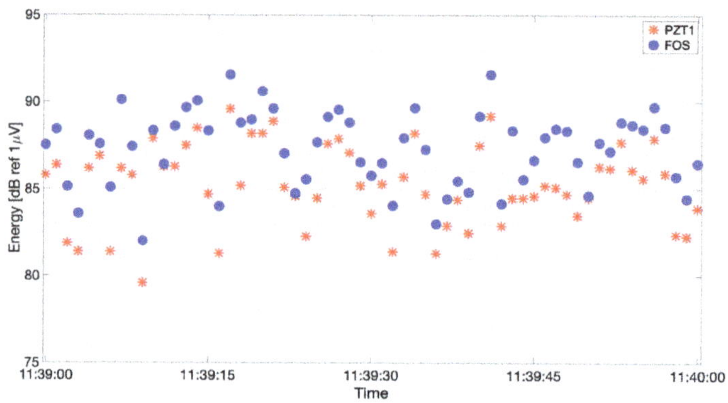

Figure 12. The evaluated energy during 1 min of the fatigue growth crack test at L = 100 mm crack length.

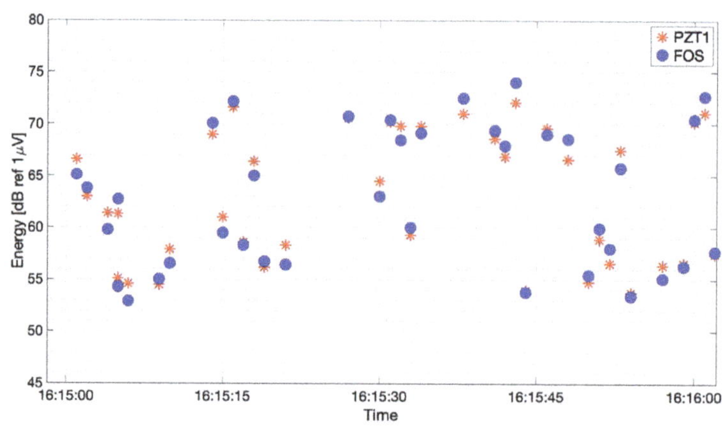

Figure 13. The evaluated energy during 1 min of the fatigue growth crack test at L = 15 mm crack length.

To further confirm this result, a *z*-test was performed on the evaluated energies from the two sensors. Table 3 reports the evaluated mean value μ and confidence interval of 95%, which gave $\alpha = 0.05$, for the FOS and the PZT, calculated over 1000 recovered energy samples, collected during the fatigue growth crack test when the crack length was both 15 and 100 mm. In spite of the resulting large confidence interval due to the AE events random nature, a good agreement was found between FOS and PZT, both in terms of the mean energy value and confidence interval, thus, confirming that the two sensors collected similar energy values.

Table 3. The mean value and critical value at 95% for the FOS- and PZT-collected energies.

Crack Length	Sensor	Mean Value μ [dB]	Confidence Interval at 95% [dB]
15 mm	FOS	64.26	13.10
	PZT	65.72	14.2
100 mm	FOS	79.36	20.57
	PZT	76.05	20.43

For the 100-mm case, the mean value difference of 3.31 dB might be ascribed to the fact that the FOS sensor was, in this case, closer, with respect to PZT1, to the crack tip as shown in Figure 9, and thus the *A*0 wave modes arrived less attenuated. For the 15 mm, the difference was lower as the FOS and PZT1 sensors were at similar distance from the crack tip. The two confidence intervals, on the contrary, were almost equal, confirming that the two sensors collected similar signal energies. The large confidence interval calculated for both types of sensor endorses the random nature of AE events, which cannot be assumed as repeatable sources.

6. Summary and Comments

In this paper, a novel fiber optic AE sensor with detection characteristics comparable with the conventional PZT sensor was presented. Two configurations were taken into consideration for the fiber-optic transducer—a solid cylindrical structure (SCT) and a stand-alone spiral (SAS)—and the latter proved to have the highest sensitivity in detecting ultracoustic signals.

The SAS configuration was chosen as a suitable FOS transducer, and thus it was further tested with PLB preliminary measurements and demonstrated a good SNR in detecting symmetric and anti-symmetric wave modes propagating in a HEM220 carbon steel beam. The time of arrivals of the two wave modes were theoretically calculated using Vallen AE-Suite Software according to the parameters of the metallic structure and were compared with the evaluated time of arrivals. Good agreement was found in the comparison of the theoretical and evaluated time of arrivals.

The FOS performance was compared to the standard PZT AE sensor that is commercially available in crack monitoring during a fatigue growth crack test, run on a 500-mm long chunk cut of a carbon steel beam. A direct relationship between PZT and FOS was found: recovered signals of the same event showed similar time durations and a comparable SNR, even though the two sensors had different frequency responses. To better compare them, an energy analysis of the detected event was performed: a good agreement was found between the PZT and FOS pulse energy behavior.

To conclude, the novel FOS sensor exhibited performances comparable to the standard commercially available PZT when tested during fatigue crack growth tests, thus, proving to be a valid alternative to PZT in AE sensing while holding all the advantages of being fiber-optic-based with a competitive cost of about half of the current PZT transducer control units. The authors believe that these early results may open up remarkable prospects for applying fiber optic sensors in AE detection. This optical fiber transducer is very simple to manufacture and can be easily multiplexed, thus, forming the basis for the construction of complex sensor networks that can be remotely installed in field applications.

Author Contributions: I.D.L., M.F. and M.C. conceived and designed the experiments and wrote the paper; I.D.L., M.F. and D.F. performed the experiments; M.M. supervised the work and contributed in writing the paper. All authors have read and agreed to the published version of the manuscript.

Funding: This research received no external funding.

Institutional Review Board Statement: Not applicable.

Informed Consent Statement: Not applicable.

Data Availability Statement: Not applicable.

Acknowledgments: This work was developed in the framework of the Saipem-Politecnico di Milano Joint Research Center, FOCA Project, under the direction of Marco Scognamiglio and Giuseppe Battista Abbate that the authors thank for the continuous stimulus and for authorizing the publication. The authors sincerely thank S. Milan for the active help given to the research. PoliNDT (Inter-Department Lab for Structural Diagnostics and Monitoring) at Politecnico di Milano is also acknowledged for its technical support in carrying out acoustic emission measurements by PZT sensors.

Conflicts of Interest: The authors declare no conflict of interest.

References

1. Beattie, A.G. Acoustic emission non-destructive testing of structures using source location techniques. *Albuq. Livermore* **2013**. [CrossRef]
2. Yan, G.; Zhou, L. An Optimal Image-Based Method for Identification of Acoustic Emission (AE) Sources in Plate-Like Structures Using a Lead Zirconium Titanate (PZT) Sensor Array. *Sensors* **2018**, *18*, 631. [CrossRef] [PubMed]
3. Sun, Z.; Rocha, B.; Wu, K.T.; Mrad, N. A Methodological Review of Piezoelectric Based Acoustic Wave Generation and Detection Techniques for Structural Health Monitoring. *Int. J. Aerosp. Eng.* **2013**, *2013*, 1–22. [CrossRef]
4. Liu, K.; Ferguson, S.M.; Measures, R.M. Fiber-optic interferometric sensor for the detection of acoustic emission within composite materials. *Opt. Lett.* **1990**, *15*, 1255–1257. [CrossRef]
5. Ansari, F. Fiber optic health monitoring of civil structures using long gage and acoustic sensors. *Smart Mater. Struct.* **2005**, *14*, S1–S7. [CrossRef]
6. Gong, J.; MacAlpine, J.M.K.; Jin, W.; Liao, Y. Locating acoustic emission with an amplitude-multiplexed acoustic sensor array based on a modified Mach–Zehnder interferometer. *Appl. Opt.* **2001**, *40*, 6199–6202. [CrossRef]
7. Liang, S.; Zhang, C.; Lin, W.; Li, L.; Li, C.; Feng, X.; Lin, B. Fiber-optic intrinsic distributed acoustic emission sensor for large structure health monitoring. *Opt. Lett.* **2009**, *34*, 1858–1860. [CrossRef] [PubMed]
8. Schenato, L.; Palmieri, L.; Gruca, G.; Iannuzzi, D.; Marcato, G.; Pasuto, A.; Galtarossa, A. Fiber optic sensors for precursory acoustic signals detection in rockfall events. *J. Eur. Opt. Soc. Rapid Public* **2012**, *7*, 12048. [CrossRef]
9. Liang, Y.; Qu, D.; Deng, H. Based on optical fiber Michelson interferometer for acoustic emission detection experimental research. In Proceedings of the International Symposium on Photoelectronic Detection and Imaging 2013: Fiber Optic Sensors and Optical Coherence Tomography, Beijing, China, 25–27 June 2013; Rao, Y., Ed.; International Society for Optics and Photonics, SPIE: Bellingham, WA, USA 2013; Volume 8914, pp. 219–226.
10. Zhou, W.; Dong, Y.; Li, H.; Xie, Y. Guided wave-based pipe damage inspection by ultrasonic fiber optic sensor. In Proceedings of the 2016 Structures Congress, Jeju Island, Korea, 28 August–1 September 2016.
11. Verstrynge, E.; Pfeiffer, H.; Wevers, M. A novel technique for acoustic emission monitoring in civil structures with global fiber optic sensors. *Smart Mater. Struct.* **2014**, *23*, 065022. [CrossRef]
12. Rajan, G.; Jinachandran, S.; Xi, J.; Li, H.; Vinod, J.S.; Moses, T.; Karekal, S.; Prusty, B. Fibre optic acoustic emission measurement technique for crack activity monitoring in civil engineering applications. In Proceedings of the 2016 IEEE Sensors Applications Symposium (SAS), Catania, Italy, 20–22 April 2016; pp. 1–6.
13. Yu, F.; Okabe, Y.; Wu, Q.; Shigeta, N. Fiber-optic sensor-based remote acoustic emission measurement of composites. *Smart Mater. Struct.* **2016**, *25*, 105033. [CrossRef]
14. Yu, F.; Okabe, Y. Regenerated fiber Bragg grating sensing system for ultrasonic detection in a 900 °C environment. *J. Nondestruct. Eval. Diagn. Progn. Eng. Syst.* **2019**, *2*, 011006. [CrossRef]
15. Wu, Q.; Yu, F.; Okabe, Y.; Kobayashi, S. Application of a novel optical fiber sensor to detection of acoustic emissions by various damages in CFRP laminates. *Smart Mater. Struct.* **2014**, *24*, 015011. [CrossRef]
16. Harish, A.V.; Varghese, B.; Rao, B.; Balasubramaniam, K.; Srinivasan, B. Dynamic interrogator for elastic wave sensing using Fabry Perot filters based on fiber Bragg gratings. *Ultrasonics* **2015**, *60*, 103–108. [CrossRef]
17. Fu, T.; Wei, P.; Han, X.; Liu, Q. Application of Fiber Bragg Grating Acoustic Emission Sensors in Thin Polymer-Bonded Explosives. *Sensors* **2018**, *18*, 3778. [CrossRef]
18. Cranch, G.A.; Johnson, L.; Algren, M.; Heerschap, S.; Miller, G.A.; Marunda, T.S.; Holtz, R.L. Crack detection in riveted lap joints using fiber laser acoustic emission sensors. *Opt. Express* **2017**, *25*, 19457–19467. [CrossRef] [PubMed]
19. Sbarufatti, C.; Beligni, A.; Gilioli, A.; Ferrario, M.; Mattarei, M.; Martinelli, M.; Giglio, M. Strain Wave Acquisition by a Fiber Optic Coherent Sensor for Impact Monitoring. *Materials* **2017**, *10*, 794. [CrossRef]
20. Lan, C.; Zhou, W.; Xie, Y. Detection of Ultrasonic Stress Waves in Structures Using 3D Shaped Optic Fiber Based on a Mach–Zehnder Interferometer. *Sensors* **2018**, *18*, 1218. [CrossRef]
21. Liguori, C.; Martinelli, M. Integral phase modulation properties of a single-mode optical fiber subjected to controlled vibration. *Appl. Opt.* **1981**, *20*, 4319–4323. [CrossRef] [PubMed]

22. Martinelli, M.; Ferrario, M. Synoptic Fiber Optic Sensor. Patent No. WO 2013/179118 Al, 5 December 2013.
23. Martinelli, M. A universal compensator for polarization changes induced by birefringence on a retracing beam. *Opt. Commun.* **1989**, *72*, 341–344. [CrossRef]
24. Ja, Y.H. Analysis of four-port optical fiber ring and loop resonators using a 3 × 3 fiber coupler and degenerate two-wave mixing. *IEEE J. Quantum Electron.* **1992**, *28*, 2749–2757. [CrossRef]
25. Boffi, P.; Cattaneo, G.; Amoriello, L.; Barberis, A.; Bucca, G.; Bocciolone, M.F.; Collina, A.; Martinelli, M. Optical Fiber Sensors to Measure Collector Performance in the Pantograph-Catenary Interaction. *IEEE Sens. J.* **2009**, *9*, 635–640. [CrossRef]
26. Wei, P.; Han, X.; Xia, D.; Liu, T.; Lang, H. Novel Fiber-Optic Ring Acoustic Emission Sensor. *Sensors* **2018**, *18*, 215. [CrossRef]
27. Viktorov, I. *Rayleigh and Lamb Waves: Physical Theory and Applications*; Ultrasonic Technology; Springer US: New York, NY, USA, 1970.
28. Hsu, N.N.; Breckenridge, F.R. Characterization and Calibration of Acoustic Emission Sensors. *Mater. Eval.* **1981**, *39*, 60–68.
29. Anderson, T. *Fracture Mechanics, Fundamentals and Applications*; CRC Press: Boca Raton, FL, USA, 2017.

sensors

MDPI

Article

Quantifying Hole-Edge Crack of Bolt Joints by Using an Embedding Triangle Eddy Current Sensing Film

Shilei Fan [1,2], Junyan Yi [3], Hu Sun [3,*] and Fenglin Yun [3]

1 College of Aerospace Engineering, Nanjing University of Aeronautics & Astronautics, Nanjing 210016, China; fanshilei@comac.cc
2 COMAC Shanghai Aircraft Design & Research Institute, Shanghai 201210, China
3 School of Aerospace Engineering, Xiamen University, Xiamen 361005, China; junyanyi@stu.xmu.edu.cn (J.Y.); yunfl@xmu.edu.cn (F.Y.)
* Correspondence: sunhu@xmu.edu.cn

Abstract: Hole-edge crack quantification of bolt joints is critical for monitoring and estimating structural integrity of aircraft. The paper proposes a new triangle eddy current sensor array for the purpose of increasing the level of quantifying hole-edge crack parameters, especially, the crack angle. The new senor array consists of triangular coils instead of planar rectangular coils. The configuration of the novel sensor array, including the excitation current directions and the excitation winding shape, is optimized by simulation. The ability of the proposed sensing film to identify the crack parameters has been verified by finite element simulations and experiments. Results shows that triangular coils with same current directions in circumferentially adjacent coils and opposite current directions in axially adjacent coils achieve better performance in sensor linearity and resolution compared to rectangular coils. In addition, it has also been proved that the sensing film has a good potential to identify the crack depth and length.

Keywords: structural health monitoring; crack quantification; triangle eddy current coils; sensing film

Citation: Fan, S.; Yi, J.; Sun, H.; Yun, F. Quantifying Hole-Edge Crack of Bolt Joints by Using an Embedding Triangle Eddy Current Sensing Film. *Sensors* **2021**, *21*, 2567. https://doi.org/10.3390/s21072567

Academic Editor: Zenghua Liu

Received: 15 March 2021
Accepted: 2 April 2021
Published: 6 April 2021

1. Introduction

Bolt joints are the most important connection types of key structures for aircraft, meanwhile, they are also the weak link due to the stress concentration caused by the opening bolt hole. Eighty percent of structural mechanical failures are caused by failure of jointed structures [1]. It is critical to monitor and prevent the damage and failure of bolt joints around the hole to ensure the structural integrity and the operational safety of advanced equipment.

In addition to the bolt-loosening failure, there are three typical hole-edge failure modes: shear failure, tensile failure and bearing failure [2,3]. According to the characteristics of structural force, the crack distribution of three failures along the circumferential angle of the bolt hole is obviously different (shear failure [30°, 60°], tensile failure [75°, 105°] and bearing failure [−15°, 15°]), which can be used to distinguish failure modes by crack angle.

In the service process of aircraft, if the damage initiation and growth in the early stage of bolt joint structure failure is monitored online and real-time by sensor network integrated with the structure, it will be helpful to monitor the failure evolution process of bolt joints and predict the hole-edge failure mode of the jointed structure, and provide technical and data support for optimizing the design of bolted joints. Structural health monitoring (SHM) technology can conduct real-time quantitative monitoring of the structure state and possible damage by using the sensor network integrated with the structure. SHM is a revolutionary innovation technology to determine the structural integrity and a key technology to improve the operation safety of advanced equipment, and moreover can improve the present design concept of damage tolerance [4]. There are some SHM technologies which have been used for the damage monitoring of bolted joints, such as ultrasonic guided wave [5–7], electromechanical impedance [8,9], comparative vacuum monitoring sensor [10], vibration [11], acoustic emission [12] and so on. However, these technologies have the following problems: (1) It is difficult to identify the cracks in the non-surface layer of a multi-layer jointed structure; (2) It is difficult to accurately quantify the angle of hole edge cracks, and then identify the failure mode.

Eddy current (EC), which is low cost and efficient for dealing with conductive bodies, has been the focus of many researchers. There are three main types of flexible EC array sensor films developed for monitoring the crack growth around the bolt hole. The first is meandering winding magnetometer arrays (MWM) proposed by JENTEK Sensors Inc. [13], which is actually designed specifically to measure the electrical conductivity and permeability of the material first and is further applied by Goldfine et al. [13–17] to detect fatigue cracks on metal structures. It is found that MWM is able to track crack growth with a resolution of 50 um. However, the output signal of the MWM is too small to be detected. The second one is the rosette-like eddy current array (RECA) sensor, which is composed of circular drive wingdings periodically deposited on the flexible film and pickup winding array distributed between the adjacent drive windings. Li et al. [18] enhanced the sensitivity of the sensor by placing a magnetic membrane washer on the surface-mounted sensor film. As found in [19], mutual interference between the channels affects the amplitude and phase of the transimpedance. Chen et al. [20] proposed a new layout of the pickup coils. Sensing channels in the outer layer and in the inner layer are misaligned to improve the angular resolution of the RECA sensor. They also discovered that exciting currents in adjacent wingdings flowing in same directions instead of opposite directions can increase the output signal and sensitivity. Fan et al. [21] further studied the effect of the excitation layout on the sensor sensitivity and the results show that the sensitivity was increased up to 43.1 times in simulation and 23.7 times in an experiment when the co-directional excitation coil layout was applied. Furthermore, research projects into the temperature compensation for RECA sensor have also been conducted in [22,23]. The RECA sensors perform well in tracking crack propagation and crack angle identifying. Nevertheless, it is hard for them to discover cracks deep in the plate due to the limit of the skin effect which is unavoidable when applying eddy current.

The third type of flexible EC array sensor is the intelligent bolt with flexible sensing film integrated on the surface of the bolt, which can detect the crack deep in thick plates or multi-layer jointed structure, as shown in Figure 1. Rakow and Chang [24] firstly proposed the concept of eddy current-based intelligent bolt and studied the capacity of the film to track flaw depth. Sun et al. [25–27] optimized the coil layout in order to offer more accurate information about the crack. First, an EC sensing array uniformly distributed along the axial direction was proposed to monitor the crack depth [25]. Furthermore, a 2D EC array-based sensing film [26] consisting of a 3×3 coil array was developed to identify the failure mode by providing the crack positions on the joint structure. However, although this proved to have the ability to identify the cracks in different sensing regions, planar rectangular coils are unable to provide a precise crack angle for the fact that there are blind zones existing in the interface between two adjacent sensing coils. Thus, this paper presents a new sensing array which is composed of triangular coils, and it also provides inspiration for sensor design in other areas.

The rest of the paper is organized as follows: the capability of the proposed sensing array to identify crack angle is demonstrated by the combination of the simulation in Section 2 where the triangular exciting coils and the rectangular exciting coils are compared and the experiment in Section 3; conclusions are drawn in Section 4.

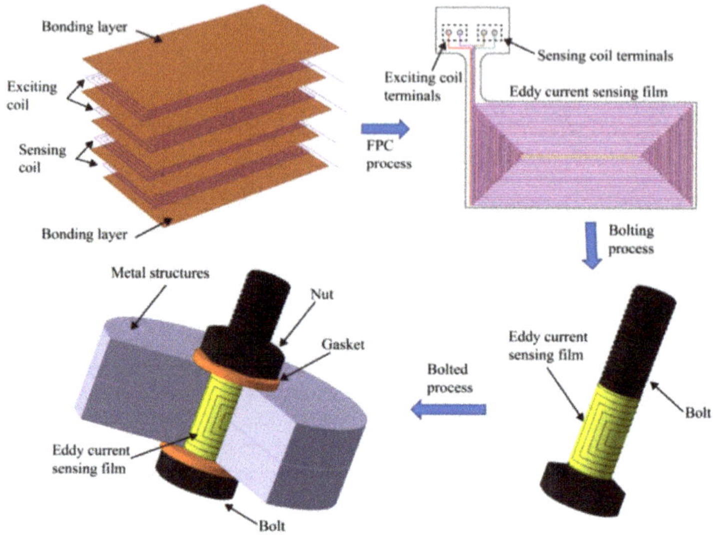

Figure 1. An eddy current (EC) sensing film-based intelligent bolt.

2. Design and Analysis of 2D-Triangular Eddy Current (EC) Sensing Film

For the purpose of improving the hole-edge crack angle identification ability, a novel 2D EC sensing array film that consists of triangular sensing coils instead of rectangular sensing coils used in [25] is proposed in this paper, and a distribution schematic of its pickup coil array is presented in Figure 2. As is shown in the picture, there are 4 triangular pickup coils in the circumferential direction and each coil covers half of the bolt hole perimeter. As depicted in the red dashed box, each pair of the circumferentially adjacent coils share a same circumferential monitoring area (a quarter of the bolt hole perimeter) and distribute separately in the axial direction after the flexible sensing film is winded by adhesive on the bolt surface. Although only one layer of the sensing coil array is plotted in the picture for the sake of brevity, the axial dimension of the sensing array is set as 3 in this paper, which should be decided by the actual demand.

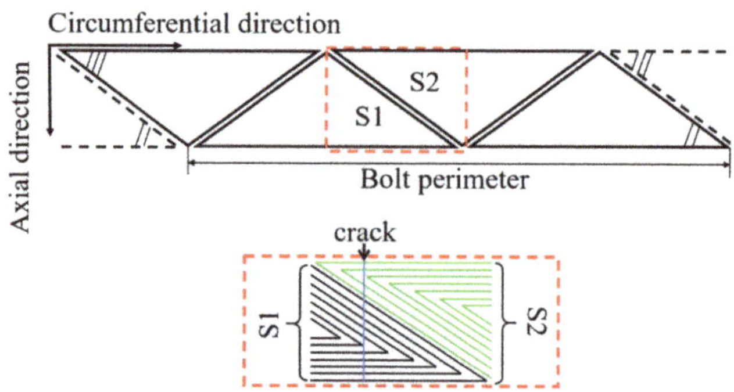

Figure 2. The simplified triangular sensing coil array configuration.

The part highlighted by a red dashed box at the lower part of Figure 2 describes half of the simplified winding model of the adjacent circumferential receiving coils. As has been demonstrated, the induced voltage of one sensing coil will change if and only if the crack reaches and extends in its area. Thus, as a result of the wingding configuration of the triangular coil, the effective coil turns feeling the change of the EC field at a specified circumferential position from the left to right decrease for S1, while increase for S2. For example, when the crack is at the blue line, there are 5 effective turns for S1 and 2 effective turns for S2. It is expected that the induced voltage of the sensing coil will vary linearly with the circumferential position of the crack.

2.1. Finite Element Simulation Model

The triangular coil winding on the bolt and the crack on the hole edge make the geometry of the structure more complex. It is difficult to obtain an analytical solution. Thus, finite element method was employed to solve the electromagnetic field equations and calculate magnetic vector potential **A** and electric scale potential φ

$$\nabla^2\varphi - \mu\varepsilon\frac{\partial^2\varphi}{\partial t^2} = -\frac{\rho}{\varepsilon} \tag{1}$$

$$\nabla^2\mathbf{A} - \mu\varepsilon\frac{\partial^2\mathbf{A}}{\partial t^2} = -\mu\mathbf{J} \tag{2}$$

where ρ denotes free charge bulk density, μ is permeability, ε is dielectric constant, and **J** is conduction current density.

The finite element simulations are conducted in Ansoft Maxwell. There are two kinds of the exciting coil considered in the paper, the rectangular one in Model 1 and the triangular one in Model 2. All of the parameters other than the exciting coils are same in both models, including the sensing coils, the joint layers and the running parameters. The joint layer with an electrical conductivity of 38 MS/m and a relative permeability of 1.00021 measures 6 mm in length, 4 mm in width and 9 mm in height, shown in Figure 3. To demonstrate the ability of the sensing coils to identify the crack angle and to track the crack growth in the radial direction, crack elements with a dimension of 1 mm × 0.2 mm × 3 mm are adopted to switch the crack position and to give rise to a mesh consistent in each crack state. Five circumferential positions marked from C1 to C5 are set from the left side to the upper middle of the model with a step of 0.55 mm. At each circumferential position, there are 3 crack elements in the radial direction, marked from R1 to R3. When the crack is at CiRj, which means the crack is at the circumferential position Ci and its radial length is jmm, the corresponding crack elements' material will be set as air.

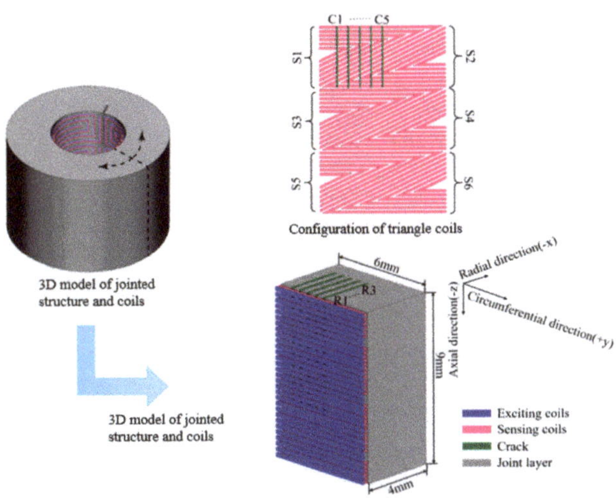

Figure 3. The simulation model.

In order to reduce the computational burden, a quarter model is selected for the simulation according to the fine symmetry in the sensing coils when it is wound on the bolt.

The turns distance and width are both 0.1 mm for the triangular coils and are set as 0.1 mm and 0.2 mm, respectively, for the rectangular exciting coils. Sensing coils are denoted as S1 to S6 from up to down. A sinusoidal exciting signal with a frequency of 1 MHz and an amplitude of 2 V is applied in the simulation. The induced voltage of each sensing coil is calculated for crack state analysis.

The exciting current direction, which has a significant influence on the crack detection, needs to be determined first. Shown in Figure 4, according to the current directions (same or opposite) between the axially/circumferentially adjacent triangular coils, there are 4 configurations compared in the simulation, numbered from configuration I to configuration IV. For example, in configuration II, the current directions of the axially adjacent coils are opposite and those of the circumferentially adjacent coils are same. The best exciting mode is determined by comparing the induced voltage variation of S1 and S2 when a crack is set at C5R1. The simulation results are plotted in Figure 5. It can be found that for both S1 and S2, the induced voltage variation is obviously greater when the circumferentially adjacent coils have same excitation current directions at the adjacent position (Configuration I, Configuration II). To explain the above result, the front view of the EC density when the test specimen is integrated and the side view when there is a crack at C5R1 are displayed in Figure 6 draw by Matlab.

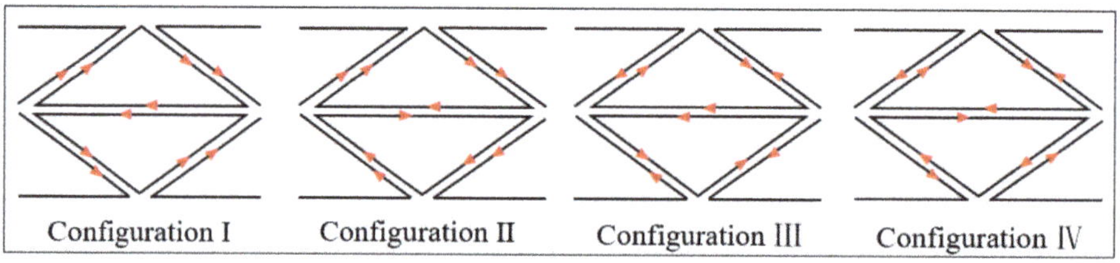

Figure 4. Four exciting configurations.

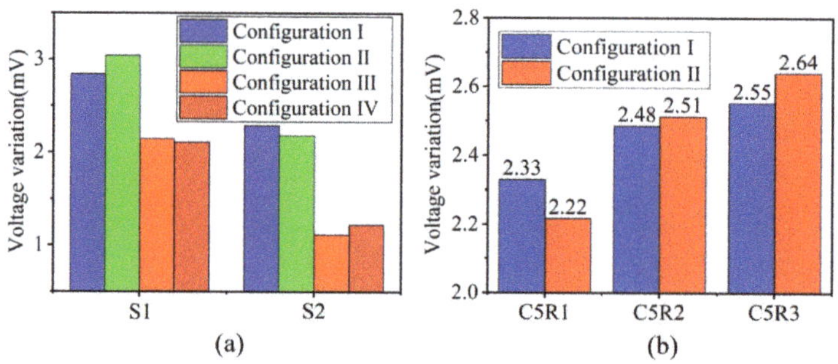

Figure 5. Simulation results for determining the best exciting configuration. (**a**) Voltage variations of the S1 and S2 when the crack is set at C5R1. (**b**) Voltage variations of S2 when the crack propagates in the radial direction at C5.

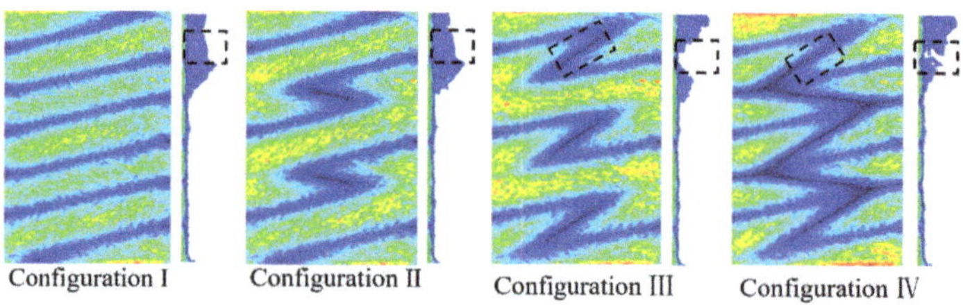

Figure 6. Current density distribution.

Figure 6 shows that the EC vanishes between the S1 and S2 (highlighted by a dashed box on the side view) in configuration III and configuration IV, which leads to less EC flowing along the radial direction (highlighted by a dashed box on the side view). Therefore, the induced voltages of the S1 and S2 in configuration III, Configuration IV change less than that of the configuration I and Configuration II.

However, the similar effects of configuration I and configuration III make it impossible to decide which is better to use to detect cracks. To address this problem, additional simulations in which the crack grows from C5R1 to C5R3 are performed to further determine the best exciting mode. The results are shown in Figure 5b where only the voltage variation of S2 is plotted because S2 is more spatially correlated with the exciting current direction between the axially adjacent coils. Although the induced voltage of configuration II changes less at first, it changes greater than that of Configuration I when the radial length of the crack is 2 mm and 3 mm, which indicates that Configuration II has better capability to track crack growth in the radial direction. Likewise, the reason for the phenomenon is given by the combination of the EC cloud picture and the EC vector on the surface of the test specimen when the crack is at C5R3, as shown in Figure 7. From the EC cloud picture, it can be seen that EC flows slightly more in the radial direction, which can be explained as follows. First of all, it is worth stressing that there are 3 modes when EC flows conductively: (1) EC flows over the crack tip when the EC flows in the same direction around the crack tip (red arrow); (2) a local EC loop is built because of the opposite EC direction under two adjacent coils (blue arrow); (3) EC flows in the radial direction, which causes a decrease of EC on the surface of the conductive test specimen (side view). EC always finds the shortest way to flow. Therefore, when the crack tip is at the middle of two adjacent coils with the same exciting direction, it is easier for EC to flow over the crack tip, leading to less EC flowing along the radial direction especially when the radial length of the crack is long enough, which results in less voltage variation in Configuration I when the crack is at C5R2 and C5R3. As for Configuration II, compared to flowing over the crack tip by overcoming the original EC direction, it seems more easier for EC to flow along the radial direction. In conclusion, Configuration II is selected as the final exciting form for Model 2. For Model 1, opposite exciting current directions are applied in axially adjacent coils, and there is no need to consider the current directions in circumferentially adjacent coils because of the space configuration of the rectangular coils.

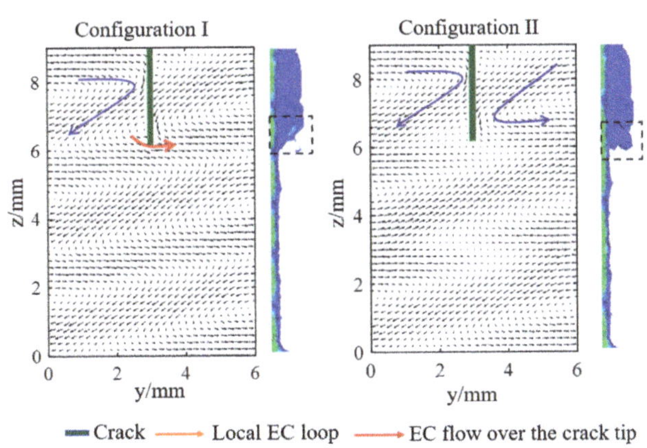

Figure 7. EC vector and cloud picture when the crack is at C5R1.

2.2. Simulation Results

2.2.1. Crack Angle Identification

Figure 8a and 8b present the crack angle identification results for Model 1 and Model 2 respectively. As shown in Figure 8a, the induced voltage variation of S1 decreases linearly and that of the S3–S6 keeps stable when the circumferential position of the crack changes from C1 to C5, which meets the expected design goal. However, compared to Model 2, the inducted voltage of S2 in Model 1 changes little until the crack is at C4 and C5, which indicates that the rectangular exciting coil deteriorates the crack angle identifying ability of the triangular sensing coil. This phenomenon can be explained by Figures 9 and 10. It can be seen from Figure 9 that a single turn can be divided into 3 parts according to the exciting direction bounded by a dashed line. It should be noted that the induced voltage of the whole single-turn coil can be represented by the sum of the 3 parts and the direction of the voltage variation caused by the crack is same for the 3 parts. Assuming the exciting direction and the induced voltage direction at a certain time is as shown in Figure 10, it is obvious that the effects of the part 1 and part 3 offset, while those of part 2 and part 3 should be summed up. As a result, the induced voltage of the S1 changes little when the crack is at C1–C3, though there is much EC flowing in the radial direction shown in the cloud picture in Figure 10, which actually contributes to the increasing induced voltage at a macroscopic level.

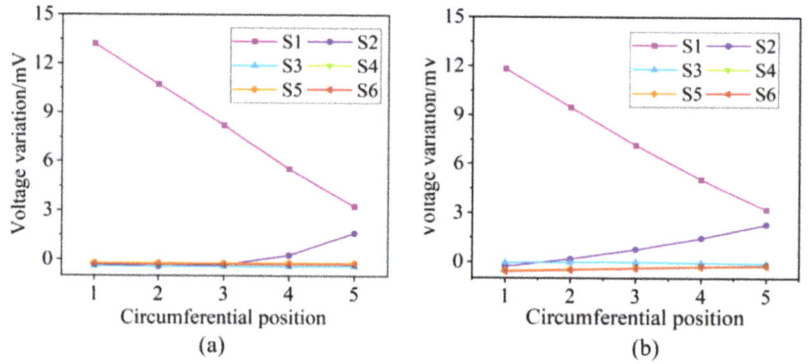

Figure 8. Simulation results of the crack angle detection. (**a**) Model 1 (rectangular exciting coils); (**b**) Model 2 (triangular exciting coils).

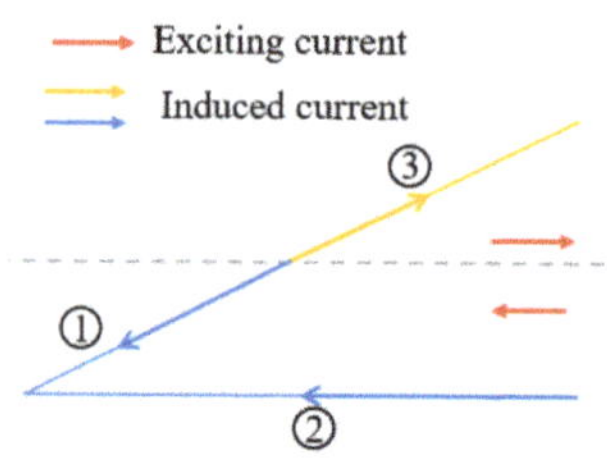

Figure 9. A schematic diagram of the inductive current directions in one triangular turn in Model 1.

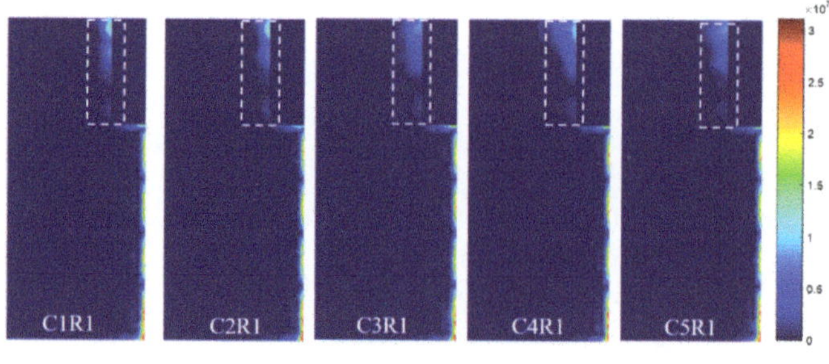

Figure 10. The side view of EC in Model 1.

The crack angle position is further represented by the difference of the induced voltage of S1 and S2, as shown in Figure 11. A fine linearity can be found in the fitting curve with a R-square of 0.9998. The simulation model consists of 3 layers in the axial direction. The first layer and the third layer are axially symmetrical. Therefore, simulations when the crack is at the second layer are also performed and the results are displayed in Figure 12. This shows that the results are similar to the first layer.

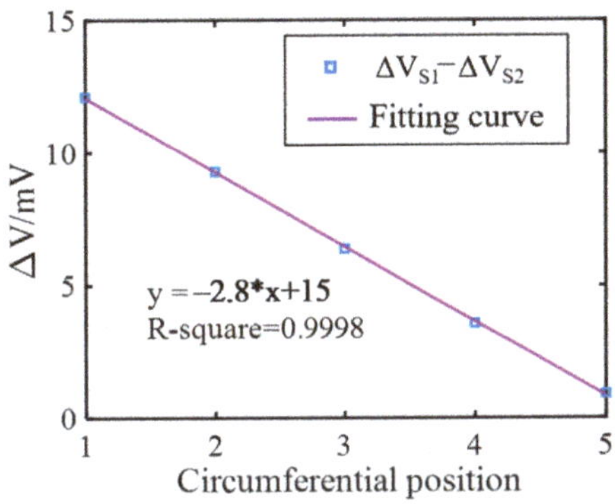

Figure 11. The difference voltage of the S1 and S2 versus the crack circumferential position.

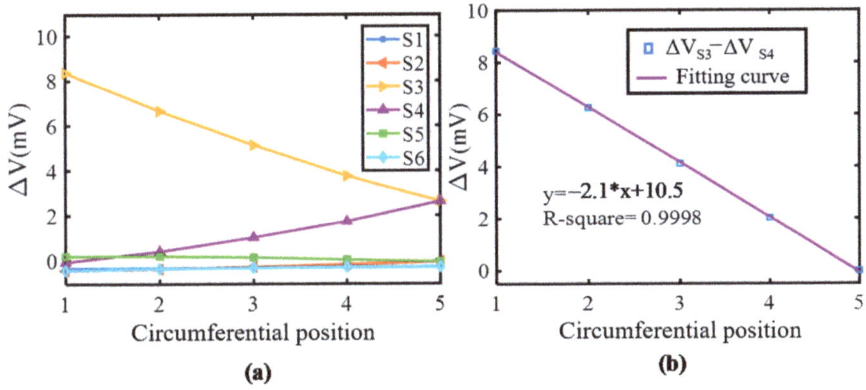

Figure 12. Simulation results for the second layer. (**a**) Voltage variation of each coil; (**b**) Difference value between adjacent coils.

In conclusion, triangular exciting coils achieve a better ability to identify the crack angle compared to rectangular exciting coils when triangular coils are applied as inductive coils. Meanwhile, triangular coils achieve good linearity with the crack angle, which is a very important property of a sensor.

2.2.2. Tracking Crack Growth in the Radial Direction

Simulations where the crack propagates along the radial direction are performed as well. Figure 13a and 13b show the voltage variation of S1 and S2 respectively. The following can be concluded from the picture: (1) the induced voltage of S1 increases gradually as the radial length of the crack increases at the circumferential position C1–C4, while that of the S2 increases only at C4 and C5; (2) the ability of the S1 to monitor the crack growth in the radial direction deteriorates as the circumferential position changes from C1 to C5, which can be attributed to the decrease of the effective coil turns; (3) S2 achieves a better effect at C5 than S1, further demonstrating the fact that the opposite exciting current direction in adjacent coils is constructive to track the radial length of the crack. Furthermore, although the voltage variation trends of the S2 shows an inability to track the crack growth at position C1–C3, S1 alone is enough as long as the circumferential position is identified.

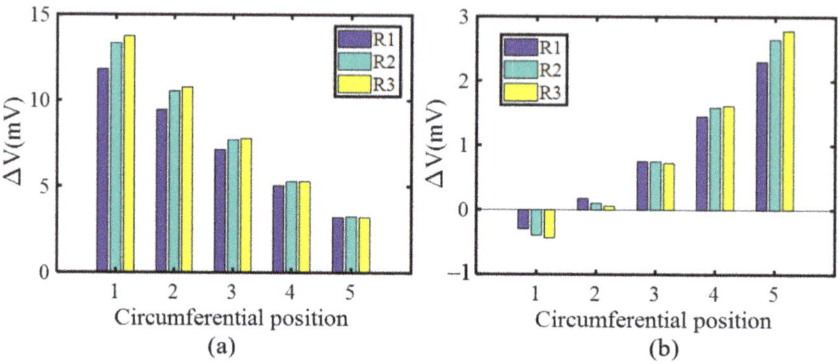

Figure 13. The induced voltage variation of S1 and S2 when the crack propagates along the radial direction at each circumferential position. (**a**) S1; (**b**) S2.

2.2.3. Tracking Crack Growth along the Axial Direction

An early crack that initiates from the hole edge of the joint layer often propagates along the axial direction at first. As a result, several simulations were carried out and the output voltage of each sensing coils were extracted and analyzed to evaluate the applicability of the triangular sensing film to monitor the crack growth in the axial direction. The crack starts from the first layer at two typical circumferential positions, C1 and C3, respectively, and propagates with a step of 1 mm along the axial direction. The results are presented in Figure 14. This shows that the axial length of the crack can be acquired by comparing the induced voltage of all sensing coils based on the following principles: (1) the induced voltage increases compared to the baseline data when the crack is at its monitoring area; (2) the induced voltage of the coil in accordance with principle (1) increases when the crack propagates in its monitoring domain; (3) when the crack has covered its axial length at a certain crack angle, the output voltage of the sensing coils in accordance with (1) and (2) will stay at a stable level, which is enough to decide that the crack exists within its axial length; (4) the space configuration of the sensing coils is settled. Taking 4 mm in Figure 14b as an example, first, it shows in Figure 14b that only the induced voltage of the S1 and S3 greatly increases which means that the axial length of the crack is between 3 mm and 6 mm. The crack angle can be obtained roughly by the nearly unchanged output value of the S2 and further deferred by S2 through the crack angle detection line in Figure 8. Second, the axial range of the crack tip can be further narrowed by the current voltage variation (about 3.2 mV) of the S2 and the maximum one (about 6 mV) when the crack length covered the S3 shown in Figure 12.

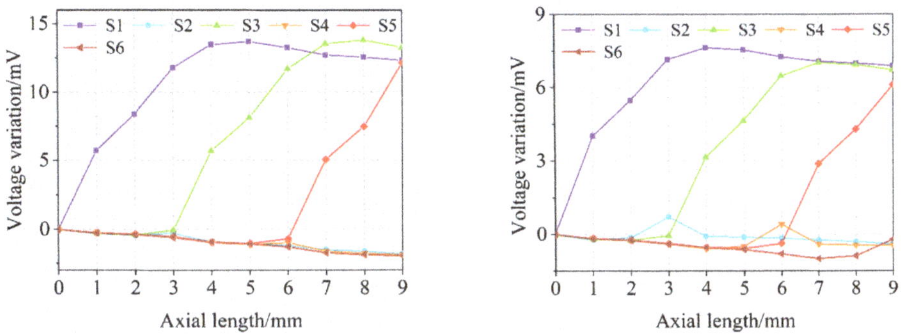

(**a**) C1 (**b**) C3

Figure 14. Voltage variation versus axial length.

3. Experiments

The performance of the EC sensing film with triangular exciting coil and triangular sensing coil array is evaluated by experiments, and the experimental setup is shown in Figure 15. The exciting signal with an amplitude of 2 V and a frequency of 1 M was generated by a signal generator made by Tektronic. The induced voltage of each coil is acquired through a channel switch and signal receiver made by National Instruments with a sample frequency of 20 M and a sample length of 20,000. Thus, the signal measurement and processing can be automatically finished nearly real-time, which shows a good monitoring efficiency similar to other SHM methods. With regard to the test specimen, three rectangular aluminum (6061) plates, each of which has dimensions 75 mm (length) × 75 mm (width) × 5 mm (thickness) and a bolt hole with diameter 13.5 mm at the center, were jointed by a steel bolt. The triangle eddy current sensing film, with dimensions 40.7 mm × 15 mm × 0.2 mm, was made by a flexible printed circuit process. In the real engineering applications, a wear-resistant coating layer of SiO_2 and Ta_2O_3 can be deposited on the surface of the sensing film via electron beam physical vapor deposition to improve the friction-resistance performance of the sensing film. As shown in Figure 14, there were 12 independent triangle coils as receiving, while similar 12 triangle coils with same shapes were connected with each other as transmitting according to Configuration II in Figure 4.

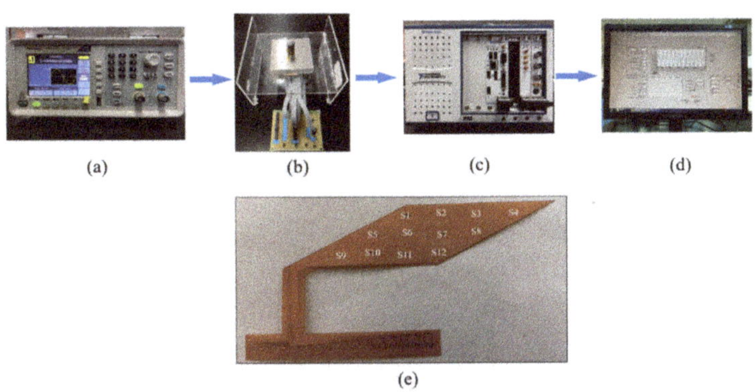

Figure 15. Experimental setup and sensor film. (**a**) Signal generator; (**b**) intelligent bolt; (**c**) signal receiver; (**d**) signal displayer; (**e**) eddy current sensing film.

3.1. Crack Angle Identification

The crack was set at the hole edge of the uppermost plate by spark erosion and the crack angle was changed by rotating the specimen from the middle of the S2 (0°) to the middle of the S4 (180°) with a step of 18°. For the sake of accuracy, the reference voltage under each crack angle was recorded. The positions of the crack represented by gray lines are shown schematically on the top of Figure 16a. As seen in Figure 16a, the voltage variation trends are consistent with the simulation: the output voltage is proportional to the effective sensing turns at a certain circumferential position. The voltage variation of S2 decreases as the crack angle changes from 0° to 90° and keeps stable when the crack is at the position of S3 and S4, while that of the S4 increases when the crack angle changes from the middle of S3 to the middle of S4, which allows the crack angle identification by the sensing film. Furthermore, the voltage difference of the S2 and S3 as well as the S3 and S4 is calculated and plotted in Figure 16b, in which both achieve a good linearity with the crack angle, showing that the proposed triangular sensing film performs well in crack angle identification.

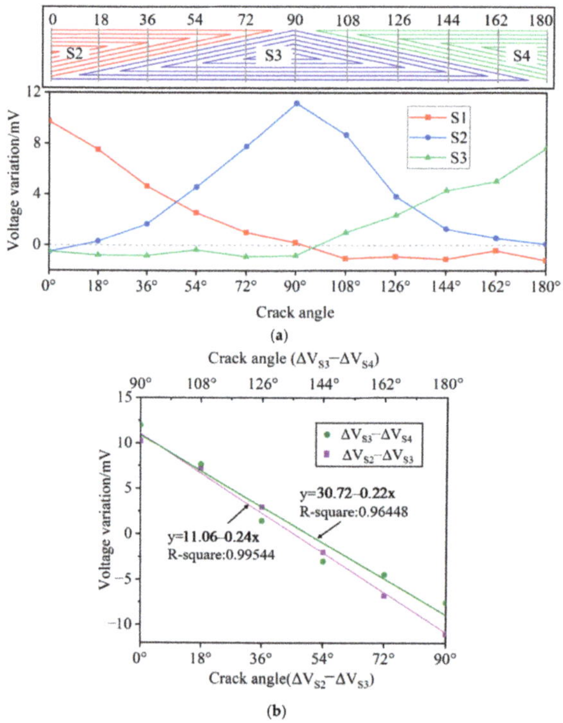

Figure 16. Crack angle detection results. (**a**) Voltage variation of the S2–S4 versus crack angle; (**b**) Voltage difference of S3 (S2) and S4 (S3).

It can be seen from the above results that when there is only a single crack in the circumferential angle monitoring area of a single crack, there is a linear relationship between the circumferential angle of the crack and the induced voltage difference of the receiving coil

$$\vartheta = k\Delta V + b \tag{3}$$

where ϑ is the circumferential angle of the crack, k is the slope of a straight line, ΔV is the difference of the induced voltage variation of the relevant receiving coil, and b is the intercept of the straight line.

In the real engineering application, the linear relationship is easy to obtain as the baseline for the crack angle quantification.

3.2. Crack Growth in the Radial Direction

Two experiments in which two typical circumferential positions were considered were conducted to verify the ability of the proposed sensing film to track crack growth in the radial direction. The crack position was set at the middle of the S2 in the first experiment and was set at the middle of two circumferentially adjacent coils (S2 and S3 in the experiment) in the second experiment. The results are given in Figure 17. First, when the crack was at the middle of the S2, only the output voltage of the S2 increased obviously when the radial length of the crack increased from 1 mm to 4 mm in a step of 1 mm. When the crack was at the middle of the S2 and S3, both the induced voltage of the S2 and S3 increased and that of the S2 increased more than that of the S2, as a result of more effective turns as shown in the crack position schematic in Figure 17b. Another point worth mentioning is that the voltage variation changes in a smaller way with the increase of the radial length of the crack, implying that the radial measuring rage of the film is not very large (but at least 4 mm in this experiment) but enough to be taken seriously in real engineering applications. It may be improved by increasing the exciting voltage amplitude, reducing the exciting frequency or increasing the equipment accuracy. Results for the coil voltage are in excellent agreement with those obtained with the finite-element method.

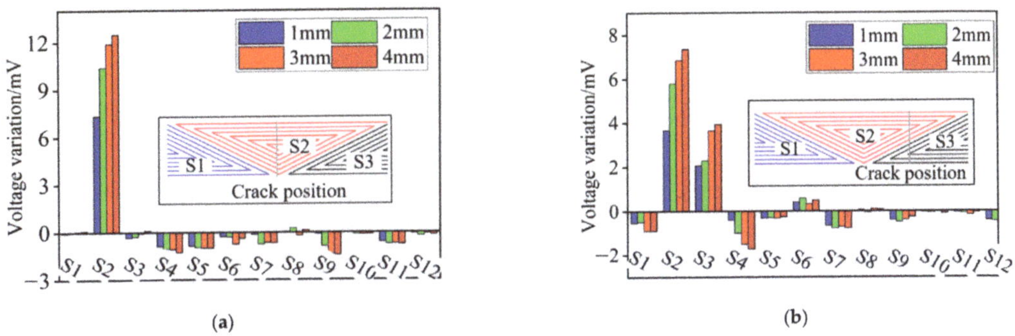

Figure 17. Experimental results when the crack grows in the radial direction at two circumferential positions. (**a**) Voltage variation when the crack position is at the middle of the S2; (**b**) voltage variation when the crack position is in the middle of the S2 and S3.

3.3. Crack Depth Identification

Three cracks, with angle at the circumferential center of Coils S2 and S1, were set at the hole edge of three jointed plates, respectively. The radial length of the crack was set as 1 mm. The results in Figure 18 show that the crack depth can be inferred by the induced voltage. When one plate has a hole-edge crack, the induced voltage of the coil at the axial location of this plate has an obvious change. For example, when one crack occurs at the location of the uppermost plate, only the induced voltages of S1 and S2 are enhanced significantly. When two cracks are present at the locations of both uppermost and middle plates, only the induced voltages of coils S1, S2, S6 and S7 change considerably, while the induced voltages of other coils change less. This indicates that the proposed sensing film has a good capability of identifying crack depth, at least, identifying which plate has a hole-edge crack for multi-layer bolted joints.

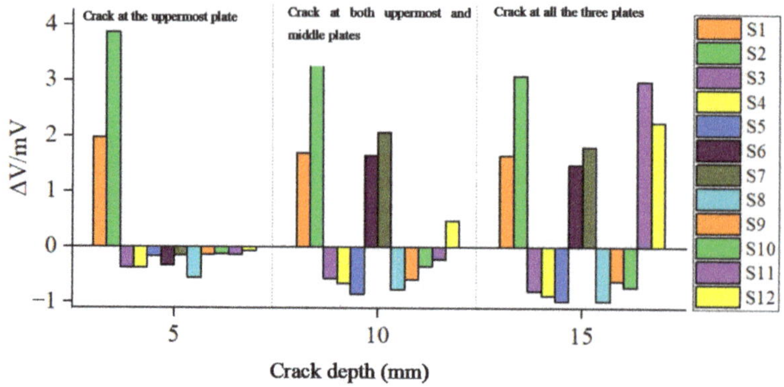

Figure 18. Experimental results as the crack grows in the axial direction.

4. Conclusions

Enhancing the damage quantification level of bolted joints has been always the research hotspot of aircraft structural health monitoring. In this paper, a novel triangle eddy current sensing film is proposed to quantitatively identify the hole-edge crack of bolted joints. The conclusions are as follows:

(1) The sensing film includes an exciting coil and multiple triangle sensing coils distributed in the axial and circumferential directions of the bolt hole. By analyzing the eddy current distribution and effect of crack quantification, five exciting-coil configurations of the sensing film, including one rectangle-coil configuration and four types of triangle-coil configuration, have been compared to give the best configuration of the eddy current sensing array.

(2) There is an approximate linear relationship between the crack angle and the difference of the induced voltage variation between two adjacent triangle sensing coils in the same layer, which can be used to quantify the crack angle with a good accuracy.

(3) By simulation and experimental analysis, the proposed sensing film has a good capacity not only to identify the crack angle, but also to quantify the crack length in the radial direction and the crack depth in the axial direction of the bolt hole.

However, some further studies will be conducted in the future for real engineering applications of this sensing technology:

(1) The configuration of the sensing film or the measuring parameters of the system should be optimized to improve the measuring range of the crack radial dimension.

(2) The robustness of the sensing film should be further verified in real engineering applications, which may include the influence of environmental variation and bolt looseness.

(3) In this paper, forward analysis between the induced voltage of eddy current coils and the crack parameters has been conducted to show the capacity of the proposed sensing film. The backward problem that the proposed sensing film is used to detect the crack will be further studied and verified in future.

Author Contributions: Conceptualization, H.S. and F.Y.; methodology, S.F. and J.Y.; validation, J.Y.; formal analysis, S.F. and J.Y.; investigation, J.Y. and F.Y.; writing—original draft preparation, H.S. and S.F.; writing—review and editing, F.Y.; project administration, H.S.; funding acquisition, H.S. All authors have read and agreed to the published version of the manuscript.

Funding: This work was funded by the projects from the Natural Science Foundation of China (Grant No. 11902280), Innovation Foundation for Young Scholar of Xiamen (Grant No. 3502Z20206042) and the Natural Science Foundation of Fujian Province (Grant No. 2018J05094).

Conflicts of Interest: The authors declare no conflict of interest.

References

1. Najafi, A.; Garg, M.; Abdi, F. Failure analysis of composite bolted joints intension. In Proceedings of the 50th AIAA/ASME/ASCE/AHS/ASC Structures, Structural Dynamics, and Materials Conference, Palm Springs, CA, USA, 4–7 May 2009.
2. Oskouei, R.H.; Keikhosravy, M.; Soutis, C. A finite element stress analysis of aircraft bolted joints loaded in tension. *Aeronaut. J.* **2016**, *114*, 315–320. [CrossRef]
3. Sun, H.-T.; Chang, F.-K.; Qing, X. The Response of Composite Joints with Bolt-Clamping Loads, Part I: Model Development. *J. Compos. Mater.* **2002**, *36*, 47–67. [CrossRef]
4. Qing, X.; Li, W.; Wang, Y.; Sun, H. Piezoelectric Transducer-Based Structural Health Monitoring for Aircraft Applications. *Sensors* **2019**, *19*, 545. [CrossRef] [PubMed]
5. Huo, L.; Wang, F.; Li, H.; Song, G. A fractal contact theory based model for bolted connection looseness monitoring using piezoceramic transducers. *Smart Mater. Struct.* **2017**, *26*, 104010. [CrossRef]
6. Qiu, L.; Yuan, S.; Chang, F.-K.; Bao, Q.; Mei, H. On-line updating Gaussian mixture model for aircraft wing spar damage evaluation under time-varying boundary condition. *Smart Mater. Struct.* **2014**, *23*, 125001. [CrossRef]
7. Kim, Y.; Lim, H.J.; Sohn, H. Nonlinear ultrasonic modulation based failure warning for aluminum plates subject to fatigue loading. *Int. J. Fatigue* **2018**, *114*, 130–137. [CrossRef]
8. Jung, H.K.; Jo, H.; Park, G.; Mascarenas, D.L.; Farrar, C.R. Relative baseline features for impedance-based structural health monitoring. *J. Intell. Mater. Syst. Struct.* **2014**, *25*, 2294–2304. [CrossRef]
9. Huo, L.; Chen, D.; Liang, Y.; Li, H.; Feng, X.; Song, G. Impedance based bolt pre-load monitoring using piezoceramic smart washer. *Smart Mater. Struct.* **2017**, *26*, 057004. [CrossRef]
10. Roach, D. Real time crack detection using mountable comparative vacuum monitoring sensors. *Smart Struct. Syst.* **2009**, *5*, 317–328. [CrossRef]
11. Argatov, I.; Butcher, E.A. On the separation of internal and boundary damage in slender bars using longitudinal vibration frequencies and equivalent linearization of damaged bolted joint response. *J. Sound Vib.* **2011**, *330*, 3245–3256. [CrossRef]
12. Hoang, T.D.; Herbelot, C.; Imad, A. Rupture and damage mechanism analysis of a bolted assembly using coupling techniques between A.E. and D.I.C. *Eng. Struct.* **2010**, *32*, 2793–2803. [CrossRef]
13. Goldfine, N.; Zilberstein, V.; Washabaugh, A.; Schlicker, D.; Shay, I.; Grundy, D. Eddy current sensor networks for aircraft fatigue monitoring. *Mater. Eval.* **2003**, *61*, 852–859.
14. Zilberstein, V.; Schlicker, D.; Walrath, K.; Weiss, V.; Goldfine, N. MWM eddy current sensors for monitoring of crack initiation and growth during fatigue tests and in service. *Int. J. Fatigue* **2001**, *23*, 477–485. [CrossRef]
15. Zilberstein, V.; Grundy, D.; Weiss, V.; Goldfine, N.; Abramovici, E.; Newman, J.; Yentzer, T. Early detection and monitoring of fatigue in high strength steels with MWM-Arrays. *Int. J. Fatigue* **2005**, *27*, 1644–1652. [CrossRef]
16. Sheiretov, Y.; Grundy, D.; Zilberstein, V.; Goldfine, N.; Maley, S. MWM-Array Sensors for In Situ Monitoring of High-Temperature Components in Power Plants. *IEEE Sens. J.* **2009**, *9*, 1527–1536. [CrossRef]
17. Goldfine, N.; Zilberstein, V.; Cargill, J.S.; Schlicker, D.; Shay, I.; Washabaugh, A.; Tsukernik, V.; Grundy, D.; Windoloski, M. Meandering winding magnetometer array eddy current sensors for detection of cracks in regions with fretting damage. *Mater. Eval.* **2002**, *60*, 870–877.
18. Li, P.; Cheng, L.; He, Y.; Jiao, S.; Du, J.; Ding, H.; Gao, J. Sensitivity boost of rosette eddy current array sensor for quantitative monitoring crack. *Sens. Actuators A Phys.* **2016**, *246*, 129–139. [CrossRef]
19. Li, P.Y.; Cheng, L.; He, Y.T.; Du, J.Q.; Jiao, S.B.; Gao, J.Y. Mutual interference effect of a rosette eddy current array sensor for quantitative crack monitoring in metallic structures. *Insight* **2018**, *60*, 495–500. [CrossRef]
20. Chen, G.; Zhang, W.; Zhang, Z.; Jin, X.; Pang, W. A new rosette-like eddy current array sensor with high sensitivity for fatigue defect around bolt hole in SHM. *NDT E Int.* **2018**, *94*, 70–78. [CrossRef]
21. Fan, X.; Chen, T.; He, Y.; Du, J.; Ma, B.; Song, Y. An excitation coil layout method for improving the sensitivity of a rosette flexible eddy current array sensor. *Smart Mater. Struct.* **2019**, *29*, 015020. [CrossRef]
22. He, Y.; Chen, T.; Du, J.; Ding, H.; Jiao, S.; Li, P. Temperature-compensated rosette eddy current array sensor (TC-RECA) using a novel temperature compensation method for quantitative monitoring crack in aluminum alloys. *Smart Mater. Struct.* **2017**, *26*, 065019. [CrossRef]
23. Chen, T.; Du, J.; He, Y.; Du, J. A structural crack monitoring gasket for aircraft bolt-jointed structures with temperature compensation. *Smart Mater. Struct.* **2018**, *27*, 115004. [CrossRef]
24. Rakow, A.; Chang, F.-K. A structural health monitoring fastener for tracking fatigue crack growth in bolted metallic joints. *Struct. Health Monit.* **2011**, *11*, 253–267. [CrossRef]
25. Sun, H.; Wang, T.; Liu, Q.; Qing, X. A novel eddy current array sensing film for quantitatively monitoring hole-edge crack growth in bolted joints. *Smart Mater. Struct.* **2018**, *28*, 015018. [CrossRef]
26. Sun, H.; Wang, T.; Liu, Q.; Wang, Y.; Qing, X. A two-dimensional eddy current array–based sensing film for estimating failure modes and tracking damage growth of bolted joints. *Struct. Health Monit.* **2019**. [CrossRef]
27. Liu, Q.; Sun, H.; Chai, Y.; Zhu, J.; Wang, T.; Qing, X. On-site monitoring of bearing failure in composite bolted joints using built-in eddy current sensing film. *J. Compos. Mater.* **2020**. [CrossRef]

MDPI
St. Alban-Anlage 66
4052 Basel
Switzerland
www.mdpi.com

Gels Editorial Office
E-mail: gels@mdpi.com
www.mdpi.com/journal/gels

www.ingramcontent.com/pod-product-compliance
Lightning Source LLC
Chambersburg PA
CBHW061451220926
51850CB00004B/63
* 9 7 8 3 7 2 5 8 0 2 2 5 8 *